Probability and Mathematical Statistics (Continued)

RUBINSTEIN • Simulation and The Monte Carlo Method
SCHEFFE • The Analysis of Variance
SEBER • Linear Regression Analysis
SEBER • Multivariate Observations
SEN • Sequential Nonparametrics: Invariance Principles and Statistical Inference
SERFLING • Approximation Theorems of Mathematical Statistics
TJUR • Probability Based on Radon Measures
WILLIAMS • Diffusions, Markov Processes, and Martingales, Volume I: Foundations
ZACKS • Theory of Statistical Inference

Applied Probability and Statistics

ABRAHAM and LEDOLTER • Statistical Methods for Forecasting
AGRESTI • Analysis of Ordinal Categorical Data
AICKIN • Linear Statistical Analysis of Discrete Data
ANDERSON, AUQUIER, HAUCK, OAKES, VANDAELE, and WEISBERG • Statistical Methods for Comparative Studies
ARTHANARI and DODGE • Mathematical Programming in Statistics
BAILEY • The Elements of Stochastic Processes with Applications to the Natural Sciences
BAILEY • Mathematics, Statistics and Systems for Health
BARNETT • Interpreting Multivariate Data
BARNETT and LEWIS • Outliers in Statistical Data
BARTHOLOMEW • Stochastic Models for Social Processes, *Third Edition*
BARTHOLOMEW and FORBES • Statistical Techniques for Manpower Planning
BECK and ARNOLD • Parameter Estimation in Engineering and Science
BELSLEY, KUH, and WELSCH • Regression Diagnostics: Identifying Influential Data and Sources of Collinearity
BHAT • Elements of Applied Stochastic Processes, *Second Edition*
BLOOMFIELD • Fourier Analysis of Time Series: An Introduction
BOX • R. A. Fisher, The Life of a Scientist
BOX and DRAPER • Evolutionary Operation: A Statistical Method for Process Improvement
BOX, HUNTER, and HUNTER • Statistics for Experimenters: An Introduction to Design, Data Analysis, and Model Building
BROWN and HOLLANDER • Statistics: A Biomedical Introduction
BROWNLEE • Statistical Theory and Methodology in Science and Engineering, *Second Edition*
CHAMBERS • Computational Methods for Data Analysis
CHATTERJEE and PRICE • Regression Analysis by Example
CHOW • Analysis and Control of Dynamic Economic Systems
CHOW • Econometric Analysis by Control Methods
COCHRAN • Sampling Techniques, *Third Edition*
COCHRAN and COX • Experimental Designs, *Second Edition*
CONOVER • Practical Nonparametric Statistics, *Second Edition*
CONOVER and IMAN • Introduction to Modern Business Statistics
CORNELL • Experiments with Mixtures: Designs, Models and The Analysis of Mixture Data
COX • Planning of Experiments
DANIEL • Biostatistics: A Foundation for Analysis in the Health Sciences, *Third Edition*
DANIEL • Applications of Statistics to Industrial Experimentation
DANIEL and WOOD • Fitting Equations to Data: Computer Analysis of Multifactor Data, *Second Edition*
DAVID • Order Statistics, *Second Edition*
DAVISON • Multidimensional Scaling
DEMING • Sample Design in Business Research
DILLON and GOLDSTEIN • Multivariate Analysis: Methods and Applications
DODGE and ROMIG • Sampling Inspection Tables, *Second Edition*
DOWDY and WEARDEN • Statistics for Research

continued on back

An Introduction to Multivariate Statistical Analysis

Also by T. W. Anderson

The Statistical Analysis of Time Series

John Wiley & Sons, 1971

An Introduction to Multivariate Statistical Analysis

Second Edition

T. W. ANDERSON
Professor of Statistics and Economics
Stanford University

JOHN WILEY & SONS
New York Chichester Brisbane Toronto Singapore

***Library of Congress Cataloging in Publication Data*:**

Anderson, T. W. (Theodore Wilbur), 1918–
An introduction to multivariate statistical analysis.

Bibliography: p.
Includes index.
1. Multivariate analysis. I. Title.

QA278.A516 1984   519.5′35   84-7334
ISBN 0-471-88987-3

Printed in the United States of America

10 9 8 7 6 5 4

To

DOROTHY

Preface to the Second Edition

Twenty-six years have passed since the first edition of this book was published. During that time great advances have been made in multivariate statistical analysis—particularly in the areas treated in that volume. This new edition purports to bring the original edition up to date by substantial revision, rewriting, and additions. The basic approach has been maintained, namely, a mathematically rigorous development of statistical methods for observations consisting of several measurements or characteristics of each subject and a study of their properties. The general outline of topics has been retained.

The method of maximum likelihood has been augmented by other considerations. In point estimation of the mean vector and covariance matrix alternatives to the maximum likelihood estimators that are better with respect to certain loss functions, such as Stein and Bayes estimators, have been introduced. In testing hypotheses likelihood ratio tests have been supplemented by other invariant procedures. New results on distributions and asymptotic distributions are given; some significance points are tabulated. Properties of these procedures, such as power functions, admissibility, unbiasedness, and monotonicity of power functions, are studied. Simultaneous confidence intervals for means and covariances are developed. A chapter on factor analysis replaces the chapter sketching miscellaneous results in the first edition. Some new topics, including simultaneous equations models and linear functional relationships, are introduced. Additional problems present further results.

It is impossible to cover all relevant material in this book; what seems most important has been included. For a comprehensive listing of papers until 1966 and books until 1970 the reader is referred to *A Bibliography of Multivariate Statistical Analysis* by Anderson, Das Gupta, and Styan (1972). Further references can be found in *Multivariate Analysis: A Selected and Abstracted Bibliography, 1957–1972* by Subrahmaniam and Subrahmaniam (1973).

I am in debt to many students, colleagues, and friends for their suggestions and assistance; they include Yasuo Amemiya, James Berger, Byoung-Seon Choi, Arthur Cohen, Margery Cruise, Somesh Das Gupta, Kai-Tai Fang, Gene Golub, Aaron Han, Takeshi Hayakawa, Jogi Henna, Huang Hsu, Fred Huffer, Mituaki Huzii, Jack Kiefer, Mark Knowles, Sue Leurgans, Alex McMillan, Masashi No, Ingram Olkin, Kartik Patel, Michael Perlman, Allen Sampson, Ashis Sen Gupta, Andrew Siegel, Charles Stein, Patrick Strout, Akimichi Takemura, Joe Verducci, Marlos Viana, and Y. Yajima. I was helped in preparing the manuscript by Dorothy Anderson, Alice Lundin, Amy Schwartz, and Pat Struse. Special thanks go to Johanne Thiffault and George P. H. Styan for their precise attention. Support was contributed by the Army Research Office, the National Science Foundation, the Office of Naval Research, and IBM Systems Research Institute.

Seven tables of significance points are given in Appendix B to facilitate carrying out test procedures. Tables 1, 5, and 7 are Tables 47, 50, and 53, respectively, of *Biometrika Tables for Statisticians*, Vol. 2, by E. S. Pearson and H. O. Hartley; permission of the Biometrika Trustees is hereby acknowledged. Table 2 is made up from three tables prepared by A. W. Davis and published in *Biometrika* (1970a), *Annals of the Institute of Statistical Mathematics* (1970b), and *Communications in Statistics, B. Simulation and Computation* (1980). Tables 3 and 4 are Tables 6.3 and 6.4, respectively, of *Concise Statistical Tables*, edited by Ziro Yamauti (1977) and published by the Japanese Standards Association; this book is a concise version of *Statistical Tables and Formulas with Computer Applications, JSA-1972*. Table 6 is Table 3 of *The Distribution of the Sphericity Test Criterion*, ARL 72-0154, by B. N. Nagarsenker and K. C. S. Pillai, Aerospace Research Laboratories (1972). The author is indebted to the authors and publishers listed above for permission to reproduce these tables.

T. W. ANDERSON

Stanford, California
June 1984

Preface to the First Edition

This book has been designed primarily as a text for a two-semester course in multivariate statistics. It is hoped that the book will also serve as an introduction to many topics in this area to statisticians who are not students and will be used as a reference by other statisticians.

For several years the book in the form of dittoed notes has been used in a two-semester sequence of graduate courses at Columbia University; the first six chapters constituted the text for the first semester, emphasizing correlation theory. It is assumed that the reader is familiar with the usual theory of univariate statistics, particularly methods based on the univariate normal distribution. A knowledge of matrix algebra is also a prerequisite; however, an appendix on this topic has been included.

It is hoped that the more basic and important topics are treated here, though to some extent the coverage is a matter of taste. Some of the more recent and advanced developments are only briefly touched on in the last chapter.

The method of maximum likelihood is used to a large extent. This leads to reasonable procedures; in some cases it can be proved that they are optimal. In many situations, however, the theory of desirable or optimum procedures is lacking.

Over the years this manuscript has been developed, a number of students and colleagues have been of considerable assistance. Allan Birnbaum, Harold Hotelling, Jacob Horowitz, Howard Levene, Ingram Olkin, Gobind Seth, Charles Stein, and Henry Teicher are to be mentioned particularly. Acknowledgments are also due to other members of the Graduate Mathematical

Statistics Society at Columbia University for aid in the preparation of the manuscript in dittoed form. The preparation of this manuscript was supported in part by the Office of Naval Research.

T. W. ANDERSON

Center for Advanced Study
in the Behavioral Sciences
Stanford, California
December 1957

Contents

CHAPTER 1
Introduction 1

1.1. Multivariate Statistical Analysis 1
1.2. The Multivariate Normal Distribution 3

CHAPTER 2
The Multivariate Normal Distribution 6

2.1. Introduction 6
2.2. Notions of Multivariate Distributions 7
2.3. The Multivariate Normal Distribution 14
2.4. The Distribution of Linear Combinations of Normally Distributed Variates; Independence of Variates; Marginal Distributions 24
2.5. Conditional Distributions and Multiple Correlation Coefficient 35
2.6. The Characteristic Function; Moments 43
Problems 50

CHAPTER 3
Estimation of the Mean Vector and the Covariance Matrix 59

3.1. Introduction 59
3.2. The Maximum Likelihood Estimators of the Mean Vector and the Covariance Matrix 60
3.3. The Distribution of the Sample Mean Vector; Inference Concerning the Mean When the Covariance Matrix Is Known 68

3.4. Theoretical Properties of Estimators of the Mean Vector 77
3.5. Improved Estimation of the Mean 86
Problems 96

CHAPTER 4
The Distributions and Uses of Sample Correlation Coefficients 102

4.1. Introduction 102
4.2. Correlation Coefficient of a Bivariate Sample 103
4.3. Partial Correlation Coefficients; Conditional Distributions 125
4.4. The Multiple Correlation Coefficient 134
Problems 149

CHAPTER 5
The Generalized T^2-Statistic 156

5.1. Introduction 156
5.2. Derivation of the Generalized T^2-Statistic and Its Distribution 157
5.3. Uses of the T^2-Statistic 164
5.4. The Distribution of T^2 Under Alternative Hypotheses; The Power Function 173
5.5. The Two-Sample Problem with Unequal Covariance Matrices 175
5.6. Some Optimal Properties of the T^2-Test 181
Problems 190

CHAPTER 6
Classification of Observations 195

6.1. The Problem of Classification 195
6.2. Standards of Good Classification 196
6.3. Procedures of Classification into One of Two Populations with Known Probability Distributions 199
6.4. Classification into One of Two Known Multivariate Normal Populations 204
6.5. Classification into One of Two Multivariate Normal Populations When the Parameters Are Estimated 208
6.6. Probabilities of Misclassification 217
6.7. Classification into One of Several Populations 224

6.8. Classification into One of Several Multivariate Normal Populations 228
6.9. An Example of Classification into One of Several Multivariate Normal Populations 231
6.10. Classification into One of Two Known Multivariate Normal Populations with Unequal Covariance Matrices 234
Problems 241

CHAPTER 7
The Distribution of the Sample Covariance Matrix and the Sample Generalized Variance 244

7.1. Introduction 244
7.2. The Wishart Distribution 245
7.3. Some Properties of the Wishart Distribution 252
7.4. Cochran's Theorem 257
7.5. The Generalized Variance 259
7.6. Distribution of the Set of Correlation Coefficients When the Population Covariance Matrix Is Diagonal 266
7.7. The Inverted Wishart Distribution and Bayes Estimation of the Covariance Matrix 268
7.8. Improved Estimation of the Covariance Matrix 273
Problems 279

CHAPTER 8
Testing the General Linear Hypothesis; Multivariate Analysis of Variance 285

8.1. Introduction 285
8.2. Estimators of Parameters in Multivariate Linear Regression 287
8.3. Likelihood Ratio Criteria for Testing Linear Hypotheses About Regression Coefficients 292
8.4. The Distribution of the Likelihood Ratio Criterion When the Hypothesis Is True 298
8.5. An Asymptotic Expansion of the Distribution of the Likelihood Ratio Criterion 311
8.6. Other Criteria for Testing the Linear Hypothesis 321
8.7. Tests of Hypotheses About Matrices of Regression Coefficients and Confidence Regions 333

8.8. Testing Equality of Means of Several Normal Distributions with Common Covariance Matrix 338
8.9. Multivariate Analysis of Variance 342
8.10. Some Optimal Properties of Tests 349
Problems 369

CHAPTER 9
Testing Independence of Sets of Variates 376

9.1. Introduction 376
9.2. The Likelihood Ratio Criterion for Testing Independence of Sets of Variates 376
9.3. The Distribution of the Likelihood Ratio Criterion When the Null Hypothesis Is True 381
9.4. An Asymptotic Expansion of the Distribution of the Likelihood Ratio Criterion 385
9.5. Other Criteria 387
9.6. Step-down Procedures 389
9.7. An Example 392
9.8. The Case of Two Sets of Variates 394
9.9. Admissibility of the Likelihood Ratio Test 397
9.10. Monotonicity of Power Functions of Tests of Independence of Sets 399
Problems 402

CHAPTER 10
Testing Hypotheses of Equality of Covariance Matrices and Equality of Mean Vectors and Covariance Matrices 404

10.1. Introduction 404
10.2. Criteria for Testing Equality of Several Covariance Matrices 405
10.3. Criteria for Testing That Several Normal Distributions Are Identical 408
10.4. Distributions of the Criteria 410
10.5. Asymptotic Expansions of the Distributions of the Criteria 419
10.6. The Case of Two Populations 422
10.7. Testing the Hypothesis That a Covariance Matrix Is Proportional to a Given Matrix; The Sphericity Test 427

10.8. Testing the Hypothesis That a Covariance Matrix Is Equal to a Given Matrix 434
10.9. Testing the Hypothesis That a Mean Vector and a Covariance Matrix Are Equal to a Given Vector and Matrix 440
10.10. Admissibility of Tests 443
Problems 446

CHAPTER 11
Principal Components 451

11.1. Introduction 451
11.2. Definition of Principal Components in the Population 452
11.3. Maximum Likelihood Estimators of the Principal Components and Their Variances 460
11.4. Computation of the Maximum Likelihood Estimates of the Principal Components 462
11.5. An Example 465
11.6. Statistical Inference 468
11.7. Testing Hypotheses about the Characteristic Roots of a Covariance Matrix 473
Problems 477

CHAPTER 12
Canonical Correlations and Canonical Variables 480

12.1. Introduction 480
12.2. Canonical Correlations and Variates in the Population 481
12.3. Estimation of Canonical Correlations and Variates 492
12.4. Statistical Inference 497
12.5. An Example 500
12.6. Linearly Related Expected Values 502
12.7. Simultaneous Equations Models 509
Problems 519

CHAPTER 13
The Distributions of Characteristic Roots and Vectors 521

13.1. Introduction 521
13.2. The Case of Two Wishart Matrices 522

13.3. The Case of One Nonsingular Wishart Matrix 532
13.4. Canonical Correlations 538
13.5. Asymptotic Distributions in the Case of One Wishart Matrix 540
13.6. Asymptotic Distributions in the Case of Two Wishart Matrices 544
Problems 548

CHAPTER 14
Factor Analysis 550

14.1. Introduction 550
14.2. The Model 551
14.3. Maximum Likelihood Estimators for Random Orthogonal Factors 557
14.4. Estimation for Fixed Factors 569
14.5. Factor Interpretation and Transformation 570
14.6. Estimation for Identification by Specified Zeros 574
14.7. Estimation of Factor Scores 575
Problems 576

APPENDIX A
Matrix Theory 579

A.1. Definition of a Matrix and Operations on Matrices 579
A.2. Characteristic Roots and Vectors 587
A.3. Partitioned Vectors and Matrices 591
A.4. Some Miscellaneous Results 596
A.5. Gram–Schmidt Orthogonalization and the Solution of Linear Equations 605

APPENDIX B
Tables 609

1. Wilks' Likelihood Criterion: Factors $C(p, m, M)$ to Adjust to χ^2_{pm} where $M = n - p + 1$ 609
2. Tables of Significance Points for the Lawley-Hotelling Trace Test 616
3. Tables of Significance Points for the Bartlett–Nanda–Pillai Trace Test 630
4. Tables of Significance Points for the Roy Maximum Root Test 634

5. Tables of Significance Points for the Modified Likelihood Ratio Test of Equality of Covariance Matrices Based on Equal Sample Sizes 638
6. Correction Factors for Significance Points for the Sphericity Test 639
7. Significance Points for the Modified Likelihood Ratio Test $\mathbf{\Sigma} = \mathbf{\Sigma}_0$ 641

References 643

Index 667

An Introduction to Multivariate Statistical Analysis

CHAPTER 1

Introduction

1.1. MULTIVARIATE STATISTICAL ANALYSIS

Multivariate statistical analysis is concerned with data that consist of sets of measurements on a number of individuals or objects. The sample data may be heights and weights of some individuals drawn randomly from a population of schoolchildren in a given city, or the statistical treatment may be made on a collection of measurements, such as lengths and widths of petals and lengths and widths of sepals of iris plants taken from two species, or one may study the scores on batteries of mental tests administered to a number of students.

The measurements made on a single individual can be assembled into a column vector. We think of the entire vector as an observation from a multivariate population or distribution. When the individual is drawn randomly, we consider the vector as a random vector with a distribution or probability law describing that population. The set of observations on all individuals in a sample constitute a sample of vectors, and the vectors set side by side make up the matrix of observations.[†] The data to be analyzed then are thought of as displayed in a matrix or in several matrices.

We shall see that it is helpful in visualizing the data and understanding the methods to think of each observation vector as constituting a point in a Euclidean space, each coordinate corresponding to a measurement or variable. Indeed, an early step in the statistical analysis is plotting the data; since most statisticians are limited to two-dimensional plots, two coordinates of the observation are plotted in turn.

[†] When data are listed on paper by individual, it is natural to print the measurements on one individual as a row of the table; then one individual corresponds to a *row* vector. Since we prefer to operate algebraically with column vectors, we have chosen to treat observations in terms of *column* vectors. (In practice, the basic data set may well be on cards, tapes, or disks.)

Characteristics of a univariate distribution of essential interest are the mean as a measure of location and the standard deviation as a measure of variability; similarly the mean and standard deviation of a univariate sample are important summary measures. In multivariate analysis, the means and variances of the separate measurements—for distributions and for samples—have corresponding relevance. An essential aspect, however, of multivariate analysis is the dependence between the different variables. The dependence between two variables may involve the covariance between them, that is, the average products of their deviations from their respective means. The covariance standardized by the corresponding standard deviations is the correlation coefficient; it serves as a measure of degree of dependence. A set of summary statistics is the mean vector (consisting of the univariate means) and the covariance matrix (consisting of the univariate variances and bivariate covariances). An alternative set of summary statistics with the same information is the mean vector, the set of standard deviations, and the correlation matrix. Similar parameter quantities describe location, variability, and dependence in the population or for a probability distribution. The multivariate *normal* distribution is completely determined by its mean vector and covariance matrix, and the sample mean vector and covariance matrix constitute a sufficient set of statistics.

The measurement and analysis of dependence between variables, between sets of variables, and between variables and sets of variables, are fundamental to multivariate analysis. The multiple correlation coefficient is an extension of the notion of correlation to the relationship of one variable to a set of variables. The partial correlation coefficient is a measure of dependence between two variables when the effects of other correlated variables have been removed. The various correlation coefficients computed from samples are used to estimate corresponding correlation coefficients of distributions. In this book tests of hypotheses of independence are developed. The properties of the estimators and test procedures are studied for sampling from the multivariate normal distribution.

A number of statistical problems arising in multivariate populations are straightforward analogs of problems arising in univariate populations; the suitable methods for handling these problems are similarly related. For example, in the univariate case we may wish to test the hypothesis that the mean of a variable is zero; in the multivariate case we may wish to test the hypothesis that the vector of the means of several variables is the zero vector. The analog of the Student t-test for the first hypothesis is the generalized T^2-test. The analysis of variance of a single variable is adapted to vector observations; in regression analysis, the dependent quantity may be a vector variable. A comparison of variances is generalized into a comparison of covariance matrices.

The test procedures of univariate statistics are generalized to the multivariate case in such ways that the dependence between variables is taken into account. These methods may not depend on the coordinate system; in other terms, the procedures are invariant with respect to linear transformations that leave the null hypothesis invariant. In some problems there may be families of tests that are invariant; then choices must be made. Optimal properties of the tests are considered.

For some other purposes, however, it may be important to select a coordinate system so that the variates have desired statistical properties. One might say that they involve characterizations of inherent properties of normal distributions and of samples. These are closely related to the algebraic problems of canonical forms of matrices. An example is finding the normalized linear combination of variables with maximum or minimum variance (finding principal components); this amounts to finding a rotation of axes that carries the covariance matrix to diagonal form. Another example is characterizing the dependence between two sets of variates (finding canonical correlations). These problems involve the characteristic roots and vectors of various matrices. The statistical properties of the corresponding sample quantities are treated.

Some statistical problems arise in models in which means and covariances are restricted. Factor analysis may be based on a model with a (population) covariance matrix that is the sum of a positive definite diagonal matrix and a positive semidefinite matrix of low rank; linear structural relationships may have a similar formulation. The simultaneous equations system of econometrics is another example of a special model.

1.2. THE MULTIVARIATE NORMAL DISTRIBUTION

The statistical methods treated in this book can be developed and evaluated in the context of the multivariate normal distribution, though many of the procedures are useful and effective when the distribution sampled is not normal. A major reason for basing statistical analysis on the normal distribution is that this probabilistic model approximates well the distribution of continuous measurements in many sampled populations. In fact, most of the methods and theory have been developed to serve statistical analysis of data. Mathematicians such as Adrian (1808), Laplace (1811), Plana (1813), Gauss (1823), and Bravais (1846) studied the bivariate normal density. Francis Galton, the geneticist, introduced the ideas of correlation, regression, and homoscedasticity in the study of pairs of measurements, one made on a parent and one in an offspring. [See, e.g., Galton (1889).] He enunciated the theory of the multivariate normal distribution as generalizations of observed properties of samples.

Karl Pearson and others carried on the development of the theory and use of different kinds of correlation coefficients† for studying problems in genetics, biology, and other fields. R. A. Fisher further developed methods for agriculture, botany, and anthropology, including the discriminant function for classification problems. In another direction, analysis of scores on mental tests led to a theory, including "factor analysis," the sampling theory of which is based on the normal distribution. In these cases, as well as in agricultural experiments, in engineering problems, in certain economic problems, and in other fields, the multivariate normal distributions have been found to be sufficiently close approximations to the populations so that statistical analyses based on these models are justified.

The univariate normal distribution arises frequently because the effect studied is the sum of many independent random effects. Similarly, the multivariate normal distribution often occurs because the multiple measurements are sums of small independent effects. Just as the central limit theorem leads to the univariate normal distribution for single variables, so does the general central limit theorem for several variables lead to the multivariate normal distribution.

Statistical theory based on the normal distribution has the advantage that the multivariate methods based on it are extensively developed and can be studied in an organized and systematic way. This is due not only to the need for such methods because they are of practical use, but also to the fact that normal theory is amenable to exact mathematical treatment. The suitable methods of analysis are mainly based on standard operations of matrix algebra; the distributions of many statistics involved can be obtained exactly or at least characterized; and in many cases optimum properties of procedures can be deduced.

The point of view in this book is to state problems of inference in terms of the multivariate normal distributions, develop efficient and often optimum

† For a detailed study of the development of the ideas of correlation, see Walker (1931).

methods in this context, and evaluate significance and confidence levels in these terms. This approach gives coherence and rigor to the exposition, but, by its very nature, cannot exhaust consideration of multivariate statistical analysis. The procedures are appropriate to many nonnormal distributions, but their adequacy may be open to question. Roughly speaking, inferences about means are robust because of the operation of the central limit theorem, but inferences about covariances are sensitive to normality, the variability of sample covariances depending on fourth-order moments. More robust methods are being developed that are suitable for distributions with "long tails." Nonparametric techniques, in particular the bootstrap, are available when nothing is known about the underlying distributions. Space does not permit inclusion of these topics as well as other considerations of "data analysis," such as treatment of outliers and transformations of variables to approximate normality and homoscedasticity.

Modern computers make multivariate methods feasible when large numbers of variables are investigated. Because hardware and software development is so explosive and programs require specialized knowledge, we are content to make a few remarks here and there about computation. Packages of statistical programs are available for most of the methods.

CHAPTER 2

The Multivariate Normal Distribution

2.1. INTRODUCTION

In this chapter we discuss the multivariate normal distribution and some of its properties. In Section 2.2 are considered the fundamental notions of multivariate distributions: the definition by means of multivariate density functions, marginal distributions, conditional distributions, expected values, and moments. In Section 2.3 the multivariate normal distribution is defined; the parameters are shown to be the means, variances, and covariances or the means, variances, and correlations of the components of the random vector. In Section 2.4 it is shown that linear combinations of normal variables are normally distributed and hence that marginal distributions are normal. In Section 2.5 we see that conditional distributions are also normal with means that are linear functions of the conditioning variables; the coefficients are regression coefficients. The variances, covariances, and correlations—called partial correlations—are constants. The multiple correlation coefficient is the maximum correlation between a scalar random variable and linear combination of other random variables; it is a measure of association between one variable and a set of others. The fact that marginal and conditional distributions of normal distributions are normal makes the treatment of this family of distributions coherent. In Section 2.6 the characteristic function, moments, and cumulants are discussed.

2.2. NOTIONS OF MULTIVARIATE DISTRIBUTIONS

2.2.1. Joint Distributions

In this section we shall consider the notions of joint distributions of several variables, derived marginal distributions of subsets of variables, and derived conditional distributions. First consider the case of two (real) random variables† X and Y. Probabilities of events defined in terms of these variables can be obtained by operations involving the *cumulative distribution function* (abbreviated as cdf),

$$F(x, y) = \Pr\{X \le x, Y \le y\}, \tag{1}$$

defined for every pair of real numbers (x, y). We are interested in cases where $F(x, y)$ is absolutely continuous; this means that the following partial derivative exists almost everywhere:

$$\frac{\partial^2 F(x, y)}{\partial x\, \partial y} = f(x, y), \tag{2}$$

and

$$F(x, y) = \int_{-\infty}^{y} \int_{-\infty}^{x} f(u, v)\, du\, dv. \tag{3}$$

The nonnegative function $f(x, y)$ is called the *density* of X and Y. The pair of random variables (X, Y) defines a random point in a plane. The probability that (X, Y) falls in a rectangle is

$$\begin{aligned} (4)\quad & \Pr\{x \le X \le x + \Delta x, y \le Y \le y + \Delta y\} \\ &= F(x + \Delta x, y + \Delta y) - F(x + \Delta x, y) - F(x, y + \Delta y) + F(x, y) \\ &= \int_{y}^{y+\Delta y} \int_{x}^{x+\Delta x} f(u, v)\, du\, dv \end{aligned}$$

$(\Delta x > 0, \Delta y > 0)$. The probability of the random point (X, Y) falling in any set E for which the following integral is defined (that is, any measurable set

† In Chapter 2 we shall distinguish between random variables and running variables by use of capital and lower-case letters, respectively. In later chapters we may be unable to hold to this convention because of other complications of notation.

E) is

$$\Pr\{(X,Y) \in E\} = \int\int_E f(x,y)\,dx\,dy. \tag{5}$$

This follows from the definition of the integral [as the limit of sums of the sort of (4)]. If $f(x, y)$ is continuous in both variables, the *probability element* $f(x, y)\Delta y \Delta x$ is approximately the probability that X falls between x and $x + \Delta x$ and Y falls between y and $y + \Delta y$ for

$$\begin{aligned} \Pr\{x \le X \le x + \Delta x, y \le Y \le y + \Delta y\} &= \int_y^{y+\Delta y} \int_x^{x+\Delta x} f(u,v)\,du\,dv \\ &= f(x_0, y_0)\,\Delta x\,\Delta y \end{aligned} \tag{6}$$

for some x_0, y_0 $(x \le x_0 \le x + \Delta x, y \le y_0 \le y + \Delta y)$ by the mean value theorem of calculus. Since $f(u, v)$ is continuous, (6) is approximately $f(x, y)\,\Delta x\,\Delta y$. In fact,

$$\begin{aligned} \lim_{\substack{\Delta x \to 0 \\ \Delta y \to 0}} \frac{1}{\Delta x\,\Delta y} |\Pr\{x \le X \le x + \Delta x, y \le Y \le y + \Delta y\} \\ - f(x,y)\,\Delta x\,\Delta y| = 0. \end{aligned} \tag{7}$$

Now we consider the case of p random variables $X_1, X_2, \dots, X_p$. The cdf is

$$F(x_1, \dots, x_p) = \Pr\{X_1 \le x_1, \dots, X_p \le x_p\} \tag{8}$$

defined for every set of real numbers $x_1, \dots, x_p$. The density function, if $F(x_1, \dots, x_p)$ is absolutely continuous, is

$$\frac{\partial^p F(x_1, \dots, x_p)}{\partial x_1 \cdots \partial x_p} = f(x_1, \dots, x_p) \tag{9}$$

(almost everywhere) and

$$F(x_1, \dots, x_p) = \int_{-\infty}^{x_p} \cdots \int_{-\infty}^{x_1} f(u_1, \dots, u_p)\,du_1 \cdots du_p. \tag{10}$$

The probability of falling in any (measurable) set R in the p-dimensional

Euclidean space is

$$\Pr\{(X_1,\dots,X_p)\in R\}=\int\underset{R}{\cdots}\int f(x_1,\dots,x_p)\,dx_1\cdots dx_p. \tag{11}$$

The probability element $f(x_1,\dots,x_p)\,\Delta x_1\cdots\Delta x_p$ is approximately the probability $\Pr\{x_1\le X_1\le x_1+\Delta x_1,\dots,x_p\le X_p\le x_p+\Delta x_p\}$ if $f(x_1,\dots,x_p)$ is continuous. The joint moments are defined as[†]

$$\mathscr{E}X_1^{h_1}\cdots X_p^{h_p}=\int_{-\infty}^{\infty}\cdots\int_{-\infty}^{\infty}x_1^{h_1}\cdots x_p^{h_p}f(x_1,\dots,x_p)\,dx_1\cdots dx_p. \tag{12}$$

2.2.2. Marginal Distributions

Given the cdf of two random variables X, Y as being $F(x,y)$, the marginal cdf of X is

$$\begin{aligned}\Pr\{X\le x\}&=\Pr\{X\le x, Y\le\infty\}\\&=F(x,\infty).\end{aligned} \tag{13}$$

Let this be $F(x)$. Clearly

$$F(x)=\int_{-\infty}^{x}\int_{-\infty}^{\infty}f(u,v)\,dv\,du. \tag{14}$$

We call

$$\int_{-\infty}^{\infty}f(u,v)\,dv=f(u), \tag{15}$$

say, the *marginal density* of X. Then (14) is

$$F(x)=\int_{-\infty}^{x}f(u)\,du. \tag{16}$$

In a similar fashion we define $G(y)$, the marginal cdf of Y, and $g(y)$, the marginal density of Y.

Now we turn to the general case. Given $F(x_1,\dots,x_p)$ as the cdf of $X_1,\dots,X_p$, we wish to find the marginal cdf of some of $X_1,\dots,X_p$, say, of $X_1,\dots,X_r$ $(r<p)$. It is

$$\begin{aligned}\Pr\{X_1\le x_1,\dots,X_r\le x_r\}&\\&=\Pr\{X_1\le x_1,\dots,X_r\le x_r,X_{r+1}\le\infty,\dots,X_p\le\infty\}\\&=F(x_1,\dots,x_r,\infty,\dots,\infty).\end{aligned} \tag{17}$$

[†] $\mathscr{E}$ will be used to denote *mathematical expectation*.

The marginal density of $X_1, \ldots, X_r$ is

$$\int_{-\infty}^{\infty} \cdots \int_{-\infty}^{\infty} f(u_1, \ldots, u_p)\, du_{r+1} \cdots du_p. \tag{18}$$

The marginal distribution and density of any other subset of $X_1, \ldots, X_p$ are obtained in the obviously similar fashion.

The joint moments of a subset of variates can be computed from the marginal distribution; for example,

(19)

$$\begin{aligned}\mathscr{E} X_1^{h_1} \cdots X_r^{h_r} &= \mathscr{E} X_1^{h_1} \cdots X_r^{h_r} X_{r+1}^0 \cdots X_p^0 \\ &= \int_{-\infty}^{\infty} \cdots \int_{-\infty}^{\infty} x_1^{h_1} \cdots x_r^{h_r} f(x_1, \ldots, x_p)\, dx_1 \cdots dx_p \\ &= \int_{-\infty}^{\infty} \cdots \int_{-\infty}^{\infty} x_1^{h_1} \cdots x_r^{h_r} \\ &\qquad \cdot \left[\int_{-\infty}^{\infty} \cdots \int_{-\infty}^{\infty} f(x_1, \ldots, x_p)\, dx_{r+1} \cdots dx_p \right] dx_1 \cdots dx_r.\end{aligned}$$

2.2.3. Statistical Independence

Two random variables X, Y with cdf $F(x, y)$ are said to be *independent* if

$$F(x, y) = F(x)G(y), \tag{20}$$

where $F(x)$ is the marginal cdf of X and $G(y)$ is the marginal cdf of Y. This implies that the density of X, Y is

$$\begin{aligned} f(x, y) &= \frac{\partial^2 F(x, y)}{\partial x\, \partial y} = \frac{\partial^2 F(x)G(y)}{\partial x\, \partial y} \\ &= \frac{dF(x)}{dx} \frac{dG(y)}{dy} \\ &= f(x)g(y). \end{aligned} \tag{21}$$

Conversely, if $f(x, y) = f(x)g(y)$, then

$$\begin{aligned} F(x, y) &= \int_{-\infty}^{y} \int_{-\infty}^{x} f(u, v)\, du\, dv = \int_{-\infty}^{y} \int_{-\infty}^{x} f(u)g(v)\, du\, dv \\ &= \int_{-\infty}^{x} f(u)\, du \int_{-\infty}^{y} g(v)\, dv = F(x)G(y). \end{aligned} \tag{22}$$

Thus an equivalent definition of independence in the case of densities existing is that $f(x, y) = f(x)g(y)$. To see the implications of statistical independence, given any $x_1 < x_2$, $y_1 < y_2$, we consider the probability

$$\begin{aligned}(23)\quad \Pr\{x_1 \le X \le x_2, y_1 \le Y \le y_2\} &= \int_{y_1}^{y_2}\int_{x_1}^{x_2} f(u,v)\,du\,dv = \int_{x_1}^{x_2} f(u)\,du \int_{y_1}^{y_2} g(v)\,dv \\ &= \Pr\{x_1 \le X \le x_2\}\Pr\{y_1 \le Y \le y_2\}.\end{aligned}$$

The probability of X falling in a given interval and Y falling in a given interval is the product of the probability of X falling in the interval and the probability of Y falling in the other interval.

If the cdf of $X_1, \dots, X_p$ is $F(x_1, \dots, x_p)$, the set of random variables is said to be *mutually independent* if

$$(24)\qquad F(x_1, \dots, x_p) = F_1(x_1) \cdots F_p(x_p),$$

where $F_i(x_i)$ is the marginal cdf of X_i, $i = 1, \dots, p$. The set $X_1, \dots, X_r$ is said to be independent of the set $X_{r+1}, \dots, X_p$ if

$$(25)\quad F(x_1, \dots, x_p) = F(x_1, \dots, x_r, \infty, \dots, \infty) \cdot F(\infty, \dots, \infty, x_{r+1}, \dots, x_p).$$

One result of independence is that joint moments factor. For example, if $X_1, \dots, X_p$ are mutually independent, then

$$\begin{aligned}(26)\quad \mathscr{E}X_1^{h_1} \cdots X_p^{h_p} &= \int_{-\infty}^{\infty} \cdots \int_{-\infty}^{\infty} x_1^{h_1} \cdots x_p^{h_p} f_1(x_1) \cdots f_p(x_p)\,dx_1 \cdots dx_p \\ &= \prod_{i=1}^{p} \int_{-\infty}^{\infty} x_i^{h_i} f_i(x_i)\,dx_i \\ &= \prod_{i=1}^{p} \{\mathscr{E}X_i^{h_i}\}.\end{aligned}$$

2.2.4. Conditional Distributions

If A and B are two events such that the probability of A and B occurring simultaneously is $P(AB)$ and the probability of B occurring is $P(B) > 0$, then the conditional probability of A occurring given that B has occurred is

$P(AB)/P(B)$. Suppose the event A is X falling in the interval $[x_1, x_2]$ and the event B is Y falling in $[y_1, y_2]$. Then the conditional probability that X fall in $[x_1, x_2]$, given that Y falls in $[y_1, y_2]$, is

$$\text{(27)} \quad \Pr\{x_1 \le X \le x_2 | y_1 \le Y \le y_2\} = \frac{\Pr\{x_1 \le X \le x_2, y_1 \le Y \le y_2\}}{\Pr\{y_1 \le Y \le y_2\}}$$

$$= \frac{\int_{x_1}^{x_2} \int_{y_1}^{y_2} f(u, v)\, dv\, du}{\int_{y_1}^{y_2} g(v)\, dv}.$$

Now let $y_1 = y$, $y_2 = y + \Delta y$. Then for a continuous density

$$\text{(28)} \qquad \int_{y}^{y+\Delta y} g(v)\, dv = g(y^*)\, \Delta y,$$

where $y \le y^* \le y + \Delta y$. Also

$$\text{(29)} \qquad \int_{y}^{y+\Delta y} f(u, v)\, dv = f[u, y^*(u)]\, \Delta y,$$

where $y \le y^*(u) \le y + \Delta y$. Therefore,

$$\text{(30)} \qquad \Pr\{x_1 \le X \le x_2 | y \le Y \le y + \Delta y\} = \int_{x_1}^{x_2} \frac{f[u, y^*(u)]}{g(y^*)}\, du.$$

It will be noticed that for fixed y and Δy (> 0), the integrand of (30) behaves as a univariate density function. Now for y such that $g(y) > 0$, we define $\Pr\{x_1 \le X \le x_2 | Y = y\}$, the probability that X lies between x_1 and x_2, given that Y is y, as the limit of (30) as $\Delta y \to 0$. Thus

$$\text{(31)} \qquad \Pr\{x_1 \le X \le x_2 | Y = y\} = \int_{x_1}^{x_2} f(u|y)\, du,$$

where $f(u|y) = f(u, y)/g(y)$. For given y, $f(u|y)$ is a density function and is called the *conditional density* of X given y. We note that if X and Y are independent, $f(x|y) = f(x)$.

In the general case of $X_1, \dots, X_p$ with cdf $F(x_1, \dots, x_p)$, the conditional density of $X_1, \dots, X_r$, given $X_{r+1} = x_{r+1}, \dots, X_p = x_p$, is

$$\text{(32)} \qquad \frac{f(x_1, \dots, x_p)}{\int_{-\infty}^{\infty} \cdots \int_{-\infty}^{\infty} f(u_1, \dots, u_r, x_{r+1}, \dots, x_p)\, du_1 \cdots du_r}.$$

For a more general discussion of conditional probabilities, the reader is referred to Chung (1974), Kolmogorov (1950), Loève (1977), (1978), and Neveu (1965).

2.2.5. Transformation of Variables

Let the density of $X_1, \ldots, X_p$ be $f(x_1, \ldots, x_p)$. Consider the p real-valued functions

$$y_i = y_i(x_1, \ldots, x_p), \qquad i = 1, \ldots, p. \tag{33}$$

We assume that the transformation from the x-space to the y-space is one-to-one;[†] the inverse transformation is

$$x_i = x_i(y_1, \ldots, y_p), \qquad i = 1, \ldots, p. \tag{34}$$

Let the random variables $Y_1, \ldots, Y_p$ be defined by

$$Y_i = y_i(X_1, \ldots, X_p), \qquad i = 1, \ldots, p. \tag{35}$$

Then the density of $Y_1, \ldots, Y_p$ is

$$g(y_1, \ldots, y_p) = f\left[x_1(y_1, \ldots, y_p), \ldots, x_p(y_1, \ldots, y_p)\right] J(y_1, \ldots, y_p), \tag{36}$$

where $J(y_1, \ldots, y_p)$ is the Jacobian

$$J(y_1, \ldots, y_p) = \operatorname{mod} \begin{vmatrix} \frac{\partial x_1}{\partial y_1} & \frac{\partial x_1}{\partial y_2} & \cdots & \frac{\partial x_1}{\partial y_p} \\ \frac{\partial x_2}{\partial y_1} & \frac{\partial x_2}{\partial y_2} & \cdots & \frac{\partial x_2}{\partial y_p} \\ \vdots & \vdots & & \vdots \\ \frac{\partial x_p}{\partial y_1} & \frac{\partial x_p}{\partial y_2} & \cdots & \frac{\partial x_p}{\partial y_p} \end{vmatrix}. \tag{37}$$

We assume the derivatives exist, and "mod" means modulus or absolute value

[†]More precisely, we assume this is true for the part of the x-space for which $f(x_1, \ldots, x_p)$ is positive.

of the expression following it. The probability that $(X_1, \dots, X_p)$ falls in a region R is given by (11); the probability that $(Y_1, \dots, Y_p)$ falls in a region S is

$$\Pr\{(Y_1, \dots, Y_p) \in S\} = \int \underset{S}{\cdots} \int g(y_1, \dots, y_p)\, dy_1 \cdots dy_p. \tag{38}$$

If S is the transform of R, that is, if each point of R transforms by (33) into a point of S and if each point of S transforms into R by (34), then (11) is equal to (38) by the usual theory of transformation of multiple integrals. From this follows the assertion that (36) is the density of $Y_1, \dots, Y_p$.

2.3. THE MULTIVARIATE NORMAL DISTRIBUTION

The univariate normal density function can be written

$$ke^{-\frac{1}{2}\alpha(x-\beta)^2} = ke^{-\frac{1}{2}(x-\beta)\alpha(x-\beta)}, \tag{1}$$

where α is positive and k is chosen so that the integral of (1) over the entire x-axis is unity. The density function of a multivariate normal distribution of $X_1, \dots, X_p$ has an analogous form. The scalar variable x is replaced by a vector

$$\boldsymbol{x} = \begin{pmatrix} x_1 \\ \vdots \\ x_p \end{pmatrix}; \tag{2}$$

the scalar constant β is replaced by a vector

$$\boldsymbol{b} = \begin{pmatrix} b_1 \\ \vdots \\ b_p \end{pmatrix}; \tag{3}$$

and the positive constant α is replaced by a positive definite (symmetric) matrix

$$\boldsymbol{A} = \begin{pmatrix} a_{11} & a_{12} & \cdots & a_{1p} \\ a_{21} & a_{22} & \cdots & a_{2p} \\ \vdots & \vdots & & \vdots \\ a_{p1} & a_{p2} & \cdots & a_{pp} \end{pmatrix}. \tag{4}$$

The square $\alpha(x-\beta)^2 = (x-\beta)\alpha(x-\beta)$ is replaced by the quadratic form

$$(x-b)'A(x-b) = \sum_{i,j=1}^{p} a_{ij}(x_i - b_i)(x_j - b_j). \tag{5}$$

Thus the density function of a p-variate normal distribution is

$$f(x_1, \dots, x_p) = Ke^{-\frac{1}{2}(x-b)'A(x-b)}, \tag{6}$$

where K (> 0) is chosen so that the integral over the entire p-dimensional Euclidean space of $x_1, \dots, x_p$ is unity.

Written in matrix notation, the similarity of the multivariate normal density (6) to the univariate density (1) is clear. Throughout this book we shall use matrix notation and operations. The reader is referred to Appendix A for a review of matrix theory and for definitions of our notation for matrix operations.

We observe that $f(x_1, \dots, x_p)$ is nonnegative. Since A is positive definite,

$$(x-b)'A(x-b) \geq 0, \tag{7}$$

and therefore the density is bounded; that is,

$$f(x_1, \dots, x_p) \leq K. \tag{8}$$

Now let us determine K so that the integral of (6) over the p-dimensional space is one. We shall evaluate

$$K^* = \int_{-\infty}^{\infty} \cdots \int_{-\infty}^{\infty} e^{-\frac{1}{2}(x-b)'A(x-b)}\, dx_p \cdots dx_1. \tag{9}$$

We use the fact (see Corollary A.1.6 of Appendix A) that if A is positive definite, there exists a nonsingular matrix C such that

$$C'AC = I, \tag{10}$$

where I denotes the identity and C' the transpose of C. Let

$$x - b = Cy, \tag{11}$$

where

$$y = \begin{pmatrix} y_1 \\ \vdots \\ y_p \end{pmatrix}. \tag{12}$$

Then

$$(13)\qquad (\boldsymbol{x}-\boldsymbol{b})'\boldsymbol{A}(\boldsymbol{x}-\boldsymbol{b}) = \boldsymbol{y}'\boldsymbol{C}'\boldsymbol{A}\boldsymbol{C}\boldsymbol{y} = \boldsymbol{y}'\boldsymbol{y}.$$

The Jacobian of the transformation is

$$(14)\qquad J = \operatorname{mod}|\boldsymbol{C}|,$$

where "$\operatorname{mod}|\boldsymbol{C}|$" indicates the absolute value of the determinant of $\boldsymbol{C}$. Thus (9) becomes

$$(15)\qquad K^* = \operatorname{mod}|\boldsymbol{C}| \int_{-\infty}^{\infty} \cdots \int_{-\infty}^{\infty} e^{-\frac{1}{2}\boldsymbol{y}'\boldsymbol{y}}\, dy_p \cdots dy_1.$$

We have

$$(16)\qquad e^{-\frac{1}{2}\boldsymbol{y}'\boldsymbol{y}} = \exp\left(-\frac{1}{2}\sum_{i=1}^{p} y_i^2\right) = \prod_{i=1}^{p} e^{-\frac{1}{2}y_i^2},$$

where $\exp(z) = e^z$. We can write (15) as

$$(17)\qquad \begin{aligned} K^* &= \operatorname{mod}|\boldsymbol{C}| \int_{-\infty}^{\infty} \cdots \int_{-\infty}^{\infty} e^{-\frac{1}{2}y_1^2} \cdots e^{-\frac{1}{2}y_p^2}\, dy_p \cdots dy_1 \\ &= \operatorname{mod}|\boldsymbol{C}| \prod_{i=1}^{p} \left\{\int_{-\infty}^{\infty} e^{-\frac{1}{2}y_i^2}\, dy_i\right\} \\ &= \operatorname{mod}|\boldsymbol{C}| \prod_{i=1}^{p} \{\sqrt{2\pi}\} \\ &= \operatorname{mod}|\boldsymbol{C}| (2\pi)^{\frac{1}{2}p} \end{aligned}$$

by virtue of

$$(18)\qquad \frac{1}{\sqrt{2\pi}} \int_{-\infty}^{\infty} e^{-\frac{1}{2}t^2}\, dt = 1.$$

Corresponding to (10) is the determinantal equation

$$(19)\qquad |\boldsymbol{C}'| \cdot |\boldsymbol{A}| \cdot |\boldsymbol{C}| = |\boldsymbol{I}|.$$

Since

$$|C'| = |C|, \tag{20}$$

and since $|I| = 1$, we deduce from (19) that

$$\operatorname{mod}|C| = 1/\sqrt{|A|}\,. \tag{21}$$

Thus

$$K = 1/K^* = \sqrt{|A|}\,(2\pi)^{-\frac{1}{2}p}. \tag{22}$$

The normal density function is

$$\frac{\sqrt{|A|}}{(2\pi)^{\frac{1}{2}p}} e^{-\frac{1}{2}(x-b)'A(x-b)}. \tag{23}$$

We shall now show the significance of b and A by finding the first and second moments of $X_1, \dots, X_p$. It will be convenient to consider these random variables as constituting a random vector

$$X = \begin{pmatrix} X_1 \\ \vdots \\ X_p \end{pmatrix}. \tag{24}$$

We shall define generally a random matrix and the expected value of a random matrix; a random vector is considered as a special case of a random matrix with one column.

DEFINITION 2.3.1. *A random matrix Z is a matrix*

$$Z = (Z_{gh}), \qquad g = 1, \dots, m; \ h = 1, \dots, n, \tag{25}$$

of random variables $Z_{11}, \dots, Z_{mn}$.

If the random variables $Z_{11}, \dots, Z_{mn}$ can take on only a finite number of values, the random matrix Z can be one of a finite number of matrices, say $Z(1), \dots, Z(q)$. If the probability of $Z = Z(i)$ is p_i, then we should like to define $\mathscr{E}Z$ as $\sum_{i=1}^{q} Z(i) p_i$. It is clear then that $\mathscr{E}Z = (\mathscr{E}Z_{gh})$. If the random variables $Z_{11}, \ldots, Z_{mn}$ have a joint density, then operating with Riemann sums we can define $\mathscr{E}\boldsymbol{Z}$ as the limit (if the limit exists) of approximating sums of the kind occurring in the discrete case; then again $\mathscr{E}\boldsymbol{Z} = (\mathscr{E}Z_{gh})$. Therefore, in general we shall use the following definition:

DEFINITION 2.3.2. *The expected value of a random matrix* $\boldsymbol{Z}$ *is*

$$\mathscr{E}\boldsymbol{Z} = \left(\mathscr{E}Z_{gh}\right), \qquad g = 1, \ldots, m;\ h = 1, \ldots, n. \tag{26}$$

In particular if $\boldsymbol{Z}$ is $\boldsymbol{X}$ defined by (24), the expected value

$$\mathscr{E}\boldsymbol{X} = \begin{pmatrix} \mathscr{E}X_1 \\ \vdots \\ \mathscr{E}X_p \end{pmatrix} \tag{27}$$

is the *mean* or *mean vector* of $\boldsymbol{X}$. We shall usually denote this mean vector by $\boldsymbol{\mu}$. If $\boldsymbol{Z}$ is $(\boldsymbol{X} - \boldsymbol{\mu})(\boldsymbol{X} - \boldsymbol{\mu})'$, the expected value is

$$\mathscr{C}(\boldsymbol{X}) = \mathscr{E}(\boldsymbol{X} - \boldsymbol{\mu})(\boldsymbol{X} - \boldsymbol{\mu})' = \left[\mathscr{E}(X_i - \mu_i)(X_j - \mu_j)\right], \tag{28}$$

the *covariance* or *covariance matrix* of $\boldsymbol{X}$. The ith diagonal element of this matrix, $\mathscr{E}(X_i - \mu_i)^2$, is the *variance* of X_i, and the i, jth off-diagonal element, $\mathscr{E}(X_i - \mu_i)(X_j - \mu_j)$, is the *covariance* of X_i and X_j, $i \neq j$. We shall usually denote the covariance matrix by $\boldsymbol{\Sigma}$. Note that

$$\mathscr{C}(\boldsymbol{X}) = \mathscr{E}\left(\boldsymbol{X}\boldsymbol{X}' - \boldsymbol{\mu}\boldsymbol{X}' - \boldsymbol{X}\boldsymbol{\mu}' + \boldsymbol{\mu}\boldsymbol{\mu}'\right) = \mathscr{E}\boldsymbol{X}\boldsymbol{X}' - \boldsymbol{\mu}\boldsymbol{\mu}'. \tag{29}$$

The operation of taking the expected value of a random matrix (or vector) satisfies certain rules which we can summarize in the following lemma:

LEMMA 2.3.1. *If* $\boldsymbol{Z}$ *is an* $m \times n$ *random matrix,* $\boldsymbol{D}$ *is an* $l \times m$ *real matrix,* $\boldsymbol{E}$ *is an* $n \times q$ *real matrix, and* $\boldsymbol{F}$ *is an* $l \times q$ *real matrix, then*

$$\mathscr{E}(\boldsymbol{D}\boldsymbol{Z}\boldsymbol{E} + \boldsymbol{F}) = \boldsymbol{D}(\mathscr{E}\boldsymbol{Z})\boldsymbol{E} + \boldsymbol{F}. \tag{30}$$

PROOF. The element in the ith row and jth column of $\mathscr{E}(\boldsymbol{D}\boldsymbol{Z}\boldsymbol{E} + \boldsymbol{F})$ is

$$\mathscr{E}\left(\sum_{h,g} d_{ih} Z_{hg} e_{gj} + f_{ij}\right) = \sum_{h,g} d_{ih}\left(\mathscr{E}Z_{hg}\right) e_{gj} + f_{ij}, \tag{31}$$

which is the element in the ith row and jth column of $\boldsymbol{D}(\mathscr{E}\boldsymbol{Z})\boldsymbol{E} + \boldsymbol{F}$. Q.E.D.

LEMMA 2.3.2. *If $\boldsymbol{Y} = \boldsymbol{D}\boldsymbol{X} + \boldsymbol{f}$, where $\boldsymbol{X}$ is a random vector, then*

$$\mathscr{E}\boldsymbol{Y} = \boldsymbol{D}\mathscr{E}\boldsymbol{X} + \boldsymbol{f}, \tag{32}$$

$$\mathscr{C}(\boldsymbol{Y}) = \boldsymbol{D}\mathscr{C}(\boldsymbol{X})\boldsymbol{D}'. \tag{33}$$

PROOF. The first assertion follows directly from Lemma 2.3.1 and the second from

$$\begin{aligned} \mathscr{C}(\boldsymbol{Y}) &= \mathscr{E}(\boldsymbol{Y} - \mathscr{E}\boldsymbol{Y})(\boldsymbol{Y} - \mathscr{E}\boldsymbol{Y})' \\ &= \mathscr{E}[\boldsymbol{D}\boldsymbol{X} + \boldsymbol{f} - (\boldsymbol{D}\mathscr{E}\boldsymbol{X} + \boldsymbol{f})][\boldsymbol{D}\boldsymbol{X} + \boldsymbol{f} - (\boldsymbol{D}\mathscr{E}\boldsymbol{X} + \boldsymbol{f})]' \\ &= \mathscr{E}[\boldsymbol{D}(\boldsymbol{X} - \mathscr{E}\boldsymbol{X})][\boldsymbol{D}(\boldsymbol{X} - \mathscr{E}\boldsymbol{X})]' \\ &= \mathscr{E}\left[\boldsymbol{D}(\boldsymbol{X} - \mathscr{E}\boldsymbol{X})(\boldsymbol{X} - \mathscr{E}\boldsymbol{X})'\boldsymbol{D}'\right], \end{aligned} \tag{34}$$

which yields the right-hand side of (33) by Lemma 2.3.1. Q.E.D.

When the transformation corresponds to (11), that is, $\boldsymbol{X} = \boldsymbol{C}\boldsymbol{Y} + \boldsymbol{b}$, then $\mathscr{E}\boldsymbol{X} = \boldsymbol{C}\mathscr{E}\boldsymbol{Y} + \boldsymbol{b}$. By the transformation theory given in Section 2.2, the density of $\boldsymbol{Y}$ is proportional to (16); that is, it is

$$\frac{1}{(2\pi)^{\frac{1}{2}p}} e^{-\frac{1}{2}y'y} = \prod_{j=1}^{p} \left\{ \frac{1}{\sqrt{2\pi}} e^{-\frac{1}{2}y_j^2} \right\}. \tag{35}$$

The expected value of the ith component of $\boldsymbol{Y}$ is

$$\begin{aligned} \mathscr{E}Y_i &= \int_{-\infty}^{\infty} \cdots \int_{-\infty}^{\infty} y_i \prod_{j=1}^{p} \left\{ \frac{1}{\sqrt{2\pi}} e^{-\frac{1}{2}y_j^2} \right\} dy_1 \cdots dy_p \\ &= \frac{1}{\sqrt{2\pi}} \int_{-\infty}^{\infty} y_i e^{-\frac{1}{2}y_i^2}\, dy_i \prod_{\substack{j=1 \\ j \neq i}}^{p} \left\{ \int_{-\infty}^{\infty} \frac{1}{\sqrt{2\pi}} e^{-\frac{1}{2}y_j^2}\, dy_j \right\} \\ &= \frac{1}{\sqrt{2\pi}} \int_{-\infty}^{\infty} y_i e^{-\frac{1}{2}y_i^2}\, dy_i \\ &= 0. \end{aligned} \tag{36}$$

The last equality follows because† $y_i e^{-\frac{1}{2}y_i^2}$ is an odd function of y_i. Thus

†Alternatively, the last equality follows because the next to last expression is the expected value of a normally distributed variable with mean 0.

$\mathscr{E}\boldsymbol{Y} = \boldsymbol{0}$. Therefore, the mean of $\boldsymbol{X}$, denoted by $\boldsymbol{\mu}$, is

$$\boldsymbol{\mu} = \mathscr{E}\boldsymbol{X} = \boldsymbol{b}. \tag{37}$$

From (33) we see that $\mathscr{C}(\boldsymbol{X}) = \boldsymbol{C}(\mathscr{E}\boldsymbol{YY}')\boldsymbol{C}'$. The i, jth element of $\mathscr{E}\boldsymbol{YY}'$ is

$$\mathscr{E}Y_iY_j = \int_{-\infty}^{\infty} \cdots \int_{-\infty}^{\infty} y_i y_j \prod_{h=1}^{p} \left\{ \frac{1}{\sqrt{2\pi}} e^{-\frac{1}{2}y_h^2} \right\} dy_1 \cdots dy_p \tag{38}$$

because the density of $\boldsymbol{Y}$ is (35). If $i = j$, we have

$$\begin{aligned} \mathscr{E}Y_i^2 &= \frac{1}{\sqrt{2\pi}} \int_{-\infty}^{\infty} y_i^2 e^{-\frac{1}{2}y_i^2}\, dy_i \prod_{\substack{h=1\\ h\neq i}}^{p} \left\{ \int_{-\infty}^{\infty} \frac{1}{\sqrt{2\pi}} e^{-\frac{1}{2}y_h^2}\, dy_h \right\} \\ &= \frac{1}{\sqrt{2\pi}} \int_{-\infty}^{\infty} y_i^2 e^{-\frac{1}{2}y_i^2}\, dy_i \\ &= 1. \end{aligned} \tag{39}$$

The last equality follows because the next to last expression is the expected value of the square of a variable normally distributed with mean 0 and variance 1. If $i \neq j$, (38) becomes

$$\begin{aligned} \mathscr{E}Y_iY_j &= \frac{1}{\sqrt{2\pi}} \int_{-\infty}^{\infty} y_i e^{-\frac{1}{2}y_i^2}\, dy_i \cdot \frac{1}{\sqrt{2\pi}} \int_{-\infty}^{\infty} y_j e^{-\frac{1}{2}y_j^2}\, dy_j \\ &\qquad \cdot \prod_{\substack{h=1\\ h\neq i,j}}^{p} \left\{ \frac{1}{\sqrt{2\pi}} \int_{-\infty}^{\infty} e^{-\frac{1}{2}y_h^2}\, dy_h \right\} \\ &= 0, \qquad i \neq j, \end{aligned} \tag{40}$$

since the first integration gives 0. We can summarize (39) and (40) as

$$\mathscr{E}\boldsymbol{YY}' = \boldsymbol{I}. \tag{41}$$

Thus

$$\mathscr{E}(\boldsymbol{X} - \boldsymbol{\mu})(\boldsymbol{X} - \boldsymbol{\mu})' = \boldsymbol{CIC}' = \boldsymbol{CC}'. \tag{42}$$

From (10) we obtain $\boldsymbol{A} = (\boldsymbol{C}')^{-1}\boldsymbol{C}^{-1}$ by multiplication by $(\boldsymbol{C}')^{-1}$ on the left and by C^{-1} on the right. Taking inverses on both sides of the equality gives us

$$CC' = A^{-1}. \tag{43}$$

Thus, the covariance matrix of X is

$$\Sigma = \mathscr{E}(X - \mu)(X - \mu)' = A^{-1}. \tag{44}$$

From (43) we see that Σ is positive definite. Let us summarize these results.

THEOREM 2.3.1. *If the density of a p-dimensional random vector X is (23), the expected value of X is b and the covariance matrix is A^{-1}. Conversely, given a vector μ and a positive definite matrix Σ, there is a multivariate normal density*

$$(2\pi)^{-\frac{1}{2}p}|\Sigma|^{-\frac{1}{2}}e^{-\frac{1}{2}(x-\mu)'\Sigma^{-1}(x-\mu)} \tag{45}$$

such that the expected value of the vector with this density is μ and the covariance matrix is Σ.

We shall denote the density (45) as $n(x|\mu, \Sigma)$ and the distribution law as $N(\mu, \Sigma)$.

The ith diagonal element of the covariance matrix, σ_{ii}, is the variance of the ith component of X; we may sometimes denote this by σ_i^2. The *correlation coefficient* between X_i and X_j is defined as

$$\rho_{ij} = \frac{\sigma_{ij}}{\sqrt{\sigma_{ii}}\sqrt{\sigma_{jj}}} = \frac{\sigma_{ij}}{\sigma_i \sigma_j}. \tag{46}$$

This measure of association is symmetric in X_i and X_j: $\rho_{ij} = \rho_{ji}$. Since

$$\begin{pmatrix} \sigma_{ii} & \sigma_{ij} \\ \sigma_{ji} & \sigma_{jj} \end{pmatrix} = \begin{pmatrix} \sigma_i^2 & \sigma_i\sigma_j\rho_{ij} \\ \sigma_i\sigma_j\rho_{ij} & \sigma_j^2 \end{pmatrix} \tag{47}$$

is positive definite (Corollary A.1.3 of Appendix A), the determinant

$$\begin{vmatrix} \sigma_i^2 & \sigma_i\sigma_j\rho_{ij} \\ \sigma_i\sigma_j\rho_{ij} & \sigma_j^2 \end{vmatrix} = \sigma_i^2\sigma_j^2(1 - \rho_{ij}^2) \tag{48}$$

is positive. Therefore, $-1 < \rho_{ij} < 1$. (For singular distributions, see Section 2.4.) The multivariate normal density can be parametrized by the means μ_i, $i = 1, \ldots, p$, the variances σ_i^2, $i = 1, \ldots, p$, and the correlations ρ_{ij}, $i < j$, $i, j = 1, \ldots, p$.

As a special case of the preceding theory, we consider the *bivariate* normal distribution. The mean vector is

$$\mathscr{E}\begin{pmatrix} X_1 \\ X_2 \end{pmatrix} = \begin{pmatrix} \mu_1 \\ \mu_2 \end{pmatrix}; \tag{49}$$

the covariance matrix may be written

$$\Sigma = \mathscr{E}\begin{pmatrix} (X_1 - \mu_1)^2 & (X_1 - \mu_1)(X_2 - \mu_2) \\ (X_2 - \mu_2)(X_1 - \mu_1) & (X_2 - \mu_2)^2 \end{pmatrix} \tag{50}$$

$$= \begin{pmatrix} \sigma_{11} & \sigma_{12} \\ \sigma_{21} & \sigma_{22} \end{pmatrix} = \begin{pmatrix} \sigma_1^2 & \sigma_1\sigma_2\rho \\ \sigma_1\sigma_2\rho & \sigma_2^2 \end{pmatrix},$$

where σ_1^2 is the variance of X_1, σ_2^2 the variance of X_2, and ρ the correlation between X_1 and X_2. The inverse of (50) is

$$\Sigma^{-1} = \frac{1}{1-\rho^2}\begin{pmatrix} \dfrac{1}{\sigma_1^2} & -\dfrac{\rho}{\sigma_1\sigma_2} \\ -\dfrac{\rho}{\sigma_1\sigma_2} & \dfrac{1}{\sigma_2^2} \end{pmatrix}. \tag{51}$$

The density function of X_1 and X_2 is

$$\frac{1}{2\pi\sigma_1\sigma_2\sqrt{1-\rho^2}} \exp\left\{ -\frac{1}{2(1-\rho^2)}\left[\frac{(x_1-\mu_1)^2}{\sigma_1^2} - 2\rho\frac{(x_1-\mu_1)(x_2-\mu_2)}{\sigma_1\sigma_2} + \frac{(x_2-\mu_2)^2}{\sigma_2^2} \right] \right\}. \tag{52}$$

THEOREM 2.3.2. *The correlation coefficient ρ of any bivariate distribution is invariant with respect to transformations $X_i^* = b_i X_i + c_i$, $b_i > 0$, $i = 1, 2$. Every function of the parameters of a bivariate normal distribution that is invariant with respect to such transformations is a function of ρ.*

PROOF. The variance of X_i^* is $b_i^2\sigma_i^2$, $i = 1, 2$, and the covariance of X_1^* and X_2^* is $b_1 b_2 \sigma_1 \sigma_2 \rho$ by Lemma 2.3.2. Insertion of these values into the definition of the correlation between X_1^* and X_2^* shows that it is ρ. If $f(\mu_1, \mu_2, \sigma_1, \sigma_2, \rho)$ is invariant with respect to such transformations, it must be $f(0, 0, 1, 1, \rho)$ by choice of $b_i = 1/\sigma_i$ and $c_i = -\mu_i/\sigma_i$, $i = 1, 2$. Q.E.D.

The correlation coefficient ρ is the natural *measure of association* between X_1 and X_2. Any function of the parameters of the bivariate normal distribution that is independent of the scale and location parameters is a function of ρ. The *standardized* variable (or standard score) is $Y_i = (X_i - \mu_i)/\sigma_i$. The mean squared difference between the two standardized variables is

$$\mathscr{E}(Y_1 - Y_2)^2 = 2(1 - \rho). \tag{53}$$

The smaller (53) is, that is, the larger ρ is, the more similar Y_1 and Y_2 are. If $\rho > 0$, X_1 and X_2 tend to be positively related, and if $\rho < 0$, they tend to be negatively related. If $\rho = 0$, the density (52) is the product of the marginal densities of X_1 and X_2; hence X_1 and X_2 are independent.

It will be noticed that the density function (45) is constant on ellipsoids

$$(\boldsymbol{x} - \boldsymbol{\mu})'\boldsymbol{\Sigma}^{-1}(\boldsymbol{x} - \boldsymbol{\mu}) = c \tag{54}$$

for every positive value of c in a p-dimensional Euclidean space. The center of each ellipsoid is at the point $\boldsymbol{\mu}$. The shape and orientation of the ellipsoid are determined by $\boldsymbol{\Sigma}$, and the size (given $\boldsymbol{\Sigma}$) is determined by c. Because (54) is a sphere if $\boldsymbol{\Sigma} = \sigma^2\boldsymbol{I}$, $n(\boldsymbol{x}|\boldsymbol{\mu}, \sigma^2\boldsymbol{I})$ is known as a *spherical* normal density.

Let us consider in detail the bivariate case of the density (52). We transform coordinates by $(x_i - \mu_i)/\sigma_i = y_i$, $i = 1, 2$, so that the centers of the loci of constant density are at the origin. These loci are defined by

$$\frac{1}{1 - \rho^2}\left(y_1^2 - 2\rho y_1 y_2 + y_2^2\right) = c. \tag{55}$$

The intercepts on the y_1-axis and y_2-axis are equal. If $\rho > 0$, the major axis of the ellipse is along the 45° line with a length of $2\sqrt{c(1 + \rho)}$ and the minor axis has a length of $2\sqrt{c(1 - \rho)}$. If $\rho < 0$, the major axis is along the 135° line with a length of $2\sqrt{c(1 - \rho)}$ and the minor axis has a length of $2\sqrt{c(1 + \rho)}$. The value of ρ determines the ratio of these lengths. In this bivariate case we can think of the density function as a surface above the plane. The contours of equal density are contours of equal altitude on a topographical map; they indicate the shape of the hill (or probability surface). If $\rho > 0$, the hill will tend to run along a line with a positive slope; most of the hill will be in the first and third quadrants. When we transform back to $x_i = \sigma_i y_i + \mu_i$, we expand each contour by a factor of σ_i in the direction of the ith axis and shift the center to (μ_1, μ_2).

The numerical values of the cdf of the univariate normal variable are obtained from tables found in most statistical texts. The numerical values of

$$F(x_1, x_2) = \Pr\{X_1 \le x_1, X_2 \le x_2\} = \Pr\left\{\frac{X_1 - \mu_1}{\sigma_1} \le y_1, \frac{X_2 - \mu_2}{\sigma_2} \le y_2\right\}, \tag{56}$$

where $y_1 = (x_1 - \mu_1)/\sigma_1$ and $y_2 = (x_2 - \mu_2)/\sigma_2$, can be found in Pearson (1931). An extensive table has been given by the National Bureau of Standards (1959). A bibliography of such tables has been given by Gupta (1963). Pearson has also shown that

$$F(x_1, x_2) = \sum_{j=0}^{\infty} \rho^j \tau_j(y_1) \tau_j(y_2), \tag{57}$$

where the so-called *tetrachoric functions* $\tau_j(y)$ are tabulated in Pearson (1930) up to $\tau_{19}(y)$. Harris and Soms (1980) have studied generalizations of (57).

2.4. THE DISTRIBUTION OF LINEAR COMBINATIONS OF NORMALLY DISTRIBUTED VARIATES; INDEPENDENCE OF VARIATES; MARGINAL DISTRIBUTIONS

One of the reasons that the study of normal multivariate distributions is so useful is that marginal distributions and conditional distributions derived from multivariate normal distributions are also normal distributions. Moreover, linear combinations of multivariate normal variates are again normally distributed. First we shall show that if we make a nonsingular linear transformation of a vector whose components have a joint distribution with a normal density, we obtain a vector whose components are jointly distributed with a normal density.

THEOREM 2.4.1. *Let* $\boldsymbol{X}$ *(with* p *components) be distributed according to* $N(\boldsymbol{\mu}, \boldsymbol{\Sigma})$. *Then*

$$\boldsymbol{Y} = \boldsymbol{C}\boldsymbol{X} \tag{1}$$

is distributed according to $N(\boldsymbol{C}\boldsymbol{\mu}, \boldsymbol{C}\boldsymbol{\Sigma}\boldsymbol{C}')$ *for* $\boldsymbol{C}$ *nonsingular.*

PROOF. The density of $\boldsymbol{Y}$ is obtained from the density of $\boldsymbol{X}$, $n(\boldsymbol{x}|\boldsymbol{\mu}, \boldsymbol{\Sigma})$, by replacing $\boldsymbol{x}$ by

$$\boldsymbol{x} = \boldsymbol{C}^{-1}\boldsymbol{y}, \tag{2}$$

and multiplying by the Jacobian of the transformation (2),

$$\text{mod}|\boldsymbol{C}^{-1}| = \frac{1}{\text{mod}|\boldsymbol{C}|} = \sqrt{\frac{1}{|\boldsymbol{C}|^2}} = \sqrt{\frac{|\boldsymbol{\Sigma}|}{|\boldsymbol{C}| \cdot |\boldsymbol{\Sigma}| \cdot |\boldsymbol{C}'|}} = \frac{|\boldsymbol{\Sigma}|^{\frac{1}{2}}}{|\boldsymbol{C}\boldsymbol{\Sigma}\boldsymbol{C}'|^{\frac{1}{2}}}. \tag{3}$$

The quadratic form in the exponent of $n(\boldsymbol{x}|\boldsymbol{\mu}, \boldsymbol{\Sigma})$ is

$$Q = (\boldsymbol{x} - \boldsymbol{\mu})'\boldsymbol{\Sigma}^{-1}(\boldsymbol{x} - \boldsymbol{\mu}). \tag{4}$$

The transformation (2) carries Q into

$$\begin{aligned} Q &= (\boldsymbol{C}^{-1}\boldsymbol{y} - \boldsymbol{\mu})'\boldsymbol{\Sigma}^{-1}(\boldsymbol{C}^{-1}\boldsymbol{y} - \boldsymbol{\mu}) \\ &= (\boldsymbol{C}^{-1}\boldsymbol{y} - \boldsymbol{C}^{-1}\boldsymbol{C}\boldsymbol{\mu})'\boldsymbol{\Sigma}^{-1}(\boldsymbol{C}^{-1}\boldsymbol{y} - \boldsymbol{C}^{-1}\boldsymbol{C}\boldsymbol{\mu}) \\ &= [\boldsymbol{C}^{-1}(\boldsymbol{y} - \boldsymbol{C}\boldsymbol{\mu})]'\boldsymbol{\Sigma}^{-1}[\boldsymbol{C}^{-1}(\boldsymbol{y} - \boldsymbol{C}\boldsymbol{\mu})] \\ &= (\boldsymbol{y} - \boldsymbol{C}\boldsymbol{\mu})'(\boldsymbol{C}^{-1})'\boldsymbol{\Sigma}^{-1}\boldsymbol{C}^{-1}(\boldsymbol{y} - \boldsymbol{C}\boldsymbol{\mu}) \\ &= (\boldsymbol{y} - \boldsymbol{C}\boldsymbol{\mu})'(\boldsymbol{C}\boldsymbol{\Sigma}\boldsymbol{C}')^{-1}(\boldsymbol{y} - \boldsymbol{C}\boldsymbol{\mu}) \end{aligned} \tag{5}$$

since $(\boldsymbol{C}^{-1})' = (\boldsymbol{C}')^{-1}$ by virtue of transposition of $\boldsymbol{C}\boldsymbol{C}^{-1} = \boldsymbol{I}$. Thus the density of $\boldsymbol{Y}$ is

$$(6)\quad n(\boldsymbol{C}^{-1}y|\boldsymbol{\mu},\boldsymbol{\Sigma})\operatorname{mod}|\boldsymbol{C}|^{-1}$$

$$= (2\pi)^{-\frac{1}{2}p}|\boldsymbol{C\Sigma C}'|^{-\frac{1}{2}}\exp\left[-\tfrac{1}{2}(y-\boldsymbol{C\mu})'(\boldsymbol{C\Sigma C}')^{-1}(y-\boldsymbol{C\mu})\right]$$

$$= n(y|\boldsymbol{C\mu},\boldsymbol{C\Sigma C}').$$

Q.E.D.

Now let us consider two sets of random variables $X_1, \dots, X_q$ and $X_{q+1}, \dots, X_p$ forming the vectors

$$(7)\qquad \boldsymbol{X}^{(1)} = \begin{pmatrix} X_1 \\ \vdots \\ X_q \end{pmatrix}, \qquad \boldsymbol{X}^{(2)} = \begin{pmatrix} X_{q+1} \\ \vdots \\ X_p \end{pmatrix}.$$

These variables form the random vector

$$(8)\qquad \boldsymbol{X} = \begin{pmatrix} \boldsymbol{X}^{(1)} \\ \boldsymbol{X}^{(2)} \end{pmatrix} = \begin{pmatrix} X_1 \\ \vdots \\ X_p \end{pmatrix}.$$

Now let us assume that the p variates have a joint normal distribution with mean vectors

$$(9)\qquad \mathscr{E}\boldsymbol{X}^{(1)} = \boldsymbol{\mu}^{(1)}, \qquad \mathscr{E}\boldsymbol{X}^{(2)} = \boldsymbol{\mu}^{(2)},$$

and covariance matrices

$$(10)\qquad \mathscr{E}(\boldsymbol{X}^{(1)} - \boldsymbol{\mu}^{(1)})(\boldsymbol{X}^{(1)} - \boldsymbol{\mu}^{(1)})' = \boldsymbol{\Sigma}_{11},$$

$$(11)\qquad \mathscr{E}(\boldsymbol{X}^{(2)} - \boldsymbol{\mu}^{(2)})(\boldsymbol{X}^{(2)} - \boldsymbol{\mu}^{(2)})' = \boldsymbol{\Sigma}_{22},$$

$$(12)\qquad \mathscr{E}(\boldsymbol{X}^{(1)} - \boldsymbol{\mu}^{(1)})(\boldsymbol{X}^{(2)} - \boldsymbol{\mu}^{(2)})' = \boldsymbol{\Sigma}_{12}.$$

We say that the random vector $\boldsymbol{X}$ has been partitioned in (8) into subvectors, that

$$(13)\qquad \boldsymbol{\mu} = \begin{pmatrix} \boldsymbol{\mu}^{(1)} \\ \boldsymbol{\mu}^{(2)} \end{pmatrix}$$

has been partitioned similarly into subvectors, and that

$$\mathbf{\Sigma} = \begin{pmatrix} \mathbf{\Sigma}_{11} & \mathbf{\Sigma}_{12} \\ \mathbf{\Sigma}_{21} & \mathbf{\Sigma}_{22} \end{pmatrix} \tag{14}$$

has been similarly partitioned into submatrices. Here $\mathbf{\Sigma}_{21} = \mathbf{\Sigma}_{12}'$. (See Appendix A, Section 3.)

We shall show that $\boldsymbol{X}^{(1)}$ and $\boldsymbol{X}^{(2)}$ are independently normally distributed if $\mathbf{\Sigma}_{12} = \mathbf{\Sigma}_{21}' = \mathbf{0}$. Then

$$\mathbf{\Sigma} = \begin{pmatrix} \mathbf{\Sigma}_{11} & \mathbf{0} \\ \mathbf{0} & \mathbf{\Sigma}_{22} \end{pmatrix}. \tag{15}$$

Its inverse is

$$\mathbf{\Sigma}^{-1} = \begin{pmatrix} \mathbf{\Sigma}_{11}^{-1} & \mathbf{0} \\ \mathbf{0} & \mathbf{\Sigma}_{22}^{-1} \end{pmatrix}. \tag{16}$$

Thus the quadratic form in the exponent of $n(\boldsymbol{x}|\boldsymbol{\mu}, \mathbf{\Sigma})$ is

$$\begin{aligned} Q &= (\boldsymbol{x} - \boldsymbol{\mu})'\mathbf{\Sigma}^{-1}(\boldsymbol{x} - \boldsymbol{\mu}) \\ &= \left[(\boldsymbol{x}^{(1)} - \boldsymbol{\mu}^{(1)})', (\boldsymbol{x}^{(2)} - \boldsymbol{\mu}^{(2)})' \right] \begin{pmatrix} \mathbf{\Sigma}_{11}^{-1} & \mathbf{0} \\ \mathbf{0} & \mathbf{\Sigma}_{22}^{-1} \end{pmatrix} \begin{pmatrix} \boldsymbol{x}^{(1)} - \boldsymbol{\mu}^{(1)} \\ \boldsymbol{x}^{(2)} - \boldsymbol{\mu}^{(2)} \end{pmatrix} \\ &= \left[(\boldsymbol{x}^{(1)} - \boldsymbol{\mu}^{(1)})'\mathbf{\Sigma}_{11}^{-1}, (\boldsymbol{x}^{(2)} - \boldsymbol{\mu}^{(2)})'\mathbf{\Sigma}_{22}^{-1} \right] \begin{pmatrix} \boldsymbol{x}^{(1)} - \boldsymbol{\mu}^{(1)} \\ \boldsymbol{x}^{(2)} - \boldsymbol{\mu}^{(2)} \end{pmatrix} \\ &= (\boldsymbol{x}^{(1)} - \boldsymbol{\mu}^{(1)})'\mathbf{\Sigma}_{11}^{-1}(\boldsymbol{x}^{(1)} - \boldsymbol{\mu}^{(1)}) + (\boldsymbol{x}^{(2)} - \boldsymbol{\mu}^{(2)})'\mathbf{\Sigma}_{22}^{-1}(\boldsymbol{x}^{(2)} - \boldsymbol{\mu}^{(2)}) \\ &= Q_1 + Q_2, \end{aligned} \tag{17}$$

say, where

$$\begin{aligned} Q_1 &= (\boldsymbol{x}^{(1)} - \boldsymbol{\mu}^{(1)})'\mathbf{\Sigma}_{11}^{-1}(\boldsymbol{x}^{(1)} - \boldsymbol{\mu}^{(1)}), \\ Q_2 &= (\boldsymbol{x}^{(2)} - \boldsymbol{\mu}^{(2)})'\mathbf{\Sigma}_{22}^{-1}(\boldsymbol{x}^{(2)} - \boldsymbol{\mu}^{(2)}). \end{aligned} \tag{18}$$

Also we note that $|\boldsymbol{\Sigma}| = |\boldsymbol{\Sigma}_{11}| \cdot |\boldsymbol{\Sigma}_{22}|$. The density of $\boldsymbol{X}$ can be written

$$
\begin{aligned}
(19)\qquad n(\boldsymbol{x}|\boldsymbol{\mu},\boldsymbol{\Sigma}) &= \frac{1}{(2\pi)^{\frac{1}{2}p}|\boldsymbol{\Sigma}|^{\frac{1}{2}}} e^{-\frac{1}{2}Q} \\
&= \frac{1}{(2\pi)^{\frac{1}{2}q}|\boldsymbol{\Sigma}_{11}|^{\frac{1}{2}}} e^{-\frac{1}{2}Q_1} \cdot \frac{1}{(2\pi)^{\frac{1}{2}(p-q)}|\boldsymbol{\Sigma}_{22}|^{\frac{1}{2}}} e^{-\frac{1}{2}Q_2} \\
&= n\left(\boldsymbol{x}^{(1)}|\boldsymbol{\mu}^{(1)},\boldsymbol{\Sigma}_{11}\right) n\left(\boldsymbol{x}^{(2)}|\boldsymbol{\mu}^{(2)},\boldsymbol{\Sigma}_{22}\right).
\end{aligned}
$$

The marginal density of $\boldsymbol{X}^{(1)}$ is given by the integral

$$
\begin{aligned}
(20)\qquad &\int_{-\infty}^{\infty} \cdots \int_{-\infty}^{\infty} n(\boldsymbol{x}|\boldsymbol{\mu},\boldsymbol{\Sigma})\, dx_{q+1} \cdots dx_p \\
&= n\left(\boldsymbol{x}^{(1)}|\boldsymbol{\mu}^{(1)},\boldsymbol{\Sigma}_{11}\right) \int_{-\infty}^{\infty} \cdots \int_{-\infty}^{\infty} n\left(\boldsymbol{x}^{(2)}|\boldsymbol{\mu}^{(2)},\boldsymbol{\Sigma}_{22}\right) dx_{q+1} \cdots dx_p \\
&= n\left(\boldsymbol{x}^{(1)}|\boldsymbol{\mu}^{(1)},\boldsymbol{\Sigma}_{11}\right).
\end{aligned}
$$

Thus the marginal distribution of $\boldsymbol{X}^{(1)}$ is $N(\boldsymbol{\mu}^{(1)},\boldsymbol{\Sigma}_{11})$; similarly the marginal distribution of $\boldsymbol{X}^{(2)}$ is $N(\boldsymbol{\mu}^{(2)},\boldsymbol{\Sigma}_{22})$. Thus the joint density of $X_1,\dots,X_p$ is the product of the marginal density of $X_1,\dots,X_q$ and the marginal density of $X_{q+1},\dots,X_p$, and, therefore, the two sets of variates are independent. Since the numbering of variates can always be done so that $\boldsymbol{X}^{(1)}$ consists of any subset of the variates, we have proved the sufficiency in the following theorem:

THEOREM 2.4.2. *If $X_1,\dots,X_p$ have a joint normal distribution, a necessary and sufficient condition that one subset of the random variables and the subset consisting of the remaining variables be independent is that each covariance of a variable from one set and a variable from the other set be 0.*

The necessity follows from the fact that if X_i is from one set and X_j from the other, then for any density (see Section 2.2.3)

$$
\begin{aligned}
(21)\qquad \sigma_{ij} &= \mathscr{E}(X_i-\mu_i)(X_j-\mu_j) \\
&= \int_{-\infty}^{\infty} \cdots \int_{-\infty}^{\infty} (x_i-\mu_i)(x_j-\mu_j) f(x_1,\dots,x_q) \\
&\qquad\qquad \cdot f(x_{q+1},\dots,x_p)\, dx_1 \cdots dx_p \\
&= \int_{-\infty}^{\infty} \cdots \int_{-\infty}^{\infty} (x_i-\mu_i) f(x_1,\dots,x_q)\, dx_1 \cdots dx_q \\
&\qquad \cdot \int_{-\infty}^{\infty} \cdots \int_{-\infty}^{\infty} (x_j-\mu_j) f(x_{q+1},\dots,x_p)\, dx_{q+1} \cdots dx_p \\
&= 0.
\end{aligned}
$$

Since $\sigma_{ij} = \sigma_i \sigma_j \rho_{ij}$, and $\sigma_i, \sigma_j \neq 0$ (we tacitly assume that $\boldsymbol{\Sigma}$ is nonsingular), the condition $\sigma_{ij} = 0$ is equivalent to $\rho_{ij} = 0$. Thus if one set of variates is uncorrelated with the remaining variates, the two sets are independent. It should be emphasized that the implication of independence by lack of correlation depends on the assumption of normality, but the converse is always true.

Let us consider the special case of the bivariate normal distribution. Then $\boldsymbol{X}^{(1)} = X_1$, $\boldsymbol{X}^{(2)} = X_2$, $\boldsymbol{\mu}^{(1)} = \mu_1$, $\boldsymbol{\mu}^{(2)} = \mu_2$, $\boldsymbol{\Sigma}_{11} = \sigma_{11} = \sigma_1^2$, $\boldsymbol{\Sigma}_{22} = \sigma_{22} = \sigma_2^2$, and $\boldsymbol{\Sigma}_{12} = \boldsymbol{\Sigma}_{21} = \sigma_{12} = \sigma_1 \sigma_2 \rho_{12}$. Thus if X_1 and X_2 have a bivariate normal distribution, they are independent if and only if they are uncorrelated. If they are uncorrelated, the marginal distribution of X_i is normal with mean μ_i and variance σ_i^2. The above discussion also proves the following corollary:

COROLLARY 2.4.1. *If $\boldsymbol{X}$ is distributed according to $N(\boldsymbol{\mu}, \boldsymbol{\Sigma})$ and if a set of components of $\boldsymbol{X}$ is uncorrelated with the other components, the marginal distribution of the set is multivariate normal with means, variances, and covariances obtained by taking the corresponding components of $\boldsymbol{\mu}$ and $\boldsymbol{\Sigma}$, respectively.*

Now let us show that the corollary holds even if the two sets are not independent. We partition $\boldsymbol{X}$, $\boldsymbol{\mu}$, and $\boldsymbol{\Sigma}$ as before. We shall make a nonsingular linear transformation to subvectors

$$\boldsymbol{Y}^{(1)} = \boldsymbol{X}^{(1)} + \boldsymbol{B}\boldsymbol{X}^{(2)}, \tag{22}$$

$$\boldsymbol{Y}^{(2)} = \boldsymbol{X}^{(2)}, \tag{23}$$

choosing $\boldsymbol{B}$ so that the components of $\boldsymbol{Y}^{(1)}$ are uncorrelated with the components of $\boldsymbol{Y}^{(2)} = \boldsymbol{X}^{(2)}$. The matrix $\boldsymbol{B}$ must satisfy the equation

$$\begin{aligned} \mathbf{0} &= \mathscr{E}(\boldsymbol{Y}^{(1)} - \mathscr{E}\boldsymbol{Y}^{(1)})(\boldsymbol{Y}^{(2)} - \mathscr{E}\boldsymbol{Y}^{(2)})' \\ &= \mathscr{E}(\boldsymbol{X}^{(1)} + \boldsymbol{B}\boldsymbol{X}^{(2)} - \mathscr{E}\boldsymbol{X}^{(1)} - \boldsymbol{B}\mathscr{E}\boldsymbol{X}^{(2)})(\boldsymbol{X}^{(2)} - \mathscr{E}\boldsymbol{X}^{(2)})' \\ &= \mathscr{E}\left[(\boldsymbol{X}^{(1)} - \mathscr{E}\boldsymbol{X}^{(1)}) + \boldsymbol{B}(\boldsymbol{X}^{(2)} - \mathscr{E}\boldsymbol{X}^{(2)})\right](\boldsymbol{X}^{(2)} - \mathscr{E}\boldsymbol{X}^{(2)})' \\ &= \boldsymbol{\Sigma}_{12} + \boldsymbol{B}\boldsymbol{\Sigma}_{22}. \end{aligned} \tag{24}$$

Thus $\boldsymbol{B} = -\boldsymbol{\Sigma}_{12}\boldsymbol{\Sigma}_{22}^{-1}$ and

$$\boldsymbol{Y}^{(1)} = \boldsymbol{X}^{(1)} - \boldsymbol{\Sigma}_{12}\boldsymbol{\Sigma}_{22}^{-1}\boldsymbol{X}^{(2)}. \tag{25}$$

The vector

$$\begin{pmatrix} Y^{(1)} \\ Y^{(2)} \end{pmatrix} = Y = \begin{pmatrix} I & -\Sigma_{12}\Sigma_{22}^{-1} \\ 0 & I \end{pmatrix} X \tag{26}$$

is a nonsingular transform of X, and therefore has a normal distribution with

$$\begin{aligned} \mathscr{E}\begin{pmatrix} Y^{(1)} \\ Y^{(2)} \end{pmatrix} &= \mathscr{E}\begin{pmatrix} I & -\Sigma_{12}\Sigma_{22}^{-1} \\ 0 & I \end{pmatrix} X \\ &= \begin{pmatrix} I & -\Sigma_{12}\Sigma_{22}^{-1} \\ 0 & I \end{pmatrix}\begin{pmatrix} \mu^{(1)} \\ \mu^{(2)} \end{pmatrix} \\ &= \begin{pmatrix} \mu^{(1)} - \Sigma_{12}\Sigma_{22}^{-1}\mu^{(2)} \\ \mu^{(2)} \end{pmatrix} = \begin{pmatrix} \nu^{(1)} \\ \nu^{(2)} \end{pmatrix} \\ &= \nu, \end{aligned} \tag{27}$$

say, and

$$\begin{aligned} \mathscr{C}(Y) &= \mathscr{E}(Y-\nu)(Y-\nu)' \\ &= \begin{pmatrix} \mathscr{E}(Y^{(1)}-\nu^{(1)})(Y^{(1)}-\nu^{(1)})' & \mathscr{E}(Y^{(1)}-\nu^{(1)})(Y^{(2)}-\nu^{(2)})' \\ \mathscr{E}(Y^{(2)}-\nu^{(2)})(Y^{(1)}-\nu^{(1)})' & \mathscr{E}(Y^{(2)}-\nu^{(2)})(Y^{(2)}-\nu^{(2)})' \end{pmatrix} \\ &= \begin{pmatrix} \Sigma_{11} - \Sigma_{12}\Sigma_{22}^{-1}\Sigma_{21} & 0 \\ 0 & \Sigma_{22} \end{pmatrix} \end{aligned} \tag{28}$$

since

$$\begin{aligned} \mathscr{E}(Y^{(1)}-\nu^{(1)})&(Y^{(1)}-\nu^{(1)})' \\ &= \mathscr{E}\left[(X^{(1)}-\mu^{(1)}) - \Sigma_{12}\Sigma_{22}^{-1}(X^{(2)}-\mu^{(2)})\right] \\ &\qquad \cdot\left[(X^{(1)}-\mu^{(1)}) - \Sigma_{12}\Sigma_{22}^{-1}(X^{(2)}-\mu^{(2)})\right]' \\ &= \Sigma_{11} - \Sigma_{12}\Sigma_{22}^{-1}\Sigma_{21} - \Sigma_{12}\Sigma_{22}^{-1}\Sigma_{21} + \Sigma_{12}\Sigma_{22}^{-1}\Sigma_{22}\Sigma_{22}^{-1}\Sigma_{21} \\ &= \Sigma_{11} - \Sigma_{12}\Sigma_{22}^{-1}\Sigma_{21}. \end{aligned} \tag{29}$$

Thus $\boldsymbol{Y}^{(1)}$ and $\boldsymbol{Y}^{(2)}$ are independent and by Corollary 2.4.1 $\boldsymbol{X}^{(2)} = \boldsymbol{Y}^{(2)}$ has the marginal distribution $N(\boldsymbol{\mu}^{(2)}, \boldsymbol{\Sigma}_{22})$. Because the numbering of the components of $\boldsymbol{X}$ is arbitrary, we can state the following theorem:

THEOREM 2.4.3. *If $\boldsymbol{X}$ is distributed according to $N(\boldsymbol{\mu}, \boldsymbol{\Sigma})$, the marginal distribution of any set of components of $\boldsymbol{X}$ is multivariate normal with means, variances, and covariances obtained by taking the corresponding components of $\boldsymbol{\mu}$ and $\boldsymbol{\Sigma}$, respectively.*

Now consider any transformation

$$\boldsymbol{Z} = \boldsymbol{D}\boldsymbol{X}, \tag{30}$$

where $\boldsymbol{Z}$ has q components and $\boldsymbol{D}$ is a $q \times p$ real matrix. The expected value of $\boldsymbol{Z}$ is

$$\mathscr{E}\boldsymbol{Z} = \boldsymbol{D}\boldsymbol{\mu}, \tag{31}$$

and the covariance matrix is

$$\mathscr{E}(\boldsymbol{Z} - \boldsymbol{D}\boldsymbol{\mu})(\boldsymbol{Z} - \boldsymbol{D}\boldsymbol{\mu})' = \boldsymbol{D}\boldsymbol{\Sigma}\boldsymbol{D}'. \tag{32}$$

The case $q = p$ and $\boldsymbol{D}$ nonsingular has been treated above. If $q \leq p$ and $\boldsymbol{D}$ is of rank q, we can find a $(p - q) \times p$ matrix $\boldsymbol{E}$ such that

$$\begin{pmatrix} \boldsymbol{Z} \\ \boldsymbol{W} \end{pmatrix} = \begin{pmatrix} \boldsymbol{D} \\ \boldsymbol{E} \end{pmatrix} \boldsymbol{X} \tag{33}$$

is a nonsingular transformation. (See Appendix A, Section 3.) Then $\boldsymbol{Z}$ and $\boldsymbol{W}$ have a joint normal distribution and $\boldsymbol{Z}$ has a marginal normal distribution by Theorem 2.4.3. Thus for $\boldsymbol{D}$ of rank q (and $\boldsymbol{X}$ having a nonsingular distribution, that is, a density) we have proved the following theorem:

THEOREM 2.4.4. *If $\boldsymbol{X}$ is distributed according to $N(\boldsymbol{\mu}, \boldsymbol{\Sigma})$, then $\boldsymbol{Z} = \boldsymbol{D}\boldsymbol{X}$ is distributed according to $N(\boldsymbol{D}\boldsymbol{\mu}, \boldsymbol{D}\boldsymbol{\Sigma}\boldsymbol{D}')$, where $\boldsymbol{D}$ is a $q \times p$ matrix of rank $q \leq p$.*

The remainder of this section is devoted to the *singular* or *degenerate* normal distribution and the extension of Theorem 2.4.4 to the case of any matrix $\boldsymbol{D}$. A singular distribution is a distribution in p-space that is concentrated on a lower dimensional set; that is, the probability associated with any set not intersecting the given set is 0. In the case of the singular normal distribution the mass is concentrated on a given linear set [that is, the intersection of a number of $(p-1)$-dimensional hyperplanes]. Let y be a set of coordinates in the linear set (the number of coordinates equaling the dimensionality of the linear set); then the "parametric" definition of the linear set can be given as $x = Ay + \lambda$, where A is a $p \times q$ matrix and λ is a p-vector. Suppose that Y is normally distributed in the q-dimensional linear set; then we say that

$$X = AY + \lambda \tag{34}$$

has a singular or degenerate normal distribution in p-space. If $\mathscr{E}Y = \nu$, then $\mathscr{E}X = A\nu + \lambda = \mu$, say. If $\mathscr{E}(Y-\nu)(Y-\nu)' = T$, then

$$\mathscr{E}(X-\mu)(X-\mu)' = \mathscr{E}A(Y-\nu)(Y-\nu)'A' = ATA' = \Sigma, \tag{35}$$

say. It should be noticed that if $p > q$ then Σ is singular and, therefore, has no inverse and thus we cannot write the normal density for X. In fact, X cannot have a density at all because the fact that the probability of any set not intersecting the q-set is 0 would imply that the density is 0 almost everywhere.

Now conversely let us see that if X has mean μ and covariance matrix Σ of rank r it can be written as (34) (except for 0 probabilities), where X has an arbitrary distribution and Y of r $(\leq p)$ components has a suitable distribution. If Σ is of rank r, there is a $p \times p$ nonsingular matrix B such that

$$B\Sigma B' = \begin{pmatrix} I_r & 0 \\ 0 & 0 \end{pmatrix}, \tag{36}$$

where the identity is of order r. (See Theorem A.4.1 of Appendix A.) The transformation

$$BX = V = \begin{pmatrix} V^{(1)} \\ V^{(2)} \end{pmatrix} \tag{37}$$

defines a random vector V with covariance matrix (36) and a mean vector

$$\mathscr{E}V = B\mu = \nu = \begin{pmatrix} \nu^{(1)} \\ \nu^{(2)} \end{pmatrix}, \tag{38}$$

say. Since the variances of the elements of $V^{(2)}$ are zero, $V^{(2)} = \nu^{(2)}$ with probability 1. Now partition

$$B^{-1} = (C\ D), \tag{39}$$

where C consists of r columns. Then (37) is equivalent to

$$X = B^{-1}V = (C\ D)\begin{pmatrix} V^{(1)} \\ V^{(2)} \end{pmatrix} = CV^{(1)} + DV^{(2)}. \tag{40}$$

Thus with probability 1

$$X = CV^{(1)} + Dv^{(2)}, \tag{41}$$

which is of the form of (34) with C as A, $V^{(1)}$ as Y, and $Dv^{(2)}$ as λ.

Now we give a formal definition of a normal distribution that includes the singular distribution.

DEFINITION 2.4.1. *A random vector X of p components with $\mathscr{E}X = \mu$ and $\mathscr{E}(X - \mu)(X - \mu)' = \Sigma$ is said to be normally distributed [or is said to be distributed according to $N(\mu, \Sigma)$] if there is a transformation* (34), *where the number of rows of A is p and the number of columns is the rank of Σ, say r, and Y (of r components) has a nonsingular normal distribution, that is, has a density*

$$ke^{-\frac{1}{2}(y-v)'T^{-1}(y-v)}. \tag{42}$$

It is clear that if Σ has rank p, then A can be taken to be I and λ to be 0; then $X = Y$ and Definition 2.4.1 agrees with Section 2.3. To avoid redundancy in Definition 2.4.1 we could take $T = I$ and $v = 0$.

THEOREM 2.4.5. *If X is distributed according to $N(\mu, \Sigma)$, then $Z = DX$ is distributed according to $N(D\mu, D\Sigma D')$.*

This theorem includes the cases where X may have a nonsingular or a singular distribution and D may be nonsingular or of rank less than q. Since X can be represented by (34), where Y has a nonsingular distribution, $N(v, T)$, we can write

$$Z = DAY + D\lambda, \tag{43}$$

where DA is $q \times r$. If the rank of DA is r, the theorem is proved. If the rank is less than r, say s, the covariance matrix of $\boldsymbol{Z}$,

$$\boldsymbol{DATA'D'} = \boldsymbol{E}, \tag{44}$$

say, is of rank s. By Theorem A.4.1 of Appendix A, there is a nonsingular matrix

$$\boldsymbol{F} = \begin{pmatrix} \boldsymbol{F}_1 \\ \boldsymbol{F}_2 \end{pmatrix} \tag{45}$$

such that

$$\begin{aligned} \boldsymbol{FEF}' &= \begin{pmatrix} \boldsymbol{F}_1\boldsymbol{EF}_1' & \boldsymbol{F}_1\boldsymbol{EF}_2' \\ \boldsymbol{F}_2\boldsymbol{EF}_1' & \boldsymbol{F}_2\boldsymbol{EF}_2' \end{pmatrix} \\ &= \begin{pmatrix} (\boldsymbol{F}_1\boldsymbol{DA})\boldsymbol{T}(\boldsymbol{F}_1\boldsymbol{DA})' & (\boldsymbol{F}_1\boldsymbol{DA})\boldsymbol{T}(\boldsymbol{F}_2\boldsymbol{DA})' \\ (\boldsymbol{F}_2\boldsymbol{DA})\boldsymbol{T}(\boldsymbol{F}_1\boldsymbol{DA})' & (\boldsymbol{F}_2\boldsymbol{DA})\boldsymbol{T}(\boldsymbol{F}_2\boldsymbol{DA})' \end{pmatrix} = \begin{pmatrix} \boldsymbol{I}_s & \boldsymbol{0} \\ \boldsymbol{0} & \boldsymbol{0} \end{pmatrix}. \end{aligned} \tag{46}$$

Thus $\boldsymbol{F}_1\boldsymbol{DA}$ is of rank s (by the converse of Theorem A.1.1 of Appendix A) and $\boldsymbol{F}_2\boldsymbol{DA} = \boldsymbol{0}$ because each diagonal element of $(\boldsymbol{F}_2\boldsymbol{DA})\boldsymbol{T}(\boldsymbol{F}_2\boldsymbol{DA})'$ is a quadratic form in a row of $\boldsymbol{F}_2\boldsymbol{DA}$ with positive definite matrix $\boldsymbol{T}$. Thus the covariance matrix of $\boldsymbol{FZ}$ is (46), and

$$\boldsymbol{FZ} = \begin{pmatrix} \boldsymbol{F}_1 \\ \boldsymbol{F}_2 \end{pmatrix} \boldsymbol{DAY} + \boldsymbol{FD}\boldsymbol{\lambda} = \begin{pmatrix} \boldsymbol{F}_1\boldsymbol{DAY} \\ \boldsymbol{0} \end{pmatrix} + \boldsymbol{FD}\boldsymbol{\lambda} = \begin{pmatrix} \boldsymbol{U}_1 \\ \boldsymbol{0} \end{pmatrix} + \boldsymbol{FD}\boldsymbol{\lambda}, \tag{47}$$

say. Clearly $\boldsymbol{U}_1$ has a nonsingular normal distribution. Let $\boldsymbol{F}^{-1} = (\boldsymbol{G}_1 \ \boldsymbol{G}_2)$. Then

$$\boldsymbol{Z} = \boldsymbol{G}_1\boldsymbol{U}_1 + \boldsymbol{D}\boldsymbol{\lambda}, \tag{48}$$

which is of the form (34). Q.E.D.

The developments in this section can be illuminated by considering the geometric interpretation put forward in the previous section. The density of $\boldsymbol{X}$ is constant on the ellipsoids (54) of Section 2.3. Since the transformation (2) is a linear transformation (i.e., a change of coordinate axes), the density of $\boldsymbol{Y}$ is constant on ellipsoids

$$(\boldsymbol{y} - \boldsymbol{C\mu})'(\boldsymbol{C\Sigma C}')^{-1}(\boldsymbol{y} - \boldsymbol{C\mu}) = k. \tag{49}$$

The marginal distribution of $X^{(1)}$ is the projection of the mass of the distribution of X onto the q-dimensional space of the first q coordinate axes. The surfaces of constant density are again ellipsoids. The projection of mass on any line is normal.

2.5. CONDITIONAL DISTRIBUTIONS AND MULTIPLE CORRELATION COEFFICIENT

2.5.1. Conditional Distributions

In this section we find that conditional distributions derived from joint normal distributions are normal. The conditional distributions are of a particularly simple nature because the means depend only linearly on the variates held fixed and the variances and covariances do not depend at all on the values of the fixed variates. The theory of partial and multiple correlation discussed in this section was originally developed by Karl Pearson (1896) for three variables and extended by Yule (1897a, 1897b).

Let X be distributed according to $N(\mu, \Sigma)$ (with Σ nonsingular). Let us partition

$$X = \begin{pmatrix} X^{(1)} \\ X^{(2)} \end{pmatrix} \tag{1}$$

as before into q- and $(p-q)$-component subvectors, respectively. We shall use the algebra developed in Section 2.4 here. The joint density of $Y^{(1)} = X^{(1)} - \Sigma_{12}\Sigma_{22}^{-1}X^{(2)}$ and $Y^{(2)} = X^{(2)}$ is

$$n\left(y^{(1)}|\mu^{(1)} - \Sigma_{12}\Sigma_{22}^{-1}\mu^{(2)}, \Sigma_{11} - \Sigma_{12}\Sigma_{22}^{-1}\Sigma_{21}\right)n\left(y^{(2)}|\mu^{(2)}, \Sigma_{22}\right).$$

The density of $X^{(1)}$ and $X^{(2)}$ then can be obtained from this expression by substituting $x^{(1)} - \Sigma_{12}\Sigma_{22}^{-1}x^{(2)}$ for $y^{(1)}$ and $x^{(2)}$ for $y^{(2)}$ (the Jacobian of this transformation being 1); the resulting density of $X^{(1)}$ and $X^{(2)}$ is

(2)

$$\begin{aligned} f(x^{(1)}, x^{(2)}) = {} & \frac{1}{(2\pi)^{\frac{1}{2}q}\sqrt{|\Sigma_{11\cdot 2}|}} \exp\left\{-\tfrac{1}{2}\left[(x^{(1)} - \mu^{(1)}) - \Sigma_{12}\Sigma_{22}^{-1}(x^{(2)} - \mu^{(2)})\right]'\right. \\ & \left.\cdot \Sigma_{11\cdot 2}^{-1}\left[(x^{(1)} - \mu^{(1)}) - \Sigma_{12}\Sigma_{22}^{-1}(x^{(2)} - \mu^{(2)})\right]\right\} \\ & \cdot \frac{1}{(2\pi)^{\frac{1}{2}(p-q)}\sqrt{|\Sigma_{22}|}} \exp\left[-\tfrac{1}{2}(x^{(2)} - \mu^{(2)})'\Sigma_{22}^{-1}(x^{(2)} - \mu^{(2)})\right], \end{aligned}$$

where

$$\boldsymbol{\Sigma}_{11\cdot 2} = \boldsymbol{\Sigma}_{11} - \boldsymbol{\Sigma}_{12}\boldsymbol{\Sigma}_{22}^{-1}\boldsymbol{\Sigma}_{21}. \tag{3}$$

This density must be $n(\boldsymbol{x}|\boldsymbol{\mu}, \boldsymbol{\Sigma})$. The conditional density of $\boldsymbol{X}^{(1)}$ given that $\boldsymbol{X}^{(2)} = \boldsymbol{x}^{(2)}$ is the quotient of (2) and the marginal density of $\boldsymbol{X}^{(2)}$ at the point $\boldsymbol{x}^{(2)}$, which is $n(\boldsymbol{x}^{(2)}|\boldsymbol{\mu}^{(2)}, \boldsymbol{\Sigma}_{22})$, the second factor of (2). The quotient is

(4)

$$f\left(\boldsymbol{x}^{(1)}|\boldsymbol{x}^{(2)}\right) = \frac{1}{(2\pi)^{\frac{1}{2}q}\sqrt{|\boldsymbol{\Sigma}_{11\cdot 2}|}}\exp\left\{-\tfrac{1}{2}\left[\left(\boldsymbol{x}^{(1)} - \boldsymbol{\mu}^{(1)}\right) - \boldsymbol{\Sigma}_{12}\boldsymbol{\Sigma}_{22}^{-1}\left(\boldsymbol{x}^{(2)} - \boldsymbol{\mu}^{(2)}\right)\right]'\right.$$

$$\left.\times\, \boldsymbol{\Sigma}_{11\cdot 2}^{-1}\left[\left(\boldsymbol{x}^{(1)} - \boldsymbol{\mu}^{(1)}\right) - \boldsymbol{\Sigma}_{12}\boldsymbol{\Sigma}_{22}^{-1}\left(\boldsymbol{x}^{(2)} - \boldsymbol{\mu}^{(2)}\right)\right]\right\}.$$

It is understood that $\boldsymbol{x}^{(2)}$ consists of $p - q$ numbers. The density $f(\boldsymbol{x}^{(1)}|\boldsymbol{x}^{(2)})$ is a q-variate normal density with mean

$$\mathscr{E}\left(\boldsymbol{X}^{(1)}|\boldsymbol{x}^{(2)}\right) = \boldsymbol{\mu}^{(1)} + \boldsymbol{\Sigma}_{12}\boldsymbol{\Sigma}_{22}^{-1}\left(\boldsymbol{x}^{(2)} - \boldsymbol{\mu}^{(2)}\right) = \boldsymbol{\nu}\left(\boldsymbol{x}^{(2)}\right), \tag{5}$$

say, and covariance matrix

$$\mathscr{E}\left\{\left[\boldsymbol{X}^{(1)} - \boldsymbol{\nu}\left(\boldsymbol{x}^{(2)}\right)\right]\left[\boldsymbol{X}^{(1)} - \boldsymbol{\nu}\left(\boldsymbol{x}^{(2)}\right)\right]'\middle|\boldsymbol{x}^{(2)}\right\} = \boldsymbol{\Sigma}_{11\cdot 2} = \boldsymbol{\Sigma}_{11} - \boldsymbol{\Sigma}_{12}\boldsymbol{\Sigma}_{22}^{-1}\boldsymbol{\Sigma}_{21}. \tag{6}$$

It should be noted that the mean of $\boldsymbol{X}^{(1)}$ given $\boldsymbol{x}^{(2)}$ is simply a linear function of $\boldsymbol{x}^{(2)}$, and the covariance matrix of $\boldsymbol{X}^{(1)}$ given $\boldsymbol{x}^{(2)}$ does not depend on $\boldsymbol{x}^{(2)}$ at all.

DEFINITION 2.5.1. *The matrix* $\boldsymbol{\beta} = \boldsymbol{\Sigma}_{12}\boldsymbol{\Sigma}_{22}^{-1}$ *is the matrix of regression coefficients of* $\boldsymbol{X}^{(1)}$ *on* $\boldsymbol{x}^{(2)}$.

The element in the ith row and $(k - q)$th column of $\boldsymbol{\beta} = \boldsymbol{\Sigma}_{12}\boldsymbol{\Sigma}_{22}^{-1}$ is often denoted by

$$\beta_{ik\cdot q+1,\dots,k-1,k+1,\dots,p}. \tag{7}$$

The vector $\boldsymbol{\mu}^{(1)} + \boldsymbol{\beta}(\boldsymbol{x}^{(2)} - \boldsymbol{\mu}^{(2)})$ is often called the *regression function*.

Let $\sigma_{ij\cdot q+1,\dots,p}$ be the i, jth element of $\boldsymbol{\Sigma}_{11\cdot 2}$. We call these *partial covariances*; $\sigma_{ii\cdot q+1,\dots,p}$ is a *partial variance*.

DEFINITION 2.5.2.

$$\rho_{ij\cdot q+1,\ldots,p} = \frac{\sigma_{ij\cdot q+1,\ldots,p}}{\sqrt{\sigma_{ii\cdot q+1,\ldots,p}}\sqrt{\sigma_{jj\cdot q+1,\ldots,p}}} \tag{8}$$

is the partial correlation between X_i and X_j holding $X_{q+1},\ldots, X_p$ fixed.

The numbering of the components of $\boldsymbol{X}$ is arbitrary and q is arbitrary. Hence, the above serves to define the conditional distribution of any q components of $\boldsymbol{X}$ given any other $p - q$ components. Indeed, we can consider the marginal distribution of any r components of $\boldsymbol{X}$ and define the conditional distribution of any q components given the other $r - q$ components.

THEOREM 2.5.1. *Let the components of $\boldsymbol{X}$ be divided into two groups composing the subvectors $\boldsymbol{X}^{(1)}$ and $\boldsymbol{X}^{(2)}$. Suppose the mean $\boldsymbol{\mu}$ is similarly divided into $\boldsymbol{\mu}^{(1)}$ and $\boldsymbol{\mu}^{(2)}$, and suppose the covariance matrix $\boldsymbol{\Sigma}$ of $\boldsymbol{X}$ is divided into $\boldsymbol{\Sigma}_{11}, \boldsymbol{\Sigma}_{12}, \boldsymbol{\Sigma}_{22}$, the covariance matrices of $\boldsymbol{X}^{(1)}$, of $\boldsymbol{X}^{(1)}$ and $\boldsymbol{X}^{(2)}$, and of $\boldsymbol{X}^{(2)}$, respectively. Then if the distribution of $\boldsymbol{X}$ is normal, the conditional distribution of $X^{(1)}$ given $\boldsymbol{X}^{(2)} = \boldsymbol{x}^{(2)}$ is normal with mean $\boldsymbol{\mu}^{(1)} + \boldsymbol{\Sigma}_{12}\boldsymbol{\Sigma}_{22}^{-1}(\boldsymbol{x}^{(2)} - \boldsymbol{\mu}^{(2)})$ and covariance matrix $\boldsymbol{\Sigma}_{11} - \boldsymbol{\Sigma}_{12}\boldsymbol{\Sigma}_{22}^{-1}\boldsymbol{\Sigma}_{21}$.*

As an example of the above considerations let us consider the bivariate normal distribution and find the conditional distribution of X_1 given $X_2 = x_2$. In this case $\boldsymbol{\mu}^{(1)} = \mu_1$, $\boldsymbol{\mu}^{(2)} = \mu_2$, $\boldsymbol{\Sigma}_{11} = \sigma_1^2$, $\boldsymbol{\Sigma}_{12} = \sigma_1\sigma_2\rho$, and $\boldsymbol{\Sigma}_{22} = \sigma_2^2$. Thus the 1×1 matrix of regression coefficients is $\boldsymbol{\Sigma}_{12}\boldsymbol{\Sigma}_{22}^{-1} = \sigma_1\rho/\sigma_2$, and the 1×1 matrix of partial covariances is

$$\boldsymbol{\Sigma}_{11\cdot 2} = \boldsymbol{\Sigma}_{11} - \boldsymbol{\Sigma}_{12}\boldsymbol{\Sigma}_{22}^{-1}\boldsymbol{\Sigma}_{21} = \sigma_1^2 - \sigma_1^2\sigma_2^2\rho^2/\sigma_2^2 = \sigma_1^2(1-\rho^2). \tag{9}$$

The density of X_1 given x_2 is $n[x_1|\mu_1 + (\sigma_1\rho/\sigma_2)(x_2 - \mu_2), \sigma_1^2(1-\rho^2)]$. The mean of this conditional distribution increases with x_2 when ρ is positive and decreases with x_2 when ρ is negative. It may be noted that when $\sigma_1 = \sigma_2$, for example, the mean of the conditional distribution of x_1 does not increase relative to μ_1 as much as x_2 increases relative to μ_2. [Galton (1889) observed that the average heights of sons whose fathers' heights were above average tended to be less than the fathers' heights; he called this effect "regression towards mediocrity."] The larger $|\rho|$ is, the smaller the variance of the conditional distribution, that is, the more information x_2 gives about x_1. This is another reason for considering ρ a measure of association between X_1 and X_2.

A geometrical interpretation of the theory is enlightening. The density $f(x_1, x_2)$ can be thought of as a surface $z = f(x_1, x_2)$ over the x_1, x_2-plane. If we intersect this surface with the plane $x_2 = c$, we obtain a curve $z = f(x_1, c)$ over the line $x_2 = c$ in the x_1, x_2-plane. The ordinate of this curve is proportional to the conditional density of X_1 given $x_2 = c$; that is, it is proportional to the ordinate of the curve of a univariate normal distribution. In the more general case it is convenient to consider the ellipsoids of constant density in the p-dimensional space. Then the surfaces of constant density of $f(x_1, \ldots, x_q | c_{q+1}, \ldots, c_p)$ are the intersections of the surfaces of constant density of $f(x_1, \ldots, x_p)$ and the hyperplanes $x_{q+1} = c_{q+1}, \ldots, x_p = c_p$; these are again ellipsoids.

Further clarification of these ideas may be had by consideration of an actual population which is idealized by a normal distribution. Consider, for example, a population of father-son pairs. If the population is reasonably homogeneous, the heights of fathers and the heights of corresponding sons have approximately a normal distribution (over a certain range). A conditional distribution may be obtained by considering sons of all fathers whose height is, say, 5 feet, 9 inches (to the accuracy of measurement); the heights of these sons will have an approximate univariate normal distribution. The mean of this normal distribution will differ from the mean of the heights of sons whose fathers' heights are 5 feet, 4 inches, say, but the variances will be about the same.

We could also consider triplets of observations, the height of a father, height of the oldest son, and height of the next oldest son. The collection of heights of two sons given that the fathers' heights are 5 feet, 9 inches is a conditional distribution of two variables; the correlation between the heights of oldest and next oldest sons is a partial correlation coefficient. Holding the fathers' heights constant eliminates the effect of heredity from fathers; however, one would expect that the partial correlation coefficient would be positive since the effect of mothers' heredity and environmental factors would tend to cause brothers' heights to vary similarly.

As we have remarked above, any conditional distribution obtained from a normal distribution is normal with the mean a linear function of the variables held fixed and the covariance matrix constant. In the case of nonnormal distributions the conditional distribution of one set of variates on another does not usually have these properties. However, one can construct nonnormal distributions such that some conditional distributions have these properties. This can be done by taking as the density of $\boldsymbol{X}$ the product $n(\boldsymbol{x}^{(1)} | \boldsymbol{\mu}^{(1)} + \boldsymbol{\beta}(\boldsymbol{x}^{(2)} - \boldsymbol{\mu}^{(2)}), \boldsymbol{\Sigma}_{11 \cdot 2}) f(\boldsymbol{x}^{(2)})$, where $f(\boldsymbol{x}^{(2)})$ is an arbitrary density.

2.5.2. The Multiple Correlation Coefficient

We again consider $\boldsymbol{X}$ partitioned into $\boldsymbol{X}^{(1)}$ and $\boldsymbol{X}^{(2)}$. We shall study some properties of $\boldsymbol{\beta X}^{(2)}$.

DEFINITION 2.5.3. *The vector* $\boldsymbol{X}^{(1\cdot 2)} = \boldsymbol{X}^{(1)} - \boldsymbol{\mu}^{(1)} - \boldsymbol{\beta}(\boldsymbol{X}^{(2)} - \boldsymbol{\mu}^{(2)})$ *is the vector of residuals of* $\boldsymbol{X}^{(1)}$ *from its regression on* $\boldsymbol{X}^{(2)}$.

THEOREM 2.5.2. *The components of* $\boldsymbol{X}^{(1\cdot 2)}$ *are uncorrelated with the components of* $\boldsymbol{X}^{(2)}$.

PROOF. The vector $\boldsymbol{X}^{(1\cdot 2)}$ is $\boldsymbol{Y}^{(1)} - \mathscr{E}\boldsymbol{Y}^{(1)}$ in (25) of Section 2.4. Q.E.D.

Let $\boldsymbol{\sigma}'_{(i)}$ be the ith row of $\boldsymbol{\Sigma}_{12}$ and $\boldsymbol{\beta}'_{(i)}$ the ith row of $\boldsymbol{\beta}$ (i.e., $\boldsymbol{\beta}'_{(i)} = \boldsymbol{\sigma}'_{(i)}\boldsymbol{\Sigma}_{22}^{-1}$). Let $\mathscr{V}(Z)$ be the variance of Z.

THEOREM 2.5.3. *For every vector* $\boldsymbol{\alpha}$

$$\mathscr{V}\left(X_i^{(1\cdot 2)}\right) \le \mathscr{V}\left(X_i - \boldsymbol{\alpha}'\boldsymbol{X}^{(2)}\right). \tag{10}$$

PROOF. By Theorem 2.5.2

$$\begin{aligned}(11)\quad \mathscr{V}&\left(X_i - \boldsymbol{\alpha}'\boldsymbol{X}^{(2)}\right)\\ &= \mathscr{E}\left[X_i - \mu_i - \boldsymbol{\alpha}'\left(\boldsymbol{X}^{(2)} - \boldsymbol{\mu}^{(2)}\right)\right]^2\\ &= \mathscr{E}\left[X_i^{(1\cdot 2)} - \mathscr{E}X_i^{(1\cdot 2)} + \left(\boldsymbol{\beta}_{(i)} - \boldsymbol{\alpha}\right)'\left(\boldsymbol{X}^{(2)} - \boldsymbol{\mu}^{(2)}\right)\right]^2\\ &= \mathscr{V}\left[X_i^{(1\cdot 2)}\right] + \left(\boldsymbol{\beta}_{(i)} - \boldsymbol{\alpha}\right)'\mathscr{E}\left(\boldsymbol{X}^{(2)} - \boldsymbol{\mu}^{(2)}\right)\left(\boldsymbol{X}^{(2)} - \boldsymbol{\mu}^{(2)}\right)'\left(\boldsymbol{\beta}_{(i)} - \boldsymbol{\alpha}\right)\\ &= \mathscr{V}\left(X_i^{(1\cdot 2)}\right) + \left(\boldsymbol{\beta}_{(i)} - \boldsymbol{\alpha}\right)'\boldsymbol{\Sigma}_{22}\left(\boldsymbol{\beta}_{(i)} - \boldsymbol{\alpha}\right).\end{aligned}$$

Since $\boldsymbol{\Sigma}_{22}$ is positive definite, the quadratic form in $(\boldsymbol{\beta}_{(i)} - \boldsymbol{\alpha})$ is nonnegative and attains its minimum of 0 at $\boldsymbol{\alpha} = \boldsymbol{\beta}_{(i)}$. Q.E.D.

Since $\mathscr{E}\boldsymbol{X}^{(1\cdot 2)} = \boldsymbol{0}$, $\mathscr{V}(X_i^{(1\cdot 2)}) = \mathscr{E}(X_i^{(1\cdot 2)})^2$. Thus $\mu_i + \boldsymbol{\beta}'_{(i)}(\boldsymbol{X}^{(2)} - \boldsymbol{\mu}^{(2)})$ is the *best linear predictor* of X_i in the sense that of all functions of $\boldsymbol{X}^{(2)}$ of the form $\boldsymbol{a}'\boldsymbol{X}^{(2)} + c$, the mean squared error of the above is a minimum.

THEOREM 2.5.4. *For every vector* $\boldsymbol{\alpha}$

$$\mathrm{Corr}\big(X_i, \boldsymbol{\beta}'_{(i)}\boldsymbol{X}^{(2)}\big) \geq \mathrm{Corr}\big(X_i, \boldsymbol{\alpha}'\boldsymbol{X}^{(2)}\big). \tag{12}$$

PROOF. Since the correlation between two variables is unchanged when either or both is multiplied by a positive constant, we can assume that $\mathscr{E}[\boldsymbol{\alpha}'(\boldsymbol{X}^{(2)} - \boldsymbol{\mu}^{(2)})]^2 = \mathscr{E}[\boldsymbol{\beta}'_{(i)}(\boldsymbol{X}^{(2)} - \boldsymbol{\mu}^{(2)})]^2$. Then the expansion of (10) is

$$\begin{aligned} \sigma_{ii} - 2\mathscr{E}(X_i - \mu_i)\boldsymbol{\beta}'_{(i)}\big(\boldsymbol{X}^{(2)} - \boldsymbol{\mu}^{(2)}\big) + \mathscr{V}\big(\boldsymbol{\beta}'_{(i)}\boldsymbol{X}^{(2)}\big) \\ \leq \sigma_{ii} - 2\mathscr{E}(X_i - \mu_i)\boldsymbol{\alpha}'\big(\boldsymbol{X}^{(2)} - \boldsymbol{\mu}^{(2)}\big) + \mathscr{V}\big(\boldsymbol{\alpha}'\boldsymbol{X}^{(2)}\big). \end{aligned} \tag{13}$$

This leads to

$$\frac{\mathscr{E}(X_i - \mu_i)\boldsymbol{\beta}'_{(i)}\big(\boldsymbol{X}^{(2)} - \boldsymbol{\mu}^{(2)}\big)}{\sqrt{\sigma_{ii}\mathscr{V}\big(\boldsymbol{\beta}'_{(i)}\boldsymbol{X}^{(2)}\big)}} \geq \frac{\mathscr{E}(X_i - \mu_i)\boldsymbol{\alpha}'\big(\boldsymbol{X}^{(2)} - \boldsymbol{\mu}^{(2)}\big)}{\sqrt{\sigma_{ii}\mathscr{V}\big(\boldsymbol{\alpha}'\boldsymbol{X}^{(2)}\big)}}. \tag{14}$$

Q.E.D.

DEFINITION 2.5.4. *The maximum correlation between* X_i *and the linear combination* $\boldsymbol{\alpha}'\boldsymbol{X}^{(2)}$ *is called the multiple correlation coefficient between* X_i *and* $\boldsymbol{X}^{(2)}$.

It follows that this is

$$\begin{aligned} \bar{R}_{i\cdot q+1,\dots,p} &= \frac{\mathscr{E}\boldsymbol{\beta}'_{(i)}\big(\boldsymbol{X}^{(2)} - \boldsymbol{\mu}^{(2)}\big)(X_i - \mu_i)}{\sqrt{\sigma_{ii}}\sqrt{\mathscr{E}\boldsymbol{\beta}'_{(i)}\big(\boldsymbol{X}^{(2)} - \boldsymbol{\mu}^{(2)}\big)\big(\boldsymbol{X}^{(2)} - \boldsymbol{\mu}^{(2)}\big)'\boldsymbol{\beta}_{(i)}}} \\ &= \frac{\boldsymbol{\sigma}'_{(i)}\boldsymbol{\Sigma}_{22}^{-1}\boldsymbol{\sigma}_{(i)}}{\sqrt{\sigma_{ii}}\sqrt{\boldsymbol{\sigma}'_{(i)}\boldsymbol{\Sigma}_{22}^{-1}\boldsymbol{\sigma}_{(i)}}} = \frac{\sqrt{\boldsymbol{\sigma}'_{(i)}\boldsymbol{\Sigma}_{22}^{-1}\boldsymbol{\sigma}_{(i)}}}{\sqrt{\sigma_{ii}}}. \end{aligned} \tag{15}$$

A useful formula is

$$1 - \bar{R}^2_{i\cdot q+1,\dots,p} = \frac{\sigma_{ii} - \boldsymbol{\sigma}'_{(i)}\boldsymbol{\Sigma}_{22}^{-1}\boldsymbol{\sigma}_{(i)}}{\sigma_{ii}} = \frac{|\boldsymbol{\Sigma}_i|}{\sigma_{ii}|\boldsymbol{\Sigma}_{22}|}, \tag{16}$$

where Theorem A.3.2 of Appendix A has been applied to

$$\Sigma_i = \begin{pmatrix} \sigma_{ii} & \sigma'_{(i)} \\ \sigma_{(i)} & \Sigma_{22} \end{pmatrix}. \tag{17}$$

Since

$$\sigma_{ii\cdot q+1,\ldots,p} = \sigma_{ii} - \sigma'_{(i)}\Sigma_{22}^{-1}\sigma_{(i)}, \tag{18}$$

it follows that

$$\sigma_{ii\cdot q+1,\ldots,p} = \left(1 - \bar{R}^2_{i\cdot q+1,\ldots,p}\right)\sigma_{ii}. \tag{19}$$

This shows incidentally that any partial variance of a component of X cannot be greater than the variance. In fact, the larger $\bar{R}_{i\cdot q+1,\ldots,p}$ is, the greater the reduction in variance by going to the conditional distribution. This fact is another reason for considering the multiple correlation coefficient a measure of association between X_i and $X^{(2)}$.

That $\beta'_{(i)}X^{(2)}$ is the best linear predictor of X_i and has the maximum correlation between X_i and linear functions of $X^{(2)}$ depends only on the covariance structure without regard to normality. Even if X does not have a normal distribution, the regression of $X^{(1)}$ on $X^{(2)}$ can be defined by $\mu^{(1)} + \Sigma_{12}\Sigma_{22}^{-1}(X^{(2)} - \mu^{(2)})$; the residuals can be defined by Definition 2.5.3; and partial covariances and correlations can be defined as the covariances and correlations of residuals yielding (3) and (8). Then these quantities do not necessarily have interpretations in terms of conditional distributions. In the case of normality $\mu_i + \beta'_{(i)}(x^{(2)} - \mu^{(2)})$ is the conditional expectation of X_i given $X^{(2)} = x^{(2)}$. Without regard to normality, $X_i - \mathscr{E}X_i|X^{(2)}$ is uncorrelated with any function of $X^{(2)}$, $\mathscr{E}X_i|X^{(2)}$ minimizes $\mathscr{E}[X_i - h(X^{(2)})]^2$ with respect to functions of $X^{(2)}$, $h(X^{(2)})$, and $\mathscr{E}X_i|X^{(2)}$ maximizes the correlation between X_i and functions of $X^{(2)}$. (See Problems 48 to 51.)

2.5.3. Some Formulas for Partial Correlations

We now consider relations between several conditional distributions obtained by holding several different sets of variates fixed. These relations are useful because they enable us to compute one set of conditional parameters from another set. A very special case is

$$\rho_{12\cdot 3} = \frac{\rho_{12} - \rho_{13}\rho_{23}}{\sqrt{1-\rho_{13}^2}\sqrt{1-\rho_{23}^2}}; \tag{20}$$

this follows from (8) when $p = 3$ and $q = 2$. We shall now find a generalization of this result. The derivation is tedious, but is given here for completeness.

Let

$$X = \begin{pmatrix} X^{(1)} \\ X^{(2)} \\ X^{(3)} \end{pmatrix}, \tag{21}$$

where $X^{(1)}$ is of p_1 components, $X^{(2)}$ of p_2 components, and $X^{(3)}$ of p_3 components. Suppose we have the conditional distribution of $X^{(1)}$ and $X^{(2)}$ given $X^{(3)} = x^{(3)}$; how do we find the conditional distribution of $X^{(1)}$ given $X^{(2)} = x^{(2)}$ and $X^{(3)} = x^{(3)}$? We use the fact that the conditional density of $X^{(1)}$ given $X^{(2)} = x^{(2)}$ and $X^{(3)} = x^{(3)}$ is

$$\begin{aligned} f(x^{(1)}|x^{(2)}, x^{(3)}) &= \frac{f(x^{(1)}, x^{(2)}, x^{(3)})}{f(x^{(2)}, x^{(3)})} \\ &= \frac{f(x^{(1)}, x^{(2)}, x^{(3)})/f(x^{(3)})}{f(x^{(2)}, x^{(3)})/f(x^{(3)})} \\ &= \frac{f(x^{(1)}, x^{(2)}|x^{(3)})}{f(x^{(2)}|x^{(3)})}. \end{aligned} \tag{22}$$

In the case of normality the conditional covariance matrix of $X^{(1)}$ and $X^{(2)}$ given $X^{(3)} = x^{(3)}$ is

$$\begin{aligned} \mathscr{C}\left[\begin{pmatrix} X^{(1)} \\ X^{(2)} \end{pmatrix}\middle| x^{(3)}\right] &= \begin{pmatrix} \Sigma_{11} & \Sigma_{12} \\ \Sigma_{21} & \Sigma_{22} \end{pmatrix} - \begin{pmatrix} \Sigma_{13} \\ \Sigma_{23} \end{pmatrix} \Sigma_{33}^{-1} (\Sigma_{31} \quad \Sigma_{32}) \\ &= \begin{pmatrix} \Sigma_{11\cdot 3} & \Sigma_{12\cdot 3} \\ \Sigma_{21\cdot 3} & \Sigma_{22\cdot 3} \end{pmatrix}, \end{aligned} \tag{23}$$

say, where

$$\Sigma = \begin{pmatrix} \Sigma_{11} & \Sigma_{12} & \Sigma_{13} \\ \Sigma_{21} & \Sigma_{22} & \Sigma_{23} \\ \Sigma_{31} & \Sigma_{32} & \Sigma_{33} \end{pmatrix}. \tag{24}$$

The conditional covariance of $\boldsymbol{X}^{(1)}$ given $\boldsymbol{X}^{(2)} = \boldsymbol{x}^{(2)}$ and $\boldsymbol{X}^{(3)} = \boldsymbol{x}^{(3)}$ is calculated from the conditional covariances of $\boldsymbol{X}^{(1)}$ and $\boldsymbol{X}^{(2)}$ given $\boldsymbol{X}^{(3)} = \boldsymbol{x}^{(3)}$ as

$$\mathscr{C}\left[\boldsymbol{X}^{(1)} | \boldsymbol{x}^{(2)}, \boldsymbol{x}^{(3)}\right] = \boldsymbol{\Sigma}_{11\cdot 3} - \boldsymbol{\Sigma}_{12\cdot 3}\left(\boldsymbol{\Sigma}_{22\cdot 3}\right)^{-1}\boldsymbol{\Sigma}_{21\cdot 3}. \tag{25}$$

This result permits the calculation of $\sigma_{ij\cdot p_1+1,\dots,p}$, $i, j = 1, \dots, p_1$, from $\sigma_{ij\cdot p_1+p_2+1,\dots,p}$, $i, j = 1, \dots, p_1 + p_2$.

In particular, for $p_1 = q$, $p_2 = 1$, and $p_3 = p - q - 1$, we obtain

$$\sigma_{ij\cdot q+1,\dots,p} = \sigma_{ij\cdot q+2,\dots,p} - \frac{\sigma_{i,q+1\cdot q+2,\dots,p}\,\sigma_{j,q+1\cdot q+2,\dots,p}}{\sigma_{q+1,q+1\cdot q+2,\dots,p}}, \qquad i, j = 1, \dots, q. \tag{26}$$

Since

$$\sigma_{ii\cdot q+1,\dots,p} = \sigma_{ii\cdot q+2,\dots,p}\left(1 - \rho^2_{i,q+1\cdot q+2,\dots,p}\right), \tag{27}$$

we obtain

$$\rho_{ij\cdot q+1,\dots,p} = \frac{\rho_{ij\cdot q+2,\dots,p} - \rho_{i,q+1\cdot q+2,\dots,p}\,\rho_{j,q+1\cdot q+2,\dots,p}}{\sqrt{1 - \rho^2_{i,q+1\cdot q+2,\dots,p}}\sqrt{1 - \rho^2_{j,q+1\cdot q+2,\dots,p}}}. \tag{28}$$

This is a useful recursion formula to compute from $\{\rho_{ij}\}$ in succession $\{\rho_{ij\cdot p}\}$, $\{\rho_{ij\cdot p-1,p}\}, \dots, \rho_{12\cdot 3,\dots,p}$.

2.6. THE CHARACTERISTIC FUNCTION; MOMENTS

2.6.1. The Characteristic Function

The characteristic function of a multivariate normal distribution has a form similar to the density function. From the characteristic function moments and cumulants can be found easily.

DEFINITION 2.6.1. *The characteristic function of a random vector $\boldsymbol{X}$ is*

$$\phi(\boldsymbol{t}) = \mathscr{E}e^{i\boldsymbol{t}'\boldsymbol{X}} \tag{1}$$

defined for every real vector $\boldsymbol{t}$.

To make this definition meaningful we need to define the expected value of a complex-valued function of a random vector.

DEFINITION 2.6.2. *Let the complex-valued function $g(\boldsymbol{x})$ be written as $g(\boldsymbol{x}) = g_1(\boldsymbol{x}) + ig_2(\boldsymbol{x})$, where $g_1(\boldsymbol{x})$ and $g_2(\boldsymbol{x})$ are real-valued. Then the expected value of $g(\boldsymbol{X})$ is*

$$\mathscr{E}g(\boldsymbol{X}) = \mathscr{E}g_1(\boldsymbol{X}) + i\mathscr{E}g_2(\boldsymbol{X}). \tag{2}$$

In particular, since $e^{i\theta} = \cos\theta + i\sin\theta$,

$$\mathscr{E}e^{i\boldsymbol{t}'\boldsymbol{X}} = \mathscr{E}\cos\boldsymbol{t}'\boldsymbol{X} + i\mathscr{E}\sin\boldsymbol{t}'\boldsymbol{X}. \tag{3}$$

To evaluate the characteristic function of a vector $\boldsymbol{X}$, it is often convenient to use the following lemma:

LEMMA 2.6.1. *Let $\boldsymbol{X}' = (\boldsymbol{X}^{(1)\prime}\ \boldsymbol{X}^{(2)\prime})$. If $\boldsymbol{X}^{(1)}$ and $\boldsymbol{X}^{(2)}$ are independent and $g(\boldsymbol{x}) = g^{(1)}(\boldsymbol{x}^{(1)})g^{(2)}(\boldsymbol{x}^{(2)})$, then*

$$\mathscr{E}g(\boldsymbol{X}) = \mathscr{E}g^{(1)}(\boldsymbol{X}^{(1)})\mathscr{E}g^{(2)}(\boldsymbol{X}^{(2)}). \tag{4}$$

PROOF. If $g(\boldsymbol{x})$ is real-valued and $\boldsymbol{X}$ has a density,

(5)

$$\begin{aligned}
\mathscr{E}g(\boldsymbol{X}) &= \int_{-\infty}^{\infty}\cdots\int_{-\infty}^{\infty} g(\boldsymbol{x})f(\boldsymbol{x})\,dx_1\cdots dx_p \\
&= \int_{-\infty}^{\infty}\cdots\int_{-\infty}^{\infty} g^{(1)}(\boldsymbol{x}^{(1)})g^{(2)}(\boldsymbol{x}^{(2)})f^{(1)}(\boldsymbol{x}^{(1)})f^{(2)}(\boldsymbol{x}^{(2)})\,dx_1\cdots dx_p \\
&= \int_{-\infty}^{\infty}\cdots\int_{-\infty}^{\infty} g^{(1)}(\boldsymbol{x}^{(1)})f^{(1)}(\boldsymbol{x}^{(1)})\,dx_1\cdots dx_q \\
&\qquad \cdot\int_{-\infty}^{\infty}\cdots\int_{-\infty}^{\infty} g^{(2)}(\boldsymbol{x}^{(2)})f^{(2)}(\boldsymbol{x}^{(2)})\,dx_{q+1}\cdots dx_p \\
&= \mathscr{E}g^{(1)}(\boldsymbol{X}^{(1)})\mathscr{E}g^{(2)}(\boldsymbol{X}^{(2)}).
\end{aligned}$$

If $g(x)$ is complex-valued,

$$g(\boldsymbol{x}) = \left[g_1^{(1)}(\boldsymbol{x}^{(1)}) + ig_2^{(1)}(\boldsymbol{x}^{(1)})\right]\left[g_1^{(2)}(\boldsymbol{x}^{(2)}) + ig_2^{(2)}(\boldsymbol{x}^{(2)})\right] \tag{6}$$
$$= g_1^{(1)}(\boldsymbol{x}^{(1)})g_1^{(2)}(\boldsymbol{x}^{(2)}) - g_2^{(1)}(\boldsymbol{x}^{(1)})g_2^{(2)}(\boldsymbol{x}^{(2)})$$
$$+ i\left[g_2^{(1)}(\boldsymbol{x}^{(1)})g_1^{(2)}(\boldsymbol{x}^{(2)}) + g_1^{(1)}(\boldsymbol{x}^{(1)})g_2^{(2)}(\boldsymbol{x}^{(2)})\right].$$

Then

$$\mathscr{E}g(\boldsymbol{X}) = \mathscr{E}\left[g_1^{(1)}(\boldsymbol{X}^{(1)})g_1^{(2)}(\boldsymbol{X}^{(2)}) - g_2^{(1)}(\boldsymbol{X}^{(1)})g_2^{(2)}(\boldsymbol{X}^{(2)})\right] \tag{7}$$
$$+ i\mathscr{E}\left[g_2^{(1)}(\boldsymbol{X}^{(1)})g_1^{(2)}(\boldsymbol{X}^{(2)}) + g_1^{(1)}(\boldsymbol{X}^{(1)})g_2^{(2)}(\boldsymbol{X}^{(2)})\right]$$
$$= \mathscr{E}g_1^{(1)}(\boldsymbol{X}^{(1)})\mathscr{E}g_1^{(2)}(\boldsymbol{X}^{(2)}) - \mathscr{E}g_2^{(1)}(\boldsymbol{X}^{(1)})\mathscr{E}g_2^{(2)}(\boldsymbol{X}^{(2)})$$
$$+ i\left[\mathscr{E}g_2^{(1)}(\boldsymbol{X}^{(1)})\mathscr{E}g_1^{(2)}(\boldsymbol{X}^{(2)}) + \mathscr{E}g_1^{(1)}(\boldsymbol{X}^{(1)})\mathscr{E}g_2^{(2)}(\boldsymbol{X}^{(2)})\right]$$
$$= \left[\mathscr{E}g_1^{(1)}(\boldsymbol{X}^{(1)}) + i\mathscr{E}g_2^{(1)}(\boldsymbol{X}^{(1)})\right]\left[\mathscr{E}g_1^{(2)}(\boldsymbol{X}^{(2)}) + i\mathscr{E}g_2^{(2)}(\boldsymbol{X}^{(2)})\right]$$
$$= \mathscr{E}g^{(1)}(\boldsymbol{X}^{(1)})\mathscr{E}g^{(2)}(\boldsymbol{X}^{(2)}).$$

Q.E.D.

By applying Lemma 2.6.1 successively to $g(\boldsymbol{X}) = e^{i\boldsymbol{t}'\boldsymbol{X}}$, we derive

LEMMA 2.6.2. *If the components of* $\boldsymbol{X}$ *are mutually independent,*

$$\mathscr{E}e^{i\boldsymbol{t}'\boldsymbol{X}} = \prod_{j=1}^{p} \mathscr{E}e^{it_jX_j}. \tag{8}$$

We now find the characteristic function of a random vector with a normal distribution.

THEOREM 2.6.1. *The characteristic function of* $\boldsymbol{X}$ *distributed according to* $N(\boldsymbol{\mu}, \boldsymbol{\Sigma})$ *is*

$$\phi(\boldsymbol{t}) = \mathscr{E}e^{i\boldsymbol{t}'\boldsymbol{X}} = e^{i\boldsymbol{t}'\boldsymbol{\mu} - \frac{1}{2}\boldsymbol{t}'\boldsymbol{\Sigma}\boldsymbol{t}} \tag{9}$$

for every real vector $\boldsymbol{t}$.

PROOF. From Corollary A.1.6 of Appendix A we know there is a nonsingular matrix $\boldsymbol{C}$ such that

$$\boldsymbol{C}'\boldsymbol{\Sigma}^{-1}\boldsymbol{C} = \boldsymbol{I}. \tag{10}$$

Thus

$$\boldsymbol{\Sigma}^{-1} = \boldsymbol{C}'^{-1}\boldsymbol{C}^{-1} = (\boldsymbol{C}\boldsymbol{C}')^{-1}. \tag{11}$$

Let

$$\boldsymbol{X} - \boldsymbol{\mu} = \boldsymbol{C}\boldsymbol{Y}. \tag{12}$$

Then $\boldsymbol{Y}$ is distributed according to $N(\boldsymbol{0}, \boldsymbol{I})$.

Now the characteristic function of $\boldsymbol{Y}$ is

$$\psi(\boldsymbol{u}) = \mathscr{E}e^{i\boldsymbol{u}'\boldsymbol{Y}} = \prod_{j=1}^{p} \mathscr{E}e^{iu_jY_j}. \tag{13}$$

Since Y_j is distributed according to $N(0,1)$,

$$\psi(\boldsymbol{u}) = \prod_{j=1}^{p} e^{-\frac{1}{2}u_j^2} = e^{-\frac{1}{2}\boldsymbol{u}'\boldsymbol{u}}. \tag{14}$$

Thus

$$\begin{aligned} \phi(\boldsymbol{t}) &= \mathscr{E}e^{i\boldsymbol{t}'\boldsymbol{X}} = \mathscr{E}e^{i\boldsymbol{t}'(\boldsymbol{C}\boldsymbol{Y}+\boldsymbol{\mu})} \\ &= e^{i\boldsymbol{t}'\boldsymbol{\mu}}\mathscr{E}e^{i\boldsymbol{t}'\boldsymbol{C}\boldsymbol{Y}} \\ &= e^{i\boldsymbol{t}'\boldsymbol{\mu}}e^{-\frac{1}{2}(\boldsymbol{t}'\boldsymbol{C})(\boldsymbol{t}'\boldsymbol{C})'} \end{aligned} \tag{15}$$

for $\boldsymbol{t}'\boldsymbol{C} = \boldsymbol{u}'$; the third equality is verified by writing both sides of this equality as integrals. But this is

$$\begin{aligned} \phi(\boldsymbol{t}) &= e^{i\boldsymbol{t}'\boldsymbol{\mu}}e^{-\frac{1}{2}\boldsymbol{t}'\boldsymbol{C}\boldsymbol{C}'\boldsymbol{t}} \\ &= e^{i\boldsymbol{t}'\boldsymbol{\mu}-\frac{1}{2}\boldsymbol{t}'\boldsymbol{\Sigma}\boldsymbol{t}} \end{aligned} \tag{16}$$

by (11). This proves the theorem.

The characteristic function of the normal distribution is very useful. For example, we can use this method of proof to demonstrate the results of Section 2.4. If $\boldsymbol{Z} = \boldsymbol{DX}$, then the characteristic function of $\boldsymbol{Z}$ is

$$(17) \qquad \mathscr{E} e^{i t' \boldsymbol{Z}} = \mathscr{E} e^{i t' \boldsymbol{DX}} = \mathscr{E} e^{i(\boldsymbol{D}' t)' \boldsymbol{X}}$$

$$= e^{i(\boldsymbol{D}'t)'\boldsymbol{\mu} - \frac{1}{2}(\boldsymbol{D}'t)'\boldsymbol{\Sigma}(\boldsymbol{D}'t)}$$

$$= e^{it'(\boldsymbol{D\mu}) - \frac{1}{2}t'(\boldsymbol{D\Sigma D}')t},$$

which is the characteristic function of $N(\boldsymbol{D\mu}, \boldsymbol{D\Sigma D}')$ (by Theorem 2.6.1).

It is interesting to use the characteristic function to show that it is only the multivariate normal distribution that has the property that every linear combination of variates is normally distributed. Consider a vector $\boldsymbol{Y}$ of p components with density $f(y)$ and characteristic function

$$(18) \qquad \psi(\boldsymbol{u}) = \mathscr{E} e^{i\boldsymbol{u}'\boldsymbol{Y}} = \int_{-\infty}^{\infty} \cdots \int_{-\infty}^{\infty} e^{i\boldsymbol{u}'y} f(y)\, dy_1 \cdots dy_p,$$

and suppose the mean of $\boldsymbol{Y}$ is $\boldsymbol{\mu}$ and the covariance matrix is $\boldsymbol{\Sigma}$. Suppose $\boldsymbol{u}'\boldsymbol{Y}$ is normally distributed for every $\boldsymbol{u}$. Then the characteristic function of such a linear combination is

$$(19) \qquad \mathscr{E} e^{it\boldsymbol{u}'\boldsymbol{Y}} = e^{it\boldsymbol{u}'\boldsymbol{\mu} - \frac{1}{2}t^2\boldsymbol{u}'\boldsymbol{\Sigma u}}.$$

Now set $t = 1$. Since the right-hand side is then the characteristic function of $N(\boldsymbol{\mu}, \boldsymbol{\Sigma})$, the result is proved (by Theorems 2.6.1 above and 2.6.3 below).

THEOREM 2.6.2. *If every linear combination of the components of a vector* $\boldsymbol{Y}$ *is normally distributed, then* $\boldsymbol{Y}$ *is normally distributed.*

It might be pointed out in passing that it is essential that *every* linear combination be normally distributed for Theorem 2.6.2 to hold. For instance, if $\boldsymbol{Y} = (Y_1, Y_2)'$ and Y_1 and Y_2 are not independent, then Y_1 and Y_2 can each have a marginal normal distribution. An example is most easily given geometrically. Let X_1, X_2 have a joint normal distribution with means 0. Move the same mass in Figure 2.1 below from rectangle A to C and from B to D. It will be seen that the resulting distribution of $\boldsymbol{Y}$ is such that the marginal distribu-

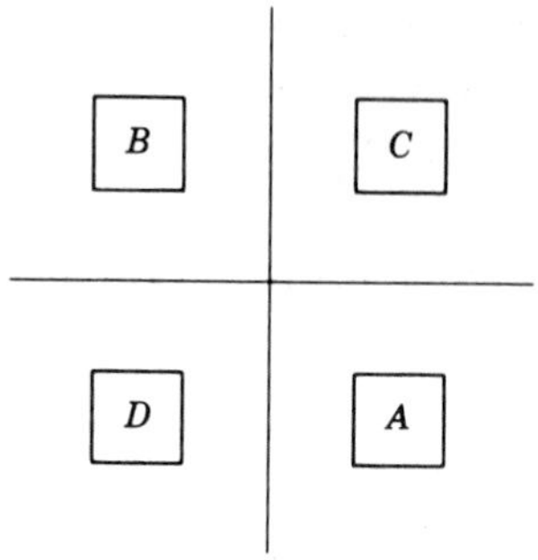

Figure 2.1

tions of Y_1 and Y_2 are the same as of X_1 and X_2, respectively, which are normal, and yet the joint distribution of Y_1 and Y_2 is not normal.

This example can be used also to demonstrate that two variables, Y_1 and Y_2, can be uncorrelated and the marginal distribution of each may be normal, but the pair need not have a joint normal distribution and need not be independent. This is done by choosing the rectangles so that for the resultant distribution the expected value of $Y_1 Y_2$ is zero. It is clear geometrically that this can be done.

For future reference we state two useful theorems concerning characteristic functions.

THEOREM 2.6.3. *If the random vector* $\boldsymbol{X}$ *has the density* $f(\boldsymbol{x})$ *and the characteristic function* $\phi(\boldsymbol{t})$, *then*

$$f(\boldsymbol{x}) = \frac{1}{(2\pi)^p} \int_{-\infty}^{\infty} \cdots \int_{-\infty}^{\infty} e^{-i\boldsymbol{t}'\boldsymbol{x}} \phi(\boldsymbol{t})\, dt_1 \cdots dt_p. \tag{20}$$

This shows that the characteristic function determines the density function uniquely. If $\boldsymbol{X}$ does not have a density, the characteristic function uniquely defines the probability of any "continuity interval." In the univariate case a continuity interval is an interval such that the cdf does not have a discontinuity at an endpoint of the interval.

THEOREM 2.6.4. *Let* $\{F_j(\boldsymbol{x})\}$ *be a sequence of* cdf's *and let* $\{\phi_j(\boldsymbol{t})\}$ *be the sequence of corresponding characteristic functions. A necessary and sufficient condition that* $F_j(\boldsymbol{x})$ *converge to a* cdf $F(\boldsymbol{x})$ *is that, for every* $\boldsymbol{t}$, $\phi_j(\boldsymbol{t})$ *converge to a limit* $\phi(\boldsymbol{t})$ *that is continuous at* $\boldsymbol{t} = \boldsymbol{0}$. *When this condition is satisfied, the limit*

$\phi(\mathbf{t})$ is identical with the characteristic function of the limiting distribution $F(\mathbf{x})$.

For the proofs of these two theorems, the reader is referred to Cramér (1946), Sections 10.6 and 10.7.

2.6.2. The Moments and Cumulants

The moments of $X_1, \dots, X_p$ with a joint normal distribution can be obtained from the characteristic function (9). The mean is

$$\mathscr{E}X_h = \frac{1}{i}\frac{\partial \phi}{\partial t_h}\bigg|_{\mathbf{t}=\mathbf{0}} = \frac{1}{i}\Big\{-\sum_j \sigma_{hj}t_j + i\mu_h\Big\}\phi(\mathbf{t})\bigg|_{\mathbf{t}=\mathbf{0}} = \mu_h. \tag{21}$$

The second moment is

$$\mathscr{E}X_hX_j = \frac{1}{i^2}\frac{\partial^2 \phi}{\partial t_h \partial t_j}\bigg|_{\mathbf{t}=\mathbf{0}} = \frac{1}{i^2}\Big\{\Big(-\sum_k \sigma_{hk}t_k + i\mu_h\Big)\Big(-\sum_k \sigma_{kj}t_k + i\mu_j\Big) - \sigma_{hj}\Big\}\phi(\mathbf{t})\bigg|_{\mathbf{t}=\mathbf{0}} = \sigma_{hj} + \mu_h\mu_j. \tag{22}$$

Thus

$$\text{Variance}(X_i) = \mathscr{E}(X_i - \mu_i)^2 = \sigma_{ii}, \tag{23}$$

$$\text{Covariance}(X_i, X_j) = \mathscr{E}(X_i - \mu_i)(X_j - \mu_j) = \sigma_{ij}. \tag{24}$$

Any third moment about the mean is

$$\mathscr{E}(X_i - \mu_i)(X_j - \mu_j)(X_k - \mu_k) = 0. \tag{25}$$

The fourth moment about the mean is

$$\mathscr{E}(X_i - \mu_i)(X_j - \mu_j)(X_k - \mu_k)(X_l - \mu_l) = \sigma_{ij}\sigma_{kl} + \sigma_{ik}\sigma_{jl} + \sigma_{il}\sigma_{jk}. \tag{26}$$

DEFINITION 2.6.3. *If all the moments of a distribution exist, then the cumulants are the coefficients* κ *in*

$$\log \phi(t) = \sum_{s_1, \ldots, s_p = 0}^{\infty} \kappa_{s_1 \cdots s_p} \frac{(it_1)^{s_1} \cdots (it_p)^{s_p}}{s_1! \cdots s_p!}. \tag{27}$$

In the case of the multivariate normal distribution $\kappa_{10 \cdots 0} = \mu_1, \ldots, \kappa_{0 \cdots 01} = \mu_p$, $\kappa_{20 \cdots 0} = \sigma_{11}, \ldots, \kappa_{0 \cdots 02} = \sigma_{pp}$, $\kappa_{110 \cdots 0} = \sigma_{12}, \ldots$. The cumulants for which $\sum s_i > 2$ are 0.

PROBLEMS

1. (Sec. 2.2) Let $f(x, y) = 1,\ 0 \le x \le 1,\ 0 \le y \le 1,$
$= 0$, otherwise.

Find:

(a) $F(x, y)$.
(b) $F(x)$.
(c) $f(x)$.
(d) $f(x|y)$. [Note: $f(x_0|y_0) = 0$ if $f(x_0, y_0) = 0$.]
(e) $\mathscr{E} X^n Y^m$.
(f) Prove X and Y are independent.

2. (Sec. 2.2) Let $f(x, y) = 2,\ 0 \le y \le x \le 1,$
$= 0$, otherwise.

Find:

(a) $F(x, y)$.
(b) $F(x)$.
(c) $f(x)$.
(d) $G(y)$.
(e) $g(y)$.
(f) $f(x|y)$.
(g) $f(y|x)$.
(h) $\mathscr{E} X^n Y^m$.
(i) Are X and Y independent?

3. (Sec. 2.2) Let $f(x, y) = C$ for $x^2 + y^2 \le k^2$ and 0 elsewhere. Prove $C = 1/(\pi k^2)$, $\mathscr{E} X = \mathscr{E} Y = 0$, $\mathscr{E} X^2 = \mathscr{E} Y^2 = k^2/4$, and $\mathscr{E} XY = 0$. Are X and Y independent?

4. (Sec. 2.2) Let $F(x_1, x_2)$ be the joint cdf of X_1, X_2 and let $F_i(x_i)$ be the marginal cdf of X_i, $i = 1, 2$. Prove that if $F_i(x_i)$ is continuous, $i = 1, 2$, then $F(x_1, x_2)$ is continuous.

5. (Sec. 2.2) Show that if the set $X_1, \ldots, X_r$ is independent of the set $X_{r+1}, \ldots, X_p$, then $\mathscr{E} g(X_1, \ldots, X_r) h(X_{r+1}, \ldots, X_p) = \mathscr{E} g(X_1, \ldots, X_r) \mathscr{E} h(X_{r+1}, \ldots, X_p)$.

6. (Sec. 2.3) Sketch the ellipses $f(x, y) = 0.06$, where $f(x, y)$ is the bivariate normal density with

(*a*) $\mu_x = 1, \mu_y = 2, \sigma_x^2 = 1, \sigma_y^2 = 1, \rho_{xy} = 0.$
(*b*) $\mu_x = 0, \mu_y = 0, \sigma_x^2 = 1, \sigma_y^2 = 1, \rho_{xy} = 0.$
(*c*) $\mu_x = 0, \mu_y = 0, \sigma_x^2 = 1, \sigma_y^2 = 1, \rho_{xy} = 0.2.$
(*d*) $\mu_x = 0, \mu_y = 0, \sigma_x^2 = 1, \sigma_y^2 = 1, \rho_{xy} = 0.8.$
(*e*) $\mu_x = 0, \mu_y = 0, \sigma_x^2 = 4, \sigma_y^2 = 1, \rho_{xy} = 0.8.$

7. (Sec. 2.3) Find $\boldsymbol{b}$ and $\boldsymbol{A}$ so that the following densities can be written in the form of (23). Also find μ_x, μ_y, σ_x, σ_y and ρ_{xy}.

(*a*) $\dfrac{1}{2\pi}\exp\{-\tfrac{1}{2}[(x-1)^2 + (y-2)^2]\}.$

(*b*) $\dfrac{1}{2.4\pi}\exp\left(-\dfrac{\dfrac{x^2}{4} - 1.6\dfrac{xy}{2} + y^2}{0.72}\right).$

(*c*) $\dfrac{1}{2\pi}\exp[-\tfrac{1}{2}(x^2 + y^2 + 4x - 6y + 13)].$

(*d*) $\dfrac{1}{2\pi}\exp[-\tfrac{1}{2}(2x^2 + y^2 + 2xy - 22x - 14y + 65)].$

8. (Sec. 2.3) For each matrix $\boldsymbol{A}$ in Problem 7 find $\boldsymbol{C}$ so that $\boldsymbol{C}'\boldsymbol{A}\boldsymbol{C} = \boldsymbol{I}$.

9. (Sec. 2.3) Let $\boldsymbol{b} = \boldsymbol{0}$,

$$\boldsymbol{A} = \begin{pmatrix} 7 & 3 & 2 \\ 3 & 4 & 1 \\ 2 & 1 & 2 \end{pmatrix}.$$

(*a*) Write the density (23).
(*b*) Find $\boldsymbol{\Sigma}$.

10. (Sec. 2.3) Prove that the principal axes of (55) of Section 2.3 are along the 45° and 135° lines with lengths $2\sqrt{c(1+\rho)}$ and $2\sqrt{c(1-\rho)}$, respectively, by transforming according to $y_1 = (z_1 + z_2)/\sqrt{2}$, $y_2 = (z_1 - z_2)/\sqrt{2}$.

11. (Sec. 2.3) Suppose the scalar random variables $X_1, \dots, X_n$ are independent and have a density which is a function only of $x_1^2 + \cdots + x_n^2$. Prove that the X_i are normally distributed with mean 0 and common variance. Indicate the mildest conditions on the density for your proof.

12. (Sec. 2.3) Show that if $\Pr\{X \geq 0, Y \geq 0\} = \alpha$ for the distribution

$$N\left[\begin{pmatrix} 0 \\ 0 \end{pmatrix}, \begin{pmatrix} 1 & \rho \\ \rho & 1 \end{pmatrix}\right],$$

then $\rho = \cos(1 - 2\alpha)\pi$. [*Hint*: Let $X = U, Y = \rho U + \sqrt{1-\rho^2}\; V$ and verify $\rho = \cos 2\pi(\frac{1}{2} - \alpha)$ geometrically.]

13. (Sec. 2.3) Prove that if $\rho_{ij} = \rho, i \neq j;\ i, j = 1, \dots, p$, then $\rho \geq -1/(p-1)$.

14. (Sec. 2.3) *Concentration ellipsoid.* Let the density of the p-component $\boldsymbol{Y}$ be $f(y) = \Gamma(\frac{1}{2}p + 1)/[(p+2)\pi]^{\frac{1}{2}p}$ for $\boldsymbol{y}'\boldsymbol{y} \leq p + 2$ and 0 elsewhere. Then $\mathscr{E}\boldsymbol{Y} = \boldsymbol{0}$ and $\mathscr{E}\boldsymbol{Y}\boldsymbol{Y}' = \boldsymbol{I}$ (Problem 4 of Chapter 7). From this result prove that if the density of $\boldsymbol{X}$ is $g(\boldsymbol{x}) = \sqrt{|\boldsymbol{A}|}\,\Gamma(\frac{1}{2}p + 1)/[(p+2)\pi]^{\frac{1}{2}p}$ for $(\boldsymbol{x} - \boldsymbol{\mu})'\boldsymbol{A}(\boldsymbol{x} - \boldsymbol{\mu}) \leq p + 2$ and 0 elsewhere then $\mathscr{E}\boldsymbol{X} = \boldsymbol{\mu}$ and $\mathscr{E}(\boldsymbol{X} - \boldsymbol{\mu})(\boldsymbol{X} - \boldsymbol{\mu})' = \boldsymbol{A}^{-1}$.

15. (Sec. 2.4) Show that when $\boldsymbol{X}$ is normally distributed the components are mutually independent if and only if the covariance matrix is diagonal.

16. (Sec. 2.4) Find necessary and sufficient conditions on $\boldsymbol{A}$ so that $\boldsymbol{A}\boldsymbol{Y} + \boldsymbol{\lambda}$ has a continuous cdf.

17. (Sec. 2.4) Which densities in Problem 7 define distributions in which X and Y are independent?

18. (Sec. 2.4) (*a*) Write the marginal density of X for each case in Problem 6.

(*b*) Indicate the marginal distribution of X for each case in Problem 7 by the notation $N(a, b)$.

(*c*) Write the marginal density of X_1 and X_2 in Problem 9.

19. (Sec. 2.4) What is the distribution of $Z = X - Y$ when X and Y have each of the densities in Problem 6?

20. (Sec. 2.4) What is the distribution of $X_1 + 2X_2 - 3X_3$ when X_1, X_2, X_3 have the distribution defined in Problem 9?

21. (Sec. 2.4) Let $\boldsymbol{X} = (X_1, X_2)'$, where $X_1 = X$ and $X_2 = aX + b$ and X has the distribution $N(0,1)$. Find the cdf of $\boldsymbol{X}$.

22. (Sec. 2.4) Let $X_1, \dots, X_N$ be independently distributed, each according to $N(\mu, \sigma^2)$.

(*a*) What is the distribution of $\boldsymbol{X} = (X_1, \dots, X_N)'$? Find the vector of means and the covariance matrix.

(*b*) Using Theorem 2.4.4, find the marginal distribution of $\bar{X} = \sum X_i/N$.

23. (Sec. 2.4) Let $X_1, \dots, X_N$ be independently distributed with X_i having distribution $N(\beta + \gamma z_i, \sigma^2)$, where z_i is a given number, $i = 1, \dots, N$, and $\sum_i z_i = 0$.

(*a*) Find the distribution of $(X_1, \dots, X_N)'$.

(*b*) Find the distribution of $\bar{X}$ and $g = \sum X_i z_i / \sum z_i^2$ for $\sum z_i^2 > 0$.

24. (Sec. 2.4) Let $(X_1, Y_1)', (X_2, Y_2)', (X_3, Y_3)'$ be independently distributed, $(X_i, Y_i)'$ according to

$$N\left[\begin{pmatrix}\mu \\ \nu\end{pmatrix}, \begin{pmatrix}\sigma_{xx} & \sigma_{xy} \\ \sigma_{xy} & \sigma_{yy}\end{pmatrix}\right], \qquad i = 1, 2, 3.$$

(*a*) Find the distribution of the six variables.

(*b*) Find the distribution of $(\bar{X}, \bar{Y})'$.

25. (Sec. 2.4) Let $\boldsymbol{X}$ have a (singular) normal distribution with mean $\mathbf{0}$ and covariance matrix

$$\boldsymbol{\Sigma} = \begin{pmatrix} 4 & 2 \\ 2 & 1 \end{pmatrix}.$$

(*a*) Prove $\boldsymbol{\Sigma}$ is of rank 1.
(*b*) Find $\boldsymbol{a}$ so $\boldsymbol{X} = \boldsymbol{a}'Y$ and Y has a nonsingular normal distribution, and give the density of Y.

26. (Sec. 2.4) Let

$$\boldsymbol{\Sigma} = \begin{pmatrix} 2 & -1 & 3 \\ -1 & 5 & -3 \\ 3 & -3 & 5 \end{pmatrix}.$$

(*a*) Find a vector $\boldsymbol{u} \neq \mathbf{0}$ so that $\boldsymbol{\Sigma u} = \mathbf{0}$. [*Hint*: Take cofactors of any column.]
(*b*) Show that any matrix of the form $\boldsymbol{G} = (\boldsymbol{H}\ \boldsymbol{u})$, where $\boldsymbol{H}$ is 3×2, has the property

$$\boldsymbol{G}'\boldsymbol{\Sigma}\boldsymbol{G} = \begin{pmatrix} \boldsymbol{H}'\boldsymbol{\Sigma}\boldsymbol{H} & \mathbf{0} \\ \mathbf{0} & 0 \end{pmatrix}.$$

(*c*) Using (*a*) and (*b*), find $\boldsymbol{B}$ to satisfy (36).
(*d*) Find $\boldsymbol{B}^{-1}$ and partition according to (39).
(*e*) Verify that $\boldsymbol{CC}' = \boldsymbol{\Sigma}$.

27. (Sec. 2.4) Prove that if the joint (marginal) distribution of X_1 and X_2 is singular (that is, degenerate), then the joint distribution of X_1, X_2, and X_3 is singular.

28. (Sec. 2.5) In each part of Problem 6, find the conditional distribution of X given $Y = y$, find the conditional distribution of Y given $X = x$, and plot each regression line on the appropriate graph in Problem 6.

29. (Sec. 2.5) Let $\boldsymbol{\mu} = \mathbf{0}$ and

$$\boldsymbol{\Sigma} = \begin{pmatrix} 1. & 0.80 & -0.40 \\ 0.80 & 1. & -0.56 \\ -0.40 & -0.56 & 1. \end{pmatrix}.$$

(*a*) Find the conditional distribution of X_1 and X_3, given $X_2 = x_2$.
(*b*) What is the partial correlation between X_1 and X_3 given X_2?

30. (Sec. 2.5) In Problem 9, find the conditional distribution of X_1 and X_2 given $X_3 = x_3$.

31. (Sec. 2.5) Verify (20) directly from Theorem 2.5.1.

32. (Sec. 2.5) (*a*) Show that finding $\boldsymbol{\alpha}$ to maximize the absolute value of the correlation between X_i and $\boldsymbol{\alpha}'\boldsymbol{X}^{(2)}$ is equivalent to maximizing $(\boldsymbol{\sigma}_{(i)}'\boldsymbol{\alpha})^2$ subject to $\boldsymbol{\alpha}'\boldsymbol{\Sigma}_{22}\boldsymbol{\alpha}$ constant. (*b*) Find $\boldsymbol{\alpha}$ by maximizing $(\boldsymbol{\sigma}_{(i)}'\boldsymbol{\alpha})^2 - \lambda(\boldsymbol{\alpha}'\boldsymbol{\Sigma}_{22}\boldsymbol{\alpha} - c)$, where c is a constant and λ is a Lagrange multiplier.

33. (Sec. 2.5) *Invariance of the multiple correlation coefficient.* Prove that $\overline{R}_{i\cdot q+1,\ldots,p}$ is an invariant characteristic of the multivariate normal distribution of X_i and $\boldsymbol{X}^{(2)}$ under the transformation $x_i^* = b_i x_i + c_i$ for $b_i \neq 0$ and $\boldsymbol{X}^{(2)*} = \boldsymbol{H}\boldsymbol{X}^{(2)} + \boldsymbol{k}$ for $\boldsymbol{H}$ nonsingular and that every function of $\mu_i, \sigma_{ii}, \boldsymbol{\sigma}_{(i)}, \boldsymbol{\mu}^{(2)}$, and $\boldsymbol{\Sigma}_{22}$ that is invariant is a function of $\overline{R}_{i\cdot q+1,\ldots,p}$.

34. (Sec. 2.5) Prove that $1 - \overline{R}^2_{i\cdot q+1,\ldots,p} = \begin{vmatrix} 1 & \rho_{ij} \\ \rho_{ki} & \rho_{kj} \end{vmatrix} \Big/ |\rho_{kj}|,\ k, j = q+1,\ldots,p.$

35. (Sec. 2.5) Find the multiple correlation coefficient between X_1 and $(X_2\ X_3)$ in Problem 29.

36. (Sec. 2.5) Prove explicitly that if $\boldsymbol{\Sigma}$ is positive definite

$$|\boldsymbol{\Sigma}| = |\boldsymbol{\Sigma}_{11} - \boldsymbol{\Sigma}_{12}\boldsymbol{\Sigma}_{22}^{-1}\boldsymbol{\Sigma}_{21}| \cdot |\boldsymbol{\Sigma}_{22}|.$$

37. (Sec. 2.5) Prove *Hadamard's inequality*

$$|\boldsymbol{\Sigma}| \leq \prod_{i=1}^{p} \sigma_{ii}.$$

[*Hint*: Using Problem 36, prove $|\boldsymbol{\Sigma}| \leq \sigma_{11}|\boldsymbol{\Sigma}_{22}|$, where $\boldsymbol{\Sigma}_{22}$ is $(p-1)\times(p-1)$ and apply induction.]

38. (Sec. 2.5) Prove equality holds in Problem 37 if and only if $\boldsymbol{\Sigma}$ is diagonal.

39. (Sec. 2.5) Prove $\beta_{12\cdot 3} = \sigma_{12\cdot 3}/\sigma_{22\cdot 3}$ and $\beta_{13\cdot 2} = \sigma_{13\cdot 2}/\sigma_{33\cdot 2}$.

40. (Sec. 2.5) Let (X_1, X_2) have the density $n(\boldsymbol{x}|\boldsymbol{0}, \boldsymbol{\Sigma}) = f(x_1, x_2)$. Let the density of X_2 given $X_1 = x_1$ be $f(x_2|x_1)$. Let the joint density of X_1, X_2, X_3 be $f(x_1, x_2)f(x_3|x_1)$. Find the covariance matrix of X_1, X_2, X_3 and the partial correlation between X_2 and X_3 for given X_1.

41. (Sec. 2.5) Prove $1 - \overline{R}^2_{1\cdot 23} = (1 - \rho_{13}^2)(1 - \rho_{12\cdot 3}^2)$. [*Hint*: Use the fact that the variance of X_1 in the conditional distribution given x_2 and x_3 is $(1 - \overline{R}^2_{1\cdot 23})\sigma_{11}$.]

42. (Sec. 2.5) If $p = 2$, can there be a difference between the simple correlation between X_1 and X_2 and the multiple correlation between X_1 and $\boldsymbol{X}^{(2)} = X_2$? Explain.

43. (Sec. 2.5) Prove

$$\beta_{ik\cdot q+1,\ldots,k-1,k+1,\ldots,p} = \frac{\sigma_{ik\cdot q+1,\ldots,k-1,k+1,\ldots,p}}{\sigma_{kk\cdot q+1,\ldots,k-1,k+1,\ldots,p}}$$

$$= \rho_{ik\cdot q+1,\ldots,k-1,k+1,\ldots,p}\frac{\sigma_{i\cdot q+1,\ldots,k-1,k+1,\ldots,p}}{\sigma_{k\cdot q+1,\ldots,k-1,k+1,\ldots,p}},$$

$i = 1,\ldots,q,\ k = q+1,\ldots,p,$ where $\sigma^2_{j\cdot q+1,\ldots,k-1,k+1,\ldots,p} = \sigma_{jj\cdot q+1,\ldots,k-1,k+1,\ldots,p}$,

$j = i, k$. [*Hint*: Prove this for the special case $k = q + 1$ by using Problem 56 with $p_1 = q, p_2 = 1, p_3 = p - q - 1$.]

44. (Sec. 2.5) Give a necessary and sufficient condition for $\overline{R}_{i \cdot q+1,\dots,p} = 0$ in terms of $\sigma_{i,q+1}, \dots, \sigma_{ip}$.

45. (Sec. 2.5) Show

$$1 - \overline{R}^2_{i \cdot q+1,\dots,p} = \left(1 - \rho^2_{ip}\right)\left(1 - \rho^2_{i,p-1 \cdot p}\right) \cdots \left(1 - \rho^2_{i,q+1 \cdot q+2,\dots,p}\right).$$

[*Hint*: Use (19) and (27) successively.]

46. (Sec. 2.5) Show

$$\rho^2_{ij \cdot q+1,\dots,p} = \beta_{ij \cdot q+1,\dots,p} \beta_{ji \cdot q+1,\dots,p}.$$

47. (Sec. 2.5) Prove

$$\rho_{12 \cdot 3 \cdots p} = \frac{-\sigma^{12}}{\sqrt{\sigma^{11}\sigma^{22}}}.$$

[*Hint*: Apply Theorem A.3.2 of Appendix A to the cofactors used to calculate σ^{ij}.]

48. (Sec. 2.5) Show that for any joint distribution for which the expectations exist and any function $h(\boldsymbol{x}^{(2)})$ that

$$\mathscr{E}\left(X_i - \mathscr{E}X_i|\boldsymbol{X}^{(2)}\right)h\left(\boldsymbol{X}^{(2)}\right) = 0.$$

[*Hint*: In the above take the expectation first with respect to X_i conditional on $\boldsymbol{X}^{(2)}$.]

49. (Sec. 2.5) Show that for any function $h(\boldsymbol{x}^{(2)})$ and any joint distribution of X_i and $\boldsymbol{X}^{(2)}$ for which the relevant expectations exist $\mathscr{E}[X_i - h(\boldsymbol{X}^{(2)})]^2 = \mathscr{E}[X_i - g(\boldsymbol{X}^{(2)})]^2 + \mathscr{E}[g(\boldsymbol{X}^{(2)}) - h(\boldsymbol{X}^{(2)})]^2$, where $g(\boldsymbol{x}^{(2)}) = \mathscr{E}X_i|\boldsymbol{x}^{(2)}$ is the conditional expectation of X_i given $\boldsymbol{X}^{(2)} = \boldsymbol{x}^{(2)}$. Hence $g(\boldsymbol{X}^{(2)})$ minimizes the mean squared error of prediction. [*Hint*: Use Problem 48.]

50. (Sec. 2.5) Show that for any function $h(\boldsymbol{x}^{(2)})$ and any joint distribution of X_i and $\boldsymbol{X}^{(2)}$ for which the relevant expectations exist, the correlation between X_i and $h(\boldsymbol{X}^{(2)})$ is not greater than the correlation between X_i and $g(\boldsymbol{X}^{(2)})$, where $g(\boldsymbol{x}^{(2)}) = \mathscr{E}X_i|\boldsymbol{x}^{(2)}$.

51. (Sec. 2.5) Show that for any vector function $\boldsymbol{h}(\boldsymbol{x}^{(2)})$

$$\mathscr{E}\left[\boldsymbol{X}^{(1)} - \boldsymbol{h}(\boldsymbol{X}^{(2)})\right]\left[\boldsymbol{X}^{(1)} - \boldsymbol{h}(\boldsymbol{X}^{(2)})\right]' - \mathscr{E}\left[\boldsymbol{X}^{(1)} - \mathscr{E}\boldsymbol{X}^{(1)}|\boldsymbol{X}^{(2)}\right]\left[\boldsymbol{X}^{(1)} - \mathscr{E}\boldsymbol{X}^{(1)}|\boldsymbol{X}^{(2)}\right]'$$

is positive semidefinite. Note this generalizes Theorem 2.5.3 and Problem 49.

52. (Sec. 2.5) Verify that $\boldsymbol{\Sigma}_{12}\boldsymbol{\Sigma}_{22}^{-1} = -\boldsymbol{\Psi}_{11}^{-1}\boldsymbol{\Psi}_{12}$, where $\boldsymbol{\Psi} = \boldsymbol{\Sigma}^{-1}$ is partitioned similarly to $\boldsymbol{\Sigma}$.

53. (Sec. 2.5) Show

$$\boldsymbol{\Sigma}^{-1} = \begin{pmatrix} \boldsymbol{\Sigma}_{11\cdot 2}^{-1} & -\boldsymbol{\Sigma}_{11\cdot 2}^{-1}\boldsymbol{\Sigma}_{12}\boldsymbol{\Sigma}_{22}^{-1} \\ -\boldsymbol{\Sigma}_{22}^{-1}\boldsymbol{\Sigma}_{21}\boldsymbol{\Sigma}_{11\cdot 2}^{-1} & \boldsymbol{\Sigma}_{22}^{-1}\boldsymbol{\Sigma}_{21}\boldsymbol{\Sigma}_{11\cdot 2}^{-1}\boldsymbol{\Sigma}_{12}\boldsymbol{\Sigma}_{22}^{-1} + \boldsymbol{\Sigma}_{22}^{-1} \end{pmatrix}.$$

$$= \begin{pmatrix} \mathbf{0} & \mathbf{0} \\ \mathbf{0} & \boldsymbol{\Sigma}_{22}^{-1} \end{pmatrix} + \begin{pmatrix} \boldsymbol{I} \\ -\boldsymbol{\beta}' \end{pmatrix} \boldsymbol{\Sigma}_{11\cdot 2}^{-1}(\boldsymbol{I} \quad -\boldsymbol{\beta}),$$

where $\boldsymbol{\beta} = \boldsymbol{\Sigma}_{12}\boldsymbol{\Sigma}_{22}^{-1}$. [*Hint*: Use Theorem A.3.3 of Appendix A and the fact that $\boldsymbol{\Sigma}^{-1}$ is symmetric.]

54. (Sec. 2.5) Use Problem 53 to show that

$$\boldsymbol{x}'\boldsymbol{\Sigma}^{-1}\boldsymbol{x} = \left(\boldsymbol{x}^{(1)} - \boldsymbol{\Sigma}_{12}\boldsymbol{\Sigma}_{22}^{-1}\boldsymbol{x}^{(2)}\right)'\boldsymbol{\Sigma}_{11\cdot 2}^{-1}\left(\boldsymbol{x}^{(1)} - \boldsymbol{\Sigma}_{12}\boldsymbol{\Sigma}_{22}^{-1}\boldsymbol{x}^{(2)}\right) + \boldsymbol{x}^{(2)\prime}\boldsymbol{\Sigma}_{22}^{-1}\boldsymbol{x}^{(2)}.$$

55. (Sec. 2.5) Show $= \mathscr{E}\{(X^{(1)}|x^{(3)})|x^{(2)}\}$

$$\mathscr{E}\left(\boldsymbol{X}^{(1)}|\boldsymbol{x}^{(2)}, \boldsymbol{x}^{(3)}\right) = \boldsymbol{\mu}^{(1)} + \boldsymbol{\Sigma}_{13}\boldsymbol{\Sigma}_{33}^{-1}\left(\boldsymbol{x}^{(3)} - \boldsymbol{\mu}^{(3)}\right)$$

$$+ \left(\boldsymbol{\Sigma}_{12} - \boldsymbol{\Sigma}_{13}\boldsymbol{\Sigma}_{33}^{-1}\boldsymbol{\Sigma}_{32}\right)\left(\boldsymbol{\Sigma}_{22} - \boldsymbol{\Sigma}_{23}\boldsymbol{\Sigma}_{33}^{-1}\boldsymbol{\Sigma}_{32}\right)^{-1}\left[\boldsymbol{x}^{(2)} - \boldsymbol{\mu}^{(2)} - \boldsymbol{\Sigma}_{23}\boldsymbol{\Sigma}_{33}^{-1}\left(\boldsymbol{x}^{(3)} - \boldsymbol{\mu}^{(3)}\right)\right].$$

56. (Sec. 2.5) Prove by matrix algebra that $= \mathscr{E}\{(X^{(1)}|x^{(3)})|x^{(2)}\}$

$$\boldsymbol{\Sigma}_{11} - (\boldsymbol{\Sigma}_{12}\ \boldsymbol{\Sigma}_{13})\begin{pmatrix} \boldsymbol{\Sigma}_{22} & \boldsymbol{\Sigma}_{23} \\ \boldsymbol{\Sigma}_{32} & \boldsymbol{\Sigma}_{33} \end{pmatrix}^{-1}\begin{pmatrix} \boldsymbol{\Sigma}_{21} \\ \boldsymbol{\Sigma}_{31} \end{pmatrix} = \boldsymbol{\Sigma}_{11} - \boldsymbol{\Sigma}_{13}\boldsymbol{\Sigma}_{33}^{-1}\boldsymbol{\Sigma}_{31}$$

$$-\left(\boldsymbol{\Sigma}_{12} - \boldsymbol{\Sigma}_{13}\boldsymbol{\Sigma}_{33}^{-1}\boldsymbol{\Sigma}_{32}\right)\left(\boldsymbol{\Sigma}_{22} - \boldsymbol{\Sigma}_{23}\boldsymbol{\Sigma}_{33}^{-1}\boldsymbol{\Sigma}_{32}\right)^{-1}\left(\boldsymbol{\Sigma}_{21} - \boldsymbol{\Sigma}_{23}\boldsymbol{\Sigma}_{33}^{-1}\boldsymbol{\Sigma}_{31}\right).$$

57. (Sec. 2.5) *Invariance of the partial correlation coefficient*. Prove that $\rho_{12\cdot 3,\ldots,p}$ is invariant under the transformations $x_i^* = a_i x_i + \boldsymbol{b}_i'\boldsymbol{x}^{(3)} + c_i$, $a_i > 0$, $i = 1, 2$, $\boldsymbol{x}^{(3)*} = \boldsymbol{C}\boldsymbol{x}^{(3)} + \boldsymbol{d}$, where $\boldsymbol{x}^{(3)} = (x_3, \ldots, x_p)'$, and that any function of $\boldsymbol{\mu}$ and $\boldsymbol{\Sigma}$ that is invariant under these transformations is a function of $\rho_{12\cdot 3,\ldots,p}$.

58. (Sec. 2.5) Suppose $\boldsymbol{X}^{(1)}$ and $\boldsymbol{X}^{(2)}$ of q and $p - q$ components, respectively, have the density

$$\frac{|\boldsymbol{A}|^{\frac{1}{2}}}{(2\pi)^{\frac{1}{2}p}} e^{-\frac{1}{2}Q},$$

where

$$Q = \left(\boldsymbol{x}^{(1)} - \boldsymbol{\mu}^{(1)}\right)'\boldsymbol{A}_{11}\left(\boldsymbol{x}^{(1)} - \boldsymbol{\mu}^{(1)}\right) + \left(\boldsymbol{x}^{(1)} - \boldsymbol{\mu}^{(1)}\right)'\boldsymbol{A}_{12}\left(\boldsymbol{x}^{(2)} - \boldsymbol{\mu}^{(2)}\right)$$

$$+ \left(\boldsymbol{x}^{(2)} - \boldsymbol{\mu}^{(2)}\right)'\boldsymbol{A}_{21}\left(\boldsymbol{x}^{(1)} - \boldsymbol{\mu}^{(1)}\right) + \left(\boldsymbol{x}^{(2)} - \boldsymbol{\mu}^{(2)}\right)'\boldsymbol{A}_{22}\left(\boldsymbol{x}^{(2)} - \boldsymbol{\mu}^{(2)}\right).$$

Show that Q can be written as $Q_1 + Q_2$, where

$$Q_1 = \left[(x^{(1)} - \mu^{(1)}) + A_{11}^{-1}A_{12}(x^{(2)} - \mu^{(2)})\right]' A_{11} \left[(x^{(1)} - \mu^{(1)}) + A_{11}^{-1}A_{12}(x^{(2)} - \mu^{(2)})\right],$$

$$Q_2 = (x^{(2)} - \mu^{(2)})'(A_{22} - A_{21}A_{11}^{-1}A_{12})(x^{(2)} - \mu^{(2)}).$$

Show that the marginal density of $X^{(2)}$ is

$$\frac{|A_{22} - A_{21}A_{11}^{-1}A_{12}|^{\frac{1}{2}}}{(2\pi)^{\frac{1}{2}(p-q)}} e^{-\frac{1}{2}Q_2}.$$

Show that the conditional density of $X^{(1)}$ given $X^{(2)} = x^{(2)}$ is

$$\frac{|A_{11}|^{\frac{1}{2}}}{(2\pi)^{\frac{1}{2}q}} e^{-\frac{1}{2}Q_1}$$

(without using Appendix A). This problem is meant to furnish an *alternative* proof of Theorems 2.4.3 and 2.5.1.

59. (Sec. 2.6) Prove Lemma 2.6.2 in detail.

60. (Sec. 2.6) Let Y be distributed according to $N(\mathbf{0}, \Sigma)$. Differentiating the characteristic function, verify (25) and (26).

61. (Sec. 2.6) Verify (25) and (26) by using the transformation $X - \mu = CY$, where $\Sigma = CC'$, and integrating the density of Y.

62. (Sec. 2.6) Let the density of (X, Y) be

$$\begin{array}{ll} 2n(x|0,1)n(y|0,1), & 0 \le y \le x < \infty, \quad 0 \le -x \le y < \infty, \\ & 0 \le -y \le -x < \infty, \quad 0 \le x \le -y < \infty, \\ 0, & \text{otherwise.} \end{array}$$

Show that $X, Y, X + Y, X - Y$ each have a marginal normal distribution.

63. (Sec. 2.6) Suppose X is distributed according to $N(\mathbf{0}, \Sigma)$. Let $\Sigma = (\sigma_1, \dots, \sigma_p)$. Prove

$$\mathscr{E}(XX' \otimes XX') = \Sigma \otimes \Sigma + \operatorname{vec}\Sigma(\operatorname{vec}\Sigma)' + \begin{bmatrix} \sigma_1\sigma_1' & \cdots & \sigma_p\sigma_1' \\ \vdots & & \vdots \\ \sigma_1\sigma_p' & \cdots & \sigma_p\sigma_p' \end{bmatrix}$$

$$= (I + K)(\Sigma \otimes \Sigma) + \operatorname{vec}\Sigma(\operatorname{vec}\Sigma)',$$

where

$$\operatorname{vec}\Sigma = \begin{bmatrix} \sigma_1 \\ \vdots \\ \sigma_p \end{bmatrix}, \qquad K = \begin{bmatrix} \varepsilon_1\varepsilon_1' & \cdots & \varepsilon_p\varepsilon_1' \\ \vdots & & \vdots \\ \varepsilon_1\varepsilon_p' & \cdots & \varepsilon_p\varepsilon_p' \end{bmatrix},$$

and ε_i is a column vector with 1 in the ith position and 0's elsewhere.

64. *Complex normal distribution*. Let $(X'\ Y')'$ have a normal distribution with mean vector $(\mu_X'\ \mu_Y')'$ and covariance matrix

$$\Sigma = \begin{pmatrix} \Gamma & -\Phi \\ \Phi & \Gamma \end{pmatrix},$$

where Γ is positive definite and $\Phi = -\Phi'$ (skew symmetric). Then $Z = X + iY$ is said to have a complex normal distribution with mean $\theta = \mu_X + i\mu_Y$ and covariance matrix $\mathscr{E}(Z-\theta)(Z-\theta)^* = P = Q + iR$, where $Z^* = X' - iY'$. Note that P is Hermitian and positive definite.

(*a*) Show $Q = 2\Gamma$ and $R = 2\Phi$.
(*b*) Show $|P|^2 = |2\Sigma|$. [*Hint*: $|\Gamma + i\Phi| = |\Gamma - i\Phi|$.]
(*c*) Show

$$P^{-1} = (Q + RQ^{-1}R)^{-1} - iQ^{-1}R(Q + RQ^{-1}R)^{-1}.$$

Note the inverse of a Hermitian matrix is Hermitian.

(*d*) Show that the density of X and Y can be written

$$\pi^{-p}|P|^{-1}e^{-(z-\theta)^*P^{-1}(z-\theta)}.$$

65. *Complex normal continued*. If Z has the complex normal distribution of Problem 64, show that $W = AZ$, where A is a nonsingular complex matrix, has the complex normal distribution with mean $A\theta$ and covariance matrix $\mathscr{C}(W) = APA^*$.

66. Show that the characteristic function of Z defined in Problem 64 is

$$\mathscr{E}e^{i\mathscr{R}(u^*Z)} = e^{i\mathscr{R}u^*\theta - u^*Pu},$$

where $\mathscr{R}(x+iy) = x$.

67. (Sec. 2.2) Show that $\int_{-a}^{a} e^{-x^2/2}\,dx/\sqrt{2\pi}$ is approximately $(1 - e^{-2a^2/\pi})^{1/2}$. [*Hint*: The probability that (X, Y) falls in a square is approximately the probability that (X, Y) falls in an approximating circle (Pólya (1949)).]

CHAPTER 3

Estimation of the Mean Vector and the Covariance Matrix

3.1. INTRODUCTION

The multivariate normal distribution is specified completely by the mean vector μ and the covariance matrix Σ. The first statistical problem is how to estimate these parameters on the basis of a sample of observations. In Section 3.2 it is shown that the maximum likelihood estimator of μ is the sample mean; the maximum likelihood estimator of Σ is proportional to the matrix of sample variances and covariances. A sample variance is a sum of squares of deviations of observations from the sample mean divided by one less than the number of observations in the sample; a sample covariance is similarly defined in terms of cross products. The sample covariance matrix is an unbiased estimator of Σ.

The distribution of the sample mean vector is given in Section 3.3, and it is shown how one can test the hypothesis that μ is a given vector when Σ is known. The case of Σ unknown will be treated in Chapter 5.

Some theoretical properties of the sample mean are given in Section 3.4, and the Bayes estimator of the population mean is derived for a normal a priori distribution. In Section 3.5 the James–Stein estimator is introduced; improvements over the sample mean for the mean squared error loss function are discussed.

3.2. THE MAXIMUM LIKELIHOOD ESTIMATORS OF THE MEAN VECTOR AND THE COVARIANCE MATRIX

Given a sample of (vector) observations from a p-variate (nondegenerate) normal distribution, we ask for estimators of the mean vector $\boldsymbol{\mu}$ and the covariance matrix $\boldsymbol{\Sigma}$ of the distribution. We shall deduce the maximum likelihood estimators.

It turns out that the method of maximum likelihood is very useful in various estimation and hypothesis testing problems concerning the multivariate normal distribution. The maximum likelihood estimators or modifications of them often have some optimum properties. In the particular case studied here, the estimators are asymptotically efficient [Cramér (1946), Sec. 33.3].

Suppose our sample of N observations on $\boldsymbol{X}$ distributed according to $N(\boldsymbol{\mu}, \boldsymbol{\Sigma})$ is $\boldsymbol{x}_1, \dots, \boldsymbol{x}_N$, where $N > p$. The likelihood function is

(1)

$$L = \prod_{\alpha=1}^{N} n(\boldsymbol{x}_\alpha | \boldsymbol{\mu}, \boldsymbol{\Sigma}) = \frac{1}{(2\pi)^{\frac{1}{2}pN} |\boldsymbol{\Sigma}|^{\frac{1}{2}N}} \exp\left[-\tfrac{1}{2} \sum_{\alpha=1}^{N} (\boldsymbol{x}_\alpha - \boldsymbol{\mu})' \boldsymbol{\Sigma}^{-1} (\boldsymbol{x}_\alpha - \boldsymbol{\mu}) \right].$$

In the likelihood function the vectors $\boldsymbol{x}_1, \dots, \boldsymbol{x}_N$ are fixed at the sample values and L is a function of $\boldsymbol{\mu}$ and $\boldsymbol{\Sigma}$. To emphasize that these quantities are variables (and not parameters) we shall denote them by $\boldsymbol{\mu}^*$ and $\boldsymbol{\Sigma}^*$. Then the logarithm of the likelihood function is

$$(2)\quad \log L = -\tfrac{1}{2} pN \log(2\pi) - \tfrac{1}{2} N \log |\boldsymbol{\Sigma}^*| - \tfrac{1}{2} \sum_{\alpha=1}^{N} (\boldsymbol{x}_\alpha - \boldsymbol{\mu}^*)' \boldsymbol{\Sigma}^{*-1} (\boldsymbol{x}_\alpha - \boldsymbol{\mu}^*).$$

Since $\log L$ is an increasing function of L, its maximum is at the same point in the space of $\boldsymbol{\mu}^*$, $\boldsymbol{\Sigma}^*$ as the maximum of L. The maximum likelihood estimators of $\boldsymbol{\mu}$ and $\boldsymbol{\Sigma}$ are the vector $\boldsymbol{\mu}^*$ and the positive definite matrix $\boldsymbol{\Sigma}^*$ that maximize $\log L$. (It remains to be seen that the supremum of $\log L$ is attained for a positive definite matrix $\boldsymbol{\Sigma}^*$.)

Let the *sample mean vector* be

$$(3)\qquad \bar{\boldsymbol{x}} = \frac{1}{N} \sum_{\alpha=1}^{N} \boldsymbol{x}_\alpha = \begin{pmatrix} \frac{1}{N} \sum_{\alpha=1}^{N} x_{1\alpha} \\ \vdots \\ \frac{1}{N} \sum_{\alpha=1}^{N} x_{p\alpha} \end{pmatrix} = \begin{pmatrix} \bar{x}_1 \\ \vdots \\ \bar{x}_p \end{pmatrix},$$

where $\boldsymbol{x}_\alpha = (x_{1\alpha}, \dots, x_{p\alpha})'$ and $\bar{x}_i = \sum_{\alpha=1}^N x_{i\alpha}/N$, and let the matrix of sums of squares and cross products of deviations about the mean be

$$(4) \qquad \boldsymbol{A} = \sum_{\alpha=1}^{N} (\boldsymbol{x}_\alpha - \bar{\boldsymbol{x}})(\boldsymbol{x}_\alpha - \bar{\boldsymbol{x}})'$$

$$= \left[\sum_{\alpha=1}^{N} (x_{i\alpha} - \bar{x}_i)(x_{j\alpha} - \bar{x}_j) \right], \qquad i, j = 1, \dots, p.$$

It will be convenient to use the following lemma:

LEMMA 3.2.1. *Let $\boldsymbol{x}_1, \dots, \boldsymbol{x}_N$ be N (p-component) vectors and let $\bar{\boldsymbol{x}}$ be defined by* (3). *Then for any vector $\boldsymbol{b}$*

$$(5) \qquad \sum_{\alpha=1}^{N} (\boldsymbol{x}_\alpha - \boldsymbol{b})(\boldsymbol{x}_\alpha - \boldsymbol{b})' = \sum_{\alpha=1}^{N} (\boldsymbol{x}_\alpha - \bar{\boldsymbol{x}})(\boldsymbol{x}_\alpha - \bar{\boldsymbol{x}})' + N(\bar{\boldsymbol{x}} - \boldsymbol{b})(\bar{\boldsymbol{x}} - \boldsymbol{b})'.$$

PROOF.

(6)

$$\sum_{\alpha=1}^{N} (\boldsymbol{x}_\alpha - \boldsymbol{b})(\boldsymbol{x}_\alpha - \boldsymbol{b})' = \sum_{\alpha=1}^{N} \left[(\boldsymbol{x}_\alpha - \bar{\boldsymbol{x}}) + (\bar{\boldsymbol{x}} - \boldsymbol{b})\right]\left[(\boldsymbol{x}_\alpha - \bar{\boldsymbol{x}}) + (\bar{\boldsymbol{x}} - \boldsymbol{b})\right]'$$

$$= \sum_{\alpha=1}^{N} \Big[(\boldsymbol{x}_\alpha - \bar{\boldsymbol{x}})(\boldsymbol{x}_\alpha - \bar{\boldsymbol{x}})' + (\boldsymbol{x}_\alpha - \bar{\boldsymbol{x}})(\bar{\boldsymbol{x}} - \boldsymbol{b})'$$

$$+ (\bar{\boldsymbol{x}} - \boldsymbol{b})(\boldsymbol{x}_\alpha - \bar{\boldsymbol{x}})' + (\bar{\boldsymbol{x}} - \boldsymbol{b})(\bar{\boldsymbol{x}} - \boldsymbol{b})'\Big]$$

$$= \sum_{\alpha=1}^{N} (\boldsymbol{x}_\alpha - \bar{\boldsymbol{x}})(\boldsymbol{x}_\alpha - \bar{\boldsymbol{x}})' + \left[\sum_{\alpha=1}^{N} (\boldsymbol{x}_\alpha - \bar{\boldsymbol{x}})\right](\bar{\boldsymbol{x}} - \boldsymbol{b})'$$

$$+ (\bar{\boldsymbol{x}} - \boldsymbol{b}) \sum_{\alpha=1}^{N} (\boldsymbol{x}_\alpha - \bar{\boldsymbol{x}})' + N(\bar{\boldsymbol{x}} - \boldsymbol{b})(\bar{\boldsymbol{x}} - \boldsymbol{b})'.$$

The second and third terms on the right-hand side are $\mathbf{0}$ because $\sum(\boldsymbol{x}_\alpha - \bar{\boldsymbol{x}}) = \sum \boldsymbol{x}_\alpha - N\bar{\boldsymbol{x}} = \mathbf{0}$ by (3). Q.E.D.

When we let $\boldsymbol{b} = \boldsymbol{\mu}^*$, we have

(7)

$$\begin{aligned}\sum_{\alpha=1}^{N} (\boldsymbol{x}_\alpha - \boldsymbol{\mu}^*)(\boldsymbol{x}_\alpha - \boldsymbol{\mu}^*)' &= \sum_{\alpha=1}^{N} (\boldsymbol{x}_\alpha - \bar{\boldsymbol{x}})(\boldsymbol{x}_\alpha - \bar{\boldsymbol{x}})' + N(\bar{\boldsymbol{x}} - \boldsymbol{\mu}^*)(\bar{\boldsymbol{x}} - \boldsymbol{\mu}^*)' \\ &= \boldsymbol{A} + N(\bar{\boldsymbol{x}} - \boldsymbol{\mu}^*)(\bar{\boldsymbol{x}} - \boldsymbol{\mu}^*)'.\end{aligned}$$

Using this result and the properties of the trace of a matrix ($\operatorname{tr} \boldsymbol{CD} = \sum c_{ij} d_{ji} = \operatorname{tr} \boldsymbol{DC}$), we have

(8)

$$\begin{aligned}\sum_{\alpha=1}^{N} (\boldsymbol{x}_\alpha - \boldsymbol{\mu}^*)'\boldsymbol{\Sigma}^{*-1}(\boldsymbol{x}_\alpha - \boldsymbol{\mu}^*) &= \operatorname{tr} \sum_{\alpha=1}^{N} (\boldsymbol{x}_\alpha - \boldsymbol{\mu}^*)'\boldsymbol{\Sigma}^{*-1}(\boldsymbol{x}_\alpha - \boldsymbol{\mu}^*) \\ &= \operatorname{tr} \sum_{\alpha=1}^{N} \boldsymbol{\Sigma}^{*-1}(\boldsymbol{x}_\alpha - \boldsymbol{\mu}^*)(\boldsymbol{x}_\alpha - \boldsymbol{\mu}^*)' \\ &= \operatorname{tr} \boldsymbol{\Sigma}^{*-1}\boldsymbol{A} + \operatorname{tr} \boldsymbol{\Sigma}^{*-1} N(\bar{\boldsymbol{x}} - \boldsymbol{\mu}^*)(\bar{\boldsymbol{x}} - \boldsymbol{\mu}^*)' \\ &= \operatorname{tr} \boldsymbol{\Sigma}^{*-1}\boldsymbol{A} + N(\bar{\boldsymbol{x}} - \boldsymbol{\mu}^*)'\boldsymbol{\Sigma}^{*-1}(\bar{\boldsymbol{x}} - \boldsymbol{\mu}^*).\end{aligned}$$

Thus we can write (2) as

$$\begin{aligned}(9)\quad \log L = {} & -\tfrac{1}{2}pN \log(2\pi) - \tfrac{1}{2}N \log|\boldsymbol{\Sigma}^*| \\ & -\tfrac{1}{2} \operatorname{tr} \boldsymbol{\Sigma}^{*-1}\boldsymbol{A} - \tfrac{1}{2}N(\bar{\boldsymbol{x}} - \boldsymbol{\mu}^*)'\boldsymbol{\Sigma}^{*-1}(\bar{\boldsymbol{x}} - \boldsymbol{\mu}^*).\end{aligned}$$

Since $\boldsymbol{\Sigma}^*$ is positive definite, $\boldsymbol{\Sigma}^{*-1}$ is positive definite and $N(\bar{\boldsymbol{x}} - \boldsymbol{\mu}^*)'\boldsymbol{\Sigma}^{*-1}(\bar{\boldsymbol{x}} - \boldsymbol{\mu}^*) \geq 0$ and is 0 if and only if $\boldsymbol{\mu}^* = \bar{\boldsymbol{x}}$. To maximize the second and third terms of (9) we use the following lemma (which is also used in later chapters):

LEMMA 3.2.2. *If $\boldsymbol{D}$ is positive definite of order p, the maximum of*

$$(10)\qquad f(\boldsymbol{G}) = -N \log|\boldsymbol{G}| - \operatorname{tr} \boldsymbol{G}^{-1}\boldsymbol{D}$$

with respect to positive definite matrices $\boldsymbol{G}$ exists, occurs at $\boldsymbol{G} = (1/N)\boldsymbol{D}$, and has the value

$$(11)\qquad f[(1/N)\boldsymbol{D}] = pN \log N - N \log|\boldsymbol{D}| - pN.$$

PROOF. Let $\boldsymbol{D} = \boldsymbol{EE}'$ and $\boldsymbol{E}'\boldsymbol{G}^{-1}\boldsymbol{E} = \boldsymbol{H}$. Then $\boldsymbol{G} = \boldsymbol{EH}^{-1}\boldsymbol{E}'$, and $|\boldsymbol{G}| = |\boldsymbol{E}| \cdot |\boldsymbol{H}^{-1}| \cdot |\boldsymbol{E}'| = |\boldsymbol{H}^{-1}| \cdot |\boldsymbol{EE}'| = |\boldsymbol{D}|/|\boldsymbol{H}|$, and $\operatorname{tr} \boldsymbol{G}^{-1}\boldsymbol{D} = \operatorname{tr} \boldsymbol{G}^{-1}\boldsymbol{EE}' =$

$\operatorname{tr} \boldsymbol{E}'\boldsymbol{G}^{-1}\boldsymbol{E} = \operatorname{tr} \boldsymbol{H}$. Then the function to be maximized (with respect to positive definite $\boldsymbol{H}$) is

$$f = -N \log|\boldsymbol{D}| + N \log|\boldsymbol{H}| - \operatorname{tr} \boldsymbol{H}. \tag{12}$$

Let $\boldsymbol{H} = \boldsymbol{TT}'$, where $\boldsymbol{T}$ is lower triangular (Corollary A.1.7). Then the maximum of

$$\begin{aligned} f &= -N \log|\boldsymbol{D}| + N \log|\boldsymbol{T}|^2 - \operatorname{tr} \boldsymbol{TT}' \\ &= -N \log|\boldsymbol{D}| + \sum_{i=1}^{p} \left(N \log t_{ii}^2 - t_{ii}^2 \right) - \sum_{i>j} t_{ij}^2 \end{aligned} \tag{13}$$

occurs at $t_{ii}^2 = N$, $t_{ij} = 0$, $i \neq j$; that is, at $\boldsymbol{H} = N\boldsymbol{I}$. Then $\boldsymbol{G} = (1/N)\boldsymbol{EE}' = (1/N)\boldsymbol{D}$. Q.E.D.

THEOREM 3.2.1. *If $\boldsymbol{x}_1, \dots, \boldsymbol{x}_N$ constitute a sample from $N(\boldsymbol{\mu}, \boldsymbol{\Sigma})$ with $p < N$, the maximum likelihood estimators of $\boldsymbol{\mu}$ and $\boldsymbol{\Sigma}$ are $\hat{\boldsymbol{\mu}} = \bar{\boldsymbol{x}} = (1/N)\sum_{\alpha=1}^{N} \boldsymbol{x}_\alpha$ and $\hat{\boldsymbol{\Sigma}} = (1/N)\sum_{\alpha=1}^{N}(\boldsymbol{x}_\alpha - \bar{\boldsymbol{x}})(\boldsymbol{x}_\alpha - \bar{\boldsymbol{x}})'$, respectively.*

Other methods of deriving the maximum likelihood estimators have been discussed by Anderson and Olkin (1979). See Problems 4, 8, and 12.

Computation of the estimate $\hat{\boldsymbol{\Sigma}}$ is made easier by the specialization of Lemma 3.2.1 ($\boldsymbol{b} = \boldsymbol{0}$)

$$\sum_{\alpha=1}^{N} (\boldsymbol{x}_\alpha - \bar{\boldsymbol{x}})(\boldsymbol{x}_\alpha - \bar{\boldsymbol{x}})' = \sum_{\alpha=1}^{N} \boldsymbol{x}_\alpha \boldsymbol{x}_\alpha' - N\bar{\boldsymbol{x}}\bar{\boldsymbol{x}}'. \tag{14}$$

An element of $\sum_{\alpha=1}^{N} \boldsymbol{x}_\alpha \boldsymbol{x}_\alpha'$ is computed as $\sum_{\alpha=1}^{N} x_{i\alpha} x_{j\alpha}$, and an element of $N\bar{\boldsymbol{x}}\bar{\boldsymbol{x}}'$ is computed as $N\bar{x}_i\bar{x}_j$ or $(\sum_{\alpha=1}^{N} x_{i\alpha})(\sum_{\alpha=1}^{N} x_{j\alpha})/N$. Rounding error may be reduced by subtracting $\sum_{\alpha=1}^{N} x_{i\alpha} \sum_{\alpha=1}^{N} x_{j\alpha}$ from $N\sum_{\alpha=1}^{N} x_{i\alpha} x_{j\alpha}$ and dividing the difference by N. If $\bar{x}_i$ is numerically large, computational error may be decreased by replacing $x_{i\alpha}$ by $x_{i\alpha} - c_i$, where c_i is a convenient approximation to $\bar{x}_i$ (in order that $x_{i\alpha} - c_i$ have a smaller number of significant figures). If N is so large that too many significant figures are lost in the differencing on the right of (14), for example, there are other methods; see Chan, Golub, and LeVeque (1981). It should be noted that if $N > p$, the probability is 1 of drawing a sample so that (14) is positive definite; see Problem 17.

The covariance matrix can be written in terms of the variances or standard deviations and correlation coefficients. These are uniquely defined by the variances and covariances. We assert that the maximum likelihood estimators of functions of the parameters are those functions of the maximum likelihood estimators of the parameters.

LEMMA 3.2.3. *Let $f(\theta)$ be a real-valued function defined on a set S and let ϕ be a single-valued function, with a single-valued inverse, on S to a set S^*; that is, to each $\theta \in S$ there corresponds a unique $\theta^* \in S^*$, and, conversely, to each $\theta^* \in S^*$ there corresponds a unique $\theta \in S$. Let*

$$g(\theta^*) = f[\phi^{-1}(\theta^*)]. \tag{15}$$

Then if $f(\theta)$ attains a maximum at $\theta = \theta_0$, $g(\theta^*)$ attains a maximum at $\theta^* = \theta_0^* = \phi(\theta_0)$. If the maximum of $f(\theta)$ at θ_0 is unique, so is the maximum of $g(\theta^*)$ at θ_0^*.

PROOF. By hypothesis $f(\theta_0) \geq f(\theta)$ for all $\theta \in S$. Then for any $\theta^* \in S^*$

$$g(\theta^*) = f[\phi^{-1}(\theta^*)] = f(\theta) \leq f(\theta_0) = g[\phi(\theta_0)] = g(\theta_0^*). \tag{16}$$

Thus $g(\theta^*)$ attains a maximum at θ_0^*. If the maximum of $f(\theta)$ at θ_0 is unique, there is strict inequality above for $\theta \neq \theta_0$, and the maximum of $g(\theta^*)$ is unique. Q.E.D.

We have the following corollary:

COROLLARY 3.2.1. *If on the basis of a given sample $\hat{\theta}_1, \ldots, \hat{\theta}_m$ are maximum likelihood estimators of the parameters $\theta_1, \ldots, \theta_m$ of a distribution, then $\phi_1(\hat{\theta}_1, \ldots, \hat{\theta}_m), \ldots, \phi_m(\hat{\theta}_1, \ldots, \hat{\theta}_m)$ are maximum likelihood estimators of $\phi_1(\theta_1, \ldots, \theta_m), \ldots, \phi_m(\theta_1, \ldots, \theta_m)$ if the transformation from $\theta_1, \ldots, \theta_m$ to $\phi_1, \ldots, \phi_m$ is one-to-one.*† *If the estimators of $\theta_1, \ldots, \theta_m$ are unique, then the estimators of $\phi_1, \ldots, \phi_m$ are unique.*

COROLLARY 3.2.2. *If $\boldsymbol{x}_1, \ldots, \boldsymbol{x}_N$ constitutes a sample from $N(\boldsymbol{\mu}, \boldsymbol{\Sigma})$, where $\sigma_{ij} = \sigma_i \sigma_j \rho_{ij}$ $(\rho_{ii} = 1)$, the maximum likelihood estimator of $\boldsymbol{\mu}$ is $\hat{\boldsymbol{\mu}} = \bar{\boldsymbol{x}} = (1/N)\sum_\alpha \boldsymbol{x}_\alpha$, the maximum likelihood estimator of σ_i^2 is $\hat{\sigma}_i^2 = (1/N)\sum_\alpha(x_{i\alpha} - \bar{x}_i)^2 = (1/N)(\sum_\alpha x_{i\alpha}^2 - N\bar{x}_i^2)$, where $x_{i\alpha}$ is the ith component of $\boldsymbol{x}_\alpha$ and $\bar{x}_i$ is*

† The assumption that the transformation is one-to-one is made so that the set $\phi_1, \ldots, \phi_m$ uniquely defines the likelihood. An alternative in case $\theta^* = \phi(\theta)$ does not have a unique inverse is to define $S(\theta^*) = \{\theta: \phi(\theta) = \theta^*\}$ and $g(\theta^*) = \sup f(\theta) | \theta \in S(\theta^*)$, which is considered the "induced likelihood" when $f(\theta)$ is the likelihood function. Then $\hat{\theta}^* = \phi(\hat{\theta})$ maximizes $g(\theta^*)$, for $g(\theta^*) = \sup f(\theta) | \theta \in S(\theta^*) \leq \sup f(\theta) | \theta \in S = f(\hat{\theta}) = g(\hat{\theta}^*)$ for all $\theta^* \in S^*$. [See, e.g., Zehna (1966).]

the ith *component of* $\bar{\mathbf{x}}$, *and the maximum likelihood estimator of* ρ_{ij} *is*

$$\hat{\rho}_{ij} = \frac{\sum_{\alpha=1}^{N}(x_{i\alpha} - \bar{x}_i)(x_{j\alpha} - \bar{x}_j)}{\sqrt{\sum_{\alpha=1}^{N}(x_{i\alpha} - \bar{x}_i)^2}\sqrt{\sum_{\alpha=1}^{N}(x_{j\alpha} - \bar{x}_j)^2}} = \frac{\sum_{\alpha=1}^{N}x_{i\alpha}x_{j\alpha} - N\bar{x}_i\bar{x}_j}{\sqrt{\sum_{\alpha=1}^{N}x_{i\alpha}^2 - N\bar{x}_i^2}\sqrt{\sum_{\alpha=1}^{N}x_{j\alpha}^2 - N\bar{x}_j^2}}. \tag{17}$$

PROOF. The set of parameters $\mu_i = \mu_i$, $\sigma_i^2 = \sigma_{ii}$, and $\rho_{ij} = \sigma_{ij}/\sqrt{\sigma_{ii}\sigma_{jj}}$ is a one-to-one transform of the set of parameters μ_i and σ_{ij}. Therefore, by Corollary 3.2.1 the estimator of μ_i is $\hat{\mu}_i$, of σ_i^2 is $\hat{\sigma}_{ii}$, and of ρ_{ij} is

$$\hat{\rho}_{ij} = \frac{\hat{\sigma}_{ij}}{\sqrt{\hat{\sigma}_{ii}\hat{\sigma}_{jj}}}. \tag{18}$$

Q.E.D.

Pearson (1896) gave a justification for this estimator of ρ_{ij}, and (17) is sometimes called the *Pearson correlation coefficient*. It is also called the *simple correlation coefficient*. It is usually denoted by r_{ij}.

A convenient geometrical interpretation of this sample $(\mathbf{x}_1, \mathbf{x}_2, \dots, \mathbf{x}_N) = \mathbf{X}$ is in terms of the rows of $\mathbf{X}$. Let

$$\mathbf{X} = \begin{pmatrix} x_{11} & \cdots & x_{1N} \\ \vdots & & \vdots \\ x_{p1} & \cdots & x_{pN} \end{pmatrix} = \begin{pmatrix} \mathbf{u}_1' \\ \vdots \\ \mathbf{u}_p' \end{pmatrix}; \tag{19}$$

that is, $\mathbf{u}_i'$ is the ith row of $\mathbf{X}$. The vector $\mathbf{u}_i$ can be considered as a vector in an N-dimensional space with the αth coordinate of one endpoint being $x_{i\alpha}$ and the other endpoint at the origin. Thus the sample is represented by p vectors in N-dimensional Euclidean space. By definition of Euclidean metric the squared length of $\mathbf{u}_i$ (that is, the squared distance of one endpoint from the other) is $\mathbf{u}_i'\mathbf{u}_i = \sum_{\alpha=1}^{N}x_{i\alpha}^2$.

Now let us show that the cosine of the angle between $\mathbf{u}_i$ and $\mathbf{u}_j$ is $\mathbf{u}_i'\mathbf{u}_j/\sqrt{\mathbf{u}_i'\mathbf{u}_i\mathbf{u}_j'\mathbf{u}_j} = \sum_{\alpha=1}^{N}x_{i\alpha}x_{j\alpha}/\sqrt{\sum_{\alpha=1}^{N}x_{i\alpha}^2\sum_{\alpha=1}^{N}x_{j\alpha}^2}$. Choose the scalar d so the vector $d\mathbf{u}_j$ is orthogonal to $\mathbf{u}_i - d\mathbf{u}_j$; that is, $0 = d\mathbf{u}_j'(\mathbf{u}_i - d\mathbf{u}_j) = d(\mathbf{u}_j'\mathbf{u}_i - d\mathbf{u}_j'\mathbf{u}_j)$. Therefore, $d = \mathbf{u}_j'\mathbf{u}_i/\mathbf{u}_j'\mathbf{u}_j$. We decompose $\mathbf{u}_i$ into $\mathbf{u}_i - d\mathbf{u}_j$ and $d\mathbf{u}_j$ ($\mathbf{u}_i = (\mathbf{u}_i - d\mathbf{u}_j) + d\mathbf{u}_j$) as indicated in Figure 3.1. The absolute value of the cosine of the angle between $\mathbf{u}_i$ and $\mathbf{u}_j$ is the length of $d\mathbf{u}_j$ divided by

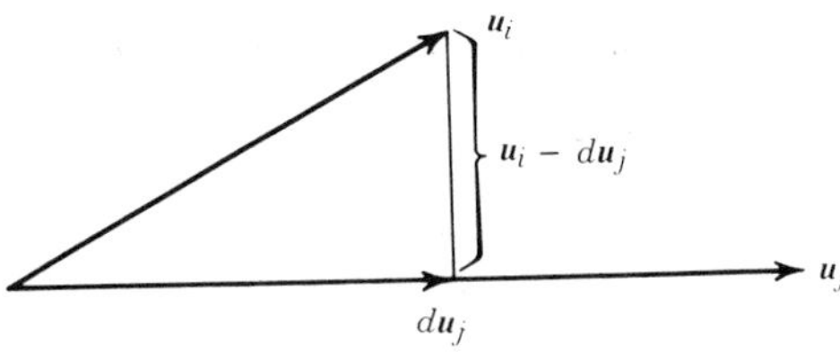

Figure 3.1

the length of $\boldsymbol{u}_i$; that is, it is $\sqrt{d\boldsymbol{u}_j'(d\boldsymbol{u}_j)/\boldsymbol{u}_i'\boldsymbol{u}_i} = \sqrt{d\boldsymbol{u}_j'\boldsymbol{u}_j d/\boldsymbol{u}_i'\boldsymbol{u}_i}$; the cosine is $\boldsymbol{u}_i'\boldsymbol{u}_j/\sqrt{\boldsymbol{u}_i'\boldsymbol{u}_i\boldsymbol{u}_j'\boldsymbol{u}_j}$. This proves the desired result.

To give a geometric interpretation of a_{ii} and $a_{ij}/\sqrt{a_{ii}a_{jj}}$, we introduce the equiangular line, which is the line going through the origin and the point $(1, 1, \ldots, 1)$. See Figure 3.2. The projection of $\boldsymbol{u}_i$ on the vector $\boldsymbol{\varepsilon} = (1, 1, \ldots, 1)'$ is $(\boldsymbol{\varepsilon}'\boldsymbol{u}_i/\boldsymbol{\varepsilon}'\boldsymbol{\varepsilon})\boldsymbol{\varepsilon} = (\Sigma_\alpha x_{i\alpha}/\Sigma_\alpha 1)\boldsymbol{\varepsilon} = \bar{x}_i\boldsymbol{\varepsilon} = (\bar{x}_i, \bar{x}_i, \ldots, \bar{x}_i)'$. Then we decompose $\boldsymbol{u}_i$ into $\bar{x}_i\boldsymbol{\varepsilon}$, the projection on the equiangular line, and $\boldsymbol{u}_i - \bar{x}_i\boldsymbol{\varepsilon}$, the projection of $\boldsymbol{u}_i$ on the plane perpendicular to the equiangular line. The squared length of $\boldsymbol{u}_i - \bar{x}_i\boldsymbol{\varepsilon}$ is $(\boldsymbol{u}_i - \bar{x}_i\boldsymbol{\varepsilon})'(\boldsymbol{u}_i - \bar{x}_i\boldsymbol{\varepsilon}) = \Sigma_\alpha(x_{i\alpha} - \bar{x}_i)^2$; this is $N\hat{\sigma}_{ii} = a_{ii}$. Translate $\boldsymbol{u}_i - \bar{x}_i\boldsymbol{\varepsilon}$ and $\boldsymbol{u}_j - \bar{x}_j\boldsymbol{\varepsilon}$, so that each vector has an endpoint at the origin; the αth coordinate of the first vector is $x_{i\alpha} - \bar{x}_i$ and of the second is $x_{j\alpha} - \bar{x}_j$. The cosine of the angle between these two vectors is

$$r_{ij} = \frac{(\boldsymbol{u}_i - \bar{x}_i\boldsymbol{\varepsilon})'(\boldsymbol{u}_j - \bar{x}_j\boldsymbol{\varepsilon})}{\sqrt{(\boldsymbol{u}_i - \bar{x}_i\boldsymbol{\varepsilon})'(\boldsymbol{u}_i - \bar{x}_i\boldsymbol{\varepsilon})(\boldsymbol{u}_j - \bar{x}_j\boldsymbol{\varepsilon})'(\boldsymbol{u}_j - \bar{x}_j\boldsymbol{\varepsilon})}} \tag{20}$$

$$= \frac{\sum_{\alpha=1}^{N}(x_{i\alpha} - \bar{x}_i)(x_{j\alpha} - \bar{x}_j)}{\sqrt{\sum_{\alpha=1}^{N}(x_{i\alpha} - \bar{x}_i)^2 \sum_{\alpha=1}^{N}(x_{j\alpha} - \bar{x}_j)^2}}.$$

As an example of the calculations consider the data in Table 3.1 and graphed in Figure 3.3, taken from Student (1908). The measurement $x_{11} = 1.9$ on the first patient is the increase in the number of hours of sleep due to the use of the sedative A, $x_{21} = 0.7$ is the increase in the number of hours due to

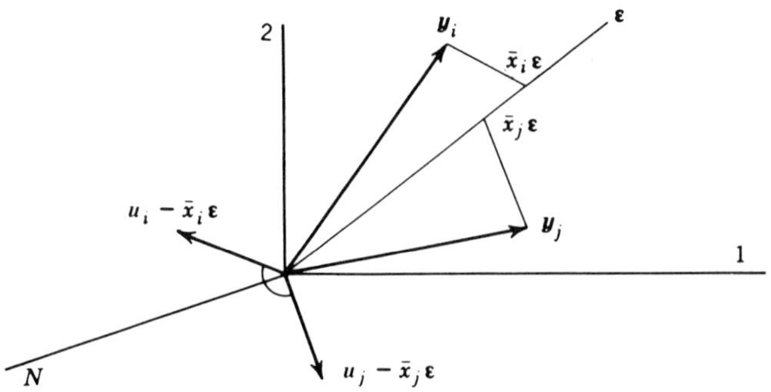

Figure 3.2

TABLE 3.1
INCREASE IN SLEEP

Patient	Drug A	Drug B
	x_1	x_2
1	1.9	0.7
2	0.8	−1.6
3	1.1	−0.2
4	0.1	−1.2
5	−0.1	−0.1
6	4.4	3.4
7	5.5	3.7
8	1.6	0.8
9	4.6	0.0
10	3.4	2.0

sedative B, and so on. Assuming that each pair (i.e., each row in the table) is an observation from $N(\boldsymbol{\mu}, \boldsymbol{\Sigma})$, we find that

$$\hat{\boldsymbol{\mu}} = \bar{\boldsymbol{x}} = \begin{pmatrix} 2.33 \\ 0.75 \end{pmatrix},$$

$$\hat{\boldsymbol{\Sigma}} = \begin{pmatrix} 3.61 & 2.56 \\ 2.56 & 2.88 \end{pmatrix}, \tag{21}$$

$$\boldsymbol{S} = \begin{pmatrix} 4.01 & 2.85 \\ 2.85 & 3.20 \end{pmatrix},$$

and $\hat{\rho}_{12} = r_{12} = 0.7952$. ($\boldsymbol{S}$ will be defined later.)

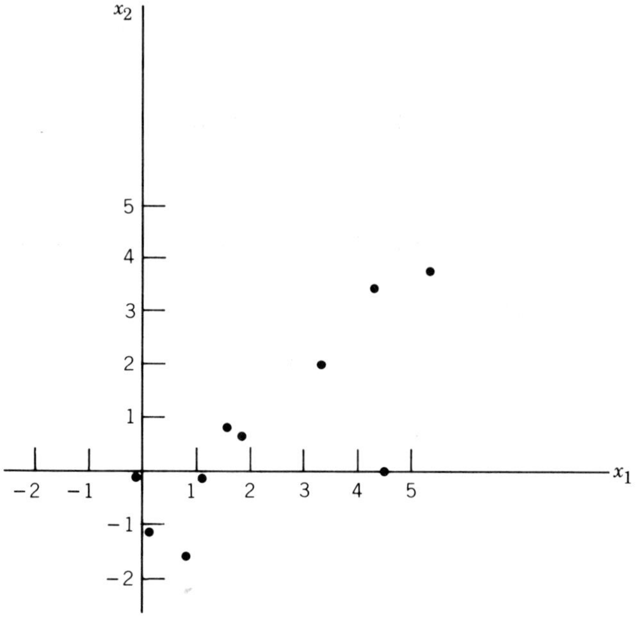

Figure 3.3. Increase in sleep.

3.3. THE DISTRIBUTION OF THE SAMPLE MEAN VECTOR; INFERENCE CONCERNING THE MEAN WHEN THE COVARIANCE MATRIX IS KNOWN

3.3.1. Distribution Theory

In the univariate case the mean of a sample is distributed normally and independently of the sample variance. Similarly, the sample mean $\bar{X}$ defined in Section 3.2 is distributed normally and independently of $\hat{\Sigma}$.

To prove this result we shall make a transformation of the set of observation vectors. Because this kind of transformation is used several times in this book, we first prove a more general theorem.

THEOREM 3.3.1. *Suppose $X_1, \dots, X_N$ are independent, where X_α is distributed according to $N(\mu_\alpha, \Sigma)$. Let $C = (c_{\alpha\beta})$ be an $N \times N$ orthogonal matrix. Then $Y_\alpha = \sum_{\beta=1}^{N} c_{\alpha\beta} X_\beta$ is distributed according to $N(\nu_\alpha, \Sigma)$, where $\nu_\alpha = \sum_{\beta=1}^{N} c_{\alpha\beta} \mu_\beta$, $\alpha = 1, \dots, N$, and $Y_1, \dots, Y_N$ are independent.*

PROOF. The set of vectors $\boldsymbol{Y}_1, \dots, \boldsymbol{Y}_N$ have a joint normal distribution because the entire set of components is a set of linear combinations of the components of $\boldsymbol{X}_1, \dots, \boldsymbol{X}_N$, which have a joint normal distribution. The expected value of $\boldsymbol{Y}_\alpha$ is

$$(1) \qquad \mathscr{E}\boldsymbol{Y}_\alpha = \mathscr{E}\sum_{\beta=1}^{N} c_{\alpha\beta}\boldsymbol{X}_\beta = \sum_{\beta=1}^{N} c_{\alpha\beta}\mathscr{E}\boldsymbol{X}_\beta = \sum_{\beta=1}^{N} c_{\alpha\beta}\boldsymbol{\mu}_\beta = \boldsymbol{\nu}_\alpha.$$

The covariance matrix between $\boldsymbol{Y}_\alpha$ and $\boldsymbol{Y}_\gamma$ is

$$(2) \qquad \begin{aligned} \mathscr{C}(\boldsymbol{Y}_\alpha, \boldsymbol{Y}_\gamma') &= \mathscr{E}(\boldsymbol{Y}_\alpha - \boldsymbol{\nu}_\alpha)(\boldsymbol{Y}_\gamma - \boldsymbol{\nu}_\gamma)' \\ &= \mathscr{E}\left[\sum_{\beta=1}^{N} c_{\alpha\beta}(\boldsymbol{X}_\beta - \boldsymbol{\mu}_\beta)\right]\left[\sum_{\varepsilon=1}^{N} c_{\gamma\varepsilon}(\boldsymbol{X}_\varepsilon - \boldsymbol{\mu}_\varepsilon)'\right] \\ &= \sum_{\beta,\varepsilon=1}^{N} c_{\alpha\beta}c_{\gamma\varepsilon}\mathscr{E}(\boldsymbol{X}_\beta - \boldsymbol{\mu}_\beta)(\boldsymbol{X}_\varepsilon - \boldsymbol{\mu}_\varepsilon)' \\ &= \sum_{\beta,\varepsilon=1}^{N} c_{\alpha\beta}c_{\gamma\varepsilon}\delta_{\beta\varepsilon}\boldsymbol{\Sigma} \\ &= \sum_{\beta=1}^{N} c_{\alpha\beta}c_{\gamma\beta}\boldsymbol{\Sigma} \\ &= \delta_{\alpha\gamma}\boldsymbol{\Sigma}, \end{aligned}$$

where $\delta_{\alpha\gamma}$ is the Kronecker delta ($= 1$ if $\alpha = \gamma$ and $= 0$ if $\alpha \neq \gamma$). This shows that $\boldsymbol{Y}_\alpha$ is independent of $\boldsymbol{Y}_\gamma$, $\alpha \neq \gamma$, and $\boldsymbol{Y}_\alpha$ has the covariance matrix $\boldsymbol{\Sigma}$. Q.E.D.

We also use the following general lemma:

LEMMA 3.3.1. *If $\boldsymbol{C} = (c_{\alpha\beta})$ is orthogonal, then $\sum_{\alpha=1}^{N} \boldsymbol{x}_\alpha \boldsymbol{x}_\alpha' = \sum_{\alpha=1}^{N} \boldsymbol{y}_\alpha \boldsymbol{y}_\alpha'$, where $\boldsymbol{y}_\alpha = \sum_{\beta=1}^{N} c_{\alpha\beta}\boldsymbol{x}_\beta$, $\alpha = 1, \dots, N$.*

PROOF.

$$\sum_{\alpha=1}^{N} y_\alpha y'_\alpha = \sum_\alpha \sum_\beta c_{\alpha\beta} \boldsymbol{x}_\beta \sum_\gamma c_{\alpha\gamma} \boldsymbol{x}'_\gamma \tag{3}$$

$$= \sum_{\beta,\gamma} \left(\sum_\alpha c_{\alpha\beta} c_{\alpha\gamma} \right) \boldsymbol{x}_\beta \boldsymbol{x}'_\gamma$$

$$= \sum_{\beta,\gamma} \delta_{\beta\gamma} \boldsymbol{x}_\beta \boldsymbol{x}'_\gamma$$

$$= \sum_\beta \boldsymbol{x}_\beta \boldsymbol{x}'_\beta. \qquad \text{Q.E.D.}$$

Let $\boldsymbol{X}_1, \dots, \boldsymbol{X}_N$ be independent, each distributed according to $N(\boldsymbol{\mu}, \boldsymbol{\Sigma})$. There exists an $N \times N$ orthogonal matrix $\boldsymbol{B} = (b_{\alpha\beta})$ with the last row

$$(1/\sqrt{N}, \dots, 1/\sqrt{N}). \tag{4}$$

(See Lemma A.4.2.) This transformation is a rotation in the N-dimensional space described in Section 3.2 with the equiangular line going into the Nth coordinate axis. Let $\boldsymbol{A} = N\hat{\boldsymbol{\Sigma}}$, defined in Section 3.2, and let

$$\boldsymbol{Z}_\alpha = \sum_{\beta=1}^{N} b_{\alpha\beta} \boldsymbol{X}_\beta. \tag{5}$$

Then

$$\boldsymbol{Z}_N = \sum_{\beta=1}^{N} b_{N\beta} \boldsymbol{X}_\beta = \sum_{\beta=1}^{N} \frac{1}{\sqrt{N}} \boldsymbol{X}_\beta = \sqrt{N}\, \bar{\boldsymbol{X}}. \tag{6}$$

By Lemma 3.3.1 we have

$$\boldsymbol{A} = \sum_{\alpha=1}^{N} \boldsymbol{X}_\alpha \boldsymbol{X}'_\alpha - N\bar{\boldsymbol{X}}\bar{\boldsymbol{X}}' \tag{7}$$

$$= \sum_{\alpha=1}^{N} \boldsymbol{Z}_\alpha \boldsymbol{Z}'_\alpha - \boldsymbol{Z}_N \boldsymbol{Z}'_N$$

$$= \sum_{\alpha=1}^{N-1} \boldsymbol{Z}_\alpha \boldsymbol{Z}'_\alpha.$$

Since $\boldsymbol{Z}_N$ is independent of $\boldsymbol{Z}_1, \dots, \boldsymbol{Z}_{N-1}$, the mean vector $\overline{\boldsymbol{X}}$ is independent of $\boldsymbol{A}$. Since

$$\mathscr{E}\boldsymbol{Z}_N = \sum_{\beta=1}^{N} b_{N\beta}\mathscr{E}\boldsymbol{X}_\beta = \sum_{\beta=1}^{N} \frac{1}{\sqrt{N}}\boldsymbol{\mu} = \sqrt{N}\,\boldsymbol{\mu}, \tag{8}$$

$\boldsymbol{Z}_N$ is distributed according to $N(\sqrt{N}\,\boldsymbol{\mu}, \boldsymbol{\Sigma})$ and $\overline{\boldsymbol{X}} = (1/\sqrt{N})\boldsymbol{Z}_N$ is distributed according to $N[\boldsymbol{\mu}, (1/N)\boldsymbol{\Sigma}]$. We note

$$\begin{aligned} \mathscr{E}\boldsymbol{Z}_\alpha &= \sum_{\beta=1}^{N} b_{\alpha\beta}\mathscr{E}\boldsymbol{X}_\beta = \sum_{\beta=1}^{N} b_{\alpha\beta}\boldsymbol{\mu} \\ &= \sum_{\beta=1}^{N} b_{\alpha\beta}b_{N\beta}\sqrt{N}\,\boldsymbol{\mu} \\ &= 0, \qquad \alpha \neq N. \end{aligned} \tag{9}$$

THEOREM 3.3.2. *The mean of a sample of size N from $N(\boldsymbol{\mu}, \boldsymbol{\Sigma})$ is distributed according to $N[\boldsymbol{\mu}, (1/N)\boldsymbol{\Sigma}]$ and independently of $\hat{\boldsymbol{\Sigma}}$, the maximum likelihood estimator of $\boldsymbol{\Sigma}$. $N\hat{\boldsymbol{\Sigma}}$ is distributed as $\sum_{\alpha=1}^{N-1}\boldsymbol{Z}_\alpha\boldsymbol{Z}_\alpha'$, where $\boldsymbol{Z}_\alpha$ is distributed according to $N(\boldsymbol{0}, \boldsymbol{\Sigma})$, $\alpha = 1, \dots, N-1$, and $\boldsymbol{Z}_1, \dots, \boldsymbol{Z}_{N-1}$ are independent.*

DEFINITION 3.3.1. *An estimator $\boldsymbol{t}$ of a parameter vector $\boldsymbol{\theta}$ is unbiased if and only if $\mathscr{E}_{\boldsymbol{\theta}}\boldsymbol{t} = \boldsymbol{\theta}$.*

Since $\mathscr{E}\overline{\boldsymbol{X}} = (1/N)\mathscr{E}\sum_{\alpha=1}^{N}\boldsymbol{X}_\alpha = \boldsymbol{\mu}$, the sample mean is an unbiased estimator of the population mean. However,

$$\mathscr{E}\hat{\boldsymbol{\Sigma}} = \frac{1}{N}\mathscr{E}\sum_{\alpha=1}^{N-1}\boldsymbol{Z}_\alpha\boldsymbol{Z}_\alpha' = \frac{N-1}{N}\boldsymbol{\Sigma}. \tag{10}$$

Thus $\hat{\boldsymbol{\Sigma}}$ is a biased estimator of $\boldsymbol{\Sigma}$. We shall, therefore, define

$$\boldsymbol{S} = \frac{1}{N-1}\boldsymbol{A} = \frac{1}{N-1}\sum_{\alpha=1}^{N}(\boldsymbol{x}_\alpha - \bar{\boldsymbol{x}})(\boldsymbol{x}_\alpha - \bar{\boldsymbol{x}})' \tag{11}$$

as the *sample covariance matrix*. It is an unbiased estimator of $\boldsymbol{\Sigma}$ and the diagonal elements are the usual (unbiased) sample variances of the components of $\boldsymbol{X}$.

3.3.2. Tests and Confidence Regions for the Mean Vector When the Covariance Matrix Is Known

A statistical problem of considerable importance is that of testing the hypothesis that the mean vector of a normal distribution is a given vector, and a related problem is that of giving a confidence region for the unknown vector of means. We now go on to study these problems under the assumption that the covariance matrix $\boldsymbol{\Sigma}$ is known. In Chapter 5 we consider these problems when the covariance matrix is unknown.

In the univariate case one bases a test or a confidence interval on the fact that the difference between the sample mean and the population mean is normally distributed with mean zero and known variance; then tables of the normal distribution can be used to set up significance points or to compute confidence intervals. In the multivariate case one uses the fact that the difference between the sample mean vector and the population mean vector is normally distributed with mean vector zero and known covariance matrix. One could set up limits for each component on the basis of the distribution, but this procedure has the disadvantages that the choice of limits is somewhat arbitrary and in the case of tests leads to tests that may be very poor against some alternatives, and, moreover, such limits are difficult to compute because tables are available only for the bivariate case. The procedures given below, however, are easily computed and furthermore can be given general intuitive and theoretical justifications.

The procedures and evaluation of their properties are based on the following theorem:

THEOREM 3.3.3. *If the m-component vector $\boldsymbol{Y}$ is distributed according to $N(\boldsymbol{\nu}, \boldsymbol{T})$ (nonsingular), then $\boldsymbol{Y}'\boldsymbol{T}^{-1}\boldsymbol{Y}$ is distributed according to the noncentral χ^2-distribution with m degrees of freedom and noncentrality parameter $\boldsymbol{\nu}'\boldsymbol{T}^{-1}\boldsymbol{\nu}$. If $\boldsymbol{\nu} = \boldsymbol{0}$, the distribution is the central χ^2-distribution.*

PROOF. Let $\boldsymbol{C}$ be a nonsingular matrix such that $\boldsymbol{C}\boldsymbol{T}\boldsymbol{C}' = \boldsymbol{I}$ and define $\boldsymbol{Z} = \boldsymbol{C}\boldsymbol{Y}$. Then $\boldsymbol{Z}$ is normally distributed with mean $\mathscr{E}\boldsymbol{Z} = \boldsymbol{C}\mathscr{E}\boldsymbol{Y} = \boldsymbol{C}\boldsymbol{\nu} = \boldsymbol{\lambda}$, say, and covariance matrix $\mathscr{E}(\boldsymbol{Z} - \boldsymbol{\lambda})(\boldsymbol{Z} - \boldsymbol{\lambda})' = \mathscr{E}\boldsymbol{C}(\boldsymbol{Y} - \boldsymbol{\nu})(\boldsymbol{Y} - \boldsymbol{\nu})'\boldsymbol{C}' = \boldsymbol{C}\boldsymbol{T}\boldsymbol{C}' = \boldsymbol{I}$. Then $\boldsymbol{Y}'\boldsymbol{T}^{-1}\boldsymbol{Y} = \boldsymbol{Z}'(\boldsymbol{C}')^{-1}\boldsymbol{T}^{-1}\boldsymbol{C}^{-1}\boldsymbol{Z} = \boldsymbol{Z}'(\boldsymbol{C}\boldsymbol{T}\boldsymbol{C}')^{-1}\boldsymbol{Z} = \boldsymbol{Z}'\boldsymbol{Z}$, which is the sum of squares of the components of $\boldsymbol{Z}$. Similarly $\boldsymbol{\nu}'\boldsymbol{T}^{-1}\boldsymbol{\nu} = \boldsymbol{\lambda}'\boldsymbol{\lambda}$. Thus $\boldsymbol{Y}'\boldsymbol{T}^{-1}\boldsymbol{Y}$ is distributed as $\sum_{i=1}^m Z_i^2$, where $Z_1, \dots, Z_m$ are independently normally distributed with means $\lambda_1, \dots, \lambda_m$, respectively, and variances 1. By definition this distribution is the noncentral χ^2-distribution with noncentrality parameter $\Sigma_{i=1}^{m}\lambda_i^2$. See Section 3.3.3. If $\lambda_1 = \cdots = \lambda_m = 0$, the distribution is central. (See Problem 5 of Chapter 7.) Q.E.D.

Since $\sqrt{N}(\overline{\mathbf{X}} - \boldsymbol{\mu})$ is distributed according to $N(\mathbf{0}, \boldsymbol{\Sigma})$, it follows from the theorem that

$$N(\overline{\mathbf{X}} - \boldsymbol{\mu})'\boldsymbol{\Sigma}^{-1}(\overline{\mathbf{X}} - \boldsymbol{\mu}) \tag{12}$$

has a (central) χ^2-distribution with p degrees of freedom. This is the fundamental fact we use in setting up tests and confidence regions concerning $\boldsymbol{\mu}$.

Let $\chi_p^2(\alpha)$ be the number such that

$$\Pr\{\chi_p^2 > \chi_p^2(\alpha)\} = \alpha. \tag{13}$$

Thus

$$\Pr\{N(\overline{\mathbf{X}} - \boldsymbol{\mu})'\boldsymbol{\Sigma}^{-1}(\overline{\mathbf{X}} - \boldsymbol{\mu}) > \chi_p^2(\alpha)\} = \alpha. \tag{14}$$

To test the hypothesis that $\boldsymbol{\mu} = \boldsymbol{\mu}_0$, where $\boldsymbol{\mu}_0$ is a specified vector, we use as our critical region

$$N(\bar{\mathbf{x}} - \boldsymbol{\mu}_0)'\boldsymbol{\Sigma}^{-1}(\bar{\mathbf{x}} - \boldsymbol{\mu}_0) > \chi_p^2(\alpha). \tag{15}$$

If we obtain a sample such that (15) is satisfied, we reject the null hypothesis. It can be seen intuitively that the probability is greater than α of rejecting the hypothesis if $\boldsymbol{\mu}$ is very much different from $\boldsymbol{\mu}_0$, since in the space of $\bar{\mathbf{x}}$ (15) defines an ellipsoid with center at $\boldsymbol{\mu}_0$, and when $\boldsymbol{\mu}$ is far from $\boldsymbol{\mu}_0$ the density of $\bar{\mathbf{x}}$ will be concentrated at a point near the edge or outside of the ellipsoid. The quantity $N(\overline{\mathbf{X}} - \boldsymbol{\mu}_0)'\boldsymbol{\Sigma}^{-1}(\overline{\mathbf{X}} - \boldsymbol{\mu}_0)$ is distributed as a noncentral χ^2 with p degrees of freedom and noncentrality parameter $N(\boldsymbol{\mu} - \boldsymbol{\mu}_0)'\boldsymbol{\Sigma}^{-1}(\boldsymbol{\mu} - \boldsymbol{\mu}_0)$ when $\overline{\mathbf{X}}$ is the mean of a sample of N from $N(\boldsymbol{\mu}, \boldsymbol{\Sigma})$ [given by Bose (1936a), (1936b)]. Pearson (1900) first proved Theorem 3.3.3 for $\boldsymbol{\nu} = \mathbf{0}$.

Now consider the following statement made on the basis of a sample with mean $\bar{\mathbf{x}}$: "The mean of the distribution satisfies

$$N(\bar{\mathbf{x}} - \boldsymbol{\mu}^*)'\boldsymbol{\Sigma}^{-1}(\bar{\mathbf{x}} - \boldsymbol{\mu}^*) \le \chi_p^2(\alpha) \tag{16}$$

as an inequality on μ^*." We see from (14) that the probability that a sample be drawn such that the above statement is true is $1 - \alpha$ because the event in (14) is equivalent to the statement being false. Thus, the set of μ^* satisfying (16) is a confidence region for μ with confidence $1 - \alpha$.

In the p-dimensional space of $\bar{x}$, (15) is the surface and exterior of an ellipsoid with center μ_0, the shape of the ellipsoid depending on Σ^{-1} and the size on $(1/N)\chi_p^2(\alpha)$ for given Σ^{-1}. In the p-dimensional space of μ^* (16) is the surface and interior of an ellipsoid with its center at $\bar{x}$. If $\Sigma^{-1} = I$, then (14) says that the probability is α that the distance between $\bar{x}$ and μ is greater than $\sqrt{\chi_p^2(\alpha)/N}$.

THEOREM 3.3.4. *If $\bar{x}$ is the mean of a sample of N drawn from $N(\mu, \Sigma)$ and Σ is known, then* (15) *gives a critical region of size α for testing the hypothesis $\mu = \mu_0$ and* (16) *gives a confidence region for μ of confidence $1 - \alpha$. $\chi_p^2(\alpha)$ is chosen to satisfy* (13).

The same technique can be used for the corresponding two sample problems. Suppose we have a sample $\{x_\alpha^{(1)}\}$, $\alpha = 1, \dots, N_1$, from the distribution $N(\mu^{(1)}, \Sigma)$ and a sample $\{x_\alpha^{(2)}\}$, $\alpha = 1, \dots, N_2$, from a second normal population $N(\mu^{(2)}, \Sigma)$ with the same covariance matrix. Then the two sample means

$$\text{(17)} \qquad \bar{x}^{(1)} = \frac{1}{N_1} \sum_{\alpha=1}^{N_1} x_\alpha^{(1)}, \qquad \bar{x}^{(2)} = \frac{1}{N_2} \sum_{\alpha=1}^{N_2} x_\alpha^{(2)}$$

are distributed independently according to $N[\mu^{(1)}, (1/N_1)\Sigma]$ and $N[\mu^{(2)}, (1/N_2)\Sigma]$, respectively. The difference of the two sample means $y = \bar{x}^{(1)} - \bar{x}^{(2)}$ is distributed according to $N\{v, [(1/N_1) + (1/N_2)]\Sigma\}$, where $v = \mu^{(1)} - \mu^{(2)}$. Thus

$$\text{(18)} \qquad \frac{N_1 N_2}{N_1 + N_2}(y - v)'\Sigma^{-1}(y - v) \le \chi_p^2(\alpha)$$

is a confidence region for the difference v of the two mean vectors, and a critical region for testing the hypothesis $\mu^{(1)} = \mu^{(2)}$ is given by

$$\text{(19)} \qquad \frac{N_1 N_2}{N_1 + N_2}(\bar{x}^{(1)} - \bar{x}^{(2)})'\Sigma^{-1}(\bar{x}^{(1)} - \bar{x}^{(2)}) > \chi_p^2(\alpha).$$

Mahalanobis (1930) suggested $(\boldsymbol{\mu}^{(1)} - \boldsymbol{\mu}^{(2)})'\boldsymbol{\Sigma}^{-1}(\boldsymbol{\mu}^{(1)} - \boldsymbol{\mu}^{(2)})$ as a measure of the *distance* squared between two populations. Let $\boldsymbol{C}$ be a matrix such that $\boldsymbol{\Sigma} = \boldsymbol{CC}'$ and let $\boldsymbol{\nu}^{(i)} = \boldsymbol{C}^{-1}\boldsymbol{\mu}^{(i)}$, $i = 1, 2$. Then the distance squared is $(\boldsymbol{\nu}^{(1)} - \boldsymbol{\nu}^{(2)})'(\boldsymbol{\nu}^{(1)} - \boldsymbol{\nu}^{(2)})$, which is the Euclidean distance squared.

3.3.3. The Noncentral χ^2-distribution; the Power Function

The power function of the test (15) of the null hypothesis that $\boldsymbol{\mu} = \boldsymbol{\mu}_0$ can be evaluated from the noncentral χ^2-distribution. The central χ^2-distribution is the distribution of the sum of squares of independent (scalar) normal variables with means 0 and variances 1; the noncentral χ^2-distribution is the generalization of this when the means may be different from 0. Let $\boldsymbol{Y}$ (of p components) be distributed according to $N(\boldsymbol{\lambda}, \boldsymbol{I})$. Let $\boldsymbol{Q}$ be an orthogonal matrix with elements of the first row being

$$q_{1i} = \frac{\lambda_i}{\sqrt{\boldsymbol{\lambda}'\boldsymbol{\lambda}}}, \qquad i = 1, \ldots, p. \tag{20}$$

Then $\boldsymbol{Z} = \boldsymbol{QY}$ is distributed according to $N(\boldsymbol{\tau}, \boldsymbol{I})$, where

$$\boldsymbol{\tau} = \begin{pmatrix} \tau \\ 0 \\ \vdots \\ 0 \end{pmatrix} \tag{21}$$

and $\tau = \sqrt{\boldsymbol{\lambda}'\boldsymbol{\lambda}}$. Let $V = \boldsymbol{Y}'\boldsymbol{Y} = \boldsymbol{Z}'\boldsymbol{Z} = \sum_{i=1}^{p} Z_i^2$. Then $W = \sum_{i=2}^{p} Z_i^2$ has a χ^2-distribution with $p - 1$ degrees of freedom (Problem 5 of Chapter 7), and Z_1 and W have as joint density

$$\begin{aligned} \frac{1}{\sqrt{2\pi}} e^{-\frac{1}{2}(z_1-\tau)^2} \frac{1}{2^{\frac{1}{2}(p-1)}\Gamma[\frac{1}{2}(p-1)]} w^{\frac{1}{2}(p-1)-1} e^{-\frac{1}{2}w} &= Ce^{-\frac{1}{2}(\tau^2+z_1^2+w)} w^{\frac{1}{2}(p-3)} e^{\tau z_1} \\ &= Ce^{-\frac{1}{2}(\tau^2+z_1^2+w)} w^{\frac{1}{2}(p-3)} \sum_{\alpha=0}^{\infty} \frac{\tau^\alpha z_1^\alpha}{\alpha!}, \end{aligned} \tag{22}$$

where $C^{-1} = 2^{\frac{1}{2}p}\sqrt{\pi}\,\Gamma[\frac{1}{2}(p-1)]$. The joint density of $V = W + Z_1^2$ and Z_1 is obtained by substituting $w = v - z_1^2$ (the Jacobian being 1),

$$Ce^{-\frac{1}{2}(\tau^2+v)}\left(v - z_1^2\right)^{\frac{1}{2}(p-3)} \sum_{\alpha=0}^{\infty} \frac{\tau^\alpha z_1^\alpha}{\alpha!}. \tag{23}$$

The joint density of V and $U = Z_1/\sqrt{V}$ is $(dz_1 = \sqrt{v}\,du)$

$$(24) \qquad Ce^{-\frac{1}{2}(\tau^2+v)}v^{\frac{1}{2}(p-2)}(1-u^2)^{\frac{1}{2}(p-3)}\sum_{\alpha=0}^{\infty}\frac{\tau^{\alpha}v^{\frac{1}{2}\alpha}u^{\alpha}}{\alpha!}.$$

The admissible range of z_1 given v is $-\sqrt{v}$ to $\sqrt{v}$, and the admissible range of u is -1 to 1. When we integrate (24) with respect to u term by term, the terms for α odd integrate to 0 since such a term is an odd function of u. In the other integrations we substitute $u = \sqrt{s}$ $(du = \frac{1}{2}ds/\sqrt{s})$ to obtain

$$(25) \qquad \begin{aligned}\int_{-1}^{1}(1-u^2)^{\frac{1}{2}(p-3)}u^{2\beta}\,du &= 2\int_0^1(1-u^2)^{\frac{1}{2}(p-3)}u^{2\beta}\,du\\ &= \int_0^1(1-s)^{\frac{1}{2}(p-3)}s^{\beta-\frac{1}{2}}\,ds\\ &= B\left[\tfrac{1}{2}(p-1),\beta+\tfrac{1}{2}\right]\\ &= \frac{\Gamma\left[\frac{1}{2}(p-1)\right]\Gamma(\beta+\frac{1}{2})}{\Gamma(\frac{1}{2}p+\beta)},\end{aligned}$$

by the usual properties of the beta and gamma functions. Thus the density of V is

$$(26) \qquad \frac{1}{2^{\frac{1}{2}p}\sqrt{\pi}}e^{-\frac{1}{2}(\tau^2+v)}v^{\frac{1}{2}p-1}\sum_{\beta=0}^{\infty}\frac{(\tau^2)^{\beta}v^{\beta}}{(2\beta)!}\frac{\Gamma(\beta+\frac{1}{2})}{\Gamma(\frac{1}{2}p+\beta)}.$$

We can use the *duplication formula for the gamma function* $\Gamma(2\beta+1) = (2\beta)!$ (Problem 40 of Chapter 7)

$$(27) \qquad \Gamma(2\beta+1) = \Gamma(\beta+\tfrac{1}{2})\Gamma(\beta+1)2^{2\beta}/\sqrt{\pi},$$

to rewrite (26) as

$$(28) \qquad \frac{1}{2^{\frac{1}{2}p}}e^{-\frac{1}{2}(\tau^2+v)}v^{\frac{1}{2}p-1}\sum_{\beta=0}^{\infty}\left(\frac{\tau^2}{4}\right)^{\beta}\frac{1}{\beta!\Gamma(\frac{1}{2}p+\beta)}v^{\beta}.$$

This is the density of the *noncentral* χ^2*-distribution* with p degrees of freedom and noncentrality parameter τ^2.

THEOREM 3.3.5. *If* $\boldsymbol{Y}$ *of* p *components is distributed according to* $N(\boldsymbol{\lambda}, \boldsymbol{I})$, *then* $V = \boldsymbol{Y}'\boldsymbol{Y}$ *has the density* (28), *where* $\tau^2 = \boldsymbol{\lambda}'\boldsymbol{\lambda}$.

To obtain the power function of the test (15), we note that $\sqrt{N}(\overline{\boldsymbol{X}} - \boldsymbol{\mu}_0)$ has the distribution $N[\sqrt{N}(\boldsymbol{\mu} - \boldsymbol{\mu}_0), \boldsymbol{\Sigma}]$. From Theorem 3.3.3 we obtain the following corollary:

COROLLARY 3.3.1. *If* $\overline{\boldsymbol{X}}$ *is the mean of a random sample of* N *drawn from* $N(\boldsymbol{\mu}, \boldsymbol{\Sigma})$, *then* $N(\overline{\boldsymbol{X}} - \boldsymbol{\mu}_0)'\boldsymbol{\Sigma}^{-1}(\overline{\boldsymbol{X}} - \boldsymbol{\mu}_0)$ *has a noncentral* χ^2*-distribution with* p *degrees of freedom and noncentrality parameter* $N(\boldsymbol{\mu} - \boldsymbol{\mu}_0)'\boldsymbol{\Sigma}^{-1}(\boldsymbol{\mu} - \boldsymbol{\mu}_0)$.

3.4. THEORETICAL PROPERTIES OF ESTIMATORS OF THE MEAN VECTOR

3.4.1. Properties of Maximum Likelihood Estimators

It was shown in Section 3.3.1 that $\bar{\boldsymbol{x}}$ and $\boldsymbol{S}$ are unbiased estimators of $\boldsymbol{\mu}$ and $\boldsymbol{\Sigma}$, respectively. In this subsection we shall show that $\bar{\boldsymbol{x}}$ and $\boldsymbol{S}$ are sufficient statistics and are complete.

Sufficiency. A statistic $\boldsymbol{T}$ is sufficient for a family of distributions of $\boldsymbol{X}$ or for a parameter $\boldsymbol{\theta}$ if the conditional distribution of $\boldsymbol{X}$ given $\boldsymbol{T} = \boldsymbol{t}$ does not depend on $\boldsymbol{\theta}$ [e.g., Cramér (1946), Section 32.4]. In this sense the statistic $\boldsymbol{T}$ gives as much information about $\boldsymbol{\theta}$ as the entire sample $\boldsymbol{X}$. (Of course, this idea depends strictly on the assumed family of distributions.)

FACTORIZATION THEOREM. *A statistic* $\boldsymbol{t}(y)$ *is sufficient for* $\boldsymbol{\theta}$ *if and only if the density* $f(y|\boldsymbol{\theta})$ *can be factored*

$$f(y|\boldsymbol{\theta}) = g[\boldsymbol{t}(y), \boldsymbol{\theta}]\,h(y), \tag{1}$$

where $g[\boldsymbol{t}(y), \boldsymbol{\theta}]$ *and* $h(y)$ *are nonnegative and* $h(y)$ *does not depend on* $\boldsymbol{\theta}$.

THEOREM 3.4.1. *If* $\boldsymbol{x}_1, \dots, \boldsymbol{x}_N$ *are observations from* $N(\boldsymbol{\mu}, \boldsymbol{\Sigma})$, *then* $\bar{\boldsymbol{x}}$ *and* $\boldsymbol{S}$ *are sufficient for* $\boldsymbol{\mu}$ *and* $\boldsymbol{\Sigma}$. *If* $\boldsymbol{\mu}$ *is given,* $\sum_{\alpha=1}^{N}(\boldsymbol{x}_\alpha - \boldsymbol{\mu})(\boldsymbol{x}_\alpha - \boldsymbol{\mu})'$ *is sufficient for* $\boldsymbol{\Sigma}$. *If* $\boldsymbol{\Sigma}$ *is given,* $\bar{\boldsymbol{x}}$ *is sufficient for* $\boldsymbol{\mu}$.

PROOF. The density of $X_1, \dots, X_N$ is

(2)
$$\prod_{\alpha=1}^{N} n(x_\alpha|\mu,\Sigma) = (2\pi)^{-\frac{1}{2}Np}|\Sigma|^{-\frac{1}{2}N}\exp\left[-\tfrac{1}{2}\operatorname{tr}\Sigma^{-1}\sum_{\alpha=1}^{N}(x_\alpha-\mu)(x_\alpha-\mu)'\right]$$
$$= (2\pi)^{-\frac{1}{2}Np}|\Sigma|^{-\frac{1}{2}N}\exp\left\{-\tfrac{1}{2}\left[N(\bar{x}-\mu)'\Sigma^{-1}(\bar{x}-\mu)+(N-1)\operatorname{tr}\Sigma^{-1}S\right]\right\}.$$

The right-hand side of (2) is in the form of (1) for $\bar{x}$, S, μ, Σ and the middle is in the form of (1) for $\sum_{\alpha=1}^{N}(x_\alpha-\mu)(x_\alpha-\mu)'$, Σ; in each case $h(x_1,\dots,x_N) = 1$. The right-hand side is in the form of (1) for $\bar{x},\mu$ with $h(x_1,\dots,x_N) = \exp\{-\frac{1}{2}(N-1)\operatorname{tr}\Sigma^{-1}S\}$. Q.E.D.

Note that if Σ is given, $\bar{x}$ is sufficient for μ, but if μ is given, S is not sufficient for Σ.

Completeness. To prove an optimality property of the T^2-test (Section 5.5), we need the result that $(\bar{x}, S)$ is a complete sufficient set of statistics for (μ, Σ).

DEFINITION 3.4.1. *A family of distributions of y indexed by θ is complete if for every real-valued function $g(y)$*

(3)
$$\mathscr{E}_\theta g(y) \equiv 0$$

identically in θ implies $g(y) = 0$ except for a set of y of probability 0 for every θ.

If the family of distributions of a sufficient set of statistics is complete, the set is called a complete sufficient set.

THEOREM 3.4.2. *The sufficient set of statistics $\bar{x}, S$ is complete for μ, Σ when the sample is drawn from $N(\mu, \Sigma)$.*

PROOF. We can define the sample in terms of $\bar{x}$ and $z_1, \dots, z_n$ as in Section 3.3 with $n = N-1$. We assume for any function $g(\bar{x}, A) = g(\bar{x}, nS)$ that

(4)
$$\int\cdots\int K|\Sigma|^{-\frac{1}{2}N} g\left(\bar{x}, \sum_{\alpha=1}^{n} z_\alpha z_\alpha'\right)$$
$$\cdot\exp\left\{-\tfrac{1}{2}\left[N(\bar{x}-\mu)'\Sigma^{-1}(\bar{x}-\mu)+\sum_{\alpha=1}^{n} z_\alpha'\Sigma^{-1}z_\alpha\right]\right\}d\bar{x}\prod_{\alpha=1}^{n}dz_\alpha \equiv 0, \forall \mu, \Sigma,$$

where $K = \sqrt{N}(2\pi)^{-\frac{1}{2}pN}$, $d\bar{x} = \prod_{i=1}^{p} d\bar{x}_i$, and $dz_\alpha = \prod_{i=1}^{p} dz_{i\alpha}$. If we let $\Sigma^{-1} = I - 2\Theta$, where $\Theta = \Theta'$ and $I - 2\Theta$ is positive definite, and let $\mu = (I - 2\Theta)^{-1}t$, then (4) is

$$(5)\quad 0 \equiv \int \cdots \int K|I - 2\Theta|^{\frac{1}{2}N} g\left(\bar{x}, \sum_{\alpha=1}^{n} z_\alpha z_\alpha'\right)$$
$$\cdot \exp\left\{-\tfrac{1}{2}\left[\operatorname{tr}(I - 2\Theta)\left(\sum_{\alpha=1}^{n} z_\alpha z_\alpha' + N\bar{x}\bar{x}'\right)\right.\right.$$
$$\left.\left. - 2Nt'\bar{x} + Nt'(I - 2\Theta)^{-1}t\right]\right\} d\bar{x} \prod_{\alpha=1}^{n} dz_\alpha$$
$$= |I - 2\Theta|^{\frac{1}{2}N} \exp\left\{-\tfrac{1}{2}Nt'(I - 2\Theta)^{-1}t\right\} \int \cdots \int g(\bar{x}, B - N\bar{x}\bar{x}')$$
$$\cdot \exp\left[\operatorname{tr}\Theta B + t'(N\bar{x})\right] n\left[\bar{x}|0, (1/N)I\right] \prod_{\alpha=1}^{n} n(z_\alpha|0, I)\, d\bar{x} \prod_{\alpha=1}^{n} dz_\alpha,$$

where $B = \sum_{\alpha=1}^{n} z_\alpha z_\alpha' + N\bar{x}\bar{x}'$. Thus

$$(6)\quad 0 \equiv \mathscr{E} g(\bar{x}, B - N\bar{x}\bar{x}')\exp\left[\operatorname{tr}\Theta B + t'(N\bar{x})\right]$$
$$= \int \cdots \int g(\bar{x}, B - N\bar{x}\bar{x}')\exp\left[\operatorname{tr}\Theta B + t'(N\bar{x})\right] h(\bar{x}, B)\, d\bar{x}\, dB,$$

where $h(\bar{x}, B)$ is the joint density of $\bar{x}$ and B and $dB = \prod_{i \le j} db_{ij}$. The right-hand side of (6) is the Laplace transform of $g(\bar{x}, B - N\bar{x}\bar{x}')h(\bar{x}, B)$. Since this is 0, $g(\bar{x}, A) = 0$ except for a set of measure 0. Q.E.D.

Efficiency. If a q-component random vector Y has mean vector $\mathscr{E}Y = \nu$ and covariance matrix $\mathscr{E}(Y - \nu)(Y - \nu)' = \Psi$, then

$$(7)\qquad (y - \nu)'\Psi^{-1}(y - \nu) = q + 2$$

is called the *concentration ellipsoid* of Y. [See Cramér (1946), p. 300.] The density defined by a uniform distribution over the interior of this ellipsoid has the same mean vector and covariance matrix as Y. (See Problem 14 of Chapter 2.) Let θ be a vector of q parameters in a distribution and let t be a vector of unbiased estimators (that is, $\mathscr{E}\boldsymbol{t} = \boldsymbol{\theta}$) based on N observations from that distribution with covariance matrix $\boldsymbol{\Psi}$. Then the ellipsoid

$$N(\boldsymbol{t}-\boldsymbol{\theta})'\mathscr{E}\left(\frac{\partial \log f}{\partial \boldsymbol{\theta}}\right)\left(\frac{\partial \log f}{\partial \boldsymbol{\theta}}\right)'(\boldsymbol{t}-\boldsymbol{\theta}) = q+2 \tag{8}$$

lies entirely within the ellipsoid of concentration of $\boldsymbol{t}$; $\partial \log f/\partial\boldsymbol{\theta}$ denotes the column vector of derivatives of the density of the distribution (or probability function) with respect to the components of $\boldsymbol{\theta}$. The discussion by Cramér (1946, p. 495) is in terms of scalar observations, but it is clear that it holds true for vector observations. If (8) is the ellipsoid of concentration of $\boldsymbol{t}$, then $\boldsymbol{t}$ is said to be efficient. In general, the ratio of the volume of (8) to that of the ellipsoid of concentration is defined as the efficiency of $\boldsymbol{t}$. In the case of the multivariate normal distribution, if $\boldsymbol{\theta} = \boldsymbol{\mu}$, then $\bar{\boldsymbol{x}}$ is efficient. If $\boldsymbol{\theta}$ includes both $\boldsymbol{\mu}$ and $\boldsymbol{\Sigma}$, then $\bar{\boldsymbol{x}}$ and $\boldsymbol{S}$ have efficiency $[(N-1)/N]^{p(p+1)/2}$. Under suitable regularity conditions, which are satisfied by the multivariate normal distribution,

$$\mathscr{E}\left(\frac{\partial \log f}{\partial \boldsymbol{\theta}}\right)\left(\frac{\partial \log f}{\partial \boldsymbol{\theta}}\right)' = -\mathscr{E}\frac{\partial^2 \log f}{\partial\boldsymbol{\theta}\,\partial\boldsymbol{\theta}'}. \tag{9}$$

This is the *information matrix* for one observation. The Cramér–Rao lower bound is that for any unbiased estimator $\boldsymbol{t}$ the matrix

$$N\mathscr{E}(\boldsymbol{t}-\boldsymbol{\theta})(\boldsymbol{t}-\boldsymbol{\theta})' - \left[-\mathscr{E}\frac{\partial^2 \log f}{\partial\boldsymbol{\theta}\,\partial\boldsymbol{\theta}'}\right]^{-1} \tag{10}$$

is positive semidefinite. (Other lower bounds can also be given.)

Consistency.

DEFINITION 3.4.2. *A sequence of vectors* $\boldsymbol{t}_n = (t_{1n}, \dots, t_{mn})'$, $n = 1, 2, \dots,$ *is a consistent estimator of* $\boldsymbol{\theta} = (\theta_1, \dots, \theta_m)'$ *if* $\operatorname{plim}_{n\to\infty} t_{in} = \theta_i$, $i = 1, \dots, m$.

By the law of large numbers each component of the sample mean $\bar{\boldsymbol{x}}$ is a consistent estimator of that component of the vector of expected values $\boldsymbol{\mu}$ if the observation vectors are independently and identically distributed with mean $\boldsymbol{\mu}$, and hence $\bar{\boldsymbol{x}}$ is a consistent estimator of $\boldsymbol{\mu}$. Normality is not involved.

An element of the sample covariance matrix is

$$s_{ij} = \frac{1}{N-1}\sum_{\alpha=1}^{N}(x_{i\alpha}-\mu_i)(x_{j\alpha}-\mu_j) - \frac{N}{N-1}(\bar{x}_i-\mu_i)(\bar{x}_j-\mu_j) \tag{11}$$

by Lemma 3.2.1 with $\boldsymbol{b} = \boldsymbol{\mu}$. The probability limit of the second term is 0. The probability limit of the first term is σ_{ij} if $\boldsymbol{x}_1, \boldsymbol{x}_2, \dots$ are independently and identically distributed with mean $\boldsymbol{\mu}$ and covariance matrix $\boldsymbol{\Sigma}$. Then $\boldsymbol{S}$ is a consistent estimator of $\boldsymbol{\Sigma}$.

Asymptotic normality. First we prove a multivariate central limit theorem.

THEOREM 3.4.3. *Let the m-component vectors* $\boldsymbol{Y}_1, \boldsymbol{Y}_2, \dots$ *be independently and identically distributed with means* $\mathscr{E}\boldsymbol{Y}_\alpha = \boldsymbol{\nu}$ *and covariance matrices* $\mathscr{E}(\boldsymbol{Y}_\alpha - \boldsymbol{\nu})(\boldsymbol{Y}_\alpha - \boldsymbol{\nu})' = \boldsymbol{T}$. *Then the limiting distribution of* $(1/\sqrt{n})\sum_{\alpha=1}^{n}(\boldsymbol{Y}_\alpha - \boldsymbol{\nu})$ *as* $n \to \infty$ *is* $N(\boldsymbol{0}, \boldsymbol{T})$.

PROOF. Let

$$\phi_n(\boldsymbol{t}, u) = \mathscr{E}\exp\left[iu\boldsymbol{t}'\frac{1}{\sqrt{n}}\sum_{\alpha=1}^{n}(\boldsymbol{Y}_\alpha - \boldsymbol{\nu})\right], \tag{12}$$

where u is a scalar and $\boldsymbol{t}$ an m-component vector. For fixed $\boldsymbol{t}$, $\phi_n(\boldsymbol{t}, u)$ can be considered as the characteristic function of $(1/\sqrt{n})\sum_{\alpha=1}^{n}(\boldsymbol{t}'\boldsymbol{Y}_\alpha - \mathscr{E}\boldsymbol{t}'\boldsymbol{Y}_\alpha)$. By the univariate central limit theorem [Cramér (1946), p. 215], the limiting distribution is $N(0, \boldsymbol{t}'\boldsymbol{T}\boldsymbol{t})$. Therefore (Theorem 2.6.4),

$$\lim_{n\to\infty}\phi_n(\boldsymbol{t}, u) = e^{-\frac{1}{2}u^2\boldsymbol{t}'\boldsymbol{T}\boldsymbol{t}} \tag{13}$$

for every u and $\boldsymbol{t}$. (For $\boldsymbol{t} = \boldsymbol{0}$ a special and obvious argument is used.) Let $u = 1$ to obtain

$$\lim_{n\to\infty}\mathscr{E}\exp\left[i\boldsymbol{t}'\frac{1}{\sqrt{n}}\sum_{\alpha=1}^{n}(\boldsymbol{Y}_\alpha - \boldsymbol{\nu})\right] = e^{-\frac{1}{2}\boldsymbol{t}'\boldsymbol{T}\boldsymbol{t}} \tag{14}$$

for every $\boldsymbol{t}$. Since $e^{-\frac{1}{2}\boldsymbol{t}'\boldsymbol{T}\boldsymbol{t}}$ is continuous at $\boldsymbol{t} = \boldsymbol{0}$, the convergence is uniform in some neighborhood of $\boldsymbol{t} = \boldsymbol{0}$. The theorem follows. Q.E.D.

Now we wish to show that the sample covariance matrix is asymptotically normally distributed as the sample size increases.

THEOREM 3.4.4. *Let* $\boldsymbol{A}(n) = \sum_{\alpha=1}^{N}(\boldsymbol{X}_\alpha - \overline{\boldsymbol{X}}_N)(\boldsymbol{X}_\alpha - \overline{\boldsymbol{X}}_N)'$, *where* $\boldsymbol{X}_1, \boldsymbol{X}_2, \dots$ *are independently distributed according to* $N(\boldsymbol{\mu}, \boldsymbol{\Sigma})$ *and* $n = N - 1$. *Then the limiting distribution of* $\boldsymbol{B}(n) = (1/\sqrt{n})[\boldsymbol{A}(n) - n\boldsymbol{\Sigma}]$ *is normal with mean* $\boldsymbol{0}$ *and covariances*

$$\mathscr{E} b_{ij}(n) b_{kl}(n) = \sigma_{ik}\sigma_{jl} + \sigma_{il}\sigma_{jk}. \tag{15}$$

PROOF. As shown earlier, $A(n)$ is distributed as $A(n) = \sum_{\alpha=1}^{n} Z_\alpha Z_\alpha'$, where $Z_1, Z_2, \dots$ are distributed independently according to $N(\mathbf{0}, \Sigma)$. We arrange the elements of $Z_\alpha Z_\alpha'$ in a vector such as

$$Y_\alpha = \begin{pmatrix} Z_{1\alpha}^2 \\ Z_{1\alpha} Z_{2\alpha} \\ \vdots \\ Z_{2\alpha}^2 \\ \vdots \\ Z_{p\alpha}^2 \end{pmatrix}. \tag{16}$$

The moments of Y_α can be deduced from the moments of Z_α as given in Section 2.6. We have $\mathscr{E} Z_{i\alpha} Z_{j\alpha} = \sigma_{ij}$, $\mathscr{E} Z_{i\alpha} Z_{j\alpha} Z_{k\alpha} Z_{l\alpha} = \sigma_{ij}\sigma_{kl} + \sigma_{ik}\sigma_{jl} + \sigma_{il}\sigma_{jk}$, $\mathscr{E}(Z_{i\alpha} Z_{j\alpha} - \sigma_{ij})(Z_{k\alpha} Z_{l\alpha} - \sigma_{kl}) = \sigma_{ik}\sigma_{jl} + \sigma_{il}\sigma_{jk}$. Thus the vectors Y_α defined by (16) satisfy the conditions of Theorem 3.4.3 with the elements of $\boldsymbol{\nu}$ being the elements of Σ arranged in vector form similar to (16) and the elements of T being given above. If the elements of $A(n)$ are arranged in vector form similar to (16), say the vector $W(n)$, then $W(n) - n\boldsymbol{\nu} = \sum_{\alpha=1}^{n}(Y_\alpha - \boldsymbol{\nu})$. By Theorem 3.4.3, $(1/\sqrt{n})[W(n) - n\boldsymbol{\nu}]$ has a limiting normal distribution with mean $\mathbf{0}$ and the covariance matrix of Y_α. Q.E.D.

The elements of $B(n)$ will have a limiting normal distribution with mean $\mathbf{0}$ if $x_1, x_2, \dots$ are independently and identically distributed with finite fourth-order moments, but the covariance structure of $B(n)$ will depend on the fourth-order moments.

3.4.2. Decision Theory

It may be enlightening to consider estimation in terms of decision theory. We review some of the concepts. An observation x is made on a random variable X (which may be a vector) whose distribution P_θ depends on a parameter θ which is an element of a set Θ. The statistician is to make a decision d in a set D. A *decision procedure* is a function $\delta(x)$ whose domain is the set of values of X and whose range is D. The *loss* in making decision d when the distribution is P_θ is a nonnegative function $L(\theta, d)$. The evaluation of a procedure $\delta(x)$ is on the basis of the *risk function*

$$R(\theta, \delta) = \mathscr{E}_\theta L[\theta, \delta(X)]. \tag{17}$$

For example, if d and θ are univariate, the loss may be squared error, $L(\theta, d) = (\theta - d)^2$ and the risk is the mean squared error $\mathscr{E}_\theta[\delta(X) - \theta]^2$.

A decision procedure $\delta(x)$ is *as good as* a procedure $\delta^*(x)$ if

$$R(\theta, \delta) \le R(\theta, \delta^*), \qquad \forall \theta; \tag{18}$$

$\delta(x)$ is *better than* $\delta^*(x)$ if (18) holds with a strict inequality for at least one value of θ. A procedure $\delta^*(x)$ is *inadmissible* if there exists another procedure $\delta(x)$ that is better than $\delta^*(x)$. A procedure is *admissible* if it is not inadmissible (i.e., if there is no procedure better than it) in terms of the given loss functions. A class of procedures is *complete* if for any procedure not in the class there is a better procedure in the class. The class is *minimal complete* if it does not contain a proper complete subclass. If a minimal complete class exists, it is identical to the class of admissible procedures. When such a class is available, there is no (mathematical) need to use a procedure outside the minimal complete class. Sometimes it is convenient to refer to an *essentially complete* class, which is a class of procedures such that for every procedure outside of the class there is one in the class that is just as good.

For a given procedure the risk function is a function of the parameter. If the parameter can be assigned an a priori distribution, say, with density $\rho(\theta)$, then the average loss from use of a decision procedure $\delta(x)$ is

$$r(\rho, \delta) = \mathscr{E}_\rho R(\theta, \delta) = \mathscr{E}_\rho \mathscr{E}_\theta L[\theta, \delta(X)]. \tag{19}$$

Given the a priori density ρ, the decision procedure $\delta(x)$ that minimizes $r(\rho, \delta)$ is the *Bayes procedure* and the resulting minimum of $r(\rho, \delta)$ is the *Bayes risk*. Under general conditions Bayes procedures are admissible and admissible procedures are Bayes or limits of Bayes procedures. If the density of X given θ is $f(x|\theta)$, the joint density of X and θ is $f(x|\theta)\rho(\theta)$ and the average risk of a procedure $\delta(x)$ is

$$\begin{aligned} r(\rho, \delta) &= \int_\Theta \int_X L[\theta, \delta(x)]\, f(x|\theta)\rho(\theta)\, dx\, d\theta \\ &= \int_X \left\{ \int_\Theta L[\theta, \delta(x)]\, g(\theta|x)\, d\theta \right\} f(x)\, dx; \end{aligned} \tag{20}$$

here

$$f(x) = \int_\Theta f(x|\theta)\rho(\theta)\, d\theta, \qquad g(\theta|x) = \frac{f(x|\theta)\rho(\theta)}{f(x)} \tag{21}$$

are the marginal density of X and the a posteriori density of θ given x. The procedure that minimizes $r(\rho, \delta)$ is one that for each x minimizes the expression in braces on the right-hand side of (20), that is, the expectation of $L[\theta, \delta(x)]$ with respect to the a posteriori distribution. If θ and d are vectors ($\boldsymbol{\theta}$ and $\boldsymbol{d}$) and $L(\boldsymbol{\theta}, \boldsymbol{d}) = (\boldsymbol{\theta} - \boldsymbol{d})'\boldsymbol{C}(\boldsymbol{\theta} - \boldsymbol{d})$, where $\boldsymbol{C}$ is positive definite, then

$$\mathscr{E}_{\boldsymbol{\theta}|x} L[\boldsymbol{\theta}, \boldsymbol{d}(x)] = \mathscr{E}_{\boldsymbol{\theta}|x}[\boldsymbol{\theta} - \mathscr{E}(\boldsymbol{\theta}|x)]'\boldsymbol{C}[\boldsymbol{\theta} - \mathscr{E}(\boldsymbol{\theta}|x)] + [\mathscr{E}(\boldsymbol{\theta}|x) - \boldsymbol{d}(x)]'\boldsymbol{C}[\mathscr{E}(\boldsymbol{\theta}|x) - \boldsymbol{d}(x)]. \tag{22}$$

The minimum occurs at $\boldsymbol{d}(x) = \mathscr{E}(\boldsymbol{\theta}|x)$, the mean of the a posteriori distribution.

THEOREM 3.4.5. *If $\boldsymbol{x}_1, \dots, \boldsymbol{x}_N$ are independently distributed, each $\boldsymbol{x}_\alpha$ according to $N(\boldsymbol{\mu}, \boldsymbol{\Sigma})$ and if $\boldsymbol{\mu}$ has an a priori distribution $N(\boldsymbol{\nu}, \boldsymbol{\Phi})$, then the a posteriori distribution of $\boldsymbol{\mu}$ given $\boldsymbol{x}_1, \dots, \boldsymbol{x}_N$ is normal with mean*

$$\boldsymbol{\Phi}\left(\boldsymbol{\Phi} + \frac{1}{N}\boldsymbol{\Sigma}\right)^{-1}\bar{\boldsymbol{x}} + \frac{1}{N}\boldsymbol{\Sigma}\left(\boldsymbol{\Phi} + \frac{1}{N}\boldsymbol{\Sigma}\right)^{-1}\boldsymbol{\nu} \tag{23}$$

and covariance matrix

$$\boldsymbol{\Phi} - \boldsymbol{\Phi}\left(\boldsymbol{\Phi} + \frac{1}{N}\boldsymbol{\Sigma}\right)^{-1}\boldsymbol{\Phi}. \tag{24}$$

PROOF. Since $\bar{\boldsymbol{x}}$ is sufficient for $\boldsymbol{\mu}$, we need only consider $\bar{\boldsymbol{x}}$, which has the distribution of $\boldsymbol{\mu} + \boldsymbol{v}$, where $\boldsymbol{v}$ has the distribution $N[\boldsymbol{0}, (1/N)\boldsymbol{\Sigma}]$ and is independent of $\boldsymbol{\mu}$. Then the joint distribution of $\boldsymbol{\mu}$ and $\bar{\boldsymbol{x}}$ is

$$N\left[\begin{pmatrix} \boldsymbol{\nu} \\ \boldsymbol{\nu} \end{pmatrix}, \begin{pmatrix} \boldsymbol{\Phi} & \boldsymbol{\Phi} \\ \boldsymbol{\Phi} & \boldsymbol{\Phi} + \frac{1}{N}\boldsymbol{\Sigma} \end{pmatrix}\right]. \tag{25}$$

The mean of the conditional distribution of $\boldsymbol{\mu}$ given $\bar{\boldsymbol{x}}$ is (by Theorem 2.5.1)

$$\boldsymbol{\nu} + \boldsymbol{\Phi}\left(\boldsymbol{\Phi} + \frac{1}{N}\boldsymbol{\Sigma}\right)^{-1}(\bar{\boldsymbol{x}} - \boldsymbol{\nu}), \tag{26}$$

which reduces to (23). Q.E.D.

COROLLARY 3.4.1. *If $\boldsymbol{x}_1, \dots, \boldsymbol{x}_N$ are independently distributed, each $\boldsymbol{x}_\alpha$ according to $N(\boldsymbol{\mu}, \boldsymbol{\Sigma})$, $\boldsymbol{\mu}$ has an a priori distribution $N(\boldsymbol{\nu}, \boldsymbol{\Phi})$, and the loss function is $(\boldsymbol{d} - \boldsymbol{\mu})'\boldsymbol{C}(\boldsymbol{d} - \boldsymbol{\mu})$, then the Bayes estimator of $\boldsymbol{\mu}$ is (23).*

The Bayes estimator of $\boldsymbol{\mu}$ is a kind of weighted average of $\bar{\boldsymbol{x}}$ and $\boldsymbol{\nu}$, the prior mean of $\boldsymbol{\mu}$. If $(1/N)\boldsymbol{\Sigma}$ is small compared to $\boldsymbol{\Phi}$ (e.g., if N is large), $\boldsymbol{\nu}$ is given little weight. Put another way, if $\boldsymbol{\Phi}$ is large, that is, the prior is relatively uninformative, a large weight is put on $\bar{\boldsymbol{x}}$. In fact, as $\boldsymbol{\Phi}$ tends to ∞ in the sense that $\boldsymbol{\Phi}^{-1} \to \boldsymbol{0}$, the estimator approaches $\bar{\boldsymbol{x}}$.

A decision procedure $\delta_0(x)$ is *minimax* if

$$\sup_\theta R(\theta, \delta_0) = \inf_\delta \sup_\theta R(\theta, \delta). \tag{27}$$

THEOREM 3.4.6. *If $\boldsymbol{x}_1, \ldots, \boldsymbol{x}_N$ are independently distributed each according to $N(\boldsymbol{\mu}, \boldsymbol{\Sigma})$ and the loss function is $(\boldsymbol{d} - \boldsymbol{\mu})'\boldsymbol{C}(\boldsymbol{d} - \boldsymbol{\mu})$, then $\bar{\boldsymbol{x}}$ is a minimax estimator.*

PROOF. This follows from a theorem in statistical decision theory that if a procedure δ_0 is *extended Bayes* [i.e., if for arbitrary ε, $r(\rho, \delta_0) \le r(\rho, \delta_\rho) + \varepsilon$ for suitable ρ, where δ_ρ is the corresponding Bayes procedure] and if $R(\theta, \delta_0)$ is constant, then δ_0 is minimax. [See, e.g., Ferguson (1967), Theorem 3 of Section 2.11.] We find

$$\begin{aligned} R(\boldsymbol{\mu}, \bar{\boldsymbol{x}}) &= \mathscr{E}(\bar{\boldsymbol{x}} - \boldsymbol{\mu})'\boldsymbol{C}(\bar{\boldsymbol{x}} - \boldsymbol{\mu}) \\ &= \mathscr{E}\operatorname{tr}\boldsymbol{C}(\bar{\boldsymbol{x}} - \boldsymbol{\mu})(\bar{\boldsymbol{x}} - \boldsymbol{\mu})' \\ &= \frac{1}{N}\operatorname{tr}\boldsymbol{C}\boldsymbol{\Sigma}. \end{aligned} \tag{28}$$

Let (23) be $\boldsymbol{d}(\bar{\boldsymbol{x}})$. Its average risk is

$$\begin{aligned} &\mathscr{E}_{\bar{x}}\mathscr{E}_{\mu}\{\operatorname{tr}\boldsymbol{C}[\boldsymbol{d}(\bar{\boldsymbol{x}}) - \boldsymbol{\mu}][\boldsymbol{d}(\bar{\boldsymbol{x}}) - \boldsymbol{\mu}]'|\bar{\boldsymbol{x}}\} \\ &\quad = \mathscr{E}_{\bar{x}}\operatorname{tr}\boldsymbol{C}\left[\boldsymbol{\Phi} - \boldsymbol{\Phi}\left(\boldsymbol{\Phi} + \frac{1}{N}\boldsymbol{\Sigma}\right)^{-1}\boldsymbol{\Phi}\right] = \operatorname{tr}\boldsymbol{C}\boldsymbol{\Phi}\left(\boldsymbol{\Phi} + \frac{1}{N}\boldsymbol{\Sigma}\right)^{-1}\frac{1}{N}\boldsymbol{\Sigma} \\ &\quad = \operatorname{tr}\boldsymbol{C}\left(\boldsymbol{I} + \frac{1}{N}\boldsymbol{\Sigma}\boldsymbol{\Phi}^{-1}\right)^{-1}\frac{1}{N}\boldsymbol{\Sigma} \to \frac{1}{N}\operatorname{tr}\boldsymbol{C}\boldsymbol{\Sigma} \end{aligned} \tag{29}$$

as $\boldsymbol{\Phi}^{-1} \to \boldsymbol{0}$. Q.E.D.

For more discussion of decision theory see Ferguson (1967), DeGroot (1970), or Berger (1980b).

3.5. IMPROVED ESTIMATION OF THE MEAN

3.5.1. Introduction

The sample mean $\bar{x}$ seems the natural estimator of the population mean μ based on a sample from $N(\mu, \Sigma)$. It is the maximum likelihood estimator, a sufficient statistic when Σ is known, and the minimum variance unbiased estimator. Moreover, it is "equivariant" in the sense that if an arbitrary vector v is added to each observation vector and to μ the error of estimation $\overline{(x+v)} - (\mu + v) = \bar{x} - \mu$ is independent of v; in other words, the error does not depend on the choice of origin. However, Stein (1956b) showed the startling fact that this conventional estimator is not admissible with respect to the loss function that is the sum of mean squared errors of the components when $\Sigma = I$ and $p \geq 3$. James and Stein (1960) produced an estimator which has a smaller sum of mean squared errors; this estimator will be studied in Section 3.5.2. Subsequent studies have shown that the phenomenon is widespread and the implications imperative.

3.5.2. The James–Stein Estimator

The loss function

$$L(\mu, m) = (m - \mu)'(m - \mu) = \sum_{i=1}^{p} (m_i - \mu_i)^2 = \|m - \mu\|^2 \tag{1}$$

defines the sum of mean squared errors of the components of the estimator. We shall show [James and Stein (1960)] that the sample mean is inadmissible by displaying an alternative estimator that has a smaller expected loss for every mean vector μ. We assume that the normal distribution sampled has covariance matrix proportional to I with the constant of proportionality known. It will be convenient to take this constant to be such that $Y = (1/N)\sum_{\alpha=1}^{N} X_\alpha = \bar{X}$ has the distribution $N(\mu, I)$. Then the expected loss or risk of the estimator Y is simply $\mathscr{E}\|Y - \mu\|^2 = \operatorname{tr} I = p$. The estimator proposed by James and Stein is (essentially)

$$m(y) = \left(1 - \frac{p-2}{\|y - v\|^2}\right)(y - v) + v, \tag{2}$$

where v is an arbitrary fixed vector and $p \geq 3$. This estimator shrinks the observed y toward the specified v. The amount of shrinkage is negligible if y is very different from v and is considerable if y is close to v. In this sense v is a favored point.

THEOREM 3.5.1. *With respect to the loss function* (1) *the risk of the estimator* (2) *is less than the risk of the estimator* $\mathbf{Y}$ *for* $p \geq 3$.

We shall show that the risk of $\mathbf{Y}$ minus the risk of (2) is positive by applying the following lemma due to Stein (1974).

LEMMA 3.5.1. *If* $f(x)$ *is a function such that*

$$f(b) - f(a) = \int_a^b f'(x)\,dx \tag{3}$$

for all a *and* b $(a < b)$ *and if*

$$\int_{-\infty}^{\infty} |f'(x)| \frac{1}{\sqrt{2\pi}} e^{-\frac{1}{2}(x-\theta)^2}\,dx < \infty, \tag{4}$$

then

$$\int_{-\infty}^{\infty} f(x)(x-\theta)\frac{1}{\sqrt{2\pi}} e^{-\frac{1}{2}(x-\theta)^2}\,dx = \int_{-\infty}^{\infty} f'(x)\frac{1}{\sqrt{2\pi}} e^{-\frac{1}{2}(x-\theta)^2}\,dx. \tag{5}$$

PROOF OF LEMMA. We write the left-hand side of (5) as

$$\begin{aligned}
(6)\quad & \int_{\theta}^{\infty} [f(x) - f(\theta)](x-\theta)\frac{1}{\sqrt{2\pi}} e^{-\frac{1}{2}(x-\theta)^2}\,dx \\
&+ \int_{-\infty}^{\theta} [f(x) - f(\theta)](x-\theta)\frac{1}{\sqrt{2\pi}} e^{-\frac{1}{2}(x-\theta)^2}\,dx \\
&= \int_{\theta}^{\infty}\int_{\theta}^{x} f'(y)(x-\theta)\frac{1}{\sqrt{2\pi}} e^{-\frac{1}{2}(x-\theta)^2}\,dy\,dx \\
&- \int_{-\infty}^{\theta}\int_{x}^{\theta} f'(y)(x-\theta)\frac{1}{\sqrt{2\pi}} e^{-\frac{1}{2}(x-\theta)^2}\,dy\,dx \\
&= \int_{\theta}^{\infty}\int_{y}^{\infty} f'(y)(x-\theta)\frac{1}{\sqrt{2\pi}} e^{-\frac{1}{2}(x-\theta)^2}\,dx\,dy \\
&- \int_{-\infty}^{\theta}\int_{-\infty}^{y} f'(y)(x-\theta)\frac{1}{\sqrt{2\pi}} e^{-\frac{1}{2}(x-\theta)^2}\,dx\,dy,
\end{aligned}$$

which yields the right-hand side of (5). Fubini's theorem justifies the interchange of order of integration. (See Problem 22.) Q.E.D.

The lemma can also be derived by integration by parts in special cases.

PROOF OF THEOREM 3.5.1. The difference in risks is

$$(7)\quad \Delta R(\boldsymbol{\mu}) = \mathscr{E}_{\boldsymbol{\mu}}\{\|\boldsymbol{Y}-\boldsymbol{\mu}\|^2 - \|\boldsymbol{m}(\boldsymbol{Y})-\boldsymbol{\mu}\|^2\}$$

$$= \mathscr{E}_{\boldsymbol{\mu}}\left\{\|\boldsymbol{Y}-\boldsymbol{\mu}\|^2 - \left\|\left(1-\frac{p-2}{\|\boldsymbol{Y}-\boldsymbol{\nu}\|^2}\right)(\boldsymbol{Y}-\boldsymbol{\nu})+\boldsymbol{\nu}-\boldsymbol{\mu}\right\|^2\right\}$$

$$= \mathscr{E}_{\boldsymbol{\mu}}\left\{\sum_{i=1}^{p}(Y_i-\mu_i)^2 - \sum_{i=1}^{p}\left[(Y_i-\mu_i)-\frac{p-2}{\|\boldsymbol{Y}-\boldsymbol{\nu}\|^2}(Y_i-\nu_i)\right]^2\right\}$$

$$= \mathscr{E}_{\boldsymbol{\mu}}\left\{2\frac{p-2}{\|\boldsymbol{Y}-\boldsymbol{\nu}\|^2}\sum_{i=1}^{p}(Y_i-\mu_i)(Y_i-\nu_i) - \frac{(p-2)^2}{\|\boldsymbol{Y}-\boldsymbol{\nu}\|^2}\right\}.$$

Now we use Lemma 3.5.1 with

$$(8)\quad f(y_i) = \frac{y_i-\nu_i}{\sum_{j=1}^{p}(y_j-\nu_j)^2},\; f'(y_i) = \frac{1}{\sum_{j=1}^{p}(y_j-\nu_j)^2} - \frac{2(y_i-\nu_i)^2}{\left[\sum_{j=1}^{p}(y_j-\nu_j)^2\right]^2}.$$

[For $p \geq 3$ condition (4) is satisfied.] Then (7) is

$$(9)\quad \Delta R(\boldsymbol{\mu}) = \mathscr{E}_{\boldsymbol{\mu}}\left\{2(p-2)\sum_{i=1}^{p}\left[\frac{1}{\|\boldsymbol{Y}-\boldsymbol{\nu}\|^2} - \frac{2(y_i-\nu_i)^2}{\|\boldsymbol{Y}-\boldsymbol{\nu}\|^4}\right] - \frac{(p-2)^2}{\|\boldsymbol{Y}-\boldsymbol{\nu}\|^2}\right\}$$

$$= (p-2)^2\mathscr{E}_{\boldsymbol{\mu}}\frac{1}{\|\boldsymbol{Y}-\boldsymbol{\nu}\|^2} > 0. \qquad \text{Q.E.D.}$$

This theorem states that $\boldsymbol{Y}$ is inadmissible for estimating $\boldsymbol{\mu}$ when $p \geq 3$ since the estimator (2) has a smaller risk for every $\boldsymbol{\mu}$ (regardless of the choice of $\boldsymbol{\nu}$).

The risk is the sum of the mean squared errors $\mathscr{E}[m_i(\boldsymbol{Y}) - \mu_i]^2$. Since $Y_1, \ldots, Y_p$ are independent and only the distribution of Y_i depends on μ_i, it is puzzling that the improved estimator uses all the Y_j's to estimate μ_i; it seems that irrelevant information is being used. Stein "explained" the phenomenon by arguing that the sample distance squared of $\boldsymbol{Y}$ from $\boldsymbol{\nu}$, that is, $\|\boldsymbol{Y} - \boldsymbol{\nu}\|^2$, overestimates the squared distance of $\boldsymbol{\mu}$ from $\boldsymbol{\nu}$ and hence that the estimator $\boldsymbol{Y}$ could be improved by bringing it nearer $\boldsymbol{\nu}$ (whatever $\boldsymbol{\nu}$ is). Berger (1980), following Brown, illustrated it by Figure 3.4. The four points $\boldsymbol{x}_1$, $\boldsymbol{x}_2$, $\boldsymbol{x}_3$, $\boldsymbol{x}_4$ represent a spherical distribution centered at $\boldsymbol{\mu}$. Consider the effects of shrinkage. The average distance of $\boldsymbol{m}(\boldsymbol{x}_1)$ and $\boldsymbol{m}(\boldsymbol{x}_3)$ from $\boldsymbol{\mu}$ is a little greater than that of $\boldsymbol{x}_1$ and $\boldsymbol{x}_3$, but $\boldsymbol{m}(\boldsymbol{x}_2)$ and $\boldsymbol{m}(\boldsymbol{x}_4)$ are a little closer to $\boldsymbol{\mu}$ than $\boldsymbol{x}_2$ and $\boldsymbol{x}_4$ if the shrinkage is a certain amount. If $p = 3$, there are two more points (not on the line $\boldsymbol{\nu}$, $\boldsymbol{\mu}$) that are shrunk closer to $\boldsymbol{\mu}$.

The risk of the estimator (2) is

$$\mathscr{E}_{\boldsymbol{\mu}}\|\boldsymbol{m}(\boldsymbol{Y}) - \boldsymbol{\mu}\|^2 = p - (p-2)^2 \mathscr{E}_{\boldsymbol{\mu}} \frac{1}{\|\boldsymbol{Y} - \boldsymbol{\nu}\|^2}, \tag{10}$$

where $\|\boldsymbol{Y} - \boldsymbol{\nu}\|^2$ has a noncentral χ^2-distribution with p degrees of freedom and noncentrality parameter $\|\boldsymbol{\mu} - \boldsymbol{\nu}\|^2$. The farther $\boldsymbol{\mu}$ is from $\boldsymbol{\nu}$, the less the improvement due to the James–Stein estimator, but there is always some improvement. The density of $\|\boldsymbol{Y} - \boldsymbol{\nu}\|^2 = V$, say, is (28) of Section 3.3.3, where

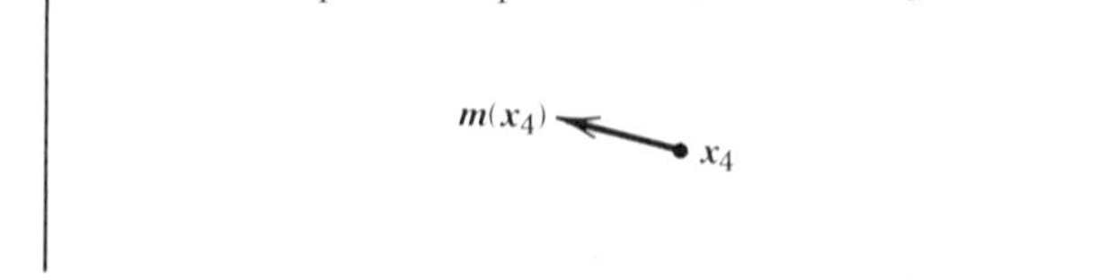

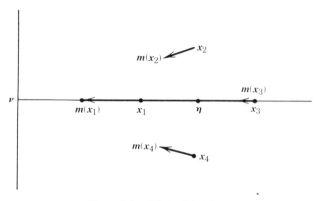

Figure 3.4. Effect of shrinkage.

$\tau^2 = \|\boldsymbol{\mu} - \boldsymbol{\nu}\|^2$. Then

$$(11)\quad \mathscr{E}_{\boldsymbol{\mu}} \frac{1}{\|\boldsymbol{Y} - \boldsymbol{\nu}\|^2} = \mathscr{E}_{\tau^2} V^{-1}$$

$$= e^{-\frac{1}{2}\tau^2} 2^{-\frac{1}{2}p} \sum_{\beta=0}^{\infty} \left(\frac{\tau^2}{4}\right)^{\beta} \frac{1}{\beta!\Gamma(\frac{1}{2}p + \beta)} \int_0^{\infty} v^{\frac{1}{2}p+\beta-2} e^{-\frac{1}{2}v}\, dv$$

$$= e^{-\frac{1}{2}\tau^2} 2^{-\frac{1}{2}p} \sum_{\beta=0}^{\infty} \left(\frac{\tau^2}{4}\right)^{\beta} \frac{\Gamma(\frac{1}{2}p + \beta - 1) 2^{\frac{1}{2}p+\beta-1}}{\beta!\Gamma(\frac{1}{2}p + \beta)}$$

$$= \tfrac{1}{2} e^{-\frac{1}{2}\tau^2} \sum_{\beta=0}^{\infty} \left(\frac{\tau^2}{2}\right)^{\beta} \frac{1}{\beta!(\frac{1}{2}p + \beta - 1)}$$

for $p \geq 3$. Note that for $\boldsymbol{\mu} = \boldsymbol{\nu}$, that is, $\tau^2 = 0$, (11) is $1/(p-2)$ and the mean squared error (10) is 2. For large p the reduction in risk is considerable.

Table 3.2 gives values of the risk for $p = 10$ and $\sigma^2 = 1$. For example, if $\tau^2 = \|\boldsymbol{\mu} - \boldsymbol{\nu}\|^2$ is 5, the mean squared error of the James–Stein estimator is 8.86 compared to 10 for the natural estimator; this is the case if $\mu_i - \nu_i = 1/\sqrt{2} = 0.707$, $i = 1, \ldots, 10$, for instance.

An obvious question in using an estimator of this class is how to choose the vector $\boldsymbol{\nu}$ toward which the observed mean vector is shrunk; any $\boldsymbol{\nu}$ yields an

TABLE 3.2†
AVERAGE MEAN SQUARED ERROR OF THE JAMES–STEIN ESTIMATOR FOR $p = 10$ AND $\sigma^2 = 1$

$\tau^2 = \|\boldsymbol{\mu} - \boldsymbol{\nu}\|^2$	$\mathscr{E}_{\boldsymbol{\mu}}\|\boldsymbol{m}(\boldsymbol{Y}) - \boldsymbol{\mu}\|^2$
0.0	2.00
0.5	4.78
1.0	6.21
2.0	7.51
3.0	8.24
4.0	8.62
5.0	8.86
6.0	9.03

†From Efron and Morris (1977).

estimator better than the natural one. However, as seen from Table 3.2, the improvement is small if $\|\mu - \nu\|$ is very large. Thus, to be effective some knowledge of the position of μ is necessary. A disadvantage of the procedure is that it is not objective; the choice of ν is up to the investigator.

A feature of the estimator we have been studying that seems disadvantageous is that for small values of $\|Y - \nu\|$, the multiplier of $Y - \nu$ is negative; that is, the estimator $m(Y)$ is in the direction from ν opposite to that of Y. This disadvantage can be overcome and the estimator improved by replacing the factor by 0 when the factor is negative.

DEFINITION 3.5.1. *For any function $g(u)$, let*

$$g^+(u) = g(u), \qquad g(u) \geq 0, \\ = 0, \qquad g(u) < 0. \tag{12}$$

LEMMA 3.5.2. *When X is distributed according to $N(\mu, I)$,*

$$\mathscr{E}_\mu\{\|g^+(\|X\|)X - \mu\|^2\} \leq \mathscr{E}_\mu\{\|g(\|X\|)X - \mu\|^2\}. \tag{13}$$

PROOF. The right-hand side of (13) minus the left-hand side is

$$\mathscr{E}_\mu\{g^2(\|X\|)\|X\|^2 - [g^+(\|X\|)]^2\|X\|^2\} \geq 0 \tag{14}$$

plus 2 times

$$\begin{aligned} \mathscr{E}_\mu \mu' X[g^+(\|X\|) - g(\|X\|)] &= \|\mu\|\int_{-\infty}^{\infty}\cdots\int_{-\infty}^{\infty} y_1[g^+(\|y\|) - g(\|y\|)] \\ &\quad \cdot \frac{1}{(2\pi)^{\frac{1}{2}p}}\exp\left\{-\tfrac{1}{2}\left[\sum_{i=1}^{p} y_i^2 - 2y_1\|\mu\| + \|\mu\|^2\right]\right\} dy, \end{aligned} \tag{15}$$

where $y' = x'P$, $(\|\mu\|, 0, \ldots, 0) = \mu' P$, and $PP' = I$. (The first column of P is $(1/\|\mu\|)\mu$.) Then (15) is $\|\mu\|$ times

$$\begin{aligned} e^{-\frac{1}{2}\|\mu\|^2}\int_{-\infty}^{\infty}\cdots\int_{-\infty}^{\infty}\int_{0}^{\infty} y_1[g^+(\|y\|) - g(\|y\|)][e^{\|\mu\|y_1} - e^{-\|\mu\|y_1}] \\ \cdot \frac{1}{(2\pi)^{\frac{1}{2}p}} e^{-\frac{1}{2}\sum_{i=1}^{p} y_i^2}\, dy_1\, dy_2 \cdots dy_p \geq 0 \end{aligned} \tag{16}$$

(by replacing y_1 by $-y_1$ for $y_1 < 0$). Q.E.D.

THEOREM 3.5.2. *The estimator*

(17) $$m^+(y) = \left(1 - \frac{p-2}{\|y-\nu\|^2}\right)^+ (y-\nu) + \nu$$

has smaller risk than $m(y)$ *defined by* (2) *and is minimax.*

PROOF. In Lemma 3.5.2 let $g(u) = 1 - (p-2)/u^2$ and $X = Y - \nu$ and replace μ by $\mu - \nu$. The second assertion in the theorem follows from Theorem 3.4.6. Q.E.D.

The theorem shows that $m(Y)$ is not admissible. However, it is known that $m^+(Y)$ is also not admissible, but it is believed that not much further improvement is possible.

This approach is easily extended to the case where one observes $x_1, \dots, x_N$ from $N(\mu, \Sigma)$ with loss function $L(\mu, m) = (m - \mu)'\Sigma^{-1}(m - \mu)$. Let $\Sigma = CC'$ for some nonsingular C, $x_\alpha = Cx_\alpha^*$, $\alpha = 1, \dots, N$, $\mu = C\mu^*$, and $L^*(m^*, \mu^*) = \|m^* - \mu^*\|^2$. Then $x_1^*, \dots, x_N^*$ are observations from $N(\mu^*, I)$ and the problem is reduced to the earlier one. Then

(18) $$\left(1 - \frac{p-2}{N(\bar{x}-\nu)'\Sigma^{-1}(\bar{x}-\nu)}\right)^+ (\bar{x}-\nu) + \nu$$

is a minimax estimator of μ.

3.5.3. Estimation for a General Known Covariance Matrix and an Arbitrary Quadratic Loss Function

Let the parent distribution be $N(\mu, \Sigma)$, where Σ is known, and let the loss function be

(19) $$L(\mu, m) = (m - \mu)'Q(m - \mu),$$

where Q is an arbitrary positive definite matrix which reflects the relative importance of errors in different directions. (If the loss function were singular, the dimensionality of x could be reduced so as to make the loss matrix nonsingular.) Then the sample mean $\bar{x}$ has the distribution $N(\mu, (1/N)\Sigma)$ and risk (expected loss)

(20) $$\mathscr{E}(\bar{x}-\mu)'Q(\bar{x}-\mu) = \mathscr{E}\,\mathrm{tr}\,Q(\bar{x}-\mu)(\bar{x}-\mu)' = \frac{1}{N}\,\mathrm{tr}\,Q\Sigma,$$

which is constant, not depending on μ.

Several estimators that improve on $\bar{x}$ have been proposed. First we take up an estimator proposed independently by Berger (1975) and Hudson (1974).

THEOREM 3.5.3. *Let $r(z)$, $0 \le z < \infty$, be a nondecreasing differentiable function such that $0 \le r(z) \le 2(p-2)$. Then for $p \ge 3$*

$$m = \left(I - \frac{r\big(N^2(\bar{x}-\nu)'\Sigma^{-1}Q^{-1}\Sigma^{-1}(\bar{x}-\nu)\big)}{N(\bar{x}-\nu)'\Sigma^{-1}Q^{-1}\Sigma^{-1}(\bar{x}-\nu)} Q^{-1}\Sigma^{-1} \right)(\bar{x}-\nu)+\nu \tag{21}$$

has smaller risk than $\bar{x}$ and is minimax.

PROOF. There exists a matrix C such that $C'QC = I$ and $(1/N)\Sigma = C\Delta C'$ where Δ is diagonal with diagonal elements $\delta_1 \ge \delta_2 \ge \cdots \ge \delta_p > 0$ (Theorem A.2.2 of Appendix A). Let $\bar{x} = Cy + \nu$ and $\mu = C\mu^* + \nu$. Then y has the distribution $N(\mu^*, \Delta)$ and the transformed loss function is

$$L^*(m^*, \mu^*) = (m^* - \mu^*)'(m^* - \mu^*) = \|m^* - \mu^*\|^2. \tag{22}$$

The estimator (21) of μ is transformed to the estimator of $\mu^* = C^{-1}(\mu - \nu)$

$$m^*(y) = \left(I - \frac{r(y'\Delta^{-2}y)}{y'\Delta^{-2}y}\Delta^{-1} \right) y. \tag{23}$$

We now proceed as in the proof of Theorem 3.5.1. The difference in risks between y and m^* is

$$\begin{aligned} \Delta R(\mu^*) &= \mathscr{E}_{\mu^*}\{\|Y-\mu^*\|^2 - \|m^*(Y)-\mu^*\|^2\} \\ &= \mathscr{E}_{\mu^*}\left\{ 2\frac{r(Y'\Delta^{-2}Y)}{Y'\Delta^{-2}Y}\sum_{i=1}^{p}\frac{1}{\delta_i}Y_i(Y_i-\mu_i^*) - \frac{r^2(Y'\Delta^{-2}Y)}{Y'\Delta^{-2}Y} \right\}. \end{aligned} \tag{24}$$

Since $r(z)$ is differentiable, we use Lemma 3.5.1 with $(x-\theta) = (y_i - \mu_i^*)/\delta_i$ and

$$f(y_i) = \frac{r(y'\Delta^{-2}y)}{y'\Delta^{-2}y} y_i, \tag{25}$$

$$f'(y_i) = \frac{r(y'\Delta^{-2}y)}{y'\Delta^{-2}y} + \frac{2r'(y'\Delta^{-2}y)}{y'\Delta^{-2}y}\frac{y_i^2}{\delta_i^2} - \frac{2r(y'\Delta^{-2}y)}{(y'\Delta^{-2}y)^2}\frac{y_i^2}{\delta_i^2}. \tag{26}$$

Then

(27)

$$\Delta R(\mu^*) = \mathscr{E}_{\mu^*}\left\{2(p-2)\frac{r(Y'\Delta^{-2}Y)}{Y'\Delta^{-2}Y} + 4r'(Y'\Delta^{-2}Y) - \frac{r^2(Y'\Delta^{-2}Y)}{Y'\Delta^{-2}Y}\right\} \geq 0$$

since $r(y'\Delta^{-2}y) \leq 2(p-2)$ and $r'(y'\Delta^{-2}y) \geq 0$. Q.E.D.

COROLLARY 3.5.1. *For $p \geq 3$*

(28)

$$\left\{I - \frac{\min\left[p-2, N^2(\bar{x}-\nu)'\Sigma^{-1}Q^{-1}\Sigma^{-1}(\bar{x}-\nu)\right]}{N(\bar{x}-\nu)'\Sigma^{-1}Q^{-1}\Sigma^{-1}(\bar{x}-\nu)}Q^{-1}\Sigma^{-1}\right\}(\bar{x}-\nu)+\nu$$

has smaller risk than $\bar{x}$ and is minimax.

PROOF. The function $r(z) = \min(p-2, z)$ is differentiable except at $z = p-2$. The function $r(z)$ can be approximated arbitrarily closely by a differentiable function. (For example, the corner at $z = p - 2$ can be smoothed by a circular arc of arbitrarily small radius.) We shall not give the details of the proof.

In canonical form y is shrunk by a scalar times a diagonal matrix. The larger the variance of a component is, the less the effect of the shrinkage.

Berger (1975) has proved these results for a more general density, that is, for a mixture of normals. Berger (1976) has also proved in the case of normality that if

(29)
$$r(z) = \frac{z\int_0^\alpha u^{\frac{1}{2}p-c+1}e^{-\frac{1}{2}uz}\,du}{\int_0^\alpha u^{\frac{1}{2}p-c}e^{-\frac{1}{2}uz}\,du}$$

for $3 - \frac{1}{2}p \leq c < 1 + \frac{1}{2}p$, where α is the smallest characteristic root of ΣQ, then the estimator m given by (21) is minimax, is admissible if $c < 2$, and is proper Bayes if $c < 1$.

Another approach to minimax estimators has been introduced by Bhattacharya (1966). Let $\boldsymbol{C}$ be such that $\boldsymbol{C}^{-1}(1/N)\boldsymbol{\Sigma}(\boldsymbol{C}^{-1})' = \boldsymbol{I}$ and $\boldsymbol{C}'\boldsymbol{Q}\boldsymbol{C} = \boldsymbol{Q}^*$, which is diagonal with diagonal elements $q_1^* \geq q_2^* \geq \cdots \geq q_p^* > 0$. Then $\boldsymbol{y} = \boldsymbol{C}^{-1}\bar{\boldsymbol{x}}$ has the distribution $N(\boldsymbol{\mu}^*, \boldsymbol{I})$, and the loss function is

$$(30) \qquad \begin{aligned} L^*(\boldsymbol{m}^*, \boldsymbol{\mu}^*) &= \sum_{i=1}^{p} q_i^*(m_i^* - \mu_i^*)^2 \\ &= \sum_{i=1}^{p} \sum_{j=i}^{p} \alpha_j (m_i^* - \mu_i^*)^2 \\ &= \sum_{j=1}^{p} \alpha_j \sum_{i=1}^{j} (m_i^* - \mu_i^*)^2 \\ &= \sum_{j=1}^{p} \alpha_j \|\boldsymbol{m}^{*(j)} - \boldsymbol{\mu}^{*(j)}\|^2, \end{aligned}$$

where $\alpha_j = q_j^* - q_{j+1}^*$, $j = 1, \ldots, p-1$, $\alpha_p = q_p^*$, $\boldsymbol{m}^{*(j)} = (m_1^*, \ldots, m_j^*)'$, and $\boldsymbol{\mu}^{*(j)} = (\mu_1^*, \ldots, \mu_j^*)'$, $j = 1, \ldots, p$. This decomposition of the loss function suggests combining minimax estimators of the vectors $\boldsymbol{\mu}^{*(j)}$, $j = 1, \ldots, p$. Let $\boldsymbol{y}^{(j)} = (y_1, \ldots, y_j)'$.

THEOREM 3.5.4. *If $\boldsymbol{h}^{(j)}(\boldsymbol{y}^{(j)}) = [h_1^{(j)}(\boldsymbol{y}^{(j)}), \ldots, h_j^{(j)}(\boldsymbol{y}^{(j)})]'$ is a minimax estimator of $\boldsymbol{\mu}^{*(j)}$ under loss function $\|\boldsymbol{m}^{*(j)} - \boldsymbol{\mu}^{*(j)}\|^2$, $j = 1, \ldots, p$, then*

$$(31) \qquad \frac{1}{q_i^*} \sum_{j=1}^{p} \alpha_j h_i^{(j)}(\boldsymbol{y}^{(j)}), \qquad i = 1, \ldots, p,$$

is a minimax estimator of $\mu_1^*, \ldots, \mu_p^*$.

PROOF. First consider the randomized estimator defined by

$$(32) \qquad \Pr\{G_i(\boldsymbol{y}) = h_i^{(j)}(\boldsymbol{y}^{(j)})\} = \frac{\alpha_j}{q_i^*}, \qquad j = i, \ldots, p,$$

for the ith component. Then the risk of this estimator is

$$(33)\quad \begin{aligned}\sum_{i=1}^{p} q_i^* \mathscr{E}_{\mu^*}\left[G_i(Y) - \mu_i^*\right]^2 &= \sum_{i=1}^{p} q_i^* \sum_{j=i}^{p} \frac{\alpha_j}{q_i^*} \mathscr{E}_{\mu^*}\left[h_i^{(j)}(Y^{(j)}) - \mu_i^*\right]^2 \\ &= \sum_{j=1}^{p} \alpha_j \sum_{i=1}^{j} \mathscr{E}_{\mu^*}\left[h_i^{(j)}(Y^{(j)}) - \mu_i^*\right]^2 \\ &= \sum_{j=1}^{p} \alpha_j \mathscr{E}_{\mu^*} \|h^{(j)}(Y^{(j)}) - \mu^{*(j)}\|^2 \\ &\le \sum_{j=1}^{p} \alpha_j j = \sum_{j=1}^{p} q_j^* \\ &= \mathscr{E}_{\mu^*} L^*(Y, \mu^*)^*,\end{aligned}$$

and hence the estimator defined by (32) is minimax.

Since the expected value of $G_i(Y)$ with respect to (32) is (31) and the loss function is convex, the risk of the estimator (31) is less than that of the randomized estimator (by Jensen's inequality). Q.E.D.

PROBLEMS

1. (Sec. 3.2) Find $\hat{\mu}$, $\hat{\Sigma}$, and $(\hat{\rho}_{ij})$ for the data given in Table 3.3, taken from Frets (1921).

2. (Sec. 3.2) Verify the numerical results of (21).

3. (Sec. 3.2) Compute $\hat{\mu}$, $\hat{\Sigma}$, S, and $\hat{\rho}$ for the following pairs of observations: (34, 55), (12, 29), (33, 75), (44, 89), (89, 62), (59, 69), (50, 41), (88, 67). Plot the observations.

4. (Sec. 3.2) Use the facts that $|C^*| = \prod \lambda_i$, $\operatorname{tr} C^* = \sum \lambda_i$, and $C^* = I$ if $\lambda_1 = \cdots = \lambda_p = 1$, where $\lambda_1, \ldots, \lambda_p$ are the characteristic roots of C^*, to prove Lemma 3.2.2. [*Hint*: Use f as given in (12).]

5. (Sec. 3.2) Let x_1 be the body weight (in kilograms) of a cat and x_2 the heart weight (in grams). [Data from Fisher (1947b).]

(a) In a sample of 47 female cats the relevant data are

$$\sum x_\alpha = \begin{pmatrix} 110.9 \\ 432.5 \end{pmatrix}, \qquad \sum x_\alpha x_\alpha' = \begin{pmatrix} 265.13 & 1029.62 \\ 1029.62 & 4064.71 \end{pmatrix}.$$

Find $\hat{\mu}$, $\hat{\Sigma}$, S, and $\hat{\rho}$.

(b) In a sample of 97 male cats the relevant data are

$$\sum x_\alpha = \begin{pmatrix} 281.3 \\ 1098.3 \end{pmatrix}, \qquad \sum x_\alpha x_\alpha' = \begin{pmatrix} 836.75 & 3275.55 \\ 3275.55 & 13056.17 \end{pmatrix}.$$

Find $\hat{\mu}$, $\hat{\Sigma}$, S, and $\hat{\rho}$.

TABLE 3.3[†]

HEAD LENGTHS AND BREADTHS OF BROTHERS

Head Length, First Son	Head Breadth, First Son	Head Length, Second Son	Head Breadth, Second Son
x_1	x_2	x_3	x_4
191	155	179	145
195	149	201	152
181	148	185	149
183	153	188	149
176	144	171	142
208	157	192	152
189	150	190	149
197	159	189	152
188	152	197	159
192	150	187	151
179	158	186	148
183	147	174	147
174	150	185	152
190	159	195	157
188	151	187	158
163	137	161	130
195	155	183	158
186	153	173	148
181	145	182	146
175	140	165	137
192	154	185	152
174	143	178	147
176	139	176	143
197	167	200	158
190	163	187	150

[†]These data, used in examples in the first edition of this book, came from Rao (1952), p. 245. Izenman (1980) has indicated some entries were apparently incorrectly copied from Frets (1921) and corrected them (p. 579).

TABLE 3.4

FOUR MEASUREMENTS ON THREE SPECIES OF IRIS (in centimeters)

Iris setosa				Iris versicolor				Iris virginica			
Sepal length	Sepal width	Petal length	Petal width	Sepal length	Sepal width	Petal length	Petal width	Sepal length	Sepal width	Petal length	Petal width
5.1	3.5	1.4	0.2	7.0	3.2	4.7	1.4	6.3	3.3	6.0	2.5
4.9	3.0	1.4	0.2	6.4	3.2	4.5	1.5	5.8	2.7	5.1	1.9
4.7	3.2	1.3	0.2	6.9	3.1	4.9	1.5	7.1	3.0	5.9	2.1
4.6	3.1	1.5	0.2	5.5	2.3	4.0	1.3	6.3	2.9	5.6	1.8
5.0	3.6	1.4	0.2	6.5	2.8	4.6	1.5	6.5	3.0	5.8	2.2
5.4	3.9	1.7	0.4	5.7	2.8	4.5	1.3	7.6	3.0	6.6	2.1
4.6	3.4	1.4	0.3	6.3	3.3	4.7	1.6	4.9	2.5	4.5	1.7
5.0	3.4	1.5	0.2	4.9	2.4	3.3	1.0	7.3	2.9	6.3	1.8
4.4	2.9	1.4	0.2	6.6	2.9	4.6	1.3	6.7	2.5	5.8	1.8
4.9	3.1	1.5	0.1	5.2	2.7	3.9	1.4	7.2	3.6	6.1	2.5
5.4	3.7	1.5	0.2	5.0	2.0	3.5	1.0	6.5	3.2	5.1	2.0
4.8	3.4	1.6	0.2	5.9	3.0	4.2	1.5	6.4	2.7	5.3	1.9
4.8	3.0	1.4	0.1	6.0	2.2	4.0	1.0	6.8	3.0	5.5	2.1
4.3	3.0	1.1	0.1	6.1	2.9	4.7	1.4	5.7	2.5	5.0	2.0
5.8	4.0	1.2	0.2	5.6	2.9	3.6	1.3	5.8	2.8	5.1	2.4
5.7	4.4	1.5	0.4	6.7	3.1	4.4	1.4	6.4	3.2	5.3	2.3
5.4	3.9	1.3	0.4	5.6	3.0	4.5	1.5	6.5	3.0	5.5	1.8
5.1	3.5	1.4	0.3	5.8	2.7	4.1	1.0	7.7	3.8	6.7	2.2
5.7	3.8	1.7	0.3	6.2	2.2	4.5	1.5	7.7	2.6	6.9	2.3
5.1	3.8	1.5	0.3	5.6	2.5	3.9	1.1	6.0	2.2	5.0	1.5
5.4	3.4	1.7	0.2	5.9	3.2	4.8	1.8	6.9	3.2	5.7	2.3
5.1	3.7	1.5	0.4	6.1	2.8	4.0	1.3	5.6	2.8	4.9	2.0
4.6	3.6	1.0	0.2	6.3	2.5	4.9	1.5	7.7	2.8	6.7	2.0
5.1	3.3	1.7	0.5	6.1	2.8	4.7	1.2	6.3	2.7	4.9	1.8
4.8	3.4	1.9	0.2	6.4	2.9	4.3	1.3	6.7	3.3	5.7	2.1
5.0	3.0	1.6	0.2	6.6	3.0	4.4	1.4	7.2	3.2	6.0	1.8
5.0	3.4	1.6	0.4	6.8	2.8	4.8	1.4	6.2	2.8	4.8	1.8
5.2	3.5	1.5	0.2	6.7	3.0	5.0	1.7	6.1	3.0	4.9	1.8
5.2	3.4	1.4	0.2	6.0	2.9	4.5	1.5	6.4	2.8	5.6	2.1
4.7	3.2	1.6	0.2	5.7	2.6	3.5	1.0	7.2	3.0	5.8	1.6
4.8	3.1	1.6	0.2	5.5	2.4	3.8	1.1	7.4	2.8	6.1	1.9
5.4	3.4	1.5	0.4	5.5	2.4	3.7	1.0	7.9	3.8	6.4	2.0
5.2	4.1	1.5	0.1	5.8	2.7	3.9	1.2	6.4	2.8	5.6	2.2
5.5	4.2	1.4	0.2	6.0	2.7	5.1	1.6	6.3	2.8	5.1	1.5
4.9	3.1	1.5	0.2	5.4	3.0	4.5	1.5	6.1	2.6	5.6	1.4

TABLE 3.4 (*Continued*)

Iris setosa				Iris versicolor				Iris virginica			
Sepal length	Sepal width	Petal length	Petal width	Sepal length	Sepal width	Petal length	Petal width	Sepal length	Sepal width	Petal length	Petal width
5.0	3.2	1.2	0.2	6.0	3.4	4.5	1.6	7.7	3.0	6.1	2.3
5.5	3.5	1.3	0.2	6.7	3.1	4.7	1.5	6.3	3.4	5.6	2.4
4.9	3.6	1.4	0.1	6.3	2.3	4.4	1.3	6.4	3.1	5.5	1.8
4.4	3.0	1.3	0.2	5.6	3.0	4.1	1.3	6.0	3.0	4.8	1.8
5.1	3.4	1.5	0.2	5.5	2.5	4.0	1.3	6.9	3.1	5.4	2.1
5.0	3.5	1.3	0.3	5.5	2.6	4.4	1.2	6.7	3.1	5.6	2.4
4.5	2.3	1.3	0.3	6.1	3.0	4.6	1.4	6.9	3.1	5.1	2.3
4.4	3.2	1.3	0.2	5.8	2.6	4.0	1.2	5.8	2.7	5.1	1.9
5.0	3.5	1.6	0.6	5.0	2.3	3.3	1.0	6.8	3.2	5.9	2.3
5.1	3.8	1.9	0.4	5.6	2.7	4.2	1.3	6.7	3.3	5.7	2.5
4.8	3.0	1.4	0.3	5.7	3.0	4.2	1.2	6.7	3.0	5.2	2.3
5.1	3.8	1.6	0.2	5.7	2.9	4.2	1.3	6.3	2.5	5.0	1.9
4.6	3.2	1.4	0.2	6.2	2.9	4.3	1.3	6.5	3.0	5.2	2.0
5.3	3.7	1.5	0.2	5.1	2.5	3.0	1.1	6.2	3.4	5.4	2.3
5.0	3.3	1.4	0.2	5.7	2.8	4.1	1.3	5.9	3.0	5.1	1.8

6. Find $\hat{\boldsymbol{\mu}}$, $\hat{\boldsymbol{\Sigma}}$, and $(\hat{\rho}_{ij})$ for *Iris setosa* from Table 3.4 of Edgar Anderson's famous Iris data [Fisher (1936)].

7. (Sec. 3.2) *Invariance of the sample correlation coefficient.* Prove that r_{12} is an invariant characteristic of the sufficient statistics $\bar{\boldsymbol{x}}$ and $\boldsymbol{S}$ of a bivariate sample under location and scale transformations ($x_{i\alpha}^* = b_i x_{i\alpha} + c_i$, $b_i > 0$, $i = 1, 2$, $\alpha = 1, \dots, N$) and that every function of $\bar{\boldsymbol{x}}$ and $\boldsymbol{S}$ that is invariant is a function of r_{12}. [*Hint*: See Theorem 2.3.2.]

8. (Sec. 3.2) Prove Lemma 3.2.2 by induction. [*Hint*: Let $\boldsymbol{H}_1 = h_{11}$,

$$\boldsymbol{H}_i = \begin{pmatrix} \boldsymbol{H}_{i-1} & \boldsymbol{h}_{(i)} \\ \boldsymbol{h}_{(i)}' & h_{ii} \end{pmatrix}, \qquad i = 2, \dots, p,$$

and use Problem 36 of Chapter 2.]

9. (Sec. 3.2) Show that

$$\frac{1}{N(N-1)} \sum_{\alpha<\beta} (\boldsymbol{x}_\alpha - \boldsymbol{x}_\beta)(\boldsymbol{x}_\alpha - \boldsymbol{x}_\beta)' = \frac{1}{N-1} \sum_{\alpha=1}^{N} (\boldsymbol{x}_\alpha - \bar{\boldsymbol{x}})(\boldsymbol{x}_\alpha - \bar{\boldsymbol{x}})'.$$

(Note: When $p = 1$, the left-hand side is the average squared differences of the observations.)

10. (Sec. 3.2) *Estimation of* $\boldsymbol{\Sigma}$ *when* $\boldsymbol{\mu}$ *is known.* Show that if $\boldsymbol{x}_1, \dots, \boldsymbol{x}_N$ constitute a sample from $N(\boldsymbol{\mu}, \boldsymbol{\Sigma})$ and $\boldsymbol{\mu}$ is known, then $(1/N)\sum_{\alpha=1}^{N}(\boldsymbol{x}_\alpha - \boldsymbol{\mu})(\boldsymbol{x}_\alpha - \boldsymbol{\mu})'$ is the maximum likelihood estimator of $\boldsymbol{\Sigma}$.

11. (Sec. 3.2) *Estimation of parameters of a complex normal distribution*. Let $z_1, \dots, z_N$ be N observations from the complex normal distribution with mean $\boldsymbol{\theta}$ and covariance matrix $\boldsymbol{P}$. (See Problem 64 of Chapter 2.)

(*a*) Show that the maximum likelihood estimators of $\boldsymbol{\theta}$ and $\boldsymbol{P}$ are

$$\hat{\boldsymbol{\theta}} = \bar{z} = \frac{1}{N}\sum_{\alpha=1}^{N} z_\alpha, \qquad \hat{\boldsymbol{P}} = \frac{1}{N}\sum_{\alpha=1}^{N}(z_\alpha - \bar{z})(z_\alpha - \bar{z})^*.$$

(*b*) Show that $\bar{z}$ has the complex normal distribution with mean $\boldsymbol{\theta}$ and covariance matrix $(1/N)\boldsymbol{P}$.

(*c*) Show that $\bar{z}$ and $\hat{\boldsymbol{P}}$ are independently distributed and that $N\hat{\boldsymbol{P}}$ has the distribution of $\sum_{\alpha=1}^{n} \boldsymbol{W}_\alpha \boldsymbol{W}_\alpha^*$, where $\boldsymbol{W}_1, \dots, \boldsymbol{W}_n$ are independently distributed, each according to the complex normal distribution with mean $\boldsymbol{0}$ and covariance matrix $\boldsymbol{P}$, and $n = N - 1$.

12. (Sec. 3.2) Prove Lemma 3.2.2 by using Lemma 3.2.3 and showing $N \log|\boldsymbol{C}| - \operatorname{tr} \boldsymbol{CD}$ has a maximum at $\boldsymbol{C} = N\boldsymbol{D}^{-1}$ by setting the derivatives of this function with respect to the elements of $\boldsymbol{C} = \boldsymbol{\Sigma}^{-1}$ equal to 0. Show that the function of $\boldsymbol{C}$ tends to $-\infty$ as $\boldsymbol{C}$ tends to a singular matrix or as one or more elements of $\boldsymbol{C}$ tend to ∞ and/or $-\infty$ (nondiagonal elements); for this latter, the equivalent of (13) can be used.

13. (Sec. 3.3) Let $\boldsymbol{X}_\alpha$ be distributed according to $N(\boldsymbol{\gamma} c_\alpha, \boldsymbol{\Sigma})$, $\alpha = 1, \dots, N$, where $\sum c_\alpha^2 > 0$. Show that the distribution of $\boldsymbol{g} = (1/\sum c_\alpha^2)\sum c_\alpha \boldsymbol{X}_\alpha$ is $N[\boldsymbol{\gamma}, (1/\sum c_\alpha^2)\boldsymbol{\Sigma}]$. Show that $\boldsymbol{E} = \sum_\alpha (\boldsymbol{X}_\alpha - \boldsymbol{g}c_\alpha)(\boldsymbol{X}_\alpha - \boldsymbol{g}c_\alpha)'$ is independently distributed as $\sum_{\alpha=1}^{N-1} \boldsymbol{Z}_\alpha \boldsymbol{Z}_\alpha'$, where $\boldsymbol{Z}_1, \dots, \boldsymbol{Z}_N$ are independent, each with distribution $N(\boldsymbol{0}, \boldsymbol{\Sigma})$. [*Hint*: Let $\boldsymbol{Z}_\alpha = \sum b_{\alpha\beta} \boldsymbol{X}_\beta$, where $b_{N\beta} = c_\beta / \sqrt{\sum c_\alpha^2}$ and $\boldsymbol{B}$ is orthogonal.]

14. (Sec. 3.3) Prove that the power of the test in (19) is a function only of p and $[N_1 N_2/(N_1 + N_2)](\boldsymbol{\mu}^{(1)} - \boldsymbol{\mu}^{(2)})' \boldsymbol{\Sigma}^{-1}(\boldsymbol{\mu}^{(1)} - \boldsymbol{\mu}^{(2)})$, given α.

15. (Sec. 3.3) *Efficiency of the mean*. Prove that $\bar{\boldsymbol{x}}$ is efficient for estimating $\boldsymbol{\mu}$.

16. (Sec. 3.3) Prove that $\bar{\boldsymbol{x}}$ and $\boldsymbol{S}$ have efficiency $[(N-1)/N]^{p(p+1)/2}$ for estimating $\boldsymbol{\mu}$ and $\boldsymbol{\Sigma}$.

17. (Sec. 3.2) Prove that $\Pr\{|\boldsymbol{A}| = 0\} = 0$ for $\boldsymbol{A}$ defined by (4) when $N > p$. [*Hint*: Argue that if $\boldsymbol{Z}_p^* = (\boldsymbol{Z}_1, \dots, \boldsymbol{Z}_p)$, then $|\boldsymbol{Z}_p^*| \neq 0$ implies $\boldsymbol{A} = \boldsymbol{Z}_p^* \boldsymbol{Z}_p^{*\prime} + \sum_{\alpha=p+1}^{N-1} \boldsymbol{Z}_\alpha \boldsymbol{Z}_\alpha'$ is positive definite. Prove $\Pr\{|\boldsymbol{Z}_j^*| = Z_{jj}|\boldsymbol{Z}_{j-1}^*| + \sum_{i=1}^{j-1} Z_{ij} \operatorname{cof}(Z_{ij}) = 0\} = 0$ by induction, $j = 2, \dots, p$.]

18. (Sec. 3.4) Prove

$$\boldsymbol{I} - \boldsymbol{\Phi}(\boldsymbol{\Phi} + \boldsymbol{\Sigma})^{-1} = \boldsymbol{\Sigma}(\boldsymbol{\Phi} + \boldsymbol{\Sigma})^{-1},$$

$$\boldsymbol{\Phi} - \boldsymbol{\Phi}(\boldsymbol{\Phi} + \boldsymbol{\Sigma})^{-1}\boldsymbol{\Phi} = (\boldsymbol{\Phi}^{-1} + \boldsymbol{\Sigma}^{-1})^{-1}.$$

19. (Sec. 3.4) Prove $(1/N)\sum_{\alpha=1}^{N}(\boldsymbol{x}_\alpha - \boldsymbol{\mu})(\boldsymbol{x}_\alpha - \boldsymbol{\mu})'$ is an unbiased estimator of $\boldsymbol{\Sigma}$ when $\boldsymbol{\mu}$ is known.

20. (Sec. 3.4) Show that

$$\Phi\left(\Phi + \frac{1}{N}\Sigma\right)^{-1} x + \frac{1}{N}\Sigma\left(\Phi + \frac{1}{N}\Sigma\right)^{-1} \nu = (\Phi^{-1} + N\Sigma^{-1})^{-1}(N\Sigma^{-1}x + \Phi^{-1}\nu).$$

21. (Sec. 3.5) Demonstrate Lemma 3.5.1 using integration by parts.

22. (Sec. 3.5) Show that

$$\int_\theta^\infty \int_y^\infty \left| f'(y)(x-\theta)\frac{1}{\sqrt{2\pi}} e^{-\frac{1}{2}(x-\theta)^2} \right| dx\,dy = \int_\theta^\infty |f'(y)| \frac{1}{\sqrt{2\pi}} e^{-\frac{1}{2}(y-\theta)^2}\,dy,$$

$$\int_{-\infty}^\theta \int_{-\infty}^y \left| f'(y)(\theta - x)\frac{1}{\sqrt{2\pi}} e^{-\frac{1}{2}(x-\theta)^2} \right| dx\,dy = \int_{-\infty}^\theta |f'(y)| \frac{1}{\sqrt{2\pi}} e^{-\frac{1}{2}(y-\theta)^2}\,dy.$$

23. Let $Z(k) = (Z_{ij}(k))$, where $i = 1, \dots, p$, $j = 1, \dots, q$, and $k = 1, 2, \dots$, be a sequence of random matrices. Let one norm of a matrix A be $N_1(A) = \max_{i,j} \operatorname{mod}(a_{ij})$ and another be $N_2(A) = \sum_{i,j} a_{ij}^2 = \operatorname{tr} AA'$. Some alternative ways of defining stochastic convergence of $Z(k)$ to B $(p \times q)$ are

(*a*) $N_1(Z(k) - B)$ converges stochastically to 0,
(*b*) $N_2(Z(k) - B)$ converges stochastically to 0, and
(*c*) $Z_{ij}(k) - b_{ij}$ converges stochastically to 0, $i = 1, \dots, p$, $j = 1, \dots, q$.

Prove that these three definitions are equivalent. Note that the definition of $X(k)$ converging stochastically to a is that for every arbitrary positive δ and ε, we can find K large enough so that for $k > K$

$$\Pr\{|X(k) - a| < \delta\} > 1 - \varepsilon.$$

24. (Sec. 3.2) *Covariance matrices with linear structure* [Anderson (1969)]. Let

$$\text{(i)} \qquad \Sigma = \sum_{g=0}^q \sigma_g G_g,$$

where $G_0, \dots, G_q$ are given symmetric matrices such that there exists at least one $(q+1)$-tuplet $\sigma_0, \sigma_1, \dots, \sigma_q$ such that (i) is positive definite. Show that the likelihood equations based on N observations are

$$\text{(ii)} \qquad -\frac{N}{2}\operatorname{tr}\Sigma^{-1}G_g + \frac{1}{2}\operatorname{tr} A\Sigma^{-1}G_g\Sigma^{-1} = 0, \qquad g = 0, 1, \dots, q.$$

Show that an iterative (scoring) method can be based on

$$\text{(iii)} \qquad \sum_{h=0}^q \operatorname{tr}\hat{\Sigma}_{i-1}^{-1}G_g\hat{\Sigma}_{i-1}^{-1}G_h\hat{\sigma}_h^{(i)} = \frac{1}{N}\operatorname{tr}\hat{\Sigma}_{i-1}^{-1}G_g\hat{\Sigma}_{i-1}^{-1}A, \qquad g = 0, 1, \dots, q,$$

where $\hat{\Sigma}_{i-1} = \sum_{g=0}^q \hat{\sigma}_g^{(i-1)} G_g$.

CHAPTER 4

The Distributions and Uses of Sample Correlation Coefficients

4.1. INTRODUCTION

In Chapter 2, in which the multivariate normal distribution was introduced, it was shown that a measure of dependence between two normal variates is the correlation coefficient $\rho_{ij} = \sigma_{ij}/\sqrt{\sigma_{ii}\sigma_{jj}}$. In a conditional distribution of $X_1, \ldots, X_q$, given $X_{q+1} = x_{q+1}, \ldots, X_p = x_p$, the partial correlation $\rho_{ij \cdot q+1, \ldots, p}$ measures the dependence between X_i and X_j. The third kind of correlation discussed was the multiple correlation which measures the relationship between one variate and a set of others. In this chapter we treat the sample equivalents of these quantities; they are point estimates of the population quantities. The distributions of the sample correlations are found. Tests of hypotheses and confidence intervals are developed.

In the cases of joint normal distributions these correlation coefficients are the natural measures of dependence. In the population they are the only parameters except for location (means) and scale (standard deviations) parameters. In the sample the correlation coefficients are derived as the reasonable estimates of the population correlations. Since the sample means and standard deviations are location and scale estimates, the sample correlations (that is, the standardized sample second moments) give all possible information about the population correlations. The sample correlations are the functions of the sufficient statistics that are invariant with respect to location and scale transformations; the population correlations are the functions of the parameters that are invariant with respect to these transformations.

In "regression theory" or least squares, one variable is considered random or "dependent" and the others fixed or "independent." In correlation theory we consider several variables as random and treat them symmetrically. If we start with a joint normal distribution and hold all variables fixed except one, we obtain the least squares model because the expected value of the random variable in the conditional distribution is a linear function of the variables held fixed. The sample regression coefficients obtained in least squares are functions of the sample variances and correlations.

In testing independence we shall see that we arrive at the same tests in either case (i.e., in the joint normal distribution or in the conditional distribution of least squares). The probability theory under the null hypothesis is the same. The distribution of the test criterion when the null hypothesis is not true differs in the two cases. If all variables may be considered random, one uses "correlation" theory as given here; if only one variable is random, one uses "least squares" theory (which is considered in some generality in Chapter 8).

In Section 4.2 we derive the distribution of the sample correlation coefficient, first when the corresponding population correlation coefficient is 0 (the two normal variables being independent) and then for any value of the population coefficient. The Fisher z-transform yields a useful approximate normal distribution. Exact and approximate confidence intervals are developed. In Section 4.3 we carry out the same program for partial correlations, that is, correlations in conditional normal distributions. In Section 4.4 the distributions and other properties of the sample multiple correlation coefficient are studied.

4.2. CORRELATION COEFFICIENT OF A BIVARIATE SAMPLE

4.2.1. The Distribution When the Population Correlation Coefficient Is Zero; Tests of the Hypothesis of Lack of Correlation

In Section 3.2 it was shown that if one has a sample (of p-component vectors) $\boldsymbol{x}_1, \dots, \boldsymbol{x}_N$ from a normal distribution the maximum likelihood estimator of the correlation between X_i and X_j (two components of the random vector $\boldsymbol{X}$) is

$$r_{ij} = \frac{\sum_{\alpha=1}^{N}(x_{i\alpha} - \bar{x}_i)(x_{j\alpha} - \bar{x}_j)}{\sqrt{\sum_{\alpha=1}^{N}(x_{i\alpha} - \bar{x}_i)^2}\sqrt{\sum_{\alpha=1}^{N}(x_{j\alpha} - \bar{x}_j)^2}}, \tag{1}$$

where $x_{i\alpha}$ is the ith component of $\boldsymbol{x}_\alpha$ and

$$\bar{x}_i = \frac{1}{N} \sum_{\alpha=1}^{N} x_{i\alpha}. \tag{2}$$

In this section we shall find the distribution of r_{ij} when the population correlation between X_i and X_j is zero, and we shall see how to use the sample correlation coefficient to test the hypothesis that the population coefficient is zero.

For convenience we shall treat r_{12}; the same theory holds for each r_{ij}. Since r_{12} depends only on the first two coordinates of each $\boldsymbol{x}_\alpha$, to find the distribution of r_{12} we need only consider the joint distribution of (x_{11}, x_{21}), $(x_{12}, x_{22}), \dots, (x_{1N}, x_{2N})$. We can reformulate the problems to be considered here, therefore, in terms of a bivariate normal distribution. Let $\boldsymbol{x}_1^*, \dots, \boldsymbol{x}_N^*$ be observation vectors from

$$N\left[\begin{pmatrix} \mu_1 \\ \mu_2 \end{pmatrix}, \begin{pmatrix} \sigma_1^2 & \sigma_1\sigma_2\rho \\ \sigma_2\sigma_1\rho & \sigma_2^2 \end{pmatrix}\right]. \tag{3}$$

We shall consider

$$r = \frac{a_{12}}{\sqrt{a_{11}}\sqrt{a_{22}}}, \tag{4}$$

where

$$a_{ij} = \sum_{\alpha=1}^{N} (x_{i\alpha} - \bar{x}_i)(x_{j\alpha} - \bar{x}_j), \qquad i, j = 1, 2, \tag{5}$$

and $\bar{x}_i$ is defined by (2), $x_{i\alpha}$ being the ith component of $\boldsymbol{x}_\alpha^*$.

From Section 3.3 we see that a_{11}, a_{12}, and a_{22} are distributed like

$$a_{ij} = \sum_{\alpha=1}^{n} z_{i\alpha} z_{j\alpha}, \qquad i, j = 1, 2, \tag{6}$$

where $n = N - 1$, $(z_{1\alpha}, z_{2\alpha})$ is distributed according to

$$N\left[\begin{pmatrix}0\\0\end{pmatrix}, \begin{pmatrix}\sigma_1^2 & \sigma_1\sigma_2\rho\\ \sigma_2\sigma_1\rho & \sigma_2^2\end{pmatrix}\right], \tag{7}$$

and the pairs $(z_{11}, z_{21}), \dots, (z_{1N}, z_{2N})$ are independently distributed.

Define the n-component vector $\boldsymbol{v}_i = (z_{i1}, \dots, z_{in})'$, $i = 1, 2$. These two vectors can be represented in an n-dimensional space; see Figure 4.1. The correlation coefficient is the cosine of the angle, say θ, between $\boldsymbol{v}_1$ and $\boldsymbol{v}_2$. (See Section 3.2.) To find the distribution of $\cos\theta$ we shall first find the distribution of $\cot\theta$. As shown in Section 3.2, if we let $b = \boldsymbol{v}_2'\boldsymbol{v}_1/\boldsymbol{v}_1'\boldsymbol{v}_1$, then $\boldsymbol{v}_2 - b\boldsymbol{v}_1$ is orthogonal to $\boldsymbol{v}_1$ and

$$\cot\theta = \frac{b\|\boldsymbol{v}_1\|}{\|\boldsymbol{v}_2 - b\boldsymbol{v}_1\|}. \tag{8}$$

If $\boldsymbol{v}_1$ is fixed, we can rotate coordinate axes so that the first coordinate axis lies along $\boldsymbol{v}_1$. Then $b\boldsymbol{v}_1$ has only the first coordinate different from zero and $\boldsymbol{v}_2 - b\boldsymbol{v}_1$ has this first coordinate equal to zero. We shall show that $\cot\theta$ is proportional to a t-variable when $\rho = 0$.

We use the following lemma.

LEMMA 4.2.1. *If $\boldsymbol{Y}_1, \dots, \boldsymbol{Y}_n$ are independently distributed, if $\boldsymbol{Y}_\alpha = (\boldsymbol{Y}_\alpha^{(1)'}, \boldsymbol{Y}_\alpha^{(2)'})$ has the density $f(\boldsymbol{y}_\alpha)$, and if the conditional density of $\boldsymbol{Y}_\alpha^{(2)}$ given $\boldsymbol{Y}_\alpha^{(1)} = \boldsymbol{y}_\alpha^{(1)}$ is $f(\boldsymbol{y}_\alpha^{(2)}|\boldsymbol{y}_\alpha^{(1)})$, $\alpha = 1, \dots, n$, then in the conditional distribution of*

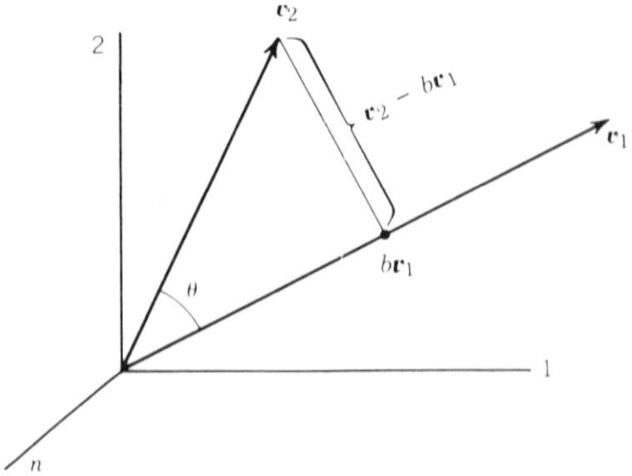

Figure 4.1

$\boldsymbol{Y}_1^{(2)}, \ldots, \boldsymbol{Y}_n^{(2)}$ *given* $\boldsymbol{Y}_1^{(1)} = \boldsymbol{y}_1^{(1)}, \ldots, \boldsymbol{Y}_n^{(1)} = \boldsymbol{y}_n^{(1)}$, *the random vectors* $\boldsymbol{Y}_1^{(2)}, \ldots, \boldsymbol{Y}_n^{(2)}$ *are independent and the density of* $\boldsymbol{Y}_\alpha^{(2)}$ *is* $f(\boldsymbol{y}_\alpha^{(2)} | \boldsymbol{y}_\alpha^{(1)})$, $\alpha = 1, \ldots, n$.

PROOF. The marginal density of $\boldsymbol{Y}_1^{(1)}, \ldots, \boldsymbol{Y}_n^{(1)}$ is $\prod_{\alpha=1}^n f_1(\boldsymbol{y}_\alpha^{(1)})$, where $f_1(\boldsymbol{y}_\alpha^{(1)})$ is the marginal density of $\boldsymbol{Y}_\alpha^{(1)}$, and the conditional density of $\boldsymbol{Y}_1^{(2)}, \ldots, \boldsymbol{Y}_n^{(2)}$ given $\boldsymbol{Y}_1^{(1)} = \boldsymbol{y}_1^{(1)}, \ldots, \boldsymbol{Y}_n^{(1)} = \boldsymbol{y}_n^{(1)}$ is

$$\frac{\prod_{\alpha=1}^n f(\boldsymbol{y}_\alpha)}{\prod_{\alpha=1}^n f_1(\boldsymbol{y}_\alpha^{(1)})} = \prod_{\alpha=1}^n \frac{f(\boldsymbol{y}_\alpha)}{f_1(\boldsymbol{y}_\alpha^{(1)})} = \prod_{\alpha=1}^n f(\boldsymbol{y}_\alpha^{(2)} | \boldsymbol{y}_\alpha^{(1)}). \tag{9}$$

Q.E.D.

Write $\boldsymbol{V}_i = (Z_{i1}, \ldots, Z_{in})'$, $i = 1, 2$, to denote random vectors. The conditional distribution of $Z_{2\alpha}$ given $Z_{1\alpha} = z_{1\alpha}$ is $N(\beta z_{1\alpha}, \sigma^2)$, where $\beta = \rho\sigma_2/\sigma_1$ and $\sigma^2 = \sigma_2^2(1 - \rho^2)$. (See Section 2.5.) The density of $\boldsymbol{V}_2$ given $\boldsymbol{V}_1 = \boldsymbol{v}_1$ is $N(\beta \boldsymbol{v}_1, \sigma^2 \boldsymbol{I})$ since the $Z_{2\alpha}$ are independent. Let $b = \boldsymbol{V}_2'\boldsymbol{v}_1/\boldsymbol{v}_1'\boldsymbol{v}_1$ $(= a_{21}/a_{11})$ so that $b\boldsymbol{v}_1'(\boldsymbol{V}_2 - b\boldsymbol{v}_1) = 0$, and let $U = (\boldsymbol{V}_2 - b\boldsymbol{v}_1)'(\boldsymbol{V}_2 - b\boldsymbol{v}_1) = \boldsymbol{V}_2'\boldsymbol{V}_2 - b^2\boldsymbol{v}_1'\boldsymbol{v}_1$ $(= a_{22} - a_{12}^2/a_{11})$. Then $\cot\theta = b\sqrt{a_{11}/U}$. The rotation of coordinate axes involves choosing an $n \times n$ orthogonal matrix $\boldsymbol{C}$ with first row $(1/c)\boldsymbol{v}_1'$, where $c^2 = \boldsymbol{v}_1'\boldsymbol{v}_1$.

We now apply Theorem 3.3.1 with $\boldsymbol{X}_\alpha = Z_{2\alpha}$. Let $Y_\alpha = \sum_\beta c_{\alpha\beta} Z_{2\beta}$, $\alpha = 1, \ldots, n$. Then $Y_1, \ldots, Y_n$ are independently normally distributed with variance σ^2 and means

$$\mathscr{E} Y_1 = \sum_{\gamma=1}^n c_{1\gamma}\beta z_{1\gamma} = \frac{\beta}{c}\sum_{\gamma=1}^n z_{1\gamma}^2 = \beta c, \tag{10}$$

$$\mathscr{E} \boldsymbol{Y}_\alpha = \sum_{\gamma=1}^n c_{\alpha\gamma}\beta z_{1\gamma} = \beta c \sum_{\gamma=1}^n c_{\alpha\gamma} c_{1\gamma} = 0, \qquad \alpha \neq 1. \tag{11}$$

We have $b = \sum_{\alpha=1}^n Z_{2\alpha} z_{1\alpha} / \sum_{\alpha=1}^n z_{1\alpha}^2 = c\sum_{\alpha=1}^n Z_{2\alpha} c_{1\alpha}/c^2 = Y_1/c$ and, from Lemma 3.3.1,

$$\begin{aligned} U &= \sum_{\alpha=1}^n Z_{2\alpha}^2 - b^2 \sum_{\alpha=1}^n z_{1\alpha}^2 = \sum_{\alpha=1}^n Y_\alpha^2 - Y_1^2 \\ &= \sum_{\alpha=2}^n Y_\alpha^2, \end{aligned} \tag{12}$$

which is independent of b. Then U/σ^2 has a χ^2-distribution with $n - 1$ degrees of freedom.

LEMMA 4.2.2. *If* $(Z_{1\alpha}, Z_{2\alpha})$, $\alpha = 1, \ldots, n$, *are independent, each pair with density* (7), *then the conditional distribution of* $b = \sum_{\alpha=1}^n Z_{2\alpha} Z_{1\alpha} / \sum_{\alpha=1}^n Z_{1\alpha}^2$ *and* $U/\sigma^2 = \sum_{\alpha=1}^n (Z_{2\alpha} - bZ_{1\alpha})^2/\sigma^2$ *given* $Z_{1\alpha} = z_{1\alpha}$, $\alpha = 1, \ldots, n$, *is* $N(\beta, \sigma^2/c^2)$ $(c^2 = \sum_{\alpha=1}^n z_{1\alpha}^2)$ *and that of* χ^2 *with* $n - 1$ *degrees of freedom, respectively, and* b *and* U *are independent.*

If $\rho = 0$, then $\beta = 0$, and b is distributed conditionally according to $N(0, \sigma^2/c^2)$, and

$$\frac{cb/\sigma}{\sqrt{\dfrac{U/\sigma^2}{n-1}}} = \frac{cb}{\sqrt{\dfrac{U}{n-1}}} \tag{13}$$

has a conditional t-distribution with $n - 1$ degrees of freedom. (See Problem 27.) However, this random variable is

$$\begin{aligned} \sqrt{n-1}\,\frac{\sqrt{a_{11}}\,a_{12}/a_{11}}{\sqrt{a_{22} - a_{12}^2/a_{11}}} &= \sqrt{n-1}\,\frac{a_{12}/\sqrt{a_{11}a_{22}}}{\sqrt{1 - [a_{12}^2/(a_{11}a_{22})]}} \\ &= \sqrt{n-1}\,\frac{r}{\sqrt{1-r^2}}. \end{aligned} \tag{14}$$

Thus $\sqrt{n-1}\,r/\sqrt{1-r^2}$ has a conditional t-distribution with $n - 1$ degrees of freedom. The density of t is

$$\frac{\Gamma(\frac{1}{2}n)}{\sqrt{n-1}\,\Gamma[\frac{1}{2}(n-1)]\sqrt{\pi}}\left(1 + \frac{t^2}{n-1}\right)^{-\frac{1}{2}n}, \tag{15}$$

and the density of $W = r/\sqrt{1-r^2}$ is

$$\frac{\Gamma(\frac{1}{2}n)}{\Gamma[\frac{1}{2}(n-1)]\sqrt{\pi}}(1 + w^2)^{-\frac{1}{2}n}. \tag{16}$$

Since $w = r(1 - r^2)^{-\frac{1}{2}}$, $dw/dr = (1 - r^2)^{-\frac{3}{2}}$. Therefore the density of r is (replacing n by $N - 1$)

$$\frac{\Gamma[\frac{1}{2}(N-1)]}{\Gamma[\frac{1}{2}(N-2)]\sqrt{\pi}}(1 - r^2)^{\frac{1}{2}(N-4)}. \tag{17}$$

It should be noted that (17) is the conditional density of r for $\boldsymbol{v}_1$ fixed. However, since (17) does not depend on $\boldsymbol{v}_1$, it is also the marginal density of r.

THEOREM 4.2.1. *Let* $\boldsymbol{X}_1, \dots, \boldsymbol{X}_N$ *be independent, each with distribution* $N(\boldsymbol{\mu}, \boldsymbol{\Sigma})$. *If* $\rho_{ij} = 0$, *the density of* r_{ij} *defined by* (1) *is* (17).

From (17) we see that the density is symmetric about the origin. For $N > 4$, it has a mode at $r = 0$ and its order of contact with the r axis at ± 1 is $\frac{1}{2}(N-5)$ for N odd and $\frac{1}{2}N - 3$ for N even. Since the density is even, the odd moments are zero; in particular, the mean is zero. The even moments are found by integration (letting $x = r^2$ and using the definition of the beta function). That $\mathscr{E} r^{2m} = \Gamma[\frac{1}{2}(N-1)]\Gamma(m+\frac{1}{2})/\{\sqrt{\pi}\,\Gamma[\frac{1}{2}(N-1)+m]\}$ and in particular that the variance is $1/(N-1)$ may be verified by the reader.

The most important use of Theorem 4.2.1 is to find significance points for testing the hypothesis that a pair of variables are not correlated. Consider the hypothesis

$$H: \rho_{ij} = 0 \tag{18}$$

for some particular pair (i, j). It would seem reasonable to reject this hypothesis if the corresponding sample correlation coefficient were very different from zero. Now how do we decide what we mean by "very different"?

Let us suppose we are interested in testing H against the alternative hypotheses $\rho_{ij} > 0$. Then we reject H if the sample correlation coefficient r_{ij} is greater than some number r_0. The probability of rejecting H when H is true is

$$\int_{r_0}^{1} k_N(r)\, dr, \tag{19}$$

where $k_N(r)$ is (17), the density of a correlation coefficient based on N observations. We choose r_0 so (19) is the desired significance level. If we test H against alternatives $\rho_{ij} < 0$, we reject H when $r_{ij} < -r_0$.

Now suppose we are interested in alternatives $\rho_{ij} \neq 0$; that is, ρ_{ij} may be either positive or negative. Then we reject the hypothesis H if $r_{ij} > r_1$ or $r_{ij} < -r_1$. The probability of rejection when H is true is

$$\int_{-1}^{-r_1} k_N(r)\, dr + \int_{r_1}^{1} k_N(r)\, dr. \tag{20}$$

The number r_1 is chosen so that (20) is the desired significance level.

The significance points r_1 are given in many books including Table VI of Fisher and Yates (1942); the index n in Table VI is equal to our $N-2$. Since $\sqrt{N-2}\,r/\sqrt{1-r^2}$ has the t-distribution with $N-2$ degrees of freedom, t-tables can also be used. Against alternatives $\rho_{ij} \neq 0$, reject H if

$$\sqrt{N-2}\,\frac{|r_{ij}|}{\sqrt{1-r_{ij}^2}} > t_{N-2}(\alpha), \tag{21}$$

where $t_{N-2}(\alpha)$ is the two-tailed significance point of the t-statistic with $N-2$ degrees of freedom for significance level α. Against alternatives $\rho_{ij} > 0$, reject H if

$$\sqrt{N-2}\,\frac{r_{ij}}{\sqrt{1-r_{ij}^2}} > t_{N-2}(2\alpha). \tag{22}$$

From (13) and (14) we see that $\sqrt{N-2}\,r/\sqrt{1-r^2}$ is the proper statistic for testing the hypothesis that the regression of V_2 on $\mathbf{v}_1$ is zero. In terms of the original observations $\{x_{i\alpha}\}$, we have

$$\sqrt{N-2}\,\frac{r}{\sqrt{1-r^2}} = \frac{b\sqrt{\sum_{\alpha=1}^N (x_{1\alpha}-\bar{x}_1)^2}}{\sqrt{\sum_{\alpha=1}^N \left[x_{2\alpha}-\bar{x}_2 - b(x_{1\alpha}-\bar{x}_1)\right]^2/(N-2)}}, \tag{23}$$

where $b = \sum_{\alpha=1}^N (x_{2\alpha}-\bar{x}_2)(x_{1\alpha}-\bar{x}_1)/\sum_{\alpha=1}^N (x_{1\alpha}-\bar{x}_1)^2$ is the least squares regression coefficient of $x_{2\alpha}$ on $x_{1\alpha}$. It is seen that the test of $\rho_{12}=0$ is equivalent to the test that the regression of X_2 on x_1 is zero (i.e., that $\rho_{12}\sigma_2/\sigma_1 = 0$).

To illustrate this procedure we consider the example given in Section 3.2. Let us test the null hypothesis that the effects of the two drugs are uncorrelated against the alternative that they are positively correlated. We shall use the 5% level of significance. For $N = 10$, the 5% significance point (r_0) is 0.5494. Our observed correlation coefficient of 0.7952 is significant; we reject the hypothesis that the effects of the two drugs are independent.

4.2.2. The Distribution When the Population Correlation Coefficient Is Nonzero; Tests of Hypotheses and Confidence Intervals

To find the distribution of the sample correlation coefficient when the population coefficient is different from zero, we shall first derive the joint density of a_{11}, a_{12}, and a_{22}. In Section 4.2.1 we saw that, conditional on $\mathbf{v}_1$

held fixed, the random variables $b = a_{12}/a_{11}$ and $U/\sigma^2 = (a_{22} - a_{12}^2/a_{11})/\sigma^2$ are distributed independently according to $N(\beta, \sigma^2/c^2)$ and the χ^2-distribution with $n-1$ degrees of freedom, respectively. Denoting the density of the χ^2-distribution by $g_{n-1}(u)$, we write the conditional density of b and U as $n(b|\beta, \sigma^2/a_{11})g_{n-1}(u/\sigma^2)/\sigma^2$. The joint density of V_1, b, and U is $n(\boldsymbol{v}_1|\boldsymbol{0}, \sigma_1^2\boldsymbol{I})n(b|\beta, \sigma^2/a_{11})g_{n-1}(u/\sigma^2)/\sigma^2$. The marginal density of $V_1'V_1/\sigma_1^2 = a_{11}/\sigma_1^2$ is $g_n(u)$; that is, the density of a_{11} is

$$\frac{1}{\sigma_1^2} g_n\left(\frac{a_{11}}{\sigma_1^2}\right) = \int_{\boldsymbol{v}_1'\boldsymbol{v}_1 = a_{11}} \cdots \int n\left(\boldsymbol{v}_1|\boldsymbol{0}, \sigma_1^2\boldsymbol{I}\right) dW, \tag{24}$$

where dW is the proper volume element.

The integration is over the sphere $\boldsymbol{v}_1'\boldsymbol{v}_1 = a_{11}$; thus, dW is an element of area on this sphere. (See Problem 1 of Chapter 7 for the use of angular coordinates in defining dW.) Thus the joint density of b, U, and a_{11} is

$$\begin{aligned}
(25)\quad & \int_{\boldsymbol{v}_1'\boldsymbol{v}_1 = a_{11}} \cdots \int n\left(b|\beta, \sigma^2/a_{11}\right) g_{n-1}\left(u/\sigma^2\right) \frac{1}{\sigma^2} n\left(\boldsymbol{v}_1|\boldsymbol{0}, \sigma_1^2\boldsymbol{I}\right) dW \\
&= g_n\left(a_{11}/\sigma_1^2\right) n\left(b|\beta, \sigma^2/a_{11}\right) g_{n-1}\left(u/\sigma^2\right)/\left(\sigma_1^2\sigma^2\right) \\
&= \frac{\left(a_{11}\right)^{\frac{1}{2}n-1}}{\left(2\sigma_1^2\right)^{\frac{1}{2}n}\Gamma\left(\frac{1}{2}n\right)} \exp\left(-\frac{a_{11}}{2\sigma_1^2}\right) \frac{\sqrt{a_{11}}}{\sqrt{2\pi\sigma^2}} \exp\left[-\frac{a_{11}}{2\sigma^2}(b-\beta)^2\right] \\
&\qquad \cdot \frac{1}{\left(2\sigma^2\right)^{\frac{1}{2}(n-1)}\Gamma\left[\frac{1}{2}(n-1)\right]} u^{\frac{1}{2}(n-3)} \exp\left(-\frac{u}{2\sigma^2}\right).
\end{aligned}$$

Now let $b = a_{12}/a_{11}$, $U = a_{22} - a_{12}^2/a_{11}$. The Jacobian is

$$\left|\frac{\partial(b, u)}{\partial(a_{12}, a_{22})}\right| = \begin{vmatrix} \dfrac{1}{a_{11}} & 0 \\ -2\dfrac{a_{12}}{a_{11}} & 1 \end{vmatrix} = \frac{1}{a_{11}}. \tag{26}$$

Thus the density of a_{11}, a_{12}, and a_{22} for $a_{11} \geq 0$, $a_{22} \geq 0$, and $a_{11}a_{22} - a_{12}^2 \geq 0$ is

$$\frac{a_{11}^{\frac{1}{2}(n-3)}\left(\dfrac{a_{11}a_{22} - a_{12}^2}{a_{11}}\right)^{\frac{1}{2}(n-3)} e^{-\frac{1}{2}Q}}{2^n\sigma_1^n\sigma_2^n(1-\rho^2)^{\frac{1}{2}n}\sqrt{\pi}\,\Gamma\left(\frac{1}{2}n\right)\Gamma\left[\frac{1}{2}(n-1)\right]}, \tag{27}$$

where

$$(28)\quad Q = \frac{a_{11}}{\sigma_1^2} + \frac{a_{11}}{\sigma^2}\left(\frac{a_{12}^2}{a_{11}^2} - 2\rho\frac{\sigma_1\sigma_2}{\sigma_1^2}\frac{a_{12}}{a_{11}} + \frac{\rho^2\sigma_1^2\sigma_2^2}{\sigma_1^4}\right) + \frac{1}{\sigma^2}\left(a_{22} - \frac{a_{12}^2}{a_{11}}\right)$$

$$= a_{11}\left[\frac{1}{\sigma_1^2} + \frac{\rho^2\sigma_1^2\sigma_2^2}{\sigma_1^4\sigma_2^2(1-\rho^2)}\right] - 2a_{12}\frac{\rho\sigma_2}{\sigma_1\sigma_2^2(1-\rho^2)} + \frac{a_{22}}{\sigma_2^2(1-\rho^2)}$$

$$= \frac{1}{1-\rho^2}\left(\frac{a_{11}}{\sigma_1^2} - 2\rho\frac{a_{12}}{\sigma_1\sigma_2} + \frac{a_{22}}{\sigma_2^2}\right).$$

The density can be written

$$(29)\qquad \frac{|\boldsymbol{A}|^{\frac{1}{2}(n-3)}e^{-\frac{1}{2}Q}}{2^n|\boldsymbol{\Sigma}|^{\frac{1}{2}n}\sqrt{\pi}\,\Gamma(\frac{1}{2}n)\Gamma[\frac{1}{2}(n-1)]}$$

for $\boldsymbol{A}$ positive definite and 0 otherwise. This is a special case of the Wishart density derived in Chapter 7.

We want to find the density of

$$(30)\qquad r = \frac{a_{12}}{\sqrt{a_{11}a_{22}}} = \frac{a_{12}/(\sigma_1\sigma_2)}{\sqrt{(a_{11}/\sigma_1^2)(a_{22}/\sigma_2^2)}} = \frac{a_{12}^*}{\sqrt{a_{11}^*a_{22}^*}},$$

where $a_{11}^* = a_{11}/\sigma_1^2$, $a_{22}^* = a_{22}/\sigma_2^2$, and $a_{12}^* = a_{12}/(\sigma_1\sigma_2)$. The transformation is equivalent to setting $\sigma_1 = \sigma_2 = 1$. Then the density of a_{11}, a_{22}, and $r = a_{12}/\sqrt{a_{11}a_{22}}$ ($da_{12} = dr\sqrt{a_{11}a_{22}}$) is

$$(31)\qquad \frac{a_{11}^{\frac{1}{2}n-1}a_{22}^{\frac{1}{2}n-1}(1-r^2)^{\frac{1}{2}(n-3)}e^{-\frac{1}{2}Q}}{2^n(1-\rho^2)^{\frac{1}{2}n}\sqrt{\pi}\,\Gamma(\frac{1}{2}n)\Gamma[\frac{1}{2}(n-1)]},$$

where

$$(32)\qquad Q = \frac{a_{11} - 2\rho r\sqrt{a_{11}}\sqrt{a_{22}} + a_{22}}{1-\rho^2}.$$

To find the density of r, we must integrate (31) with respect to a_{11} and a_{22} over the range 0 to ∞. There are various ways of carrying out the integration, which result in different expressions for the density. The method we shall indicate here is straightforward. We expand part of the exponential

$$\exp\left[\frac{\rho r\sqrt{a_{11}}\sqrt{a_{22}}}{(1-\rho^2)}\right] = \sum_{\alpha=0}^{\infty} \frac{\left(\rho r\sqrt{a_{11}}\sqrt{a_{22}}\right)^{\alpha}}{\alpha!(1-\rho^2)^{\alpha}}. \tag{33}$$

Then the density (31) is

$$\frac{(1-r^2)^{\frac{1}{2}(n-3)}}{(1-\rho^2)^{\frac{1}{2}n}2^n\sqrt{\pi}\,\Gamma(\frac{1}{2}n)\Gamma[\frac{1}{2}(n-1)]} \sum_{\alpha=0}^{\infty} \frac{(\rho r)^{\alpha}}{\alpha!(1-\rho^2)^{\alpha}} \tag{34}$$

$$\cdot\left\{\exp\left[-\frac{a_{11}}{2(1-\rho^2)}\right]a_{11}^{(n+\alpha)/2-1}\right\}\left\{\exp\left[-\frac{a_{22}}{2(1-\rho^2)}\right]a_{22}^{(n+\alpha)/2-1}\right\}.$$

Since

$$\int_0^{\infty} a^{\frac{1}{2}(n+\alpha)-1}\exp\left[-\frac{a}{2(1-\rho^2)}\right]da = \Gamma[\tfrac{1}{2}(n+\alpha)][2(1-\rho^2)]^{\frac{1}{2}(n+\alpha)}, \tag{35}$$

the integral of (34) (term by term integration is permissible) is

$$\frac{(1-r^2)^{\frac{1}{2}(n-3)}}{(1-\rho^2)^{\frac{1}{2}n}2^n\sqrt{\pi}\,\Gamma(\frac{1}{2}n)\Gamma[\frac{1}{2}(n-1)]} \sum_{\alpha=0}^{\infty} \frac{(\rho r)^{\alpha}}{\alpha!(1-\rho^2)^{\alpha}} \tag{36}$$

$$\cdot\Gamma^2[\tfrac{1}{2}(n+\alpha)]2^{n+\alpha}(1-\rho^2)^{n+\alpha}$$

$$= \frac{(1-\rho^2)^{\frac{1}{2}n}(1-r^2)^{\frac{1}{2}(n-3)}}{\sqrt{\pi}\,\Gamma(\frac{1}{2}n)\Gamma[\frac{1}{2}(n-1)]} \sum_{\alpha=0}^{\infty} \frac{(2\rho r)^{\alpha}}{\alpha!}\Gamma^2[\tfrac{1}{2}(n+\alpha)].$$

The *duplication formula for the gamma function* is

$$\Gamma(2z) = \frac{2^{2z-1}\Gamma(z)\Gamma(z+\frac{1}{2})}{\sqrt{\pi}}. \tag{37}$$

It can be used to modify the constant in (36).

THEOREM 4.2.2. *The correlation coefficient in a sample of N from a bivariate normal distribution with correlation ρ is distributed with density*

$$\frac{2^{n-2}(1-\rho^2)^{\frac{1}{2}n}(1-r^2)^{\frac{1}{2}(n-3)}}{(n-2)!\pi}\sum_{\alpha=0}^{\infty}\frac{(2\rho r)^{\alpha}}{\alpha!}\Gamma^2\left[\tfrac{1}{2}(n+\alpha)\right], \qquad -1 \le r \le 1, \tag{38}$$

where $n = N - 1$.

The distribution of r was first found by Fisher (1915). He also gave as another form of the density

$$\frac{(1-\rho^2)^{\frac{1}{2}n}(1-r^2)^{\frac{1}{2}(n-3)}}{\pi(n-2)!}\left[\frac{d^{n-1}}{dx^{n-1}}\left\{\frac{\cos^{-1}(-x)}{\sqrt{1-x^2}}\right\}\bigg|_{x=r\rho}\right]. \tag{39}$$

See Problem 24.

Hotelling (1953) has made an exhaustive study of the distribution of r. He has recommended the following form:

$$\frac{n-1}{\sqrt{2\pi}}\frac{\Gamma(n)}{\Gamma(n+\frac{1}{2})}(1-\rho^2)^{\frac{1}{2}n}(1-r^2)^{\frac{1}{2}(n-3)} \cdot(1-\rho r)^{-n+\frac{1}{2}}F\left(\tfrac{1}{2},\tfrac{1}{2};n+\tfrac{1}{2};\frac{1+\rho r}{2}\right), \tag{40}$$

where

$$F(a,b;c;x) = \sum_{j=0}^{\infty}\frac{\Gamma(a+j)}{\Gamma(a)}\frac{\Gamma(b+j)}{\Gamma(b)}\frac{\Gamma(c)}{\Gamma(c+j)}\frac{x^j}{j!} \tag{41}$$

is a *hypergeometric function*. (See Problem 25.) The series in (40) converges more rapidly than the one in (38). Hotelling discusses methods of integrating the density and also calculates moments of r.

The cumulative distribution of r,

$$\Pr\{r \le r^*\} = F(r^*|N,\rho), \tag{42}$$

has been tabulated by David (1938) for† $\rho = 0(.1).9$, $N = 3(1)25, 50, 100, 200, 400$, and $r^* = -1(.05)1$. (David's n is our N.) It is clear from the density (38) that $F(r^*|N,\rho) = 1 - F(-r^*|N,-\rho)$ because the density for r, ρ is equal to

†$\rho = 0(.1).9$ means $\rho = 0, 0.1, 0.2, \ldots, 0.9$.

the density for $-r$, $-\rho$. These tables can be used for a number of statistical procedures.

First, we consider the problem of using a sample to test the hypothesis

$$H: \rho = \rho_0. \tag{43}$$

If the alternatives are $\rho > \rho_0$, we reject the hypothesis if the sample correlation coefficient is greater than r_0 where r_0 is chosen so $1 - F(r_0|N, \rho_0) = \alpha$, the significance level. If the alternatives are $\rho < \rho_0$, we reject the hypothesis if the sample correlation coefficient is less than r_0', where r_0' is chosen so $F(r_0'|N, \rho_0) = \alpha$. If the alternatives are $\rho \neq \rho_0$, the region of rejection is $r > r_1$ and $r < r_1'$ where r_1 and r_1' are chosen so $[1 - F(r_1|N, \rho_0)] + F(r_1'|N, \rho_0) = \alpha$. David suggests that r_1 and r_1' be chosen so $1 - F(r_1|N, \rho_0) = F(r_1'|N, \rho_0) = \frac{1}{2}\alpha$. She has shown (1937) that for $N \geq 10$, $|\rho| \leq 0.8$ this critical region is nearly the region of an unbiased test of H, that is, a test whose power function has its minimum at ρ_0.

It should be pointed out that any test based on r is invariant under transformations of location and scale, that is, $x_{i\alpha}^* = b_i x_{i\alpha} + c_i$, $b_i > 0$, $i = 1, 2$, $\alpha = 1, \ldots, N$; and r is essentially the only invariant of the sufficient statistics (Problem 7 of Chapter 3). The above procedure for testing $H: \rho = \rho_0$ against alternatives $\rho > \rho_0$ is uniformly most powerful among all invariant tests. (See Problems 16, 17, and 18.)

As an example suppose one wishes to test the hypothesis that $\rho = 0.5$ against alternatives $\rho \neq 0.5$ at the 5% level of significance using the correlation observed in a sample of 15. In David's tables we find (by interpolation) that $F(0.027|15, 0.5) = 0.025$ and $F(0.805|15, 0.5) = 0.975$. Hence we reject the hypothesis if our sample r is less than 0.027 or greater than 0.805.

Secondly, we can use David's tables to compute the power function of a test of correlation. If the region of rejection of H is $r > r_1$ and $r < r_1'$, the power of the test is a function of the true correlation ρ, $[1 - F(r_1|N, \rho)] + [F(r_1'|N, \rho)]$; this is the probability of rejecting the null hypothesis when the population correlation is ρ.

As an example consider finding the power function of the test for $\rho = 0$ considered in the preceding section. The rejection region (one-sided) is $r \geq 0.5494$ at the 5% significance level. The probabilities of rejection are given in Table 4.1. The graph of the power function is illustrated in Figure 4.2.

Thirdly, David's computations lead to confidence regions for ρ. For given N, r_1' (defining a significance point) is a function of ρ, say $f_1(\rho)$, and r_1 is another

TABLE 4.1
A POWER FUNCTION

ρ	Probability
−1.0	0.0000
−0.8	0.0000
−0.6	0.0004
−0.4	0.0032
−0.2	0.0147
0.0	0.0500
0.2	0.1376
0.4	0.3215
0.6	0.6235
0.8	0.9279
1.0	1.0000

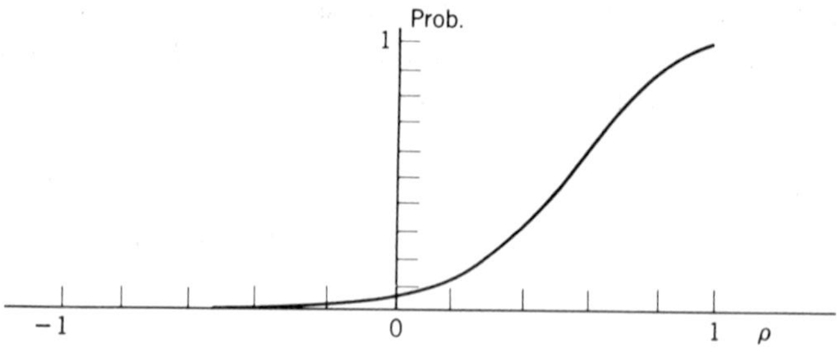

Figure 4.2. A power function.

function of ρ, say $f_2(\rho)$, such that

(44) $$\Pr\{f_1(\rho) < r < f_2(\rho)|\rho\} = 1 - \alpha.$$

Clearly, $f_1(\rho)$ and $f_2(\rho)$ are monotonically increasing functions of ρ if r_1 and r_1' are chosen so $1 - F(r_1|N, \rho) = \frac{1}{2}\alpha = F(r_1'|N, \rho)$. If $\rho = f_i^{-1}(r)$ is the inverse of $r = f_i(\rho)$, $i = 1, 2$, then the inequality $f_1(\rho) < r$ is equivalent to[†] $\rho < f_1^{-1}(r)$ and $r < f_2(\rho)$ is equivalent to $f_2^{-1}(r) < \rho$. Thus (44) can be written

(45) $$\Pr\{f_2^{-1}(r) < \rho < f_1^{-1}(r)|\rho\} = 1 - \alpha.$$

[†] The point $(f_1(\rho), \rho)$ on the first curve is to the left of (r, ρ), and the point $(r, f_1^{-1}(r))$ is above (r, ρ).

This equation says that the probability is $1 - \alpha$ that we draw a sample such that the interval $(f_2^{-1}(r), f_1^{-1}(r))$ cover the parameter ρ. Thus this interval is a confidence interval for ρ with confidence coefficient $1 - \alpha$. For a given N and α the curves $r = f_1(\rho)$ and $r = f_2(\rho)$ appear as in Figure 4.3. In testing the hypothesis $\rho = \rho_0$, the intersection of the line $\rho = \rho_0$ and two curves gives the significance points r_1 and r_1'. In setting up a confidence region for ρ on the basis of a sample correlation r^*, we find the limits $f_2^{-1}(r^*)$ and $f_1^{-1}(r^*)$ by the intersection of the line $r = r^*$ with the two curves. David gives these curves for $\alpha = 0.1, 0.05, 0.02$, and 0.01 for various values of N. One-sided confidence regions can be obtained by using only one inequality above.

The tables of $F(r|N, \rho)$ can also be used instead of the curves for finding the confidence interval. Given the sample value r^*, $f_1^{-1}(r^*)$ is the value of ρ such that $\frac{1}{2}\alpha = \Pr\{r \le r^*|\rho\} = F(r^*|N, \rho)$ and similarly $f_2^{-1}(r^*)$ is the value of ρ such that $\frac{1}{2}\alpha = \Pr\{r \ge r^*|\rho\} = 1 - F(r^*|N, \rho)$. The interval between these two values of ρ, $(f_2^{-1}(r^*), f_1^{-1}(r^*))$, is the confidence interval.

As an example, consider the confidence interval with confidence coefficient 0.95 based on the correlation of 0.7952 observed in a sample of 10. Using Graph II of David, we find the two limits are 0.34 and 0.94. Hence we state that $0.34 < \rho < 0.94$ with confidence 95%.

DEFINITION 4.2.1. *Let $L(\boldsymbol{x}, \boldsymbol{\theta})$ be the likelihood function of the observation vector $\boldsymbol{x}$ and the parameter vector $\boldsymbol{\theta} \in \Omega$. Let a null hypothesis be defined by a*

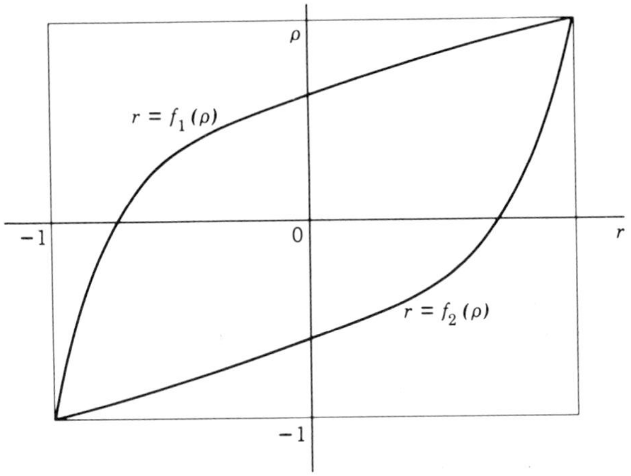

Figure 4.3

proper subset ω of Ω. The likelihood ratio criterion is

$$\lambda(\boldsymbol{x}) = \frac{\sup_{\boldsymbol{\theta} \in \omega} L(\boldsymbol{x}, \boldsymbol{\theta})}{\sup_{\boldsymbol{\theta} \in \Omega} L(\boldsymbol{x}, \boldsymbol{\theta})}. \tag{46}$$

The likelihood ratio test is the procedure of rejecting the null hypothesis when $\lambda(\boldsymbol{x})$ is less than a predetermined constant.

Intuitively, one rejects the null hypothesis if the density of the observations under the most favorable choice of parameters in the null hypothesis is much less than the density under the most favorable unrestricted choice of the parameters. Likelihood ratio tests have some desirable features; see Lehmann (1959), for example. Wald (1943) has proved some favorable asymptotic properties. For most hypotheses concerning the multivariate normal distribution, likelihood ratio tests are appropriate and often are optimal.

Let us consider the likelihood ratio test of the hypothesis that $\rho = \rho_0$ based on a sample $\boldsymbol{x}_1, \dots, \boldsymbol{x}_N$ from the bivariate normal distribution. The set Ω consists of μ_1, μ_2, σ_1, σ_2, and ρ such that $\sigma_1 > 0$, $\sigma_2 > 0$, $-1 < \rho < 1$. The set ω is the subset for which $\rho = \rho_0$. The likelihood maximized in Ω is (by Lemmas 3.2.2 and 3.2.3)

$$\max_{\Omega} L = \frac{N^N e^{-N}}{(2\pi)^N (1 - r^2)^{\frac{1}{2}N} a_{11}^{N/2} a_{22}^{N/2}}. \tag{47}$$

Under the null hypothesis the likelihood function is

$$\frac{1}{(2\pi)^N (1 - \rho_0^2)^{\frac{1}{2}N} (\sigma^2)^N} \exp\left[- \frac{a_{11}/\tau + \tau a_{22} - 2\rho_0 a_{12}}{2\sigma^2 (1 - \rho_0^2)} \right], \tag{48}$$

where $\sigma^2 = \sigma_1 \sigma_2$ and $\tau = \sigma_1 / \sigma_2$. The maximum of (48) with respect to τ occurs at $\hat{\tau} = \sqrt{a_{11}} / \sqrt{a_{22}}$. The concentrated likelihood is

$$\frac{1}{(2\pi)^N (1 - \rho_0^2)^{\frac{1}{2}N} (\sigma^2)^N} \exp\left[- \frac{\sqrt{a_{11}}\sqrt{a_{22}}\,(1 - \rho_0 r)}{\sigma^2 (1 - \rho_0^2)} \right]; \tag{49}$$

the maximum of (49) occurs at

$$\hat{\sigma}^2 = \frac{a_{11}^{\frac{1}{2}} a_{22}^{\frac{1}{2}} (1 - \rho_0 r)}{N(1 - \rho_0^2)}. \tag{50}$$

The likelihood ratio criterion is, therefore,

$$\text{(51)} \qquad \frac{\max_{\omega} L}{\max_{\Omega} L} = \frac{\left(1-\rho_0^2\right)^{\frac{1}{2}N}\left(1-r^2\right)^{\frac{1}{2}N}}{\left(1-\rho_0 r\right)^N} = \left[\frac{\left(1-\rho_0^2\right)\left(1-r^2\right)}{\left(1-\rho_0 r\right)^2}\right]^{\frac{1}{2}N}.$$

The likelihood ratio test is $(1-\rho_0^2)(1-r^2)(1-\rho_0 r)^{-2} < c$, where c is chosen so the probability of the inequality when samples are drawn from normal populations with correlation ρ_0 is the prescribed significance level. The critica' region can be written equivalently as

$$\text{(52)} \qquad \left(\rho_0^2 c - \rho_0^2 + 1\right) r^2 - 2\rho_0 c r + c - 1 + \rho_0^2 > 0,$$

or

$$\text{(53)} \qquad \begin{aligned} r &> \frac{\rho_0 c + \left(1-\rho_0^2\right)\sqrt{1-c}}{\rho_0^2 c + 1 - \rho_0^2}, \\ r &< \frac{\rho_0 c - \left(1-\rho_0^2\right)\sqrt{1-c}}{\rho_0^2 c + 1 - \rho_0^2}. \end{aligned}$$

Thus the likelihood ratio test of $H: \rho = \rho_0$ against alternatives $\rho \neq \rho_0$ has a rejection region of the form $r > r_1$ and $r < r_1'$; but r_1 and r_1' are not chosen so that the probability of each inequality is $\alpha/2$ when H is true, but they are taken to be of the form given in (53), where c is chosen so that the probability of the two inequalities is α.

Unbiased Estimation. Since r is not an unbiased estimator of ρ (which follows from the density of r), we ask for a function, say $G(r)$, that is unbiased. From (38) we see that $\mathscr{E}G(r) = \rho$ is equivalent to

$$\text{(54)} \qquad \frac{2^{n-2}}{(n-2)!\pi} \sum_{\alpha=0}^{\infty} \frac{\Gamma^2\left[\frac{1}{2}(n+\alpha)\right](2\rho)^{\alpha}}{\alpha!} \int_{-1}^{1} G(r)\left(1-r^2\right)^{\frac{1}{2}(n-3)} r^{\alpha}\, dr$$

$$= \rho\left(1-\rho^2\right)^{-\frac{1}{2}n} = \sum_{j=0}^{\infty} \frac{\Gamma\left(\frac{1}{2}n+j\right)}{\Gamma\left(\frac{1}{2}n\right)} \frac{\rho^{2j+1}}{j!}$$

as an identity in ρ. The terms on the left-hand side for α even are 0 (because the right-hand side has only odd powers of ρ), confirming that $G(r)$ is an odd function $[G(-r) = -G(r)]$. Then for $\alpha = 2j + 1$ we have (from the coefficient of ρ^{2j+1})

$$(55)\quad 2\int_0^1 G(r)(1 - r^2)^{\frac{1}{2}(n-3)} r^{2j+1}\, dr$$
$$= \frac{\pi\Gamma(n-1)\Gamma(2j+2)\Gamma(\frac{1}{2}n + j)}{2^{n+2j-1}\Gamma^2[\frac{1}{2}(n+2j+1)]\Gamma(\frac{1}{2}n)\Gamma(j+1)}, \quad j = 0, 1, \ldots .$$

Use of the duplication formula for the gamma function and substitution of $r = e^{-\frac{1}{2}y}$ yields

$$(56)\quad \int_0^\infty \left[G(e^{-\frac{1}{2}y})(1 - e^{-y})^{\frac{1}{2}(n-3)} e^{-y}\right] e^{-jy}\, dy$$
$$= \frac{\Gamma[\frac{1}{2}(n-1)]\Gamma(j + 3/2)\Gamma(\frac{1}{2}n + j)}{\Gamma^2[\frac{1}{2}(n+1) + j]}, \qquad j = 0, 1, \ldots .$$

The right-hand side of (56) as a function of j is the unilateral Laplace transform [Erdélyi (1954), 12.4.4] of

$$(57)\qquad e^{-\frac{3}{2}y}(1 - e^{-y})^{\frac{1}{2}(n-3)} F(\tfrac{1}{2}, \tfrac{1}{2}; \tfrac{1}{2}(n-1); 1 - e^{-y}).$$

The uniqueness of the transform (except for a set of measure 0) follows from the uniqueness of the solution to the moment problem [e.g., Widder (1941), p. 60]. The left-hand side of (55) is the jth moment of $G(r)(1 - r^2)^{\frac{1}{2}(n-3)}$ as a function of r^2. Therefore, the unbiased estimator of ρ is

$$(58)\qquad G(r) = rF(\tfrac{1}{2}, \tfrac{1}{2}; \tfrac{1}{2}(n-1); 1 - r^2)$$
$$= r\left[1 + \frac{1 - r^2}{2(n-1)} + \frac{9}{8}\frac{(1 - r^2)^2}{(n-1)(n+1)} + \cdots\right].$$

It can be verified that $G(r)$ is increasing in r, while $G(r)/r$ is decreasing in r^2. The series is convergent and asymptotic in the sense that if the term with $(1 - r^2)^j$ is retained, the error is $\mathcal{O}(n^{-j-1})$. Olkin and Pratt (1958) give tables of $G(r)$ and $G(r)/r$. They assert that the first two terms of the right-hand side of (58) divided by r agree with $G(r)/r$ by an amount less than 0.01 for $n \geq 14$ and 0.001 for $n \geq 36$; the first three terms agree by an amount less than 0.01 for $n \geq 10$ and 0.001 for $n \geq 18$.

Harley (1956), (1957) has shown that $\mathscr{E} \sin^{-1} r = \sin^{-1} \rho$. A short geometrical proof was given by Daniels and Kendall (1958), and a proof using characteristic functions was presented by Khatri (1968). The only function $g(r)$ such that $\mathscr{E} g(r) = g(\rho)$ is $g(r) = \sin^{-1} r$.

4.2.3. The Asymptotic Distribution of a Sample Correlation Coefficient and Fisher's z

In this section we shall show that as the sample size increases, a sample correlation coefficient tends to be normally distributed. The distribution of a particular function of a sample correlation, Fisher's z [Fisher (1921)], which has a variance approximately independent of the population correlation, tends to normality faster.

We are particularly interested in the sample correlation coefficient

$$r(n) = \frac{A_{ij}(n)}{\sqrt{A_{ii}(n) A_{jj}(n)}} \tag{59}$$

for some i and j, $i \neq j$. This can also be written

$$r(n) = \frac{C_{ij}(n)}{\sqrt{C_{ii}(n) C_{jj}(n)}}, \tag{60}$$

where $C_{gh}(n) = A_{gh}(n)/\sqrt{\sigma_{gg}\sigma_{hh}}$. The set $C_{ii}(n)$, $C_{jj}(n)$, and $C_{ij}(n)$ is distributed like the distinct elements of the matrix

$$\sum_{\alpha=1}^{n} \begin{pmatrix} Z_{i\alpha}^{*} \\ Z_{j\alpha}^{*} \end{pmatrix} \begin{pmatrix} Z_{i\alpha}^{*}, Z_{j\alpha}^{*} \end{pmatrix} = \sum_{\alpha=1}^{n} \begin{pmatrix} Z_{i\alpha}/\sqrt{\sigma_{ii}} \\ Z_{j\alpha}/\sqrt{\sigma_{jj}} \end{pmatrix} \begin{pmatrix} Z_{i\alpha}/\sqrt{\sigma_{ii}}, Z_{j\alpha}/\sqrt{\sigma_{jj}} \end{pmatrix}, \tag{61}$$

where the $(Z_{i\alpha}^{*}, Z_{j\alpha}^{*})$ are independent, each with distribution $N\left[\begin{pmatrix} 0 \\ 0 \end{pmatrix}, \begin{pmatrix} 1 & \rho \\ \rho & 1 \end{pmatrix}\right]$ and $\rho = \sigma_{ij}/\sqrt{\sigma_{ii}\sigma_{jj}}$. Let

$$\boldsymbol{U}(n) = \frac{1}{n} \begin{pmatrix} C_{ii}(n) \\ C_{jj}(n) \\ C_{ij}(n) \end{pmatrix}, \tag{62}$$

$$\boldsymbol{b} = \begin{pmatrix} 1 \\ 1 \\ \rho \end{pmatrix}. \tag{63}$$

Then by Theorem 3.4.4 the vector $\sqrt{n}\,[\boldsymbol{U}(n) - \boldsymbol{b}]$ has a limiting normal distribution with mean $\mathbf{0}$ and covariance matrix

$$\begin{pmatrix} 2 & 2\rho^2 & 2\rho \\ 2\rho^2 & 2 & 2\rho \\ 2\rho & 2\rho & 1+\rho^2 \end{pmatrix}. \tag{64}$$

Now we need the general theorem:

THEOREM 4.2.3. *Let $\{\boldsymbol{U}(n)\}$ be a sequence of m-component random vectors and $\boldsymbol{b}$ a fixed vector such that $\sqrt{n}\,[\boldsymbol{U}(n) - \boldsymbol{b}]$ has the limiting distribution $N(\mathbf{0}, \mathbf{T})$ as $n \to \infty$. Let $\boldsymbol{f}(\boldsymbol{u})$ be a vector-valued function of $\boldsymbol{u}$ such that each component $f_j(\boldsymbol{u})$ has a nonzero differential at $\boldsymbol{u} = \boldsymbol{b}$, and let $\partial f_j(\boldsymbol{u})/\partial u_i|\boldsymbol{u} = \boldsymbol{b}$ be the i, jth component of $\boldsymbol{\Phi}_b$. Then $\sqrt{n}\,\{\boldsymbol{f}[\boldsymbol{u}(n)] - \boldsymbol{f}(\boldsymbol{b})\}$ has the limiting distribution $N(\mathbf{0}, \boldsymbol{\Phi}_b'\mathbf{T}\boldsymbol{\Phi}_b)$.*

PROOF. See Serfling (1980), Section 3.3, or Rao (1973), Section 6a.2. A function $g(\boldsymbol{u})$ is said to have a differential at $\boldsymbol{b}$ or to be totally differentiable at $\boldsymbol{b}$ if the partial derivatives $\partial g(\boldsymbol{u})/\partial u_i$ exist at $\boldsymbol{u} = \boldsymbol{b}$ and for every $\varepsilon > 0$ there exists a neighborhood $N_\varepsilon(\boldsymbol{b})$ such that

$$\left| g(\boldsymbol{u}) - g(\boldsymbol{b}) - \sum_{i=1}^{m} \frac{\partial g(\boldsymbol{u})}{\partial u_i}(u_i - b_i) \right| \le \varepsilon \|\boldsymbol{u} - \boldsymbol{b}\| \quad \text{for all } \boldsymbol{u} \in N_\varepsilon(\boldsymbol{b}). \tag{65}$$

Q.E.D.

It is clear that $\boldsymbol{U}(n)$ defined by (62) with $\boldsymbol{b}$ and $\mathbf{T}$ defined by (63) and (64), respectively, satisfy the conditions of the theorem. The function

$$r = \frac{u_3}{\sqrt{u_1 u_2}} = u_3 u_1^{-\frac{1}{2}} u_2^{-\frac{1}{2}} \tag{66}$$

satisfies the conditions; the elements of $\boldsymbol{\Phi}_b$ are

$$\begin{aligned} \left.\frac{\partial r}{\partial u_1}\right|_{\boldsymbol{u}=\boldsymbol{b}} &= -\tfrac{1}{2} u_3 u_1^{-\frac{3}{2}} u_2^{-\frac{1}{2}}\Big|_{\boldsymbol{u}=\boldsymbol{b}} = -\tfrac{1}{2}\rho, \\ \left.\frac{\partial r}{\partial u_2}\right|_{\boldsymbol{u}=\boldsymbol{b}} &= -\tfrac{1}{2} u_3 u_1^{-\frac{1}{2}} u_2^{-\frac{3}{2}}\Big|_{\boldsymbol{u}=\boldsymbol{b}} = -\tfrac{1}{2}\rho, \\ \left.\frac{\partial r}{\partial u_3}\right|_{\boldsymbol{u}=\boldsymbol{b}} &= u_1^{-\frac{1}{2}} u_2^{-\frac{1}{2}}\Big|_{\boldsymbol{u}=\boldsymbol{b}} = 1, \end{aligned} \tag{67}$$

and $f(\boldsymbol{b}) = \rho$. The variance of the limiting distribution of $\sqrt{n}\,[r(n) - \rho]$ is

$$(68)\quad \left(-\tfrac{1}{2}\rho, -\tfrac{1}{2}\rho, 1\right)\begin{pmatrix} 2 & 2\rho^2 & 2\rho \\ 2\rho^2 & 2 & 2\rho \\ 2\rho & 2\rho & 1+\rho^2 \end{pmatrix}\begin{pmatrix} -\tfrac{1}{2}\rho \\ -\tfrac{1}{2}\rho \\ 1 \end{pmatrix}$$

$$= \left(\rho - \rho^3, \rho - \rho^3, 1 - \rho^2\right)\begin{pmatrix} -\tfrac{1}{2}\rho \\ -\tfrac{1}{2}\rho \\ 1 \end{pmatrix}$$

$$= 1 - 2\rho^2 + \rho^4$$

$$= \left(1 - \rho^2\right)^2.$$

Thus we obtain the following:

THEOREM 4.2.4. *If $r(n)$ is the sample correlation coefficient of a sample of N $(= n + 1)$ from a normal distribution with correlation ρ, then $\sqrt{n}\,[r(n) - \rho]/(1 - \rho^2)$ [or $\sqrt{N}\,[r(n) - \rho]/(1 - \rho^2)$] has the limiting distribution $N(0, 1)$.*

It is clear from Theorem 4.2.3 that if $f(x)$ is differentiable at $x = \rho$, then $\sqrt{n}\,[f(r) - f(\rho)]$ is asymptotically normally distributed with mean zero and variance

$$\left(\left.\frac{\partial f}{\partial x}\right|_{x=\rho}\right)^2 \left(1 - \rho^2\right)^2.$$

A useful function to consider is one whose asymptotic variance is constant (here unity) independent of the parameter ρ. This function satisfies the equation

$$(69)\qquad f'(\rho) = \frac{1}{1-\rho^2} = \frac{1}{2}\left(\frac{1}{1+\rho} + \frac{1}{1-\rho}\right).$$

Thus $f(\rho)$ is taken as $\tfrac{1}{2}[\log(1 + \rho) - \log(1 - \rho)] = \tfrac{1}{2}\log[(1 + \rho)/(1 - \rho)]$. The so-called Fisher's z is

$$(70)\qquad z = \tfrac{1}{2}\log\frac{1+r}{1-r}.$$

Let

$$(71)\qquad \zeta = \tfrac{1}{2}\log\frac{1+\rho}{1-\rho}.$$

THEOREM 4.2.5. *Let z be defined by* (70), *where r is the correlation coefficient of a sample of N $(= n + 1)$ from a bivariate normal distribution with correlation ρ; let ζ be defined by* (71). *Then $\sqrt{n}(z - \zeta)$ has a limiting normal distribution with mean* 0 *and variance* 1.

It can be shown that to a closer approximation

$$\mathscr{E}z \sim \zeta + \frac{\rho}{2n}, \tag{72}$$

$$\mathscr{E}(z - \zeta)^2 \sim \frac{1}{n-2} \sim \mathscr{E}\left(z - \zeta - \frac{\rho}{2n}\right)^2. \tag{73}$$

The latter follows from

$$\mathscr{E}(z - \zeta)^2 = \frac{1}{n} + \frac{8 - \rho^2}{4n^2} + \cdots \tag{74}$$

and holds good for ρ^2/n^2 small. Hotelling (1953) gives moments of z to order n^{-3}. An important property of Fisher's z is that the approach to normality is much more rapid than for r. David (1938) makes some comparisons between the tabulated probabilities and the probabilities computed by assuming z is normally distributed. She recommends that for $N > 25$ one take z as normally distributed with mean and variance given by (72) and (73). Konishi (1978a, 1978b, 1979) has also studied z. [Ruben (1966) has suggested an alternative approach, which is more complicated, but possibly more accurate.]

We shall now indicate how Theorem 4.2.5 can be used.

(*a*) Suppose we wish to test the hypothesis $\rho = \rho_0$ on the basis of a sample of N against the alternatives $\rho \neq \rho_0$. We compute r and then z by (70). Let

$$\zeta_0 = \tfrac{1}{2}\log\frac{1+\rho_0}{1-\rho_0}. \tag{75}$$

Then a region of rejection at the 5% significance level is

$$\sqrt{N-3}\,|z - \zeta_0| > 1.96. \tag{76}$$

A better region is

$$\sqrt{N-3}\,|z - \zeta_0 - \tfrac{1}{2}\rho_0/(N-1)| > 1.96. \tag{77}$$

(*b*) Suppose we have a sample of N_1 from one population and a sample of N_2 from a second population. How do we test the hypothesis that the two correlation coefficients are equal, $\rho_1 = \rho_2$? From Theorem 4.2.5 we know that if the null hypothesis is true then $z_1 - z_2$ [where z_1 and z_2 are defined by (70) for the two sample correlation coefficients] is asymptotically normally distributed with mean 0 and variance $1/(N_1 - 3) + 1/(N_2 - 3)$. As a critical region of size 5%, we use

$$\frac{|z_1 - z_2|}{\sqrt{1/(N_1 - 3) + 1/(N_2 - 3)}} > 1.96. \tag{78}$$

(*c*) Under the conditions of (*b*) assume that $\rho_1 = \rho_2 = \rho$. How do we use the results of both samples to give a joint estimate of ρ? Since z_1 and z_2 have variances $1/(N_1 - 3)$ and $1/(N_2 - 3)$, respectively, we can estimate ζ by

$$\frac{(N_1 - 3)z_1 + (N_2 - 3)z_2}{N_1 + N_2 - 6} \tag{79}$$

and convert this to an estimate of ρ by the inverse of (70).

(*d*) Let r be the sample correlation from N observations. How do we obtain a confidence interval for ρ? We know that approximately

$$\Pr\{-1.96 \le \sqrt{N-3}\,(z - \zeta) \le 1.96\} = 0.95. \tag{80}$$

From this we deduce that $[-1.96/\sqrt{N-3} + z, 1.96/\sqrt{N-3} + z]$ is a confidence interval for ζ. From this we obtain an interval for ρ using the fact $\rho = \tanh \zeta = (e^{\zeta} - e^{-\zeta})/(e^{\zeta} + e^{-\zeta})$, which is a monotonic transformation. Thus a 95% confidence interval is

$$\tanh\left(z - 1.96/\sqrt{N-3}\right) \le \rho \le \tanh\left(z + 1.96/\sqrt{N-3}\right). \tag{81}$$

Recently, the bootstrap method has been developed to assess the variability of a sample quantity. See Efron (1982). We shall illustrate the method on the sample correlation coefficient, but it can be applied to other quantities studied in this book.

Suppose $\boldsymbol{x}_1, \dots, \boldsymbol{x}_N$ is a sample from some bivariate population not necessarily normal. The approach of the bootstrap is to consider these N vectors as a finite population of size N; a random vector $\boldsymbol{X}$ has the (discrete) probability

$$\Pr\{\boldsymbol{X} = \boldsymbol{x}_\alpha\} = \frac{1}{N}, \qquad \alpha = 1, \dots, N. \tag{82}$$

A random sample of size N drawn from this finite population has a probability distribution, and the correlation coefficient calculated from such a sample has a (discrete) probability distribution, say $p_N(r)$. The bootstrap proposes to use this distribution in place of the unobtainable distribution of the correlation coefficient of random samples from the parent population. However, it is prohibitively expensive to compute; instead $p_N(r)$ is estimated by the empirical distribution of r calculated from a large number of random samples from (82). Diaconis and Efron (1983) have given an example of $N = 15$; they find the empirical distribution closely resembles the actual distribution of r (essentially obtainable in this special case). An advantage of this approach is that it is not necessary to assume knowledge of the parent population; a disadvantage is the massive computation.

4.3. PARTIAL CORRELATION COEFFICIENTS; CONDITIONAL DISTRIBUTIONS

4.3.1. Estimation of Partial Correlation Coefficients

Partial correlation coefficients in normal distributions are correlation coefficients in conditional distributions. It was shown in Section 2.5 that if $\boldsymbol{X}$ is distributed according to $N(\boldsymbol{\mu}, \boldsymbol{\Sigma})$, where

$$(1) \qquad \boldsymbol{X} = \begin{pmatrix} \boldsymbol{X}^{(1)} \\ \boldsymbol{X}^{(2)} \end{pmatrix}, \qquad \boldsymbol{\mu} = \begin{pmatrix} \boldsymbol{\mu}^{(1)} \\ \boldsymbol{\mu}^{(2)} \end{pmatrix}, \qquad \boldsymbol{\Sigma} = \begin{pmatrix} \boldsymbol{\Sigma}_{11} & \boldsymbol{\Sigma}_{12} \\ \boldsymbol{\Sigma}_{21} & \boldsymbol{\Sigma}_{22} \end{pmatrix},$$

then the conditional distribution of $\boldsymbol{X}^{(1)}$ given $\boldsymbol{X}^{(2)} = \boldsymbol{x}^{(2)}$ is $N[\boldsymbol{\mu}^{(1)} + \boldsymbol{\beta}(\boldsymbol{x}^{(2)} - \boldsymbol{\mu}^{(2)}), \boldsymbol{\Sigma}_{11\cdot 2}]$, where

$$(2) \qquad \boldsymbol{\beta} = \boldsymbol{\Sigma}_{12}\boldsymbol{\Sigma}_{22}^{-1},$$

$$(3) \qquad \boldsymbol{\Sigma}_{11\cdot 2} = \boldsymbol{\Sigma}_{11} - \boldsymbol{\Sigma}_{12}\boldsymbol{\Sigma}_{22}^{-1}\boldsymbol{\Sigma}_{21}.$$

The partial correlations of $\boldsymbol{X}^{(1)}$ given $\boldsymbol{x}^{(2)}$ are the correlations calculated in the usual way from $\boldsymbol{\Sigma}_{11\cdot 2}$. In this section we are interested in statistical problems concerning these correlation coefficients.

First we consider the problem of estimation on the basis of a sample of N from $N(\boldsymbol{\mu}, \boldsymbol{\Sigma})$. What are the maximum likelihood estimators of the partial correlations of $\boldsymbol{X}^{(1)}$ (of q components), $\rho_{ij\cdot q+1,\ldots,p}$? We know that the maximum likelihood estimator of $\boldsymbol{\Sigma}$ is $(1/N)\boldsymbol{A}$, where

$$\boldsymbol{A} = \sum_{\alpha=1}^{N} (\boldsymbol{x}_\alpha - \bar{\boldsymbol{x}})(\boldsymbol{x}_\alpha - \bar{\boldsymbol{x}})' \tag{4}$$

$$= \sum_{\alpha=1}^{N} \begin{pmatrix} \boldsymbol{x}_\alpha^{(1)} - \bar{\boldsymbol{x}}^{(1)} \\ \boldsymbol{x}_\alpha^{(2)} - \bar{\boldsymbol{x}}^{(2)} \end{pmatrix} \left(\boldsymbol{x}_\alpha^{(1)\prime} - \bar{\boldsymbol{x}}^{(1)\prime}, \ \boldsymbol{x}_\alpha^{(2)\prime} - \bar{\boldsymbol{x}}^{(2)\prime} \right)$$

$$= \begin{pmatrix} \boldsymbol{A}_{11} & \boldsymbol{A}_{12} \\ \boldsymbol{A}_{21} & \boldsymbol{A}_{22} \end{pmatrix}$$

and $\bar{\boldsymbol{x}} = (1/N)\Sigma_{\alpha=1}^{N}\boldsymbol{x}_\alpha = (\bar{\boldsymbol{x}}^{(1)\prime}\ \bar{\boldsymbol{x}}^{(2)\prime})'$. The correspondence between $\boldsymbol{\Sigma}$ and $\boldsymbol{\Sigma}_{11\cdot 2}$, $\boldsymbol{\beta}$, and $\boldsymbol{\Sigma}_{22}$ is one-to-one by virtue of (2) and (3) and

$$\boldsymbol{\Sigma}_{12} = \boldsymbol{\beta}\boldsymbol{\Sigma}_{22}, \tag{5}$$

$$\boldsymbol{\Sigma}_{11} = \boldsymbol{\Sigma}_{11\cdot 2} + \boldsymbol{\beta}\boldsymbol{\Sigma}_{22}\boldsymbol{\beta}'. \tag{6}$$

We can now apply Corollary 3.2.1 to the effect that maximum likelihood estimators of functions of parameters are those functions of the maximum likelihood estimators of those parameters.

THEOREM 4.3.1. *Let* $\boldsymbol{x}_1, \ldots, \boldsymbol{x}_N$ *be a sample from* $N(\boldsymbol{\mu}, \boldsymbol{\Sigma})$, *where* $\boldsymbol{\mu}$ *and* $\boldsymbol{\Sigma}$ *are partitioned as in* (1). *Define* $\boldsymbol{A}$ *by* (4) *and* $(\bar{\boldsymbol{x}}^{(1)\prime}\ \bar{\boldsymbol{x}}^{(2)\prime}) = (1/N)\Sigma_{\alpha=1}^{N}(\boldsymbol{x}_\alpha^{(1)\prime}\ \boldsymbol{x}_\alpha^{(2)\prime})$. *Then the maximum likelihood estimators of* $\boldsymbol{\mu}^{(1)}$, $\boldsymbol{\mu}^{(2)}$, $\boldsymbol{\beta}$, $\boldsymbol{\Sigma}_{11\cdot 2}$, *and* $\boldsymbol{\Sigma}_{22}$ *are* $\hat{\boldsymbol{\mu}}^{(1)} = \bar{\boldsymbol{x}}^{(1)}$, $\hat{\boldsymbol{\mu}}^{(2)} = \bar{\boldsymbol{x}}^{(2)}$,

$$\hat{\boldsymbol{\beta}} = \boldsymbol{A}_{12}\boldsymbol{A}_{22}^{-1}, \qquad \hat{\boldsymbol{\Sigma}}_{11\cdot 2} = \frac{1}{N}\left(\boldsymbol{A}_{11} - \boldsymbol{A}_{12}\boldsymbol{A}_{22}^{-1}\boldsymbol{A}_{21}\right), \tag{7}$$

and $\hat{\boldsymbol{\Sigma}}_{22} = (1/N)\boldsymbol{A}_{22}$, *respectively.*

In turn, Corollary 3.2.1 can be used to obtain the maximum likelihood estimators of $\boldsymbol{\mu}^{(1)}$, $\boldsymbol{\mu}^{(2)}$, $\boldsymbol{\beta}$, $\boldsymbol{\Sigma}_{22}$, $\sigma_{ii\cdot q+1,\ldots,p}$, $i = 1,\ldots,q$, and $\rho_{ij\cdot q+1,\ldots,p}$, $i, j = 1,\ldots,q$. It follows that the maximum likelihood estimators of the partial correlation coefficients are

$$\hat{\rho}_{ij\cdot q+1,\ldots,p} = \frac{\hat{\sigma}_{ij\cdot q+1,\ldots,p}}{\sqrt{\hat{\sigma}_{ii\cdot q+1,\ldots,p}\hat{\sigma}_{jj\cdot q+1,\ldots,p}}}, \qquad i, j = 1,\ldots,q, \tag{8}$$

where $\hat{\sigma}_{ij\cdot q+1,\ldots,p}$ is the i, jth element of $\hat{\boldsymbol{\Sigma}}_{11\cdot 2}$.

THEOREM 4.3.2. *Let $x_1, \ldots, x_N$ be a sample of N from $N(\mu, \Sigma)$. The maximum likelihood estimators of $\rho_{ij \cdot q+1, \ldots, p}$, the partial correlations of the first q components conditional on the last $p - q$ components, are given by*

$$\hat{\rho}_{ij \cdot q+1, \ldots, p} = \frac{a_{ij \cdot q+1, \ldots, p}}{\sqrt{a_{ii \cdot q+1, \ldots, p} a_{jj \cdot q+1, \ldots, p}}}, \qquad i, j = 1, \ldots, q, \tag{9}$$

where

$$(a_{ij \cdot q+1, \ldots, p}) = A_{11} - A_{12} A_{22}^{-1} A_{21} = A_{11 \cdot 2}. \tag{10}$$

The estimator $\hat{\rho}_{ij \cdot q+1, \ldots, p}$, denoted by $r_{ij \cdot q+1, \ldots, p}$, is called the *sample partial correlation coefficient between X_i and X_j holding $X_{q+1}, \ldots, X_p$ fixed*. It is also called the sample partial correlation coefficient between X_i and X_j having taken account of $X_{q+1}, \ldots, X_p$. Note that the calculations can be done in terms of (r_{ij}).

The matrix $A_{11 \cdot 2}$ can also be represented as

$$\begin{aligned} A_{11 \cdot 2} &= \sum_{\alpha=1}^{N} \left[x_\alpha^{(1)} - \bar{x}^{(1)} - \hat{\beta}\left(x_\alpha^{(2)} - \bar{x}^{(2)} \right) \right] \left[x_\alpha^{(1)} - \bar{x}^{(1)} - \hat{\beta}\left(x_\alpha^{(2)} - \bar{x}^{(2)} \right) \right]' \\ &= A_{11} - \hat{\beta} A_{22} \hat{\beta}'. \end{aligned} \tag{11}$$

The vector $x_\alpha^{(1)} - \bar{x}^{(1)} - \hat{\beta}(x_\alpha^{(2)} - \bar{x}^{(2)})$ is the residual of $x_\alpha^{(1)}$ from its regression on $x_\alpha^{(2)}$ and 1. The partial correlations are simple correlations between these residuals. The definition can be used also when the distributions involved are not normal.

Two geometric interpretations of the above theory can be given. In p-dimensional space $x_1, \ldots, x_N$ represent N points. The sample regression function

$$x^{(1)} = \bar{x}^{(1)} + \hat{\beta}(x^{(2)} - \bar{x}^{(2)}) \tag{12}$$

is a $(p-q)$-dimensional hyperplane which is the intersection of q $(p-1)$-dimensional hyperplanes,

$$x_i = \bar{x}_i + \sum_{j=q+1}^{p} \hat{\beta}_{ij}(x_j - \bar{x}_j), \qquad i = 1, \ldots, q, \tag{13}$$

where x_i, x_j are running variables. Here $\hat{\beta}_{ij}$ is an element of $\hat{\beta} = \hat{\Sigma}_{12} \hat{\Sigma}_{22}^{-1} = A_{12} A_{22}^{-1}$. The ith row of $\hat{\beta}$ is $(\hat{\beta}_{i, q+1}, \ldots, \hat{\beta}_{ip})$. Each right-hand side of (13) is the least squares "regression" function of x_i on $x_{q+1}, \ldots, x_p$; that is, if we project the points $\boldsymbol{x}_1, \dots, \boldsymbol{x}_N$ on the coordinate hyperplane of $x_i, x_{q+1}, \dots, x_p$, then (13) is the regression plane. The point with coordinates

$$(14) \qquad \begin{aligned} x_i &= \bar{x}_i + \sum_{j=q+1}^{p} \hat{\beta}_{ij}(x_{j\alpha} - \bar{x}_j), & i &= 1, \dots, q, \\ x_j &= x_{j\alpha}, & j &= q+1, \dots, p, \end{aligned}$$

is on the hyperplane (13). The difference in the ith coordinate of $\boldsymbol{x}_\alpha$ and the point (14) is $y_{i\alpha} = x_{i\alpha} - [\bar{x}_i + \sum_{j=q+1}^{p} \hat{\beta}_{ij}(x_{j\alpha} - \bar{x}_j)]$ for $i = 1, \dots, q$ and 0 for the other coordinates. Let $\boldsymbol{y}'_\alpha = (y_{1\alpha}, \dots, y_{q\alpha})$. These points can be represented as N points in a q-dimensional space. Then $\boldsymbol{A}_{11 \cdot 2} = \sum_{\alpha=1}^{N} \boldsymbol{y}_\alpha \boldsymbol{y}'_\alpha$.

We can also interpret the sample as p points in N-space (Figure 4.4). Let $\boldsymbol{u}_j = (x_{j1}, \dots, x_{jN})'$ be the jth point, and let $\boldsymbol{\varepsilon} = (1, \dots, 1)'$ be another point. The point with coordinates $\bar{x}_i, \dots, \bar{x}_i$ is $\bar{x}_i \boldsymbol{\varepsilon}$. The projection of $\boldsymbol{u}_i$ on the hyperplane spanned by $\boldsymbol{u}_{q+1}, \dots, \boldsymbol{u}_p, \boldsymbol{\varepsilon}$ is

$$(15) \qquad \hat{\boldsymbol{u}}_i = \bar{x}_i \boldsymbol{\varepsilon} + \sum_{j=q+1}^{p} \hat{\beta}_{ij}(\boldsymbol{u}_j - \bar{x}_j \boldsymbol{\varepsilon});$$

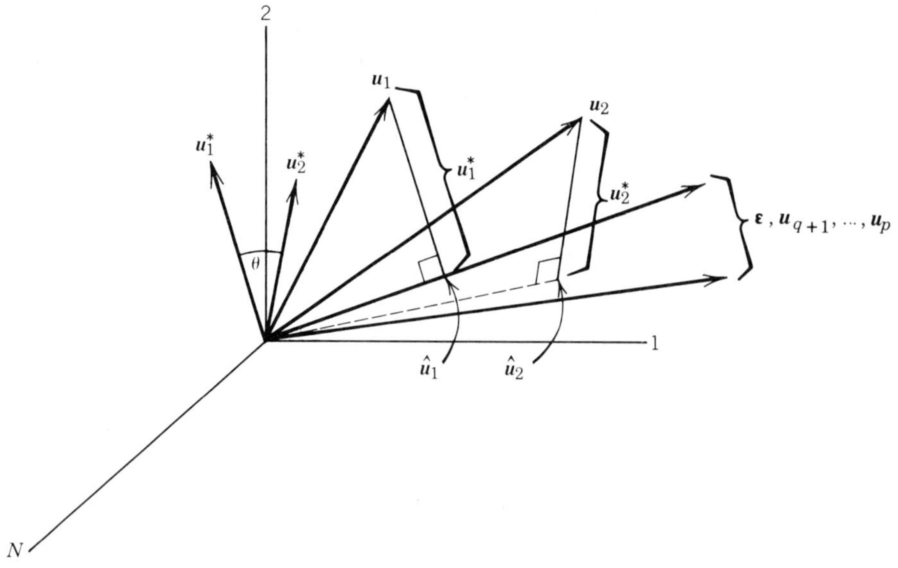

Figure 4.4

this is the point on the hyperplane that is at a minimum distance from $\boldsymbol{u}_i$. Let $\boldsymbol{u}_i^*$ be the vector from $\hat{\boldsymbol{u}}_i$ to $\boldsymbol{u}_i$, that is, $\boldsymbol{u}_i - \hat{\boldsymbol{u}}_i$, or, equivalently, this vector translated so that one endpoint is at the origin. The set of vectors $\boldsymbol{u}_1^*, \ldots, \boldsymbol{u}_q^*$ are the projections of $\boldsymbol{u}_1, \ldots, \boldsymbol{u}_q$ on the hyperplane orthogonal to $\boldsymbol{u}_{q+1}, \ldots, \boldsymbol{u}_p, \boldsymbol{\varepsilon}$. Then $\boldsymbol{u}_i^{*\prime}\boldsymbol{u}_i^* = a_{ii\cdot q+1,\ldots,p}$, the length squared of $\boldsymbol{u}_i^*$ (i.e., the square of the distance of $\boldsymbol{u}$ from $\hat{\boldsymbol{u}}_i$). Then $\boldsymbol{u}_i^{*\prime}\boldsymbol{u}_j^* / \sqrt{\boldsymbol{u}_i^{*\prime}\boldsymbol{u}_i^*\boldsymbol{u}_j^{*\prime}\boldsymbol{u}_j^*} = r_{ij\cdot q+1,\ldots,p}$ is the cosine of the angle between $\boldsymbol{u}_i^*$ and $\boldsymbol{u}_j^*$.

As an example of the use of partial correlations we consider some data [Hooker (1907)] on yield of hay (X_1) in hundredweights per acre, spring rainfall (X_2) in inches, and accumulated temperature above 42°F in the spring (X_3) for an English area over 20 years. The estimates of μ_i, σ_i $(= \sqrt{\sigma_{ii}})$, and ρ_{ij} are

$$\hat{\mu} = \bar{x} = \begin{pmatrix} 28.02 \\ 4.91 \\ 594 \end{pmatrix},$$

(16)
$$\begin{pmatrix} \hat{\sigma}_1 \\ \hat{\sigma}_2 \\ \hat{\sigma}_3 \end{pmatrix} = \begin{pmatrix} 4.42 \\ 1.10 \\ 85 \end{pmatrix},$$

$$\begin{pmatrix} 1 & \hat{\rho}_{12} & \hat{\rho}_{13} \\ \hat{\rho}_{21} & 1 & \hat{\rho}_{23} \\ \hat{\rho}_{31} & \hat{\rho}_{32} & 1 \end{pmatrix} = \begin{pmatrix} 1.00 & 0.80 & -0.40 \\ 0.80 & 1.00 & -0.56 \\ -0.40 & -0.56 & 1.00 \end{pmatrix}.$$

From the correlations we observe that yield and rainfall are positively related, yield and temperature are negatively related, and rainfall and temperature are negatively related. What interpretation is to be given to the apparent negative relation between yield and temperature? Does high temperature tend to cause low yield or is high temperature associated with low rainfall and hence with low yield? To answer this question we consider the correlation between yield and temperature when rainfall is held fixed; that is, we use the data given above to estimate the partial correlation between X_1 and X_3 with X_2 held fixed. It is[†]

(17)
$$r_{13\cdot 2} = \frac{\hat{\sigma}_{13\cdot 2}}{\sqrt{\hat{\sigma}_{11\cdot 2}\hat{\sigma}_{33\cdot 2}}} = 0.097.$$

[†] We compute with $\hat{\boldsymbol{\Sigma}}$ as if it were $\boldsymbol{\Sigma}$.

Thus, if the effect of rainfall is removed, yield and temperature are positively correlated. The conclusion is that both high rainfall and high temperature increase hay yield, but in most years high rainfall occurs with low temperature and vice versa.

4.3.2. The Distribution of the Sample Partial Correlation Coefficient

In order to test a hypothesis about a population partial correlation coefficient we want the distribution of the sample partial correlation coefficient. The partial correlations are computed from $\boldsymbol{A}_{11\cdot 2} = \boldsymbol{A}_{11} - \boldsymbol{A}_{12}\boldsymbol{A}_{22}^{-1}\boldsymbol{A}_{21}$ (as indicated in Theorem 4.3.1) in the same way that correlations are computed from $\boldsymbol{A}$. To obtain the distribution of a simple correlation we showed that $\boldsymbol{A}$ was distributed as $\sum_{\alpha=1}^{N-1}\boldsymbol{Z}_\alpha \boldsymbol{Z}_\alpha'$, where $\boldsymbol{Z}_1, \dots, \boldsymbol{Z}_{N-1}$ are distributed independently according to $N(\boldsymbol{0}, \boldsymbol{\Sigma})$ and independently of $\bar{\boldsymbol{X}}$ (Theorem 3.3.2). Here we want to show that $\boldsymbol{A}_{11\cdot 2}$ is distributed as $\sum_{\alpha=1}^{N-1-(p-q)}\boldsymbol{U}_\alpha \boldsymbol{U}_\alpha'$, where $\boldsymbol{U}_1, \dots, \boldsymbol{U}_{N-1-(p-q)}$ are distributed independently according to $N(\boldsymbol{0}, \boldsymbol{\Sigma}_{11\cdot 2})$ and independently of $\hat{\boldsymbol{\beta}}$. The distribution of a partial correlation coefficient will follow from this characterization of the distribution of $\boldsymbol{A}_{11\cdot 2}$. We state the theorem in a general form; it will be used in Chapter 8, where we treat regression in detail. The following corollary applies it to $\boldsymbol{A}_{11\cdot 2}$, expressed in terms of residuals.

THEOREM 4.3.3. *Suppose* $\boldsymbol{Y}_1, \dots, \boldsymbol{Y}_m$ *are independent with* $\boldsymbol{Y}_\alpha$ *distributed according to* $N(\boldsymbol{\Gamma} \boldsymbol{w}_\alpha, \boldsymbol{\Phi})$, *where* $\boldsymbol{w}_\alpha$ *is an r-component vector. Let* $\boldsymbol{H} = \sum_{\alpha=1}^{m}\boldsymbol{w}_\alpha \boldsymbol{w}_\alpha'$, *assumed nonsingular,* $\boldsymbol{G} = \sum_{\alpha=1}^{m}\boldsymbol{Y}_\alpha \boldsymbol{w}_\alpha' \boldsymbol{H}^{-1}$, *and*

$$\boldsymbol{C} = \sum_{\alpha=1}^{m}(\boldsymbol{Y}_\alpha - \boldsymbol{G}\boldsymbol{w}_\alpha)(\boldsymbol{Y}_\alpha - \boldsymbol{G}\boldsymbol{w}_\alpha)' = \sum_{\alpha=1}^{m}\boldsymbol{Y}_\alpha \boldsymbol{Y}_\alpha' - \boldsymbol{G}\boldsymbol{H}\boldsymbol{G}'. \tag{18}$$

Then $\boldsymbol{C}$ *is distributed as* $\sum_{\alpha=1}^{m-r}\boldsymbol{U}_\alpha \boldsymbol{U}_\alpha'$, *where* $\boldsymbol{U}_1, \dots, \boldsymbol{U}_{m-r}$ *are independently distributed according to* $N(\boldsymbol{0}, \boldsymbol{\Phi})$ *and independently of* $\boldsymbol{G}$.

PROOF. The rows of $\boldsymbol{Y} = (\boldsymbol{Y}_1, \dots, \boldsymbol{Y}_m)$ are random vectors in an m-dimensional space, and the rows of $\boldsymbol{W} = (\boldsymbol{w}_1, \dots, \boldsymbol{w}_m)$ are fixed vectors in that space. The idea of the proof is to rotate coordinate axes so that the last r coordinate axes are in the space spanned by the rows of $\boldsymbol{W}$. Let $\boldsymbol{E}_2 = \boldsymbol{F}\boldsymbol{W}$, where $\boldsymbol{F}$ is a square matrix such that $\boldsymbol{F}\boldsymbol{H}\boldsymbol{F}' = \boldsymbol{I}$. Then

$$\begin{aligned} \boldsymbol{E}_2 \boldsymbol{E}_2' &= \boldsymbol{F}\boldsymbol{W}\boldsymbol{W}'\boldsymbol{F}' = \boldsymbol{F}\sum_{\alpha=1}^{m}\boldsymbol{w}_\alpha \boldsymbol{w}_\alpha' \boldsymbol{F}' \\ &= \boldsymbol{F}\boldsymbol{H}\boldsymbol{F}' = \boldsymbol{I}. \end{aligned} \tag{19}$$

Thus the m-component rows of $\boldsymbol{E}_2$ are orthogonal and of unit length. It is possible to find an $(m-r)\times m$ matrix $\boldsymbol{E}_1$ such that

$$\boldsymbol{E} = \begin{pmatrix} \boldsymbol{E}_1 \\ \boldsymbol{E}_2 \end{pmatrix} \tag{20}$$

is orthogonal. (See Appendix A, Lemma A.4.2.) Now let $\boldsymbol{U} = \boldsymbol{YE}'$ (i.e., $\boldsymbol{U}_\alpha = \sum_{\beta=1}^m e_{\alpha\beta}\boldsymbol{Y}_\beta$). By Theorem 3.3.1 the columns of $\boldsymbol{U} = (\boldsymbol{U}_1, \dots, \boldsymbol{U}_m)$ are independently and normally distributed, each with covariance matrix $\boldsymbol{\Phi}$. The means are given by

$$\begin{aligned} \mathscr{E}\boldsymbol{U} &= \mathscr{E}\boldsymbol{YE}' = \boldsymbol{\Gamma WE}' \\ &= \boldsymbol{\Gamma F}^{-1}\boldsymbol{E}_2(\boldsymbol{E}_1'\ \boldsymbol{E}_2') \\ &= (\boldsymbol{0}\ \boldsymbol{\Gamma F}^{-1}) \end{aligned} \tag{21}$$

by orthogonality of $\boldsymbol{E}$. To complete the proof we need to show that $\boldsymbol{C}$ transforms to $\sum_{\alpha=1}^{m-r}\boldsymbol{U}_\alpha\boldsymbol{U}_\alpha'$. We have

$$\sum_{\alpha=1}^m \boldsymbol{Y}_\alpha\boldsymbol{Y}_\alpha' = \boldsymbol{YY}' = \boldsymbol{UEE}'\boldsymbol{U}' = \boldsymbol{UU}' = \sum_{\alpha=1}^m \boldsymbol{U}_\alpha\boldsymbol{U}_\alpha'. \tag{22}$$

Note that

$$\begin{aligned} \boldsymbol{G} &= \boldsymbol{YW}'\boldsymbol{H}^{-1} = \boldsymbol{UEE}_2'(\boldsymbol{F}^{-1})'\boldsymbol{F}'\boldsymbol{F} \\ &= \boldsymbol{U}\begin{pmatrix} \boldsymbol{E}_1 \\ \boldsymbol{E}_2 \end{pmatrix}\boldsymbol{E}_2'\boldsymbol{F} \\ &= \boldsymbol{U}\begin{pmatrix} \boldsymbol{0} \\ \boldsymbol{I} \end{pmatrix}\boldsymbol{F} = \boldsymbol{U}^{(2)}\boldsymbol{F}, \end{aligned} \tag{23}$$

where $\boldsymbol{U}^{(2)} = (\boldsymbol{U}_{m-r+1}, \dots, \boldsymbol{U}_m)$. Then

$$\boldsymbol{GHG}' = \boldsymbol{U}^{(2)}\boldsymbol{FHF}'\boldsymbol{U}^{(2)\prime} = \boldsymbol{U}^{(2)}\boldsymbol{U}^{(2)\prime} = \sum_{\alpha=m-r+1}^{m} \boldsymbol{U}_\alpha\boldsymbol{U}_\alpha'. \tag{24}$$

Thus $\boldsymbol{C}$ is

$$\sum_{\alpha=1}^m \boldsymbol{Y}_\alpha\boldsymbol{Y}_\alpha' - \boldsymbol{GHG}' = \sum_{\alpha=1}^m \boldsymbol{U}_\alpha\boldsymbol{U}_\alpha' - \sum_{\alpha=m-r+1}^{m} \boldsymbol{U}_\alpha\boldsymbol{U}_\alpha' = \sum_{\alpha=1}^{m-r} \boldsymbol{U}_\alpha\boldsymbol{U}_\alpha'. \tag{25}$$

This proves the theorem. Q.E.D.

It follows from the above considerations that when $\Gamma = \mathbf{0}$, then $\mathscr{E}\boldsymbol{U} = \mathbf{0}$, and we obtain the following:

COROLLARY 4.3.1. *If* $\Gamma = \mathbf{0}$, *the matrix* $\boldsymbol{GHG}'$ *defined in Theorem* 4.3.3 *is distributed as* $\sum_{\alpha=m-r+1}^{m} \boldsymbol{U}_\alpha \boldsymbol{U}_\alpha'$, *where* $\boldsymbol{U}_{m-r+1}, \dots, \boldsymbol{U}_m$ *are independently distributed, each according to* $N(\mathbf{0}, \boldsymbol{\Phi})$.

We now find the distribution of $\boldsymbol{A}_{11 \cdot 2}$ in the same form. It was shown in Theorem 3.3.1 that $\boldsymbol{A}$ is distributed as $\sum_{\alpha=1}^{N-1} \boldsymbol{Z}_\alpha \boldsymbol{Z}_\alpha'$, where $\boldsymbol{Z}_1, \dots, \boldsymbol{Z}_{N-1}$ are independent, each with distribution $N(\mathbf{0}, \boldsymbol{\Sigma})$. Let $\boldsymbol{Z}_\alpha$ be partitioned into two subvectors of q and $p-q$ components, respectively,

$$\boldsymbol{Z}_\alpha = \begin{pmatrix} \boldsymbol{Z}_\alpha^{(1)} \\ \boldsymbol{Z}_\alpha^{(2)} \end{pmatrix}. \tag{26}$$

Then $\boldsymbol{A}_{ij} = \sum_{\alpha=1}^{N-1} \boldsymbol{Z}_\alpha^{(i)} \boldsymbol{Z}_\alpha^{(j)\prime}$. By Lemma 4.2.1 conditionally on $\boldsymbol{Z}_1^{(2)} = \boldsymbol{z}_1^{(2)}, \dots, \boldsymbol{Z}_{N-1}^{(2)} = \boldsymbol{z}_{N-1}^{(2)}$, the random vectors $\boldsymbol{Z}_1^{(1)}, \dots, \boldsymbol{Z}_{N-1}^{(1)}$ are independently distributed, with $\boldsymbol{Z}_\alpha^{(1)}$ distributed according to $N(\boldsymbol{\beta} \boldsymbol{z}_\alpha^{(2)}, \boldsymbol{\Sigma}_{11 \cdot 2})$, where $\boldsymbol{\beta} = \boldsymbol{\Sigma}_{12} \boldsymbol{\Sigma}_{22}^{-1}$ and $\boldsymbol{\Sigma}_{11 \cdot 2} = \boldsymbol{\Sigma}_{11} - \boldsymbol{\Sigma}_{12} \boldsymbol{\Sigma}_{22}^{-1} \boldsymbol{\Sigma}_{21}$. Now we apply Theorem 4.3.3 with $\boldsymbol{Z}_\alpha^{(1)} = \boldsymbol{Y}_\alpha$, $\boldsymbol{z}_\alpha^{(2)} = \boldsymbol{w}_\alpha$, $N-1 = m$, $p - q = r$, $\boldsymbol{\beta} = \Gamma$, $\boldsymbol{\Sigma}_{11 \cdot 2} = \boldsymbol{\Phi}$, $\boldsymbol{A}_{11} = \sum_{\alpha=1}^{N-1} \boldsymbol{Y}_\alpha \boldsymbol{Y}_\alpha'$, $\boldsymbol{A}_{12} \boldsymbol{A}_{22}^{-1} = \boldsymbol{G}$, $\boldsymbol{A}_{22} = \boldsymbol{H}$. We find that the conditional distribution of $\boldsymbol{A}_{11} - (\boldsymbol{A}_{12} \boldsymbol{A}_{22}^{-1}) \boldsymbol{A}_{22} (\boldsymbol{A}_{22}^{-1} \boldsymbol{A}_{12}') = \boldsymbol{A}_{11 \cdot 2}$ given $\boldsymbol{Z}_\alpha^{(2)} = \boldsymbol{z}_\alpha^{(2)}$, $\alpha = 1, \dots, N-1$, is that of $\sum_{\alpha=1}^{N-1-(p-q)} \boldsymbol{U}_\alpha \boldsymbol{U}_\alpha'$, where $\boldsymbol{U}_1, \dots, \boldsymbol{U}_{N-1-(p-q)}$ are independent, each with distribution $N(\mathbf{0}, \boldsymbol{\Sigma}_{11 \cdot 2})$. Since this distribution does not depend on $\{\boldsymbol{z}_\alpha^{(2)}\}$, we obtain the following theorem:

THEOREM 4.3.4. *The matrix* $\boldsymbol{A}_{11 \cdot 2} = \boldsymbol{A}_{11} - \boldsymbol{A}_{12} \boldsymbol{A}_{22}^{-1} \boldsymbol{A}_{21}$ *is distributed as* $\sum_{\alpha=1}^{N-1-(p-q)} \boldsymbol{U}_\alpha \boldsymbol{U}_\alpha'$, *where* $\boldsymbol{U}_1, \dots, \boldsymbol{U}_{N-1-(p-q)}$ *are independently distributed, each according to* $N(\mathbf{0}, \boldsymbol{\Sigma}_{11 \cdot 2})$, *and independently of* $\boldsymbol{A}_{12}$ *and* $\boldsymbol{A}_{22}$.

COROLLARY 4.3.2. *If* $\boldsymbol{\Sigma}_{12} = \mathbf{0}$ $(\boldsymbol{\beta} = \mathbf{0})$, *then* $\boldsymbol{A}_{11 \cdot 2}$ *is distributed as* $\sum_{\alpha=1}^{N-1-(p-q)} \boldsymbol{U}_\alpha \boldsymbol{U}_\alpha'$ *and* $\boldsymbol{A}_{12} \boldsymbol{A}_{22}^{-1} \boldsymbol{A}_{21}$ *is distributed as* $\sum_{\alpha=N-(p-q)}^{N-1} \boldsymbol{U}_\alpha \boldsymbol{U}_\alpha'$, *where* $\boldsymbol{U}_1, \dots, \boldsymbol{U}_{N-1}$ *are independently distributed, each according to* $N(\mathbf{0}, \boldsymbol{\Sigma}_{11 \cdot 2})$.

Now it follows that the distribution of $r_{ij \cdot q+1, \dots, p}$ based on N observations is the same as that of a simple correlation coefficient based on $N - (p - q)$ observations with a corresponding population correlation value of $\rho_{ij \cdot q+1, \dots, p}$.

THEOREM 4.3.5. *If the* cdf *of* r_{ij} *based on a sample of* N *from a normal distribution with correlation* ρ_{ij} *is denoted by* $F(r|N, \rho_{ij})$, *then the* cdf *of the sample partial correlation* $r_{ij\cdot q+1,\dots,p}$ *based on a sample of* N *from a normal distribution with partial correlation coefficient* $\rho_{ij\cdot q+1,\dots,p}$ *is* $F[r|N - (p-q), \rho_{ij\cdot q+1,\dots,p}]$.

This distribution was derived by Fisher (1924).

4.3.3. Tests of Hypotheses and Confidence Regions for Partial Correlation Coefficients

Since the distribution of a sample partial correlation $r_{ij\cdot q+1,\dots,p}$ based on a sample of N from a distribution with population correlation $\rho_{ij\cdot q+1,\dots,p}$ equal to a certain value, ρ, say, is the same as the distribution of a simple correlation r based on a sample of size $N - (p-q)$ from a distribution with the corresponding population correlation of ρ, all statistical inference procedures for the simple population correlation can be used for the partial correlation. The procedure for the partial correlation is exactly the same except that N is replaced by $N - (p-q)$. To illustrate this rule we give two examples.

Example 1. Suppose that on the basis of a sample of size N we wish to obtain a confidence interval for $\rho_{ij\cdot q+1,\dots,p}$. The sample partial correlation is $r_{ij\cdot q+1,\dots,p}$. The procedure is to use David's charts for $N - (p-q)$. In the example at the end of Section 4.3.1, we might want to find a confidence interval for $\rho_{12\cdot 3}$ with confidence coefficient 0.95. The sample partial correlation is $r_{12\cdot 3} = 0.759$. We use the chart (or table) for $N - (p-q) = 20 - 1 = 19$. The interval is $0.50 < \rho_{12\cdot 3} < 0.88$.

Example 2. Suppose that on the basis of a sample of size N we use Fisher's z for an approximate significance test of $\rho_{ij\cdot q+1,\dots,p} = \rho_0$ against two-sided alternatives. We let

$$z = \tfrac{1}{2}\log\frac{1 + r_{ij\cdot q+1,\dots,p}}{1 - r_{ij\cdot q+1,\dots,p}},$$

$$\zeta_0 = \tfrac{1}{2}\log\frac{1+\rho_0}{1-\rho_0}. \tag{27}$$

Then $\sqrt{N - (p-q) - 3}\,(z - \zeta_0)$ is compared with the significance points of the standardized normal distribution. In the example at the end of Section 4.3.1, we might wish to test the hypothesis $\rho_{13\cdot 2} = 0$ at the 0.05 level. Then $\zeta_0 = 0$ and $\sqrt{20-1-3}\,(0.0973) = 0.3892$. This value is clearly nonsignificant ($|0.3892| < 1.96$), and hence the data do not indicate rejection of the null hypothesis.

To answer the question whether two variables x_1 and x_2 are related when both may be related to a vector $\boldsymbol{x}^{(2)} = (x_3, \dots, x_p)$, two approaches may be used. One is to consider the regression of x_1 on x_2 and $\boldsymbol{x}^{(2)}$ and test whether the regression of x_1 on x_2 is 0. Another is to test whether $\rho_{12\cdot 3,\dots,p} = 0$. Problems 43–47 show that these approaches lead to exactly the same test.

4.4. THE MULTIPLE CORRELATION COEFFICIENT

4.4.1. Estimation of the Multiple Correlation Coefficient

The population multiple correlation between one variate and a set of variates was defined in Section 2.5. For the sake of convenience in this section we shall treat the case of the multiple correlation between X_1 and the vector $\boldsymbol{X}^{(2)} = (X_2, \dots, X_p)'$; we shall not need subscripts on R. The variables can always be numbered so that the desired multiple correlation is this one (any irrelevant variables being omitted). Then the multiple correlation in the population is

$$\bar{R} = \frac{\boldsymbol{\beta}'\boldsymbol{\Sigma}_{22}\boldsymbol{\beta}}{\sqrt{\sigma_{11}\boldsymbol{\beta}'\boldsymbol{\Sigma}_{22}\boldsymbol{\beta}}} = \sqrt{\frac{\boldsymbol{\beta}'\boldsymbol{\Sigma}_{22}\boldsymbol{\beta}}{\sigma_{11}}} = \sqrt{\frac{\boldsymbol{\sigma}_{(1)}'\boldsymbol{\Sigma}_{22}^{-1}\boldsymbol{\sigma}_{(1)}}{\sigma_{11}}}, \tag{1}$$

where $\boldsymbol{\beta}$, $\boldsymbol{\sigma}_{(1)}$, and $\boldsymbol{\Sigma}_{22}$ are defined by

$$\boldsymbol{\Sigma} = \begin{pmatrix} \sigma_{11} & \boldsymbol{\sigma}_{(1)}' \\ \boldsymbol{\sigma}_{(1)} & \boldsymbol{\Sigma}_{22} \end{pmatrix}, \tag{2}$$

$$\boldsymbol{\beta} = \boldsymbol{\Sigma}_{22}^{-1}\boldsymbol{\sigma}_{(1)}. \tag{3}$$

Given a sample $\boldsymbol{x}_1, \dots, \boldsymbol{x}_N$ $(N > p)$ we estimate $\boldsymbol{\Sigma}$ by $\boldsymbol{S} = [N/(N-1)]\hat{\boldsymbol{\Sigma}}$ or

$$\hat{\boldsymbol{\Sigma}} = \frac{1}{N}\boldsymbol{A} = \frac{1}{N}\sum_{\alpha=1}^{N}(\boldsymbol{x}_\alpha - \bar{\boldsymbol{x}})(\boldsymbol{x}_\alpha - \bar{\boldsymbol{x}})' = \begin{pmatrix} \hat{\sigma}_{11} & \hat{\boldsymbol{\sigma}}_{(1)}' \\ \hat{\boldsymbol{\sigma}}_{(1)} & \hat{\boldsymbol{\Sigma}}_{22} \end{pmatrix}, \tag{4}$$

and we estimate $\boldsymbol{\beta}$ by $\hat{\boldsymbol{\beta}} = \hat{\boldsymbol{\Sigma}}_{22}^{-1}\hat{\boldsymbol{\sigma}}_{(1)} = \boldsymbol{A}_{22}^{-1}\boldsymbol{a}_{(1)}$. We define the *sample multiple correlation coefficient* by

$$(5) \qquad R = \sqrt{\frac{\hat{\boldsymbol{\beta}}'\hat{\boldsymbol{\Sigma}}_{22}\hat{\boldsymbol{\beta}}}{\hat{\sigma}_{11}}} = \sqrt{\frac{\hat{\boldsymbol{\sigma}}_{(1)}'\hat{\boldsymbol{\Sigma}}_{22}^{-1}\hat{\boldsymbol{\sigma}}_{(1)}}{\hat{\sigma}_{11}}} = \sqrt{\frac{\boldsymbol{a}_{(1)}'\boldsymbol{A}_{22}^{-1}\boldsymbol{a}_{(1)}}{a_{11}}}.$$

That this is the maximum likelihood estimator of $\overline{R}$ is justified by Corollary 3.2.1 since we can define $\overline{R}$, $\boldsymbol{\sigma}_{(1)}$, $\boldsymbol{\Sigma}_{22}$ as a one-to-one transformation of $\boldsymbol{\Sigma}$. Another expression for R [see (16) of Section 2.5] follows from

$$(6) \qquad 1 - R^2 = \frac{|\hat{\boldsymbol{\Sigma}}|}{\hat{\sigma}_{11}|\hat{\boldsymbol{\Sigma}}_{22}|} = \frac{|\boldsymbol{A}|}{a_{11}|\boldsymbol{A}_{22}|}.$$

The quantities R and $\hat{\boldsymbol{\beta}}$ have properties in the sample that are similar to those $\overline{R}$ and $\boldsymbol{\beta}$ have in the population. We have analogs of Theorems 2.5.2, 2.5.3, and 2.5.4. Let $\hat{x}_{1\alpha} = \bar{x}_1 + \hat{\boldsymbol{\beta}}'(\boldsymbol{x}_\alpha^{(2)} - \bar{\boldsymbol{x}}^{(2)})$ and $x_{1\alpha}^* = x_{1\alpha} - \hat{x}_{1\alpha}$ be the residual.

THEOREM 4.4.1. *The residuals $x_{1\alpha}^*$ are uncorrelated in the sample with the components of $\boldsymbol{x}_\alpha^{(2)}$, $\alpha = 1, \dots, N$. For every vector $\boldsymbol{a}$*

$$(7) \qquad \sum_{\alpha=1}^{N}\left[x_{1\alpha} - \bar{x}_1 - \hat{\boldsymbol{\beta}}'\left(\boldsymbol{x}_\alpha^{(2)} - \bar{\boldsymbol{x}}^{(2)}\right)\right]^2 \le \sum_{\alpha=1}^{N}\left[x_{1\alpha} - \bar{x}_1 - \boldsymbol{a}'\left(\boldsymbol{x}_\alpha^{(2)} - \bar{\boldsymbol{x}}^{(2)}\right)\right]^2.$$

The sample correlation between $x_{1\alpha}$ and $\boldsymbol{a}'\boldsymbol{x}_\alpha^{(2)}$, $\alpha = 1, \dots, N$, is maximized for $\boldsymbol{a} = \hat{\boldsymbol{\beta}}$ and that maximum correlation is R.

PROOF. Since the sample mean of the residuals is 0, the vector of sample covariances between $x_{1\alpha}^*$ and $\boldsymbol{x}_\alpha^{(2)}$ is proportional to

$$(8) \qquad \sum_{\alpha=1}^{N}\left[(x_{1\alpha} - \bar{x}_1) - \hat{\boldsymbol{\beta}}'\left(\boldsymbol{x}_\alpha^{(2)} - \bar{\boldsymbol{x}}^{(2)}\right)\right]\left(\boldsymbol{x}_\alpha^{(2)} - \bar{\boldsymbol{x}}^{(2)}\right)' = \boldsymbol{a}_{(1)}' - \hat{\boldsymbol{\beta}}'\boldsymbol{A}_{22} = \boldsymbol{0}.$$

The right-hand side of (7) can be written as the left-hand side plus

$$(9) \qquad \sum_{\alpha=1}^{N}\left[(\hat{\boldsymbol{\beta}} - \boldsymbol{a})'\left(\boldsymbol{x}_\alpha^{(2)} - \bar{\boldsymbol{x}}^{(2)}\right)\right]^2$$

$$= (\hat{\boldsymbol{\beta}} - \boldsymbol{a})'\sum_{\alpha=1}^{N}\left(\boldsymbol{x}_\alpha^{(2)} - \bar{\boldsymbol{x}}^{(2)}\right)\left(\boldsymbol{x}_\alpha^{(2)} - \bar{\boldsymbol{x}}^{(2)}\right)'(\hat{\boldsymbol{\beta}} - \boldsymbol{a}),$$

which is 0 if and only if $\boldsymbol{a} = \hat{\boldsymbol{\beta}}$. To prove the third assertion we consider the vector $\boldsymbol{a}$ for which $\sum_{\alpha=1}^N [\boldsymbol{a}'(\boldsymbol{x}_\alpha^{(2)} - \bar{\boldsymbol{x}}^{(2)})]^2 = \sum_{\alpha=1}^N [\hat{\boldsymbol{\beta}}'(\boldsymbol{x}_\alpha^{(2)} - \bar{\boldsymbol{x}}^{(2)})]^2$ since the correlation is unchanged when the linear function is multiplied by a positive constant. From (7) we obtain

$$(10) \quad a_{11} - 2\sum_{\alpha=1}^N (x_{1\alpha} - \bar{x}_1)\hat{\boldsymbol{\beta}}'(\boldsymbol{x}_\alpha^{(2)} - \bar{\boldsymbol{x}}^{(2)}) + \sum_{\alpha=1}^N [\hat{\boldsymbol{\beta}}'(\boldsymbol{x}_\alpha^{(2)} - \bar{\boldsymbol{x}}^{(2)})]^2$$
$$\leq a_{11} - 2\sum_{\alpha=1}^N (x_{1\alpha} - \bar{x}_1)\boldsymbol{a}'(\boldsymbol{x}_\alpha^{(2)} - \bar{\boldsymbol{x}}^{(2)}) + \sum_{\alpha=1}^N [\boldsymbol{a}'(\boldsymbol{x}_\alpha^{(2)} - \bar{\boldsymbol{x}}^{(2)})]^2,$$

from which we deduce

$$(11) \quad \frac{\sum_{\alpha=1}^N (x_{1\alpha} - \bar{x}_1)(\boldsymbol{x}_\alpha^{(2)} - \bar{\boldsymbol{x}}^{(2)})'\boldsymbol{a}}{\sqrt{a_{11}}\sqrt{\sum_{\alpha=1}^N [\boldsymbol{a}'(\boldsymbol{x}_\alpha^{(2)} - \bar{\boldsymbol{x}}^{(2)})]^2}} \leq \frac{\sum_{\alpha=1}^N (x_{1\alpha} - \bar{x}_1)(\boldsymbol{x}_\alpha^{(2)} - \bar{\boldsymbol{x}}^{(2)})'\hat{\boldsymbol{\beta}}}{\sqrt{a_{11}}\sqrt{\sum_{\alpha=1}^N [\hat{\boldsymbol{\beta}}'(\boldsymbol{x}_\alpha^{(2)} - \bar{\boldsymbol{x}}^{(2)})]^2}}$$
$$= \frac{\boldsymbol{a}'_{(1)}\hat{\boldsymbol{\beta}}}{\sqrt{a_{11}}\sqrt{\hat{\boldsymbol{\beta}}'\boldsymbol{A}_{22}\hat{\boldsymbol{\beta}}}},$$

which is (5). Q.E.D.

Thus $\bar{x}_1 + \hat{\boldsymbol{\beta}}'(\boldsymbol{x}_\alpha^{(2)} - \bar{\boldsymbol{x}}^{(2)})$ is the best linear predictor of $x_{1\alpha}$ in the sample, and $\hat{\boldsymbol{\beta}}'\boldsymbol{x}_\alpha^{(2)}$ is the linear function of $\boldsymbol{x}_\alpha^{(2)}$ that has maximum sample correlation with $x_{1\alpha}$. The minimum sum of squares of deviations [the left-hand side of (7)] is

$$(12) \quad \sum_{\alpha=1}^N [(x_{1\alpha} - \bar{x}_1) - \hat{\boldsymbol{\beta}}'(\boldsymbol{x}_\alpha^{(2)} - \bar{\boldsymbol{x}}^{(2)})]^2 = a_{11} - \hat{\boldsymbol{\beta}}'\boldsymbol{A}_{22}\hat{\boldsymbol{\beta}}$$
$$= a_{11} - \boldsymbol{a}'_{(1)}\boldsymbol{A}_{22}^{-1}\boldsymbol{a}_{(1)}$$
$$= a_{11\cdot 2}$$

as defined in Section 4.3 with $q = 1$. The maximum likelihood estimator of $\sigma_{11\cdot 2}$ is $\hat{\sigma}_{11\cdot 2} = a_{11\cdot 2}/N$. It follows that

$$(13) \quad \hat{\sigma}_{11\cdot 2} = (1 - R^2)\hat{\sigma}_{11}.$$

Thus $1 - R^2$ measures the proportional reduction in the variance by using residuals. We can say that R^2 is the fraction of the variance "explained" by $\boldsymbol{x}^{(2)}$. The larger R^2 is, the more the variance is decreased by use of the explanatory variables in $\boldsymbol{x}^{(2)}$.

In p-dimensional space $\boldsymbol{x}_1, \dots, \boldsymbol{x}_N$ represent N points. The sample regression function, $x_1 = \bar{x}_1 + \hat{\boldsymbol{\beta}}'(\boldsymbol{x}^{(2)} - \bar{\boldsymbol{x}}^{(2)})$ is the $(p-1)$-dimensional hyperplane that minimizes the squared deviations of the points from the hyperplane, the deviations calculated in the x_1 direction. The hyperplane goes through the point $\bar{\boldsymbol{x}}$.

In N-dimensional space the rows of $(\boldsymbol{x}_1, \dots, \boldsymbol{x}_N)$ represent p points. The N-component vector with αth component $x_{i\alpha} - \bar{x}_i$ is the projection of the vector with αth component $x_{i\alpha}$ on the plane orthogonal to the equiangular line. We have p such vectors. $\boldsymbol{a}'(\boldsymbol{x}_\alpha^{(2)} - \bar{\boldsymbol{x}}^{(2)})$ is the αth component of a vector in the hyperplane spanned by the last $p-1$ vectors. Since the right-hand side of (7) is the squared distance between the first vector and the linear combination of the last $p-1$ vectors, $\hat{\boldsymbol{\beta}}'(\boldsymbol{x}_\alpha^{(2)} - \bar{\boldsymbol{x}}^{(2)})$ is a component of the vector which minimizes this squared distance. The interpretation of (8) is that the vector with αth component $(x_{1\alpha} - \bar{x}_1) - \hat{\boldsymbol{\beta}}'(\boldsymbol{x}_\alpha^{(2)} - \bar{\boldsymbol{x}}^{(2)})$ is orthogonal to each of the last $p-1$ vectors. Thus the vector with αth component $\hat{\boldsymbol{\beta}}'(\boldsymbol{x}_\alpha^{(2)} - \bar{\boldsymbol{x}}^{(2)})$ is the projection of the first vector on the hyperplane. See Figure 4.5. The length squared of the projection vector is

$$\sum_{\alpha=1}^{N} \left[\hat{\boldsymbol{\beta}}'\left(\boldsymbol{x}_\alpha^{(2)} - \bar{\boldsymbol{x}}^{(2)}\right)\right]^2 = \hat{\boldsymbol{\beta}}'\boldsymbol{A}_{22}\hat{\boldsymbol{\beta}} = \boldsymbol{a}_{(1)}'\boldsymbol{A}_{22}^{-1}\boldsymbol{a}_{(1)}, \tag{14}$$

and the length squared of the first vector is $\sum_{\alpha=1}^{N}(x_{1\alpha} - \bar{x}_1)^2 = a_{11}$. Thus R is the cosine of the angle between the first vector and its projection.

In Section 3.2 we saw that the simple correlation coefficient is the cosine of the angle between the two vectors involved (in the plane orthogonal to the

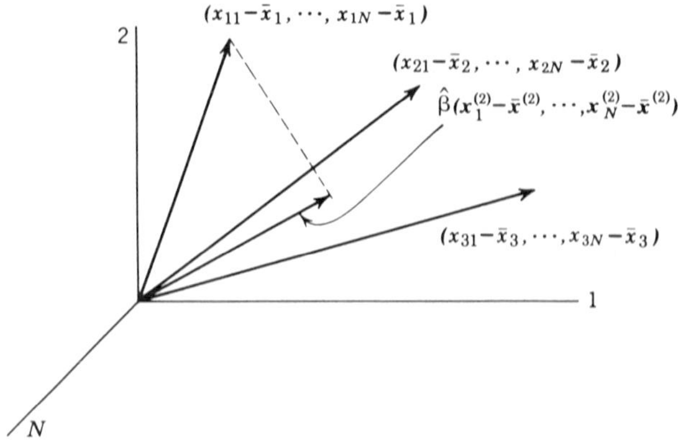

Figure 4.5

equiangular line). The property of R that it is the maximum correlation between $x_{1\alpha}$ and linear combinations of the components of $\boldsymbol{x}_\alpha^{(2)}$ corresponds to the geometric property that R is the cosine of the smallest angle between the vector with components $(x_{1\alpha} - \bar{x}_1)$ and a vector in the hyperplane spanned by the other $p-1$ vectors.

The geometric interpretations are in terms of the vectors in the $(N-1)$-dimensional hyperplane orthogonal to the equiangular line. It was shown in Section 3.3 that the vector $(x_{i1} - \bar{x}_i, \dots, x_{iN} - \bar{x}_i)$ in this hyperplane can be designated as $(z_{i1}, \dots, z_{i,N-1})$, where the $z_{i\alpha}$ are the coordinates referred to an $(N-1)$-dimensional coordinate system in the hyperplane. It was shown that the new coordinates are obtained from the old by the transformation $z_{i\alpha} = \sum_{\beta=1}^{N} b_{\alpha\beta} x_{i\beta}$, $\alpha = 1, \dots, N$, where $\boldsymbol{B} = (b_{\alpha\beta})$ is an orthogonal matrix with last row $(1/\sqrt{N}, \dots, 1/\sqrt{N})$. Then

$$a_{ij} = \sum_{\alpha=1}^{N} (x_{i\alpha} - \bar{x}_i)(x_{j\alpha} - \bar{x}_j) = \sum_{\alpha=1}^{N-1} z_{i\alpha} z_{j\alpha}. \tag{15}$$

It will be convenient to refer to the multiple correlation defined in terms of $z_{i\alpha}$ as the "multiple correlation without subtracting the means."

The computation of R involves taking the square root of the ratio of $\boldsymbol{a}_{(1)}' \boldsymbol{A}_{22}^{-1} \boldsymbol{a}_{(1)}$ to a_{11}. Since $\boldsymbol{A}$ is computed directly from the observations, only the calculation of $\boldsymbol{a}_{(1)}' \boldsymbol{A}_{22}^{-1} \boldsymbol{a}_{(1)}$ requires special techniques. This kind of computation is considered in Section 5.3.1.

The population multiple correlation $\bar{R}$ is essentially the only function of the parameters $\boldsymbol{\mu}$ and $\boldsymbol{\Sigma}$ that is invariant under changes of location, changes of scale of X_1, and nonsingular linear transformations of $\boldsymbol{X}^{(2)}$, that is, transformations $X_1^* = cX_1 + d$, $\boldsymbol{X}^{(2)*} = \boldsymbol{C}\boldsymbol{X}^{(2)} + \boldsymbol{d}$. Similarly, the sample multiple correlation coefficient R is essentially the only function of $\bar{\boldsymbol{x}}$ and $\hat{\boldsymbol{\Sigma}}$, the sufficient set of statistics for $\boldsymbol{\mu}$ and $\boldsymbol{\Sigma}$, that is invariant under these transformations. Just as the simple correlation r is a measure of association between two scalar variables in a sample, the multiple correlation R is a measure of association between a scalar variable and a vector variable in a sample.

4.4.2. Distribution of the Sample Multiple Correlation Coefficient When the Population Multiple Correlation Coefficient Is Zero

From (5) we have

$$R^2 = \frac{\boldsymbol{a}_{(1)}' \boldsymbol{A}_{22}^{-1} \boldsymbol{a}_{(1)}}{a_{11}}; \tag{16}$$

then

$$1 - R^2 = 1 - \frac{\boldsymbol{a}'_{(1)}\boldsymbol{A}_{22}^{-1}\boldsymbol{a}_{(1)}}{a_{11}} = \frac{a_{11} - \boldsymbol{a}'_{(1)}\boldsymbol{A}_{22}^{-1}\boldsymbol{a}_{(1)}}{a_{11}} = \frac{a_{11\cdot 2}}{a_{11}}, \tag{17}$$

and

$$\frac{R^2}{1 - R^2} = \frac{\boldsymbol{a}'_{(1)}\boldsymbol{A}_{22}^{-1}\boldsymbol{a}_{(1)}}{a_{11\cdot 2}}. \tag{18}$$

For $q = 1$, Corollary 4.3.2 states that when $\boldsymbol{\beta} = \boldsymbol{0}$, that is, when $\overline{R} = 0$, $a_{11\cdot 2}$ is distributed as $\sum_{\alpha=1}^{N-p} V_\alpha^2$ and $\boldsymbol{a}'_{(1)}\boldsymbol{A}_{22}^{-1}\boldsymbol{a}_{(1)}$ is distributed as $\sum_{\alpha=N-p+1}^{N-1} V_\alpha^2$, where $V_1, \dots, V_{N-1}$ are independent, each with distribution $N(0, \sigma_{11\cdot 2})$. Then $a_{11\cdot 2}/\sigma_{11\cdot 2}$ and $\boldsymbol{a}'_{(1)}\boldsymbol{A}_{22}^{-1}\boldsymbol{a}_{(1)}/\sigma_{11\cdot 2}$ are distributed independently as χ^2-variables with $N - p$ and $p - 1$ degrees of freedom, respectively. Thus

$$\begin{aligned} \frac{R^2}{1 - R^2} \cdot \frac{N - p}{p - 1} &= \frac{\boldsymbol{a}'_{(1)}\boldsymbol{A}_{22}^{-1}\boldsymbol{a}_{(1)}/\sigma_{11\cdot 2}}{a_{11\cdot 2}/\sigma_{11\cdot 2}} \cdot \frac{N - p}{p - 1} \\ &= \frac{\chi^2_{p-1}}{\chi^2_{N-p}} \cdot \frac{N - p}{p - 1} \\ &= F_{p-1, N-p} \end{aligned} \tag{19}$$

has the F-distribution with $p - 1$ and $N - p$ degrees of freedom. The density of F is

(20)

$$\frac{\Gamma[\frac{1}{2}(N-1)]}{\Gamma[\frac{1}{2}(p-1)]\Gamma[\frac{1}{2}(N-p)]}\left(\frac{p-1}{N-p}\right)^{\frac{1}{2}(p-1)} f^{\frac{1}{2}(p-1)-1}\left(1 + \frac{p-1}{N-p}f\right)^{-\frac{1}{2}(N-1)}.$$

Thus the density of

$$R = \sqrt{\frac{\frac{p-1}{N-p}F_{p-1,N-p}}{1 + \frac{p-1}{N-p}F_{p-1,N-p}}} \tag{21}$$

is

$$2\frac{\Gamma[\frac{1}{2}(N-1)]}{\Gamma[\frac{1}{2}(p-1)]\Gamma[\frac{1}{2}(N-p)]} R^{p-2}(1 - R^2)^{\frac{1}{2}(N-p)-1}, \quad 0 \le R \le 1. \tag{22}$$

THEOREM 4.4.2. *Let R be the sample multiple correlation coefficient [defined by (5)] between X_1 and $\boldsymbol{X}^{(2)\prime} = (X_2, \dots, X_p)$ based on a sample of N from $N(\boldsymbol{\mu}, \boldsymbol{\Sigma})$. If $\overline{R} = 0$ [that is, if $(\sigma_{12}, \dots, \sigma_{1p})' = \mathbf{0} = \boldsymbol{\beta}$], then $[R^2/(1 - R^2)] \cdot [(N-p)/(p-1)]$ is distributed as F with $p-1$ and $N-p$ degrees of freedom.*

It should be noticed that $p - 1$ is the number of components of $\boldsymbol{X}^{(2)}$ and $N - p = N - (p-1) - 1$. If the multiple correlation is between a component X_i and q other components, the numbers are q and $N - q - 1$.

It might be observed that $R^2/(1 - R^2)$ is the quantity that arises in regression (or least squares) theory for testing the hypothesis that the regression of X_1 on $X_2, \dots, X_p$ is zero.

If $\overline{R} \neq 0$, the distribution of R is much more difficult to derive. This distribution will be obtained in Section 4.4.3.

Now let us consider the statistical problem of testing the hypothesis $H: \overline{R} = 0$ on the basis of a sample of N from $N(\boldsymbol{\mu}, \boldsymbol{\Sigma})$. [$\overline{R}$ is the population multiple correlation between X_1 and $(X_2, \dots, X_p)$.] Since $\overline{R} \geq 0$, the alternatives considered are $\overline{R} > 0$.

Let us derive the likelihood ratio test of this hypothesis. The likelihood function is

$$(23) \qquad L(\boldsymbol{\mu}^*, \boldsymbol{\Sigma}^*) = \frac{1}{(2\pi)^{\frac{1}{2}pN} |\boldsymbol{\Sigma}^*|^{\frac{1}{2}N}} \exp\left[-\tfrac{1}{2} \sum_{\alpha=1}^{N} (\boldsymbol{x}_\alpha - \boldsymbol{\mu}^*)' \boldsymbol{\Sigma}^{*-1} (\boldsymbol{x}_\alpha - \boldsymbol{\mu}^*) \right].$$

The observations are given; L is a function of the indeterminates $\boldsymbol{\mu}^*$, $\boldsymbol{\Sigma}^*$. Let ω be the region in the parameter space Ω specified by the null hypothesis. The likelihood ratio criterion is

$$(24) \qquad \lambda = \frac{\max_{\boldsymbol{\mu}^*, \boldsymbol{\Sigma}^* \in \omega} L(\boldsymbol{\mu}^*, \boldsymbol{\Sigma}^*)}{\max_{\boldsymbol{\mu}^*, \boldsymbol{\Sigma}^* \in \Omega} L(\boldsymbol{\mu}^*, \boldsymbol{\Sigma}^*)}.$$

Here Ω is the space of $\boldsymbol{\mu}^*$, $\boldsymbol{\Sigma}^*$ positive definite and ω is the region in this space where $\overline{R} = \sqrt{\boldsymbol{\sigma}_{(1)}' \boldsymbol{\Sigma}_{22}^{-1} \boldsymbol{\sigma}_{(1)}} \big/ \sqrt{\sigma_{11}} = 0$; that is, where $\boldsymbol{\sigma}_{(1)}' \boldsymbol{\Sigma}_{22}^{-1} \boldsymbol{\sigma}_{(1)} = 0$. Because $\boldsymbol{\Sigma}_{22}^{-1}$ is positive definite, this condition is equivalent to $\boldsymbol{\sigma}_{(1)} = \mathbf{0}$. The maximum of $L(\boldsymbol{\mu}^*, \boldsymbol{\Sigma}^*)$ over Ω occurs at $\boldsymbol{\mu}^* = \hat{\boldsymbol{\mu}} = \bar{\boldsymbol{x}}$ and $\boldsymbol{\Sigma}^* = \hat{\boldsymbol{\Sigma}} = (1/N)\boldsymbol{A} = (1/N)\sum_{\alpha=1}^{N} (\boldsymbol{x}_\alpha - \bar{\boldsymbol{x}})(\boldsymbol{x}_\alpha - \bar{\boldsymbol{x}})'$ and is

$$(25) \qquad \max_{\boldsymbol{\mu}^*, \boldsymbol{\Sigma}^* \in \Omega} L(\boldsymbol{\mu}^*, \boldsymbol{\Sigma}^*) = \frac{N^{\frac{1}{2}pN} e^{-\frac{1}{2}pN}}{(2\pi)^{\frac{1}{2}pN} |\boldsymbol{A}|^{\frac{1}{2}N}}.$$

In ω the likelihood function is

$$L(\boldsymbol{\mu}^*, \boldsymbol{\Sigma}^* | \boldsymbol{\sigma}_{(1)}^* = \mathbf{0}) = \frac{1}{(2\pi)^{\frac{1}{2}N} \sigma_{11}^{*\frac{1}{2}N}} \exp\left[-\tfrac{1}{2} \sum_{\alpha=1}^{N} (x_{1\alpha} - \mu_1^*)^2 / \sigma_{11}^* \right] \tag{26}$$

$$\cdot \frac{1}{(2\pi)^{\frac{1}{2}(p-1)N} |\boldsymbol{\Sigma}_{22}^*|^{\frac{1}{2}N}} \exp\left[-\tfrac{1}{2} \sum_{\alpha=1}^{N} (\boldsymbol{x}_\alpha^{(2)} - \boldsymbol{\mu}^{(2)*})' \boldsymbol{\Sigma}_{22}^{*-1} (\boldsymbol{x}_\alpha^{(2)} - \boldsymbol{\mu}^{(2)*}) \right].$$

The first factor is maximized at $\mu_1^* = \hat{\mu}_1 = \bar{x}_1$ and $\sigma_{11}^* = \hat{\sigma}_{11} = (1/N)a_{11}$, and the second factor is maximized at $\boldsymbol{\mu}^{(2)*} = \hat{\boldsymbol{\mu}}^{(2)} = \bar{\boldsymbol{x}}^{(2)}$ and $\boldsymbol{\Sigma}_{22}^* = \hat{\boldsymbol{\Sigma}}_{22} = (1/N)\boldsymbol{A}_{22}$. The value of the maximized function is

$$\max_{\boldsymbol{\mu}^*, \boldsymbol{\Sigma}^* \in \omega} L(\boldsymbol{\mu}^*, \boldsymbol{\Sigma}^*) = \frac{N^{\frac{1}{2}N} e^{-\frac{1}{2}N}}{(2\pi)^{\frac{1}{2}N} a_{11}^{\frac{1}{2}N}} \cdot \frac{N^{\frac{1}{2}(p-1)N} e^{-\frac{1}{2}(p-1)N}}{(2\pi)^{\frac{1}{2}(p-1)N} |\boldsymbol{A}_{22}|^{\frac{1}{2}N}}. \tag{27}$$

Thus the likelihood ratio criterion is [see (6)]

$$\lambda = \frac{|\boldsymbol{A}|^{\frac{1}{2}N}}{a_{11}^{\frac{1}{2}N} |\boldsymbol{A}_{22}|^{\frac{1}{2}N}} = (1 - R^2)^{\frac{1}{2}N}. \tag{28}$$

The likelihood ratio test consists of the critical region $\lambda < \lambda_0$, where λ_0 is chosen so the probability of this inequality when $\bar{R} = 0$ is the significance level α. An equivalent test is

$$1 - \lambda^{2/N} = R^2 > 1 - \lambda_0^{2/N}. \tag{29}$$

Since $[R^2/(1 - R^2)][(N - p)/(p - 1)]$ is a monotonic function of R, an equivalent test involves this ratio being larger than a constant. When $\bar{R} = 0$, this ratio has an $F_{p-1, N-p}$-distribution. Hence, the critical region is

$$\frac{R^2}{1 - R^2} \cdot \frac{N - p}{p - 1} > F_{p-1, N-p}(\alpha), \tag{30}$$

where $F_{p-1, N-p}(\alpha)$ is the (upper) significance point corresponding to the α significance level.

THEOREM 4.4.3. *Given a sample $\boldsymbol{x}_1, \dots, \boldsymbol{x}_N$ from $N(\boldsymbol{\mu}, \boldsymbol{\Sigma})$, the likelihood ratio test at significance level α for the hypothesis $\bar{R} = 0$, where $\bar{R}$ is the* *population multiple correlation coefficient between* X_1 *and* $(X_2, \ldots, X_p)$, *is given by* (30), *where* R *is the sample multiple correlation coefficient defined by* (5).

As an example consider the data given at the end of Section 4.3.1. The sample multiple correlation coefficient is found from

$$(31)\quad 1 - R^2 = \frac{\begin{vmatrix} 1 & r_{12} & r_{13} \\ r_{12} & 1 & r_{23} \\ r_{13} & r_{23} & 1 \end{vmatrix}}{\begin{vmatrix} 1 & r_{23} \\ r_{23} & 1 \end{vmatrix}} = \frac{\begin{vmatrix} 1.00 & 0.80 & -0.40 \\ 0.80 & 1.00 & -0.56 \\ -0.40 & -0.56 & 1.00 \end{vmatrix}}{\begin{vmatrix} 1.00 & -0.56 \\ -0.56 & 1.00 \end{vmatrix}} = 0.357.$$

Thus R is 0.802. If we wish to test the hypothesis at the 0.01 level that hay yield is independent of spring rainfall and temperature, we compare the observed $[R^2/(1 - R^2)][(20 - 3)/(3 - 1)] = 15.3$ with $F_{2,17}(0.01) = 6.11$ and find the result significant; that is, we reject the null hypothesis.

The test of independence between X_1 and $(X_2, \ldots, X_p) = \boldsymbol{X}^{(2)\prime}$ is equivalent to the test that if the regression of X_1 on $\boldsymbol{x}^{(2)}$ (that is, the conditional expected value of X_1 given $X_2 = x_2, \ldots, X_p = x_p$) is $\mu_1 + \boldsymbol{\beta}'(\boldsymbol{x}^{(2)} - \boldsymbol{\mu}^{(2)})$, the vector of regression coefficients is $\boldsymbol{0}$. Here $\hat{\boldsymbol{\beta}} = \boldsymbol{A}_{22}^{-1}\boldsymbol{a}_{(1)}$ is the usual least squares estimate of $\boldsymbol{\beta}$ with expected value $\boldsymbol{\beta}$ and covariance matrix $\sigma_{11\cdot 2}\boldsymbol{A}_{22}^{-1}$ (when the $\boldsymbol{X}_\alpha^{(2)}$ are fixed) and $a_{11\cdot 2}/(N - p)$ is the usual estimate of $\sigma_{11\cdot 2}$. Thus [see (18)]

$$(32)\qquad \frac{R^2}{1 - R^2} \cdot \frac{N - p}{p - 1} = \frac{\hat{\boldsymbol{\beta}}'\boldsymbol{A}_{22}\hat{\boldsymbol{\beta}}}{a_{11\cdot 2}} \cdot \frac{N - p}{p - 1}$$

is the usual F-statistic for testing the hypothesis that the regression of X_1 on $x_2, \ldots, x_p$ is 0. In this book we are primarily interested in the multiple correlation coefficient as a measure of association between one variable and a vector of variables when both are random. We shall not treat problems of univariate regression. In Chapter 8 we study regression when the dependent variable is a vector.

Adjusted Multiple Correlation Coefficient. The expression (17) is the ratio of $a_{11\cdot 2}$, the sum of squared deviations from the fitted regression, to a_{11}, the sum of squared deviations around the mean. To obtain unbiased estimators of σ_{11} when $\boldsymbol{\beta} = \boldsymbol{0}$ we would divide these quantities by their numbers of degrees of freedom, $N - p$ and $N - 1$, respectively. Accordingly we can define an "adjusted multiple correlation coefficient" R^* by

$$1 - R^{*2} = \frac{a_{11\cdot 2}/(N-p)}{a_{11}/(N-1)} = \frac{N-1}{N-p}(1 - R^2), \tag{33}$$

which is equivalent to

$$R^{*2} = R^2 - \frac{p-1}{N-p}(1 - R^2). \tag{34}$$

This quantity is smaller than R^2 (unless $p = 1$ or $R^2 = 1$). A possible merit to it is that it takes account of p; the idea is that the larger p is relative to N, the greater the tendency of R^2 to be large by chance.

4.4.3. Distribution of the Sample Multiple Correlation Coefficient When the Population Multiple Correlation Coefficient Is Not Zero

In this section we shall find the distribution of R when the null hypothesis $\overline{R} = 0$ is not true. We shall find that the distribution depends only on the population multiple correlation coefficient $\overline{R}$.

First let us consider the conditional distribution of $R^2/(1 - R^2) = \boldsymbol{a}'_{(1)}\boldsymbol{A}_{22}^{-1}\boldsymbol{a}_{(1)}/a_{11\cdot 2}$ given $\boldsymbol{Z}_\alpha^{(2)} = \boldsymbol{z}_\alpha^{(2)}$, $\alpha = 1, \ldots, n$. Under these conditions $Z_{11}, \ldots, Z_{1n}$ are independently distributed, $Z_{1\alpha}$ according to $N(\boldsymbol{\beta}'\boldsymbol{z}_\alpha^{(2)}, \sigma_{11\cdot 2})$, where $\boldsymbol{\beta} = \boldsymbol{\Sigma}_{22}^{-1}\boldsymbol{\sigma}_{(1)}$ and $\sigma_{11\cdot 2} = \sigma_{11} - \boldsymbol{\sigma}'_{(1)}\boldsymbol{\Sigma}_{22}^{-1}\boldsymbol{\sigma}_{(1)}$. The conditions are those of Theorem 4.3.3 with $\boldsymbol{Y}_\alpha = Z_{1\alpha}$, $\boldsymbol{\Gamma} = \boldsymbol{\beta}'$, $\boldsymbol{w}_\alpha = \boldsymbol{z}_\alpha^{(2)}$, $r = p - 1$, $\boldsymbol{\Phi} = \sigma_{11\cdot 2}$, $m = n$. Then $a_{11\cdot 2} = a_{11} - \boldsymbol{a}'_{(1)}\boldsymbol{A}_{22}^{-1}\boldsymbol{a}_{(1)}$ corresponds to $\sum_{\alpha=1}^m \boldsymbol{Y}_\alpha\boldsymbol{Y}'_\alpha - \boldsymbol{GHG}'$, and $a_{11\cdot 2}/\sigma_{11\cdot 2}$ has a χ^2-distribution with $n - (p-1)$ degrees of freedom. $\boldsymbol{a}'_{(1)}\boldsymbol{A}_{22}^{-1}\boldsymbol{a}_{(1)} = (\boldsymbol{A}_{22}^{-1}\boldsymbol{a}_{(1)})'\boldsymbol{A}_{22}(\boldsymbol{A}_{22}^{-1}\boldsymbol{a}_{(1)})$ corresponds to $\boldsymbol{GHG}'$ and is distributed as $\sum_\alpha U_\alpha^2$, $\alpha = n - (p-1) + 1, \ldots, n$, where $\operatorname{Var}(U_\alpha) = \sigma_{11\cdot 2}$ and

$$\mathscr{E}(U_{n-p+2}, \ldots, U_n) = \boldsymbol{\Gamma F}^{-1}, \tag{35}$$

where $\boldsymbol{FHF}' = \boldsymbol{I}$ $[\boldsymbol{H} = \boldsymbol{F}^{-1}(\boldsymbol{F}')^{-1}]$. Then $\boldsymbol{a}'_{(1)}\boldsymbol{A}_{22}^{-1}\boldsymbol{a}_{(1)}/\sigma_{11\cdot 2}$ is distributed as $\sum_\alpha (U_\alpha/\sqrt{\sigma_{11\cdot 2}})^2$, where $\operatorname{Var}(U_\alpha/\sqrt{\sigma_{11\cdot 2}}) = 1$ and

$$\sum_{\alpha=n-p+2}^{n} \left(\frac{\mathscr{E}U_\alpha}{\sqrt{\sigma_{11\cdot 2}}}\right)^2 = \frac{1}{\sigma_{11\cdot 2}}\boldsymbol{\Gamma F}^{-1}(\boldsymbol{\Gamma F}^{-1})' = \frac{\boldsymbol{\Gamma H \Gamma}'}{\sigma_{11\cdot 2}} = \frac{\boldsymbol{\beta}'\boldsymbol{A}_{22}\boldsymbol{\beta}}{\sigma_{11\cdot 2}}. \tag{36}$$

Thus (conditionally) $\boldsymbol{a}'_{(1)}\boldsymbol{A}_{22}^{-1}\boldsymbol{a}_{(1)}/\sigma_{11\cdot 2}$ has a noncentral χ^2-distribution with

$p-1$ degrees of freedom and noncentrality parameter $\boldsymbol{\beta}'\boldsymbol{A}_{22}\boldsymbol{\beta}/\sigma_{11\cdot 2}$. (See Theorem 5.4.1.) We are led to the following theorem:

THEOREM 4.4.4. *Let R be the sample multiple correlation coefficient between $X_{(1)}$ and $\boldsymbol{X}^{(2)\prime} = (X_2, \dots, X_p)$ based on N observations $(x_{11}, \boldsymbol{x}_1^{(2)}), \dots, (x_{1N}, \boldsymbol{x}_N^{(2)})$. The conditional distribution of $[R^2/(1-R^2)][(N-p)/(p-1)]$ given $\boldsymbol{x}_\alpha^{(2)}$ fixed is noncentral F with $p-1$ and $N-p$ degrees of freedom and noncentrality parameter $\boldsymbol{\beta}'\boldsymbol{A}_{22}\boldsymbol{\beta}/\sigma_{11\cdot 2}$.*

The conditional density (from Theorem 5.4.1) of $F = [R^2/(1-R^2)][(N-p)/(p-1)]$ is

$$\text{(37)}\qquad \frac{(p-1)\exp\left[-\frac{1}{2}\boldsymbol{\beta}'\boldsymbol{A}_{22}\boldsymbol{\beta}/\sigma_{11\cdot 2}\right]}{(N-p)\Gamma\left[\frac{1}{2}(N-p)\right]} \cdot \sum_{\alpha=0}^{\infty} \frac{\left(\dfrac{\boldsymbol{\beta}'\boldsymbol{A}_{22}\boldsymbol{\beta}}{2\sigma_{11\cdot 2}}\right)^{\alpha} \left[\dfrac{(p-1)f}{N-p}\right]^{\frac{1}{2}(p-1)+\alpha-1} \Gamma\left[\frac{1}{2}(N-1)+\alpha\right]}{\alpha!\,\Gamma\left[\frac{1}{2}(p-1)+\alpha\right]\left[1+\dfrac{(p-1)f}{N-p}\right]^{\frac{1}{2}(N-1)+\alpha}},$$

and the conditional density of $W = R^2$ is $(df = [(N-p)/(p-1)](1-w)^{-2}\,dw)$

$$\text{(38)}\qquad \frac{\exp\left[-\frac{1}{2}\boldsymbol{\beta}'\boldsymbol{A}_{22}\boldsymbol{\beta}/\sigma_{11\cdot 2}\right]}{\Gamma\left[\frac{1}{2}(N-p)\right]}(1-w)^{\frac{1}{2}(N-p)-1} \cdot \sum_{\alpha=0}^{\infty} \frac{\left(\dfrac{\boldsymbol{\beta}'\boldsymbol{A}_{22}\boldsymbol{\beta}}{2\sigma_{11\cdot 2}}\right)^{\alpha} w^{\frac{1}{2}(p-1)+\alpha-1}\Gamma\left[\frac{1}{2}(N-1)+\alpha\right]}{\alpha!\,\Gamma\left[\frac{1}{2}(p-1)+\alpha\right]}.$$

To obtain the unconditional density we need to multiply (38) by the density of $\boldsymbol{Z}_1^{(2)}, \dots, \boldsymbol{Z}_n^{(2)}$ to obtain the joint density of W and $\boldsymbol{Z}_1^{(2)}, \dots, \boldsymbol{Z}_n^{(2)}$ and then integrate with respect to the latter set to obtain the marginal density of W. We have

$$\text{(39)}\qquad \frac{\boldsymbol{\beta}'\boldsymbol{A}_{22}\boldsymbol{\beta}}{\sigma_{11\cdot 2}} = \frac{\boldsymbol{\beta}'\sum_{\alpha=1}^{n} \boldsymbol{z}_\alpha^{(2)}\boldsymbol{z}_\alpha^{(2)\prime}\boldsymbol{\beta}}{\sigma_{11\cdot 2}} = \sum_{\alpha=1}^{n}\left(\frac{\boldsymbol{\beta}'\boldsymbol{z}_\alpha^{(2)}}{\sqrt{\sigma_{11\cdot 2}}}\right)^2.$$

Since the distribution of $\boldsymbol{Z}_\alpha^{(2)}$ is $N(\mathbf{0}, \boldsymbol{\Sigma}_{22})$, the distribution of $\boldsymbol{\beta}'\boldsymbol{Z}_\alpha^{(2)}/\sqrt{\sigma_{11\cdot 2}}$ is normal with mean zero and variance

$$
(40) \qquad \mathscr{E}\left(\frac{\boldsymbol{\beta}'\boldsymbol{Z}_\alpha^{(2)}}{\sqrt{\sigma_{11\cdot 2}}}\right)^2 = \frac{\mathscr{E}\boldsymbol{\beta}'\boldsymbol{Z}_\alpha^{(2)}\boldsymbol{Z}_\alpha^{(2)\prime}\boldsymbol{\beta}}{\sigma_{11\cdot 2}}
$$

$$
= \frac{\boldsymbol{\beta}'\boldsymbol{\Sigma}_{22}\boldsymbol{\beta}}{\sigma_{11} - \boldsymbol{\beta}'\boldsymbol{\Sigma}_{22}\boldsymbol{\beta}} = \frac{\boldsymbol{\beta}'\boldsymbol{\Sigma}_{22}\boldsymbol{\beta}/\sigma_{11}}{1 - \boldsymbol{\beta}'\boldsymbol{\Sigma}_{22}\boldsymbol{\beta}/\sigma_{11}}
$$

$$
= \frac{\overline{R}^2}{1 - \overline{R}^2}.
$$

Thus $(\boldsymbol{\beta}'\boldsymbol{A}_{22}\boldsymbol{\beta}/\sigma_{11\cdot 2})/[\overline{R}^2/(1-\overline{R}^2)]$ has a χ^2-distribution with n degrees of freedom. Let $\overline{R}^2/(1-\overline{R}^2) = \phi$. Then $\boldsymbol{\beta}'\boldsymbol{A}_{22}\boldsymbol{\beta}/\sigma_{11\cdot 2} = \phi\chi_n^2$. We compute

$$
(41) \qquad \mathscr{E}e^{-\frac{1}{2}\phi\chi_n^2}\left(\frac{\phi\chi_n^2}{2}\right)^\alpha
$$

$$
= \frac{\phi^\alpha}{2^\alpha}\int_0^\infty u^\alpha e^{-\frac{1}{2}\phi u}\frac{1}{2^{\frac{1}{2}n}\Gamma\left(\frac{1}{2}n\right)}u^{\frac{1}{2}n-1}e^{-\frac{1}{2}u}\,du
$$

$$
= \frac{\phi^\alpha}{2^\alpha}\int_0^\infty \frac{1}{2^{\frac{1}{2}n}\Gamma\left(\frac{1}{2}n\right)}u^{\frac{1}{2}n+\alpha-1}e^{-\frac{1}{2}(1+\phi)u}\,du
$$

$$
= \frac{\phi^\alpha}{(1+\phi)^{\frac{1}{2}n+\alpha}}\frac{\Gamma\left(\frac{1}{2}n+\alpha\right)}{\Gamma\left(\frac{1}{2}n\right)}\int_0^\infty \frac{1}{2^{\frac{1}{2}n+\alpha}\Gamma\left(\frac{1}{2}n+\alpha\right)}v^{\frac{1}{2}n+\alpha-1}e^{-\frac{1}{2}v}\,dv
$$

$$
= \frac{\phi^\alpha}{(1+\phi)^{\frac{1}{2}n+\alpha}}\frac{\Gamma\left(\frac{1}{2}n+\alpha\right)}{\Gamma\left(\frac{1}{2}n\right)}.
$$

Applying this result to (38), we obtain as the density of R^2

$$
(42) \qquad \frac{(1-R^2)^{\frac{1}{2}(n-p-1)}(1-\overline{R}^2)^{\frac{1}{2}n}}{\Gamma\left[\frac{1}{2}(n-p+1)\right]\Gamma\left(\frac{1}{2}n\right)}\sum_{\mu=0}^{\infty}\frac{(\overline{R}^2)^\mu(R^2)^{\frac{1}{2}(p-1)+\mu-1}\Gamma^2\left(\frac{1}{2}n+\mu\right)}{\mu!\,\Gamma\left[\frac{1}{2}(p-1)+\mu\right]}.
$$

Fisher (1928) found this distribution. It can also be written

$$(43)\qquad \frac{\Gamma(\frac{1}{2}n)(1-\bar{R}^2)^{\frac{1}{2}n}}{\Gamma[\frac{1}{2}(n-p+1)]\Gamma[\frac{1}{2}(p-1)]}(R^2)^{\frac{1}{2}(p-3)}(1-R^2)^{\frac{1}{2}(n-p-1)} \cdot F\left[\tfrac{1}{2}n, \tfrac{1}{2}n; \tfrac{1}{2}(p-1); R^2\bar{R}^2\right],$$

where F is the hypergeometric function defined in (41) of Section 4.2.

Another form of the density can be obtained when $n-p+1$ is even. We have

$$(44)\qquad \sum_{\mu=0}^{\infty}\frac{(R^2\bar{R}^2)^{\mu}}{\mu!}\frac{\Gamma^2(\frac{1}{2}n+\mu)}{\Gamma[\frac{1}{2}(p-1)+\mu]}$$

$$= \sum_{\mu=0}^{\infty}\frac{(R^2\bar{R}^2)^{\mu}}{\mu!}\Gamma(\tfrac{1}{2}n+\mu)\left(\frac{\partial}{\partial t}\right)^{\frac{1}{2}(n-p+1)} t^{\frac{1}{2}n+\mu-1}\Bigg]_{t=1}$$

$$= \left(\frac{\partial}{\partial t}\right)^{\frac{1}{2}(n-p+1)} t^{\frac{1}{2}n-1}\sum_{\mu=0}^{\infty}\frac{(t\bar{R}^2R^2)^{\mu}}{\mu!}\frac{\Gamma(\frac{1}{2}n+\mu)}{\Gamma(\frac{1}{2}n)}\Bigg]_{t=1}\Gamma(\tfrac{1}{2}n)$$

$$= \Gamma(\tfrac{1}{2}n)\left(\frac{\partial}{\partial t}\right)^{\frac{1}{2}(n-p+1)} t^{\frac{1}{2}n-1}(1-tR^2\bar{R}^2)^{-\frac{1}{2}n}\Bigg]_{t=1}.$$

The density is therefore

$$(45)\qquad \frac{(1-\bar{R}^2)^{\frac{1}{2}n}(R^2)^{\frac{1}{2}(p-3)}(1-R^2)^{\frac{1}{2}(n-p-1)}}{\Gamma[\frac{1}{2}(n-p+1)]} \cdot \left(\frac{\partial}{\partial t}\right)^{\frac{1}{2}(n-p+1)} t^{\frac{1}{2}n-1}(1-tR^2\bar{R}^2)^{-\frac{1}{2}n}\Bigg]_{t=1}.$$

THEOREM 4.4.5. *The density of the square of the multiple correlation coefficient, R^2, between X_1 and $X_2, \dots, X_p$ based on a sample of $N = n+1$ is given by* (42) *or* (43) [*or* (45) *in the case of $n-p+1$ even*], *where $\bar{R}^2$ is the corresponding population multiple correlation coefficient.*

The moments of R are

$$(46)\quad \mathscr{E}R^h = \frac{(1-\bar{R}^2)^{\frac{1}{2}n}}{\Gamma[\frac{1}{2}(n-p+1)]\Gamma(\frac{1}{2}n)} \sum_{\mu=0}^{\infty} \frac{(\bar{R}^2)^{\mu}\Gamma^2(\frac{1}{2}n+\mu)}{\Gamma[\frac{1}{2}(p-1)+\mu]\mu!}$$

$$\cdot \int_0^1 (1-R^2)^{\frac{1}{2}(n-p+1)-1}(R^2)^{\frac{1}{2}(p+h-1)+\mu-1}\,d(R^2)$$

$$= \frac{(1-\bar{R}^2)^{\frac{1}{2}n}}{\Gamma(\frac{1}{2}n)} \sum_{\mu=0}^{\infty} \frac{(\bar{R}^2)^{\mu}\Gamma^2(\frac{1}{2}n+\mu)\Gamma[\frac{1}{2}(p+h-1)+\mu]}{\mu!\Gamma[\frac{1}{2}(p-1)+\mu]\Gamma[\frac{1}{2}(n+h)+\mu]}.$$

The multiple correlation coefficient is defined to be nonnegative. In fact, the probability is 1 that A_{22} is positive definite and hence that $R > 0$, even if $\bar{R} = 0$. It seems reasonable that any function of R, say $h(R)$, to be used as an estimator of $\bar{R}$ should be in the closed interval $[0, 1]$, the range of possible values of $\bar{R}$. Since R has a density that is positive on $[0, 1]$, $\mathscr{E}h(R) = 0$ for $\bar{R} = 0$ implies that $h(R) = 0$ almost everywhere and hence $\mathscr{E}h(R) = 0$ for every $\bar{R}$. Olkin and Pratt (1958) have given as the unique unbiased estimator of $\bar{R}^2$ based on R^2

$$(47)\qquad 1 - \frac{N-3}{N-p}(1-R^2)F(1,1;\tfrac{1}{2}(N-p+2);1-R^2),$$

but the value is $-(p-1)/(N-p-2)$ for $R = 0$. It has the disadvantage that the estimator is negative for small values of R, but the parameter being estimated is nonnegative.

The sample multiple correlation tends to overestimate the population multiple correlation. The sample multiple correlation is the maximum sample correlation between x_1 and linear combinations of $\boldsymbol{x}^{(2)}$ and hence is greater than the sample correlation between x_1 and $\boldsymbol{\beta}'\boldsymbol{x}^{(2)}$; however, the latter is the simple sample correlation corresponding to the simple population correlation between x_1 and $\boldsymbol{\beta}'\boldsymbol{x}^{(2)}$, which is $\bar{R}$, the population multiple correlation.

Suppose R_1 is the multiple correlation in the first sample of two samples and $\hat{\boldsymbol{\beta}}_1$ is the estimate of $\boldsymbol{\beta}$; then the simple correlation between x_1 and $\hat{\boldsymbol{\beta}}_1'\boldsymbol{x}^{(2)}$ in the second sample will tend to be less than R_1 and in particular will be less than R_2, the multiple correlation in the second sample. This has been called "the shrinkage of the multiple correlation."

Kramer (1963) and Lee (1972) have given tables of the upper significance points of R. Gajjar (1967), Gurland (1968), Gurland and Milton (1970), Khatri (1966), and Lee (1971b) have suggested approximations to the distributions of $R^2/(1-R^2)$ and obtained large-sample results.

4.4.4. Some Optimal Properties of the Multiple Correlation Test

THEOREM 4.4.6. *Given the observations $\boldsymbol{x}_1, \dots, \boldsymbol{x}_N$ from $N(\boldsymbol{\mu}, \boldsymbol{\Sigma})$, of all tests of $\overline{R} = 0$ at a given significance level based on $\bar{\boldsymbol{x}}$ and $\boldsymbol{A} = \sum_{\alpha=1}^{N}(\boldsymbol{x}_\alpha - \bar{\boldsymbol{x}})(\boldsymbol{x}_\alpha - \bar{\boldsymbol{x}})'$ that are invariant with respect to transformations*

$$\begin{aligned} &\bar{x}_1^* = c\bar{x}_1 + d, \qquad \bar{\boldsymbol{x}}^{(2)*} = \boldsymbol{C}\bar{\boldsymbol{x}}^{(2)} + \boldsymbol{d}, \\ &a_{11}^* = c^2 a_{11}, \qquad \boldsymbol{a}_{(1)}^* = c\boldsymbol{C}\boldsymbol{a}_{(1)}, \qquad \boldsymbol{A}_{22}^* = \boldsymbol{C}\boldsymbol{A}_{22}\boldsymbol{C}', \end{aligned} \tag{48}$$

any critical rejection region given by R greater than a constant is uniformly most powerful.

PROOF. The multiple correlation coefficient R is invariant under the transformation and any function of the sufficient statistics that is invariant is a function of R. (See Problem 34.) Therefore, any invariant test must be based on R. The Neyman–Pearson fundamental lemma applied to testing the null hypothesis $\overline{R} = 0$ against a specific alternative $\overline{R} = \overline{R}_0 > 0$ tells us the most powerful test at a given level of significance is based on the ratio of the density of R for $\overline{R} = \overline{R}_0$, which is (42) times $2R$ [because (42) is the density of R^2] to the density for $R = 0$, which is (22). The ratio is a positive constant times

$$\sum_{\mu=0}^{\infty} \frac{\left(\overline{R}_0^2\right)^{\mu} \Gamma^2\left(\frac{1}{2}n + \mu\right)}{\mu!\,\Gamma\left[\frac{1}{2}(p-1) + \mu\right]} R^{p-2+2\mu}. \tag{49}$$

Since (49) is an increasing function of R for $R \geq 0$, the set of R for which (49) is greater than a constant is an interval of R greater than a constant. Q.E.D.

THEOREM 4.4.7. *On the basis of observations $\boldsymbol{x}_1, \dots, \boldsymbol{x}_N$ from $N(\boldsymbol{\mu}, \boldsymbol{\Sigma})$, of all tests of $\overline{R} = 0$ at a given significance level with power depending only on $\overline{R}$, the test with critical region given by R greater than a constant is uniformly most powerful.*

Theorem 4.4.7 follows from Theorem 4.4.6 in the same way that Theorem 5.6.4 follows from Theorem 5.6.1.

PROBLEMS

1. (Sec. 4.2.1) Sketch

$$k_N(r) = \frac{\Gamma[\frac{1}{2}(N-1)]}{\Gamma(\frac{1}{2}N-1)\sqrt{\pi}}(1-r^2)^{\frac{1}{2}(N-4)}$$

for (*a*) $N = 3$, (*b*) $N = 4$, (*c*) $N = 5$, and (*d*) $N = 10$.

2. (Sec. 4.2.1) Using the data of Problem 1 of Chapter 3, test the hypothesis that X_1 and X_2 are independent against all alternatives of dependence at significance level 0.01.

3. (Sec. 4.2.1) Suppose a sample correlation of 0.65 is observed in a sample of 10. Test the hypothesis of independence against the alternatives of positive correlation at significance level 0.05.

4. (Sec. 4.2.2) Suppose a sample correlation of 0.65 is observed in a sample of 20. Test the hypothesis that the population correlation is 0.4 against the alternatives that the population correlation is greater than 0.4 at significance level 0.05.

5. (Sec. 4.2.1) Find the significance points for testing $\rho = 0$ at the 0.01 level with $N = 15$ observations against alternatives (*a*) $\rho \neq 0$, (*b*) $\rho > 0$, and (*c*) $\rho < 0$.

6. (Sec. 4.2.2) Find significance points for testing $\rho = 0.6$ at the 0.01 level with $N = 20$ observations against alternatives (*a*) $\rho \neq 0.6$, (*b*) $\rho > 0.6$, and (*c*) $\rho < 0.6$.

7. (Sec. 4.2.2) Table the power function at $\rho = -1(0.2)1$ for the tests in Problem 5. Sketch the graph of each power function.

8. (Sec. 4.2.2) Table the power function at $\rho = -1(0.2)1$ for the tests in Problem 6. Sketch the graph of each power function.

9. (Sec. 4.2.2) Using the data of Problem 1 of Chapter 3 find a (two-sided) confidence interval for ρ_{12} with confidence coefficient 0.99.

10. (Sec. 4.2.2) Suppose $N = 10$, $r = 0.795$. Find a one-sided confidence interval for ρ [of the form $(r_0, 1)$] with confidence coefficient 0.95.

11. (Sec. 4.2.3) Use Fisher's z to test the hypothesis $\rho = 0.7$ against alternatives $\rho \neq 0.7$ at the 0.05 level with $r = 0.5$ and $N = 50$.

12. (Sec. 4.2.3) Use Fisher's z to test the hypothesis $\rho_1 = \rho_2$ against the alternatives $\rho_1 \neq \rho_2$ at the 0.01 level with $r_1 = 0.5$, $N_1 = 40$, $r_2 = 0.6$, $N_2 = 40$.

13. (Sec. 4.2.3) Use Fisher's z to estimate ρ based on sample correlations of -0.7 ($N = 30$) and of -0.6 ($N = 40$).

14. (Sec. 4.2.3) Use Fisher's z to obtain a confidence interval for ρ with confidence 0.95 based on a sample correlation of 0.65 and a sample size of 25.

15. (Sec. 4.2.2) Prove that when $N = 2$ and $\rho = 0$, $\Pr\{r = 1\} = \Pr\{r = -1\} = \frac{1}{2}$.

16. (Sec. 4.2) Let $k_N(r, \rho)$ be the density of the sample correlation coefficient r for a given value of ρ and N. Prove that r has a monotone likelihood ratio; that is, show that if $\rho_1 > \rho_2$, then $k_N(r, \rho_1)/k_N(r, \rho_2)$ is monotonically increasing in r. [*Hint*: Using (40) prove that if

$$F\left[\tfrac{1}{2}, \tfrac{1}{2}; n + \tfrac{1}{2}; \tfrac{1}{2}(1 + \rho r)\right] = \sum_{\alpha=0}^{\infty} c_\alpha (1 + \rho r)^\alpha = g(r, \rho)$$

has a monotone ratio, then $k_N(r, \rho)$ does. Show

$$\frac{\partial^2}{\partial\rho\,\partial r} \log g(r, \rho) = \frac{\sum_{\alpha,\beta=0}^{\infty} c_\alpha c_\beta \left[(\alpha - \beta)^2 r\rho + (\alpha + \beta)\right](1 + r\rho)^{\alpha+\beta-2}}{2\left[\sum_{\alpha=0}^{\infty} c_\alpha (1 + r\rho)^\alpha\right]^2};$$

if $(\partial^2/\partial\rho\,\partial r)\log g(r, \rho) > 0$, then $g(r, \rho)$ has a monotone ratio. Show the numerator of the above expression is positive by showing that for each α the sum on β is positive; use the fact that $c_{\alpha+1} < \frac{1}{2} c_\alpha$.]

17. (Sec. 4.2) Show that of all tests of ρ_0 against a specific ρ_1 $(> \rho_0)$ based on r, the procedures for which $r > c$ implies rejection are the best. [*Hint*: This follows from Problem 16.]

18. (Sec. 4.2) Show that of all tests of $\rho = \rho_0$ against $\rho > \rho_0$ based on r, a procedure for which $r > c$ implies rejection is uniformly most powerful.

19. (Sec. 4.2) Prove r has a monotone likelihood ratio for $r > 0$, $\rho > 0$ by proving $h(r) = k_N(r, \rho_1)/k_N(r, \rho_2)$ is monotonically increasing for $\rho_1 > \rho_2$. $h(r)$ is a constant times $(\sum_{\alpha=0}^{\infty} c_\alpha \rho_1^\alpha r^\alpha)/(\sum_{\alpha=0}^{\infty} c_\alpha \rho_2^\alpha r^\alpha)$. In the numerator of $h'(r)$, show that the coefficient of r^β is positive.

20. (Sec. 4.2) Prove that if $\mathbf{\Sigma}$ is diagonal then the sets r_{ij} and a_{ii} are independently distributed. [*Hint*: Use the facts that r_{ij} is invariant under scale transformations and that the density of the observations depends only on the a_{ii}.]

21. (Sec. 4.2.1) Prove that if $\rho = 0$

$$\mathscr{E} r^{2m} = \frac{\Gamma\left[\frac{1}{2}(N-1)\right]\Gamma\left(m + \frac{1}{2}\right)}{\sqrt{\pi}\,\Gamma\left[\frac{1}{2}(N-1) + m\right]}.$$

22. (Sec. 4.2.2) Prove $f_1(\rho)$ and $f_2(\rho)$ are monotonically increasing functions of ρ.

23. (Sec. 4.2.2) Prove that the density of the sample correlation r [given by (38)] is

$$\frac{n-1}{\pi}(1 - \rho^2)^{\frac{1}{2}n}(1 - r^2)^{\frac{1}{2}(n-3)} \int_0^1 \frac{x^{n-1}\,dx}{(1 - \rho r x)^n \sqrt{1 - x^2}}.$$

[*Hint*: Expand $(1 - \rho r x)^{-n}$ in a power series, integrate, and use the duplication formula for the gamma function.]

24. (Sec. 4.2) Prove that (39) is the density of r. [*Hint*: From Problem 12 of Chapter 2 show

$$\int_0^\infty \int_0^\infty e^{-\frac{1}{2}(y^2 - 2xyz + z^2)}\, dy\, dz = \frac{\cos^{-1}(-x)}{\sqrt{1-x^2}}.$$

Then argue

$$\int_0^\infty \int_0^\infty (yz)^{n-1} e^{-\frac{1}{2}(y^2 - 2xyz + z^2)}\, dy\, dz = \frac{d^{n-1}}{dx^{n-1}} \frac{\cos^{-1}(-x)}{\sqrt{1-x^2}}.$$

Finally show that the integral of (31) with respect to $a_{11}(= y^2)$ and $a_{22}(= z^2)$ is (39).]

25. (Sec. 4.2) Prove that (40) is the density of r. [*Hint*: In (31) let $a_{11} = ue^{-v}$ and $a_{22} = ue^{v}$; show that the density of v $(0 \le v < \infty)$ and r $(-1 \le r \le 1)$ is

$$\frac{n-1}{\pi\sqrt{2}}(1-\rho^2)^{\frac{1}{2}n}(1-\rho r)^{-n+\frac{1}{2}}(1-r^2)^{\frac{1}{2}(n-3)} v^{-\frac{1}{2}}(1-v)^{n-1}\left[1-\tfrac{1}{2}(1+\rho r)v\right]^{-\frac{1}{2}}.$$

Use the expansion

$$(1-y)^{-\frac{1}{2}} = \sum_{j=0}^{\infty} \frac{\Gamma(j+\frac{1}{2})}{\Gamma(\frac{1}{2})\, j!} y^j.$$

Show that the integral is (40).]

26. (Sec. 4.2) Prove for integer h

$$\mathscr{E} r^{2h+1} = \frac{(1-\rho^2)^{\frac{1}{2}n}}{\sqrt{\pi}\,\Gamma(\frac{1}{2}n)} \sum_{\beta=0}^{\infty} \frac{(2\rho)^{2\beta+1}}{(2\beta+1)!} \frac{\Gamma^2[\frac{1}{2}(n+1)+\beta]\Gamma(h+\beta+\frac{3}{2})}{\Gamma(\frac{1}{2}n+h+\beta+1)},$$

$$\mathscr{E} r^{2h} = \frac{(1-\rho^2)^{\frac{1}{2}n}}{\sqrt{\pi}\,\Gamma(\frac{1}{2}n)} \sum_{\beta=0}^{\infty} \frac{(2\rho)^{2\beta}}{(2\beta)!} \frac{\Gamma^2(\frac{1}{2}n+\beta)\Gamma(h+\beta+\frac{1}{2})}{\Gamma(\frac{1}{2}n+h+\beta)}.$$

27. (Sec. 4.2) *The t-distribution*. Prove that if X and Y are independently distributed, X having the distribution $N(0,1)$ and Y having the χ^2-distribution with m degrees of freedom, then $W = X/\sqrt{Y/m}$ has the density

$$\frac{\Gamma[\frac{1}{2}(m+1)]}{\sqrt{m}\sqrt{\pi}\,\Gamma(\frac{1}{2}m)}\left(1+\frac{t^2}{m}\right)^{-\frac{1}{2}(m+1)}.$$

[*Hint*: In the joint density of X and Y, let $x = tw^{\frac{1}{2}}m^{-\frac{1}{2}}$ and integrate out w.]

28. (Sec. 4.2) Prove

$$\mathscr{E} r = \frac{(1-\rho^2)^{\frac{1}{2}n}}{\Gamma(\frac{1}{2}n)} \sum_{\beta=0}^{\infty} \frac{\rho^{2\beta+1}\Gamma^2[\frac{1}{2}(n+1)+\beta]}{\beta!\Gamma[\frac{1}{2}n+\beta+1]}.$$

[*Hint*: Use Problem 26 and the duplication formula for the gamma function.]

29. (Sec. 4.2) Show that $\sqrt{n}(r_{ij} - \rho_{ij})$, $(i, j) = (1,2), (1,3), (2,3)$, have a joint limiting distribution with variances $(1-\rho_{ij}^2)^2$ and covariances of r_{ij} and r_{ik}, $j \neq k$, being $\frac{1}{2}(2\rho_{jk} - \rho_{ij}\rho_{jk})(1 - \rho_{ij}^2 - \rho_{ik}^2 - \rho_{jk}^2) + \rho_{jk}^2$.

30. (Sec. 4.3.2) Find a confidence interval for $\rho_{13\cdot 2}$ with confidence 0.95 based on $r_{13\cdot 2} = 0.097$ and $N = 20$.

31. (Sec. 4.3.2) Use Fisher's z to test the hypothesis $\rho_{12\cdot 34} = 0$ against alternatives $\rho_{12\cdot 34} \neq 0$ at significance level 0.01 with $r_{12\cdot 34} = 0.14$ and $N = 40$.

32. (Sec. 4.3) Show that the inequality $r_{12\cdot 3}^2 \leq 1$ is the same as the inequality $|r_{ij}| \geq 0$, where $|r_{ij}|$ denotes the determinant of the 3×3 correlation matrix.

33. (Sec. 4.3) *Invariance of the sample partial correlation coefficient*. Prove that $r_{12\cdot 3,\ldots,p}$ is invariant under the transformations $x_{i\alpha}^* = a_i x_{i\alpha} + \boldsymbol{b}_i' \boldsymbol{x}_\alpha^{(3)} + c_i$, $a_i > 0$, $i = 1,2$, $\boldsymbol{x}_\alpha^{(3)*} = \boldsymbol{C}\boldsymbol{x}_\alpha^{(3)} + \boldsymbol{b}$, $\alpha = 1, \ldots, N$, where $\boldsymbol{x}_\alpha^{(3)} = (x_{3\alpha}, \ldots, x_{p\alpha})'$, and that any function of $\bar{\boldsymbol{x}}$ and $\hat{\boldsymbol{\Sigma}}$ that is invariant under these transformations is a function of $r_{12\cdot 3,\ldots,p}$.

34. (Sec. 4.4) *Invariance of the sample multiple correlation coefficient*. Prove that R is a function of the sufficient statistics $\bar{\boldsymbol{x}}$ and $\boldsymbol{S}$ that is invariant under changes of location and scale of $x_{1\alpha}$ and nonsingular linear transformations of $\boldsymbol{x}_\alpha^{(2)}$ (that is, $x_{1\alpha}^* = c x_{1\alpha} + d$, $\boldsymbol{x}_\alpha^{(2)*} = \boldsymbol{C}\boldsymbol{x}_\alpha^{(2)} + \boldsymbol{d}$, $\alpha = 1, \ldots, N$) and that every function of $\bar{\boldsymbol{x}}$ and $\boldsymbol{S}$ that is invariant is a function of R.

35. (Sec. 4.4) Prove that conditional on $Z_{1\alpha} = z_{1\alpha}$, $\alpha = 1, \ldots, n$, $R^2/(1-R^2)$ is distributed like $T^2/(N^*-1)$, where $T^2 = N^* \bar{\boldsymbol{x}}' \boldsymbol{S}^{-1} \bar{\boldsymbol{x}}$ based on $N^* = n$ observations on a vector $\boldsymbol{X}$ with $p^* = p - 1$ components, with mean vector $(c/\sigma_{11})\boldsymbol{\sigma}_{(1)}'$ $(nc^2 = \Sigma z_{1\alpha}^2)$ and covariance matrix $\boldsymbol{\Sigma}_{22\cdot 1} = \boldsymbol{\Sigma}_{22} - (1/\sigma_{11})\boldsymbol{\sigma}_{(1)}\boldsymbol{\sigma}_{(1)}'$. [*Hint*: The conditional distribution of $\boldsymbol{Z}_\alpha^{(2)}$ given $Z_{1\alpha} = z_{1\alpha}$ is $N[(1/\sigma_{11})\boldsymbol{\sigma}_{(1)} z_{1\alpha}, \boldsymbol{\Sigma}_{22\cdot 1}]$. There is an $n \times n$ orthogonal matrix $\boldsymbol{B}$ which carries $(z_{11}, \ldots, z_{1n})$ into $(c, \ldots, c)$ and $(Z_{i1}, \ldots, Z_{in})$ into $(Y_{i1}, \ldots, Y_{in})$, $i = 2, \ldots, p$. Let the new $\boldsymbol{X}_\alpha'$ be $(Y_{2\alpha}, \ldots, Y_{p\alpha})$.]

36. (Sec. 4.4) Prove that the noncentrality parameter in the distribution in Problem 35 is $(a_{11}/\sigma_{11})\bar{R}^2/(1-\bar{R}^2)$.

37. (Sec. 4.4) Find the distribution of $R^2/(1-R^2)$ by multiplying the density of Problem 35 by the density of a_{11} and integrating with respect to a_{11}.

38. (Sec. 4.4) Show that the density of r^2 derived from (38) of Section 4.2 is identical with (42) in Section 4.4 for $p = 2$. [*Hint*: Use the duplication formula for the gamma function.]

39. (Sec. 4.4) Prove that (30) is the uniformly most powerful test of $\bar{R} = 0$ based on R. [*Hint*: Use the Neyman–Pearson fundamental lemma.]

40. (Sec. 4.4) Prove that (47) is the unique unbiased estimator of $\overline{R}^2$ based on R^2.

41. The estimates of $\boldsymbol{\mu}$ and $\boldsymbol{\Sigma}$ in Problem 1 of Chapter 3 are

$$\bar{\boldsymbol{x}} = (185.72 \quad 151.12 \qquad 183.84 \quad 149.24)'$$

$$\boldsymbol{S} = \left(\begin{array}{cc|cc} 95.2933 & 52.8683 & 69.6617 & 46.1117 \\ 52.8683 & 54.3600 & 51.3117 & 35.0533 \\ \hline 69.6617 & 51.3117 & 100.8067 & 56.5400 \\ 46.1117 & 35.0533 & 56.5400 & 45.0233 \end{array}\right).$$

(*a*) Find the estimates of the parameters of the conditional distribution of (x_3, x_4) given (x_1, x_2); that is, find $\boldsymbol{S}_{21}\boldsymbol{S}_{11}^{-1}$ and $\boldsymbol{S}_{22\cdot 1} = \boldsymbol{S}_{22} - \boldsymbol{S}_{21}\boldsymbol{S}_{11}^{-1}\boldsymbol{S}_{12}$.

(*b*) Find the partial correlation $r_{34\cdot 12}$.

(*c*) Use Fisher's z to find a confidence interval for $\rho_{34\cdot 12}$ with confidence 0.95.

(*d*) Find the sample multiple correlation coefficients between x_3 and (x_1, x_2) and between x_4 and (x_1, x_2).

(*e*) Test the hypotheses that x_3 is independent of (x_1, x_2) and x_4 is independent of (x_1, x_2) at significance level 0.05.

42. Let the components of $\boldsymbol{X}$ correspond to scores on tests in arithmetic speed (X_1), arithmetic power (X_2), memory for words (X_3), memory for meaningful symbols (X_4), and memory for meaningless symbols (X_5). The observed correlations in a sample of 140 are [Kelley (1928)]

$$\begin{pmatrix} 1.0000 & 0.4248 & 0.0420 & 0.0215 & 0.0573 \\ 0.4248 & 1.0000 & 0.1487 & 0.2489 & 0.2843 \\ 0.0420 & 0.1487 & 1.0000 & 0.6693 & 0.4662 \\ 0.0215 & 0.2489 & 0.6693 & 1.0000 & 0.6915 \\ 0.0573 & 0.2843 & 0.4662 & 0.6915 & 1.0000 \end{pmatrix}.$$

(*a*) Find the partial correlation between X_4 and X_5, holding X_3 fixed.

(*b*) Find the partial correlation between X_1 and X_2, holding X_3, X_4, and X_5 fixed.

(*c*) Find the multiple correlation between X_1 and the set X_3, X_4, and X_5.

(*d*) Test the hypothesis at the 1% significance level that arithmetic speed is independent of the three memory scores.

43. (Sec. 4.3) Prove that if $\rho_{ij\cdot q+1,\dots,p} = 0$, then $\sqrt{N-2-(p-q)}\, r_{ij\cdot q+1,\dots,p} / \sqrt{1 - r_{ij\cdot q+1,\dots,p}^2}$ is distributed according to the t-distribution with $N-2-(p-q)$ degrees of freedom.

44. (Sec. 4.3) Let $\boldsymbol{X}' = (X_1, X_2, \boldsymbol{X}^{(2)\prime})$ have the distribution $N(\boldsymbol{\mu}, \boldsymbol{\Sigma})$. The conditional distribution of X_1 given $X_2 = x_2$ and $\boldsymbol{X}^{(2)} = \boldsymbol{x}^{(2)}$ is

$$N\left[\mu_1 + \gamma_2(x_2 - \mu_2) + \boldsymbol{\gamma}'(\boldsymbol{x}^{(2)} - \boldsymbol{\mu}^{(2)}), \sigma_{11\cdot 2,\dots,p}\right],$$

where

$$\begin{pmatrix} \sigma_{22} & \boldsymbol{\sigma}'_{(2)} \\ \boldsymbol{\sigma}_{(2)} & \boldsymbol{\Sigma}_{22} \end{pmatrix} \begin{pmatrix} \gamma_2 \\ \boldsymbol{\gamma} \end{pmatrix} = \begin{pmatrix} \sigma_{12} \\ \boldsymbol{\sigma}_{(1)} \end{pmatrix}.$$

The estimators of γ_2 and $\boldsymbol{\gamma}$ are defined by

$$\begin{pmatrix} a_{22} & \boldsymbol{a}'_{(2)} \\ \boldsymbol{a}_{(2)} & \boldsymbol{A}_{22} \end{pmatrix} \begin{pmatrix} c_2 \\ \boldsymbol{c} \end{pmatrix} = \begin{pmatrix} a_{12} \\ \boldsymbol{a}_{(1)} \end{pmatrix}.$$

Show $c_2 = a_{12 \cdot 3, \dots, p} / a_{22 \cdot 3, \dots, p}$. [*Hint*: Solve for $\boldsymbol{c}$ in terms of c_2 and the a's and substitute.]

45. (Sec. 4.3) In the notation of Problem 44 prove

$$\begin{aligned} a_{11 \cdot 2, \dots, p} &= a_{11} - \boldsymbol{a}'_{(1)} \boldsymbol{A}_{22}^{-1} \boldsymbol{a}_{(1)} - c_2^2 \left(a_{22} - \boldsymbol{a}'_{(2)} \boldsymbol{A}_{22}^{-1} \boldsymbol{a}_{(2)} \right) \\ &= a_{11 \cdot 3, \dots, p} - c_2^2 a_{22 \cdot 3, \dots, p}. \end{aligned}$$

Hint: Use

$$a_{11 \cdot 2, \dots, p} = a_{11} - \begin{pmatrix} c_2 & \boldsymbol{c}' \end{pmatrix} \begin{pmatrix} a_{22} & \boldsymbol{a}'_{(2)} \\ \boldsymbol{a}_{(2)} & \boldsymbol{A}_{22} \end{pmatrix} \begin{pmatrix} c_2 \\ \boldsymbol{c} \end{pmatrix}.$$

46. (Sec. 4.3) Prove that $1/a_{22 \cdot 3, \dots, p}$ is the element in the upper left-hand corner of

$$\begin{pmatrix} a_{22} & \boldsymbol{a}'_{(2)} \\ \boldsymbol{a}_{(2)} & \boldsymbol{A}_{22} \end{pmatrix}^{-1}.$$

47. (Sec. 4.3) Using the results in Problems 43–46 prove that the test for $\rho_{12 \cdot 3, \dots, p} = 0$ is equivalent to the usual t-test that $\gamma_2 = 0$.

48. *Missing observations*. Let $\boldsymbol{X} = (\boldsymbol{Y}' \ \boldsymbol{Z}')'$, where $\boldsymbol{Y}$ has p components and $\boldsymbol{Z}$ has q components, be distributed according to $N(\boldsymbol{\mu}, \boldsymbol{\Sigma})$, where

$$\boldsymbol{\mu} = \begin{pmatrix} \boldsymbol{\mu}_y \\ \boldsymbol{\mu}_z \end{pmatrix}, \quad \boldsymbol{\Sigma} = \begin{pmatrix} \boldsymbol{\Sigma}_{yy} & \boldsymbol{\Sigma}_{yz} \\ \boldsymbol{\Sigma}_{zy} & \boldsymbol{\Sigma}_{zz} \end{pmatrix}.$$

Let M observations be made on $\boldsymbol{X}$ and $N - M$ additional observations be made on $\boldsymbol{Y}$. Find the maximum likelihood estimates of $\boldsymbol{\mu}$ and $\boldsymbol{\Sigma}$. [*Hint*: Express the likelihood function in terms of the marginal density of $\boldsymbol{Y}$ and the conditional density of $\boldsymbol{Z}$ given $\boldsymbol{Y}$.]

49. Suppose $\boldsymbol{X}$ is distributed according to $N(\mathbf{0}, \boldsymbol{\Sigma})$, where

$$\boldsymbol{\Sigma} = \begin{pmatrix} 1 & \rho & \rho^2 \\ \rho & 1 & \rho \\ \rho^2 & \rho & 1 \end{pmatrix}.$$

Show that on the basis of one observation, $\boldsymbol{x}' = (x_1, x_2, x_3)$, we can obtain a confidence interval for ρ (with confidence coefficient $1 - \alpha$) by using as endpoints of the interval the solutions in t of

$$\left(x_2^2 + \chi_3^2(\alpha)\right)t^2 - 2(x_1 x_2 + x_2 x_3)t + x_1^2 + x_2^2 + x_3^2 - \chi_3^2(\alpha) = 0,$$

where $\chi_3^2(\alpha)$ is the significance point of the χ^2-distribution with three degrees of freedom at significance level α.

CHAPTER 5

The Generalized T^2-Statistic

5.1. INTRODUCTION

One of the most important groups of problems in univariate statistics relates to questions concerning the mean of a given distribution when the variance of the distribution is unknown. On the basis of a sample one may wish to decide whether the mean is equal to a number specified in advance, or one may wish to give an interval within which the mean lies. The statistic usually used in univariate statistics is the difference between the mean of the sample $\bar{x}$ and the hypothetical population mean μ divided by the sample standard deviation s. If the distribution sampled is $N(\mu, \sigma^2)$, then

$$t = \sqrt{N}\,\frac{\bar{x} - \mu}{s} \tag{1}$$

has the well-known t-distribution with $N - 1$ degrees of freedom, where N is the number of observations in the sample. On the basis of this fact, one can set up a test of the hypothesis $\mu = \mu_0$, where μ_0 is specified, or one can set up a confidence interval for the unknown parameter μ.

The multivariate analogue of the square of t given in (1) is

$$T^2 = N(\bar{\boldsymbol{x}} - \boldsymbol{\mu})'\boldsymbol{S}^{-1}(\bar{\boldsymbol{x}} - \boldsymbol{\mu}), \tag{2}$$

where $\bar{\boldsymbol{x}}$ is the mean vector of a sample of N and $\boldsymbol{S}$ is the sample covariance matrix. It will be shown how this statistic can be used for testing hypotheses about the mean vector μ of the population and for obtaining confidence regions for the unknown μ. The distribution of T^2 will be obtained when μ in (2) is the mean of the distribution sampled and when μ is different from the population mean. Hotelling (1931) proposed the T^2-statistic for two samples and derived the distribution when μ is the population mean.

In Section 5.3 various uses of the T^2-statistic are presented, including simultaneous confidence intervals for all linear combinations of the mean vector. A James–Stein type estimator is given when Σ is unknown. The power function of the T^2-test is treated in Section 5.4 and the multivariate Behrens–Fisher problem in Section 5.5. In the last section optimum properties of the T^2-test are considered, both in terms of invariance and admissibility. Stein's criterion for admissibility in the general exponential family is proved and applied.

5.2. DERIVATION OF THE GENERALIZED T^2-STATISTIC AND ITS DISTRIBUTION

5.2.1. Derivation of the T^2-Statistic As a Function of the Likelihood Ratio Criterion

Although the T^2-statistic has many uses, we shall begin our discussion by showing that the likelihood ratio test of the hypothesis $H: \mu = \mu_0$ on the basis of a sample from $N(\mu, \Sigma)$ is based on the T^2-statistic given in (2) of Section 5.1. Suppose we have N observations $x_1, \dots, x_N$ $(N > p)$. The likelihood function is

$$(1) \qquad L(\mu, \Sigma) = (2\pi)^{-\frac{1}{2}pN} |\Sigma|^{-\frac{1}{2}N} \exp\left[-\tfrac{1}{2} \sum_{\alpha=1}^{N} (x_\alpha - \mu)' \Sigma^{-1} (x_\alpha - \mu) \right].$$

The observations are given; L is a function of the indeterminates μ, Σ. (We shall not distinguish in notation between the indeterminates and the parameters.) The likelihood ratio criterion is

$$(2) \qquad \lambda = \frac{\max_{\Sigma} L(\mu_0, \Sigma)}{\max_{\mu, \Sigma} L(\mu, \Sigma)},$$

that is, the numerator is the maximum of the likelihood function for $\boldsymbol{\mu}$, $\boldsymbol{\Sigma}$ in the parameter space restricted by the null hypothesis ($\boldsymbol{\mu} = \boldsymbol{\mu}_0$, $\boldsymbol{\Sigma}$ positive definite), and the denominator is the maximum over the entire parameter space ($\boldsymbol{\Sigma}$ positive definite). When the parameters are unrestricted, the maximum occurs when $\boldsymbol{\mu}$, $\boldsymbol{\Sigma}$ are defined by the maximum likelihood estimators (Section 3.2) of $\boldsymbol{\mu}$ and $\boldsymbol{\Sigma}$

$$(3) \qquad \hat{\boldsymbol{\mu}}_\Omega = \bar{\boldsymbol{x}},$$

$$(4) \qquad \hat{\boldsymbol{\Sigma}}_\Omega = \frac{1}{N} \sum_{\alpha=1}^{N} (\boldsymbol{x}_\alpha - \bar{\boldsymbol{x}})(\boldsymbol{x}_\alpha - \bar{\boldsymbol{x}})'.$$

When $\boldsymbol{\mu} = \boldsymbol{\mu}_0$, the likelihood function is maximized at

$$(5) \qquad \hat{\boldsymbol{\Sigma}}_\omega = \frac{1}{N} \sum_{\alpha=1}^{N} (\boldsymbol{x}_\alpha - \boldsymbol{\mu}_0)(\boldsymbol{x}_\alpha - \boldsymbol{\mu}_0)'$$

by Lemma 3.2.2. Furthermore, by Lemma 3.2.2

$$(6) \qquad \max_{\boldsymbol{\Sigma}, \boldsymbol{\mu}} L(\boldsymbol{\mu}, \boldsymbol{\Sigma}) = \frac{1}{(2\pi)^{\frac{1}{2}pN} |\hat{\boldsymbol{\Sigma}}_\Omega|^{\frac{1}{2}N}} e^{-\frac{1}{2}pN},$$

$$(7) \qquad \max_{\boldsymbol{\Sigma}} L(\boldsymbol{\mu}_0, \boldsymbol{\Sigma}) = \frac{1}{(2\pi)^{\frac{1}{2}pN} |\hat{\boldsymbol{\Sigma}}_\omega|^{\frac{1}{2}N}} e^{-\frac{1}{2}pN}.$$

Thus the likelihood ratio criterion is

$$(8) \qquad \lambda = \frac{|\hat{\boldsymbol{\Sigma}}_\Omega|^{\frac{1}{2}N}}{|\hat{\boldsymbol{\Sigma}}_\omega|^{\frac{1}{2}N}} = \frac{|\sum (\boldsymbol{x}_\alpha - \bar{\boldsymbol{x}})(\boldsymbol{x}_\alpha - \bar{\boldsymbol{x}})'|^{\frac{1}{2}N}}{|\sum (\boldsymbol{x}_\alpha - \boldsymbol{\mu}_0)(\boldsymbol{x}_\alpha - \boldsymbol{\mu}_0)'|^{\frac{1}{2}N}}$$

$$= \frac{|\boldsymbol{A}|^{\frac{1}{2}N}}{|\boldsymbol{A} + N(\bar{\boldsymbol{x}} - \boldsymbol{\mu}_0)(\bar{\boldsymbol{x}} - \boldsymbol{\mu}_0)'|^{\frac{1}{2}N}},$$

where

$$(9) \qquad \boldsymbol{A} = \sum_{\alpha=1}^{N} (\boldsymbol{x}_\alpha - \bar{\boldsymbol{x}})(\boldsymbol{x}_\alpha - \bar{\boldsymbol{x}})' = (N-1)\boldsymbol{S}.$$

Application of Corollary A.3.1 of Appendix A shows

$$\lambda^{2/N} = \frac{|A|}{\left|A + \left[\sqrt{N}(\bar{x} - \mu_0)\right]\left[\sqrt{N}(\bar{x} - \mu_0)\right]'\right|} \tag{10}$$

$$= \frac{1}{1 + N(\bar{x} - \mu_0)'A^{-1}(\bar{x} - \mu_0)}$$

$$= \frac{1}{1 + T^2/(N-1)},$$

where

$$T^2 = N(\bar{x} - \mu_0)'S^{-1}(\bar{x} - \mu_0) = (N-1)N(\bar{x} - \mu_0)'A^{-1}(\bar{x} - \mu_0). \tag{11}$$

The likelihood ratio test is defined by the critical region (region of rejection)

$$\lambda \le \lambda_0, \tag{12}$$

where λ_0 is chosen so that the probability of (12) when the null hypothesis is true is equal to the significance level. If we take the $\frac{1}{2}N$th root of both sides of (12) and invert, subtract 1, and multiply by $N - 1$, we obtain

$$T^2 \ge T_0^2, \tag{13}$$

where

$$T_0^2 = (N-1)\left(\lambda_0^{-2/N} - 1\right). \tag{14}$$

THEOREM 5.2.1. *The likelihood ratio test of the hypothesis $\mu = \mu_0$ for the distribution $N(\mu, \Sigma)$ is given by* (13), *where T^2 is defined by* (11), *$\bar{x}$ is the mean of a sample of N from $N(\mu, \Sigma)$, S is the covariance matrix of the sample, and T_0^2 is chosen so that the probability of* (13) *under the null hypothesis is equal to the chosen significance level.*

The Student t-test has the property that when testing $\mu = 0$ it is invariant with respect to scale transformations. If the scalar random variable X is distributed according to $N(\mu, \sigma^2)$, then $X^* = cX$ is distributed according to $N(c\mu, c^2\sigma^2)$ which is in the same class of distributions, and the hypothesis

$\mathscr{E}X = 0$ is equivalent to $\mathscr{E}X^* = \mathscr{E}cX = 0$. If the observations x_α are transformed similarly, $x_\alpha^* = cx_\alpha$, then, for $c > 0$, t^* computed from x_α^* is the same as t computed from x_α. Thus, whatever the unit of measurement the statistical result is the same.

The generalized T^2-test has a similar property. If the vector random variable $\mathbf{X}$ is distributed according to $N(\boldsymbol{\mu}, \boldsymbol{\Sigma})$, then $\mathbf{X}^* = \mathbf{C}\mathbf{X}$ (for $|\mathbf{C}| \neq \mathbf{0}$) is distributed according to $N(\mathbf{C}\boldsymbol{\mu}, \mathbf{C}\boldsymbol{\Sigma}\mathbf{C}')$ which is in the same class of distributions. The hypothesis $\mathscr{E}\mathbf{X} = \mathbf{0}$ is equivalent to the hypothesis $\mathscr{E}\mathbf{X}^* = \mathscr{E}\mathbf{C}\mathbf{X} = \mathbf{0}$. If the observations $\mathbf{x}_\alpha$ are transformed in the same way, $\mathbf{x}_\alpha^* = \mathbf{C}\mathbf{x}_\alpha$, then T^{*2} computed on the basis of $\mathbf{x}_\alpha^*$ is the same as T^2 computed on the basis of $\mathbf{x}_\alpha$. This follows from the facts that $\bar{\mathbf{x}}^* = \mathbf{C}\bar{\mathbf{x}}$ and $\mathbf{A}^* = \mathbf{C}\mathbf{A}\mathbf{C}'$ and the following lemma:

LEMMA 5.2.1. *For any $p \times p$ nonsingular matrices $\mathbf{C}$ and $\mathbf{H}$ and any vector $\mathbf{k}$,*

$$\mathbf{k}'\mathbf{H}^{-1}\mathbf{k} = (\mathbf{C}\mathbf{k})'(\mathbf{C}\mathbf{H}\mathbf{C}')^{-1}(\mathbf{C}\mathbf{k}). \tag{15}$$

PROOF. The right-hand side of (15) is

$$\begin{aligned}(\mathbf{C}\mathbf{k})'(\mathbf{C}\mathbf{H}\mathbf{C}')^{-1}(\mathbf{C}\mathbf{k}) &= \mathbf{k}'\mathbf{C}'(\mathbf{C}')^{-1}\mathbf{H}^{-1}\mathbf{C}^{-1}\mathbf{C}\mathbf{k} \\ &= \mathbf{k}'\mathbf{H}^{-1}\mathbf{k}.\end{aligned} \tag{16}$$

Q.E.D.

We shall show in Section 5.6 that of all tests invariant with regard to such transformations, (13) is the uniformly most powerful.

We can give a geometric interpretation of the $\frac{1}{2}N$th root of the likelihood ratio criterion

$$\lambda^{2/N} = \frac{\left|\sum_{\alpha=1}^{N}(\mathbf{x}_\alpha - \bar{\mathbf{x}})(\mathbf{x}_\alpha - \bar{\mathbf{x}})'\right|}{\left|\sum_{\alpha=1}^{N}(\mathbf{x}_\alpha - \boldsymbol{\mu}_0)(\mathbf{x}_\alpha - \boldsymbol{\mu}_0)'\right|} \tag{17}$$

in terms of parallelotopes. (See Section 7.5.) In the p-dimensional representation the numerator of $\lambda^{2/N}$ is the sum of squares of volumes of all parallelotopes with principal edges p vectors, each with one endpoint at $\bar{\mathbf{x}}$ and the other at an $\mathbf{x}_\alpha$. The denominator is the sum of squares of volumes of all parallelotopes with principal edges p vectors, each with one endpoint at $\boldsymbol{\mu}_0$ and the other at $\mathbf{x}_\alpha$. If the sum of squared volumes involving vectors emanating from $\bar{x}$, the "center" of the x_α, is much less than that involving vectors emanating from μ_0, then we reject the hypothesis that μ_0 is the mean of the distribution.

There is also an interpretation in the N-dimensional representation. Let $y_i = (x_{i1}, \dots, x_{iN})'$ be the ith vector. Then

$$\sqrt{N}\,\bar{x}_i = \sum_{\alpha=1}^{N} \frac{1}{\sqrt{N}} x_{i\alpha} \tag{18}$$

is the distance from the origin of the projection of y_i on the equiangular line (with direction cosines $1/\sqrt{N}, \dots, 1/\sqrt{N}$). The coordinates of the projection are $(\bar{x}_i, \dots, \bar{x}_i)$. Then $(x_{i1} - \bar{x}_i, \dots, x_{iN} - \bar{x}_i)$ is the projection of y_i on the plane through the origin perpendicular to the equiangular line. The numerator of $\lambda^{2/N}$ is the square of the p-dimensional volume of the parallelotope with principal edges, the vectors $(x_{i1} - \bar{x}_i, \dots, x_{iN} - \bar{x}_i)$. A point $(x_{i1} - \mu_{0i}, \dots, x_{iN} - \mu_{0i})$ is obtained from y_i by translation parallel to the equiangular line (by a distance $\sqrt{N}\,\mu_{0i}$). The denominator of $\lambda^{2/N}$ is the square of the volume of the parallelotope with principal edges these vectors. Then $\lambda^{2/N}$ is the ratio of these squared volumes.

5.2.2. The Distribution of T^2

In this section we will find the distribution of T^2 under general conditions, including the case when the null hypothesis is not true. Let $T^2 = Y'S^{-1}Y$ where Y is distributed according to $N(\nu, \Sigma)$ and nS is distributed independently as $\sum_{\alpha=1}^{n} Z_\alpha Z_\alpha'$ with $Z_1, \dots, Z_n$ independent, each with distribution $N(0, \Sigma)$. The T^2 defined in Section 5.2.1 is a special case of this with $Y = \sqrt{N}(\bar{x} - \mu_0)$ and $\nu = \sqrt{N}(\mu - \mu_0)$ and $n = N - 1$. Let D be a nonsingular matrix such that $D\Sigma D' = I$, and define

$$Y^* = DY, \qquad S^* = DSD', \qquad \nu^* = D\nu. \tag{19}$$

Then $T^2 = Y^{*\prime}S^{*-1}Y^*$ (by Lemma 5.2.1), where Y^* is distributed according to $N(\nu^*, I)$ and nS^* is distributed independently as $\sum_{\alpha=1}^{n} Z_\alpha^* Z_\alpha^{*\prime} = \sum_{\alpha=1}^{n} DZ_\alpha(DZ_\alpha)'$ with the $Z_\alpha^* = DZ_\alpha$ independent, each with distribution $N(0, I)$. We note $\nu'\Sigma^{-1}\nu = \nu^{*\prime}(I)^{-1}\nu^* = \nu^{*\prime}\nu^*$ by Lemma 5.2.1.

Let the first row of a $p \times p$ orthogonal matrix Q be defined by

$$q_{1i} = \frac{Y_i^*}{\sqrt{Y^{*\prime}Y^*}}, \qquad i = 1, \dots, p; \tag{20}$$

this is permissible since $\sum_{i=1}^{p} q_{1i}^2 = 1$. The other $p-1$ rows can be defined by some arbitrary rule (Lemma A.4.2 of Appendix A). Since $\boldsymbol{Q}$ depends on $\boldsymbol{Y}^*$, it is a random matrix. Now let

$$(21) \qquad \boldsymbol{U} = \boldsymbol{Q}\boldsymbol{Y}^*, \qquad \boldsymbol{B} = \boldsymbol{Q}n\boldsymbol{S}^*\boldsymbol{Q}'.$$

From the way $\boldsymbol{Q}$ was defined

$$(22) \qquad \begin{aligned} U_1 &= \sum q_{1i} Y_i^* = \sqrt{\boldsymbol{Y}^{*\prime}\boldsymbol{Y}^*}, \\ U_j &= \sum q_{ji} Y_i^* = \sqrt{\boldsymbol{Y}^{*\prime}\boldsymbol{Y}^*} \sum q_{ji} q_{1i} = 0, \qquad j \neq 1. \end{aligned}$$

Then

$$(23) \qquad \begin{aligned} \frac{T^2}{n} = \boldsymbol{U}'\boldsymbol{B}^{-1}\boldsymbol{U} &= (U_1, 0, \ldots, 0) \begin{pmatrix} b^{11} & b^{12} & \cdots & b^{1p} \\ b^{21} & b^{22} & \cdots & b^{2p} \\ \vdots & \vdots & & \vdots \\ b^{p1} & b^{p2} & \cdots & b^{pp} \end{pmatrix} \begin{pmatrix} U_1 \\ 0 \\ \vdots \\ 0 \end{pmatrix} \\ &= U_1^2 b^{11}, \end{aligned}$$

where $(b^{ij}) = \boldsymbol{B}^{-1}$. By Theorem A.3.3 of Appendix A, $1/b^{11} = b_{11} - \boldsymbol{b}_{(1)}'\boldsymbol{B}_{22}^{-1}\boldsymbol{b}_{(1)} = b_{11 \cdot 2, \ldots, p}$, where

$$(24) \qquad \boldsymbol{B} = \begin{pmatrix} b_{11} & \boldsymbol{b}_{(1)}' \\ \boldsymbol{b}_{(1)} & \boldsymbol{B}_{22} \end{pmatrix},$$

and $T^2/n = U_1^2/b_{11 \cdot 2, \ldots, p} = \boldsymbol{Y}^{*\prime}\boldsymbol{Y}^*/b_{11 \cdot 2, \ldots, p}$. The conditional distribution of $\boldsymbol{B}$ given $\boldsymbol{Q}$ is that of $\sum_{\alpha=1}^{n} \boldsymbol{V}_\alpha \boldsymbol{V}_\alpha'$, where conditionally the $\boldsymbol{V}_\alpha = \boldsymbol{Q}\boldsymbol{Z}_\alpha^*$ are independent, each with distribution $N(\boldsymbol{0}, \boldsymbol{I})$. By Theorem 4.3.3 $b_{11 \cdot 2, \ldots, p}$ is conditionally distributed as $\sum_{\alpha=1}^{n-(p-1)} W_\alpha^2$, where conditionally the W_α are independent, each with the distribution $N(0,1)$; that is, $b_{11 \cdot 2, \ldots, p}$ is conditionally distributed as χ^2 with $n-(p-1)$ degrees of freedom. Since the conditional distribution of $b_{11 \cdot 2, \ldots, p}$ does not depend on $\boldsymbol{Q}$, it is unconditionally distributed as χ^2. The quantity $\boldsymbol{Y}^{*\prime}\boldsymbol{Y}^*$ has a noncentral χ^2-distribution with p degrees of freedom and noncentrality parameter $\boldsymbol{\nu}^{*\prime}\boldsymbol{\nu}^* = \boldsymbol{\nu}'\boldsymbol{\Sigma}^{-1}\boldsymbol{\nu}$. Then T^2/n is distributed as the ratio of a noncentral χ^2 and an independent χ^2.

THEOREM 5.2.2. *Let $T^2 = \boldsymbol{Y}'\boldsymbol{S}^{-1}\boldsymbol{Y}$, where $\boldsymbol{Y}$ is distributed according to $N(\boldsymbol{\nu}, \boldsymbol{\Sigma})$ and $n\boldsymbol{S}$ is independently distributed as $\sum_{\alpha=1}^{n}\boldsymbol{Z}_\alpha\boldsymbol{Z}_\alpha'$ with $\boldsymbol{Z}_1, \dots, \boldsymbol{Z}_n$ independent, each with distribution $N(\boldsymbol{0}, \boldsymbol{\Sigma})$. Then $(T^2/n)[(n-p+1)/p]$ is distributed as a noncentral F with p and $n-p+1$ degrees of freedom and noncentrality parameter $\boldsymbol{\nu}'\boldsymbol{\Sigma}^{-1}\boldsymbol{\nu}$. If $\boldsymbol{\nu} = \boldsymbol{0}$, the distribution is central F.*

We shall call this the T^2-distribution with n degrees of freedom.

COROLLARY 5.2.1. *Let $\boldsymbol{x}_1, \dots, \boldsymbol{x}_N$ be a sample from $N(\boldsymbol{\mu}, \boldsymbol{\Sigma})$, and let $T^2 = N(\bar{\boldsymbol{x}} - \boldsymbol{\mu}_0)'\boldsymbol{S}^{-1}(\bar{\boldsymbol{x}} - \boldsymbol{\mu}_0)$. The distribution of $[T^2/(N-1)][(N-p)/p]$ is noncentral F with p and $N-p$ degrees of freedom and noncentrality parameter $N(\boldsymbol{\mu} - \boldsymbol{\mu}_0)'\boldsymbol{\Sigma}^{-1}(\boldsymbol{\mu} - \boldsymbol{\mu}_0)$. If $\boldsymbol{\mu} = \boldsymbol{\mu}_0$, then the F-distribution is central.*

The above derivation of the T^2-distribution is due to Bowker (1960). The noncentral F-density and tables of the distribution are discussed in Section 5.4.

For large samples the distribution of T^2 given by Corollary 5.2.1 is approximately valid even if the parent distribution is not normal; in this sense the T^2-test is a robust procedure.

THEOREM 5.2.3. *Let $\{\boldsymbol{X}_\alpha\}$, $\alpha = 1, 2, \dots,$ be a sequence of independently identically distributed random vectors with mean vector $\boldsymbol{\mu}$ and covariance matrix $\boldsymbol{\Sigma}$; let $\bar{\boldsymbol{X}}_N = (1/N)\sum_{\alpha=1}^{N}\boldsymbol{X}_\alpha$, $\boldsymbol{S}_N = [1/(N-1)]\sum_{\alpha=1}^{N}(\boldsymbol{X}_\alpha - \bar{\boldsymbol{X}}_N)(\boldsymbol{X}_\alpha - \bar{\boldsymbol{X}}_N)'$, and $T_N^2 = N(\bar{\boldsymbol{X}}_N - \boldsymbol{\mu}_0)'\boldsymbol{S}_N^{-1}(\bar{\boldsymbol{X}}_N - \boldsymbol{\mu}_0)$. Then the limiting distribution of T_N^2 as $N \to \infty$ is the χ^2-distribution with p degrees of freedom if $\boldsymbol{\mu} = \boldsymbol{\mu}_0$.*

PROOF. By the central limit theorem (Theorem 4.2.3) the limiting distribution of $\sqrt{N}(\bar{\boldsymbol{X}}_N - \boldsymbol{\mu})$ is $N(\boldsymbol{0}, \boldsymbol{\Sigma})$. The sample covariance matrix converges stochastically to $\boldsymbol{\Sigma}$. Then the limiting distribution of T_N^2 is the distribution of $\boldsymbol{Y}'\boldsymbol{\Sigma}^{-1}\boldsymbol{Y}$, when $\boldsymbol{Y}$ has the distribution $N(\boldsymbol{0}, \boldsymbol{\Sigma})$. The theorem follows from Theorem 3.3.3. Q.E.D.

When the null hypothesis is true, T^2/n is distributed as χ_p^2/χ_{n-p+1}^2 and $\lambda^{2/N}$ given by (10) has the distribution of $\chi_{n-p+1}^2/(\chi_{n-p+1}^2 + \chi_p^2)$. The density of $V = \chi_a^2/(\chi_a^2 + \chi_b^2)$, when χ_a^2 and χ_b^2 are independent is

$$\frac{\Gamma[\frac{1}{2}(a+b)]}{\Gamma(\frac{1}{2}a)\Gamma(\frac{1}{2}b)} v^{\frac{1}{2}a-1}(1-v)^{\frac{1}{2}b-1} = \beta(v; \tfrac{1}{2}a, \tfrac{1}{2}b); \tag{25}$$

this is the density of the *beta distribution* with parameters $\frac{1}{2}a$ and $\frac{1}{2}b$ (Problem 27). Thus the distribution of $\lambda^{2/N} = (1 + T^2/n)^{-1}$ is the beta distribution with parameters $\frac{1}{2}p$ and $\frac{1}{2}(n-p+1)$.

5.3. USES OF THE T^2-STATISTIC

5.3.1. Testing the Hypothesis That the Mean Vector Is a Given Vector

The likelihood ratio test of the hypothesis $\mu = \mu_0$ on the basis of a sample of N from $N(\mu, \Sigma)$ is equivalent to

$$T^2 \geq T_0^2 \tag{1}$$

as given in Section 5.2.1. If the significance level is α, then the $100\alpha\%$ point of the F-distribution is taken, that is,

$$T_0^2 = \frac{(N-1)p}{N-p} F_{p,N-p}(\alpha) = T_{p,N-1}^2(\alpha), \tag{2}$$

say. The choice of significance level may depend on the power of the test. We shall discuss this in Section 5.4.

The statistic T^2 is easily computed from the sample. Let

$$\boldsymbol{A}^{-1}(\bar{\boldsymbol{x}} - \mu_0) = \boldsymbol{b}. \tag{3}$$

This vector is the solution of

$$\boldsymbol{A}\boldsymbol{b} = \bar{\boldsymbol{x}} - \mu_0. \tag{4}$$

If the vector $\boldsymbol{b}$ is obtained by solving (4), then

$$\frac{T^2}{N-1} = N(\bar{\boldsymbol{x}} - \mu_0)'\boldsymbol{b}. \tag{5}$$

Thus it is unnecessary to compute $\boldsymbol{A}^{-1}$ or $\boldsymbol{S}^{-1}$. In fact, if one uses a method of Gaussian elimination for the solution of (4), one need only proceed through the forward solution. In Section A.5 of Appendix A it is shown that the forward solution amounts to multiplying (4) on the left by a matrix $\boldsymbol{F}$ (such that $\boldsymbol{FA}$ is triangular) and this result by $\boldsymbol{D}^{-2}$, where $\boldsymbol{D}^2$ is a diagonal matrix with diagonal elements equal to the corresponding diagonal elements of $\boldsymbol{FA}$. On the right-hand side one has $\boldsymbol{F}(\bar{\boldsymbol{x}} - \mu_0)$ and $\boldsymbol{D}^{-2}\boldsymbol{F}(\bar{\boldsymbol{x}} - \mu_0)$, respectively.

Then $[F(\bar{x} - \mu_0)]'[D^{-2}F(\bar{x} - \mu_0)] = (\bar{x} - \mu_0)'F'D^{-2}F(\bar{x} - \mu_0) = (\bar{x} - \mu_0)'A^{-1}(\bar{x} - \mu_0)$ because $F'D^{-2}F = A^{-1}$. The details of the procedure and proof are given in Section A.5.

It is of interest to note that $T^2/(N-1)$ is the nonzero root of

$$|N(\bar{x} - \mu_0)(\bar{x} - \mu_0)' - \lambda A| = 0. \tag{6}$$

LEMMA 5.3.1. *If v is a vector of p components and if B is a nonsingular $p \times p$ matrix, then $v'B^{-1}v$ is the nonzero root of*

$$|vv' - \lambda B| = 0. \tag{7}$$

PROOF. The nonzero root, say λ_1, of (7) is associated with a characteristic vector β satisfying

$$vv'\beta = \lambda_1 B\beta. \tag{8}$$

Since $\lambda_1 \neq 0$, $v'\beta \neq 0$. Multiplying on the left by $v'B^{-1}$, we obtain

$$(v'B^{-1}v)(v'\beta) = \lambda_1(v'\beta); \tag{9}$$

this proves the lemma. In the case above $v = \sqrt{N}(\bar{x} - \mu_0)$ and $B = A$.

5.3.2. A Confidence Region for the Mean Vector

If μ is the mean of $N(\mu, \Sigma)$, the probability is $1 - \alpha$ of drawing a sample of N with mean $\bar{x}$ and covariance matrix S such that

$$N(\bar{x} - \mu)'S^{-1}(\bar{x} - \mu) \leq T^2_{p,N-1}(\alpha). \tag{10}$$

Thus, if we compute (10) for a particular sample, we have confidence $1 - \alpha$ that (10) is a true statement concerning μ. The inequality

$$N(\bar{x} - m)'S^{-1}(\bar{x} - m) \leq T^2_{p,N-1}(\alpha) \tag{11}$$

is the interior and boundary of an ellipsoid in the p-dimensional space of m with center at $\bar{x}$ and with size and shape depending on S^{-1} and α. See Figure 5.1. We state that μ lies within this ellipsoid with confidence $1 - \alpha$. Over random samples (11) is a random ellipsoid.

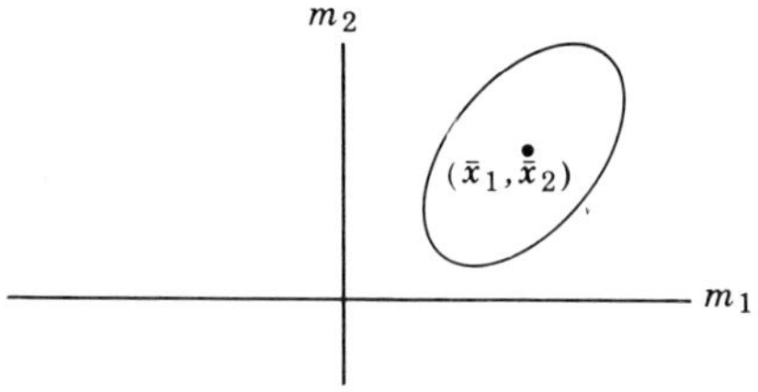

Figure 5.1. A confidence ellipse.

5.3.3. Simultaneous Confidence Intervals for All Linear Combinations of the Mean Vector

From the confidence region (11) for $\boldsymbol{\mu}$ we can obtain confidence intervals for linear functions $\boldsymbol{\gamma}'\boldsymbol{\mu}$ that hold simultaneously with a given confidence coefficient.

LEMMA 5.3.2. *For a positive definite matrix* $\boldsymbol{S}$

$$(\boldsymbol{\gamma}'\boldsymbol{y})^2 \le \boldsymbol{\gamma}'\boldsymbol{S}\boldsymbol{\gamma}\, \boldsymbol{y}'\boldsymbol{S}^{-1}\boldsymbol{y}. \tag{12}$$

PROOF. Let $b = \boldsymbol{\gamma}'\boldsymbol{y}/\boldsymbol{\gamma}'\boldsymbol{S}\boldsymbol{\gamma}$. Then

$$\begin{aligned} 0 &\le (\boldsymbol{y} - b\boldsymbol{S}\boldsymbol{\gamma})'\boldsymbol{S}^{-1}(\boldsymbol{y} - b\boldsymbol{S}\boldsymbol{\gamma}) \\ &= \boldsymbol{y}'\boldsymbol{S}^{-1}\boldsymbol{y} - b\boldsymbol{\gamma}'\boldsymbol{S}\boldsymbol{S}^{-1}\boldsymbol{y} - \boldsymbol{y}'\boldsymbol{S}^{-1}\boldsymbol{S}\boldsymbol{\gamma}b + b^2\boldsymbol{\gamma}'\boldsymbol{S}\boldsymbol{S}^{-1}\boldsymbol{S}\boldsymbol{\gamma} \\ &= \boldsymbol{y}'\boldsymbol{S}^{-1}\boldsymbol{y} - \frac{(\boldsymbol{\gamma}'\boldsymbol{y})^2}{\boldsymbol{\gamma}'\boldsymbol{S}\boldsymbol{\gamma}}, \end{aligned} \tag{13}$$

which yields (12). Note this is a generalization of the Cauchy–Schwarz inequality.

When $\boldsymbol{y} = \bar{\boldsymbol{x}} - \boldsymbol{\mu}$, then (12) implies

$$\begin{aligned} |\boldsymbol{\gamma}'(\bar{\boldsymbol{x}} - \boldsymbol{\mu})| &\le \sqrt{\boldsymbol{\gamma}'\boldsymbol{S}\boldsymbol{\gamma}(\bar{\boldsymbol{x}} - \boldsymbol{\mu})'\boldsymbol{S}^{-1}(\bar{\boldsymbol{x}} - \boldsymbol{\mu})} \\ &\le \sqrt{\boldsymbol{\gamma}'\boldsymbol{S}\boldsymbol{\gamma}}\sqrt{T^2_{p,N-1}(\alpha)/N} \end{aligned} \tag{14}$$

holds for all $\boldsymbol{\gamma}$ with probability $1 - \alpha$. Thus we can assert with confidence $1 - \alpha$ that the unknown parameter vector satisfies simultaneously for all $\boldsymbol{\gamma}$ the inequalities

$$|\gamma'\bar{x} - \gamma' m| \leq \sqrt{\gamma' S \gamma}\sqrt{T^2_{p,N-1}(\alpha)/N}. \tag{15}$$

The confidence region (11) can be explored by setting γ in (15) equal to simple vectors such as $(1, 0, \ldots, 0)'$ to obtain m_1, $(1, -1, 0, \ldots, 0)$ to yield $m_1 - m_2$, and so on. It should be noted that if only one linear function $\gamma'\mu$ were of interest, $\sqrt{T^2_{p,N-1}(\alpha)} = \sqrt{npF_{p,n-p+1}(\alpha)/(n-p+1)}$ would be replaced by $t_n(\alpha)$.

5.3.4. Two-Sample Problems

Another situation in which the T^2-statistic is used is one in which the null hypothesis is that the mean of one normal population is equal to the mean of the other where the covariance matrices are assumed equal but unknown. Suppose $y_1^{(i)}, \ldots, y_{N_i}^{(i)}$ is a sample from $N(\mu^{(i)}, \Sigma)$, $i = 1, 2$. We wish to test the null hypothesis $\mu^{(1)} = \mu^{(2)}$. The vector $\bar{y}^{(i)}$ is distributed according to $N[\mu^{(i)}, (1/N_i)\Sigma]$. Consequently, $\sqrt{N_1 N_2/(N_1 + N_2)}\,(\bar{y}^{(1)} - \bar{y}^{(2)})$ is distributed according to $N(\mathbf{0}, \Sigma)$ under the null hypothesis. If we let

$$S = \frac{1}{N_1 + N_2 - 2}\left\{\sum_{\alpha=1}^{N_1}(y_\alpha^{(1)} - \bar{y}^{(1)})(y_\alpha^{(1)} - \bar{y}^{(1)})' + \sum_{\alpha=1}^{N_2}(y_\alpha^{(2)} - \bar{y}^{(2)})(y_\alpha^{(2)} - \bar{y}^{(2)})'\right\}, \tag{16}$$

then $(N_1 + N_2 - 2)S$ is distributed as $\sum_{\alpha=1}^{N_1+N_2-2} Z_\alpha Z_\alpha'$, where Z_α is distributed according to $N(\mathbf{0}, \Sigma)$. Thus

$$T^2 = \frac{N_1 N_2}{N_1 + N_2}(\bar{y}^{(1)} - \bar{y}^{(2)})' S^{-1}(\bar{y}^{(1)} - \bar{y}^{(2)}) \tag{17}$$

is distributed as T^2 with $N_1 + N_2 - 2$ degrees of freedom. The critical region is

$$T^2 > \frac{(N_1 + N_2 - 2)p}{N_1 + N_2 - p - 1} F_{p, N_1+N_2-p-1}(\alpha) \tag{18}$$

with significance level α.

A confidence region for $\boldsymbol{\mu}^{(1)} - \boldsymbol{\mu}^{(2)}$ with confidence level $1 - \alpha$ is the set of vectors $\boldsymbol{m}$ satisfying

(19)

$$(\bar{\boldsymbol{y}}^{(1)} - \bar{\boldsymbol{y}}^{(2)} - \boldsymbol{m})'\boldsymbol{S}^{-1}(\bar{\boldsymbol{y}}^{(1)} - \bar{\boldsymbol{y}}^{(2)} - \boldsymbol{m}) \le \frac{N_1 + N_2}{N_1 N_2} T^2_{p,N_1+N_2-2}(\alpha)$$

$$= \frac{N_1 + N_2}{N_1 N_2} \frac{(N_1 + N_2 - 2)p}{N_1 + N_2 - p - 1} F_{p,N_1+N_2-p-1}(\alpha).$$

Simultaneous confidence intervals are

$$|\boldsymbol{\gamma}'(\bar{\boldsymbol{y}}^{(1)} - \bar{\boldsymbol{y}}^{(2)}) - \boldsymbol{\gamma}'\boldsymbol{m}| \le \sqrt{\boldsymbol{\gamma}'\boldsymbol{S}\boldsymbol{\gamma}} \sqrt{\frac{N_1 + N_2}{N_1 N_2} T^2_{p,N_1+N_2-2}(\alpha)}\,. \tag{20}$$

An example may be taken from Fisher (1936). Let x_1 = sepal length, x_2 = sepal width, x_3 = petal length, x_4 = petal width. Fifty observations are taken from the population Iris versicolor (1) and 50 from the population Iris setosa (2). See Table 3.4. The data may be summarized (in centimeters) as

$$\bar{\boldsymbol{x}}^{(1)} = \begin{pmatrix} 5.936 \\ 2.770 \\ 4.260 \\ 1.326 \end{pmatrix}, \tag{21}$$

$$\bar{\boldsymbol{x}}^{(2)} = \begin{pmatrix} 5.006 \\ 3.428 \\ 1.462 \\ 0.246 \end{pmatrix}, \tag{22}$$

$$98\boldsymbol{S} = \begin{pmatrix} 19.1434 & 9.0356 & 9.7634 & 3.2394 \\ 9.0356 & 11.8658 & 4.6232 & 2.4746 \\ 9.7634 & 4.6232 & 12.2978 & 3.8794 \\ 3.2394 & 2.4746 & 3.8794 & 2.4604 \end{pmatrix}. \tag{23}$$

The value of $T^2/98$ is 26.334 and $T^2/98 \times 95/4 = 625.5$. This value is highly significant compared to the F-value for 4 and 95 degrees of freedom of 3.52 at the 0.01 significance level.

Simultaneous confidence intervals for the differences of component means $\mu_i^{(1)} - \mu_i^{(2)}$, $i = 1, 2, 3, 4$, are 0.930 ± 0.337, -0.658 ± 0.265, -2.798 ± 0.270, and 1.080 ± 0.121. In each case 0 does not lie in the interval. (Since $t_{98}(.01) < T_{4,98}(.01)$, a univariate test on any component would lead to rejection of the null hypothesis.) The last two components show the most significant differences from 0.

5.3.5. A Problem of Several Samples

After considering the above example, Fisher considers a third sample drawn from a population assumed to have the same covariance matrix. He treats the same measurements on 50 Iris virginica (Table 3.4). There is a theoretical reason for believing the gene structures of these three species to be such that the mean vectors of the three populations are related as

$$3\boldsymbol{\mu}^{(1)} = \boldsymbol{\mu}^{(3)} + 2\boldsymbol{\mu}^{(2)}, \tag{24}$$

where $\boldsymbol{\mu}^{(3)}$ is the mean vector of the third population.

This is a special case of the following general problem. Let $\{\boldsymbol{x}_\alpha^{(i)}\}$, $\alpha = 1,\dots,N_i$; $i = 1,\dots,q$, be samples from $N(\boldsymbol{\mu}^{(i)}, \boldsymbol{\Sigma})$, $i = 1,\dots,q$, respectively. Let us test the hypothesis

$$H\colon \sum_{i=1}^{q} \beta_i \boldsymbol{\mu}^{(i)} = \boldsymbol{\mu}, \tag{25}$$

where $\beta_1,\dots,\beta_q$ are given scalars and $\boldsymbol{\mu}$ is a given vector. The criterion is

$$T^2 = c\left(\sum_{i=1}^{q} \beta_i \bar{\boldsymbol{x}}^{(i)} - \boldsymbol{\mu}\right)' \boldsymbol{S}^{-1} \left(\sum_{i=1}^{q} \beta_i \bar{\boldsymbol{x}}^{(i)} - \boldsymbol{\mu}\right), \tag{26}$$

where

$$\bar{\boldsymbol{x}}^{(i)} = \frac{1}{N_i} \sum_{\alpha=1}^{N_i} \boldsymbol{x}_\alpha^{(i)}, \tag{27}$$

$$\left(\sum_{i=1}^{q} N_i - q\right)\boldsymbol{S} = \sum_{i=1}^{q} \sum_{\alpha=1}^{N_i} \left(\boldsymbol{x}_\alpha^{(i)} - \bar{\boldsymbol{x}}^{(i)}\right)\left(\boldsymbol{x}_\alpha^{(i)} - \bar{\boldsymbol{x}}^{(i)}\right)', \tag{28}$$

$$\frac{1}{c} = \sum_{i=1}^{q} \frac{\beta_i^2}{N_i}. \tag{29}$$

This T^2 has the T^2-distribution with $\sum_{i=1}^{q} N_i - q$ degrees of freedom.

Fisher actually assumes in his example that the covariance matrices of the three populations may be different. Hence he uses the technique described in Section 5.5.

5.3.6. A Problem of Symmetry

Consider testing the hypothesis $H: \mu_1 = \mu_2 = \cdots = \mu_p$ on the basis of a sample $\boldsymbol{x}_1, \dots, \boldsymbol{x}_N$ from $N(\boldsymbol{\mu}, \boldsymbol{\Sigma})$, where $\boldsymbol{\mu}' = (\mu_1, \dots, \mu_p)$. Let $\boldsymbol{C}$ be any $(p-1) \times p$ matrix of rank $p-1$ such that

$$\boldsymbol{C}\boldsymbol{\varepsilon} = \boldsymbol{0}, \tag{30}$$

where $\boldsymbol{\varepsilon}' = (1, \dots, 1)$. Then

$$\boldsymbol{y}_\alpha = \boldsymbol{C}\boldsymbol{x}_\alpha, \qquad \alpha = 1, \dots, N, \tag{31}$$

has mean $\boldsymbol{C}\boldsymbol{\mu}$ and covariance matrix $\boldsymbol{C}\boldsymbol{\Sigma}\boldsymbol{C}'$. The hypothesis H is $\boldsymbol{C}\boldsymbol{\mu} = \boldsymbol{0}$. The statistic to be used is

$$T^2 = N\bar{\boldsymbol{y}}'\boldsymbol{S}^{-1}\bar{\boldsymbol{y}}, \tag{32}$$

where

$$\bar{\boldsymbol{y}} = \frac{1}{N}\sum_{\alpha=1}^{N} \boldsymbol{y}_\alpha = \boldsymbol{C}\bar{\boldsymbol{x}}, \tag{33}$$

$$\begin{aligned} \boldsymbol{S} &= \frac{1}{N-1}\sum_{\alpha=1}^{N}(\boldsymbol{y}_\alpha - \bar{\boldsymbol{y}})(\boldsymbol{y}_\alpha - \bar{\boldsymbol{y}})' \\ &= \frac{1}{N-1}\boldsymbol{C}\sum_{\alpha=1}^{N}(\boldsymbol{x}_\alpha - \bar{\boldsymbol{x}})(\boldsymbol{x}_\alpha - \bar{\boldsymbol{x}})'\boldsymbol{C}'. \end{aligned} \tag{34}$$

This statistic has the T^2-distribution with $N-1$ degrees of freedom for a $(p-1)$-dimensional distribution. This T^2-statistic is invariant under any linear transformation in the $p-1$ dimensions orthogonal to $\boldsymbol{\varepsilon}$. Hence the statistic is independent of the choice of $\boldsymbol{C}$.

An example of this sort has been given by Rao (1948b). Let N be the amount of cork in a boring from the north into a cork tree; let E, S, and W be defined similarly. The set of amounts in four borings on one tree is considered as an observation from a 4-variate normal distribution. The question is whether the cork trees have the same amount of cork on each side. We make a transformation

$$\begin{aligned} y_1 &= N - E - W + S, \\ y_2 &= S - W, \\ y_3 &= N - S. \end{aligned} \tag{35}$$

The number of observations is 28. The vector of means is

$$\bar{y} = \begin{pmatrix} 8.86 \\ 4.50 \\ 0.86 \end{pmatrix}; \tag{36}$$

the covariance matrix for y is

$$S = \begin{pmatrix} 128.72 & 61.41 & -21.02 \\ 61.41 & 56.93 & -28.30 \\ -21.02 & -28.30 & 63.53 \end{pmatrix}. \tag{37}$$

The value of $T^2/(N-1)$ is 0.768. The statistic $0.768 \times 25/3 = 6.402$ is to be compared with the F-significance point with 3 and 25 degrees of freedom. It is significant at the 1% level.

5.3.7. Improved Estimation of the Mean

In Section 3.5 we considered estimation of the mean when the covariance matrix was known and showed that the Stein-type estimation based on this knowledge yielded lower quadratic risks than did the sample mean. In particular, if the loss is $(\boldsymbol{m}-\boldsymbol{\mu})'\boldsymbol{\Sigma}^{-1}(\boldsymbol{m}-\boldsymbol{\mu})$ then

$$\left(1 - \frac{p-2}{N(\bar{\boldsymbol{x}}-\boldsymbol{\nu})'\boldsymbol{\Sigma}^{-1}(\bar{\boldsymbol{x}}-\boldsymbol{\nu})}\right)^{+}(\bar{\boldsymbol{x}}-\boldsymbol{\nu})+\boldsymbol{\nu} \tag{38}$$

is a minimax estimator of $\boldsymbol{\mu}$ for any $\boldsymbol{\nu}$ and has a smaller risk than $\bar{\boldsymbol{x}}$ when $p \geq 3$. When $\boldsymbol{\Sigma}$ is unknown, we consider replacing it by an estimator, namely, a multiple of $\boldsymbol{A} = n\boldsymbol{S}$.

THEOREM 5.3.1. *When the loss is* $(\boldsymbol{m}-\boldsymbol{\mu})'\boldsymbol{\Sigma}^{-1}(\boldsymbol{m}-\boldsymbol{\mu})$, *the estimator for* $p \geq 3$

$$\left(1 - \frac{a}{N(\bar{\boldsymbol{x}}-\boldsymbol{\nu})'\boldsymbol{A}^{-1}(\bar{\boldsymbol{x}}-\boldsymbol{\nu})}\right)(\bar{\boldsymbol{x}}-\boldsymbol{\nu})+\boldsymbol{\nu} \tag{39}$$

has smaller risk than $\bar{\boldsymbol{x}}$ *and is minimax for* $0 < a < 2(p-2)/(n-p+3)$, *and the risk is minimized for* $a = (p-2)/(n-p+3)$.

PROOF. As in the case when $\boldsymbol{\Sigma}$ is known (Section 3.5.2), we can make a transformation that carries $(1/N)\boldsymbol{\Sigma}$ to $\boldsymbol{I}$. Then the problem is to estimate $\boldsymbol{\mu}$

based on $\boldsymbol{Y}$ with the distribution $N(\boldsymbol{\mu}, \boldsymbol{I})$ and $\boldsymbol{A} = \sum_{\alpha=1}^{n} \boldsymbol{Z}_\alpha \boldsymbol{Z}_\alpha'$, where $\boldsymbol{Z}_1, \dots, \boldsymbol{Z}_n$ are independently distributed, each according to $N(\boldsymbol{0}, \boldsymbol{I})$, and the loss is $(\boldsymbol{m} - \boldsymbol{\mu})'(\boldsymbol{m} - \boldsymbol{\mu})$. (We have dropped a factor of N.) The difference in risks is

(40)

$$\Delta R(\boldsymbol{\mu}) = \mathscr{E}_{\boldsymbol{\mu}}\left\{ \|\boldsymbol{Y} - \boldsymbol{\mu}\|^2 - \left\| \left(1 - \frac{a}{(\boldsymbol{Y} - \boldsymbol{\nu})'\boldsymbol{A}^{-1}(\boldsymbol{Y} - \boldsymbol{\nu})}\right)(\boldsymbol{Y} - \boldsymbol{\nu}) + \boldsymbol{\nu} - \boldsymbol{\mu} \right\|^2 \right\}$$

$$= \mathscr{E}_{\boldsymbol{\mu}}\left\{ 2\frac{a}{(\boldsymbol{Y} - \boldsymbol{\nu})'\boldsymbol{A}^{-1}(\boldsymbol{Y} - \boldsymbol{\nu})} \sum_{i=1}^{p} (Y_i - \mu_i)(Y_i - \nu_i) \right.$$

$$\left. - \frac{a^2}{\left[(\boldsymbol{Y} - \boldsymbol{\nu})'\boldsymbol{A}^{-1}(\boldsymbol{Y} - \boldsymbol{\nu})\right]^2} \|\boldsymbol{Y} - \boldsymbol{\nu}\|^2 \right\}.$$

The proof of Theorem 5.2.2 shows that $(\boldsymbol{Y} - \boldsymbol{\nu})'\boldsymbol{A}^{-1}(\boldsymbol{Y} - \boldsymbol{\nu})$ is distributed as $\|\boldsymbol{Y} - \boldsymbol{\nu}\|^2/\chi^2_{n-p+1}$, where the χ^2_{n-p+1} is independent of $\boldsymbol{Y}$. Then the difference in risks is

$$(41) \quad \Delta R(\boldsymbol{\mu}) = \mathscr{E}_{\boldsymbol{\mu}}\left\{ \frac{2a\chi^2_{n-p+1}}{\|\boldsymbol{Y} - \boldsymbol{\nu}\|^2} \sum_{i=1}^{p} (Y_i - \mu_i)(Y_i - \nu_i) - \frac{a^2\left(\chi^2_{n-p+1}\right)^2}{\|\boldsymbol{Y} - \boldsymbol{\nu}\|^2} \right\}$$

$$= \mathscr{E}_{\boldsymbol{\mu}}\left\{ \frac{2a(p-2)\chi^2_{n-p+1}}{\|\boldsymbol{Y} - \boldsymbol{\nu}\|^2} - \frac{a^2\left(\chi^2_{n-p+1}\right)^2}{\|\boldsymbol{Y} - \boldsymbol{\nu}\|^2} \right\}$$

$$= \{2(p-2)(n-p+1)a$$

$$-\left[2(n-p+1) + (n-p+1)^2\right]a^2\} \mathscr{E}_{\boldsymbol{\mu}} \frac{1}{\|\boldsymbol{Y} - \boldsymbol{\nu}\|^2}.$$

The term in braces is $n - p + 1$ times $2(p-2)a - (n-p+3)a^2$ which is positive for $0 < a < 2(p-2)/(n-p+3)$ and is maximized for $a = (p-2)/(n-p+3)$. Q.E.D.

The improvement over the risk of $\boldsymbol{Y}$ is $(n-p+1)(p-2)^2/(n-p+3)$ $\mathscr{E}_{\boldsymbol{\mu}}\|\boldsymbol{Y} - \boldsymbol{\nu}\|^{-2}$ as compared to the improvement of $\boldsymbol{m}(y)$ of Section 3.5 when $\boldsymbol{\Sigma}$ is known of $(p-2)^2\mathscr{E}_{\boldsymbol{\mu}}\|\boldsymbol{Y} - \boldsymbol{\nu}\|^{-2}$.

COROLLARY 5.3.1. *The estimator for* $p \geq 3$

$$\left(1 - \frac{a}{N(\bar{\mathbf{x}} - \mathbf{v})'\mathbf{A}^{-1}(\bar{\mathbf{x}} - \mathbf{v})}\right)^{+} (\bar{\mathbf{x}} - \mathbf{v}) + \mathbf{v} \tag{42}$$

has smaller risk than (39) *and is minimax for* $0 < a < 2(p-2)/(n-p+3)$.

PROOF. This corollary follows from Theorem 5.3.1 and Lemma 3.5.2.

The risk of (42) is not necessarily minimized at $a = (p-2)/(n-p+3)$, but that value seems like a good choice. This is the estimator (18) of Section 3.5 with $\boldsymbol{\Sigma}$ replaced by $[1/(n-p+3)]\mathbf{A}$.

When the loss function is $(\mathbf{m} - \boldsymbol{\mu})'\mathbf{Q}(\mathbf{m} - \boldsymbol{\mu})$, where $\mathbf{Q}$ is an arbitrary positive definite matrix, it is harder to present a uniformly improved estimator that is attractive. The estimators of Section 3.5 can be used with $\boldsymbol{\Sigma}$ replaced by an estimate.

5.4. THE DISTRIBUTION OF T^2 UNDER ALTERNATIVE HYPOTHESES; THE POWER FUNCTION

In Section 5.2.2 we showed that $(T^2/n)(N-p)/p$ has a noncentral F-distribution. In this section we shall discuss the noncentral F-distribution, its tabulation, and applications to procedures based on T^2.

The noncentral F-distribution is defined as the distribution of the ratio of a noncentral χ^2 and an independent χ^2 divided by the ratio of corresponding degrees of freedom. Let V have the noncentral χ^2-distribution with p degrees of freedom and noncentrality parameter τ^2 (as given in Theorem 3.3.5), and let W be independently distributed as χ^2 with m degrees of freedom. We shall find the density of $F = (V/p)/(W/m)$, which is the noncentral F with noncentrality parameter τ^2. The joint density of V and W is (28) of Section 3.3 multiplied by the density of W, which is $2^{-\frac{1}{2}m}\Gamma^{-1}(\frac{1}{2}m)w^{\frac{1}{2}m-1}e^{-\frac{1}{2}w}$. The joint density of F and W ($dv = pw\,df/m$) is

$$\frac{e^{-\frac{1}{2}\tau^2}}{2^{\frac{1}{2}(p+m)}\Gamma(\frac{1}{2}m)} e^{-\frac{1}{2}w(1+pf/m)} \tag{1}$$

$$\cdot \frac{p}{m} \sum_{\beta=0}^{\infty} \left(\frac{\tau^2}{4}\right)^{\beta} \frac{1}{\beta!\,\Gamma(\frac{1}{2}p+\beta)} \left(\frac{pf}{m}\right)^{\frac{1}{2}p+\beta-1} w^{\frac{1}{2}(p+m)+\beta-1}.$$

The marginal density, obtained by integrating (1) with respect to w from 0 to ∞, is

$$(2)\qquad \frac{pe^{-\frac{1}{2}\tau^2}}{m\Gamma(\frac{1}{2}m)} \sum_{\beta=0}^{\infty} \frac{(\tau^2/2)^\beta (pf/m)^{\frac{1}{2}p+\beta-1}\Gamma[\frac{1}{2}(p+m)+\beta]}{\beta!\Gamma(\frac{1}{2}p+\beta)(1+pf/m)^{\frac{1}{2}(p+m)+\beta}}.$$

THEOREM 5.4.1. *If V has a noncentral χ^2-distribution with p degrees of freedom and noncentrality parameter τ^2 and W has an independent χ^2-distribution with m degrees of freedom, then $F = (V/p)/(W/m)$ has the density* (2).

If $T^2 = N(\bar{x} - \mu_0)'S^{-1}(\bar{x} - \mu_0)$ is based on a sample of N from $N(\mu, \Sigma)$, then $(T^2/n)(N-p)/p$ has the noncentral F-distribution with p and $N-p$ degrees of freedom and noncentrality parameter $N(\mu - \mu_0)'\Sigma^{-1}(\mu - \mu_0) = \tau^2$. From (2) we find that the density of T^2 is

$$(3)\qquad \frac{e^{-\frac{1}{2}\tau^2}}{(N-1)\Gamma[\frac{1}{2}(N-p)]} \sum_{\beta=0}^{\infty} \frac{(\tau^2/2)^\beta [t^2/(N-1)]^{\frac{1}{2}p+\beta-1}\Gamma(\frac{1}{2}N+\beta)}{\beta!\Gamma(\frac{1}{2}p+\beta)[1+t^2/(N-1)]^{\frac{1}{2}N+\beta}}$$

$$= \frac{\Gamma(\frac{1}{2}N)}{(N-1)\Gamma[\frac{1}{2}(N-p)]\Gamma(\frac{1}{2}p)} \left(\frac{t^2}{N-1}\right)^{\frac{1}{2}p-1} \left(1+\frac{t^2}{N-1}\right)^{-\frac{1}{2}N}$$

$$\cdot e^{-\frac{1}{2}\tau^2}\,{}_1F_1\left(\tfrac{1}{2}N; \tfrac{1}{2}p; \frac{\tau^2 t^2}{2(N-1)}\right),$$

where

$$(4)\qquad {}_1F_1(a;b;x) = \sum_{\beta=0}^{\infty} \frac{\Gamma(a+\beta)\Gamma(b)x^\beta}{\Gamma(a)\Gamma(b+\beta)\beta!}.$$

Tables have been given by Tang (1938) of the probability of accepting the null hypothesis (that is, the probability of Type II error) for various values of τ^2 and for significance levels 0.05 and 0.01. His number of degrees of freedom f_1 is our p [1(1)8], his f_2 is our $n - p + 1$ [2, 4(1)30, 60, ∞] and his noncentrality parameter ϕ is related to our τ^2 by

$$(5)\qquad \phi = \frac{\tau}{\sqrt{p+1}}$$

[1($\frac{1}{2}$)3(1)8]. His accompanying tables of significance points are for $T^2/(T^2 + N - 1)$.

As an example, suppose $p = 4$, $n - p + 1 = 20$, and consider testing the null hypothesis $\boldsymbol{\mu} = \mathbf{0}$ at the 1% level of significance. We would like to know the probability, say, that we accept the null hypothesis when $\phi = 2.5$ ($\tau^2 = 31.25$). It is 0.227. If we think the disadvantage of accepting the null hypothesis when N, $\boldsymbol{\mu}$, and $\boldsymbol{\Sigma}$ are such that $\tau^2 = 31.25$ is less than the disadvantage of rejecting the null hypothesis when it is true, then we may find it reasonable to conduct the test as assumed. However, if the disadvantage of one type of error is about equal to that of the other, it would seem reasonable to bring down the probability of a Type II error. Thus, if we used a significance level of 5%, the probability of Type II error (for $\phi = 2.5$) is only 0.043.

Lehmer (1944) has computed tables of ϕ for given significance level and given probability of Type II error. Her tables can be used to see what value of τ^2 is needed to make the probability of acceptance of the null hypothesis sufficiently low when $\boldsymbol{\mu} \neq \mathbf{0}$. For instance, if we want to be able to reject the hypothesis $\boldsymbol{\mu} = \mathbf{0}$ on the basis of a sample for a given $\boldsymbol{\mu}$ and $\boldsymbol{\Sigma}$, we may be able to choose N so that $N\boldsymbol{\mu}'\boldsymbol{\Sigma}^{-1}\boldsymbol{\mu} = \tau^2$ is sufficiently large. Of course, the difficulty with these considerations is that we usually do not know exactly the values of $\boldsymbol{\mu}$ and $\boldsymbol{\Sigma}$ (and hence of τ^2) for which we want the probability of rejection at a certain value.

The distribution of T^2 when the null hypothesis is not true was derived by different methods by Hsu (1938) and Bose and Roy (1938).

5.5. THE TWO-SAMPLE PROBLEM WITH UNEQUAL COVARIANCE MATRICES

If the covariance matrices are not the same, the T^2-test for equality of mean vectors has a probability of rejection under the null hypothesis that depends on these matrices. If the difference between the matrices is small or if the sample sizes are large, there is no practical effect. However, if the covariance matrices are quite different and/or the sample sizes are relatively small, the nominal significance level may be distorted. Hence we develop a procedure with assigned significance level. Let $\{\boldsymbol{x}_\alpha^{(i)}\}$, $\alpha = 1, \dots, N_i$, be samples from $N(\boldsymbol{\mu}^{(i)}, \boldsymbol{\Sigma}_i)$, $i = 1, 2$. We wish to test the hypothesis H: $\boldsymbol{\mu}^{(1)} = \boldsymbol{\mu}^{(2)}$. The mean $\bar{\boldsymbol{x}}^{(1)}$ of the first sample is normally distributed with expected value

$$\mathscr{E}\bar{\boldsymbol{x}}^{(1)} = \boldsymbol{\mu}^{(1)} \tag{1}$$

and covariance matrix

(2) $$\mathscr{E}(\bar{\mathbf{x}}^{(1)} - \boldsymbol{\mu}^{(1)})(\bar{\mathbf{x}}^{(1)} - \boldsymbol{\mu}^{(1)})' = \frac{1}{N_1}\boldsymbol{\Sigma}_1.$$

Similarly, the mean $\bar{\mathbf{x}}^{(2)}$ of the second sample is normally distributed with expected value

(3) $$\mathscr{E}\bar{\mathbf{x}}^{(2)} = \boldsymbol{\mu}^{(2)}$$

and covariance matrix

(4) $$\mathscr{E}(\bar{\mathbf{x}}^{(2)} - \boldsymbol{\mu}^{(2)})(\bar{\mathbf{x}}^{(2)} - \boldsymbol{\mu}^{(2)})' = \frac{1}{N_2}\boldsymbol{\Sigma}_2.$$

Thus $\bar{\mathbf{x}}^{(1)} - \bar{\mathbf{x}}^{(2)}$ has mean $\boldsymbol{\mu}^{(1)} - \boldsymbol{\mu}^{(2)}$ and covariance matrix $(1/N_1)\boldsymbol{\Sigma}_1 + (1/N_2)\boldsymbol{\Sigma}_2$. We cannot use the technique of Section 5.2, however, because

(5) $$\sum_{\alpha=1}^{N_1} (\mathbf{x}_\alpha^{(1)} - \bar{\mathbf{x}}^{(1)})(\mathbf{x}_\alpha^{(1)} - \bar{\mathbf{x}}^{(1)})' + \sum_{\alpha=1}^{N_2} (\mathbf{x}_\alpha^{(2)} - \bar{\mathbf{x}}^{(2)})(\mathbf{x}_\alpha^{(2)} - \bar{\mathbf{x}}^{(2)})'$$

does not have the Wishart distribution with covariance matrix a multiple of $(1/N_1)\boldsymbol{\Sigma}_1 + (1/N_2)\boldsymbol{\Sigma}_2$.

If $N_1 = N_2 = N$, say, we can use the T^2-test in an obvious way. Let $\mathbf{y}_\alpha = \mathbf{x}_\alpha^{(1)} - \mathbf{x}_\alpha^{(2)}$ (assuming the numbering of the observations in the two samples is independent of the observations themselves). Then $\mathbf{y}_\alpha$ is normally distributed with mean $\boldsymbol{\mu}^{(1)} - \boldsymbol{\mu}^{(2)}$ and covariance matrix $\boldsymbol{\Sigma}_1 + \boldsymbol{\Sigma}_2$, and $\mathbf{y}_1, \dots, \mathbf{y}_N$ are independent. Let $\bar{\mathbf{y}} = (1/N)\sum_{\alpha=1}^N \mathbf{y}_\alpha = \bar{\mathbf{x}}^{(1)} - \bar{\mathbf{x}}^{(2)}$, and define $\mathbf{S}$ by

(6) $$\begin{aligned}(N-1)\mathbf{S} &= \sum_{\alpha=1}^{N} (\mathbf{y}_\alpha - \bar{\mathbf{y}})(\mathbf{y}_\alpha - \bar{\mathbf{y}})' \\ &= \sum_{\alpha=1}^{N} (\mathbf{x}_\alpha^{(1)} - \mathbf{x}_\alpha^{(2)} - \bar{\mathbf{x}}^{(1)} + \bar{\mathbf{x}}^{(2)})(\mathbf{x}_\alpha^{(1)} - \mathbf{x}_\alpha^{(2)} - \bar{\mathbf{x}}^{(1)} + \bar{\mathbf{x}}^{(2)})'.\end{aligned}$$

Then

(7) $$T^2 = N\bar{\mathbf{y}}'\mathbf{S}^{-1}\bar{\mathbf{y}}$$

is suitable for testing the hypothesis $\boldsymbol{\mu}^{(1)} - \boldsymbol{\mu}^{(2)} = \mathbf{0}$, and has the T^2-distribution with $N-1$ degrees of freedom. It should be observed that if we had known $\boldsymbol{\Sigma}_1 = \boldsymbol{\Sigma}_2$, we would have used a T^2-statistic with $2N-2$ degrees of freedom; thus we have lost $N-1$ degrees of freedom in constructing a test which is independent of the two covariance matrices. If $N_1 = N_2 = 50$ as in the example in Section 5.3.4, $T_{4,49}^2(.01) = 15.93$ as compared to $T_{4,98}^2(.01) = 14.52$.

Now let us turn our attention to the case of $N_1 \neq N_2$. For convenience, let $N_1 < N_2$. Then we define

$$(8)\qquad \mathbf{y}_\alpha = \mathbf{x}_\alpha^{(1)} - \sqrt{\frac{N_1}{N_2}}\,\mathbf{x}_\alpha^{(2)} + \frac{1}{\sqrt{N_1 N_2}}\sum_{\beta=1}^{N_1}\mathbf{x}_\beta^{(2)} - \frac{1}{N_2}\sum_{\gamma=1}^{N_2}\mathbf{x}_\gamma^{(2)}, \qquad \alpha = 1,\ldots,N_1.$$

The expected value of $\mathbf{y}_\alpha$ is

$$(9)\qquad \begin{aligned} \mathscr{E}\mathbf{y}_\alpha &= \boldsymbol{\mu}^{(1)} - \sqrt{\frac{N_1}{N_2}}\,\boldsymbol{\mu}^{(2)} + \frac{N_1}{\sqrt{N_1 N_2}}\boldsymbol{\mu}^{(2)} - \frac{N_2}{N_2}\boldsymbol{\mu}^{(2)} \\ &= \boldsymbol{\mu}^{(1)} - \boldsymbol{\mu}^{(2)}. \end{aligned}$$

The covariance matrix of $\mathbf{y}_\alpha$ and $\mathbf{y}_\beta$ is

$$(10)\qquad \begin{aligned} &\mathscr{E}(\mathbf{y}_\alpha - \mathscr{E}\mathbf{y}_\alpha)(\mathbf{y}_\beta - \mathscr{E}\mathbf{y}_\beta)' \\ &\quad = \mathscr{E}\Bigg[\left(\mathbf{x}_\alpha^{(1)} - \boldsymbol{\mu}^{(1)}\right) - \sqrt{\frac{N_1}{N_2}}\left(\mathbf{x}_\alpha^{(2)} - \boldsymbol{\mu}^{(2)}\right) \\ &\qquad\qquad + \frac{1}{\sqrt{N_1 N_2}}\sum_{\gamma=1}^{N_1}\left(\mathbf{x}_\gamma^{(2)} - \boldsymbol{\mu}^{(2)}\right) - \frac{1}{N_2}\sum_{\gamma=1}^{N_2}\left(\mathbf{x}_\gamma^{(2)} - \boldsymbol{\mu}^{(2)}\right)\Bigg] \\ &\qquad \times \Bigg[\left(\mathbf{x}_\beta^{(1)} - \boldsymbol{\mu}^{(1)}\right)' - \sqrt{\frac{N_1}{N_2}}\left(\mathbf{x}_\beta^{(2)} - \boldsymbol{\mu}^{(2)}\right)' \\ &\qquad\qquad + \frac{1}{\sqrt{N_1 N_2}}\sum_{\gamma=1}^{N_1}\left(\mathbf{x}_\gamma^{(2)} - \boldsymbol{\mu}^{(2)}\right)' - \frac{1}{N_2}\sum_{\gamma=1}^{N_2}\left(\mathbf{x}_\gamma^{(2)} - \boldsymbol{\mu}^{(2)}\right)'\Bigg] \\ &\quad = \delta_{\alpha\beta}\boldsymbol{\Sigma}_1 + \frac{N_1}{N_2}\delta_{\alpha\beta}\boldsymbol{\Sigma}_2 \\ &\qquad + \boldsymbol{\Sigma}_2\left(-2\frac{1}{N_2} + \frac{2}{N_2}\sqrt{\frac{N_1}{N_2}} + \frac{N_1}{N_1 N_2} - 2\frac{N_1}{\sqrt{N_1 N_2}\,N_2} + \frac{N_2}{N_2^2}\right) \\ &\quad = \delta_{\alpha\beta}\left(\boldsymbol{\Sigma}_1 + \frac{N_1}{N_2}\boldsymbol{\Sigma}_2\right). \end{aligned}$$

Thus a suitable statistic for testing $\boldsymbol{\mu}^{(1)} - \boldsymbol{\mu}^{(2)} = \mathbf{0}$, which has the T^2-distribution with $N_1 - 1$ degrees of freedom, is

$$T^2 = N_1 \bar{\boldsymbol{y}}' \boldsymbol{S}^{-1} \bar{\boldsymbol{y}}, \tag{11}$$

where

$$\bar{\boldsymbol{y}} = \frac{1}{N_1} \sum_{\alpha=1}^{N_1} \boldsymbol{y}_\alpha = \bar{\boldsymbol{x}}^{(1)} - \bar{\boldsymbol{x}}^{(2)} \tag{12}$$

and

$$\begin{aligned}(N_1 - 1)\boldsymbol{S} &= \sum_{\alpha=1}^{N_1} (\boldsymbol{y}_\alpha - \bar{\boldsymbol{y}})(\boldsymbol{y}_\alpha - \bar{\boldsymbol{y}})' \\ &= \sum_{\alpha=1}^{N_1} \left[\boldsymbol{x}_\alpha^{(1)} - \bar{\boldsymbol{x}}^{(1)} - \sqrt{\frac{N_1}{N_2}} \left(\boldsymbol{x}_\alpha^{(2)} - \frac{1}{N_1} \sum_{\beta=1}^{N_1} \boldsymbol{x}_\beta^{(2)} \right) \right] \\ &\quad \times \left[\boldsymbol{x}_\alpha^{(1)} - \bar{\boldsymbol{x}}^{(1)} - \sqrt{\frac{N_1}{N_2}} \left(\boldsymbol{x}_\alpha^{(2)} - \frac{1}{N_1} \sum_{\beta=1}^{N_1} \boldsymbol{x}_\beta^{(2)} \right) \right]'.\end{aligned} \tag{13}$$

In terms of $\boldsymbol{u}_\alpha = \boldsymbol{x}_\alpha^{(1)} - \sqrt{N_1/N_2}\, \boldsymbol{x}_\alpha^{(2)}$, $\alpha = 1, \dots, N_1$, this last equation may be written

$$(N_1 - 1)\boldsymbol{S} = \sum_{\alpha=1}^{N_1} (\boldsymbol{u}_\alpha - \bar{\boldsymbol{u}})(\boldsymbol{u}_\alpha - \bar{\boldsymbol{u}})', \tag{14}$$

where $\bar{\boldsymbol{u}} = (1/N_1)\sum_{\alpha=1}^{N_1} \boldsymbol{u}_\alpha$.

This procedure has been suggested by Scheffé (1943) in the univariate case. Scheffé has shown that in the univariate case this technique gives the shortest confidence intervals obtained by using the t-distribution. The advantage of the method is that $\bar{\boldsymbol{x}}^{(1)} - \bar{\boldsymbol{x}}^{(2)}$ is used, and this statistic is most relevant to $\boldsymbol{\mu}^{(1)} - \boldsymbol{\mu}^{(2)}$. The sacrifice of observations in estimating a covariance matrix is not so important. Bennett (1951) gave the extension of the procedure to the multivariate case.

This approach can be used for more general cases. Let $\{\boldsymbol{x}_\alpha^{(i)}\}$, $\alpha = 1, \dots, N_i$; $i = 1, \dots, q$, be samples from $N(\boldsymbol{\mu}^{(i)}, \boldsymbol{\Sigma}_i)$, $i = 1, \dots, q$, respectively. Consider testing the hypothesis

$$H\colon \sum_{i=1}^{q} \beta_i \boldsymbol{\mu}^{(i)} = \boldsymbol{\mu}, \tag{15}$$

where $\beta_1, \dots, \beta_q$ are given scalars and $\boldsymbol{\mu}$ is a given vector. If the N_i are unequal, take N_1 to be the smallest. Let

$$\boldsymbol{y}_\alpha = \beta_1 \boldsymbol{x}_\alpha^{(1)} + \sum_{i=2}^{q} \beta_i \sqrt{\frac{N_1}{N_i}} \left(\boldsymbol{x}_\alpha^{(i)} - \frac{1}{N_1} \sum_{\beta=1}^{N_1} \boldsymbol{x}_\beta^{(i)} + \frac{1}{\sqrt{N_1 N_i}} \sum_{\gamma=1}^{N_i} \boldsymbol{x}_\gamma^{(i)} \right). \tag{16}$$

Then

$$\begin{aligned} \mathscr{E} \boldsymbol{y}_\alpha &= \beta_1 \boldsymbol{\mu}^{(1)} + \sum_{i=2}^{q} \beta_i \sqrt{\frac{N_1}{N_i}} \left(\boldsymbol{\mu}^{(i)} - \frac{1}{N_1} N_1 \boldsymbol{\mu}^{(i)} + \frac{N_i}{\sqrt{N_1 N_i}} \boldsymbol{\mu}^{(i)} \right) \\ &= \sum_{i=1}^{q} \beta_i \boldsymbol{\mu}^{(i)}, \end{aligned} \tag{17}$$

$$\mathscr{E}(\boldsymbol{y}_\alpha - \mathscr{E}\boldsymbol{y}_\alpha)(\boldsymbol{y}_\beta - \mathscr{E}\boldsymbol{y}_\beta)' = \delta_{\alpha\beta} \left(\sum_{i=1}^{q} \frac{\beta_i^2 N_1}{N_i} \boldsymbol{\Sigma}_i \right). \tag{18}$$

Let $\bar{\boldsymbol{y}}$ and $\boldsymbol{S}$ be defined by

$$\bar{\boldsymbol{y}} = \frac{1}{N_1} \sum_{\alpha=1}^{N_1} \boldsymbol{y}_\alpha = \sum_{i=1}^{q} \beta_i \bar{\boldsymbol{x}}^{(i)}, \tag{19}$$

where

$$\begin{aligned} \bar{\boldsymbol{x}}^{(i)} &= \frac{1}{N_i} \sum_{\beta=1}^{N_i} \boldsymbol{x}_\beta^{(i)}, \\ (N_1 - 1)\boldsymbol{S} &= \sum_{\alpha=1}^{N_1} (\boldsymbol{y}_\alpha - \bar{\boldsymbol{y}})(\boldsymbol{y}_\alpha - \bar{\boldsymbol{y}})'. \end{aligned} \tag{20}$$

Then

$$T^2 = N_1(\bar{y} - \mu)'S^{-1}(\bar{y} - \mu) \tag{21}$$

is suitable for testing H, and when the hypothesis is true, this statistic has the T^2-distribution for dimension p with $N_1 - 1$ degrees of freedom. If we let

$$u_\alpha = \sum_{i=1}^{q} \beta_i \sqrt{\frac{N_1}{N_i}}\, x_\alpha^{(i)}, \qquad \alpha = 1, \dots, N_1, \tag{22}$$

then S can be defined as

$$(N_1 - 1)S = \sum_{\alpha=1}^{N_1} (u_\alpha - \bar{u})(u_\alpha - \bar{u})'. \tag{23}$$

Another problem that is amenable to this kind of treatment is testing the hypothesis that two subvectors have equal means. Let

$$x = \begin{pmatrix} x^{(1)} \\ x^{(2)} \end{pmatrix} \tag{24}$$

be distributed normally with mean

$$\mu = \begin{pmatrix} \mu^{(1)} \\ \mu^{(2)} \end{pmatrix} \tag{25}$$

and covariance matrix

$$\Sigma = \begin{pmatrix} \Sigma_{11} & \Sigma_{12} \\ \Sigma_{21} & \Sigma_{22} \end{pmatrix}. \tag{26}$$

We assume that $x^{(1)}$ and $x^{(2)}$ are each of q components. Then $x^{(1)} - x^{(2)}$ is distributed normally with mean $\mu^{(1)} - \mu^{(2)}$ and covariance matrix

$$\mathscr{E}\big[(x^{(1)} - \mu^{(1)}) - (x^{(2)} - \mu^{(2)})\big]\big[(x^{(1)} - \mu^{(1)}) - (x^{(2)} - \mu^{(2)})\big]' = \Sigma_{11} - \Sigma_{21} - \Sigma_{12} + \Sigma_{22}. \tag{27}$$

To test the hypothesis $\boldsymbol{\mu}^{(1)} = \boldsymbol{\mu}^{(2)}$ we use a T^2-statistic

$$N(\bar{\boldsymbol{x}}^{(1)} - \bar{\boldsymbol{x}}^{(2)})'(\boldsymbol{S}_{11} - \boldsymbol{S}_{21} - \boldsymbol{S}_{12} + \boldsymbol{S}_{22})^{-1}(\bar{\boldsymbol{x}}^{(1)} - \bar{\boldsymbol{x}}^{(2)}), \tag{28}$$

where the mean vector and covariance matrix of the sample are partitioned similarly to $\boldsymbol{\mu}$ and $\boldsymbol{\Sigma}$.

5.6. SOME OPTIMAL PROPERTIES OF THE T^2-TEST

5.6.1. Optimal Invariant Tests

In this section we shall indicate that the T^2-test is the best in certain classes of tests and sketch briefly the proofs of these results.

The hypothesis $\boldsymbol{\mu} = \boldsymbol{0}$ is to be tested on the basis of the N observations $\boldsymbol{x}_1, \dots, \boldsymbol{x}_N$ from $N(\boldsymbol{\mu}, \boldsymbol{\Sigma})$. First we consider the class of tests based on the statistics $\boldsymbol{A} = \Sigma(\boldsymbol{x}_\alpha - \bar{\boldsymbol{x}})(\boldsymbol{x}_\alpha - \bar{\boldsymbol{x}})'$ and $\bar{\boldsymbol{x}}$ which are invariant with respect to the transformations $\boldsymbol{A}^* = \boldsymbol{C}\boldsymbol{A}\boldsymbol{C}'$ and $\bar{\boldsymbol{x}}^* = \boldsymbol{C}\bar{\boldsymbol{x}}$, where $\boldsymbol{C}$ is nonsingular. The transformation $\boldsymbol{x}_\alpha^* = \boldsymbol{C}\boldsymbol{x}_\alpha$ leaves the problem invariant; that is, in terms of $\boldsymbol{x}_\alpha^*$ we test the hypothesis $\mathscr{E}\boldsymbol{x}_\alpha^* = \boldsymbol{0}$ given that $\boldsymbol{x}_1^*, \dots, \boldsymbol{x}_N^*$ are N observations from a multivariate normal population. It seems reasonable that we require a solution that is also invariant with respect to these transformations; that is, we look for a critical region that is not changed by a nonsingular linear transformation. (The definition of the region is the same in different coordinate systems.)

THEOREM 5.6.1. *Given the observations $\boldsymbol{x}_1, \dots, \boldsymbol{x}_N$ from $N(\boldsymbol{\mu}, \boldsymbol{\Sigma})$, of all tests of $\boldsymbol{\mu} = \boldsymbol{0}$ based on $\bar{\boldsymbol{x}}$ and $\boldsymbol{A} = \Sigma(\boldsymbol{x}_\alpha - \bar{\boldsymbol{x}})(\boldsymbol{x}_\alpha - \bar{\boldsymbol{x}})'$ that are invariant with respect to transformations $\bar{\boldsymbol{x}}^* = \boldsymbol{C}\bar{\boldsymbol{x}}$, $\boldsymbol{A}^* = \boldsymbol{C}\boldsymbol{A}\boldsymbol{C}'$ ($\boldsymbol{C}$ nonsingular), the T^2-test is uniformly most powerful.*

PROOF. First, as we have seen in Section 5.2.1, any test based on T^2 is invariant. Second, this function is essentially the only invariant, for if $f(\bar{\boldsymbol{x}}, \boldsymbol{A})$ is invariant, then $f(\bar{\boldsymbol{x}}, \boldsymbol{A}) = f(\bar{\boldsymbol{x}}^*, \boldsymbol{I})$, where only the first coordinate of $\bar{\boldsymbol{x}}^*$ is different from zero and it is $\sqrt{\bar{\boldsymbol{x}}'\boldsymbol{A}^{-1}\bar{\boldsymbol{x}}}$. (There is a matrix $\boldsymbol{C}$ such that $\boldsymbol{C}\bar{\boldsymbol{x}} = \bar{\boldsymbol{x}}^*$ and $\boldsymbol{C}\boldsymbol{A}\boldsymbol{C}' = \boldsymbol{I}$.) Thus $f(\bar{\boldsymbol{x}}, \boldsymbol{A})$ depends only on $\bar{\boldsymbol{x}}'\boldsymbol{A}^{-1}\bar{\boldsymbol{x}}$. Thus an invariant test must be based on $\bar{\boldsymbol{x}}'\boldsymbol{A}^{-1}\bar{\boldsymbol{x}}$. Third, we can apply the Neyman–Pearson fundamental lemma to the distribution of T^2 [(3) of Section 5.4] to find the uniformly most powerful test based on T^2 against a simple alternative $\tau^2 = N\mu'\Sigma^{-1}\mu$. The most powerful test of $\tau^2 = 0$ is based on the ratio of (3) of Section 5.4 to (3) with $\tau^2 = 0$. The critical region is

(1)

$$c < e^{-\frac{1}{2}\tau^2} \sum_{\alpha=0}^{\infty} \frac{(\tau^2/2)^{\alpha}(t^2/n)^{\frac{1}{2}p+\alpha-1}(1+t^2/n)^{-\frac{1}{2}(n+1)-\alpha}\Gamma[\frac{1}{2}(n+1)+\alpha]}{\alpha!\Gamma(\frac{1}{2}p+\alpha)} \Bigg/ \frac{(t^2/n)^{\frac{1}{2}p-1}(1+t^2/n)^{-\frac{1}{2}(n+1)}\Gamma[\frac{1}{2}(n+1)]}{\Gamma(\frac{1}{2}p)}$$

$$= \frac{\Gamma(\frac{1}{2}p)}{\Gamma[\frac{1}{2}(n+1)]} e^{-\frac{1}{2}\tau^2} \sum_{\alpha=0}^{\infty} \frac{(\tau^2/2)^{\alpha}\Gamma[\frac{1}{2}(n+1)+\alpha]}{\alpha!\Gamma(\frac{1}{2}p+\alpha)} \left(\frac{t^2/n}{1+t^2/n}\right)^{\alpha}.$$

The right-hand side of (1) is a strictly increasing function of $(t^2/n)/(1 + t^2/n)$, hence of t^2. Thus the inequality is equivalent to $t^2 > k$ for k suitably chosen. Since this does not depend on the alternative τ^2, the test is uniformly most powerful invariant. Q.E.D.

DEFINITION 5.6.1. *A critical function $\psi(\bar{x}, A)$ is a function with values between 0 and 1 (inclusive) such that $\mathscr{E}\psi(\bar{x}, A) = \varepsilon$, the significance level, when $\mu = 0$.*

A randomized test consists of rejecting the hypothesis with probability $\psi(x, B)$ when $\bar{x} = x$ and $A = B$. A nonrandomized test is defined when $\psi(\bar{x}, A)$ takes on only the values 0 and 1. Using the form of the Neyman–Pearson lemma appropriate for critical functions we obtain the following corollary:

COROLLARY 5.6.1. *On the basis of observations $x_1, \dots, x_N$ from $N(\mu, \Sigma)$, of all randomized tests based on $\bar{x}$ and A that are invariant with respect to transformations $\bar{x}^* = C\bar{x}$, $A^* = CAC'$ (C nonsingular), the T^2-test is uniformly most powerful.*

THEOREM 5.6.2. *On the basis of observations $x_1, \dots, x_N$ from $N(\mu, \Sigma)$, of all tests of $\mu = 0$ that are invariant with respect to transformations $x_\alpha^* = Cx_\alpha$ (C nonsingular), the T^2-test is a uniformly most powerful test; that is, the T^2-test is at least as powerful as any other invariant test.*

PROOF. Let $\psi(\boldsymbol{x}_1, \ldots, \boldsymbol{x}_N)$ be the critical function of an invariant test. Then

$$\mathscr{E}[\psi(\boldsymbol{x}_1, \ldots, \boldsymbol{x}_N)] = \mathscr{E}_{\bar{\boldsymbol{x}}, \boldsymbol{A}}\{\mathscr{E}[\psi(\boldsymbol{x}_1, \ldots, \boldsymbol{x}_N)|\bar{\boldsymbol{x}}, \boldsymbol{A}]\}. \tag{2}$$

Since $\bar{\boldsymbol{x}}$, $\boldsymbol{A}$ are sufficient statistics for $\boldsymbol{\mu}$, $\boldsymbol{\Sigma}$, $\mathscr{E}[\psi(\boldsymbol{x}_1, \ldots, \boldsymbol{x}_N)|\bar{\boldsymbol{x}}, \boldsymbol{A}]$ depends only on $\bar{\boldsymbol{x}}$, $\boldsymbol{A}$. It is invariant and has the same power as $\psi(\boldsymbol{x}_1, \ldots, \boldsymbol{x}_N)$. Thus each test in this larger class can be replaced by one in the smaller class (depending only on $\bar{\boldsymbol{x}}$ and $\boldsymbol{A}$) that has identical power. Corollary 5.6.1 completes the proof. Q.E.D.

THEOREM 5.6.3. *Given observations $\boldsymbol{x}_1, \ldots, \boldsymbol{x}_N$ from $N(\boldsymbol{\mu}, \boldsymbol{\Sigma})$, of all tests of $\boldsymbol{\mu} = \boldsymbol{0}$ based on $\bar{\boldsymbol{x}}$ and $\boldsymbol{A} = \sum(\boldsymbol{x}_\alpha - \bar{\boldsymbol{x}})(\boldsymbol{x}_\alpha - \bar{\boldsymbol{x}})'$ with power depending only on $N\boldsymbol{\mu}'\boldsymbol{\Sigma}^{-1}\boldsymbol{\mu}$, the T^2-test is uniformly most powerful.*

PROOF. We wish to reduce this theorem to Theorem 5.6.1 by identifying the class of tests with power depending on $N\boldsymbol{\mu}'\boldsymbol{\Sigma}^{-1}\boldsymbol{\mu}$ with the class of invariant tests. We need the following definition:

DEFINITION 5.6.2. *A test $\psi(\boldsymbol{x}_1, \ldots, \boldsymbol{x}_N)$ is said to be almost invariant if*

$$\psi(\boldsymbol{x}_1, \ldots, \boldsymbol{x}_N) = \psi(\boldsymbol{C}\boldsymbol{x}_1, \ldots, \boldsymbol{C}\boldsymbol{x}_N) \tag{3}$$

for all $\boldsymbol{x}_1, \ldots, \boldsymbol{x}_N$ except for a set of $\boldsymbol{x}_1, \ldots, \boldsymbol{x}_N$ of Lebesgue measure zero; this exception set may depend on $\boldsymbol{C}$.

It is clear that Theorems 5.6.1 and 5.6.2 hold if we extend the definition of invariant test to mean that (3) holds except for a fixed set of $\boldsymbol{x}_1, \ldots, \boldsymbol{x}_N$ of measure 0 (the set not depending on $\boldsymbol{C}$). It has been shown by Hunt and Stein [Lehmann (1959)] that in our problem almost invariance implies invariance (in the broad sense).

Now we wish to argue that if $\psi(\bar{\boldsymbol{x}}, \boldsymbol{A})$ has power depending only on $N\boldsymbol{\mu}'\boldsymbol{\Sigma}^{-1}\boldsymbol{\mu}$, it is almost invariant. Since the power of $\psi(\bar{\boldsymbol{x}}, \boldsymbol{A})$ depends only on $N\boldsymbol{\mu}'\boldsymbol{\Sigma}^{-1}\boldsymbol{\mu}$, the power is

$$\begin{aligned}\mathscr{E}_{\boldsymbol{\mu}, \boldsymbol{\Sigma}}\psi(\bar{\boldsymbol{x}}, \boldsymbol{A}) &\equiv \mathscr{E}_{\boldsymbol{C}^{-1}\boldsymbol{\mu}, \boldsymbol{C}^{-1}\boldsymbol{\Sigma}(\boldsymbol{C}^{-1})'}\psi(\bar{\boldsymbol{x}}, \boldsymbol{A}) \\ &\equiv \mathscr{E}_{\boldsymbol{\mu}, \boldsymbol{\Sigma}}\psi(\boldsymbol{C}\bar{\boldsymbol{x}}, \boldsymbol{C}\boldsymbol{A}\boldsymbol{C}').\end{aligned} \tag{4}$$

The second and third terms of (4) are merely different ways of writing the same integral. Thus

$$\mathscr{E}_{\mu,\Sigma}[\psi(\bar{x}, A) - \psi(C\bar{x}, CAC')] \equiv 0, \tag{5}$$

identically in μ, Σ. Since $\bar{x}$, A are a complete sufficient set of statistics for μ, Σ (Theorem 3.4.2), $f(\bar{x}, A) = \psi(\bar{x}, A) - \psi(C\bar{x}, CAC') = 0$ almost everywhere. Theorem 5.6.3 follows. Q.E.D.

As Theorem 5.6.2 follows from Theorem 5.6.1, so does the following theorem from Theorem 5.6.3:

THEOREM 5.6.4. *On the basis of observations $x_1, \dots, x_N$ from $N(\mu, \Sigma)$, of all tests of $\mu = \mathbf{0}$ with power depending only on $N\mu'\Sigma^{-1}\mu$, the T^2-test is a uniformly most powerful test.*

Theorem 5.6.4 was first proved by Simaika (1941). The results and proofs given in this section follow Lehmann (1959). Hsu (1945) has proved an optimal property of the T^2-test that involves averaging the power over μ and Σ.

5.6.2. Admissible Tests

We now turn to the question of whether the T^2-test is a good test compared to all possible tests; the comparison in the previous section was to the restricted class of invariant tests. The main result is that the T^2-test is admissible in the class of all tests; that is, there is no other procedure that is better.

DEFINITION 5.6.3. *A test T^* of the null hypothesis H_0: $\omega \in \Omega_0$ against the alternative $\omega \in \Omega_1$ (disjoint from Ω_0) is admissible if there exists no other test T such that*

$$\Pr\{\text{Reject } H_0 | T, \omega\} \le \Pr\{\text{Reject } H_0 | T^*, \omega\}, \qquad \omega \in \Omega_0, \tag{6}$$

$$\Pr\{\text{Reject } H_0 | T, \omega\} \ge \Pr\{\text{Reject } H_0 | T^*, \omega\}, \qquad \omega \in \Omega_1, \tag{7}$$

with strict inequality for at least one ω.

The admissibility of the T^2-test follows from a theorem of Stein (1956a) that applies to any exponential family of distributions.

An *exponential family* of distributions $(\mathscr{Y}, \mathscr{B}, m, \Omega, P)$ consists of a finite-dimensional Euclidean space $\mathscr{Y}$, a measure m on the σ-algebra $\mathscr{B}$ of all ordinary Borel sets of $\mathscr{Y}$, a subset Ω of the adjoint space $\mathscr{Y}'$ (the linear space of all real-valued linear functions on $\mathscr{Y}$) such that

$$\psi(\omega) = \int_{\mathscr{Y}} e^{\omega' y}\, dm(y) < \infty, \qquad \omega \in \Omega, \tag{8}$$

and P, the function on Ω to the set of probability measures on $\mathscr{B}$ given by

$$P_\omega(A) = \frac{1}{\psi(\omega)} \int_A e^{\omega' y}\, dm(y), \qquad A \in \mathscr{B}.$$

The family of normal distributions $N(\mu, \Sigma)$ constitutes an exponential family, for the density can be written

$$n(x|\mu, \Sigma) = \frac{e^{-\frac{1}{2}\mu'\Sigma^{-1}\mu}}{(2\pi)^{\frac{1}{2}p}|\Sigma|^{\frac{1}{2}}} e^{(\mu'\Sigma^{-1})x + \operatorname{tr}(-\frac{1}{2}\Sigma^{-1})xx'}. \tag{9}$$

We map from $\mathscr{X}$ to $\mathscr{Y}$; the vector $y = (y^{(1)\prime}, y^{(2)\prime})'$ is composed of $y^{(1)} = x$ and $y^{(2)} = (x_1^2, 2x_1x_2, \dots, 2x_1x_p, x_2^2, \dots, x_p^2)'$. The vector $\omega = (\omega^{(1)\prime}, \omega^{(2)\prime})'$ is composed of $\omega^{(1)} = \Sigma^{-1}\mu$ and $\omega^{(2)} = -\frac{1}{2}(\sigma^{11}, \sigma^{12}, \dots, \sigma^{1p}, \sigma^{22}, \dots, \sigma^{pp})'$, where $(\sigma^{ij}) = \Sigma^{-1}$; the transformation of parameters is one to one. The measure $m(A)$ of a set $A \in \mathscr{B}$ is the ordinary Lebesgue measure of the set of x that maps into the set A. (Note the probability measure in $\mathscr{Y}$ is not defined by a density.)

THEOREM 5.6.5 (Stein). *Let $(\mathscr{Y}, \mathscr{B}, m, \Omega, P)$ be an exponential family and Ω_0 a nonempty proper subset of Ω. (i) Let A be a subset of $\mathscr{Y}$ that is closed and convex. (ii) Suppose that for every vector $\omega \in \mathscr{Y}'$ and real c for which $\{y|\omega' y > c\}$ and A are disjoint, there exists $\omega_1 \in \Omega$ such that for arbitrarily large λ the vector $\omega_1 + \lambda\omega \in \Omega - \Omega_0$. Then the test with acceptance region A is admissible for testing the hypothesis that $\omega \in \Omega_0$ against the alternative $\omega \in \Omega - \Omega_0$.*

The conditions of the theorem are illustrated in Figure 5.2, which is drawn simultaneously in the space $\mathscr{Y}$ and the set Ω.

PROOF. The critical function of the test with acceptance region A is $\phi_A(y) = 0$, $y \in A$, and $\phi_A(y) = 1$, $y \notin A$. Suppose $\phi(y)$ is the critical

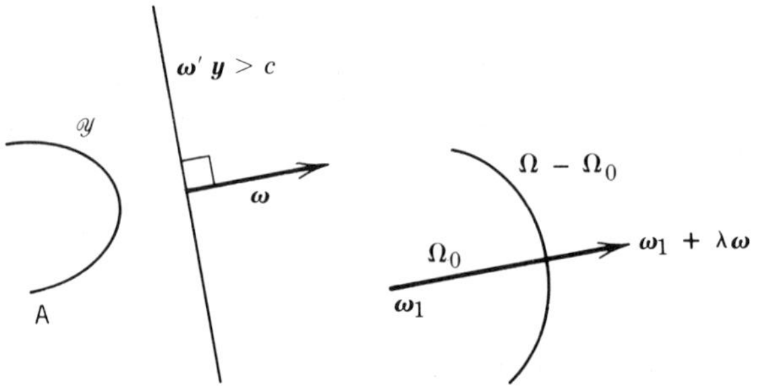

Figure 5.2

function of a better test, that is,

(10) $$\int \phi(y)\, dP_\omega(y) \leq \int \phi_A(y)\, dP_\omega(y), \qquad \omega \in \Omega_0,$$

(11) $$\int \phi(y)\, dP_\omega(y) \geq \int \phi_A(y)\, dP_\omega(y), \qquad \omega \in \Omega - \Omega_0,$$

with strict inequality for some ω; we shall show that this assumption leads to a contradiction. Let $B = \{y|\phi(y) < 1\}$. (If the competing test is nonrandomized, B is its acceptance region.) Then

(12) $$\{y \mid \phi_A(y) - \phi(y) > 0\} = \bar{A} \cap B,$$

where $\bar{A}$ is the complement of A. The m-measure of the set (12) is positive; otherwise $\phi_A(y) = \phi(y)$ almost everywhere, and (10) and (11) would hold with equality for all ω. Since A is convex, there exists an ω and a c such that the intersection of $\bar{A} \cap B$ and $\{y|\omega' y > c\}$ has positive m-measure. (Since A is closed, $\bar{A}$ is open and it can be covered with a denumerable collection of open spheres, for example, with rational radii and centers with rational coordinates. Because there is a hyperplane separating A and each sphere, there exists a denumerable collection of open half-spaces H_j disjoint from A that covers $\bar{A}$. Then at least one half-space has an intersection with $\bar{A} \cap B$ with positive m-measure.) By hypothesis there exists $\omega_1 \in \Omega$ and an arbitrarily large λ such that

(13) $$\omega_\lambda = \omega_1 + \lambda\omega \in \Omega - \Omega_0.$$

Then

$$
\begin{aligned}
(14)\quad & \int [\phi_A(y) - \phi(y)]\, dP_{\omega_\lambda}(y) \\
&= \frac{1}{\psi(\omega_\lambda)} \int [\phi_A(y) - \phi(y)]\, e^{\omega'_\lambda y}\, dm(y) \\
&= \frac{\psi(\omega_1)}{\psi(\omega_\lambda)} \int [\phi_A(y) - \phi(y)]\, e^{\lambda\omega' y}\, dP_{\omega_1}(y) \\
&= \frac{\psi(\omega_1)}{\psi(\omega_\lambda)} e^{\lambda c} \int [\phi_A(y) - \phi(y)]\, e^{\lambda(\omega' y - c)}\, dP_{\omega_1}(y) \\
&= \frac{\psi(\omega_1)}{\psi(\omega_\lambda)} e^{\lambda c} \left\{ \int_{\omega' y > c} [\phi_A(y) - \phi(y)]\, e^{\lambda(\omega' y - c)}\, dP_{\omega_1}(y) \right. \\
&\qquad \left. + \int_{\omega' y \le c} [\phi_A(y) - \phi(y)]\, e^{\lambda(\omega' y - c)}\, dP_{\omega_1}(y) \right\}.
\end{aligned}
$$

For $\omega' y > c$, $\phi_A(y) = 1$, $\phi_A(y) - \phi(y) \ge 0$ and $\{y \mid \phi_A(y) - \phi(y) > 0\}$ has positive measure; therefore, the first integral in the braces approaches ∞ as $\lambda \to \infty$. The second integral is bounded because the integrand is bounded by 1, and hence the last expression is positive for sufficiently large λ. This contradicts (11). Q.E.D.

This proof was given by Stein (1956a). It is a generalization of a theorem of Birnbaum (1955).

COROLLARY 5.6.2. *If the conditions of Theorem* 5.6.5 *hold except that A is not necessarily closed, but the boundary of A has m-measure* 0, *then the conclusion of Theorem* 5.6.5 *holds.*

PROOF. The closure of A is convex (Problem 18), and the test with acceptance region closure of A differs from A by a set of probability 0 for all $\omega \in \Omega$. Furthermore,

$$
\begin{aligned}
(15)\quad & A \cap \{ y \mid \omega' y > c \} = \varnothing \Rightarrow A \subset \{ y \mid \omega' y \le c \} \\
&\Rightarrow \text{closure } A \subset \{ y \mid \omega' y \le c \}.
\end{aligned}
$$

Then Theorem 5.6.5 holds with A replaced by the closure of A. Q.E.D.

THEOREM 5.6.6. *Based on observations* $\mathbf{x}_1, \dots, \mathbf{x}_N$ *from* $N(\boldsymbol{\mu}, \boldsymbol{\Sigma})$, *Hotelling's* T^2*-test is admissible for testing the hypothesis* $\boldsymbol{\mu} = \mathbf{0}$.

PROOF. To apply Theorem 5.6.5 we put the distribution of the observations into the form of an exponential family. By Theorems 3.3.1 and 3.3.2 we can transform $\mathbf{x}_1, \dots, \mathbf{x}_N$ to $\mathbf{z}_\alpha = \sum_{\beta=1}^{N} c_{\alpha\beta}\mathbf{x}_\beta$, where $(c_{\alpha\beta})$ is orthogonal and $\mathbf{z}_N = \sqrt{N}\,\bar{\mathbf{x}}$. Then the density of $\mathbf{z}_1, \dots, \mathbf{z}_N$ (with respect to Lebesgue measure) is

$$\frac{e^{-\frac{1}{2}N\boldsymbol{\mu}'\boldsymbol{\Sigma}^{-1}\boldsymbol{\mu}}}{(2\pi)^{\frac{1}{2}pN}|\boldsymbol{\Sigma}|^{\frac{1}{2}N}} \exp\left[\sqrt{N}\,\boldsymbol{\mu}'\boldsymbol{\Sigma}^{-1}\mathbf{z}_N + \operatorname{tr}\left(-\tfrac{1}{2}\boldsymbol{\Sigma}^{-1}\right)\sum_{\alpha=1}^{N}\mathbf{z}_\alpha\mathbf{z}_\alpha'\right]. \tag{16}$$

The vector $\mathbf{y} = (\mathbf{y}^{(1)\prime}, \mathbf{y}^{(2)\prime})'$ is composed of $\mathbf{y}^{(1)} = \mathbf{z}_N$ $(= \sqrt{N}\,\bar{\mathbf{x}})$ and $\mathbf{y}^{(2)} = (b_{11}, 2b_{12}, \dots, 2b_{1p}, b_{22}, \dots, b_{pp})$, where

$$\mathbf{B} = \sum_{\alpha=1}^{N}\mathbf{z}_\alpha\mathbf{z}_\alpha' \left(= \sum_{\alpha=1}^{N}\mathbf{x}_\alpha\mathbf{x}_\alpha' \right). \tag{17}$$

The vector $\boldsymbol{\omega} = (\boldsymbol{\omega}^{(1)\prime}, \boldsymbol{\omega}^{(2)\prime})'$ is composed of $\boldsymbol{\omega}^{(1)} = \sqrt{N}\,\boldsymbol{\Sigma}^{-1}\boldsymbol{\mu}$ and $\boldsymbol{\omega}^{(2)} = -\frac{1}{2}(\sigma^{11}, \sigma^{12}, \dots, \sigma^{1p}, \sigma^{22}, \dots, \sigma^{pp})'$. The measure $m(A)$ is the Lebesgue measure of the set of $\mathbf{z}_1, \dots, \mathbf{z}_N$ that maps into the set A.

LEMMA 5.6.1. *Let* $\mathbf{B} = \mathbf{A} + N\bar{\mathbf{x}}\bar{\mathbf{x}}'$. *Then*

$$N\bar{\mathbf{x}}'\mathbf{A}^{-1}\bar{\mathbf{x}} = \frac{N\bar{\mathbf{x}}'\mathbf{B}^{-1}\bar{\mathbf{x}}}{1 - N\bar{\mathbf{x}}'\mathbf{B}^{-1}\bar{\mathbf{x}}}. \tag{18}$$

PROOF OF LEMMA. If we let $\mathbf{B} = \mathbf{A} + \sqrt{N}\,\bar{\mathbf{x}}\sqrt{N}\,\bar{\mathbf{x}}'$ in (10) of Section 5.2, we obtain by Corollary A.3.1

$$\begin{aligned} \frac{1}{1 + T^2/(N-1)} &= \lambda^{2/N} = \frac{|\mathbf{B} - \sqrt{N}\,\bar{\mathbf{x}}\sqrt{N}\,\bar{\mathbf{x}}'|}{|\mathbf{B}|} \\ &= 1 - N\bar{\mathbf{x}}'\mathbf{B}^{-1}\bar{\mathbf{x}}. \end{aligned} \tag{19}$$

Q.E.D.

Thus the acceptance region of a T^2-test is

$$A = \left\{\mathbf{z}_N, \mathbf{B} \,|\, \mathbf{z}_N'\mathbf{B}^{-1}\mathbf{z}_N \le k, \mathbf{B} \text{ positive definite}\right\} \tag{20}$$

for a suitable k.

The function $z_N' B^{-1} z_N$ is convex in (z, B) for B positive definite (Problem 17). Therefore, the set $z_N' B^{-1} z_N \le k$ is convex. This shows that the set A is convex. Furthermore, the closure of A is convex (Problem 18) and the probability of the boundary of A is 0.

Now consider the other condition of Theorem 5.6.5. Suppose A is disjoint with the half-space

$$c < \omega' y = \nu' z_N - \tfrac{1}{2} \operatorname{tr} \Lambda B, \tag{21}$$

where Λ is a symmetric matrix and B is positive semidefinite. We shall take $\Lambda_1 = I$. We want to show that $\omega_1 + \lambda \omega \in \Omega - \Omega_0$; that is, that $\nu_1 + \lambda \nu \neq 0$ (which is trivial) and $\Lambda_1 + \lambda \Lambda$ is positive definite for $\lambda > 0$. This is the case when Λ is positive semidefinite. Now we shall show that a half-space (21) disjoint with A and Λ not positive semidefinite implies a contradiction. If Λ is not positive semidefinite, it can be written (by Corollary A.4.1 of Appendix A)

$$\Lambda = D \begin{bmatrix} I & 0 & 0 \\ 0 & -I & 0 \\ 0 & 0 & 0 \end{bmatrix} D', \tag{22}$$

where D is nonsingular. If Λ is not positive semidefinite, $-I$ is not vacuous because its order is the number of negative characteristic roots of Λ. Let $z_N = (1/\gamma) z_0$ and

$$B = (D')^{-1} \begin{bmatrix} I & 0 & 0 \\ 0 & \gamma I & 0 \\ 0 & 0 & I \end{bmatrix} D^{-1}. \tag{23}$$

Then

$$\omega' y = \frac{1}{\gamma} \nu' z_0 + \tfrac{1}{2} \operatorname{tr} \begin{bmatrix} -I & 0 & 0 \\ 0 & \gamma I & 0 \\ 0 & 0 & 0 \end{bmatrix}, \tag{24}$$

which is greater than c for sufficiently large γ. On the other hand

$$z_N' B^{-1} z_N = \frac{1}{\gamma^2} z_0' D \begin{bmatrix} I & 0 & 0 \\ 0 & \gamma^{-1} I & 0 \\ 0 & 0 & I \end{bmatrix} D' z_0, \tag{25}$$

which is less than k for sufficiently large γ. This contradicts the fact that (20) and (21) are disjoint. Thus the conditions of Theorem 5.6.5 are satisfied and the theorem is proved. Q.E.D. (This proof is due to Stein.)

An alternative proof of admissibility is to show that the T^2-test is a proper Bayes procedure. Suppose an arbitrary random vector $\boldsymbol{X}$ has density $f(\boldsymbol{x}|\boldsymbol{\omega})$ for $\boldsymbol{\omega} \in \Omega$. Consider testing the null hypothesis $H_0: \boldsymbol{\omega} \in \Omega_0$ against the alternative $H_1: \boldsymbol{\omega} \in \Omega - \Omega_0$. Let Π_0 be a prior finite measure on Ω_0 and Π_1 a prior finite measure on Ω_1. Then the Bayes procedure (with 0–1 loss function) is to reject H_0 if

$$\frac{\int f(\boldsymbol{x}|\boldsymbol{\omega})\Pi_1(d\boldsymbol{\omega})}{\int f(\boldsymbol{x}|\boldsymbol{\omega})\Pi_0(d\boldsymbol{\omega})} \geq c \tag{26}$$

for some c $(0 \leq c \leq \infty)$. If equality in (26) occurs with probability 0 for all $\boldsymbol{\omega} \in \Omega_0$, then the Bayes procedure is unique and hence admissible. Since the measures are finite, they can be normed to be probability measures. For the T^2-test of H_0: $\boldsymbol{\mu} = \boldsymbol{0}$ a pair of measures is suggested in Problem 15. (This pair is not unique.) The reader can verify that with these measures (26) reduces to the complement of (20).

Among invariant tests it was shown that the T^2-test is uniformly most powerful; that is, it is most powerful against every value of $\boldsymbol{\mu}'\boldsymbol{\Sigma}^{-1}\boldsymbol{\mu}$ among invariant tests of the specified significance level. We can ask whether the T^2-test is "best" against a specified value of $\boldsymbol{\mu}'\boldsymbol{\Sigma}^{-1}\boldsymbol{\mu}$ among all tests. Here "best" can be taken to mean admissible "minimax"; and "minimax" means maximizing with respect to procedures the minimum with respect to parameter values of the power. This property was shown in the simplest case of $p = 2$ and $N = 3$ by Giri, Kiefer, and Stein (1963). The property for general p and N was announced by Šalaevskiĭ (1968). He has furnished a proof for the case of $p = 2$ [Šalaevskiĭ (1971)], but has not given a proof for $p > 2$.

Giri and Kiefer (1964) have proved the T^2-test is locally minimax (as $\boldsymbol{\mu}'\boldsymbol{\Sigma}^{-1}\boldsymbol{\mu} \to 0$) and asymptotically (logarithmically) minimax as $\boldsymbol{\mu}'\boldsymbol{\Sigma}^{-1}\boldsymbol{\mu} \to \infty$.

PROBLEMS

1. (Sec. 5.2) Let $\boldsymbol{x}_\alpha$ be distributed according to $N(\boldsymbol{\mu} + \boldsymbol{\beta}(z_\alpha - \bar{z}), \boldsymbol{\Sigma})$, $\alpha = 1, \dots, N$, where $\bar{z} = (1/N)\Sigma z_\alpha$. Let $\boldsymbol{b} = [1/\Sigma(z_\alpha - \bar{z})^2]\Sigma \boldsymbol{x}_\alpha(z_\alpha - \bar{z})$, $(N-2)\boldsymbol{S} = \Sigma[\boldsymbol{x}_\alpha - \bar{\boldsymbol{x}} - \boldsymbol{b}(z_\alpha - \bar{z})][\boldsymbol{x}_\alpha - \bar{\boldsymbol{x}} - \boldsymbol{b}(z_\alpha - \bar{z})]'$, and $T^2 = \Sigma(z_\alpha - \bar{z})^2 \boldsymbol{b}'\boldsymbol{S}^{-1}\boldsymbol{b}$. Show that T^2 has the T^2-distribution with $N-2$ degrees of freedom. [*Hint*: See Problem 13 of Chapter 3.]

TABLE 5.1

Section 5.2	Section 4.4
$x_{0\alpha} = 1/\sqrt{N}$	$z_{1\alpha}$
$\boldsymbol{x}_\alpha$	$\boldsymbol{z}_\alpha^{(2)}$
$\sqrt{N}\bar{\boldsymbol{x}}$	$\boldsymbol{a}_{(1)} = \sum z_{1\alpha}\boldsymbol{z}_\alpha^{(2)}$
$\boldsymbol{B} = \sum \boldsymbol{x}_\alpha \boldsymbol{x}_\alpha'$	$\boldsymbol{A}_{22} = \sum \boldsymbol{z}_\alpha^{(2)}\boldsymbol{z}_\alpha^{(2)\prime}$
$1 = \sum x_{0\alpha}^2$	$a_{11} = \sum z_{1\alpha}^2$
$\dfrac{T^2}{N-1}$	$\dfrac{R^2}{1-R^2}$
p	$p-1$
N	n

2. (Sec. 5.2.2) Show that $T^2/(N-1)$ can be written as $R^2/(1-R^2)$ with the correspondences given in Table 5.1.

3. (Sec. 5.2.2) Let

$$\frac{R^2}{1-R^2} = \frac{\sum u_\alpha \boldsymbol{x}_\alpha' \left(\sum \boldsymbol{x}_\alpha \boldsymbol{x}_\alpha'\right)^{-1} \sum u_\alpha \boldsymbol{x}_\alpha}{\sum u_\alpha^2 - \sum u_\alpha \boldsymbol{x}_\alpha' \left(\sum \boldsymbol{x}_\alpha \boldsymbol{x}_\alpha'\right)^{-1} \sum u_\alpha \boldsymbol{x}_\alpha},$$

where $u_1, \dots, u_N$ are N numbers and $\boldsymbol{x}_1, \dots, \boldsymbol{x}_N$ are independent, each with the distribution $N(\boldsymbol{0}, \boldsymbol{\Sigma})$. Prove that the distribution of $R^2/(1-R^2)$ is independent of $u_1, \dots, u_N$. [*Hint*: There is an orthogonal $N \times N$ matrix $\boldsymbol{C}$ that carries $(u_1, \dots, u_N)$ into a vector proportional to $(1/\sqrt{N}, \dots, 1/\sqrt{N})$.]

4. (Sec. 5.2.2) Use Problems 2 and 3 to show that $[T^2/(N-1)][(N-p)/p]$ has the $F_{p, N-p}$-distribution (under the null hypothesis). [*Note*: This is the analysis that corresponds to Hotelling's geometric proof (1931).]

5. (Sec. 5.2.2) Let $T^2 = N\bar{\boldsymbol{x}}'\boldsymbol{S}^{-1}\bar{\boldsymbol{x}}$, where $\bar{\boldsymbol{x}}$ and $\boldsymbol{S}$ are the mean vector and covariance matrix of a sample of N from $N(\boldsymbol{\mu}, \boldsymbol{\Sigma})$. Show that T^2 is distributed the same when $\boldsymbol{\mu}$ is replaced by $\boldsymbol{\lambda} = (\tau, 0, \dots, 0)'$, where $\tau^2 = \boldsymbol{\mu}'\boldsymbol{\Sigma}^{-1}\boldsymbol{\mu}$, and $\boldsymbol{\Sigma}$ is replaced by $\boldsymbol{I}$.

6. (Sec. 5.2.2) Let $u = [T^2/(N-1)]/[1 + T^2/(N-1)]$. Show that $u = \boldsymbol{\gamma}\boldsymbol{V}'(\boldsymbol{V}\boldsymbol{V}')^{-1}\boldsymbol{V}\boldsymbol{\gamma}'$, where $\boldsymbol{\gamma} = (1/\sqrt{N}, \dots, 1/\sqrt{N})$ and

$$\boldsymbol{V} = \begin{pmatrix} \boldsymbol{v}_1 \\ \vdots \\ \boldsymbol{v}_p \end{pmatrix} = \begin{pmatrix} x_{11} & \cdots & x_{1N} \\ \vdots & & \vdots \\ x_{p1} & \cdots & x_{pN} \end{pmatrix}.$$

7. (Sec. 5.2.2) Let

$$\boldsymbol{v}_1^* = \boldsymbol{v}_1,$$

$$\boldsymbol{v}_i^* = \boldsymbol{v}_i - \frac{\boldsymbol{v}_i\boldsymbol{v}_1'}{\boldsymbol{v}_1\boldsymbol{v}_1'}\boldsymbol{v}_1 = \boldsymbol{v}_i\left(\boldsymbol{I} - \frac{1}{\boldsymbol{v}_1\boldsymbol{v}_1'}\boldsymbol{v}_1'\boldsymbol{v}_1\right), \qquad i \neq 1,$$

$$\boldsymbol{\gamma}^* = \boldsymbol{\gamma} - \frac{\boldsymbol{\gamma}\boldsymbol{v}_1'}{\boldsymbol{v}_1\boldsymbol{v}_1'}\boldsymbol{v}_1,$$

$$\boldsymbol{V}^* = \begin{pmatrix} \boldsymbol{v}_1^* \\ \vdots \\ \boldsymbol{v}_p^* \end{pmatrix}.$$

Prove that $U = s + (1 - s)w$, where

$$s = \frac{(\boldsymbol{\gamma}\boldsymbol{v}_1^{*\prime})^2}{\boldsymbol{v}_1^*\boldsymbol{v}_1^{*\prime}} = \frac{(\boldsymbol{\gamma}\boldsymbol{v}_1')^2}{\boldsymbol{v}_1\boldsymbol{v}_1'},$$

$$w = \frac{1}{\boldsymbol{\gamma}^*\boldsymbol{\gamma}^{*\prime}}\boldsymbol{\gamma}^*\begin{pmatrix} \boldsymbol{v}_2^* \\ \vdots \\ \boldsymbol{v}_p^* \end{pmatrix}'\begin{pmatrix} \boldsymbol{v}_2^*\boldsymbol{v}_2^{*\prime} & \cdots & \boldsymbol{v}_2^*\boldsymbol{v}_p^{*\prime} \\ \vdots & & \vdots \\ \boldsymbol{v}_p^*\boldsymbol{v}_2^{*\prime} & \cdots & \boldsymbol{v}_p^*\boldsymbol{v}_p^{*\prime} \end{pmatrix}^{-1}\begin{pmatrix} \boldsymbol{v}_2^* \\ \vdots \\ \boldsymbol{v}_p^* \end{pmatrix}\boldsymbol{\gamma}^{*\prime}.$$

Hint: $\boldsymbol{E}\boldsymbol{V} = \boldsymbol{V}^*$, where

$$\boldsymbol{E} = \begin{pmatrix} 1 & 0 & \cdots & 0 \\ -\dfrac{\boldsymbol{v}_2\boldsymbol{v}_1'}{\boldsymbol{v}_1\boldsymbol{v}_1'} & 1 & \cdots & 0 \\ \vdots & \vdots & & \vdots \\ -\dfrac{\boldsymbol{v}_p\boldsymbol{v}_1'}{\boldsymbol{v}_1\boldsymbol{v}_1'} & 0 & \cdots & 1 \end{pmatrix}.$$

8. (Sec. 5.2.2) Prove that w has the distribution of the square of a multiple correlation between one vector and $p - 1$ vectors in $(N - 1)$-space without subtracting means; that is, has density

$$\frac{\Gamma[\frac{1}{2}(N-1)]}{\Gamma[\frac{1}{2}(N-p)]\Gamma[\frac{1}{2}(p-1)]} w^{\frac{1}{2}(p-1)-1}(1-w)^{\frac{1}{2}(N-p)-1}.$$

[*Hint*: The transformation of Problem 7 is a projection of $\boldsymbol{v}_2, \ldots, \boldsymbol{v}_p$, $\boldsymbol{\gamma}$ on the $(N - 1)$-space orthogonal to $\boldsymbol{v}_1$.]

9. (Sec. 5.2.2) Verify that $r = s/(1 - s)$ multiplied by $(N - 1)/1$ has the noncentral F-distribution with 1 and $N - 1$ degrees of freedom and noncentrality parameter $N\tau^2$.

10. (Sec. 5.2.2) From Problems 5–9, verify Corollary 5.2.1.

11. (Sec. 5.3) Use the data in Section 3.2 to test the hypothesis that neither drug has a soporific effect at significance level 0.01.

12. (Sec. 5.3) Using the data in Section 3.2, give a confidence region for $\boldsymbol{\mu}$ with confidence coefficient 0.95.

13. (Sec. 5.3) Prove the statement in Section 5.3.6 that the T^2-statistic is independent of the choice of $\boldsymbol{C}$.

14. (Sec. 5.5) Use the data of Problem 41 of Chapter 4 to test the hypothesis that the mean head length and breadth of first sons is equal to those of second sons at significance level 0.01.

15. (Sec. 5.6.2) *T^2-test as a Bayes procedure.* [Kiefer and Schwartz (1965)] Let $\boldsymbol{x}_1, \dots, \boldsymbol{x}_N$ be independently distributed, each according to $N(\boldsymbol{\mu}, \boldsymbol{\Sigma})$. Let Π_0 be defined by $[\boldsymbol{\mu}, \boldsymbol{\Sigma}] = [\boldsymbol{0}, (\boldsymbol{I} + \boldsymbol{\eta}\boldsymbol{\eta}')^{-1}]$ with $\boldsymbol{\eta}$ having a density proportional to $|\boldsymbol{I} + \boldsymbol{\eta}\boldsymbol{\eta}'|^{-\frac{1}{2}N}$, and let Π_1 be defined by $[\boldsymbol{\mu}, \boldsymbol{\Sigma}] = [(\boldsymbol{I} + \boldsymbol{\eta}\boldsymbol{\eta}')^{-1}\boldsymbol{\eta}, (\boldsymbol{I} + \boldsymbol{\eta}\boldsymbol{\eta}')^{-1}]$ with $\boldsymbol{\eta}$ having a density proportional to

$$|\boldsymbol{I} + \boldsymbol{\eta}\boldsymbol{\eta}'|^{-\frac{1}{2}N} \exp\left[\tfrac{1}{2}N\boldsymbol{\eta}'(\boldsymbol{I} + \boldsymbol{\eta}\boldsymbol{\eta}')^{-1}\boldsymbol{\eta}\right].$$

(*a*) Show that the measures are finite for $N > p$ by showing $\boldsymbol{\eta}'(\boldsymbol{I} + \boldsymbol{\eta}\boldsymbol{\eta}')^{-1}\boldsymbol{\eta} \le 1$ and verifying that the integral of $|\boldsymbol{I} + \boldsymbol{\eta}\boldsymbol{\eta}'|^{-\frac{1}{2}N} = (1 + \boldsymbol{\eta}'\boldsymbol{\eta})^{-\frac{1}{2}N}$ is finite.

(*b*) Show that the inequality (26) is equivalent to $N\bar{\boldsymbol{x}}'(\sum_{\alpha=1}^{N}\boldsymbol{x}_\alpha\boldsymbol{x}_\alpha')^{-1}\bar{\boldsymbol{x}} \ge k$. Hence the T^2-test is Bayes and thus admissible.

16. (Sec. 5.6.2) Let $g(t) = f[t\boldsymbol{y}_1 + (1-t)\boldsymbol{y}_2]$, where $f(\boldsymbol{y})$ is a real-valued function of the vector $\boldsymbol{y}$. Prove that if $g(t)$ is convex, then $f(\boldsymbol{y})$ is convex.

17. (Sec. 5.6.2) Show that $\boldsymbol{z}'\boldsymbol{B}^{-1}\boldsymbol{z}$ is a convex function of $(\boldsymbol{z}, \boldsymbol{B})$, where $\boldsymbol{B}$ is a positive definite matrix. [*Hint*: Use Problem 16.]

18. (Sec. 5.6.2) Prove that if the set A is convex, then the closure of A is convex.

19. (Sec. 5.3) Let $\bar{\boldsymbol{x}}$ and $\boldsymbol{S}$ be based on N observations from $N(\boldsymbol{\mu}, \boldsymbol{\Sigma})$ and let $\boldsymbol{x}$ be an additional observation from $N(\boldsymbol{\mu}, \boldsymbol{\Sigma})$. Show that $\boldsymbol{x} - \bar{\boldsymbol{x}}$ is distributed according to

$$N[\boldsymbol{0}, (1 + 1/N)\boldsymbol{\Sigma}].$$

Verify that $[N/(N+1)](\boldsymbol{x} - \bar{\boldsymbol{x}})'\boldsymbol{S}^{-1}(\boldsymbol{x} - \bar{\boldsymbol{x}})$ has the T^2-distribution with $N-1$ degrees of freedom. Show how this statistic can be used to give a prediction region for $\boldsymbol{x}$ based on $\bar{\boldsymbol{x}}$ and $\boldsymbol{S}$ (i.e., a region such that one has a given confidence that the next observation will fall into it).

20. (Sec. 5.3) Let $\boldsymbol{x}_\alpha^{(i)}$ be observations from $N(\boldsymbol{\mu}^{(i)}, \boldsymbol{\Sigma}_i)$, $\alpha = 1, \dots, N_i$, $i = 1, 2$. Find the likelihood ratio criterion for testing the hypothesis $\boldsymbol{\mu}^{(1)} = \boldsymbol{\mu}^{(2)}$.

21. (Sec. 5.4) Prove that $\boldsymbol{\mu}'\boldsymbol{\Sigma}^{-1}\boldsymbol{\mu}$ is larger for $\boldsymbol{\mu}' = (\mu_1, \mu_2)$ than for $\boldsymbol{\mu} = \mu_1$ by verifying

$$\frac{1}{1-\rho^2}\left(\frac{\mu_1^2}{\sigma_1^2} - 2\rho\frac{\mu_1\mu_2}{\sigma_1\sigma_2} + \frac{\mu_2^2}{\sigma_2^2}\right) = \frac{\mu_1^2}{\sigma_1^2} + \frac{(\mu_2 - \rho\sigma_2\mu_1/\sigma_1)^2}{(1-\rho^2)\sigma_2^2}.$$

Discuss the power of the test $\mu_1 = 0$ compared to the power of the test $\mu_1 = 0$, $\mu_2 = 0$.

22. (Sec. 5.3) (*a*) Using the data of Section 5.3.3 test the hypothesis $\mu_1^{(1)} = \mu_1^{(2)}$.
(*b*) Test the hypothesis $\mu_1^{(1)} = \mu_1^{(2)}$, $\mu_2^{(1)} = \mu_2^{(2)}$.

23. (Sec. 5.4) Let

$$\boldsymbol{\mu} = \begin{pmatrix} \boldsymbol{\mu}^{(1)} \\ \boldsymbol{\mu}^{(2)} \end{pmatrix}, \qquad \boldsymbol{\Sigma} = \begin{pmatrix} \boldsymbol{\Sigma}_{11} & \boldsymbol{\Sigma}_{12} \\ \boldsymbol{\Sigma}_{21} & \boldsymbol{\Sigma}_{22} \end{pmatrix}.$$

Prove $\boldsymbol{\mu}'\boldsymbol{\Sigma}^{-1}\boldsymbol{\mu} \geq \boldsymbol{\mu}^{(1)\prime}\boldsymbol{\Sigma}_{11}^{-1}\boldsymbol{\mu}^{(1)}$. Give a condition for strict inequality to hold. [*Hint*: This is the vector analogue of Problem 21.]

24. Let $\boldsymbol{X}^{(i)\prime} = (\boldsymbol{Y}^{(i)\prime}, \boldsymbol{Z}^{(i)\prime})$, $i = 1, 2$, where $\boldsymbol{Y}^{(i)}$ has p components and $\boldsymbol{Z}^{(i)}$ has q components, be distributed according to $N(\boldsymbol{\mu}^{(i)}, \boldsymbol{\Sigma})$, where

$$\boldsymbol{\mu}^{(i)} = \begin{pmatrix} \boldsymbol{\mu}_y^{(i)} \\ \boldsymbol{\mu}_z^{(i)} \end{pmatrix}, \qquad \boldsymbol{\Sigma} = \begin{pmatrix} \boldsymbol{\Sigma}_{yy} & \boldsymbol{\Sigma}_{yz} \\ \boldsymbol{\Sigma}_{zy} & \boldsymbol{\Sigma}_{zz} \end{pmatrix}, \qquad i = 1, 2.$$

Find the likelihood ratio criterion (or equivalent T^2-criterion) for testing $\boldsymbol{\mu}_z^{(1)} = \boldsymbol{\mu}_z^{(2)}$ given $\boldsymbol{\mu}_y^{(1)} = \boldsymbol{\mu}_y^{(2)}$ on the basis of a sample of N_i on $\boldsymbol{X}^{(i)}$, $i = 1, 2$. [*Hint*: Express the likelihood in terms of the marginal density of $\boldsymbol{Y}^{(i)}$ and the conditional density of $\boldsymbol{Z}^{(i)}$ given $\boldsymbol{Y}^{(i)}$.]

25. Find the distribution of the criterion in the preceding problem under the null hypothesis.

26. (Sec. 5.5) Suppose $\boldsymbol{x}_\alpha^{(g)}$ is an observation from $N(\boldsymbol{\mu}^{(g)}, \boldsymbol{\Sigma}_g)$, $\alpha = 1, \dots, N_g$, $g = 1, \dots, q$. (a) Show that the hypothesis $\boldsymbol{\mu}^{(1)} = \cdots = \boldsymbol{\mu}^{(q)}$ is equivalent to $\mathscr{E}\boldsymbol{y}_\alpha^{(i)} = \boldsymbol{0}$, $i = 1, \dots, q-1$, where

$$\boldsymbol{y}_\alpha^{(i)} = a_1^{(i)}\boldsymbol{x}_\alpha^{(1)} + \sum_{g=2}^{q} a_g^{(i)} \left(\frac{N_1}{N_g} \right)^{\frac{1}{2}} \left[\boldsymbol{x}_\alpha^{(g)} - \frac{1}{N_1} \sum_{\beta=1}^{N_1} \boldsymbol{x}_\beta^{(g)} + \frac{1}{(N_1 N_g)^{\frac{1}{2}}} \sum_{\beta=1}^{N_g} \boldsymbol{x}_\beta^{(g)} \right],$$
$$\alpha = 1, \dots, N_1, \; i = 1, \dots, q-1,$$

$N_1 \leq N_g$, $g = 2, \dots, q$, and $(a_1^{(i)}, \dots, a_q^{(i)})$, $i = 1, \dots, q-1$, are linearly independent. (b) Show how to construct a T^2-test of the hypothesis using $(\bar{\boldsymbol{y}}^{(1)\prime}, \dots, \bar{\boldsymbol{y}}^{(q-1)\prime})'$ yielding an F-statistic with $(q-1)p$ and $N - (q-1)p$ degrees of freedom [Anderson (1963b)].

27. (Sec. 5.2) Prove (25) is the density of $V = \chi_a^2/(\chi_a^2 + \chi_b^2)$. [*Hint*: In the joint density of $U = \chi_a^2$ and $W = \chi_b^2$ make the transformation $u = vw(1-v)^{-1}$, $w = w$ and integrate out w.]

CHAPTER 6

Classification of Observations

6.1. THE PROBLEM OF CLASSIFICATION

The problem of classification arises when an investigator makes a number of measurements on an individual and wishes to classify the individual into one of several categories on the basis of these measurements. The investigator cannot identify the individual with a category directly but must use these measurements. In many cases it can be assumed that there are a finite number of categories or populations from which the individual may have come and each population is characterized by a probability distribution of the measurements. Thus an individual is considered as a random observation from this population. The question is: Given an individual with certain measurements, from which population did the person arise?

The problem of classification may be considered as a problem of "statistical decision functions." We have a number of hypotheses: Each hypothesis is that the distribution of the observation is a given one. We must accept one of these hypotheses and reject the others. If only two populations are admitted, we have an elementary problem of testing one hypothesis of a specified distribution against another.

In some instances, the categories are specified beforehand in the sense that the probability distributions of the measurements are assumed completely known. In other cases, the form of each distribution may be known, but the parameters of the distribution must be estimated from a sample from that population.

Let us give an example of a problem of classification. Prospective students applying for admission into college are given a battery of tests; the vector of scores is a set of measurements $\boldsymbol{x}$. The prospective student may be a member of one population consisting of those students who will successfully complete college training or, rather, have potentialities for successfully completing training, or the student may be a member of the other population, those who will not complete the college course successfully. The problem is to classify a student applying for admission on the basis of his scores on the entrance examination.

In this chapter we shall develop the theory of classification in general terms and then apply it to cases involving the normal distribution. In Section 6.2 the problem of classification with two populations is defined in terms of decision theory, and in Section 6.3 Bayes and admissible solutions are obtained. In Section 6.4 the theory is applied to two known normal populations, differing with respect to means, yielding the population linear discriminant function. When the parameters are unknown, they are replaced by estimates (Section 6.5). An alternative procedure is maximum likelihood. In Section 6.6 the probabilities of misclassification by the two methods are evaluated in terms of asymptotic expansions of the distributions. Then these developments are carried out for several populations. Finally, in Section 6.10 linear procedures for the two populations are studied when the covariance matrices are different and the parameters are known.

6.2. STANDARDS OF GOOD CLASSIFICATION

6.2.1. Preliminary Considerations

In constructing a procedure of classification, it is desired to minimize the probability of misclassification, or, more specifically, it is desired to minimize on the average the bad effects of misclassification. Now let us make this notion precise. For convenience we shall now consider the case of only two categories. Later we shall treat the more general case. This section develops the ideas of Section 3.4 in more detail for the problem of two decisions.

Suppose an individual is an observation from either population π_1 or population π_2. The classification of an observation depends on the vector of measurements $\boldsymbol{x}' = (x_1, \ldots, x_p)$ on that individual. We set up a rule that if an individual is characterized by certain sets of values of $x_1, \ldots, x_p$ that person will be classified as from π_1, if other values, as from π_2.

We can think of an observation as a point in a p-dimensional space. We divide this space into two regions. If the observation falls in R_1, we classify it

TABLE 6.1

		Statistician's Decision	
		π_1	π_2
Population	π_1	0	$C(2\|1)$
	π_2	$C(1\|2)$	0

as coming from population π_1, and if it falls in R_2 we classify it as coming from population π_2.

In following a given classification procedure, the statistician can make two kinds of errors in classification. If the individual is actually from π_1 the statistician can classify that person as coming from population π_2; or if that person is from π_2 the statistician may classify that individual as from π_1. We need to know the relative undesirability of these two kinds of misclassification. Let the "cost" of the first type of misclassification be $C(2|1)$ (> 0) and let the cost of misclassifying an individual from π_2 as from π_1 be $C(1|2)$ (> 0). These costs may be measured in any kind of units. As we shall see later, it is only the ratio of the two costs that is important. The statistician may not know these costs in each case, but will often have at least a rough idea of them.

Table 6.1 indicates the costs of correct and incorrect classification. Clearly, a good classification procedure is one that minimizes in some sense or other the cost of misclassification.

6.2.2. Two Cases of Two Populations

We shall consider ways of defining "minimum cost" in two cases. In one case we shall suppose that we have a priori probabilities of the two populations. Let the probability that an observation comes from population π_1 be q_1 and from population π_2 be q_2 ($q_1 + q_2 = 1$). The probability properties of population π_1 are specified by a distribution function. For convenience we shall treat only the case where the distribution has a density, although the case of discrete probabilities lends itself to almost the same treatment. Let the density of population π_1 be $p_1(\boldsymbol{x})$ and that of π_2 be $p_2(\boldsymbol{x})$. If we have a region R_1 of classification as from π_1, the probability of correctly classifying an observation that actually is drawn from population π_1 is

$$(1) \qquad P(1|1, R) = \int_{R_1} p_1(\boldsymbol{x})\, d\boldsymbol{x},$$

where $dx = dx_1 \cdots dx_p$, and the probability of misclassification of an observation from π_1 is

$$P(2|1, R) = \int_{R_2} p_1(x)\, dx. \tag{2}$$

Similarly, the probability of correctly classifying an observation from π_2 is

$$P(2|2, R) = \int_{R_2} p_2(x)\, dx, \tag{3}$$

and the probability of misclassifying such an observation is

$$P(1|2, R) = \int_{R_1} p_2(x)\, dx. \tag{4}$$

Since the probability of drawing an observation from π_1 is q_1, the probability of drawing an observation from π_1 and correctly classifying it is $q_1 P(1|1, R)$; that is, this is the probability of the situation in the upper left-hand corner of Table 6.1. Similarly, the probability of drawing an observation from π_1 and misclassifying it is $q_1 P(2|1, R)$. The probability associated with the lower left-hand corner of Table 6.1 is $q_2 P(1|2, R)$ and with the lower right-hand corner is $q_2 P(2|2, R)$.

What is the average or expected loss from costs of misclassification? It is the sum of the products of costs of each misclassification multiplied by the probability of its occurrence:

$$C(2|1) P(2|1, R) q_1 + C(1|2) P(1|2, R) q_2. \tag{5}$$

It is this average loss that we wish to minimize. That is, we want to divide our space into regions R_1 and R_2 such that the expected loss is as small as possible. A procedure that minimizes (5) for given q_1 and q_2 is called a *Bayes procedure*.

In the example of admission of students, the undesirability of misclassification is, in one instance, the expense of teaching a student who will not complete the course successfully and is, in the other instance, the undesirability of excluding from college a potentially good student.

The other case we shall treat is that in which there are no known a priori probabilities. In this case the expected loss if the observation is from π_1 is

$$C(2|1) P(2|1, R) = r(1, R); \tag{6}$$

the expected loss if the observation is from π_2 is

$$C(1|2)P(1|2, R) = r(2, R). \tag{7}$$

We do not know whether the observation is from π_1 or from π_2, and we do not know probabilities of these two instances.

A procedure R is at least as good as a procedure R^* if $r(1, R) \leq r(1, R^*)$ and $r(2, R) \leq r(2, R^*)$; R is better than R^* if at least one of these inequalities is a strict inequality. Usually there is no one procedure that is better than all other procedures or is at least as good as all other procedures. A procedure R is called *admissible* if there is no procedure better than R; we shall be interested in the entire class of admissible procedures. It will be shown that under certain conditions this class is the same as the class of Bayes procedures. A class of procedures is *complete* if for every procedure outside the class there is one in the class which is better; a class is called *essentially complete* if for every procedure outside the class there is one in the class which is at least as good. A *minimal complete class* (if it exists) is a complete class such that no proper subset is a complete class; a similar definition holds for a minimal essentially complete class. Under certain conditions we shall show that the admissible class is minimal complete. To simplify the discussion we shall consider procedures the same if they only differ on sets of probability zero. In fact, throughout the next section we shall make statements which are meant to hold "except for sets of probability zero" without saying so explicitly.

A principle that usually leads to a unique procedure is the minimax principle. A procedure is *minimax* if the maximum expected loss, $r(i, R)$, is a minimum. From a conservative point of view, this may be considered an optimum procedure. For a general discussion of the concepts in this section and the next see Wald (1950), Blackwell and Girshick (1954), Ferguson (1967), DeGroot (1970), and Berger (1980b).

6.3. PROCEDURES OF CLASSIFICATION INTO ONE OF TWO POPULATIONS WITH KNOWN PROBABILITY DISTRIBUTIONS

6.3.1. The Case When A Priori Probabilities Are Known

We now turn to the problem of choosing regions R_1 and R_2 so as to minimize (5) of Section 6.2. Since we have a priori probabilities, we can define joint probabilities of the population and the observed set of variables. The probability that an observation comes from π_1 and that each variate is less than the corresponding component in $\boldsymbol{y}$ is

$$(1) \qquad \int_{-\infty}^{y_p} \cdots \int_{-\infty}^{y_1} q_1 p_1(\boldsymbol{x})\, dx_1 \cdots dx_p.$$

We can also define the conditional probability that an observation came from a certain population given the values of the observed variates. For instance, the conditional probability of coming from population π_1, given an observation $\boldsymbol{x}$, is

$$(2) \qquad \frac{q_1 p_1(\boldsymbol{x})}{q_1 p_1(\boldsymbol{x}) + q_2 p_2(\boldsymbol{x})}.$$

Suppose for a moment that $C(1|2) = C(2|1) = 1$. Then the expected loss is

$$(3) \qquad q_1 \int_{R_2} p_1(\boldsymbol{x})\, d\boldsymbol{x} + q_2 \int_{R_1} p_2(\boldsymbol{x})\, d\boldsymbol{x}.$$

This is also the probability of a misclassification; hence we wish to minimize the probability of misclassification.

For a given observed point $\boldsymbol{x}$ we minimize the probability of a misclassification by assigning the population that has the higher conditional probability. If

$$(4) \qquad \frac{q_1 p_1(\boldsymbol{x})}{q_1 p_1(\boldsymbol{x}) + q_2 p_2(\boldsymbol{x})} \geq \frac{q_2 p_2(\boldsymbol{x})}{q_1 p_1(\boldsymbol{x}) + q_2 p_2(\boldsymbol{x})},$$

we choose population π_1. Otherwise we choose population π_2. Since we minimize the probability of misclassification at each point, we minimize it over the whole space. Thus the rule is

$$(5) \qquad \begin{aligned} R_1&: q_1 p_1(\boldsymbol{x}) \geq q_2 p_2(\boldsymbol{x}), \\ R_2&: q_1 p_1(\boldsymbol{x}) < q_2 p_2(\boldsymbol{x}). \end{aligned}$$

If $q_1 p_1(\boldsymbol{x}) = q_2 p_2(\boldsymbol{x})$, the point could be classified as either from π_1 or π_2; we have arbitrarily put it into R_1. If $q_1 p_1(\boldsymbol{x}) + q_2 p_2(\boldsymbol{x}) = 0$ for a given $\boldsymbol{x}$, that point also may go into either region.

Now let us prove formally that (5) is the best procedure. For any procedure $R^* = (R_1^*, R_2^*)$, the probability of misclassification is

$$(6) \quad q_1 \int_{R_2^*} p_1(\boldsymbol{x})\, d\boldsymbol{x} + q_2 \int_{R_1^*} p_2(\boldsymbol{x})\, d\boldsymbol{x}$$

$$= \int_{R_2^*} [q_1 p_1(\boldsymbol{x}) - q_2 p_2(\boldsymbol{x})]\, d\boldsymbol{x} + q_2 \int p_2(\boldsymbol{x})\, d\boldsymbol{x}.$$

On the right-hand side the second term is a given number; the first term is minimized if R_2^* includes the points $\boldsymbol{x}$ such that $q_1 p_1(\boldsymbol{x}) - q_2 p_2(\boldsymbol{x}) < 0$ and excludes the points for which $q_1 p_1(\boldsymbol{x}) - q_2 p_2(\boldsymbol{x}) > 0$. If

$$(7) \qquad \Pr\left\{ \frac{p_1(\boldsymbol{x})}{p_2(\boldsymbol{x})} = \frac{q_2}{q_1} \,\middle|\, \pi_i \right\} = 0, \qquad i = 1, 2,$$

then the Bayes procedure is unique except for sets of probability zero.

Now we notice that mathematically the problem was: given nonnegative constants q_1 and q_2 and nonnegative functions $p_1(\boldsymbol{x})$ and $p_2(\boldsymbol{x})$, choose regions R_1 and R_2 so as to minimize (3). The solution is (5). If we wish to minimize (5) of Section 6.2, which can be written

$$(8) \qquad [C(2|1)q_1] \int_{R_2} p_1(\boldsymbol{x})\, d\boldsymbol{x} + [C(1|2)q_2] \int_{R_1} p_2(\boldsymbol{x})\, d\boldsymbol{x},$$

we choose R_1 and R_2 according to

$$(9) \qquad \begin{aligned} R_1&: [C(2|1)q_1]\, p_1(\boldsymbol{x}) \geq [C(1|2)q_2]\, p_2(\boldsymbol{x}), \\ R_2&: [C(2|1)q_1]\, p_1(\boldsymbol{x}) < [C(1|2)q_2]\, p_2(\boldsymbol{x}), \end{aligned}$$

since $C(2|1)q_1$ and $C(1|2)q_2$ are nonnegative constants. Another way of writing (9) is

$$(10) \qquad \begin{aligned} R_1&: \frac{p_1(\boldsymbol{x})}{p_2(\boldsymbol{x})} \geq \frac{C(1|2)q_2}{C(2|1)q_1}, \\ R_2&: \frac{p_1(\boldsymbol{x})}{p_2(\boldsymbol{x})} < \frac{C(1|2)q_2}{C(2|1)q_1}. \end{aligned}$$

THEOREM 6.3.1. *If q_1 and q_2 are a priori probabilities of drawing an observation from population π_1 with density $p_1(\boldsymbol{x})$ and π_2 with density $p_2(\boldsymbol{x})$,* *respectively, and if the cost of misclassifying an observation from π_1 as from π_2 is $C(2|1)$ and an observation from π_2 as from π_1 is $C(1|2)$, then the regions of classification R_1 and R_2, defined by* (10), *minimize the expected cost. If*

$$\Pr\left\{\frac{p_1(x)}{p_2(x)} = \frac{q_2 C(1|2)}{q_1 C(2|1)}\bigg|\pi_i\right\} = 0, \qquad i = 1, 2, \tag{11}$$

then the procedure is unique except for sets of probability zero.

6.3.2. The Case When No A Priori Probabilities Are Known

In many instances of classification the statistician cannot assign a priori probabilities to the two populations. In this case we shall look for the class of admissible procedures, that is, the set of procedures that cannot be improved upon.

First, let us prove that a Bayes procedure is admissible. Let $R = (R_1, R_2)$ be a Bayes procedure for a given q_1, q_2; is there a procedure $R^* = (R_1^*, R_2^*)$ such that $P(1|2, R^*) \le P(1|2, R)$ and $P(2|1, R^*) \le P(2|1, R)$ with at least one strict inequality? Since R is a Bayes procedure,

$$q_1 P(2|1, R) + q_2 P(1|2, R) \le q_1 P(2|1, R^*) + q_2 P(1|2, R^*). \tag{12}$$

This inequality can be written

$$q_1[P(2|1, R) - P(2|1, R^*)] \le q_2[P(1|2, R^*) - P(1|2, R)]. \tag{13}$$

Suppose $0 < q_1 < 1$. Then if $P(1|2, R^*) < P(1|2, R)$, the right-hand side of (13) is less than zero and therefore $P(2|1, R) < P(2|1, R^*)$. Then $P(2|1, R^*) < P(2|1, R)$ similarly implies $P(1|2, R) < P(1|2, R^*)$. Thus R^* is not better than R, and R is admissible. If $q_1 = 0$, then (13) implies $0 \le P(1|2, R^*) - P(1|2, R)$. For a Bayes procedure, R_1 includes only points for which $p_2(x) = 0$. Therefore, $P(1|2, R) = 0$ and if R^* is to be better $P(1|2, R^*) = 0$. If $\Pr\{p_2(x) = 0|\pi_1\} = 0$, then $P(2|1, R) = \Pr\{p_2(x) > 0|\pi_1\} = 1$. If $P(1|2, R^*) = 0$, then R_1^* contains only points for which $p_2(x) = 0$. Then $P(2|1, R^*) = \Pr\{R_2^*|\pi_1\} = \Pr\{p_2(x) > 0|\pi_1\} = 1$, and R^* is not better than R.

THEOREM 6.3.2. *If* $\Pr\{p_2(x) = 0|\pi_1\} = 0$ *and* $\Pr\{p_1(x) = 0|\pi_2\} = 0$, *then every Bayes procedure is admissible.*

Now let us prove the converse, namely, that every admissible procedure is a Bayes procedure. We assume†

$$(14) \qquad \Pr\left\{ \frac{p_1(x)}{p_2(x)} = k \,\middle|\, \pi_i \right\} = 0, \qquad i = 1, 2; 0 \leq k \leq \infty.$$

Then for any q_1 the Bayes procedure is unique. Moreover, the cdf of $p_1(x)/p_2(x)$ for π_1 and π_2 is continuous.

Let R be an admissible procedure. Then there exists a k such that

$$(15) \qquad P(2|1, R) = \Pr\left\{ \frac{p_1(x)}{p_2(x)} \leq k \,\middle|\, \pi_1 \right\} = P(2|1, R^*),$$

where R^* is the Bayes procedure corresponding to $q_2/q_1 = k$ [i.e., $q_1 = 1/(1 + k)$]. Since R is admissible, $P(1|2, R) \leq P(1|2, R^*)$. However, since by Theorem 6.3.2 R^* is admissible, $P(1|2, R) \geq P(1|2, R^*)$; that is, $P(1|2, R) = P(1|2, R^*)$. Therefore, R is also a Bayes procedure; by the uniqueness of Bayes procedures R is the same as R^*.

THEOREM 6.3.3. *If* (14) *holds, then every admissible procedure is a Bayes procedure.*

The proof of Theorem 6.3.3 shows that the class of Bayes procedures is complete. For if R is any procedure outside the class, we construct a Bayes procedure R^* so that $P(2|1, R) = P(2|1, R^*)$. Then, since R^* is admissible, $P(1|2, R) \geq P(1|2, R^*)$. Furthermore, the class of Bayes procedures is minimal complete since it is identical with the class of admissible procedures.

THEOREM 6.3.4. *If* (14) *holds, the class of Bayes procedures is minimal complete.*

Finally, let us consider the minimax procedure. Let $P(i|j, q_1) = P(i|j, R)$, where R is the Bayes procedure corresponding to q_1. $P(i|j, q_1)$ is a continuous function of q_1. $P(2|1, q_1)$ varies from 1 to 0 as q_1 goes from 0 to 1; $P(1|2, q_1)$

†$p_1(x)/p_2(x) = \infty$ means $p_2(x) = 0$.

varies from 0 to 1. Thus there is a value of q_1, say q_1^*, such that $P(2|1, q_1^*) = P(1|2, q_1^*)$. This is the minimax solution, for if there were another procedure R^* such that $\max\{P(2|1, R^*), P(1|2, R^*)\} \le P(2|1, q_1^*) = P(1|2, q_1^*)$ this would contradict the fact that every Bayes solution is admissible.

6.4. CLASSIFICATION INTO ONE OF TWO KNOWN MULTIVARIATE NORMAL POPULATIONS

Now we shall use the general procedure outlined above in the case of two multivariate normal populations with equal covariance matrices, namely, $N(\boldsymbol{\mu}^{(1)}, \boldsymbol{\Sigma})$ and $N(\boldsymbol{\mu}^{(2)}, \boldsymbol{\Sigma})$, where $\boldsymbol{\mu}^{(i)\prime} = (\mu_1^{(i)}, \dots, \mu_p^{(i)})$ is the vector of means of the ith population, $i = 1, 2$, and $\boldsymbol{\Sigma}$ is the matrix of variances and covariances of each population. [The approach was first used by Wald (1944).] Then the ith density is

$$(1) \qquad p_i(\boldsymbol{x}) = \frac{1}{(2\pi)^{\frac{1}{2}p}|\boldsymbol{\Sigma}|^{\frac{1}{2}}} \exp\left[-\tfrac{1}{2}(\boldsymbol{x} - \boldsymbol{\mu}^{(i)})'\boldsymbol{\Sigma}^{-1}(\boldsymbol{x} - \boldsymbol{\mu}^{(i)})\right].$$

The ratio of densities is

$$(2) \qquad \begin{aligned} \frac{p_1(\boldsymbol{x})}{p_2(\boldsymbol{x})} &= \frac{\exp\left[-\frac{1}{2}(\boldsymbol{x} - \boldsymbol{\mu}^{(1)})'\boldsymbol{\Sigma}^{-1}(\boldsymbol{x} - \boldsymbol{\mu}^{(1)})\right]}{\exp\left[-\frac{1}{2}(\boldsymbol{x} - \boldsymbol{\mu}^{(2)})'\boldsymbol{\Sigma}^{-1}(\boldsymbol{x} - \boldsymbol{\mu}^{(2)})\right]} \\ &= \exp\Big\{-\tfrac{1}{2}\Big[(\boldsymbol{x} - \boldsymbol{\mu}^{(1)})'\boldsymbol{\Sigma}^{-1}(\boldsymbol{x} - \boldsymbol{\mu}^{(1)}) \\ &\qquad\qquad -(\boldsymbol{x} - \boldsymbol{\mu}^{(2)})'\boldsymbol{\Sigma}^{-1}(\boldsymbol{x} - \boldsymbol{\mu}^{(2)})\Big]\Big\}. \end{aligned}$$

The region of classification into π_1, R_1, is the set of $\boldsymbol{x}$'s for which (2) is greater than or equal to k (for k suitably chosen). Since the logarithmic function is monotonically increasing, the inequality can be written in terms of the logarithm of (2) as

$$(3) \qquad -\tfrac{1}{2}\left[(\boldsymbol{x} - \boldsymbol{\mu}^{(1)})'\boldsymbol{\Sigma}^{-1}(\boldsymbol{x} - \boldsymbol{\mu}^{(1)}) - (\boldsymbol{x} - \boldsymbol{\mu}^{(2)})'\boldsymbol{\Sigma}^{-1}(\boldsymbol{x} - \boldsymbol{\mu}^{(2)})\right] \ge \log k.$$

The left-hand side of (3) can be expanded as

$$(4) \qquad \begin{aligned} -\tfrac{1}{2}\big[&\boldsymbol{x}'\boldsymbol{\Sigma}^{-1}\boldsymbol{x} - \boldsymbol{x}'\boldsymbol{\Sigma}^{-1}\boldsymbol{\mu}^{(1)} - \boldsymbol{\mu}^{(1)\prime}\boldsymbol{\Sigma}^{-1}\boldsymbol{x} + \boldsymbol{\mu}^{(1)\prime}\boldsymbol{\Sigma}^{-1}\boldsymbol{\mu}^{(1)} \\ &-\boldsymbol{x}'\boldsymbol{\Sigma}^{-1}\boldsymbol{x} + \boldsymbol{x}'\boldsymbol{\Sigma}^{-1}\boldsymbol{\mu}^{(2)} + \boldsymbol{\mu}^{(2)\prime}\boldsymbol{\Sigma}^{-1}\boldsymbol{x} - \boldsymbol{\mu}^{(2)\prime}\boldsymbol{\Sigma}^{-1}\boldsymbol{\mu}^{(2)}\big]. \end{aligned}$$

By rearrangement of the terms we obtain

$$x'\Sigma^{-1}(\mu^{(1)} - \mu^{(2)}) - \tfrac{1}{2}(\mu^{(1)} + \mu^{(2)})'\Sigma^{-1}(\mu^{(1)} - \mu^{(2)}). \tag{5}$$

The first term is the well-known *discriminant function*. It is a linear function of the components of the observation vector.

The following theorem is now a direct consequence of Theorem 6.3.1.

THEOREM 6.4.1. *If π_i has the density* (1), $i = 1, 2$, *the best regions of classification are given by*

$$\begin{aligned} R_1&: x'\Sigma^{-1}(\mu^{(1)} - \mu^{(2)}) - \tfrac{1}{2}(\mu^{(1)} + \mu^{(2)})'\Sigma^{-1}(\mu^{(1)} - \mu^{(2)}) \geq \log k, \\ R_2&: x'\Sigma^{-1}(\mu^{(1)} - \mu^{(2)}) - \tfrac{1}{2}(\mu^{(1)} + \mu^{(2)})'\Sigma^{-1}(\mu^{(1)} - \mu^{(2)}) < \log k. \end{aligned} \tag{6}$$

If a priori probabilities q_1 and q_2 are known, then k is given by

$$k = \frac{q_2 C(1|2)}{q_1 C(2|1)}. \tag{7}$$

In the particular case of the two populations being equally likely and the costs being equal, $k = 1$ and $\log k = 0$. Then the region of classification into π_1 is

$$R_1: x'\Sigma^{-1}(\mu^{(1)} - \mu^{(2)}) \geq \tfrac{1}{2}(\mu^{(1)} + \mu^{(2)})'\Sigma^{-1}(\mu^{(1)} - \mu^{(2)}). \tag{8}$$

If we do not have a priori probabilities, we may select $\log k = c$, say, on the basis of making the expected losses due to misclassification equal. Let X be a random observation. Then we wish to find the distribution of

$$U = X'\Sigma^{-1}(\mu^{(1)} - \mu^{(2)}) - \tfrac{1}{2}(\mu^{(1)} + \mu^{(2)})'\Sigma^{-1}(\mu^{(1)} - \mu^{(2)}) \tag{9}$$

on the assumption that X is distributed according to $N(\mu^{(1)}, \Sigma)$ and then on the assumption that X is distributed according to $N(\mu^{(2)}, \Sigma)$. When X is distributed according to $N(\mu^{(1)}, \Sigma)$, U is normally distributed with mean

$$\begin{aligned} \mathscr{E}_1 U &= \mu^{(1)\prime}\Sigma^{-1}(\mu^{(1)} - \mu^{(2)}) - \tfrac{1}{2}(\mu^{(1)} + \mu^{(2)})'\Sigma^{-1}(\mu^{(1)} - \mu^{(2)}) \\ &= \tfrac{1}{2}(\mu^{(1)} - \mu^{(2)})'\Sigma^{-1}(\mu^{(1)} - \mu^{(2)}) \end{aligned} \tag{10}$$

and variance

$$(11)\quad \begin{aligned} \operatorname{Var}_1(U) &= \mathscr{E}_1(\boldsymbol{\mu}^{(1)} - \boldsymbol{\mu}^{(2)})'\boldsymbol{\Sigma}^{-1}(\boldsymbol{X} - \boldsymbol{\mu}^{(1)})(\boldsymbol{X} - \boldsymbol{\mu}^{(1)})'\boldsymbol{\Sigma}^{-1}(\boldsymbol{\mu}^{(1)} - \boldsymbol{\mu}^{(2)}) \\ &= (\boldsymbol{\mu}^{(1)} - \boldsymbol{\mu}^{(2)})'\boldsymbol{\Sigma}^{-1}(\boldsymbol{\mu}^{(1)} - \boldsymbol{\mu}^{(2)}). \end{aligned}$$

The Mahalanobis squared distance between $N(\boldsymbol{\mu}^{(1)}, \boldsymbol{\Sigma})$ and $N(\boldsymbol{\mu}^{(2)}, \boldsymbol{\Sigma})$ is

$$(12)\qquad (\boldsymbol{\mu}^{(1)} - \boldsymbol{\mu}^{(2)})'\boldsymbol{\Sigma}^{-1}(\boldsymbol{\mu}^{(1)} - \boldsymbol{\mu}^{(2)}) = \Delta^2,$$

say. Then U is distributed according to $N(\frac{1}{2}\Delta^2, \Delta^2)$ if $\boldsymbol{X}$ is distributed according to $N(\boldsymbol{\mu}^{(1)}, \boldsymbol{\Sigma})$. If $\boldsymbol{X}$ is distributed according to $N(\boldsymbol{\mu}^{(2)}, \boldsymbol{\Sigma})$, then

$$(13)\quad \begin{aligned} \mathscr{E}_2 U &= \boldsymbol{\mu}^{(2)\prime}\boldsymbol{\Sigma}^{-1}(\boldsymbol{\mu}^{(1)} - \boldsymbol{\mu}^{(2)}) - \tfrac{1}{2}(\boldsymbol{\mu}^{(1)} + \boldsymbol{\mu}^{(2)})'\boldsymbol{\Sigma}^{-1}(\boldsymbol{\mu}^{(1)} - \boldsymbol{\mu}^{(2)}) \\ &= \tfrac{1}{2}(\boldsymbol{\mu}^{(2)} - \boldsymbol{\mu}^{(1)})'\boldsymbol{\Sigma}^{-1}(\boldsymbol{\mu}^{(1)} - \boldsymbol{\mu}^{(2)}) \\ &= -\tfrac{1}{2}\Delta^2. \end{aligned}$$

The variance is the same as when $\boldsymbol{X}$ is distributed according to $N(\boldsymbol{\mu}^{(1)}, \boldsymbol{\Sigma})$ because it depends only on the second-order moments of $\boldsymbol{X}$. Thus U is distributed according to $N(-\frac{1}{2}\Delta^2, \Delta^2)$.

The probability of misclassification if the observation is from π_1 is

$$(14)\quad P(2|1) = \int_{-\infty}^{c} \frac{1}{\sqrt{2\pi}\,\Delta} e^{-\frac{1}{2}(z - \frac{1}{2}\Delta^2)^2/\Delta^2}\, dz = \int_{-\infty}^{(c - \frac{1}{2}\Delta^2)/\Delta} \frac{1}{\sqrt{2\pi}} e^{-\frac{1}{2}y^2}\, dy,$$

and the probability of misclassification if the observation is from π_2 is

$$(15)\quad P(1|2) = \int_{c}^{\infty} \frac{1}{\sqrt{2\pi}\,\Delta} e^{-\frac{1}{2}(z + \frac{1}{2}\Delta^2)^2/\Delta}\, dz = \int_{(c + \frac{1}{2}\Delta^2)/\Delta}^{\infty} \frac{1}{\sqrt{2\pi}} e^{-\frac{1}{2}y^2}\, dy.$$

Figure 6.1 indicates the two probabilities as the shaded portion in the tails. For the minimax solution we choose c so that

$$(16)\quad C(1|2)\int_{(c + \frac{1}{2}\Delta^2)/\Delta}^{\infty} \frac{1}{\sqrt{2\pi}} e^{-\frac{1}{2}y^2}\, dy = C(2|1)\int_{-\infty}^{(c - \frac{1}{2}\Delta^2)/\Delta} \frac{1}{\sqrt{2\pi}} e^{-\frac{1}{2}y^2}\, dy.$$

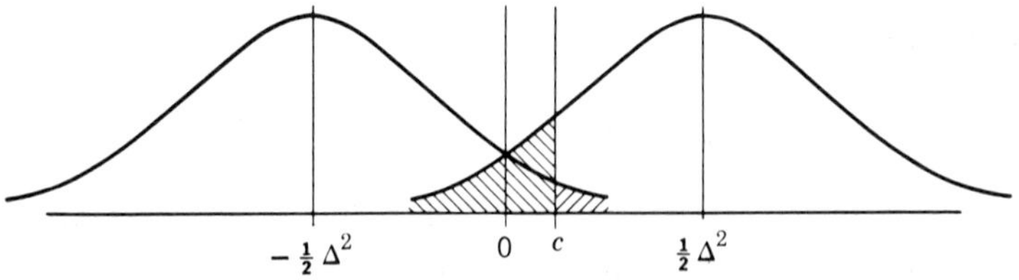

Figure 6.1

THEOREM 6.4.2. *If the* π_i *have densities* (1), $i = 1, 2$, *the minimax regions of classification are given by* (6) *where* $c = \log k$ *is chosen by the condition* (16) *with* $C(i|j)$ *the two costs of misclassification.*

It should be noted that if the costs of misclassification are equal, $c = 0$ and the probability of misclassification is

$$\int_{\Delta/2}^{\infty} \frac{1}{\sqrt{2\pi}} e^{-\frac{1}{2}y^2}\, dy. \tag{17}$$

In case the costs of misclassification are unequal, c could be determined to sufficient accuracy by a trial-and-error method with the normal tables.

Both terms in (5) involve the vector

$$\boldsymbol{\delta} = \boldsymbol{\Sigma}^{-1}\left(\boldsymbol{\mu}^{(1)} - \boldsymbol{\mu}^{(2)}\right). \tag{18}$$

This is obtained as the solution of

$$\boldsymbol{\Sigma\delta} = \left(\boldsymbol{\mu}^{(1)} - \boldsymbol{\mu}^{(2)}\right) \tag{19}$$

by an efficient computing method. The *discriminant function* $\boldsymbol{x}'\boldsymbol{\delta}$ is the linear function that maximizes

$$\frac{\left[\mathscr{E}_1(\boldsymbol{X}'\boldsymbol{d}) - \mathscr{E}_2(\boldsymbol{X}'\boldsymbol{d})\right]^2}{\operatorname{Var}(\boldsymbol{X}'\boldsymbol{d})} \tag{20}$$

for all choices of $\boldsymbol{d}$. The numerator of (20) is

$$\left[\boldsymbol{\mu}^{(1)\prime}\boldsymbol{d} - \boldsymbol{\mu}^{(2)\prime}\boldsymbol{d}\right]^2 = \boldsymbol{d}'\left[\left(\boldsymbol{\mu}^{(1)} - \boldsymbol{\mu}^{(2)}\right)\left(\boldsymbol{\mu}^{(1)} - \boldsymbol{\mu}^{(2)}\right)'\right]\boldsymbol{d}; \tag{21}$$

the denominator is

$$(22) \qquad \boldsymbol{d}'\mathscr{E}(\boldsymbol{X} - \mathscr{E}\boldsymbol{X})(\boldsymbol{X} - \mathscr{E}\boldsymbol{X})'\boldsymbol{d} = \boldsymbol{d}'\boldsymbol{\Sigma}\boldsymbol{d}.$$

We wish to maximize (21) with respect to $\boldsymbol{d}$, holding (22) constant. If λ is a Lagrange multiplier, we ask for the maximum of

$$(23) \qquad \boldsymbol{d}'\left[(\boldsymbol{\mu}^{(1)} - \boldsymbol{\mu}^{(2)})(\boldsymbol{\mu}^{(1)} - \boldsymbol{\mu}^{(2)})'\right]\boldsymbol{d} - \lambda(\boldsymbol{d}'\boldsymbol{\Sigma}\boldsymbol{d} - 1).$$

The derivatives of (23) with respect to the components of $\boldsymbol{d}$ are set equal to zero to obtain

$$(24) \qquad 2\left[(\boldsymbol{\mu}^{(1)} - \boldsymbol{\mu}^{(2)})(\boldsymbol{\mu}^{(1)} - \boldsymbol{\mu}^{(2)})'\right]\boldsymbol{d} = 2\lambda\boldsymbol{\Sigma}\boldsymbol{d}.$$

Since $(\boldsymbol{\mu}^{(1)} - \boldsymbol{\mu}^{(2)})'\boldsymbol{d}$ is a scalar, say ν, we can write (24) as

$$(25) \qquad \boldsymbol{\mu}^{(1)} - \boldsymbol{\mu}^{(2)} = \frac{\lambda}{\nu}\boldsymbol{\Sigma}\boldsymbol{d}.$$

Thus the solution is proportional to $\boldsymbol{\delta}$.

We may finally note that if we have a sample of N from either π_1 or π_2, we use the mean of the sample and classify it as from $N[\boldsymbol{\mu}^{(1)}, (1/N)\boldsymbol{\Sigma}]$ or $N[\boldsymbol{\mu}^{(2)}, (1/N)\boldsymbol{\Sigma}]$.

6.5. CLASSIFICATION INTO ONE OF TWO MULTIVARIATE NORMAL POPULATIONS WHEN THE PARAMETERS ARE ESTIMATED

6.5.1. The Criterion of Classification

Thus far we have assumed that the two populations are known exactly. In most applications of this theory the populations are not known, but must be inferred from samples, one from each population. We shall now treat the case in which we have a sample from each of two normal populations and we wish to use that information in classifying another observation as coming from one of the two populations.

Suppose that we have a sample $\boldsymbol{x}_1^{(1)}, \dots, \boldsymbol{x}_{N_1}^{(1)}$ from $N(\boldsymbol{\mu}^{(1)}, \boldsymbol{\Sigma})$ and a sample $\boldsymbol{x}_1^{(2)}, \dots, \boldsymbol{x}_{N_2}^{(2)}$ from $N(\boldsymbol{\mu}^{(2)}, \boldsymbol{\Sigma})$. In one terminology these are "training samples."

On the basis of this information we wish to classify the observation $\boldsymbol{x}$ as coming from π_1 or π_2. Clearly, our best estimate of $\boldsymbol{\mu}^{(1)}$ is $\bar{\boldsymbol{x}}^{(1)} = \sum_1^{N_1} \boldsymbol{x}_\alpha^{(1)} / N_1$, of $\boldsymbol{\mu}^{(2)}$ is $\bar{\boldsymbol{x}}^{(2)} = \sum_1^{N_2} \boldsymbol{x}_\alpha^{(2)} / N_2$, and of $\boldsymbol{\Sigma}$ is $\boldsymbol{S}$ defined by

$$(1) \quad (N_1 + N_2 - 2)\boldsymbol{S} = \sum_{\alpha=1}^{N_1} \left(\boldsymbol{x}_\alpha^{(1)} - \bar{\boldsymbol{x}}^{(1)}\right)\left(\boldsymbol{x}_\alpha^{(1)} - \bar{\boldsymbol{x}}^{(1)}\right)' + \sum_{\alpha=1}^{N_2} \left(\boldsymbol{x}_\alpha^{(2)} - \bar{\boldsymbol{x}}^{(2)}\right)\left(\boldsymbol{x}_\alpha^{(2)} - \bar{\boldsymbol{x}}^{(2)}\right)'.$$

We substitute these estimates for the parameters in (5) of Section 6.4 to obtain

$$(2) \qquad W(\boldsymbol{x}) = \boldsymbol{x}'\boldsymbol{S}^{-1}\left(\bar{\boldsymbol{x}}^{(1)} - \bar{\boldsymbol{x}}^{(2)}\right) - \tfrac{1}{2}\left(\bar{\boldsymbol{x}}^{(1)} + \bar{\boldsymbol{x}}^{(2)}\right)'\boldsymbol{S}^{-1}\left(\bar{\boldsymbol{x}}^{(1)} - \bar{\boldsymbol{x}}^{(2)}\right).$$

The first term of (2) is the discriminant function based on two samples [suggested by Fisher (1936)]. It is the linear function that has greatest "variance between samples" relative to the "variance within samples" (Problem 12). We propose that (2) be used as the criterion of classification in the same way that (5) of Section 6.4 is used.

When the populations are known, we can argue that the classification criterion is the best in the sense that its use minimizes the expected loss in the case of known a priori probabilities and generates the class of admissible procedures when a priori probabilities are not known. We cannot justify the use of (2) in the same way. However, it seems intuitively reasonable that (2) should give good results. Another criterion is indicated in Section 6.5.5.

Suppose we have a sample $\boldsymbol{x}_1, \dots, \boldsymbol{x}_N$ from either π_1 or π_2, and we wish to classify the sample as a whole. Then we define $\boldsymbol{S}$ by

$$(3) \quad (N_1 + N_2 + N - 3)\boldsymbol{S} = \sum_{\alpha=1}^{N_1} \left(\boldsymbol{x}_\alpha^{(1)} - \bar{\boldsymbol{x}}^{(1)}\right)\left(\boldsymbol{x}_\alpha^{(1)} - \bar{\boldsymbol{x}}^{(1)}\right)' + \sum_{\alpha=1}^{N_2} \left(\boldsymbol{x}_\alpha^{(2)} - \bar{\boldsymbol{x}}^{(2)}\right)\left(\boldsymbol{x}_\alpha^{(2)} - \bar{\boldsymbol{x}}^{(2)}\right)' + \sum_{\alpha=1}^{N} (\boldsymbol{x}_\alpha - \bar{\boldsymbol{x}})(\boldsymbol{x}_\alpha - \bar{\boldsymbol{x}})',$$

where

$$(4) \qquad \bar{\boldsymbol{x}} = \frac{1}{N} \sum_{\alpha=1}^{N} \boldsymbol{x}_\alpha.$$

Then the criterion is

$$[\bar{x} - \tfrac{1}{2}(\bar{x}^{(1)} + \bar{x}^{(2)})]'S^{-1}(\bar{x}^{(1)} - \bar{x}^{(2)}). \tag{5}$$

The larger N is, the smaller are the probabilities of misclassification.

6.5.2. On the Distribution of the Criterion

Let

$$\begin{aligned} W &= X'S^{-1}(\bar{X}^{(1)} - \bar{X}^{(2)}) - \tfrac{1}{2}(\bar{X}^{(1)} + \bar{X}^{(2)})'S^{-1}(\bar{X}^{(1)} - \bar{X}^{(2)}) \\ &= [X - \tfrac{1}{2}(\bar{X}^{(1)} + \bar{X}^{(2)})]'S^{-1}(\bar{X}^{(1)} - \bar{X}^{(2)}) \end{aligned} \tag{6}$$

for random X, $\bar{X}^{(1)}$, $\bar{X}^{(2)}$, and S.

The distribution of W is extremely complicated. It depends on the sample sizes and the unknown Δ^2. Let

$$Y_1 = c_1[X - (N_1 + N_2)^{-1}(N_1\bar{X}^{(1)} + N_2\bar{X}^{(2)})], \tag{7}$$

$$Y_2 = c_2(\bar{X}^{(1)} - \bar{X}^{(2)}), \tag{8}$$

where $c_1 = \sqrt{(N_1 + N_2)/(N_1 + N_2 + 1)}$ and $c_2 = \sqrt{N_1 N_2/(N_1 + N_2)}$. Then Y_1 and Y_2 are independently normally distributed with covariance matrix Σ. The expected value of Y_2 is $c_2(\mu^{(1)} - \mu^{(2)})$, and the expected value of Y_1 is $c_1[N_2/(N_1 + N_2)](\mu^{(1)} - \mu^{(2)})$ if X is from π_1 and $-c_1[N_1/(N_1 + N_2)](\mu^{(1)} - \mu^{(2)})$ if X is from π_2. Let $Y = (Y_1 \;\; Y_2)$ and

$$M = Y'S^{-1}Y = \begin{pmatrix} m_{11} & m_{12} \\ m_{21} & m_{22} \end{pmatrix}. \tag{9}$$

Then

$$W = \sqrt{\frac{N_1 + N_2 + 1}{N_1 N_2}}\, m_{12} + \frac{N_1 - N_2}{2N_1 N_2} m_{22}. \tag{10}$$

The density of M has been given by Sitgreaves (1952). Anderson (1951a) and Wald (1944) have also studied the distribution of W.

If $N_1 = N_2$, the distribution of W for X from π_1 is the same as that of $-W$ for X from π_2. Thus, if $W \geq 0$ is the region of classification as π_1, then the probability of misclassifying $\boldsymbol{X}$ when it is from π_1 is equal to the probability of misclassifying it when it is from π_2.

6.5.3. The Asymptotic Distribution of the Criterion

In the case of large samples from $N(\boldsymbol{\mu}^{(1)}, \boldsymbol{\Sigma})$ and $N(\boldsymbol{\mu}^{(2)}, \boldsymbol{\Sigma})$, we can apply limiting distribution theory. Since $\overline{\boldsymbol{X}}^{(1)}$ is the mean of a sample of N_1 independent observations from $N(\boldsymbol{\mu}^{(1)}, \boldsymbol{\Sigma})$, we know that

$$\operatorname*{plim}_{N_1 \to \infty} \overline{\boldsymbol{X}}^{(1)} = \boldsymbol{\mu}^{(1)}. \tag{11}$$

The explicit definition of (11) is as follows: Given arbitrary positive δ and ε, we can find N large enough so that for $N_1 \geq N$

$$\Pr\left\{ |\overline{X}_i^{(1)} - \mu_i^{(1)}| < \delta, i = 1, \dots, p \right\} > 1 - \varepsilon. \tag{12}$$

(See Problem 23 of Chapter 3.) This can be proved by using the Tchebycheff inequality. Similarly,

$$\operatorname*{plim}_{N_2 \to \infty} \overline{\boldsymbol{X}}^{(2)} = \boldsymbol{\mu}^{(2)}, \tag{13}$$

$$\operatorname{plim} \boldsymbol{S} = \boldsymbol{\Sigma}, \tag{14}$$

as $N_1 \to \infty$, $N_2 \to \infty$ or as both $N_1, N_2 \to \infty$. From (14) we obtain

$$\operatorname{plim} \boldsymbol{S}^{-1} = \boldsymbol{\Sigma}^{-1}, \tag{15}$$

since the probability limits of sums, differences, products, and quotients of random variables are the sums, differences, products, and quotients of the probability limits as long as the probability limit of each denominator is different from zero [Cramér (1946), p. 254]. Furthermore,

$$\operatorname*{plim}_{N_1, N_2 \to \infty} \boldsymbol{S}^{-1}\left(\overline{\boldsymbol{X}}^{(1)} - \overline{\boldsymbol{X}}^{(2)}\right) = \boldsymbol{\Sigma}^{-1}\left(\boldsymbol{\mu}^{(1)} - \boldsymbol{\mu}^{(2)}\right), \tag{16}$$

(17)

$$\operatorname*{plim}_{N_1, N_2 \to \infty} \left(\overline{\boldsymbol{X}}^{(1)} + \overline{\boldsymbol{X}}^{(2)}\right)' \boldsymbol{S}^{-1}\left(\overline{\boldsymbol{X}}^{(1)} - \overline{\boldsymbol{X}}^{(2)}\right) = \left(\boldsymbol{\mu}^{(1)} + \boldsymbol{\mu}^{(2)}\right)' \boldsymbol{\Sigma}^{-1}\left(\boldsymbol{\mu}^{(1)} - \boldsymbol{\mu}^{(2)}\right).$$

It follows then that the limiting distribution of W is the distribution of U. For sufficiently large samples from π_1 and π_2 we can use the criterion as if we knew the populations exactly and we make only a small error. [The result was first given by Wald (1944).]

THEOREM 6.5.1. *Let W be given by* (6) *with $\bar{\mathbf{X}}^{(1)}$ the mean of a sample of N_1 from $N(\boldsymbol{\mu}^{(1)}, \boldsymbol{\Sigma})$, $\bar{\mathbf{X}}^{(2)}$ the mean of a sample of N_2 from $N(\boldsymbol{\mu}^{(2)}, \boldsymbol{\Sigma})$, and $\mathbf{S}$ the estimate of $\boldsymbol{\Sigma}$ based on the pooled sample. The limiting distribution of W as $N_1 \to \infty$ and $N_2 \to \infty$ is $N(\frac{1}{2}\Delta^2, \Delta^2)$ if $\mathbf{X}$ is distributed according to $N(\boldsymbol{\mu}^{(1)}, \boldsymbol{\Sigma})$ and is $N(-\frac{1}{2}\Delta^2, \Delta^2)$ if $\mathbf{X}$ is distributed according to $N(\boldsymbol{\mu}^{(2)}, \boldsymbol{\Sigma})$.*

6.5.4. Another Derivation of the Criterion

A convenient mnemonic derivation of the criterion is the use of regression of a dummy variate [given by Fisher (1936)]. Let

$$(18)\qquad y_\alpha^{(1)} = \frac{N_2}{N_1 + N_2}, \quad \alpha = 1, \dots, N_1, \quad y_\alpha^{(2)} = \frac{-N_1}{N_1 + N_2}, \qquad \alpha = 1, \dots, N_2.$$

Then formally find the regression on the variates $\mathbf{x}_\alpha^{(i)}$ by choosing $\mathbf{b}$ to minimize

$$(19)\qquad \sum_{i=1}^{2} \sum_{\alpha=1}^{N_i} \left[y_\alpha^{(i)} - \mathbf{b}'\left(\mathbf{x}_\alpha^{(i)} - \bar{\mathbf{x}}\right)\right]^2,$$

where

$$(20)\qquad \bar{\mathbf{x}} = \left(N_1 \bar{\mathbf{x}}^{(1)} + N_2 \bar{\mathbf{x}}^{(2)}\right) / \left(N_1 + N_2\right).$$

The "normal equations" are

$$(21)\qquad \sum_{i=1}^{2} \sum_{\alpha=1}^{N_i} \left(\mathbf{x}_\alpha^{(i)} - \bar{\mathbf{x}}\right)\left(\mathbf{x}_\alpha^{(i)} - \bar{\mathbf{x}}\right)' \mathbf{b} = \sum_{i=1}^{2} \sum_{\alpha=1}^{N_i} y_\alpha^{(i)}\left(\mathbf{x}_\alpha^{(i)} - \bar{\mathbf{x}}\right)$$

$$= \frac{N_1 N_2}{N_1 + N_2}\left[\left(\bar{\mathbf{x}}^{(1)} - \bar{\mathbf{x}}\right) - \left(\bar{\mathbf{x}}^{(2)} - \bar{\mathbf{x}}\right)\right]$$

$$= \frac{N_1 N_2}{N_1 + N_2}\left(\bar{\mathbf{x}}^{(1)} - \bar{\mathbf{x}}^{(2)}\right).$$

The matrix multiplying $\boldsymbol{b}$ can be written as

$$(22)\quad \sum_{i=1}^{2}\sum_{\alpha=1}^{N_i}\left(\boldsymbol{x}_\alpha^{(i)}-\bar{\boldsymbol{x}}\right)\left(\boldsymbol{x}_\alpha^{(i)}-\bar{\boldsymbol{x}}\right)'$$

$$=\sum_{i=1}^{2}\sum_{\alpha=1}^{N_i}\left(\boldsymbol{x}_\alpha^{(i)}-\bar{\boldsymbol{x}}^{(i)}\right)\left(\boldsymbol{x}_\alpha^{(i)}-\bar{\boldsymbol{x}}^{(i)}\right)'$$

$$+N_1\left(\bar{\boldsymbol{x}}^{(1)}-\bar{\boldsymbol{x}}\right)\left(\bar{\boldsymbol{x}}^{(1)}-\bar{\boldsymbol{x}}\right)'+N_2\left(\bar{\boldsymbol{x}}^{(2)}-\bar{\boldsymbol{x}}\right)\left(\bar{\boldsymbol{x}}^{(2)}-\bar{\boldsymbol{x}}\right)'$$

$$=\sum_{i=1}^{2}\sum_{\alpha=1}^{N_i}\left(\boldsymbol{x}_\alpha^{(i)}-\bar{\boldsymbol{x}}^{(i)}\right)\left(\boldsymbol{x}_\alpha^{(i)}-\bar{\boldsymbol{x}}^{(i)}\right)'$$

$$+\frac{N_1N_2}{N_1+N_2}\left(\bar{\boldsymbol{x}}^{(1)}-\bar{\boldsymbol{x}}^{(2)}\right)\left(\bar{\boldsymbol{x}}^{(1)}-\bar{\boldsymbol{x}}^{(2)}\right)'.$$

Thus (21) can be written as

$$(23)\qquad \boldsymbol{A}\boldsymbol{b}=\left(\bar{\boldsymbol{x}}^{(1)}-\bar{\boldsymbol{x}}^{(2)}\right)\left[\frac{N_1N_2}{N_1+N_2}-\frac{N_1N_2}{N_1+N_2}\left(\bar{\boldsymbol{x}}^{(1)}-\bar{\boldsymbol{x}}^{(2)}\right)'\boldsymbol{b}\right],$$

where

$$(24)\qquad \boldsymbol{A}=\sum_{i=1}^{2}\sum_{\alpha=1}^{N_i}\left(\boldsymbol{x}_\alpha^{(i)}-\bar{\boldsymbol{x}}^{(i)}\right)\left(\boldsymbol{x}_\alpha^{(i)}-\bar{\boldsymbol{x}}^{(i)}\right)'.$$

Since $(\bar{\boldsymbol{x}}^{(1)}-\bar{\boldsymbol{x}}^{(2)})'\boldsymbol{b}$ is a scalar, we see that the solution $\boldsymbol{b}$ of (23) is proportional to $\boldsymbol{S}^{-1}(\bar{\boldsymbol{x}}^{(1)}-\bar{\boldsymbol{x}}^{(2)})$.

6.5.5. The Likelihood Ratio Criterion

Another criterion which can be used in classification is the likelihood ratio criterion. Consider testing the composite null hypothesis that $\boldsymbol{x}, \boldsymbol{x}_1^{(1)}, \dots, \boldsymbol{x}_{N_1}^{(1)}$ are drawn from $N(\boldsymbol{\mu}^{(1)}, \boldsymbol{\Sigma})$ and $\boldsymbol{x}_1^{(2)}, \dots, \boldsymbol{x}_{N_2}^{(2)}$ are drawn from $N(\boldsymbol{\mu}^{(2)}, \boldsymbol{\Sigma})$ against the composite alternative hypothesis that $\boldsymbol{x}_1^{(1)}, \dots, \boldsymbol{x}_{N_1}^{(1)}$ are drawn from $N(\boldsymbol{\mu}^{(1)}, \boldsymbol{\Sigma})$ and $\boldsymbol{x}, \boldsymbol{x}_1^{(2)}, \dots, \boldsymbol{x}_{N_2}^{(2)}$ are drawn from $N(\boldsymbol{\mu}^{(2)}, \boldsymbol{\Sigma})$, with $\boldsymbol{\mu}^{(1)}$, $\boldsymbol{\mu}^{(2)}$, and $\boldsymbol{\Sigma}$ unspecified. Under the first hypothesis the maximum likelihood estimators of

$\boldsymbol{\mu}^{(1)}$, $\boldsymbol{\mu}^{(2)}$, and $\boldsymbol{\Sigma}$ are

(25)

$$\hat{\boldsymbol{\mu}}_1^{(1)} = \left(N_1 \bar{\boldsymbol{x}}^{(1)} + \boldsymbol{x}\right)/(N_1 + 1),$$

$$\hat{\boldsymbol{\mu}}_1^{(2)} = \bar{\boldsymbol{x}}^{(2)},$$

$$\hat{\boldsymbol{\Sigma}}_1 = \frac{1}{N_1 + N_2 + 1}\left[\sum_{\alpha=1}^{N_1}\left(\boldsymbol{x}_\alpha^{(1)} - \hat{\boldsymbol{\mu}}_1^{(1)}\right)\left(\boldsymbol{x}_\alpha^{(1)} - \hat{\boldsymbol{\mu}}_1^{(1)}\right)' + \left(\boldsymbol{x} - \hat{\boldsymbol{\mu}}_1^{(1)}\right)\left(\boldsymbol{x} - \hat{\boldsymbol{\mu}}_1^{(1)}\right)' + \sum_{\alpha=1}^{N_2}\left(\boldsymbol{x}_\alpha^{(2)} - \hat{\boldsymbol{\mu}}_1^{(2)}\right)\left(\boldsymbol{x}_\alpha^{(2)} - \hat{\boldsymbol{\mu}}_1^{(2)}\right)'\right].$$

Since

$$(26)\quad \sum_{\alpha=1}^{N_1}\left(\boldsymbol{x}_\alpha^{(1)} - \hat{\boldsymbol{\mu}}_1^{(1)}\right)\left(\boldsymbol{x}_\alpha^{(1)} - \hat{\boldsymbol{\mu}}_1^{(1)}\right)' + \left(\boldsymbol{x} - \hat{\boldsymbol{\mu}}_1^{(1)}\right)\left(\boldsymbol{x} - \hat{\boldsymbol{\mu}}_1^{(1)}\right)'$$

$$= \sum_{\alpha=1}^{N_1}\left(\boldsymbol{x}_\alpha^{(1)} - \bar{\boldsymbol{x}}^{(1)}\right)\left(\boldsymbol{x}_\alpha^{(1)} - \bar{\boldsymbol{x}}^{(1)}\right)' + N_1\left(\bar{\boldsymbol{x}}^{(1)} - \hat{\boldsymbol{\mu}}_1^{(1)}\right)\left(\bar{\boldsymbol{x}}^{(1)} - \hat{\boldsymbol{\mu}}_1^{(1)}\right)' + \left(\boldsymbol{x} - \hat{\boldsymbol{\mu}}_1^{(1)}\right)\left(\boldsymbol{x} - \hat{\boldsymbol{\mu}}_1^{(1)}\right)'$$

$$= \sum_{\alpha=1}^{N_1}\left(\boldsymbol{x}_\alpha^{(1)} - \bar{\boldsymbol{x}}^{(1)}\right)\left(\boldsymbol{x}_\alpha^{(1)} - \bar{\boldsymbol{x}}^{(1)}\right)' + \frac{N_1}{N_1 + 1}(\boldsymbol{x} - \bar{\boldsymbol{x}}^{(1)})(\boldsymbol{x} - \bar{\boldsymbol{x}}^{(1)})',$$

we can write $\hat{\boldsymbol{\Sigma}}_1$ as

$$(27)\qquad \hat{\boldsymbol{\Sigma}}_1 = \frac{1}{N_1 + N_2 + 1}\left[\boldsymbol{A} + \frac{N_1}{N_1 + 1}(\boldsymbol{x} - \bar{\boldsymbol{x}}^{(1)})(\boldsymbol{x} - \bar{\boldsymbol{x}}^{(1)})'\right],$$

where $\boldsymbol{A}$ is given by (24). Under the assumptions of the alternative hypothesis we find (by considerations of symmetry) that the maximum likelihood estimators of the parameters are

$$\hat{\boldsymbol{\mu}}_2^{(1)} = \bar{\boldsymbol{x}}^{(1)},$$

$$(28) \qquad \hat{\boldsymbol{\mu}}_2^{(2)} = \big(N_2 \bar{\boldsymbol{x}}^{(2)} + \boldsymbol{x}\big)/(N_2 + 1),$$

$$\hat{\boldsymbol{\Sigma}}_2 = \frac{1}{N_1 + N_2 + 1}\left[\boldsymbol{A} + \frac{N_2}{N_2 + 1}(\boldsymbol{x} - \bar{\boldsymbol{x}}^{(2)})(\boldsymbol{x} - \bar{\boldsymbol{x}}^{(2)})'\right].$$

The likelihood ratio criterion is, therefore, the $(N_1 + N_2 + 1)/2$th power of

$$(29) \qquad \frac{|\hat{\boldsymbol{\Sigma}}_2|}{|\hat{\boldsymbol{\Sigma}}_1|} = \frac{\left|\boldsymbol{A} + \frac{N_2}{N_2 + 1}(\boldsymbol{x} - \bar{\boldsymbol{x}}^{(2)})(\boldsymbol{x} - \bar{\boldsymbol{x}}^{(2)})'\right|}{\left|\boldsymbol{A} + \frac{N_1}{N_1 + 1}(\boldsymbol{x} - \bar{\boldsymbol{x}}^{(1)})(\boldsymbol{x} - \bar{\boldsymbol{x}}^{(1)})'\right|}.$$

This ratio can also be written (Corollary A.3.1)

$$(30) \qquad \frac{1 + \frac{N_2}{N_2 + 1}(\boldsymbol{x} - \bar{\boldsymbol{x}}^{(2)})'\boldsymbol{A}^{-1}(\boldsymbol{x} - \bar{\boldsymbol{x}}^{(2)})}{1 + \frac{N_1}{N_1 + 1}(\boldsymbol{x} - \bar{\boldsymbol{x}}^{(1)})'\boldsymbol{A}^{-1}(\boldsymbol{x} - \bar{\boldsymbol{x}}^{(1)})}$$

$$= \frac{n + \frac{N_2}{N_2 + 1}(\boldsymbol{x} - \bar{\boldsymbol{x}}^{(2)})'\boldsymbol{S}^{-1}(\boldsymbol{x} - \bar{\boldsymbol{x}}^{(2)})}{n + \frac{N_1}{N_1 + 1}(\boldsymbol{x} - \bar{\boldsymbol{x}}^{(1)})'\boldsymbol{S}^{-1}(\boldsymbol{x} - \bar{\boldsymbol{x}}^{(1)})},$$

where $n = N_1 + N_2 - 2$. The region of classification into π_1 consists of those points for which the ratio (30) is greater than or equal to a given number K_n. It can be written

$$(31) \qquad R_1\colon n + \frac{N_2}{N_2 + 1}(\boldsymbol{x} - \bar{\boldsymbol{x}}^{(2)})'\boldsymbol{S}^{-1}(\boldsymbol{x} - \bar{\boldsymbol{x}}^{(2)})$$

$$\geq K_n\left[n + \frac{N_1}{N_1 + 1}(\boldsymbol{x} - \bar{\boldsymbol{x}}^{(1)})'\boldsymbol{S}^{-1}(\boldsymbol{x} - \bar{\boldsymbol{x}}^{(1)})\right].$$

If $K_n = 1 + 2c/n$ and N_1 and N_2 are large, the region (31) is approximately $W(\boldsymbol{x}) \geq c$.

If we take $K_n = 1$, the rule is to classify as π_1 if (30) is greater than 1 and as π_2 if (30) is less than 1. This is the *maximum likelihood* rule. Let

$$Z = \frac{1}{2}\left[\frac{N_2}{N_2+1}(\boldsymbol{x} - \bar{\boldsymbol{x}}^{(2)})'\boldsymbol{S}^{-1}(\boldsymbol{x} - \bar{\boldsymbol{x}}^{(2)}) - \frac{N_1}{N_1+1}(\boldsymbol{x} - \bar{\boldsymbol{x}}^{(1)})'\boldsymbol{S}^{-1}(\boldsymbol{x} - \bar{\boldsymbol{x}}^{(1)})\right]. \tag{32}$$

Then the maximum likelihood rule is to classify as π_1 if $Z > 0$ and π_2 if $Z < 0$. Roughly speaking, assign $\boldsymbol{x}$ to π_1 or π_2 according to whether the distance to $\bar{\boldsymbol{x}}^{(1)}$ is less or greater than the distance to $\bar{\boldsymbol{x}}^{(2)}$. The difference between W and Z is

$$W - Z = \frac{1}{2}\left[\frac{1}{N_2+1}(\boldsymbol{x} - \bar{\boldsymbol{x}}^{(2)})'\boldsymbol{S}^{-1}(\boldsymbol{x} - \bar{\boldsymbol{x}}^{(2)}) - \frac{1}{N_1+1}(\boldsymbol{x} - \bar{\boldsymbol{x}}^{(1)})'\boldsymbol{S}^{-1}(\boldsymbol{x} - \bar{\boldsymbol{x}}^{(1)})\right], \tag{33}$$

which has the probability limit 0 as $N_1, N_2 \to \infty$. The probabilities of misclassification with W are equivalent asymptotically to those with Z for large samples.

Note that for $N_1 = N_2$, $Z = [N_1/(N_1+1)]W$. Then the symmetric test based on the cutoff $c = 0$ is the same for Z and W.

6.5.6. Invariance

The classification problem is invariant with respect to transformations

$$\begin{aligned} \boldsymbol{x}_\alpha^{(1)*} &= \boldsymbol{B}\boldsymbol{x}_\alpha^{(1)} + \boldsymbol{c}, & \alpha &= 1, \ldots, N_1, \\ \boldsymbol{x}_\alpha^{(2)*} &= \boldsymbol{B}\boldsymbol{x}_\alpha^{(2)} + \boldsymbol{c}, & \alpha &= 1, \ldots, N_2, \\ \boldsymbol{x}^* &= \boldsymbol{B}\boldsymbol{x} + \boldsymbol{c}, \end{aligned} \tag{34}$$

where $\boldsymbol{B}$ is nonsingular and $\boldsymbol{c}$ is a vector. This transformation induces the following transformation on the sufficient statistics:

$$(35) \qquad \bar{\boldsymbol{x}}^{(1)*} = \boldsymbol{B}\bar{\boldsymbol{x}}^{(1)} + \boldsymbol{c}, \qquad \bar{\boldsymbol{x}}^{(2)*} = \boldsymbol{B}\bar{\boldsymbol{x}}^{(2)} + \boldsymbol{c}, \qquad \boldsymbol{x}^* = \boldsymbol{B}\boldsymbol{x} + \boldsymbol{c},$$

$$\boldsymbol{S}^* = \boldsymbol{B}\boldsymbol{S}\boldsymbol{B}',$$

with the same transformations on the parameters, $\boldsymbol{\mu}^{(1)}$, $\boldsymbol{\mu}^{(2)}$, and $\boldsymbol{\Sigma}$. (Note $\mathscr{E}\boldsymbol{x} = \boldsymbol{\mu}^{(1)}$ or $\boldsymbol{\mu}^{(2)}$.) Any invariant of the parameters is a function of $\Delta^2 = (\boldsymbol{\mu}^{(1)} - \boldsymbol{\mu}^{(2)})'\boldsymbol{\Sigma}^{-1}(\boldsymbol{\mu}^{(1)} - \boldsymbol{\mu}^{(2)})$. There exists a matrix $\boldsymbol{B}$ and a vector $\boldsymbol{c}$ such that

$$(36) \qquad \boldsymbol{\mu}^{(1)*} = \boldsymbol{B}\boldsymbol{\mu}^{(1)} + \boldsymbol{c} = \boldsymbol{0}, \qquad \boldsymbol{\mu}^{(2)*} = \boldsymbol{B}\boldsymbol{\mu}^{(2)} + \boldsymbol{c} = (\Delta, 0, \ldots, 0)',$$

$$\boldsymbol{\Sigma}^* = \boldsymbol{B}\boldsymbol{\Sigma}\boldsymbol{B}' = \boldsymbol{I}.$$

Therefore, Δ^2 is the minimal invariant of the parameters. The elements of $\boldsymbol{M}$ defined by (9) are invariant and are the minimal invariants of the sufficient statistics. Thus invariant procedures depend only on $\boldsymbol{M}$, and the distribution of $\boldsymbol{M}$ depends only on Δ^2. The statistics W and Z are invariant.

6.6. PROBABILITIES OF MISCLASSIFICATION

6.6.1. Asymptotic Expansions of the Probabilities of Misclassification Using W

We may like to know the probabilities of misclassification before we draw the two samples for determining the classification rule and we may like to know the (conditional) probabilities of misclassification after drawing the samples. As observed earlier, the exact distributions of W and Z are very difficult to calculate. Therefore, we treat asymptotic expansions of their probabilities as N_1 and N_2 increase. The background is that the limiting distribution of W and Z is $N(\frac{1}{2}\Delta^2, \Delta^2)$ if $\boldsymbol{x}$ is from π_1 and is $N(-\frac{1}{2}\Delta^2, \Delta^2)$ if $\boldsymbol{x}$ is from π_2.

Okamoto (1963) has obtained the asymptotic expansion of the distribution of W to terms of order n^{-2} and Siotani and Wang (1975), (1977) to terms of order n^{-3}. [Bowker and Sitgreaves (1961) treated the case of $N_1 = N_2$.] Let $\Phi(\cdot)$ and $\phi(\cdot)$ be the cdf and density of $N(0,1)$, respectively.

THEOREM 6.6.1. *As* $N_1 \to \infty$, $N_2 \to \infty$, *and* $N_1/N_2 \to$ *a positive limit* ($n = N_1 + N_2 - 2$)

$$
\begin{aligned}
(1)\quad \Pr\left\{\frac{W - \frac{1}{2}\Delta^2}{\Delta} \le u \middle| \pi_1\right\}
&= \Phi(u) - \phi(u)\left\{\frac{1}{2N_1\Delta^2}\left[u^3 + (p-3)u - p\Delta\right]\right. \\
&\quad + \frac{1}{2N_2\Delta^2}\left[u^3 + 2\Delta u^2 + (p - 3 + \Delta^2)u + (p-2)\Delta\right] \\
&\quad \left. + \frac{1}{4n}\left[4u^3 + 4\Delta u^2 + (6p - 6 + \Delta^2)u + 2(p-1)\Delta\right]\right\} \\
&\quad + O(n^{-2})
\end{aligned}
$$

and $\Pr\{-(W + \frac{1}{2}\Delta^2)/\Delta \le u|\pi_2\}$ *is* (1) *with* N_1 *and* N_2 *interchanged.*

The rule using W is to assign the observation $\boldsymbol{x}$ to π_1 if $W(\boldsymbol{x}) > c$ and to π_2 if $W(\boldsymbol{x}) \le c$. The probabilities of misclassification are given by Theorem 6.6.1 with $u = (c - \frac{1}{2}\Delta^2)/\Delta$ and $u = -(c + \frac{1}{2}\Delta^2)/\Delta$, respectively. For $c = 0$, $u = -\frac{1}{2}\Delta$. If $N_1 = N_2$, this defines an exact minimax procedure [Das Gupta (1965)].

COROLLARY 6.6.1.

$$
\begin{aligned}
(2)\quad \Pr\left\{W \le 0|\pi_1,\ \lim_{n\to\infty}\frac{N_1}{N_2} = 1\right\}
&= \Phi(-\tfrac{1}{2}\Delta) + \frac{1}{n}\phi(\tfrac{1}{2}\Delta)\left[\frac{p-1}{\Delta} + \frac{p}{4}\Delta\right] + o(n^{-1}) \\
&= \Pr\left\{W \ge 0|\pi_2,\ \lim_{n\to\infty}\frac{N_1}{N_2} = 1\right\}.
\end{aligned}
$$

Note that the correction term is positive, as far as this correction goes; that is, the probability of misclassification is greater than the value of the normal approximation. The correction term (to order n^{-1}) increases with p for given Δ and decreases with Δ for given p.

Since Δ is usually unknown, it is relevant to "Studentize" W. The sample Mahalanobis squared distance

$$D^2 = (\bar{\boldsymbol{x}}^{(1)} - \bar{\boldsymbol{x}}^{(2)})' \boldsymbol{S}^{-1} (\bar{\boldsymbol{x}}^{(1)} - \bar{\boldsymbol{x}}^{(2)}) \tag{3}$$

is an estimator of the population Mahalanobis squared distance Δ^2. The expectation of D^2 is

$$\mathscr{E} D^2 = \frac{n}{n-p-1}\left[\Delta^2 + p\left(\frac{1}{N_1} + \frac{1}{N_2}\right)\right]. \tag{4}$$

See Problem 14. If N_1 and N_2 are large, this is approximately Δ^2.

Anderson (1973b) has shown the following:

THEOREM 6.6.2. *If $N_1/N_2 \to$ a positive limit as $n \to \infty$,*

$$\Pr\left\{\frac{W - \frac{1}{2}D^2}{D} \le u \,\middle|\, \pi_1\right\} \tag{5}$$

$$= \Phi(u) - \phi(u)\left\{\frac{1}{N_1}\left(\frac{u}{2} - \frac{p-1}{\Delta}\right) + \frac{1}{n}\left[\frac{u^3}{4} + \left(p - \frac{3}{4}\right)u\right]\right\} + O(n^{-2}),$$

$$\Pr\left\{-\frac{W + \frac{1}{2}D^2}{D} \le u \,\middle|\, \pi_2\right\} \tag{6}$$

$$= \Phi(u) - \phi(u)\left\{\frac{1}{N_2}\left(\frac{u}{2} - \frac{p-1}{\Delta}\right) + \frac{1}{n}\left[\frac{u^3}{4} + \left(p - \frac{3}{4}\right)u\right]\right\} + O(n^{-2}).$$

Usually, one is interested in $u \le 0$ (small probabilities of error). Then the correction term is positive; that is, the normal approximation underestimates the probability of misclassification.

One may want to choose the cutoff point c so that one probability of misclassification is controlled. Let α be the desired $\Pr\{W < c|\pi_1\}$. Anderson (1973b, 1973c) has derived the following theorem:

THEOREM 6.6.3. *Let u_0 be such that $\Phi(u_0) = \alpha$, and let*

$$u = u_0 - \frac{1}{N_1}\left[\frac{p-1}{D} - \frac{1}{2}u_0\right] + \frac{1}{n}\left[\left(p - \frac{3}{4}\right)u_0 + \frac{1}{4}u_0^3\right]. \tag{7}$$

Then as $N_1 \to \infty$, $N_2 \to \infty$, and $N_1/N_2 \to$ a positive limit

$$\Pr\left\{\frac{W - \frac{1}{2}D^2}{D} \le u \,\middle|\, \pi_1\right\} = \alpha + O(n^{-2}). \tag{8}$$

Then $c = Du + \frac{1}{2}D^2$ will attain the desired probability α to within $O(n^{-2})$.

We now turn to evaluating the probabilities of misclassification after the two samples have been drawn. Conditional on $\bar{\boldsymbol{x}}^{(1)}$, $\bar{\boldsymbol{x}}^{(2)}$, and $\boldsymbol{S}$, W is normally distributed with conditional mean

$$(9) \qquad \mathscr{E}\left(W|\pi_i, \bar{\boldsymbol{x}}^{(1)}, \bar{\boldsymbol{x}}^{(2)}, \boldsymbol{S}\right) = \left[\boldsymbol{\mu}^{(i)} - \tfrac{1}{2}\left(\bar{\boldsymbol{x}}^{(1)} + \bar{\boldsymbol{x}}^{(2)}\right)\right]'\boldsymbol{S}^{-1}\left(\bar{\boldsymbol{x}}^{(1)} - \bar{\boldsymbol{x}}^{(2)}\right) = \mu^{(i)}\left(\bar{\boldsymbol{x}}^{(1)}, \bar{\boldsymbol{x}}^{(2)}, \boldsymbol{S}\right),$$

when $\boldsymbol{x}$ is from π_i, $i = 1, 2$, and conditional variance

$$(10) \qquad \mathscr{V}\left(W|\bar{\boldsymbol{x}}^{(1)}, \bar{\boldsymbol{x}}^{(2)}, \boldsymbol{S}\right) = \left(\bar{\boldsymbol{x}}^{(1)} - \bar{\boldsymbol{x}}^{(2)}\right)'\boldsymbol{S}^{-1}\boldsymbol{\Sigma}\boldsymbol{S}^{-1}\left(\bar{\boldsymbol{x}}^{(1)} - \bar{\boldsymbol{x}}^{(2)}\right) = \sigma^2\left(\bar{\boldsymbol{x}}^{(1)}, \bar{\boldsymbol{x}}^{(2)}, \boldsymbol{S}\right).$$

Note that these means and variance are functions of the samples with probability limits

$$(11) \qquad \underset{N_1, N_2 \to \infty}{\text{plim}} \ \mu^{(i)}\left(\bar{\boldsymbol{x}}^{(1)}, \bar{\boldsymbol{x}}^{(2)}, \boldsymbol{S}\right) = (-1)^{i-1}\tfrac{1}{2}\Delta^2, \qquad \underset{N_1, N_2 \to \infty}{\text{plim}} \ \sigma^2\left(\bar{\boldsymbol{x}}^{(1)}, \bar{\boldsymbol{x}}^{(2)}, \boldsymbol{S}\right) = \Delta^2.$$

For large N_1 and N_2 the conditional probabilities of misclassification are close to the limiting normal probabilities (with high probability relative to $\bar{\boldsymbol{x}}^{(1)}$, $\bar{\boldsymbol{x}}^{(2)}$, and $\boldsymbol{S}$).

When c is the cutoff point, probabilities of misclassification conditional on $\bar{\boldsymbol{x}}^{(1)}$, $\bar{\boldsymbol{x}}^{(2)}$, and $\boldsymbol{S}$ are

$$(12) \qquad P\left(2|1, c, \bar{\boldsymbol{x}}^{(1)}, \bar{\boldsymbol{x}}^{(2)}, \boldsymbol{S}\right) = \Phi\left[\frac{c - \mu^{(1)}\left(\bar{\boldsymbol{x}}^{(1)}, \bar{\boldsymbol{x}}^{(2)}, \boldsymbol{S}\right)}{\sigma\left(\bar{\boldsymbol{x}}^{(1)}, \bar{\boldsymbol{x}}^{(2)}, \boldsymbol{S}\right)}\right],$$

$$(13) \qquad P\left(1|2, c, \bar{\boldsymbol{x}}^{(1)}, \bar{\boldsymbol{x}}^{(2)}, \boldsymbol{S}\right) = 1 - \Phi\left[\frac{c - \mu^{(2)}\left(\bar{\boldsymbol{x}}^{(1)}, \bar{\boldsymbol{x}}^{(2)}, \boldsymbol{S}\right)}{\sigma\left(\bar{\boldsymbol{x}}^{(1)}, \bar{\boldsymbol{x}}^{(2)}, \boldsymbol{S}\right)}\right].$$

In (12) write c as $Du_1 + \frac{1}{2}D^2$. Then the argument of $\Phi(\cdot)$ in (12) is $u_1 D/\sigma + (\bar{\boldsymbol{x}}^{(1)} - \bar{\boldsymbol{x}}^{(2)})'\boldsymbol{S}^{-1}(\bar{\boldsymbol{x}}^{(1)} - \boldsymbol{\mu}^{(1)})/\sigma$; the first term converges in probability to u_1, the second term tends to 0 as $N_1 \to \infty$, $N_2 \to \infty$, and (12) to $\Phi(u_1)$. In (13) write c as $Du_2 - \frac{1}{2}D^2$. Then the argument of $\Phi(\cdot)$ in (13) is $u_2 D/\sigma + (\bar{\boldsymbol{x}}^{(1)} - \bar{\boldsymbol{x}}^{(2)})'\boldsymbol{S}^{-1}(\bar{\boldsymbol{x}}^{(2)} - \boldsymbol{\mu}^{(2)})/\sigma$. The first term converges in probability to u_2 and the second term to 0; (13) converges to $1 - \Phi(u_2)$.

For given $\bar{\boldsymbol{x}}^{(1)}$, $\bar{\boldsymbol{x}}^{(2)}$, and $\boldsymbol{S}$ the (conditional) probabilities of misclassification (12) and (13) are functions of the parameters $\boldsymbol{\mu}^{(1)}$, $\boldsymbol{\mu}^{(2)}$, $\boldsymbol{\Sigma}$ and can be estimated. Consider them when $c = 0$. Then (12) and (13) converge in probability to $\Phi(-\frac{1}{2}\Delta)$; that suggests $\Phi(-\frac{1}{2}D)$ as an estimator of (12) and (13). A better estimator is $\Phi(-\frac{1}{2}\overline{D})$, where $\overline{D}^2 = (n-p-1)D^2/n$, which is closer to being an unbiased estimator of Δ^2. [See (4).] McLachlan (1973, 1974a, 1974b, 1974c) has given an estimator of (12) whose bias is of order n^{-2}; it is

$$(14) \qquad \Phi(-\tfrac{1}{2}D) + \phi(\tfrac{1}{2}D)\left\{\frac{p-D}{N_1 D} + \frac{1}{32n}\left[-D^3 + 4(4p-1)D\right]\right\}.$$

[McLachlan gave (14) to terms of order n^{-2}]. McLachlan has explored the properties of these and other estimators as have Lachenbruch and Mickey (1968).

Now consider (12) with $c = Du_1 + \frac{1}{2}D^2$; u_1 might be chosen to control $P(2|1)$ conditional on $\bar{\boldsymbol{x}}^{(1)}, \bar{\boldsymbol{x}}^{(2)}, \boldsymbol{S}$. This conditional probability as a function of $\bar{\boldsymbol{x}}^{(1)}, \bar{\boldsymbol{x}}^{(2)}, \boldsymbol{S}$ is a random variable whose distribution may be approximated. McLachlan has shown the following

THEOREM 6.6.4. *As $N_1 \to \infty$, $N_2 \to \infty$, and $N_1/N_2 \to$ a positive limit,*

$$(15) \quad \Pr\left\{\sqrt{n}\,\frac{P(2|1, Du_1 + \frac{1}{2}D^2, \bar{\boldsymbol{x}}^{(1)}, \bar{\boldsymbol{x}}^{(2)}, \boldsymbol{S}) - \Phi(u_1)}{\phi(u_1)\left[\frac{1}{2}u_1^2 + n/N_1\right]^{\frac{1}{2}}} \leq x\right\}$$

$$= \Phi\left[x - \frac{(p-1)n/N_1 - (p - \frac{3}{4} + n/N_1)u_1 - u_1^3/4}{\sqrt{n}\left[\frac{1}{2}u_1^2 + n/N_1\right]^{\frac{1}{2}}}\right] + O(n^{-2}).$$

McLachlan (1977) has given a method of selecting u_1 so that the probability of one misclassification is less than a preassigned δ with a preassigned confidence level $1 - \varepsilon$.

6.6.2. Asymptotic Expansions of the Probabilities of Misclassification Using Z

We now turn our attention to Z defined by (32) of Section 6.5. The results are parallel to those for W. Memon and Okamoto (1971) expanded the distribution of Z to terms of order n^{-2} and Siotani and Wang (1975), (1977) to terms of order n^{-3}.

THEOREM 6.6.5. *As* $N_1 \to \infty$, $N_2 \to \infty$, *and* N_1/N_2 *approaches a positive limit,*

$$\text{(16)} \quad \Pr\left\{ \frac{Z - \frac{1}{2}\Delta^2}{\Delta} \le u \middle| \pi_1 \right\}$$

$$= \Phi(u) - \phi(u)\left\{ \frac{1}{2N_1\Delta^2}\left[u^3 + \Delta u^2 + (p-3)u - \Delta\right] \right.$$

$$+ \frac{1}{2N_2\Delta^2}\left[u^3 + \Delta u^2 + (p - 3 - \Delta^2)u - \Delta^3 - \Delta\right]$$

$$\left. + \frac{1}{4n}\left[4u^3 + 4\Delta u^2 + (6p - 6 + \Delta^2)u + 2(p-1)\Delta\right]\right\}$$

$$+ O(n^{-2})$$

and $\Pr\{-(Z + \frac{1}{2}\Delta^2)/\Delta \le u|\pi_2\}$ *is* (16) *with* N_1 *and* N_2 *interchanged.*

When $c = 0$, then $u = -\frac{1}{2}\Delta$. If $N_1 = N_2$, the rule with Z is identical to the rule with W and the probability of misclassification is given by (2).

Fujikoshi and Kanazawa (1976) have proved

THEOREM 6.6.6.

$$\text{(17)} \quad \Pr\left\{ \frac{Z - \frac{1}{2}D^2}{D} \le u \middle| \pi_1 \right\}$$

$$= \Phi(u) - \phi(u)\left\{ \frac{1}{2N_1\Delta}\left[u^2 + \Delta u - (p-1)\right] \right.$$

$$- \frac{1}{2N_2\Delta}\left[u^2 + 2\Delta u + p - 1 + \Delta^2\right]$$

$$\left. + \frac{1}{4n}\left[u^3 + (4p-3)u\right]\right\} + O(n^{-2}),$$

$$\text{(18)} \quad \Pr\left\{ -\frac{Z + \frac{1}{2}D^2}{D} \le u \middle| \pi_2 \right\}$$

$$= \Phi(u) - \phi(u)\left\{ -\frac{1}{2N_1\Delta}\left[u^2 + 2\Delta u + p - 1 + \Delta^2\right] \right.$$

$$\left. + \frac{1}{2N_2\Delta}\left[u^2 + \Delta u - (p-1)\right] + \frac{1}{4n}\left[u^3 + (4p-3)u\right]\right\} + O(n^{-2}).$$

Kanazawa (1979) has shown the following:

THEOREM 6.6.7. *Let u_0 be such that $\Phi(u_0) = \alpha$, and let*

$$(19)\quad u = u_0 + \frac{1}{2N_1 D}\left[u_0^2 + Du_0 - (p-1)\right] - \frac{1}{2N_2 D}\left[u_0^2 + Du_0 + (p-1) - D^2\right] + \frac{1}{4n}\left[u_0^3 + (4p-5)u_0\right].$$

Then as $N_1 \to \infty$, $N_2 \to \infty$, and $N_1/N_2 \to$ a positive limit

$$(20)\quad \Pr\left\{\frac{Z - \frac{1}{2}D^2}{D} \le u\right\} = \alpha + O(n^{-2}).$$

Now consider the probabilities of misclassification after the samples have been drawn. The conditional distribution of Z is not normal because Z is quadratic in $\boldsymbol{x}$ unless $N_1 = N_2$. We do not have expressions equivalent to (12) and (13). Siotani (1980) has shown the following:

THEOREM 6.6.8. *As $N_1 \to \infty$, $N_2 \to \infty$, and $N_1/N_2 \to$ a positive limit*

$$(21)\quad \Pr\left\{2\sqrt{\frac{N_1 N_2}{N_1 + N_2}}\,\frac{P(2|1, 0, \bar{\boldsymbol{x}}^{(1)}, \bar{\boldsymbol{x}}^{(2)}, \boldsymbol{S}) - \Phi(-\frac{1}{2}\Delta)}{\phi(\frac{1}{2}\Delta)} \le x\right\}$$
$$= \Phi\left[x - 2\sqrt{\frac{N_1 N_2}{N_1 + N_2}}\left\{\frac{1}{16N_1\Delta}\left[4(p-1) - \Delta^2\right] + \frac{1}{16N_2}\left[4(p-1) + 3\Delta^2\right] - \frac{(p-1)\Delta}{4n}\right\}\right] + O(n^{-2}).$$

It is also possible to obtain a similar expression for $P(2|1, Du_1 + \frac{1}{2}D^2, \bar{\boldsymbol{x}}^{(1)}, \bar{\boldsymbol{x}}^{(2)}, \boldsymbol{S})$ for Z and a confidence interval. See Siotani (1980).

6.7. CLASSIFICATION INTO ONE OF SEVERAL POPULATIONS

Let us now consider the problem of classifying an observation into one of several populations. We shall extend the consideration of the previous sections to the cases of more than two populations. Let $\pi_1, \dots, \pi_m$ be m populations with density functions $p_1(\boldsymbol{x}), \dots, p_m(\boldsymbol{x})$ respectively. We wish to divide the space of observations into m mutually exclusive and exhaustive regions $R_1, \dots, R_m$. If an observation falls into R_i we shall say that it comes from π_i. Let the cost of misclassifying an observation from π_i as coming from π_j be $C(j|i)$. The probability of this misclassification is

$$P(j|i, R) = \int_{R_j} p_i(\boldsymbol{x})\, d\boldsymbol{x}. \tag{1}$$

Suppose we have a priori probabilities of the populations, $q_1, \dots, q_m$. Then the expected loss is

$$\sum_{i=1}^{m} q_i \left\{ \sum_{\substack{j=1 \\ j \neq i}}^{m} C(j|i) P(j|i, R) \right\}. \tag{2}$$

We should like to choose $R_1, \dots, R_m$ to make this a minimum.

Since we have a priori probabilities for the populations, we can define the conditional probability of an observation coming from a population given the values of the components of the vector $\boldsymbol{x}$. The conditional probability of the observation coming from π_i is

$$\frac{q_i p_i(\boldsymbol{x})}{\sum_{k=1}^{m} q_k p_k(\boldsymbol{x})}. \tag{3}$$

If we classify the observation as from π_j, the expected loss is

$$\sum_{\substack{i=1 \\ i \neq j}}^{m} \frac{q_i p_i(\boldsymbol{x})}{\sum_{k=1}^{m} q_k p_k(\boldsymbol{x})} C(j|i). \tag{4}$$

We minimize the expected loss at this point if we choose j so as to minimize (4); that is, we consider

$$\sum_{\substack{i=1\\ i\neq j}}^{m} q_i p_i(\mathbf{x}) C(j|i) \tag{5}$$

for all j and select that j that gives the minimum. (If two different indices give the minimum, it is irrelevant which index is selected.) This procedure assigns the point $\mathbf{x}$ to one of the R_j. Following this procedure for each $\mathbf{x}$, we define our regions $R_1, \dots, R_m$. The classification procedure, then, is to classify an observation as coming from π_j if it falls in R_j.

THEOREM 6.7.1. *If q_i is the a priori probability of drawing an observation from population π_i with density $p_i(\mathbf{x})$, $i = 1, \dots, m$, and if the cost of misclassifying an observation from π_i as from π_j is $C(j|i)$, then the regions of classification, $R_1, \dots, R_m$, that minimize the expected cost are defined by assigning $\mathbf{x}$ to R_k if*

$$\sum_{\substack{i=1\\ i\neq k}}^{m} q_i p_i(\mathbf{x}) C(k|i) < \sum_{\substack{i=1\\ i\neq j}}^{m} q_i p_i(\mathbf{x}) C(j|i), \qquad j = 1, \dots, m, \ j \neq k. \tag{6}$$

[*If* (6) *holds for all j ($j \neq k$) except for h indices and the inequality is replaced by equality for those indices, then this point can be assigned to any of the $h + 1$ π's.*] *If the probability of equality between right-hand and left-hand sides of* (6) *is zero for each k and j under π_i (each i), then the minimizing procedure is unique except for sets of probability zero.*

PROOF. We now verify this result. Let

$$h_j(\mathbf{x}) = \sum_{\substack{i=1\\ i\neq j}}^{m} q_i p_i(\mathbf{x}) C(j|i). \tag{7}$$

Then the expected loss of a procedure R is

$$\sum_{j=1}^{m} \int_{R_j} h_j(\mathbf{x})\, d\mathbf{x} = \int h(\mathbf{x}|R)\, d\mathbf{x}, \tag{8}$$

where $h(\boldsymbol{x}|R) = h_j(\boldsymbol{x})$ for $\boldsymbol{x}$ in R_j. For the Bayes procedure R^* described in the theorem $h(\boldsymbol{x}|R)$ is $h(\boldsymbol{x}|R^*) = \min_i h_i(\boldsymbol{x})$. Thus the difference between the expected loss for any procedure R and for R^* is

$$(9) \qquad \int [h(\boldsymbol{x}|R) - h(\boldsymbol{x}|R^*)]\, d\boldsymbol{x} = \sum_j \int_{R_j} [h_j(\boldsymbol{x}) - \min_i h_i(\boldsymbol{x})]\, d\boldsymbol{x}$$
$$\geq 0.$$

Equality can hold only if $h_j(\boldsymbol{x}) = \min_i h_i(\boldsymbol{x})$ for $\boldsymbol{x}$ in R_j except for sets of probability zero. Q.E.D.

Let us see how this method applies when $C(j|i) = 1$ for all i and j, $i \neq j$. Then in R_k

$$(10) \qquad \sum_{\substack{i=1 \\ i\neq k}}^{m} q_i p_i(\boldsymbol{x}) < \sum_{\substack{i=1 \\ i\neq j}}^{m} q_i p_i(\boldsymbol{x}), \qquad j \neq k.$$

Subtracting $\sum_{i=1, i\neq k, j}^{m} q_i p_i(\boldsymbol{x})$ from both sides of (10), we obtain

$$(11) \qquad q_j p_j(\boldsymbol{x}) < q_k p_k(\boldsymbol{x}), \qquad j \neq k.$$

In this case the point $\boldsymbol{x}$ is in R_k if k is the index for which $q_i p_i(\boldsymbol{x})$ is a maximum; that is, π_k is the most probable population.

Now suppose that we do not have a priori probabilities. Then we cannot define an unconditional expected loss for a classification procedure. However, we can define an expected loss on the condition that the observation comes from a given population. The conditional expected loss if the observation is from π_i is

$$(12) \qquad \sum_{\substack{j=1 \\ j\neq i}}^{m} C(j|i) P(j|i, R) = r(i, R).$$

A procedure R is *at least as good* as R^* if $r(i, R) \leq r(i, R^*)$, $i = 1, \ldots, m$; R is *better* if at least one inequality is strict. R is *admissible* if there is no procedure R^* that is better. A class of procedures is *complete* if for every procedure R outside the class there is a procedure R^* in the class that is better.

Now let us show that a Bayes procedure is admissible. Let R be a Bayes procedure; let R^* be another procedure. Since R is Bayes,

$$\sum_{i=1}^{m} q_i r(i, R) \le \sum_{i=1}^{m} q_i r(i, R^*). \tag{13}$$

Suppose $q_1 > 0, q_2 > 0, r(2, R^*) < r(2, R)$, and $r(i, R^*) \le r(i, R), i = 3, \dots, m$. Then

$$q_1[r(1, R) - r(1, R^*)] \le \sum_{i=2}^{m} q_i[r(i, R^*) - r(i, R)] < 0, \tag{14}$$

and $r(1, R) < r(1, R^*)$. Thus R^* is not better than R.

THEOREM 6.7.2. *If $q_i > 0, i = 1, \dots, m$, then a Bayes procedure is admissible.*

We shall now assume that $C(i|j) = 1, i \neq j$, and $\Pr\{p_i(x) = 0|\pi_j\} = 0$. The latter condition implies that all $p_i(x)$ are positive on the same set (except for a set of measure 0). Suppose $q_i = 0$ for $i = 1, \dots, t$, and $q_i > 0$ for $i = t + 1, \dots, m$. Then for the Bayes solution R_i, $i = 1, \dots, t$, is empty (except for a set of probability 0) as seen from (11) (that is, $p_m(x) = 0$ for x in R_i). It follows that $r(i, R) = \sum_{j \neq i} P(j|i, R) = 1 - P(i|i, R) = 1$ for $i = 1, \dots, t$. Then $(R_{t+1}, \dots, R_m)$ is a Bayes solution for the problem involving $p_{t+1}(x), \dots, p_m(x)$ and $q_{t+1}, \dots, q_m$. It follows from Theorem 6.7.2 that no procedure R^* for which $P(i|i, R^*) = 0$, $i = 1, \dots, t$, can be better than the Bayes procedure. Now consider a procedure R^* such that R_1^* includes a set of positive probability so that $P(1|1, R^*) > 0$. For R^* to be better than R,

$$\begin{aligned} P(i|i, R) &= \int_{R_i} p_i(x)\,dx \\ &\le P(i|i, R^*) = \int_{R_i^*} p_i(x)\,dx, \qquad i = 2, \dots, m. \end{aligned} \tag{15}$$

In such a case a procedure R^{**} where R_i^{**} is empty, $i = 1, \dots, t$, $R_i^{**} = R_i^*$, $i = t + 1, \dots, m - 1$, and $R_m^{**} = R_m^* \cup R_1^* \cup \cdots \cup R_t^*$ would give risks such that

$$
\begin{aligned}
&P(i|i, R^{**}) = 0, && i = 1, \ldots, t, \\
&P(i|i, R^{**}) = P(i|i, R^*) \geq P(i|i, R), && i = t+1, \ldots, m-1, \qquad (16) \\
&P(m|m, R^{**}) > P(m|m, R^*) \geq P(m|m, R).
\end{aligned}
$$

Then $(R_{t+1}^{**}, \ldots, R_m^{**})$ would be better than $(R_{t+1}, \ldots, R_m)$ for the $(m-t)$-decision problem which contradicts the preceding discussion.

THEOREM 6.7.3. *If* $C(i|j) = 1, i \neq j$, *and* $\Pr\{p_i(\mathbf{x}) = 0 | \pi_j\} = 0$, *then a Bayes procedure is admissible.*

The converse is true without conditions (except that the parameter space is finite).

THEOREM 6.7.4. *Every admissible procedure is a Bayes procedure.*

We shall not prove this theorem. It is Theorem 1 of Section 2.10 of Ferguson (1967), for example. The class of Bayes procedures is minimal complete if each Bayes procedure is unique (for the specified probabilities).

The minimax procedure is the Bayes procedure for which the risks are equal.

There are now available general treatments of statistical decision procedures by Wald (1950), Blackwell and Girshick (1954), Ferguson (1967), DeGroot (1970), Berger (1980b), and others.

6.8. CLASSIFICATION INTO ONE OF SEVERAL MULTIVARIATE NORMAL POPULATIONS

We shall now apply the theory of Section 6.7 to the case in which each population has a normal distribution. [See von Mises (1945).] We assume that the means are different and the covariance matrices are alike. Let $N(\boldsymbol{\mu}^{(i)}, \boldsymbol{\Sigma})$ be the distribution of π_i. The density is given by (1) of Section 6.4. At the outset the parameters are assumed known. For general costs with known a priori probabilities we can form the m functions (5) of Section 6.7 and define the region R_j as consisting of points $\mathbf{x}$ such that the jth function is minimum.

In the remainder of our discussion we shall assume that the costs of misclassification are equal. Then we use the functions

$$(1)\qquad u_{jk}(\boldsymbol{x}) = \log\frac{p_j(\boldsymbol{x})}{p_k(\boldsymbol{x})} = \left[\boldsymbol{x} - \tfrac{1}{2}\left(\boldsymbol{\mu}^{(j)} + \boldsymbol{\mu}^{(k)}\right)\right]'\boldsymbol{\Sigma}^{-1}\left(\boldsymbol{\mu}^{(j)} - \boldsymbol{\mu}^{(k)}\right).$$

If a priori probabilities are known, the region R_j is defined by those $\boldsymbol{x}$ satisfying

$$(2)\qquad R_j\colon u_{jk}(\boldsymbol{x}) > \log\frac{q_k}{q_j}, \qquad k = 1,\dots,m;\ k \neq j.$$

THEOREM 6.8.1. *If q_i is the a priori probability of drawing an observation from $\pi_i = N(\boldsymbol{\mu}^{(i)}, \boldsymbol{\Sigma})$, $i = 1,\dots,m$, and if the costs of misclassification are equal, then the regions of classification, $R_1,\dots,R_m$, that minimize the expected cost are defined by* (2), *where $u_{jk}(\boldsymbol{x})$ is given by* (1).

It should be noted that each $u_{jk}(\boldsymbol{x})$ is the classification function related to the jth and kth populations, and $u_{jk}(\boldsymbol{x}) = -u_{kj}(\boldsymbol{x})$. Since these are linear functions, the region R_i is bounded by hyperplanes. If the means span an $(m-1)$-dimensional hyperplane (for example, if the vectors $\boldsymbol{\mu}^{(i)}$ are linearly independent and $p \geq m-1$), then R_i is bounded by $m-1$ hyperplanes.

In the case of no a priori probabilities known, the region R_j is defined by inequalities

$$(3)\qquad u_{jk}(\boldsymbol{x}) \geq c_j - c_k, \qquad k = 1,\dots,m,\ k \neq j.$$

The constants c_k can be taken nonnegative. These sets of regions form the class of admissible procedures. For the minimax procedure these constants are determined so all $P(i|i, R)$ are equal.

We now show how to evaluate the probabilities of correct classification. If $\boldsymbol{X}$ is a random observation, we consider the random variables

$$(4)\qquad U_{ji} = \left[\boldsymbol{X} - \tfrac{1}{2}\left(\boldsymbol{\mu}^{(i)} + \boldsymbol{\mu}^{(j)}\right)\right]'\boldsymbol{\Sigma}^{-1}\left(\boldsymbol{\mu}^{(j)} - \boldsymbol{\mu}^{(i)}\right).$$

Here $U_{ji} = -U_{ij}$. Thus we use $m(m-1)/2$ classification functions if the means span an $(m-1)$-dimensional hyperplane. If $\boldsymbol{X}$ is from π_j, then U_{ji} is distributed according to $N(\frac{1}{2}\Delta_{ji}^2, \Delta_{ji}^2)$ where

$$(5)\qquad \Delta_{ji}^2 = \left(\boldsymbol{\mu}^{(j)} - \boldsymbol{\mu}^{(i)}\right)'\boldsymbol{\Sigma}^{-1}\left(\boldsymbol{\mu}^{(j)} - \boldsymbol{\mu}^{(i)}\right).$$

The covariance of U_{ji} and U_{jk} is

$$\Delta_{jk,ji} = \left(\boldsymbol{\mu}^{(j)} - \boldsymbol{\mu}^{(k)}\right)'\boldsymbol{\Sigma}^{-1}\left(\boldsymbol{\mu}^{(j)} - \boldsymbol{\mu}^{(i)}\right). \tag{6}$$

To determine the constants c_j we consider the integrals

$$P(j|j,R) = \int_{c_j - c_m}^{\infty} \cdots \int_{c_j - c_1}^{\infty} f_j \, du_{j1} \cdots du_{j,j-1} \, du_{j,j+1} \cdots du_{jm}, \tag{7}$$

where f_j is the density of U_{ji}, $i = 1, 2, \ldots, m$, $i \neq j$.

THEOREM 6.8.2. *If π_i is $N(\boldsymbol{\mu}^{(i)}, \boldsymbol{\Sigma})$ and the costs of misclassification are equal, the regions of classification, $R_1, \ldots, R_m$, that minimize the maximum conditional expected loss are defined by* (3), *where $u_{jk}(\boldsymbol{x})$ is given by* (1). *The constants c_j are determined so that the integrals* (7) *are equal.*

As an example consider the case of $m = 3$. There is no loss of generality in taking $p = 2$, for the density for higher p can be projected on the two-dimensional plane determined by the means of the three populations if they are not collinear (i.e., we can transform the vector $\boldsymbol{x}$ into u_{12}, u_{13} and $p - 2$ other coordinates, where these last $p - 2$ components are distributed independently of u_{12} and u_{13} and with zero means). The regions R_j are determined by three half-lines as shown in Figure 6.2. If this procedure is minimax, we cannot move the line between R_1 and R_2 nearer $(\mu_1^{(1)}, \mu_2^{(1)})$, the line between R_2 and R_3 nearer $(\mu_1^{(2)}, \mu_2^{(2)})$, and the line between R_3 and R_1 nearer $(\mu_1^{(3)}, \mu_2^{(3)})$ and still retain the equality $P(1|1, R) = P(2|2, R) = P(3|3, R)$ without leaving a triangle that is not included in any region. Thus, since the regions must exhaust the space, the lines must meet in a point, and the equality of probabilities determines $c_i - c_j$ uniquely.

To do this in a specific case in which we have numerical values for the components of the vectors $\boldsymbol{\mu}^{(1)}, \boldsymbol{\mu}^{(2)}, \boldsymbol{\mu}^{(3)}$, and the matrix $\boldsymbol{\Sigma}$, we would consider the three ($\leq p + 1$) joint distributions, each of 2 U_{ij}'s ($j \neq i$). We could try the values of $c_i = 0$, and using tables [Pearson (1931)] of the bivariate normal distribution, compute $P(i|i, R)$. By a trial-and-error method we could obtain c_i to approximate the above condition.

The preceding theory has been given on the assumption that the parameters are known. If they are not known and if a sample from each population is available, the estimators of the parameters can be substituted in the definition

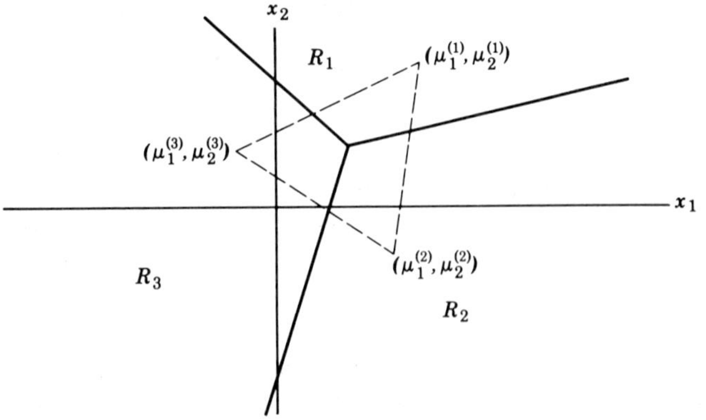

Figure 6.2. Classification regions.

of $u_{ij}(\boldsymbol{x})$. Let the observations be $\boldsymbol{x}_1^{(i)}, \dots, \boldsymbol{x}_{N_i}^{(i)}$ from $N(\boldsymbol{\mu}^{(i)}, \boldsymbol{\Sigma})$, $i = 1, \dots, m$. We estimate $\boldsymbol{\mu}^{(i)}$ by

$$\bar{\boldsymbol{x}}^{(i)} = \frac{1}{N_i} \sum_{\alpha=1}^{N_i} \boldsymbol{x}_\alpha^{(i)} \tag{8}$$

and $\boldsymbol{\Sigma}$ by $\boldsymbol{S}$ defined by

$$\left(\sum_{i=1}^{m} N_i - m \right) \boldsymbol{S} = \sum_{i=1}^{m} \sum_{\alpha=1}^{N_i} \left(\boldsymbol{x}_\alpha^{(i)} - \bar{\boldsymbol{x}}^{(i)} \right) \left(\boldsymbol{x}_\alpha^{(i)} - \bar{\boldsymbol{x}}^{(i)} \right)'. \tag{9}$$

Then, the analog of $u_{ij}(\boldsymbol{x})$ is

$$w_{ij}(\boldsymbol{x}) = \left[\boldsymbol{x} - \tfrac{1}{2}\left(\bar{\boldsymbol{x}}^{(i)} + \bar{\boldsymbol{x}}^{(j)} \right) \right]' \boldsymbol{S}^{-1} \left(\bar{\boldsymbol{x}}^{(i)} - \bar{\boldsymbol{x}}^{(j)} \right). \tag{10}$$

If the variables above are random, the distributions are different from those of U_{ij}. However, as $N_i \to \infty$, the joint distributions approach those of U_{ij}. Hence, for sufficiently large samples one can use the theory given above.

6.9. AN EXAMPLE OF CLASSIFICATION INTO ONE OF SEVERAL MULTIVARIATE NORMAL POPULATIONS

Rao (1948a) considers three populations consisting of the Brahmin caste (π_1), the Artisan caste (π_2), and the Korwa caste (π_3) of India. The measurements for each individual of a caste are stature (x_1), sitting height (x_2), nasal

TABLE 6.2

	Brahmin (π_1)	Artisan (π_2)	Korwa (π_3)
Stature (x_1)	164.51	160.53	158.17
Sitting height (x_2)	86.43	81.47	81.16
Nasal depth (x_3)	25.49	23.84	21.44
Nasal height (x_4)	51.24	48.62	46.72

depth (x_3), and nasal height (x_4). The means of these variables in the three populations are given in Table 6.2. The matrix of correlations for all the populations is

$$\text{(1)} \qquad \begin{bmatrix} 1.0000 & 0.5849 & 0.1774 & 0.1974 \\ 0.5849 & 1.0000 & 0.2094 & 0.2170 \\ 0.1774 & 0.2094 & 1.0000 & 0.2910 \\ 0.1974 & 0.2170 & 0.2910 & 1.0000 \end{bmatrix}.$$

The standard deviations are $\sigma_1 = 5.74, \sigma_2 = 3.20, \sigma_3 = 1.75, \sigma_4 = 3.50$. We assume that each population is normal. Our problem is to divide the space of the four variables x_1, x_2, x_3, x_4 into three regions of classification. We assume that the costs of misclassification are equal. We shall find (i) a set of regions under the assumption that drawing a new observation from each population is equally likely ($q_1 = q_2 = q_3 = \frac{1}{3}$), and (ii) a set of regions such that the largest probability of misclassification is minimized (the minimax solution).

We first compute the coefficients of $\mathbf{\Sigma}^{-1}(\boldsymbol{\mu}^{(1)} - \boldsymbol{\mu}^{(2)})$ and $\mathbf{\Sigma}^{-1}(\boldsymbol{\mu}^{(1)} - \boldsymbol{\mu}^{(3)})$. Then $\mathbf{\Sigma}^{-1}(\boldsymbol{\mu}^{(2)} - \boldsymbol{\mu}^{(3)}) = \mathbf{\Sigma}^{-1}(\boldsymbol{\mu}^{(1)} - \boldsymbol{\mu}^{(3)}) - \mathbf{\Sigma}^{-1}(\boldsymbol{\mu}^{(1)} - \boldsymbol{\mu}^{(2)})$. Then we calculate $\frac{1}{2}(\boldsymbol{\mu}^{(i)} + \boldsymbol{\mu}^{(j)})'\mathbf{\Sigma}^{-1}(\boldsymbol{\mu}^{(i)} - \boldsymbol{\mu}^{(j)})$. We obtain the discriminant functions†

$$\begin{aligned} u_{12}(\boldsymbol{x}) &= -0.0708x_1 + 0.4990x_2 + 0.3373x_3 + 0.0887x_4 - 43.13, \\ \text{(2)} \quad u_{13}(\boldsymbol{x}) &= 0.0003x_1 + 0.3550x_2 + 1.1063x_3 + 0.1375x_4 - 62.49, \\ u_{23}(\boldsymbol{x}) &= 0.0711x_1 - 0.1440x_2 + 0.7690x_3 + 0.0488x_4 - 19.36. \end{aligned}$$

The other three functions are $u_{21}(\boldsymbol{x}) = -u_{12}(\boldsymbol{x})$, $u_{31}(\boldsymbol{x}) = -u_{13}(\boldsymbol{x})$, and

†Due to an error in computations, Rao's discriminant functions are incorrect. I am indebted to Mr. Peter Frank for assistance in the computations.

TABLE 6.3

Population of x	u	Means	Standard Deviation	Correlation
π_1	u_{12}	1.491	1.727	0.8658
	u_{13}	3.487	2.641	
π_2	u_{21}	1.491	1.727	−0.3894
	u_{23}	1.031	1.436	
π_3	u_{31}	3.487	2.641	0.7983
	u_{32}	1.031	1.436	

$u_{32}(\boldsymbol{x}) = -u_{23}(\boldsymbol{x})$. If there are a priori probabilities and they are equal, the best set of regions of classification are R_1: $u_{12}(\boldsymbol{x}) \geq 0, u_{13}(\boldsymbol{x}) \geq 0$; R_2: $u_{21}(\boldsymbol{x}) \geq 0, u_{23}(\boldsymbol{x}) \geq 0$; and R_3: $u_{31}(\boldsymbol{x}) \geq 0, u_{32}(\boldsymbol{x}) \geq 0$. For example, if we obtain an individual with measurements $\boldsymbol{x}$ such that $u_{12}(\boldsymbol{x}) \geq 0$ and $u_{13}(\boldsymbol{x}) \geq 0$, we classify him as a Brahmin.

To find the probabilities of misclassification when an individual is drawn from population π_g we need the means, variances, and covariances of the proper pairs of u's. They are given in Table 6.3.†

The probabilities of misclassification are then obtained by use of the tables for the bivariate normal distribution. These probabilities are 0.21 for π_1, 0.42 for π_2, and 0.25 for π_3. For example, if measurements are made on a Brahmin, the probability that he is classified as an Artisan or Korwa is 0.21.

The minimax solution is obtained by finding the constants c_1, c_2, and c_3 for (3) of Section 6.8 so that the probabilities of misclassification are equal. The regions of classification are

$$\begin{aligned} &R_1': u_{12}(\boldsymbol{x}) \geq 0.54, \qquad u_{13}(\boldsymbol{x}) \geq 0.29; \\ &R_2': u_{21}(\boldsymbol{x}) \geq -0.54, \qquad u_{23}(\boldsymbol{x}) \geq -0.25; \qquad (3) \\ &R_3': u_{31}(\boldsymbol{x}) \geq -0.29, \qquad u_{32}(\boldsymbol{x}) \geq 0.25. \end{aligned}$$

The common probability of misclassification (to two decimal places) is 0.30.

†Some numerical errors in T. W. Anderson (1951a) are corrected in Table 6.3 and (3).

Thus the maximum probability of misclassification has been reduced from 0.42 to 0.30.

6.10. CLASSIFICATION INTO ONE OF TWO KNOWN MULTIVARIATE NORMAL POPULATIONS WITH UNEQUAL COVARIANCE MATRICES

6.10.1. Likelihood Procedures

Let π_1 and π_2 be $N(\boldsymbol{\mu}^{(1)}, \boldsymbol{\Sigma}_1)$ and $N(\boldsymbol{\mu}^{(2)}, \boldsymbol{\Sigma}_2)$ with $\boldsymbol{\mu}^{(1)} \neq \boldsymbol{\mu}^{(2)}$ and $\boldsymbol{\Sigma}_1 \neq \boldsymbol{\Sigma}_2$. When the parameters are known, the likelihood ratio is

$$(1) \qquad \frac{p_1(\boldsymbol{x})}{p_2(\boldsymbol{x})} = \frac{|\boldsymbol{\Sigma}_2|^{\frac{1}{2}}\exp\left[-\frac{1}{2}(\boldsymbol{x}-\boldsymbol{\mu}^{(1)})'\boldsymbol{\Sigma}_1^{-1}(\boldsymbol{x}-\boldsymbol{\mu}^{(1)})\right]}{|\boldsymbol{\Sigma}_1|^{\frac{1}{2}}\exp\left[-\frac{1}{2}(\boldsymbol{x}-\boldsymbol{\mu}^{(2)})'\boldsymbol{\Sigma}_2^{-1}(\boldsymbol{x}-\boldsymbol{\mu}^{(2)})\right]}$$

$$= |\boldsymbol{\Sigma}_2|^{\frac{1}{2}}|\boldsymbol{\Sigma}_1|^{-\frac{1}{2}}\exp\left[\tfrac{1}{2}(\boldsymbol{x}-\boldsymbol{\mu}^{(2)})'\boldsymbol{\Sigma}_2^{-1}(\boldsymbol{x}-\boldsymbol{\mu}^{(2)}) - \tfrac{1}{2}(\boldsymbol{x}-\boldsymbol{\mu}^{(1)})'\boldsymbol{\Sigma}_1^{-1}(\boldsymbol{x}-\boldsymbol{\mu}^{(1)})\right].$$

The logarithm of (1) is quadratic in $\boldsymbol{x}$. The probabilities of misclassification are difficult to compute. [One can make a linear transformation of $\boldsymbol{x}$ so that its covariance matrix is $\boldsymbol{I}$ and the matrix of the quadratic form is diagonal; then the logarithm of (1) has the distribution of a linear combination of noncentral χ^2-variables plus a constant.]

When the parameters are unknown, we consider the problem as testing the hypothesis that $\boldsymbol{x}, \boldsymbol{x}_1^{(1)}, \dots, \boldsymbol{x}_{N_1}^{(1)}$ are observations from $N(\boldsymbol{\mu}^{(1)}, \boldsymbol{\Sigma}_1)$ and $\boldsymbol{x}_1^{(2)}, \dots, \boldsymbol{x}_{N_2}^{(2)}$ are observations from $N(\boldsymbol{\mu}^{(2)}, \boldsymbol{\Sigma}_2)$ against the alternative that $\boldsymbol{x}_1^{(1)}, \dots, \boldsymbol{x}_{N_1}^{(1)}$ are observations from $N(\boldsymbol{\mu}^{(1)}, \boldsymbol{\Sigma}_1)$ and $\boldsymbol{x}, \boldsymbol{x}_1^{(2)}, \dots, \boldsymbol{x}_{N_2}^{(2)}$ are observations from $N(\boldsymbol{\mu}^{(2)}, \boldsymbol{\Sigma}_2)$. Under the first hypothesis the maximum likelihood estimators are $\hat{\boldsymbol{\mu}}_1^{(1)} = (N_1\bar{\boldsymbol{x}}^{(1)} + \boldsymbol{x})/(N_1+1)$, $\hat{\boldsymbol{\mu}}_1^{(2)} = \bar{\boldsymbol{x}}^{(2)}$,

$$(2) \qquad \begin{aligned} \hat{\boldsymbol{\Sigma}}_1(1) &= \frac{1}{N_1+1}\left[\boldsymbol{A}_1 + \frac{N_1}{N_1+1}(\boldsymbol{x}-\bar{\boldsymbol{x}}^{(1)})(\boldsymbol{x}-\bar{\boldsymbol{x}}^{(1)})'\right], \\ \hat{\boldsymbol{\Sigma}}_2(1) &= \frac{1}{N_2}\boldsymbol{A}_2, \end{aligned}$$

where $\boldsymbol{A}_i = \sum_{\alpha=1}^{N_i}(\boldsymbol{x}_\alpha^{(i)} - \bar{\boldsymbol{x}}^{(i)})(\boldsymbol{x}_\alpha^{(i)} - \bar{\boldsymbol{x}}^{(i)})'$, $i = 1, 2$. (See Section 6.5.5.) Under the second hypothesis the maximum likelihood estimators are $\hat{\mu}_2^{(1)} = \bar{x}^{(1)}$, $\hat{\mu}_2^{(2)} = (N_2\bar{x}^{(2)} + x)/(N_2 + 1)$,

$$\hat{\Sigma}_1(2) = \frac{1}{N_1}A_1,$$

(3)

$$\hat{\Sigma}_2(2) = \frac{1}{N_2+1}\left[A_2 + \frac{N_2}{N_2+1}(x - \bar{x}^{(2)})(x - \bar{x}^{(2)})'\right].$$

The likelihood ratio criterion is

(4)
$$\frac{|\hat{\Sigma}_1(2)|^{\frac{1}{2}N_1}|\hat{\Sigma}_2(2)|^{\frac{1}{2}(N_2+1)}}{|\hat{\Sigma}_1(1)|^{\frac{1}{2}(N_1+1)}|\hat{\Sigma}_2(1)|^{\frac{1}{2}N_2}} = \frac{\left[1 + (x - \bar{x}^{(2)})'A_2^{-1}(x - \bar{x}^{(2)})\right]^{\frac{1}{2}(N_2+1)}}{\left[1 + (x - \bar{x}^{(1)})'A_1^{-1}(x - \bar{x}^{(1)})\right]^{\frac{1}{2}(N_1+1)}}$$

$$\times \frac{(N_1+1)^{\frac{1}{2}(N_1+1)p}N_2^{\frac{1}{2}N_2 p}|A_2|^{\frac{1}{2}}}{N_1^{\frac{1}{2}N_1 p}(N_2+1)^{\frac{1}{2}(N_2+1)p}|A_1|^{\frac{1}{2}}}.$$

The observation x is classified into π_1 if (4) is greater than 1 and into π_2 if (4) is less than 1.

An alternative criterion is to plug estimates into the logarithm of (1). Use

(5)
$$\tfrac{1}{2}\left[(x - \bar{x}^{(2)})'S_2^{-1}(x - \bar{x}^{(2)}) - (x - \bar{x}^{(1)})'S_1^{-1}(x - \bar{x}^{(1)})\right],$$

to classify into π_1 if (5) is large and into π_2 if (5) is small. Again it is difficult to evaluate the probabilities of misclassification.

6.10.2. Linear Procedures

The best procedures when $\Sigma_1 \neq \Sigma_2$ are not linear; when the parameters are known, the best procedures are based on a quadratic function of the vector observation x. The procedure depends very much on the assumed normality. For example, in the case of $p = 1$, the region for classification with one population is an interval and for the other is the complement of the interval, that is, if the observation is sufficiently large or sufficiently small. In the bivariate case the regions are defined by conic sections; for example, the region of classification into one population might be the interior of an ellipse or the region between two hyperbolas. In general, the regions are defined by means of a quadratic function of the observations which is not necessarily a positive definite quadratic form. These procedures depend very much on the assumption of normality and especially on the shape of the normal distribution relatively far from its center. For instance, in the univariate case cited above the region of classification into the first population is a finite interval because the density of the first population falls off in either direction more rapidly than the density of the second since its standard deviation is smaller.

One may want to use a classification procedure in a situation where the two populations are centered around different points and have different patterns of scatter, and where one considers multivariate normal distributions to be reasonably good approximations for these two populations near their centers and between their two centers (though not far from the centers, where the densities are small). In such a case one may want to divide the sample space into the two regions of classification by some simple curve or surface. The simplest is a line or hyperplane; the procedure may then be termed linear.

Let $\boldsymbol{b}$ ($\neq \mathbf{0}$) be a vector (of p components) and c a scalar. An observation $\boldsymbol{x}$ is classified as from the first population if $\boldsymbol{b}'\boldsymbol{x} \geq c$ and as from the second if $\boldsymbol{b}'\boldsymbol{x} < c$. We are primarily interested in situations where the important difference between the two populations is the difference between the centers; we assume $\boldsymbol{\mu}^{(1)} \neq \boldsymbol{\mu}^{(2)}$ as well as $\boldsymbol{\Sigma}_1 \neq \boldsymbol{\Sigma}_2$, and that $\boldsymbol{\Sigma}_1$ and $\boldsymbol{\Sigma}_2$ are nonsingular.

When sampling from the ith population, $\boldsymbol{b}'\boldsymbol{x}$ has a univariate normal distribution with mean $\mathscr{E}(\boldsymbol{b}'\boldsymbol{x}|\pi_i) = \boldsymbol{b}'\boldsymbol{\mu}^{(i)}$ and variance

$$\mathscr{E}\left(\boldsymbol{b}'\boldsymbol{x} - \boldsymbol{b}'\boldsymbol{\mu}^{(i)}\right)^2|\pi_i = \mathscr{E}\boldsymbol{b}'\left(\boldsymbol{x} - \boldsymbol{\mu}^{(i)}\right)\left(\boldsymbol{x} - \boldsymbol{\mu}^{(i)}\right)'\boldsymbol{b}|\pi_i = \boldsymbol{b}'\boldsymbol{\Sigma}_i\boldsymbol{b}. \tag{6}$$

The probability of misclassifying an observation when it comes from the first population is

$$\begin{aligned} P(2|1) = \Pr\left\{\boldsymbol{b}'\boldsymbol{x} < c|\pi_1\right\} &= \Pr\left\{\frac{\boldsymbol{b}'\boldsymbol{x} - \boldsymbol{b}'\boldsymbol{\mu}^{(1)}}{\left(\boldsymbol{b}'\boldsymbol{\Sigma}_1\boldsymbol{b}\right)^{\frac{1}{2}}} < \frac{c - \boldsymbol{b}'\boldsymbol{\mu}^{(1)}}{\left(\boldsymbol{b}'\boldsymbol{\Sigma}_1\boldsymbol{b}\right)^{\frac{1}{2}}}\Bigg|\pi_1\right\} \\ &= \Phi\left(\frac{c - \boldsymbol{b}'\boldsymbol{\mu}^{(1)}}{\left(\boldsymbol{b}'\boldsymbol{\Sigma}_1\boldsymbol{b}\right)^{\frac{1}{2}}}\right) = 1 - \Phi\left(\frac{\boldsymbol{b}'\boldsymbol{\mu}^{(1)} - c}{\left(\boldsymbol{b}'\boldsymbol{\Sigma}_1\boldsymbol{b}\right)^{\frac{1}{2}}}\right). \end{aligned} \tag{7}$$

The probability of misclassifying an observation when it comes from the second population is

$$(8) \qquad P(1|2) = \Pr\{ \boldsymbol{b}'\boldsymbol{x} \geq c | \pi_2 \} = \Pr\left\{ \frac{\boldsymbol{b}'\boldsymbol{x} - \boldsymbol{b}'\boldsymbol{\mu}^{(2)}}{(\boldsymbol{b}'\boldsymbol{\Sigma}_2\boldsymbol{b})^{\frac{1}{2}}} \geq \frac{c - \boldsymbol{b}'\boldsymbol{\mu}^{(2)}}{(\boldsymbol{b}'\boldsymbol{\Sigma}_2\boldsymbol{b})^{\frac{1}{2}}} \middle| \pi_2 \right\}$$

$$= 1 - \Phi\left(\frac{c - \boldsymbol{b}'\boldsymbol{\mu}^{(2)}}{(\boldsymbol{b}'\boldsymbol{\Sigma}_2\boldsymbol{b})^{\frac{1}{2}}} \right).$$

It is desired to make these probabilities small or, equivalently, to make the arguments

$$(9) \qquad y_1 = \frac{\boldsymbol{b}'\boldsymbol{\mu}^{(1)} - c}{(\boldsymbol{b}'\boldsymbol{\Sigma}_1\boldsymbol{b})^{\frac{1}{2}}}, \qquad y_2 = \frac{c - \boldsymbol{b}'\boldsymbol{\mu}^{(2)}}{(\boldsymbol{b}'\boldsymbol{\Sigma}_2\boldsymbol{b})^{\frac{1}{2}}}$$

large. We shall consider making y_1 large for given y_2.

When we eliminate c from (9), we obtain

$$(10) \qquad y_1 = \left[\boldsymbol{b}'\boldsymbol{\gamma} - y_2(\boldsymbol{b}'\boldsymbol{\Sigma}_2\boldsymbol{b})^{\frac{1}{2}} \right] (\boldsymbol{b}'\boldsymbol{\Sigma}_1\boldsymbol{b})^{-\frac{1}{2}},$$

where $\boldsymbol{\gamma} = \boldsymbol{\mu}^{(1)} - \boldsymbol{\mu}^{(2)}$. To maximize y_1 for given y_2 we differentiate y_1 with respect to $\boldsymbol{b}$ to obtain

$$(11) \qquad \frac{\partial y_1}{\partial \boldsymbol{b}} = \left[\boldsymbol{\gamma} - y_2(\boldsymbol{b}'\boldsymbol{\Sigma}_2\boldsymbol{b})^{-\frac{1}{2}}\boldsymbol{\Sigma}_2\boldsymbol{b} \right] (\boldsymbol{b}'\boldsymbol{\Sigma}_1\boldsymbol{b})^{-\frac{1}{2}}$$

$$- \left[\boldsymbol{b}'\boldsymbol{\gamma} - y_2(\boldsymbol{b}'\boldsymbol{\Sigma}_2\boldsymbol{b})^{\frac{1}{2}} \right] (\boldsymbol{b}'\boldsymbol{\Sigma}_1\boldsymbol{b})^{-\frac{3}{2}}\boldsymbol{\Sigma}_1\boldsymbol{b}.$$

If we let

$$(12) \qquad t_1 = \frac{\boldsymbol{b}'\boldsymbol{\gamma} - y_2(\boldsymbol{b}'\boldsymbol{\Sigma}_2\boldsymbol{b})^{\frac{1}{2}}}{\boldsymbol{b}'\boldsymbol{\Sigma}_1\boldsymbol{b}},$$

$$(13) \qquad t_2 = \frac{y_2}{\sqrt{\boldsymbol{b}'\boldsymbol{\Sigma}_2\boldsymbol{b}'}},$$

then (11) set equal to $\mathbf{0}$ is

$$(14) \qquad (t_1\boldsymbol{\Sigma}_1 + t_2\boldsymbol{\Sigma}_2)\boldsymbol{b} = \boldsymbol{\gamma}.$$

Note that (13) and (14) imply (12). If there is a pair t_1, t_2, and a vector $\boldsymbol{b}$ satisfying (12) and (13), then c is obtained from (9) as

$$c = y_2\sqrt{\boldsymbol{b}'\boldsymbol{\Sigma}_2\boldsymbol{b}} + \boldsymbol{b}'\boldsymbol{\mu}^{(2)} = t_2\boldsymbol{b}'\boldsymbol{\Sigma}_2\boldsymbol{b} + \boldsymbol{b}'\boldsymbol{\mu}^{(2)}. \tag{15}$$

Then from (9), (12), and (13)

$$y_1 = \frac{\boldsymbol{b}'\boldsymbol{\mu}^{(1)} - \left(t_2\boldsymbol{b}'\boldsymbol{\Sigma}_2\boldsymbol{b} + \boldsymbol{b}'\boldsymbol{\mu}^{(2)}\right)}{\sqrt{\boldsymbol{b}'\boldsymbol{\Sigma}_1\boldsymbol{b}}} = t_1\sqrt{\boldsymbol{b}'\boldsymbol{\Sigma}_1\boldsymbol{b}}\,. \tag{16}$$

Now consider (14) as a function of t ($0 \le t \le 1$). Let $t_1 = t$ and $t_2 = 1 - t$; then $\boldsymbol{b} = (t_1\boldsymbol{\Sigma}_1 + t_2\boldsymbol{\Sigma}_2)^{-1}\boldsymbol{\gamma}$. Define $v_1 = t_1\sqrt{\boldsymbol{b}'\boldsymbol{\Sigma}_1\boldsymbol{b}}$ and $v_2 = t_2\sqrt{\boldsymbol{b}'\boldsymbol{\Sigma}_2\boldsymbol{b}}$. The derivative of v_1^2 with respect to t is

$$\begin{aligned}
(17)\quad \frac{d}{dt}t^2\boldsymbol{\gamma}'&\left[t\boldsymbol{\Sigma}_1 + (1-t)\boldsymbol{\Sigma}_2\right]^{-1}\boldsymbol{\Sigma}_1\left[t\boldsymbol{\Sigma}_1 + (1-t)\boldsymbol{\Sigma}_2\right]^{-1}\boldsymbol{\gamma} \\
&= 2t\boldsymbol{\gamma}'\left[t\boldsymbol{\Sigma}_1 + (1-t)\boldsymbol{\Sigma}_2\right]^{-1}\boldsymbol{\Sigma}_1\left[t\boldsymbol{\Sigma}_1 + (1-t)\boldsymbol{\Sigma}_2\right]^{-1}\boldsymbol{\gamma} \\
&\quad - t^2\boldsymbol{\gamma}'\left[t\boldsymbol{\Sigma}_1 + (1-t)\boldsymbol{\Sigma}_2\right]^{-1}(\boldsymbol{\Sigma}_1 - \boldsymbol{\Sigma}_2)\left[t\boldsymbol{\Sigma}_1 + (1-t)\boldsymbol{\Sigma}_2\right]^{-1} \\
&\qquad\qquad \times \boldsymbol{\Sigma}_1\left[t\boldsymbol{\Sigma}_1 + (1-t)\boldsymbol{\Sigma}_2\right]^{-1}\boldsymbol{\gamma} \\
&\quad - t^2\boldsymbol{\gamma}'\left[t\boldsymbol{\Sigma}_1 + (1-t)\boldsymbol{\Sigma}_2\right]^{-1}\boldsymbol{\Sigma}_1\left[t\boldsymbol{\Sigma}_1 + (1-t)\boldsymbol{\Sigma}_2\right]^{-1} \\
&\qquad\qquad \times (\boldsymbol{\Sigma}_1 - \boldsymbol{\Sigma}_2)\left[t\boldsymbol{\Sigma}_1 + (1-t)\boldsymbol{\Sigma}_2\right]^{-1}\boldsymbol{\gamma} \\
&= t\boldsymbol{\gamma}'\left[t\boldsymbol{\Sigma}_1 + (1-t)\boldsymbol{\Sigma}_2\right]^{-1}\left\{\boldsymbol{\Sigma}_2\left[t\boldsymbol{\Sigma}_1 + (1-t)\boldsymbol{\Sigma}_2\right]^{-1}\boldsymbol{\Sigma}_1\right. \\
&\quad \left. + \boldsymbol{\Sigma}_1\left[t\boldsymbol{\Sigma}_1 + (1-t)\boldsymbol{\Sigma}_2\right]^{-1}\boldsymbol{\Sigma}_2\right\}\left[t\boldsymbol{\Sigma}_1 + (1-t)\boldsymbol{\Sigma}_2\right]^{-1}\boldsymbol{\gamma} \\
&> 0
\end{aligned}$$

by the following lemma.

LEMMA 6.10.1. *If $\boldsymbol{\Sigma}_1$ and $\boldsymbol{\Sigma}_2$ are positive definite and $t_1 > 0$, $t_2 > 0$, then*

$$\boldsymbol{\Sigma}_2\left[t_1\boldsymbol{\Sigma}_1 + t_2\boldsymbol{\Sigma}_2\right]^{-1}\boldsymbol{\Sigma}_1 \tag{18}$$

is positive definite.

PROOF. The matrix (18) is

$$\Sigma_2\left[\Sigma_1\left(t_1\Sigma_2^{-1}+t_2\Sigma_1^{-1}\right)\Sigma_2\right]^{-1}\Sigma_1=\left(t_1\Sigma_2^{-1}+t_2\Sigma_1^{-1}\right)^{-1}. \tag{19}$$ Q.E.D.

Similarly $dv_2^2/dt < 0$. Since $v_1 \geq 0, v_2 \geq 0$, we see that v_1 increases with t from 0 at $t = 0$ to $\sqrt{\gamma'\Sigma_1^{-1}\gamma}$ at $t = 1$ and v_2 decreases with t from $\sqrt{\gamma'\Sigma_2^{-1}\gamma}$ at $t = 0$ to 0 at $t = 1$. The coordinates v_1 and v_2 are continuous functions of t. For given $y_2, 0 \leq y_2 \leq \sqrt{\gamma'\Sigma_2^{-1}\gamma}$, there is a t such that $y_2 = v_2 = t_2\sqrt{b'\Sigma_2 b}$ and b satisfies (14) for $t_1 = t$ and $t_2 = 1 - t$. Then $y_1 = v_1 = t_1\sqrt{b'\Sigma_1 b}$ maximizes y_1 for that value of y_2. Similarly given $y_1, 0 \leq y_1 \leq \sqrt{\gamma'\Sigma_1^{-1}\gamma}$, there is a t such that $y_1 = v_1 = t_1\sqrt{b'\Sigma_1 b}$ and b satisfies (14) for $t_1 = t$ and $t_2 = 1 - t$, and $y_2 = v_2 = t_2\sqrt{b'\Sigma_2 b}$ maximizes y_2. Note that $y_1 \geq 0, y_2 \geq 0$ implies the errors of misclassification are not greater than $\frac{1}{2}$.

We now argue that the set of y_1, y_2 defined this way correspond to admissible linear procedures. Let x_1, x_2 be in this set and suppose another procedure defined by z_1, z_2 were better than x_1, x_2, that is, $x_1 \leq z_1, x_2 \leq z_2$ with at least one strict inequality. For $y_1 = z_1$ let y_2^* be the maximum y_2 among linear procedures; then $z_1 = y_1, z_2 \leq y_2^*$ and hence $x_1 \leq y_1, x_2 \leq y_2^*$. However, this is possible only if $x_1 = y_1, x_2 = y_2^*$ because $dy_1/dy_2 < 0$. Now we have a contradiction to the assumption that z_1, z_2 was better than x_1, x_2. Thus x_1, x_2 corresponds to an admissible linear procedure.

Use of Admissible Linear Procedures. Given t_1 and t_2 (so $t_1\Sigma_1 + t_2\Sigma_2$ is positive definite), one would compute the optimum b by solving the linear equations (15) and then compute c by one of (9). Usually t_1 and t_2 are not given, but a desired solution is specified in another way. We consider three ways.

Minimization of one probability of misclassification for a specified probability of the other. Suppose we are given y_2 (or, equivalently, the probability of misclassification when sampling from the second distribution) and we want to maximize y_1 (or, equivalently, minimize the probability of misclassification when sampling from the first distribution). Suppose $y_2 > 0$ (i.e., the given probability of misclassification is less than $\frac{1}{2}$). Then if the maximum $y_1 \geq 0$ we want to find $t_2 = 1 - t_1$, so $y_2 = t_2(b'\Sigma_2 b)^{\frac{1}{2}}$, where $b = [t_1\Sigma_1 + t_2\Sigma_2]^{-1}\gamma$. The solution can be approximated by trial and error since y_2 is an increasing function of t_2. For $t_2 = 0$, $y_2 = 0$ and for $t_2 = 1$, $y_2 = (b'\Sigma_2 b)^{\frac{1}{2}} = (b'\gamma)^{\frac{1}{2}} =$

$(\boldsymbol{\gamma}'\boldsymbol{\Sigma}_2^{-1}\boldsymbol{\gamma})$, where $\boldsymbol{\Sigma}_2\boldsymbol{b} = \boldsymbol{\gamma}$. One could try other values of t_2 successively by solving (14) and inserting in $\boldsymbol{b}'\boldsymbol{\Sigma}_2\boldsymbol{b}$ until $t_2(\boldsymbol{b}'\boldsymbol{\Sigma}_2\boldsymbol{b})^{\frac{1}{2}}$ agrees closely enough with the desired y_2. ($y_1 > 0$ if the specified $y_2 < (\boldsymbol{\gamma}'\boldsymbol{\Sigma}_2^{-1}\boldsymbol{\gamma})^{\frac{1}{2}}$.)

The minimax procedure. The minimax procedure is the admissible procedure for which $y_1 = y_2$. Since for this procedure both probabilities of correct classification are greater than $\frac{1}{2}$, $y_1 = y_2 > 0$ and $t_1 > 0, t_2 > 0$. We want to find t $(= t_1 = 1 - t_2)$ so that

$$\begin{aligned} (20)\qquad 0 = y_1^2 - y_2^2 &= t^2\boldsymbol{b}'\boldsymbol{\Sigma}_1\boldsymbol{b} - (1-t)^2\boldsymbol{b}'\boldsymbol{\Sigma}_2\boldsymbol{b} \\ &= \boldsymbol{b}'\left[t^2\boldsymbol{\Sigma}_1 - (1-t)^2\boldsymbol{\Sigma}_2\right]\boldsymbol{b}. \end{aligned}$$

Since y_1^2 increases with t and y_2^2 decreases with t, there is one and only one solution to (20) and this can be approximated by trial and error by guessing a value of t $(0 < t < 1)$, solving (14) for $\boldsymbol{b}$, and computing the quadratic form on the right of (20). Then another t can be tried.

An alternative approach is to set $y_1 = y_2$ in (9) and solve for c. Then the common value of $y_1 = y_2$ is

$$(21)\qquad \frac{\boldsymbol{b}'\boldsymbol{\gamma}}{(\boldsymbol{b}'\boldsymbol{\Sigma}_1\boldsymbol{b})^{\frac{1}{2}} + (\boldsymbol{b}'\boldsymbol{\Sigma}_2\boldsymbol{b})^{\frac{1}{2}}}$$

and we want to find $\boldsymbol{b}$ to maximize this, where $\boldsymbol{b}$ is of the form

$$(22)\qquad [t\boldsymbol{\Sigma}_1 + (1-t)\boldsymbol{\Sigma}_2]^{-1}\boldsymbol{\gamma}$$

with $0 < t < 1$.

When $\boldsymbol{\Sigma}_1 = \boldsymbol{\Sigma}_2$, twice the maximum of (21) is the squared Mahalanobis distance between the populations. This suggests that when $\boldsymbol{\Sigma}_1$ may be unequal to $\boldsymbol{\Sigma}_2$, twice the maximum of (21) might be called the distance between the populations.

Welch and Wimpress (1961) have programmed the minimax procedure and applied it to the recognition of spoken sounds.

Case of a priori probabilities. Suppose we are given a priori probabilities, q_1 and q_2, of the first and second populations, respectively. Then the probability of a misclassification is

$$(23)\qquad q_1[1 - \Phi(y_1)] + q_2[1 - \Phi(y_2)] = 1 - [q_1\Phi(y_1) + q_2\Phi(y_2)],$$

which we want to minimize. The solution will be an admissible linear procedure. If we know it involves $y_1 \geq 0$ and $y_2 \geq 0$, we can substitute $y_1 = t(b'\Sigma_1 b)^{\frac{1}{2}}$ and $y_2 = (1-t)(b'\Sigma_2 b)^{\frac{1}{2}}$, where $b = [t\Sigma_1 + (1-t)\Sigma_2]^{-1}\gamma$, into (23) and set the derivative of (23) with respect to t equal to 0, obtaining

$$q_1\phi(y_1)\frac{dy_1}{dt} + q_2\phi(y_2)\frac{dy_2}{dt} = 0, \tag{24}$$

where $\phi(u) = (2\pi)^{-\frac{1}{2}}e^{-\frac{1}{2}u^2}$. There does not seem to be any easy or direct way of solving (24) for t. The left-hand side of (24) is not necessarily monotonic. In fact, there may be several roots to (24). If there are, the absolute minimum will be found by putting the solution into (23). (We remind the reader that the curve of admissible error probabilities is not necessarily convex.)

Anderson and Bahadur (1962) have studied these linear procedures in general, including $y_1 < 0$ and $y_2 < 0$. Clunies-Ross and Riffenburgh (1960) have approached the problem from a more geometric point of view.

PROBLEMS

1. (Sec. 6.3) Let π_i be $N(\mu, \Sigma_i)$, $i = 1, 2$. Find the form of the admissible classification procedures.

2. (Sec. 6.3) Prove that every complete class of procedures includes the class of admissible procedures.

3. (Sec. 6.3) Prove that if the class of admissible procedures is complete it is minimal complete.

4. (Sec. 6.3) The *Neyman–Pearson fundamental lemma* states that of all tests at a given significance level of the null hypothesis that x is drawn from $p_1(x)$ against alternative that it is drawn from $p_2(x)$ the most powerful test has the critical region $p_1(x)/p_2(x) < k$. Show that the discussion in Section 6.3 proves this result.

5. (Sec. 6.3) When $p(x) = n(x|\mu, \Sigma)$ find the best test of $\mu = \mathbf{0}$ against $\mu = \mu^*$ at significance level ε. Show that this test is uniformly most powerful against all alternatives $\mu = c\mu^*$, $c > 0$. Prove that there is no uniformly most powerful test against $\mu = \mu^{(1)}$ and $\mu = \mu^{(2)}$ unless $\mu^{(1)} = c\mu^{(2)}$ for some $c > 0$.

6. (Sec. 6.4) Let $P(2|1)$ and $P(1|2)$ be defined by (14) and (15). Prove if $-\frac{1}{2}\Delta^2 < c < \frac{1}{2}\Delta^2$, then $P(2|1)$ and $P(1|2)$ are decreasing functions of Δ.

7. (Sec. 6.4) Let $x' = (x^{(1)\prime}, x^{(2)\prime})$. Using Problem 23 of Chapter 5 and Problem 6 of this chapter, prove that the class of classification procedures based on x is uniformly as good as the class of procedures based on $x^{(1)}$.

8. (Sec. 6.5.1) Find the criterion for classifying Iris as Iris setosa or Iris versicolor on the basis of data given in Section 5.3.4. Classify a random sample of 5 Iris Virginica in Table 3.4.

9. (Sec. 6.5.1) Let $W(\boldsymbol{x})$ be the classification criterion given by (2). Show that the T^2-criterion for testing $N(\boldsymbol{\mu}^{(1)}, \boldsymbol{\Sigma}) = N(\boldsymbol{\mu}^{(2)}, \boldsymbol{\Sigma})$ is proportional to $W(\bar{\boldsymbol{x}}^{(1)})$ and $W(\bar{\boldsymbol{x}}^{(2)})$.

10. (Sec. 6.5.1) Show that the probabilities of misclassification of $\boldsymbol{x}_1, \ldots, \boldsymbol{x}_N$ (all assumed to be from either π_1 or π_2) decrease as N increases.

11. (Sec. 6.5) Show that the elements of $\boldsymbol{M}$ are invariant under the transformation (34) and that any function of the sufficient statistics that is invariant is a function of $\boldsymbol{M}$.

12. (Sec. 6.5) Consider $\boldsymbol{d}'\boldsymbol{x}^{(i)}$. Prove that the ratio

$$\frac{[\boldsymbol{d}'\bar{\boldsymbol{x}}^{(1)} - \boldsymbol{d}'\bar{\boldsymbol{x}}^{(2)}]^2}{\sum_{\alpha=1}^{N_1}\left(\boldsymbol{d}'\boldsymbol{x}_\alpha^{(1)} - \boldsymbol{d}'\bar{\boldsymbol{x}}^{(1)}\right)^2 + \sum_{\alpha=1}^{N_2}\left(\boldsymbol{d}'\boldsymbol{x}_\alpha^{(2)} - \boldsymbol{d}'\bar{\boldsymbol{x}}^{(2)}\right)^2}$$

is maximized by $\boldsymbol{d} = \boldsymbol{S}^{-1}(\bar{\boldsymbol{x}}^{(1)} - \bar{\boldsymbol{x}}^{(2)})$.

13. (Sec. 6.6) Show that the derivative of (2) to terms of order n^{-1} is

$$-\phi\left(\tfrac{1}{2}\Delta\right)\left\{\frac{1}{2} + \frac{1}{n}\left[\frac{p-1}{\Delta^2} + \frac{p-2}{4} - \frac{p}{8}\Delta^2\right]\right\}.$$

14. (Sec. 6.6) Show $\mathscr{E}D^2$ is (4). [*Hint:* Let $\boldsymbol{\Sigma} = \boldsymbol{I}$ and show that $\mathscr{E}(\boldsymbol{S}^{-1}|\boldsymbol{\Sigma} = \boldsymbol{I}) = [n/(n-p-1)]\boldsymbol{I}$.]

15. (Sec. 6.6.2) Show

$$\Pr\left\{\frac{Z - \frac{1}{2}D^2}{D} \le u \middle| \pi_1\right\} - \Pr\left\{\frac{Z - \frac{1}{2}\Delta^2}{\Delta} \le u \middle| \pi_1\right\}$$

$$= \phi(u)\left\{\frac{1}{2N_1\Delta^2}\left[u^3 + (p-3)u - \Delta^2 u + p\Delta\right]\right.$$

$$+ \frac{1}{2N_2\Delta^2}\left[u^3 + 2\Delta u^2 + (p - 3 + \Delta^2)u - \Delta^3 + p\Delta\right]$$

$$\left. + \frac{1}{4n}\left[3u^3 + 4\Delta u^2 + (2p - 3 + \Delta^2)u + 2(p-1)\Delta\right]\right\}$$

$$+ O(n^{-2}).$$

16. (Sec. 6.8) Let π_i be $N(\boldsymbol{\mu}^{(i)}, \boldsymbol{\Sigma}), i = 1, \ldots, m$. If the $\boldsymbol{\mu}^{(i)}$ are on a line (i.e., $\boldsymbol{\mu}^{(i)} = \boldsymbol{\mu} + \nu_i \boldsymbol{\beta}$), show that for admissible procedures the R_i are defined by parallel planes. Thus show that only one discriminant function $u_{jk}(\boldsymbol{x})$ need be used.

17. (Sec. 6.8) In Section 8.8 data are given on samples from four populations of skulls. Consider the first two measurements and the first three samples. Construct the classification functions $u_{ij}(\boldsymbol{x})$. Find the procedure for $q_i = N_i/(N_1 + N_2 + N_3)$. Find the minimax procedure.

18. (Sec. 6.10) Show that $\boldsymbol{b}'\boldsymbol{x} = c$ is the equation of a plane that is tangent to an ellipsoid of constant density of π_1 and to an ellipsoid of constant density of π_2 at a common point.

19. (Sec. 6.8) Let $\boldsymbol{x}_1^{(i)}, \dots, \boldsymbol{x}_{N_i}^{(i)}$ be observations from $N(\boldsymbol{\mu}^{(i)}, \boldsymbol{\Sigma})$, $i = 1, 2, 3$, and let $\boldsymbol{x}$ be an observation to be classified. Give explicitly the maximum likelihood rule.

20. (Sec. 6.5) Verify (33).

CHAPTER 7

The Distribution of the Sample Covariance Matrix and the Sample Generalized Variance

7.1. INTRODUCTION

The sample covariance matrix, $S = [1/(N-1)]\sum_{\alpha=1}^{n}(x_\alpha - \bar{x})(x_\alpha - \bar{x})'$, is an estimator of the population covariance matrix Σ. In Section 4.2 we found the density of $A = (N-1)S$ in the case of a 2×2 matrix. In Section 7.2 this result will be generalized to the case of a matrix A of any order. When $\Sigma = I$, this distribution is in a sense a generalization of the χ^2-distribution. The distribution of A (or S), often called the Wishart distribution, is fundamental to multivariate statistical analysis. In Sections 7.3 and 7.4 we discuss some properties of the Wishart distribution.

The generalized variance of the sample is defined as $|S|$ in Section 7.5; it is a kind of measure of the scatter of the sample. Its distribution is characterized. The density of the set of all correlation coefficients when the components of the observed vector are independent is obtained in Section 7.6.

The inverted Wishart distribution is introduced in Section 7.7 and is used as an a priori distribution of Σ to obtain a Bayes estimator of the covariance matrix. In Section 7.8 we consider improving on S as an estimator of Σ with respect to two loss functions.

7.2. THE WISHART DISTRIBUTION

We shall obtain the distribution of $\boldsymbol{A} = \sum_{\alpha=1}^{N}(\boldsymbol{X}_\alpha - \overline{\boldsymbol{X}})(\boldsymbol{X}_\alpha - \overline{\boldsymbol{X}})'$, where $\boldsymbol{X}_1, \dots, \boldsymbol{X}_N$ $(N > p)$ are independent, each with the distribution $N(\boldsymbol{\mu}, \boldsymbol{\Sigma})$. As was shown in Section 3.3, $\boldsymbol{A}$ is distributed as $\sum_{\alpha=1}^{n}\boldsymbol{Z}_\alpha \boldsymbol{Z}_\alpha'$, where $n = N - 1$ and $\boldsymbol{Z}_1, \dots, \boldsymbol{Z}_n$ are independent, each with the distribution $N(\boldsymbol{0}, \boldsymbol{\Sigma})$. We shall show that the density of $\boldsymbol{A}$ for $\boldsymbol{A}$ positive definite is

$$(1) \qquad \frac{|\boldsymbol{A}|^{\frac{1}{2}(n-p-1)}\exp\left(-\frac{1}{2}\operatorname{tr}\boldsymbol{\Sigma}^{-1}\boldsymbol{A}\right)}{2^{\frac{1}{2}np}\pi^{p(p-1)/4}|\boldsymbol{\Sigma}|^{\frac{1}{2}n}\prod_{i=1}^{p}\Gamma\left[\frac{1}{2}(n+1-i)\right]}.$$

We shall first consider the case of $\boldsymbol{\Sigma} = \boldsymbol{I}$. Let

$$(2) \qquad (\boldsymbol{Z}_1, \dots, \boldsymbol{Z}_n) = \begin{pmatrix} \boldsymbol{v}_1' \\ \vdots \\ \boldsymbol{v}_p' \end{pmatrix}.$$

Then the elements of $\boldsymbol{A} = (a_{ij})$ are inner products of these n-component vectors, $a_{ij} = \boldsymbol{v}_i'\boldsymbol{v}_j$. The vectors $\boldsymbol{v}_1, \dots, \boldsymbol{v}_p$ are independently distributed, each according to $N(\boldsymbol{0}, \boldsymbol{I}_n)$. It will be convenient to transform to new coordinates according to the Gram–Schmidt orthogonalization. Let $\boldsymbol{w}_1 = \boldsymbol{v}_1$,

$$(3) \qquad \boldsymbol{w}_i = \boldsymbol{v}_i - \sum_{j=1}^{i-1}\frac{\boldsymbol{v}_i'\boldsymbol{w}_j}{\boldsymbol{w}_j'\boldsymbol{w}_j}\boldsymbol{w}_j, \qquad i = 2, \dots, p.$$

We can prove by induction that $\boldsymbol{w}_k$ is orthogonal to $\boldsymbol{w}_i$, $k < i$. Assume $\boldsymbol{w}_k'\boldsymbol{w}_h = 0$, $k \neq h$, $k, h = 1, \dots, i-1$; then take the inner product of $\boldsymbol{w}_k$ and (3) to obtain $\boldsymbol{w}_k'\boldsymbol{w}_i = 0$, $k = 1, \dots, i-1$. (Note $\Pr\{\|\boldsymbol{w}_i\| = 0\} = 0$.)

Define

$$(4) \qquad t_{ii} = \|\boldsymbol{w}_i\| = \sqrt{\boldsymbol{w}_i'\boldsymbol{w}_i}, \qquad i = 1, \dots, p,$$

$$(5) \qquad t_{ij} = \frac{\boldsymbol{v}_i'\boldsymbol{w}_j}{\|\boldsymbol{w}_j\|}, \qquad j = 1, \dots, i-1,\ i = 2, \dots, p.$$

Since $\boldsymbol{v}_i = \sum_{j=1}^{i}(t_{ij}/\|\boldsymbol{w}_j\|)\boldsymbol{w}_j$,

$$(6) \qquad a_{hi} = \boldsymbol{v}_h'\boldsymbol{v}_i = \sum_{j=1}^{\min(h,i)} t_{hj}t_{ij}.$$

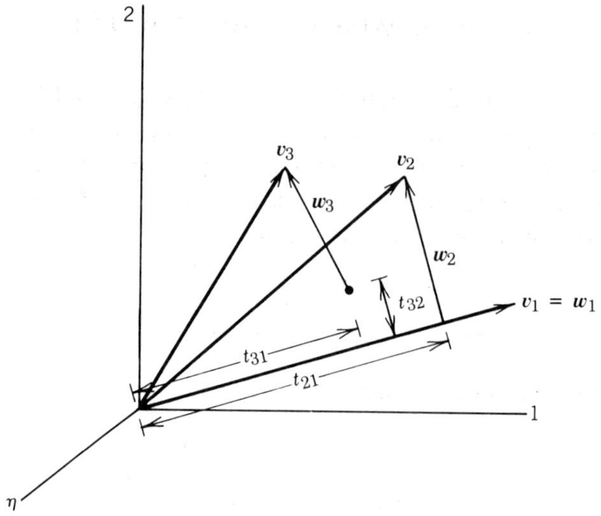

Figure 7.1. Transformation of coordinates.

If we define the lower triangular matrix $T = (t_{ij})$ with $t_{ij} = 0$, $i < j$, then

$$A = TT'. \tag{7}$$

Note that t_{ij}, $j = 1, \ldots, i-1$, are the first $i-1$ coordinates of v_i in the coordinate system with $w_1, \ldots, w_{i-1}$ as the first $i-1$ coordinate axes. (See Figure 7.1.) The sum of the other $n-i+1$ coordinates squared is $\|v_i\|^2 - \sum_{j=1}^{i-1} t_{ij}^2 = t_{ii}^2 = \|w_i\|^2$; w_i is the vector from v_i to its projection on $w_1, \ldots, w_{i-1}$ (or equivalently on $v_1, \ldots, v_{i-1}$).

LEMMA 7.2.1. *Conditional on* $w_1, \ldots, w_{i-1}$ (*or equivalently on* $v_1, \ldots, v_{i-1}$), $t_{i1}, \ldots, t_{i,i-1}$ *and* t_{ii}^2 *are independently distributed*; t_{ij} *is distributed according to* $N(0,1)$; *and* t_{ii}^2 *has the* χ^2-*distribution with* $n-i+1$ *degrees of freedom*.

PROOF OF LEMMA 7.2.1. The coordinates of v_i referred to the new orthogonal coordinates with $v_1, \ldots, v_{i-1}$ defining the first coordinate axes are independently normally distributed with means 0 and variances 1 (Theorem 3.3.1). t_{ii}^2 is the sum of the coordinates squared omitting the first $i-1$. Q.E.D.

Since the conditional distribution of $t_{i1}, \ldots, t_{ii}$ does not depend on $v_1, \ldots, v_{i-1}$, they are distributed independently of $t_{11}, t_{21}, t_{22}, \ldots, t_{i-1,i-1}$.

COROLLARY 7.2.1. *Let $\boldsymbol{Z}_1, \dots, \boldsymbol{Z}_n$ $(n \geq p)$ be independently distributed, each according to $N(\boldsymbol{0}, \boldsymbol{I})$, let $\boldsymbol{A} = \sum_{\alpha=1}^{n} \boldsymbol{Z}_\alpha \boldsymbol{Z}_\alpha' = \boldsymbol{T}\boldsymbol{T}'$, where $t_{ij} = 0$, $i < j$, and $t_{ii} > 0$. Then $t_{11}, t_{21}, \dots, t_{pp}$ are independently distributed; t_{ij} is distributed according to $N(0,1)$; and t_{ii}^2 has the χ^2-distribution with $n - i + 1$ degrees of freedom.*

Since t_{ii} has density $2^{-\frac{1}{2}(n-i-1)} t^{n-i} e^{-\frac{1}{2}t^2} / \Gamma[\frac{1}{2}(n+1-i)]$, the joint density of t_{ij}, $j = 1, \dots, i$; $i = 1, \dots, p$, is

$$(8)\quad \prod_{i=1}^{p} \frac{t_{ii}^{n-i} \exp\left(-\frac{1}{2}\sum_{j=1}^{i} t_{ij}^2\right)}{\pi^{\frac{1}{2}(i-1)} 2^{\frac{1}{2}n-1} \Gamma\left[\frac{1}{2}(n+1-i)\right]}$$

$$= \frac{\prod_{i=1}^{p} t_{ii}^{n-i} \exp\left(-\frac{1}{2}\sum_{i=1}^{p}\sum_{j=1}^{i} t_{ij}^2\right)}{2^{\frac{1}{2}p(n-2)} \pi^{p(p-1)/4} \prod_{i=1}^{p} \Gamma\left[\frac{1}{2}(n+1-i)\right]}.$$

Let $\boldsymbol{C}$ be a lower triangular matrix ($c_{ij} = 0$, $i < j$) such that $\boldsymbol{\Sigma} = \boldsymbol{C}\boldsymbol{C}'$ and $c_{ii} > 0$. The linear transformation $\boldsymbol{T}^* = \boldsymbol{C}\boldsymbol{T}$, that is,

$$(9)\quad \begin{aligned} t_{ij}^* &= \sum_{k=j}^{i} c_{ik} t_{kj}, && i \geq j, \\ &= 0, && i < j, \end{aligned}$$

can be written

$$(10)\quad \begin{bmatrix} t_{11}^* \\ t_{21}^* \\ t_{22}^* \\ \vdots \\ t_{p1}^* \\ \vdots \\ t_{pp}^* \end{bmatrix} = \begin{bmatrix} c_{11} & 0 & 0 & \cdots & 0 & \cdots & 0 \\ x & c_{22} & 0 & \cdots & 0 & \cdots & 0 \\ x & x & c_{22} & \cdots & 0 & \cdots & 0 \\ \vdots & \vdots & \vdots & & \vdots & & \vdots \\ x & x & x & \cdots & c_{pp} & \cdots & 0 \\ \vdots & \vdots & \vdots & & \vdots & & \vdots \\ x & x & x & \cdots & x & \cdots & c_{pp} \end{bmatrix} \begin{bmatrix} t_{11} \\ t_{21} \\ t_{22} \\ \vdots \\ t_{p1} \\ \vdots \\ t_{pp} \end{bmatrix}$$

where x denotes an element, possibly nonzero. Since the matrix of the transformation is triangular, its determinant is the product of the diagonal elements, namely, $\prod_{i=1}^{p} c_{ii}^{i}$. The Jacobian of the transformation from $\boldsymbol{T}$ to $\boldsymbol{T}^*$ is the reciprocal of the determinant. The density of $\boldsymbol{T}^*$ is obtained by substituting into (8) $t_{ii} = t_{ii}^*/c_{ii}$ and

$$\sum_{i=1}^{p}\sum_{j=1}^{i} t_{ij}^2 = \operatorname{tr} \boldsymbol{T}\boldsymbol{T}' \tag{11}$$

$$= \operatorname{tr} \boldsymbol{C}^{-1}\boldsymbol{T}^*\boldsymbol{T}^{*\prime}(\boldsymbol{C}^{-1})'$$

$$= \operatorname{tr} \boldsymbol{T}^*\boldsymbol{T}^{*\prime}\boldsymbol{C}'^{-1}\boldsymbol{C}^{-1}$$

$$= \operatorname{tr} \boldsymbol{T}^*\boldsymbol{T}^{*\prime}\boldsymbol{\Sigma}^{-1} = \operatorname{tr} \boldsymbol{T}^{*\prime}\boldsymbol{\Sigma}^{-1}\boldsymbol{T}^*,$$

and using $\prod_{i=1}^{p} c_{ii}^2 = |\boldsymbol{C}||\boldsymbol{C}'| = |\boldsymbol{\Sigma}|$.

THEOREM 7.2.1. *Let* $\boldsymbol{Z}_1, \dots, \boldsymbol{Z}_n$ $(n \geq p)$ *be independently distributed, each according to* $N(\boldsymbol{0}, \boldsymbol{\Sigma})$; *let* $\boldsymbol{A} = \sum_{\alpha=1}^{n} \boldsymbol{Z}_\alpha \boldsymbol{Z}_\alpha'$; *and let* $\boldsymbol{A} = \boldsymbol{T}^*\boldsymbol{T}^{*\prime}$, *where* $t_{ij}^* = 0$, $i < j$, *and* $t_{ii}^* > 0$. *Then the density of* $\boldsymbol{T}^*$ *is*

$$\frac{\prod_{i=1}^{p} t_{ii}^{*\,n-i} e^{-\frac{1}{2}\operatorname{tr} \boldsymbol{\Sigma}^{-1}\boldsymbol{T}^*\boldsymbol{T}^{*\prime}}}{2^{\frac{1}{2}p(n-2)} \pi^{p(p-1)/4} |\boldsymbol{\Sigma}|^{\frac{1}{2}n} \prod_{i=1}^{p} \Gamma\left[\frac{1}{2}(n+1-i)\right]}. \tag{12}$$

We can write (6) as $a_{hi} = \sum_{j=1}^{i} t_{hj}^* t_{ij}^*$ for $h \geq i$. Then

$$\frac{\partial a_{hi}}{\partial t_{kl}^*} = 0, \qquad k > h \tag{13}$$

$$= 0, \qquad k = h, l > i;$$

that is, $\partial a_{hi}/\partial t_{kl}^* = 0$ if k, l is beyond h, i in the lexicographic ordering. The Jacobian of the transformation from $\boldsymbol{A}$ to $\boldsymbol{T}^*$ is the determinant of the lower triangular matrix with diagonal elements

$$\frac{\partial a_{hh}}{\partial t_{hh}^*} = 2t_{hh}^*, \tag{14}$$

$$\frac{\partial a_{hi}}{\partial t_{hi}^*} = t_{ii}^*, \qquad h > i. \tag{15}$$

The Jacobian is, therefore, $2^p \prod_{i=1}^p t_{ii}^{*p+1-i}$. The Jacobian of the transformation from $\boldsymbol{T}^*$ to $\boldsymbol{A}$ is the reciprocal.

THEOREM 7.2.2. *Let $\boldsymbol{Z}_1, \dots, \boldsymbol{Z}_n$ be independently distributed, each according to $N(\boldsymbol{0}, \boldsymbol{\Sigma})$. The density of $\boldsymbol{A} = \sum_{\alpha=1}^n \boldsymbol{Z}_\alpha \boldsymbol{Z}_\alpha'$ is*

$$\frac{|\boldsymbol{A}|^{\frac{1}{2}(n-p-1)} e^{-\frac{1}{2}\operatorname{tr}\boldsymbol{\Sigma}^{-1}\boldsymbol{A}}}{2^{\frac{1}{2}pn} \pi^{p(p-1)/4} |\boldsymbol{\Sigma}|^{\frac{1}{2}n} \prod_{i=1}^p \Gamma\left[\frac{1}{2}(n+1-i)\right]} \tag{16}$$

for $\boldsymbol{A}$ positive definite and 0 otherwise.

COROLLARY 7.2.2. *Let $\boldsymbol{X}_1, \dots, \boldsymbol{X}_N$ $(N > p)$ be independently distributed, each according to $N(\boldsymbol{\mu}, \boldsymbol{\Sigma})$. Then the density of $\boldsymbol{A} = \sum_{\alpha=1}^N (\boldsymbol{X}_\alpha - \overline{\boldsymbol{X}})(\boldsymbol{X}_\alpha - \overline{\boldsymbol{X}})'$ is (16) for $n = N - 1$.*

The density (16) will be denoted by $w(\boldsymbol{A}|\boldsymbol{\Sigma}, n)$, and the associated distribution will be termed $W(\boldsymbol{\Sigma}, n)$. If $n < p$, $\boldsymbol{A}$ does not have a density, but its distribution is, nevertheless, defined and we shall refer to it as $W(\boldsymbol{\Sigma}, n)$.

COROLLARY 7.2.3. *Let $\boldsymbol{X}_1, \dots, \boldsymbol{X}_N$ $(N > p)$ be independently distributed, each according to $N(\boldsymbol{\mu}, \boldsymbol{\Sigma})$. The distribution of $\boldsymbol{S} = (1/n)\sum_{\alpha=1}^N (\boldsymbol{X}_\alpha - \overline{\boldsymbol{X}})(\boldsymbol{X}_\alpha - \overline{\boldsymbol{X}})'$ is $W[(1/n)\boldsymbol{\Sigma}, n]$, where $n = N - 1$.*

PROOF. $\boldsymbol{S}$ has the distribution of $\sum_{\alpha=1}^n [(1/\sqrt{n})\boldsymbol{Z}_\alpha][(1/\sqrt{n})\boldsymbol{Z}_\alpha]'$, where $(1/\sqrt{n})\boldsymbol{Z}_1, \dots, (1/\sqrt{n})\boldsymbol{Z}_N$ are independently distributed, each according to $N(\boldsymbol{0}, (1/n)\boldsymbol{\Sigma})$. Theorem 7.2.2 implies this corollary. Q.E.D.

The Wishart distribution for $p = 2$ as given in Section 4.2.1 was derived by Fisher (1915). The distribution for arbitrary p was obtained by Wishart (1928) by a geometric argument using $\boldsymbol{v}_1, \dots, \boldsymbol{v}_p$ defined above. As noted in Section 3.2, the ith diagonal element of $\boldsymbol{A}$ is the squared length of the ith vector, $a_{ii} = \boldsymbol{v}_i'\boldsymbol{v}_i = \|\boldsymbol{v}_i\|^2$, and the i, jth off-diagonal element of $\boldsymbol{A}$ is the product of the lengths of $\boldsymbol{v}_i$ and $\boldsymbol{v}_j$ and the cosine of the angle between them. The matrix $\boldsymbol{A}$ specifies the lengths and configuration of the vectors.

We shall give a geometric interpretation† of the derivation of the density of the rectangular coordinates t_{ij}, $i \geq j$, when $\boldsymbol{\Sigma} = \boldsymbol{I}$. The probability element of

†In the first edition of this book, the derivation of the Wishart distribution and its geometric interpretation were in terms of the nonorthogonal vectors $\boldsymbol{v}_1, \dots, \boldsymbol{v}_p$.

t_{11} is approximately the probability that $\|v_1\|$ lies in the interval $t_{11} < \|v_1\| < t_{11} + dt_{11}$. This is the probability that v_1 falls in a spherical shell in n dimensions with inner radius t_{11} and thickness dt_{11}. In this region, the density $(2\pi)^{-\frac{1}{2}n}\exp(-\frac{1}{2}v_1'v_1)$ is approximately constant, that is, $(2\pi)^{-\frac{1}{2}n}\exp(-\frac{1}{2}t_{11}^2)$. The surface area of the unit sphere in n dimensions is $C(n) = 2\pi^{\frac{1}{2}n}/\Gamma(\frac{1}{2}n)$ (Problems 1–3), and the volume of the spherical shell is approximately $C(n)t_{11}^{n-1}dt_{11}$. The probability element is the product of the volume and approximate density, namely,

$$(17) \qquad 2^{-(\frac{1}{2}n-1)}t_{11}^{n-1}\exp\left(-\tfrac{1}{2}t_{11}^2\right)dt_{11}/\Gamma\left(\tfrac{1}{2}n\right).$$

The probability element of $t_{i1},\dots,t_{i,i-1},t_{ii}$ given $v_1,\dots,v_{i-1}$ (i.e., given $w_1,\dots,w_{i-1}$) is approximately the probability that v_i falls in the region for which $t_{i1} < v_i'w_1/\|w_1\| < t_{i1} + dt_{i1},\dots,t_{i,i-1} < v_i'w_{i-1}/\|w_{i-1}\| < t_{i,i-1} + dt_{i,i-1}$, and $t_{ii} < \|w_i\| < t_{ii} + dt_{ii}$, where w_i is the projection of v_i on the $(n-i+1)$-dimensional space orthogonal to $w_1,\dots,w_{i-1}$. Each of the first $i-1$ pairs of inequalities defines the region between two hyperplanes (the different pairs being orthogonal). The last pair of inequalities defines a cylindrical shell whose intersection with the $(i-1)$-dimensional hyperplane spanned by $v_1,\dots,v_{i-1}$ is a spherical shell in $n-i+1$ dimensions with inner radius t_{ii}. In this region the density $(2\pi)^{-\frac{1}{2}n}\exp(-\frac{1}{2}v_i'v_i)$ is approximately constant, that is, $(2\pi)^{-\frac{1}{2}n}\exp(-\frac{1}{2}\Sigma_{j=1}^{i}t_{ij}^2)$. The volume of the region is approximately $dt_{i1}\cdots dt_{i,i-1}\,C(n-i+1)t_{ii}^{n-i}dt_{ii}$. The probability element is

$$(18) \quad 2^{-(\frac{1}{2}n-1)}\pi^{-\frac{1}{2}(i-1)}t_{ii}^{n-i}\exp\left(-\frac{1}{2}\sum_{j=1}^{i}t_{ij}^2\right)\Bigg/\Gamma\left[\tfrac{1}{2}(n+1-i)\right]dt_{i1}\cdots dt_{ii}.$$

Then the product of (17) and (18) for $i = 2,\dots,p$ is (8) times $dt_{11}\cdots dt_{pp}$.

The analysis that exactly parallels the geometric derivation by Wishart [and later by Mahalanobis, Bose, and Roy (1937)] has been given by Sverdrup (1947) [and by Fog (1948) for $p = 3$]. Another method has been used by Madow (1938), who draws on the distribution of correlation coefficients (for $\boldsymbol{\Sigma} = \boldsymbol{I}$) obtained by Hotelling by considering certain partial correlation coefficients. Hsu (1939b) has given an inductive proof, and Rasch (1948) has given a method involving the use of a functional equation. A different method is to obtain the characteristic function and invert it, as was done by Ingham (1933) and by Wishart and Bartlett (1933).

Cramér (1946) verifies that the Wishart distribution has the characteristic function of $\boldsymbol{A}$. By means of alternative matrix transformations Elfving (1947), Mauldon (1955), and Olkin and Roy (1954) derived the Wishart distribution via the "Bartlett decomposition"; Kshirsagar (1959) based his derivation on random orthogonal tranformations. Narain (1948a), (1948b) and Ogawa (1953) used a regression approach. James (1954), Khatri and Ramachandran (1958), and Khatri (1963) applied different methods. Giri (1977) used invariance. Wishart (1948) surveyed the derivations up to that date. Some of these methods are indicated in the problems.

The relation $\boldsymbol{A} = \boldsymbol{TT}'$ is known as the *Bartlett decomposition* [Bartlett (1939)] and the (nonzero) elements of $\boldsymbol{T}$ were termed *rectangular coordinates* by Mahalanobis, Bose, and Roy (1937).

COROLLARY 7.2.4.

$$\int_{\boldsymbol{B}>\boldsymbol{0}}\cdots\int |\boldsymbol{B}|^{t-\frac{1}{2}(p+1)} e^{-\operatorname{tr}\boldsymbol{B}}\, d\boldsymbol{B} = \pi^{p(p-1)/4}\prod_{i=1}^{p}\Gamma\left[t-\tfrac{1}{2}(i-1)\right]. \tag{19}$$

PROOF. Here $\boldsymbol{B} > \boldsymbol{0}$ denotes $\boldsymbol{B}$ positive definite. Since (16) is a density, its integral for $\boldsymbol{A} > \boldsymbol{0}$ is 1. Let $\boldsymbol{\Sigma} = \boldsymbol{I}$, $\boldsymbol{A} = 2\boldsymbol{B}$ ($d\boldsymbol{A} = 2d\boldsymbol{B}$), and $n = 2t$. Then the fact that the integral is 1 is identical to (19) for t being a half integer. However, if we derive (16) from (8), we can let n be any real number greater than $p-1$. In fact (19) holds for complex t such that $\mathscr{R}t > p-1$. ($\mathscr{R}t$ means the real part of t.) Q.E.D.

DEFINITION 7.2.1. *The multivariate gamma function is*

$$\Gamma_p(t) = \pi^{p(p-1)/4}\prod_{i=1}^{p}\Gamma\left[t-\tfrac{1}{2}(i-1)\right]. \tag{20}$$

The Wishart density can be written

$$w(\boldsymbol{A}|\boldsymbol{\Sigma},n) = \frac{|\boldsymbol{A}|^{\frac{1}{2}(n-p-1)}e^{-\frac{1}{2}\operatorname{tr}\boldsymbol{\Sigma}^{-1}\boldsymbol{A}}}{2^{\frac{1}{2}pn}|\boldsymbol{\Sigma}|^{\frac{1}{2}n}\Gamma_p\left(\frac{1}{2}n\right)}. \tag{21}$$

7.3. SOME PROPERTIES OF THE WISHART DISTRIBUTION

7.3.1. The Characteristic Function

The characteristic function of the Wishart distribution can be obtained directly from the distribution of the observations. Suppose $Z_1, \ldots, Z_n$ are distributed independently, each with density

$$\text{(1)} \qquad \frac{1}{(2\pi)^{\frac{1}{2}p}|\Sigma|^{\frac{1}{2}}} \exp\left(-\tfrac{1}{2}z'\Sigma^{-1}z\right).$$

Let

$$\text{(2)} \qquad A = \sum_{\alpha=1}^{n} Z_\alpha Z_\alpha'.$$

Introduce the $p \times p$ matrix $\Theta = (\theta_{ij})$ with $\theta_{ij} = \theta_{ji}$. The characteristic function of $A_{11}, A_{22}, \ldots, A_{pp}, 2A_{12}, 2A_{13}, \ldots, 2A_{p-1,p}$ is

$$\text{(3)} \qquad \begin{aligned} \mathscr{E}\exp[i\operatorname{tr}(A\Theta)] &= \mathscr{E}\exp\left(i\operatorname{tr}\sum_{\alpha=1}^{n} Z_\alpha Z_\alpha'\Theta\right) \\ &= \mathscr{E}\exp\left(i\operatorname{tr}\sum_{\alpha=1}^{n} Z_\alpha'\Theta Z_\alpha\right) \\ &= \mathscr{E}\exp\left(i\sum_{\alpha=1}^{n} Z_\alpha'\Theta Z_\alpha\right). \end{aligned}$$

It follows from Lemma 2.6.1 that

$$\text{(4)} \qquad \begin{aligned} \mathscr{E}\exp\left(i\sum_{\alpha=1}^{n} Z_\alpha'\Theta Z_\alpha\right) &= \prod_{\alpha=1}^{n} \mathscr{E}\exp\left(iZ_\alpha'\Theta Z_\alpha\right) \\ &= \left[\mathscr{E}\exp(iZ'\Theta Z)\right]^n, \end{aligned}$$

where Z has the density (1). For Θ real, there is a real nonsingular matrix B such that

$$\text{(5)} \qquad B'\Sigma^{-1}B = I,$$

$$\text{(6)} \qquad B'\Theta B = D,$$

where $\boldsymbol{D}$ is a real diagonal matrix (Theorem A.2.2 of Appendix A). If we let

$$z = \boldsymbol{B}y, \tag{7}$$

then

$$\begin{aligned}\mathscr{E}\exp(i\boldsymbol{Z}'\boldsymbol{\Theta}\boldsymbol{Z}) &= \mathscr{E}\exp(i\boldsymbol{Y}'\boldsymbol{D}\boldsymbol{Y}) \\ &= \mathscr{E}\prod_{j=1}^{p}\exp\left(id_{jj}Y_j^2\right) \\ &= \prod_{j=1}^{p}\mathscr{E}\exp\left(id_{jj}Y_j^2\right)\end{aligned} \tag{8}$$

by Lemma 2.6.2. The jth term in the product is $\mathscr{E}\exp(id_{jj}Y_j^2)$, where Y_j has the distribution $N(0,1)$; this is the characteristic function of the χ^2-distribution with one degree of freedom, namely $(1-2id_{jj})^{-\frac{1}{2}}$ (and can be proved easily by expanding $\exp(id_{jj}y_j^2)$ in a power series and integrating term by term). Thus

$$\mathscr{E}\exp(i\boldsymbol{Z}'\boldsymbol{\Theta}\boldsymbol{Z}) = \prod_{j=1}^{p}\left(1-2id_{jj}\right)^{-\frac{1}{2}} = |\boldsymbol{I}-2i\boldsymbol{D}|^{-\frac{1}{2}} \tag{9}$$

since $\boldsymbol{I}-2i\boldsymbol{D}$ is a diagonal matrix. From (5) and (6) we see that

$$\begin{aligned}|\boldsymbol{I}-2i\boldsymbol{D}| &= |\boldsymbol{B}'\boldsymbol{\Sigma}^{-1}\boldsymbol{B}-2i\boldsymbol{B}'\boldsymbol{\Theta}\boldsymbol{B}| \\ &= |\boldsymbol{B}'(\boldsymbol{\Sigma}^{-1}-2i\boldsymbol{\Theta})\boldsymbol{B}| \\ &= |\boldsymbol{B}'|\cdot|\boldsymbol{\Sigma}^{-1}-2i\boldsymbol{\Theta}|\cdot|\boldsymbol{B}| \\ &= |\boldsymbol{B}|^2\cdot|\boldsymbol{\Sigma}^{-1}-2i\boldsymbol{\Theta}|,\end{aligned} \tag{10}$$

$|\boldsymbol{B}'|\cdot|\boldsymbol{\Sigma}^{-1}|\cdot|\boldsymbol{B}| = |\boldsymbol{I}| = 1$, and $|\boldsymbol{B}|^2 = 1/|\boldsymbol{\Sigma}^{-1}|$. Combining the above results, we obtain

$$\mathscr{E}\exp[i\,\mathrm{tr}(\boldsymbol{A}\boldsymbol{\Theta})] = \frac{|\boldsymbol{\Sigma}^{-1}|^{\frac{1}{2}n}}{|\boldsymbol{\Sigma}^{-1}-2i\boldsymbol{\Theta}|^{\frac{1}{2}n}} = |\boldsymbol{I}-2i\boldsymbol{\Theta}\boldsymbol{\Sigma}|^{-\frac{1}{2}n}. \tag{11}$$

It can be shown that the result is valid provided $(\mathscr{R}(\sigma^{jk}-2i\theta_{jk}))$ is positive definite. In particular, it is true for all real $\boldsymbol{\Theta}$. It also holds for $\boldsymbol{\Sigma}$ singular.

THEOREM 7.3.1. *If* $Z_1, \dots, Z_n$ *are independent, each with distribution* $N(\mathbf{0}, \Sigma)$, *the characteristic function of* $A_{11}, \dots, A_{pp}, 2A_{12}, \dots, 2A_{p-1,p}$, *where* $(A_{ij}) = A = \sum_{\alpha=1}^{n} Z_\alpha Z'_\alpha$, *is given by* (11).

7.3.2. The Sum of Wishart Matrices

Suppose the A_i, $i = 1, 2$, are distributed independently according to $W(\Sigma, n_i)$, respectively. Then A_1 is distributed as $\sum_{\alpha=1}^{n_1} Z_\alpha Z'_\alpha$, and A_2 is distributed as $\sum_{\alpha=n_1+1}^{n_1+n_2} Z_\alpha Z'_\alpha$, where $Z_1, \dots, Z_{n_1+n_2}$ are independent, each with distribution $N(\mathbf{0}, \Sigma)$. Then

$$A = A_1 + A_2 \tag{12}$$

is distributed as $\sum_{\alpha=1}^{n} Z_\alpha Z'_\alpha$, where $n = n_1 + n_2$. Thus A is distributed according to $W(\Sigma, n)$. Similarly, the sum of q matrices distributed independently, each according to a Wishart distribution with covariance Σ, has a Wishart distribution with covariance matrix Σ and number of degrees of freedom equal to the sum of the numbers of degrees of freedom of the component matrices.

THEOREM 7.3.2. *If* $A_1, \dots, A_q$ *are independently distributed with* A_i *distributed according to* $W(\Sigma, n_i)$, *then*

$$A = \sum_{i=1}^{q} A_i \tag{13}$$

is distributed according to $W(\Sigma, \sum_{i=1}^{q} n_i)$.

7.3.3. A Certain Linear Transformation

We shall frequently make the transformation

$$A = CBC', \tag{14}$$

where C is a nonsingular $p \times p$ matrix. If A is distributed according to $W(\Sigma, n)$, then B is distributed according to $W(\Phi, n)$ where

$$\Phi = C^{-1}\Sigma C'^{-1}. \tag{15}$$

This is proved by the following argument:

$$A = \sum_{\alpha=1}^{n} Z_\alpha Z'_\alpha, \tag{16}$$

where $Z_1, \dots, Z_n$ are independently distributed, each according to $N(\mathbf{0}, \Sigma)$. Then

$$Y_\alpha = C^{-1} Z_\alpha \tag{17}$$

is distributed according to $N(\mathbf{0}, \Phi)$. However,

$$B = \sum_{\alpha=1}^{n} Y_\alpha Y_\alpha' = C^{-1} \sum_{\alpha=1}^{n} Z_\alpha Z_\alpha' C'^{-1} = C^{-1} A C'^{-1} \tag{18}$$

is distributed according to $W(\Phi, n)$. $\left|\dfrac{\partial(A)}{\partial(B)}\right|$, the Jacobian of the transformation (14), is

$$\left|\frac{\partial(A)}{\partial(B)}\right| = \frac{w(B, \Phi, n)}{w(A, \Sigma, n)} = \frac{|B|^{\frac{1}{2}(n-p-1)} |\Sigma|^{\frac{1}{2}n}}{|A|^{\frac{1}{2}(n-p-1)} |\Phi|^{\frac{1}{2}n}} = \operatorname{mod}|C|^{p+1}. \tag{19}$$

THEOREM 7.3.3. *The Jacobian of the transformation* (14) *from A to B, where A and B are symmetric, is* $\operatorname{mod}|C|^{p+1}$.

7.3.4. Marginal Distributions

If A is distributed according to $W(\Sigma, n)$, the marginal distribution of any arbitrary set of the elements of A may be awkward to obtain. However, the marginal distribution of some sets of elements can be found easily. We give some of these in the following two theorems.

THEOREM 7.3.4. *Let A and Σ be partitioned into q and $p - q$ rows and columns*

$$A = \begin{pmatrix} A_{11} & A_{12} \\ A_{21} & A_{22} \end{pmatrix}, \qquad \Sigma = \begin{pmatrix} \Sigma_{11} & \Sigma_{12} \\ \Sigma_{21} & \Sigma_{22} \end{pmatrix}. \tag{20}$$

If A is distributed according to $W(\Sigma, n)$, then A_{11} is distributed according to $W(\Sigma_{11}, n)$.

PROOF. A is distributed as $\sum_{\alpha=1}^{n} Z_\alpha Z_\alpha'$, where the Z_α are independent, each with the distribution $N(\mathbf{0}, \Sigma)$. Partition Z_α into subvectors of q and $p - q$ components,

$$Z_\alpha = \begin{pmatrix} Z_\alpha^{(1)} \\ Z_\alpha^{(2)} \end{pmatrix}. \tag{21}$$

Then $Z_1^{(1)}, \dots, Z_n^{(1)}$ are independent, each with the distribution $N(\mathbf{0}, \Sigma_{11})$, and A_{11} is distributed as $\sum_{\alpha=1}^n Z_\alpha^{(1)} Z_\alpha^{(1)\prime}$, which has the distribution $W(\Sigma_{11}, n)$. Q.E.D.

THEOREM 7.3.5. *Let A and Σ be partitioned into $p_1, p_2, \dots, p_q$ rows and columns* ($p_1 + \cdots + p_q = p$)

$$A = \begin{pmatrix} A_{11} & \cdots & A_{1q} \\ \vdots & & \vdots \\ A_{q1} & \cdots & A_{qq} \end{pmatrix}, \qquad \Sigma = \begin{pmatrix} \Sigma_{11} & \cdots & \Sigma_{1q} \\ \vdots & & \vdots \\ \Sigma_{q1} & \cdots & \Sigma_{qq} \end{pmatrix}. \tag{22}$$

If $\Sigma_{ij} = \mathbf{0}$ for $i \neq j$ and if A is distributed according to $W(\Sigma, n)$, then $A_{11}, A_{22}, \dots, A_{qq}$ are independently distributed and A_{jj} is distributed according to $W(\Sigma_{jj}, n)$.

PROOF. A is distributed as $\sum_{\alpha=1}^n Z_\alpha Z_\alpha'$, where $Z_1, \dots, Z_n$ are independently distributed, each according to $N(\mathbf{0}, \Sigma)$. Let Z_α be partitioned

$$Z_\alpha = \begin{pmatrix} Z_\alpha^{(1)} \\ \vdots \\ Z_\alpha^{(q)} \end{pmatrix} \tag{23}$$

as A and Σ have been partitioned. Since $\Sigma_{ij} = \mathbf{0}$, the sets $Z_1^{(1)}, \dots, Z_n^{(1)}, \dots, Z_1^{(q)}, \dots, Z_n^{(q)}$ are independent. Then $A_{11} = \sum_{\alpha=1}^n Z_\alpha^{(1)} Z_\alpha^{(1)\prime}, \dots, A_{qq} = \sum_{\alpha=1}^n Z_\alpha^{(q)} Z_\alpha^{(q)\prime}$ are independent. The rest of Theorem 7.3.5 follows from Theorem 7.3.4.

7.3.5. Conditional Distributions

In Section 4.3 we considered estimation of the parameters of the conditional distribution of $X^{(1)}$ given $X^{(2)} = x^{(2)}$. Application of Theorem 7.2.2 to Theorem 4.3.3 yields the following theorem:

THEOREM 7.3.6. *Let A and Σ be partitioned into q and $p - q$ rows and columns as in* (20). *If A is distributed according to $W(\Sigma, n)$, the distribution of $A_{11 \cdot 2} = A_{11} - A_{12} A_{22}^{-1} A_{21}$ is $W(\Sigma_{11 \cdot 2}, n - p + q)$.*

Note that Theorem 7.3.6 implies that $A_{11\cdot 2}$ is independent of A_{22} and $A_{12}A_{22}^{-1}$ regardless of Σ.

7.4. COCHRAN'S THEOREM

Cochran's theorem [Cochran (1934)] is useful in proving that certain "vector quadratic forms" are distributed as sums of "vector squares." It is a statistical statement of an algebraic theorem, which we shall give as a lemma.

LEMMA 7.4.1. *If the $N \times N$ symmetric matrix C_i has rank r_i, $i = 1, \ldots, m$, and*

$$\sum_{i=1}^{m} C_i = I_N, \tag{1}$$

then

$$\sum_{i=1}^{m} r_i = N \tag{2}$$

is a necessary and sufficient condition such that there exists an $N \times N$ orthogonal matrix P such that for $i = 1, \ldots, m$

$$PC_iP' = \begin{pmatrix} 0 & 0 & 0 \\ 0 & I & 0 \\ 0 & 0 & 0 \end{pmatrix}, \tag{3}$$

where I is of order r_i, the upper left-hand 0 is square of order $\sum_{j=1}^{i-1} r_j$ (which is 0 for $i = 1$), and the lower right-hand 0 is square of order $\sum_{j=i+1}^{m} r_j$ (which is 0 for $i = m$).

PROOF. The necessity follows from the fact that (1) implies the sum of (3) over $i = 1, \ldots, m$ is I_N. Now let us prove the sufficiency; we assume (2). There exists an orthogonal matrix P_j such that $P_jC_jP_j'$ is diagonal with diagonal elements the characteristic roots of C_j. The number of nonzero roots is r_j, the rank of C_j, and the number of 0 roots is $N - r_j$. We write

$$P_jC_jP_j' = \begin{pmatrix} 0 & 0 & 0 \\ 0 & \Delta_j & 0 \\ 0 & 0 & 0 \end{pmatrix}, \tag{4}$$

where the partitioning is according to (3) and $\boldsymbol{\Delta}_j$ is diagonal of order r_j. This is possible in view of (2). Then

$$(5) \qquad \boldsymbol{P}_j \sum_{\substack{i=1 \\ i \neq j}}^{m} \boldsymbol{C}_i \boldsymbol{P}_j' = \boldsymbol{P}_j (\boldsymbol{I} - \boldsymbol{C}_j) \boldsymbol{P}_j' = \begin{pmatrix} \boldsymbol{I} & \boldsymbol{0} & \boldsymbol{0} \\ \boldsymbol{0} & \boldsymbol{I} - \boldsymbol{\Delta}_j & \boldsymbol{0} \\ \boldsymbol{0} & \boldsymbol{0} & \boldsymbol{I} \end{pmatrix}.$$

Since the rank of (5) is not greater than $\sum_{i=1}^m r_i - r_j = N - r_j$, which is the sum of the orders of the upper left-hand and lower right-hand $\boldsymbol{I}$'s in (5), the rank of $\boldsymbol{I} - \boldsymbol{\Delta}_j$ is 0 and $\boldsymbol{\Delta}_j = \boldsymbol{I}$. (Thus the r_j nonzero roots of $\boldsymbol{C}_j$ are 1, and $\boldsymbol{C}_j$ is positive semidefinite.) From (4) we obtain

$$(6) \qquad \begin{aligned} \boldsymbol{C}_j &= \boldsymbol{P}_j' \begin{pmatrix} \boldsymbol{0} & \boldsymbol{0} & \boldsymbol{0} \\ \boldsymbol{0} & \boldsymbol{I} & \boldsymbol{0} \\ \boldsymbol{0} & \boldsymbol{0} & \boldsymbol{0} \end{pmatrix} \boldsymbol{P}_j \\ &= \boldsymbol{B}_j \boldsymbol{B}_j', \end{aligned}$$

where $\boldsymbol{B}_j$ consists of the r_j columns of $\boldsymbol{P}_j'$ corresponding to $\boldsymbol{I}$ in (6). From (1) we obtain

$$(7) \qquad \boldsymbol{I} = \sum_{j=1}^{m} \boldsymbol{B}_j \boldsymbol{B}_j' = (\boldsymbol{B}_1, \boldsymbol{B}_2, \dots, \boldsymbol{B}_m) \begin{pmatrix} \boldsymbol{B}_1' \\ \boldsymbol{B}_2' \\ \vdots \\ \boldsymbol{B}_m' \end{pmatrix} = \boldsymbol{P}'\boldsymbol{P},$$

where $\boldsymbol{P} = (\boldsymbol{B}_1, \boldsymbol{B}_2, \dots, \boldsymbol{B}_m)'$. Q.E.D.

We now state a multivariate analogue to Cochran's theorem.

THEOREM 7.4.1. *Suppose* $\boldsymbol{Y}_1, \dots, \boldsymbol{Y}_N$ *are independently distributed, each according to* $N(\boldsymbol{0}, \boldsymbol{\Sigma})$. *Suppose the matrix* $(c_{\alpha\beta}^i) = \boldsymbol{C}_i$ *used in forming*

$$(8) \qquad \boldsymbol{Q}_i = \sum_{\alpha, \beta=1}^{N} c_{\alpha\beta}^i \boldsymbol{Y}_\alpha \boldsymbol{Y}_\beta', \qquad i = 1, \dots, m,$$

is of rank r_i, *and suppose*

$$(9) \qquad \sum_{i=1}^{m} \boldsymbol{Q}_i = \sum_{\alpha=1}^{N} \boldsymbol{Y}_\alpha \boldsymbol{Y}_\alpha'.$$

Then (2) *is a necessary and sufficient condition that* $Q_1, \ldots, Q_m$ *are independently distributed with* Q_i *having the distribution* $W(\Sigma, r_i)$.

This theorem is useful in generalizing results from the univariate analysis of variance. (See Chapter 8.) As an example of the use of this theorem, let us prove that the mean of a sample of size N times its transpose and a multiple of the sample covariance matrix are independently distributed with a singular and a nonsingular Wishart distribution, respectively. Let $Y_1, \ldots, Y_N$ be independently distributed, each according to $N(\mathbf{0}, \Sigma)$. We shall use the matrices $C_1 = (c_{\alpha\beta}^{(1)}) = (1/N)$ and $C_2 = (c_{\alpha\beta}^{(2)}) = [\delta_{\alpha\beta} - (1/N)]$. Then

$$Q_1 = \sum_{\alpha,\beta=1}^{N} \frac{1}{N} Y_\alpha Y_\beta' = N\bar{Y}\bar{Y}', \tag{10}$$

$$\begin{aligned} Q_2 &= \sum_{\alpha,\beta=1}^{N} \left(\delta_{\alpha\beta} - \frac{1}{N}\right) Y_\alpha Y_\beta' \\ &= \sum_{\alpha=1}^{N} Y_\alpha Y_\alpha' - N\bar{Y}\bar{Y}' \\ &= \sum_{\alpha=1}^{N} (Y_\alpha - \bar{Y})(Y_\alpha - \bar{Y})', \end{aligned} \tag{11}$$

and (9) is satisfied. The matrix C_1 is of rank 1; the matrix C_2 is of rank $N - 1$ (since the rank of the sum of two matrices is less than or equal to the sum of the ranks of the matrices and the rank of the second matrix is less than N). The conditions of the theorem are satisfied; therefore Q_1 is distributed as ZZ', where Z is distributed according to $N(\mathbf{0}, \Sigma)$, and Q_2 is distributed independently according to $W(\Sigma, N - 1)$.

Anderson and Styan (1982) have given a survey of proofs and extensions of Cochran's theorem.

7.5. THE GENERALIZED VARIANCE

7.5.1. Definition of the Generalized Variance

One multivariate analogue of the variance σ^2 of a univariate distribution is the covariance matrix Σ. Another multivariate analogue is the scalar $|\Sigma|$, which is called the *generalized variance* of the multivariate distribution [Wilks (1932);

see also Frisch (1929)]. Similarly, the generalized variance of the sample of vectors $x_1, \dots, x_N$ is

$$|S| = \left| \frac{1}{N-1} \sum_{\alpha=1}^{N} (x_\alpha - \bar{x})(x_\alpha - \bar{x})' \right|. \tag{1}$$

In some sense each of these is a measure of spread. We consider them here because the sample generalized variance will recur in many likelihood ratio criteria for testing hypotheses.

A geometric interpretation of the sample generalized variance comes from considering the p rows of $X = (x_1, \dots, x_N)$ as p vectors in N-dimensional space. In Section 3.2 it was shown that the rows of

$$(x_1 - \bar{x}, \dots, x_N - \bar{x}) = X - \bar{x}\varepsilon', \tag{2}$$

where $\varepsilon = (1, \dots, 1)'$, are orthogonal to the equiangular line (through the origin and ε); see Figure 3.2. Then the entries of

$$A = (X - \bar{x}\varepsilon')(X - \bar{x}\varepsilon')' \tag{3}$$

are the inner products of rows of $X - \bar{x}\varepsilon'$.

We now define a *parallelotope* determined by p vectors $v_1, \dots, v_p$ in an n-dimensional space ($n \geq p$). If $p = 1$, the "parallelotope" is the line segment v_1. If $p = 2$, the "parallelotope" is the parallelotope with v_1 and v_2 as principal edges; that is, its sides are v_1, v_2, v_1 translated so its initial endpoint is at v_2, and v_2 translated so its initial endpoint is at v_1. See Figure 7.2. If $p = 3$, the "parallelotope" is the conventional parallelotope with v_1, v_2, and v_3 as principal edges. In general, the "parallelotope" is the figure defined by the principal edges $v_1, \dots, v_p$. It is cut out by p pairs of parallel $(p-1)$-dimensional hyperplanes, one hyperplane of a pair being spanned by $p-1$ of $v_1, \dots, v_p$ and the other hyperplane going through the endpoint of the remaining vector.

THEOREM 7.5.1. *If $V = (v_1, \dots, v_p)$, then the square of the p-dimensional volume of the parallelotope with $v_1, \dots, v_p$ as principal edges is $|V'V|$.*

PROOF. If $p = 1$, $|V'V| = v_1'v_1 = \|v_1\|^2$, which is the square of the one-dimensional volume of v_1. If two k-dimensional parallelotopes have bases

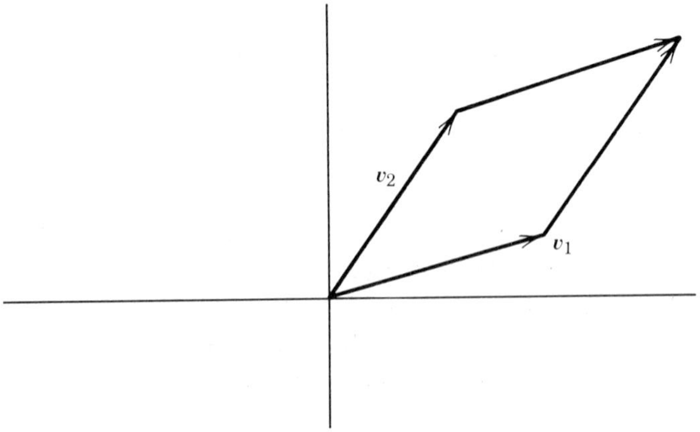

Figure 7.2. A parallelogram.

consisting of $(k-1)$-dimensional parallelotopes of equal $(k-1)$-dimensional volumes and equal altitudes, their k-dimensional volumes are equal [since the k-dimensional volume is the integral of the $(k-1)$-dimensional volumes]. In particular, the volume of a k-dimensional parallelotope is equal to the volume of a parallelotope with the same base (in $k-1$ dimensions) and same altitude with sides in the kth direction orthogonal to the first $k-1$ directions. Thus the volume of the parallelotope with principal edges $\boldsymbol{v}_1, \dots, \boldsymbol{v}_k$, say P_k, is equal to the volume of the parallelotope with principal edges $\boldsymbol{v}_1, \dots, \boldsymbol{v}_{k-1}$, say P_{k-1}, times the altitude of P_k over P_{k-1}; that is,

$$\text{(4)} \qquad \operatorname{Vol}(P_k) = \operatorname{Vol}(P_{k-1}) \times \operatorname{Alt}(P_k | P_{k-1}).$$

It follows (by induction) that

$$\text{(5)} \qquad \operatorname{Vol}(P_p) = \operatorname{Vol}(P_1) \times \operatorname{Alt}(P_2|P_1) \times \cdots \times \operatorname{Alt}(P_p|P_{p-1}).$$

By the construction in Section 7.2 the altitude of P_k over P_{k-1} is $t_{kk} = \|\boldsymbol{w}_k\|$; that is, t_{kk} is the distance of $\boldsymbol{v}_k$ from the $(k-1)$-dimensional space spanned by $\boldsymbol{v}_1, \dots, \boldsymbol{v}_{k-1}$ (or $\boldsymbol{w}_1, \dots, \boldsymbol{w}_{k-1}$). Hence $\operatorname{Vol}(P_p) = \prod_{k=1}^{p} t_{kk}$. Since $|\boldsymbol{V}'\boldsymbol{V}| = |\boldsymbol{T}\boldsymbol{T}'| = \prod_{i=1}^{p} t_{ii}^2$, the theorem is proved. Q.E.D.

We now apply this theorem to the parallelotope having the rows of (2) as principal edges. The dimensionality in Theorem 7.5.1 is arbitrary (but at least p).

COROLLARY 7.5.1. *The square of the p-dimensional volume of the parallelotope with the rows of* (2) *as principal edges is* $|\boldsymbol{A}|$, *where* $\boldsymbol{A}$ *is given by* (3).

We shall see later that many multivariate statistics can be given an interpretation in terms of these volumes. These volumes are analogous to distances that arise in special cases when $p = 1$.

We now consider a geometric interpretation of $|\boldsymbol{A}|$ in terms of N points in p-space. Let the columns of the matrix (2) be $\boldsymbol{y}_1, \dots, \boldsymbol{y}_N$, representing N points in p-space. When $p = 1$, $|\boldsymbol{A}| = \sum_\alpha y_{1\alpha}^2$, which is the sum of squares of the distances from the points to the origin. In general $|\boldsymbol{A}|$ is the sum of squares of the volumes of all parallelotopes formed by taking as principal edges p vectors from the set $\boldsymbol{y}_1, \dots, \boldsymbol{y}_N$.

We see that

$$
(6) \qquad |\boldsymbol{A}| = \begin{vmatrix} \sum_\alpha y_{1\alpha}^2 & \cdots & \sum_\alpha y_{1\alpha} y_{p-1,\alpha} & \sum_\beta y_{1\beta} y_{p\beta} \\ \vdots & & \vdots & \vdots \\ \sum_\alpha y_{p-1,\alpha} y_{1\alpha} & \cdots & \sum_\alpha y_{p-1,\alpha}^2 & \sum_\beta y_{p-1,\beta} y_{p\beta} \\ \sum_\alpha y_{p\alpha} y_{1\alpha} & \cdots & \sum_\alpha y_{p\alpha} y_{p-1,\alpha} & \sum_\beta y_{p\beta}^2 \end{vmatrix}
$$

$$
= \sum_\beta \begin{vmatrix} \sum_\alpha y_{1\alpha}^2 & \cdots & \sum_\alpha y_{1\alpha} y_{p-1,\alpha} & y_{1\beta} y_{p\beta} \\ \vdots & & \vdots & \vdots \\ \sum_\alpha y_{p-1,\alpha} y_{1\alpha} & \cdots & \sum_\alpha y_{p-1,\alpha}^2 & y_{p-1,\beta} y_{p\beta} \\ \sum_\alpha y_{p\alpha} y_{1\alpha} & \cdots & \sum_\alpha y_{p\alpha} y_{p-1,\alpha} & y_{p\beta}^2 \end{vmatrix}
$$

by the rule for expanding determinants. [See (24) of Section A.1 of Appendix A.] In (6) the matrix $\boldsymbol{A}$ has been partitioned into $p - 1$ and 1 columns. Applying the rule successively to the columns, we find

$$
(7) \qquad |\boldsymbol{A}| = \sum_{\alpha_1, \dots, \alpha_p = 1}^{N} |y_{i\alpha_j} y_{j\alpha_j}|.
$$

By Theorem 7.5.1 the square of the volume of the parallelotope with $y_{\gamma_1}, \dots, y_{\gamma_p}$, $\gamma_1 < \cdots < \gamma_p$, as principal edges is

$$(8) \qquad V^2_{\gamma_1, \dots, \gamma_p} = \left| \sum_\beta y_{i\beta} y_{j\beta} \right|,$$

where the sum on β is over $(\gamma_1, \dots, \gamma_p)$. If we now expand this determinant in the manner used for $|A|$, we obtain

$$(9) \qquad V^2_{\gamma_1, \dots, \gamma_p} = \sum |y_{i\beta_j} y_{j\beta_j}|,$$

where the sum is for each β_j over the range $(\gamma_1, \dots, \gamma_p)$. Summing (9) over all different sets $(\gamma_1 < \cdots < \gamma_p)$ we obtain (7). ($|y_{i\beta_j} y_{j\beta_j}| = 0$ if two or more β_j are equal.) Thus $|A|$ is the sum of volumes squared of all different parallelotopes formed by sets of p of the vectors y_α as principal edges. If we replace y_α by $x_\alpha - \bar{x}$, we can state the following theorem:

THEOREM 7.5.2. *Let $|S|$ be defined by* (1), *where $x_1, \dots, x_N$ are the N vectors of a sample. Then $|S|$ is proportional to the sum of squares of the volumes of all the different parallelotopes formed by using as principal edges p vectors with p of $x_1, \dots, x_N$ as one set of endpoints and $\bar{x}$ as the other, and the factor of proportionality is* $1/(N-1)^p$.

The population analogue of $|S|$ is $|\Sigma|$, which can also be given a geometric interpretation. From Section 3.3 we know that

$$(10) \qquad \Pr\left\{ X'\Sigma^{-1}X \le \chi_p^2(\alpha) \right\} = 1 - \alpha$$

if X is distributed according to $N(\mathbf{0}, \Sigma)$; that is, the probability is $1 - \alpha$ that X fall inside the ellipsoid

$$(11) \qquad x'\Sigma^{-1}x = \chi_p^2(\alpha).$$

The volume of this ellipsoid is $C(p)|\Sigma|^{\frac{1}{2}}[\chi_p^2(\alpha)]^{\frac{1}{2}p}/p$, where $C(p)$ is defined in Problem 3.

7.5.2. Distribution of the Sample Generalized Variance

The distribution of $|S|$ is the same as the distribution of $|A|/(N-1)^p$, where $A = \sum_{\alpha=1}^n Z_\alpha Z_\alpha'$ and $Z_1, \dots, Z_n$ are distributed independently, each according to $N(\mathbf{0}, \Sigma)$, and $n = N - 1$. Let $Z_\alpha = CY_\alpha$, $\alpha = 1, \dots, n$, where $CC' = \Sigma$. Then $Y_1, \dots, Y_n$ are independently distributed, each with distribution $N(\mathbf{0}, I)$. Let

$$B = \sum_{\alpha=1}^{n} Y_\alpha Y_\alpha' = \sum_{\alpha=1}^{n} C^{-1} Z_\alpha Z_\alpha' (C^{-1})' = C^{-1} A (C^{-1})'; \tag{12}$$

then $|A| = |C| \cdot |B| \cdot |C'| = |B| \cdot |\Sigma|$. By the development in Section 7.2 we see that $|B|$ has the distribution of $\prod_{i=1}^{p} t_{ii}^2$ and $t_{11}^2, \dots, t_{pp}^2$ are independently distributed with χ^2-distributions.

THEOREM 7.5.3. *The distribution of the generalized variance $|S|$ of a sample $X_1, \dots, X_N$ from $N(\mu, \Sigma)$ is the same as the distribution of $|\Sigma|/(N-1)^p$ times the product of p independent factors, the distribution of the* ith *factor being the χ^2-distribution with $N - i$ degrees of freedom.*

If $p = 1$, $|S|$ has the distribution of $|\Sigma| \cdot \chi_{N-1}^2/(N-1)$. If $p = 2$, $|S|$ has the distribution of $|\Sigma| \chi_{N-1}^2 \cdot \chi_{N-2}^2/(N-1)^2$. It follows from Problem 15 or 37 that when $p = 2$ $|S|$ has the distribution of $|\Sigma| (\chi_{2N-4}^2)^2/(2N-2)^2$. We can write

$$|A| = |\Sigma| \times \chi_{N-1}^2 \times \chi_{N-2}^2 \times \cdots \times \chi_{N-p}^2. \tag{13}$$

If $p = 2r$, then $|A|$ is distributed as

$$|\Sigma| \left(\chi_{2N-4}^2 \times \chi_{2N-8}^2 \times \cdots \times \chi_{2N-4r}^2 \right)^2 / 2^{2r}. \tag{14}$$

Since the hth moment of a χ^2-variable with m degrees of freedom is $2^h \Gamma(\frac{1}{2}m + h)/\Gamma(\frac{1}{2}m)$ and the moment of a product of independent variables is the product of the moments of the variables, the hth moment of $|A|$ is

$$\begin{aligned} |\Sigma|^h \prod_{i=1}^{p} \left\{ 2^h \frac{\Gamma[\frac{1}{2}(N-i) + h]}{\Gamma[\frac{1}{2}(N-i)]} \right\} &= 2^{hp} |\Sigma|^h \frac{\prod_{i=1}^{p} \Gamma[\frac{1}{2}(N-i) + h]}{\prod_{i=1}^{p} \Gamma[\frac{1}{2}(N-i)]} \\ &= 2^{hp} |\Sigma|^h \frac{\Gamma_p[\frac{1}{2}(N-1) + h]}{\Gamma_p[\frac{1}{2}(N-1)]}. \end{aligned} \tag{15}$$

Thus

$$(16)\qquad \mathscr{E}|\boldsymbol{A}| = |\boldsymbol{\Sigma}|\prod_{i=1}^{p}(N-i),$$

$$(17)\qquad \mathscr{V}(|\boldsymbol{A}|) = |\boldsymbol{\Sigma}|^2\prod_{i=1}^{p}(N-i)\left[\prod_{j=1}^{p}\Gamma(N-j+2) - \prod_{j=1}^{p}\Gamma(N-j)\right],$$

where $\mathscr{V}(|\boldsymbol{A}|)$ is the variance of $|\boldsymbol{A}|$.

Hoel (1937) has suggested as an approximate density for the pth root of $|\boldsymbol{A}|/|\boldsymbol{\Sigma}|$

$$(18)\qquad \frac{c^{\frac{1}{2}p(N-p)}y^{\frac{1}{2}p(N-p)-1}e^{-cy}}{\Gamma\left[\frac{1}{2}p(N-p)\right]},$$

where

$$(19)\qquad c = \frac{p}{2}\left(1 - \frac{(p-1)(p-2)}{2N}\right)^{1/p}.$$

[See also Mathai (1972).]

7.5.3. The Asymptotic Distribution of the Sample Generalized Variance

Let $|\boldsymbol{B}|/n^p = V_1(n)\times V_2(n)\times\cdots\times V_p(n)$, where the V's are independently distributed and $nV_i(n) = \chi^2_{n-p+i}$. Since χ^2_{n-p+i} is distributed as $\sum_{\alpha=1}^{n-p+i}W_\alpha^2$, where the W_α are independent, each with distribution $N(0,1)$, the central limit theorem (applied to W_α^2) states that

$$(20)\qquad \frac{nV_i(n) - (n-p+i)}{\sqrt{2(n-p+i)}} = \sqrt{n}\,\frac{V_i(n) - 1 + \dfrac{p-i}{n}}{\sqrt{2}\sqrt{1-\dfrac{p-i}{n}}}$$

is asymptotically distributed according to $N(0,1)$. Then $\sqrt{n}\,[V_i(n)-1]$ is asymptotically distributed according to $N(0,2)$. We now apply Theorem 4.2.3. We have

$$(21)\qquad \boldsymbol{U}(n) = \begin{pmatrix} V_1(n) \\ \vdots \\ V_p(n)\end{pmatrix}, \qquad \boldsymbol{b} = \begin{pmatrix} 1 \\ \vdots \\ 1\end{pmatrix},$$

$|\boldsymbol{B}|/n^p = w = f(u_1, \ldots, u_p) = u_1 u_2 \cdots u_p$, $\mathbf{T} = 2\boldsymbol{I}$, $\partial f/\partial u_i|_{\boldsymbol{u}=\boldsymbol{b}} = 1$, and $\boldsymbol{\phi}_b' \mathbf{T} \boldsymbol{\phi}_b = 2p$. Thus

$$\sqrt{n}\left(\frac{|\boldsymbol{B}|}{n^p} - 1\right) \tag{22}$$

is asymptotically distributed according to $N(0, 2p)$.

THEOREM 7.5.4. *Let $\boldsymbol{S}$ be a $p \times p$ sample covariance matrix with n degrees of freedom. Then $\sqrt{n}\,(|\boldsymbol{S}|/|\boldsymbol{\Sigma}| - 1)$ is asymptotically normally distributed with mean 0 and variance $2p$.*

7.6. DISTRIBUTION OF THE SET OF CORRELATION COEFFICIENTS WHEN THE POPULATION COVARIANCE MATRIX IS DIAGONAL

In Section 4.2.1 we found the distribution of a single sample correlation when the corresponding population correlation was zero. Here we shall find the density of the set r_{ij}, $i < j$, $i, j = 1, \ldots, p$, when $\rho_{ij} = 0$, $i < j$.

We start with the distribution of $\boldsymbol{A}$ when $\boldsymbol{\Sigma}$ is diagonal; that is, $W[(\sigma_{ii}\delta_{ij}), n]$. The density of $\boldsymbol{A}$ is

$$\frac{|a_{ij}|^{\frac{1}{2}(n-p-1)} \exp\left(-\frac{1}{2}\sum_{i=1}^{p} \frac{a_{ii}}{\sigma_{ii}}\right)}{2^{\frac{1}{2}np} \prod_{i=1}^{p} \sigma_{ii}^{\frac{1}{2}n} \Gamma_p\left(\frac{1}{2}n\right)}, \tag{1}$$

since

$$|\boldsymbol{\Sigma}| = \begin{vmatrix} \sigma_{11} & 0 & \cdots & 0 \\ 0 & \sigma_{22} & \cdots & 0 \\ \vdots & \vdots & & \vdots \\ 0 & 0 & \cdots & \sigma_{pp} \end{vmatrix} = \prod_{i=1}^{p} \sigma_{ii}. \tag{2}$$

We make the transformation

$$a_{ij} = \sqrt{a_{ii}}\sqrt{a_{jj}}\, r_{ij}, \qquad i \neq j, \tag{3}$$

$$a_{ii} = a_{ii}. \tag{4}$$

The Jacobian is the product of the Jacobian of (4) and that of (3) for a_{ii} fixed. The Jacobian of (3) is the determinant of a $p(p-1)/2$-order diagonal matrix with diagonal elements $\sqrt{a_{ii}}\sqrt{a_{jj}}$. Since each particular subscript k, say, appears in the set r_{ij}, $i<j$, $p-1$ times, the Jacobian is

$$J = \prod_{i=1}^{p} a_{ii}^{\frac{1}{2}(p-1)}. \tag{5}$$

If we substitute from (3) and (4) into $w[\boldsymbol{A}|(\sigma_{ii}\delta_{ij}), n]$ and multiply by (5) we obtain as the joint density of $\{a_{ii}\}$ and $\{r_{ij}\}$

$$\frac{|\sqrt{a_{ii}}\sqrt{a_{jj}}\,r_{ij}|^{\frac{1}{2}(n-p-1)}\exp\left(-\frac{1}{2}\sum_{i=1}^{p}\frac{a_{ii}}{\sigma_{ii}}\right)}{2^{\frac{1}{2}np}\prod_{i=1}^{p}\sigma_{ii}^{\frac{1}{2}n}\Gamma_p\left(\frac{1}{2}n\right)}\prod_{i=1}^{p}a_{ii}^{\frac{1}{2}(p-1)} \tag{6}$$

$$= \frac{|r_{ij}|^{\frac{1}{2}(n-p-1)}}{\Gamma_p\left(\frac{1}{2}n\right)}\prod_{i=1}^{p}\left\{\frac{a_{ii}^{\frac{1}{2}n-1}\exp\left(-\frac{1}{2}\frac{a_{ii}}{\sigma_{ii}}\right)}{2^{\frac{1}{2}n}\sigma_{ii}^{\frac{1}{2}n}}\right\}$$

since

$$|\sqrt{a_{ii}}\sqrt{a_{jj}}\,r_{ij}| = \left(\prod_{i=1}^{p}a_{ii}\right)|r_{ij}|, \tag{7}$$

where $r_{ii}=1$. In the ith term of the product on the right-hand side of (6), let $a_{ii}/(2\sigma_{ii}) = u_i$; then the integral of this term is

$$\int_0^\infty \frac{a_{ii}^{\frac{1}{2}n-1}\exp\left(-\frac{1}{2}\frac{a_{ii}}{\sigma_{ii}}\right)}{2^{\frac{1}{2}n}\sigma_{ii}^{\frac{1}{2}n}}\,da_{ii} = \int_0^\infty u_i^{\frac{1}{2}n-1}e^{-u_i}\,du_i = \Gamma\left(\tfrac{1}{2}n\right) \tag{8}$$

by definition of the gamma function (or by the fact that a_{ii}/σ_{ii} has the χ^2-density with n degrees of freedom). Hence the density of r_{ij} is

$$\frac{\Gamma^p\left(\frac{1}{2}n\right)|r_{ij}|^{\frac{1}{2}(n-p-1)}}{\Gamma_p\left(\frac{1}{2}n\right)}. \tag{9}$$

THEOREM 7.6.1. *If $X_1, \dots, X_N$ are independent, each with distribution $N[\mu, (\sigma_{ii}\delta_{ij})]$, then the density of the sample correlation coefficients is given by* (9) *where* $n = N - 1$.

7.7. THE INVERTED WISHART DISTRIBUTION AND BAYES ESTIMATION OF THE COVARIANCE MATRIX

7.7.1. The Inverted Wishart Distribution

As indicated in Section 3.4.2, Bayes estimators are usually admissible. The calculation of Bayes estimators is facilitated when the prior distributions of the parameter is chosen conveniently. When there is a sufficient statistic there will exist a family of prior distributions for the parameter such that the posterior distribution is a member of this family; such a family is called a *conjugate family of distributions.* In Section 3.4.2 we saw that the normal family of priors is conjugate to the normal family of distributions when the covariance matrix is given. In Section 7.7 we shall consider Bayesian estimation of the covariance matrix and estimation of the mean vector and the covariance matrix.

THEOREM 7.7.1. *If $\boldsymbol{A}$ has the distribution $W(\Sigma, m)$, then $\boldsymbol{B} = \boldsymbol{A}^{-1}$ has the density*

$$\frac{|\boldsymbol{\Psi}|^{\frac{1}{2}m}|\boldsymbol{B}|^{-\frac{1}{2}(m+p+1)}e^{-\frac{1}{2}\operatorname{tr}\boldsymbol{\Psi}\boldsymbol{B}^{-1}}}{2^{\frac{1}{2}mp}\Gamma_p\left(\frac{1}{2}m\right)} \tag{1}$$

for $\boldsymbol{B}$ positive definite and 0 *elsewhere, where $\boldsymbol{\Psi} = \Sigma^{-1}$.*

PROOF. By Theorem A.4.6 of Appendix A the Jacobian of the transformation $\boldsymbol{A} = \boldsymbol{B}^{-1}$ is $|\boldsymbol{B}|^{-(p+1)}$. Substitution of $\boldsymbol{B}^{-1}$ for $\boldsymbol{A}$ in (16) of Section 7.2 and multiplication by $|\boldsymbol{B}|^{-(p+1)}$ yields (1). Q.E.D.

We shall call (1) the density of the *inverted Wishart distribution with m degrees of freedom*† and denote the distribution by $W^{-1}(\boldsymbol{\Psi}, m)$ and the density by $w^{-1}(\boldsymbol{B}|\boldsymbol{\Psi}, m)$. We shall call $\boldsymbol{\Psi}$ the *precision matrix.*

† The definition of the number of degrees of freedom differs from that of Giri (1977), p. 104, and Muirhead (1982), p. 113.

7.7.2. Bayes Estimation of the Covariance Matrix

The covariance matrix of a sample of size N from $N(\mu, \Sigma)$ has the distribution of $(1/n)A$, where A has the distribution $W(\Sigma, n)$ and $n = N - 1$. We shall now show that if Σ is assigned an inverted Wishart distribution, then the conditional distribution of Σ given A is an inverted Wishart distribution. In other words, the family of inverted Wishart distributions for Σ is conjugate to the family of Wishart distributions.

THEOREM 7.7.2. *If A has the distribution $W(\Sigma, n)$ and Σ has the a priori distribution $W^{-1}(\Psi, m)$, then the conditional distribution of Σ is $W^{-1}(A + \Psi, n + m)$.*

PROOF. The joint density of A and Σ is

$$\frac{|\Psi|^{\frac{1}{2}m}|\Sigma|^{-\frac{1}{2}(n+m+p+1)}|A|^{\frac{1}{2}(n-p-1)}e^{-\frac{1}{2}\operatorname{tr}(A+\Psi)\Sigma^{-1}}}{2^{\frac{1}{2}(n+m)p}\Gamma_p\left(\frac{1}{2}n\right)\Gamma_p\left(\frac{1}{2}m\right)} \tag{2}$$

for A and Σ positive definite. The marginal density of A is the integral of (2) over the set of Σ positive definite. Since the integral of (1) with respect to B is 1 identical in Ψ, the integral of (2) with respect to Σ is

$$\frac{\Gamma_p\left[\frac{1}{2}(n+m)\right]|\Psi|^{\frac{1}{2}m}|A|^{\frac{1}{2}(n-p-1)}|A+\Psi|^{-\frac{1}{2}(n+m)}}{\Gamma_p\left(\frac{1}{2}n\right)\Gamma_p\left(\frac{1}{2}m\right)} \tag{3}$$

for A positive definite. The conditional density of Σ given A is the ratio of (2) to (3), namely,

$$\frac{|A+\Psi|^{\frac{1}{2}(n+m)}|\Sigma|^{-\frac{1}{2}(n+m+p+1)}e^{-\frac{1}{2}\operatorname{tr}(A+\Psi)\Sigma^{-1}}}{2^{\frac{1}{2}(n+m)p}\Gamma_p\left[\frac{1}{2}(n+m)\right]}, \tag{4}$$

which is $w^{-1}(\Sigma|A + \Psi, n + m)$. Q.E.D.

COROLLARY 7.7.1. *If nS has the distribution $W(\Sigma, n)$ and Σ has the a priori distribution $W^{-1}(\Psi, m)$, then the conditional distribution of Σ given S is $W^{-1}(nS + \Psi, n + m)$.*

COROLLARY 7.7.2. *If nS has the distribution $W(\Sigma, n)$, Σ has the a priori distribution $W^{-1}(\Psi, m)$, and the loss function is $\operatorname{tr}(D - \Sigma)G(D - \Sigma)H$, where*

$\boldsymbol{G}$ and $\boldsymbol{H}$ are positive definite, then the Bayes estimator for $\boldsymbol{\Sigma}$ is

$$\frac{1}{n+m-p-1}(n\boldsymbol{S}+\boldsymbol{\Psi}). \tag{5}$$

PROOF. It follows from Section 3.4.2 that the Bayes estimator for $\boldsymbol{\Sigma}$ is $\mathscr{E}(\boldsymbol{\Sigma}|\boldsymbol{S})$. From Theorem 7.7.2 we see that $\boldsymbol{\Sigma}^{-1}$ has the a posteriori distribution $W[(n\boldsymbol{S}+\boldsymbol{\Psi})^{-1}, n+m]$. The theorem results from the following lemma.

LEMMA 7.7.1. *If $\boldsymbol{A}$ has the distribution $W(\boldsymbol{\Sigma}, n)$, then*

$$\mathscr{E}\boldsymbol{A}^{-1} = \frac{1}{n-p-1}\boldsymbol{\Sigma}^{-1}. \tag{6}$$

PROOF. If $\boldsymbol{C}$ is a nonsingular matrix such that $\boldsymbol{\Sigma} = \boldsymbol{C}\boldsymbol{C}'$, then $\boldsymbol{A}$ has the distribution of $\boldsymbol{C}\boldsymbol{B}\boldsymbol{C}'$, where $\boldsymbol{B}$ has the distribution $W(\boldsymbol{I}, n)$, and $\mathscr{E}\boldsymbol{A}^{-1} = (\boldsymbol{C}')^{-1}(\mathscr{E}\boldsymbol{B}^{-1})\boldsymbol{C}^{-1}$. By symmetry the diagonal elements of $\mathscr{E}\boldsymbol{B}^{-1}$ are the same and the off-diagonal elements are the same; that is, $\mathscr{E}\boldsymbol{B}^{-1} = k_1\boldsymbol{I} + k_2\boldsymbol{\varepsilon}\boldsymbol{\varepsilon}'$. For every orthogonal matrix $\boldsymbol{Q}$, $\boldsymbol{Q}\boldsymbol{B}\boldsymbol{Q}'$ has the distribution $W(\boldsymbol{I}, n)$ and hence $\mathscr{E}(\boldsymbol{Q}\boldsymbol{B}\boldsymbol{Q}')^{-1} = \boldsymbol{Q}\mathscr{E}\boldsymbol{B}^{-1}\boldsymbol{Q}' = \mathscr{E}\boldsymbol{B}^{-1}$. Thus $k_2 = 0$. A diagonal element of $\boldsymbol{B}^{-1}$ has the distribution of $(\chi^2_{n-p+1})^{-1}$. (See, e.g., the proof of Theorem 5.2.2.) Since $\mathscr{E}(\chi^2_{n-p+1})^{-1} = (n-p-1)^{-1}$, $\mathscr{E}\boldsymbol{B}^{-1} = (n-p-1)^{-1}\boldsymbol{I}$. Then (6) follows. Q.E.D.

We note that $(n-p-1)\boldsymbol{A}^{-1} = [(n-p-1)/(n-1)]\boldsymbol{S}^{-1}$ is an unbiased estimator of the precision $\boldsymbol{\Sigma}^{-1}$.

If $\boldsymbol{\mu}$ is known, the unbiased estimator of $\boldsymbol{\Sigma}$ is $(1/N)\sum_{\alpha=1}^{N}(\boldsymbol{x}_\alpha - \boldsymbol{\mu})(\boldsymbol{x}_\alpha - \boldsymbol{\mu})'$. The above can be applied with n replaced by N. Note that if n (or N) is large, (5) is approximately $\boldsymbol{S}$.

THEOREM 7.7.3. *Let $\boldsymbol{x}_1, \dots, \boldsymbol{x}_N$ be observations from $N(\boldsymbol{\mu}, \boldsymbol{\Sigma})$. Suppose $\boldsymbol{\mu}$ and $\boldsymbol{\Sigma}$ have the a priori density $n(\boldsymbol{\mu}|\boldsymbol{\nu}, (1/K)\boldsymbol{\Sigma}) \times w^{-1}(\boldsymbol{\Sigma}|\boldsymbol{\Psi}, m)$. Then the a posteriori density of $\boldsymbol{\mu}$ and $\boldsymbol{\Sigma}$ given $\bar{\boldsymbol{x}} = (1/N)\sum_{\alpha=1}^{N}\boldsymbol{x}_\alpha$ and $\boldsymbol{S} = (1/n)\sum_{\alpha=1}^{N}(\boldsymbol{x}_\alpha - \bar{\boldsymbol{x}})(\boldsymbol{x}_\alpha - \bar{\boldsymbol{x}})'$ is*

$$n\left(\boldsymbol{\mu}\,\middle|\,\frac{1}{N+K}(N\bar{\boldsymbol{x}} + K\boldsymbol{\nu}), \frac{1}{N+K}\boldsymbol{\Sigma}\right) \tag{7}$$

$$\times w^{-1}\left(\boldsymbol{\Sigma}\,\middle|\,\boldsymbol{\Psi} + n\boldsymbol{S} + \frac{NK}{N+K}(\bar{\boldsymbol{x}} - \boldsymbol{\nu})(\bar{\boldsymbol{x}} - \boldsymbol{\nu})', N+m\right).$$

PROOF. Since $\bar{x}$ and $nS = A$ are a sufficient set of statistics, we can consider the joint density of $\bar{x}$, A, μ, and Σ, which is

$$(8)\quad \frac{K^{\frac{1}{2}p} N^{\frac{1}{2}p} |\Psi|^{\frac{1}{2}m} |\Sigma|^{-\frac{1}{2}(N+m+p+2)} |A|^{\frac{1}{2}(N-p-2)}}{2^{\frac{1}{2}(N+m+1)p} \pi^p \Gamma_p[\frac{1}{2}(N-1)] \Gamma_p(\frac{1}{2}m)} \times \exp\{-\tfrac{1}{2}[N(\bar{x}-\mu)'\Sigma^{-1}(\bar{x}-\mu) + \operatorname{tr} A\Sigma^{-1} + K(\mu-\nu)'\Sigma^{-1}(\mu-\nu) + \operatorname{tr}\Psi\Sigma^{-1}]\}.$$

The marginal density of $\bar{x}$ and A is the integral of (8) with respect to μ and Σ. The exponential in (8) is $-\frac{1}{2}$ times

$$(9)\quad \begin{aligned} &(N+K)\mu'\Sigma^{-1}\mu - 2(N\bar{x}+K\nu)\Sigma^{-1}\mu + N\bar{x}'\Sigma^{-1}\bar{x} + K\nu'\Sigma^{-1}\nu + \operatorname{tr}(A+\Psi)\Sigma^{-1} \\ &= (N+K)\left[\mu - \frac{1}{N+K}(N\bar{x}+K\nu)\right]'\Sigma^{-1}\left[\mu - \frac{1}{N+K}(N\bar{x}+K\nu)\right] \\ &\quad + \frac{NK}{N+K}(\bar{x}-\nu)'\Sigma^{-1}(\bar{x}-\nu) + \operatorname{tr}(A+\Psi)\Sigma^{-1}. \end{aligned}$$

The integral of (8) with respect to μ is

$$(10)\quad \frac{K^{\frac{1}{2}p} N^{\frac{1}{2}p} |\Psi|^{\frac{1}{2}m} |\Sigma|^{-\frac{1}{2}(N+m+p+1)} |A|^{\frac{1}{2}(N-p-2)}}{(N+K)^{\frac{1}{2}p} 2^{\frac{1}{2}(N+m)p} \pi^{\frac{1}{2}p} \Gamma_p[\frac{1}{2}(N-1)] \Gamma_p(\frac{1}{2}m)} \times \exp\left\{-\frac{1}{2}\left[\operatorname{tr} A\Sigma^{-1} + \frac{NK}{N+K}(\bar{x}-\nu)'\Sigma^{-1}(\bar{x}-\nu) + \operatorname{tr}\Psi\Sigma^{-1}\right]\right\}.$$

In turn, the integral of (10) with respect to Σ is

$$(11)\quad \frac{K^{\frac{1}{2}p} N^{\frac{1}{2}} \prod_{i=1}^{p} \Gamma[\frac{1}{2}(N+m+1-i)]}{\pi^{\frac{1}{2}p} \Gamma_p[\frac{1}{2}(N-1)] \Gamma_p(\frac{1}{2}m)(N+K)^{\frac{1}{2}p}} \times |A|^{\frac{1}{2}(N-p-2)} |\Psi|^{\frac{1}{2}m} \left|\Psi + A + \frac{NK}{N+K}(\bar{x}-\nu)(\bar{x}-\nu)'\right|^{-\frac{1}{2}(N+m)}.$$

The conditional density of $\boldsymbol{\mu}$ and $\boldsymbol{\Sigma}$ given $\bar{\boldsymbol{x}}$ and $\boldsymbol{A}$ is the ratio of (8) to (11), namely,

$$(12) \qquad \frac{(N+K)^{\frac{1}{2}p}|\boldsymbol{\Sigma}|^{-\frac{1}{2}(N+m+p+2)}\left|\boldsymbol{\Psi}+\boldsymbol{A}+\frac{NK}{N+K}(\bar{\boldsymbol{x}}-\boldsymbol{\nu})(\bar{\boldsymbol{x}}-\boldsymbol{\nu})'\right|^{\frac{1}{2}(N+m)}}{2^{\frac{1}{2}(N+m+1)p}\pi^{\frac{1}{2}p}\Gamma_p\left[\frac{1}{2}(N+m)\right]}$$

$$\times \exp\left\{-\frac{1}{2}\left[(N+K)\left[\boldsymbol{\mu}-\frac{1}{N+K}(N\bar{\boldsymbol{x}}+K\boldsymbol{\nu})\right]'\boldsymbol{\Sigma}^{-1}\left[\boldsymbol{\mu}-\frac{1}{N+K}(N\bar{\boldsymbol{x}}+K\boldsymbol{\nu})\right]\right.\right.$$
$$\left.\left.+\operatorname{tr}\left[\boldsymbol{\Psi}+\boldsymbol{A}+\frac{NK}{N+K}(\bar{\boldsymbol{x}}-\boldsymbol{\nu})(\bar{\boldsymbol{x}}-\boldsymbol{\nu})'\right]\boldsymbol{\Sigma}^{-1}\right]\right\}$$

Then (12) can be written as (7). Q.E.D.

COROLLARY 7.7.2. *If $\boldsymbol{x}_1, \dots, \boldsymbol{x}_N$ are observations from $N(\boldsymbol{\mu}, \boldsymbol{\Sigma})$, if $\boldsymbol{\mu}$ and $\boldsymbol{\Sigma}$ have the a priori density $n[\boldsymbol{\mu}, (1/K)\boldsymbol{\Sigma}] \times w^{-1}(\boldsymbol{\Sigma}|\boldsymbol{\Psi}, m)$, and if the loss function is $(\boldsymbol{d}-\boldsymbol{\mu})'\boldsymbol{J}(\boldsymbol{d}-\boldsymbol{\mu}) + \operatorname{tr}(\boldsymbol{D}-\boldsymbol{\Sigma})\boldsymbol{G}(\boldsymbol{D}-\boldsymbol{\Sigma})\boldsymbol{H}$, then the Bayes estimators of $\boldsymbol{\mu}$ and $\boldsymbol{\Sigma}$ are*

$$(13) \qquad \frac{1}{N+K}(N\bar{\boldsymbol{x}}+K\boldsymbol{\nu})$$

and

$$(14) \qquad \frac{1}{N+m-p-1}\left[n\boldsymbol{S}+\boldsymbol{\Psi}+\frac{NK}{N+K}(\bar{\boldsymbol{x}}-\boldsymbol{\nu})(\bar{\boldsymbol{x}}-\boldsymbol{\nu})'\right],$$

respectively.

The estimator of $\boldsymbol{\mu}$ is a weighted average of the sample mean $\bar{\boldsymbol{x}}$ and the a priori mean $\boldsymbol{\nu}$. If N is large, the a priori mean has relatively little weight. The estimator of $\boldsymbol{\Sigma}$ is a weighted average of the sample covariance $\boldsymbol{S}$, $\boldsymbol{\Psi}$ and a term deriving from the difference between the sample mean and the a priori mean. If N is large, the estimator is close to the sample covariance matrix.

THEOREM 7.7.4. *If $\boldsymbol{x}_1, \dots, \boldsymbol{x}_N$ are observations from $N(\boldsymbol{\mu}, \boldsymbol{\Sigma})$ and if $\boldsymbol{\mu}$ and $\boldsymbol{\Sigma}$ have the a priori density $n[\boldsymbol{\mu}|\boldsymbol{\nu}, (1/K)\boldsymbol{\Sigma}] \times w^{-1}(\boldsymbol{\Sigma}|\boldsymbol{\Psi}, m)$, then the marginal a posteriori density of $\boldsymbol{\mu}$ given $\bar{\boldsymbol{x}}$ and $\boldsymbol{S}$ is*

$$(15) \qquad \frac{(N+K)^{\frac{1}{2}p}\Gamma\left[\frac{1}{2}(N+m+1)\right]|\boldsymbol{B}|^{-\frac{1}{2}}}{\pi^{\frac{1}{2}p}\Gamma\left[\frac{1}{2}(N+m+1-p)\right]\left[1+(N+K)(\boldsymbol{\mu}-\boldsymbol{\mu}^*)'\boldsymbol{B}^{-1}(\boldsymbol{\mu}-\boldsymbol{\mu}^*)\right]^{\frac{1}{2}(N+m+1)}},$$

where $\boldsymbol{\mu}^*$ is (13) and $\boldsymbol{B}$ is $(N+m-p-1)$ times (14).

PROOF. The exponent in (12) is $-\frac{1}{2}$ times

$$\operatorname{tr}\left[\boldsymbol{B} + (N+K)(\boldsymbol{\mu} - \boldsymbol{\mu}^*)(\boldsymbol{\mu} - \boldsymbol{\mu}^*)'\right]\boldsymbol{\Sigma}^{-1}. \tag{16}$$

Then the integral of (12) with respect to $\boldsymbol{\Sigma}$ is

$$\frac{(N+K)^{\frac{1}{2}p}\Gamma_p\left[\frac{1}{2}(N+m+1)\right]|\boldsymbol{B}|^{\frac{1}{2}(N+m)}}{\pi^{\frac{1}{2}p}\Gamma_p\left[\frac{1}{2}(N+m)\right]|\boldsymbol{B} + (N+K)(\boldsymbol{\mu} - \boldsymbol{\mu}^*)(\boldsymbol{\mu} - \boldsymbol{\mu}^*)'|^{\frac{1}{2}(N+m+1)}}. \tag{17}$$

Since $|\boldsymbol{B} + \boldsymbol{x}\boldsymbol{x}'| = |\boldsymbol{B}|(1 + \boldsymbol{x}'\boldsymbol{B}^{-1}\boldsymbol{x})$ (Corollary A.3.1), (15) follows. Q.E.D.

The density (15) is known as the *multivariate t-distribution* with mean vector $\boldsymbol{\mu}^*$, covariance matrix $[(N+m+1-p)/(N+m-1-p)]\boldsymbol{B}$, and $N+m+1-p$ degrees of freedom.

7.8. IMPROVED ESTIMATION OF THE COVARIANCE MATRIX

Just as the sample mean $\bar{\boldsymbol{x}}$ can be improved on as an estimator of the population mean $\boldsymbol{\mu}$ when the loss function is quadratic, so can the sample covariance $\boldsymbol{S}$ be improved on as an estimator of the population covariance $\boldsymbol{\Sigma}$ for certain loss functions. The loss function for estimation of the location parameter $\boldsymbol{\mu}$ was invariant with respect to translation ($\boldsymbol{x} \to \boldsymbol{x} + \boldsymbol{a}, \boldsymbol{\mu} \to \boldsymbol{\mu} + \boldsymbol{a}$), and the risk of the sample mean (which is the unique unbiased function of the sufficient statistic when $\boldsymbol{\Sigma}$ is known) does not depend on the parameter value. The natural group of transformations of covariance matrices is multiplication on the left by a nonsingular matrix and on the right by its transpose ($\boldsymbol{x} \to \boldsymbol{C}\boldsymbol{x}$, $\boldsymbol{S} \to \boldsymbol{C}\boldsymbol{S}\boldsymbol{C}'$, $\boldsymbol{\Sigma} \to \boldsymbol{C}\boldsymbol{\Sigma}\boldsymbol{C}'$). We consider two loss functions which are invariant with respect to such transformations.

One loss function is quadratic:

$$\begin{aligned} L_q(\boldsymbol{\Sigma}, \boldsymbol{G}) &= \operatorname{tr}(\boldsymbol{G} - \boldsymbol{\Sigma})\boldsymbol{\Sigma}^{-1}(\boldsymbol{G} - \boldsymbol{\Sigma})\boldsymbol{\Sigma}^{-1} \\ &= \operatorname{tr}(\boldsymbol{G}\boldsymbol{\Sigma}^{-1} - \boldsymbol{I})^2, \end{aligned} \tag{1}$$

where $\boldsymbol{G}$ is a positive definite matrix. The other is based on the form of the likelihood function:

$$L_l(\boldsymbol{\Sigma}, \boldsymbol{G}) = \operatorname{tr}\boldsymbol{G}\boldsymbol{\Sigma}^{-1} - \log|\boldsymbol{G}\boldsymbol{\Sigma}^{-1}| - p. \tag{2}$$

(See Lemma 3.2.2 and alternative proofs in Problems 4, 8, and 12 of Chapter 3.) Each of these is 0 when $\boldsymbol{G} = \boldsymbol{\Sigma}$ and is positive when $\boldsymbol{G} \neq \boldsymbol{\Sigma}$. The second loss function approaches ∞ as $\boldsymbol{G}$ approaches a singular matrix or when one or more elements (or one or more characteristic roots) of $\boldsymbol{G}$ approaches ∞. (See proof of Lemma 3.2.2.) Each is invariant with respect to transformations $\boldsymbol{G}^* = \boldsymbol{C}\boldsymbol{G}\boldsymbol{C}'$, $\boldsymbol{\Sigma}^* = \boldsymbol{C}\boldsymbol{\Sigma}\boldsymbol{C}'$. We can see some properties of the loss functions from $L_q(\boldsymbol{I}, \boldsymbol{D}) = \sum_{i=1}^p (d_{ii} - 1)^2$ and $L_l(\boldsymbol{I}, \boldsymbol{D}) = \sum_{i=1}^p (d_{ii} - \log d_{ii} - 1)$, where $\boldsymbol{D}$ is diagonal. (By Theorem A.2.2 of Appendix A for arbitrary positive definite $\boldsymbol{\Sigma}$ and symmetric $\boldsymbol{G}$, there exists a nonsingular $\boldsymbol{C}$ such that $\boldsymbol{C}\boldsymbol{\Sigma}\boldsymbol{C}' = \boldsymbol{I}$ and $\boldsymbol{C}\boldsymbol{G}\boldsymbol{C}' = \boldsymbol{D}$.) If we let $\boldsymbol{g} = (g_{11}, \dots, g_{pp}, g_{12}, \dots, g_{p-1,p})'$, $\boldsymbol{s} = (s_{11}, \dots, s_{pp}, s_{12}, \dots, s_{p-1,p})'$, $\boldsymbol{\sigma} = (\sigma_{11}, \dots, \sigma_{pp}, \sigma_{12}, \dots, \sigma_{p-1,p})'$, and $\boldsymbol{\Phi} = \mathscr{E}(\boldsymbol{s} - \boldsymbol{\sigma})(\boldsymbol{s} - \boldsymbol{\sigma})'$, then $L_q(\boldsymbol{\Sigma}, \boldsymbol{G})$ is a constant multiple of $(\boldsymbol{g} - \boldsymbol{\sigma})'\boldsymbol{\Phi}^{-1}(\boldsymbol{g} - \boldsymbol{\sigma})$. (See Problem 33.)

The maximum likelihood estimator $\hat{\boldsymbol{\Sigma}}$ and the unbiased estimator $\boldsymbol{S}$ are of the form $a\boldsymbol{A}$, where $\boldsymbol{A}$ has the distribution $W(\boldsymbol{\Sigma}, n)$ and $n = N - 1$.

THEOREM 7.8.1. *The quadratic risk of $a\boldsymbol{A}$ is minimized at $a = 1/(n + p + 1)$ and its value is $p(p + 1)/(n + p + 1)$. The likelihood risk of $a\boldsymbol{A}$ is minimized at $a = 1/n\boldsymbol{A}$* (i.e., $a\boldsymbol{A} = \boldsymbol{S}$) *and its value is $p \log n - \sum_{i=1}^p \mathscr{E} \log \chi^2_{n+1-i}$.*

PROOF. By the invariance of the loss function

$$
\begin{aligned}
(3) \qquad \mathscr{E}_{\boldsymbol{\Sigma}} L_q(\boldsymbol{\Sigma}, a\boldsymbol{A}) &= \mathscr{E}_{\boldsymbol{I}} L_q(\boldsymbol{I}, a\boldsymbol{A}^*) \\
&= \mathscr{E}_{\boldsymbol{I}} \operatorname{tr}(a\boldsymbol{A}^* - \boldsymbol{I})^2 \\
&= \mathscr{E}_{\boldsymbol{I}} \left(a^2 \sum_{i,j=1}^{p} a_{ij}^{*2} - 2a \sum_{i=1}^{p} a_{ii}^* + p \right) \\
&= a^2 \left[(2n + n^2)p + np(p - 1) \right] - 2anp + p \\
&= p\left[n(n + p + 1)a^2 - 2na + 1 \right],
\end{aligned}
$$

which has its minimum at $a = 1/(n + p + 1)$. Similarly

$$
\begin{aligned}
(4) \qquad \mathscr{E}_{\boldsymbol{\Sigma}} L_l(\boldsymbol{\Sigma}, a\boldsymbol{A}) &= \mathscr{E}_{\boldsymbol{I}} L_l(\boldsymbol{I}, a\boldsymbol{A}^*) \\
&= \mathscr{E}_{\boldsymbol{I}} \left\{ a \operatorname{tr} \boldsymbol{A}^* - \log|\boldsymbol{A}^*| - p \log a - p \right\} \\
&= p[na - \log a - 1] - \mathscr{E}_{\boldsymbol{I}} \log|\boldsymbol{A}^*|,
\end{aligned}
$$

which is minimized at $a = 1/n$. Q.E.D.

Although the minimum risk of the estimator of the form $a\boldsymbol{A}$ is constant for its respective loss function, the estimator is not minimax. We shall now consider estimators $\boldsymbol{G}(\boldsymbol{A})$ such that

$$\boldsymbol{G}(\boldsymbol{HAH}') = \boldsymbol{HG}(\boldsymbol{A})\boldsymbol{H}' \tag{5}$$

for lower triangular matrices $\boldsymbol{H}$. The two loss functions are invariant with respect to transformations $\boldsymbol{G}^* = \boldsymbol{HGH}'$, $\boldsymbol{\Sigma}^* = \boldsymbol{H\Sigma H}'$.

Let $\boldsymbol{A} = \boldsymbol{I}$ and $\boldsymbol{H}$ be the diagonal matrix $\boldsymbol{D}_i$ with -1 as the ith diagonal element and 1 as each other diagonal element. Then $\boldsymbol{HAH}' = \boldsymbol{I}$ and the i,jth component of (5) is

$$g_{ij}(\boldsymbol{I}) = -g_{ij}(\boldsymbol{I}), \qquad j \neq i. \tag{6}$$

Hence, $g_{ij}(\boldsymbol{I}) = 0$, $i \neq j$, and $\boldsymbol{G}(\boldsymbol{I})$ is diagonal, say $\boldsymbol{D}$. Since $\boldsymbol{A} = \boldsymbol{TT}'$ for $\boldsymbol{T}$ lower triangular, we have

$$\begin{aligned} \boldsymbol{G}(\boldsymbol{A}) &= \boldsymbol{G}(\boldsymbol{TIT}') \\ &= \boldsymbol{TG}(\boldsymbol{I})\boldsymbol{T}' \\ &= \boldsymbol{TDT}', \end{aligned} \tag{7}$$

where $\boldsymbol{D}$ is a diagonal matrix not depending on $\boldsymbol{A}$. We note in passing that if (5) holds for *all* nonsingular $\boldsymbol{H}$ then $\boldsymbol{D} = a\boldsymbol{I}$ for some a. ($\boldsymbol{H}$ can be taken as a permutation matrix.)

If $\boldsymbol{\Sigma} = \boldsymbol{KK}'$, where $\boldsymbol{K}$ is lower triangular, then

$$\begin{aligned} \mathscr{E}_{\boldsymbol{\Sigma}} L[\boldsymbol{\Sigma}, \boldsymbol{G}(\boldsymbol{A})] &= \int L[\boldsymbol{\Sigma}, \boldsymbol{G}(\boldsymbol{A})] C(p,n) |\boldsymbol{\Sigma}|^{-\frac{1}{2}n} |\boldsymbol{A}|^{\frac{1}{2}(n-p-1)} e^{-\frac{1}{2}\operatorname{tr}\boldsymbol{\Sigma}^{-1}\boldsymbol{A}}\, d\boldsymbol{A} \\ &= \int L[\boldsymbol{KK}', \boldsymbol{G}(\boldsymbol{A})] C(p,n) |\boldsymbol{KK}'|^{-\frac{1}{2}n} |\boldsymbol{A}|^{\frac{1}{2}(n-p-1)} \\ &\qquad \times e^{-\frac{1}{2}\operatorname{tr}\boldsymbol{K}'^{-1}\boldsymbol{K}^{-1}\boldsymbol{A}}\, d\boldsymbol{A} \\ &= \int L[\boldsymbol{KK}', \boldsymbol{G}(\boldsymbol{KA}^*\boldsymbol{K}')] C(p,n) |\boldsymbol{A}^*|^{\frac{1}{2}(n-p-1)} e^{-\frac{1}{2}\operatorname{tr}\boldsymbol{A}^*}\, d\boldsymbol{A}^* \\ &= \mathscr{E}_{\boldsymbol{I}} L[\boldsymbol{KK}', \boldsymbol{KG}(\boldsymbol{A}^*)\boldsymbol{K}'] \\ &= \mathscr{E}_{\boldsymbol{I}} L[\boldsymbol{I}, \boldsymbol{G}(\boldsymbol{A}^*)] \end{aligned} \tag{8}$$

by invariance of the loss function. The risk does not depend on $\boldsymbol{\Sigma}$.

For the quadratic loss function we calculate

$$
\begin{aligned}
(9)\qquad \mathscr{E}_{\boldsymbol{I}} L_q[\boldsymbol{I}, \boldsymbol{G}(\boldsymbol{A})] &= \mathscr{E}_{\boldsymbol{I}} L_q[\boldsymbol{I}, \boldsymbol{TDT}'] \\
&= \mathscr{E}_{\boldsymbol{I}} \operatorname{tr}(\boldsymbol{TDT}' - \boldsymbol{I})^2 \\
&= \mathscr{E}_{\boldsymbol{I}} \operatorname{tr}(\boldsymbol{TDT}'\boldsymbol{TDT}' - 2\boldsymbol{TDT}' + \boldsymbol{I}) \\
&= \mathscr{E}_{\boldsymbol{I}} \sum_{i,j,k,l=1}^{p} t_{ij} d_j t_{kj} t_{kl} d_l t_{il} - 2\mathscr{E}_{\boldsymbol{I}} \sum_{i,j=1}^{p} t_{ij}^2 d_j + p.
\end{aligned}
$$

The expectations can be evaluated by using the fact that the (nonzero) elements of $\boldsymbol{T}$ are independent, t_{ii}^2 has the χ^2-distribution with $n + 1 - i$ degrees of freedom, and t_{ij}, $i > j$, has the distribution $N(0,1)$. Then

$$
(10)\qquad \mathscr{E}_{\boldsymbol{I}} L_q[\boldsymbol{I}, \boldsymbol{G}(\boldsymbol{A})] = \boldsymbol{d}'\boldsymbol{F}\boldsymbol{d} - 2\boldsymbol{f}'\boldsymbol{d} + p,
$$

where $\boldsymbol{F} = (f_{ij})$, $\boldsymbol{f} = (f_i)$,

$$
\begin{aligned}
(11)\qquad f_{ii} &= (n + p - 2i + 1)(n + p - 2i + 3), \\
f_{ij} &= n + p - 2j + 1, \qquad\qquad i < j, \\
f_i &= n + p - 2i + 1,
\end{aligned}
$$

and $\boldsymbol{d} = (d_1, \dots, d_p)'$. Since $\boldsymbol{d}'\boldsymbol{F}\boldsymbol{d} = \mathscr{E} \operatorname{tr}(\boldsymbol{TDT}')^2 > 0$, $\boldsymbol{F}$ is positive definite and (10) has a unique minimum. It is attained at $\boldsymbol{d} = \boldsymbol{F}^{-1}\boldsymbol{f}$ and the minimum is $p - \boldsymbol{f}'\boldsymbol{F}^{-1}\boldsymbol{f}$.

THEOREM 7.8.2. *With respect to the quadratic loss function the best estimator invariant with respect to linear transformations* $\boldsymbol{\Sigma} \to \boldsymbol{H}\boldsymbol{\Sigma}\boldsymbol{H}'$, $\boldsymbol{A} \to \boldsymbol{HAH}'$, *where* $\boldsymbol{H}$ *is lower triangular, is* $\boldsymbol{G}(\boldsymbol{A}) = \boldsymbol{TDT}'$, *where* $\boldsymbol{D}$ *is the diagonal matrix whose diagonal elements compose* $\boldsymbol{d} = \boldsymbol{F}^{-1}\boldsymbol{f}$, $\boldsymbol{F}$ *and* $\boldsymbol{f}$ *are defined by* (11) *and* $\boldsymbol{A} = \boldsymbol{TT}'$ *with* $\boldsymbol{T}$ *lower triangular.*

Since $\boldsymbol{d} = \boldsymbol{F}^{-1}\boldsymbol{f}$ is not proportional to $\boldsymbol{\varepsilon} = (1, \dots, 1)'$, that is, $\boldsymbol{F}\boldsymbol{\varepsilon}$ is not proportional to $\boldsymbol{f}$ (see Problem 28), this estimator has a smaller (quadratic) loss than any estimator of the form $a\boldsymbol{A}$ (which is the only type of estimator invariant under the full linear group). Kiefer (1957) has shown that if an estimator is minimax in the class of estimators invariant with respect to a group of transformations satisfying certain conditions,† then it is minimax with respect to all estimators. In this problem the group of triangular linear transformations satisfies the conditions while the group of all linear transformations does not.

The definition of this estimator depends on the coordinate system and on the numbering of the coordinates. These properties are intuitively unappealing.

THEOREM 7.8.3. *The estimator* $\boldsymbol{G}(\boldsymbol{A})$ *defined in Theorem* 7.8.2 *is minimax with respect to the quadratic loss function.*

In the case of $p = 2$

(12)
$$d_1 = \frac{(n+1)^2 - (n-1)}{(n+1)^2(n+3) - (n-1)}, \qquad d_2 = \frac{(n+1)(n+2)}{(n+1)^2(n+3) - (n-1)}.$$

The risk is

(13)
$$2\frac{3n^2 + 5n + 4}{n^3 + 5n^2 + 6n + 4}.$$

The difference between the risks of the best estimator $a\boldsymbol{A}$ and the best estimator $\boldsymbol{TDT}'$ is

(14)
$$\frac{6}{n+3} - \frac{6n^2 + 10n + 8}{n^3 + 5n^2 + 6n + 4} = \frac{2n(n-1)}{(n+3)(n^3 + 5n^2 + 6n + 4)}.$$

The difference is $1/55$ for $n = 2$ (relative to $6/5$) and $1/47$ for $n = 3$ (relative to 1); it is of the order $2/n^2$; the improvement due to using the estimator $\boldsymbol{TDT}'$ is not great, at least for $p = 2$.

For the likelihood loss function we calculate

(15)
$$\begin{aligned}
\mathscr{E}_{\boldsymbol{I}} L_l[\boldsymbol{I}, \boldsymbol{G}(\boldsymbol{A})] &= \mathscr{E}_{\boldsymbol{I}} L_l[\boldsymbol{I}, \boldsymbol{TDT}'] \\
&= \mathscr{E}_{\boldsymbol{I}}\left[\operatorname{tr} \boldsymbol{TDT}' - \log|\boldsymbol{TDT}'| - p\right] \\
&= \mathscr{E}_{\boldsymbol{I}}\left[\sum_{i,j=1}^{p} t_{ij}^2 d_j - \sum_{i=1}^{p} \log t_{ii}^2\right] - \sum_{i=1}^{p} \log d_i - p \\
&= \sum_{j=1}^{p} (n + p - 2j + 1) d_j - \sum_{j=1}^{p} \log d_j - \sum_{j=1}^{p} \mathscr{E} \log \chi_{n+1-j}^2 - p.
\end{aligned}$$

† The essential condition is that the group is solvable. See Kiefer (1966) and Kudo (1955).

The minimum of (15) occurs at $d_j = 1/(n+p-2j+1)$, $j = 1, \ldots, p$.

THEOREM 7.8.4. *With respect to the likelihood loss function, the best estimator invariant with respect to linear transformations* $\Sigma \to H\Sigma H'$, $A \to HAH'$, *where* H *is lower triangular, is* $G(A) = TDT'$, *where the* j*th diagonal element of the diagonal matrix* D *is* $1/(n+p-2j+1)$, $j = 1, \ldots, p$, *and* $A = TT'$, *with* T *lower triangular. The minimum risk is*

$$\mathscr{E}_{\Sigma} L[\Sigma, G(A)] = \sum_{j=1}^{p} \log(n+p-2j+1) - \sum_{j=1}^{p} \mathscr{E} \log \chi^2_{n+1-j}. \tag{16}$$

THEOREM 7.8.5. *The estimator* $G(A)$ *defined in Theorem 7.8.4 is minimax with respect to the likelihood loss function.*

James and Stein (1961) gave this estimator. Note that the reciprocals of the weights $1/(n+p-1), 1/(n+p-3), \ldots, 1/(n-p+1)$ are symmetrically distributed about the reciprocal of $1/n$.

If $p = 2$,

$$G(A) = \frac{1}{n+1} A + \begin{pmatrix} 0 & 0 \\ 0 & \dfrac{2}{n^2-1} \end{pmatrix} \frac{|A|}{a_{11}}, \tag{17}$$

$$\mathscr{E} G(A) = \frac{n}{n+1} \Sigma + \frac{2}{n+1} \begin{pmatrix} 0 & 0 \\ 0 & \dfrac{|\Sigma|}{\sigma_{11}} \end{pmatrix}. \tag{18}$$

The difference between the risks of the best estimator aA and the best estimator TDT' is

$$p \log n - \sum_{j=1}^{p} \log(n+p-2j+1) = -\sum_{j=1}^{p} \log\left(1 + \frac{p-2j+1}{n}\right). \tag{19}$$

If $p = 2$, the improvement is

$$\begin{aligned} -\log\left(1+\frac{1}{n}\right) - \log\left(1-\frac{1}{n}\right) &= -\log\left(1-\frac{1}{n^2}\right) \\ &= \frac{1}{n^2} + \frac{1}{2n^4} + \frac{1}{3n^6} + \cdots, \end{aligned} \tag{20}$$

which is 0.288 for $n = 2$, 0.118 for $n = 3$, 0.065 for $n = 4$, etc. The risk (19) is $O(1/n^2)$ for any p. (See Problem 31.)

An obvious disadvantage of these estimators is that they depend on the coordinate system. Let $\boldsymbol{P}_i$ be the ith permutation matrix, $i = 1, \ldots, p!$, and let $\boldsymbol{P}_i \boldsymbol{A} \boldsymbol{P}_i' = \boldsymbol{T}_i \boldsymbol{T}_i'$, where $\boldsymbol{T}_i$ is lower triangular and $t_{jj} > 0$, $j = 1, \ldots, p$. Then a randomized estimator that does not depend on the numbering of coordinates is to let the estimator be $\boldsymbol{P}_i' \boldsymbol{T}_i \boldsymbol{D} \boldsymbol{T}_i' \boldsymbol{P}_i$ with probability $1/p!$; this estimator has the same risk as the estimator for the original numbering of coordinates. Since the loss functions are convex, $(1/p!)\Sigma_i \boldsymbol{P}_i' \boldsymbol{T}_i \boldsymbol{D} \boldsymbol{T}_i' \boldsymbol{P}_i$ will have at least as good a risk function; in this case the risk will depend on $\boldsymbol{\Sigma}$.

Haff (1980) has shown that $\boldsymbol{G}(\boldsymbol{A}) = [1/(n+p+1)](\boldsymbol{A} + \gamma u \boldsymbol{C})$, where γ is constant, $0 \le \gamma \le 2(p-1)/(n-p+3)$, $u = 1/\mathrm{tr}(\boldsymbol{A}^{-1}\boldsymbol{C})$, and $\boldsymbol{C}$ is an arbitrary positive definite matrix, has a smaller quadratic risk than $[1/(n+p+1)]\boldsymbol{A}$. $\boldsymbol{G}(\boldsymbol{A}) = (1/n)[\boldsymbol{A} + ut(u)\boldsymbol{C}]$, where $t(u)$ is an absolutely continuous, nonincreasing function, $0 \le t(u) \le 2(p-1)/n$ has a smaller likelihood risk than $\boldsymbol{S}$.

PROBLEMS

1. (Sec. 7.2) A transformation from rectangular to *polar coordinates* is

$$
\begin{aligned}
y_1 &= w \sin\theta_1, \\
y_2 &= w \cos\theta_1 \sin\theta_2, \\
y_3 &= w \cos\theta_1 \cos\theta_2 \sin\theta_3, \\
&\vdots \\
y_{n-1} &= w \cos\theta_1 \cos\theta_2 \cdots \cos\theta_{n-2} \sin\theta_{n-1}, \\
y_n &= w \cos\theta_1 \cos\theta_2 \cdots \cos\theta_{n-2} \cos\theta_{n-1},
\end{aligned}
$$

where $-\frac{1}{2}\pi < \theta_i \le \frac{1}{2}\pi$, $i = 1, \ldots, n-2$, and $-\pi < \theta_{n-1} \le \pi$.

(*a*) Prove $w^2 = \Sigma y_\alpha^2$. [*Hint*: Compute in turn $y_n^2 + y_{n-1}^2$, $(y_n^2 + y_{n-1}^2) + y_{n-2}^2$, and so forth.]

(*b*) Show that the Jacobian is $w^{n-1}\cos^{n-2}\theta_1\cos^{n-3}\theta_2 \cdots \cos\theta_{n-2}$. *Hint*: Prove

$$\left|\frac{\partial(y_1,\ldots,y_n)}{\partial(\theta_1,\ldots,\theta_{n-1},w)}\right| \cdot \begin{vmatrix} \cos\theta_1 & 0 & \cdots & 0 & 0 \\ 0 & \cos\theta_2 & \cdots & 0 & 0 \\ \vdots & \vdots & & \vdots & \vdots \\ 0 & 0 & \cdots & \cos\theta_{n-1} & 0 \\ w\sin\theta_1 & w\sin\theta_2 & \cdots & w\sin\theta_{n-1} & 1 \end{vmatrix}$$

$$= \begin{vmatrix} w & x & \cdots & x & x \\ 0 & w\cos\theta_1 & \cdots & x & x \\ \vdots & \vdots & & \vdots & \vdots \\ 0 & 0 & \cdots & w\cos\theta_1\cdots\cos\theta_{n-2} & x \\ 0 & 0 & \cdots & 0 & \cos\theta_1\cdots\cos\theta_{n-1} \end{vmatrix},$$

where x denotes elements whose explicit values are not needed.

2. (Sec. 7.2) Prove that

$$\int_{-\pi/2}^{\pi/2} \cos^{h-1}\theta\, d\theta = \frac{\Gamma(\frac{1}{2}h)\Gamma(\frac{1}{2})}{\Gamma[\frac{1}{2}(h+1)]}.$$

[*Hint*: Let $\cos^2\theta = u$, and use the definition of $B(p,q)$.]

3. (Sec. 7.2) Use Problems 1 and 2 to prove that the *surface area of a sphere* of unit radius in n dimensions is

$$C(n) = \frac{2\pi^{\frac{1}{2}n}}{\Gamma(\frac{1}{2}n)}.$$

4. (Sec. 7.2) Use Problems 1, 2, and 3 to prove that if the density of $y' = (y_1,\ldots,y_n)$ is $f(y'y)$, then the density of $u = y'y$ is $\frac{1}{2}C(n)f(u)u^{\frac{1}{2}n-1}$.

5. (Sec. 7.2) *Chi-square distribution*. Use Problem 4 to show that if $y_1,\ldots,y_n$ are independently distributed, each according to $N(0,1)$, then $U = \sum_{\alpha=1}^n y_\alpha^2$ has the density $u^{\frac{1}{2}n-1}e^{-\frac{1}{2}u}/[2^{\frac{1}{2}n}\Gamma(\frac{1}{2}n)]$, which is the chi-square density with n degrees of freedom.

6. (Sec. 7.2) Use (9) of Section 7.6 to derive the distribution of $\boldsymbol{A}$.

7. (Sec. 7.2) Use the proof of Theorem 7.2.1 to demonstrate $\Pr\{|\boldsymbol{A}| = 0\} = 0$.

8. (Sec. 7.2) *Independence of estimators of the parameters of the complex normal distribution*. Let $z_1,\ldots,z_N$ be N observations from the complex normal distribution with mean $\boldsymbol{\theta}$ and covariance matrix $\boldsymbol{P}$. (See Problem 64 of Chapter 2.) Show that $\bar{\boldsymbol{Z}}$ and $\boldsymbol{A} = \sum_{\alpha=1}^N(\boldsymbol{Z}_\alpha - \bar{\boldsymbol{Z}})(\boldsymbol{Z}_\alpha - \bar{\boldsymbol{Z}})^*$ are independently distributed and show that $\boldsymbol{A}$ has the distribution of $\sum_{\alpha=1}^n \boldsymbol{W}_\alpha \boldsymbol{W}_\alpha^*$, where $\boldsymbol{W}_1,\ldots,\boldsymbol{W}_N$ are independently distributed, each according to the complex normal distribution with mean $\boldsymbol{0}$ and covariance matrix $\boldsymbol{P}$.

9. (Sec. 7.2) *The complex Wishart distribution*. Let $\boldsymbol{W}_1, \dots, \boldsymbol{W}_n$ be independently distributed, each according to the complex normal distribution with mean $\boldsymbol{0}$ and covariance matrix $\boldsymbol{P}$. (See Problem 64 of Chapter 2). Show that the density of $\boldsymbol{B} = \sum_{\alpha=1}^n \boldsymbol{W}_\alpha \boldsymbol{W}_\alpha^*$ is

$$\frac{|\boldsymbol{B}|^{n-p} e^{-\frac{1}{2}\operatorname{tr} \boldsymbol{B}\boldsymbol{P}^{-1}}}{|\boldsymbol{P}|^n \pi^{\frac{1}{2}p(p-1)} \prod_{i=1}^p \Gamma(n+1-i)}.$$

10. (Sec. 7.3) Find the characteristic function of $\boldsymbol{A}$ from $W(\boldsymbol{\Sigma}, n)$. [*Hint*: From $\int w(\boldsymbol{A}|\boldsymbol{\Sigma}, n)\, d\boldsymbol{A} = 1$, one derives

$$\int \frac{|\boldsymbol{A}|^{\frac{1}{2}(n-p-1)} \exp\left(-\frac{1}{2}\operatorname{tr} \boldsymbol{\Phi}^{-1}\boldsymbol{A}\right) d\boldsymbol{A}}{2^{\frac{1}{2}pn}\Gamma_p\left(\frac{1}{2}n\right)} = |\boldsymbol{\Phi}|^{\frac{1}{2}n}$$

as an identity in $\boldsymbol{\Phi}$.] Note that comparison of this result with that of Section 7.3.1 is a proof of the Wishart distribution.

11. (Sec. 7.3.2) Prove Theorem 7.3.2 by use of characteristic functions.

12. (Sec. 7.3.1) Find the first two moments of the elements of $\boldsymbol{A}$ by differentiating the characteristic function (11).

13. (Sec. 7.3) Let $\boldsymbol{Z}_1, \dots, \boldsymbol{Z}_n$ be independently distributed, each according to $N(\boldsymbol{0}, \boldsymbol{I})$. Let $\boldsymbol{W} = \sum_{\alpha,\beta=1}^n b_{\alpha\beta} \boldsymbol{Z}_\alpha \boldsymbol{Z}_\beta'$. Prove that if $\boldsymbol{a}'\boldsymbol{W}\boldsymbol{a} = \chi_m^2$ for all $\boldsymbol{a}$ such that $\boldsymbol{a}'\boldsymbol{a} = 1$, then $\boldsymbol{W}$ is distributed according to $W(\boldsymbol{I}, m)$. [*Hint*: Use the characteristic function of $\boldsymbol{a}'\boldsymbol{W}\boldsymbol{a}$.]

14. (Sec. 7.4) Let $\boldsymbol{x}_\alpha$ be an observation from $N(\boldsymbol{\beta} z_\alpha, \boldsymbol{\Sigma})$, $\alpha = 1, \dots, N$, where z_α is a scalar. Let $\boldsymbol{b} = \sum_\alpha z_\alpha \boldsymbol{x}_\alpha / \sum_\alpha z_\alpha^2$. Use Theorem 7.4.1 to show that $\sum_\alpha \boldsymbol{x}_\alpha \boldsymbol{x}_\alpha' - \boldsymbol{b}\boldsymbol{b}' \sum_\alpha z_\alpha^2$ and $\boldsymbol{b}\boldsymbol{b}'$ are independent.

15. (Sec. 7.4) Show that

$$\mathscr{E}\left(\chi_{N-1}^2 \chi_{N-2}^2\right)^h = \mathscr{E}\left(\chi_{2N-4}^2/4\right)^h, \qquad h \ge 0,$$

by use of the duplication formula for the gamma function; χ_{N-1}^2 and χ_{N-2}^2 are independent. Hence show that the distribution of $\chi_{N-1}^2 \chi_{N-2}^2$ is the distribution of $\chi_{2N-4}^2/4$.

16. (Sec. 7.4) Verify that Theorem 7.4.1 follows from Lemma 7.4.1. [*Hint*: Prove that $\boldsymbol{Q}_i$ having the distribution $W(\boldsymbol{\Sigma}, r_i)$ implies the existence of (6) where $\boldsymbol{I}$ is of order r_i and that the independence of the $\boldsymbol{Q}_i$'s implies that the $\boldsymbol{I}$'s in (6) do not overlap.]

17. (Sec. 7.5) Find $\mathscr{E}|\boldsymbol{A}|^h$ directly from $W(\boldsymbol{\Sigma}, n)$. [*Hint*: The fact that

$$\int w(\boldsymbol{A}|\boldsymbol{\Sigma}, n)\, d\boldsymbol{A} \equiv 1$$

shows

$$\int |A|^{\frac{1}{2}(n-p-1)} \exp\left(-\tfrac{1}{2}\operatorname{tr}\Sigma^{-1}A\right) dA = 2^{\frac{1}{2}np}|\Sigma|^{\frac{1}{2}n}\Gamma_p\left(\tfrac{1}{2}n\right)$$

as an identity in n.]

18. (Sec. 7.5) Consider the confidence region for μ given by

$$N(\bar{x}-\mu^*)'S^{-1}(\bar{x}-\mu^*) \le \frac{(N-1)p}{N-p} F_{p,N-p}(\varepsilon),$$

where $\bar{x}$ and S are based on a sample of N from $N(\mu, \Sigma)$. Find the expected value of the volume of the confidence region.

19. (Sec. 7.6) Prove that if $\Sigma = I$ the joint density of $r_{ij\cdot p}$, $i, j = 1, \dots, p-1$, and $r_{1p}, \dots, r_{p-1,p}$ is

$$\frac{\Gamma^{p-1}\left[\frac{1}{2}(n-1)\right]|R_{11\cdot p}|^{\frac{1}{2}(n-p-1)}}{\pi^{(p-1)(p-2)/4}\prod_{i=1}^{p-1}\Gamma\left[\frac{1}{2}(n-i)\right]} \prod_{i=1}^{p-1} \frac{\Gamma\left(\frac{1}{2}n\right)}{\pi^{\frac{1}{2}}\Gamma\left[\frac{1}{2}(n-1)\right]}\left(1-r_{ip}^2\right)^{\frac{1}{2}(n-3)},$$

where $R_{11\cdot p} = (r_{ij\cdot p})$. [*Hint*: $r_{ij\cdot p} = (r_{ij} - r_{ip}r_{jp})/(\sqrt{1-r_{ip}^2}\sqrt{1-r_{jp}^2})$ and $|r_{ij}| = |\sqrt{1-r_{ip}^2}\sqrt{1-r_{jp}^2}\, r_{ij\cdot p}|$. Use (9).]

20. (Sec. 7.6) Prove that the joint density of $r_{12\cdot 3,\dots,p}$, $r_{13\cdot 4,\dots,p}$, $r_{23\cdot 4,\dots,p}, \dots$, $r_{1p}, \dots, r_{p-1,p}$ is

$$\frac{\Gamma\left\{\frac{1}{2}[n-(p-2)]\right\}}{\pi^{\frac{1}{2}}\Gamma\left\{\frac{1}{2}[n-(p-1)]\right\}}\left(1-r_{12\cdot 3,\dots,p}^2\right)^{\frac{1}{2}[n-(p+1)]}$$

$$\times \prod_{i=1}^{2} \frac{\Gamma\left\{\frac{1}{2}[n-(p-3)]\right\}}{\pi^{\frac{1}{2}}\Gamma\left\{\frac{1}{2}[n-(p-2)]\right\}}\left(1-r_{i3\cdot 4,\dots,p}^2\right)^{\frac{1}{2}(n-p)}$$

$$\cdots \times \prod_{i=1}^{p-2} \frac{\Gamma\left[\frac{1}{2}(n-1)\right]}{\pi^{\frac{1}{2}}\Gamma\left[\frac{1}{2}(n-2)\right]}\left(1-r_{i,p-1\cdot p}^2\right)^{\frac{1}{2}(n-4)}$$

$$\times \prod_{i=1}^{p-1} \frac{\Gamma\left(\frac{1}{2}n\right)}{\pi^{\frac{1}{2}}\Gamma\left[\frac{1}{2}(n-1)\right]}\left(1-r_{ip}^2\right)^{\frac{1}{2}(n-3)}.$$

[*Hint*: Use the result of Problem 19 inductively.]

21. (Sec. 7.6) Prove (without the use of Problem 20) that if $\Sigma = I$, then $r_{1p}, \dots, r_{p-1,p}$ are independently distributed. [*Hint*: $r_{ip} = a_{ip}/(\sqrt{a_{ii}}\sqrt{a_{pp}})$. Prove that the pairs $(a_{1p}, a_{11}), \dots, (a_{p-1,p}, a_{p-1,p-1})$ are independent when $(z_{1p}, \dots, z_{np})$ are fixed, and note from Section 4.2.1 that the marginal distribution of r_{ip}, conditional on $z_{\alpha p}$, does not depend on $z_{\alpha p}$.]

22. (Sec. 7.6) Prove (without the use of Problems 19 and 20) that if $\Sigma = I$, then the set $r_{1p}, \dots, r_{p-1,p}$ is independent of the set $r_{ij\cdot p}$, $i, j = 1, \dots, p-1$. [*Hint*: From Section 4.3.2 $a_{pp}, (a_{ip})$ are independent of $(a_{ij\cdot p})$. Prove that a_{pp}, (a_{ip}) and a_{ii}, $i = 1, \dots, p-1$, are independent of $(r_{ij\cdot p})$ by proving that $a_{ii\cdot p}$ are independent of $(r_{ij\cdot p})$. See Problem 21 of Chapter 4.]

23. (Sec. 7.6) Prove the conclusion of Problem 20 by using Problems 21 and 22.

24. (Sec. 7.6) Reverse the steps in Problem 20 to derive (9) of Section 7.6.

25. (Sec. 7.6) Show that when $p = 3$ and Σ is diagonal r_{12}, r_{13}, r_{23} are not mutually independent.

26. (Sec. 7.6) Show that when Σ is diagonal the set r_{ij} are pairwise independent.

27. (Sec. 7.7) *Multivariate t-distribution*. Let y and u be independently distributed according to $N(\mathbf{0}, \Sigma)$ and the χ_n^2-distribution, respectively, and let $\sqrt{n/u}\, y = x - \mu$.

(*a*) Show that the density of x is

$$\frac{\Gamma[\frac{1}{2}(n+p)]}{\Gamma(\frac{1}{2}n)\, n^{\frac{1}{2}p} \pi^{\frac{1}{2}p} |\Sigma|^{\frac{1}{2}} \left[1 + \frac{1}{n}(x-\mu)'\Sigma^{-1}(x-\mu)\right]^{\frac{1}{2}(n+p)}}.$$

(*b*) Show that $\mathscr{E} x = \mu$ and

$$\mathscr{E}(x-\mu)(x-\mu)' = \frac{n}{n-2}\Sigma.$$

28. (Sec. 7.8) Prove that $F\varepsilon$ is not proportional to f by calculating $F\varepsilon$.

29. (Sec. 7.8) Prove for $p = 2$

$$TDT' = d_1 A + (d_2 - d_1)\begin{pmatrix} 0 & 0 \\ 0 & \dfrac{|A|}{a_{11}} \end{pmatrix}.$$

30. (Sec. 7.8) Verify (17) and (18). [*Hint*: To verify (18) let $\Sigma = KK'$, $A = KA^*K'$, and $A^* = T^*T^{*\prime}$, where K and T^* are lower triangular.]

31. (Sec. 7.8) Prove for optimal D

$$\mathscr{E}_I L_l(I, S) - \mathscr{E}_I L_l(I, TDT') = -\sum_{i=1}^{\frac{1}{2}p} \log\left[1 - \left(\frac{p-2i+1}{n}\right)^2\right], \qquad p \text{ even},$$

$$= -\sum_{i=1}^{\frac{1}{2}(p-1)} \log\left[1 - \left(\frac{p-2i+1}{n}\right)^2\right], \qquad p \text{ odd}.$$

32. (Sec. 7.8) Prove $L_q(\boldsymbol{\Sigma}, \boldsymbol{G})$ and $L_l(\boldsymbol{\Sigma}, \boldsymbol{G})$ are invariant with respect to transformations $\boldsymbol{G}^* = \boldsymbol{CGC}'$, $\boldsymbol{\Sigma}^* = \boldsymbol{C\Sigma C}'$ for $\boldsymbol{C}$ nonsingular.

33. (Sec. 7.8) Prove $L_q(\boldsymbol{\Sigma}, \boldsymbol{G})$ is a multiple of $(\boldsymbol{g} - \boldsymbol{\sigma})'\boldsymbol{\Phi}^{-1}(\boldsymbol{g} - \boldsymbol{\sigma})$. [*Hint*: Transform so $\boldsymbol{\Sigma} = \boldsymbol{I}$. Then show

$$\boldsymbol{\Phi} = \frac{1}{n}\begin{pmatrix} 2\boldsymbol{I} & \boldsymbol{0} \\ \boldsymbol{0} & \boldsymbol{I} \end{pmatrix}.\Bigg]$$

34. (Sec. 7.8) Verify (11).

35. Let the density of $\boldsymbol{Y}$ be $f(y) = K$ for $y'y \le p + 2$ and 0 elsewhere. Prove that $K = \Gamma(\frac{1}{2}p + 1)/[(p+2)\pi]^{\frac{1}{2}p}$ and show that $\mathscr{E}\boldsymbol{Y} = \boldsymbol{0}$ and $\mathscr{E}\boldsymbol{YY}' = \boldsymbol{I}$.

36. (Sec. 7.2) *Dirichlet distribution*. Let $Y_1, \dots, Y_m$ be independently distributed as χ^2-variables with $p_1, \dots, p_m$ degrees of freedom, respectively. Define $Z_i = Y_i / \sum_{j=1}^m Y_j$, $i = 1, \dots, m$. Show that the density of $Z_1, \dots, Z_{m-1}$ is

$$\frac{\Gamma\left(\frac{1}{2}\sum_{i=1}^m p_i\right)}{\prod_{i=1}^m \Gamma\left(\frac{1}{2}p_i\right)} z_1^{p_1 - 1} \cdots z_m^{p_m - 1}, \qquad z_m \equiv 1 - \sum_{i=1}^{m-1} z_i,$$

for $z_i \ge 0$, $i = 1, \dots, m$.

37. (Sec. 7.5) Show that if χ_{N-1}^2 and χ_{N-2}^2 are independently distributed, then $\chi_{N-1}^2 \cdot \chi_{N-2}^2$ is distributed as $(\chi_{2N-4}^2)^2/4$. [*Hint*: In the joint density of $x = \chi_{N-1}^2$ and $y = \chi_{N-2}^2$ substitute $z = 2\sqrt{xy}$, $x = x$ and express the marginal density of z as $z^{N-3}h(z)$, where $h(z)$ is an integral with respect to x. Find $h'(z)$ and solve the differential equation. Srivastava and Khatri (1979), Chapter 3.]

CHAPTER 8

Testing the General Linear Hypothesis; Multivariate Analysis of Variance

8.1. INTRODUCTION

In this chapter we generalize the univariate least squares theory (i.e., "regression analysis") and the analysis of variance to vector variates. The algebra of the multivariate case is essentially the same as that of the univariate case. This leads to distribution theory that is analogous to that of the univariate case and to test criteria that are analogs of F-statistics. In fact, given a univariate test, we shall be able to write down immediately a corresponding multivariate test. Since the analysis of variance based on the model of fixed effects can be obtained from least squares theory, we obtain directly a multivariate analysis of variance theory. However, in the multivariate case there is more latitude in the choice of tests of significance.

In univariate least squares we consider scalar dependent variates $x_1, \dots, x_N$ drawn from populations with expected values $\boldsymbol{\beta}'z_1, \dots, \boldsymbol{\beta}'z_N$, respectively, where $\boldsymbol{\beta}$ is a column vector of q components and each of the z_α is a column vector of q known components. Under the assumption that the variances in the populations are the same, the least squares estimators of $\boldsymbol{\beta}'$ is

(1)
$$\boldsymbol{b}' = \left(\sum_{\alpha=1}^{N} x_\alpha z_\alpha' \right) \left(\sum_{\alpha=1}^{N} z_\alpha z_\alpha' \right)^{-1}.$$

If the populations are normal, the vector is the maximum likelihood estimator of $\boldsymbol{\beta}$. The unbiased estimator of the common variance σ^2 is

$$s^2 = \sum_{\alpha=1}^{N} \left(x_\alpha - \boldsymbol{b}'\boldsymbol{z}_\alpha\right)^2 / (N-q), \tag{2}$$

and under the assumption of normality, the maximum likelihood estimator of σ^2 is $\hat{\sigma}^2 = (N-q)s^2/N$.

In the multivariate case $\boldsymbol{x}_\alpha$ is a vector, $\boldsymbol{\beta}'$ is replaced by a matrix $\mathbf{B}$, and σ^2 is replaced by a covariance matrix Σ. The estimators of $\mathbf{B}$ and Σ, given in Section 8.2, are matric analogues of (1) and (2).

To test a hypothesis concerning $\boldsymbol{\beta}$, say the hypothesis $\boldsymbol{\beta} = \mathbf{0}$, we use an F-test. A criterion equivalent to the F-ratio is

$$\frac{1}{[q/(N-q)]F + 1} = \frac{\hat{\sigma}^2}{\hat{\sigma}_0^2}, \tag{3}$$

where $\hat{\sigma}_0^2$ is the maximum likelihood estimator of σ^2 under the null hypothesis. We shall find that the likelihood ratio criterion for the corresponding multivariate hypothesis, say $\mathbf{B} = \mathbf{0}$, is the above with the variances replaced by generalized variances. The distribution of the likelihood ratio criterion under the null hypothesis is characterized, the moments are found, and some specific distributions obtained. Satisfactory approximations are given as well as tables of significance points (in Appendix B).

The hypothesis testing problem is invariant under several groups of linear transformations. Other invariant criteria are treated, including the Lawley–Hotelling trace, the Bartlett–Nanda–Pillai trace, and the Roy maximum root criteria. Some comparison of power is made.

Confidence regions or simultaneous confidence intervals for elements of $\mathbf{B}$ can be based on the likelihood ratio test, the Lawley–Hotelling trace test, and the Roy maximum root test. Procedures are given explicitly for several problems of the analysis of variance. Finally optimal properties of admissibility, unbiasedness, and monotonicity of power functions are studied.

8.2. ESTIMATORS OF PARAMETERS IN MULTIVARIATE LINEAR REGRESSION

8.2.1. The Maximum Likelihood Estimators

Suppose $x_1, \dots, x_N$ are a set of N independent observations, x_α being drawn from $N(\boldsymbol{\beta} z_\alpha, \Sigma)$. Ordinarily the vectors z_α (with q components) are known vectors and the $p \times p$ matrix Σ and the $p \times q$ matrix $\boldsymbol{\beta}$ are unknown. We assume $N \geq p + q$ and the rank of

$$Z = (z_1, \dots, z_N) \tag{1}$$

is q. We shall estimate Σ and $\boldsymbol{\beta}$ by the method of maximum likelihood. The likelihood function is

$$L = (2\pi)^{-\frac{1}{2}Np} |\Sigma^*|^{-\frac{1}{2}N} \exp\left[-\frac{1}{2} \sum_{\alpha=1}^{N} (x_\alpha - \boldsymbol{\beta}^* z_\alpha)' \Sigma^{*-1} (x_\alpha - \boldsymbol{\beta}^* z_\alpha) \right]. \tag{2}$$

In (2) the elements of Σ^* and $\boldsymbol{\beta}^*$ are indeterminates. The method of maximum likelihood specifies the estimators of Σ and $\boldsymbol{\beta}$ based on the given sample $x_1, z_1, \dots, x_N, z_N$ as the Σ^* and $\boldsymbol{\beta}^*$ that maximize (2). It is convenient to use the following lemma.

LEMMA 8.2.1. *Let*

$$B = \sum_{\alpha=1}^{N} x_\alpha z_\alpha' \left(\sum_{\alpha=1}^{N} z_\alpha z_\alpha' \right)^{-1}. \tag{3}$$

Then for any $p \times q$ matrix F

$$\sum_{\alpha=1}^{N} (x_\alpha - F z_\alpha)(x_\alpha - F z_\alpha)' = \sum_{\alpha=1}^{N} (x_\alpha - B z_\alpha)(x_\alpha - B z_\alpha)' + (B - F) \sum_{\alpha=1}^{N} z_\alpha z_\alpha' (B - F)'. \tag{4}$$

PROOF. The left-hand side of (4) is

$$\sum_{\alpha=1}^{N} [(x_\alpha - B z_\alpha) + (B - F) z_\alpha][(x_\alpha - B z_\alpha) + (B - F) z_\alpha]', \tag{5}$$

which is equal to the right-hand side of (4) because

$$(B - F) \sum_{\alpha=1}^{N} z_\alpha (x_\alpha - Bz_\alpha)' = 0 \tag{6}$$

by virtue of (3). Q.E.D.

The exponential in L is $-\frac{1}{2}$ times

$$\begin{aligned} \operatorname{tr} \Sigma^{*-1} \sum_{\alpha=1}^{N} (x_\alpha - \boldsymbol{\beta}^* z_\alpha)(x_\alpha - \boldsymbol{\beta}^* z_\alpha)' &= \operatorname{tr} \Sigma^{*-1} \sum_{\alpha=1}^{N} (x_\alpha - Bz_\alpha)(x_\alpha - Bz_\alpha)' \\ &\quad + \operatorname{tr} \Sigma^{*-1} (B - \boldsymbol{\beta}^*) A (B - \boldsymbol{\beta}^*)', \end{aligned} \tag{7}$$

where

$$A = \sum_{\alpha=1}^{N} z_\alpha z_\alpha'. \tag{8}$$

The likelihood is maximized with respect to $\boldsymbol{\beta}^*$ by minimizing the last term in (7).

LEMMA 8.2.2. *If A and G are positive definite,* $\operatorname{tr} FAF'G > 0$ *for* $F \neq 0$.

PROOF. Let $A = HH'$, $G = KK'$. Then

$$\begin{aligned} \operatorname{tr} FAF'G &= \operatorname{tr} FHH'F'KK' = \operatorname{tr} K'FHH'F'K \\ &= \operatorname{tr}(K'FH)(K'FH)' > 0 \end{aligned} \tag{9}$$

for $F \neq 0$ because then $K'FH \neq 0$ since H and K are nonsingular. Q.E.D.

It follows from (7) and the lemma that L is maximized with respect to $\boldsymbol{\beta}^*$ by $\hat{\boldsymbol{\beta}}^* = B$, that is,

$$\hat{\boldsymbol{\beta}} = CA^{-1}, \tag{10}$$

where

$$C = \sum_{\alpha=1}^{N} x_\alpha z_\alpha'. \tag{11}$$

Then by Lemma 3.2.2 L is maximized with respect to Σ^* at

$$\hat{\Sigma} = \frac{1}{N} \sum_{\alpha=1}^{N} \left(x_\alpha - \hat{\boldsymbol{\beta}} z_\alpha \right)\left(x_\alpha - \hat{\boldsymbol{\beta}} z_\alpha \right)'. \tag{12}$$

This is the multivariate analogue of $\hat{\sigma}^2 = (N-q)s^2/N$ defined by (2) of Section 8.1.

THEOREM 8.2.1. *If x_α is an observation from $N(\boldsymbol{\beta} z_\alpha, \Sigma)$, $\alpha = 1, \dots, N$, with $(z_1, \dots, z_N)$ of rank q, the maximum likelihood estimator of $\boldsymbol{\beta}$ is given by* (10), *where $C = \sum_\alpha x_\alpha z_\alpha'$ and $A = \sum_\alpha z_\alpha z_\alpha'$. The maximum likelihood estimator of Σ is given by* (12).

A useful algebraic result follows from (12) and (4) with $F = 0$:

$$N\hat{\Sigma} = \sum_{\alpha=1}^{N} x_\alpha x_\alpha' - \hat{\boldsymbol{\beta}} A \hat{\boldsymbol{\beta}}' = \sum_{\alpha=1}^{N} x_\alpha x_\alpha' - CA^{-1}C'. \tag{13}$$

Now let us consider a geometric interpretation of the estimation procedure. Let the ith row of $(x_1, \dots, x_N)$ be x_i^* (with N components) and the ith row of $(z_1, \dots, z_N)$ be z_i^* (with N components). Then $\sum_j \hat{\beta}_{ij} z_j^*$, being a linear combination of the vectors $z_1^*, \dots, z_q^*$, is a vector in the q-space spanned by $z_1^*, \dots, z_q^*$, and is in fact of all such vectors the one nearest to x_i^*; hence, it is the projection of x_i^* on the q-space. Thus $x_i^* - \sum_j \hat{\beta}_{ij} z_j^*$ is the vector orthogonal to the q-space going from the projection of x_i^* on the q-space to x_i^*. Translate this vector so that one endpoint of this vector is at the origin. Then the set of p vectors, $x_1^* - \sum_j \hat{\beta}_{1j} z_j^*, \dots, x_p^* - \sum_j \hat{\beta}_{pj} z_j^*$, is a set of vectors emanating from the origin. $N\hat{\sigma}_{ii} = (x_i^* - \sum_j \hat{\beta}_{ij} z_j^*)(x_i^* - \sum_j \hat{\beta}_{ij} z_j^*)'$ is the square of the length of the ith such vector, and $N\hat{\sigma}_{ij} = (x_i^* - \sum_h \hat{\beta}_{ih} z_h^*)(x_j^* - \sum_g \hat{\beta}_{jg} z_g^*)'$ is the product of the length of the ith vector, the length of the jth vector, and the cosine of the angle between them.

The equations defining the maximum likelihood estimator of $\boldsymbol{\beta}$, namely, $AB' = C'$, consist of p sets of q linear equations in q unknowns. Each set is solved by the method of pivotal condensation or successive elimination (Section A.5 of Appendix A). The forward solutions are the same (except the right-hand sides) for all sets. Use of (13) to compute $N\hat{\Sigma}$ involves an efficient computation of $\hat{\boldsymbol{\beta}} A \hat{\boldsymbol{\beta}}'$.

8.2.2. Distribution of $\hat{\boldsymbol{\beta}}$ and $\hat{\Sigma}$

Now let us find the joint distribution of $\hat{\beta}_{ig}$ $(i = 1,\dots,p;\ g = 1,\dots,q)$. Clearly, the joint distribution is normal since the $\hat{\beta}_{ig}$ are linear combinations of the $X_{i\alpha}$. From (10) we see that

$$\text{(14)} \qquad \begin{aligned} \mathscr{E}\hat{\boldsymbol{\beta}} &= \mathscr{E}\sum_{\alpha=1}^{N} \boldsymbol{X}_\alpha \boldsymbol{z}_\alpha' \boldsymbol{A}^{-1} \\ &= \sum_{\alpha=1}^{N} \boldsymbol{\beta} \boldsymbol{z}_\alpha \boldsymbol{z}_\alpha' \boldsymbol{A}^{-1} = \boldsymbol{\beta}\boldsymbol{A}\boldsymbol{A}^{-1} \\ &= \boldsymbol{\beta}. \end{aligned}$$

The covariance between $\hat{\boldsymbol{\beta}}_i'$ and $\hat{\boldsymbol{\beta}}_j'$, two rows of $\hat{\boldsymbol{\beta}}$, is

(15)

$$\begin{aligned} \mathscr{E}(\hat{\boldsymbol{\beta}}_i - \boldsymbol{\beta}_i)(\hat{\boldsymbol{\beta}}_j - \boldsymbol{\beta}_j)' &= \boldsymbol{A}^{-1}\mathscr{E}\sum_{\alpha=1}^{N} (X_{i\alpha} - \mathscr{E}X_{i\alpha})\boldsymbol{z}_\alpha \sum_{\gamma=1}^{N} (X_{j\gamma} - \mathscr{E}X_{j\gamma})\boldsymbol{z}_\gamma' \boldsymbol{A}^{-1} \\ &= \boldsymbol{A}^{-1} \sum_{\alpha,\gamma=1}^{N} \mathscr{E}(X_{i\alpha} - \mathscr{E}X_{i\alpha})(X_{j\gamma} - \mathscr{E}X_{j\gamma})\boldsymbol{z}_\alpha \boldsymbol{z}_\gamma' \boldsymbol{A}^{-1} \\ &= \boldsymbol{A}^{-1} \sum_{\alpha,\gamma=1}^{N} \delta_{\alpha\gamma}\sigma_{ij}\boldsymbol{z}_\alpha \boldsymbol{z}_\gamma' \boldsymbol{A}^{-1} \\ &= \boldsymbol{A}^{-1} \sum_{\alpha=1}^{N} \sigma_{ij}\boldsymbol{z}_\alpha \boldsymbol{z}_\alpha' \boldsymbol{A}^{-1} \\ &= \sigma_{ij}\boldsymbol{A}^{-1}\boldsymbol{A}\boldsymbol{A}^{-1} \\ &= \sigma_{ij}\boldsymbol{A}^{-1}. \end{aligned}$$

To summarize, the vector of pq components $(\hat{\boldsymbol{\beta}}_1',\dots,\hat{\boldsymbol{\beta}}_p')' = \operatorname{vec}\hat{\boldsymbol{\beta}}'$ is normally distributed with mean $(\boldsymbol{\beta}_1',\dots,\boldsymbol{\beta}_p')' = \operatorname{vec}\boldsymbol{\beta}'$ and covariance matrix

$$\text{(16)} \qquad \begin{pmatrix} \sigma_{11}\boldsymbol{A}^{-1} & \sigma_{12}\boldsymbol{A}^{-1} & \cdots & \sigma_{1p}\boldsymbol{A}^{-1} \\ \sigma_{21}\boldsymbol{A}^{-1} & \sigma_{22}\boldsymbol{A}^{-1} & \cdots & \sigma_{2p}\boldsymbol{A}^{-1} \\ \vdots & \vdots & & \vdots \\ \sigma_{p1}\boldsymbol{A}^{-1} & \sigma_{p2}\boldsymbol{A}^{-1} & \cdots & \sigma_{pp}\boldsymbol{A}^{-1} \end{pmatrix}.$$

The matrix (16) is the Kronecker (or direct) product of the matrices Σ and A^{-1}, denoted by $\Sigma \otimes A^{-1}$.

From Theorem 4.3.3 it follows that $N\hat{\Sigma} = \sum_{\alpha=1}^{N} x_\alpha x_\alpha' - \hat{\boldsymbol{\beta}} A \hat{\boldsymbol{\beta}}'$ is distributed according to $W(\Sigma, N-q)$. From this we see that an unbiased estimator of Σ is $S = [N/(N-q)]\hat{\Sigma}$.

THEOREM 8.2.2. *The maximum likelihood estimator* $\hat{\boldsymbol{\beta}}$ *based on a set of* N *observations, the* αth *from* $N(\boldsymbol{\beta} z_\alpha, \Sigma)$, *is normally distributed with mean* $\boldsymbol{\beta}$, *and the covariance matrix of the* ith *and* jth *rows of* $\hat{\boldsymbol{\beta}}$ *is* $\sigma_{ij}A^{-1}$, *where* $A = \sum_\alpha z_\alpha z_\alpha'$. *The maximum likelihood estimator* $\hat{\Sigma}$ *multiplied by* N *is independently distributed according to* $W(\Sigma, N-q)$, *where* q *is the number of components of* z_α.

The density then can be written (by virtue of (4))

$$\frac{1}{(2\pi)^{\frac{1}{2}pN}|\Sigma|^{\frac{1}{2}N}} \exp\left(-\frac{1}{2}\operatorname{tr}\left\{\Sigma^{-1}\left[(\hat{\boldsymbol{\beta}}-\boldsymbol{\beta})A(\hat{\boldsymbol{\beta}}-\boldsymbol{\beta})' + N\hat{\Sigma}\right]\right\}\right). \tag{17}$$

This proves the following:

COROLLARY 8.2.1. $\hat{\boldsymbol{\beta}}$ *and* $\hat{\Sigma}$ *form a sufficient set of statistics for* $\boldsymbol{\beta}$ *and* Σ.

A useful theorem is the following.

THEOREM 8.2.3. *Let* X_α *be distributed according to* $N(\boldsymbol{\beta} z_\alpha, \Sigma)$, $\alpha = 1, \ldots, N$, *and suppose* $X_1, \ldots, X_N$ *are independent.*

(*a*) *If* $w_\alpha = Hz_\alpha$ *and* $\Gamma = \boldsymbol{\beta} H^{-1}$, *then* X_α *is distributed according to* $N(\Gamma w_\alpha, \Sigma)$.

(*b*) *The maximum likelihood estimator of* Γ *based on observations* x_α *on* X_α, $\alpha = 1, \ldots, N$, *is* $\hat{\Gamma} = \hat{\boldsymbol{\beta}} H^{-1}$, *where* $\hat{\boldsymbol{\beta}}$ *is the maximum likelihood estimator of* $\boldsymbol{\beta}$.

(*c*) $\hat{\Gamma}(\sum_\alpha w_\alpha w_\alpha')\hat{\Gamma}' = \hat{\boldsymbol{\beta}} A \hat{\boldsymbol{\beta}}'$, *where* $A = \sum_\alpha z_\alpha z_\alpha'$, *and the maximum likelihood estimator of* $N\Sigma$ *is* $N\hat{\Sigma} = \sum_\alpha x_\alpha x_\alpha' - \hat{\Gamma}(\sum_\alpha w_\alpha w_\alpha')\hat{\Gamma}' = \sum_\alpha x_\alpha x_\alpha' - \hat{\boldsymbol{\beta}} A \hat{\boldsymbol{\beta}}'$.

(*d*) $\hat{\Gamma}$ *and* $\hat{\Sigma}$ *are independently distributed.*

(*e*) $\hat{\Gamma}$ *is normally distributed with mean* Γ *and the covariance matrix of the* ith *and* jth *rows of* $\hat{\Gamma}$ *is* $\sigma_{ij}(HAH')^{-1} = \sigma_{ij}H'^{-1}A^{-1}H^{-1}$.

The proof is left to the reader.

8.3. LIKELIHOOD RATIO CRITERIA FOR TESTING LINEAR HYPOTHESES ABOUT REGRESSION COEFFICIENTS

8.3.1. Likelihood Ratio Criteria

Suppose we partition

$$\boldsymbol{\beta} = (\boldsymbol{\beta}_1 \ \boldsymbol{\beta}_2) \tag{1}$$

such that $\boldsymbol{\beta}_1$ has q_1 columns and $\boldsymbol{\beta}_2$ has q_2 columns. We shall derive the likelihood ratio criterion for testing the hypothesis

$$H: \boldsymbol{\beta}_1 = \boldsymbol{\beta}_1^*, \tag{2}$$

where $\boldsymbol{\beta}_1^*$ is a given matrix. The maximum of the likelihood function L for the sample $\boldsymbol{x}_1, \ldots, \boldsymbol{x}_N$ is

$$\max_{\boldsymbol{\beta}, \boldsymbol{\Sigma}} L = (2\pi)^{-\frac{1}{2}pN} |\hat{\boldsymbol{\Sigma}}_\Omega|^{-\frac{1}{2}N} e^{-\frac{1}{2}pN}, \tag{3}$$

where $\hat{\boldsymbol{\Sigma}}_\Omega$ is given by (12) or (13) of Section 8.2.

To find the maximum of the likelihood function for the parameters restricted to ω defined by (2) we let

$$\boldsymbol{y}_\alpha = \boldsymbol{x}_\alpha - \boldsymbol{\beta}_1^* \boldsymbol{z}_\alpha^{(1)}, \qquad \alpha = 1, \ldots, N, \tag{4}$$

where

$$\boldsymbol{z}_\alpha = \begin{pmatrix} \boldsymbol{z}_\alpha^{(1)} \\ \boldsymbol{z}_\alpha^{(2)} \end{pmatrix}, \qquad \alpha = 1, \ldots, N, \tag{5}$$

is partitioned in a manner corresponding to the partitioning of $\boldsymbol{\beta}$. Then $\boldsymbol{y}_\alpha$ can be considered as an observation from $N(\boldsymbol{\beta}_2 \boldsymbol{z}_\alpha^{(2)}, \boldsymbol{\Sigma})$. The estimator of $\boldsymbol{\beta}_2$ is obtained by the procedure of Section 8.2 as

$$\begin{aligned} \hat{\boldsymbol{\beta}}_{2\omega} &= \sum_{\alpha=1}^{N} \boldsymbol{y}_\alpha \boldsymbol{z}_\alpha^{(2)\prime} \boldsymbol{A}_{22}^{-1} = \sum_{\alpha=1}^{N} \left(\boldsymbol{x}_\alpha - \boldsymbol{\beta}_1^* \boldsymbol{z}_\alpha^{(1)}\right) \boldsymbol{z}_\alpha^{(2)\prime} \boldsymbol{A}_{22}^{-1} \\ &= \left(\boldsymbol{C}_2 - \boldsymbol{\beta}_1^* \boldsymbol{A}_{12}\right) \boldsymbol{A}_{22}^{-1} \end{aligned} \tag{6}$$

with C and A being partitioned in the manner corresponding to the partitioning of $\boldsymbol{\beta}$ and z_α,

$$C = (C_1 \; C_2), \tag{7}$$

$$A = \begin{pmatrix} A_{11} & A_{12} \\ A_{21} & A_{22} \end{pmatrix}. \tag{8}$$

The estimator of $\boldsymbol{\Sigma}$ is given by

$$\begin{aligned} N\hat{\boldsymbol{\Sigma}}_\omega &= \sum_{\alpha=1}^{N} \left(y_\alpha - \hat{\boldsymbol{\beta}}_{2\omega} z_\alpha^{(2)} \right)\left(y_\alpha - \hat{\boldsymbol{\beta}}_{2\omega} z_\alpha^{(2)} \right)' \\ &= \sum_{\alpha=1}^{N} y_\alpha y_\alpha' - \hat{\boldsymbol{\beta}}_{2\omega} A_{22} \hat{\boldsymbol{\beta}}_{2\omega}' \\ &= \sum_{\alpha=1}^{N} \left(x_\alpha - \boldsymbol{\beta}_1^* z_\alpha^{(1)} \right)\left(x_\alpha - \boldsymbol{\beta}_1^* z_\alpha^{(1)} \right)' - \hat{\boldsymbol{\beta}}_{2\omega} A_{22} \hat{\boldsymbol{\beta}}_{2\omega}'. \end{aligned} \tag{9}$$

Thus the maximum of the likelihood function over ω is

$$\max_{\boldsymbol{\beta}_2, \boldsymbol{\Sigma}} L = (2\pi)^{-\frac{1}{2}pN} |\hat{\boldsymbol{\Sigma}}_\omega|^{-\frac{1}{2}N} e^{-\frac{1}{2}pN}. \tag{10}$$

The likelihood ratio criterion for testing H is (10) divided by (3), namely,

$$\lambda = \frac{|\hat{\boldsymbol{\Sigma}}_\Omega|^{\frac{1}{2}N}}{|\hat{\boldsymbol{\Sigma}}_\omega|^{\frac{1}{2}N}}. \tag{11}$$

In testing H, one rejects the hypothesis if $\lambda < \lambda_0$, where λ_0 is a suitably chosen number.

A special case of this problem led to Hotelling's T^2-criterion. If $q = q_1 = 1$ ($q_2 = 0$), $z_\alpha = 1$, $\alpha = 1, \dots, N$, $\boldsymbol{\beta} = \boldsymbol{\beta}_1 = \boldsymbol{\mu}$, then the T^2-criterion for testing the hypothesis $\boldsymbol{\mu} = \boldsymbol{\mu}_0$ is a monotonic function of (11) for $\boldsymbol{\beta}_1^* = \boldsymbol{\mu}_0$.

The hypothesis $\boldsymbol{\mu} = \mathbf{0}$ and the T^2-statistic are invariant with respect to the transformations $X^* = DX$ and $x_\alpha^* = Dx_\alpha$, $\alpha = 1, \dots, N$, for nonsingular D. Similarly, in this problem the null hypothesis $\boldsymbol{\beta}_1 = \mathbf{0}$ and the likelihood ratio criterion for testing it are invariant with respect to nonsingular linear transformations.

THEOREM 8.3.1. *The likelihood ratio criterion* (11) *for testing the null hypothesis* $\boldsymbol{\beta}_1 = \mathbf{0}$ *is invariant with respect to transformations* $\boldsymbol{x}_\alpha^* = \boldsymbol{D}\boldsymbol{x}_\alpha$, $\alpha = 1, \dots, N$, *for nonsingular* $\boldsymbol{D}$.

PROOF. The estimators in terms of $\boldsymbol{x}_\alpha^*$ are

$$\hat{\boldsymbol{\beta}}^* = \boldsymbol{D}\boldsymbol{C}\boldsymbol{A}^{-1} = \boldsymbol{D}\hat{\boldsymbol{\beta}}, \tag{12}$$

$$\hat{\boldsymbol{\Sigma}}_\Omega^* = \frac{1}{N}\sum_{\alpha=1}^{N}\left(\boldsymbol{D}\boldsymbol{x}_\alpha - \boldsymbol{D}\hat{\boldsymbol{\beta}}\boldsymbol{z}_\alpha\right)\left(\boldsymbol{D}\boldsymbol{x}_\alpha - \boldsymbol{D}\hat{\boldsymbol{\beta}}\boldsymbol{z}_\alpha\right)' = \boldsymbol{D}\hat{\boldsymbol{\Sigma}}_\Omega\boldsymbol{D}', \tag{13}$$

$$\hat{\boldsymbol{\beta}}_{2\omega}^* = \boldsymbol{D}\boldsymbol{C}_2\boldsymbol{A}_{22}^{-1} = \boldsymbol{D}\hat{\boldsymbol{\beta}}_{2\omega}, \tag{14}$$

$$\hat{\boldsymbol{\Sigma}}_\omega^* = \frac{1}{N}\sum_{\alpha=1}^{N}\left(\boldsymbol{D}\boldsymbol{x}_\alpha - \boldsymbol{D}\hat{\boldsymbol{\beta}}_{2\omega}\boldsymbol{z}_\alpha^{(2)}\right)\left(\boldsymbol{D}\boldsymbol{x}_\alpha - \boldsymbol{D}\hat{\boldsymbol{\beta}}_{2\omega}\boldsymbol{z}_\alpha^{(2)}\right)' = \boldsymbol{D}\hat{\boldsymbol{\Sigma}}_\omega\boldsymbol{D}'. \tag{15}$$

Q.E.D.

8.3.2. Geometric Interpretation

An insight into the algebra developed here can be given in terms of a geometric interpretation. It will be convenient to use the following lemma:

LEMMA 8.3.1.

$$\hat{\boldsymbol{\beta}}_{2\omega} - \hat{\boldsymbol{\beta}}_{2\Omega} = \left(\hat{\boldsymbol{\beta}}_{1\Omega} - \boldsymbol{\beta}_1^*\right)\boldsymbol{A}_{12}\boldsymbol{A}_{22}^{-1}. \tag{16}$$

PROOF. The normal equation $\hat{\boldsymbol{\beta}}_\Omega\boldsymbol{A} = \boldsymbol{C}$ is written in partitioned form

$$\left(\hat{\boldsymbol{\beta}}_{1\Omega}\boldsymbol{A}_{11} + \hat{\boldsymbol{\beta}}_{2\Omega}\boldsymbol{A}_{21}, \hat{\boldsymbol{\beta}}_{1\Omega}\boldsymbol{A}_{12} + \hat{\boldsymbol{\beta}}_{2\Omega}\boldsymbol{A}_{22}\right) = \left(\boldsymbol{C}_1, \boldsymbol{C}_2\right). \tag{17}$$

Thus $\hat{\boldsymbol{\beta}}_{2\Omega} = \boldsymbol{C}_2\boldsymbol{A}_{22}^{-1} - \hat{\boldsymbol{\beta}}_{1\Omega}\boldsymbol{A}_{12}\boldsymbol{A}_{22}^{-1}$. The lemma follows by comparison with (6). Q.E.D.

We can now write

$$\begin{aligned} \boldsymbol{X} - \boldsymbol{\beta}\boldsymbol{Z} &= \left(\boldsymbol{X} - \hat{\boldsymbol{\beta}}_\Omega\boldsymbol{Z}\right) + \left(\hat{\boldsymbol{\beta}}_{2\Omega} - \boldsymbol{\beta}_2\right)\boldsymbol{Z}_2 + \left(\hat{\boldsymbol{\beta}}_{1\Omega} - \boldsymbol{\beta}_1^*\right)\boldsymbol{Z}_1 \\ &= \left(\boldsymbol{X} - \hat{\boldsymbol{\beta}}_\Omega\boldsymbol{Z}\right) + \left(\hat{\boldsymbol{\beta}}_{2\omega} - \boldsymbol{\beta}_2\right)\boldsymbol{Z}_2 \\ &\qquad - \left(\hat{\boldsymbol{\beta}}_{2\omega} - \hat{\boldsymbol{\beta}}_{2\Omega}\right)\boldsymbol{Z}_2 + \left(\hat{\boldsymbol{\beta}}_{1\Omega} - \boldsymbol{\beta}_1^*\right)\boldsymbol{Z}_1 \\ &= \left(\boldsymbol{X} - \hat{\boldsymbol{\beta}}_\Omega\boldsymbol{Z}\right) + \left(\hat{\boldsymbol{\beta}}_{2\omega} - \boldsymbol{\beta}_2\right)\boldsymbol{Z}_2 \\ &\qquad + \left(\hat{\boldsymbol{\beta}}_{1\Omega} - \boldsymbol{\beta}_1^*\right)\left(\boldsymbol{Z}_1 - \boldsymbol{A}_{12}\boldsymbol{A}_{22}^{-1}\boldsymbol{Z}_2\right) \end{aligned} \tag{18}$$

as an identity; here $X = (x_1, \ldots, x_N)$, $Z_1 = (z_1^{(1)}, \ldots, z_N^{(1)})$, and $Z_2 = (z_1^{(2)}, \ldots, z_N^{(2)})$. The rows of $Z = \begin{pmatrix} Z_1 \\ Z_2 \end{pmatrix}$ span a q-dimensional subspace in N-space. Each row of $\boldsymbol{\beta} Z$ is a vector in the q-space, and hence each row of $X - \boldsymbol{\beta} Z$ is a vector from a vector in the q-space to the corresponding row vector of X. Each row vector of $X - \boldsymbol{\beta} Z$ is expressed above as the sum of three row vectors. The first matrix on the right of (18) has as its ith row a vector orthogonal to the q-space and leading to the ith row vector of X (as shown in the preceding section). The row vectors of $(\hat{\boldsymbol{\beta}}_{2\omega} - \boldsymbol{\beta}_2)Z_2$ are vectors in the q_2-space spanned by the rows of Z_2 (since they are linear combinations of the rows of Z_2). The row vectors of $(\hat{\boldsymbol{\beta}}_{1\Omega} - \boldsymbol{\beta}_1^*)(Z_1 - A_{12}A_{22}^{-1}Z_2)$ are vectors in the q_1-space of $Z_1 - A_{12}A_{22}^{-1}Z_2$ and this space is in the q-space of Z, but orthogonal to the q_2-space of Z_2 [since $(Z_1 - A_{12}A_{22}^{-1}Z_2)Z_2' = \mathbf{0}$]. Thus each row of $X - \boldsymbol{\beta} Z$ is indicated in Figure 8.1 as the sum of three orthogonal vectors: one vector is in the space orthogonal to Z, one is in the space of Z_2, and one is in the subspace of Z that is orthogonal to Z_2.

From the orthogonality relations we have

$$
\begin{aligned}
(19)\quad (X - \boldsymbol{\beta} Z)(X - \boldsymbol{\beta} Z)'
&= (X - \hat{\boldsymbol{\beta}}_\Omega Z)(X - \hat{\boldsymbol{\beta}}_\Omega Z)' + (\hat{\boldsymbol{\beta}}_{2\omega} - \boldsymbol{\beta}_2)Z_2Z_2'(\hat{\boldsymbol{\beta}}_{2\omega} - \boldsymbol{\beta}_2)' \\
&\quad + (\hat{\boldsymbol{\beta}}_{1\Omega} - \boldsymbol{\beta}_1^*)(Z_1 - A_{12}A_{22}^{-1}Z_2)(Z_1 - A_{12}A_{22}^{-1}Z_2)'(\hat{\boldsymbol{\beta}}_{1\Omega} - \boldsymbol{\beta}_1^*)' \\
&= N\hat{\boldsymbol{\Sigma}}_\Omega + (\hat{\boldsymbol{\beta}}_{2\omega} - \boldsymbol{\beta}_2)A_{22}(\hat{\boldsymbol{\beta}}_{2\omega} - \boldsymbol{\beta}_2)' \\
&\qquad + (\hat{\boldsymbol{\beta}}_{1\Omega} - \boldsymbol{\beta}_1^*)(A_{11} - A_{12}A_{22}^{-1}A_{21})(\hat{\boldsymbol{\beta}}_{1\Omega} - \boldsymbol{\beta}_1^*)'.
\end{aligned}
$$

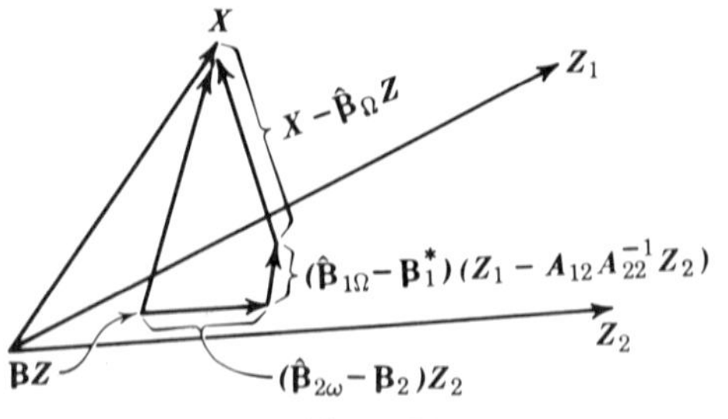

Figure 8.1

If we subtract $(\hat{\boldsymbol{\beta}}_{2\omega} - \boldsymbol{\beta}_2)\boldsymbol{Z}_2$ from both sides of (18) we have

$$(20)\quad \boldsymbol{X} - \boldsymbol{\beta}_1^*\boldsymbol{Z}_1 - \hat{\boldsymbol{\beta}}_{2\omega}\boldsymbol{Z}_2 = \left(\boldsymbol{X} - \hat{\boldsymbol{\beta}}_\Omega\boldsymbol{Z}\right) + \left(\hat{\boldsymbol{\beta}}_{1\Omega} - \boldsymbol{\beta}_1^*\right)\left(\boldsymbol{Z}_1 - \boldsymbol{A}_{12}\boldsymbol{A}_{22}^{-1}\boldsymbol{Z}_2\right).$$

From this we obtain

$$\begin{aligned}(21)\quad N\hat{\boldsymbol{\Sigma}}_\omega &= \left(\boldsymbol{X} - \boldsymbol{\beta}_1^*\boldsymbol{Z}_1 - \hat{\boldsymbol{\beta}}_{2\omega}\boldsymbol{Z}_2\right)\left(\boldsymbol{X} - \boldsymbol{\beta}_1^*\boldsymbol{Z}_1 - \hat{\boldsymbol{\beta}}_{2\omega}\boldsymbol{Z}_2\right)' \\ &= \left(\boldsymbol{X} - \hat{\boldsymbol{\beta}}_\Omega\boldsymbol{Z}\right)\left(\boldsymbol{X} - \hat{\boldsymbol{\beta}}_\Omega\boldsymbol{Z}\right)' \\ &\quad + \left(\hat{\boldsymbol{\beta}}_{1\Omega} - \boldsymbol{\beta}_1^*\right)\left(\boldsymbol{Z}_1 - \boldsymbol{A}_{12}\boldsymbol{A}_{22}^{-1}\boldsymbol{Z}_2\right)\left(\boldsymbol{Z}_1 - \boldsymbol{A}_{12}\boldsymbol{A}_{22}^{-1}\boldsymbol{Z}_2\right)'\left(\hat{\boldsymbol{\beta}}_{1\Omega} - \boldsymbol{\beta}_1^*\right)' \\ &= N\hat{\boldsymbol{\Sigma}}_\Omega + \left(\hat{\boldsymbol{\beta}}_{1\Omega} - \boldsymbol{\beta}_1^*\right)\left(\boldsymbol{A}_{11} - \boldsymbol{A}_{12}\boldsymbol{A}_{22}^{-1}\boldsymbol{A}_{21}\right)\left(\hat{\boldsymbol{\beta}}_{1\Omega} - \boldsymbol{\beta}_1^*\right)'.\end{aligned}$$

The determinant $|\hat{\boldsymbol{\Sigma}}_\Omega| = (1/N^p)|(\boldsymbol{X} - \hat{\boldsymbol{\beta}}_\Omega\boldsymbol{Z})(\boldsymbol{X} - \hat{\boldsymbol{\beta}}_\Omega\boldsymbol{Z})'|$ is proportional to the volume squared of the parallelotope spanned by the row vectors of $\boldsymbol{X} - \hat{\boldsymbol{\beta}}_\Omega\boldsymbol{Z}$ (translated to the origin). The determinant $|\hat{\boldsymbol{\Sigma}}_\omega| = (1/N^p)|(\boldsymbol{X} - \boldsymbol{\beta}_1^*\boldsymbol{Z}_1 - \hat{\boldsymbol{\beta}}_{2\omega}\boldsymbol{Z}_2)(\boldsymbol{X} - \boldsymbol{\beta}_1^*\boldsymbol{Z}_1 - \hat{\boldsymbol{\beta}}_{2\omega}\boldsymbol{Z}_2)'|$ is proportional to the volume squared of the parallelotope spanned by the row vectors of $\boldsymbol{X} - \boldsymbol{\beta}_1^*\boldsymbol{Z}_1 - \hat{\boldsymbol{\beta}}_{2\omega}\boldsymbol{Z}_2$ (translated to the origin); each of these vectors is the part of the vector of $\boldsymbol{X} - \boldsymbol{\beta}_1^*\boldsymbol{Z}_1$ that is orthogonal to $\boldsymbol{Z}_2$. Thus the test based on the likelihood ratio criterion depends on the ratio of volumes of parallelotopes. One parallelotope involves vectors orthogonal to $\boldsymbol{Z}$ and the other involves vectors orthogonal to $\boldsymbol{Z}_2$.

From (15) we see that the density of $\boldsymbol{x}_1, \ldots, \boldsymbol{x}_N$ can be written as

$$\begin{aligned}(22)\quad \frac{1}{(2\pi)^{\frac{1}{2}pN}|\boldsymbol{\Sigma}|^{\frac{1}{2}N}} \exp\Big(-\tfrac{1}{2}\operatorname{tr}\Big\{\boldsymbol{\Sigma}^{-1}\Big[N\hat{\boldsymbol{\Sigma}} + \left(\hat{\boldsymbol{\beta}}_{2\omega} - \boldsymbol{\beta}_2\right)\boldsymbol{A}_{22}\left(\hat{\boldsymbol{\beta}}_{2\omega} - \boldsymbol{\beta}_2\right)' \\ + \left(\hat{\boldsymbol{\beta}}_{1\Omega} - \boldsymbol{\beta}_1^*\right)\left(\boldsymbol{A}_{11} - \boldsymbol{A}_{12}\boldsymbol{A}_{22}^{-1}\boldsymbol{A}_{21}\right)\left(\hat{\boldsymbol{\beta}}_{1\Omega} - \boldsymbol{\beta}_1^*\right)'\Big]\Big\}\Big).\end{aligned}$$

Thus $\hat{\boldsymbol{\Sigma}}$, $\hat{\boldsymbol{\beta}}_{1\Omega}$, and $\hat{\boldsymbol{\beta}}_{2\omega}$ form a sufficient set of statistics for $\boldsymbol{\Sigma}$, $\boldsymbol{\beta}_1$, and $\boldsymbol{\beta}_2$.

Wilks (1932) first gave the likelihood ratio criterion for testing the equality of mean vectors from several populations (Section 8.8). Wilks (1934) and Bartlett (1934) extended its use to regression coefficients.

8.3.3. The Canonical Form

In studying the distributions of criteria it will be convenient to put the distribution of the observations in canonical form. This amounts to picking a coordinate system in the N-dimensional space so that the first q_1 coordinate axes are in the space of $\boldsymbol{Z}$ that is orthogonal to $\boldsymbol{Z}_2$, the next q_2 coordinate axes are in the space of $\boldsymbol{Z}_2$, and the last n $(=N-q)$ coordinate axes are orthogonal to the $\boldsymbol{Z}$ space.

Let $\boldsymbol{P}_2$ be a $q_2 \times q_2$ matrix such that

$$\boldsymbol{I} = \boldsymbol{P}_2 \boldsymbol{A}_{22} \boldsymbol{P}_2' = (\boldsymbol{P}_2 \boldsymbol{Z}_2)(\boldsymbol{P}_2 \boldsymbol{Z}_2)', \tag{23}$$

and let $\boldsymbol{P}_1$ be a $q_1 \times q_1$ matrix such that $(\boldsymbol{A}_{11\cdot 2} = \boldsymbol{A}_{11} - \boldsymbol{A}_{12}\boldsymbol{A}_{22}^{-1}\boldsymbol{A}_{21})$

$$\boldsymbol{I} = \boldsymbol{P}_1 \boldsymbol{A}_{11\cdot 2} \boldsymbol{P}_1' = \left[\boldsymbol{P}_1(\boldsymbol{Z}_1 - \boldsymbol{A}_{12}\boldsymbol{A}_{22}^{-1}\boldsymbol{Z}_2)\right]\left[\boldsymbol{P}_1(\boldsymbol{Z}_1 - \boldsymbol{A}_{12}\boldsymbol{A}_{22}^{-1}\boldsymbol{Z}_2)\right]'. \tag{24}$$

Then define the $N \times N$ orthogonal matrix $\boldsymbol{Q}$ as

$$\boldsymbol{Q} = \begin{pmatrix} \boldsymbol{Q}_1 \\ \boldsymbol{Q}_2 \\ \boldsymbol{Q}_3 \end{pmatrix} = \begin{pmatrix} \boldsymbol{P}_1(\boldsymbol{Z}_1 - \boldsymbol{A}_{12}\boldsymbol{A}_{22}^{-1}\boldsymbol{Z}_2) \\ \boldsymbol{P}_2 \boldsymbol{Z}_2 \\ \boldsymbol{Q}_3 \end{pmatrix}, \tag{25}$$

where $\boldsymbol{Q}_3$ is any $n \times N$ matrix making $\boldsymbol{Q}$ orthogonal. Then the columns of

$$\boldsymbol{W} = (\boldsymbol{W}_1 \ \boldsymbol{W}_2 \ \boldsymbol{W}_3) = \boldsymbol{X}\boldsymbol{Q}' = \boldsymbol{X}(\boldsymbol{Q}_1' \ \boldsymbol{Q}_2' \ \boldsymbol{Q}_3') \tag{26}$$

are independently normally distributed with covariance matrix $\boldsymbol{\Sigma}$ (Theorem 3.3.1). Then

$$\begin{aligned} \mathscr{E}\boldsymbol{W}_1 &= \mathscr{E}\boldsymbol{X}\boldsymbol{Q}_1' = (\boldsymbol{\beta}_1\boldsymbol{Z}_1 + \boldsymbol{\beta}_2\boldsymbol{Z}_2)(\boldsymbol{Z}_1 - \boldsymbol{A}_{12}\boldsymbol{A}_{22}^{-1}\boldsymbol{Z}_2)'\boldsymbol{P}_1' \\ &= \boldsymbol{\beta}_1\boldsymbol{A}_{11\cdot 2}\boldsymbol{P}_1' = \boldsymbol{\beta}_1\boldsymbol{P}_1^{-1}, \end{aligned} \tag{27}$$

$$\begin{aligned} \mathscr{E}\boldsymbol{W}_2 &= \mathscr{E}\boldsymbol{X}\boldsymbol{Q}_2' = (\boldsymbol{\beta}_1\boldsymbol{Z}_1 + \boldsymbol{\beta}_2\boldsymbol{Z}_2)\boldsymbol{Z}_2'\boldsymbol{P}_2' \\ &= (\boldsymbol{\beta}_1\boldsymbol{A}_{12} + \boldsymbol{\beta}_2\boldsymbol{A}_{22})\boldsymbol{P}_2', \end{aligned} \tag{28}$$

$$\mathscr{E}\boldsymbol{W}_3 = \mathscr{E}\boldsymbol{X}\boldsymbol{Q}_3' = \boldsymbol{\beta}\boldsymbol{Z}\boldsymbol{Q}_3' = \boldsymbol{0}. \tag{29}$$

Let

$$\boldsymbol{\Gamma}_1 = (\boldsymbol{\gamma}_1, \dots, \boldsymbol{\gamma}_{q_1}) = \boldsymbol{\beta}_1\boldsymbol{A}_{11\cdot 2}\boldsymbol{P}_1' = \boldsymbol{\beta}_1\boldsymbol{P}_1^{-1}, \tag{30}$$

$$\boldsymbol{\Gamma}_2 = (\boldsymbol{\gamma}_{q_1+1}, \dots, \boldsymbol{\gamma}_q) = (\boldsymbol{\beta}_1\boldsymbol{A}_{12} + \boldsymbol{\beta}_2\boldsymbol{A}_{22})\boldsymbol{P}_2', \tag{31}$$

$$\boldsymbol{W} = (\boldsymbol{W}_1 \ \boldsymbol{W}_2 \ \boldsymbol{W}_3) = (\boldsymbol{w}_1, \dots, \boldsymbol{w}_{q_1}, \boldsymbol{w}_{q_1+1}, \dots, \boldsymbol{w}_q, \boldsymbol{w}_{q+1}, \dots, \boldsymbol{w}_N). \tag{32}$$

Then $w_1, \dots, w_N$ are independently normally distributed with covariance matrix Σ and $\mathscr{E} w_\alpha = \gamma_\alpha$, $\alpha = 1, \dots, q$, and $\mathscr{E} w_\alpha = \mathbf{0}$, $\alpha = q+1, \dots, N$.

The hypothesis $\boldsymbol{\beta}_1 = \boldsymbol{\beta}_1^*$ can be transformed to $\boldsymbol{\beta}_1 = \mathbf{0}$ by subtraction; that is, by letting $x_\alpha - \boldsymbol{\beta}_1^* z_\alpha^{(1)} = y_\alpha$, as in Section 8.3.1. In canonical form then, the hypothesis is $\Gamma_1 = \mathbf{0}$. We can study problems in the canonical form, if we wish, and transform solutions back to terms of X and Z.

In (17), which is the partitioned form of $\hat{\boldsymbol{\beta}}_\Omega A = C$, eliminate $\hat{\boldsymbol{\beta}}_{2\Omega}$ to obtain

$$\hat{\boldsymbol{\beta}}_{1\Omega}\left(A_{11} - A_{12}A_{22}^{-1}A_{21}\right) = C_1 - C_2 A_{22}^{-1}A_{21} \qquad (33)$$

$$= X\left(Z_1' - Z_2' A_{22}^{-1}A_{21}\right)$$

$$= W_1 P_1'^{-1};$$

that is, $W_1 = \hat{\boldsymbol{\beta}}_{1\Omega} A_{11\cdot 2} P_1' = \hat{\boldsymbol{\beta}}_{1\Omega} P_1^{-1}$ and $\Gamma_1 = \boldsymbol{\beta}_1 P_1^{-1}$. Similarly, from (6) we obtain

$$\hat{\boldsymbol{\beta}}_{2\omega} A_{22} + \boldsymbol{\beta}_1^* A_{12} = C_2 = XZ_2' = W_2 P_2'^{-1}; \qquad (34)$$

that is, $W_2 = (\hat{\boldsymbol{\beta}}_{2\omega} A_{22} + \boldsymbol{\beta}_1^* A_{12}) P_2' = \boldsymbol{\beta}_{2\omega} P_2^{-1} + \boldsymbol{\beta}_1^* A_{12} P_2^{-1}$ and $\Gamma_2 = \boldsymbol{\beta}_2 P_2^{-1} + \boldsymbol{\beta}_1 A_{12} P_2^{-1}$.

8.4. THE DISTRIBUTION OF THE LIKELIHOOD RATIO CRITERION WHEN THE HYPOTHESIS IS TRUE

8.4.1. Characterization of the Distribution

The likelihood ratio criterion is the $\frac{1}{2}N$th power of

$$U = \lambda^{2/N} = \frac{|\hat{\Sigma}_\Omega|}{|\hat{\Sigma}_\omega|} = \frac{|N\hat{\Sigma}_\Omega|}{|N\hat{\Sigma}_\Omega + (\hat{\boldsymbol{\beta}}_{1\Omega} - \boldsymbol{\beta}_1^*) A_{11\cdot 2} (\hat{\boldsymbol{\beta}}_{1\Omega} - \boldsymbol{\beta}_1^*)'|}, \qquad (1)$$

where $A_{11\cdot 2} = A_{11} - A_{12}A_{22}^{-1}A_{21}$. We shall study the distribution and the moments of U when $\boldsymbol{\beta}_1 = \boldsymbol{\beta}_1^*$. It has been shown in Section 8.2 that $N\hat{\Sigma}_\Omega$ is distributed according to $W(\Sigma, n)$, where $n = N - q$, and the elements of $\hat{\boldsymbol{\beta}}_\Omega - \boldsymbol{\beta}$ have a joint normal distribution independent of $N\hat{\Sigma}_\Omega$.

From (33) of Section 8.3, we have

$$(\hat{\boldsymbol{\beta}}_{1\Omega} - \boldsymbol{\beta}_1^*) A_{11\cdot 2} (\hat{\boldsymbol{\beta}}_{1\Omega} - \boldsymbol{\beta}_1^*)' = (W_1 - \Gamma_1) P_1 A_{11\cdot 2} P_1' (W_1 - \Gamma_1)' \qquad (2)$$

$$= (W_1 - \Gamma_1)(W_1 - \Gamma_1)',$$

by (24) of Section 8.3; the columns of $W_1 - \Gamma_1$ are independently distributed, each according to $N(\mathbf{0}, \Sigma)$.

LEMMA 8.4.1. *$(\hat{\boldsymbol{\beta}}_{1\Omega} - \boldsymbol{\beta}_1^*)A_{11\cdot 2}(\hat{\boldsymbol{\beta}}_{1\Omega} - \boldsymbol{\beta}_1^*)'$ is distributed according to $W(\boldsymbol{\Sigma}, q_1)$.*

LEMMA 8.4.2. *The criterion U has the distribution of*

$$U = \frac{|\boldsymbol{G}|}{|\boldsymbol{G} + \boldsymbol{H}|}, \tag{3}$$

where $\boldsymbol{G}$ is distributed according to $W(\boldsymbol{\Sigma}, n)$, $\boldsymbol{H}$ is distributed according to $W(\boldsymbol{\Sigma}, m)$, where $m = q_1$, and $\boldsymbol{G}$ and $\boldsymbol{H}$ are independent.

Let

$$\boldsymbol{G} = N\hat{\boldsymbol{\Sigma}}_\Omega = \boldsymbol{X}\boldsymbol{X}' - \boldsymbol{X}\boldsymbol{Z}'(\boldsymbol{Z}\boldsymbol{Z}')^{-1}\boldsymbol{Z}\boldsymbol{X}', \tag{4}$$

$$\begin{aligned} \boldsymbol{G} + \boldsymbol{H} &= N\hat{\boldsymbol{\Sigma}}_\Omega + (\hat{\boldsymbol{\beta}}_{1\Omega} - \boldsymbol{\beta}_1^*)A_{11\cdot 2}(\hat{\boldsymbol{\beta}}_{1\Omega} - \boldsymbol{\beta}_1^*)' \\ &= N\hat{\boldsymbol{\Sigma}}_\omega = \boldsymbol{Y}\boldsymbol{Y}' - \boldsymbol{Y}\boldsymbol{Z}_2'(\boldsymbol{Z}_2\boldsymbol{Z}_2')^{-1}\boldsymbol{Z}_2\boldsymbol{Y}', \end{aligned} \tag{5}$$

where $\boldsymbol{Y} = \boldsymbol{X} - \boldsymbol{\beta}_1^*\boldsymbol{Z}_1 = \boldsymbol{X} - (\boldsymbol{\beta}_1^* \ \boldsymbol{0})\boldsymbol{Z}$. Then

$$\boldsymbol{G} = \boldsymbol{Y}\boldsymbol{Y}' - \boldsymbol{Y}\boldsymbol{Z}'(\boldsymbol{Z}\boldsymbol{Z}')^{-1}\boldsymbol{Z}\boldsymbol{Y}'. \tag{6}$$

We shall denote this criterion as $U_{p,m,n}$, where p is the dimensionality, $m = q_1$ is the number of columns of $\boldsymbol{\beta}_1$, and $n = N - q$ is the number of degrees of freedom of $\boldsymbol{G}$.

We now proceed to characterize the distribution of U as the product of beta variables (Section 5.2). Write the criterion U as

$$U = V_1 V_2 \cdots V_p, \tag{7}$$

where $V_1 = g_{11}/(g_{11} + h_{11})$,

$$V_i = \frac{|\boldsymbol{G}_i|}{|\boldsymbol{G}_{i-1}|} \bigg/ \frac{|\boldsymbol{G}_i + \boldsymbol{H}_i|}{|\boldsymbol{G}_{i-1} + \boldsymbol{H}_{i-1}|}, \qquad i = 2, \ldots, p, \tag{8}$$

and $\boldsymbol{G}_i$ and $\boldsymbol{H}_i$ are the submatrices of $\boldsymbol{G}$ and $\boldsymbol{H}$, respectively, of the first i rows and columns. Correspondingly, let $\boldsymbol{y}_\alpha^{(i)}$ consist of the first i components of $\boldsymbol{y}_\alpha = \boldsymbol{x}_\alpha - \boldsymbol{\beta}_1^*\boldsymbol{z}_\alpha^{(1)}$, $\alpha = 1, \ldots, N$. We shall show that V_i is the ratio of the length squared of the vector from $\boldsymbol{y}_i^* = (y_{i1}, \ldots, y_{iN})$ to its projection on $\boldsymbol{Z}$ and $\boldsymbol{Y}_{i-1} = (\boldsymbol{y}_1^{(i-1)}, \ldots, \boldsymbol{y}_N^{(i-1)})$ and the length squared of the vector from $\boldsymbol{y}_i^*$ to its projection on $\boldsymbol{Z}_2$ and $\boldsymbol{Y}_{i-1}$.

LEMMA 8.4.3. *Let $\boldsymbol{y}$ be an N-component row vector and $\boldsymbol{U}$ an $r \times N$ matrix. Then the sum of squares of the residuals of $\boldsymbol{y}$ from its regression on $\boldsymbol{U}$ is*

$$\frac{\begin{vmatrix} \boldsymbol{y}\boldsymbol{y}' & \boldsymbol{y}\boldsymbol{U}' \\ \boldsymbol{U}\boldsymbol{y}' & \boldsymbol{U}\boldsymbol{U}' \end{vmatrix}}{|\boldsymbol{U}\boldsymbol{U}'|}. \tag{9}$$

PROOF. By Corollary A.3.1 of Appendix A, (9) is $\boldsymbol{y}\boldsymbol{y}' - \boldsymbol{y}\boldsymbol{U}'(\boldsymbol{U}\boldsymbol{U}')^{-1}\boldsymbol{U}\boldsymbol{y}'$ which is the sum of squares of residuals as indicated in (13) of Section 8.2. Q.E.D.

LEMMA 8.4.4. *V_i defined by (8) is the ratio of the sum of squares of the residuals of $y_{i1}, \dots, y_{iN}$ from their regression on $\boldsymbol{y}_1^{(i-1)}, \dots, \boldsymbol{y}_N^{(i-1)}$ and $\boldsymbol{Z}$ to the sum of squares of residuals of $y_{i1}, \dots, y_{iN}$ from their regression on $\boldsymbol{y}_1^{(i-1)}, \dots, \boldsymbol{y}_N^{(i-1)}$ and $\boldsymbol{Z}_2$.*

PROOF. The numerator of V_i can be written [from (13) of Section 8.2]

$$\begin{aligned}
\frac{|\boldsymbol{G}_i|}{|\boldsymbol{G}_{i-1}|} &= \frac{|\boldsymbol{Y}_i\boldsymbol{Y}_i' - (\boldsymbol{Y}_i\boldsymbol{Z}'\boldsymbol{Z}\boldsymbol{Z}')^{-1}\boldsymbol{Z}\boldsymbol{Y}_i'|}{|\boldsymbol{Y}_{i-1}\boldsymbol{Y}_{i-1}' - \boldsymbol{Y}_{i-1}\boldsymbol{Z}'(\boldsymbol{Z}\boldsymbol{Z}')^{-1}\boldsymbol{Z}\boldsymbol{Y}_{i-1}'|} \\
&= \frac{\begin{vmatrix} \boldsymbol{Y}_i\boldsymbol{Y}_i' & \boldsymbol{Y}_i\boldsymbol{Z}' \\ \boldsymbol{Z}\boldsymbol{Y}_i' & \boldsymbol{Z}\boldsymbol{Z}' \end{vmatrix} \Big/ |\boldsymbol{Z}\boldsymbol{Z}'|}{\begin{vmatrix} \boldsymbol{Y}_{i-1}\boldsymbol{Y}_{i-1}' & \boldsymbol{Y}_{i-1}\boldsymbol{Z}' \\ \boldsymbol{Z}\boldsymbol{Y}_{i-1}' & \boldsymbol{Z}\boldsymbol{Z}' \end{vmatrix} \Big/ |\boldsymbol{Z}\boldsymbol{Z}'|} \\
&= \frac{\begin{vmatrix} \boldsymbol{Y}_{i-1}\boldsymbol{Y}_{i-1}' & \boldsymbol{Y}_i\boldsymbol{y}_i^{*\prime} & \boldsymbol{Y}_{i-1}\boldsymbol{Z}' \\ \boldsymbol{y}_i^*\boldsymbol{Y}_{i-1}' & \boldsymbol{y}_i^*\boldsymbol{y}_i^{*\prime} & \boldsymbol{y}_i^*\boldsymbol{Z}' \\ \boldsymbol{Z}\boldsymbol{Y}_{i-1}' & \boldsymbol{Z}\boldsymbol{y}_i^{*\prime} & \boldsymbol{Z}\boldsymbol{Z}' \end{vmatrix}}{\begin{vmatrix} \boldsymbol{Y}_{i-1}\boldsymbol{Y}_{i-1}' & \boldsymbol{Y}_{i-1}\boldsymbol{Z}' \\ \boldsymbol{Z}\boldsymbol{Y}_{i-1}' & \boldsymbol{Z}\boldsymbol{Z}' \end{vmatrix}} \\
&= \frac{\begin{vmatrix} \boldsymbol{y}_i^*\boldsymbol{y}_i^{*\prime} & \boldsymbol{y}_i^*\begin{bmatrix} \boldsymbol{Y}_{i-1}' & \boldsymbol{Z}' \end{bmatrix} \\ \begin{bmatrix} \boldsymbol{Y}_{i-1} \\ \boldsymbol{Z} \end{bmatrix}\boldsymbol{y}_i^* & \begin{bmatrix} \boldsymbol{Y}_{i-1} \\ \boldsymbol{Z} \end{bmatrix}\begin{bmatrix} \boldsymbol{Y}_{i-1}' & \boldsymbol{Z}' \end{bmatrix} \end{vmatrix}}{\left| \begin{bmatrix} \boldsymbol{Y}_{i-1} \\ \boldsymbol{Z} \end{bmatrix}\begin{bmatrix} \boldsymbol{Y}_{i-1}' & \boldsymbol{Z}' \end{bmatrix} \right|}
\end{aligned} \tag{10}$$

by Corollary A.3.1. Application of Lemma 8.4.3 shows that the right-hand side of (10) is the sum of squares of the residuals of y_i^* on Y_{i-1} and Z. The denominator is evaluated similarly with Z replaced by Z_2. Q.E.D.

The ratio V_i is the $2/N$th power of the likelihood ratio criterion for testing the hypothesis that the regression of $y_i^* = x_i^* - \beta_{i1}^* Z_1$ on Z_1 is $\mathbf{0}$ (in the presence of regression on Y_{i-1} and Z_2); here β_{i1}^* is the ith row of $\boldsymbol{\beta}_1^*$. For $i = 1$, g_{11} is the sum of squares of the residuals of $y_1^* = (y_{11}, \dots, y_{1N})$ from its regression on Z, and $g_{11} + h_{11}$ is the sum of squares of the residuals from Z_2. The ratio $V_1 = g_{11}/(g_{11} + h_{11})$, which is appropriate to test the hypothesis that regression of y_1^* on Z_1 is $\mathbf{0}$, is distributed as $\chi_n^2/(\chi_n^2 + \chi_m^2)$ (by Lemma 8.4.2) and has the beta distribution $\beta(v; \frac{1}{2}n, \frac{1}{2}m)$. (See Section 5.2, for example.) Thus V_i has the beta-density

(11)

$$\beta\left[v; \tfrac{1}{2}(n+1-i), \tfrac{1}{2}m\right] = \frac{\Gamma\left[\frac{1}{2}(n+m+1-i)\right]}{\Gamma\left[\frac{1}{2}(n+1-i)\right]\Gamma\left(\frac{1}{2}m\right)} v^{\frac{1}{2}(n+1-i)-1}(1-v)^{\frac{1}{2}m-1},$$

for $0 \le v \le 1$ and 0 for v outside this interval. Since this distribution does not depend on Y_{i-1}, we see that the ratio V_i is independent of Y_{i-1}, and hence independent of $V_1, \dots, V_{i-1}$. Then $V_1, \dots, V_p$ are independent.

THEOREM 8.4.1. *The distribution of U defined by* (3) *is the distribution of the product $\prod_{i=1}^p V_i$, where $V_1, \dots, V_p$ are independent and V_i has the density* (11).

The cdf of U can be found by integrating the joint density of $V_1, \dots, V_p$ over the range

$$\prod_{i=1}^{p} V_i \le u. \tag{12}$$

We shall now show that for given $N - q_2$ the indices p and q_1 can be interchanged; that is, the distributions of $U_{p, q_1, N-q_2-q_1} = U_{p,m,n}$ and of $U_{q_1, p, N-q_2-p} = U_{m,p,n+m-p}$ are the same. The joint density of G and W_1 defined in Section 8.2 when $\Sigma = I$ and $\boldsymbol{\beta}_1 = \mathbf{0}$ is

$$\frac{|G|^{\frac{1}{2}(n-p-1)} e^{-\frac{1}{2}\operatorname{tr} G - \frac{1}{2}\operatorname{tr} W_1 W_1'}}{2^{\frac{1}{2}np} \pi^{p(p-1)/4} \prod_{i=1}^p \Gamma\left[\frac{1}{2}(n+1-i)\right] (2\pi)^{\frac{1}{2}mp}}. \tag{13}$$

Let $G + W_1W_1' = J = CC'$ and let $W_1 = CU$. Then

$$U_{p,m,n} = \frac{|G|}{|G + W_1W_1'|} = \frac{|CC' - CUU'C'|}{|CC'|} = |I_p - UU'| \tag{14}$$

$$= \begin{vmatrix} I_p & U \\ U' & I_m \end{vmatrix} = \begin{vmatrix} I_m & U' \\ U & I_p \end{vmatrix} = |I_m - U'U|;$$

the fourth and sixth equalities follow from Theorem A.3.2 of Appendix A and the fifth from permutation of rows and columns. Since the Jacobian of $W_1 = CU$ is $\operatorname{mod}|C|^m = |J|^{\frac{1}{2}m}$, the joint density of J and U is

$$\frac{|J|^{\frac{1}{2}(n+m-p-1)}e^{-\frac{1}{2}\operatorname{tr} J}}{2^{\frac{1}{2}(n+m)p}\pi^{p(p-1)/4}\prod_{i=1}^{p}\Gamma[\frac{1}{2}(n+m+1-i)]} \tag{15}$$

$$\cdot \prod_{i=1}^{p}\left\{\frac{\Gamma[\frac{1}{2}(n+m+1-i)]}{\Gamma[\frac{1}{2}(n+1-i)]}\right\}\frac{|I_p - UU'|^{\frac{1}{2}(n-p-1)}}{\pi^{\frac{1}{2}mp}}$$

for J and $I_p - UU'$ positive definite and 0 otherwise. Thus J and U are independently distributed; the density of J is the first term in (15), namely, $w(J|I_p, n+m)$, and the density of U is the second term, namely, of the form

$$K|I_p - UU'|^{\frac{1}{2}(n-p-1)} \tag{16}$$

for $I_p - UU'$ positive definite and 0 otherwise. Let $U_* = U'$, $p^* = m$, $m^* = p$, and $n^* = n + m - p$. Then the density of U_* is

$$K|I_p - U_*'U_*|^{\frac{1}{2}(n-p-1)} \tag{17}$$

for $I_p - U_*'U_*$ positive definite and 0 otherwise. By (14) $|I_p - U_*'U_*| = |I_m - U_*U_*'|$ and hence the density of U_* is

$$K|I_{p^*} - U_*U_*'|^{\frac{1}{2}(n^*-p^*-1)}, \tag{18}$$

which is of the form of (16) with p replaced by $p^* = m$, m replaced by $m^* = p$, and $n - p - 1$ replaced by $n^* - p^* - 1 = n - p - 1$. Finally we note that $U_{p,m,n}$ given by (14) is $|I_m - U_*U_*'| = U_{m,p,n+m-p}$.

THEOREM 8.4.2. *When the hypothesis is true, the distribution of* $U_{p,q_1,N-q_1-q_2}$ *is the same as that of* $U_{q_1,p,N-p-q_2}$ *(i.e., that of* $U_{p,m,n}$ *is that of* $U_{m,p,n+m-p}$*).*

8.4.2. Moments

Since (11) is a density and hence integrates to 1, by change of notation

$$\int_0^1 v^{a-1}(1-v)^{b-1}\,dv = B(a,b) = \frac{\Gamma(a)\Gamma(b)}{\Gamma(a+b)}. \tag{19}$$

From this fact we see that the hth moment of V_i is

$$\begin{aligned} \mathscr{E}V_i^h &= \int_0^1 \frac{\Gamma[\frac{1}{2}(n+m+1-i)]}{\Gamma[\frac{1}{2}(n+1-i)]\Gamma(\frac{1}{2}m)} v^{\frac{1}{2}(n+1-i)+h-1}(1-v)^{\frac{1}{2}m-1}\,dv \\ &= \frac{\Gamma[\frac{1}{2}(n+1-i)+h]\Gamma[\frac{1}{2}(n+m+1-i)]}{\Gamma[\frac{1}{2}(n+1-i)]\Gamma[\frac{1}{2}(n+m+1-i)+h]}. \end{aligned} \tag{20}$$

Since $V_1, \dots, V_p$ are independent, $\mathscr{E}U^h = \mathscr{E}\prod_{i=1}^p V_i^h = \prod_{i=1}^p \mathscr{E}V_i^h$. We obtain the following theorem:

THEOREM 8.4.3. *The* hth *moment of* U [*if* $h > -\frac{1}{2}(n+1-p)$] *is*

$$\begin{aligned} \mathscr{E}U^h &= \prod_{i=1}^p \frac{\Gamma[\frac{1}{2}(n+1-i)+h]\Gamma[\frac{1}{2}(n+m+1-i)]}{\Gamma[\frac{1}{2}(n+1-i)]\Gamma[\frac{1}{2}(n+m+1-i)+h]} \\ &= \prod_{i=1}^p \frac{\Gamma[\frac{1}{2}(N-q_1-q_2+1-i)+h]\Gamma[\frac{1}{2}(N-q_2+1-i)]}{\Gamma[\frac{1}{2}(N-q_1-q_2+1-i)]\Gamma[\frac{1}{2}(N-q_2+1-i)+h]}. \end{aligned} \tag{21}$$

In the first expression p can be replaced by m, m by p, and n by $n+m-p$.

Suppose p is even, that is, $p = 2r$. We use the fact that

$$\Gamma(\alpha+\tfrac{1}{2})\Gamma(\alpha+1) = \frac{\sqrt{\pi}\,\Gamma(2\alpha+1)}{2^{2\alpha}}. \tag{22}$$

Then the hth moment of $U_{2r,m,n}$ is

(23)

$$\begin{aligned} \mathscr{E}U_{2r,m,n}^h &= \prod_{j=1}^r \left\{ \frac{\Gamma[\frac{1}{2}(m+n+2)-j]}{\Gamma[\frac{1}{2}(m+n+2)-j+h]} \frac{\Gamma[\frac{1}{2}(m+n+1)-j]}{\Gamma[\frac{1}{2}(m+n+1)-j+h]} \right. \\ &\qquad \left. \cdot \frac{\Gamma[\frac{1}{2}(n+2)-j+h]\Gamma[\frac{1}{2}(n+1)-j+h]}{\Gamma[\frac{1}{2}(n+2)-j]\Gamma[\frac{1}{2}(n+1)-j]} \right\} \\ &= \prod_{j=1}^r \left\{ \frac{\Gamma(m+n+1-2j)\Gamma(n+1-2j+2h)}{\Gamma(m+n+1-2j+2h)\Gamma(n+1-2j)} \right\}. \end{aligned}$$

It is clear from the definition of the beta function that (23) is

$$\prod_{j=1}^{r}\left\{\int_0^1 \frac{\Gamma(m+n+1-2j)}{\Gamma(n+1-2j)\Gamma(m)} y^{(n+1-2j)+2h-1}(1-y)^{m-1}\,dy\right\} \tag{24}$$

$$= \prod_{j=1}^{r} \mathscr{E} Y_j^{2h}$$

$$= \mathscr{E}\left(\prod_{j=1}^{r} Y_j^2\right)^h,$$

where the Y_j are independent and Y_j has density $\beta(y; n+1-2j, m)$.

Suppose p is odd; that is, $p = 2s+1$. Then

$$\mathscr{E} U_{2s+1,m,n}^h = \mathscr{E}\left(\prod_{i=1}^{s} Z_i^2 Z_{s+1}\right)^h, \tag{25}$$

where the Z_i are independent and Z_i has density $\beta(z; n+1-2i, m)$ for $i = 1, \ldots, s$ and Z_{s+1} is distributed with density $\beta[z; (n+1-p)/2, m/2]$.

Theorem 8.4.4. *$U_{2r,m,n}$ is distributed as $\prod_{i=1}^r Y_i^2$, where $Y_1, \ldots, Y_r$ are independent and Y_i has density $\beta\ (y; n+1-2i, m)$; $U_{2s+1,m,n}$ is distributed as $\prod_{i=1}^s Z_i^2 Z_{s+1}$, where the Z_i $(i = 1, \ldots, s)$ are independent and Z_i has density $\beta(z; n+1-2i, m)$ and Z_{s+1} is independently distributed with density $\beta[z; \frac{1}{2}(n+1-p), \frac{1}{2}m]$.*

8.4.3. Some Special Distributions

$p = 1$. From the preceding characterization we see that the density of $U_{1,m,n}$ is

$$\frac{\Gamma[\frac{1}{2}(n+m)]}{\Gamma(\frac{1}{2}n)\Gamma(\frac{1}{2}m)} u^{\frac{1}{2}n-1}(1-u)^{\frac{1}{2}m-1}. \tag{26}$$

Another way of writing $U_{1,m,n}$ is

$$U_{1,m,n} = \frac{1}{1+\sum_{i=1}^m Y_i^2/g_{11}} = \frac{1}{1+(m/n)F_{m,n}}, \tag{27}$$

where g_{11} is the one element of $\boldsymbol{G} = N\hat{\boldsymbol{\Sigma}}_\Omega$ and $F_{m,n}$ is an F-statistic. Thus

$$\frac{1 - U_{1,m,n}}{U_{1,m,n}} \cdot \frac{n}{m} = F_{m,n}. \tag{28}$$

THEOREM 8.4.5. *The distribution of* $(1 - U_{1,m,n})/U_{1,m,n} \cdot n/m$ *is the F-distribution with m and n degrees of freedom; the distribution of* $(1 - U_{p,1,n})/U_{p,1,n} \cdot (n+1-p)/p$ *is the F-distribution with p and* $n+1-p$ *degrees of freedom.*

$p = 2$. From Theorem 8.4.4, we see that the density of $\sqrt{U_{2,m,n}}$ is

$$\frac{\Gamma(n+m-1)}{\Gamma(n-1)\Gamma(m)} x^{n-2}(1-x)^{m-1}, \tag{29}$$

and thus the density of $U_{2,m,n}$ is

$$\frac{\Gamma(n+m-1)}{2\Gamma(n-1)\Gamma(m)} u^{\frac{1}{2}(n-3)}(1-\sqrt{u})^{m-1}. \tag{30}$$

From (29) it follows that

$$\frac{1 - \sqrt{U_{2,m,n}}}{\sqrt{U_{2,m,n}}} \cdot \frac{n-1}{m} = F_{2m,2(n-1)}. \tag{31}$$

THEOREM 8.4.6. *The distribution of* $(1 - \sqrt{U_{2,m,n}})/\sqrt{U_{2,m,n}} \cdot (n-1)/m$ *is the F-distribution with* $2m$ *and* $2(n-1)$ *degrees of freedom; the distribution of* $(1 - \sqrt{U_{p,2,n}})/\sqrt{U_{p,2,n}} \cdot (n+1-p)/p$ *is the F-distribution with* $2p$ *and* $2(n+1-p)$ *degrees of freedom.*

p even. Wald and Brookner (1941) gave a method for finding the distribution of $U_{p,m,n}$ for p or m even. We shall present the method of Schatzoff (1966a). It will be convenient first to consider $U_{p,m,n}$ for $m = 2r$. We can write the event $\prod_{i=1}^{p} V_i \le u$ as

$$Y_1 + \cdots + Y_p \ge -\log u, \tag{32}$$

where $Y_1, \dots, Y_p$ are independent and $Y_i = -\log V_i$ has the density

$$K_i e^{-\frac{1}{2}(n+1-i)y}(1-e^{-y})^{r-1} = K_i \sum_{j=0}^{r-1} (-1)^j \binom{r-1}{j} e^{-[\frac{1}{2}(n+1-i)+j]y} \tag{33}$$

for $0 \le y < \infty$ and 0 otherwise and

$$(34)\qquad K_i = \frac{\Gamma\left[\frac{1}{2}(n+1-i)+r\right]}{\Gamma\left[\frac{1}{2}(n+1-i)\right]\Gamma(r)} = \frac{1}{(r-1)!}\prod_{j=0}^{r-1}\frac{n+1-i+2j}{2}.$$

The joint density of $Y_1, \ldots, Y_p$ is then a linear combination of terms $\exp[-\sum_{i=1}^{p} a_i y_i]$. The density of $W_j = \sum_{i=1}^{j} Y_i$ can be obtained inductively from the density of $W_{j-1} = \sum_{i=1}^{j-1} Y_i$ and Y_j, $j = 2, \ldots, p$, which is a linear combination of terms $w_{j-1}^k e^{cw_{j-1}+a_j y_j}$. The density of W_j consists of linear combinations of

$$(35)\qquad e^{a_j w_j}\int_0^{w_j} w^k e^{(c-a_j)w}\,dw = e^{a_j w_j}\cdot\frac{w_j^{k+1}}{k+1} \qquad \text{if } a_j = c,$$

$$= e^{cw_j}\sum_{h=0}^{k}(-1)^h\frac{k!}{(k-h)!}\frac{w_j^{k-h}}{(c-a_j)^{h+1}}$$

$$+(-1)^{k+1}e^{a_j w_j}\frac{k!}{(c-a_j)^{k+1}} \qquad \text{if } a_j \ne c.$$

The evaluation involves integration by parts.

THEOREM 8.4.7. *If p is even or if m is even, the density of $U_{p,m,n}$ can be expressed as a linear combination of terms $(-\log u)^k u^l$, where k is an integer and l is a half-integer.*

From (35) we see that the cumulative distribution function of $-\log U$ is a linear combination of terms $w^k e^{-lw}$ and hence the cumulative distribution function of U is a linear combination of terms $(-\log u)^k u^l$. The values of k and l and the coefficients depend on p, m, and n. They can be obtained by inductively carrying out the procedure leading to Theorem 8.4.7. Schatzoff (1966a) reports that a recursive algorithm has been programmed for these integrations and used for computing tables. Pillai and Gupta (1969) have used Theorem 8.4.3.

An alternative approach is to use Theorem 8.4.4. The complement to the cumulative distribution function $U_{2r,m,n}$ is

(36)

$$\Pr\{U_{2r,m,n} \geq u\} = \Pr\left\{\prod_{i=1}^{r} Y_i > \sqrt{u}\right\}$$

$$= \int_{\sqrt{u}}^{1} \int_{\frac{\sqrt{u}}{y_1}}^{1} \cdots \int_{\frac{\sqrt{u}}{\prod_{i=1}^{r-1} y_i}}^{1} \prod_{i=1}^{r} \beta(y_i | n+1-2i, m)\, dy_r \ldots dy_2\, dy_1.$$

In the density $(1-y_i)^{m-1}$ can be expanded by the binomial theorem. Then all integrations are expressed as integrations of powers of the variables.

As an example, consider $r = 2$. The density of Y_1 and Y_2 is

$$(37) \qquad C y_1^{n-2} y_2^{n-4} (1-y_1)^{m-1} (1-y_2)^{m-1}$$

$$= C \sum_{i,j=0}^{m-1} \frac{(m-1)!^2 (-1)^{i+j}}{(m-i-1)!(m-j-1)!i!j!} y_1^{n-2+i} y_2^{n-4+j},$$

where

$$(38) \qquad C = \frac{\Gamma(n+m-1)\Gamma(n+m-3)}{\Gamma(n-1)\Gamma(n-3)\Gamma^2(m)}.$$

The complement to the cdf of $U_{4,m,n}$ is

$$(39) \quad \Pr\{U_{4,m,n} \geq u\} = C \sum_{i,j=0}^{m-1} \frac{(m-1)!^2 (-1)^{i+j}}{(m-i-1)!(m-j-1)!i!j!}$$

$$\cdot \int_{\sqrt{u}}^{1} \int_{\sqrt{u}/y_1}^{1} y_1^{n-2+i} y_2^{n-4+j}\, dy_2\, dy_1$$

$$= C \sum_{i,j=0}^{m-1} \frac{(m-1)!^2 (-1)^{i+j}}{(m-i-1)!(m-j-1)!i!j!(n-3+j)}$$

$$\cdot \int_{\sqrt{u}}^{1} \left[y_1^{n-2+i} - u^{\frac{1}{2}(n-3+j)} y_1^{1+i-j} \right] dy_1.$$

The last step of the integration yields powers of $\sqrt{u}$ and products of powers of $\sqrt{u}$ and $\log u$ (for $1 + i - j = -1$).

Wilks (1935) gives explicitly the distributions of U for $p = 1$, $p = 2$, $p = 3$ with $m = 3$; $p = 3$ with $m = 4$; and $p = 4$ with $m = 4$. Wilks' formula for $p = 3$ with $m = 4$ appears to be incorrect; see the first edition of this book. Consul (1966) gives many distributions for special cases. See also Mathai (1971).

8.4.4. The Likelihood Ratio Procedure

Let $u_{p,m,n}(\alpha)$ be the α significance point for $U_{p,m,n}$; that is,

$$\Pr\{U_{p,m,n} \le u_{p,m,n}(\alpha) | H \text{ true}\} = \alpha. \tag{40}$$

It is shown in Section 8.5 that $-[n - \frac{1}{2}(p - m + 1)]\log U_{p,m,n}$ has a limiting χ^2-distribution with pm degrees of freedom. Let $\chi^2_{pm}(\alpha)$ denote the α significance point of χ^2_{pm}, and let

$$C_{p,m,n-p+1}(\alpha) = \frac{-\left[n - \frac{1}{2}(p - m + 1)\right]\log u_{p,m,n}(\alpha)}{\chi^2_{pm}(\alpha)}. \tag{41}$$

Table 1 of Appendix B gives values of $C_{p,m,M}(\alpha)$ for $\alpha = .10$ and $.05$, $p = 1(1)10$, various even values of m, and $M = n - p + 1 = 1(1)10(2)20$, 24, 30, 40, 60, 120.

To test a null hypothesis one computes $U_{p,m,n}$ and rejects the null hypothesis at significance level α if

$$-\left[n - \frac{1}{2}(p - m + 1)\right]\log U_{p,m,n} > C_{p,m,n-p+1}(\alpha)\chi^2_{pm}(\alpha). \tag{42}$$

Since $C_{p,m,n}(\alpha) > 1$, the hypothesis is accepted if the left-hand side of (42) is less than $\chi^2_{pm}(\alpha)$.

The purpose of tabulating $C_{p,m,M}(\alpha)$ is that linear interpolation is relatively accurate because the entries decrease monotonically and smoothly to 1 as M increases. Schatzoff (1966a) has recommended interpolation for odd p by using adjacent even values of p and displays some examples. The table also indicates how accurate the χ^2-approximation is. The table has been extended by Pillai and Gupta (1969). The table is also given in Pearson and Hartley (1972).

8.4.5. A Step-down Procedure

The criterion U has been expressed in (7) as the product of independent beta variables $V_1, V_2, \ldots, V_p$. The ratio V_i is a least squares criterion for testing the null hypothesis that in the regression of $\boldsymbol{x}_i^* - \boldsymbol{\beta}_{i1}^* \boldsymbol{Z}_1$ on $\boldsymbol{Z} = (\boldsymbol{Z}_1' \ \boldsymbol{Z}_2')'$ and $\boldsymbol{X}_{i-1}$ the coefficient of $\boldsymbol{Z}_1$ is $\boldsymbol{0}$. The null hypothesis that the regression of $\boldsymbol{X}$ on $\boldsymbol{Z}_1$ is $\boldsymbol{\beta}_1^*$, which is equivalent to the hypothesis that the regression of $\boldsymbol{X} - \boldsymbol{\beta}_1^* \boldsymbol{Z}_1$ on $\boldsymbol{Z}_1$ is $\boldsymbol{0}$, is composed of the hypotheses that the regression of $\boldsymbol{x}_i^* - \boldsymbol{\beta}_{i1}^* \boldsymbol{Z}_1$ on $\boldsymbol{Z}_1$ is $\boldsymbol{0}$, $i = 1, \ldots, p$. Hence the null hypothesis $\boldsymbol{\beta}_1 = \boldsymbol{\beta}_1^*$ can be tested by use of $V_1, \ldots, V_p$.

Since V_i has the beta-density (11) under the hypothesis $\boldsymbol{\beta}_{i1} = \boldsymbol{\beta}_{i1}^*$,

$$\frac{1 - V_i}{V_i} \frac{n - i + 1}{m} \tag{43}$$

has the F-distribution with m and $n - i + 1$ degrees of freedom. The step-down testing procedure is to compare (42) for $i = 1$ with the significance point $F_{m,n}(\varepsilon_1)$; if (42) for $i = 1$ is larger, reject the null hypothesis that the regression of $\boldsymbol{x}_1^* - \boldsymbol{\beta}_{i1}^* \boldsymbol{Z}_1$ on $\boldsymbol{Z}_1$ is $\boldsymbol{0}$ and hence reject the null hypothesis that $\boldsymbol{\beta}_1 = \boldsymbol{\beta}_1^*$. If this first component null hypothesis is accepted, compare (42) for $i = 2$ with $F_{m,n-1}(\varepsilon_2)$. In sequence, the component null hypotheses are tested. If one is rejected, the sequence is stopped and the hypothesis $\boldsymbol{\beta}_1 = \boldsymbol{\beta}_1^*$ is rejected. If all component null hypotheses are accepted, the composite hypothesis is accepted. When the hypothesis $\boldsymbol{\beta}_1 = \boldsymbol{\beta}_1^*$ is true, the probability of accepting it is $\prod_{i=1}^p (1 - \varepsilon_i)$. Hence the significance level of the step-down test is $1 - \prod_{i=1}^p (1 - \varepsilon_i)$.

In the step-down procedure the investigator usually has a choice of the ordering of the variables[†] (i.e., the numbering of the components of $\boldsymbol{X}$) and a selection of component significance levels. It seems reasonable to order the variables in descending order of importance. The choice of significance levels will affect the power. If ε_i is a very small number, it will take a relatively large deviation from the ith null hypothesis to lead to rejection. In the absence of any other reason, the component significance levels can be taken equal. This procedure, of course, is not invariant with respect to linear transformation of the dependent vector variable. However, before carrying out a step-down procedure, a linear transformation can be used to determine the p variables.

[†] In some cases the ordering of variables may be imposed; for example, x_1 might be an observation at the first time point, x_2 at the second time point, and so on.

The factors can be grouped. For example, group $x_1, \dots, x_k$ into one set and $x_{k+1}, \dots, x_p$ into another set. Then $U_{k,m,n} = \prod_{i=1}^{k} V_i$ can be used to test the null hypothesis that the first k rows of $\mathbf{B}_1$ are the first k rows of $\mathbf{B}_1^*$. Subsequently $\prod_{i=k+1}^{p} V_i$ is used to test the hypothesis that the last $p-k$ rows of $\mathbf{B}_1$ are those of $\mathbf{B}_1^*$; this latter criterion has the distribution under the null hypothesis of $U_{p-k,m,n-k}$.

The investigator may test the null hypothesis $\mathbf{B}_1 = \mathbf{B}_1^*$ by the likelihood ratio procedure. If the hypothesis is rejected, he may look at the factors $V_1, \dots, V_p$ to try to determine which rows of $\mathbf{B}_1$ might be different from $\mathbf{B}_1^*$.

The factors can also be used to obtain confidence regions for $\beta_{11}, \dots, \beta_{p1}$. Let $v_i(\varepsilon_i)$ be defined by

$$\frac{1 - v_i(\varepsilon_i)}{v_i(\varepsilon_i)} \cdot \frac{n-i+1}{m} = F_{m,n-i+1}(\varepsilon_i). \tag{44}$$

Then a confidence region for β_{i1} of confidence $1 - \varepsilon_i$ is

$$\frac{\begin{vmatrix} x_i^* x_i^{*\prime} & x_i^* X_{i-1}' & x_i^* Z' \\ X_{i-1} x_i^{*\prime} & X_{i-1} X_{i-1}' & X_{i-1} Z' \\ Z x_i^{*\prime} & Z X_{i-1}' & Z Z' \end{vmatrix}}{\begin{vmatrix} (x_i^* - \bar{\beta}_{i1} Z_1)(x_i^* - \bar{\beta}_{i1} Z_1)' & (x_i^* - \bar{\beta}_{i1} Z_1) X_{i-1}' & (x_i^* - \bar{\beta}_{i1} Z_1) Z_2' \\ X_{i-1}(x_i^* - \bar{\beta}_{i1} Z_1)' & X_{i-1} X_{i-1}' & X_{i-1} Z_2' \\ Z_2 (x_i^* - \bar{\beta}_{i1} Z_1)' & Z_2 X_{i-1}' & Z_2 Z_2' \end{vmatrix}} \cdot \frac{\begin{vmatrix} X_{i-1} X_{i-1}' & X_{i-1} Z_2' \\ Z_2 X_{i-1}' & Z_2 Z_2' \end{vmatrix}}{\begin{vmatrix} X_{i-1} X_{i-1}' & X_{i-1} Z' \\ Z X_{i-1}' & Z Z' \end{vmatrix}} \geq v_i(\varepsilon_i). \tag{45}$$

8.5. AN ASYMPTOTIC EXPANSION OF THE DISTRIBUTION OF THE LIKELIHOOD RATIO CRITERION

8.5.1. General Theory of Asymptotic Expansions

In this section we develop a large sample distribution theory for the criterion studied in this chapter. First we develop a general asymptotic expansion of the distribution of a random variable whose moments are certain functions of gamma functions [Box (1949)]. Then we apply it to the case of the likelihood ratio criterion for the linear hypothesis.

We consider a random variable W $(0 \le W \le 1)$ with hth moment[†]

$$(1) \qquad \mathscr{E}W^h = K\left(\frac{\prod_{j=1}^b y_j^{y_j}}{\prod_{k=1}^a x_k^{x_k}}\right)^h \frac{\prod_{k=1}^a \Gamma[x_k(1+h)+\xi_k]}{\prod_{j=1}^b \Gamma[y_j(1+h)+\eta_j]}, \qquad h = 0, 1, \dots,$$

where K is a constant such that $\mathscr{E}W^0 = 1$ and

$$(2) \qquad \sum_{k=1}^a x_k = \sum_{j=1}^b y_j.$$

It will be observed that the hth moment of $\lambda = U_{p,q_1,n}^{\frac{1}{2}N}$ is of this form where $x_k = \frac{1}{2}N = y_j$, $\xi_k = \frac{1}{2}(-q+1-k)$, $\eta_j = \frac{1}{2}(-q_2+1-j)$, $a = b = p$. We treat a more general case here because applications later in this book require it.

If we let

$$(3) \qquad M = -2\log W,$$

the characteristic function of ρM $(0 \le \rho < 1)$ is

$$\begin{aligned}(4) \qquad \phi(t) &= \mathscr{E}e^{it\rho M} \\ &= \mathscr{E}W^{-2it\rho} \\ &= K\left(\frac{\prod_{j=1}^b y_j^{y_j}}{\prod_{k=1}^a x_k^{x_k}}\right)^{-2it\rho} \frac{\prod_{k=1}^a \Gamma[x_k(1-2it\rho)+\xi_k]}{\prod_{j=1}^b \Gamma[y_j(1-2it\rho)+\eta_j]}.\end{aligned}$$

Here ρ is arbitrary; later it will depend on N. If $a = b$, $x_k = y_k$, $\xi_k \le \eta_k$, (1) is the hth moment of the product of powers of variables with beta distributions, and then (1) holds for all h for which the gamma functions exist. In this case (4) is valid for all real t. We shall assume here that (4) holds for all real t, and in each case where we apply the result we shall verify this assumption.

[†] In all cases where we apply this result the parameters x_k, ξ_k, y_j, and η_j will be such that there is a distribution with such moments.

Let

$$\Phi(t) = \log \phi(t) = g(t) - g(0), \tag{5}$$

where

$$g(t) = 2it\rho\left(\sum_{k=1}^{a} x_k \log x_k - \sum_{j=1}^{b} y_j \log y_j\right) + \sum_{k=1}^{a} \log\Gamma[\rho x_k(1-2it) + \beta_k + \xi_k] - \sum_{j=1}^{b} \log\Gamma[\rho y_j(1-2it) + \varepsilon_j + \eta_j],$$

where $\beta_k = (1-\rho)x_k$ and $\varepsilon_j = (1-\rho)y_j$. The form $g(t) - g(0)$ makes $\Phi(0) = 0$ which agrees with the fact that K is such that $\phi(0) = 1$. We make use of an expansion formula for the gamma function [Barnes (1899), p. 64] which is asymptotic in x for bounded h:

$$\log\Gamma(x+h) = \log\sqrt{2\pi} + (x + h - \tfrac{1}{2})\log x - x - \sum_{r=1}^{m} (-1)^r \frac{B_{r+1}(h)}{r(r+1)x^r} + R_{m+1}(x), \tag{6}$$

where† $R_{m+1}(x) = O(x^{-(m+1)})$ and $B_r(h)$ is the Bernoulli polynomial of degree r and order unity defined by‡

$$\frac{\tau e^{h\tau}}{e^{\tau} - 1} = \sum_{r=0}^{\infty} \frac{\tau^r}{r!} B_r(h). \tag{7}$$

The first three polynomials are $[B_0(h) = 1]$

$$\begin{aligned} B_1(h) &= h - \tfrac{1}{2}, \\ B_2(h) &= h^2 - h + \tfrac{1}{6}, \\ B_3(h) &= h^3 - \tfrac{3}{2}h^2 + \tfrac{1}{2}h. \end{aligned} \tag{8}$$

†$R_{m+1}(x) = O(x^{-(m+1)})$ means $|x^{m+1}R_{m+1}(x)|$ is bounded as $|x| \to \infty$.

‡This definition differs slightly from that of Whittaker and Watson [(1943), p. 126] who expand $\tau(e^{h\tau} - 1)/(e^{\tau} - 1)$. If $B_r^*(h)$ is this second type of polynomial, $B_1(h) = B_1^*(h) - \frac{1}{2}$, $B_{2r}(h) = B_{2r}^*(h) + (-1)^{r+1}B_r$, where B_r is the rth Bernoulli number, and $B_{2r+1}(h) = B_{2r+1}^*(h)$.

Taking $x = \rho x_k(1-2it)$, $\rho y_j(1-2it)$ and $h = \beta_k + \xi_k$, $\varepsilon_j + \eta_j$ in turn, we obtain

$$\Phi(t) = Q - g(0) - \tfrac{1}{2}f\log(1-2it) + \sum_{r=1}^{m}\omega_r(1-2it)^{-r} + \sum O\left(x_k^{-(m+1)}\right) + \sum O\left(y_j^{-(m+1)}\right), \tag{9}$$

where

$$f = -2\left\{\sum\xi_i - \sum\eta_j - \tfrac{1}{2}(a-b)\right\}, \tag{10}$$

$$\omega_r = \frac{(-1)^{r+1}}{r(r+1)}\left\{\sum_k \frac{B_{r+1}(\beta_k+\xi_k)}{(\rho x_k)^r} - \sum_j \frac{B_{r+1}(\varepsilon_j+\eta_j)}{(\rho y_j)^r}\right\}, \tag{11}$$

$$Q = \tfrac{1}{2}(a-b)\log 2\pi - \tfrac{1}{2}f\log\rho + \sum_k\left(x_k+\xi_k-\tfrac{1}{2}\right)\log x_k - \sum_j\left(y_j+\eta_j-\tfrac{1}{2}\right)\log y_j. \tag{12}$$

One resulting form for $\phi(t)$ (which we shall not use here) is

$$\phi(t) = e^{\Phi(t)} = e^{Q-g(0)}(1-2it)^{-\frac{1}{2}f}\sum_{v=0}^{m}a_v(1-2it)^{-v} + R^*_{m+1}, \tag{13}$$

where $\sum_{v=0}^{m}a_v z^{-m}$ are the first $m+1$ terms in the series expansion of $\exp(-\sum_{r=0}^{m}\omega_r z^{-r})$ and R^*_{m+1} is a remainder term. Alternatively

$$\Phi(t) = -\tfrac{1}{2}f\log(1-2it) + \sum_{r=1}^{m}\omega_r\left[(1-2it)^{-r}-1\right] + R'_{m+1}, \tag{14}$$

where

$$R'_{m+1} = \sum O\left(x_k^{-(m+1)}\right) + \sum O\left(y_j^{-(m+1)}\right). \tag{15}$$

In (14) we have expanded $g(0)$ in the same way we had expanded $g(t)$ and have collected similar terms.

Then

$$\begin{aligned}(16)\quad \phi(t) &= e^{\Phi(t)}\\
&= (1-2it)^{-\frac{1}{2}f}\exp\left[\sum_{r=1}^{m}\omega_r(1-2it)^{-r} - \sum_{r=1}^{m}\omega_r + R'_{m+1}\right]\\
&= (1-2it)^{-\frac{1}{2}f}\left\{\prod_{r=1}^{m}\left[1+\omega_r(1-2it)^{-r} + \frac{1}{2!}\omega_r^2(1-2it)^{-2r}\cdots\right]\right.\\
&\qquad\left.\times\prod_{r=1}^{m}\left(1-\omega_r+\frac{1}{2!}\omega_r^2-\cdots\right) + R''_{m+1}\right\}\\
&= (1-2it)^{-\frac{1}{2}f}\left[1+T_1(t)+T_2(t)+\cdots+T_m(t)+R'''_{m+1}\right],\end{aligned}$$

where $T_r(t)$ is the term in the expansion with terms $\omega_1^{s_1}\cdots\omega_r^{s_r}$, $\Sigma i s_i = r$; for example,

$$(17)\qquad T_1(t) = \omega_1\left[(1-2it)^{-1}-1\right],$$

(18)

$$T_2(t) = \omega_2\left[(1-2it)^{-2}-1\right] + \tfrac{1}{2}\omega_1^2\left[(1-2it)^{-2} - 2(1-2it)^{-1}+1\right].$$

In most applications, we will have $x_k = c_k\theta$ and $y_j = d_j\theta$, where c_k and d_j will be constant and θ will vary (i.e., will grow with the sample size). In this case if ρ is chosen so $(1-\rho)x_k$ and $(1-\rho)y_j$ have limits, then R'''_{m+1} is $O(\theta^{-(m+1)})$. We collect in (16) all terms $\omega_1^{s_1}\cdots\omega_r^{s_r}$, $\Sigma i s_i = r$, because these terms are $O(\theta^{-r})$.

It will be observed that $T_r(t)$ is a polynomial of degree r in $(1-2it)^{-1}$ and each term of $(1-2it)^{-\frac{1}{2}f}T_r(t)$ is a constant times $(1-2it)^{-\frac{1}{2}v}$ for an integral v. We know that $(1-2it)^{-\frac{1}{2}v}$ is the characteristic function of the χ^2-density with v degrees of freedom; that is,

$$\begin{aligned}(19)\qquad g_v(z) &= \frac{1}{2^{\frac{1}{2}v}\Gamma\left(\frac{1}{2}v\right)}z^{\frac{1}{2}v-1}e^{-\frac{1}{2}z}\\
&= \int_{-\infty}^{\infty}\frac{1}{2\pi}(1-2it)^{-\frac{1}{2}v}e^{-itz}\,dt.\end{aligned}$$

Let

$$\begin{aligned}S_r(z) &= \int_{-\infty}^{\infty}\frac{1}{2\pi}(1-2it)^{-\frac{1}{2}f}T_r(t)e^{-itz}\,dt,\\
(20)\qquad & \\
R^{\text{iv}}_{m+1} &= \int_{-\infty}^{\infty}\frac{1}{2\pi}(1-2it)^{-\frac{1}{2}f}R'''_{m+1}e^{-itz}\,dt.\end{aligned}$$

Then the density of ρM is

$$(21)\quad \int_{-\infty}^{\infty} \frac{1}{2\pi}\phi(t)e^{-itz}\,dt = \sum_{r=0}^{m} S_r(z) + R_{m+1}^{\text{iv}}$$
$$= g_f(z) + \omega_1\big[g_{f+2}(z) - g_f(z)\big]$$
$$+\Big\{\omega_2\big[g_{f+4}(z) - g_f(z)\big]$$
$$+\frac{\omega_1^2}{2}\big[g_{f+4}(z) - 2g_{f+2}(z) + g_f(z)\big]\Big\}$$
$$+\cdots + S_m(z) + R_{m+1}^{\text{iv}}.$$

Let

$$(22)\quad \begin{aligned} U_r(z_0) &= \int_0^{z_0} S_r(z)\,dz, \\ R_{m+1}^{\text{v}} &= \int_0^{z_0} R_{m+1}^{\text{iv}}\,dz. \end{aligned}$$

The cdf of M is written in terms of the cdf of ρM which is the integral of the density, namely,

$$(23)\quad \Pr\{M \le M_0\}$$
$$= \Pr\{\rho M \le \rho M_0\}$$
$$= \sum_{r=0}^{m} U_r(\rho M_0) + R_{m+1}^{\text{v}}$$
$$= \Pr\{\chi_f^2 \le \rho M_0\} + \omega_1\big(\Pr\{\chi_{f+2}^2 \le \rho M_0\} - \Pr\{\chi_f^2 \le \rho M_0\}\big)$$
$$+\Big[\omega_2\big(\Pr\{\chi_{f+4}^2 \le \rho M_0\} - \Pr\{\chi_f^2 \le \rho M_0\}\big)$$
$$+\frac{\omega_1^2}{2}\big(\Pr\{\chi_{f+4}^2 \le \rho M_0\} - 2\Pr\{\chi_{f+2}^2 \le \rho M_0\}$$
$$+\Pr\{\chi_f^2 \le \rho M_0\}\big)\Big] + \cdots + U_m(\rho M_0) + R_{m+1}^{\text{v}}.$$

The remainder R_{m+1}^{v} is $O(\theta^{-(m+1)})$; this last statement can be verified by following the remainder terms along. (In fact, to make the proof rigorous one needs to verify that each remainder is of the proper order in a uniform sense.)

In many cases it is desirable to choose ρ so that $\omega_1 = 0$. In such a case using only the first term of (23) gives an error of order θ^{-2}.

Further details of the expansion can be found in Box's paper (1949).

THEOREM 8.5.1. *Suppose that* $\mathscr{E}W^h$ *is given by* (1) *for all purely imaginary* h, *with* (2) *holding. Then the* cdf *of* $-2\rho\log W$ *is given by* (23). *The error,* R^{v}_{m+1}, *is* $O(\theta^{-(m+1)})$ *if* $x_k \geq c_k\theta$, $y_j \geq d_j\theta$ ($c_k > 0, d_j > 0$), *and* $(1-\rho)x_k$, $(1-\rho)y_j$ *have limits, where* ρ *may depend on* θ.

Box also considers approximating the distribution of $-2\rho\log W$ by an F-distribution. He finds that the error in this approximation can be made to be of order θ^{-3}.

8.5.2. Asymptotic Distributions of Criterion

We now apply theorem 8.5.1 to the distribution of $-2\log\lambda$, the likelihood ratio criterion developed in Section 8.3. We let $W = \lambda$. The hth moment of λ is

$$\mathscr{E}\lambda^h = K\frac{\prod_{k=1}^{p}\Gamma\left[\frac{1}{2}(N-q+1-k+Nh)\right]}{\prod_{j=1}^{p}\Gamma\left[\frac{1}{2}(N-q_2+1-j+Nh)\right]}, \tag{24}$$

and this holds for all h for which the gamma functions exist including purely imaginary h. We let $a = b = p$,

$$\begin{aligned} x_k &= \tfrac{1}{2}N, & \xi_k &= \tfrac{1}{2}(-q+1-k), & \beta_k &= \tfrac{1}{2}(1-\rho)N, \\ y_j &= \tfrac{1}{2}N, & \eta_j &= \tfrac{1}{2}(-q_2+1-j), & \varepsilon_j &= \tfrac{1}{2}(1-\rho)N. \end{aligned} \tag{25}$$

We observe that

(26)

$$\begin{aligned} 2\omega_1 &= \sum_{k=1}^{p}\left\{\frac{\left\{\frac{1}{2}[(1-\rho)N-q+1-k]\right\}^2 - \frac{1}{2}[(1-\rho)N-q+1-k]}{\frac{1}{2}\rho N}\right. \\ &\qquad \left. - \frac{\left\{\frac{1}{2}[(1-\rho)N-q_2+1-k]\right\}^2 - \frac{1}{2}[(1-\rho)N-q_2+1-k]}{\frac{1}{2}\rho N}\right\} \\ &= \frac{2}{\rho N}\sum_{k=1}^{p}\left[\frac{-2[(1-\rho)N-q_2+1-k]q_1+q_1^2}{4}+\frac{q_1}{2}\right] \\ &= \frac{pq_1}{2\rho N}\left[-2(1-\rho)N+2q_2-2+(p+1)+q_1+2\right]. \end{aligned}$$

To make this zero, we require that

$$\rho = \frac{N - q_2 - \frac{1}{2}(p + q_1 + 1)}{N}. \tag{27}$$

Then

$$\begin{aligned}
\Pr\left\{-2\frac{k}{N}\log\lambda \le z\right\} &= \Pr\{-k\log U_{p,q_1,N-q} \le z\} \\
&= \Pr\{\chi^2_{pq_1} \le z\} \\
&\quad + \frac{\gamma_2}{k^2}\left(\Pr\{\chi^2_{pq_1+4} \le z\} - \Pr\{\chi^2_{pq_1} \le z\}\right) \\
&\quad + \frac{1}{k^4}\Big[\gamma_4\left(\Pr\{\chi^2_{pq_1+8} \le z\} - \Pr\{\chi^2_{pq_1} \le z\}\right) \\
&\quad -\gamma_2^2\left(\Pr\{\chi^2_{pq_1+4} \le z\} - \Pr\{\chi^2_{pq_1} \le z\}\right)\Big] + R_5^{\mathrm{v}},
\end{aligned} \tag{28}$$

where

$$k = \rho N = N - q_2 - \tfrac{1}{2}(p + q_1 + 1) = n - \tfrac{1}{2}(p - q_1 + 1), \tag{29}$$

$$\gamma_2 = \frac{pq_1(p^2 + q_1^2 - 5)}{48}, \tag{30}$$

$$\gamma_4 = \frac{\gamma_2^2}{2} + \frac{pq_1}{1920}\left[3p^4 + 3q_1^4 + 10p^2q_1^2 - 50(p^2 + q_1^2) + 159\right]. \tag{31}$$

Since $\lambda = U_{p,q_1,n}^{\frac{1}{2}N}$, where $n = N - q$, (28) gives $\Pr\{-k\log U_{p,q_1,n} \le z\}$.

Theorem 8.5.2. *The cdf of* $-k\log U_{p,q_1,n}$ *is given by* (28) *with* $k = n - \frac{1}{2}(p - q_1 + 1)$, *and* γ_2 *and* γ_4 *given by* (30) *and* (31), *respectively. The remainder term is* $O(N^{-6})$.

If the first term of (28) is used, the error is of the order N^{-2}; if the second, N^{-4}; and if the third,† N^{-6}. The second term is always negative and is

† Box has shown that the term of order N^{-5} is 0 and gives the coefficients to be used in the term of order N^{-6}.

numerically maximum for $z = \sqrt{(pq_1 + 2)(pq_1)}$ ($= pq_1 + 1$, approximately). For $p \geq 3$, $q_1 \geq 3$, $\gamma_2/k^2 \leq [(p^2 + q_1^2)/k]^2/96$ and the contribution of the second term lies between $-0.005[(p^2 + q_1^2)/k]^2$ and 0. For $p \geq 3$, $q_1 \geq 3$, $\gamma_4 \leq \gamma_2^2$ and the contribution of the third term is numerically less than $(\gamma_2/k^2)^2$. A rough rule that may be followed is that use of the first term is accurate to three decimal places if $p^2 + q_1^2 \leq k/3$.

As an example of the calculation, consider the case of $p = 3$, $q_1 = 6$, $N - q_2 = 24$, and $z = 26.0$ (the 10% significance point of χ^2_{18}, $18 = pq_1$). In this case $\gamma_2/k^2 = 0.048$ and the second term is -0.007; $\gamma_4/k^4 = 0.0015$ and the third term is -0.0001. Thus the probability of $-19 \log U_{3,6,18} \leq 26.0$ is 0.893 to three decimal places.

Since

$$-\left[n - \tfrac{1}{2}(p - m + 1)\right] \log u_{p,m,n}(\alpha) = C_{p,m,n-p+1}(\alpha)\chi^2_{pm}(\alpha), \tag{32}$$

the proportional error in approximating the left-hand side by $\chi^2_{pm}(\alpha)$ is $C_{p,m,n-p+1} - 1$. The proportional error increases slowly with p and m.

8.5.3. A Normal Approximation

Mudholkar and Trivedi (1980), (1981) have developed a normal approximation to the distribution of $-\log U_{p,m,n}$ which is asymptotic as p and/or $m \to \infty$. It is related to the Wilson–Hilferty normal approximation for the χ^2-distribution.

First, we give the background of the approximation. Suppose $\{Y_k\}$ is a sequence of nonnegative random variables such that $(Y_k - \mu_k)/\sigma_k \xrightarrow{\mathcal{L}} N(0,1)$ as $k \to \infty$, where $\mathscr{E} Y_k = \mu_k$ and $\mathscr{V}(Y_k) = \sigma_k^2$. Suppose also that $\mu_k \to \infty$ and σ_k^2/μ_k is bounded as $k \to \infty$. Let $Z_k = (Y_k/\mu_k)^h$. Then

$$\frac{Z_k - 1}{h(\sigma_k/\mu_k)} = \frac{\mu_k(Z_k - 1)}{h\sigma_k} \xrightarrow{\mathcal{L}} N(0,1) \tag{33}$$

by Theorem 4.2.3. The approach to normality may be accelerated by choosing h to make the distribution of Z_k relatively symmetric as measured by its third cumulant. The normal distribution is to be used as an approximation and is justified by its accuracy in practice. However, it will be convenient to develop the ideas in terms of limits although rigor is not necessary.

By a Taylor expansion we express the hth moment of Y_k/μ_k as

$$(34)\quad \mathscr{E}Z_k = \mathscr{E}\left(\frac{Y_k}{\mu_k}\right)^h = 1 + \frac{h(h-1)}{2}\frac{\sigma_k^2}{\mu_k} + \frac{h(h-1)(h-2)}{24}\frac{4\phi_k + 3(h-3)(\sigma_k^2/\mu_k)^2}{\mu_k^2} + O(\mu_k^{-3}),$$

where $\phi_k = \mathscr{E}(Y_k - \mu_k)^3/\mu_k$, assumed bounded. The rth moment of Z_k is expressed by replacement of h by rh in (34). The central moments of Z_k are

(35)

$$\mathscr{E}(Z_k - 1)^2 = h^2\frac{\sigma_k^2/\mu_k}{\mu_k} + \frac{h^2(h-1)}{2}\frac{2\phi_k + (3h-5)(\sigma_k^2/\mu_k)^2}{\mu_k^2} + O(\mu_k^{-3}).$$

$$(36)\qquad \mathscr{E}(Z_k - 1)^3 = h^3\frac{\phi_k + 3(h-1)(\sigma_k^2/\mu_k)^2}{\mu_k^2} + O(\mu_k^{-3}).$$

To make the third moment approximately 0 we take h to be

$$(37)\qquad h_0 = 1 - \frac{\mathscr{E}(Y_k - \mu_k)^3\mu_k}{3\sigma_k^4}.$$

Then $Z_k = (Y_k/\mu_k)^{h_0}$ is treated as normally distributed with mean and variance given by (34) and (35), respectively, with $h = h_0$.

Now we consider $-\log U_{p,m,n} = -\sum_{i=1}^p \log V_i$, where $V_1, \dots, V_p$ are independent and V_i has the density $\beta(x; (n+1-i)/2, m/2)$, $i = 1, \dots, p$. As $n \to \infty$ and $m \to \infty$, $-\log V_i$ tends to normality. If V has the density $\beta(x; a/2, b/2)$, the moment generating function of $-\log V$ is

$$(38)\qquad \mathscr{E}e^{-t\log V} = \frac{\Gamma[(a+b)/2]\Gamma(a/2 - t)}{\Gamma(a/2)\Gamma[(a+b)/2 - t]}.$$

Its logarithm is the cumulant generating function. Differentiation of the last yields as the rth cumulant of V

$$(39)\quad C_r = (-1)^r\left\{\psi^{(r-1)}(a/2) - \psi^{(r-1)}[(a+b)/2]\right\},\quad r = 1, 2, \dots,$$

where $\psi(w) = d \log \Gamma(w)/dw$. [See Abramovitz and Stegun (1972), p. 258, for example.] From $\Gamma(w+1) = w\Gamma(w)$ we obtain the recursion relation $\psi(w+1) = \psi(w) + 1/w$. This yields for $s = 0$ and l an integer

$$(40) \qquad \psi^{(s)}(w+l) - \psi^{(s)}(w) = (-1)^{s+1} s! \sum_{j=0}^{l-1} \frac{1}{(w+j)^{s+1}}.$$

The validity of (40) for $s = 1, 2, \ldots$ is verified by differentiation. (The expression for $\psi'(Z)$ in the first line of page 223 of Mudholkar and Trivedi (1981) is incorrect.) Thus for $b = 2l$

$$(41) \qquad C_r = (r-1)! \sum_{j=0}^{l-1} \frac{1}{\left(\frac{a}{2} + j\right)^r}.$$

From these results we obtain as the rth cumulant of $-\log U_{p,2l,n}$

$$(42) \qquad \kappa_r(-\log U_{p,2l,n}) = 2^r (r-1)! \sum_{i=1}^{p} \sum_{j=0}^{l-1} \frac{1}{(n-i+1+2j)^r}.$$

As $l \to \infty$ the series diverges for $r = 1$ and converges for $r = 2, 3$, and hence $\kappa_r/\kappa_1 \to 0$, $r = 2, 3$. The same is true as $p \to \infty$ (if n/p approaches a positive constant).

Given n, p, and l, the first three cumulants are calculated from (42). Then h_0 is determined from (37) and $(-\log U_{p,2l,n})^{h_0}$ is treated as approximately normally distributed with mean and variance calculated from (34) and (35) for $h = h_0$.

Mudholkar and Trivedi (1980) have calculated the error of approximation for significance levels of .01 and .05 for n from 4 to 66, $p = 3, 7$, and $q = 2, 6, 10$. The maximum error is less than .0007; in most cases the error is considerably less. The error for the χ^2-approximation is much larger, especially for small values of n.

In case of m odd the rth cumulant can be approximated by

$$(43) \qquad 2^r (r-1)! \sum_{i=1}^{p} \left[\sum_{j=0}^{\frac{1}{2}(m-3)} \frac{1}{(n-i+1+2j)^r} + \frac{1}{2} \frac{1}{(n-i+m)^r} \right].$$

Davis (1933, 1935) has given tables of $\psi(w)$ and its derivatives.

8.5.4. An F-Approximation

Rao (1951) has used the expansion of Section 8.5.2 to develop an expansion of the distribution of another function of $U_{p,m,n}$ in terms of beta distributions. The constants can be adjusted so that the term after the leading one is of order m^{-4}. A good approximation is to consider

$$\frac{1 - U^{1/s}}{U^{1/s}} \cdot \frac{ks - r}{pm} \tag{44}$$

as F with pm and $ks - r$ degrees of freedom, where

$$s = \sqrt{\frac{p^2m^2 - 4}{p^2 + m^2 - 5}}, \qquad r = \frac{pm}{2} - 1, \tag{45}$$

and k is $n - \frac{1}{2}(p - m + 1)$. For $p = 1$ or 2 or $m = 1$ or 2 the F-distribution is exactly as given in Section 8.4. If $ks - r$ is not an integer, interpolation between two integer values can be used. For smaller values of m this approximation is more accurate than the χ^2 approximation.

8.6. OTHER CRITERIA FOR TESTING THE LINEAR HYPOTHESIS

8.6.1. Functions of Roots

Thus far the only test of the linear hypothesis we have considered is the likelihood ratio test. In this section we consider other test procedures.

Let $\hat{\Sigma}_\Omega$, $\hat{\boldsymbol{\beta}}_{1\Omega}$, and $\hat{\boldsymbol{\beta}}_{2\omega}$ be the estimates of the parameters in $N(\boldsymbol{\beta}z, \Sigma)$, based on a sample of N observations. These are a sufficient set of statistics, and we shall base test procedures on them. As was shown in Section 8.3, if the hypothesis is $\boldsymbol{\beta}_1 = \boldsymbol{\beta}_1^*$, one can reformulate the hypothesis as $\boldsymbol{\beta}_1 = \mathbf{0}$ (by replacing $\boldsymbol{x}_\alpha$ by $\boldsymbol{x}_\alpha - \boldsymbol{\beta}_1^* z_\alpha^{(1)}$). Moreover,

$$\begin{aligned} \boldsymbol{\beta}z_\alpha &= \boldsymbol{\beta}_1 z_\alpha^{(1)} + \boldsymbol{\beta}_2 z_\alpha^{(2)} \\ &= \boldsymbol{\beta}_1\left(z_\alpha^{(1)} - A_{12}A_{22}^{-1}z_\alpha^{(2)}\right) + \left(\boldsymbol{\beta}_2 + \boldsymbol{\beta}_1 A_{12}A_{22}^{-1}\right)z_\alpha^{(2)} \\ &= \boldsymbol{\beta}_1 z_\alpha^{*(1)} + \boldsymbol{\beta}_2^* z_\alpha^{(2)}, \end{aligned} \tag{1}$$

where $\sum_\alpha z_\alpha^{*(1)} z_\alpha^{(2)\prime} = \mathbf{0}$ and $\sum_\alpha z_\alpha^{*(1)} z_\alpha^{*(1)\prime} = A_{11\cdot 2}$. Then $\hat{\boldsymbol{\beta}}_1 = \hat{\boldsymbol{\beta}}_{1\Omega}$ and $\hat{\boldsymbol{\beta}}_2^* = \hat{\boldsymbol{\beta}}_{2\omega}$.

We shall use the principle of invariance to reduce the set of tests to be considered. First, if we make the transformation $X_\alpha^* = X_\alpha + \Gamma z_\alpha^{(2)}$, we leave the null hypothesis invariant since $\mathscr{E}X_\alpha^* = \boldsymbol{\beta}_1 z_\alpha^{*(1)} + (\boldsymbol{\beta}_2^* + \Gamma)z_\alpha^{(2)}$ and $\boldsymbol{\beta}_2^* + \Gamma$ is unspecified. The only invariants of the sufficient statistics are $\hat{\Sigma}$ and $\hat{\boldsymbol{\beta}}_1$ (since for each $\hat{\boldsymbol{\beta}}_2^*$, there is a Γ that transforms it to $\mathbf{0}$, i.e., $-\hat{\boldsymbol{\beta}}_2^*$).

Second, the null hypothesis is invariant under the transformation $z_\alpha^{**(1)} = Cz_\alpha^{*(1)}$ (C nonsingular); the transformation carries $\boldsymbol{\beta}_1$ to $\boldsymbol{\beta}_1 C^{-1}$. Under this transformation $\hat{\Sigma}$ and $\hat{\boldsymbol{\beta}}_1 A_{11\cdot 2}\hat{\boldsymbol{\beta}}_1'$ are invariant; we consider $A_{11\cdot 2}$ as information relevant to inference. But these are the only invariants. For consider a function of $\hat{\boldsymbol{\beta}}_1$ and $A_{11\cdot 2}$, say $f(\hat{\boldsymbol{\beta}}_1, A_{11\cdot 2})$. Then there is a C^* that carries this into $f(\hat{\boldsymbol{\beta}}_1 C^{*-1}, I)$ and a further orthogonal transformation carries this into $f(T, I)$, where $t_{iv} = 0$, $i < v$, $t_{ii} \geq 0$. (If each row of T is considered a vector in q_1-space, the rotation of coordinate axes can be done so the first vector is along the first coordinate axis, the second vector is in the plane determined by the first two coordinate axes, and so forth.) But T is a function of $TT' = \hat{\boldsymbol{\beta}}_1 A_{11\cdot 2}\hat{\boldsymbol{\beta}}_1'$; that is, the elements of T are uniquely determined by this equation and the preceding restrictions. Thus our tests will depend on $\hat{\Sigma}$ and $\hat{\boldsymbol{\beta}}_1 A_{11\cdot 2}\hat{\boldsymbol{\beta}}_1'$. Let $N\hat{\Sigma} = G$ and $\hat{\boldsymbol{\beta}}_1 A_{11\cdot 2}\hat{\boldsymbol{\beta}}_1' = H$.

Third, the null hypothesis is invariant when x_α is replaced by Kx_α, for Σ and $\boldsymbol{\beta}_2^*$ are unspecified. This transforms G to KGK' and H to KHK'. The only invariants of G and H under such transformations are the roots of

$$|H - lG| = 0. \tag{2}$$

It is clear the roots are invariant, for

$$\begin{aligned} 0 &= |KHK' - lKGK'| \\ &= |K(H - lG)K'| \\ &= |K| \cdot |H - lG| \cdot |K'|. \end{aligned} \tag{3}$$

On the other hand, these are the only invariants, for given G and H there is a K such that $KGK' = I$ and

$$KHK' = L = \begin{pmatrix} l_1 & 0 & \cdots & 0 \\ 0 & l_2 & \cdots & 0 \\ \vdots & \vdots & & \vdots \\ 0 & 0 & \cdots & l_p \end{pmatrix}, \tag{4}$$

where $l_1 \geq \cdots \geq l_p$ are the roots of (2). (See Theorem A.2.2 of Appendix A.)

THEOREM 8.6.1. *Let $\boldsymbol{x}_\alpha$ be an observation from $N(\boldsymbol{\beta}_1 z_\alpha^{*(1)} + \boldsymbol{\beta}_2^* z_\alpha^{(2)}, \Sigma)$, where $\sum_\alpha z_\alpha^{*(1)} z_\alpha^{(2)\prime} = \mathbf{0}$ and $\sum_\alpha z_\alpha^{*(1)} z_\alpha^{*(1)\prime} = \boldsymbol{A}_{11\cdot 2}$. The only functions of the sufficient statistics and $\boldsymbol{A}_{11\cdot 2}$ invariant under the transformations $\boldsymbol{x}_\alpha^* = \boldsymbol{x}_\alpha + \Gamma z_\alpha^{(2)}$, $z_\alpha^{**(1)} = \boldsymbol{C} z_\alpha^{*(1)}$, and $\boldsymbol{x}_\alpha^* = \boldsymbol{K}\boldsymbol{x}_\alpha$ are the roots of (2), where $\boldsymbol{G} = N\hat{\boldsymbol{\Sigma}}$ and $\boldsymbol{H} = \hat{\boldsymbol{\beta}}_1 \boldsymbol{A}_{11\cdot 2} \hat{\boldsymbol{\beta}}_1'$.*

The likelihood ratio criterion is a function of

$$U = \frac{|\boldsymbol{G}|}{|\boldsymbol{G}+\boldsymbol{H}|} = \frac{|\boldsymbol{K}\boldsymbol{G}\boldsymbol{K}'|}{|\boldsymbol{K}\boldsymbol{G}\boldsymbol{K}' + \boldsymbol{K}\boldsymbol{H}\boldsymbol{K}'|} = \frac{|\boldsymbol{I}|}{|\boldsymbol{I}+\boldsymbol{L}|} \tag{5}$$

$$= \prod_{i=1}^{p} (1 + l_i)^{-1},$$

which is clearly invariant under the transformations.

Intuitively it would appear that good tests should reject the null hypothesis when the roots in some sense are large, for if $\boldsymbol{\beta}_1$ is very different from $\mathbf{0}$, $\hat{\boldsymbol{\beta}}_1$ will tend to be large and so will $\boldsymbol{H}$. Some other criteria that have been suggested are (*a*) $\sum l_i$, (*b*) $\sum l_i/(1+l_i)$, (*c*) $\max l_i$, and (*d*) $\min l_i$. In each case we reject the null hypothesis if the criterion exceeds some specified number.

8.6.2. The Lawley–Hotelling Trace Criterion

Let $\boldsymbol{K}$ be the matrix such that $\boldsymbol{K}\boldsymbol{G}\boldsymbol{K}' = \boldsymbol{I}$ [$\boldsymbol{G} = \boldsymbol{K}^{-1}(\boldsymbol{K}')^{-1}$ or $\boldsymbol{G}^{-1} = \boldsymbol{K}'\boldsymbol{K}$] and so (4) holds. Then the sum of the roots can be written

$$\sum_{i=1}^{p} l_i = \operatorname{tr} \boldsymbol{L} = \operatorname{tr} \boldsymbol{K}\boldsymbol{H}\boldsymbol{K}' \tag{6}$$

$$= \operatorname{tr} \boldsymbol{H}\boldsymbol{K}'\boldsymbol{K} = \operatorname{tr} \boldsymbol{H}\boldsymbol{G}^{-1}.$$

This criterion was suggested by Lawley (1938), Bartlett (1939), and Hotelling (1947), (1951). The test procedure is to reject the hypothesis if (6) is greater than a constant depending on p, m, and n.

The general distribution† of $\operatorname{tr} \boldsymbol{H}\boldsymbol{G}^{-1}$ cannot be characterized as easily as that of $U_{p,m,n}$. In the case of $p = 2$, Hotelling (1951) obtained an explicit expression for the distribution of $\operatorname{tr} \boldsymbol{H}\boldsymbol{G}^{-1} = l_1 + l_2$. A slightly different form

†Lawley (1938) purported to derive the exact distribution, but the result is in error.

of this distribution is obtained from the density of the two roots l_1 and l_2 in Chapter 13. It is

$$(7)\quad \Pr\{\operatorname{tr}\boldsymbol{HG}^{-1} \le w\} = I_{w/(2+w)}(m-1, n-1)$$
$$-\frac{\sqrt{\pi}\,\Gamma[\frac{1}{2}(m+n-1)]}{\Gamma(\frac{1}{2}m)\Gamma(\frac{1}{2}n)}(1+w)^{-\frac{1}{2}(n-1)} I_{w^2/(2+w)^2}[\tfrac{1}{2}(m-1), \tfrac{1}{2}(n-1)],$$

where $I_x(a,b)$ is the *incomplete beta function*, that is, the integral of $\beta(y; a, b)$ from 0 to x.

Constantine (1966) expressed the density of $\operatorname{tr}\boldsymbol{HG}^{-1}$ as an infinite series in generalized Laguerre polynomials and as an infinite series in zonal polynomials; these series, however, converge only for $\operatorname{tr}\boldsymbol{HG}^{-1} < 1$. Davis (1968) showed that the analytic continuation of these series satisfies a system of linear homogeneous differential equations of order p. Davis (1970) used a solution to compute tables as given in Appendix B.

Under the null hypothesis $\boldsymbol{G}$ is distributed as $\sum_{\alpha=1}^{n}\boldsymbol{Z}_\alpha\boldsymbol{Z}_\alpha'$ $(n = N - q)$ and $\boldsymbol{H}$ is distributed as $\sum_{\upsilon=1}^{q_1}\boldsymbol{Y}_\upsilon\boldsymbol{Y}_\upsilon'$, where the $\boldsymbol{Z}_\alpha$ and $\boldsymbol{Y}_\upsilon$ are independent, each with distribution $N(\boldsymbol{0}, \boldsymbol{\Sigma})$. Since the roots are invariant under the previously specified linear transformation, we can choose $\boldsymbol{K}$ so that $\boldsymbol{K\Sigma K}' = \boldsymbol{I}$ and let $\boldsymbol{G}^* = \boldsymbol{KGK}'[= \sum(\boldsymbol{KZ}_\alpha)(\boldsymbol{KZ}_\alpha)']$ and $\boldsymbol{H}^* = \boldsymbol{KHK}'$. This is equivalent to assuming at the outset that $\boldsymbol{\Sigma} = \boldsymbol{I}$.

Now

$$(8)\qquad \operatorname*{plim}_{N\to\infty}\frac{1}{N}\boldsymbol{G} = \operatorname*{plim}_{n\to\infty}\frac{n}{n+q}\,\frac{1}{n}\sum_{\alpha=1}^{n}\boldsymbol{Z}_\alpha\boldsymbol{Z}_\alpha' = \boldsymbol{I}.$$

This result follows applying the (weak) law of large numbers to each element of $(1/n)\boldsymbol{G}$,

$$(9)\qquad \operatorname*{plim}_{n\to\infty}\frac{1}{n}\sum_{\alpha=1}^{n} Z_{i\alpha}Z_{j\alpha} = \mathscr{E}Z_{i\alpha}Z_{j\alpha} = \delta_{ij}.$$

THEOREM 8.6.2. *Let $f(\boldsymbol{H})$ be a function whose discontinuities form a set of probability zero when $\boldsymbol{H}$ is distributed as $\sum_{\upsilon=1}^{q_1}\boldsymbol{Y}_\upsilon\boldsymbol{Y}_\upsilon'$ with the $\boldsymbol{Y}_\upsilon$ independent, each with distribution $N(\boldsymbol{0}, \boldsymbol{I})$. Then the limiting distribution of $f(N\boldsymbol{HG}^{-1})$ is the distribution of $f(\boldsymbol{H})$.*

PROOF. This is a straightforward application of a general theorem [for example, Theorem 2, Chernoff (1956)] to the effect that if the cdf of $\boldsymbol{X}_n$

converges to that of $\boldsymbol{X}$ (in every continuity point of the latter) and if $g(\boldsymbol{x})$ is a function whose discontinuities form a set of probability 0 according to the distribution of $\boldsymbol{X}$, then the cdf of $g(\boldsymbol{X}_n)$ converges to that of $g(\boldsymbol{X})$. In our case $\boldsymbol{X}_n$ consists of the components of $\boldsymbol{H}$ and $\boldsymbol{G}$, and $\boldsymbol{X}$ consists of the components of $\boldsymbol{H}$ and $\boldsymbol{I}$. Q.E.D.

COROLLARY 8.6.1. *The limiting distribution of $N \operatorname{tr} \boldsymbol{HG}^{-1}$ or $n \operatorname{tr} \boldsymbol{HG}^{-1}$ is the χ^2-distribution with pq_1 degrees of freedom.*

This follows from Theorem 8.6.2 because

$$\operatorname{tr} \boldsymbol{H} = \sum_{i=1}^{p} h_{ii} = \sum_{i=1}^{p} \sum_{v=1}^{q_1} Y_{iv}^2. \tag{10}$$

Ito (1956), (1960) developed asymptotic formulas and Fujikoshi (1973) extended them. Let $w_{p,m,n}(\alpha)$ be the α-significance point of $\operatorname{tr} \boldsymbol{HG}^{-1}$; that is,

$$\Pr\left\{\operatorname{tr} \boldsymbol{HG}^{-1} \geq w_{p,m,n}(\alpha)\right\} = \alpha, \tag{11}$$

and let $\chi_k^2(\alpha)$ be the α-significance point of the χ^2-distribution with k degrees of freedom. Then

$$nw_{p,m,n}(\alpha) = \chi_{pm}^2(\alpha) + \frac{1}{2n}\left[\frac{p+m+1}{pm+2}\chi_{pm}^4(\alpha) + (p-m+1)\chi_{pm}^2(\alpha)\right] + O(n^{-2}). \tag{12}$$

Ito also gives the term of order n^{-2}. See also Muirhead (1970). Davis (1970) evaluated the accuracy of the approximation (12). Ito also found

$$\Pr\{n \operatorname{tr} \boldsymbol{HG}^{-1} \leq z\} = G_{pm}(z) - \frac{1}{2n}\left[\frac{p+m+1}{pm+2}z^2 + (p-m+1)g_{pm}(z)\right] + O(n^{-2}), \tag{13}$$

where $G_k(z) = \Pr\{\chi_k^2 \leq z\}$ and $g_k(z) = (d/dz)G_k(z)$. Pillai (1956) has suggested another approximation to $nw_{p,m,n}(\alpha)$, and Pillai and Samson (1959) have given moments of $\operatorname{tr} \boldsymbol{H}\boldsymbol{G}^{-1}$. Pillai and Young (1971) and Krishnaiah and Chang (1972) have evaluated the Laplace transform of $\operatorname{tr} \boldsymbol{H}\boldsymbol{G}^{-1}$ and shown how to invert the transform. Khatri and Pillai (1966) suggest an approximate distribution based on moments. Pillai and Young (1971) suggest approximate distributions based on the first three moments.

Tables of the significance points have been given by Grubbs (1954) for $p = 2$ and by Davis (1970a) for $p = 3$ and 4, Davis (1970b) for $p = 5$, and Davis (1980) for $p = 6(1)10$; approximate significance points have been given by Pillai (1960). Davis's tables are reproduced in Appendix B, Table 2.

8.6.3. The Bartlett–Nanda–Pillai Trace Criterion

Another criterion, proposed by Bartlett (1939), Nanda (1950), and Pillai (1955), is

$$
\begin{aligned}
(14) \qquad V &= \sum_{i=1}^{p} \frac{l_i}{1 + l_i} = \operatorname{tr} \boldsymbol{L}(\boldsymbol{I} + \boldsymbol{L})^{-1} \\
&= \operatorname{tr} \boldsymbol{K}\boldsymbol{H}\boldsymbol{K}'(\boldsymbol{K}\boldsymbol{G}\boldsymbol{K}' + \boldsymbol{K}\boldsymbol{H}\boldsymbol{K}')^{-1} \\
&= \operatorname{tr} \boldsymbol{H}\boldsymbol{K}'\left[\boldsymbol{K}(\boldsymbol{G} + \boldsymbol{H})\boldsymbol{K}'\right]^{-1}\boldsymbol{K} \\
&= \operatorname{tr} \boldsymbol{H}(\boldsymbol{G} + \boldsymbol{H})^{-1},
\end{aligned}
$$

where as before $\boldsymbol{K}$ is such that $\boldsymbol{K}\boldsymbol{G}\boldsymbol{K}' = \boldsymbol{I}$ and (4) holds. In terms of the roots $f_i = l_i/(1 + l_i)$, $i = 1, \dots, p$, of

$$
(15) \qquad |\boldsymbol{H} - f(\boldsymbol{H} + \boldsymbol{G})| = 0
$$

the criterion is $\sum_{i=1}^{p} f_i$. In principle, the cdf, density, and moments under the null hypothesis can be found from the density of the roots (Sec. 13.2.3)

$$
(16) \qquad C \prod_{i=1}^{p} f_i^{\frac{1}{2}(m-p-1)} \prod_{i=1}^{p} (1 - f_i)^{\frac{1}{2}(n-p-1)} \prod_{i<j} (f_i - f_j),
$$

where

$$
(17) \qquad C = \frac{\pi^{\frac{1}{2}p^2} \Gamma_p\left[\frac{1}{2}(m + n)\right]}{\Gamma_p\left(\frac{1}{2}n\right)\Gamma_p\left(\frac{1}{2}m\right)\Gamma_p\left(\frac{1}{2}p\right)}
$$

for $1 > f_1 > \cdots > f_p > 0$ and 0 otherwise. If $m - p$ and $n - p$ are odd, the density is a polynomial in $f_1, \ldots, f_p$. Then the density and cdf of the sum of the roots are polynomials.

Many authors have written about the moments, Laplace transforms, densities, and cdf's, using various approaches. Nanda (1950) derived the distribution for $p = 2, 3, 4$ and $m = p + 1$. Pillai (1954), (1956), (1960) and Pillai and Mijares (1959) calculated the first four moments of V and proposed approximating the distribution by a beta-distribution based on the first four moments. Pillai and Jayachandran (1970) show how to evaluate the moment generating function as a weighted sum of determinants whose elements are incomplete gamma functions; they derive exact densities for some special cases and use them for a table of significance points. Krishnaiah and Chang (1972) express the distributions as linear combinations of inverse Laplace transforms of the products of certain double integrals and have further developed this technique for finding the distribution. Davis (1970) has shown that the distribution satisfies a differential equation and shown the nature of the solution. Khatri and Pillai (1968) have obtained the (nonnull) distributions in series forms. The characteristic function (under the null hypothesis) has been given by James (1964). Pillai and Jayachandran (1967) found the nonnull distribution for $p = 2$ and computed power functions. For an extensive bibliography see Krishnaiah (1978).

We now turn to the asymptotic theory. It follows from Theorem 8.6.2 that nV or NV has a limiting χ^2-distribution with pm degrees of freedom.

Let $v_{p,m,n}(\alpha)$ be defined by

$$\Pr\left\{\operatorname{tr} \boldsymbol{H}(\boldsymbol{H} + \boldsymbol{G})^{-1} \geq v_{p,m,n}(\alpha)\right\} = \alpha. \tag{18}$$

Then Davis (1970), Fujikoshi (1973), and Rothenberg (1977) have shown that

$$nv_{p,m,n}(\alpha) = \chi^2_{pm}(\alpha) + \frac{1}{2n}\left[-\frac{p+m+1}{pm+2}\chi^4_{pm}(\alpha) + (p-m+1)\chi^2_{pm}(\alpha)\right] + O(n^{-2}). \tag{19}$$

Since we can write (for the likelihood ratio test)

$$nu_{p,m,n}(\alpha) = \chi^2_{pm}(\alpha) + \frac{1}{2n}(p-m+1)\chi^2_{pm}(\alpha) + O(n^{-2}), \tag{20}$$

we have the comparison

$$(21)\quad nw_{p,m,n}(\alpha) = nu_{p,m,n}(\alpha) + \frac{1}{2n}\cdot\frac{p+m+1}{pm+2}\chi^4_{pm}(\alpha) + O(n^{-2}),$$

$$(22)\quad nv_{p,m,n}(\alpha) = nu_{p,m,n}(\alpha) - \frac{1}{2n}\cdot\frac{p+m+1}{pm+2}\chi^4_{pm}(\alpha) + O(n^{-2}).$$

An asymptotic expansion [Muirhead (1970), Fujikoshi (1973)] is

$$(23)\quad \Pr\{nV \le z\} = G_{pm}(z) + \frac{pm}{4n}\big[(m-p-1)G_{pm}(z) + 2(p+1)G_{pm+2}(z) - (p+m+1)G_{pm+4}(z)\big] + O(n^{-2}).$$

Higher-order terms are given by Muirhead and Fujikoshi.

Tables. Pillai (1960) tabulated 1% and 5% significance points of V for $p = 2(1)8$ based on fitting Pearson curves (i.e., beta distributions with adjusted ranges) to the first four moments. Mijares (1964) extended the tables to $p = 50$. The table in Appendix B of the significance points of $(n+m)V/m = \operatorname{tr}(1/m)\boldsymbol{H}\{[1/(n+m)](\boldsymbol{G}+\boldsymbol{H})\}^{-1}$ is taken from *Concise Statistical Tables*, which was computed on the same basis as Pillai. Schuurman, Krishnaiah, and Chattopodhyay (1975) gave exact significance points of V for $p = 2(1)5$; a more extensive table is in their technical report (ARL 73-0008). A comparison of some values with those of Appendix B shows a maximum difference of 3 in the third decimal place.

8.6.4. The Roy Maximum Root Criterion

Any characteristic root of $\boldsymbol{HG}^{-1}$ can be used as a test criterion. Roy (1953) proposed l_1, the maximum characteristic root of $\boldsymbol{HG}^{-1}$, on the basis of his "union-intersection principle." The test procedure is to reject the null hypothesis if l_1 is greater than a number or equivalently $f_1 = l_1/(1+l_1) = R$ is greater than a number $r_{p,m,n}(\alpha)$, which satisfies

$$(24)\qquad \Pr\{R \ge r_{p,m,n}(\alpha)\} = \alpha.$$

The density of the roots $f_1, \dots, f_p$ for $p \le m$ under the null hypothesis are given in (16). The cdf of $R = f_1$, $\Pr\{f_1 \le f^*\}$, can be obtained from the joint density by integration over the range $0 \le f_p \le \cdots \le f_1 \le f^*$. If $m - p$ and $n - p$ are both odd, the density of $f_1, \dots, f_p$ is a polynomial; then the cdf of f_1 is a polynomial in f^* and the density of f_1 is a polynomial. The only difficulty in carrying out the integration is keeping track of the different terms.

Roy (1945), (1957), Appendix 9, developed a method of integration that results in a cdf that is a linear combination of products of univariate beta densities and beta cdf's. The cdf of f_1 for $p = 2$ is

$$(25)\quad \Pr\{f_1 \le f\} = I_f(m-1, n-1)$$
$$-\frac{\sqrt{\pi}\,\Gamma\left[\frac{1}{2}(m+n-1)\right]}{\Gamma\left(\frac{1}{2}m\right)\Gamma\left(\frac{1}{2}n\right)} f^{\frac{1}{2}(m-1)}(1-f)^{\frac{1}{2}(n-1)} I_f\left[\tfrac{1}{2}(m-1), \tfrac{1}{2}(n-1)\right].$$

This is derived in Section 13.5. Roy (1957), Chapter 8, gives the cdf's for $p = 3$ and 4 also.

By Theorem 8.6.2 the limiting distribution of the largest characteristic root of $n\boldsymbol{H}\boldsymbol{G}^{-1}$, $N\boldsymbol{H}\boldsymbol{G}^{-1}$, $n\boldsymbol{H}(\boldsymbol{H}+\boldsymbol{G})^{-1}$ or $N\boldsymbol{H}(\boldsymbol{H}+\boldsymbol{G})^{-1}$ is the distribution of the largest characteristic root of $\boldsymbol{H}$ having the distribution $W(\boldsymbol{I}, m)$. The density of the roots of $\boldsymbol{H}$ are given in Section 13.3. In principle, the marginal density of the largest root can be obtained from the joint density by integration, but in actual fact the integration is more difficult than that for the density of the roots of $\boldsymbol{H}\boldsymbol{G}^{-1}$ or $\boldsymbol{H}(\boldsymbol{H}+\boldsymbol{G})^{-1}$.

The literature on this subject is too extensive to summarize here. Nanda (1948) obtained the distribution for $p = 2$, 3, 4, and 5. Pillai (1954, 1956, 1965, 1967) has treated the distribution under the null hypothesis. Other results were obtained by Sugiyama and Fukutomi (1966) and Sugiyama (1967). Pillai (1967) derived an appropriate distribution as a linear combination of incomplete beta functions. Davis (1972a) showed that the density of a single ordered root satisfies a differential equation and derived [Davis (1972b)] a recurrence relation for it. Hayakawa (1967), Khatri and Pillai (1968), Pillai and Sugiyama (1969), and Khatri (1972) have treated the noncentral case. See Krishnaiah (1978) for more references.

Tables. Tables of the percentage points have been calculated by Nanda (1951) and Foster and Rees (1957) for $p = 2$, Foster (1957) for $p = 3$, Foster (1958) for $p = 4$, and Pillai (1960) for $p = 2(1)6$ on the basis of an approximation. [See also Pillai (1956, 1960, 1964, 1965, 1967).] Heck (1960) presented charts of the significance points for $p = 2(1)6$. The table of significance points of nl_1/m in Appendix B is from *Concise Statistical Tables*, based on the approximation by Pillai (1967).

8.6.5. Comparison of Powers

The four tests that have been given most consideration are those based on Wilks' U, the Lawley–Hotelling W, the Bartlett–Nanda–Pillai V, and Roy's R. To guide in the choice of one of these four, we would like to compare power functions. The first three have been compared by Rothenberg on the basis of the asymptotic expansions of their distributions in the nonnull case.

Let $\nu_1^N, \dots, \nu_p^N$ be the roots of

$$|(\boldsymbol{\beta}_1 - \boldsymbol{\beta}_1^*)A_{11\cdot 2}(\boldsymbol{\beta}_1 - \boldsymbol{\beta}_1^*)' - \nu\Sigma| = 0. \tag{26}$$

The distribution of

$$\operatorname{tr}(\hat{\boldsymbol{\beta}}_{1\Omega} - \boldsymbol{\beta}_1^*)A_{11\cdot 2}(\hat{\boldsymbol{\beta}}_{1\Omega} - \boldsymbol{\beta}_1^*)'\Sigma^{-1} \tag{27}$$

is the noncentral χ^2-distribution with pm degrees of freedom and noncentrality parameter $\sum_{i=1}^p \nu_i^N$. As $N \to \infty$, the quantity $(1/n)\boldsymbol{G}$ or $(1/N)\boldsymbol{G}$ approaches $\boldsymbol{\Sigma}$ with probability one. If we let $N \to \infty$ and $\boldsymbol{A}_{11\cdot 2}$ is unbounded, the noncentrality parameter grows indefinitely and the power approaches 1. It is more informative to consider a sequence of alternatives such that the powers of the different tests are different. Suppose $\boldsymbol{\beta}_1 = \boldsymbol{\beta}_1^N$ is a sequence of matrices such that as $N \to \infty$, $(\boldsymbol{\beta}_1^N - \boldsymbol{\beta}_1^*)A_{11\cdot 2}(\boldsymbol{\beta}_1^N - \boldsymbol{\beta}_1^*)'$ approaches a limit and hence that $\nu_1^N, \dots, \nu_p^N$ approach some limiting values $\nu_1, \dots, \nu_p$, respectively. Then the limiting distribution of $N \operatorname{tr} \boldsymbol{HG}^{-1}$, $n \operatorname{tr} \boldsymbol{HG}^{-1}$, $N \operatorname{tr} \boldsymbol{H}(\boldsymbol{H} + \boldsymbol{G})^{-1}$, and $n \operatorname{tr} \boldsymbol{H}(\boldsymbol{H} + \boldsymbol{G})^{-1}$ is the noncentral χ^2-distribution with pm degrees of freedom and noncentrality parameter $\sum_{i=1}^p \nu_i$. Similarly for $-N \log U$ and $-n \log U$.

Rothenberg (1977) has shown under the above conditions that

(28)
$$\Pr\{U \le u_{p,m,n}(\alpha)\} = 1 - G_{pm}\left[\chi^2_{pm}(\alpha)\middle|\sum_{i=1}^{p}\nu_i\right] - \frac{1}{2n}\left\{(p+m+1)\sum_{i=1}^{p}\nu_i g_{pm+4}\left[\chi^2_{pm}(\alpha)\right] + \sum_{i=1}^{p}\nu_i^2 g_{pm+6}\left[\chi^2_{pm}(\alpha)\right]\right\} + o\left(\frac{1}{n}\right),$$

(29)
$$\Pr\{\operatorname{tr}\boldsymbol{H}\boldsymbol{G}^{-1} \ge w_{p,m,n}(\alpha)\}$$
$$= 1 - G_{pm}\left[\chi^2_{pm}(\alpha)\middle|\sum_{i=1}^{p}\nu_i\right] - \frac{1}{2n}\left\{(p+m+1)\sum_{i=1}^{p}\nu_i g_{pm+4}\left[\chi^2_{pm}(\alpha)\right] + \sum_{i=1}^{p}\nu_i^2 g_{pm+6}\left[\chi^2_{pm}(\alpha)\right] - \left[\sum_{i=1}^{p}\nu_i^2 - \frac{p+m+1}{pm+2}\left(\sum_{i=1}^{p}\nu_i\right)^2\right] g_{pm+8}\left[\chi^2_{pm}(\alpha)\right]\right\} + o\left(\frac{1}{n}\right),$$

(30)
$$\Pr\{\operatorname{tr}\boldsymbol{H}(\boldsymbol{H}+\boldsymbol{G})^{-1} \ge v_{p,m,n}(\alpha)\}$$
$$= 1 - G_{pm}\left[\chi^2_{pm}(\alpha)\middle|\sum_{i=1}^{p}\nu_i\right] - \frac{1}{2n}\left\{(p+m+1)\sum_{i=1}^{p}\nu_i g_{pm+4}\left[\chi^2_{pm}(\alpha)\right] + \sum_{i=1}^{p}\nu_i^2 g_{pm+6}\left[\chi^2_{pm}(\alpha)\right] + \left[\sum_{i=1}^{p}\nu_i^2 - \frac{p+m+1}{pm+2}\left(\sum_{i=1}^{p}\nu_i\right)^2\right] g_{pm+8}\left[\chi^2_{pm}(\alpha)\right]\right\} + o\left(\frac{1}{n}\right),$$

where $G_f(x|y)$ is the noncentral χ^2-distribution with f degrees of freedom and noncentrality parameter y and $g_f(x)$ is the (central) χ^2-density with f degrees of freedom. The leading terms are the noncentral χ^2-distribution; the power functions of the three tests agree to this order. The power functions of the two trace tests differ from that of the likelihood ratio test by $\pm g_{pm+8}[\chi^2_{pm}(\alpha)]/(2n)$ times

$$(31)\qquad \sum_{i=1}^{p} \nu_i^2 - \frac{p+m+1}{pm+2}\left(\sum_{i=1}^{p} \nu_i\right)^2 = \sum_{i=1}^{p} (\nu_i - \bar{\nu})^2 - \frac{p(p-1)(p+2)}{pm+2}\bar{\nu}^2,$$

where $\bar{\nu} = \sum_{i=1}^{p} \nu_i/p$. This is positive if

$$(32)\qquad \frac{\sigma_\nu}{\bar{\nu}} > \sqrt{\frac{(p-1)(p+2)}{pm+2}},$$

where $\sigma_\nu^2 = \sum_{i=1}^{p} (\nu_i - \bar{\nu})^2/p$ is the (population) variance of $\nu_1, \dots, \nu_p$; the left-hand side of (32) is the coefficient of variation. If the ν_i's are relatively variable in the sense that (32) holds, the power of the Lawley–Hotelling trace test is greater than that of the likelihood ratio test, which in turn is greater than that of the Bartlett–Nanda–Pillai trace tests (to order $1/n$); if the inequality (32) is reversed, the ordering of power is reversed.

The differences between the powers decrease as n increases for fixed $\nu_1, \dots, \nu_p$. (However, this comparison is not very meaningful because increasing n decreases $\boldsymbol{\beta}_1^N - \boldsymbol{\beta}_1^*$ and increases $\boldsymbol{Z}'\boldsymbol{Z}$.)

A number of numerical comparisons have been made. Schatzoff (1966b) and Olson (1974) have used Monte Carlo methods; Mikhail (1965), Pillai and Jayachandran (1967), and Lee (1971a) have used asymptotic expansions of distributions. All of these results agree with Rothenberg's. Among these three procedures, the Bartlett–Nanda–Pillai trace test is to be preferred if the roots are roughly equal in the alternative, and the Lawley–Hotelling trace is more powerful when the roots are substantially unequal. Wilks' likelihood ratio test seems to come in second best; in a sense it is maximin.

As noted in Section 8.6.4, the Roy largest root has a limiting distribution which is not a χ^2-distribution under the null hypothesis and a noncentral χ^2-distribution under sequence of the alternative hypotheses. Hence the comparison of Rothenberg cannot be extended to this case. In fact, the distributions in the nonnull case are difficult to evaluate. However, the Monte Carlo results of Schatzoff (1966b) and Olson (1974) are clear-cut. The maximum root test has greatest power if the alternative is one-dimensional, that is, if $\nu_2 = \cdots = \nu_p = 0$. On the other hand, if the alternative is not one-dimensional, then the maximum root test is inferior.

These test procedures tend to be robust. Under the null hypothesis the limiting distribution of $\hat{\boldsymbol{\beta}}_1 - \boldsymbol{\beta}_1^*$ suitably normalized is normal with mean $\mathbf{0}$ and covariances the same as if $\boldsymbol{X}$ were normal as long as its distribution satisfies some condition such as bounded fourth-order moments. Then $\hat{\boldsymbol{\Sigma}}_\Omega = (1/N)\boldsymbol{G}$ converges with probability one. The limiting distribution of each criterion suitably normalized is the same as if $\boldsymbol{X}$ were normal. Olson (1974) studied the robustness under departures from covariance homogeneity as well as departures from normality. His conclusion was that the two trace tests and the likelihood ratio test were rather robust and the maximum root test least robust. See also Pillai and Hsu (1979).

Berndt and Savin (1977) have noted that

$$\text{(33)} \qquad \operatorname{tr} \boldsymbol{H}(\boldsymbol{H}+\boldsymbol{G})^{-1} \le \log U^{-1} \le \operatorname{tr} \boldsymbol{H}\boldsymbol{G}^{-1}.$$

(See Problem 19.) If the χ^2 significance point is used, then a larger criterion may lead to rejection while a smaller one may not.

8.7. TESTS OF HYPOTHESES ABOUT MATRICES OF REGRESSION COEFFICIENTS AND CONFIDENCE REGIONS

8.7.1. Testing Hypotheses

Suppose we are given a set of vector observations $\boldsymbol{x}_1, \dots, \boldsymbol{x}_N$ with accompanying fixed vectors $z_1, \dots, z_N$, where $\boldsymbol{x}_\alpha$ is an observation from $N(\boldsymbol{\beta} z_\alpha, \Sigma)$. We let $\boldsymbol{\beta} = (\boldsymbol{\beta}_1 \ \boldsymbol{\beta}_2)$ and $z_\alpha' = (z_\alpha^{(1)\prime}, z_\alpha^{(2)\prime})$, where $\boldsymbol{\beta}_1$ and $z_\alpha^{(1)\prime}$ have $q_1 \ (= q - q_2)$ columns. The null hypothesis is

$$\text{(1)} \qquad H: \boldsymbol{\beta}_1 = \boldsymbol{\beta}_1^*,$$

where $\boldsymbol{\beta}_1^*$ is a specified matrix. Suppose the desired significance level is α. A test procedure is to compute

$$\text{(2)} \qquad U = \frac{|N\hat{\boldsymbol{\Sigma}}_\Omega|}{|N\hat{\boldsymbol{\Sigma}}_\omega|}$$

and compare this number with $U_{p,q_1,n}(\alpha)$, the α significance point of the $U_{p,q_1,n}$-distribution. For $p = 2,\dots,10$ and even m Table 1 in Appendix B can be used. For $m = 2,\dots,10$ and even p Table 1, Appendix B, can be used with m replaced by p and p replaced by m. (M as given in the table remains unchanged.) For p and m both odd interpolation between even values of either p or m will give sufficient accuracy for most purposes. For reasonably large n, the asymptotic theory can be used. An equivalent procedure is to calculate $\Pr\{U_{p,m,n} \le U\}$; if this is less than α, the null hypothesis is rejected.

Alternatively one an use the Lawley–Hotelling trace criterion

$$W = \operatorname{tr}(N\hat{\Sigma}_\omega - N\hat{\Sigma}_\Omega)(N\hat{\Sigma}_\Omega)^{-1} \tag{3}$$
$$= \operatorname{tr}\big(\hat{\mathbf{B}}_{1\Omega} - \mathbf{B}_1^*\big)A_{11\cdot 2}\big(\hat{\mathbf{B}}_{1\Omega} - \mathbf{B}_1^*\big)'(N\hat{\Sigma}_\Omega)^{-1},$$

the Pillai trace criterion

$$V = \operatorname{tr}(N\hat{\Sigma}_\omega - N\hat{\Sigma}_\Omega)(N\hat{\Sigma}_\omega)^{-1} \tag{4}$$
$$= \operatorname{tr}\big(\hat{\mathbf{B}}_{1\Omega} - \mathbf{B}_1^*\big)A_{11\cdot 2}\big(\hat{\mathbf{B}}_{1\Omega} - \mathbf{B}_1^*\big)'(N\hat{\Sigma}_\omega)^{-1},$$

or the Roy maximum root criterion R, where R is the maximum root of

$$|N\hat{\Sigma}_\omega - N\hat{\Sigma}_\Omega - rN\hat{\Sigma}_\Omega| = \big|\big(\hat{\mathbf{B}}_{1\Omega} - \mathbf{B}_1^*\big)A_{11\cdot 2}\big(\hat{\mathbf{B}}_{1\Omega} - \mathbf{B}_1^*\big)' - rN\hat{\Sigma}_\Omega\big| = 0. \tag{5}$$

These criteria can be referred to the appropriate tables in Appendix B.

In most studies a computer package will be used. We only outline an approach to computing the criterion. If we let $y_\alpha = x_\alpha - \mathbf{B}_1^* z_\alpha^{(1)}$, then y_α can be considered as an observation from $N(\Delta z_\alpha, \Sigma)$, where $\Delta = (\Delta_1 \;\; \Delta_2) = (\mathbf{B}_1 - \mathbf{B}_1^* \;\; \mathbf{B}_2)$. Then the null hypothesis is $H\colon \Delta_1 = \mathbf{0}$. Then

$$\sum y_\alpha y_\alpha' = \sum x_\alpha x_\alpha' - \mathbf{B}_1^* C_1' - C_1 \mathbf{B}_1^{*\prime} + \mathbf{B}_1^* A_{11} \mathbf{B}_1^{*\prime}, \tag{6}$$

$$\sum y_\alpha z_\alpha' = C - \mathbf{B}_1^*(A_{11} \;\; A_{12}). \tag{7}$$

Thus the problem of testing the hypothesis $\mathbf{B}_1 = \mathbf{B}_1^*$ is equivalent to testing the hypothesis $\Delta_1 = \mathbf{0}$ where $\mathscr{E} y_\alpha = \Delta z_\alpha$. Hence let us suppose the problem is testing the hypothesis $\mathbf{B}_1 = \mathbf{0}$. Then $N\hat{\Sigma}_\omega = \Sigma x_\alpha x_\alpha' - \hat{\mathbf{B}}_{2\omega} A_{22} \hat{\mathbf{B}}_{2\omega}'$ and $N\hat{\Sigma}_\Omega = \Sigma x_\alpha x_\alpha' - \hat{\mathbf{B}}_\Omega A \hat{\mathbf{B}}_\Omega'$. We have discussed in Section 8.2.2 the computation of $\hat{\mathbf{B}}_\Omega A \hat{\mathbf{B}}_\Omega'$, and hence $N\hat{\Sigma}_\Omega$. Then $\hat{\mathbf{B}}_{2\omega} A_{22} \hat{\mathbf{B}}_{2\omega}'$ can be computed in a similar manner. If the method is laid out

$$(8)\qquad \begin{pmatrix} A_{22} & A_{21} \\ A_{12} & A_{11} \end{pmatrix} \begin{pmatrix} \hat{\boldsymbol{\beta}}'_{2\Omega} \\ \hat{\boldsymbol{\beta}}'_{1\Omega} \end{pmatrix} = \begin{pmatrix} C'_2 \\ C'_1 \end{pmatrix},$$

the first q_2 rows and columns of A^* and of A^{**} are the same as the result of applying the forward solution to the left-hand side of

$$(9)\qquad A_{22}\hat{\boldsymbol{\beta}}'_{2\omega} = C'_2$$

and the first q_2 rows of $\bar{C}^*$ and $\bar{C}^{**}$ are the same as the result of applying the forward solution to the right-hand side of (9). Thus $\hat{\boldsymbol{\beta}}_{2\omega}A_{22}\hat{\boldsymbol{\beta}}'_{2\omega} = \bar{C}_2^*\bar{C}_2^{**\prime}$, where $\bar{C}^{*\prime} = (\bar{C}_2^{*\prime}\ \bar{C}_1^{*\prime})$ and $\bar{C}^{**\prime} = (\bar{C}_2^{**\prime}\ \bar{C}_1^{**\prime})$.

It might be noticed that the method implies a method for computing a determinant. In Section A.5 of Appendix A it is shown that the result of the forward solution is $\boldsymbol{F}\boldsymbol{A} = \boldsymbol{A}^*$. Thus $|\boldsymbol{F}| \cdot |\boldsymbol{A}| = |\boldsymbol{A}^*|$. Since the determinant of a triangular matrix is the product of its diagonal elements, $|\boldsymbol{F}| = 1$ and $|\boldsymbol{A}| = |\boldsymbol{A}^*| = \prod_{i=1}^{q} a_{ii}^*$. This result holds for any positive definite matrix in place of $\boldsymbol{A}$ (with a suitable modification of $\boldsymbol{F}$) and hence can be used to compute $|N\hat{\Sigma}_\Omega|$ and $|N\hat{\Sigma}_\omega|$.

8.7.2. Confidence Regions Based on U

We have considered tests of hypotheses $\boldsymbol{\beta}_1 = \boldsymbol{\beta}_1^*$, where $\boldsymbol{\beta}_1^*$ is specified. In the usual way we can deduce from the family of tests a confidence region for $\boldsymbol{\beta}_1$. From the theory given before, we know that the probability is $1 - \alpha$ of drawing a sample so that

$$(10)\qquad \frac{|N\hat{\Sigma}_\Omega|}{|N\hat{\Sigma}_\Omega + (\hat{\boldsymbol{\beta}}_{1\Omega} - \boldsymbol{\beta}_1)A_{11\cdot 2}(\hat{\boldsymbol{\beta}}_{1\Omega} - \boldsymbol{\beta}_1)'|} \geq u_{p,q_1,n}(\alpha).$$

Thus if we make the confidence region statement that $\boldsymbol{\beta}_1$ satisfies

$$(11)\qquad \frac{|N\hat{\Sigma}_\Omega|}{|N\hat{\Sigma}_\Omega + (\hat{\boldsymbol{\beta}}_{1\Omega} - \bar{\boldsymbol{\beta}}_1)A_{11\cdot 2}(\hat{\boldsymbol{\beta}}_{1\Omega} - \bar{\boldsymbol{\beta}}_1)'|} \geq u_{p,q_1,n}(\alpha),$$

where (11) is interpreted as an inequality on $\boldsymbol{\beta}_1 = \bar{\boldsymbol{\beta}}_1$, the probability is $1 - \alpha$ of drawing a sample such that the statement is true.

THEOREM 8.7.1. *The region* (11) *in the* $\overline{\boldsymbol{\beta}}_1$ *space is a confidence region for* $\boldsymbol{\beta}_1$ *with confidence coefficient* $1 - \alpha$.

Usually the set of $\overline{\boldsymbol{\beta}}_1$ satisfying (11) is difficult to visualize. However, the inequality can be used to determine whether trial matrices are included in the region.

8.7.3. Simultaneous Confidence Intervals Based on the Lawley–Hotelling Trace

Each test procedure implies a set of confidence regions. The Lawley–Hotelling trace criterion can be used to develop simultaneous confidence intervals for linear combinations of elements of $\boldsymbol{\beta}_1$. A confidence region with confidence coefficient $1 - \alpha$ is

$$\text{(12)} \qquad \operatorname{tr}\left(\hat{\boldsymbol{\beta}}_{1\Omega} - \overline{\boldsymbol{\beta}}_1'\right) A_{11\cdot 2}\left(\hat{\boldsymbol{\beta}}_{1\Omega} - \overline{\boldsymbol{\beta}}_1'\right)'\left(N\hat{\Sigma}_\Omega\right)^{-1} \le w_{p,m,n}(\alpha).$$

To derive the confidence bounds we generalize Lemma 5.3.2.

LEMMA 8.7.1. *For positive definite matrices* A *and* G

$$\text{(13)} \qquad |\operatorname{tr} \boldsymbol{\Phi}'Y| \le \sqrt{\operatorname{tr} A^{-1}\boldsymbol{\Phi}'G\boldsymbol{\Phi}}\sqrt{\operatorname{tr} A Y'G^{-1}Y}.$$

PROOF. Let $b = \operatorname{tr} \boldsymbol{\Phi}'Y / \operatorname{tr} A^{-1}\boldsymbol{\Phi}'G\boldsymbol{\Phi}$. Then

$$\begin{aligned}
\text{(14)} \qquad 0 &\le \operatorname{tr} A(Y - bG\boldsymbol{\Phi}A^{-1})'G^{-1}(Y - bG\boldsymbol{\Phi}A^{-1}) \\
&= \operatorname{tr} A Y'G^{-1}Y - b \operatorname{tr} \boldsymbol{\Phi}'Y - b \operatorname{tr} Y'\boldsymbol{\Phi} + b^2 \operatorname{tr} \boldsymbol{\Phi}'G\boldsymbol{\Phi}A^{-1} \\
&= \operatorname{tr} A Y'G^{-1}Y - \frac{(\operatorname{tr} \boldsymbol{\Phi}'Y)^2}{\operatorname{tr} A^{-1}\boldsymbol{\Phi}'G\boldsymbol{\Phi}},
\end{aligned}$$

which yields (13). Q.E.D.

Now (12) and (13) imply that

$$\begin{aligned}
\text{(15)} \quad |\operatorname{tr} \boldsymbol{\Phi}'\hat{\boldsymbol{\beta}}_{1\Omega} - \operatorname{tr} \boldsymbol{\Phi}'\overline{\boldsymbol{\beta}}_1| &= |\operatorname{tr} \boldsymbol{\Phi}'(\hat{\boldsymbol{\beta}}_{1\Omega} - \overline{\boldsymbol{\beta}}_1)| \\
&\le \sqrt{\operatorname{tr} A_{11\cdot 2}^{-1}\boldsymbol{\Phi}'N\hat{\Sigma}_\Omega\boldsymbol{\Phi} \cdot \operatorname{tr} A_{11\cdot 2}\left(\hat{\boldsymbol{\beta}}_{1\Omega} - \overline{\boldsymbol{\beta}}_1\right)'\left(N\hat{\Sigma}_\Omega\right)^{-1}\left(\hat{\boldsymbol{\beta}}_{1\Omega} - \overline{\boldsymbol{\beta}}_1\right)} \\
&\le \sqrt{\operatorname{tr} A_{11\cdot 2}^{-1}\boldsymbol{\Phi}'N\hat{\Sigma}_\Omega\boldsymbol{\Phi}}\sqrt{w_{p,m,n}(\alpha)}
\end{aligned}$$

holds for all $p \times m$ matrices $\mathbf{\Phi}$. We assert that

$$(16)\quad \operatorname{tr}\mathbf{\Phi}'\hat{\boldsymbol{\beta}}_{1\Omega} - \sqrt{N \operatorname{tr} A_{11\cdot 2}^{-1}\mathbf{\Phi}'\hat{\mathbf{\Sigma}}_{\Omega}\mathbf{\Phi}}\sqrt{w_{p,m,n}(\alpha)} \le \operatorname{tr}\mathbf{\Phi}'\overline{\boldsymbol{\beta}}_1$$

$$\le \operatorname{tr}\mathbf{\Phi}'\hat{\boldsymbol{\beta}}_{1\Omega} + \sqrt{N \operatorname{tr} A_{11\cdot 2}^{-1}\mathbf{\Phi}'\hat{\mathbf{\Sigma}}_{\Omega}\mathbf{\Phi}}\sqrt{w_{p,m,n}(\alpha)}$$

holds for all $\mathbf{\Phi}$ with confidence $1-\alpha$.

The confidence region (12) can be explored by use of (16) for various $\mathbf{\Phi}$. If $\phi_{ik} = 1$ for some pair (I, K) and 0 for other elements, then (16) gives an interval for β_{IK}. If $\phi_{ik} = 1$ for a pair (I, K), -1 for (I, L), and 0 otherwise, the interval pertains to $\beta_{IK} - \beta_{IL}$, the difference of coefficients of two independent variables. If $\phi_{ik} = 1$ for a pair (I, K), -1 for (J, K), and 0 otherwise, one obtains an interval for $\beta_{IK} - \beta_{JK}$, the difference of coefficients for two dependent variables.

8.7.4. Simultaneous Confidence Intervals based on the Roy Maximum Root Criterion

A confidence region with confidence $1-\alpha$ based on the maximum root criteria is

$$(17)\qquad \operatorname{ch}_1\left(\hat{\boldsymbol{\beta}}_{1\Omega} - \overline{\boldsymbol{\beta}}_1\right)A_{11\cdot 2}\left(\hat{\boldsymbol{\beta}}_{1\Omega} - \overline{\boldsymbol{\beta}}_1\right)'\left(N\hat{\mathbf{\Sigma}}_{\Omega}\right)^{-1} \le r_{p,m,n}(\alpha),$$

where $\operatorname{ch}_1(C)$ denotes the largest characteristic root of C. We can derive simultaneous confidence bounds from (17). From Lemma 5.3.2, we find for any vectors a and b

(18)

$$\left[a'\left(\hat{\boldsymbol{\beta}}_{1\Omega} - \boldsymbol{\beta}_1\right)b\right]^2 = \left\{\left[\left(\hat{\boldsymbol{\beta}}_{1\Omega} - \boldsymbol{\beta}_1\right)'a\right]'b\right\}^2$$

$$\le \left[\left(\hat{\boldsymbol{\beta}}_{1\Omega} - \boldsymbol{\beta}_1\right)'a\right]'A_{11\cdot 2}\left[\left(\hat{\boldsymbol{\beta}}_{1\Omega} - \boldsymbol{\beta}_1\right)'a\right]\cdot b'A_{11\cdot 2}^{-1}b$$

$$= \frac{a'\left(\hat{\boldsymbol{\beta}}_{1\Omega} - \boldsymbol{\beta}_1\right)A_{11\cdot 2}\left(\hat{\boldsymbol{\beta}}_{1\Omega} - \boldsymbol{\beta}_1\right)'a}{a'Ga}\cdot a'Ga\cdot b'A_{11\cdot 2}^{-1}b$$

$$\le \operatorname{ch}_1\left[\left(\hat{\boldsymbol{\beta}}_{1\Omega} - \boldsymbol{\beta}_1\right)A_{11\cdot 2}\left(\hat{\boldsymbol{\beta}}_{1\Omega} - \boldsymbol{\beta}_1\right)'G^{-1}\right]\cdot a'Ga\cdot b'A_{11\cdot 2}^{-1}b$$

$$\le r_{p,m,n}(\alpha)\cdot a'Ga\cdot b'A_{11\cdot 2}^{-1}b$$

with probability $1 - \alpha$; the second inequality follows from Theorem A.2.4 of Appendix A. Then a set of confidence intervals on all linear combinations $\boldsymbol{a}'\boldsymbol{\beta}_1\boldsymbol{b}$ holding with confidence $1 - \alpha$ is

$$(19)\quad \boldsymbol{a}'\hat{\boldsymbol{\beta}}_{1\Omega}\boldsymbol{b} - \sqrt{r_{p,m,n}(\alpha)\cdot \boldsymbol{a}'\boldsymbol{G}\boldsymbol{a}\cdot \boldsymbol{b}'\boldsymbol{A}_{11\cdot 2}^{-1}\boldsymbol{b}} \le \boldsymbol{a}'\boldsymbol{\beta}_1\boldsymbol{b}$$

$$\le \boldsymbol{a}'\hat{\boldsymbol{\beta}}_{1\Omega}\boldsymbol{b} + \sqrt{r_{p,m,n}(\alpha)\cdot \boldsymbol{a}'\boldsymbol{G}\boldsymbol{a}\cdot \boldsymbol{b}'\boldsymbol{A}_{11\cdot 2}^{-1}\boldsymbol{b}}\,.$$

The linear combinations are $\boldsymbol{a}'\boldsymbol{\beta}_1\boldsymbol{b} = \sum_{i=1}^{p}\sum_{h=1}^{m} a_i\beta_{ih}b_h$. If $a_1 = 1$, $a_i = 0$, $i \ne 1$ and $b_1 = 1$, $b_h = 0$, $h \ne 1$, the linear combination is simply β_{11}. If $a_1 = 1$, $a_i = 0$, $i \ne 1$ and $b_1 = 1$, $b_2 = -1$, $b_h = 0$, $h \ne 1, 2$, the linear combination is $\beta_{11} - \beta_{12}$.

We can compare these intervals with (16) for $\boldsymbol{\Phi} = \boldsymbol{a}\boldsymbol{b}'$, which is of rank 1. The term subtracted and added to $\operatorname{tr}\boldsymbol{\Phi}'\hat{\boldsymbol{\beta}}_{1\Omega} = \boldsymbol{a}'\hat{\boldsymbol{\beta}}_{1\Omega}\boldsymbol{b}$ is the square root of

$$(20)\qquad w_{p,m,n}(\alpha)\cdot \operatorname{tr}\boldsymbol{A}_{11\cdot 2}^{-1}\boldsymbol{b}\boldsymbol{a}'\boldsymbol{G}\boldsymbol{a}\boldsymbol{b}' = w_{p,m,n}(\alpha)\cdot \boldsymbol{a}'\boldsymbol{G}\boldsymbol{a}\cdot \boldsymbol{b}'\boldsymbol{A}_{11\cdot 2}^{-1}\boldsymbol{b}.$$

This is greater than the term subtracted and added to $\boldsymbol{a}'\hat{\boldsymbol{\beta}}_{1\Omega}\boldsymbol{b}$ in (19) because $w_{p,m,n}(\alpha)$, pertaining to the sum of the roots, is greater than $r_{p,m,n}(\alpha)$, relating to one root. The bounds (16) hold for all $p \times m$ matrices $\boldsymbol{\Phi}$, while (19) holds only for matrices $\boldsymbol{a}\boldsymbol{b}'$ of rank 1.

Mudholkar (1966) gives a very general method of constructing simultaneous confidence intervals based on symmetric gauge functions. Gabriel (1969) relates confidence bounds with simultaneous test procedures. Wijsman (1979) has shown that under certain conditions the confidence sets based on the maximum root are smallest. [See also Wijsman (1980).]

8.8. TESTING EQUALITY OF MEANS OF SEVERAL NORMAL DISTRIBUTIONS WITH COMMON COVARIANCE MATRIX

In univariate analysis it is well known that many hypotheses can be put in the form of hypotheses concerning regression coefficients. The same is true for the corresponding multivariate cases. As an example we consider testing the hypothesis that the means of, say, q normal distributions with a common covariance matrix are equal.

Let $y_\alpha^{(i)}$ be an observation from $N(\boldsymbol{\mu}^{(i)}, \boldsymbol{\Sigma})$, $\alpha = 1, \dots, N_i$; $i = 1, \dots, q$. The null hypothesis is

$$H: \boldsymbol{\mu}^{(1)} = \cdots = \boldsymbol{\mu}^{(q)}. \tag{1}$$

To put the problem in the form considered earlier in this chapter, let

$$\boldsymbol{X} = (\boldsymbol{x}_1\ \boldsymbol{x}_2\ \cdots\ \boldsymbol{x}_{N_1}\ \boldsymbol{x}_{N_1+1}\ \cdots\ \boldsymbol{x}_N) = \left(\boldsymbol{y}_1^{(1)}\ \boldsymbol{y}_2^{(1)}\ \cdots\ \boldsymbol{y}_{N_1}^{(1)}\ \boldsymbol{y}_1^{(2)}\ \cdots\ \boldsymbol{y}_{N_q}^{(q)}\right) \tag{2}$$

with $N = N_1 + \cdots + N_q$. Let

$$\begin{aligned} \boldsymbol{Z} &= (z_1\ z_2\ \cdots\ z_{N_1}\ z_{N_1+1}\ \cdots\ z_N) \\ &= \begin{pmatrix} 1 & 1 & \cdots & 1 & 0 & \cdots & 0 \\ 0 & 0 & \cdots & 0 & 1 & \cdots & 0 \\ 0 & 0 & \cdots & 0 & 0 & \cdots & 0 \\ \vdots & \vdots & & \vdots & \vdots & & \vdots \\ 0 & 0 & \cdots & 0 & 0 & \cdots & 0 \\ 1 & 1 & \cdots & 1 & 1 & \cdots & 1 \end{pmatrix}; \end{aligned} \tag{3}$$

that is, $z_{i\alpha} = 1$ if $N_1 + \cdots + N_{i-1} < \alpha \le N_1 + \cdots + N_i$, and $z_{i\alpha} = 0$ otherwise, for $i = 1, \dots, q-1$, and $z_{q\alpha} = 1$ (all α). Let $\boldsymbol{\beta} = (\boldsymbol{\beta}_1\ \boldsymbol{\beta}_2)$, where

$$\begin{aligned} \boldsymbol{\beta}_1 &= \left(\boldsymbol{\mu}^{(1)} - \boldsymbol{\mu}^{(q)}, \dots, \boldsymbol{\mu}^{(q-1)} - \boldsymbol{\mu}^{(q)}\right), \\ \boldsymbol{\beta}_2 &= \boldsymbol{\mu}^{(q)}. \end{aligned} \tag{4}$$

Then $\boldsymbol{x}_\alpha$ is an observation from $N(\boldsymbol{\beta} z_\alpha, \boldsymbol{\Sigma})$ and the null hypothesis is $\boldsymbol{\beta}_1 = \mathbf{0}$. Thus we can use the above theory for finding the criterion for testing the hypothesis.

We have

$$\boldsymbol{A} = \sum_{\alpha=1}^{N} z_\alpha z_\alpha' = \begin{pmatrix} N_1 & 0 & \cdots & 0 & N_1 \\ 0 & N_2 & \cdots & 0 & N_2 \\ \vdots & \vdots & & \vdots & \vdots \\ 0 & 0 & \cdots & N_{q-1} & N_{q-1} \\ N_1 & N_2 & \cdots & N_{q-1} & N \end{pmatrix}, \tag{5}$$

$$\boldsymbol{C} = \sum_{\alpha=1}^{N} \boldsymbol{x}_\alpha z_\alpha' = \left(\sum_\alpha \boldsymbol{y}_\alpha^{(1)}\ \sum_\alpha \boldsymbol{y}_\alpha^{(2)}\ \cdots\ \sum_\alpha \boldsymbol{y}_\alpha^{(q-1)}\ \sum_{i,\alpha} \boldsymbol{y}_\alpha^{(i)}\right). \tag{6}$$

Here $\boldsymbol{A}_{22} = N$ and $\boldsymbol{C}_2 = \Sigma_{i,\alpha}\boldsymbol{y}_\alpha^{(i)}$. Thus $\hat{\boldsymbol{\beta}}_{2\omega} = \Sigma_{i,\alpha}\boldsymbol{y}_\alpha^{(i)} \cdot (1/N) = \bar{\boldsymbol{y}}$, say, and

$$N\hat{\boldsymbol{\Sigma}}_\omega = \sum_\alpha \boldsymbol{x}_\alpha \boldsymbol{x}_\alpha' - \bar{\boldsymbol{y}}N\bar{\boldsymbol{y}}' \tag{7}$$

$$= \sum_{i,\alpha} \boldsymbol{y}_\alpha^{(i)}\boldsymbol{y}_\alpha^{(i)\prime} - N\bar{\boldsymbol{y}}\bar{\boldsymbol{y}}'$$

$$= \sum_{i,\alpha}\left(\boldsymbol{y}_\alpha^{(i)} - \bar{\boldsymbol{y}}\right)\left(\boldsymbol{y}_\alpha^{(i)} - \bar{\boldsymbol{y}}\right)'.$$

For $\hat{\boldsymbol{\Sigma}}_\Omega$, we use the formula $N\hat{\boldsymbol{\Sigma}}_\Omega = \Sigma \boldsymbol{x}_\alpha \boldsymbol{x}_\alpha' - \hat{\boldsymbol{\beta}}_\Omega \boldsymbol{A}\hat{\boldsymbol{\beta}}_\Omega' = \Sigma \boldsymbol{x}_\alpha \boldsymbol{x}_\alpha' - \boldsymbol{C}\boldsymbol{A}^{-1}\boldsymbol{C}'$. Let

$$\boldsymbol{D} = \begin{pmatrix} 1 & 0 & \cdots & 0 & 0 \\ 0 & 1 & \cdots & 0 & 0 \\ \vdots & \vdots & & \vdots & \vdots \\ 0 & 0 & \cdots & 1 & 0 \\ -1 & -1 & \cdots & -1 & 1 \end{pmatrix}; \tag{8}$$

then

$$\boldsymbol{D}^{-1} = \begin{pmatrix} 1 & 0 & \cdots & 0 & 0 \\ 0 & 1 & \cdots & 0 & 0 \\ \vdots & \vdots & & \vdots & \vdots \\ 0 & 0 & \cdots & 1 & 0 \\ 1 & 1 & \cdots & 1 & 1 \end{pmatrix}. \tag{9}$$

Thus

$$\boldsymbol{C}\boldsymbol{A}^{-1}\boldsymbol{C}' = \boldsymbol{C}\boldsymbol{D}'\boldsymbol{D}^{-1\prime}\boldsymbol{A}^{-1}\boldsymbol{D}^{-1}\boldsymbol{D}\boldsymbol{C}' \tag{10}$$

$$= \boldsymbol{C}\boldsymbol{D}'(\boldsymbol{D}\boldsymbol{A}\boldsymbol{D}')^{-1}\boldsymbol{D}\boldsymbol{C}'$$

$$= \left(\sum_\alpha \boldsymbol{y}_\alpha^{(1)} \cdots \sum_\alpha \boldsymbol{y}_\alpha^{(q)}\right)\begin{pmatrix} N_1 & 0 & \cdots & 0 \\ 0 & N_2 & \cdots & 0 \\ \vdots & \vdots & & \vdots \\ 0 & 0 & \cdots & N_q \end{pmatrix}^{-1}\begin{pmatrix} \sum_\alpha \boldsymbol{y}_\alpha^{(1)\prime} \\ \vdots \\ \sum_\alpha \boldsymbol{y}_\alpha^{(q)\prime} \end{pmatrix}$$

$$= \sum_i \left(\sum_\alpha \boldsymbol{y}_\alpha^{(i)} \frac{1}{N_i} \sum_\gamma \boldsymbol{y}_\gamma^{(i)\prime}\right)$$

$$= \sum_i N_i \bar{\boldsymbol{y}}^{(i)}\bar{\boldsymbol{y}}^{(i)\prime},$$

where $\bar{\mathbf{y}}^{(i)} = \frac{1}{N_i}\sum_\alpha \mathbf{y}_\alpha^{(i)}$. Thus

$$N\hat{\boldsymbol{\Sigma}}_\Omega = \sum_{i,\alpha} \mathbf{y}_\alpha^{(i)}\mathbf{y}_\alpha^{(i)\prime} - \sum_i N_i\bar{\mathbf{y}}^{(i)}\bar{\mathbf{y}}^{(i)\prime} = \sum_{i,\alpha}\left(\mathbf{y}_\alpha^{(i)} - \bar{\mathbf{y}}^{(i)}\right)\left(\mathbf{y}_\alpha^{(i)} - \bar{\mathbf{y}}^{(i)}\right)'. \tag{11}$$

It will be seen that $\hat{\boldsymbol{\Sigma}}_\omega$ is the estimator of $\boldsymbol{\Sigma}$ when $\boldsymbol{\mu}^{(1)} = \cdots = \boldsymbol{\mu}^{(q)}$ and $\hat{\boldsymbol{\Sigma}}_\Omega$ is the weighted average of the estimators of $\boldsymbol{\Sigma}$ based on the separate samples.

When the null hypothesis is true, $|N\hat{\boldsymbol{\Sigma}}_\Omega|/|N\hat{\boldsymbol{\Sigma}}_\omega|$ is distributed as $U_{p,q-1,n}$, where $n = N - q$. Therefore, the rejection region at the α significance level is

$$\frac{|N\hat{\boldsymbol{\Sigma}}_\Omega|}{|N\hat{\boldsymbol{\Sigma}}_\omega|} < u_{p,q-1,n}(\alpha). \tag{12}$$

We observe that

$$N\hat{\boldsymbol{\Sigma}}_\omega - N\hat{\boldsymbol{\Sigma}}_\Omega = \sum_{i,\alpha}\mathbf{y}_\alpha^{(i)}\mathbf{y}_\alpha^{(i)\prime} - N\bar{\mathbf{y}}\bar{\mathbf{y}}' - \left(\sum_{i,\alpha}\mathbf{y}_\alpha^{(i)}\mathbf{y}_\alpha^{(i)\prime} - \sum_i N_i\bar{\mathbf{y}}^{(i)}\bar{\mathbf{y}}^{(i)\prime}\right) = \sum N_i\left(\bar{\mathbf{y}}^{(i)} - \bar{\mathbf{y}}\right)\left(\bar{\mathbf{y}}^{(i)} - \bar{\mathbf{y}}\right)'. \tag{13}$$

It will be seen that when $p = 1$, this test reduces to the usual F-test

$$\frac{\sum N_i\left(\bar{y}^{(i)} - \bar{y}\right)^2}{\sum\left(y_\alpha^{(i)} - \bar{y}^{(i)}\right)^2}\cdot\frac{n}{q-1} > F_{q-1,n}(\alpha). \tag{14}$$

We give an example of the analysis. The data are taken from Barnard's study of Egyptian skulls (1935). The 4 ($= q$) populations are Late Predynastic ($i = 1$), Sixth to Twelfth ($i = 2$), Twelfth to Thirteenth ($i = 3$), and Ptolemaic Dynasties ($i = 4$). The 4 ($= p$) measurements (i.e., components of $y_\alpha^{(i)}$) are maximum breadth, basialveolar length, nasal height, and basibregmatic height. The numbers of observations are $N_1 = 91$, $N_2 = 162$, $N_3 = 70$, $N_4 = 75$. The data are summarized as

$$\left(\bar{\mathbf{y}}^{(1)}\ \bar{\mathbf{y}}^{(2)}\ \bar{\mathbf{y}}^{(3)}\ \bar{\mathbf{y}}^{(4)}\right) = \begin{pmatrix} 133.582\,418 & 134.265\,432 & 134.371\,429 & 135.306\,667 \\ 98.307\,692 & 96.462\,963 & 95.857\,143 & 95.040\,000 \\ 50.835\,165 & 51.148\,148 & 50.100\,000 & 52.093\,333 \\ 133.000\,000 & 134.882\,716 & 133.642\,857 & 131.466\,667 \end{pmatrix}, \tag{15}$$

(16) $N\hat{\boldsymbol{\Sigma}}_\Omega$

$$= \begin{pmatrix} 9661.997\,470 & 445.573\,301 & 1130.623\,900 & 2148.584\,210 \\ 445.573\,301 & 9073.115\,027 & 1239.211\,990 & 2255.812\,722 \\ 1130.623\,900 & 1239.211\,990 & 3938.320\,351 & 1271.054\,662 \\ 2148.584\,210 & 2255.812\,722 & 1271.054\,662 & 8741.508\,829 \end{pmatrix}.$$

From these data we find

(17) $N\hat{\boldsymbol{\Sigma}}_\omega$

$$= \begin{pmatrix} 9785.178\,098 & 214.197\,666 & 1217.929\,248 & 2019.820\,216 \\ 214.197\,666 & 9559.460\,890 & 1131.716\,372 & 2381.126\,040 \\ 1217.929\,248 & 1131.716\,372 & 4088.731\,856 & 1133.473\,898 \\ 2019.820\,216 & 2381.126\,040 & 1133.473\,898 & 9382.242\,720 \end{pmatrix}.$$

We shall use the likelihood ratio test. The ratio of determinants is

$$U = \frac{|N\hat{\boldsymbol{\Sigma}}_\Omega|}{|N\hat{\boldsymbol{\Sigma}}_\omega|} = \frac{2.426\,905\,4 \times 10^5}{2.954\,477\,5 \times 10^5} = 0.821\,434\,4. \tag{18}$$

Here $N = 398$, $n = 394$, $p = 4$, and $q = 4$. Thus $k = 393$. Since n is very large, we may assume $-k \log U_{4,3,394}$ is distributed as χ^2 with 12 degrees of freedom (when the null hypothesis is true). Here $-k \log U = 77.30$. Since the 1% point of the χ^2_{12}-distribution is 26.2, the hypothesis of $\boldsymbol{\mu}^{(1)} = \boldsymbol{\mu}^{(2)} = \boldsymbol{\mu}^{(3)} = \boldsymbol{\mu}^{(4)}$ is rejected.†

8.9. MULTIVARIATE ANALYSIS OF VARIANCE

The univariate analysis of variance has a direct generalization for vector variables leading to an analysis of vector sums of squares (i.e., sums such as $\sum \boldsymbol{x}_\alpha \boldsymbol{x}_\alpha'$). In fact, in the preceding section this generalization was considered for an analysis of variance problem involving a single classification.

As another example consider a two-way layout. Suppose that we are interested in the question whether the column effects are zero. We shall review the analysis for a scalar variable and then show the analysis for a vector variable. Let Y_{ij}, $i = 1, \dots, r$; $j = 1, \dots, c$, be a set of rc random variables.

† The above computations have been given by Bartlett (1947).

We assume that

$$\mathscr{E}Y_{ij} = \mu + \lambda_i + \nu_j, \qquad i = 1,\dots,r;\ j = 1,\dots,c, \tag{1}$$

with the restrictions

$$\sum_{i=1}^{r} \lambda_i = \sum_{j=1}^{c} \nu_j = 0, \tag{2}$$

that the variance of Y_{ij} is σ^2, and that the Y_{ij} are independently normally distributed. To test that column effects are zero is to test that

$$\nu_j = 0, \qquad j = 1,\dots,c. \tag{3}$$

This problem can be treated as a problem of regression by the introduction of dummy fixed variates. Let

$$\begin{aligned} z_{00,ij} &= 1, & & \\ z_{k0,ij} &= 1, & k &= i, \\ &= 0, & k &\neq i, \\ z_{0k,ij} &= 1, & k &= j, \\ &= 0, & k &\neq j. \end{aligned} \tag{4}$$

Then (1) can be written

$$\mathscr{E}Y_{ij} = \mu z_{00,ij} + \sum_{k=1}^{r} \lambda_k z_{k0,ij} + \sum_{k=1}^{c} \nu_k z_{0k,ij}. \tag{5}$$

The hypothesis is that the coefficients of $z_{0k,ij}$ are zero. Since the matrix of fixed variates here

$$\begin{pmatrix} z_{00,11} & \cdots & z_{00,rc} \\ z_{10,11} & \cdots & z_{10,rc} \\ z_{20,11} & \cdots & z_{20,rc} \\ \vdots & & \vdots \\ z_{0c,11} & \cdots & z_{0c,rc} \end{pmatrix} \tag{6}$$

is singular (for example, row 00 is the sum of rows $10, 20, \dots, r0$), one must elaborate the regression theory. When one does, one finds that the test criterion indicated by the regression theory is the usual F-test of analysis of variance.

Let

$$
\begin{aligned}
Y_{\cdot\cdot} &= \frac{1}{rc}\sum_{i,j} Y_{ij}, \\
Y_{i\cdot} &= \frac{1}{c}\sum_{j} Y_{ij}, \qquad (7) \\
Y_{\cdot j} &= \frac{1}{r}\sum_{i} Y_{ij},
\end{aligned}
$$

and let

$$
\begin{aligned}
a &= \sum_{i,j}\left(Y_{ij} - Y_{i\cdot} - Y_{\cdot j} + Y_{\cdot\cdot}\right)^2 \\
&= \sum_{i,j} Y_{ij}^2 - c\sum_{i} Y_{i\cdot}^2 - r\sum_{j} Y_{\cdot j}^2 + rcY_{\cdot\cdot}^2, \qquad (8) \\
b &= r\sum_{j}\left(Y_{\cdot j} - Y_{\cdot\cdot}\right)^2 \\
&= r\sum_{j} Y_{\cdot j}^2 - rcY_{\cdot\cdot}^2 .
\end{aligned}
$$

Then the F-statistic is given by

$$
F = \frac{b}{a}\cdot\frac{(c-1)(r-1)}{c-1}. \qquad (9)
$$

Under the null hypothesis, this has the F-distribution with $c - 1$ and $(r-1)(c-1)$ degrees of freedom. The likelihood ratio criterion for the hypothesis is the $rc/2$ power of

$$
\frac{a}{a+b} = \frac{1}{1 + (c-1)/[(r-1)(c-1)]\,F}. \qquad (10)
$$

Now let us turn to the multivariate analysis of variance. We have a set of p-dimensional random vectors $\boldsymbol{Y}_{ij}$, $i = 1,\dots,r$; $j = 1,\dots,c$, with expected values (1), where $\boldsymbol{\mu}$, $\boldsymbol{\lambda}$'s, and $\boldsymbol{\nu}$'s are vectors, and covariance matrix $\boldsymbol{\Sigma}$, and they are independently normally distributed. Then the same algebra may be used to reduce this problem to the regression problem. We define $\mathbf{Y}_{\cdot\cdot}, \mathbf{Y}_{i\cdot}, \mathbf{Y}_{\cdot j}$ by (7) and

$$
\begin{aligned}
(11) \qquad \mathbf{A} &= \sum_{i,j} (\mathbf{Y}_{ij} - \mathbf{Y}_{i\cdot} - \mathbf{Y}_{\cdot j} + \mathbf{Y}_{\cdot\cdot})(\mathbf{Y}_{ij} - \mathbf{Y}_{i\cdot} - \mathbf{Y}_{\cdot j} + \mathbf{Y}_{\cdot\cdot})' \\
&= \sum_{i,j} \mathbf{Y}_{ij}\mathbf{Y}_{ij}' - c\sum_{i} \mathbf{Y}_{i\cdot}\mathbf{Y}_{i\cdot}' - r\sum_{j} \mathbf{Y}_{\cdot j}\mathbf{Y}_{\cdot j}' + rc\mathbf{Y}_{\cdot\cdot}\mathbf{Y}_{\cdot\cdot}', \\
\mathbf{B} &= r\sum_{j} (\mathbf{Y}_{\cdot j} - \mathbf{Y}_{\cdot\cdot})(\mathbf{Y}_{\cdot j} - \mathbf{Y}_{\cdot\cdot})' \\
&= r\sum_{j} \mathbf{Y}_{\cdot j}\mathbf{Y}_{\cdot j}' - rc\mathbf{Y}_{\cdot\cdot}\mathbf{Y}_{\cdot\cdot}'.
\end{aligned}
$$

A statistic analogous to (10) is

$$
(12) \qquad \frac{|\mathbf{A}|}{|\mathbf{A} + \mathbf{B}|}.
$$

Under the null hypothesis, this has the distribution of U for $p, n = (r-1)(c-1)$ and $q_1 = c - 1$ given in Section 8.4. In order for $\mathbf{A}$ to be nonsingular (with probability 1), we must require $p \le (r-1)(c-1)$.

As an example we use data first published by Immer, Hayes, and Powers (1934), and later used by Fisher (1947a), by Yates and Cochran (1938), and by Tukey (1949). The first component of the observation vector is a barley yield in a given year; the second component is the same measurement made the following year. Column indices run over the varieties of barley and row indices over the locations. The data are given in Table 8.1 [e.g., $\begin{pmatrix} 81 \\ 81 \end{pmatrix}$ in the upper left-hand corner indicates a yield of 81 in each year of variety M in location UF.] The numbers along the borders are sums.

We consider the square of (147, 100) to be

$$
\begin{pmatrix} 147 \\ 100 \end{pmatrix} (147 \; 100) = \begin{pmatrix} 21609 & 14700 \\ 14700 & 10000 \end{pmatrix}.
$$

TABLE 8.1

Location	Varieties					Sums
	M	*S*	*V*	*T*	*P*	
UF	81	105	120	110	98	514
	81	82	80	87	84	414
W	147	142	151	192	146	778
	100	116	112	148	108	584
M	82	77	78	131	90	458
	103	105	117	140	130	595
C	120	121	124	141	125	631
	99	62	96	126	76	459
GR	99	89	69	89	104	450
	66	50	97	62	80	355
D	87	77	79	102	96	441
	68	67	67	92	94	388
Sums	616	611	621	765	659	3272
	517	482	569	655	572	2795

Then

(13) $$\sum_{i,j} \boldsymbol{Y}_{ij}\boldsymbol{Y}_{ij}' = \begin{pmatrix} 380{,}944 & 315{,}381 \\ 315{,}381 & 277{,}625 \end{pmatrix},$$

(14) $$\sum_{j} (6\boldsymbol{Y}_{\cdot j})(6\boldsymbol{Y}_{\cdot j})' = \begin{pmatrix} 2{,}157{,}924 & 1{,}844{,}346 \\ 1{,}844{,}346 & 1{,}579{,}583 \end{pmatrix},$$

(15) $$\sum_{i} (5\boldsymbol{Y}_{i\cdot})(5\boldsymbol{Y}_{i\cdot})' = \begin{pmatrix} 1{,}874{,}386 & 1{,}560{,}145 \\ 1{,}560{,}145 & 1{,}353{,}727 \end{pmatrix},$$

(16) $$(30\boldsymbol{Y}_{\cdot\cdot})(30\boldsymbol{Y}_{\cdot\cdot})' = \begin{pmatrix} 10{,}705{,}984 & 9{,}145{,}240 \\ 9{,}145{,}240 & 7{,}812{,}025 \end{pmatrix}.$$

Then the "error sum of squares" is

(17) $$\boldsymbol{A} = \begin{pmatrix} 3279 & 802 \\ 802 & 4017 \end{pmatrix},$$

the "row sum of squares" is

$$5\sum_i (Y_{i\cdot} - Y_{\cdot\cdot})(Y_{i\cdot} - Y_{\cdot\cdot})' = \begin{pmatrix} 18{,}011 & 7{,}188 \\ 7{,}188 & 10{,}345 \end{pmatrix}, \tag{18}$$

and the "column sum of squares" is

$$\boldsymbol{B} = \begin{pmatrix} 2788 & 2550 \\ 2550 & 2863 \end{pmatrix}. \tag{19}$$

The test criterion is

$$\frac{|\boldsymbol{A}|}{|\boldsymbol{A} + \boldsymbol{B}|} = \frac{\begin{vmatrix} 3279 & 802 \\ 802 & 4017 \end{vmatrix}}{\begin{vmatrix} 6067 & 3352 \\ 3352 & 6880 \end{vmatrix}} = 0.4107. \tag{20}$$

This result is to be compared with the significant point for $U_{2,4,20}$. Using the result of Section 8.4, we see that

$$\frac{1 - \sqrt{0.4107}}{\sqrt{0.4107}} \cdot \frac{19}{4} = 2.66$$

is to be compared with the significance point of $F_{8,38}$. This is significant at the 5% level. Our data show that there are differences between varieties.

Now let us see that each F-test in the univariate analysis of variance has analogous tests in the multivariate analysis of variance. In the linear hypothesis model for the univariate analysis of variance, one assumes that the random variables $Y_1, \dots, Y_N$ have expected values that are linear combinations of unknown parameters

$$\mathscr{E} Y_\alpha = \sum_g \beta_g z_{g\alpha}, \tag{21}$$

where the β's are the parameters and the z's are the known coefficients. The variables $\{Y_\alpha\}$ are assumed to be normally and independently distributed with common variance σ^2. In this model there are a set of linear combinations, say $\sum_{\alpha=1}^{N} \gamma_{i\alpha} Y_\alpha$, where the γ's are known, such that

$$a = \sum_{i=1}^{n} \left(\sum_{\alpha=1}^{N} \gamma_{i\alpha} Y_\alpha \right)^2 = \sum_{\alpha,\beta=1}^{N} d_{\alpha\beta} Y_\alpha Y_\beta \tag{22}$$

is distributed as $\sigma^2\chi^2$ with n degrees of freedom. There is another set of linear combinations, say $\sum_\alpha \phi_{g\alpha} Y_\alpha$, where the ϕ's are known, such that

$$b = \sum_{g=1}^{m} \left(\sum_{\alpha=1}^{N} \phi_{g\alpha} Y_\alpha \right)^2 = \sum_{\alpha,\beta=1}^{N} c_{\alpha\beta} Y_\alpha Y_\beta \tag{23}$$

is distributed as $\sigma^2\chi^2$ with m degrees of freedom when the null hypothesis is true and as σ^2 times a noncentral χ^2 when the null hypothesis is not true; and in either case b is distributed independently of a. Then

$$\frac{b}{a} \cdot \frac{n}{m} = \frac{\sum c_{\alpha\beta} Y_\alpha Y_\beta}{\sum d_{\alpha\beta} Y_\alpha Y_\beta} \cdot \frac{n}{m} \tag{24}$$

has the F-distribution with m and n degrees of freedom, respectively, when the null hypothesis is true. The null hypothesis is that certain β's are zero.

In the multivariate analysis of variance, $\boldsymbol{Y}_1, \dots, \boldsymbol{Y}_N$ are vector variables with p components. The expected value of $\boldsymbol{Y}_\alpha$ is given by (21) where $\boldsymbol{\beta}_g$ is a vector of p parameters. We assume that the $\{\boldsymbol{Y}_\alpha\}$ are normally and independently distributed with common covariance matrix $\boldsymbol{\Sigma}$. The linear combinations $\sum \gamma_{i\alpha} \boldsymbol{Y}_\alpha$ can be formed for the vectors. Then

$$\boldsymbol{A} = \sum_{i=1}^{n} \left(\sum_{\alpha=1}^{N} \gamma_{i\alpha} \boldsymbol{Y}_\alpha \right) \left(\sum_{\alpha=1}^{N} \gamma_{i\alpha} \boldsymbol{Y}_\alpha \right)' = \sum_{\alpha,\beta=1}^{N} d_{\alpha\beta} \boldsymbol{Y}_\alpha \boldsymbol{Y}_\beta' \tag{25}$$

has the distribution $W(\boldsymbol{\Sigma}, n)$. When the null hypothesis is true,

$$\boldsymbol{B} = \sum_{g=1}^{m} \left(\sum_{\alpha=1}^{N} \phi_{g\alpha} \boldsymbol{Y}_\alpha \right) \left(\sum_{\alpha=1}^{N} \phi_{g\alpha} \boldsymbol{Y}_\alpha \right)' = \sum_{\alpha,\beta=1}^{N} c_{\alpha\beta} \boldsymbol{Y}_\alpha \boldsymbol{Y}_\beta' \tag{26}$$

has the distribution $W(\boldsymbol{\Sigma}, m)$, and $\boldsymbol{B}$ is independent of $\boldsymbol{A}$. Then

$$\frac{|\boldsymbol{A}|}{|\boldsymbol{A} + \boldsymbol{B}|} = \frac{|\sum d_{\alpha\beta} \boldsymbol{Y}_\alpha \boldsymbol{Y}_\beta'|}{|\sum d_{\alpha\beta} \boldsymbol{Y}_\alpha \boldsymbol{Y}_\beta' + \sum c_{\alpha\beta} \boldsymbol{Y}_\alpha \boldsymbol{Y}_\beta'|} \tag{27}$$

has the $U_{p,m,n}$-distribution.

The argument for the distribution of a and b involves showing that $\mathscr{E} \sum_\alpha \gamma_{i\alpha} Y_\alpha = 0$ and $\mathscr{E} \sum_\alpha \phi_{g\alpha} Y_\alpha = 0$ when certain β's are equal to zero as specified by the null hypothesis (as identities in the unspecified β's). Clearly this argument holds for the vector case as well. Secondly, one argues, in the univariate case, that there is an orthogonal matrix $\Psi = (\psi_{\alpha\beta})$ such that when the transformation $Y_\beta = \sum_\alpha \psi_{\beta\alpha} Z_\alpha$ is made

$$(28)\qquad \begin{aligned} a &= \sum_{\alpha,\beta,\gamma,\delta} d_{\alpha\beta}\psi_{\alpha\gamma}\psi_{\beta\delta}Z_\gamma Z_\delta = \sum_{\alpha=1}^{n} Z_\alpha^2, \\ b &= \sum_{\alpha,\beta,\gamma,\delta} c_{\alpha\beta}\psi_{\alpha\gamma}\psi_{\beta\delta}Z_\gamma Z_\delta = \sum_{\alpha=n+1}^{n+m} Z_\alpha^2. \end{aligned}$$

Because the transformation is orthogonal, the $\{Z_\alpha\}$ are independently and normally distributed with common variance σ^2. Since the Z_α, $\alpha = 1, \dots, n$, must be linear combinations of $\sum_\alpha \gamma_{i\alpha} Y_\alpha$ and Z_α, $\alpha = n+1, \dots, n+m$, must be linear combinations of $\sum_\alpha \phi_{g\alpha} Y_\alpha$, they must have means zero (under the null hypothesis). Thus a/σ^2 and b/σ^2 have the stated independent χ^2-distributions.

In the multivariate case the transformation $\boldsymbol{Y}_\beta = \sum_\alpha \psi_{\beta\alpha}\boldsymbol{Z}_\alpha$ is used, where $\boldsymbol{Y}_\beta$ and $\boldsymbol{Z}_\alpha$ are vectors. Then

$$(29)\qquad \begin{aligned} \boldsymbol{A} &= \sum_{\alpha,\beta,\gamma,\delta} d_{\alpha\beta}\psi_{\alpha\gamma}\psi_{\beta\delta}\boldsymbol{Z}_\gamma \boldsymbol{Z}_\delta' = \sum_{\alpha=1}^{n} \boldsymbol{Z}_\alpha \boldsymbol{Z}_\alpha', \\ \boldsymbol{B} &= \sum_{\alpha,\beta,\gamma,\delta} c_{\alpha\beta}\psi_{\alpha\gamma}\psi_{\beta\delta}\boldsymbol{Z}_\gamma \boldsymbol{Z}_\delta' = \sum_{\alpha=n+1}^{n+m} \boldsymbol{Z}_\alpha \boldsymbol{Z}_\alpha' \end{aligned}$$

because it follows from (28) that $\sum_{\alpha,\beta} d_{\alpha\beta}\psi_{\alpha\gamma}\psi_{\beta\delta} = 1$, $\gamma = \delta \le n$, $= 0$, otherwise, and $\sum_{\alpha,\beta} c_{\alpha\beta}\psi_{\alpha\gamma}\psi_{\beta\delta} = 1$, $n+1 \le \gamma = \delta \le n+m$, $= 0$, otherwise. Since Ψ is orthogonal, the $\{\boldsymbol{Z}_\alpha\}$ are independently normally distributed with covariance matrix $\boldsymbol{\Sigma}$. The same argument shows $\mathscr{E}\boldsymbol{Z}_\alpha = \boldsymbol{0}$, $\alpha = 1, \dots, n+m$, under the null hypothesis. Thus $\boldsymbol{A}$ and $\boldsymbol{B}$ are independently distributed according to $W(\boldsymbol{\Sigma}, n)$ and $W(\boldsymbol{\Sigma}, m)$, respectively.

8.10. SOME OPTIMAL PROPERTIES OF TESTS

8.10.1. Admissibility of Invariant Tests

In this chapter we have considered several tests of a linear hypothesis which are invariant with respect to transformations that leave the null hypothesis invariant. We raise the question of which invariant tests are good tests. In particular we ask for admissible procedures, that is, procedures that cannot be improved on in the sense of smaller probabilities of Type I or Type II error. The competing tests are not necessarily invariant. Clearly, if an invariant test is admissible in the class of all tests, it is admissible in the class of invariant tests.

Testing the general linear hypothesis as treated here is a generalization of testing the hypothesis concerning one mean vector as treated in Chapter 5. The invariant procedures in Chapter 8 are generalizations of the T^2-test. One way of showing a procedure is admissible is to display a prior distribution on the parameters such that the Bayes procedure is a given test procedure. This approach requires some ingenuity in constructing the prior, but the verification of the property given the prior is straightforward. Problems 26 and 27 show that the Bartlett–Nanda–Pillai trace criterion V and Wilks' likelihood ratio criterion U yield admissible tests. The disadvantage of this approach to admissibility is that one must invent a prior distribution for each procedure; a general theorem does not cover many cases.

The other approach to admissibility is to apply Stein's theorem (Theorem 5.6.5), which yields general results. The invariant tests can be stated in terms of the roots of the determinantal equation

$$|\boldsymbol{H} - \lambda(\boldsymbol{H} + \boldsymbol{G})| = 0, \tag{1}$$

where $\boldsymbol{H} = \hat{\boldsymbol{\beta}}_1 A_{11\cdot 2} \hat{\boldsymbol{\beta}}_1' = W_1 W_1'$ and $\boldsymbol{G} = N\hat{\boldsymbol{\Sigma}}_\Omega = W_3 W_3'$. There is also a matrix $\hat{\boldsymbol{\beta}}_2$ (or W_2) associated with the nuisance parameters $\boldsymbol{\beta}_2$. For convenience, we define the canonical form in the following notation. Let $W_1 = \boldsymbol{X}\ (p \times m)$, $W_2 = \boldsymbol{Y}\ (p \times r)$, $W_3 = \boldsymbol{Z}\ (p \times n)$, $\mathscr{E}\boldsymbol{X} = \Xi$, $\mathscr{E}\boldsymbol{Y} = \mathrm{H}$, and $\mathscr{E}\boldsymbol{Z} = \boldsymbol{0}$; the columns are independently normally distributed with covariance matrix $\boldsymbol{\Sigma}$. The null hypothesis is $\Xi = \boldsymbol{0}$, and the alternative hypothesis is $\Xi \neq \boldsymbol{0}$.

The usual tests are given in terms of the (nonzero) roots of

$$|\boldsymbol{XX}' - \lambda(\boldsymbol{ZZ}' + \boldsymbol{XX}')| = |\boldsymbol{XX}' - \lambda(\boldsymbol{U} - \boldsymbol{YY}')| = 0, \tag{2}$$

where $\boldsymbol{U} = \boldsymbol{XX}' + \boldsymbol{YY}' + \boldsymbol{ZZ}'$. Except for roots that are identically zero, the roots of (2) coincide with the nonzero characteristic roots of $\boldsymbol{X}'(\boldsymbol{U} - \boldsymbol{YY}')^{-1}\boldsymbol{X}$. Let $\boldsymbol{V} = (\boldsymbol{X}, \boldsymbol{Y}, \boldsymbol{U})$ and

$$\boldsymbol{M}(\boldsymbol{V}) = \boldsymbol{X}'(\boldsymbol{U} - \boldsymbol{YY}')^{-1}\boldsymbol{X}. \tag{3}$$

The vector of ordered characteristic roots of $\boldsymbol{M}(\boldsymbol{V})$ is denoted by

$$(\lambda_1, \dots, \lambda_m)' = \lambda(\boldsymbol{M}(\boldsymbol{V})), \tag{4}$$

where $\lambda_1 \geq \cdots \geq \lambda_m \geq 0$. Since the inclusion of zero roots (when $m > p$) causes no trouble in the sequel, we assume that the tests depend on $\boldsymbol{\lambda}(\boldsymbol{M}(\boldsymbol{V}))$.

The admissibility of these tests can be stated in terms of the geometric characteristics of the acceptance regions. Let

$$
\begin{aligned}
R_{<}^{m} &= \{\boldsymbol{\lambda} \in R^m | \lambda_1 \geq \lambda_2 \geq \cdots \geq \lambda_m \geq 0\}, \\
R_{+}^{m} &= \{\boldsymbol{\lambda} \in R^m | \lambda_1 \geq 0, \ldots, \lambda_m \geq 0\}.
\end{aligned} \tag{5}
$$

It seems reasonable that if a set of sample roots leads to acceptance of the null hypothesis, then a set of smaller roots would as well (Figure 8.2).

DEFINITION 8.10.1. *A region* $A \subset R_{<}^{m}$ *is monotone if* $\boldsymbol{\lambda} \in A$, $\boldsymbol{\nu} \in R_{<}^{m}$, $\nu_i \leq \lambda_i$, $i = 1, \ldots, m$, *imply* $\boldsymbol{\nu} \in A$.

DEFINITION 8.10.2. *For* $A \subset R_{<}^{m}$ *the extended region* A^* *is*

$$
A^* = \bigcup_{\pi} \left\{ (x_{\pi(1)}, \ldots, x_{\pi(m)})' | \boldsymbol{x} \in A \right\}, \tag{6}
$$

where π *ranges over all permutations of* $(1, \ldots, m)$.

The main result, first proved by Schwartz (1967), is the following theorem:

THEOREM 8.10.1. *If the region* $A \subset R_{<}^{m}$ *is monotone and if the extended region* A^* *is closed and convex, then* A *is the acceptance region of an admissible test.*

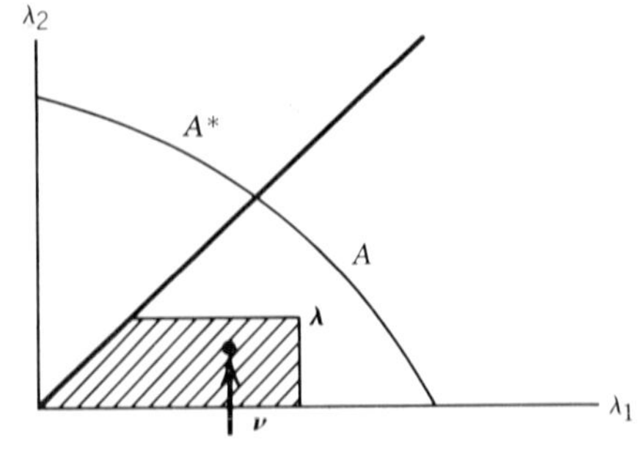

Figure 8.2. A monotone acceptance region.

Another characterization of admissible tests is given in terms of *majorization*.

DEFINITION 8.10.3. *A vector* $\boldsymbol{\lambda} = (\lambda_1, \dots, \lambda_m)'$ *weakly majorizes a vector* $\boldsymbol{\nu} = (\nu_1, \dots, \nu_m)'$ *if*

(7)

$$\lambda_{[1]} \geq \nu_{[1]}, \lambda_{[1]} + \lambda_{[2]} \geq \nu_{[1]} + \nu_{[2]}, \dots, \lambda_{[1]} + \cdots + \lambda_{[m]} \geq \nu_{[1]} + \cdots + \nu_{[m]},$$

where $\lambda_{[i]}$ *and* $\nu_{[i]}$, $i = 1, \dots, m$, *are the coordinates rearranged in nonascending order.*

We use the notation $\boldsymbol{\lambda} \succ_w \boldsymbol{\nu}$ or $\boldsymbol{\nu} \prec_w \boldsymbol{\lambda}$ if $\boldsymbol{\lambda}$ weakly majorizes $\boldsymbol{\nu}$. If $\boldsymbol{\lambda}, \boldsymbol{\nu} \in R^m_<$ then $\boldsymbol{\lambda} \succ_w \boldsymbol{\nu}$ is simply

$$(8) \qquad \lambda_1 \geq \nu_1, \lambda_1 + \lambda_2 \geq \nu_1 + \nu_2, \dots, \lambda_1 + \cdots + \lambda_m \geq \nu_1 + \cdots + \nu_m.$$

If the last inequality in (7) is replaced by an equality, we say simply that $\boldsymbol{\lambda}$ *majorizes* $\boldsymbol{\nu}$ and denote this by $\boldsymbol{\lambda} \succ \boldsymbol{\nu}$ or $\boldsymbol{\nu} \prec \boldsymbol{\lambda}$. The theory of majorization and the related inequalities are developed in detail in Marshall and Olkin (1979).

DEFINITION 8.10.4. *A region* $A \subset R^m_<$ *is monotone in majorization if* $\boldsymbol{\lambda} \in A$, $\boldsymbol{\nu} \in R^m_<$, $\boldsymbol{\nu} \prec_w \boldsymbol{\lambda}$ *imply* $\boldsymbol{\nu} \in A$. (See Figure 8.3.)

THEOREM 8.10.2. *If a region* $A \subset R^m_<$ *is closed, convex, and monotone in majorization, then* A *is the acceptance region of an admissible test.*

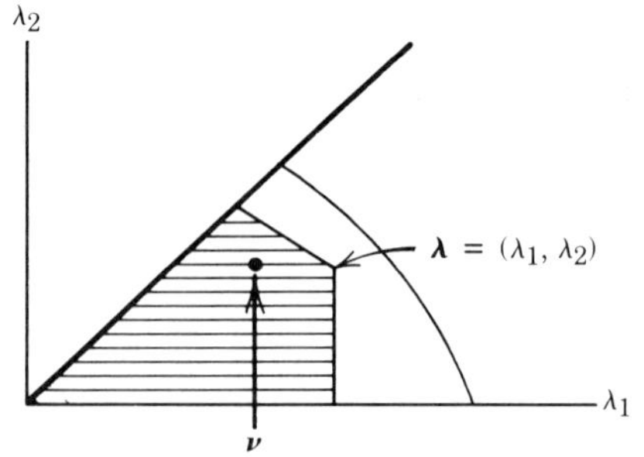

Figure 8.3. A region monotone in majorization.

Theorems 8.10.1 and 8.10.2 are equivalent; it will be convenient to prove Theorem 8.10.2 first. Then an argument about the extreme points of a certain convex set (Lemma 8.10.11) establishes the equivalence of the two theorems.

Theorem 5.6.5 (Stein's theorem) will be used because we can write the distribution of $(\boldsymbol{X}, \boldsymbol{Y}, \boldsymbol{Z})$ in exponential form. Let $\boldsymbol{U} = \boldsymbol{XX}' + \boldsymbol{YY}' + \boldsymbol{ZZ}' = (u_{ij})$ and $\boldsymbol{\Sigma}^{-1} = (\sigma^{ij})$. For a general matrix $\boldsymbol{C} = (\boldsymbol{c}_1, \dots, \boldsymbol{c}_k)$, let $\operatorname{vec}(\boldsymbol{C}) = (\boldsymbol{c}_1', \dots, \boldsymbol{c}_k')'$. The density of $(\boldsymbol{X}, \boldsymbol{Y}, \boldsymbol{Z})$ can be written as

$$(9)\quad f(\boldsymbol{X}, \boldsymbol{Y}, \boldsymbol{Z}) = K(\Xi, \mathrm{H}, \Sigma)\exp\{\operatorname{tr}\Xi'\Sigma^{-1}\boldsymbol{X} + \operatorname{tr}\mathrm{H}'\Sigma^{-1}\boldsymbol{Y} - \tfrac{1}{2}\operatorname{tr}\Sigma^{-1}\boldsymbol{U}\}$$

$$= K(\Xi, \mathrm{H}, \Sigma)\exp\{\boldsymbol{\omega}_{(1)}'\boldsymbol{y}_{(1)} + \boldsymbol{\omega}_{(2)}'\boldsymbol{y}_{(2)} + \boldsymbol{\omega}_{(3)}'\boldsymbol{y}_{(3)}\},$$

where $K(\Xi, \mathrm{H}, \Sigma)$ is a constant,

$$(10)\qquad \boldsymbol{\omega}_{(1)} = \operatorname{vec}(\Sigma^{-1}\Xi), \qquad \boldsymbol{\omega}_{(2)} = \operatorname{vec}(\Sigma^{-1}\mathrm{H}),$$

$$\boldsymbol{\omega}_{(3)} = -\tfrac{1}{2}(\sigma^{11}, 2\sigma^{12}, \dots, 2\sigma^{1p}, \sigma^{22}, \dots, \sigma^{pp})',$$

$$\boldsymbol{y}_{(1)} = \operatorname{vec}(\boldsymbol{X}), \qquad \boldsymbol{y}_{(2)} = \operatorname{vec}(\boldsymbol{Y}),$$

$$\boldsymbol{y}_{(3)} = (u_{11}, u_{12}, \dots, u_{1p}, u_{22}, \dots, u_{pp})'.$$

If we denote the mapping $(\boldsymbol{X}, \boldsymbol{Y}, \boldsymbol{Z}) \to \boldsymbol{y} = (\boldsymbol{y}_{(1)}', \boldsymbol{y}_{(2)}', \boldsymbol{y}_{(3)}')'$ by g, $\boldsymbol{y} = g(\boldsymbol{X}, \boldsymbol{Y}, \boldsymbol{Z})$, then the measure of a set A in the space of $\boldsymbol{y}$ is $m(A) = \mu(g^{-1}(A))$, where μ is the ordinary Lebesgue measure on $R^{p(m+r+n)}$. We note that $(\boldsymbol{X}, \boldsymbol{Y}, \boldsymbol{U})$ is a sufficient statistic and so is $\boldsymbol{y} = (\boldsymbol{y}_{(1)}', \boldsymbol{y}_{(2)}', \boldsymbol{y}_{(3)}')'$. Because a test that is admissible with respect to the class of tests based on a sufficient statistic is admissible in the whole class of tests, we consider only tests based on a sufficient statistic. Then the acceptance regions of these tests are subsets in the space of $\boldsymbol{y}$. The density of $\boldsymbol{y}$ given by the right-hand side of (9) is of the form of the exponential family and therefore we can apply Stein's theorem. Furthermore, since the transformation $(\boldsymbol{X}, \boldsymbol{Y}, \boldsymbol{U}) \to \boldsymbol{y}$ is linear, we prove the convexity of an acceptance region in terms of $(\boldsymbol{X}, \boldsymbol{Y}, \boldsymbol{U})$. The acceptance region of an invariant test is given in terms of $\boldsymbol{\lambda}(\boldsymbol{M}(\boldsymbol{V})) = (\lambda_1, \dots, \lambda_m)'$. Therefore, in order to prove the admissibility of these tests we have to check that the inverse image of A, namely, $\tilde{A} = \{V | \lambda(M(V)) \in A\}$ satisfies the conditions of Stein's theorem, namely, is convex.

Suppose $V_i = (X_i, Y_i, U_i) \in \tilde{A}$, $i = 1, 2$, that is, $\lambda[M(V_i)] \in A$. By the convexity of A, $p\lambda[M(V_1)] + q\lambda[M(V_2)] \in A$ for $0 \le p = 1 - q \le 1$. To show $pV_1 + qV_2 \in \tilde{A}$, that is, $\lambda[M(pV_1 + qV_2)] \in A$, we use the property of monotonicity of majorization of A and the following theorem.

THEOREM 8.10.3.

$$\lambda[M(pV_1 + qV_2)] \prec_w p\lambda[M(V_1)] + q\lambda[M(V_2)]. \tag{11}$$

The proof of Theorem 8.10.3 (Figure 8.4) follows from the pair of majorizations

$$\begin{aligned} \lambda[M(pV_1 + qV_2)] &\prec_w \lambda[pM(V_1) + qM(V_2)] \\ &\prec_w p\lambda[M(V_1)] + q\lambda[M(V_2)]. \end{aligned} \tag{12}$$

The second majorization in (12) is a special case of the following lemma.

LEMMA 8.10.1. *For* A *and* B *symmetric,*

$$\lambda(A + B) \prec_w \lambda(A) + \lambda(B). \tag{13}$$

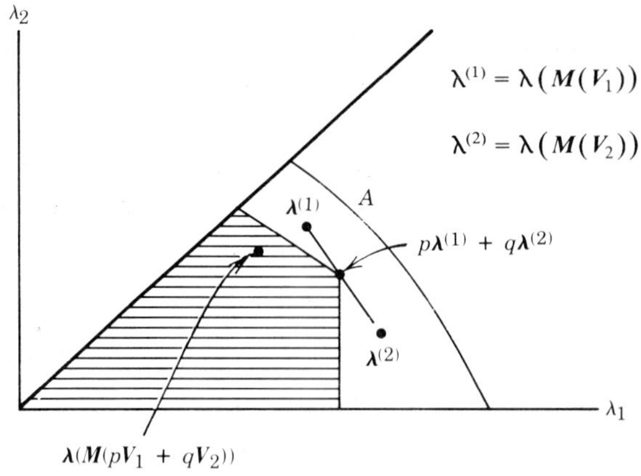

Figure 8.4. Theorem 8.10.3.

PROOF. By Corollary A.4.2 of Appendix A

$$\begin{aligned}(14)\quad \sum_{i=1}^{k}\lambda_i(A+B) &= \max_{R'R=I_k}\operatorname{tr}R'(A+B)R \\ &\le \max_{R'R=I_k}\operatorname{tr}R'AR+\max_{R'R=I_k}\operatorname{tr}R'BR \\ &= \sum_{i=1}^{k}\lambda_i(A)+\sum_{i=1}^{k}\lambda_i(B) \\ &= \sum_{i=1}^{k}\{\lambda_i(A)+\lambda_i(B)\}, \qquad k=1,\dots,p. \quad \text{Q.E.D.}\end{aligned}$$

Let $A > B$ mean $A - B$ is positive definite and $A \ge B$ mean $A - B$ is positive semidefinite.

The first majorization in (12) follows from several lemmas.

LEMMA 8.10.2.

$$\begin{aligned}(15)\quad pU_1+qU_2-(pY_1+qY_2)(pY_1+qY_2)' \\ \ge p(U_1-Y_1Y_1')+q(U_2-Y_2Y_2').\end{aligned}$$

PROOF. The left-hand side minus the right-hand side is

$$\begin{aligned}(16)\quad pY_1Y_1'+qY_2Y_2'-p^2Y_1Y_1'-q^2Y_2Y_2'-pq(Y_1Y_2'+Y_2Y_1') \\ = p(1-p)Y_1Y_1'+q(1-q)Y_2Y_2'-pq(Y_1Y_2'+Y_2Y_1') \\ = pq(Y_1-Y_2)(Y_1-Y_2)'\ge \mathbf{0}. \qquad \text{Q.E.D.}\end{aligned}$$

LEMMA 8.10.3. *If* $A \ge B > \mathbf{0}$, *then* $A^{-1} \le B^{-1}$.

PROOF. See Problem 31.

LEMMA 8.10.4. *If* $A > \mathbf{0}$, *then* $f(x, A) = x'A^{-1}x$ *is convex in* (x, A).

PROOF. See Problem 17 of Chapter 5.

LEMMA 8.10.5. *If* $A_1 > 0$, $A_2 > 0$, *then*

$$(pB_1 + qB_2)'(pA_1 + qA_2)^{-1}(pB_1 + qB_2) \le pB_1'A_1^{-1}B_1 + qB_2'A_2^{-1}B_2. \tag{17}$$

PROOF. From Lemma 8.10.4 we have for all y

$$\begin{aligned} py'B_1'A_1^{-1}B_1y + qy'B_2'A_2^{-1}B_2y & \\ -y'(pB_1 + qB_2)'(pA_1 + qA_2)^{-1}(pB_1 + qB_2)y & \\ = p(B_1y)'A_1^{-1}(B_1y) + q(B_2y)'A_2^{-1}(B_2y) & \\ -(pB_1y + qB_2y)'(pA_1 + qA_2)^{-1}(pB_1y + qB_2y) & \\ \ge 0. & \end{aligned} \tag{18}$$

Thus the matrix of the quadratic form in y is positive semidefinite. Q.E.D.

The relation as in (17) is sometimes called *matrix convexity*. [See Marshall and Olkin (1979).]

LEMMA 8.10.6.

$$M(pV_1 + qV_2) \le pM(V_1) + qM(V_2), \tag{19}$$

where $V_1 = (X_1, Y_1, U_1)$, $V_2 = (X_2, Y_2, U_2)$, $U_1 - Y_1Y_1' > 0$, $U_2 - Y_2Y_2' > 0$, $0 \le p = 1 - q \le 1$.

PROOF. Lemmas 8.10.2 and 8.10.3 show that

$$\begin{aligned} \left[pU_1 + qU_2 - (pY_1 + qY_2)(pY_1 + qY_2)'\right]^{-1} & \\ \le \left[p(U_1 - Y_1Y_1') + q(U_2 - Y_2Y_2')\right]^{-1}. & \end{aligned} \tag{20}$$

This implies

$$\begin{aligned} M(pV_1 + qV_2) & \\ \le (pX_1 + qX_2)'\left[p(U_1 - Y_1Y_1') + q(U_2 - Y_2Y_2')\right]^{-1}(pX_1 + qX_2). & \end{aligned} \tag{21}$$

Then Lemma 8.10.5 implies that the right-hand side of (21) is less than or equal to

$$pX_1'(U_1 - Y_1Y_1')^{-1}X_1 + qX_2'(U_2 - Y_2Y_2')^{-1}X_2 = pM(V_1) + qM(V_2). \tag{22}$$

Q.E.D.

LEMMA 8.10.7. *If* $A \leq B$, *then* $\lambda(A) \prec_w \lambda(B)$.

PROOF. From Corollary A.4.2 of Appendix A

$$\sum_{i=1}^{k} \lambda_i(A) = \max_{R'R=I_k} \operatorname{tr} R'AR \leq \max_{R'R=I_k} \operatorname{tr} R'BR = \sum_{i=1}^{k} \lambda_i(B), \tag{23}$$

$k = 1, \dots, p$. Q.E.D.

From Lemma 8.10.7 we obtain the first majorization in (12) and hence Theorem 8.10.3, which in turn implies the convexity of $\tilde{A}$. Thus the acceptance region satisfies condition (i) of Stein's theorem.

LEMMA 8.10.8. *For the acceptance region* A *of Theorem* 8.10.1 *or Theorem* 8.10.2 *the condition* (*ii*) *of Stein's theorem is satisfied.*

PROOF. Let ω correspond to (Φ, Ψ, Θ), then

$$\begin{aligned} \omega' y &= \omega_{(1)}' y_{(1)} + \omega_{(2)}' y_{(2)} + \omega_{(3)}' y_{(3)} \\ &= \operatorname{tr} \Phi' X + \operatorname{tr} \Psi' Y - \tfrac{1}{2} \operatorname{tr} \Theta U, \end{aligned} \tag{24}$$

where Θ is symmetric. Suppose that $\{y|\omega' y > c\}$ is disjoint from $\tilde{A} = \{V|\lambda(M(V)) \in A\}$. We want to show that in this case Θ is positive semidefinite. If this were not true, then

$$\Theta = D \begin{bmatrix} I & 0 & 0 \\ 0 & -I & 0 \\ 0 & 0 & 0 \end{bmatrix} D', \tag{25}$$

where D is nonsingular and $-I$ is not vacuous. Let $X = (1/\gamma)X_0$, $Y = (1/\gamma)Y_0$,

$$U = (D')^{-1} \begin{pmatrix} I & 0 & 0 \\ 0 & \gamma I & 0 \\ 0 & 0 & I \end{pmatrix} D^{-1}, \tag{26}$$

and $\boldsymbol{V} = (\boldsymbol{X}, \boldsymbol{Y}, \boldsymbol{U})$, where $\boldsymbol{X}_0, \boldsymbol{Y}_0$ are fixed matrices and γ is a positive number. Then

$$(27) \qquad \boldsymbol{\omega}'\boldsymbol{y} = \frac{1}{\gamma}\operatorname{tr}\boldsymbol{\Phi}'\boldsymbol{X}_0 + \frac{1}{\gamma}\operatorname{tr}\boldsymbol{\Psi}'\boldsymbol{Y}_0 + \tfrac{1}{2}\operatorname{tr}\begin{pmatrix} -\boldsymbol{I} & \mathbf{0} & \mathbf{0} \\ \mathbf{0} & \gamma\boldsymbol{I} & \mathbf{0} \\ \mathbf{0} & \mathbf{0} & \mathbf{0} \end{pmatrix} > c$$

for sufficiently large γ. On the other hand,

$$(28) \quad \lambda(\boldsymbol{M}(\boldsymbol{V})) = \lambda\{\boldsymbol{X}'(\boldsymbol{U} - \boldsymbol{Y}\boldsymbol{Y}')^{-1}\boldsymbol{X}\}$$

$$= \frac{1}{\gamma^2}\lambda\left\{\boldsymbol{X}_0'\left[(\boldsymbol{D}')^{-1}\begin{pmatrix} \boldsymbol{I} & \mathbf{0} & \mathbf{0} \\ \mathbf{0} & \gamma\boldsymbol{I} & \mathbf{0} \\ \mathbf{0} & \mathbf{0} & \boldsymbol{I} \end{pmatrix}\boldsymbol{D}^{-1} - \frac{1}{\gamma^2}\boldsymbol{Y}_0\boldsymbol{Y}_0'\right]^{-1}\boldsymbol{X}_0\right\}$$

$$\to \mathbf{0}$$

as $\gamma \to \infty$. Therefore, $\boldsymbol{V} \in \tilde{A}$ for sufficiently large γ. This is a contradiction. Hence $\boldsymbol{\Theta}$ is positive semidefinite.

Now let $\boldsymbol{\omega}_1$ correspond to $(\boldsymbol{\Phi}_1, \mathbf{0}, \boldsymbol{I})$, where $\boldsymbol{\Phi}_1 \neq \mathbf{0}$. Then $\boldsymbol{I} + \lambda\boldsymbol{\Theta}$ is positive definite and $\boldsymbol{\Phi}_1 + \lambda\boldsymbol{\Phi} \neq \mathbf{0}$ for sufficiently large λ. Hence $\boldsymbol{\omega}_1 + \lambda\boldsymbol{\omega} \in \Omega - \Omega_0$ for sufficiently large λ. Q.E.D.

The preceding proof was suggested by Charles Stein.

By Theorem 5.6.5, Theorem 8.10.3 and Lemma 8.10.8 now imply Theorem 8.10.2.

To obtain Theorem 8.10.1 from Theorem 8.10.2, we use the following lemmas.

Lemma 8.10.9. *$A \subset R^m_<$ is convex and monotone in majorization if and only if A is monotone and A^* is convex.*

Proof. *Necessity.* If A is monotone in majorization, then it is obviously monotone. A^* is convex (see Problem 35).

Sufficiency. For $\lambda \in R^m_<$ let

$$(29) \qquad \begin{aligned} C(\lambda) &= \{x \mid x \in R^m_+,\, x \prec_w \lambda\}, \\ D(\lambda) &= \{x \mid x \in R^m_<,\, x \prec_w \lambda\}. \end{aligned}$$

It will be proved in Lemma 8.10.10, Lemma 8.10.11, and its corollary that monotonicity of A and convexity of A^* implies $C(\lambda) \subset A^*$. Then $D(\lambda) = C(\lambda) \cap R^m_< \subset A^* \cap R^m_< = A$. Now suppose $\nu \in R^m_<$ and $\nu \prec_w \lambda$. Then $\nu \in D(\lambda) \subset A$. This shows that A is monotone in majorization. Furthermore, if A^* is convex, then $A = R^m_< \cap A^*$ is convex. (See Figure 8.5.) Q.E.D.

LEMMA 8.10.10. *Let C be compact and convex and let D be convex. If the extreme points of C are contained in D, then $C \subset D$.*

PROOF. Obvious.

LEMMA 8.10.11. *Every extreme point of $C(\lambda)$ is of the form*

$$(30) \qquad \left(\delta_{\pi(1)}\lambda_{\pi(1)}, \ldots, \delta_{\pi(m)}\lambda_{\pi(m)}\right),$$

where π is a permutation of $(1, \ldots, m)$ and $\delta_1 = \cdots = \delta_k = 1$, $\delta_{k+1} = \cdots = \delta_m = 0$ for some k.

PROOF. $C(\lambda)$ is convex. (See Problem 34.) Now note that $C(\lambda)$ is permutation symmetric, that is, if $(x_1, \ldots, x_m)' \in C(\lambda)$, then $(x_{\pi(1)}, \ldots, x_{\pi(m)})' \in C(\lambda)$ for any permutation π. Therefore, for any permutation π, $\pi(C(\lambda)) = \{(x_{\pi(1)}, \ldots, x_{\pi(m)})' | x \in C(\lambda)\}$ coincides with $C(\lambda)$. This implies that if $(x_1, \ldots, x_m)'$ is an extreme point of $C(\lambda)$, then $(x_{\pi(1)}, \ldots, x_{\pi(m)})'$ is also an extreme point. In particular, $(x_{[1]}, \ldots, x_{[m]}) \in R^m_<$ is an extreme point. Conversely, if $(x_1, \ldots, x_m) \in R^m_<$ is an extreme point of $C(\lambda)$, then $(x_{\pi(1)}, \ldots, x_{\pi(m)})'$ is an extreme point.

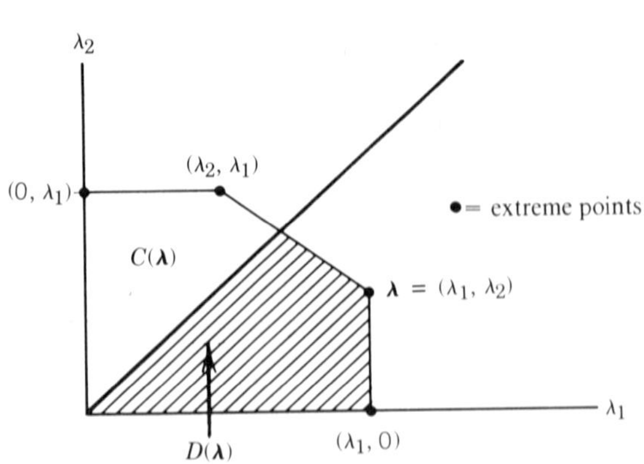

Figure 8.5

We see that once we enumerate the extreme points of $C(\lambda)$ in $R^m_<$, the rest of the extreme points can be obtained by permutation.

Suppose $\boldsymbol{x} \in R^m_<$. An extreme point being the intersection of m hyperplanes has to satisfy m or more of the following $2m$ equations:

$$(31)\qquad \begin{aligned} &E_1 : x_1 = 0, && F_1 : x_1 = \lambda_1, \\ &E_2 : x_2 = 0, && F_2 : x_1 + x_2 = \lambda_1 + \lambda_2, \\ &\vdots && \vdots \\ &E_m : x_m = 0, && F_m : x_1 + \cdots + x_m = \lambda_1 + \cdots + \lambda_m. \end{aligned}$$

Suppose that k is the first index such that E_k holds. Then $\boldsymbol{x} \in R^m_<$ implies $0 = x_k \geq x_{k+1} \geq \cdots \geq x_m \geq 0$. Therefore, $E_k, \ldots, E_m$ hold. The remaining $k - 1 = m - (m - k + 1)$ or more equations are among F's. We order them as $F_{i_1}, \ldots, F_{i_l}$, where $i_1 < \cdots < i_l$, $l \geq k - 1$. Now $i_1 < \cdots < i_l$ implies $i_l \geq l$ with equality if and only if $i_1 = 1, \ldots, i_l = l$. In this case $F_1, \ldots, F_{k-1}$ hold ($l \geq k - 1$). Now suppose $i_l > l$. Since $x_k = \cdots = x_m = 0$,

$$(32)\qquad F_{i_l} : x_1 + \cdots + x_{k-1} = \lambda_1 + \cdots + \lambda_{k-1} + \cdots + \lambda_{i_l}.$$

But $x_1 + \cdots + x_{k-1} \leq \lambda_1 + \cdots + \lambda_{k-1}$ and we have $\lambda_k + \cdots + \lambda_{i_l} = 0$. Therefore, $0 = \lambda_k + \cdots + \lambda_{i_l} \geq \lambda_k \geq \cdots \geq \lambda_m \geq 0$. In this case $F_{k-1}, \ldots, F_m$ reduce to the same equation $x_1 + \cdots + x_{k-1} = \lambda_1 + \cdots + \lambda_{k-1}$. It follows that $\boldsymbol{x}$ satisfies $k - 2$ more equations which have to be $F_1, \ldots, F_{k-2}$. We have shown that in either case $E_k, \ldots, E_m, F_1, \ldots, F_{k-1}$ hold and this gives the point $\boldsymbol{\beta} = (\lambda_1, \ldots, \lambda_{k-1}, 0, \ldots, 0)$, which is in $R^m_< \cap C(\lambda)$. Therefore, $\boldsymbol{\beta}$ is an extreme point. Q.E.D.

COROLLARY 8.10.1. $C(\lambda) \subset A^*$.

PROOF. If A is monotone, then A^* is monotone in the sense that if $\lambda = (\lambda_1, \ldots, \lambda_m)' \in A^*$, $\boldsymbol{\nu} = (\nu_1, \ldots, \nu_m)'$, $\nu_i \leq \lambda_i$, $i = 1, \ldots, m$, then $\boldsymbol{\nu} \in A^*$. (See Problem 35). Now the extreme points of $C(\lambda)$ given by (30) are in A^* because of permutation symmetry and monotonicity of A^*. Hence by Lemma 8.10.10 $C(\lambda) \subset A^*$. Q.E.D.

PROOF OF THEOREM 8.10.1. Immediate from Theorem 8.10.2 and Lemma 8.10.9. Q.E.D.

Application of the theory of Schur-convex functions yields several corollaries to Theorem 8.10.2.

COROLLARY 8.10.2. *Let g be continuous, nondecreasing, and convex in* $[0,1)$. *Let*

$$f(\boldsymbol{\lambda}) = f(\lambda_1, \ldots, \lambda_m) = \sum_{i=1}^{m} g(\lambda_i). \tag{33}$$

Then a test with the acceptance region $A = \{\boldsymbol{\lambda} | f(\boldsymbol{\lambda}) \leq c\}$ *is admissible.*

PROOF. Being a sum of convex functions f is convex and hence A is convex. A is closed because f is continuous. We want to show that if $f(\boldsymbol{x}) \leq c$ and $\boldsymbol{y} \prec_w \boldsymbol{x}$ ($\boldsymbol{x}, \boldsymbol{y} \in R^m_<$) then $f(\boldsymbol{y}) \leq c$. Let $\tilde{x}_k = \sum_{i=1}^k x_i$, $\tilde{y}_k = \sum_{i=1}^k y_i$. Then $\boldsymbol{y} \prec_w \boldsymbol{x}$ if and only if $\tilde{x}_k \geq \tilde{y}_k$, $k = 1, \ldots, m$. Let $f(\boldsymbol{x}) = h(\tilde{x}_1, \ldots, \tilde{x}_m) = g(\tilde{x}_1) + \sum_{i=2}^m g(\tilde{x}_i - \tilde{x}_{i-1})$. It suffices to show that $h(\tilde{x}_1, \ldots, \tilde{x}_m)$ is increasing in each $\tilde{x}_i$. For $i \leq m-1$ the convexity of g implies that

$$\begin{aligned} h(\tilde{x}_1, \ldots, \tilde{x}_i + \varepsilon, \ldots, \tilde{x}_m) &- h(\tilde{x}_1, \ldots, \tilde{x}_i, \ldots, \tilde{x}_m) \\ &= g(x_i + \varepsilon) - g(x_i) - \{g(x_{i+1}) - g(x_{i+1} - \varepsilon)\} \geq 0. \end{aligned} \tag{34}$$

For $i = m$ the monotonicity of g implies

$$h(\tilde{x}_1, \ldots, \tilde{x}_m + \varepsilon) - h(\tilde{x}_1, \ldots, \tilde{x}_m) = g(x_m + \varepsilon) - g(x_m) \geq 0. \tag{35}$$

Q.E.D.

Setting $g(\lambda) = -\log(1-\lambda)$, $g(\lambda) = \lambda/(1-\lambda)$, $g(\lambda) = \lambda$, respectively, shows that Wilks' likelihood ratio test, the Lawley–Hotelling trace test, and the Bartlett–Nanda–Pillai test are admissible. Admissibility of Roy's maximum root test $A: \lambda_1 \leq c$ follows directly from Theorem 8.10.1 or Theorem 8.10.2. On the contrary, the minimum root test $\lambda_t \leq c$, where $t = \min(m, p)$, does not satisfy the convexity condition. The following theorem shows that this test is actually inadmissible.

THEOREM 8.10.4. *A necessary condition that an invariant test be admissible is that the extended region in the space of* $\sqrt{\lambda_1}, \ldots, \sqrt{\lambda_t}$ *is convex and monotone.*

We shall only sketch the proof of this theorem [following Schwartz (1967)]. Let $\sqrt{\lambda_i} = d_i$, $i = 1, \ldots, t$, and let the density of $d_1, \ldots, d_t$ be $f(\boldsymbol{d}|\boldsymbol{\nu})$, where $\boldsymbol{\nu} = (\nu_1, \ldots, \nu_t)'$ is defined in Section 8.6.5 and $f(\boldsymbol{d}|\boldsymbol{\nu})$ is given in Chapter 13.

The ratio $f(\boldsymbol{d}|\boldsymbol{\nu})/f(\boldsymbol{d}|\boldsymbol{0})$ can be extended symmetrically to the unit cube $(0 \le d_i \le 1,\ i = 1, \ldots, t)$. The extended ratio is then a convex function and is strictly increasing in each d_i. A proper Bayes procedure has an acceptance region

$$\int \frac{f(\boldsymbol{d}|\boldsymbol{\nu})}{f(\boldsymbol{d}|\boldsymbol{0})}\, d\Pi(\boldsymbol{\nu}) \le c, \tag{36}$$

where $\Pi(\boldsymbol{\nu})$ is a finite measure on the space of $\boldsymbol{\nu}$'s. Then the symmetric extension of the set of $\boldsymbol{d}$ satisfying (36) is convex and monotone [as shown by Birnbaum (1955)]. The closure (in the weak * topology) of the set of Bayes procedures forms an essentially complete class [Wald (1950)]. In this case the limit of the convex monotone acceptance regions is convex and monotone. The exposition of admissibility here was developed by Anderson and Takemura (1982).

8.10.2. Unbiasedness of Tests and Monotonicity of Power Functions

A test T is called *unbiased* if the power achieves its minimum at the null hypothesis. When there is a natural parametrization and a notion of "distance" in the parameter space, the power function is *monotone* if the power increases as the distance between the alternative hypothesis and the null hypothesis increases. Note that monotonicity implies unbiasedness. In this section we shall show that the power functions of many of the invariant tests of the general linear hypothesis are monotone in the invariants of the parameters, namely, the roots; these can be considered as measures of distance.

To introduce the approach, we consider the acceptance interval $(-a, a)$ for testing the null hypothesis $\mu = 0$ against the alternative $\mu \ne 0$ on the basis of an observation from $N(\mu, \sigma^2)$. In Figure 8.6 the probabilities of acceptance are

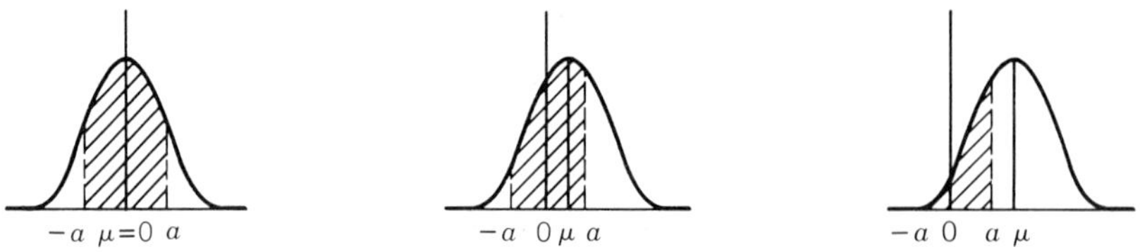

Figure 8.6. Three acceptance regions.

represented by the shaded regions for three values of μ. It is clear that the probability of acceptance decreases monotonically (or equivalently the power increases monotonically) as μ moves away from zero. In fact, this property depends only on the density function being unimodal and symmetric.

In higher dimensions we generalize the interval by a symmetric convex set and we ask that the density function be symmetric and unimodal in the sense that every contour of constant density surrounds a convex set. In Figure 8.7 we illustrate that in this case the probability of acceptance decreases monotonically. The following theorem is due to Anderson (1955b).

THEOREM 8.10.5. *Let E be a convex set in n-space, symmetric about the origin. Let $f(\boldsymbol{x}) \geq 0$ be a function such that (i) $f(\boldsymbol{x}) = f(-\boldsymbol{x})$, (ii) $\{\boldsymbol{x} | f(\boldsymbol{x}) \geq u\} = K_u$ is convex for every u $(0 < u < \infty)$, and (iii) $\int_E f(\boldsymbol{x})\, d\boldsymbol{x} < \infty$. Then*

$$\int_E f(\boldsymbol{x} + k\boldsymbol{y})\, d\boldsymbol{x} \geq \int_E f(\boldsymbol{x} + \boldsymbol{y})\, d\boldsymbol{x} \tag{37}$$

for $0 \leq k \leq 1$.

The proof of Theorem 8.10.5 is based on the following lemma.

LEMMA 8.10.12. *Let E, F be convex and symmetric about the origin. Then*

$$V\{(E + k\boldsymbol{y}) \cap F\} \geq V\{(E + \boldsymbol{y}) \cap F\}, \tag{38}$$

where $0 \leq k \leq 1$ and V denotes the n-dimensional volume.

PROOF. Consider the set $\alpha(E + \boldsymbol{y}) + (1 - \alpha)(E - \boldsymbol{y}) = \alpha E + (1 - \alpha)E + (2\alpha - 1)\boldsymbol{y}$ which consists of points $\alpha(\boldsymbol{x} + \boldsymbol{y}) + (1 - \alpha)(\boldsymbol{z} - \boldsymbol{y})$ with $\boldsymbol{x}$,

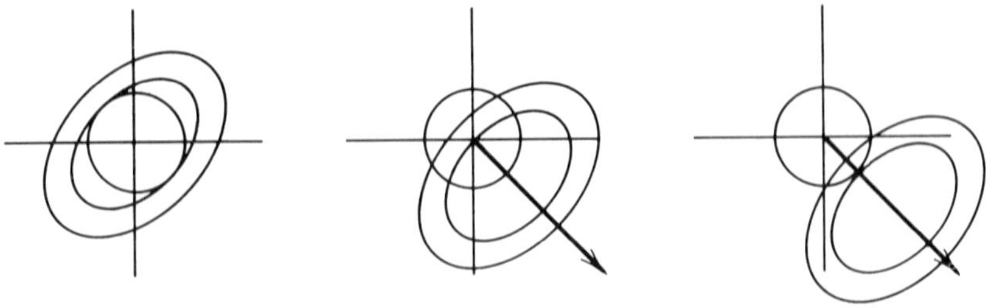

Figure 8.7. Acceptance regions.

$z \in E$. Let $\alpha_0 = (k+1)/2$ so that $2\alpha_0 - 1 = k$. Then by convexity of E we have

$$\alpha_0(E + y) + (1 - \alpha_0)(E - y) \subset E + ky. \tag{39}$$

Hence by convexity of F

$$\alpha_0[(E + y) \cap F] + (1 - \alpha_0)[(E - y) \cap F] \subset [(E + ky) \cap F]$$

and

(40)

$$V\{\alpha_0[(E + y) \cap F] + (1 - \alpha_0)[(E - y) \cap F]\} \le V\{(E + ky) \cap F\}.$$

Now by the Brunn–Minkowski inequality [e.g., Bonnesen and Fenchel (1948), Section 48], we have

$$\begin{aligned} (41) \quad & V^{1/n}\{\alpha_0[(E + y) \cap F] + (1 - \alpha_0)[(E - y) \cap F]\} \\ & \ge \alpha_0 V^{1/n}\{(E + y) \cap F\} + (1 - \alpha_0) V^{1/n}\{(E - y) \cap F\} \\ & = \alpha_0 V^{1/n}\{(E + y) \cap F\} + (1 - \alpha_0) V^{1/n}\{(-E + y) \cap (-F)\} \\ & = V^{1/n}\{(E + y) \cap F\}. \end{aligned}$$

The last equality follows from the symmetry of E and F. Q.E.D.

Proof of Theorem 8.10.5. Let

$$H(u) = V\{(E + ky) \cap K_u\}, \tag{42}$$

$$H^*(u) = V\{(E + y) \cap K_u\}. \tag{43}$$

Then

$$\begin{aligned} (44) \quad \int_E f(x + y)\, dx &= \int_{E+y} f(x)\, dx \\ &= \int_{E+y} \int_0^\infty I_{\{0 \le u \le f(x)\}}(u)\, du\, dx \\ &= \int_0^\infty \int_{E+y} I_{\{0 \le u \le f(x)\}}(u)\, dx\, du \\ &= \int_0^\infty H^*(u)\, du. \end{aligned}$$

Similarly,

$$\int_E f(x + ky)\, dx = \int_0^\infty H(u)\, du. \tag{45}$$

By Lemma 8.10.12, $H(u) \geq H^*(u)$. Hence Theorem 8.10.5 follows from (44) and (45). Q.E.D.

We start with the canonical form given in Section 8.10.1. We further simplify the problem as follows. Let $t = \min(m, p)$ and let $\nu_1, \ldots, \nu_t$, $(\nu_1 \geq \nu_2 \geq \cdots \geq \nu_t)$ be the nonzero characteristic roots of $\Xi' \Sigma^{-1} \Xi$, where $\Xi = \mathscr{E} \boldsymbol{X}$.

LEMMA 8.10.13. *There exist matrices* $\boldsymbol{B}$ $(p \times p)$ *and* $\boldsymbol{F}$ $(m \times m)$ *such that*

$$\boldsymbol{B} \Sigma \boldsymbol{B}' = \boldsymbol{I}_p, \qquad \boldsymbol{F}\boldsymbol{F}' = \boldsymbol{I}_m,$$

$$\boldsymbol{B} \Xi \boldsymbol{F}' = \begin{cases} \left(\boldsymbol{D}_\nu^{\frac{1}{2}}, \boldsymbol{0}\right), & p \leq m, \\ \begin{pmatrix} \boldsymbol{D}_\nu^{\frac{1}{2}} \\ \boldsymbol{0} \end{pmatrix}, & p > m, \end{cases} \tag{46}$$

where $\boldsymbol{D}_\nu = \operatorname{diag}(\nu_1, \ldots, \nu_t)$.

PROOF. We prove this for the case $p \leq m$ and $\nu_p > 0$. Other cases can be proved similarly. By Theorem A.2.2 of Appendix A there is a matrix $\boldsymbol{B}$ such that

$$\boldsymbol{B} \Sigma \boldsymbol{B}' = \boldsymbol{I}, \qquad \boldsymbol{B} \Xi \Xi' \boldsymbol{B}' = \boldsymbol{D}_\nu. \tag{47}$$

Let

$$\boldsymbol{F}_1 = \boldsymbol{D}_\nu^{-\frac{1}{2}} \boldsymbol{B} \Xi \qquad (p \times m). \tag{48}$$

Then

$$\boldsymbol{F}_1 \boldsymbol{F}_1' = \boldsymbol{I}_p. \tag{49}$$

Let $\boldsymbol{F}' = (\boldsymbol{F}_1', \boldsymbol{F}_2')$ be a full $m \times m$ orthogonal matrix. Then

$$\boldsymbol{B} \Xi \boldsymbol{F}_2' = \boldsymbol{D}_\nu^{\frac{1}{2}} \boldsymbol{F}_1 \boldsymbol{F}_2' = \boldsymbol{0} \tag{50}$$

and

$$\mathbf{B\Xi F}' = \mathbf{B\Xi}(\mathbf{F}_1', \mathbf{F}_2') = \mathbf{B\Xi}(\mathbf{\Xi}'\mathbf{B}'\mathbf{D}_\nu^{-\frac{1}{2}}, \mathbf{F}_2') = (\mathbf{D}_\nu^{\frac{1}{2}}, \mathbf{0}). \qquad \text{Q.E.D.} \tag{51}$$

Now let

$$\mathbf{U} = \mathbf{BXF}', \qquad \mathbf{V} = \mathbf{BZ}. \tag{52}$$

Then the columns of $\mathbf{U}, \mathbf{V}$ are independently normally distributed with covariance matrix $\mathbf{I}$ and means when $p \leq m$

$$\begin{aligned} \mathscr{E}\mathbf{U} &= (\mathbf{D}_\nu^{\frac{1}{2}}, \mathbf{0}), \\ \mathscr{E}\mathbf{V} &= \mathbf{0}. \end{aligned} \tag{53}$$

Invariant tests are given in terms of characteristic roots $l_1, \ldots, l_t$ ($l_1 \geq \cdots \geq l_t$) of $\mathbf{U}'(\mathbf{VV}')^{-1}\mathbf{U}$. Note that for the admissibility we used the characteristic roots λ_i of $\mathbf{U}'(\mathbf{UU}' + \mathbf{VV}')^{-1}\mathbf{U}$ rather than $l_i = \lambda_i/(1 - \lambda_i)$. Here it is more natural to use l_i, which corresponds to the parameter value ν_i. The following theorem is given by Das Gupta, Anderson, and Mudholkar (1964).

THEOREM 8.10.6. *If the acceptance region of an invariant test is convex in the space of each column vector of* $\mathbf{U}$ *for each set of fixed values of* $\mathbf{V}$ *and of the other column vectors of* $\mathbf{U}$, *then the power of the test increases monotonically in each* ν_i.

PROOF. Since $\mathbf{UU}'$ is unchanged when any column vector of $\mathbf{U}$ is multiplied by -1, the acceptance region is symmetric about the origin in each of the column vectors of $\mathbf{U}$. Now the density of $\mathbf{U} = (u_{ij})$, $\mathbf{V} = (v_{ij})$ is

$$f(\mathbf{U}, \mathbf{V}) = (2\pi)^{-\frac{1}{2}(n+m)p} \exp\left[-\tfrac{1}{2}\left\{ \operatorname{tr} \mathbf{VV}' + \sum_{i=1}^{t} \left(u_{ii} - \sqrt{\nu_i}\right)^2 + \sum_{i=1}^{p} \sum_{\substack{j=1 \\ j \neq i}}^{m} u_{ij}^2 \right\}\right]. \tag{54}$$

Applying Theorem 8.10.5 to (54) we see that the power increases monotonically in each $\sqrt{\nu_i}$. Q.E.D.

Since the section of a convex set is convex, we have the following corollary.

COROLLARY 8.10.3. *If the acceptance region A of an invariant test is convex in $\boldsymbol{U}$ for each fixed $\boldsymbol{V}$, then the power of the test increases monotonically in each ν_i.*

From this we see that Roy's maximum root test $A: l_1 \le K$ and the Lawley–Hotelling's trace test $A: \operatorname{tr} \boldsymbol{U}'(\boldsymbol{V}\boldsymbol{V}')^{-1}\boldsymbol{U} \le K$ have power functions that are monotonically increasing in each ν_i.

To see that the acceptance region of the likelihood ratio test

$$A: \prod_{i=1}^{t}(1 + l_i) \le K \tag{55}$$

satisfies the condition of Theorem 8.10.6 let

$$(\boldsymbol{V}\boldsymbol{V}')^{-1} = \boldsymbol{T}'\boldsymbol{T}, \qquad \boldsymbol{T}: p \times p \tag{56}$$

$$\boldsymbol{U}^* = (\boldsymbol{u}_1^*, \dots, \boldsymbol{u}_m^*) = \boldsymbol{T}\boldsymbol{U}.$$

Then

$$\begin{aligned} \prod_{i=1}^{t}(1 + l_i) &= |\boldsymbol{U}'(\boldsymbol{V}\boldsymbol{V}')^{-1}\boldsymbol{U} + \boldsymbol{I}| = |\boldsymbol{U}^{*\prime}\boldsymbol{U}^* + \boldsymbol{I}| \\ &= |\boldsymbol{U}^*\boldsymbol{U}^{*\prime} + \boldsymbol{I}| = |\boldsymbol{u}_1^*\boldsymbol{u}_1^{*\prime} + \boldsymbol{B}| \\ &= (\boldsymbol{u}_1^{*\prime}\boldsymbol{B}^{-1}\boldsymbol{u}_1^* + 1)|\boldsymbol{B}| \\ &= (\boldsymbol{u}_1'\boldsymbol{T}'\boldsymbol{B}^{-1}\boldsymbol{T}\boldsymbol{u}_1 + 1)|\boldsymbol{B}|, \end{aligned} \tag{57}$$

where $\boldsymbol{B} = \boldsymbol{u}_2^*\boldsymbol{u}_2^{*\prime} + \cdots + \boldsymbol{u}_m^*\boldsymbol{u}_m^{*\prime} + \boldsymbol{I}$. Since $\boldsymbol{T}'\boldsymbol{B}^{-1}\boldsymbol{T}$ is positive definite, (55) is convex in $\boldsymbol{u}_1$. Therefore, the likelihood ratio test has a power function which is monotone increasing in each ν_i.

The Bartlett–Nanda–Pillai trace test

$$A: \operatorname{tr} \boldsymbol{U}'(\boldsymbol{U}\boldsymbol{U}' + \boldsymbol{V}\boldsymbol{V}')^{-1}\boldsymbol{U} = \sum_{i=1}^{t} \frac{l_i}{1 + l_i} \le K \tag{58}$$

has an acceptance region that is an ellipsoid if $K < 1$ and is convex in each column $\boldsymbol{u}_i$ of $\boldsymbol{U}$ *provided* $K \le 1$. (See Problem 36.) For $K > 1$ (58) may not be convex in each column of $\boldsymbol{U}$. The reader can work out an example for $p = 2$.

Eaton and Perlman (1974) have shown that if an invariant test is convex in $\boldsymbol{U}$ and $\boldsymbol{W} = \boldsymbol{V}\boldsymbol{V}'$ then the power at $(\nu_1^0, \dots, \nu_t^0)$ is greater than at $(\nu_1, \dots, \nu_t)$ if $(\sqrt{\nu_1}, \dots, \sqrt{\nu_t}) \prec_w (\sqrt{\nu_1^0}, \dots, \sqrt{\nu_t^0})$. We shall not prove this result. Roy's maximum root test and the Lawley–Hotelling trace test satisfy the condition, but the likelihood ratio and the Bartlett–Nanda–Pillai trace test do not.

Takemura has shown that if the acceptance region is convex in $\boldsymbol{U}$ and $\boldsymbol{W}$, the set of $\sqrt{\nu_1}, \dots, \sqrt{\nu_t}$ for which the power is not greater than a constant is monotone and convex.

It is enlightening to consider the contours of the power function, $\Pi(\sqrt{\nu_1}, \dots, \sqrt{\nu_t})$. Theorem 8.10.6 does not exclude the case (a) of Figure 8.8 and similarly the Eaton–Perlman result does not exclude (b). The last result guarantees that the contour looks like (c) for the Roy's maximum root test and the Lawley–Hotelling trace test. These results relate to the fact that these two tests are more likely to detect alternative hypotheses where few ν_i's are far from zero. In contrast with this the likelihood ratio test and the Bartlett–Nanda–Pillai trace test are sensitive for the overall departure from the null hypothesis. It might be noted that the convexity in $\sqrt{\nu}$-space cannot be translated into the convexity in ν-space.

By using the noncentral density of l_i's, which depends on the parameter values $\nu_1, \dots, \nu_t$, Perlman and Olkin (1980) showed that any invariant test with monotone acceptance region (in the space of roots) is unbiased. Note that this result covers all the standard tests considered earlier.

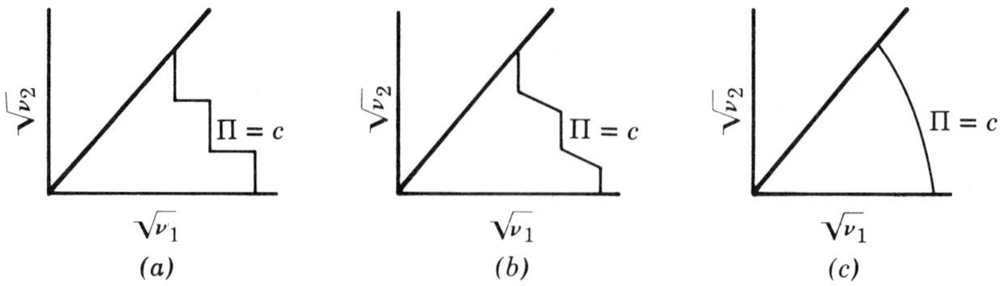

Figure 8.8. Contours of power functions.

PROBLEMS

1. (Sec. 8.2.2) Consider the following sample (for $N = 8$):

Weight of grain	40	17	9	15	6	12	5	9
Weight of straw	53	19	10	29	13	27	19	30
Amount of fertilizer	24	11	5	12	7	14	11	18

Let $z_{2\alpha} = 1$, and let $z_{1\alpha}$ be the amount of fertilizer on the αth plot. Estimate $\boldsymbol{\beta}$ for this sample. Test the hypothesis $\boldsymbol{\beta}_1 = \mathbf{0}$ at the 0.01 significance level.

2. (Sec. 8.2) Show that Theorem 3.2.1 is a special case of Theorem 8.2.1. [*Hint*: Let $q = 1$, $z_\alpha = 1$, $\boldsymbol{\beta} = \boldsymbol{\mu}$.]

3. (Sec. 8.2) Prove Theorem 8.2.3.

4. (Sec. 8.2) Show that $\hat{\boldsymbol{\beta}}$ minimizes the generalized variance

$$\left| \sum_{\alpha=1}^{N} (\boldsymbol{x}_\alpha - \boldsymbol{\beta} \boldsymbol{z}_\alpha)(\boldsymbol{x}_\alpha - \boldsymbol{\beta} \boldsymbol{z}_\alpha)' \right|.$$

5. (Sec. 8.3) In the following data [Woltz, Reid, and Colwell (1948), used by R. L. Anderson and Bancroft (1952)] the variables are x_1, rate of cigarette burn; x_2, the per cent nicotine; z_1, per cent of nitrogen; z_2, per cent of chlorine; z_3 of potassium; z_4 of phosphorus; z_5 of calcium; and z_6 of magnesium; and $z_7 = 1$, and $N = 25$.

$$\sum_{\alpha=1}^{N} \boldsymbol{x}_\alpha = \begin{pmatrix} 42.20 \\ 54.03 \end{pmatrix}, \qquad \sum_{\alpha=1}^{N} \boldsymbol{z}_\alpha = \begin{pmatrix} 53.92 \\ 62.02 \\ 56.00 \\ 12.25 \\ 89.79 \\ 24.10 \\ 25 \end{pmatrix},$$

$$\sum_{\alpha=1}^{N} (\boldsymbol{x}_\alpha - \bar{\boldsymbol{x}})(\boldsymbol{x}_\alpha - \bar{\boldsymbol{x}})' = \begin{pmatrix} 0.6690 & 0.4527 \\ 0.4527 & 6.5921 \end{pmatrix},$$

$$\sum_{\alpha=1}^{N} (\boldsymbol{z}_\alpha - \bar{\boldsymbol{z}})(\boldsymbol{z}_\alpha - \bar{\boldsymbol{z}})'$$

$$= \begin{pmatrix} 1.8311 & -0.3589 & -0.0125 & -0.0244 & 1.6379 & 0.5057 & 0 \\ -0.3589 & 8.8102 & -0.3469 & 0.0352 & 0.7920 & 0.2173 & 0 \\ -0.0125 & -0.3469 & 1.5818 & -0.0415 & -1.4278 & -0.4753 & 0 \\ -0.0244 & 0.0352 & -0.0415 & 0.0258 & 0.0043 & 0.0154 & 0 \\ 1.6379 & 0.7920 & -1.4278 & 0.0043 & 3.7248 & 0.9120 & 0 \\ 0.5057 & 0.2173 & -0.4753 & 0.0154 & 0.9120 & 0.3828 & 0 \\ 0 & 0 & 0 & 0 & 0 & 0 & 0 \end{pmatrix},$$

$$\sum_{\alpha=1}^{N} (z_\alpha - \bar{z})(x_\alpha - \bar{x})' = \begin{pmatrix} 0.2501 & 2.6691 \\ -1.5136 & -2.0617 \\ 0.5007 & -0.9503 \\ -0.0421 & -0.0187 \\ -0.1914 & 3.4020 \\ -0.1586 & 1.1663 \\ 0 & 0 \end{pmatrix}$$

(*a*) Estimate the regression of x_1 and x_2 on z_1, z_5, z_6, and z_7.
(*b*) Estimate the regression on all seven variables.
(*c*) Test the hypothesis that the regression on z_2, z_3, and z_4 is 0.

6. (Sec. 8.3) Let $q = 2$, $z_{1\alpha} = w_\alpha$ (scalar), $z_{2\alpha} = 1$. Show that the U-statistic for testing the hypothesis $\boldsymbol{\beta}_1 = \mathbf{0}$ is a monotonic function of a T^2-statistic and give the T^2-statistic in a simple form. (See Problem 1 of Chapter 5.)

7. (Sec. 8.3) Let $z_{q\alpha} = 1$, let $q_2 = 1$, and let

$$A^* = \left[\sum_\alpha (z_{i\alpha} - \bar{z}_i)(z_{j\alpha} - \bar{z}_j)\right], \qquad i, j = 1, \dots, q_1 = q - 1.$$

Prove that

$$(\hat{\boldsymbol{\beta}}_{1\Omega} - \boldsymbol{\beta}_1)(A_{11} - A_{12}A_{22}^{-1}A_{21})(\hat{\boldsymbol{\beta}}_{1\Omega} - \boldsymbol{\beta}_1)' = (\hat{\boldsymbol{\beta}}_{1\Omega} - \boldsymbol{\beta}_1)A^*(\hat{\boldsymbol{\beta}}_{1\Omega} - \boldsymbol{\beta}_1)'.$$

8. (Sec. 8.3) Let $q_1 = q_2$. How do you test the hypothesis $\boldsymbol{\beta}_1 = \boldsymbol{\beta}_2$?

9. (Sec. 8.3) Prove

$$\hat{\boldsymbol{\beta}}_{1\Omega} = \sum_\alpha x_\alpha \left(z_\alpha^{(1)} - A_{12}A_{22}^{-1}z_\alpha^{(2)}\right)' \left[\sum_\alpha \left(z_\alpha^{(1)} - A_{12}A_{22}^{-1}z_\alpha^{(2)}\right)\left(z_\alpha^{(1)} - A_{12}A_{22}^{-1}z_\alpha^{(2)}\right)'\right]^{-1}$$

$$= \left(C_1 - C_2A_{22}^{-1}A_{21}\right)\left(A_{11} - A_{12}A_{22}^{-1}A_{21}\right)^{-1}$$

10. (Sec. 8.4) By comparing Theorem 8.2.2 and Problem 9, prove Lemma 8.4.1.

11. (Sec. 8.4) Prove Lemma 8.4.1 by showing that the density of $\hat{\boldsymbol{\beta}}_{1\Omega}$ and $\hat{\boldsymbol{\beta}}_{2\omega}$ is

$$K_1 \exp\left[-\tfrac{1}{2}\operatorname{tr}\boldsymbol{\Sigma}^{-1}(\hat{\boldsymbol{\beta}}_{1\Omega} - \boldsymbol{\beta}_1^*)A_{11\cdot 2}(\hat{\boldsymbol{\beta}}_{1\Omega} - \boldsymbol{\beta}_1^*)'\right]$$

$$\times K_2 \exp\left[-\tfrac{1}{2}\operatorname{tr}\boldsymbol{\Sigma}^{-1}(\hat{\boldsymbol{\beta}}_{2\omega} - \boldsymbol{\beta}_2)A_{22}(\hat{\boldsymbol{\beta}}_{2\omega} - \boldsymbol{\beta}_2)'\right].$$

12. (Sec. 8.4) Show that the cdf of $U_{3,3,n}$ is

$$I_u\left(\tfrac{1}{2}n-1,\tfrac{3}{2}\right)+\frac{\Gamma(n+2)\Gamma\left[\frac{1}{2}(n+1)\right]}{\Gamma(n-1)\Gamma\left(\frac{1}{2}n-1\right)\sqrt{\pi}}\left\{\frac{2u^{\frac{1}{2}n-1}\sqrt{1-u}}{n(n-1)}\right.$$

$$+\frac{u^{\frac{1}{2}(n-1)}}{n-1}\left[\arcsin(2u-1)-\tfrac{1}{2}\pi\right]+\frac{2u^{\frac{1}{2}n}}{n}\log\left(\frac{1+\sqrt{1-u}}{\sqrt{u}}\right)$$

$$\left.+\frac{2u^{\frac{1}{2}n-1}(1-u)^{\frac{3}{2}}}{3(n+1)}\right\}.$$

[*Hint*: Use Theorem 8.4.4. The region $0 \le z_1 \le 1, 0 \le z_2 \le 1, z_1^2 z_2 \le u$ is the union of $0 \le z_1 \le 1, 0 \le z_2 \le u$ and $0 \le z_1 \le u/z_2,\ u \le z_2 \le 1$.]

13. (Sec. 8.4) Find $\Pr\{U_{4,3,n} \ge u\}$.

14. (Sec. 8.4) Find $\Pr\{U_{4,4,n} \ge u\}$.

15. (Sec. 8.4) For $p \le m$ find $\mathscr{E}U^h$ from the density of G and H.

16. (Sec. 8.4) (a) Show that when p is even, the characteristic function of $Y = \log U_{p,m,n}$, say $\phi(t) = \mathscr{E}e^{itY}$, is the reciprocal of a polynomial. (b) Sketch a method of inverting the characteristic function of Y by the method of residues. (c) Show that the resulting density of U is a polynomial in $\sqrt{u}$ and $\log u$ with possibly a factor of $u^{-\frac{1}{2}}$.

17. (Sec. 8.5) Use the asymptotic expansion of the distribution to compute $\Pr\{-k\log U_{3,3,n} \le M^*\}$ for

(a) $n = 8,\ M^* = 14.7$,
(b) $n = 8,\ M^* = 21.7$,
(c) $n = 16,\ M^* = 14.7$,
(d) $n = 16,\ M^* = 21.7$.

(Either compute to the third decimal place or use the expansion to the k^{-4} term.)

18. (Sec. 8.5) In case $p = 3$, $q_1 = 4$, and $n = N - q = 20$, find the 5% significance point for $k\log U$ using (a) $-2\log\lambda$ as χ^2 and (b) using $-k\log U$ as χ^2. Using more terms of this expansion, evaluate the exact significance levels for your answers to (a) and (b).

19. (Sec. 8.6.5) Prove for $l_i \ge 0,\ i = 1,\dots,p$,

$$\sum_{i=1}^{p}\frac{l_i}{1+l_i} \le \log\prod_{i=1}^{p}(1+l_i) \le \sum_{i=1}^{p} l_i.$$

Comment: The inequalities imply an ordering of the values of the Bartlett–Nanda–Pillai trace, the negative logarithm of the likelihood ratio criterion, and the Lawley–Hotelling trace.

20. (Sec. 8.6) *The multivariate beta density*. Let $\boldsymbol{H}$ and $\boldsymbol{G}$ be independently distributed according to $W(\boldsymbol{\Sigma}, m)$ and $W(\boldsymbol{\Sigma}, n)$, respectively. Let $\boldsymbol{C}$ be a matrix such that $\boldsymbol{CC}' = \boldsymbol{H} + \boldsymbol{G}$ and let

$$\boldsymbol{L} = \boldsymbol{C}^{-1}\boldsymbol{H}\boldsymbol{C}'^{-1}.$$

Show that the density of $\boldsymbol{L}$ is

$$\frac{\Gamma_p\left[\frac{1}{2}(m+n)\right]}{\Gamma_p\left(\frac{1}{2}m\right)\Gamma_p\left(\frac{1}{2}n\right)}|\boldsymbol{L}|^{\frac{1}{2}(m-p-1)}|\boldsymbol{I}-\boldsymbol{L}|^{\frac{1}{2}(n-p-1)}$$

for $\boldsymbol{L}$ and $\boldsymbol{I} - \boldsymbol{L}$ positive definite and 0 otherwise.

21. (Sec. 8.9) Let $\boldsymbol{Y}_{ij}$ (a p-component vector) be distributed according to $N(\boldsymbol{\mu}_{ij}, \boldsymbol{\Sigma})$, where $\mathscr{E}\boldsymbol{Y}_{ij} = \boldsymbol{\mu}_{ij} = \boldsymbol{\mu} + \boldsymbol{\lambda}_i + \boldsymbol{\nu}_j + \boldsymbol{\gamma}_{ij}$, $\sum_i \boldsymbol{\lambda}_i = \boldsymbol{0} = \sum_j \boldsymbol{\nu}_j = \sum_i \boldsymbol{\gamma}_{ij} = \sum_j \boldsymbol{\gamma}_{ij}$; the $\boldsymbol{\gamma}_{ij}$ are the interactions. If m observations are made on each $\boldsymbol{Y}_{ij}$ (say $\boldsymbol{y}_{ij1}, \dots, \boldsymbol{y}_{ijm}$), how do you test the hypothesis $\boldsymbol{\lambda}_i = \boldsymbol{0}$, $i = 1, \dots, r$? How do you test the hypothesis $\boldsymbol{\gamma}_{ij} = \boldsymbol{0}$, $i = 1, \dots, r$; $j = 1, \dots, c$?

22. (Sec. 8.9) *The Latin Square*. Let $\boldsymbol{Y}_{ij}$, $i, j = 1, \dots, r$, be distributed according to $N(\boldsymbol{\mu}_{ij}, \boldsymbol{\Sigma})$, where $\mathscr{E}\boldsymbol{Y}_{ij} = \boldsymbol{\mu}_{ij} = \boldsymbol{\gamma} + \boldsymbol{\lambda}_i + \boldsymbol{\nu}_j + \boldsymbol{\mu}_k$ and $k = j - i + 1 \pmod r$ with $\sum \boldsymbol{\lambda}_i = \sum \boldsymbol{\nu}_i = \sum \boldsymbol{\mu}_i = 0$.

(*a*) Give the univariate analysis of variance table for main effects and error (including sums of squares, numbers of degrees of freedom, and mean squares).
(*b*) Give the table for the vector case.
(*c*) Indicate in the vector case how to test the hypothesis $\boldsymbol{\lambda}_i = \boldsymbol{0}$, $i = 1, \dots, r$.

23. (Sec. 8.9) Let x_1 be the yield of a process and x_2 a quality measure. Let $z_1 = 1$, $z_2 = \pm 10°$ (temperature relative to average), $z_3 = \pm 0.75$ (relative measure of flow of one agent), and $z_4 = \pm 1.50$ (relative measure of flow of another agent). [See Anderson (1955a) for details.] Three observations were made on x_1 and x_2 for each possible triplet of values of z_2, z_3, and z_4. The estimate of $\boldsymbol{\beta}$ is

$$\hat{\boldsymbol{\beta}} = \begin{pmatrix} 58.529 & -0.3829 & -5.050 & 2.308 \\ 98.675 & 0.1558 & 4.144 & -0.700 \end{pmatrix};$$

$s_1 = 3.090$, $s_2 = 1.619$, and $r = -0.6632$ can be used to compute $\boldsymbol{S}$ or $\hat{\boldsymbol{\Sigma}}$.

(*a*) Formulate an analysis of variance model for this situation.
(*b*) Find a confidence region for the effects of temperature (i.e., β_{12}, β_{22}).
(*c*) Test the hypothesis that the two agents have no effect on the yield and quantity.

24. (Sec. 8.6) Interpret the transformations referred to in Theorem 8.6.1 in the original terms; that is, $H: \boldsymbol{\beta}_1 = \boldsymbol{\beta}_1^*$ and $z_\alpha^{(1)}$.

25. (Sec. 8.6) Find the cdf of $\operatorname{tr} \boldsymbol{H}\boldsymbol{G}^{-1}$ for $p = 2$. [*Hint*: Use the distribution of the roots given in Chapter 13.]

26. (Sec. 8.10.1) *Bartlett–Nanda–Pillai V test as a Bayes procedure*. Let $\boldsymbol{w}_1, \boldsymbol{w}_2, \ldots, \boldsymbol{w}_{m+n}$ be independently normally distributed with covariance matrix $\boldsymbol{\Sigma}$ and means $\mathscr{E}\boldsymbol{w}_i = \boldsymbol{\gamma}_i$, $i = 1, \ldots, m$, $\mathscr{E}\boldsymbol{w}_i = \boldsymbol{0}$, $i = m+1, \ldots, m+n$. Let Π_0 be defined by $[\boldsymbol{\Gamma}_1, \boldsymbol{\Sigma}] = [\boldsymbol{0}, (\boldsymbol{I} + \boldsymbol{C}\boldsymbol{C}')^{-1}]$, where the $p \times m$ matrix $\boldsymbol{C}$ has a density proportional to $|\boldsymbol{I} + \boldsymbol{C}\boldsymbol{C}'|^{-\frac{1}{2}(n+m)}$, and $\boldsymbol{\Gamma}_1 = (\boldsymbol{\gamma}_1, \ldots, \boldsymbol{\gamma}_m)$; let Π_1 be defined by $[\boldsymbol{\Gamma}_1, \boldsymbol{\Sigma}] = [(\boldsymbol{I} + \boldsymbol{C}\boldsymbol{C}')^{-1}\boldsymbol{C}, (\boldsymbol{I} + \boldsymbol{C}\boldsymbol{C}')^{-1}]$ where $\boldsymbol{C}$ has a density proportional to $|\boldsymbol{I} + \boldsymbol{C}\boldsymbol{C}'|^{-\frac{1}{2}(n+m)} e^{\frac{1}{2}\operatorname{tr} \boldsymbol{C}'(\boldsymbol{I}+\boldsymbol{C}\boldsymbol{C}')^{-1}\boldsymbol{C}}$.

(*a*) Show that the measures are finite for $n \ge p$ by showing $\operatorname{tr} \boldsymbol{C}'(\boldsymbol{I} + \boldsymbol{C}\boldsymbol{C}')^{-1}\boldsymbol{C} < m$ and verifying that the integral of $|\boldsymbol{I} + \boldsymbol{C}\boldsymbol{C}'|^{-\frac{1}{2}(n+m)}$ is finite. [*Hint*: Let $\boldsymbol{C} = (\boldsymbol{c}_1, \ldots, \boldsymbol{c}_m)$, $\boldsymbol{D}_j = \boldsymbol{I} + \sum_{i=1}^{j} \boldsymbol{c}_i\boldsymbol{c}_i' = \boldsymbol{E}_j\boldsymbol{E}_j'$, $\boldsymbol{c}_j = \boldsymbol{E}_{j-1}\boldsymbol{d}_j$, $j = 1, \ldots, m$ $(\boldsymbol{E}_0 = \boldsymbol{I})$. Show $|\boldsymbol{D}_j| = |\boldsymbol{D}_{j-1}|(1 + \boldsymbol{d}_j'\boldsymbol{d}_j)$ and hence $|\boldsymbol{D}_m| = \prod_{j=1}^{m}(1 + \boldsymbol{d}_j'\boldsymbol{d}_j)$. Then refer to Problem 15 of Chapter 5.]

(*b*) Show that the inequality (26) of Section 5.6 is equivalent to

$$\operatorname{tr}\left(\sum_{i=1}^{m+n} \boldsymbol{w}_i\boldsymbol{w}_i'\right)^{-1} \sum_{i=1}^{m} \boldsymbol{w}_i\boldsymbol{w}_i' \ge k.$$

Hence the Bartlett–Nanda–Pillai V test is Bayes and thus admissible.

27. (Sec. 8.10.1) *Likelihood ratio test as a Bayes procedure*. Let $\boldsymbol{w}_1, \ldots, \boldsymbol{w}_{m+n}$ be independently normally distributed with covariance matrix $\boldsymbol{\Sigma}$ and means $\mathscr{E}\boldsymbol{w}_i = \boldsymbol{\gamma}_i$, $i = 1, \ldots, m$, $\mathscr{E}\boldsymbol{w}_i = \boldsymbol{0}$, $i = m+1, \ldots, m+n$, with $n \ge m + p$. Let Π_0 be defined by $[\boldsymbol{\Gamma}_1, \boldsymbol{\Sigma}] = [\boldsymbol{0}, (\boldsymbol{I} + \boldsymbol{C}\boldsymbol{C}')^{-1}]$, where the $p \times m$ matrix $\boldsymbol{C}$ has a density proportional to $|\boldsymbol{I} + \boldsymbol{C}\boldsymbol{C}'|^{-\frac{1}{2}(n+m)}$ and $\boldsymbol{\Gamma}_1 = (\boldsymbol{\gamma}_1, \ldots, \boldsymbol{\gamma}_m)$; let Π_1 be defined by

$$[\boldsymbol{\Gamma}_1, \boldsymbol{\Sigma}] = \left[(\boldsymbol{I} + \boldsymbol{C}\boldsymbol{C}')^{-1}\boldsymbol{C}\boldsymbol{D}, (\boldsymbol{I} + \boldsymbol{C}\boldsymbol{C}')^{-1}\right],$$

where the m columns of $\boldsymbol{D}$ are conditionally independently normally distributed with means $\boldsymbol{0}$ and covariance matrix $[\boldsymbol{I} - \boldsymbol{C}'(\boldsymbol{I} + \boldsymbol{C}\boldsymbol{C}')^{-1}\boldsymbol{C}]^{-1}$ and $\boldsymbol{C}$ has the (marginal) density proportional to

$$|\boldsymbol{I} + \boldsymbol{C}\boldsymbol{C}'|^{-\frac{1}{2}(n+m)} |\boldsymbol{I} - \boldsymbol{C}'(\boldsymbol{I} + \boldsymbol{C}\boldsymbol{C}')^{-1}\boldsymbol{C}|^{-\frac{1}{2}m}.$$

(*a*) Show the measures are finite. [*Hint*: See Problem 26.]

(*b*) Show that the inequality (26) of Section 5.6 is equivalent to

$$\frac{\left|\sum_{i=1}^{m+n} \boldsymbol{w}_i\boldsymbol{w}_i'\right|}{\left|\sum_{i=m+1}^{m+n} \boldsymbol{w}_i\boldsymbol{w}_i'\right|} \ge k.$$

Hence the likelihood ratio test is Bayes and thus admissible.

28. (Sec. 8.10.1) *Admissibility of the likelihood ratio test*. Show that the acceptance region $|ZZ'|/|ZZ' + XX'| \geq c$ satisfies the conditions of Theorem 8.10.1. [*Hint*: The acceptance region can be written $\prod_{i=1}^{t} m_i > c$, where $m_i = 1 - \lambda_i$, $i = 1, \ldots, t$.]

29. (Sec. 8.10.1) *Admissibility of the Lawley–Hotelling test*. Show that the acceptance region $\operatorname{tr} XX'(ZZ')^{-1} \leq c$ satisfies the conditions of Theorem 8.10.1.

30. (Sec. 8.10.1) *Admissibility of the Bartlett–Nanda–Pillai trace test*. Show that the acceptance region $\operatorname{tr} X'(ZZ' + XX')^{-1}X \leq c$ satisfies the conditions of Theorem 8.10.1.

31. (Sec. 8.10.1) Show that if A and B are positive definite and $A - B$ is positive semidefinite, then $B^{-1} - A^{-1}$ is positive semidefinite. [*Hint*: There exists a nonsingular matrix F and a diagonal matrix D such that $A = FDF'$ and $B = FF'$ (Theorem A.2.2 of Appendix A).]

32. (Sec. 8.10.1) Show that the boundary of $\tilde{A}$ has m-measure 0. [*Hint*: Show that closure of $\tilde{A} \subset \tilde{A} \cup C$, where $C = \{V | U - YY' \text{ is singular}\}$.]

33. (Sec. 8.10.1) Show that if $A \subset R_{<}^{m}$ is convex and monotone in majorization, then A^* is convex. [*Hint*: Show

$$(px + qy)_{\downarrow} \prec_w px_{\downarrow} + qy_{\downarrow},$$

where

$$z_{\downarrow} = (z_{[1]}, \ldots, z_{[m]})' \in R_{<}^{m}.]$$

34. (Sec. 8.10.1) Show that $C(\lambda)$ is convex. [*Hint*: Follow the solution of Problem 33 to show $(px + qy) \prec_w \lambda$ if $x \prec_w \lambda$ and $y \prec_w \lambda$.]

35. (Sec. 8.10.1) Show that if A is monotone, then A^* is monotone. [*Hint*: Use the fact that

$$x_{[k]} = \max_{i_1, \ldots, i_k} \{\min(x_{i_1}, \ldots, x_{i_k})\}.]$$

36. (Sec. 8.10.2) *Monotonicity of the power function of the Bartlett–Nanda–Pillai trace test*. Show that

$$\operatorname{tr}(uu' + B)(uu' + B + W)^{-1} \leq K$$

is convex in u for fixed positive semidefinite B and positive definite $B + W$ if $0 \leq K \leq 1$. [*Hint*: Verify

$$(uu' + B + W)^{-1}$$
$$= (B + W)^{-1} - \frac{1}{1 + u'(B + W)^{-1}u}(B + W)^{-1}uu'(B + W)^{-1}.$$

The resulting quadratic form in $\boldsymbol{u}$ involves the matrix $(\operatorname{tr}\boldsymbol{A})\boldsymbol{I} - \boldsymbol{A}$ for $\boldsymbol{A} = (\boldsymbol{B} + \boldsymbol{W})^{-\frac{1}{2}}\boldsymbol{B}(\boldsymbol{B} + \boldsymbol{W})^{-\frac{1}{2}}$; show that this matrix is positive semidefinite by diagonalizing $\boldsymbol{A}$.]

37. (Sec. 8.8) Let $\boldsymbol{x}_\alpha^{(\nu)}$, $\alpha = 1, \dots, N_\nu$, be observations from $N(\boldsymbol{\mu}^{(\nu)}, \boldsymbol{\Sigma})$, $\nu = 1, \dots, q$. What criterion may be used to test the hypothesis that

$$\boldsymbol{\mu}^{(\nu)} = \sum_{h=1}^{m} \boldsymbol{\gamma}_h c_{h\nu} + \boldsymbol{\mu},$$

where $c_{h\nu}$ are given numbers and $\boldsymbol{\gamma}_\nu, \boldsymbol{\mu}$ are unknown vectors? [*Note*: This hypothesis (that the means lie on an m-dimensional hyperplane with ratios of distances known) can be put in the form of the general linear hypothesis.]

38. (Sec. 8.2) Let $\boldsymbol{x}_\alpha$ be an observation from $N(\boldsymbol{\beta} \boldsymbol{z}_\alpha, \boldsymbol{\Sigma})$, $\alpha = 1, \dots, N$. Suppose there is a known fixed vector $\boldsymbol{\gamma}$ such that $\boldsymbol{\beta\gamma} = \boldsymbol{0}$. How do you estimate $\boldsymbol{\beta}$?

39. (Sec. 8.8) What is the largest group of transformations on $y_\alpha^{(i)}$, $\alpha = 1, \dots, N_i$, $i = 1, \dots, q$, that leaves (1) invariant? Prove the test (12) is invariant under this group.

CHAPTER 9

Testing Independence of Sets of Variates

9.1. INTRODUCTION

In this section we divide a set of p variates with a joint normal distribution into q subsets and ask whether the q subsets are mutually independent; this is equivalent to testing the hypothesis that each variable in one subset is uncorrelated with each variable in the others. We find the likelihood ratio criterion for this hypothesis, the moments of the criterion under the null hypothesis, some particular distributions, and an asymptotic expansion of the distribution.

The likelihood ratio criterion is invariant under linear transformations within sets; another such criterion is developed. Alternative test procedures are step-down procedures, which are not invariant, but are flexible. In the case of two sets, independence of the two sets is equivalent to the regression of one on the other being 0; the criteria for Chapter 8 are available. Some optimal properties of the likelihood ratio test are treated.

9.2. THE LIKELIHOOD RATIO CRITERION FOR TESTING INDEPENDENCE OF SETS OF VARIATES

Let the p-component vector $\boldsymbol{X}$ be distributed according to $N(\boldsymbol{\mu}, \boldsymbol{\Sigma})$. We partition $\boldsymbol{X}$ into q subvectors with $p_1, p_2, \dots, p_q$ components, respectively; that is,

$$\boldsymbol{X} = \begin{pmatrix} \boldsymbol{X}^{(1)} \\ \boldsymbol{X}^{(2)} \\ \vdots \\ \boldsymbol{X}^{(q)} \end{pmatrix}. \tag{1}$$

The vector of means $\boldsymbol{\mu}$ and the covariance matrix $\boldsymbol{\Sigma}$ are partitioned similarly,

$$\boldsymbol{\mu} = \begin{pmatrix} \boldsymbol{\mu}^{(1)} \\ \boldsymbol{\mu}^{(2)} \\ \vdots \\ \boldsymbol{\mu}^{(q)} \end{pmatrix}, \tag{2}$$

$$\boldsymbol{\Sigma} = \begin{pmatrix} \boldsymbol{\Sigma}_{11} & \boldsymbol{\Sigma}_{12} & \cdots & \boldsymbol{\Sigma}_{1q} \\ \boldsymbol{\Sigma}_{21} & \boldsymbol{\Sigma}_{22} & \cdots & \boldsymbol{\Sigma}_{2q} \\ \vdots & \vdots & & \vdots \\ \boldsymbol{\Sigma}_{q1} & \boldsymbol{\Sigma}_{q2} & \cdots & \boldsymbol{\Sigma}_{qq} \end{pmatrix}. \tag{3}$$

The null hypothesis we wish to test is that the subvectors $\boldsymbol{X}^{(1)}, \ldots, \boldsymbol{X}^{(q)}$ are mutually independently distributed, that is, that the density of $\boldsymbol{X}$ factors into the densities of $\boldsymbol{X}^{(1)}, \ldots, \boldsymbol{X}^{(q)}$. It is

$$H: n(\boldsymbol{x}|\boldsymbol{\mu}, \boldsymbol{\Sigma}) = \prod_{i=1}^{q} n(\boldsymbol{x}^{(i)}|\boldsymbol{\mu}^{(i)}, \boldsymbol{\Sigma}_{ii}). \tag{4}$$

If $\boldsymbol{X}^{(1)}, \ldots, \boldsymbol{X}^{(q)}$ are independent subvectors,

$$\mathscr{E}(\boldsymbol{X}^{(i)} - \boldsymbol{\mu}^{(i)})(\boldsymbol{X}^{(j)} - \boldsymbol{\mu}^{(j)})' = \boldsymbol{\Sigma}_{ij} = \boldsymbol{0}, \qquad i \neq j. \tag{5}$$

(See Section 2.4.) Conversely, if (5) holds, then (4) is true. Thus the null hypothesis is equivalently $H: \boldsymbol{\Sigma}_{ij} = \boldsymbol{0}$, $i \neq j$. This can be stated alternatively as the hypothesis that $\boldsymbol{\Sigma}$ is of the form

$$\boldsymbol{\Sigma}_0 = \begin{pmatrix} \boldsymbol{\Sigma}_{11} & \boldsymbol{0} & \cdots & \boldsymbol{0} \\ \boldsymbol{0} & \boldsymbol{\Sigma}_{22} & \cdots & \boldsymbol{0} \\ \vdots & \vdots & & \vdots \\ \boldsymbol{0} & \boldsymbol{0} & \cdots & \boldsymbol{\Sigma}_{qq} \end{pmatrix}. \tag{6}$$

Given a sample $\boldsymbol{x}_1, \ldots, \boldsymbol{x}_N$ of N observations on $\boldsymbol{X}$, the likelihood ratio criterion is

$$\lambda = \frac{\max_{\boldsymbol{\mu}, \boldsymbol{\Sigma}_0} L(\boldsymbol{\mu}, \boldsymbol{\Sigma}_0)}{\max_{\boldsymbol{\mu}, \boldsymbol{\Sigma}} L(\boldsymbol{\mu}, \boldsymbol{\Sigma})}, \tag{7}$$

where

$$L(\boldsymbol{\mu}, \boldsymbol{\Sigma}) = \prod_{\alpha=1}^{N} \frac{1}{(2\pi)^{\frac{1}{2}p} |\boldsymbol{\Sigma}|^{\frac{1}{2}}} e^{-\frac{1}{2}(\boldsymbol{x}_\alpha - \boldsymbol{\mu})' \boldsymbol{\Sigma}^{-1} (\boldsymbol{x}_\alpha - \boldsymbol{\mu})} \tag{8}$$

and $L(\boldsymbol{\mu}, \boldsymbol{\Sigma}_0)$ is $L(\boldsymbol{\mu}, \boldsymbol{\Sigma})$ with $\boldsymbol{\Sigma}_{ij} = \mathbf{0}$, $i \neq j$, and the maximum is taken with respect to all vectors $\boldsymbol{\mu}$ and positive definite $\boldsymbol{\Sigma}$ and $\boldsymbol{\Sigma}_0$ (i.e., $\boldsymbol{\Sigma}_{ii}$). As derived in Section 5.2, Equation (6),

$$\max_{\boldsymbol{\mu}, \boldsymbol{\Sigma}} L(\boldsymbol{\mu}, \boldsymbol{\Sigma}) = \frac{1}{(2\pi)^{\frac{1}{2}pN} |\hat{\boldsymbol{\Sigma}}_\Omega|^{\frac{1}{2}N}} e^{-\frac{1}{2}pN}, \tag{9}$$

where

$$\hat{\boldsymbol{\Sigma}}_\Omega = \frac{1}{N} \boldsymbol{A} = \frac{1}{N} \sum_{\alpha=1}^{N} (\boldsymbol{x}_\alpha - \bar{\boldsymbol{x}})(\boldsymbol{x}_\alpha - \bar{\boldsymbol{x}})'. \tag{10}$$

Under the null hypothesis

$$L(\boldsymbol{\mu}, \boldsymbol{\Sigma}_0) = \prod_{i=1}^{q} L_i(\boldsymbol{\mu}^{(i)}, \boldsymbol{\Sigma}_{ii}), \tag{11}$$

where

$$L_i(\boldsymbol{\mu}^{(i)}, \boldsymbol{\Sigma}_{ii}) = \prod_{\alpha=1}^{N} \frac{1}{(2\pi)^{\frac{1}{2}p_i} |\boldsymbol{\Sigma}_{ii}|^{\frac{1}{2}}} e^{-\frac{1}{2}(\boldsymbol{x}_\alpha^{(i)} - \boldsymbol{\mu}^{(i)})' \boldsymbol{\Sigma}_{ii}^{-1} (\boldsymbol{x}_\alpha^{(i)} - \boldsymbol{\mu}^{(i)})}. \tag{12}$$

Clearly

$$\begin{aligned} \max_{\boldsymbol{\mu}, \boldsymbol{\Sigma}_0} L(\boldsymbol{\mu}, \boldsymbol{\Sigma}_0) &= \prod_{i=1}^{q} \max_{\boldsymbol{\mu}^{(i)}, \boldsymbol{\Sigma}_{ii}} L_i(\boldsymbol{\mu}^{(i)}, \boldsymbol{\Sigma}_{ii}) \\ &= \prod_{i=1}^{q} \frac{1}{(2\pi)^{\frac{1}{2}p_i N} |\hat{\boldsymbol{\Sigma}}_{ii\omega}|^{\frac{1}{2}N}} e^{-\frac{1}{2}p_i N} \\ &= \frac{1}{(2\pi)^{\frac{1}{2}pN} \prod_{i=1}^{q} |\hat{\boldsymbol{\Sigma}}_{ii\omega}|^{\frac{1}{2}N}} e^{-\frac{1}{2}pN}, \end{aligned} \tag{13}$$

where

$$\hat{\boldsymbol{\Sigma}}_{ii\omega} = \frac{1}{N} \sum_{\alpha=1}^{N} (\boldsymbol{x}_\alpha^{(i)} - \bar{\boldsymbol{x}}^{(i)})(\boldsymbol{x}_\alpha^{(i)} - \bar{\boldsymbol{x}}^{(i)})'. \tag{14}$$

If we partition $\boldsymbol{A}$ and $\hat{\boldsymbol{\Sigma}}_\Omega$ as we have $\boldsymbol{\Sigma}$,

$$(15)\quad \boldsymbol{A} = \begin{pmatrix} \boldsymbol{A}_{11} & \boldsymbol{A}_{12} & \cdots & \boldsymbol{A}_{1q} \\ \boldsymbol{A}_{21} & \boldsymbol{A}_{22} & \cdots & \boldsymbol{A}_{2q} \\ \vdots & \vdots & & \vdots \\ \boldsymbol{A}_{q1} & \boldsymbol{A}_{q2} & \cdots & \boldsymbol{A}_{qq} \end{pmatrix}, \qquad \hat{\boldsymbol{\Sigma}}_\Omega = \begin{pmatrix} \hat{\boldsymbol{\Sigma}}_{11} & \hat{\boldsymbol{\Sigma}}_{12} & \cdots & \hat{\boldsymbol{\Sigma}}_{1q} \\ \hat{\boldsymbol{\Sigma}}_{21} & \hat{\boldsymbol{\Sigma}}_{22} & \cdots & \hat{\boldsymbol{\Sigma}}_{2q} \\ \vdots & \vdots & & \vdots \\ \hat{\boldsymbol{\Sigma}}_{q1} & \hat{\boldsymbol{\Sigma}}_{q2} & \cdots & \hat{\boldsymbol{\Sigma}}_{qq} \end{pmatrix},$$

we see that $\hat{\boldsymbol{\Sigma}}_{ii\omega} = \hat{\boldsymbol{\Sigma}}_{ii} = (1/N)\boldsymbol{A}_{ii}$.

The likelihood ratio criterion is

$$(16)\quad \lambda = \frac{\max_{\boldsymbol{\mu},\boldsymbol{\Sigma}_0} L(\boldsymbol{\mu},\boldsymbol{\Sigma}_0)}{\max_{\boldsymbol{\mu},\boldsymbol{\Sigma}} L(\boldsymbol{\mu},\boldsymbol{\Sigma})} = \frac{|\hat{\boldsymbol{\Sigma}}_\Omega|^{\frac{1}{2}N}}{\prod_{i=1}^q |\hat{\boldsymbol{\Sigma}}_{ii}|^{\frac{1}{2}N}} = \frac{|\boldsymbol{A}|^{\frac{1}{2}N}}{\prod_{i=1}^q |\boldsymbol{A}_{ii}|^{\frac{1}{2}N}}.$$

The critical region of the likelihood ratio test is

$$(17)\quad \lambda \le \lambda(\varepsilon),$$

where $\lambda(\varepsilon)$ is a number such that the probability of (17) is ε when $\boldsymbol{\Sigma} = \boldsymbol{\Sigma}_0$. (It remains to show that such a number can be found.) Let

$$(18)\quad V = \frac{|\boldsymbol{A}|}{\prod_{i=1}^q |\boldsymbol{A}_{ii}|}.$$

Then $\lambda = V^{\frac{1}{2}N}$ is a monotonic increasing function of V. The critical region (17) can be equivalently written as

$$(19)\quad V \le V(\varepsilon).$$

THEOREM 9.2.1. *Let $\boldsymbol{x}_1, \dots, \boldsymbol{x}_N$ be a sample of N observations drawn from $N(\boldsymbol{\mu}, \boldsymbol{\Sigma})$, where $\boldsymbol{x}_\alpha$, $\boldsymbol{\mu}$, and $\boldsymbol{\Sigma}$ are partitioned into $p_1, \dots, p_q$ rows (and columns in the case of $\boldsymbol{\Sigma}$) as indicated in* (1), (2), *and* (3). *The likelihood ratio criterion that the q sets of components are mutually independent is given by* (16), *where $\boldsymbol{A}$ is defined by* (10) *and partitioned according to* (15). *The likelihood ratio test is given by* (17) *and equivalently by* (19), *where V is defined by* (18) *and $\lambda(\varepsilon)$ or $V(\varepsilon)$ is chosen to obtain the significance level ε.*

Since $r_{ij} = a_{ij}/\sqrt{a_{ii}a_{jj}}$, we have

$$|A| = |R| \prod_{i=1}^{p} a_{ii}, \tag{20}$$

where

$$R = (r_{ij}) = \begin{pmatrix} R_{11} & R_{12} & \cdots & R_{1q} \\ R_{21} & R_{22} & \cdots & R_{2q} \\ \vdots & \vdots & & \vdots \\ R_{q1} & R_{q2} & \cdots & R_{qq} \end{pmatrix} \tag{21}$$

and

$$|A_{ii}| = |R_{ii}| \prod_{j=p_1+\cdots+p_{i-1}+1}^{p_1+\cdots+p_i} a_{jj}. \tag{22}$$

Thus

$$V = \frac{|A|}{\prod |A_{ii}|} = \frac{|R|}{\prod |R_{ii}|}. \tag{23}$$

That is, V can be expressed entirely in terms of sample correlation coefficients.

We can interpret the criterion V in terms of generalized variance. Each set $(x_{i1}, \ldots, x_{iN})$ can be considered as a vector in N-space; the set $(x_{i1} - \bar{x}_i, \ldots, x_{iN} - \bar{x}_i) = z_i$, say, is the projection on the plane orthogonal to the equiangular line. The determinant $|A|$ is the p-dimensional volume squared of the parallelotope with $z_1, \ldots, z_p$ as principal edges. The determinant $|A_{ii}|$ is the p_i-dimensional volume squared of the parallelotope having as principal edges the ith set of vectors. If each set of vectors is orthogonal to each other set (i.e., $R_{ij} = 0$, $i \neq j$), then the volume squared $|A|$ is the product of the volumes squared $|A_{ii}|$. For example, if $p = 2$, $p_1 = p_2 = 1$, this statement is that the area of a parallelogram is the product of the lengths of the sides if the sides are at right angles. If the sets are almost orthogonal, $|A|$ is almost $\prod |A_{ii}|$ and V is almost 1.

The criterion has an invariance property. Let C_i be an arbitrary nonsingular matrix of order p_i and let

$$C = \begin{pmatrix} C_1 & 0 & \cdots & 0 \\ 0 & C_2 & \cdots & 0 \\ \vdots & \vdots & & \vdots \\ 0 & 0 & \cdots & C_q \end{pmatrix}. \tag{24}$$

Let $Cx_\alpha + d = x_\alpha^*$. Then the criterion for independence in terms of x_α^* is identical to the criterion in terms of x_α. Let $A^* = \sum_\alpha (x_\alpha^* - \bar{x}^*)(x_\alpha^* - \bar{x}^*)'$ be partitioned into submatrices A_{ij}^*. Then

$$(25)\qquad \begin{aligned} A_{ij}^* &= \sum_{\alpha} \left(x_{\alpha}^{*(i)} - \bar{x}^{*(i)} \right) \left(x_{\alpha}^{*(j)} - \bar{x}^{*(j)} \right)' \\ &= C_i \sum_{\alpha} \left(x_{\alpha}^{(i)} - \bar{x}^{(i)} \right) \left(x_{\alpha}^{(j)} - \bar{x}^{(j)} \right)' C_j' \\ &= C_i A_{ij} C_j' \end{aligned}$$

and $A^* = CAC'$. Thus

$$(26)\qquad \begin{aligned} V^* &= \frac{|A^*|}{\prod |A_{ii}^*|} = \frac{|CAC'|}{\prod |C_i A_{ii} C_i'|} \\ &= \frac{|C| \cdot |A| \cdot |C'|}{\prod |C_i| \cdot |A_{ii}| \cdot |C_i'|} = \frac{|A|}{\prod |A_{ii}|} = V, \end{aligned}$$

for $|C| = \prod |C_i|$. Thus the test is invariant with respect to linear transformations within each set.

Narain (1950) has shown that the test based on V is strictly unbiased; that is, the probability of rejecting the null hypothesis is greater than the significance level if the hypothesis is not true. [See also Daly (1940).]

9.3. THE DISTRIBUTION OF THE LIKELIHOOD RATIO CRITERION WHEN THE NULL HYPOTHESIS IS TRUE

9.3.1. Characterization of the Distribution

We shall show that under the null hypothesis the distribution of the criterion V is the distribution of a product of independent variables, each of which has the distribution of a criterion U for the linear hypothesis (Section 8.4).

Let

$$(1)\qquad V_i = \frac{\begin{vmatrix} A_{11} & \cdots & A_{1,i-1} & A_{1,i} \\ \vdots & & \vdots & \vdots \\ A_{i-1,1} & \cdots & A_{i-1,i-1} & A_{i-1,i} \\ A_{i1} & \cdots & A_{i,i-1} & A_{ii} \end{vmatrix}}{\begin{vmatrix} A_{11} & \cdots & A_{1,i-1} \\ \vdots & & \vdots \\ A_{i-1,1} & \cdots & A_{i-1,i-1} \end{vmatrix} \cdot |A_{ii}|}, \qquad i = 2, \ldots, q.$$

Then $V = V_2 V_3 \cdots V_q$. Note that V_i is the $N/2$th root of the likelihood ratio criterion for testing the null hypothesis

$$H_i : \boldsymbol{\Sigma}_{1,i} = \mathbf{0}, \dots, \boldsymbol{\Sigma}_{i-1,i} = \mathbf{0}, \tag{2}$$

that is, that $\boldsymbol{X}^{(i)}$ is independent of $(\boldsymbol{X}^{(1)\prime}, \dots, \boldsymbol{X}^{(i-1)\prime})'$. The null hypothesis H is the intersection of these hypotheses.

THEOREM 9.3.1. *When H_i is true, V_i has the distribution of $U_{p_i, \bar{p}_i, n-\bar{p}_i}$, where $n = N - 1$ and $\bar{p}_i = p_1 + \cdots + p_{i-1}$, $i = 2, \dots, q$.*

PROOF. The matrix $\boldsymbol{A}$ has the distribution of $\sum_{\alpha=1}^{n} \boldsymbol{Z}_\alpha \boldsymbol{Z}_\alpha'$, where $\boldsymbol{Z}_1, \dots, \boldsymbol{Z}_n$ are independently distributed according to $N(\mathbf{0}, \boldsymbol{\Sigma})$ and $\boldsymbol{Z}_\alpha$ is partitioned as $(\boldsymbol{Z}_\alpha^{(1)\prime}, \dots, \boldsymbol{Z}_\alpha^{(q)\prime})'$. Then conditional on $\boldsymbol{Z}_\alpha^{(1)} = \boldsymbol{z}_\alpha^{(1)}, \dots, \boldsymbol{Z}_\alpha^{(i-1)} = \boldsymbol{z}_\alpha^{(i-1)}$, $\alpha = 1, \dots, n$, the subvectors $\boldsymbol{Z}_1^{(i)}, \dots, \boldsymbol{Z}_n^{(i)}$ are independently distributed, $\boldsymbol{Z}_\alpha^{(i)}$ having a normal distribution with mean

$$\boldsymbol{\beta}_i \begin{pmatrix} \boldsymbol{z}_\alpha^{(1)} \\ \vdots \\ \boldsymbol{z}_\alpha^{(i-1)} \end{pmatrix} \tag{3}$$

and covariance matrix

$$\boldsymbol{\Sigma}_{ii} - \boldsymbol{\beta}_i \begin{pmatrix} \boldsymbol{\Sigma}_{11} & \cdots & \boldsymbol{\Sigma}_{1,i-1} \\ \vdots & & \vdots \\ \boldsymbol{\Sigma}_{i-1,1} & \cdots & \boldsymbol{\Sigma}_{i-1,i-1} \end{pmatrix} \boldsymbol{\beta}_i', \tag{4}$$

where

$$\boldsymbol{\beta}_i = (\boldsymbol{\Sigma}_{i1} \cdots \boldsymbol{\Sigma}_{i,i-1}) \begin{pmatrix} \boldsymbol{\Sigma}_{11} & \cdots & \boldsymbol{\Sigma}_{1,i-1} \\ \vdots & & \vdots \\ \boldsymbol{\Sigma}_{i-1,1} & \cdots & \boldsymbol{\Sigma}_{i-1,i-1} \end{pmatrix}^{-1}. \tag{5}$$

When the null hypothesis is not assumed, the estimator of $\boldsymbol{\beta}_i$ is (5) with $\boldsymbol{\Sigma}_{jk}$ replaced by $\boldsymbol{A}_{jk}$, and the estimator of (4) is (4) with $\boldsymbol{\Sigma}_{jk}$ replaced by $(1/n)\boldsymbol{A}_{jk}$ and $\boldsymbol{\beta}_i$ replaced by its estimator. Under $H_i : \boldsymbol{\beta}_i = \mathbf{0}$ and the covariance matrix (4) is $\boldsymbol{\Sigma}_{ii}$, which is estimated by $(1/n)\boldsymbol{A}_{ii}$. The $N/2$th root of the likelihood ratio criterion for H_i is

$$
(6)\quad \frac{\left| \boldsymbol{A}_{ii} - (\boldsymbol{A}_{i1}, \dots, \boldsymbol{A}_{i,i-1}) \begin{pmatrix} \boldsymbol{A}_{11} & \cdots & \boldsymbol{A}_{1,i-1} \\ \vdots & & \vdots \\ \boldsymbol{A}_{i-1,1} & \cdots & \boldsymbol{A}_{i-1,i-1} \end{pmatrix}^{-1} \begin{pmatrix} \boldsymbol{A}_{1i} \\ \vdots \\ \boldsymbol{A}_{i-1,i} \end{pmatrix} \right|}{|\boldsymbol{A}_{ii}|}
$$

$$
= \frac{\begin{vmatrix} \boldsymbol{A}_{11} & \cdots & \boldsymbol{A}_{1,i-1} & \boldsymbol{A}_{1i} \\ \vdots & & \vdots & \vdots \\ \boldsymbol{A}_{i-1,1} & \cdots & \boldsymbol{A}_{i-1,i-1} & \boldsymbol{A}_{i-1,i} \\ \boldsymbol{A}_{i1} & \cdots & \boldsymbol{A}_{i,i-1} & \boldsymbol{A}_{ii} \end{vmatrix}}{\begin{vmatrix} \boldsymbol{A}_{11} & \cdots & \boldsymbol{A}_{1,i-1} \\ \vdots & & \vdots \\ \boldsymbol{A}_{i-1,1} & \cdots & \boldsymbol{A}_{i-1,i-1} \end{vmatrix} \cdot |\boldsymbol{A}_{ii}|}
$$

which is V_i. This is the U-statistic for p_i dimensions, $\bar{p}_i$ components of the conditioning vector and $n - \bar{p}_i$ degrees of freedom in the estimator of the covariance matrix. Q.E.D.

THEOREM 9.3.2. *The distribution of V under the null hypothesis is the distribution of $V_2 V_3 \cdots V_q$, where $V_2, \dots, V_q$ are independently distributed with V_i having the distribution of $U_{p_i, \bar{p}_i, n-\bar{p}_i}$, where $\bar{p}_i = p_1 + \cdots + p_{i-1}$.*

PROOF. From the proof of Theorem 9.3.1, we see that the distribution of V_i is that of $U_{p_i, \bar{p}_i, n-\bar{p}_i}$ not depending on the conditioning $z_\alpha^{(k)}$, $k = 1, \dots, i-1$, $\alpha = 1, \dots, n$. Hence the distribution of V_i does not depend on $V_2, \dots, V_{i-1}$. Q.E.D.

THEOREM 9.3.3. *Under the null hypothesis V is distributed as $\prod_{i=2}^{q} \prod_{j=1}^{p_i} X_{ij}$, where the X_{ij}'s are independent and X_{ij} has the density $\beta[x|\frac{1}{2}(n - \bar{p}_i + 1 - j), \frac{1}{2}p_i]$.*

PROOF. This theorem follows from Theorems 9.3.2 and 8.4.1. Q.E.D.

9.3.2. Moments

THEOREM 9.3.4. *When the null hypothesis is true, the* hth *moment of the criterion is*

$$\mathscr{E}V^h = \prod_{i=2}^{q}\left\{\prod_{j=1}^{p_i}\frac{\Gamma\left[\frac{1}{2}(n-\bar{p}_i+1-j)+h\right]\Gamma\left[\frac{1}{2}(n+1-j)\right]}{\Gamma\left[\frac{1}{2}(n-\bar{p}_i+1-j)\right]\Gamma\left[\frac{1}{2}(n+1-j)+h\right]}\right\}. \tag{7}$$

PROOF. Because $V_2, \dots, V_q$ are independent

$$\mathscr{E}V^h = \mathscr{E}V_2^h \mathscr{E}V_3^h \cdots \mathscr{E}V_q^h. \tag{8}$$

Theorem 9.3.2 implies $\mathscr{E}V_i^h = \mathscr{E}U^h_{p_i, \bar{p}_i, n-\bar{p}_i}$. Then the theorem follows by substituting from Theorem 8.4.3. Q.E.D.

If the p_i are even, say $p_i = 2r_i$, $i > 1$, then by using the duplication formula $\Gamma(\alpha + \frac{1}{2})\Gamma(\alpha + 1) = \sqrt{\pi}\,\Gamma(2\alpha + 1)2^{-2\alpha}$ for the gamma function we can reduce the hth moment of V to

$$\begin{aligned}\mathscr{E}V^h &= \prod_{i=2}^{q}\left\{\prod_{k=1}^{r_i}\frac{\Gamma(n+1-\bar{p}_i-2k+2h)\Gamma(n+1-2k)}{\Gamma(n+1-\bar{p}_i-2k)\Gamma(n+1-2k+2h)}\right\} \\ &= \prod_{i=2}^{q}\left\{\prod_{k=1}^{r_i} B^{-1}(n+1-\bar{p}_i-2k, \bar{p}_i) \right. \\ &\qquad \left. \times \int_0^1 x^{n+1-\bar{p}_i-2k+2h-1}(1-x)^{\bar{p}_i-1}\,dx\right\}.\end{aligned} \tag{9}$$

Thus V is distributed as $\prod_{i=2}^{q}\{\prod_{k=1}^{r_i} Y_{ik}^2\}$ where the Y_{ik} are independent, and Y_{ik} has density $\beta(y; n+1-\bar{p}_i-2k, \bar{p}_i)$.

In general, the duplication formula for the gamma function can be used to reduce the moments as indicated in Section 8.4.

9.3.3. Some Special Distributions

If $q = 2$, V is distributed as $U_{p_2, p_1, n-p_1}$. Special cases have been treated in Section 8.4, and references to the literature given. The distribution for $p_1 = p_2 = p_3 = 1$ is given in Problem 2 and for $p_1 = p_2 = p_3 = 2$ in Problem 3. Wilks

(1935) has given the distributions for $p_1 = p_2 = 1$, $p_3 = p - 2$,† for $p_1 = 1$, $p_2 = p_3 = 2$, for $p_1 = 1$, $p_2 = 2$, $p_3 = 3$, for $p_1 = 1$, $p_2 = 2$, $p_3 = 4$, and for $p_1 = p_2 = 2$, $p_3 = 3$. Consul (1967a) has treated the case $p_1 = 2$, $p_2 = 3$, p_3 even.

Wald and Brookner (1941) have given a method for deriving the distribution if not more than one p_i is odd. It can be seen that the same result can be obtained by integration of products of beta functions after using the duplication formula to reduce the moments.

Mathai and Saxena (1973) have given the exact distribution for the general case. Mathai and Katiyar (1979) have given exact significance points for $p = 3(1)10$ and $n = 3(1)20$ for significance levels of 5% and 1% (of $-k \log V$ of Section 9.4).

9.4. AN ASYMPTOTIC EXPANSION OF THE DISTRIBUTION OF THE LIKELIHOOD RATIO CRITERION

The hth moment of $\lambda = V^{\frac{1}{2}N}$ is

$$\mathscr{E}\lambda^h = K \frac{\prod_{i=1}^{p} \Gamma\{\frac{1}{2}[N(1+h) - i]\}}{\prod_{i=1}^{q}\left\{\prod_{j=1}^{p_i} \Gamma\{\frac{1}{2}[N(1+h) - j]\}\right\}}, \tag{1}$$

where K is chosen so that $\mathscr{E}\lambda^0 = 1$. This is of the form of (1) of Section 8.5 with

$$a = p, \qquad b = p, \qquad x_k = \frac{N}{2}, \qquad \xi_k = \frac{-k}{2}, \quad k = 1, \dots, p,$$

$$y_j = \frac{N}{2}, \qquad \eta_j = \frac{-j + p_1 + \cdots + p_{i-1}}{2}, \tag{2}$$

$$j = p_1 + \cdots + p_{i-1} + 1, \dots, p_1 + \cdots + p_i; \; i = 1, \dots, q.$$

Then $f = \frac{1}{2}[p(p+1) - \Sigma p_i(p_i + 1)] = \frac{1}{2}(p^2 - \Sigma p_i^2)$, $\beta_k = \varepsilon_j = \frac{1}{2}(1-\rho)N$.

In order to make the second term in the expansion vanish we take ρ as

$$\rho = 1 - \frac{2(p^3 - \Sigma p_i^3) + 9(p^2 - \Sigma p_i^2)}{6N(p^2 - \Sigma p_i^2)}. \tag{3}$$

†In Wilks' formula $\Gamma[\frac{1}{2}(N - 2 - i)]$ should be $\Gamma[\frac{1}{2}(n - 2 - i)]$.

Let

$$k = \rho N = N - \frac{3}{2} - \frac{p^3 - \Sigma p_i^3}{3(p^2 - \Sigma p_i^2)}. \tag{4}$$

Then $\omega_2 = \gamma_2/k^2$, where [as shown by Box (1949)]

$$\gamma_2 = \frac{p^4 - \Sigma p_i^4}{48} - \frac{5(p^2 - \Sigma p_i^2)}{96} - \frac{(p^3 - \Sigma p_i^3)^2}{72(p^2 - \Sigma p_i^2)}. \tag{5}$$

We obtain from Section 8.5 the following expansion:

$$\Pr\{-k \log V \le v\} = \Pr\{\chi_f^2 \le v\} + \frac{\gamma_2}{k^2}\left[\Pr\{\chi_{f+4}^2 \le v\} - \Pr\{\chi_f^2 \le v\}\right] + O(k^{-3}). \tag{6}$$

If $q = 2$, we obtain further terms in the expansion by using the results of Section 8.5.

If $p_i = 1$, we have

$$\begin{aligned} f &= \tfrac{1}{2}p(p-1), \\ k &= N - \frac{2p+11}{6}, \\ \gamma_2 &= \frac{p(p-1)}{288}(2p^2 - 2p - 13), \\ \gamma_3 &= \frac{p(p-1)}{3240}(p-2)(2p-1)(p+1); \end{aligned} \tag{7}$$

other terms are given by Box (1949). If $p_i = 2$ $(p = 2q)$

$$\begin{aligned} f &= 2q(q-1), \\ k &= N - \frac{4q+13}{6}, \\ \gamma_2 &= \frac{q(q-1)}{72}(8q^2 - 8q - 7). \end{aligned} \tag{8}$$

Table 9.1 gives an indication of the order of approximation of (6) for $p_i = 1$. In each case v is chosen so that the first term is 0.95.

TABLE 9.1

p	f	v	γ_2	N	k	γ_2/k^2	Second Term
4	6	12.592	$\frac{11}{24}$	15	$\frac{71}{6}$	0.0033	-0.0007
5	10	18.307	$\frac{15}{8}$	15	$\frac{69}{6}$	0.0142	-0.0021
6	15	24.996	$\frac{235}{48}$	15	$\frac{67}{6}$	0.0393	-0.0043
				16	$\frac{73}{6}$	0.0331	-0.0036

If $q = 2$, the approximate distributions given in Sections 8.5.3 and 8.5.4 are available. [See also Nagao (1973c).]

9.5. OTHER CRITERIA

In case $q = 2$, the criteria considered in Section 8.6 can be used with $\boldsymbol{G} + \boldsymbol{H}$ replaced by $\boldsymbol{A}_{11}$ and $\boldsymbol{H}$ replaced by $\boldsymbol{A}_{12}\boldsymbol{A}_{22}^{-1}\boldsymbol{A}_{21}$ or $\boldsymbol{G} + \boldsymbol{H}$ replaced by $\boldsymbol{A}_{22}$ and $\boldsymbol{H}$ replaced by $\boldsymbol{A}_{21}\boldsymbol{A}_{11}^{-1}\boldsymbol{A}_{12}$.

The null hypothesis of independence is that $\boldsymbol{\Sigma} - \boldsymbol{\Sigma}_0 = \boldsymbol{0}$, where $\boldsymbol{\Sigma}_0$ is defined in (6) of Section 9.2. An appropriate test procedure will reject the null hypothesis if the elements of $\boldsymbol{A} - \boldsymbol{A}_0$ are large compared to the elements of the diagonal blocks of $\boldsymbol{A}_0$ (where $\boldsymbol{A}_0$ is composed of diagonal blocks $\boldsymbol{A}_{ii}$ and off-diagonal blocks of $\boldsymbol{0}$). Let the nonsingular matrix $\boldsymbol{B}_{ii}$ be such that $\boldsymbol{B}_{ii}\boldsymbol{A}_{ii}\boldsymbol{B}_{ii}' = \boldsymbol{I}$, that is, $\boldsymbol{A}_{ii}^{-1} = \boldsymbol{B}_{ii}'\boldsymbol{B}_{ii}$, and let $\boldsymbol{B}_0$ be the matrix with $\boldsymbol{B}_{ii}$ as the ith diagonal block and $\boldsymbol{0}$'s as off-diagonal blocks. Then $\boldsymbol{B}_0\boldsymbol{A}_0\boldsymbol{B}_0' = \boldsymbol{I}$ and

$$(1)\quad \boldsymbol{B}_0(\boldsymbol{A} - \boldsymbol{A}_0)\boldsymbol{B}_0' = \begin{pmatrix} \boldsymbol{0} & \boldsymbol{B}_{11}\boldsymbol{A}_{12}\boldsymbol{B}_{22}' & \cdots & \boldsymbol{B}_{11}\boldsymbol{A}_{1q}\boldsymbol{B}_{qq}' \\ \boldsymbol{B}_{22}\boldsymbol{A}_{21}\boldsymbol{B}_{11}' & \boldsymbol{0} & & \boldsymbol{B}_{22}\boldsymbol{A}_{2q}\boldsymbol{B}_{qq}' \\ \vdots & \vdots & & \vdots \\ \boldsymbol{B}_{qq}\boldsymbol{A}_{q1}\boldsymbol{B}_{11}' & \boldsymbol{B}_{qq}\boldsymbol{A}_{q2}\boldsymbol{B}_{22}' & \cdots & \boldsymbol{0} \end{pmatrix}.$$

This matrix is invariant with respect to transformations (24) of Section 9.2 operating on $\boldsymbol{A}$. A different choice of $\boldsymbol{B}_{ii}$ amounts to multiplying (1) on the left by $\boldsymbol{Q}_0$ and on the right by $\boldsymbol{Q}_0'$, where $\boldsymbol{Q}_0$ is a matrix with orthogonal diagonal blocks and off-diagonal blocks of $\boldsymbol{0}$'s. A test procedure should reject the null hypothesis if some measure of the numerical values of the elements of (1) is too large. The likelihood ratio criterion is the $N/2$ power of $|\boldsymbol{B}_0(\boldsymbol{A} - \boldsymbol{A}_0)\boldsymbol{B}_0' + \boldsymbol{I}| = |\boldsymbol{B}_0 \boldsymbol{A} \boldsymbol{B}_0'|$.

Another measure, suggested by Nagao (1973a), is

$$(2) \quad \tfrac{1}{2}\operatorname{tr}\left[\boldsymbol{B}_0(\boldsymbol{A} - \boldsymbol{A}_0)\boldsymbol{B}_0'\right]^2 = \tfrac{1}{2}\operatorname{tr}\left[(\boldsymbol{A} - \boldsymbol{A}_0)\boldsymbol{A}_0^{-1}\right]^2 = \tfrac{1}{2}\operatorname{tr}\left(\boldsymbol{A}\boldsymbol{A}_0^{-1} - \boldsymbol{I}\right)^2$$

$$= \frac{1}{2} \sum_{\substack{i,j=1 \\ i \neq j}}^{q} \operatorname{tr} \boldsymbol{A}_{ij} \boldsymbol{A}_{jj}^{-1} \boldsymbol{A}_{ji} \boldsymbol{A}_{ii}^{-1}.$$

For $q = 2$ this measure is the average of the Bartlett–Nanda–Pillai trace criterion with $\boldsymbol{G} + \boldsymbol{H}$ replaced by $\boldsymbol{A}_{11}$ and $\boldsymbol{H}$ replaced by $\boldsymbol{A}_{12}\boldsymbol{A}_{22}^{-1}\boldsymbol{A}_{21}$ and the criterion with $\boldsymbol{G} + \boldsymbol{H}$ replaced by $\boldsymbol{A}_{22}$ and $\boldsymbol{H}$ replaced by $\boldsymbol{A}_{21}\boldsymbol{A}_{11}^{-1}\boldsymbol{A}_{12}$.

This criterion multiplied by n or N has a limiting χ^2-distribution with number of degrees of freedom $f = \frac{1}{2}(p^2 - \Sigma_{i=1}^{q} p_i^2)$, which is the same number as for $-N \log V$. Nagao has obtained an asymptotic expansion of the distribution:

$$(3) \quad \Pr\left\{\tfrac{1}{2} n \operatorname{tr}\left(\boldsymbol{A}\boldsymbol{A}_0^{-1} - \boldsymbol{I}\right)^2 \le x\right\}$$

$$= \Pr\left\{\chi_f^2 \le x\right\}$$

$$+ \frac{1}{n}\left[\frac{1}{12}\left(p^3 - 3p\sum_{i=1}^{q} p_i^2 + 2\sum_{i=1}^{q} p_i^3\right)\Pr\left\{\chi_{f+6}^2 \le x\right\}\right.$$

$$+ \frac{1}{8}\left(-2p^3 + 4p\sum_{i=1}^{q} p_i^2 - 2\sum_{i=1}^{q} p_i^3 - p^2 + \sum_{i=1}^{q} p_i^2\right)\Pr\left\{\chi_{f+4}^2 \le x\right\}$$

$$+ \frac{1}{4}\left(p^3 - p\sum_{i=1}^{q} p_i^2 + p^2 - \sum_{i=1}^{q} p_i^3\right)\Pr\left\{\chi_{f+2}^2 \le x\right\}$$

$$\left. - \frac{1}{24}\left(2p^3 - 2\sum_{i=1}^{q} p_i^3 + 3p^2 - 3\sum_{i=1}^{q} p_i^2\right)\Pr\left\{\chi_f^2 \le x\right\}\right] + O(n^{-2}).$$

9.6. STEP-DOWN PROCEDURES

9.6.1. Step-down by Blocks

It was shown in Section 9.3 that the $N/2$th root of the likelihood ratio criterion, namely V, is the product of $q-1$ of these criteria, that is, $V_2, \dots, V_q$. The ith subcriterion V_i provides a likelihood ratio test of the hypothesis H_i [(2) of Section 9.3] that the ith subvector is independent of the preceding $i-1$ subvectors. Under the null hypothesis H [$= \cap_{i=2}^{q} H_i$], these $q-1$ criteria are independent (Theorem 9.3.2). A step-down testing procedure is to accept the null hypothesis if

$$V_i \geq v_i(\varepsilon_i), \qquad i = 2, \dots, q, \tag{1}$$

and reject the null hypothesis if $V_i < v_i(\varepsilon_i)$ for any i. Here $v_i(\varepsilon_i)$ is the number such that the probability of (1) when H_i is true is $1 - \varepsilon_i$. The significance level of the procedure is ε satisfying

$$1 - \varepsilon = \prod_{i=2}^{q} (1 - \varepsilon_i). \tag{2}$$

The subtests can be done sequentially, say, in the order $2, \dots, q$. As soon as a subtest calls for rejection, the procedure is terminated; if no subtest leads to rejection, H is accepted. The ordering of the subvectors is at the discretion of the investigator as well as the ordering of the tests.

Suppose, for example, that measurements on an individual are grouped into physiological measurements, measurements of intelligence, and measurements of emotional characteristics. One could test that intelligence is independent of physiology and then that emotions are independent of physiology and intelligence, or the order of these could be reversed. Alternatively, one could test that intelligence is independent of emotions and then that physiology is independent of these two aspects, or the order reversed. There is a third pair of procedures.

Other criteria for the linear hypothesis discussed in Section 8.6 can be used to test the component hypotheses $H_2, \dots, H_q$ in a similar fashion. When H_i is true, the criterion is distributed independently of $\boldsymbol{X}_\alpha^{(1)}, \dots, \boldsymbol{X}_\alpha^{(i-1)}$, $\alpha = 1, \dots, N$, and hence independently of the criteria for $H_2, \dots, H_{i-1}$.

9.6.2. Step-down by Components

In Section 8.4.5 we discussed a componentwise step-down procedure for testing that a submatrix of regression coefficients was a specified matrix. We adapt this procedure to test the null hypothesis H_i cast in the form

$$(3)\quad H_i: (\boldsymbol{\Sigma}_{i1}\ \boldsymbol{\Sigma}_{i2}\ \cdots\ \boldsymbol{\Sigma}_{i,i-1}) \begin{pmatrix} \boldsymbol{\Sigma}_{11} & \boldsymbol{\Sigma}_{12} & \cdots & \boldsymbol{\Sigma}_{1,i-1} \\ \boldsymbol{\Sigma}_{21} & \boldsymbol{\Sigma}_{22} & \cdots & \boldsymbol{\Sigma}_{2,i-1} \\ \vdots & \vdots & & \vdots \\ \boldsymbol{\Sigma}_{i-1,1} & \boldsymbol{\Sigma}_{i-1,2} & \cdots & \boldsymbol{\Sigma}_{i-1,i-1} \end{pmatrix}^{-1} = \mathbf{0},$$

where $\mathbf{0}$ is of order $p_i \times \bar{p}_i$. The matrix in (3) consists of the coefficients of the regression of $\boldsymbol{X}^{(i)}$ on $(\boldsymbol{X}^{(1)\prime}, \dots, \boldsymbol{X}^{(i-1)\prime})'$.

For $i = 2$, we test in sequence whether the regression of X_{p_1+1} on $\boldsymbol{X}^{(1)} = (X_1, \dots, X_{p_1})'$ is $\mathbf{0}$, whether the regression of X_{p_1+2} on $\boldsymbol{X}^{(1)}$ is $\mathbf{0}$ in the regression of X_{p_1+2} on $\boldsymbol{X}^{(1)}$ and $X_{p_1+1}, \dots,$ whether the regression of $X_{p_1+p_2}$ on $\boldsymbol{X}^{(1)}$ is $\mathbf{0}$ in the regression of $X_{p_1+p_2}$ on $\boldsymbol{X}^{(1)}$, $X_{p_1+1}, \dots, X_{p_1+p_2-1}$. These hypotheses are equivalently that the first, second, ..., and p_2th rows of the matrix in (3) for $i = 2$ are $\mathbf{0}$ vectors.

Let $\boldsymbol{A}_{ii}^{(k)}$ be the $k \times k$ matrix in the upper left-hand corner of $\boldsymbol{A}_{ii}$, let $\boldsymbol{A}_{ij}^{(k)}$ consist of the upper k rows of $\boldsymbol{A}_{ij}$, and let $\boldsymbol{A}_{ji}^{(k)}$ consist of the first k columns of $\boldsymbol{A}_{ji}$, $k = 1, \dots, p_i$. Then the criterion for testing that the first row of (3) is $\mathbf{0}$ is

(4)

$$X_{i1} = \left| \boldsymbol{A}_{ii}^{(1)} - \left(\boldsymbol{A}_{i1}^{(1)} \cdots \boldsymbol{A}_{i,i-1}^{(1)}\right) \begin{pmatrix} \boldsymbol{A}_{11} & \cdots & \boldsymbol{A}_{1,i-1} \\ \vdots & & \vdots \\ \boldsymbol{A}_{i-1,1} & \cdots & \boldsymbol{A}_{i-1,i-1} \end{pmatrix}^{-1} \begin{pmatrix} \boldsymbol{A}_{1i}^{(1)} \\ \vdots \\ \boldsymbol{A}_{i-1,i}^{(1)} \end{pmatrix} \right| \div \boldsymbol{A}_{ii}^{(1)}$$

$$= \frac{\begin{vmatrix} \boldsymbol{A}_{11} & \cdots & \boldsymbol{A}_{1,i-1} & \boldsymbol{A}_{1i}^{(1)} \\ \vdots & & \vdots & \vdots \\ \boldsymbol{A}_{i-1,1} & \cdots & \boldsymbol{A}_{i-1,i-1} & \boldsymbol{A}_{i-1,i}^{(1)} \\ \boldsymbol{A}_{i1}^{(1)} & \cdots & \boldsymbol{A}_{i,i-1}^{(1)} & \boldsymbol{A}_{ii}^{(1)} \end{vmatrix}}{\begin{vmatrix} \boldsymbol{A}_{11} & \cdots & \boldsymbol{A}_{1,i-1} \\ \vdots & & \vdots \\ \boldsymbol{A}_{i-1,1} & \cdots & \boldsymbol{A}_{i-1,i-1} \end{vmatrix} \boldsymbol{A}_{ii}^{(1)}}.$$

For $k > 1$, the criterion for testing that the kth row of the matrix in (3) is $\mathbf{0}$ is [see (8) in Section 8.4]

(5)

$$X_{ik} = \frac{\left| \boldsymbol{A}_{ii}^{(k)} - \left(\boldsymbol{A}_{i1}^{(k)} \cdots \boldsymbol{A}_{i,i-1}^{(k)} \right) \begin{pmatrix} \boldsymbol{A}_{11} & \cdots & \boldsymbol{A}_{1,i-1} \\ \vdots & & \vdots \\ \boldsymbol{A}_{i-1,1} & \cdots & \boldsymbol{A}_{i-1,i-1} \end{pmatrix}^{-1} \begin{pmatrix} \boldsymbol{A}_{1i}^{(k)} \\ \vdots \\ \boldsymbol{A}_{1,i-1}^{(k)} \end{pmatrix} \right|}{\left| \boldsymbol{A}_{ii}^{(k-1)} - \left(\boldsymbol{A}_{i1}^{(k-1)} \cdots \boldsymbol{A}_{i,i-1}^{(k-1)} \right) \begin{pmatrix} \boldsymbol{A}_{11} & \cdots & \boldsymbol{A}_{1,i-1} \\ \vdots & & \vdots \\ \boldsymbol{A}_{i-1,1} & \cdots & \boldsymbol{A}_{i-1,i-1} \end{pmatrix}^{-1} \begin{pmatrix} \boldsymbol{A}_{1i}^{(k-1)} \\ \vdots \\ \boldsymbol{A}_{1,i-1}^{(k-1)} \end{pmatrix} \right|}$$

$$\div \frac{|\boldsymbol{A}_{ii}^{(k)}|}{|\boldsymbol{A}_{ii}^{(k-1)}|}$$

$$= \frac{\begin{vmatrix} \boldsymbol{A}_{11} & \cdots & \boldsymbol{A}_{1,i-1} & \boldsymbol{A}_{1i}^{(k)} \\ \vdots & & \vdots & \vdots \\ \boldsymbol{A}_{i-1,1} & \cdots & \boldsymbol{A}_{i-1,i-1} & \boldsymbol{A}_{i-1,i}^{(k)} \\ \boldsymbol{A}_{i1}^{(k)} & \cdots & \boldsymbol{A}_{i,i-1}^{(k)} & \boldsymbol{A}_{ii}^{(k)} \end{vmatrix}}{\begin{vmatrix} \boldsymbol{A}_{11} & \cdots & \boldsymbol{A}_{1,i-1} & \boldsymbol{A}_{1i}^{(k-1)} \\ \vdots & & \vdots & \vdots \\ \boldsymbol{A}_{i-1,1} & \cdots & \boldsymbol{A}_{i-1,i-1} & \boldsymbol{A}_{i-1,i}^{(k-1)} \\ \boldsymbol{A}_{i1}^{(k-1)} & \cdots & \boldsymbol{A}_{i,i-1}^{(k-1)} & \boldsymbol{A}_{ii}^{(k-1)} \end{vmatrix}} \cdot \frac{|\boldsymbol{A}_{ii}^{(k-1)}|}{|\boldsymbol{A}_{ii}^{(k)}|},$$

$$k = 2, \ldots, p_i, \qquad i = 2, \ldots, q.$$

Under the null hypothesis the criterion has the beta-density $\beta[x; \frac{1}{2}(n - \bar{p}_i + 1 - j), \frac{1}{2}p_i]$. For given i, the criteria $X_{i1}, \ldots, X_{ip_i}$ are independent (Theorem 8.4.1). The sets for different i are independent by the argument in Section 9.6.1.

A step-down procedure consists of a sequence of tests based on $X_{21}, \ldots, X_{2p_2}, X_{31}, \ldots, X_{qp_q}$. A particular component test leads to rejection if

$$(6) \qquad \frac{1 - X_{ij}}{X_{ij}} \frac{n - \bar{p}_i + 1 - j}{p_i} > F_{p_i, n - \bar{p}_i + 1 - j}(\varepsilon_{ij}).$$

The significance level is ε, where

$$(7) \qquad 1 - \varepsilon = \prod_{i=2}^{q} \prod_{j=1}^{p_i} (1 - \varepsilon_{ij}).$$

The sequence of subvectors and the sequence of components within each subvector is at the discretion of the investigator.

The criterion V_i for testing H_i is $V_i = \prod_{k=1}^{p_i} X_{ik}$, and the criterion

$$(8) \qquad V = \prod_{i=2}^{q} V_i = \prod_{i=2}^{q} \prod_{k=1}^{p_i} X_{ik}.$$

These are the random variables described in Theorem 9.3.3.

9.7. AN EXAMPLE

We take the following example from an industrial time study [Abruzzi (1950)]. The purpose of the study was to investigate the length of time taken by various operators in a garment factory to do several elements of a pressing operation. The entire pressing operation was divided into the following six elements:

1. Pick up and position garment.
2. Press and repress short dart.
3. Reposition garment on ironing board.
4. Press three quarters of length of long dart.
5. Press balance of long dart.
6. Hang garment on rack.

In this case $\boldsymbol{x}_\alpha$ is the vector of measurements on individual α. The component $x_{i\alpha}$ is the time taken to do the ith element of the operation. N is 76. The data (in seconds) are summarized in the sample mean vector and covariance matrix:

$$\bar{\boldsymbol{x}} = \begin{pmatrix} 9.47 \\ 25.56 \\ 13.25 \\ 31.44 \\ 27.29 \\ 8.80 \end{pmatrix}, \tag{1}$$

$$\boldsymbol{S} = \begin{pmatrix} 2.57 & 0.85 & 1.56 & 1.79 & 1.33 & 0.42 \\ 0.85 & 37.00 & 3.34 & 13.47 & 7.59 & 0.52 \\ 1.56 & 3.34 & 8.44 & 5.77 & 2.00 & 0.50 \\ 1.79 & 13.47 & 5.77 & 34.01 & 10.50 & 1.77 \\ 1.33 & 7.59 & 2.00 & 10.50 & 23.01 & 3.43 \\ 0.42 & 0.52 & 0.50 & 1.77 & 3.43 & 4.59 \end{pmatrix}. \tag{2}$$

The sample standard deviations are (1.604, 6.041, 2.903, 5.832, 4.798, 2.141). The sample correlation matrix is

$$\boldsymbol{R} = \begin{pmatrix} 1.000 & 0.088 & 0.334 & 0.191 & 0.173 & 0.123 \\ 0.088 & 1.000 & 0.186 & 0.384 & 0.262 & 0.040 \\ 0.334 & 0.186 & 1.000 & 0.343 & 0.144 & 0.080 \\ 0.191 & 0.384 & 0.343 & 1.000 & 0.375 & 0.142 \\ 0.173 & 0.262 & 0.144 & 0.375 & 1.000 & 0.334 \\ 0.123 & 0.040 & 0.080 & 0.142 & 0.334 & 1.000 \end{pmatrix}. \tag{3}$$

The investigators are interested in testing the hypothesis that the six variates are mutually independent. It often happens in time studies that a new operation is proposed in which the elements are combined in a different way; the new operation may use some of the elements several times and some elements may be omitted. If the times for the different elements in the operation for which data are available are independent, it may reasonably be assumed that they will be independent in a new operation. Then the distribution of time for the new operation can be estimated by using the means and variances of the individual items.

In this problem the criterion V is $V = |\boldsymbol{R}| = 0.472$. Since the sample size is large we can use asymptotic theory: $k = 433/6$, $f = 15$, and $-k \log V = 54.1$. Since the significance point for the χ^2-distribution with 15 degrees of freedom is 30.6 at the 0.01 significance level, we find the result significant. We reject the hypothesis of independence; we cannot consider the times of the elements independent.

9.8. THE CASE OF TWO SETS OF VARIATES

In the case of two sets of variates ($q = 2$), the random vector $\boldsymbol{X}$, the observation vector $\boldsymbol{x}_\alpha$, the mean vector $\boldsymbol{\mu}$, and the covariance matrix $\boldsymbol{\Sigma}$ are partitioned as follows:

$$(1) \qquad \begin{aligned} \boldsymbol{X} &= \begin{pmatrix} \boldsymbol{X}^{(1)} \\ \boldsymbol{X}^{(2)} \end{pmatrix}, & \boldsymbol{x}_\alpha &= \begin{pmatrix} \boldsymbol{x}_\alpha^{(1)} \\ \boldsymbol{x}_\alpha^{(2)} \end{pmatrix}, \\ \boldsymbol{\mu} &= \begin{pmatrix} \boldsymbol{\mu}^{(1)} \\ \boldsymbol{\mu}^{(2)} \end{pmatrix}, & \boldsymbol{\Sigma} &= \begin{pmatrix} \boldsymbol{\Sigma}_{11} & \boldsymbol{\Sigma}_{12} \\ \boldsymbol{\Sigma}_{21} & \boldsymbol{\Sigma}_{22} \end{pmatrix}. \end{aligned}$$

The null hypothesis of independence specifies that $\boldsymbol{\Sigma}_{12} = \mathbf{0}$; that is, that $\boldsymbol{\Sigma}$ is of the form

$$(2) \qquad \boldsymbol{\Sigma}_0 = \begin{pmatrix} \boldsymbol{\Sigma}_{11} & \mathbf{0} \\ \mathbf{0} & \boldsymbol{\Sigma}_{22} \end{pmatrix}.$$

The test criterion is

$$(3) \qquad V = \frac{|\boldsymbol{A}|}{|\boldsymbol{A}_{11}| \cdot |\boldsymbol{A}_{22}|}.$$

It was shown in Section 9.3 that when the null hypothesis is true, this criterion is distributed as $U_{p_1, p_2, N-1-p_2}$, the criterion for testing a hypothesis about regression coefficients (Chapter 8). We now wish to study further the relationship between testing the hypothesis of independence of two sets and testing the hypothesis that regression of one set on the other is zero.

The conditional distribution of $\boldsymbol{X}_\alpha^{(1)}$ given $\boldsymbol{X}_\alpha^{(2)} = \boldsymbol{x}_\alpha^{(2)}$ is $N[\boldsymbol{\mu}^{(1)} + \boldsymbol{\beta}(\boldsymbol{x}_\alpha^{(2)} - \boldsymbol{\mu}^{(2)}), \boldsymbol{\Sigma}_{11\cdot 2}] = N[\boldsymbol{\beta}(\boldsymbol{x}_\alpha^{(2)} - \bar{\boldsymbol{x}}^{(2)}) + \boldsymbol{\nu}, \boldsymbol{\Sigma}_{11\cdot 2}]$, where $\boldsymbol{\beta} = \boldsymbol{\Sigma}_{12}\boldsymbol{\Sigma}_{22}^{-1}$, $\boldsymbol{\Sigma}_{11\cdot 2} = \boldsymbol{\Sigma}_{11} - \boldsymbol{\Sigma}_{12}\boldsymbol{\Sigma}_{22}^{-1}\boldsymbol{\Sigma}_{21}$, and $\boldsymbol{\nu} = \boldsymbol{\mu}^{(1)} + \boldsymbol{\beta}(\bar{\boldsymbol{x}}^{(2)} - \boldsymbol{\mu}^{(2)})$. Let $\boldsymbol{X}_\alpha^* = \boldsymbol{X}_\alpha^{(1)}$, $\boldsymbol{z}_\alpha^{*\prime} = [(\boldsymbol{x}_\alpha^{(2)} - \bar{\boldsymbol{x}}^{(2)})' \ 1]$, $\boldsymbol{\beta}^* = (\boldsymbol{\beta} \ \boldsymbol{\nu})$, and $\boldsymbol{\Sigma}^* = \boldsymbol{\Sigma}_{11\cdot 2}$. Then the conditional distribution of $\boldsymbol{X}_\alpha^*$ is $N(\boldsymbol{\beta}^*\boldsymbol{z}_\alpha^*, \boldsymbol{\Sigma}^*)$. This is exactly the distribution studied in Chapter 8.

The null hypothesis that $\boldsymbol{\Sigma}_{12} = \mathbf{0}$ is equivalent to the null hypothesis $\boldsymbol{\beta} = \mathbf{0}$. Considering $\boldsymbol{x}_\alpha^{(2)}$ fixed, we know from Chapter 8 that the criterion (based on the likelihood ratio criterion) for testing this hypothesis is

$$(4) \qquad U = \frac{\left|\sum \left(\boldsymbol{x}_\alpha^* - \hat{\boldsymbol{\beta}}_\Omega^* \boldsymbol{z}_\alpha^*\right)\left(\boldsymbol{x}_\alpha^* - \hat{\boldsymbol{\beta}}_\Omega^* \boldsymbol{z}_\alpha^*\right)'\right|}{\left|\sum \left(\boldsymbol{x}_\alpha^* - \hat{\boldsymbol{\beta}}_{2\omega}^* \boldsymbol{z}_\alpha^{*(2)}\right)\left(\boldsymbol{x}_\alpha^* - \hat{\boldsymbol{\beta}}_{2\omega}^* \boldsymbol{z}_\alpha^{*(2)}\right)'\right|},$$

where

$$z_\alpha^{*(2)} = 1,$$

(5) $$\hat{\boldsymbol{\beta}}_{2\omega}^* = \hat{\nu} = \bar{x}^* = \bar{x}^{(1)},$$

$$\hat{\boldsymbol{\beta}}_\Omega^* = \left(\hat{\boldsymbol{\beta}}_{1\Omega}^* \ \hat{\boldsymbol{\beta}}_{2\Omega}^*\right)$$

$$= \left(\sum x_\alpha^* z_\alpha^{*(1)\prime} \ \sum x_\alpha^* z_\alpha^{*(2)\prime}\right) \begin{pmatrix} \sum z_\alpha^{*(1)} z_\alpha^{*(1)\prime} & \sum z_\alpha^{*(1)} z_\alpha^{*(2)\prime} \\ \sum z_\alpha^{*(2)} z_\alpha^{*(1)\prime} & \sum z_\alpha^{*(2)} z_\alpha^{*(2)\prime} \end{pmatrix}^{-1}$$

$$= \left(A_{12} \ N\bar{x}^{(1)}\right) \begin{pmatrix} A_{22} & \mathbf{0} \\ \mathbf{0} & N \end{pmatrix}^{-1}$$

$$= \left(A_{12} A_{22}^{-1} \ \bar{x}^{(1)}\right).$$

The matrix in the denominator of U is

(6) $$\sum_{\alpha=1}^{N} \left(x_\alpha^{(1)} - \bar{x}^{(1)}\right)\left(\bar{x}_\alpha^{(1)} - \bar{x}^{(1)}\right)' = A_{11}.$$

The matrix in the numerator is

(7) $$\sum_{\alpha=1}^{N} \left[x_\alpha^{(1)} - \bar{x}^{(1)} - A_{12} A_{22}^{-1}\left(x_\alpha^{(2)} - \bar{x}^{(2)}\right)\right]\left[x_\alpha^{(1)} - \bar{x}^{(1)} - A_{12} A_{22}^{-1}\left(x_\alpha^{(2)} - \bar{x}^{(2)}\right)\right]'$$
$$= A_{11} - A_{12} A_{22}^{-1} A_{21}.$$

Therefore,

(8) $$U = \frac{|A_{11} - A_{12} A_{22}^{-1} A_{21}|}{|A_{11}|} = \frac{|A|}{|A_{11}| \cdot |A_{22}|},$$

which is exactly V.

Now let us see why it is that when the null hypothesis is true the distribution of $U = V$ does not depend on whether the $X_\alpha^{(2)}$ are held fixed. It was shown in Chapter 8 that when the null hypothesis is true the distribution of U depends only on p, q_1 and $N - q_2$, but not on z_α. Thus the conditional distribution of V given $\boldsymbol{X}_\alpha^{(2)} = \boldsymbol{x}_\alpha^{(2)}$ does not depend on $\boldsymbol{x}_\alpha^{(2)}$; the joint distribution of V and $\boldsymbol{X}_\alpha^{(2)}$ is the product of the distribution of V and the distribution of $\boldsymbol{X}_\alpha^{(2)}$, and the marginal distribution of V is this conditional distribution. This shows that the distribution of V (under the null hypothesis) does not depend on whether the $\boldsymbol{X}_\alpha^{(2)}$ are fixed or have any distribution (normal or not).

We can extend this result to show that if $q > 2$, the distribution of V under the null hypothesis of independence does not depend on the distribution of one set of variates, say $\boldsymbol{X}_\alpha^{(1)}$. We have $V = V_2 \cdots V_q$, where V_i is defined in (1) of Section 9.3. When the null hypothesis is true, V_q is distributed independently of $\boldsymbol{X}_\alpha^{(1)}, \dots, \boldsymbol{X}_\alpha^{(q-1)}$ by the previous result. In turn we argue that V_j is distributed independently of $\boldsymbol{X}_\alpha^{(1)}, \dots, \boldsymbol{X}_\alpha^{(j-1)}$. Thus $V_2 \cdots V_q$ is distributed independently of $\boldsymbol{X}_\alpha^{(1)}$.

THEOREM 9.8.1. *Under the null hypothesis of independence, the distribution of V is that given earlier in this chapter if $q-1$ sets are jointly normally distributed, even though one set is not normally distributed.*

In the case of two sets of variates, we may be interested in a measure of association between the two sets which is a generalization of the correlation coefficient. The square of the correlation between two scalars X_1 and X_2 can be considered as the ratio of the variance of the regression of X_1 on X_2 to the variance of X_1; this is $\mathscr{V}(\beta X_2)/\mathscr{V}(X_1) = \beta^2\sigma_{22}/\sigma_{11} = (\sigma_{12}^2/\sigma_{22})/\sigma_{11} = \rho_{12}^2$. A corresponding measure for vectors $\boldsymbol{X}^{(1)}$ and $\boldsymbol{X}^{(2)}$ is the ratio of the generalized variance of the regression of $\boldsymbol{X}^{(1)}$ on $\boldsymbol{X}^{(2)}$ to the generalized variance of $\boldsymbol{X}^{(1)}$, namely,

$$\frac{|\mathscr{E}\mathbf{B}\boldsymbol{X}^{(2)}(\mathbf{B}\boldsymbol{X}^{(2)})'|}{|\Sigma_{11}|} = \frac{|\mathbf{B}\Sigma_{22}\mathbf{B}'|}{|\Sigma_{11}|} = \frac{|\Sigma_{12}\Sigma_{22}^{-1}\Sigma_{21}|}{|\Sigma_{11}|} = (-1)^{p_1}\frac{\begin{vmatrix} \mathbf{0} & \Sigma_{12} \\ \Sigma_{21} & \Sigma_{22}\end{vmatrix}}{|\Sigma_{11}|}. \tag{9}$$

If $p_1 = p_2$, the measure is

$$\frac{|\Sigma_{12}|^2}{|\Sigma_{11}|\cdot|\Sigma_{22}|}. \tag{10}$$

In a sense this measure shows how well $\boldsymbol{X}^{(1)}$ can be predicted from $\boldsymbol{X}^{(2)}$.

In the case of two scalar variables X_1 and X_2 the coefficient of alienation is $\sigma_{1\cdot 2}^2/\sigma_1^2$, where $\sigma_{1\cdot 2}^2 = \mathscr{E}(X_1 - \beta X_2)^2$ is the variance of X_1 about its regression on X_2 when $\mathscr{E}X_1 = \mathscr{E}X_2 = 0$ and $\mathscr{E}(X_1|X_2) = \beta X_2$. In the case of two vectors $\boldsymbol{X}^{(1)}$ and $\boldsymbol{X}^{(2)}$, the regression matrix is $\boldsymbol{\beta} = \Sigma_{12}\Sigma_{22}^{-1}$, and the generalized variance of $\boldsymbol{X}^{(1)}$ about its regression on $\boldsymbol{X}^{(2)}$ is

$$(11)\qquad \left|\mathscr{E}\left\{(\boldsymbol{X}^{(1)} - \boldsymbol{\beta}\boldsymbol{X}^{(2)})(\boldsymbol{X}^{(1)} - \boldsymbol{\beta}\boldsymbol{X}^{(2)})'\right\}\right| = |\Sigma_{11} - \Sigma_{12}\Sigma_{22}^{-1}\Sigma_{21}| = \frac{|\Sigma|}{|\Sigma_{22}|}.$$

Since the generalized variance of $\boldsymbol{X}^{(1)}$ is $|\mathscr{E}\boldsymbol{X}^{(1)}\boldsymbol{X}^{(1)\prime}| = |\Sigma_{11}|$, the *vector coefficient of alienation* is

$$(12)\qquad \frac{|\Sigma_{11} - \Sigma_{12}\Sigma_{22}^{-1}\Sigma_{21}|}{|\Sigma_{11}|} = \frac{|\Sigma|}{|\Sigma_{11}|\cdot|\Sigma_{22}|}.$$

The sample equivalent of (12) is simply V.

A measure of association is 1 minus the coefficient of alienation. Either of these two measures of association can be modified to take account of the number of components. In the first case, one can take the p_1th root of (9); in the second case, one can subtract the p_1th root of the coefficient of alienation from 1. Another measure of association is

$$(13)\qquad \frac{\operatorname{tr}\mathscr{E}\left[\boldsymbol{\beta}\boldsymbol{X}^{(2)}(\boldsymbol{\beta}\boldsymbol{X}^{(2)})'\right](\mathscr{E}\boldsymbol{X}^{(1)}\boldsymbol{X}^{(1)\prime})^{-1}}{p} = \frac{\operatorname{tr}\Sigma_{12}\Sigma_{22}^{-1}\Sigma_{21}\Sigma_{11}^{-1}}{p}.$$

This measure of association ranges between 0 and 1. If $\boldsymbol{X}^{(1)}$ can be predicted exactly from $\boldsymbol{X}^{(2)}$ for $p_1 \leq p_2$ (i.e., $\Sigma_{11\cdot 2} = \boldsymbol{0}$), then this measure is 1. If no linear combination of $\boldsymbol{X}^{(1)}$ can be predicted exactly, this measure is 0.

9.9. ADMISSIBILITY OF THE LIKELIHOOD RATIO TEST

The admissibility of the likelihood ratio test in the case of the $0-1$ loss function can be proved by showing that it is the Bayes procedure with respect to an appropriate a priori distribution of the parameters. (See Section 5.6.)

THEOREM 9.9.1. *The likelihood ratio test of the hypothesis that Σ is of the form* (6) *of Section* 9.2 *is Bayes and admissible if $N > p + 1$.*

PROOF. We shall show that the likelihood ratio test is equivalent to rejection of the hypothesis when

$$\frac{\int f(\boldsymbol{x}|\boldsymbol{\theta})\Pi_1(d\boldsymbol{\theta})}{\int f(\boldsymbol{x}|\boldsymbol{\theta})\Pi_0(d\boldsymbol{\theta})} \geq c, \tag{1}$$

where $\boldsymbol{x}$ represents the sample, $\boldsymbol{\theta}$ represents the parameters ($\boldsymbol{\mu}$ and $\boldsymbol{\Sigma}$). $f(\boldsymbol{x}|\boldsymbol{\theta})$ is the density, and Π_1 and Π_0 are proportional to probability measures of $\boldsymbol{\theta}$ under the alternative and null hypotheses, respectively. Specifically, the left-hand side is to be proportional to the square root of $\prod_{i=1}^{q}|\boldsymbol{A}_{ii}|/|\boldsymbol{A}|$.

To define Π_1, let

$$\boldsymbol{\mu} = (\boldsymbol{I} + \boldsymbol{V}\boldsymbol{V}')^{-1}\boldsymbol{V}Y, \qquad \boldsymbol{\Sigma} = (\boldsymbol{I} + \boldsymbol{V}\boldsymbol{V}')^{-1}, \tag{2}$$

where the p-component random vector $\boldsymbol{V}$ has the density proportional to $(1 + \boldsymbol{v}'\boldsymbol{v})^{-\frac{1}{2}n}$, $n = N - 1$, and the conditional distribution of Y given $\boldsymbol{V} = \boldsymbol{v}$ is $N[0, (1 + \boldsymbol{v}'\boldsymbol{v})/N]$. Note that the integral of $(1 + \boldsymbol{v}'\boldsymbol{v})^{-\frac{1}{2}n}$ is finite if $n > p$ (Problem 15 of Chapter 5). The numerator of (1) is then

$$\begin{aligned}
(3) \quad &\text{const}\int_{-\infty}^{\infty} \cdots \int_{-\infty}^{\infty} |\boldsymbol{I} + \boldsymbol{v}\boldsymbol{v}'|^{\frac{1}{2}N} \\
&\times \exp\left\{-\frac{1}{2}\sum_{\alpha=1}^{N}\left[\boldsymbol{x}_\alpha - (\boldsymbol{I} + \boldsymbol{v}\boldsymbol{v}')^{-1}\boldsymbol{v}y\right]'(\boldsymbol{I} + \boldsymbol{v}\boldsymbol{v}')\left[\boldsymbol{x}_\alpha - (\boldsymbol{I} + \boldsymbol{v}\boldsymbol{v}')^{-1}\boldsymbol{v}y\right]\right\} \\
&\times (1 + \boldsymbol{v}'\boldsymbol{v})^{-\frac{1}{2}n} \times (1 + \boldsymbol{v}'\boldsymbol{v})^{-\frac{1}{2}} \exp\left\{\frac{-\frac{1}{2}Ny^2}{(1 + \boldsymbol{v}'\boldsymbol{v})}\right\} d\boldsymbol{v}\, dy.
\end{aligned}$$

The exponent in the integrand of (3) is -2 times

$$\begin{aligned}
(4) \quad &\sum_{\alpha=1}^{N} \boldsymbol{x}_\alpha'(\boldsymbol{I} + \boldsymbol{v}\boldsymbol{v}')\boldsymbol{x}_\alpha - 2y\boldsymbol{v}'\sum_{\alpha=1}^{N}\boldsymbol{x}_\alpha + Ny^2\boldsymbol{v}'(\boldsymbol{I} + \boldsymbol{v}\boldsymbol{v}')^{-1}\boldsymbol{v} + \frac{Ny^2}{(1 + \boldsymbol{v}'\boldsymbol{v})} \\
&= \sum_{\alpha=1}^{N}\boldsymbol{x}_\alpha'\boldsymbol{x}_\alpha + \boldsymbol{v}'\sum_{\alpha=1}^{N}\boldsymbol{x}_\alpha\boldsymbol{x}_\alpha'\boldsymbol{v} - 2y\boldsymbol{v}'N\bar{\boldsymbol{x}} + Ny^2 \\
&= \operatorname{tr}\boldsymbol{A} + \boldsymbol{v}'\boldsymbol{A}\boldsymbol{v} + N\bar{\boldsymbol{x}}'\bar{\boldsymbol{x}} + N(y - \bar{\boldsymbol{x}}'\boldsymbol{v})^2,
\end{aligned}$$

where $A = \sum_{\alpha=1}^{N} x_\alpha x_\alpha' - N\bar{x}\bar{x}'$. We have used $v'(I + vv')^{-1}v + (1 + v'v)^{-1} = 1$ [from $(I + vv')^{-1} = I - (1 + v'v)^{-1}vv'$]. Using $|I + vv'| = 1 + v'v$ (Corollary A.3.1), we write (3) as

$$(5) \quad \text{const}\, e^{-\frac{1}{2}\operatorname{tr} A - \frac{1}{2}N\bar{x}'\bar{x}} \int_{-\infty}^{\infty} \cdots \int_{-\infty}^{\infty} e^{-\frac{1}{2}v'Av}\, dv = \text{const}\, |A|^{-\frac{1}{2}} e^{-\frac{1}{2}\operatorname{tr} A - \frac{1}{2}N\bar{x}'\bar{x}}.$$

To define Π_0 let Σ have the form of (6) of Section 9.2. Let

$$(6) \qquad \left[\mu^{(i)}, \Sigma_{ii}\right] = \left[(I + V^{(i)}V^{(i)\prime})^{-1}V^{(i)}Y_i, \qquad (I + V^{(i)}V^{(i)\prime})^{-1}\right], \qquad i = 1, \dots, q,$$

where the p_i-component random vector $V^{(i)}$ has the density proportional to $(1 + v^{(i)\prime}v^{(i)})^{-\frac{1}{2}n}$, and the conditional distribution of Y_i given $V^{(i)} = v^{(i)}$ is $N[0, (1 + v^{(i)\prime}v^{(i)})/N]$, and let $(V_1, Y_1), \dots, (V_q, Y_q)$ be mutually independent. Then the denominator of (1) is

$$(7) \quad \prod_{i=1}^{q} \text{const}|A_{ii}|^{-\frac{1}{2}} \exp\left[-\tfrac{1}{2}\left(\operatorname{tr} A_{ii} + N\bar{x}^{(i)\prime}\bar{x}^{(i)}\right)\right]$$

$$= \text{const}\left(\prod_{i=1}^{q} |A_{ii}|^{-\frac{1}{2}}\right) \exp\left[-\tfrac{1}{2}(\operatorname{tr} A + N\bar{x}'\bar{x})\right].$$

The left-hand side of (1) is then proportional to the square root of $\prod_{i=1}^{q}|A_{ii}|/|A|$. Q.E.D.

This proof has been adapted from that of Kiefer and Schwartz (1965).

9.10. MONOTONICITY OF POWER FUNCTIONS OF TESTS OF INDEPENDENCE OF SETS

Let $Z_\alpha = [Z_\alpha^{(1)\prime}, Z_\alpha^{(2)\prime}]'$, $\alpha = 1, \dots, n$, be distributed according to

$$(1) \qquad N\left[\begin{pmatrix} \mathbf{0} \\ \mathbf{0} \end{pmatrix}, \begin{pmatrix} \Sigma_{11} & \Sigma_{12} \\ \Sigma_{21} & \Sigma_{22} \end{pmatrix}\right].$$

We want to test $H: \Sigma_{12} = \mathbf{0}$. We suppose $p_1 \le p_2$ without loss of generality.

Let $\rho_1, \dots, \rho_{p_1}$ $(\rho_1 \geq \cdots \geq \rho_{p_1})$ be the (population) canonical correlation coefficients. (The ρ_i^2's are the characteristic roots of $\boldsymbol{\Sigma}_{11}^{-1}\boldsymbol{\Sigma}_{12}\boldsymbol{\Sigma}_{22}^{-1}\boldsymbol{\Sigma}_{21}$, Chapter 12.) Let $\boldsymbol{R} = \operatorname{diag}(\rho_1, \dots, \rho_{p_1})$ and $\boldsymbol{\Delta} = [\boldsymbol{R}, \boldsymbol{0}]$ $(p_1 \times p_2)$.

LEMMA 9.10.1. *There exist matrices $\boldsymbol{B}_1$ $(p_1 \times p_1)$, $\boldsymbol{B}_2$ $(p_2 \times p_2)$ such that*

$$\boldsymbol{B}_1\boldsymbol{\Sigma}_{11}\boldsymbol{B}_1' = \boldsymbol{I}_{p_1}, \qquad \boldsymbol{B}_2\boldsymbol{\Sigma}_{22}\boldsymbol{B}_2' = \boldsymbol{I}_{p_2}, \qquad \boldsymbol{B}_1\boldsymbol{\Sigma}_{12}\boldsymbol{B}_2' = \boldsymbol{\Delta}. \tag{2}$$

PROOF. Let $m = p_2$, $\boldsymbol{B} = \boldsymbol{B}_1$, $\boldsymbol{F}' = \boldsymbol{\Sigma}_{22}^{\frac{1}{2}}\boldsymbol{B}_2'$, $\boldsymbol{\Xi} = \boldsymbol{\Sigma}_{12}\boldsymbol{\Sigma}_{22}^{-\frac{1}{2}}$ in Lemma 8.10.13. Then $\boldsymbol{F}'\boldsymbol{F} = \boldsymbol{B}_2\boldsymbol{\Sigma}_{22}\boldsymbol{B}_2' = \boldsymbol{I}_{p_2}$, $\boldsymbol{B}_1\boldsymbol{\Sigma}_{12}\boldsymbol{B}_2' = \boldsymbol{B}_1\boldsymbol{\Xi}\boldsymbol{F} = \boldsymbol{\Delta}$. Q.E.D.

(This lemma is also contained in Section 12.2.)

Let $\boldsymbol{x}_\alpha = \boldsymbol{B}_1\boldsymbol{Z}_\alpha^{(1)}$, $\boldsymbol{y}_\alpha = \boldsymbol{B}_2\boldsymbol{Z}_\alpha^{(2)}$, $\alpha = 1, \dots, n$, and $\boldsymbol{X} = (\boldsymbol{x}_1, \dots, \boldsymbol{x}_n)$, $\boldsymbol{Y} = (\boldsymbol{y}_1, \dots, \boldsymbol{y}_n)$. Then $(\boldsymbol{x}_\alpha', \boldsymbol{y}_\alpha')'$, $\alpha = 1, \dots, n$, are independently distributed according to

$$N\left[\begin{pmatrix}\boldsymbol{0}\\ \boldsymbol{0}\end{pmatrix}, \begin{pmatrix}\boldsymbol{I}_p & \boldsymbol{\Delta}\\ \boldsymbol{\Delta}' & \boldsymbol{I}_q\end{pmatrix}\right]. \tag{3}$$

The hypothesis $H: \boldsymbol{\Sigma}_{12} = \boldsymbol{0}$ is equivalent to $H: \boldsymbol{\Delta} = \boldsymbol{0}$ (i.e., all the canonical correlation coefficients $\rho_1, \dots, \rho_{p_1}$ are zero). Now given $\boldsymbol{Y}$, the vectors $\boldsymbol{x}_\alpha$, $\alpha = 1, \dots, n$, are conditionally independently distributed according to $N(\boldsymbol{\Delta}\boldsymbol{y}_\alpha, \boldsymbol{I} - \boldsymbol{\Delta}\boldsymbol{\Delta}') = N(\boldsymbol{\Delta}\boldsymbol{y}_\alpha, \boldsymbol{I} - \boldsymbol{R}^2)$. Then $\boldsymbol{x}_\alpha^* = (\boldsymbol{I}_{p_1} - \boldsymbol{R}^2)^{-\frac{1}{2}}\boldsymbol{x}_\alpha$ is distributed according to $N(\boldsymbol{M}\boldsymbol{y}_\alpha, \boldsymbol{I}_{p_1})$ where

$$\begin{aligned} \boldsymbol{M} &= (\boldsymbol{D}, \boldsymbol{0}), \\ \boldsymbol{D} &= \operatorname{diag}(\delta_1, \dots, \delta_{p_1}), \\ \delta_i &= \rho_i/(1 - \rho_i^2)^{\frac{1}{2}}, \qquad i = 1, \dots, p_1. \end{aligned} \tag{4}$$

Note that δ_i^2 is a characteristic root of $\boldsymbol{\Sigma}_{12}\boldsymbol{\Sigma}_{22}^{-1}\boldsymbol{\Sigma}_{21}\boldsymbol{\Sigma}_{11\cdot 2}^{-1}$ where $\boldsymbol{\Sigma}_{11\cdot 2} = \boldsymbol{\Sigma}_{11} - \boldsymbol{\Sigma}_{12}\boldsymbol{\Sigma}_{22}^{-1}\boldsymbol{\Sigma}_{21}$.

Invariant tests depend only on the (sample) canonical correlation coefficients $r_i = \sqrt{c_i}$, where

$$c_i = \lambda_i\left[(\boldsymbol{X}^*\boldsymbol{X}^{*\prime})^{-1}(\boldsymbol{X}^*\boldsymbol{Y}')(\boldsymbol{Y}\boldsymbol{Y}')^{-1}(\boldsymbol{Y}\boldsymbol{X}^{*\prime})\right]. \tag{5}$$

Let

$$(6)\qquad \begin{aligned} \boldsymbol{S}_h &= \boldsymbol{X}^*\boldsymbol{Y}'(\boldsymbol{Y}\boldsymbol{Y}')^{-1}\boldsymbol{Y}\boldsymbol{X}^{*\prime}, \\ \boldsymbol{S}_e &= \boldsymbol{X}^*\boldsymbol{X}^{*\prime} - \boldsymbol{S}_h = \boldsymbol{X}^*\left[\boldsymbol{I} - \boldsymbol{Y}'(\boldsymbol{Y}\boldsymbol{Y}')^{-1}\boldsymbol{Y}\right]\boldsymbol{X}^{*\prime}. \end{aligned}$$

Then

$$(7)\qquad \lambda_i\left[\boldsymbol{S}_h\boldsymbol{S}_e^{-1}\right] = \frac{c_i}{1-c_i}.$$

Now given $\boldsymbol{Y}$, the problem reduces to the MANOVA problem and we can apply Theorem 8.10.6 as follows. There is an orthogonal transformation (Sec. 8.3.3) that carries $\boldsymbol{X}^*$ to $(\boldsymbol{U}, \boldsymbol{V})$ such that $\boldsymbol{S}_h = \boldsymbol{U}\boldsymbol{U}'$, $\boldsymbol{S}_e = \boldsymbol{V}\boldsymbol{V}'$, $\boldsymbol{U} = (\boldsymbol{u}_1, \dots, \boldsymbol{u}_{p_2})$, $\boldsymbol{V}$ is $p_1 \times (n - p_2)$, $\boldsymbol{u}_i$ has the distribution $N(\delta_i \boldsymbol{\varepsilon}_i, \boldsymbol{I})$, $i = 1, \dots, p_1$ ($\boldsymbol{\varepsilon}_i$ being the ith column of $\boldsymbol{I}$), and $N(\boldsymbol{0}, \boldsymbol{I})$, $i = p_1 + 1, \dots, p_2$, and the columns of $\boldsymbol{V}$ are independently distributed according to $N(\boldsymbol{0}, \boldsymbol{I})$. Then $c_1, \dots, c_{p_1}$ are the characteristic roots of $\boldsymbol{U}\boldsymbol{U}'(\boldsymbol{V}\boldsymbol{V}')^{-1}$, and their distribution depends on the characteristic roots of $\boldsymbol{M}\boldsymbol{Y}\boldsymbol{Y}'\boldsymbol{M}'$, say, $\tau_1^2, \dots, \tau_{p_1}^2$. Now from Theorem 8.10.6, we obtain the following lemma.

LEMMA 9.10.2. *If the acceptance region of an invariant test is convex in each column of $\boldsymbol{U}$ given $\boldsymbol{V}$ and the other columns of $\boldsymbol{U}$, then the conditional power given $\boldsymbol{Y}$ increases in each τ_i^2, where τ_i^2 is a characteristic root of $\boldsymbol{M}\boldsymbol{Y}\boldsymbol{Y}'\boldsymbol{M}'$.*

LEMMA 9.10.3. *If $\boldsymbol{A} \geq \boldsymbol{B}$, then $\lambda_i(\boldsymbol{A}) \geq \lambda_i(\boldsymbol{B})$.*

PROOF. By the minimaxity property of the characteristic roots [see, e.g., Courant and Hilbert (1953)],

$$(8)\qquad \lambda_i(\boldsymbol{A}) = \max_{S_i} \min_{x \in S_i} \frac{\boldsymbol{x}'\boldsymbol{A}\boldsymbol{x}}{\boldsymbol{x}'\boldsymbol{x}} \geq \max_{S_i} \min_{x \in S_i} \frac{\boldsymbol{x}'\boldsymbol{B}\boldsymbol{x}}{\boldsymbol{x}'\boldsymbol{x}} = \lambda_i(\boldsymbol{B})$$

where S_i ranges over i-dimensional subspaces. Q.E.D.

Now Lemma 9.10.3 applied to $\boldsymbol{M}\boldsymbol{Y}\boldsymbol{Y}'\boldsymbol{M}'$ shows that for every j, τ_j^2 is an increasing function of $\delta_i = \rho_i/(1-\rho_i^2)^{\frac{1}{2}}$ and hence of ρ_i. Since the marginal distribution of $\boldsymbol{Y}$ does not depend on ρ_i's, by taking the unconditional power we obtain the following theorem.

THEOREM 9.10.1. *An invariant test for which the acceptance region is convex in each column of $\boldsymbol{U}$ for each set of fixed $\boldsymbol{V}$ and other columns of $\boldsymbol{U}$ has a power function that is monotonically increasing in each ρ_i.*

PROBLEMS

1. (Sec. 9.3) Prove

$$\mathscr{E}V^h = \frac{\prod_{i=1}^{p}\Gamma[\frac{1}{2}(n+1-i)+h]\prod_{i=1}^{q}\left\{\prod_{j=1}^{p_i}\Gamma[\frac{1}{2}(n+1-j)]\right\}}{\prod_{i=1}^{p}\Gamma[\frac{1}{2}(n+1-i)]\prod_{i=1}^{q}\left\{\prod_{j=1}^{p_i}\Gamma[\frac{1}{2}(n+1-j)+h]\right\}}$$

by integration of $V^h w(\boldsymbol{A}|\boldsymbol{\Sigma}_0, n)$. [*Hint*: Show

$$\mathscr{E}V^h = \frac{K(\boldsymbol{\Sigma}_0, n)}{K(\boldsymbol{\Sigma}_0, n+2h)}\int\cdots\int\prod_{i=1}^{q}|\boldsymbol{A}_{ii}|^{-h}w(\boldsymbol{A}, \boldsymbol{\Sigma}_0, n+2h)\,d\boldsymbol{A}$$

where $K(\boldsymbol{\Sigma}, n)$ is defined by $w(\boldsymbol{A}|\boldsymbol{\Sigma}, n) = K(\boldsymbol{\Sigma}, n)|\boldsymbol{A}|^{\frac{1}{2}(n-p-1)}e^{-\frac{1}{2}\operatorname{tr}\boldsymbol{\Sigma}^{-1}\boldsymbol{A}}$. Use Theorem 7.3.5 to show

$$\mathscr{E}V^h = \frac{K(\boldsymbol{\Sigma}_0, n)}{K(\boldsymbol{\Sigma}_0, n+2h)}\prod_{i=1}^{q}\frac{K(\boldsymbol{\Sigma}_{ii}, n+2h)}{K(\boldsymbol{\Sigma}_{ii}, n)}\int\cdots\int w(\boldsymbol{A}_{ii}|\boldsymbol{\Sigma}_{ii}, n)\,d\boldsymbol{A}_{ii}.]$$

2. (Sec. 9.3) Prove that if $p_1 = p_2 = p_3 = 1$ [Wilks (1935)]

$$\Pr\{V \le v\} = I_v[\tfrac{1}{2}(n-1), \tfrac{1}{2}] + 2B^{-1}[\tfrac{1}{2}(n-1), \tfrac{1}{2}]\sin^{-1}\sqrt{1-v}\,.$$

[*Hint*: Use Theorem 9.3.3 and $\Pr\{V \le v\} = 1 - \Pr\{v \le V\}$.]

3. (Sec. 9.3) Prove that if $p_1 = p_2 = p_3 = 2$ [Wilks (1935)]

$$\begin{aligned}\Pr\{V \le v\} = {} & I_{\sqrt{v}}(n-5, 4) + B^{-1}(n-5, 4)v^{\frac{1}{2}(n-5)} \\ & \times\{n/6 - (3/2)(n-1)\sqrt{v} - (3/2)(n-4)v \\ & + (17n/6 - 15/2)v^{3/2} \\ & - (3/2)(n-2)v\log v - \tfrac{1}{2}(n-3)v^{3/2}\log v\}.\end{aligned}$$

[*Hint*: Use (9).]

4. (Sec. 9.3) Derive some of the distributions obtained by Wilks (1935) and referred to at the end of Section 9.3.3. [*Hint*: In addition to the results for Problems 2 and 3, use those of Section 9.3.2.]

5. (Sec. 9.4) For the case $p_i = 2$, express k and γ_2. Compute the second term of (6) when v is chosen so that the first term is 0.95 for $p = 4$ and 6 and $N = 15$.

6. (Sec. 9.5) Prove that if $\boldsymbol{BAB}' = \boldsymbol{CAC}' = \boldsymbol{I}$ for $\boldsymbol{A}$ positive definite and $\boldsymbol{B}$ and $\boldsymbol{C}$ nonsingular, then $\boldsymbol{B} = \boldsymbol{QC}$ where $\boldsymbol{Q}$ is orthogonal.

7. (Sec. 9.5) Prove N times (2) has a limiting χ^2-distribution with f degrees of freedom under the null hypothesis.

8. (Sec. 9.8) Give the sample vector coefficient of alienation and the vector correlation coefficient.

9. (Sec. 9.8) If y is the sample vector coefficient of alienation and z the square of the vector correlation coefficient, find $\mathscr{E} y^g z^h$ when $\boldsymbol{\Sigma}_{12} = \boldsymbol{0}$.

10. (Sec. 9.9) Prove

$$\int_{-\infty}^{\infty} \cdots \int_{-\infty}^{\infty} \frac{1}{\left(1 + \sum_{i=1}^{p} v_i^2\right)^{\frac{1}{2}r}} \, dv_1 \cdots dv_p < \infty$$

if $p < n$. [*Hint*: Let $y_j = w_j \sqrt{1 + \sum_{i=j+1}^{p} y_i^2}$, $j = 1, \dots, p-1$, in turn.]

11. Let x_1 = arithmetic speed, x_2 = arithmetic power, x_3 = intellectual interest, x_4 = social interest, x_5 = activity interest. Kelley (1928) observed the following correlations between batteries of tests identified as above, based on 109 pupils:

$$\begin{pmatrix} 1.0000 & 0.4249 & -0.0552 & -0.0031 & 0.1927 \\ 0.4249 & 1.0000 & -0.0416 & 0.0495 & 0.0687 \\ -0.0552 & -0.0416 & 1.0000 & 0.7474 & 0.1691 \\ -0.0031 & 0.0495 & 0.7474 & 1.0000 & 0.2653 \\ 0.1927 & 0.0687 & 0.1691 & 0.2653 & 1.0000 \end{pmatrix}.$$

Let $\boldsymbol{x}^{(1)\prime} = (x_1, x_2)$ and $\boldsymbol{x}^{(2)\prime} = (x_3, x_4, x_5)$. Test the hypothesis that $\boldsymbol{x}^{(1)}$ is independent of $\boldsymbol{x}^{(2)}$ at the 1% significance level.

12. Carry out the same exercise on the data in Problem 42 of Chapter 3.

13. Another set of time-study data [Abruzzi (1950)] is summarized by the correlation matrix based on 188 observations:

$$\begin{pmatrix} 1.00 & -0.27 & 0.06 & 0.07 & 0.02 \\ -0.27 & 1.00 & -0.01 & -0.02 & -0.02 \\ 0.06 & -0.01 & 1.00 & -0.07 & -0.04 \\ 0.07 & -0.02 & -0.07 & 1.00 & -0.10 \\ 0.02 & -0.02 & -0.04 & -0.10 & 1.00 \end{pmatrix}.$$

Test the hypothesis that $\sigma_{ij} = 0$ $(i \neq j)$ at the 5% significance level.

CHAPTER 10

Testing Hypotheses of Equality of Covariance Matrices and Equality of Mean Vectors and Covariance Matrices

10.1. INTRODUCTION

In this chapter we study the problems of testing hypotheses of equality of covariance matrices and equality of both covariance matrices and mean vectors. In each case (except one) the problem and tests considered are multivariate generalizations of a univariate problem and test. Many of the tests are likelihood ratio tests or modifications of likelihood ratio tests. Invariance considerations lead to other test procedures.

First, we consider equality of covariance matrices and equality of covariance matrices and mean vectors of several populations without specifying the common covariance matrix or the common covariance matrix and mean vector. The multivariate analysis of variance with random factors is considered in this context. Later we treat the equality of a covariance matrix to a given matrix and also simultaneous equality of a covariance matrix to a given matrix and equality of a mean vector to a given vector. One other hypothesis considered, the equality of a covariance matrix to a given matrix except for a proportionality factor, has only a trivial corresponding univariate hypothesis.

In each case the class of tests for a class of hypotheses leads to a confidence region. Families of simultaneous confidence intervals for covariances and for ratios of covariances are given.

10.2. CRITERIA FOR TESTING EQUALITY OF SEVERAL COVARIANCE MATRICES

In this section we study several normal distributions and consider using a set of samples, one from each population, to test the hypothesis that the covariance matrices of these populations are equal. Let $x_\alpha^{(g)}$, $\alpha = 1, \dots, N_g$; $g = 1, \dots, q$, be an observation from the gth population $N(\mu^{(g)}, \Sigma_g)$. We wish to test the hypothesis

$$H_1: \Sigma_1 = \cdots = \Sigma_q. \tag{1}$$

Let

$$\sum_{g=1}^{q} N_g = N,$$

$$A_g = \sum_{\alpha=1}^{N_g} \left(x_\alpha^{(g)} - \bar{x}^{(g)}\right)\left(x_\alpha^{(g)} - \bar{x}^{(g)}\right)', \qquad g = 1, \dots, q, \tag{2}$$

$$A = \sum_{g=1}^{q} A_g.$$

First we shall obtain the likelihood ratio criterion. The likelihood function is

$$L = \prod_{g=1}^{q} \frac{1}{(2\pi)^{\frac{1}{2}pN_g}|\Sigma_g|^{\frac{1}{2}N_g}} \exp\left[-\frac{1}{2}\sum_{\alpha=1}^{N_g}\left(x_\alpha^{(g)} - \mu^{(g)}\right)'\Sigma_g^{-1}\left(x_\alpha^{(g)} - \mu^{(g)}\right)\right]. \tag{3}$$

The space Ω is the parameter space in which each Σ_g is positive definite and $\mu^{(g)}$ any vector. The space ω is the parameter space in which $\Sigma_1 = \Sigma_2 = \cdots = \Sigma_q$ (positive definite) and $\mu^{(g)}$ any vector. The maximum likelihood estimators of $\mu^{(g)}$ and Σ_g in Ω are given by

$$\hat{\mu}_\Omega^{(g)} = \bar{x}^{(g)}, \qquad \hat{\Sigma}_{g\Omega} = \frac{1}{N_g} A_g, \qquad g = 1, \dots, q. \tag{4}$$

The maximum likelihood estimators of $\boldsymbol{\mu}^{(g)}$ in ω are given by (4), $\hat{\boldsymbol{\mu}}_\omega^{(g)} = \bar{\boldsymbol{x}}^{(g)}$, since the maximizing values of $\boldsymbol{\mu}^{(g)}$ are the same regardless of $\boldsymbol{\Sigma}_g$. The function to be maximized with respect to $\boldsymbol{\Sigma}_1 = \cdots = \boldsymbol{\Sigma}_q = \boldsymbol{\Sigma}$, say, is

$$(5)\qquad \frac{1}{(2\pi)^{\frac{1}{2}pN}|\boldsymbol{\Sigma}|^{\frac{1}{2}N}} \exp\left[-\frac{1}{2}\sum_{g=1}^{q}\sum_{\alpha=1}^{N_g}\left(\boldsymbol{x}_\alpha^{(g)} - \bar{\boldsymbol{x}}^{(g)}\right)'\boldsymbol{\Sigma}^{-1}\left(\boldsymbol{x}_\alpha^{(g)} - \bar{\boldsymbol{x}}^{(g)}\right)\right].$$

By Lemma 3.2.2, the maximizing value of $\boldsymbol{\Sigma}$ is

$$(6)\qquad \hat{\boldsymbol{\Sigma}}_\omega = \frac{1}{N}\boldsymbol{A},$$

and the maximum of the likelihood function is

$$(7)\qquad \frac{1}{(2\pi)^{\frac{1}{2}pN}|\hat{\boldsymbol{\Sigma}}_\omega|^{\frac{1}{2}N}} e^{-\frac{1}{2}pN}.$$

The likelihood ratio criterion for testing (1) is

$$(8)\qquad \lambda_1 = \frac{\prod_{g=1}^{q}|\hat{\boldsymbol{\Sigma}}_{g\Omega}|^{\frac{1}{2}N_g}}{|\hat{\boldsymbol{\Sigma}}_\omega|^{\frac{1}{2}N}}$$

$$= \frac{\prod_{g=1}^{q}|\boldsymbol{A}_g|^{\frac{1}{2}N_g}}{|\boldsymbol{A}|^{\frac{1}{2}N}} \cdot \frac{N^{\frac{1}{2}pN}}{\prod_{g=1}^{q}N_g^{\frac{1}{2}pN_g}}.$$

The critical region is

$$(9)\qquad \lambda_1 \le \lambda_1(\varepsilon),$$

where $\lambda_1(\varepsilon)$ is defined so that (9) holds with probability ε when (1) is true.

Bartlett (1937a) has suggested modifying λ_1 in the univariate case by replacing sample numbers by the numbers of degrees of freedom of the $\boldsymbol{A}_g$. Except for a numerical constant, the statistic he proposes is

$$(10)\qquad V_1 = \frac{\prod_{g=1}^{q}|\boldsymbol{A}_g|^{\frac{1}{2}n_g}}{|\boldsymbol{A}|^{\frac{1}{2}n}},$$

where $n_g = N_g - 1$ and $n = \sum_{g=1}^{q} n_g = N - q$. The numerator is proportional to a power of a weighted geometric mean of the sample generalized variances, and the denominator is proportional to a power of the determinant of a weighted arithmetic mean of the sample covariance matrices.

In the scalar case ($p = 1$) of two samples the criterion (10) is

$$\frac{(n_1)^{\frac{1}{2}n_1}(n_2)^{\frac{1}{2}n_2}(s_1^2)^{\frac{1}{2}n_1}(s_2^2)^{\frac{1}{2}n_2}}{(n_1 s_1^2 + n_2 s_2^2)^{\frac{1}{2}(n_1+n_2)}} = \frac{(n_1)^{\frac{1}{2}n_1}(n_2)^{\frac{1}{2}n_2}F^{\frac{1}{2}n_1}}{(n_1 F + n_2)^{\frac{1}{2}(n_1+n_2)}}, \tag{11}$$

where s_1^2 and s_2^2 are the usual unbiased estimators of σ_1^2 and σ_2^2 (the two population variances) and

$$F = \frac{s_1^2}{s_2^2}. \tag{12}$$

Thus the critical region

$$V_1 \le V_1(\varepsilon) \tag{13}$$

is based on the F-statistic with n_1 and n_2 degrees of freedom, and the inequality (13) implies a particular method of choosing $F_1(\varepsilon)$ and $F_2(\varepsilon)$ for the critical region

$$\begin{aligned} F &\le F_1(\varepsilon), \\ F &\ge F_2(\varepsilon). \end{aligned} \tag{14}$$

Brown (1939) and Scheffé (1942) have shown that (14) yields an unbiased test.

Bartlett gave a more intuitive argument for the use of V_1 in place of λ_1. He argues that if N_1, say, is small, A_1 is given too much weight in λ_1, and other effects may be missed. Perlman (1980) has shown that the test based on V_1 is unbiased.

If one assumes

$$\mathscr{E} X_\alpha^{(g)} = \boldsymbol{\beta}_g z_\alpha^{(g)}, \tag{15}$$

where $z_\alpha^{(g)}$ consists of k_g components, and if one estimates the matrix $\boldsymbol{\beta}_g$, defining

$$A_g = \sum_{\alpha=1}^{N_g} \left(x_\alpha^{(g)} - \hat{\boldsymbol{\beta}}_g z_\alpha^{(g)}\right)\left(x_\alpha - \hat{\boldsymbol{\beta}}_g z_\alpha^{(g)}\right)', \tag{16}$$

one uses (10) with $n_g = N_g - k_g$.

The statistical problem (parameter space Ω and null hypothesis ω) is invariant with respect to changes of location within populations and a common linear transformation

$$X^{*(g)} = CX^{(g)} + \nu^{(g)}, \qquad g = 1, \ldots, q, \tag{17}$$

where C is nonsingular. Each matrix A_g is invariant under change of location and the modified criterion (10) is invariant:

$$V_1^* = \frac{\prod_{g=1}^{q} |A_g^*|^{\frac{1}{2}n_g}}{|A^*|^{\frac{1}{2}n}} = \frac{\prod_{g=1}^{q} |CA_gC'|^{\frac{1}{2}n_g}}{|CAC'|^{\frac{1}{2}n}} = \frac{\prod_{g=1}^{q} |A_g|^{\frac{1}{2}n_g}}{|A|^{\frac{1}{2}n}} = V_1. \tag{18}$$

Similarly the likelihood ratio criterion (8) is invariant.

An alternative invariant test procedure [Nagao (1973a)] is based on the criterion

$$\frac{1}{2} \sum_{g=1}^{q} n_g \operatorname{tr}\left(S_g S^{-1} - I\right)^2 = \frac{1}{2} \sum_{g=1}^{q} n_g \operatorname{tr}(S_g - S)S^{-1}(S_g - S)S^{-1}, \tag{19}$$

where $S_g = (1/n_g)A_g$ and $S = (1/n)A$. (See Section 7.8.)

10.3. CRITERIA FOR TESTING THAT SEVERAL NORMAL DISTRIBUTIONS ARE IDENTICAL

In Section 8.8 we considered testing the equality of mean vectors when we assumed the covariance matrices were the same; that is, we tested

$$H_2: \mu^{(1)} = \mu^{(2)} = \cdots = \mu^{(q)} \text{ given } \Sigma_1 = \Sigma_2 \cdots = \Sigma_q. \tag{1}$$

The test of the assumption in H_2 was considered in Section 10.2. Now let us consider the hypothesis that both means and covariances are the same; this is a combination of H_1 and H_2. We test

$$H: \mu^{(1)} = \mu^{(2)} = \cdots = \mu^{(q)}, \quad \Sigma_1 = \Sigma_2 = \cdots = \Sigma_q. \tag{2}$$

As in Section 10.2, let $x_\alpha^{(g)}$, $\alpha = 1, \ldots, N_g$, be an observation from $N(\mu^{(g)}, \Sigma_g)$, $g = 1, \ldots, q$. Then Ω is the unrestricted parameter space of $\{\mu^{(g)}, \Sigma_g\}$, $g =$

$1, \dots, q$, where $\boldsymbol{\Sigma}_g$ is positive definite, and ω^* consists of the space restricted by (2).

The likelihood function is given by (3) of Section 10.2. The hypothesis H_1 of Section 10.2 is that the parameter point falls in ω; the hypothesis H_2 of Section 8.8 is that the parameter point falls in ω^* given it falls in $\omega \supset \omega^*$; and the hypothesis H here is that the parameter point falls in ω^* given that it is in Ω.

We use the following lemma:

LEMMA 10.3.1. *Let $\boldsymbol{y}$ be an observation vector on a random vector with density $f(z, \boldsymbol{\theta})$, where $\boldsymbol{\theta}$ is a parameter vector in a space Ω. Let H_a be the hypothesis $\boldsymbol{\theta} \in \Omega_a \subset \Omega$, let H_b be the hypothesis $\boldsymbol{\theta} \in \Omega_b \subset \Omega_a$, given $\boldsymbol{\theta} \in \Omega_a$, and let H_{ab} be the hypothesis $\boldsymbol{\theta} \in \Omega_b$, given $\boldsymbol{\theta} \in \Omega$. If λ_a, the likelihood ratio criterion for testing H_a, λ_b for H_b, and λ_{ab} for H_{ab} are uniquely defined for the observation vector $\boldsymbol{y}$, then*

$$\lambda_{ab} = \lambda_a \lambda_b. \tag{3}$$

PROOF. The lemma follows from the definitions:

$$\lambda_a = \frac{\max_{\boldsymbol{\theta} \in \Omega_a} f(\boldsymbol{y}, \boldsymbol{\theta})}{\max_{\boldsymbol{\theta} \in \Omega} f(\boldsymbol{y}, \boldsymbol{\theta})}, \tag{4}$$

$$\lambda_b = \frac{\max_{\boldsymbol{\theta} \in \Omega_b} f(\boldsymbol{y}, \boldsymbol{\theta})}{\max_{\boldsymbol{\theta} \in \Omega_a} f(\boldsymbol{y}, \boldsymbol{\theta})}, \tag{5}$$

$$\lambda_{ab} = \frac{\max_{\boldsymbol{\theta} \in \Omega_b} f(\boldsymbol{y}, \boldsymbol{\theta})}{\max_{\boldsymbol{\theta} \in \Omega} f(\boldsymbol{y}, \boldsymbol{\theta})}. \qquad \text{Q.E.D.} \tag{6}$$

Thus the likelihood ratio criterion for the hypothesis H is the product of the likelihood ratio criteria for H_1 and H_2,

$$\lambda = \lambda_1 \lambda_2 = \left(\prod_{g=1}^{q} \frac{|\boldsymbol{A}_g|^{\frac{1}{2}N_g}}{N_g^{\frac{1}{2}pN_g}} \right) \frac{N^{\frac{1}{2}pN}}{|\boldsymbol{B}|^{\frac{1}{2}N}}, \tag{7}$$

where

$$\begin{aligned} \boldsymbol{B} &= \sum_{g=1}^{q} \sum_{\alpha=1}^{N_g} \left(\boldsymbol{x}_\alpha^{(g)} - \bar{\boldsymbol{x}} \right) \left(\boldsymbol{x}_\alpha^{(g)} - \bar{\boldsymbol{x}} \right)' \\ &= \boldsymbol{A} + \sum_{g=1}^{q} N_g \left(\bar{\boldsymbol{x}}^{(g)} - \bar{\boldsymbol{x}} \right) \left(\bar{\boldsymbol{x}}^{(g)} - \bar{\boldsymbol{x}} \right)'. \end{aligned} \tag{8}$$

The critical region is defined by

$$\lambda \le \lambda(\varepsilon), \tag{9}$$

where $\lambda(\varepsilon)$ is chosen so that the probability of (9) under H is ε.

Let

$$V_2 = \frac{|A|^{\frac{1}{2}n}}{|B|^{\frac{1}{2}n}} = \lambda_2^{n/N}; \tag{10}$$

this is clearly equivalent to λ_2 for testing H_2. We might consider

$$V = V_1 V_2 = \frac{\prod_{g=1}^{q} |A_g|^{\frac{1}{2}n_g}}{|B|^{\frac{1}{2}n}}. \tag{11}$$

However, Perlman (1980) has shown that the likelihood ratio test is unbiased.

10.4. DISTRIBUTIONS OF THE CRITERIA

10.4.1. Characterization of the Distributions

First let us consider V_1 given by (10) of Section 10.2. If

$$V_{1g} = \frac{|A_1 + \cdots + A_{g-1}|^{\frac{1}{2}(n_1 + \cdots + n_{g-1})} |A_g|^{\frac{1}{2}n_g}}{|A_1 + \cdots + A_g|^{\frac{1}{2}(n_1 + \cdots + n_g)}}, \qquad g = 2, \dots, q, \tag{1}$$

then

$$V_1 = \prod_{g=2}^{q} V_{1g}. \tag{2}$$

THEOREM 10.4.1. *$V_{12}, V_{13}, \dots, V_{1q}$ defined by* (1) *are independent when $\Sigma_1 = \cdots = \Sigma_q$ and $n_g \ge p$, $g = 1, \dots, q$.*

The theorem is a consequence of the following lemma:

LEMMA 10.4.1. *If A and B are independently distributed according to $W(\Sigma, m)$ and $W(\Sigma, n)$, respectively, $n \ge p$, $m \ge p$, and C is such that $C(A + B)C' = I$, then $A + B$ and CAC' are independently distributed; $A + B$ has the Wishart* *distribution with $m + n$ degrees of freedom, and $\boldsymbol{CAC}'$ has the multivariate beta distribution with n and m degrees of freedom.*

PROOF OF LEMMA. The density of $\boldsymbol{D} = \boldsymbol{A} + \boldsymbol{B}$ and $\boldsymbol{E} = \boldsymbol{CAC}'$ is found by replacing $\boldsymbol{A}$ and $\boldsymbol{B}$ in their joint density by $\boldsymbol{C}^{-1}\boldsymbol{E}\boldsymbol{C}'^{-1}$ and $\boldsymbol{D} - \boldsymbol{C}^{-1}\boldsymbol{E}\boldsymbol{C}'^{-1} = \boldsymbol{C}^{-1}(\boldsymbol{I} - \boldsymbol{E})\boldsymbol{C}'^{-1}$, respectively, and multiplying by the Jacobian which is $\operatorname{mod}|\boldsymbol{C}|^{-(p+1)} = |\boldsymbol{D}|^{\frac{1}{2}(p+1)}$ to obtain

$$(3) \qquad K(\boldsymbol{\Sigma}, m)K(\boldsymbol{\Sigma}, n)|\boldsymbol{C}^{-1}\boldsymbol{E}\boldsymbol{C}'^{-1}|^{\frac{1}{2}(m-p-1)} \cdot|\boldsymbol{C}^{-1}(\boldsymbol{I} - \boldsymbol{E})\boldsymbol{C}'^{-1}|^{\frac{1}{2}(n-p-1)}e^{-\frac{1}{2}\operatorname{tr}\boldsymbol{\Sigma}^{-1}\boldsymbol{D}}|\boldsymbol{D}|^{\frac{1}{2}(p+1)}$$

$$= K(\boldsymbol{\Sigma}, m + n)|\boldsymbol{D}|^{\frac{1}{2}(m+n-p-1)}e^{-\frac{1}{2}\operatorname{tr}\boldsymbol{\Sigma}^{-1}\boldsymbol{D}} \cdot \frac{\Gamma_p\left[\frac{1}{2}(m+n)\right]}{\Gamma_p\left(\frac{1}{2}m\right)\Gamma_p\left(\frac{1}{2}n\right)}|\boldsymbol{E}|^{\frac{1}{2}(m-p-1)}|\boldsymbol{I} - \boldsymbol{E}|^{\frac{1}{2}(n-p-1)}$$

for $\boldsymbol{D}$, $\boldsymbol{E}$, and $\boldsymbol{I} - \boldsymbol{E}$ positive definite. Q.E.D.

PROOF OF THEOREM. If we let $\boldsymbol{A}_1 + \cdots + \boldsymbol{A}_g = \boldsymbol{D}_g$ and $\boldsymbol{C}_g(\boldsymbol{A}_1 + \cdots + \boldsymbol{A}_{g-1})\,\boldsymbol{C}_g' = \boldsymbol{E}_g$, where $\boldsymbol{C}_g\boldsymbol{D}_g\boldsymbol{C}_g' = \boldsymbol{I}$, $g = 2, \dots, q$, then

$$(4) \qquad V_{1g} = \frac{|\boldsymbol{C}_g^{-1}\boldsymbol{E}_g\boldsymbol{C}_g'^{-1}|^{\frac{1}{2}(n_1 + \cdots + n_{g-1})}|\boldsymbol{C}_g^{-1}(\boldsymbol{I} - \boldsymbol{E}_g)\boldsymbol{C}_g'^{-1}|^{\frac{1}{2}n_g}}{|\boldsymbol{C}_g^{-1}\boldsymbol{C}_g'^{-1}|^{\frac{1}{2}(n_1 + \cdots + n_g)}}$$

$$= |\boldsymbol{E}_g|^{\frac{1}{2}(n_1 + \cdots + n_{g-1})}|\boldsymbol{I} - \boldsymbol{E}_g|^{\frac{1}{2}n_g}, \qquad g = 2, \dots, q,$$

and $\boldsymbol{E}_2, \dots, \boldsymbol{E}_q$ are independent by Lemma 10.4.1. Q.E.D.

We shall now find a characterization of the distribution of V_{1g}. A statistic V_{1g} is of the form

$$(5) \qquad \frac{|\boldsymbol{B}|^b|\boldsymbol{C}|^c}{|\boldsymbol{B} + \boldsymbol{C}|^{b+c}}.$$

Let $\boldsymbol{B}_i$ and $\boldsymbol{C}_i$ be the upper left-hand square submatrices of $\boldsymbol{B}$ and $\boldsymbol{C}$, respectively, of order i. Define $\boldsymbol{b}_{(i)}$ and $\boldsymbol{c}_{(i)}$ by

$$(6) \qquad \boldsymbol{B}_i = \begin{pmatrix} \boldsymbol{B}_{i-1} & \boldsymbol{b}_{(i)} \\ \boldsymbol{b}_{(i)}' & b_{ii} \end{pmatrix}, \qquad \boldsymbol{C}_i = \begin{pmatrix} \boldsymbol{C}_{i-1} & \boldsymbol{c}_{(i)} \\ \boldsymbol{c}_{(i)}' & c_{ii} \end{pmatrix}, \qquad i = 2, \dots, p.$$

Then (5) is ($\boldsymbol{B}_0 = \boldsymbol{C}_0 = \boldsymbol{I}$, $\boldsymbol{b}_{(1)} = \boldsymbol{c}_{(1)} = \boldsymbol{0}$).

(7)

$$\frac{|\boldsymbol{B}|^b|\boldsymbol{C}|^c}{|\boldsymbol{B}+\boldsymbol{C}|^{b+c}} = \prod_{i=1}^{p} \frac{|\boldsymbol{B}_i|^b|\boldsymbol{C}_i|^c}{|\boldsymbol{B}_{i-1}|^b|\boldsymbol{C}_{i-1}|^c} \cdot \frac{|\boldsymbol{B}_{i-1}+\boldsymbol{C}_{i-1}|^{b+c}}{|\boldsymbol{B}_i+\boldsymbol{C}_i|^{b+c}}$$

$$= \prod_{i=1}^{p} \frac{\left(b_{ii} - \boldsymbol{b}'_{(i)}\boldsymbol{B}_{i-1}^{-1}\boldsymbol{b}_{(i)}\right)^b \left(c_{ii} - \boldsymbol{c}'_{(i)}\boldsymbol{C}_{i-1}^{-1}\boldsymbol{c}_{(i)}\right)^c}{\left[b_{ii} + c_{ii} - \left(\boldsymbol{b}_{(i)} + \boldsymbol{c}_{(i)}\right)'\left(\boldsymbol{B}_{i-1} + \boldsymbol{C}_{i-1}\right)^{-1}\left(\boldsymbol{b}_{(i)} + \boldsymbol{c}_{(i)}\right)\right]^{b+c}}$$

$$= \prod_{i=1}^{p} \left\{ \frac{b_{ii\cdot i-1}^{b}c_{ii\cdot i-1}^{c}}{\left(b_{ii\cdot i-1} + c_{ii\cdot i-1}\right)^{b+c}} \right.$$

$$\left. \cdot \frac{\left(b_{ii\cdot i-1} + c_{ii\cdot i-1}\right)^{b+c}}{\left[b_{ii\cdot i-1} + c_{ii\cdot i-1} + \boldsymbol{b}'_{(i)}\boldsymbol{B}_{i-1}^{-1}\boldsymbol{b}_{(i)} + \boldsymbol{c}'_{(i)}\boldsymbol{C}_{i-1}^{-1}\boldsymbol{c}_{(i)} - \left(\boldsymbol{b}_{(i)} + \boldsymbol{c}_{(i)}\right)'\left(\boldsymbol{B}_{i-1} + \boldsymbol{C}_{i-1}\right)^{-1}\left(\boldsymbol{b}_{(i)} + \boldsymbol{c}_{(i)}\right)\right]^{b+c}} \right\},$$

where $b_{ii\cdot i-1} = b_{ii} - \boldsymbol{b}'_{(i)}\boldsymbol{B}_{i-1}^{-1}\boldsymbol{b}_{(i)}$ and $c_{ii\cdot i-1} = c_{ii} - \boldsymbol{c}'_{(i)}\boldsymbol{C}_{i-1}^{-1}\boldsymbol{c}_{(i)}$. The second term for $i = 1$ is defined as 1.

Now we want to argue that the ratios on the right-hand side of (7) are statistically independent when $\boldsymbol{B}$ and $\boldsymbol{C}$ are independently distributed according to $W(\boldsymbol{\Sigma}, m)$ and $W(\boldsymbol{\Sigma}, n)$, respectively. It follows from Theorem 4.3.3 that for $\boldsymbol{B}_{i-1}$ fixed $\boldsymbol{b}_{(i)}$ and $b_{ii\cdot i-1}$ are independently distributed according to $N(\boldsymbol{\beta}_{(i)}, \sigma_{ii\cdot i-1}\boldsymbol{B}_{i-1}^{-1})$ and $\sigma_{ii\cdot i-1}\chi^2$ with $m - (i-1)$ degrees of freedom, respectively. Lemma 10.4.1 implies that the first term (which is a function of $b_{ii\cdot i-1}/c_{ii\cdot i-1}$) is independent of $b_{ii\cdot i-1} + c_{ii\cdot i-1}$.

We apply the following lemma:

LEMMA 10.4.2. *For $\boldsymbol{B}_{i-1}$ and $\boldsymbol{C}_{i-1}$ positive definite*

$$(8)\quad \boldsymbol{b}'_{(i)}\boldsymbol{B}_{i-1}^{-1}\boldsymbol{b}_{(i)} + \boldsymbol{c}'_{(i)}\boldsymbol{C}_{i-1}^{-1}\boldsymbol{c}_{(i)} - \left(\boldsymbol{b}_{(i)} + \boldsymbol{c}_{(i)}\right)'\left(\boldsymbol{B}_{i-1} + \boldsymbol{C}_{i-1}\right)^{-1}\left(\boldsymbol{b}_{(i)} + \boldsymbol{c}_{(i)}\right)$$

$$= \left(\boldsymbol{B}_{i-1}^{-1}\boldsymbol{b}_{(i)} - \boldsymbol{C}_{i-1}^{-1}\boldsymbol{c}_{(i)}\right)'\left(\boldsymbol{B}_{i-1}^{-1} + \boldsymbol{C}_{i-1}^{-1}\right)^{-1}\left(\boldsymbol{B}_{i-1}^{-1}\boldsymbol{b}_{(i)} - \boldsymbol{C}_{i-1}^{-1}\boldsymbol{c}_{(i)}\right).$$

PROOF. Use of $(B^{-1}+C^{-1})^{-1}=[C^{-1}(B+C)B^{-1}]^{-1}=B(B+C)^{-1}C$ shows the left-hand side of (8) is (omitting i and $i-1$)

(9)

$$\begin{aligned} b'B^{-1}(B^{-1}+C^{-1})^{-1}(B^{-1}+C^{-1})b + c'(B^{-1}+C^{-1})(B^{-1}+C^{-1})^{-1}C^{-1}c \\ -(b+c)'B^{-1}(B^{-1}+C^{-1})^{-1}C^{-1}(b+c) \\ = b'B^{-1}(B^{-1}+C^{-1})^{-1}B^{-1}b + c'C^{-1}(B^{-1}+C^{-1})^{-1}C^{-1}c \\ -b'B^{-1}(B^{-1}+C^{-1})^{-1}C^{-1}c - c'C^{-1}(B^{-1}+C^{-1})B^{-1}b, \end{aligned}$$

which is the right-hand side of (8). Q.E.D.

The denominator of the ith second term in (7) is the numerator plus (8). The conditional distribution of $B_{i-1}^{-1}b_{(i)} - C_{i-1}^{-1}c_{(i)}$ is normal with mean $B_{i-1}^{-1}\beta_{(i)} - C_{i-1}^{-1}\gamma_{(i)}$ and covariance matrix $\sigma_{ii\cdot i-1}(B_{i-1}^{-1}+C_{i-1}^{-1})$. The covariance matrix is $\sigma_{ii\cdot i-1}$ times the inverse of the second matrix on the right-hand side of (8). Thus (8) is distributed as $\sigma_{ii\cdot i-1}\chi^2$ with $i-1$ degrees of freedom, independent of B_{i-1}, C_{i-1}, $b_{ii\cdot i-1}$, and $c_{ii\cdot i-1}$.

Then

$$(10)\qquad \frac{b_{ii\cdot i-1}^{b}c_{ii\cdot i-1}^{c}}{(b_{ii\cdot i-1}+c_{ii\cdot i-1})^{b+c}} = \left(\frac{b_{ii\cdot i-1}}{b_{ii\cdot i-1}+c_{ii\cdot i-1}}\right)^{b}\left(\frac{c_{ii\cdot i-1}}{b_{ii\cdot i-1}+c_{ii\cdot i-1}}\right)^{c}$$

is distributed as $X_i^b(1-X_i)^c$, where X_i has the $\beta[\frac{1}{2}(m-i+1),\frac{1}{2}(n-i+1)]$ distribution, $i=1,\dots,p$. Also

$$(11)\qquad \left[\frac{b_{ii\cdot i-1}+c_{ii\cdot i-1}}{b_{ii\cdot i-1}+c_{ii\cdot i-1}+(8)}\right]^{b+c}, \qquad i=2,\dots,p,$$

is distributed as Y_i^{b+c} where Y_i has the $\beta[\frac{1}{2}(m+n)-i+1,\frac{1}{2}(i-1)]$ distribution. Then (5) is distributed as $\prod_{i=1}^p X_i^b(1-X_i)^c\prod_{i=2}^p Y_i^{b+c}$ and the factors are mutually independent.

THEOREM 10.4.2.

$$V_1 = \prod_{g=2}^{q} \left\{ \prod_{i=1}^{p} X_{ig}^{\frac{1}{2}(n_1 + \cdots + n_{g-1})} (1 - X_{ig})^{\frac{1}{2}n_g} \cdot \prod_{i=2}^{p} Y_{ig}^{\frac{1}{2}(n_1 + \cdots + n_g)} \right\}, \tag{12}$$

where the X*'s and* Y*'s are independent,* X_{ig} *has the* $\beta[\frac{1}{2}(n_1 + \cdots + n_{g-1} - i + 1), \frac{1}{2}(n_g - i + 1)]$ *distribution, and* Y_{ig} *has the* $\beta[\frac{1}{2}(n_1 + \cdots + n_g) - i + 1, \frac{1}{2}(i - 1)]$ *distribution.*

PROOF. The factors $V_{12}, \dots, V_{1q}$ are independent by Theorem 10.4.1. Each term V_{1g} is decomposed according to (7) and the factors are independent. Q.E.D.

The factors of V_1 can be interpreted as test criteria for subhypotheses. The term depending on X_{i2} is the criterion for testing the hypothesis that $\sigma_{ii \cdot i-1}^{(1)} = \sigma_{ii \cdot i-1}^{(2)}$, and the term depending on Y_{i2} is the criterion for testing $\boldsymbol{\sigma}_{(i)}^{(1)} = \boldsymbol{\sigma}_{(i)}^{(2)}$ given $\sigma_{ii \cdot i-1}^{(1)} = \sigma_{ii \cdot i-1}^{(2)}$, and $\boldsymbol{\Sigma}_{i-1,1} = \boldsymbol{\Sigma}_{i-1,2}$. The terms depending on X_{ig} and Y_{ig} similarly furnish criteria for testing $\boldsymbol{\Sigma}_1 = \boldsymbol{\Sigma}_g$ given $\boldsymbol{\Sigma}_1 = \cdots = \boldsymbol{\Sigma}_{g-1}$.

Now consider the likelihood ratio criterion λ given by (7) of Section 10.3 for testing the hypothesis $\boldsymbol{\mu}^{(1)} = \cdots = \boldsymbol{\mu}^{(q)}$ and $\boldsymbol{\Sigma}_1 = \cdots = \boldsymbol{\Sigma}_q$. It is equivalent to the criterion

$$W = \frac{\prod_{g=1}^{q} |\boldsymbol{A}_g|^{\frac{1}{2}N_g}}{|\boldsymbol{A}_1 + \cdots + \boldsymbol{A}_q|^{\frac{1}{2}(N_1 + \cdots + N_g)}} \times \frac{|\boldsymbol{A}_1 + \cdots + \boldsymbol{A}_q|^{\frac{1}{2}N}}{|\boldsymbol{A}_1 + \cdots + \boldsymbol{A}_q + \sum_{g=1}^{q} N_g (\bar{\boldsymbol{x}}^{(g)} - \bar{\boldsymbol{x}})(\bar{\boldsymbol{x}}^{(g)} - \bar{\boldsymbol{x}})'|^{\frac{1}{2}N}}. \tag{13}$$

The two factors of (13) are independent because the first factor is independent of $\boldsymbol{A}_1 + \cdots + \boldsymbol{A}_q$ (by Lemma 10.4.1 and the proof of Theorem 10.4.1) and of $\bar{\boldsymbol{x}}^{(1)}, \dots, \bar{\boldsymbol{x}}^{(q)}$.

THEOREM 10.4.3.

$$W = \prod_{g=2}^{q} \left\{ \prod_{i=1}^{p} X_{ig}^{\frac{1}{2}(N_1 + \cdots + N_{g-1})} (1 - X_{ig})^{\frac{1}{2}N_g} \prod_{i=2}^{p} Y_{ig}^{\frac{1}{2}(N_1 + \cdots + N_g)} \right\} \prod_{i=1}^{p} Z_i^{\frac{1}{2}N}, \tag{14}$$

where the X*'s,* Y*'s, and* Z*'s are independent,* X_{ig} *has the* $\beta[\frac{1}{2}(n_1 + \cdots + n_{g-1} - i + 1), \frac{1}{2}(n_g - i + 1)]$ *distribution,* Y_{ig} *has the* $\beta[\frac{1}{2}(n_1 + \cdots + n_g) - i + 1, \frac{1}{2}(i-1)]$ *distribution, and* Z_i *has the* $\beta[\frac{1}{2}(n+1-i), \frac{1}{2}(q-1)]$ *distribution.*

PROOF. The characterization of the first factor in (13) corresponds to that of V_1 with the exponents of X_{ig} and $(1 - X_{ig})$ modified by replacing n_g by N_g. The second term is $U_{p,q-1,n}^{\frac{1}{2}N}$ and its characterization follows from Theorem 8.4.1. Q.E.D.

10.4.2. Moments of the Distributions

We now find the moments of V_1 and of W. Since $0 \le V_1 \le 1$ and $0 \le W \le 1$, the moments determine the distributions uniquely. The hth moment of V_1 we find from the characterization of the distribution in Theorem 10.4.2.

$$(15)\quad \mathscr{E}V_1^h = \prod_{g=2}^{q}\left\{\prod_{i=1}^{p}\mathscr{E}X_{ig}^{\frac{1}{2}(n_1+\cdots+n_{g-1})h}(1-X_{ig})^{\frac{1}{2}n_g h}\prod_{i=2}^{p}\mathscr{E}Y_{ig}^{\frac{1}{2}(n_1+\cdots+n_g)h}\right\}$$

$$= \prod_{g=2}^{q}\left\{\prod_{i=1}^{p}\frac{\Gamma\left[\frac{1}{2}(n_1+\cdots+n_{g-1})(1+h)-\frac{1}{2}(i-1)\right]}{\Gamma\left[\frac{1}{2}(n_1+\cdots+n_{g-1}-i+1)\right]}\right.$$

$$\times\frac{\Gamma\left[\frac{1}{2}n_g(1+h)-\frac{1}{2}(i-1)\right]\Gamma\left[\frac{1}{2}(n_1+\cdots+n_g)-i+1\right]}{\Gamma\left[\frac{1}{2}(n_g-i+1)\right]\Gamma\left[\frac{1}{2}(n_1+\cdots+n_g)(1+h)-i+1\right]}$$

$$\left.\times\prod_{i=2}^{p}\frac{\Gamma\left[\frac{1}{2}(n_1+\cdots+n_g)(1+h)-i+1\right]\Gamma\left[\frac{1}{2}(n_1+\cdots+n_g-i+1)\right]}{\Gamma\left[\frac{1}{2}(n_1+\cdots+n_g)-i+1\right]\Gamma\left[\frac{1}{2}(n_1+\cdots+n_g)(1+h)-\frac{1}{2}(i-1)\right]}\right\}$$

$$= \prod_{i=1}^{p}\left\{\frac{\Gamma\left[\frac{1}{2}(n+1-i)\right]}{\Gamma\left[\frac{1}{2}(n+hn+1-i)\right]}\prod_{g=1}^{q}\frac{\Gamma\left[\frac{1}{2}(n_g+hn_g+1-i)\right]}{\Gamma\left[\frac{1}{2}(n_g+1-i)\right]}\right\}$$

$$= \frac{\Gamma_p\left(\frac{1}{2}n\right)}{\Gamma_p\left[\frac{1}{2}(n+hn)\right]}\prod_{g=1}^{q}\frac{\Gamma_p\left[\frac{1}{2}(n_g+hn_g)\right]}{\Gamma_p\left(\frac{1}{2}n_g\right)}.$$

The hth moment of W can be found from its representation in Theorem 10.4.3. We have

(16)

$$\mathscr{E}W^h = \prod_{g=2}^{q}\prod_{i=1}^{p}\mathscr{E}X_{ig}^{\frac{1}{2}(N_1+\cdots+N_{g-1})h}\left(1-X_{ig}\right)^{\frac{1}{2}N_g h}\prod_{i=2}^{p}\mathscr{E}Y_{ig}^{\frac{1}{2}(N_1+\cdots+N_g)h}\mathscr{E}U_{p,q-1,n}^{\frac{1}{2}Nh}$$

$$= \prod_{g=2}^{q}\left\{\prod_{i=1}^{p}\frac{\Gamma\left[\frac{1}{2}(n_1+\cdots+n_{g-1}+1-i)+\frac{1}{2}h(N_1+\cdots+N_{g-1})\right]}{\Gamma\left[\frac{1}{2}(n_1+\cdots+n_{g-1}+1-i)\right]\Gamma\left[\frac{1}{2}(n_g+1-i)\right]}\right.$$

$$\times\frac{\Gamma\left[\frac{1}{2}(n_g+1-i+N_g h)\right]\Gamma\left[\frac{1}{2}(n_1+\cdots+n_g)-i+1\right]}{\Gamma\left[\frac{1}{2}(n_1+\cdots+n_g)+\frac{1}{2}h(N_1+\cdots+N_g)+1-i\right]}$$

$$\times\prod_{i=2}^{p}\frac{\Gamma\left[\frac{1}{2}(n_1+\cdots+n_g)+\frac{1}{2}h(N_1+\cdots+N_g)+1-i\right]}{\Gamma\left[\frac{1}{2}(n_1+\cdots+n_g)+1-i\right]}$$

$$\left.\times\frac{\Gamma\left[\frac{1}{2}(n_1+\cdots+n_g+1-i)\right]}{\Gamma\left[\frac{1}{2}(n_1+\cdots+n_g+1-i)+\frac{1}{2}h(N_1+\cdots+N_g)\right]}\right\}$$

$$\times\prod_{i=1}^{p}\frac{\Gamma\left[\frac{1}{2}(n+1-i+hN)\right]\Gamma\left[\frac{1}{2}(N-i)\right]}{\Gamma\left[\frac{1}{2}(n+1-i)\right]\Gamma\left[\frac{1}{2}(N+hN-i)\right]}$$

$$= \prod_{i=1}^{p}\left\{\prod_{g=1}^{q}\frac{\Gamma\left[\frac{1}{2}(N_g+hN_g-i)\right]}{\Gamma\left[\frac{1}{2}(N_g-i)\right]}\right\}\frac{\Gamma\left[\frac{1}{2}(N-i)\right]}{\Gamma\left[\frac{1}{2}(N+hN-i)\right]}$$

$$= \frac{\Gamma_p\left(\frac{1}{2}n\right)}{\Gamma_p\left(\frac{1}{2}n+\frac{1}{2}hN\right)}\prod_{g=1}^{q}\frac{\Gamma_p\left[\frac{1}{2}(n_g+hN_g)\right]}{\Gamma_p\left(\frac{1}{2}n_g\right)}.$$

We summarize in the following theorem:

THEOREM 10.4.4. *Let V_1 be the criterion defined by* (10) *of Section* 10.2 *for testing the hypothesis that H_1: $\boldsymbol{\Sigma}_1 = \cdots = \boldsymbol{\Sigma}_q$, where $\boldsymbol{A}_g$ is n_g times the sample covariance matrix and $n_g + 1$ is the size of the sample from the* gth *population; let W be the criterion defined by* (13) *for testing the hypothesis H: $\boldsymbol{\mu}_1 = \cdots = \boldsymbol{\mu}_q$*

and H_1, *where* $\boldsymbol{B} = \boldsymbol{A} + \Sigma_g N_g(\bar{\boldsymbol{x}}^{(g)} - \bar{\boldsymbol{x}})(\bar{\boldsymbol{x}}^{(g)} - \bar{\boldsymbol{x}})'$. *The* hth *moment of* V_1 *when* H_1 *is true is given by* (15). *The* hth *moment of* W, *the criterion for testing* H, *is given by* (16).

This theorem was first proved by Wilks (1932a). See Problem 5 for an alternative approach.

If p is even, say $p = 2r$, we can use the duplication formula for the gamma function $[\Gamma(\alpha + \frac{1}{2})\Gamma(\alpha + 1) = \sqrt{\pi}\,\Gamma(2\alpha + 1)2^{-2\alpha}]$. Then

$$(17)\quad \mathscr{E}V_1^h = \prod_{j=1}^{r}\left\{\left[\prod_{g=1}^{q}\frac{\Gamma(n_g + hn_g + 1 - 2j)}{\Gamma(n_g + 1 - 2j)}\right]\frac{\Gamma(n + 1 - 2j)}{\Gamma(n + hn + 1 - 2j)}\right\}$$

and

$$(18)\quad \mathscr{E}W^h = \prod_{j=1}^{r}\left\{\left[\prod_{g=1}^{q}\frac{\Gamma(n_g + hN_g + 1 - 2j)}{\Gamma(n_g + 1 - 2j)}\right]\frac{\Gamma(N - 2j)}{\Gamma(N + hN - 2j)}\right\}.$$

In principle the distributions of the factors can be integrated to obtain the distributions of V_1 and W. In Section 10.6 we consider V_1 when $p = 2$, $q = 2$ (the case of $p = 1$, $q = 2$ being a function of an F-statistic). In other cases, the integrals become unmanageable. To find probabilities we use the asymptotic expansion given in the next section. Box (1949) has given some other approximate distributions.

10.4.3. Step-down Tests

The characterizations of the distributions of the criteria in terms of independent factors suggests testing the hypotheses H_1 and H by testing component hypotheses sequentially. First, we consider testing H_1: $\Sigma_1 = \Sigma_2$ for $q = 2$. Let

$$(19)\quad \boldsymbol{X}_{(i)}^{(g)} = \begin{pmatrix} \boldsymbol{X}_{(i-1)}^{(g)} \\ X_i^{(g)} \end{pmatrix}, \qquad \boldsymbol{\mu}_{(i)}^{(g)} = \begin{pmatrix} \boldsymbol{\mu}_{(i-1)}^{(g)} \\ \mu_i^{(g)} \end{pmatrix}, \qquad \Sigma_i^{(g)} = \begin{bmatrix} \Sigma_{(i-1)}^{(g)} & \boldsymbol{\sigma}_{(i)}^{(g)\prime} \\ \boldsymbol{\sigma}_{(i)}^{(g)} & \sigma_{ii}^{(g)} \end{bmatrix},$$

$$i = 2, \ldots, p, \quad g = 1, 2.$$

The conditional distribution of $X_i^{(g)}$ given $\boldsymbol{X}_{(i-1)}^{(g)} = \boldsymbol{x}_{(i-1)}^{(g)}$ is

$$N\left[\mu_i^{(g)} + \boldsymbol{\sigma}_{(i)}^{(g)\prime}\boldsymbol{\Sigma}_{i-1}^{-1}\left(\boldsymbol{x}_{(i-1)}^{(g)} - \boldsymbol{\mu}_{(i-1)}^{(g)}\right), \sigma_{ii\cdot i-1}^{(g)}\right], \tag{20}$$

where $\sigma_{ii\cdot i-1}^{(g)} = \sigma_{ii}^{(g)} - \boldsymbol{\sigma}_{(i)}^{(g)\prime}\boldsymbol{\Sigma}_{i-1}^{(g)-1}\boldsymbol{\sigma}_{(i)}^{(g)}$. It is assumed that the components of $\boldsymbol{X}$ have been numbered in descending order of importance. At the ith step the component hypothesis $\sigma_{ii\cdot i-1}^{(1)} = \sigma_{ii\cdot i-1}^{(2)}$ is tested at significance level ε_i by means of an F-test based on $s_{ii\cdot i-1}^{(1)}/s_{ii\cdot i-1}^{(2)}$; $\boldsymbol{S}_1$ and $\boldsymbol{S}_2$ are partitioned like $\boldsymbol{\Sigma}^{(1)}$ and $\boldsymbol{\Sigma}^{(2)}$. If that hypothesis is accepted, then the hypothesis $\boldsymbol{\sigma}_{(i)}^{(1)} = \boldsymbol{\sigma}_{(i)}^{(2)}$ (or $\boldsymbol{\Sigma}_{i-1}^{(1)-1}\boldsymbol{\sigma}_{(i)}^{(1)} = \boldsymbol{\Sigma}_{i-1}^{(2)-1}\boldsymbol{\sigma}_{(i)}^{(2)}$) is tested at significance level δ_i on the assumption that $\boldsymbol{\Sigma}_{i-1}^{(1)} = \boldsymbol{\Sigma}_{i-1}^{(2)}$ (a hypothesis previously accepted). The criterion is

$$\frac{\left(\boldsymbol{S}_{i-1}^{(1)-1}\boldsymbol{s}_{(i)}^{(1)} - \boldsymbol{S}_{i-1}^{(2)-1}\boldsymbol{s}_{(i)}^{(2)}\right)'\left(\boldsymbol{S}_{i-1}^{(1)-1} + \boldsymbol{S}_{i-1}^{(2)-1}\right)^{-1}\left(\boldsymbol{S}_{i-1}^{(1)-1}\boldsymbol{s}_{(i)}^{(1)} - \boldsymbol{S}_{i-1}^{(2)-1}\boldsymbol{s}_{(i)}^{(2)}\right)}{(i-1)s_{ii\cdot i-1}}, \tag{21}$$

where $(n_1 + n_2 - 2i + 2)s_{ii\cdot i-1} = (n_1 - i + 1)s_{ii\cdot i-1}^{(1)} + (n_2 - i + 1)s_{ii\cdot i-1}^{(2)}$. Under the null hypothesis (21) has the F-distribution with $i - 1$ and $n_1 + n_2 - 2i + 2$ degrees of freedom. If this hypothesis is accepted, the $(i + 1)$st step is taken. The overall hypothesis $\boldsymbol{\Sigma}_1 = \boldsymbol{\Sigma}_2$ is accepted if the $2p - 1$ component hypotheses are accepted. (At the first step, $\boldsymbol{\sigma}_{(1)}^{(g)}$ is vacuous.) The overall significance level is

$$1 - \prod_{i=1}^{p}(1 - \varepsilon_i)\prod_{i=2}^{p}(1 - \delta_i). \tag{22}$$

If any component null hypothesis is rejected, the overall hypothesis is rejected.

If $q > 2$, the null hypotheses H_1: $\boldsymbol{\Sigma}_1 = \cdots = \boldsymbol{\Sigma}_q$ is broken down into a sequence of hypotheses $[1/(g-1)](\boldsymbol{\Sigma}_1 + \cdots + \boldsymbol{\Sigma}_{g-1}) = \boldsymbol{\Sigma}_g$ and tested sequentially. Each such matrix hypothesis is tested as $\boldsymbol{\Sigma}_1 = \boldsymbol{\Sigma}_2$ with $\boldsymbol{S}_2$ replaced by $\boldsymbol{S}_g$ and $\boldsymbol{S}_1$ replaced by $[1/(n_1 + \cdots + n_{g-1})](\boldsymbol{A}_1 + \cdots + \boldsymbol{A}_{g-1})$.

In the case of the hypothesis H, consider first $q = 2$, $\boldsymbol{\Sigma}_1 = \boldsymbol{\Sigma}_2$, and $\boldsymbol{\mu}^{(1)} = \boldsymbol{\mu}^{(2)}$. One can test $\boldsymbol{\Sigma}_1 = \boldsymbol{\Sigma}_2$. The steps for testing $\boldsymbol{\mu}^{(1)} = \boldsymbol{\mu}^{(2)}$ consists of t-tests for $\mu_i^{(1)} = \mu_i^{(2)}$ based on the conditional distribution of $X_i^{(1)}$ and $X_i^{(2)}$ given $\bar{\boldsymbol{x}}_{(i-1)}^{(1)}$ and $\bar{\boldsymbol{x}}_{(i-1)}^{(2)}$. Alternatively one can test in sequence the equality of the conditional distributions of $X_i^{(1)}$ and $X_i^{(2)}$ given $\boldsymbol{x}_{(i-1)}^{(1)}$ and $\boldsymbol{x}_{(i-1)}^{(2)}$.

For $q > 2$, the hypothesis $\boldsymbol{\Sigma}_1 = \cdots = \boldsymbol{\Sigma}_q$ can be tested, and then $\boldsymbol{\mu}_1 = \cdots = \boldsymbol{\mu}_q$. Alternatively, one can test $[1/(g-1)](\boldsymbol{\Sigma}_1 + \cdots + \boldsymbol{\Sigma}_{g-1}) = \boldsymbol{\Sigma}_g$ and $[1/(g-1)](\boldsymbol{\mu}^{(1)} + \cdots + \boldsymbol{\mu}^{(g-1)}) = \boldsymbol{\mu}^{(g)}$.

10.5. ASYMPTOTIC EXPANSIONS OF THE DISTRIBUTIONS OF THE CRITERIA

Again we make use of Theorem 8.5.1 to obtain asymptotic expansions of the distributions of V_1 and of λ. We assume that $n_g = k_g n$, where $\sum_{g=1}^q k_g = 1$. The asymptotic expansion is in terms of n increasing with $k_1, \dots, k_q$ fixed. (We could assume only $\lim n_g/n = k_g > 0$.)

The hth moment of

$$(1) \quad \lambda_1^* = V_1 \cdot \frac{n^{\frac{1}{2}pn}}{\prod_{g=1}^q n_g^{\frac{1}{2}pn_g}} = V_1 \cdot \prod_{g=1}^q \left(\frac{n}{n_g}\right)^{\frac{1}{2}pn_g} = \left[\prod_{g=1}^q \left(\frac{1}{k_g}\right)^{k_g}\right]^{\frac{1}{2}pn} V_1$$

is

$$(2) \quad \mathscr{E}\lambda_1^{*h} = K\left(\frac{\prod_{j=1}^p \left(\frac{1}{2}n\right)^{\frac{1}{2}n}}{\prod_{g=1}^q \prod_{i=1}^p \left(\frac{1}{2}n_g\right)^{\frac{1}{2}n_g}}\right)^h \frac{\prod_{g=1}^q \prod_{i=1}^p \Gamma\left[\frac{1}{2}n_g(1+h) + \frac{1}{2}(1-i)\right]}{\prod_{j=1}^p \Gamma\left[\frac{1}{2}n(1+h) + \frac{1}{2}(1-j)\right]}.$$

This is of the form of (1) of Section 8.6 with

$$(3) \quad \begin{aligned} &b = p, && y_j = \tfrac{1}{2}n, && \eta_j = \tfrac{1}{2}(1-j), && j = 1, \dots, p, \\ &a = pq, && x_k = \tfrac{1}{2}n_g, && k = (g-1)p+1, \dots, gp, && g = 1, \dots, q, \\ &\xi_k = \tfrac{1}{2}(1-i), && && k = i, p+i, \dots, (q-1)p+i, && i = 1, \dots, p. \end{aligned}$$

Then

$$(4) \quad \begin{aligned} f &= -2\left[\sum \xi_k - \sum \eta_j - \tfrac{1}{2}(a-b)\right] \\ &= -\left[q\sum_{i=1}^p (1-i) - \sum_{j=1}^p (1-j) - (qp-p)\right] \\ &= -\left[-q\tfrac{1}{2}p(p-1) + \tfrac{1}{2}p(p-1) - (q-1)p\right] \\ &= \tfrac{1}{2}(q-1)p(p+1), \end{aligned}$$

$\varepsilon_j = \frac{1}{2}(1-\rho)n$, $j = 1, \dots, p$, and $\beta_k = \frac{1}{2}(1-\rho)n_g = \frac{1}{2}(1-\rho)k_g n$, $k = (g-1)p+1, \dots, gp$.

In order to make the second term in the expansion vanish, we take ρ as

$$\rho = 1 - \left(\sum_{g=1}^{q} \frac{1}{n_g} - \frac{1}{n} \right) \frac{2p^2 + 3p - 1}{6(p+1)(q-1)}. \tag{5}$$

Then

$$\omega_2 = \frac{p(p+1)\left[(p-1)(p+2)\left(\sum_{g=1}^{q} \frac{1}{n_g^2} - \frac{1}{n^2} \right) - 6(q-1)(1-\rho)^2 \right]}{48\rho^2}. \tag{6}$$

Thus

$$\begin{aligned} \Pr\{-2\rho \log \lambda_1^* \le z\} = \Pr\{\chi_f^2 \le z\} &+ \omega_2\left[\Pr\{\chi_{f+4}^2 \le z\} \right. \\ &\left. - \Pr\{\chi_f^2 \le z\}\right] + O(n^{-3}). \end{aligned} \tag{7}$$

Let $\lambda = W N^{\frac{1}{2}pN} \prod_{g=1}^{q} N_g^{-\frac{1}{2}pN_g}$. The hth moment is

$$\mathscr{E}\lambda^h = K \left[\frac{\prod_{j=1}^{p} \left(\frac{1}{2}N\right)^{\frac{1}{2}N}}{\prod_{g=1}^{q} \prod_{i=1}^{p} \left(\frac{1}{2}N_g\right)^{\frac{1}{2}N_g}} \right]^h \frac{\prod_{g=1}^{q} \prod_{i=1}^{p} \Gamma\left[\frac{1}{2}N_g(1+h) - \frac{1}{2}i\right]}{\prod_{j=1}^{p} \Gamma\left[\frac{1}{2}N(1+h) - \frac{1}{2}j\right]}. \tag{8}$$

This is the form (1) of Section 8.5 with

$$\begin{aligned} &b = p, && y_j = \tfrac{1}{2}N = \tfrac{1}{2}\sum_{g=1}^{q} N_g, && \eta_j = -\tfrac{1}{2}j, \qquad j = 1, \dots, p, \\ &a = pq, && x_k = \tfrac{1}{2}N_g, && k = (g-1)p + 1, \dots, gp,\ g = 1, \dots, q, \\ &\xi_k = -\tfrac{1}{2}i, && && k = i, p+i, \dots, (q-1)p + i,\ i = 1, \dots, p. \end{aligned} \tag{9}$$

The basic number of degrees of freedom is $f = \frac{1}{2}p(p+3)(q-1)$. We use (11) of Section 8.5 with $\beta_k = (1-\rho)x_k$ and $\varepsilon_j = (1-\rho)y_j$. To make $\omega_1 = 0$, we take

$$\rho = 1 - \left(\sum_{g=1}^{q} \frac{1}{N_g} - \frac{1}{N} \right) \frac{2p^2 + 9p + 11}{6(q-1)(p+3)}. \tag{10}$$

Then

(11)
$$\omega_2 = \frac{p(p+3)}{48\rho^2}\left[\sum_{g=1}^{q}\left(\frac{1}{N_g^2} - \frac{1}{N^2}\right)(p+1)(p+2) - 6(1-\rho)^2(q-1)\right].$$

The asymptotic expansion of the distribution of $-2\rho\log\lambda$ is

(12)
$$\Pr\{-2\rho\log\lambda \le z\} = \Pr\{\chi_f^2 \le z\} + \omega_2\left[\Pr\{\chi_{f+4}^2 \le z\} - \Pr\{\chi_f^2 \le z\}\right] + O(n^{-3}).$$

Box (1949) has considered the case of λ_1^* in considerable detail. In addition to this expansion he has considered the use of (13) of Section 8.6. He also has given an F-approximation.

As an example, we use one given by E. S. Pearson and Wilks (1933). The measurements are made on tensile strength (X_1) and hardness (X_2) of aluminum die-castings. There are 12 observations in each of five samples. The observed sums of squares and cross-products in the five samples are

(13)
$$A_1 = \begin{pmatrix} 78.948 & 214.18 \\ 214.18 & 1247.18 \end{pmatrix},$$
$$A_2 = \begin{pmatrix} 223.695 & 657.62 \\ 657.62 & 2519.31 \end{pmatrix},$$
$$A_3 = \begin{pmatrix} 57.448 & 190.63 \\ 190.63 & 1241.78 \end{pmatrix},$$
$$A_4 = \begin{pmatrix} 187.618 & 375.91 \\ 375.91 & 1473.44 \end{pmatrix},$$
$$A_5 = \begin{pmatrix} 88.456 & 259.18 \\ 259.18 & 1171.73 \end{pmatrix},$$

and the sum of these is

(14)
$$\sum A_i = \begin{pmatrix} 636.165 & 1697.52 \\ 1697.52 & 7653.44 \end{pmatrix}.$$

Then $-\log\lambda_1^*$ is 5.399. To use the asymptotic expansion we find $\rho = 152/165$

$= 0.9212$ and $\omega_2 = 0.0022$. Since ω_2 is small, we can consider $-2\rho\log\lambda_1^*$ as χ^2 with 12 degrees of freedom. Our observed criterion is, therefore, clearly not significant.

Table 5 [due to Korin (1969)] of Appendix B gives 5% significance points for $-2\log\lambda_1^*$ for $N_1 = \cdots = N_q$ for various q, small values of N_g, and $p = 2(1)6$.

The limiting distribution of the criterion (19) of Section 10.1 is also χ_f^2. An asymptotic expansion of the distribution has been given by Nagao (1973b) to terms of order $1/n$ involving χ^2-distributions with f, $f+2$, $f+4$, and $f+6$ degrees of freedom.

10.6. THE CASE OF TWO POPULATIONS

10.6.1. Invariant Tests

When $q = 2$, the null hypothesis H_1 is $\Sigma_1 = \Sigma_2$. It is invariant with respect to transformations

$$x^{*(1)} = Cx^{(1)} + \nu^{(1)}, \qquad x^{*(2)} = Cx^{(2)} + \nu^{(2)}, \tag{1}$$

where C is nonsingular. The maximal invariant of the parameters under the transformation of locations ($C = I$) is the pair of covariance matrices Σ_1, Σ_2, and the maximal invariant of the sufficient statistics $\bar{x}^{(1)}$, S_1, $\bar{x}^{(2)}$, S_2 is the pair of matrices S_1, S_2 (or equivalently A_1, A_2). The transformation (1) induces the transformations $\Sigma_1^* = C\Sigma_1 C'$, $\Sigma_2^* = C\Sigma_2 C'$, $S_1^* = CS_1C'$, and $S_2^* = CS_2C'$. The roots $\lambda_1 \geq \lambda_2 \geq \cdots \geq \lambda_p$ of

$$|\Sigma_1 - \lambda\Sigma_2| = 0 \tag{2}$$

are invariant under these transformations since

$$|\Sigma_1^* - \lambda\Sigma_2^*| = |C\Sigma_1C' - \lambda C\Sigma_2C'| = |CC'|\,|\Sigma_1 - \lambda\Sigma_2|. \tag{3}$$

Moreover, the roots are the only invariants because there exists a nonsingular matrix C such that

$$C\Sigma_1C' = \Lambda, \qquad C\Sigma_2C' = I, \tag{4}$$

where Λ is the diagonal matrix with λ_i as the ith diagonal element, $i = 1, \ldots, p$.

(See Theorem A.2.2 of Appendix A.) Similarly the maximal invariants of S_1 and S_2 are the roots $l_1 \geq l_2 \geq \cdots \geq l_p$ of

$$|S_1 - lS_2| = 0. \tag{5}$$

THEOREM 10.6.1. *The maximal invariant of the parameters of* $N(\mu^{(1)}, \Sigma_1)$ *and* $N(\mu^{(2)}, \Sigma_2)$ *under the transformation* (1) *is the set of roots* $\lambda_1 \geq \cdots \geq \lambda_p$ *of* (2). *The maximal invariant of the sufficient statistics* $\bar{x}^{(1)}$, S_1, $\bar{x}^{(2)}$, S_2 *is the set of roots* $l_1 \geq \cdots \geq l_p$ *of* (5).

Any invariant test criterion can be expressed in terms of the roots $l_1, \dots, l_p$. The criterion V_1 is $n_1^{\frac{1}{2}pn_1} n_2^{\frac{1}{2}pn_2}$ times

$$\frac{|S_1|^{\frac{1}{2}n_1} |S_2|^{\frac{1}{2}n_2}}{|n_1 S_1 + n_2 S_2|^{\frac{1}{2}n}} = \frac{|L|^{\frac{1}{2}n_1} |I|^{\frac{1}{2}n_2}}{|n_1 L + n_2 I|^{\frac{1}{2}n}} = \prod_{i=1}^{p} \frac{l_i^{\frac{1}{2}n_1}}{(n_1 l_i + n_2)^{\frac{1}{2}n}}, \tag{6}$$

where L is the diagonal matrix with l_i as the ith diagonal element. The null hypothesis is rejected if the smaller roots are too small or if the larger roots are too large, or both.

The null hypothesis is that $\lambda_1 = \cdots = \lambda_p = 1$. Any useful invariant test of the null hypothesis has a rejection region in the space of $l_1, \dots, l_p$ that includes the points that in some sense are far from $l_1 = \cdots = l_p = 1$. The power of an invariant test depends on the parameters through the roots $\lambda_1, \dots, \lambda_p$.

The criterion (19) of Section 10.2 is (with $nS = n_1 S_1 + n_2 S_2$)

$$\begin{aligned}
(7)\quad & \tfrac{1}{2}n_1 \operatorname{tr}[(S_1 - S)S^{-1}]^2 + \tfrac{1}{2}n_2 \operatorname{tr}[(S_2 - S)S^{-1}]^2 \\
&= \tfrac{1}{2}n_1 \operatorname{tr}\left[C(S_1 - S)C'(CSC')^{-1}\right]^2 \\
&\qquad + \tfrac{1}{2}n_2 \operatorname{tr}\left[C(S_2 - S)C'(CSC')^{-1}\right]^2 \\
&= \tfrac{1}{2}n_1 \operatorname{tr}\left[\left\{L - \left(\frac{n_1}{n}L + \frac{n_2}{n}I\right)\right\}\left(\frac{n_1}{n}L + \frac{n_2}{n}I\right)^{-1}\right]^2 \\
&\qquad + \tfrac{1}{2}n_2 \operatorname{tr}\left[\left\{I - \left(\frac{n_1}{n}L + \frac{n_2}{n}I\right)\right\}\left(\frac{n_1}{n}L + \frac{n_2}{n}I\right)^{-1}\right]^2 \\
&= \tfrac{1}{2}n_1 n_2 n \sum_{i=1}^{p} \frac{(l_i - 1)^2}{(n_1 l_i + n_2)^2}.
\end{aligned}$$

This criterion is a measure of how close $l_1, \dots, l_p$ are to 1; the hypothesis is rejected if the measure is too large. Under the null hypothesis, (7) has the χ^2-distribution with $f = \frac{1}{2}p(p+1)$ degrees of freedom as $n_1 \to \infty$, $n_2 \to \infty$ and n_1/n_2 approaches a positive constant. Nagao (1973b) has given an asymptotic expansion of this distribution to terms of order $1/n$.

Roy (1953) suggested a test based on the largest and smallest roots, l_1 and l_p. The procedure is to reject the null hypothesis if $l_1 > k_1$ or if $l_p < k_p$, where k_1 and k_p are chosen so that the probability of rejection when $\Lambda = \boldsymbol{I}$ is the desired significance level. Roy (1957) proposed determining k_1 and k_p so that the test is locally unbiased; that is, that the power functions have a relative minimum at $\Lambda = \boldsymbol{I}$. Since it is hard to determine k_1 and k_p on this basis, other proposals have been made. The limit k_1 can be determined so that $\Pr\{l_1 > k_1 | H_1\}$ is one-half of the significance level, or $\Pr\{l_p < k_p | H_1\}$ is one-half of the significance level, or $k_1 + k_p = 2$, or $k_1 k_p = 1$. In principle k_1 and k_p can be determined from the distribution of the roots, given in Section 13.2. Schuurmann, Waikar, and Krishnaiah (1975) and Chu and Pillai (1979) have given some exact values of k_1 and k_p for small values of p. Chu and Pillai (1979) have made some power comparisons of several test procedures.

In the case of $p = 1$ the only invariant of the sufficient statistics is S_1/S_2, which is the usual F-statistic with n_1 and n_2 degrees of freedom. The criterion V_1 is $(A_1/A_2)^{\frac{1}{2}n_1}\,[1 + (A_1/A_2)]^{-\frac{1}{2}n}$; the critical region V_1 less than a constant is equivalent to a two-tailed critical region for the F-statistic. The quantity $n(B - A)/A$ has an independent F-distribution with 1 and n degrees of freedom. (See Section 10.3.)

In the case of $p = 2$, the hth moment of V_1 is from (15) of Section 10.4

$$\mathscr{E}V_1^h = \frac{\Gamma(n_1 + hn_1 - 1)\Gamma(n_2 + hn_2 - 1)\Gamma(n - 1)}{\Gamma(n_1 - 1)\Gamma(n_2 - 1)\Gamma(n + hn - 1)} \tag{8}$$

$$= \mathscr{E}\left[X_1^{n_1}(1 - X_1)^{n_2} X_2^{n_1 + n_2}\right]^h,$$

where X_1 and X_2 are independently distributed according to $\beta(x|n_1 - 1, n_2 - 1)$ and $\beta(x|n_1 + n_2 - 2, 1)$, respectively. Then $\Pr\{V_1 \le v\}$ can be found by integration. (See Problems 8 and 9.)

Anderson (1965a) has shown that a confidence interval for $\boldsymbol{a}'\Sigma_1\boldsymbol{a}/\boldsymbol{a}'\Sigma_2\boldsymbol{a}$ for all $\boldsymbol{a}$ with confidence coefficient ε is given by $(l_p/U, l_1/L)$, where $\Pr\{(n_2 - p + 1)L \le n_2 F_{n_1, n_2 - p + 1}\} \times \Pr\{(n_1 - p + 1)F_{n_1 - p + 1, n_2} \le n_1 U\} = 1 - \varepsilon$.

10.6.2. Components of Variance

In Section 8.8 we considered what is equivalent to the one-way analysis of variance with fixed effects. We can write the model in the balanced case $(N_1 = N_2 = \cdots = N_q)$ as

$$(9)\qquad \begin{aligned} \boldsymbol{X}_\alpha^{(g)} &= \boldsymbol{\mu}^{(g)} + \boldsymbol{U}_\alpha^{(g)} \\ &= \boldsymbol{\mu} + \boldsymbol{\nu}_g + \boldsymbol{U}_\alpha^{(g)}, \qquad \alpha = 1, \dots, M,\ g = 1, \dots, q, \end{aligned}$$

where $\mathscr{E}\boldsymbol{U}^{(g)} = \boldsymbol{0}$ and $\mathscr{E}\boldsymbol{U}^{(g)}\boldsymbol{U}^{(g)\prime} = \boldsymbol{\Sigma}$, $\boldsymbol{\nu}_g = \boldsymbol{\mu}^{(g)} - \boldsymbol{\mu}$ and $\boldsymbol{\mu} = (1/q)\Sigma_{g=1}^{q}\boldsymbol{\mu}^{(g)}$ $(\Sigma_{g=1}^{q}\boldsymbol{\nu}_g = \boldsymbol{0})$. The null hypothesis of no effects is $\boldsymbol{\nu}_1 = \cdots = \boldsymbol{\nu}_q = \boldsymbol{0}$. Let $\bar{\boldsymbol{x}}^{(g)} = (1/M)\Sigma_{\alpha=1}^{M}\boldsymbol{x}_\alpha^{(g)}$ and $\bar{\boldsymbol{x}} = (1/q)\Sigma_{g=1}^{q}\bar{\boldsymbol{x}}^{(g)}$. The analysis of variance table is

Source	Sum of Squares	Degrees of Freedom
Effect	$\boldsymbol{H} = M\sum_{g=1}^{q}(\bar{\boldsymbol{x}}^{(g)} - \bar{\boldsymbol{x}})(\bar{\boldsymbol{x}}^{(g)} - \bar{\boldsymbol{x}})'$	$q - 1$
Error	$\boldsymbol{G} = \sum_{g=1}^{q}\sum_{\alpha=1}^{M}(\boldsymbol{x}_\alpha^{(g)} - \bar{\boldsymbol{x}}^{(g)})(\boldsymbol{x}_\alpha^{(g)} - \bar{\boldsymbol{x}}^{(g)})'$	$q(M - 1)$
Total	$\sum_{g=1}^{q}\sum_{\alpha=1}^{M}(\boldsymbol{x}_\alpha^{(g)} - \bar{\boldsymbol{x}})(\boldsymbol{x}_\alpha^{(g)} - \bar{\boldsymbol{x}})'$	$qM - 1$

Invariant tests of the null hypothesis of no effects are based on the roots of $|\boldsymbol{H} - m\boldsymbol{G}| = 0$ or of $|\boldsymbol{S}_h - l\boldsymbol{S}_e| = 0$, where $\boldsymbol{S}_h = (1/(q-1))\boldsymbol{H}$ and $\boldsymbol{S}_e = (1/q(M-1))\boldsymbol{G}$. The null hypothesis is rejected if one or more of the roots is too large. The error matrix $\boldsymbol{G}$ has the distribution $W(\boldsymbol{\Sigma}, q(M-1))$. The effects matrix $\boldsymbol{H}$ has the distribution $W(\boldsymbol{\Sigma}, q-1)$ when the null hypothesis is true and has the noncentral Wishart distribution when the null hypothesis is not true; its expected value is

$$(10)\qquad \begin{aligned} \mathscr{E}\boldsymbol{H} &= (q-1)\boldsymbol{\Sigma} + M\sum_{g=1}^{q}(\boldsymbol{\mu}^{(g)} - \boldsymbol{\mu})(\boldsymbol{\mu}^{(g)} - \boldsymbol{\mu})' \\ &= (q-1)\boldsymbol{\Sigma} + M\sum_{g=1}^{q}\boldsymbol{\nu}_g\boldsymbol{\nu}_g'. \end{aligned}$$

The MANOVA model with random effects is

$$(11)\qquad \boldsymbol{X}_\alpha^{(g)} = \boldsymbol{\mu} + \boldsymbol{V}_g + \boldsymbol{U}_\alpha^{(g)}, \qquad \alpha = 1,\dots,M,\ g = 1,\dots,q,$$

where $\boldsymbol{V}_g$ has the distribution $N(\boldsymbol{0}, \boldsymbol{\Theta})$. Then $\boldsymbol{X}_\alpha^{(g)}$ has the distribution $N(\boldsymbol{\mu}, \boldsymbol{\Sigma} + \boldsymbol{\Theta})$. The null hypothesis of no effects is

$$(12)\qquad \boldsymbol{\Theta} = \boldsymbol{0}.$$

In this model $\boldsymbol{G}$ again has the distribution $W(\boldsymbol{\Sigma}, q(M-1))$. Since $\bar{\boldsymbol{X}}^{(g)} = \boldsymbol{\mu} + \boldsymbol{V}_g + \bar{\boldsymbol{U}}^{(g)}$ has the distribution $N(\boldsymbol{\mu}, (1/M)\boldsymbol{\Sigma} + \boldsymbol{\Theta})$, $\boldsymbol{H}$ has the distribution $W(\boldsymbol{\Sigma} + M\boldsymbol{\Theta}, q-1)$. The null hypothesis (12) is equivalent to the equality of the covariance matrices in these two Wishart distributions; that is, $\boldsymbol{\Sigma} = \boldsymbol{\Sigma} + M\boldsymbol{\Theta}$. The matrices $\boldsymbol{G}$ and $\boldsymbol{H}$ correspond to $\boldsymbol{A}_1$ and $\boldsymbol{A}_2$ in Section 10.6.1. However, here the alternative to the null hypothesis is that $(\boldsymbol{\Sigma} + M\boldsymbol{\Theta}) - \boldsymbol{\Sigma}$ is positive semidefinite, rather than $\boldsymbol{\Sigma}_1 \neq \boldsymbol{\Sigma}_2$. The null hypothesis is to be rejected if $\boldsymbol{H}$ is too large relative to $\boldsymbol{G}$. Any of the criteria presented in Section 10.2 can be used to test the null hypothesis here, and its distribution under the null hypothesis is the same as given there.

The likelihood ratio criterion for testing $\boldsymbol{\Theta} = \boldsymbol{0}$ must take into account the fact that $\boldsymbol{\Theta}$ is positive semidefinite; that is, the maximum likelihood estimators of $\boldsymbol{\Sigma}$ and $\boldsymbol{\Sigma} + M\boldsymbol{\Theta}$ under Ω must be such that the estimator of $\boldsymbol{\Theta}$ is positive semidefinite. Let $l_1 > l_2 > \cdots > l_p$ be the roots of

$$(13)\qquad \left| \boldsymbol{H} - l\frac{1}{M-1}\boldsymbol{G} \right| = 0.$$

(Note $\{1/[q(M-1)]\}\boldsymbol{G}$ and $(1/q)\boldsymbol{H}$ maximize the likelihood without regard to $\boldsymbol{\Theta}$ being positive definite.) Let $l_i^* = l_i$ if $l_i > 1$ and let $l_i^* = 1$ if $l_i \leq 1$. Then the likelihood ratio criterion for testing the hypothesis $\boldsymbol{\Theta} = \boldsymbol{0}$ against the alternative $\boldsymbol{\Theta}$ positive semidefinite and $\boldsymbol{\Theta} \neq \boldsymbol{0}$ is

$$(14)\qquad M^{\frac{1}{2}qMp} \prod_{i=1}^{p} \frac{l_i^{*\frac{1}{2}q}}{\left(l_i^* + M - 1\right)^{\frac{1}{2}qM}} = M^{\frac{1}{2}qMk} \prod_{i=1}^{k} \frac{l_i^{\frac{1}{2}q}}{\left(l_i + M - 1\right)^{\frac{1}{2}qM}},$$

where k is the number of roots of (13) greater than 1. [See Anderson (1946b), Morris and Olkin (1964), Klotz and Putter (1969), and Anderson (1984a).]

10.7. TESTING THE HYPOTHESIS THAT A COVARIANCE MATRIX IS PROPORTIONAL TO A GIVEN MATRIX; THE SPHERICITY TEST

10.7.1. The Hypothesis

In many statistical analyses that are considered univariate, the assumption is made that a set of random variables are independent and have a common variance. In this section we consider a test of these assumptions based on repeated sets of observations.

More precisely, we use a sample of p-component vectors $\boldsymbol{x}_1, \dots, \boldsymbol{x}_N$ from $N(\boldsymbol{\mu}, \boldsymbol{\Sigma})$ to test the hypothesis H: $\boldsymbol{\Sigma} = \sigma^2 \boldsymbol{I}$, where σ^2 is not specified. The hypothesis can be given an algebraic interpretation in terms of the characteristic roots of $\boldsymbol{\Sigma}$, that is, the roots of

$$|\boldsymbol{\Sigma} - \phi \boldsymbol{I}| = 0. \tag{1}$$

The hypothesis is true, if and only if, all the roots of (1) are equal.† Another way of putting it is that the arithmetic mean of roots $\phi_1, \dots, \phi_p$ is equal to the geometric mean, that is,

$$\frac{\prod_{i=1}^{p} \phi_i^{1/p}}{\frac{1}{p}\sum_{i=1}^{p} \phi_i} = \frac{|\boldsymbol{\Sigma}|^{1/p}}{(1/p)\operatorname{tr}\boldsymbol{\Sigma}}. \tag{2}$$

The lengths squared of the principal axes of the ellipsoids of constant density are proportional to the roots ϕ_i (see Chapter 11); the hypothesis specifies that these are equal, that is, that the ellipsoids are spheres.

The hypothesis H is equivalent to the more general form, $\boldsymbol{\Psi} = \sigma^2 \boldsymbol{\Psi}_0$, with $\boldsymbol{\Psi}_0$ specified, having observation vectors $\boldsymbol{y}_1, \dots, \boldsymbol{y}_N$ from $N(\boldsymbol{\nu}, \boldsymbol{\Psi})$. Let $\boldsymbol{C}$ be a matrix such that

$$\boldsymbol{C}\boldsymbol{\Psi}_0\boldsymbol{C}' = \boldsymbol{I}, \tag{3}$$

and let $\boldsymbol{\mu}^* = \boldsymbol{C}\boldsymbol{\nu}$, $\boldsymbol{\Sigma}^* = \boldsymbol{C}\boldsymbol{\Psi}\boldsymbol{C}'$, $\boldsymbol{x}_\alpha^* = \boldsymbol{C}\boldsymbol{y}_\alpha$. Then $\boldsymbol{x}_1^*, \dots, \boldsymbol{x}_N^*$ are observations from $N(\boldsymbol{\mu}^*, \boldsymbol{\Sigma}^*)$ and the hypothesis is transformed into H: $\boldsymbol{\Sigma}^* = \sigma^2 \boldsymbol{I}$.

†This follows from the fact that $\boldsymbol{\Sigma} = \boldsymbol{O}'\boldsymbol{\Phi}\boldsymbol{O}$, where $\boldsymbol{\Phi}$ is a diagonal matrix with roots as diagonal elements and $\boldsymbol{O}$ is an orthogonal matrix.

10.7.2. The Criterion

In the canonical form the hypothesis H is a combination of the hypothesis H_1: $\boldsymbol{\Sigma}$ is diagonal or the components of $\boldsymbol{X}$ are independent and H_2: The diagonal elements of $\boldsymbol{\Sigma}$ are equal given $\boldsymbol{\Sigma}$ is diagonal or the variances of the components of $\boldsymbol{X}$ are equal given the components are independent. Thus by Lemma 10.3.1 the likelihood ratio criterion λ for H is the product of the criterion λ_1 for H_1 and λ_2 for H_2. From Section 9.2 we see that the criterion for H_1 is

$$\lambda_1 = \frac{|\boldsymbol{A}|^{\frac{1}{2}N}}{\prod a_{ii}^{\frac{1}{2}N}} = |r_{ij}|^{\frac{1}{2}N}, \tag{4}$$

where

$$\boldsymbol{A} = \sum_{\alpha=1}^{N} (\boldsymbol{x}_\alpha - \bar{\boldsymbol{x}})(\boldsymbol{x}_\alpha - \bar{\boldsymbol{x}})' = (a_{ij}) \tag{5}$$

and $r_{ij} = a_{ij}/\sqrt{a_{ii}a_{jj}}$. We use the results of Section 10.2 to obtain λ_2 by considering the ith component of $\boldsymbol{x}_\alpha$ as the αth observation from the ith population. (p here is q in Section 10.2, N here is N_g there, pN here is N there.) Thus

$$\begin{aligned}\lambda_2 &= \frac{\prod_i \left[\sum_\alpha (x_{i\alpha} - \bar{x}_i)^2\right]^{\frac{1}{2}N}}{\left[\sum_{i,\alpha} (x_{i\alpha} - \bar{x}_i)^2/p\right]^{\frac{1}{2}pN}} \\ &= \frac{\prod a_{ii}^{\frac{1}{2}N}}{(\operatorname{tr} \boldsymbol{A}/p)^{\frac{1}{2}pN}}.\end{aligned} \tag{6}$$

Thus the criterion for H is

$$\lambda = \lambda_1\lambda_2 = \frac{|\boldsymbol{A}|^{\frac{1}{2}N}}{(\operatorname{tr} \boldsymbol{A}/p)^{\frac{1}{2}pN}}. \tag{7}$$

It will be observed that λ resembles (2). If $l_1, \dots, l_p$ are the roots of

$$|\boldsymbol{S} - l\boldsymbol{I}| = 0, \tag{8}$$

where $\boldsymbol{S} = (1/n)\boldsymbol{A}$, the criterion is a power of the ratio of the geometric mean and the arithmetic mean

$$\lambda = \left(\frac{\prod l_i^{1/p}}{\sum l_i/p} \right)^{\frac{1}{2}pN}. \tag{9}$$

Now let us go back to the hypothesis $\boldsymbol{\Psi} = \sigma^2 \boldsymbol{\Psi}_0$, given observation vectors $\boldsymbol{y}_1, \dots, \boldsymbol{y}_N$ from $N(\boldsymbol{\nu}, \boldsymbol{\Psi})$. In the transformed variables $\{\boldsymbol{x}_\alpha^*\}$ the criterion is $|\boldsymbol{A}^*|^{\frac{1}{2}N}(\operatorname{tr} \boldsymbol{A}^*/p)^{-\frac{1}{2}pN}$, where

$$\begin{aligned} \boldsymbol{A}^* &= \sum_{\alpha=1}^{N} (\boldsymbol{x}_\alpha^* - \bar{\boldsymbol{x}}^*)(\boldsymbol{x}_\alpha^* - \bar{\boldsymbol{x}}^*)' \\ &= \boldsymbol{C} \sum_{\alpha=1}^{N} (\boldsymbol{y}_\alpha - \bar{\boldsymbol{y}})(\boldsymbol{y}_\alpha - \bar{\boldsymbol{y}})' \boldsymbol{C}' \\ &= \boldsymbol{CBC}', \end{aligned} \tag{10}$$

where

$$\boldsymbol{B} = \sum_{\alpha=1}^{N} (\boldsymbol{y}_\alpha - \bar{\boldsymbol{y}})(\boldsymbol{y}_\alpha - \bar{\boldsymbol{y}})'. \tag{11}$$

From (3) we have $\boldsymbol{\Psi}_0 = \boldsymbol{C}^{-1}(\boldsymbol{C}')^{-1} = (\boldsymbol{C}'\boldsymbol{C})^{-1}$. Thus

$$\begin{aligned} |\boldsymbol{A}^*| &= \frac{|\boldsymbol{B}|}{|\boldsymbol{\Psi}_0|} = |\boldsymbol{B}\boldsymbol{\Psi}_0^{-1}|, \\ \operatorname{tr} \boldsymbol{A}^* &= \operatorname{tr} \boldsymbol{CBC}' = \operatorname{tr} \boldsymbol{BC}'\boldsymbol{C} \\ &= \operatorname{tr} \boldsymbol{B}\boldsymbol{\Psi}_0^{-1}. \end{aligned} \tag{12}$$

The results can be summarized.

THEOREM 10.7.1. *Given a set of p-component observation vectors $\boldsymbol{y}_1, \dots, \boldsymbol{y}_N$ from $N(\boldsymbol{\nu}, \boldsymbol{\Psi})$, the likelihood ratio criterion for testing the hypothesis H: $\boldsymbol{\Psi} = \sigma^2 \boldsymbol{\Psi}_0$, where $\boldsymbol{\Psi}_0$ is specified and σ^2 is not specified, is*

$$\frac{|\boldsymbol{B}\boldsymbol{\Psi}_0^{-1}|^{\frac{1}{2}N}}{(\operatorname{tr} \boldsymbol{B}\boldsymbol{\Psi}_0^{-1}/p)^{\frac{1}{2}pN}}. \tag{13}$$

Mauchly (1940) gives this criterion and its moments under the null hypothesis.

The maximum likelihood estimator of σ^2 under the null hypothesis is $\operatorname{tr} \boldsymbol{B\Psi}_0^{-1}/(pN)$, which is $\operatorname{tr} \boldsymbol{A}/(pN)$ in canonical form; an unbiased estimator is $\operatorname{tr} \boldsymbol{B\Psi}_0^{-1}/[p(N-1)]$ or $\operatorname{tr} \boldsymbol{A}/[p(N-1)]$ in canonical form [Hotelling (1951)]. Then $\operatorname{tr} \boldsymbol{B\Psi}_0^{-1}/\sigma^2$ has the χ^2-distribution with $p(N-1)$ degrees of freedom.

10.7.3. The Distribution and Moments of the Criterion

The distribution of the likelihood ratio criterion under the null hypothesis can be characterized by the facts that $\lambda = \lambda_1 \lambda_2$ and λ_1 and λ_2 are independent and the characterizations of λ_1 and λ_2. As was observed in Section 7.6, when $\boldsymbol{\Sigma}$ is diagonal the correlation coefficients $\{r_{ij}\}$ are distributed independently of the variances $\{a_{ii}/(N-1)\}$. Since λ_1 depends only on $\{r_{ij}\}$ and λ_2 depends only on $\{a_{ii}\}$, they are independently distributed when the null hypothesis is true. Let $W = \lambda^{2/N}$, $W_1 = \lambda_1^{2/N}$, $W_2 = \lambda_2^{2/N}$. From Theorem 9.3.3, we see that W_1 is distributed as $\prod_{i=2}^p X_i$, where $X_2, \dots, X_p$ are independent and X_i has the density $\beta(x|\frac{1}{2}(n-i+1), \frac{1}{2})$, where $n = N-1$. From Theorem 10.4.2 with $W_2 = p^p V_1^{2/n}$, we find that W_2 is distributed as $p^p \prod_{j=2}^p Y_j^{j-1}(1 - Y_j)$, where $Y_2, \dots, Y_p$ are independent and Y_j has the density $\beta(y|\frac{1}{2}n(j-1), \frac{1}{2}n)$. Then W is distributed as $W_1 W_2$, where W_1 and W_2 are independent.

The moments of W can be found from this characterization or from Theorems 9.3.4 and 10.4.4. We have

$$\mathscr{E} W_1^h = \frac{\Gamma^p(\frac{1}{2}n)}{\Gamma^p(\frac{1}{2}n + h)} \frac{\Gamma_p(\frac{1}{2}n + h)}{\Gamma_p(\frac{1}{2}n)}, \tag{14}$$

$$\mathscr{E} W_2^h = p^{hp} \frac{\Gamma^p(\frac{1}{2}n + h)\Gamma(\frac{1}{2}pn)}{\Gamma^p(\frac{1}{2}n)\Gamma(\frac{1}{2}pn + ph)}. \tag{15}$$

It follows that

$$\mathscr{E} W^h = p^{hp} \frac{\Gamma(\frac{1}{2}pn)}{\Gamma(\frac{1}{2}pn + ph)} \frac{\Gamma_p(\frac{1}{2}n + h)}{\Gamma_p(\frac{1}{2}n)}. \tag{16}$$

For $p = 2$ we have

$$\mathscr{E}W^h = 4^h \frac{\Gamma(n)}{\Gamma(n+2h)} \prod_{i=1}^{2} \frac{\Gamma[\frac{1}{2}(n+1-i)+h]}{\Gamma[\frac{1}{2}(n+1-i)]} \tag{17}$$

$$= \frac{\Gamma(n)\Gamma(n-1+2h)}{\Gamma(n+2h)\Gamma(n-1)} = \frac{n-1}{n-1+2h}$$

$$= (n-1)\int_0^1 z^{n-2+2h}\,dz,$$

by use of the duplication formula for the gamma function. Thus W is distributed as Z^2 where Z has the density $(n-1)z^{n-2}$ and W has the density $\frac{1}{2}(n-1)w^{\frac{1}{2}(n-3)}$. The cdf is

$$\Pr\{W \le w\} = F(w) = w^{\frac{1}{2}(n-1)}. \tag{18}$$

This result can also be found from the joint distribution of l_1, l_2, the roots of (8). The density for $p = 3$, 4, and 6 has been obtained by Consul (1967b). See also Pillai and Nagarsenkar (1971).

10.7.4. Asymptotic Expansion of the Distribution

From (16) we see that the rth moment of $W^{\frac{1}{2}n} = Z$, say, is

$$\mathscr{E}Z^r = Kp^{\frac{1}{2}npr} \frac{\prod_{i=1}^{p}\Gamma[\frac{1}{2}n(1+r)+\frac{1}{2}(1-i)]}{\Gamma[\frac{1}{2}pn(1+r)]}. \tag{19}$$

This is of the form of (1), Section 8.5, with

$$\begin{aligned} a &= p, & x_k &= \tfrac{1}{2}n, & \xi_k &= \tfrac{1}{2}(1-k), & k &= 1,\ldots,p, \\ b &= 1, & y_1 &= \tfrac{1}{2}np, & \eta_1 &= 0. \end{aligned} \tag{20}$$

Thus the expansion of Section 8.5 is valid with $f = \frac{1}{2}p(p+1) - 1$. To make the second term in the expansion zero we take ρ so

$$1 - \rho = \frac{2p^2 + p + 2}{6pn}. \tag{21}$$

Then

$$\omega_2 = \frac{(p+2)(p-1)(p-2)(2p^3+6p^2+3p+2)}{288p^2n^2\rho^2}. \tag{22}$$

Thus the cdf of W is found from

$$\begin{aligned}\Pr\{-2\rho\log Z \le z\} &= \Pr\{-n\rho\log W \le z\} \\ &= \Pr\{\chi_f^2 \le z\} + \omega_2\big(\Pr\{\chi_{f+4}^2 \le z\} \\ &\quad - \Pr\{\chi_f^2 \le z\}\big) + O(n^{-3}).\end{aligned} \tag{23}$$

Table 6 in Appendix B gives factors, say $c(n, p, \varepsilon)$, such that

$$\Pr\left\{-n\rho\log W \le c(n,p,\varepsilon)\chi^2_{\frac{1}{2}p(p+1)-1}(\varepsilon)\right\} = \varepsilon. \tag{24}$$

Nagarsenkar and Pillai (1973a) have tables for W.

10.7.5. Invariant Tests

The null hypothesis H: $\boldsymbol{\Sigma} = \sigma^2\boldsymbol{I}$ is invariant with respects to transformations $\boldsymbol{X}^* = c\boldsymbol{Q}\boldsymbol{X} + \boldsymbol{\nu}$, where c is a scalar and $\boldsymbol{Q}$ is an orthogonal matrix. The invariant of the sufficient statistic under shift of location is $\boldsymbol{A}$, the invariants of $\boldsymbol{A}$ under orthogonal transformations are the characteristic roots $l_1, \dots, l_p$, and the invariants of the roots under scale transformations are functions that are homogeneous of degree 0, such as the ratios of roots, say $l_1/l_2, \dots, l_{p-1}/l_p$. Invariant tests are based on such functions; the likelihood ratio criterion is such a function.

Nagao (1973a) has proposed the criterion

$$\begin{aligned}&\tfrac{1}{2}n\operatorname{tr}\left(\boldsymbol{S} - \frac{\operatorname{tr}\boldsymbol{S}}{p}\boldsymbol{I}\right)\frac{p}{\operatorname{tr}\boldsymbol{S}}\left(\boldsymbol{S} - \frac{\operatorname{tr}\boldsymbol{S}}{p}\boldsymbol{I}\right)\frac{p}{\operatorname{tr}\boldsymbol{S}} \\ &\quad = \tfrac{1}{2}n\operatorname{tr}\left(\frac{p}{\operatorname{tr}\boldsymbol{S}}\boldsymbol{S} - \boldsymbol{I}\right)^2 = \tfrac{1}{2}n\left[\frac{p^2}{(\operatorname{tr}\boldsymbol{S})^2}\operatorname{tr}\boldsymbol{S}^2 - p\right] \\ &\quad = \tfrac{1}{2}n\left[\frac{p^2}{\left(\sum_{i=1}^p l_i\right)^2}\sum_{i=1}^p l_i^2 - p\right] = \tfrac{1}{2}n\frac{\sum_{i=1}^p (l_i - \bar{l})^2}{\bar{l}^2},\end{aligned} \tag{25}$$

where $\bar{l} = \sum_{i=1}^p l_i/p$. The left-hand side of (25) is based on the loss function

$L_q(\boldsymbol{\Sigma}, \boldsymbol{G})$ of Section 7.8; the right-hand side shows it is proportional to the square of the coefficient of variation of the characteristic roots of the sample covariance matrix $\boldsymbol{S}$. Another criterion is l_1/l_p. Percentage points have been given by Krishnaiah and Schuurmann (1974).

10.7.6. Confidence Regions

Given observations $y_1, \ldots, y_N$ from $N(\boldsymbol{\nu}, \boldsymbol{\Psi})$, we can test $\boldsymbol{\Psi} = \sigma^2 \boldsymbol{\Psi}_0$ for any specified $\boldsymbol{\Psi}_0$. From this family of tests we can set up a confidence region for $\boldsymbol{\Psi}$. If any matrix is in the confidence region, all multiples of it are. This kind of confidence region is of interest if all components of y_α are measured in the same unit, but the investigator wants a region independent of this common unit. The confidence region of confidence $1 - \varepsilon$ consists of all matrices $\boldsymbol{\Psi}^*$ satisfying

$$\frac{|\boldsymbol{B}\boldsymbol{\Psi}^{*-1}|}{\left[(\operatorname{tr} \boldsymbol{B}\boldsymbol{\Psi}^{*-1})/p\right]^p} \geq \lambda^{2/N}(\varepsilon), \tag{26}$$

where $\lambda(\varepsilon)$ is the ε significance level for the criterion.

Consider the case of $p = 2$. If the common unit of measurement is irrelevant, the investigator is interested in $\tau = \psi_{11}/\psi_{22}$ and $\rho = \psi_{12}/\sqrt{\psi_{11}\psi_{22}}$. In this case

$$\boldsymbol{\Psi}^{-1} = \frac{1}{\psi_{11}\psi_{22}(1-\rho^2)} \begin{pmatrix} \psi_{22} & -\rho\sqrt{\psi_{11}\psi_{22}} \\ -\rho\sqrt{\psi_{11}\psi_{22}} & \psi_{11} \end{pmatrix} \tag{27}$$

$$= \frac{1}{\psi_{11}(1-\rho^2)} \begin{pmatrix} 1 & -\rho\sqrt{\tau} \\ -\rho\sqrt{\tau} & \tau \end{pmatrix}.$$

The region in terms of τ and ρ is

$$4\frac{\left(b_{11}b_{22} - b_{12}^2\right)(1-\rho^2)^\tau}{\left(b_{11} + \tau b_{22} - 2\rho\sqrt{\tau}\, b_{12}\right)^2} \geq \lambda^{2/N}(\varepsilon). \tag{28}$$

Hickman (1953) has given an example of such a confidence region.

10.8. TESTING THE HYPOTHESIS THAT A COVARIANCE MATRIX IS EQUAL TO A GIVEN MATRIX

10.8.1. The Criteria

If Y is distributed according to $N(\boldsymbol{\nu}, \boldsymbol{\Psi})$, we wish to test H_1 that $\boldsymbol{\Psi} = \boldsymbol{\Psi}_0$, where $\boldsymbol{\Psi}_0$ is a given positive definite matrix. By the argument of the preceding section we see that this is equivalent to testing the hypothesis H_1: $\boldsymbol{\Sigma} = \boldsymbol{I}$, where $\boldsymbol{\Sigma}$ is the covariance matrix of a vector X distributed according to $N(\boldsymbol{\mu}, \boldsymbol{\Sigma})$. Given a sample $\boldsymbol{x}_1, \dots, \boldsymbol{x}_N$, the likelihood ratio criterion is

$$\lambda_1 = \frac{\max_{\boldsymbol{\mu}} L(\boldsymbol{\mu}, \boldsymbol{I})}{\max_{\boldsymbol{\mu}, \boldsymbol{\Sigma}} L(\boldsymbol{\mu}, \boldsymbol{\Sigma})}, \tag{1}$$

where the likelihood function is

$$L(\boldsymbol{\mu}, \boldsymbol{\Sigma}) = (2\pi)^{-\frac{1}{2}pN} |\boldsymbol{\Sigma}|^{-\frac{1}{2}N} \exp\left[-\tfrac{1}{2} \sum_{\alpha=1}^{N} (\boldsymbol{x}_\alpha - \boldsymbol{\mu})' \boldsymbol{\Sigma}^{-1} (\boldsymbol{x}_\alpha - \boldsymbol{\mu})\right]. \tag{2}$$

Results in Chapter 3 show that

$$\begin{aligned} \lambda_1 &= \frac{(2\pi)^{-\frac{1}{2}pN} \exp\left[-\frac{1}{2}\sum_{\alpha=1}^{N} (\boldsymbol{x}_\alpha - \bar{\boldsymbol{x}})'(\boldsymbol{x}_\alpha - \bar{\boldsymbol{x}})\right]}{(2\pi)^{-\frac{1}{2}pN} \left|\frac{1}{N}\boldsymbol{A}\right|^{-\frac{1}{2}N} e^{-\frac{1}{2}pN}} \\ &= \left(\frac{e}{N}\right)^{\frac{1}{2}pN} |\boldsymbol{A}|^{\frac{1}{2}N} e^{-\frac{1}{2}\operatorname{tr}\boldsymbol{A}}, \end{aligned} \tag{3}$$

where

$$\boldsymbol{A} = \sum_{\alpha} (\boldsymbol{x}_\alpha - \bar{\boldsymbol{x}})(\boldsymbol{x}_\alpha - \bar{\boldsymbol{x}})'. \tag{4}$$

Sugiura and Nagao (1968) have shown that the likelihood ratio test is biased, but the modified likelihood ratio test based on

$$\begin{aligned} \lambda_1^* &= \left(\frac{e}{n}\right)^{\frac{1}{2}pn} |\boldsymbol{A}|^{\frac{1}{2}n} e^{-\frac{1}{2}\operatorname{tr}\boldsymbol{A}} \\ &= e^{\frac{1}{2}pn} \left(|\boldsymbol{S}| e^{-\operatorname{tr}\boldsymbol{S}}\right)^{\frac{1}{2}n}, \end{aligned} \tag{5}$$

where $\boldsymbol{S} = (1/n)\boldsymbol{A}$ is unbiased. Note that

$$(6)\qquad -\frac{2}{n}\log\lambda_1^* = \operatorname{tr}\boldsymbol{S} - \log|\boldsymbol{S}| - p$$
$$= L_l(\boldsymbol{I}, \boldsymbol{S}),$$

where $L_l(\boldsymbol{I}, \boldsymbol{S})$ is the loss function for estimating $\boldsymbol{I}$ by $\boldsymbol{S}$ defined in (2) of Section 7.8. In terms of the characteristic roots of $\boldsymbol{S}$ the criterion (6) is a constant plus

$$(7)\qquad \sum_{i=1}^{p} l_i - \log\prod_{i=1}^{p} l_i - p = \sum_{i=1}^{p}(l_i - \log l_i - 1);$$

for each i the minimum of (7) is at $l_i = 1$.

Using the algebra of the preceding section we see that given $\boldsymbol{y}_1, \dots, \boldsymbol{y}_N$ as observation vectors of p components from $N(\boldsymbol{\nu}, \boldsymbol{\Psi})$, the modified likelihood ratio criterion for testing the hypothesis H_1: $\boldsymbol{\Psi} = \boldsymbol{\Psi}_0$, where $\boldsymbol{\Psi}_0$ is specified, is

$$(8)\qquad \lambda_1^* = \left(\frac{e}{n}\right)^{\frac{1}{2}pn} |\boldsymbol{B}\boldsymbol{\Psi}_0^{-1}|^{\frac{1}{2}n} e^{-\frac{1}{2}\operatorname{tr}\boldsymbol{B}\boldsymbol{\Psi}_0^{-1}},$$

where

$$(9)\qquad \boldsymbol{B} = \sum_{\alpha=1}^{N}(\boldsymbol{y}_\alpha - \bar{\boldsymbol{y}})(\boldsymbol{y}_\alpha - \bar{\boldsymbol{y}})'.$$

10.8.2. The Distribution and Moments of the Modified Likelihood Ratio Criterion

The null hypothesis H_1: $\boldsymbol{\Sigma} = \boldsymbol{I}$ is the intersection of the null hypothesis of Section 10.7, H: $\boldsymbol{\Sigma} = \sigma^2\boldsymbol{I}$, and the null hypothesis $\sigma^2 = 1$ given $\boldsymbol{\Sigma} = \sigma^2\boldsymbol{I}$. The likelihood ratio criterion for H_1 given by (3) is the product of (7) of Section 10.7 and

$$(10)\qquad \left(\frac{\operatorname{tr}\boldsymbol{A}}{pN}\right)^{\frac{1}{2}pN} e^{-\frac{1}{2}\operatorname{tr}\boldsymbol{A} + \frac{1}{2}pN},$$

which is the likelihood ratio criterion for testing the hypothesis $\sigma^2 = 1$ given $\boldsymbol{\Sigma} = \sigma^2\boldsymbol{I}$. The modified criterion λ_1^* is the product of $|\boldsymbol{A}|^{\frac{1}{2}n}/(\operatorname{tr}\boldsymbol{A}/p)^{\frac{1}{2}pn}$ and

$$(11)\qquad \left(\frac{\operatorname{tr}\boldsymbol{A}}{pn}\right)^{\frac{1}{2}pn} e^{-\frac{1}{2}\operatorname{tr}\boldsymbol{A} + \frac{1}{2}pn};$$

these two factors are independent (Lemma 10.4.1). The characterization of the distribution of the modified criterion can be obtained from Section 10.7.3. The quantity $\operatorname{tr} \boldsymbol{A}$ has the χ^2-distribution with np degrees of freedom under the null hypothesis.

Instead of obtaining the moments and characteristic function of λ_1^* [defined by (5)] from the preceding characterization, we shall find them by use of the fact that $\boldsymbol{A}$ has the distribution $W(\boldsymbol{\Sigma}, n)$. We shall calculate

$$\text{(12)} \qquad \mathscr{E}\lambda_1^{*h} = \int \cdots \int \left(\frac{e^{\frac{1}{2}pn}}{n^{\frac{1}{2}pn}} |\boldsymbol{A}|^{\frac{1}{2}n} e^{-\frac{1}{2}\operatorname{tr} \boldsymbol{A}} \right)^h w(\boldsymbol{A}|\boldsymbol{\Sigma}, n)\, d\boldsymbol{A}$$

$$= \frac{e^{\frac{1}{2}pnh}}{n^{\frac{1}{2}pnh}} \int \cdots \int |\boldsymbol{A}|^{\frac{1}{2}nh} e^{-\frac{1}{2}h \operatorname{tr} \boldsymbol{A}} w(\boldsymbol{A}|\boldsymbol{\Sigma}, n)\, d\boldsymbol{A}.$$

Since

(13)

$$|\boldsymbol{A}|^{\frac{1}{2}nh} e^{-\frac{1}{2}h \operatorname{tr} \boldsymbol{A}} w(\boldsymbol{A}|\boldsymbol{\Sigma}, n) = \frac{|\boldsymbol{A}|^{\frac{1}{2}(n+nh-p-1)} e^{-\frac{1}{2}(\operatorname{tr} \boldsymbol{\Sigma}^{-1}\boldsymbol{A} + \operatorname{tr} h\boldsymbol{A})}}{2^{\frac{1}{2}pn} |\boldsymbol{\Sigma}|^{\frac{1}{2}n} \Gamma_p\left(\frac{1}{2}n\right)}$$

$$= \frac{2^{\frac{1}{2}pnh} \Gamma_p\left[\frac{1}{2}n(1+h)\right]}{|\boldsymbol{\Sigma}^{-1} + h\boldsymbol{I}|^{\frac{1}{2}(n+nh)} |\boldsymbol{\Sigma}|^{\frac{1}{2}n} \Gamma_p\left(\frac{1}{2}n\right)}$$

$$\cdot \frac{|\boldsymbol{\Sigma}^{-1} + h\boldsymbol{I}|^{\frac{1}{2}(n+nh)} |\boldsymbol{A}|^{\frac{1}{2}(n+nh-p-1)} e^{-\frac{1}{2}\operatorname{tr}(\boldsymbol{\Sigma}^{-1} + h\boldsymbol{I})\boldsymbol{A}}}{2^{\frac{1}{2}p(n+nh)} \Gamma_p\left[\frac{1}{2}n(1+h)\right]}$$

$$= \frac{2^{\frac{1}{2}pnh} |\boldsymbol{\Sigma}|^{\frac{1}{2}nh} \Gamma_p\left[\frac{1}{2}n(1+h)\right]}{|\boldsymbol{I} + h\boldsymbol{\Sigma}|^{\frac{1}{2}(n+nh)} \Gamma_p\left(\frac{1}{2}n\right)}$$

$$\times w\left(\boldsymbol{A}|(\boldsymbol{\Sigma}^{-1} + h\boldsymbol{I})^{-1}, n + nh\right),$$

the hth moment of λ_1^* is

$$\text{(14)} \qquad \mathscr{E}\lambda_1^{*h} = \left(\frac{2e}{n}\right)^{\frac{1}{2}phn} \frac{|\boldsymbol{\Sigma}|^{\frac{1}{2}nh} \prod_{j=1}^p \Gamma\left[\frac{1}{2}(n+nh+1-j)\right]}{|\boldsymbol{I} + h\boldsymbol{\Sigma}|^{\frac{1}{2}(n+nh)} \prod_{j=1}^p \Gamma\left[\frac{1}{2}(n+1-j)\right]}.$$

Then the characteristic function of $-2\log\lambda_1^*$ is

$$(15)\quad \mathscr{E}e^{-2it\log\lambda_1^*} = \mathscr{E}\lambda_1^{*\,-2it}$$

$$= \left(\frac{2e}{n}\right)^{-ipnt} \frac{|\mathbf{\Sigma}|^{-int}}{|\mathbf{I} - 2it\mathbf{\Sigma}|^{\frac{1}{2}n-int}} \cdot \prod_{j=1}^{p} \frac{\Gamma\left[\frac{1}{2}(n+1-j) - int\right]}{\Gamma\left[\frac{1}{2}(n+1-j)\right]}.$$

When the null hypothesis is true, $\mathbf{\Sigma} = \mathbf{I}$, and

(16)

$$\mathscr{E}e^{-2it\log\lambda_1^*} = \left(\frac{2e}{n}\right)^{-ipnt}(1-2it)^{-\frac{1}{2}p(n-2int)} \prod_{j=1}^{p} \frac{\Gamma\left[\frac{1}{2}(n+1-j) - int\right]}{\Gamma\left[\frac{1}{2}(n+1-j)\right]}.$$

This characteristic function is the product of p terms such as

$$(17)\quad \phi_j(t) = \left(\frac{2e}{n}\right)^{-int}(1-2it)^{-\frac{1}{2}(n-2int)} \frac{\Gamma\left[\frac{1}{2}(n+1-j) - int\right]}{\Gamma\left[\frac{1}{2}(n+1-j)\right]}.$$

Thus $-2\log\lambda_1^*$ is distributed as the sum of p independent variates, the characteristic function of the jth being (17). Using Stirling's approximation for the gamma function, we have

$$(18)\quad \phi_j(t) \sim 2^{-int}e^{-int}n^{int}(1-2it)^{\frac{1}{2}(2int-n)}$$

$$\cdot \frac{e^{-[\frac{1}{2}(n+1-j)-int]}\left[\frac{1}{2}(n+1-j) - int\right]^{\frac{1}{2}(n-j)-int}}{e^{-[\frac{1}{2}(n+1-j)]}\left[\frac{1}{2}(n-j+1)\right]^{\frac{1}{2}(n-j)}}$$

$$= (1-2it)^{-\frac{1}{2}j}\left(1 - \frac{it(j-1)}{\frac{1}{2}(n-j+1)(1-2it)}\right)^{\frac{1}{2}(n+1-j)-\frac{1}{2}}$$

$$\cdot\left(1 - \frac{2j-1}{n(1-2it)}\right)^{-int}.$$

As $n \to \infty$, $\phi_j(t) \to (1-2it)^{-\frac{1}{2}j}$, the characteristic function of χ_j^2, χ^2 with j degrees of freedom. Thus $-2\log\lambda_1^*$ is asymptotically distributed as $\sum_{j=1}^p \chi_j^2$, which is χ^2 with $\sum_{j=1}^p j = \frac{1}{2}p(p+1)$ degrees of freedom. The distribution of

λ_1^* can be further expanded [Korin (1968), Davis (1971)] as

$$\Pr\{-2\rho \log \lambda_1^* \le z\} = \Pr\{\chi_f^2 \le z\} + \frac{\gamma_2}{\rho^2 N^2}\left(\Pr\{\chi_{f+4}^2 \le z\} - \Pr\{\chi_f^2 \le z\}\right) + O(N^{-3}), \tag{19}$$

where

$$\rho = 1 - \frac{2p^2 + 3p - 1}{6N(p+1)}, \tag{20}$$

$$\gamma_2 = \frac{p(2p^4 + 6p^3 + p^2 - 12p - 13)}{288(p+1)}. \tag{21}$$

Nagarsenker and Pillai (1973b) have found exact distributions and tabulated 5% and 1% significant points as have Davis and Field (1971) for $p = 2(1)10$ and $n = 6(1)30(5)50, 60, 120$. In Table 7 [due to Korin (1968)] of Appendix B are given some 5% and 1% significance points of $-2\log\lambda_1^*$ for small values of n and $p = 2(1)10$.

10.8.3. Invariant Tests

The null hypothesis H: $\boldsymbol{\Sigma} = \boldsymbol{I}$ is invariant with respect to transformations $\boldsymbol{X}^* = \boldsymbol{Q}\boldsymbol{X} + \nu$, where $\boldsymbol{Q}$ is an orthogonal matrix. The invariants of the sufficient statistics are the characteristic roots $l_1, \dots, l_p$ of $\boldsymbol{S}$, and the invariants of the parameters are the characteristic roots of $\boldsymbol{\Sigma}$. Invariant tests are based on the roots of $\boldsymbol{S}$; the modified likelihood ratio criterion is one of them. Nagao (1973a) has suggested the criterion

$$\tfrac{1}{2} n \operatorname{tr}(\boldsymbol{S} - \boldsymbol{I})^2 = \tfrac{1}{2} n \sum_{i=1}^{p} (l_i - 1)^2. \tag{22}$$

Under the null hypothesis this criterion has a limiting χ^2-distribution with $\frac{1}{2}p(p+1)$ degrees of freedom.

Roy (1957), Section 6.4, proposed a test based on the largest and smallest characteristic roots l_1 and l_p. Reject the null hypothesis if

$$l_p < l \quad \text{or} \quad l_1 > u \tag{23}$$

where

$$\Pr\{l < l_p,\, l_1 < u | \Sigma = I\} = 1 - \varepsilon \tag{24}$$

and ε is the significance level. Clemm, Krishnaiah, and Waikar (1973) have given tables of $u = 1/l$. See also Schuurman and Waikar (1973).

10.8.4. Confidence Bounds for Quadratic Forms

The test procedure based on the smallest and largest characteristic roots can be inverted to give confidence bounds on quadratic forms in Σ. Suppose nS has the distribution $W(\Sigma, n)$. Let C be a nonsingular matrix such that $\Sigma = C'C$. Then $nS^* = nC'^{-1}SC^{-1}$ has the distribution $W(I, n)$. Since $l_p^* \leq a'S^*a/a'a < l_1^*$ for all a, where l_p^* and l_1^* are the smallest and largest characteristic roots of S^* (Section 11.2 and Appendix A.2),

$$\Pr\left\{ l \leq \frac{a'S^*a}{a'a} \leq u,\ \forall a \neq 0 \right\} = 1 - \varepsilon, \tag{25}$$

where

$$\Pr\{l \leq l_p^* \leq l_1^* \leq u\} = 1 - \varepsilon. \tag{26}$$

Let $a = Cb$. Then $a'a = b'C'Cb = b'\Sigma b$ and $a'S^*a = b'C'S^*Cb = b'Sb$. Thus (25) is

$$1 - \varepsilon = \Pr\left\{ l \leq \frac{b'Sb}{b'\Sigma b} \leq u,\ \forall b \neq 0 \right\} \tag{27}$$

$$= \Pr\left\{ \frac{b'Sb}{u} \leq b'\Sigma b \leq \frac{b'Sb}{l},\ \forall b \right\}.$$

Given an observed S, one can assert

$$\frac{b'Sb}{u} \leq b'\Sigma b \leq \frac{b'Sb}{l}, \qquad \forall b, \tag{28}$$

with confidence $1 - \varepsilon$.

If b has 1 in the ith position and 0's elsewhere, (28) is $s_{ii}/u \leq \sigma_{ii} \leq s_{ii}/l$. If b has 1 in the ith position, -1 in the jth position, $i \neq j$, and 0's elsewhere, then (28) is

$$(s_{ii} + s_{jj} - 2s_{ij})/u \leq \sigma_{ii} + \sigma_{jj} - 2\sigma_{ij} \leq (s_{ii} + s_{jj} - 2s_{ij})/l. \tag{29}$$

Manipulation of these inequalities yields

$$\frac{s_{ij}}{l} - \frac{s_{ii}+s_{jj}}{2}\left(\frac{1}{l}-\frac{1}{u}\right) \le \sigma_{ij} \le \frac{s_{ij}}{u} + \frac{s_{ii}+s_{jj}}{2}\left(\frac{1}{l}-\frac{1}{u}\right), \qquad i \ne j. \tag{30}$$

We can obtain simultaneously confidence intervals on all elements of Σ.

From (27) we can obtain

$$\begin{aligned} 1-\varepsilon &= \Pr\left\{\frac{1}{u}\frac{b'Sb}{b'b} \le \frac{b'\Sigma b}{b'b} \le \frac{1}{l}\frac{b'Sb}{b'b}, \forall b\right\} \\ &\le \Pr\left\{\frac{1}{u}\min_a \frac{a'Sa}{a'a} \le \frac{b'\Sigma b}{b'b} \le \frac{1}{l}\max_a \frac{a'Sa}{a'a}, \forall b\right\} \\ &= \Pr\left\{\frac{1}{u}l_p \le \lambda_p \le \lambda_1 \le \frac{1}{l}l_1\right\}, \end{aligned} \tag{31}$$

where l_1 and l_p are the largest and smallest characteristic roots of S and λ_1 and λ_p are the largest and smallest characteristic roots of Σ. Then

$$\frac{1}{u}l_p \le \lambda(\Sigma) \le \frac{1}{l}l_1 \tag{32}$$

is a confidence interval for all characteristic roots of Σ with confidence at least $1-\varepsilon$. In Section 11.6 we give tighter bounds on $\lambda(\Sigma)$ with exact confidence.

10.9. TESTING THE HYPOTHESIS THAT A MEAN VECTOR AND A COVARIANCE MATRIX ARE EQUAL TO A GIVEN VECTOR AND MATRIX

In Chapter 3 we pointed out that if Ψ is known, $(\bar{y}-\nu_0)'\Psi_0^{-1}(\bar{y}-\nu_0)$ is suitable for testing

$$H_2\colon \nu = \nu_0, \qquad \text{given } \Psi \equiv \Psi_0. \tag{1}$$

Now let us combine H_1 of Section 10.8 and H_2 and test

$$H\colon \nu = \nu_0, \qquad \Psi = \Psi_0, \tag{2}$$

on the basis of a sample $y_1, \dots, y_N$ from $N(\nu, \Psi)$.

Let

$$X = C(Y - \nu_0), \tag{3}$$

where

$$C\Psi_0 C' = I. \tag{4}$$

Then $x_1, \dots, x_N$ constitutes a sample from $N(\mu, \Sigma)$ and the hypothesis is

$$H: \mu = 0, \qquad \Sigma = I. \tag{5}$$

The likelihood ratio criterion for H_2: $\mu = 0$, given $\Sigma = I$, is

$$\lambda_2 = e^{-\frac{1}{2}N\bar{x}'\bar{x}}. \tag{6}$$

The likelihood ratio criterion for H is (by Lemma 10.3.1)

$$\begin{aligned} \lambda = \lambda_1\lambda_2 &= \left(\frac{e}{N}\right)^{\frac{1}{2}pN} |A|^{\frac{1}{2}N} e^{-\frac{1}{2}\operatorname{tr} A} e^{-\frac{1}{2}N\bar{x}'\bar{x}} \\ &= \left(\frac{e}{N}\right)^{\frac{1}{2}pN} |A|^{\frac{1}{2}N} e^{-\frac{1}{2}\operatorname{tr}(A + N\bar{x}\bar{x}')} \\ &= \left(\frac{e}{N}\right)^{\frac{1}{2}pN} |A|^{\frac{1}{2}N} e^{-\frac{1}{2}\Sigma x_\alpha' x_\alpha}. \end{aligned} \tag{7}$$

The likelihood ratio test (rejecting H if λ is less than a suitable constant) is unbiased [Srivastava and Khatri (1979), Theorem 10.4.5]. The two factors λ_1 and λ_2 are independent because λ_1 is a function of A and λ_2 is a function of $\bar{x}$, and A and $\bar{x}$ are independent. Since

$$\begin{aligned} \mathscr{E}\lambda_2^h &= \mathscr{E}e^{-\frac{1}{2}hN\Sigma\bar{x}_i^2} = \mathscr{E}e^{-\frac{1}{2}h\chi_p^2} \\ &= (1+h)^{-\frac{1}{2}p}, \end{aligned} \tag{8}$$

the hth moment of λ is

$$\mathscr{E}\lambda^h = \mathscr{E}\lambda_1^h \mathscr{E}\lambda_2^h = \left(\frac{2e}{N}\right)^{\frac{1}{2}pNh} \frac{1}{(1+h)^{\frac{1}{2}pN(1+h)}} \frac{\Gamma_p\left[\frac{1}{2}(n+Nh)\right]}{\Gamma_p\left(\frac{1}{2}n\right)} \tag{9}$$

under the null hypothesis. Clearly

$$-2\log\lambda = -2\log\lambda_1 - 2\log\lambda_2 \tag{10}$$

has asymptotically the χ^2-distribution with $f = p(p+1)/2 + p$ degrees of freedom. In fact, an asymptotic expansion of the distribution [Davis (1971)] of

$-2\rho\log\lambda$ is

$$\Pr\{-2\rho\log\lambda\le z\}=\Pr\{\chi_f^2\le z\}+\frac{\gamma_2}{\rho^2N^2}\left(\Pr\{\chi_{f+4}^2\le z\}-\Pr\{\chi_f^2\le z\}\right)+O(N^{-3}),\tag{11}$$

where

$$\rho=1-\frac{2p^2+9p+11}{6N(p+3)},\tag{12}$$

$$\gamma_2=\frac{p(2p^4+18p^3+49p^2+36p-13)}{288(p+3)}.\tag{13}$$

Nagarsenker and Pillai (1974) have used the moments to derive exact distributions and have tabulated the 5% and 1% significance points for $p = 2(1)6$ and $N = 4(1)20(2)40(5)100$.

Now let us return to the observations $y_1,\dots,y_N$. Then

$$\begin{aligned}\sum_\alpha x_\alpha'x_\alpha&=\sum_\alpha(y_\alpha-\nu_0)'C'C(y_\alpha-\nu_0)\\&=\sum_\alpha(y_\alpha-\nu_0)'\Psi_0^{-1}(y_\alpha-\nu_0)\\&=\operatorname{tr}A+N\bar{x}'\bar{x}\\&=\operatorname{tr}(B\Psi_0^{-1})+N(\bar{y}-\nu_0)'\Psi_0^{-1}(\bar{y}-\nu_0)\end{aligned}\tag{14}$$

and

$$|A|=|B\Psi_0^{-1}|.\tag{15}$$

THEOREM 10.9.1. *Given the p-component observation vectors $y_1,\dots,y_N$ from $N(\nu,\Psi)$, the likelihood ratio criterion for testing the hypothesis H: $\nu=\nu_0$, $\Psi=\Psi_0$, is*

$$\lambda=\left(\frac{e}{N}\right)^{\frac{1}{2}pN}|B\Psi_0^{-1}|^{\frac{1}{2}N}e^{-\frac{1}{2}[\operatorname{tr}B\Psi_0^{-1}+N(\bar{y}-\nu_0)'\Psi_0^{-1}(\bar{y}-\nu_0)]}.\tag{16}$$

When the null hypothesis is true, $-2\log\lambda$ is asymptotically distributed as χ^2 with $\frac{1}{2}p(p+1)+p$ degrees of freedom.

10.10. ADMISSIBILITY OF TESTS

We shall consider some Bayes solutions to the problem of testing the hypothesis

(1) $$\Sigma_1 = \cdots = \Sigma_q$$

as in Section 10.2. Under the alternative hypothesis, let

(2) $$\left[\mu^{(g)}, \Sigma_g\right] = \left[\left(I + C_g C_g'\right)^{-1} C_g y^{(g)}, \left(I + C_g C_g\right)^{-1}\right], \qquad g = 1, \ldots, q,$$

where the $p \times r_g$ matrix C_g has the density proportional to $|I + C_g C_g'|^{-\frac{1}{2}n_g}$, $n_g = N_g - 1$, the r_g-component random vector $y^{(g)}$ has the conditional normal distribution with mean $\mathbf{0}$ and covariance matrix $(1/N_g)\,[I_{r_g} - C_g'(I_p + C_g C_g')^{-1} C_g]^{-1}$ given C_g, and $(C_1, y^{(1)}), \ldots, (C_q, y^{(q)})$ are independently distributed. As we shall see, we need to choose suitable integers $r_1, \ldots, r_q$. Note that the integral of $|I + C_g C_g'|^{-\frac{1}{2}n_g}$ is finite if $n_g \geq p + r_g$. Then the numerator of the Bayes ratio is

(3)
$$\begin{aligned}
\text{const} \prod_{g=1}^{q} \int_{-\infty}^{\infty} \cdots \int_{-\infty}^{\infty} & |I + C_g C_g'|^{\frac{1}{2}N_g} \exp\left\{ -\frac{1}{2} \sum_{\alpha=1}^{N_g} \left[x_\alpha^{(g)} - \left(I + C_g C_g'\right)^{-1} C_g y^{(g)} \right]' \right. \\
& \left. \times \left(I + C_g C_g'\right) \left[x_\alpha^{(g)} - \left(I + C_g C_g'\right)^{-1} C_g y^{(g)} \right] \right\} \times |I + C_g C_g'|^{-\frac{1}{2}n_g} \\
& \times |I - C_g'\left(I + C_g C_g'\right)^{-1} C_g|^{\frac{1}{2}} \\
& \times \exp\left\{ -\tfrac{1}{2} N_g y^{(g)\prime} \left[I - C_g'\left(I + C_g C_g'\right)^{-1} C_g^{(g)} \right] y^{(g)} \right\} dy^{(g)}\, dC_g \\
= \text{const} \prod_{g=1}^{q} \int_{-\infty}^{\infty} \cdots \int_{-\infty}^{\infty} & \exp\left\{ -\frac{1}{2} \left[\sum_{\alpha=1}^{N_g} x_\alpha^{(g)\prime} \left(I + C_g C_g'\right) x_\alpha^{(g)} \right. \right. \\
& \left. \left. - 2 y^{(g)\prime} C_g' \sum_{\alpha=1}^{N_g} x_\alpha^{(g)} + N_g y^{(g)\prime} y^{(g)} \right] \right\} dy^{(g)}\, dC_g \\
= \text{const} \prod_{g=1}^{q} & \exp\left\{ -\tfrac{1}{2} \sum_{\alpha=1}^{N_g} x_\alpha^{(g)\prime} x_\alpha^{(g)} \right\} \\
& \times \int_{-\infty}^{\infty} \cdots \int_{-\infty}^{\infty} \exp\left\{ -\tfrac{1}{2} N_g \left(y^{(g)} - C_g' \bar{x}^{(g)} \right)' \left(y^{(g)} - C_g' \bar{x}^{(g)} \right) \right. \\
& \left. - \tfrac{1}{2} \operatorname{tr} C_g' A_g C_g \right\} dy^{(g)}\, dC_g \\
= \text{const} \prod_{g=1}^{q} & \exp\left\{ -\tfrac{1}{2} \left[\operatorname{tr} A_g + N_g \bar{x}^{(g)\prime} \bar{x}^{(g)} \right] \right\} |A_g|^{-\frac{1}{2} r_g}.
\end{aligned}$$

Under the null hypothesis let

$$(4)\qquad \left[\mu^{(g)}, \Sigma_g\right] = \left[(I + CC')^{-1}Cy^{(g)}, (I + CC')^{-1}\right],$$

where the $p \times r$ matrix C has the density proportional to $|I + CC'|^{-\frac{1}{2}n}$, $n = \sum_{g=1}^{q} n_g$, the r-component vector $y^{(g)}$ has the conditional normal distribution with mean $\mathbf{0}$ and covariance matrix $(1/N_g)[I_r - C'(I_p + CC')^{-1}C]^{-1}$ given C, and $y^{(1)}, \ldots, y^{(g)}$ are conditionally independent. Note that the integral of $|I + CC'|^{-\frac{1}{2}n}$ is finite if $n \geq p + r$. The denominator of the Bayes ratio is

(5)

$$\int_{-\infty}^{\infty} \cdots \int_{-\infty}^{\infty} \prod_{g=1}^{q} \left[|I + CC'|^{\frac{1}{2}N_g} \exp\left\{ -\frac{1}{2} \sum_{\alpha=1}^{N_g} \left[x_\alpha^{(g)} - (I + CC')^{-1} C y^{(g)} \right]' \right.\right.$$
$$\times (I + CC')\left[x_\alpha^{(g)} - (I + CC')^{-1} C y^{(g)} \right] \Bigg\} \times |I + CC'|^{-\frac{1}{2}n_g}$$
$$\times |I - C'(I + CC')^{-1}C|^{\frac{1}{2}}$$
$$\left. \times \exp\left\{ -\tfrac{1}{2} N_g y^{(g)\prime}\left[I - C'(I + CC')^{-1} C \right] y^{(g)} \right\} dy^{(g)} \right] dC$$
$$= \text{const} \int_{-\infty}^{\infty} \cdots \int_{-\infty}^{\infty} \prod_{g=1}^{q} \times \exp\left\{ -\frac{1}{2}\left[\sum_{\alpha=1}^{N_g} x_\alpha^{(g)\prime}(I + CC')x_\alpha^{(g)} \right.\right.$$
$$\left.\left. -2y^{(g)\prime}C' \sum_{\alpha=1}^{N_g} x_\alpha^{(g)} + N_g y^{(g)\prime} y^{(g)} \right]\right\} dy^{(g)}\, dC$$
$$= \text{const} \exp\left\{ -\frac{1}{2} \sum_{g=1}^{q} \sum_{\alpha=1}^{N_g} x_\alpha^{(g)\prime} x_\alpha^{(g)} \right\} \int_{-\infty}^{\infty} \cdots \int_{-\infty}^{\infty}$$
$$\times \exp\left\{ -\frac{1}{2} \sum_{g=1}^{q} N_g \left(y^{(g)} - C'\bar{x}^{(g)} \right)' \left(y^{(g)} - C'\bar{x}^{(g)} \right) \right.$$
$$\left. -\tfrac{1}{2} \operatorname{tr} C'AC \right\} \prod_{g=1}^{q} dy^{(g)}\, dC$$
$$= \text{const} \exp\left\{ -\frac{1}{2}\left(\operatorname{tr} A + \sum_{g=1}^{q} N_g \bar{x}^{(g)\prime} \bar{x}^{(g)} \right) \right\} |A|^{-\frac{1}{2}r}.$$

The Bayes test procedure is to reject the hypothesis if

$$(6) \qquad \frac{|A|^{\frac{1}{2}r}}{\prod_{g=1}^{q}|A_g|^{\frac{1}{2}r_g}} \geq c.$$

For invariance we want $\sum_{g=1}^{q} r_g = r$.

The binding constraint on the choice of $r_1, \ldots, r_q$ is $r_g \leq n_g - p$, $g = 1, \ldots, q$. It is possible in some special cases to choose $r_1, \ldots, r_q$ so that $(r_1, \ldots, r_q)$ is proportional to $(N_1, \ldots, N_q)$ and hence yield the likelihood ratio test or proportional to $(n_1, \ldots, n_q)$ to yield the modified likelihood ratio test, but since $r_1, \ldots, r_q$ have to be integers it may not be possible to choose them in either such way. Next we consider an extension of this approach that involves the choice of numbers $t_1, \ldots, t_q$, and t as well as $r_1, \ldots, r_q$, and r.

Suppose $2(p-1) < n_g$, $g = 1, \ldots, q$, and take $r_g \geq p$. Let t_g be a real number such that $2p - 1 < r_g + t_g + p < n_g + 1$, and let t be a real number such that $2p - 1 < r + t + p < n + 1$. Under the alternative hypothesis let the marginal density of C_g be proportional to $|C_g C_g'|^{\frac{1}{2}t_g}|I + C_g C_g'|^{-\frac{1}{2}n_g}$, $g = 1, \ldots, q$, and under the null hypothesis let the marginal density of C be proportional to $|CC'|^{\frac{1}{2}t}|I + CC'|^{-\frac{1}{2}n}$. (The conditions on $t_1, \ldots, t_q$, and t ensure that the purported densities have finite integrals; see Problem 18.) Then the Bayes procedure is to reject the null hypothesis if

$$(7) \qquad \frac{|A|^{\frac{1}{2}(r+t)}}{\prod_{g=1}^{q}|A_g|^{\frac{1}{2}(r_g+t_g)}} \geq c.$$

For invariance we want $t = \sum_{g=1}^{q} t_g$. If $t_1, \ldots, t_q$ are taken so $r_g + t_g = kN_g$ and $p - 1 < kN_g < N_g - p$, $g = 1, \ldots, q$, for some k, then (7) is the likelihood ratio test; if $r_g + t_g = kn_g$ and $p - 1 < kn_g < n_g + 1 - p$, $g = 1, \ldots, q$, for some k then (7) is the modified test [i.e., $(p-1)/\min_g N_g < k < 1 - p/\min_g N_g$].

THEOREM 10.10.1. *If $2p < N_g + 1$, $g = 1, \ldots, q$, then the likelihood ratio test and the modified likelihood ratio test of the null hypothesis* (1) *are admissible.*

Now consider the hypothesis

$$(8) \qquad \mu^{(1)} = \cdots = \mu^{(q)}, \qquad \Sigma_1 = \cdots = \Sigma_q.$$

The alternative hypothesis is treated before. For the null hypothesis let

$$\left[\mu^{(g)}, \Sigma_g\right] = \left[(I + CC')Cy, (I + CC')^{-1}\right], \tag{9}$$

where the $p \times r$ matrix C has the density proportional to $|I + CC'|^{-\frac{1}{2}(N-1)}$ and the r-component vector y has the conditional normal distribution with mean $\mathbf{0}$ and covariance matrix $(1/N)[I - C'(I + CC')^{-1}C]^{-1}$ given C. Then the Bayes procedure is to reject the null hypothesis (8) if

$$\frac{\left|\sum_{g=1}^{q} A_g + \sum_{g=1}^{q} N_g(\bar{y}^{(g)} - \bar{y})(\bar{y}^{(g)} - \bar{y})'\right|^{\frac{1}{2}r}}{\prod_{g=1}^{q} |A_g|^{\frac{1}{2}r_g}} \geq c. \tag{10}$$

If $2p < N_g + 1$, $g = 1, \ldots, q$, the prior distributions can be modified as before to obtain the likelihood ratio test and modified likelihood ratio test.

THEOREM 10.10.2. *If $2p < N_g + 1$, $g = 1, \ldots, q$, the likelihood ratio test and modified likelihood ratio test of the null hypothesis* (8) *are admissible.*

For more details see Kiefer and Schwartz (1965).

PROBLEMS

1. (Sec. 10.2) Sums of squares and cross-products of deviations from the means of four measurements are given below (from Table 3.4). The populations are Iris versicolor (1), Iris setosa (2), and Iris virginica (3); each sample consists of 50 observations.

$$A_1 = \begin{pmatrix} 13.0552 & 4.1740 & 8.9620 & 2.7332 \\ 4.1740 & 4.8250 & 4.0500 & 2.0190 \\ 8.9620 & 4.0500 & 10.8200 & 3.5820 \\ 2.7332 & 2.0190 & 3.5820 & 1.9162 \end{pmatrix},$$

$$A_2 = \begin{pmatrix} 6.0882 & 4.8616 & 0.8014 & 0.5062 \\ 4.8616 & 7.0408 & 0.5732 & 0.4556 \\ 0.8014 & 0.5732 & 1.4778 & 0.2974 \\ 0.5062 & 0.4556 & 0.2974 & 0.5442 \end{pmatrix},$$

$$A_3 = \begin{pmatrix} 19.8128 & 4.5944 & 14.8612 & 2.4056 \\ 4.5944 & 5.0962 & 3.4976 & 2.3338 \\ 14.8612 & 3.4976 & 14.9248 & 2.3924 \\ 2.4056 & 2.3338 & 2.3924 & 3.6962 \end{pmatrix}.$$

(*a*) Test the hypothesis $\Sigma_1 = \Sigma_2$ at the 5% significance level.
(*b*) Test the hypothesis $\Sigma_1 = \Sigma_2 = \Sigma_3$ at the 5% significance level.

2. (Sec. 10.2) (*a*) Let $\boldsymbol{Y}^{(g)}$, $g = 1, \dots, q$, be a set of random vectors each with p components. Suppose

$$\mathscr{E}\boldsymbol{Y}^{(g)} = \mathbf{0}, \qquad \mathscr{E}\boldsymbol{Y}^{(g)}\boldsymbol{Y}^{(h)\prime} = \delta_{gh}\boldsymbol{\Sigma}_g.$$

Let $\boldsymbol{C}$ be an orthogonal matrix of order q such that each element of the last row is

$$c_{qh} = 1/\sqrt{q}\,.$$

Define

$$\boldsymbol{Z}^{(g)} = \sum_{h=1}^{q} c_{gh}\boldsymbol{Y}^{(h)}, \qquad g = 1, \dots, q.$$

Show that

$$\mathscr{E}\boldsymbol{Z}^{(q)}\boldsymbol{Z}^{(g)\prime} = \mathbf{0}, \qquad g = 1, \dots, q-1,$$

if and only if

$$\boldsymbol{\Sigma}_1 = \boldsymbol{\Sigma}_2 = \cdots = \boldsymbol{\Sigma}_q.$$

(*b*) Let $\boldsymbol{X}_\alpha^{(g)}, \alpha = 1, \dots, N$, be a random sample from $N(\boldsymbol{\mu}^{(g)}, \boldsymbol{\Sigma}_g), g = 1, \dots, q$. Use the result from (*a*) to construct a test of the hypothesis

$$H\colon \boldsymbol{\Sigma}_1 = \cdots = \boldsymbol{\Sigma}_q,$$

based on a test of independence of $\boldsymbol{Z}^{(q)}$ and the set $\boldsymbol{Z}^{(1)}, \dots, \boldsymbol{Z}^{(q-1)}$. Find the exact distribution of the criterion for the case $p = 2$.

3. (Sec 10.2) *Unbiasedness of the modified likelihood ratio test of* $\sigma_1^2 = \sigma_2^2$. Show that (14) is unbiased. [*Hint*: Let $G = n_1 F/n_2$, $r = \sigma_1^2/\sigma_2^2$, and $c_1 < c_2$ be the solutions to $G^{\frac{1}{2}n_1}(1 + G)^{-\frac{1}{2}(n_1+n_2)} = k$, the critical value for the modified likelihood ratio criterion. Then

$$\Pr\{\text{Acceptance}|\sigma_1^2/\sigma_2^2 = r\} = \text{const}\int_{c_1}^{c_2} r^{\frac{1}{2}n_1} G^{\frac{1}{2}n_1 - 1}(1 + rG)^{-\frac{1}{2}(n_1+n_2)}\,dG$$

$$= \text{const}\int_{rc_1}^{rc_2} H^{\frac{1}{2}n_1 - 1}(1 + H)^{-\frac{1}{2}(n_1+n_2)}\,dH.$$

Show that the derivative of the above with respect to r is positive for $0 < r < 1$, 0 for $r = 1$, and negative for $r > 1$.]

4. (Sec. 10.2) Prove that the limiting distribution of (19) is χ_f^2, where $f = \frac{1}{2}p(p+1)(q-1)$. [*Hint*: Let $\boldsymbol{\Sigma} = \boldsymbol{I}$. Show that the limiting distribution of (19) is the limiting distribution of

$$\tfrac{1}{2}\sum_{i=1}^{p}\sum_{g=1}^{q} n_g\left(s_{ii}^{(g)} - s_{ii}\right)^2 + \sum_{i<j}\sum_{g=1}^{q} n_g\left(s_{ij}^{(g)} - s_{ij}\right)^2,$$

where $\boldsymbol{S}^{(g)} = (s_{ij}^{(g)})$, $\boldsymbol{S} = (s_{ij})$, and the $\sqrt{n_g}(s_{ij}^{(g)} - \delta_{ij})$, $i \le j$, are independent in the limiting distribution, the limiting distribution of $\sqrt{n_g}(s_{ii}^{(g)} - 1)$ is $N(0,2)$, and the limiting distribution of $\sqrt{n_g}\, s_{ij}^{(g)}$, $i<j$, is $N(0,1)$.]

5. (Sec. 10.4) Prove (15) by integration of Wishart densities. [*Hint*: $\mathscr{E}V_i^h = \mathscr{E}\prod_{g=1}^{q}|\boldsymbol{A}_g|^{\frac{1}{2}n_g}|\boldsymbol{A}|^{-\frac{1}{2}n}$ can be written as the integral of a constant times $|\boldsymbol{A}|^{-\frac{1}{2}n}\prod_{g=1}^{q} w(\boldsymbol{A}_g|\boldsymbol{\Sigma}, n_g + hn_g)$. Integration over $\sum_{g=1}^{q}\boldsymbol{A}_g = \boldsymbol{A}$ gives a constant times $w(\boldsymbol{A}|\boldsymbol{\Sigma}, n)$.]

6. (Sec. 10.4) Prove (15) by integration of Wishart and normal densities. [*Hint*: $\sum_{g=1}^{q} N_g(\bar{\boldsymbol{x}}^{(g)} - \bar{\boldsymbol{x}})(\bar{\boldsymbol{x}}^{(g)} - \bar{\boldsymbol{x}})'$ is distributed as $\sum_{f=1}^{q-1}\boldsymbol{y}_f\boldsymbol{y}_f'$. Use the hint of Problem 5.]

7. (Sec. 10.6) Let $\boldsymbol{x}_1^{(\nu)}, \dots, \boldsymbol{x}_N^{(\nu)}$ be observations from $N(\boldsymbol{\mu}^{(\nu)}, \boldsymbol{\Sigma}_\nu)$, $\nu = 1,2$, and let $\boldsymbol{A}_\nu = \sum(\boldsymbol{x}_\alpha^{(\nu)} - \bar{\boldsymbol{x}}^{(\nu)})(\boldsymbol{x}_\alpha^{(\nu)} - \bar{\boldsymbol{x}}^{(\nu)})'$.

(*a*) Prove that the likelihood ratio test for H: $\boldsymbol{\Sigma}_1 = \boldsymbol{\Sigma}_2$ is equivalent to rejecting H if

$$T = \frac{|\boldsymbol{A}_1| \cdot |\boldsymbol{A}_2|}{|\boldsymbol{A}_1 + \boldsymbol{A}_2|^2} \le C.$$

(*b*) Let $d_1^2, d_2^2, \dots, d_p^2$ be the roots of $|\boldsymbol{\Sigma}_1 - \lambda\boldsymbol{\Sigma}_2| = 0$, and let

$$\boldsymbol{D} = \begin{pmatrix} d_1 & 0 & \cdots & 0 \\ 0 & d_2 & \cdots & 0 \\ \vdots & \vdots & & \vdots \\ 0 & 0 & \cdots & d_p \end{pmatrix}.$$

Show that T is distributed as $|\boldsymbol{B}_1| \cdot |\boldsymbol{B}_2| / |\boldsymbol{B}_1 + \boldsymbol{B}_2|^2$, where $\boldsymbol{B}_1$ is distributed according to $W(\boldsymbol{D}^2, N-1)$ and $\boldsymbol{B}_2$ is distributed according to $W(\boldsymbol{I}, N-1)$. Show that T is distributed as $|\boldsymbol{D}\boldsymbol{C}_1\boldsymbol{D}| \cdot |\boldsymbol{C}_2| / |\boldsymbol{D}\boldsymbol{C}_1\boldsymbol{D} + \boldsymbol{C}_2|^2$, where $\boldsymbol{C}_i$ is distributed according to $W(\boldsymbol{I}, N-1)$.

8. (Sec. 10.6) For $p = 2$ show

$$\begin{aligned}\Pr\{V_1 \le v\} = {} & I_a(n_1 - 1, n_2 - 1) \\ & + B^{-1}(n_1 - 1, n_2 - 1)\, v^{(n_1+n_2-2)/n}\int_a^b x^{-2n_2/n}(1 - x_1)^{-n_1/n}\, dx_1 \\ & + 1 - I_b(n_1 - 1, n_2 - 1),\end{aligned}$$

where $a < b$ are the two roots of $x_1^{n_1}(1 - x_1)^{n_2} = v \le n_1^{n_1} n_2^{n_2}/n^n$. [*Hint*: This follows from integrating the density defined by (8).]

9. (Sec. 10.6) For $p = 2$ and $n_1 = n_2 = m$, say, show

$$\Pr\{V_1 \le v\}$$

$$= 2I_a(m-1, m-1) + 2B^{-1}(m-1, m-1)v^{1-(1/m)}\log\frac{1+\sqrt{1-4v^{1/m}}}{1-\sqrt{1-4v^{1/m}}},$$

where $a = \frac{1}{2}[1 - \sqrt{1-4v^{1/m}}]$.

10. (Sec. 10.7) Find the distribution of W for $p = 2$ under the null hypothesis (a) directly from the distribution of $\boldsymbol{A}$ and (b) from the distribution of the characteristic roots (Chapter 13).

11. (Sec. 10.7) Let $\boldsymbol{x}_1, \dots, \boldsymbol{x}_N$ be a sample from $N(\boldsymbol{\mu}, \boldsymbol{\Sigma})$. What is the likelihood ratio criterion for testing the hypothesis $\boldsymbol{\mu} = k\boldsymbol{\mu}_0$, $\boldsymbol{\Sigma} = k^2\boldsymbol{\Sigma}_0$, where $\boldsymbol{\mu}_0$ and $\boldsymbol{\Sigma}_0$ are specified and k is unspecified?

12. (Sec. 10.7) Let $\boldsymbol{x}_1^{(1)}, \dots, \boldsymbol{x}_{N_1}^{(1)}$ be a sample from $N(\boldsymbol{\mu}^{(1)}, \boldsymbol{\Sigma}_1)$ and $\boldsymbol{x}_1^{(2)}, \dots, \boldsymbol{x}_{N_2}^{(2)}$ be a sample from $N(\boldsymbol{\mu}^{(2)}, \boldsymbol{\Sigma}_2)$. What is the likelihood ratio criterion for testing the hypothesis that $\boldsymbol{\Sigma}_1 = k^2\boldsymbol{\Sigma}_2$, where k is unspecified? What is the likelihood ratio criterion for testing the hypothesis that $\boldsymbol{\mu}^{(1)} = k\boldsymbol{\mu}^{(2)}$ and $\boldsymbol{\Sigma}_1 = k^2\boldsymbol{\Sigma}_2$, where k is unspecified?

13. (Sec. 10.7) Let $\boldsymbol{x}_\alpha$ of p components, $\alpha = 1, \dots, N$, be observations from $N(\boldsymbol{\mu}, \boldsymbol{\Sigma})$. We define the following hypotheses

$$H\colon \boldsymbol{\mu} = \boldsymbol{0}, \qquad \boldsymbol{\Sigma} = k^2\boldsymbol{\Sigma}_0,$$

$$H_1\colon \boldsymbol{\Sigma} = k^2\boldsymbol{\Sigma}_0,$$

$$H_2\colon \boldsymbol{\mu} = \boldsymbol{0}, \qquad \text{given that } \boldsymbol{\Sigma} = k^2\boldsymbol{\Sigma}_0.$$

In each case k^2 is unspecified, but $\boldsymbol{\Sigma}_0$ is specified. Find the likelihood ratio criterion λ_2 for testing H_2. Give the asymptotic distribution of $-2\log\lambda_2$ under H_2. Obtain the exact distribution of a suitable monotonic function of λ_2 under H_2.

14. (Sec. 10.7) Find the likelihood ratio criterion λ for testing H of Problem 13 (given $\boldsymbol{x}_1, \dots, \boldsymbol{x}_N$). What is the asymptotic distribution of $-2\log\lambda$ under H?

15. (Sec. 10.7) Show that $\lambda = \lambda_1\lambda_2$, where λ is defined in Problem 14, λ_2 is defined in Problem 13, and λ_1 is the likelihood ratio criterion for H_1 in Problem 13. Are λ_1 and λ_2 independently distributed under H? Prove your answer.

16. (Sec. 10.7) Verify that $\operatorname{tr}\boldsymbol{B}\boldsymbol{\Psi}_0^{-1}$ has the χ^2-distribution with $p(N-1)$ degrees of freedom.

17. (Sec. 10.7.1) *Admissibility of sphericity test*. Prove that the likelihood ratio test of sphericity is admissible. [*Hint*: Under the null hypothesis let $\boldsymbol{\Sigma} = [1/(1+\eta^2)]\boldsymbol{I}$, and let η have the density $(1+\eta^2)^{-\frac{1}{2}np}(\eta^2)^{p-\frac{1}{2}}$.]

18. (Sec. 10.10.1) Show that for $r \geq p$

$$\int_{-\infty}^{\infty} \cdots \int_{-\infty}^{\infty} \left| \sum_{i=1}^{r} \boldsymbol{x}_i \boldsymbol{x}_i' \right|^{\frac{1}{2}t} \left| \boldsymbol{I} + \sum_{i=1}^{r} \boldsymbol{x}_i \boldsymbol{x}_i' \right|^{-\frac{1}{2}n} \prod_{i=1}^{r} d\boldsymbol{x}_i < \infty$$

if $2p - 1 < t + r + p < n + 1$. [*Hint*: $|\boldsymbol{A}|/|\boldsymbol{I} + \boldsymbol{A}| \leq 1$ if $\boldsymbol{A}$ is positive semidefinite. Also $|\sum_{i=1}^{r} \boldsymbol{x}_i \boldsymbol{x}_i'|$ has the distribution of $\chi_r^2 \cdot \chi_{r-1}^2 \cdot \cdots \cdot \chi_{r-p+1}^2$ if $\boldsymbol{x}_1, \ldots, \boldsymbol{x}_r$ are independently distributed according to $N(\boldsymbol{0}, \boldsymbol{I})$.]

19. (Sec. 10.10.1) Show

$$\int_{-\infty}^{\infty} \cdots \int_{-\infty}^{\infty} |\boldsymbol{C}\boldsymbol{C}'|^{\frac{1}{2}t} e^{-\frac{1}{2}\operatorname{tr} \boldsymbol{C}'\boldsymbol{A}\boldsymbol{C}} \, d\boldsymbol{C} = \text{const} |\boldsymbol{A}|^{-\frac{1}{2}(t+r)}$$

where $\boldsymbol{C}$ is $p \times r$. [*Hint*: $\boldsymbol{C}\boldsymbol{C}'$ has the distribution $W(\boldsymbol{A}^{-1}, r)$ if $\boldsymbol{C}$ has a density proportional to $e^{-\frac{1}{2}\operatorname{tr} \boldsymbol{C}'\boldsymbol{A}\boldsymbol{C}}$].

20. (Sec. 10.10.1) Using Problem 18, complete the proof of Theorem 10.10.1.

CHAPTER 11

Principal Components

11.1. INTRODUCTION

Principal components are linear combinations of random or statistical variables which have special properties in terms of variances. For example, the first principal component is the normalized linear combination (the sum of squares of the coefficients being one) with maximum variance. In effect, transforming the original vector variable to the vector of principal components amounts to a rotation of coordinate axes to a new coordinate system that has inherent statistical properties. This choosing of a coordinate system is to be contrasted with the many problems treated previously where the coordinate system is irrelevant.

The principal components turn out to be the characteristic vectors of the covariance matrix. Thus the study of principal components can be considered as putting into statistical terms the usual developments of characteristic roots and vectors (for positive semidefinite matrices).

From the point of view of statistical theory, the set of principal components yields a convenient set of coordinates, and the accompanying variances of the components characterize their statistical properties. In statistical practice, the method of principal components is used to find the linear combinations with large variance. In many exploratory studies the number of variables under consideration is too large to handle. Since it is the deviations in these studies that are of interest, a way of reducing the number of variables to be treated is to discard the linear combinations which have small variances and study only those with large variances. For example, a physical anthropologist may make dozens of measurements of lengths and breadths of each of a number of individuals, such measurements as ear length, ear breadth, facial length, facial breadth, and so forth. He may be interested in describing and analyzing how individuals differ in these kinds of physiological characteristics. Eventually he will want to "explain" these differences, but first he wants to know what measurements or combinations of measurements show considerable variation; that is, which should have further study. The principal components give a new set of linearly combined measurements. It may be that most of the variation from individual to individual resides in three linear combinations; then the anthropologist can direct his study to these three quantities; the other linear combinations vary so little from one person to the next that study of them will tell little of individual variation.

Hotelling (1933), who developed many of these ideas, gave a rather thorough discussion.

In Section 11.2 we define principal components in the population to have the properties described above; they define an orthogonal transformation to a diagonal covariance matrix. The maximum likelihood estimators have similar properties in the sample (Section 11.3). A brief discussion of computation is given in Section 11.4, and a numerical example is carried out in Section 11.5. Asymptotic distributions of the coefficients of the sample principal components and the sample variances are derived and applied to obtain large-sample tests and confidence intervals for individual parameters (Section 11.6); exact confidence bounds are found for the characteristic roots of a covariance matrix. In the last section we consider other tests of hypotheses about these roots.

11.2. DEFINITION OF PRINCIPAL COMPONENTS IN THE POPULATION

Suppose the random vector $\boldsymbol{X}$ of p components has the covariance matrix $\boldsymbol{\Sigma}$. Since we shall be interested only in variances and covariances in this chapter, we shall assume that the mean vector is $\mathbf{0}$. Moreover, in developing the ideas and algebra here, the actual distribution of $\boldsymbol{X}$ is irrelevant except for the covariance matrix; however, if $\boldsymbol{X}$ is normally distributed, more meaning can be given to the principal components.

In the following treatment we shall not use the usual theory of characteristic roots and vectors; as a matter of fact, that theory will be implicitly derived. The treatment will include the cases where $\boldsymbol{\Sigma}$ is singular (i.e., positive semidefinite) and where $\boldsymbol{\Sigma}$ has multiple roots.

Let $\boldsymbol{\beta}$ be a p-component column vector such that $\boldsymbol{\beta}'\boldsymbol{\beta} = 1$. The variance of $\boldsymbol{\beta}'\boldsymbol{X}$ is

$$\mathscr{E}(\boldsymbol{\beta}'\boldsymbol{X})^2 = \mathscr{E}\boldsymbol{\beta}'\boldsymbol{X}\boldsymbol{X}'\boldsymbol{\beta} = \boldsymbol{\beta}'\boldsymbol{\Sigma}\boldsymbol{\beta}. \tag{1}$$

To determine the normalized linear combination $\boldsymbol{\beta}'\boldsymbol{X}$ with maximum variance, we must find a vector $\boldsymbol{\beta}$ satisfying $\boldsymbol{\beta}'\boldsymbol{\beta} = 1$ which maximizes (1). Let

$$\phi = \boldsymbol{\beta}'\boldsymbol{\Sigma}\boldsymbol{\beta} - \lambda(\boldsymbol{\beta}'\boldsymbol{\beta} - 1) = \sum_{i,j} \beta_i \sigma_{ij} \beta_j - \lambda\left(\sum_i \beta_i^2 - 1\right), \tag{2}$$

where λ is a Lagrange multiplier. The vector of partial derivatives $(\partial\phi/\partial\beta_i)$ is

$$\frac{\partial\phi}{\partial\boldsymbol{\beta}} = 2\boldsymbol{\Sigma}\boldsymbol{\beta} - 2\lambda\boldsymbol{\beta} \tag{3}$$

(by Theorem A.4.3 of Appendix A). Since $\boldsymbol{\beta}'\boldsymbol{\Sigma}\boldsymbol{\beta}$ and $\boldsymbol{\beta}'\boldsymbol{\beta}$ have derivatives everywhere in a region containing $\boldsymbol{\beta}'\boldsymbol{\beta} = 1$, a vector $\boldsymbol{\beta}$ maximizing $\boldsymbol{\beta}'\boldsymbol{\Sigma}\boldsymbol{\beta}$ must satisfy the expression (3) set equal to $\mathbf{0}$; that is

$$(\boldsymbol{\Sigma} - \lambda\boldsymbol{I})\boldsymbol{\beta} = \mathbf{0}. \tag{4}$$

In order to get a solution of (4) with $\boldsymbol{\beta}'\boldsymbol{\beta} = 1$ we must have $\boldsymbol{\Sigma} - \lambda\boldsymbol{I}$ singular; in other words, λ must satisfy

$$|\boldsymbol{\Sigma} - \lambda\boldsymbol{I}| = 0. \tag{5}$$

The function $|\boldsymbol{\Sigma} - \lambda\boldsymbol{I}|$ is a polynomial in λ of degree p. Therefore (5) has p roots; let these be $\lambda_1 \geq \lambda_2 \geq \cdots \geq \lambda_p$. ($\boldsymbol{\beta}'$ complex conjugate in (6) proves λ real.) If we multiply (4) on the left by $\boldsymbol{\beta}'$ we obtain

$$\boldsymbol{\beta}'\boldsymbol{\Sigma}\boldsymbol{\beta} = \lambda\boldsymbol{\beta}'\boldsymbol{\beta} = \lambda. \tag{6}$$

This shows that if $\boldsymbol{\beta}$ satisfies (4) (and $\boldsymbol{\beta}'\boldsymbol{\beta} = 1$), then the variance of $\boldsymbol{\beta}'\boldsymbol{X}$ [given by (1)] is λ. Thus for the maximum variance we should use in (4) the largest root λ_1. Let $\boldsymbol{\beta}^{(1)}$ be a normalized solution of $(\boldsymbol{\Sigma} - \lambda_1\boldsymbol{I})\boldsymbol{\beta} = \mathbf{0}$. Then $U_1 = \boldsymbol{\beta}^{(1)\prime}\boldsymbol{X}$ is a normalized linear combination with maximum variance. [If $\boldsymbol{\Sigma} - \lambda_1\boldsymbol{I}$ is of rank $p - 1$, then there is only one solution to $(\boldsymbol{\Sigma} - \lambda_1\boldsymbol{I})\boldsymbol{\beta} = \mathbf{0}$ and $\boldsymbol{\beta}'\boldsymbol{\beta} = 1$.]

Now let us find a normalized combination $\boldsymbol{\beta}'\boldsymbol{X}$ that has maximum variance of all linear combinations uncorrelated with U_1. Lack of correlation is

$$(7) \qquad \begin{aligned} 0 &= \mathscr{E}\boldsymbol{\beta}'\boldsymbol{X}U_1 = \mathscr{E}\boldsymbol{\beta}'\boldsymbol{X}\boldsymbol{X}'\boldsymbol{\beta}^{(1)} \\ &= \boldsymbol{\beta}'\boldsymbol{\Sigma}\boldsymbol{\beta}^{(1)} = \lambda_1\boldsymbol{\beta}'\boldsymbol{\beta}^{(1)}, \end{aligned}$$

since $\boldsymbol{\Sigma}\boldsymbol{\beta}^{(1)} = \lambda_1\boldsymbol{\beta}^{(1)}$. Thus $\boldsymbol{\beta}'\boldsymbol{X}$ is orthogonal to U in both the statistical sense (of lack of correlation) and in the geometric sense (of the inner product of the vectors $\boldsymbol{\beta}$ and $\boldsymbol{\beta}^{(1)}$ being zero). (That is, $\lambda_1\boldsymbol{\beta}'\boldsymbol{\beta}^{(1)} = 0$ only if $\boldsymbol{\beta}'\boldsymbol{\beta}^{(1)} = 0$ when $\lambda_1 \neq 0$, and $\lambda_1 \neq 0$ if $\boldsymbol{\Sigma} \neq \mathbf{0}$; the case of $\boldsymbol{\Sigma} = \mathbf{0}$ is trivial and is not treated.) We now want to maximize

$$(8) \qquad \phi_2 = \boldsymbol{\beta}'\boldsymbol{\Sigma}\boldsymbol{\beta} - \lambda(\boldsymbol{\beta}'\boldsymbol{\beta} - 1) - 2\nu_1\boldsymbol{\beta}'\boldsymbol{\Sigma}\boldsymbol{\beta}^{(1)},$$

where λ and ν_1 are Lagrange multipliers. The vector of partial derivatives is

$$(9) \qquad \frac{\partial\phi_2}{\partial\boldsymbol{\beta}} = 2\boldsymbol{\Sigma}\boldsymbol{\beta} - 2\lambda\boldsymbol{\beta} - 2\nu_1\boldsymbol{\Sigma}\boldsymbol{\beta}^{(1)},$$

and we set this equal to $\mathbf{0}$. From (9) we obtain by multiplying on the left by $\boldsymbol{\beta}^{(1)\prime}$

$$(10) \qquad \begin{aligned} 0 &= 2\boldsymbol{\beta}^{(1)\prime}\boldsymbol{\Sigma}\boldsymbol{\beta} - 2\lambda\boldsymbol{\beta}^{(1)\prime}\boldsymbol{\beta} - 2\nu_1\boldsymbol{\beta}^{(1)\prime}\boldsymbol{\Sigma}\boldsymbol{\beta}^{(1)} \\ &= -2\nu_1\lambda_1, \end{aligned}$$

by (7). Therefore, $\nu_1 = 0$ and $\boldsymbol{\beta}$ must satisfy (4), and therefore λ must satisfy (5). Let $\lambda_{(2)}$ be the maximum of $\lambda_1, \ldots, \lambda_p$ such that there is a vector $\boldsymbol{\beta}$ satisfying $(\boldsymbol{\Sigma} - \lambda_{(2)}\boldsymbol{I})\boldsymbol{\beta} = \mathbf{0}$, $\boldsymbol{\beta}'\boldsymbol{\beta} = 1$, and (7); call this vector $\boldsymbol{\beta}^{(2)}$ and the corresponding linear combination $U_2 = \boldsymbol{\beta}^{(2)\prime}\boldsymbol{X}$. (It will be shown eventually that $\lambda_{(2)} = \lambda_2$. We define $\lambda_{(1)} = \lambda_1$.)

This procedure is continued; at the $(r+1)$st step, we want to find a vector $\boldsymbol{\beta}$ such that $\boldsymbol{\beta}'\boldsymbol{X}$ has maximum variance of all normalized linear combinations which are uncorrelated with $U_1, \ldots, U_r$, that is, such that

$$(11) \qquad \begin{aligned} 0 &= \mathscr{E}\boldsymbol{\beta}'\boldsymbol{X}U_i = \mathscr{E}\boldsymbol{\beta}'\boldsymbol{X}\boldsymbol{X}'\boldsymbol{\beta}^{(i)} \\ &= \boldsymbol{\beta}'\boldsymbol{\Sigma}\boldsymbol{\beta}^{(i)} = \lambda_{(i)}\boldsymbol{\beta}'\boldsymbol{\beta}^{(i)}, \qquad i = 1, \ldots, r. \end{aligned}$$

We want to maximize

$$(12) \qquad \phi_{r+1} = \boldsymbol{\beta}'\boldsymbol{\Sigma}\boldsymbol{\beta} - \lambda(\boldsymbol{\beta}'\boldsymbol{\beta} - 1) - 2\sum_{i=1}^{r}\nu_i\boldsymbol{\beta}'\boldsymbol{\Sigma}\boldsymbol{\beta}^{(i)},$$

where λ and $\nu_1, \ldots, \nu_r$ are Lagrange multipliers. The vector of partial derivatives is

$$(13) \qquad \frac{\partial \phi_{r+1}}{\partial \boldsymbol{\beta}} = 2\boldsymbol{\Sigma}\boldsymbol{\beta} - 2\lambda\boldsymbol{\beta} - 2\sum_{i=1}^{r} \nu_i \boldsymbol{\Sigma}\boldsymbol{\beta}^{(i)},$$

and we set this equal to $\mathbf{0}$. Multiplying (13) on the left by $\boldsymbol{\beta}^{(j)\prime}$, we obtain

$$(14) \qquad 0 = 2\boldsymbol{\beta}^{(j)\prime}\boldsymbol{\Sigma}\boldsymbol{\beta} - 2\lambda\boldsymbol{\beta}^{(j)\prime}\boldsymbol{\beta} - 2\nu_j\boldsymbol{\beta}^{(j)\prime}\boldsymbol{\Sigma}\boldsymbol{\beta}^{(j)}.$$

If $\lambda_{(j)} \neq 0$, this gives $-2\nu_j\lambda_{(j)} = 0$ and $\nu_j = 0$. If $\lambda_{(j)} = 0$, $\boldsymbol{\Sigma}\boldsymbol{\beta}^{(j)} = \lambda_{(j)}\boldsymbol{\beta}^{(j)} = \mathbf{0}$ and the jth term in the sum in (13) vanishes. Thus $\boldsymbol{\beta}$ must satisfy (4) and therefore λ must satisfy (5).

Let $\lambda_{(r+1)}$ be the maximum of $\lambda_1, \ldots, \lambda_p$ such that there is a vector $\boldsymbol{\beta}$ satisfying $(\boldsymbol{\Sigma} - \lambda_{(r+1)}\boldsymbol{I})\boldsymbol{\beta} = \mathbf{0}$, $\boldsymbol{\beta}'\boldsymbol{\beta} = 1$ and (11); call this vector $\boldsymbol{\beta}^{(r+1)}$, and the corresponding linear combination $U_{r+1} = \boldsymbol{\beta}^{(r+1)\prime}\boldsymbol{X}$. If $\lambda_{(r+1)} = 0$ and $\lambda_{(j)} = 0$, $j \neq r+1$, then $\boldsymbol{\beta}^{(j)\prime}\boldsymbol{\Sigma}\boldsymbol{\beta}^{(r+1)} = 0$ does not imply $\boldsymbol{\beta}^{(j)\prime}\boldsymbol{\beta}^{(r+1)} = 0$. However, $\boldsymbol{\beta}^{(r+1)}$ can be replaced by a linear combination of $\boldsymbol{\beta}^{(r+1)}$ and the $\boldsymbol{\beta}^{(j)}$'s with $\lambda_{(j)}$'s being 0 so that the new $\boldsymbol{\beta}^{(r+1)}$ is orthogonal to all $\boldsymbol{\beta}^{(j)}$, $j = 1, \ldots, r$. This procedure is carried on until at the $(m+1)$st stage one cannot find a vector $\boldsymbol{\beta}$ satisfying $\boldsymbol{\beta}'\boldsymbol{\beta} = 1$, (4), and (11). Either $m = p$ or $m < p$ since $\boldsymbol{\beta}^{(1)}, \ldots, \boldsymbol{\beta}^{(m)}$ must be linearly independent.

We shall now show that the inequality $m < p$ leads to a contradiction. If $m < p$ there exist $p - m$ vectors, say $\boldsymbol{e}_{m+1}, \ldots, \boldsymbol{e}_p$, such that $\boldsymbol{\beta}^{(i)\prime}\boldsymbol{e}_j = 0$, $\boldsymbol{e}_i'\boldsymbol{e}_j = \delta_{ij}$. (This follows from Lemma A.4.2 in Appendix A.) Let $(\boldsymbol{e}_{m+1}, \ldots, \boldsymbol{e}_p) = \boldsymbol{E}$. Now we shall show that there exists a $(p-m)$-component vector $\boldsymbol{c}$ and a number θ such that $\boldsymbol{E}\boldsymbol{c} = \sum c_i \boldsymbol{e}_i$ is a solution to (4) with $\lambda = \theta$. Consider a root of $|\boldsymbol{E}'\boldsymbol{\Sigma}\boldsymbol{E} - \theta\boldsymbol{I}| = 0$ and a corresponding vector $\boldsymbol{c}$ satisfying $\boldsymbol{E}'\boldsymbol{\Sigma}\boldsymbol{E}\boldsymbol{c} = \theta\boldsymbol{c}$. The vector $\boldsymbol{\Sigma}\boldsymbol{E}\boldsymbol{c}$ is orthogonal to $\boldsymbol{\beta}^{(1)}, \ldots, \boldsymbol{\beta}^{(m)}$ (since $\boldsymbol{\beta}^{(i)\prime}\boldsymbol{\Sigma}\boldsymbol{E}\boldsymbol{c} = \lambda_{(i)}\boldsymbol{\beta}^{(i)\prime}\sum c_j\boldsymbol{e}_j = \lambda_{(i)}\sum c_j\boldsymbol{\beta}^{(i)\prime}\boldsymbol{e}_j = 0$) and therefore is a vector in the space spanned by $\boldsymbol{e}_{m+1}, \ldots, \boldsymbol{e}_p$ and can be written as $\boldsymbol{E}\boldsymbol{g}$ (where $\boldsymbol{g}$ is a $(p-m)$-component vector). Multiplying $\boldsymbol{\Sigma}\boldsymbol{E}\boldsymbol{c} = \boldsymbol{E}\boldsymbol{g}$ on the left by $\boldsymbol{E}'$, we obtain $\boldsymbol{E}'\boldsymbol{\Sigma}\boldsymbol{E}\boldsymbol{c} = \boldsymbol{E}'\boldsymbol{E}\boldsymbol{g} = \boldsymbol{g}$. Thus $\boldsymbol{g} = \theta\boldsymbol{c}$ and we have $\boldsymbol{\Sigma}(\boldsymbol{E}\boldsymbol{c}) = \theta(\boldsymbol{E}\boldsymbol{c})$. Then $(\boldsymbol{E}\boldsymbol{c})'\boldsymbol{X}$ is uncorrelated with $\boldsymbol{\beta}^{(j)\prime}\boldsymbol{X}$, $j = 1, \ldots, m$, and thus leads to a new $\boldsymbol{\beta}^{(m+1)}$. Since this contradicts the assumption that $m < p$, we must have $m = p$.

Let $\mathbf{B} = (\boldsymbol{\beta}^{(1)} \cdots \boldsymbol{\beta}^{(p)})$ and

$$(15) \qquad \boldsymbol{\Lambda} = \begin{pmatrix} \lambda_{(1)} & 0 & \cdots & 0 \\ 0 & \lambda_{(2)} & \cdots & 0 \\ \vdots & \vdots & & \vdots \\ 0 & 0 & \cdots & \lambda_{(p)} \end{pmatrix}.$$

The equations $\Sigma\beta^{(r)} = \lambda_{(r)}\beta^{(r)}$ can be written in matrix form as

$$\Sigma\mathbf{\beta} = \mathbf{\beta}\Lambda, \tag{16}$$

and the equations $\beta^{(r)\prime}\beta^{(r)} = 1$ and $\beta^{(r)\prime}\beta^{(s)} = 0$, $r \neq s$, can be written as

$$\mathbf{\beta}'\mathbf{\beta} = \boldsymbol{I}. \tag{17}$$

From (16) and (17) we obtain

$$\mathbf{\beta}'\Sigma\mathbf{\beta} = \Lambda. \tag{18}$$

From the fact that

$$\begin{aligned} |\Sigma - \lambda \boldsymbol{I}| &= |\mathbf{\beta}'|\cdot|\Sigma - \lambda\boldsymbol{I}|\cdot|\mathbf{\beta}| \\ &= |\mathbf{\beta}'\Sigma\mathbf{\beta} - \lambda\mathbf{\beta}'\mathbf{\beta}| = |\Lambda - \lambda\boldsymbol{I}| \\ &= \prod(\lambda_{(i)} - \lambda) \end{aligned} \tag{19}$$

we see that the roots of (19) are the diagonal elements of Λ; that is, $\lambda_{(1)} = \lambda_1$, $\lambda_{(2)} = \lambda_2, \ldots, \lambda_{(p)} = \lambda_p$.

We have proved the following theorem:

THEOREM 11.2.1. *Let the p-component random vector* $\boldsymbol{X}$ *have* $\mathscr{E}\boldsymbol{X} = \mathbf{0}$ *and* $\mathscr{E}\boldsymbol{X}\boldsymbol{X}' = \Sigma$. *Then there exists an orthogonal linear transformation*

$$\boldsymbol{U} = \mathbf{\beta}'\boldsymbol{X} \tag{20}$$

such that the covariance matrix of $\boldsymbol{U}$ *is* $\mathscr{E}\boldsymbol{U}\boldsymbol{U}' = \Lambda$ *and*

$$\Lambda = \begin{pmatrix} \lambda_1 & 0 & \cdots & 0 \\ 0 & \lambda_2 & \cdots & 0 \\ \vdots & \vdots & & \vdots \\ 0 & 0 & \cdots & \lambda_p \end{pmatrix}, \tag{21}$$

where $\lambda_1 \geq \lambda_2 \geq \cdots \geq \lambda_p \geq 0$ *are the roots of* (5). *The* rth *column of* $\mathbf{\beta}$, $\beta^{(r)}$, *satisfies* $(\Sigma - \lambda_r\boldsymbol{I})\beta = \mathbf{0}$. *The* rth *component of* $\boldsymbol{U}$, $U_r = \beta^{(r)\prime}\boldsymbol{X}$, *has maximum variance of all normalized linear combinations uncorrelated with* $U_1, \ldots, U_{r-1}$.

The vector $\boldsymbol{U}$ is defined as the vector of principal components of $\boldsymbol{X}$. It will be observed that we have proved Theorem A.2.1 of Appendix A for $\boldsymbol{B}$ positive semidefinite, and indeed, the proof holds for any symmetric $\boldsymbol{B}$. It might be noted that once the transformation to $U_1, \ldots, U_p$ has been made, it is obvious that U_1 is the normalized linear combination with maximum variance, for if $U^* = \Sigma c_i U_i$, where $\Sigma c_i^2 = 1$ (U^* also being a normalized linear combination of the X's), then $\operatorname{Var}(U^*) = \Sigma c_i^2 \lambda_i = \lambda_1 + \Sigma_{i=2}^p c_i^2(\lambda_i - \lambda_1)$ (since $c_1^2 = 1 - \Sigma_2^p c_i^2$), which is clearly maximum for $c_i^2 = 0$, $i = 2, \ldots, p$. Similarly, U_2 is the normalized linear combination uncorrelated with U_1 which has maximum variance ($U^* = \Sigma c_i U_i$ being uncorrelated with U_1 implying $c_1 = 0$); in turn the maximal properties of $U_3, \ldots, U_p$ are verified.

Some other consequences can be derived.

COROLLARY 11.2.1. *Suppose* $\lambda_{r+1} = \cdots = \lambda_{r+m} = \nu$ *(i.e.,* ν *is a root of multiplicity* m*); then* $(\boldsymbol{\Sigma} - \nu \boldsymbol{I})$ *is of rank* $p - m$. *Furthermore* $\boldsymbol{\beta}^* = (\boldsymbol{\beta}^{(r+1)} \cdots \boldsymbol{\beta}^{(r+m)})$ *is uniquely determined except for multiplication on the right by an orthogonal matrix.*

PROOF. From the derivation of the theorem we have $(\boldsymbol{\Sigma} - \nu \boldsymbol{I})\boldsymbol{\beta}^{(i)} = \mathbf{0}$, $i = r+1, \ldots, r+m$; that is, $\boldsymbol{\beta}^{(r+1)}, \ldots, \boldsymbol{\beta}^{(r+m)}$ are m linearly independent solutions of $(\boldsymbol{\Sigma} - \nu \boldsymbol{I})\boldsymbol{\beta} = \mathbf{0}$. To show that there cannot be another linearly independent solution, take $\Sigma_{i=1}^p x_i \boldsymbol{\beta}^{(i)}$, where the x_i are scalars. If it is a solution we have $\nu \Sigma x_i \boldsymbol{\beta}^{(i)} = \boldsymbol{\Sigma}(\Sigma x_i \boldsymbol{\beta}^{(i)}) = \Sigma x_i \boldsymbol{\Sigma} \boldsymbol{\beta}^{(i)} = \Sigma x_i \lambda_i \boldsymbol{\beta}^{(i)}$. Since $\nu x_i = \lambda_i x_i$, we must have $x_i = 0$ unless $i = r+1, \ldots, r+m$. Thus the rank is $p - m$.

If $\boldsymbol{\beta}^*$ is one set of solutions to $(\boldsymbol{\Sigma} - \nu \boldsymbol{I})\boldsymbol{\beta} = \mathbf{0}$, then any other set of solutions are linear combinations of the others, that is, are $\boldsymbol{\beta}^* \boldsymbol{A}$ for $\boldsymbol{A}$ nonsingular. However, the orthogonality conditions $\boldsymbol{\beta}^{*\prime}\boldsymbol{\beta}^* = \boldsymbol{I}$ applied to the linear combinations give $\boldsymbol{I} = (\boldsymbol{\beta}^* \boldsymbol{A})'(\boldsymbol{\beta}^* \boldsymbol{A}) = \boldsymbol{A}'\boldsymbol{\beta}^{*\prime}\boldsymbol{\beta}^* \boldsymbol{A} = \boldsymbol{A}'\boldsymbol{A}$, and thus $\boldsymbol{A}$ must be orthogonal. Q.E.D.

THEOREM 11.2.2. *An orthogonal transformation* $\boldsymbol{V} = \boldsymbol{C}\boldsymbol{X}$ *of a random vector* $\boldsymbol{X}$ *leaves invariant the generalized variance and the sum of the variances of the components.*

PROOF. Let $\mathscr{E}\boldsymbol{X} = \mathbf{0}$ and $\mathscr{E}\boldsymbol{X}\boldsymbol{X}' = \boldsymbol{\Sigma}$. Then $\mathscr{E}\boldsymbol{V} = \mathbf{0}$ and $\mathscr{E}\boldsymbol{V}\boldsymbol{V}' = \boldsymbol{C}\boldsymbol{\Sigma}\boldsymbol{C}'$. The generalized variance of $\boldsymbol{V}$ is

$$|\boldsymbol{C}\boldsymbol{\Sigma}\boldsymbol{C}'| = |\boldsymbol{C}| \cdot |\boldsymbol{\Sigma}| \cdot |\boldsymbol{C}'| = |\boldsymbol{\Sigma}| \cdot |\boldsymbol{C}\boldsymbol{C}'| = |\boldsymbol{\Sigma}|, \tag{22}$$

which is the generalized variance of $\boldsymbol{X}$. The sum of the variances of the components of $\boldsymbol{V}$ is

$$(23) \qquad \sum \mathscr{E} V_i^2 = \operatorname{tr}(\boldsymbol{C\Sigma C'}) = \operatorname{tr}(\boldsymbol{\Sigma C'C}) = \operatorname{tr}(\boldsymbol{\Sigma I}) = \operatorname{tr} \boldsymbol{\Sigma} = \sum \mathscr{E} X_i^2.$$

COROLLARY 11.2.2. *The generalized variance of the vector of principal components is the generalized variance of the original vector, and the sum of the variances of the principal components is the sum of the variances of the original variates.*

Another approach to the above theory can be based on the surfaces of constant density of the normal distribution with mean vector $\mathbf{0}$ and covariance matrix $\boldsymbol{\Sigma}$ (nonsingular). The density is

$$(24) \qquad \frac{1}{(2\pi)^{\frac{1}{2}p}|\boldsymbol{\Sigma}|^{\frac{1}{2}}} e^{-\frac{1}{2}\boldsymbol{x}'\boldsymbol{\Sigma}^{-1}\boldsymbol{x}},$$

and surfaces of constant density are ellipsoids

$$(25) \qquad \boldsymbol{x}'\boldsymbol{\Sigma}^{-1}\boldsymbol{x} = C.$$

A principal axis of this ellipsoid is defined as the line from $-\boldsymbol{y}$ to $\boldsymbol{y}$, where $\boldsymbol{y}$ is a point on the ellipsoid where the squared distance $\boldsymbol{x}'\boldsymbol{x}$ has a stationary point. Using the method of Lagrange multipliers, we determine the stationary points by considering

$$(26) \qquad \psi = \boldsymbol{x}'\boldsymbol{x} - \lambda \boldsymbol{x}'\boldsymbol{\Sigma}^{-1}\boldsymbol{x},$$

where λ is a Lagrange multiplier. We differentiate ψ with respect to the components of $\boldsymbol{x}$, and the derivatives set equal to 0 are

$$(27) \qquad \frac{\partial \psi}{\partial \boldsymbol{x}} = 2\boldsymbol{x} - 2\lambda\boldsymbol{\Sigma}^{-1}\boldsymbol{x} = \mathbf{0},$$

or

$$(28) \qquad \boldsymbol{x} = \lambda\boldsymbol{\Sigma}^{-1}\boldsymbol{x}.$$

Multiplication by $\boldsymbol{\Sigma}$ gives

$$\boldsymbol{\Sigma} \boldsymbol{x} = \lambda \boldsymbol{x}. \tag{29}$$

This equation is the same as (4) and the same algebra can be developed. Thus the vectors $\boldsymbol{\beta}^{(1)}, \dots, \boldsymbol{\beta}^{(p)}$ give the principal axes of the ellipsoid. The transformation $\boldsymbol{u} = \mathbf{B}'\boldsymbol{x}$ is a rotation of the coordinate axes so that the new axes are in the direction of the principal axes of the ellipsoid. In the new coordinates the ellipsoid is

$$\boldsymbol{u}'\boldsymbol{\Lambda}^{-1}\boldsymbol{u} = \sum \frac{u_i^2}{\lambda_i} = C. \tag{30}$$

Thus the length of the ith principal axis is $2\sqrt{\lambda_i C}$.

A third approach to the same results is in terms of "planes of closest fit" [Pearson (1901)]. Consider a plane through the origin $\boldsymbol{\alpha}'\boldsymbol{x} = 0$, where $\boldsymbol{\alpha}'\boldsymbol{\alpha} = 1$. The distance of a point $\boldsymbol{x}$ from this plane is $\boldsymbol{\alpha}'\boldsymbol{x}$. Let us find the coefficients of a plane such that the expected distance squared of a random point $\boldsymbol{X}$ from the plane is a minimum, where $\mathscr{E}\boldsymbol{X} = \boldsymbol{0}$ and $\mathscr{E}\boldsymbol{X}\boldsymbol{X}' = \boldsymbol{\Sigma}$. Thus we wish to minimize $\mathscr{E}(\boldsymbol{\alpha}'\boldsymbol{X})^2 = \mathscr{E}\boldsymbol{\alpha}'\boldsymbol{X}\boldsymbol{X}'\boldsymbol{\alpha} = \boldsymbol{\alpha}'\boldsymbol{\Sigma}\boldsymbol{\alpha}$, subject to the restriction $\boldsymbol{\alpha}'\boldsymbol{\alpha} = 1$. Comparison with the first approach immediately shows that the solution is $\boldsymbol{\alpha} = \boldsymbol{\beta}^{(p)}$.

Analysis into principal components is most suitable when all the components of $\boldsymbol{X}$ are measured in the same units. If they are not measured in the same units, the rationale of maximizing $\boldsymbol{\beta}'\boldsymbol{\Sigma}\boldsymbol{\beta}$ relative to $\boldsymbol{\beta}'\boldsymbol{\beta}$ is questionable; in fact, the analysis will depend on the various units of measurement. Suppose $\boldsymbol{\Delta}$ is a diagonal matrix and let $\boldsymbol{Y} = \boldsymbol{\Delta}\boldsymbol{X}$. For example, one component of $\boldsymbol{X}$ may be measured in inches and the corresponding component of $\boldsymbol{Y}$ may be measured in feet; another component of $\boldsymbol{X}$ may be in pounds and the corresponding one of $\boldsymbol{Y}$ in ounces. The covariance matrix of $\boldsymbol{Y}$ is $\mathscr{E}\boldsymbol{Y}\boldsymbol{Y}' = \mathscr{E}\boldsymbol{\Delta}\boldsymbol{X}\boldsymbol{X}'\boldsymbol{\Delta} = \boldsymbol{\Delta}\boldsymbol{\Sigma}\boldsymbol{\Delta} = \boldsymbol{\Psi}$, say. Then analysis of $\boldsymbol{Y}$ into principal components involves maximizing $\mathscr{E}(\boldsymbol{\gamma}'\boldsymbol{Y})^2 = \boldsymbol{\gamma}'\boldsymbol{\Psi}\boldsymbol{\gamma}$ relative to $\boldsymbol{\gamma}'\boldsymbol{\gamma}$ and leads to the equation $\boldsymbol{0} = (\boldsymbol{\Psi} - \nu\boldsymbol{I})\boldsymbol{\gamma} = (\boldsymbol{\Delta}\boldsymbol{\Sigma}\boldsymbol{\Delta} - \nu\boldsymbol{I})\boldsymbol{\gamma}$, where ν must satisfy $|\boldsymbol{\Psi} - \nu\boldsymbol{I}| = 0$. Multiplication on the left by $\boldsymbol{\Delta}^{-1}$ gives

$$\boldsymbol{0} = (\boldsymbol{\Sigma} - \nu\boldsymbol{\Delta}^{-2})(\boldsymbol{\Delta}\boldsymbol{\gamma}). \tag{31}$$

Let $\boldsymbol{\Delta}\boldsymbol{\gamma} = \boldsymbol{\alpha}$; that is, $\boldsymbol{\gamma}'\boldsymbol{Y} = \boldsymbol{\gamma}'\boldsymbol{\Delta}\boldsymbol{X} = \boldsymbol{\alpha}'\boldsymbol{X}$. Then (31) results from maximizing $\mathscr{E}(\boldsymbol{\alpha}'\boldsymbol{X})^2 = \boldsymbol{\alpha}'\boldsymbol{\Sigma}\boldsymbol{\alpha}$ relative to $\boldsymbol{\alpha}'\boldsymbol{\Delta}^{-2}\boldsymbol{\alpha}$. This last quadratic form is a weighted sum of squares, the weights being the diagonal elements of $\boldsymbol{\Delta}^{-2}$.

It might be noted that if $\boldsymbol{\Delta}^{-2}$ is taken to be the matrix

$$\boldsymbol{\Delta}^{-2} = \begin{pmatrix} \sigma_{11} & 0 & \cdots & 0 \\ 0 & \sigma_{22} & \cdots & 0 \\ \vdots & \vdots & & \vdots \\ 0 & 0 & \cdots & \sigma_{pp} \end{pmatrix}, \tag{32}$$

then $\boldsymbol{\Psi}$ is the matrix of correlations.

11.3. MAXIMUM LIKELIHOOD ESTIMATORS OF THE PRINCIPAL COMPONENTS AND THEIR VARIANCES

A primary problem of statistical inference in principal component analysis is to estimate the vectors $\boldsymbol{\beta}^{(1)}, \ldots, \boldsymbol{\beta}^{(p)}$ and the scalars $\lambda_1, \ldots, \lambda_p$. We apply the algebra of the preceding section to an estimate of the covariance matrix.

THEOREM 11.3.1. *Let $\boldsymbol{x}_1, \ldots, \boldsymbol{x}_N$ be N $(> p)$ observations from $N(\boldsymbol{\mu}, \boldsymbol{\Sigma})$, where $\boldsymbol{\Sigma}$ is a matrix with p different characteristic roots. Then a set of maximum likelihood estimators of $\lambda_1, \ldots, \lambda_p$ and $\boldsymbol{\beta}^{(1)}, \ldots, \boldsymbol{\beta}^{(p)}$ defined in Theorem* 11.2.1 *are the roots $k_1 > \cdots > k_p$ of*

$$|\hat{\boldsymbol{\Sigma}} - k\boldsymbol{I}| = 0, \tag{1}$$

and a set of corresponding vectors $\boldsymbol{b}^{(1)}, \ldots, \boldsymbol{b}^{(p)}$ satisfying

$$(\hat{\boldsymbol{\Sigma}} - k_i\boldsymbol{I})\boldsymbol{b}^{(i)} = \boldsymbol{0}, \tag{2}$$

$$\boldsymbol{b}^{(i)\prime}\boldsymbol{b}^{(i)} = 1, \tag{3}$$

where $\hat{\boldsymbol{\Sigma}}$ is the maximum likelihood estimate of $\boldsymbol{\Sigma}$.

PROOF. When the roots of $|\boldsymbol{\Sigma} - \lambda\boldsymbol{I}| = 0$ are different, each vector $\boldsymbol{\beta}^{(i)}$ is uniquely defined except that $\boldsymbol{\beta}^{(i)}$ can be rcplaced by $-\boldsymbol{\beta}^{(i)}$. If we require that the first nonzero component of $\boldsymbol{\beta}^{(i)}$ be positive, then $\boldsymbol{\beta}^{(i)}$ is uniquely defined, and $\boldsymbol{\mu}$, $\boldsymbol{\Lambda}$, $\boldsymbol{\beta}$ is a single-valued function of $\boldsymbol{\mu}$, $\boldsymbol{\Sigma}$. By Corollary 3.2.1, the set of maximum likelihood estimates of $\boldsymbol{\mu}$, $\boldsymbol{\Lambda}$, $\boldsymbol{\beta}$ is the same function of $\hat{\boldsymbol{\mu}}$, $\hat{\boldsymbol{\Sigma}}$. This function is defined by (1), (2), and (3) with the corresponding restriction that the first nonzero component of $\boldsymbol{b}^{(i)}$ be positive. [It can be shown that if $|\Sigma| \neq 0$, the probability is 1 that the roots of (1) are different because the conditions on $\hat{\Sigma}$ for the roots to have multiplicities higher than 1 determine a region in the space of $\hat{\Sigma}$ of dimensionality less than $\frac{1}{2}p(p+1)$; see Okamoto (1973).] From (18) of Section 11.2 we see that

$$\Sigma = \boldsymbol{\beta}\Lambda\boldsymbol{\beta}' = \sum \lambda_i \boldsymbol{\beta}^{(i)}\boldsymbol{\beta}^{(i)\prime} \tag{4}$$

and by the same algebra

$$\hat{\Sigma} = \sum k_i \boldsymbol{b}^{(i)}\boldsymbol{b}^{(i)\prime}. \tag{5}$$

Replacing $\boldsymbol{b}^{(i)}$ by $-\boldsymbol{b}^{(i)}$ clearly does not change $\sum k_i \boldsymbol{b}^{(i)}\boldsymbol{b}^{(i)\prime}$. Since the likelihood function depends only on $\hat{\Sigma}$ (see Section 3.2), the maximum of the likelihood function is attained by taking any set of solutions to (2) and (3). Q.E.D.

It is possible to assume explicitly arbitrary multiplicities of roots of Σ. If these multiplicities are not all unity, the maximum likelihood estimates are not defined as in Theorem 11.3.1. [See Anderson (1963a).] As an example suppose that we assume that the equation $|\Sigma - \lambda \boldsymbol{I}| = 0$ has one root of multiplicity p. Let this root be λ_1. Then by Corollary 11.2.1, $\Sigma - \lambda_1 \boldsymbol{I}$ is of rank 0; that is, $\Sigma - \lambda_1 \boldsymbol{I} = \boldsymbol{0}$ or $\Sigma = \lambda_1 \boldsymbol{I}$. If $\boldsymbol{X}$ is distributed according to $N(\mu, \Sigma) = N(\mu, \lambda_1 \boldsymbol{I})$, the components of $\boldsymbol{X}$ are independently distributed with variance λ_1. Thus the maximum likelihood estimator of λ_1 is

$$\hat{\lambda}_1 = \frac{1}{pN} \sum_{i=1}^{p} \sum_{\alpha=1}^{N} (x_{i\alpha} - \bar{x}_i)^2, \tag{6}$$

and $\hat{\Sigma} = \hat{\lambda}_1 \boldsymbol{I}$, and $\hat{\boldsymbol{\beta}}$ can be any orthogonal matrix. It might be pointed out that in Section 10.7 we considered a test of the hypothesis that $\Sigma = \lambda_1 \boldsymbol{I}$ (with λ_1 unspecified); that is, the hypothesis is that Σ has one characteristic root of multiplicity p.

In most applications of principal component analysis it can be assumed that the roots of Σ are different. It might also be pointed out that in some uses of this method the algebra is applied to the matrix of correlation coefficients rather than to the covariance matrix. In general this leads to different roots and vectors.

11.4. COMPUTATION OF THE MAXIMUM LIKELIHOOD ESTIMATES OF THE PRINCIPAL COMPONENTS

There are several ways of computing the characteristic roots and characteristic vectors (principal components) of a matrix $\boldsymbol{\Sigma}$ or $\hat{\boldsymbol{\Sigma}}$. We shall indicate several of these methods.

One method for small p involves expanding the determinantal equation

$$0 = |\boldsymbol{\Sigma} - \lambda \boldsymbol{I}| \tag{1}$$

and solving the resulting pth degree equation in λ (e.g., by Newton's method or the secant method) for the roots $\lambda_1 > \lambda_2 > \cdots > \lambda_p$. Then $\boldsymbol{\Sigma} - \lambda_i \boldsymbol{I}$ is of rank $p - 1$, and a solution of $(\boldsymbol{\Sigma} - \lambda_i \boldsymbol{I})\boldsymbol{\beta}^{(i)} = \boldsymbol{0}$ can be obtained by taking $\beta_j^{(i)}$ as the cofactor of the element in the first (or any other fixed) column and jth row of $\boldsymbol{\Sigma} - \lambda_i \boldsymbol{I}$.

The second method iterates using the equation for a characteristic root and the corresponding characteristic vector

$$\boldsymbol{\Sigma x} = \lambda \boldsymbol{x}, \tag{2}$$

where we have written the equation for the population. Let $\boldsymbol{x}_{(0)}$ be any vector not orthogonal to the first characteristic vector, and define

$$\boldsymbol{x}_{(i)} = \boldsymbol{\Sigma y}_{(i-1)}, \qquad i = 1, 2, \ldots, \tag{3}$$

$$\boldsymbol{y}_{(i)} = \frac{1}{\sqrt{\boldsymbol{x}_{(i)}' \boldsymbol{x}_{(i)}}} \boldsymbol{x}_{(i)}, \qquad i = 0, 1, 2, \ldots . \tag{4}$$

It can be shown (Problem 12) that

$$\lim_{i \to \infty} \boldsymbol{y}_{(i)} = \pm \boldsymbol{\beta}^{(1)}, \tag{5}$$

$$\lim_{i \to \infty} \boldsymbol{x}_{(i)}' \boldsymbol{x}_{(i)} = \lambda_1^2. \tag{6}$$

The rate of convergence depends on the ratio λ_2/λ_1; the closer this ratio is to 1, the slower the rate of convergence.

To find the second root and vector we define

$$\boldsymbol{\Sigma}_2 = \boldsymbol{\Sigma} - \lambda_1 \boldsymbol{\beta}^{(1)} \boldsymbol{\beta}^{(1)\prime}. \tag{7}$$

Then

$$\text{(8)} \qquad \boldsymbol{\Sigma}_2\boldsymbol{\beta}^{(i)} = \boldsymbol{\Sigma}\boldsymbol{\beta}^{(i)} - \lambda_1\boldsymbol{\beta}^{(1)}\boldsymbol{\beta}^{(1)\prime}\boldsymbol{\beta}^{(i)}$$

$$= \boldsymbol{\Sigma}\boldsymbol{\beta}^{(i)} = \lambda_i\boldsymbol{\beta}^{(i)}$$

if $i \neq 1$ and

$$\text{(9)} \qquad \boldsymbol{\Sigma}_2\boldsymbol{\beta}^{(1)} = \mathbf{0}.$$

Thus λ_2 is the largest root of $\boldsymbol{\Sigma}_2$ and $\boldsymbol{\beta}^{(2)}$ is the corresponding vector. The iteration process is now applied to $\boldsymbol{\Sigma}_2$ to find λ_2 and $\boldsymbol{\beta}^{(2)}$. Defining $\boldsymbol{\Sigma}_3 = \boldsymbol{\Sigma}_2 - \lambda_2\boldsymbol{\beta}^{(2)}\boldsymbol{\beta}^{(2)\prime}$, we can find λ_3 and $\boldsymbol{\beta}^{(3)}$, and so forth.

There are several ways in which the labor of the iteration procedure may be reduced. One is to raise $\boldsymbol{\Sigma}$ to a power before proceeding with the iteration. Thus one can use $\boldsymbol{\Sigma}^2$ defining

$$\text{(10)} \qquad \boldsymbol{x}_{(i)} = \boldsymbol{\Sigma}^2\boldsymbol{y}_{(i-1)}, \qquad i = 1,2,\ldots,$$

$$\text{(11)} \qquad \boldsymbol{y}_{(i)} = \frac{\boldsymbol{x}_{(i)}}{\sqrt{\boldsymbol{x}_{(i)}'\boldsymbol{x}_{(i)}}}, \qquad i = 0,1,2,\ldots.$$

This procedure will give twice as rapid convergence as the use of (3) and (4). Using $\boldsymbol{\Sigma}^4 = \boldsymbol{\Sigma}^2\boldsymbol{\Sigma}^2$ will lead to convergence four times as rapid, and so on. It should be noted that since $\boldsymbol{\Sigma}^2$ is symmetric, there are only $p(p+1)/2$ elements to be found.

Another acceleration procedure has been suggested by Aitken (1937). Suppose one requires accuracy to q significant figures. If the components of $\boldsymbol{y}_{(i)}$ agree with those of $\boldsymbol{y}_{(i-1)}$ to almost q significant figures, define $\boldsymbol{z}^{(0)} = \boldsymbol{y}_{(i)}$, $\boldsymbol{z}^{(1)} = \boldsymbol{\Sigma}\boldsymbol{z}^{(0)}$, $\boldsymbol{z}^{(2)} = \boldsymbol{\Sigma}\boldsymbol{z}^{(1)}$, $\boldsymbol{z}^{(3)} = \boldsymbol{\Sigma}\boldsymbol{z}^{(2)}$,

$$\text{(12)} \qquad r_i^{(k)} = \frac{z_i^{(k)}}{z_i^{(k-1)}}, \qquad i = 1,\ldots,p;\ k = 1,2,3.$$

Then

$$\text{(13)} \qquad R_i = \frac{r_i^{(1)}r_i^{(3)} - \left[r_i^{(2)}\right]^2}{r_i^{(1)} + r_i^{(3)} - 2r_i^{(2)}}, \qquad i = 1,\ldots,p,$$

should agree to the desired number of significant figures. Take the common value of the R_i as the estimate of $\lambda_{(1)}$. The components of the estimate of $\boldsymbol{\beta}^{(1)}$ are proportional to

$$(14) \qquad Z_i = \frac{z_i^{(1)} z_i^{(3)} - \left[z_i^{(2)}\right]^2}{R_i z_i^{(1)} + z_i^{(3)}/R_i - 2z_i^{(2)}}, \qquad i = 1, \dots, p.$$

Efficient computation, however, uses other methods. One method is the *QR* or *QL* algorithm. Let $\boldsymbol{\Sigma}_0 = \boldsymbol{\Sigma}$. Define recursively the orthogonal $\boldsymbol{Q}_i$ and lower triangular $\boldsymbol{L}_i$ by $\boldsymbol{\Sigma}_i = \boldsymbol{Q}_i \boldsymbol{L}_i$ and $\boldsymbol{\Sigma}_{i+1} = \boldsymbol{L}_i \boldsymbol{Q}_i$ $(= \boldsymbol{Q}_i' \boldsymbol{\Sigma}_i \boldsymbol{Q}_i)$, $i = 1, 2, \dots$. (The Gram–Schmidt orthogonalization is a way of finding $\boldsymbol{Q}_i$ and $\boldsymbol{L}_i$; the *QR* method replaces a lower triangular matrix $\boldsymbol{L}$ by an upper triangular matrix $\boldsymbol{R}$.) If the characteristic roots of $\boldsymbol{\Sigma}$ are distinct, $\lim_{i \to \infty} \boldsymbol{\Sigma}_{i+1} = \boldsymbol{\Lambda}^*$, where $\boldsymbol{\Lambda}^*$ is the diagonal matrix with the roots usually ordered in ascending order. The characteristic vectors are the columns of $\lim_{i \to \infty} \boldsymbol{Q}_i' \boldsymbol{Q}_{i-1}' \cdots \boldsymbol{Q}_1'$ (which is computed recursively).

A more efficient algorithm (for the symmetric $\boldsymbol{\Sigma}$) uses a sequence of Householder transformations to carry $\boldsymbol{\Sigma}$ to tridiagonal form. A *Householder matrix* is $\boldsymbol{H} = \boldsymbol{I} - 2\boldsymbol{\alpha}\boldsymbol{\alpha}'$, where $\boldsymbol{\alpha}'\boldsymbol{\alpha} = 1$. Such a matrix is orthogonal and symmetric. A Householder transformation of the symmetric matrix $\boldsymbol{\Sigma}$ is $\boldsymbol{H}\boldsymbol{\Sigma}\boldsymbol{H}$. It is symmetric and has the same characteristic roots as $\boldsymbol{\Sigma}$; its characteristic vectors are $\boldsymbol{H}$ times those of $\boldsymbol{\Sigma}$.

A *tridiagonal matrix* is one with all entries 0 except on the main diagonal, the first superdiagonal, and the first subdiagonal. A sequence of $p-2$ Householder transformations carries the symmetric $\boldsymbol{\Sigma}$ to tridiagonal form. (The first one inserts 0's into the last $p-2$ entries of the first column and row of $\boldsymbol{H}\boldsymbol{\Sigma}\boldsymbol{H}$, etc. See Problem 13.)

The *QL* method is applied to the tridiagonal form. At the ith step let the tridiagonal matrix be $\boldsymbol{T}_0^{(i)}$, let $\boldsymbol{P}_j^{(i)}$ be a block diagonal matrix (Givens matrix)

$$(15) \qquad \boldsymbol{P}_j^{(i)} = \begin{bmatrix} \boldsymbol{I} & \boldsymbol{0} & \boldsymbol{0} & \boldsymbol{0} \\ \boldsymbol{0} & \cos\theta_j & -\sin\theta_j & \boldsymbol{0} \\ \boldsymbol{0} & \sin\theta_j & \cos\theta_j & \boldsymbol{0} \\ \boldsymbol{0} & \boldsymbol{0} & \boldsymbol{0} & \boldsymbol{I} \end{bmatrix},$$

where $\cos\theta_j$ is the jth and $j+1$st diagonal element, and let $\boldsymbol{T}_j^{(i)} = \boldsymbol{P}_{p-j}^{(i)} \boldsymbol{T}_{j-1}^{(i)}$, $j = 1, \dots, p-1$. θ_j is chosen so that the element in position $j, j+1$ in $\boldsymbol{T}_j$ is 0.

Then $P^{(i)} = P_1^{(i)} P_2^{(i)} \ldots P_{p-1}^{(i)}$ is orthogonal and $P^{(i)} T_0^{(i)} = L^{(i)}$ is lower triangular. Then $T_0^{(i+1)} = L^{(i)} P^{(i)\prime}$ $(= P^{(i)} T_0^{(i)} P^{(i)\prime})$ is symmetric and tridiagonal. It converges to Λ^* (if the roots are all different). For more details see Chapters II/2 and II/3 of Wilkinson and Reinsch (1971) and Chapter 5 of Wilkinson (1965).

11.5. AN EXAMPLE

In Table 3.4 we presented three samples of observations on varieties of iris [Fisher (1936)]; as an example of principal component analysis we use one of those samples, namely Iris versicolor. There are 50 observations ($N = 50$, $n = N - 1 = 49$). Each observation consists of four measurements on a plant; x_1 is sepal length, x_2 is sepal width, x_3 is petal length, and x_4 is petal width. The observed sums of squares and cross products of deviations from means are

(1)

$$A = \sum_{\alpha=1}^{50} (x_\alpha - \bar{x})(x_\alpha - \bar{x})' = \begin{pmatrix} 13.0552 & 4.1740 & 8.9620 & 2.7332 \\ 4.1740 & 4.8250 & 4.0500 & 2.0190 \\ 8.9620 & 4.0500 & 10.8200 & 3.5820 \\ 2.7332 & 2.0190 & 3.5820 & 1.9162 \end{pmatrix},$$

and an estimate of Σ is

$$S = \tfrac{1}{49} A = \begin{pmatrix} 0.266{,}433 & 0.085{,}184 & 0.182{,}899 & 0.055{,}780 \\ 0.085{,}184 & 0.098{,}469 & 0.082{,}653 & 0.041{,}204 \\ 0.182{,}899 & 0.082{,}653 & 0.220{,}816 & 0.073{,}102 \\ 0.055{,}780 & 0.041{,}204 & 0.073{,}102 & 0.039{,}106 \end{pmatrix}. \tag{2}$$

We use the iterative procedure to find the first principal component, by computing in turn $z^{(j)} = Sz^{(j-1)}$. As an initial approximation, we use $z^{(0)\prime} = (1, 0, 1, 0)$. It is not necessary to normalize the vector at each iteration, but to compare successive vectors, we compute $z_i^{(j)}/z_i^{(j-1)} = r_i^{(j)}$, which is an approximation to l_1, the largest root of S. After seven iterations, $r_i^{(7)}$ agree to within two units in the fifth decimal place (fifth significant figure). This vector is normalized, and S is applied to the normalized vector. The ratios, $r_i^{(8)}$, agree to within two units in the sixth place; the value of l_1 is (nearly accurate to the sixth place) $l_1 = 0.487{,}875$. The normalized eighth iterated vector is our estimate of $\boldsymbol{\beta}^{(1)}$, namely,

$$\boldsymbol{b}^{(1)} = \begin{pmatrix} 0.686,724,4 \\ 0.305,346,3 \\ 0.623,662,8 \\ 0.214,983,7 \end{pmatrix}. \tag{3}$$

This vector agrees with the normalized seventh iterate to about one unit in the sixth place. It should be pointed out that l_1 and $\boldsymbol{b}^{(1)}$ have to be calculated more accurately than l_2 and $\boldsymbol{b}^{(2)}$, and so forth. The trace of $\boldsymbol{S}$ is 0.624,824, which is the sum of the roots. Thus l_1 is more than three times the sum of the other roots.

We next compute

$$\boldsymbol{S}_2 = \boldsymbol{S} - l_1 \boldsymbol{b}^{(1)}\boldsymbol{b}^{(1)\prime} \tag{4}$$

$$= \begin{pmatrix} 0.036,355,9 & -0.017,117,9 & -0.026,050,2 & -0.016,247,2 \\ -0.017,117,9 & 0.052,981,3 & -0.010,254,6 & 0.009,177,7 \\ -0.026,050,2 & -0.010,254,6 & 0.031,054,4 & 0.007,689,0 \\ -0.016,247,2 & 0.009,177,7 & 0.007,689,0 & 0.016,557,4 \end{pmatrix},$$

and iterate $z^{(j)} = \boldsymbol{S}_2 z^{(j-1)}$, using $z^{(0)\prime} = (0, 1, 0, 0)$. (In the actual computation $\boldsymbol{S}_2$ was multiplied by 10 and the first row and column were multiplied by -1.) In this case the iteration does not proceed as rapidly; as will be seen, the ratio of l_2 to l_3 is approximately 1.32. After 15 iterations, $r_i^{(15)}$ agree to three decimal places or two significant figures. On iterations 15, 16, and 17, the $r_i^{(j)}$ are computed to ten significant figures. The R_i for the acceleration can then be computed to about six figures, and the accelerated vector to about three places. (In this case the number of valid significant figures in the accelerated vector is only one-third the number in the $r_i^{(j)}$.) The accelerated vector is iterated once, normalized, and iterated again. On this last iteration, the ratios agree to within four units in the fifth significant figure. We obtain $l_2 = 0.072,382,8$ and

$$\boldsymbol{b}^{(2)} = \begin{pmatrix} -0.669,033 \\ 0.567,484 \\ 0.343,309 \\ 0.335,307 \end{pmatrix}. \tag{5}$$

We find

$$\boldsymbol{S}_3 = \boldsymbol{S}_2 - l_2 \boldsymbol{b}^{(2)}\boldsymbol{b}^{(2)\prime} \tag{6}$$

$$= \begin{pmatrix} 0.003,955 & 0.010,363 & -0.009,425 & -0.000,010 \\ 0.010,363 & 0.029,671 & -0.024,356 & -0.004,595 \\ -0.009,425 & -0.024,356 & 0.022,523 & -0.000,643 \\ -0.000,010 & -0.004,595 & -0.000,643 & 0.008,419 \end{pmatrix}.$$

Using an initial approximation of $z^{(0)\prime} = (0, 1, 1, 0)$, we iterate six times to obtain $l_3 = 0.054{,}775$ and

$$(7) \qquad b^{(3)} = \begin{pmatrix} -0.265{,}105 \\ -0.729{,}589 \\ 0.627{,}178 \\ 0.063{,}676 \end{pmatrix}.$$

Next we compute

$$(8) \qquad S_4 = S_3 - l_3 b^{(3)} b^{(3)\prime}$$

$$= \begin{pmatrix} 0.000{,}105 & -0.000{,}232 & -0.000{,}318 & 0.000{,}915 \\ -0.000{,}232 & 0.000{,}514 & 0.000{,}708 & -0.002{,}050 \\ -0.000{,}318 & 0.000{,}708 & 0.000{,}977 & -0.002{,}831 \\ 0.000{,}915 & -0.002{,}050 & -0.002{,}831 & 0.008{,}197 \end{pmatrix}.$$

This matrix is approximately of rank one; the characteristic vector is proportional to any column. We find $l_4 = 0.009{,}793$ and

$$(9) \qquad b^{(4)} = \begin{pmatrix} 0.102{,}31 \\ -0.228{,}90 \\ -0.316{,}02 \\ 0.915{,}02 \end{pmatrix}.$$

We compute $S_4 - l_4 b^{(4)} b^{(4)\prime}$ as a check, and we find that the entries of this matrix are within 0.000,003 of being zero. Because the entries in S_4 are so small, the elements of $b^{(4)}$ can be correct to only four places and probably are correct to only three places. Since the other vectors are probably accurate to four or five places, it would likely be more accurate to compute $b^{(4)}$ to be orthogonal to $b^{(1)}$, $b^{(2)}$, and $b^{(3)}$.

The results may be summarized as follows:

$$(10) \qquad (l_1, l_2, l_3, l_4) = (0.4879 \quad 0.0724 \quad 0.0548 \quad 0.0098),$$

$$(11) \qquad B = \begin{pmatrix} 0.6867 & -0.6690 & -0.2651 & 0.1023 \\ 0.3053 & 0.5675 & -0.7296 & -0.2289 \\ 0.6237 & 0.3433 & 0.6272 & -0.3160 \\ 0.2150 & 0.3353 & 0.0637 & 0.9150 \end{pmatrix}.$$

It will be observed that the first component accounts for 78% of the total variance in the four measurements; the last component accounts for a little more than 1% of the total variance. In fact, the variance of $0.7x_1 + 0.3x_2 +$

$0.6x_3 + 0.2x_4$ (an approximation to the first principal component) is 0.478, which is almost 77% of the total variance. If one is interested in studying the variations in conditions that lead to variations of (x_1, x_2, x_3, x_4), one can look for variations in conditions that lead to variations of $0.7x_1 + 0.3x_2 + 0.6x_3 + 0.2x_4$. It is not very important if the other variations in (x_1, x_2, x_3, x_4) are neglected in exploratory investigations.

11.6. STATISTICAL INFERENCE

11.6.1. Asymptotic Distributions

In Section 13.3 we shall derive the exact distribution of the sample characteristic roots and vectors when the population covariance matrix is $\boldsymbol{I}$ or proportional to $\boldsymbol{I}$, that is, in the case of all population roots equal. The exact distribution of roots and vectors when the population roots are not all equal involves a multiply infinite series of zonal polynomials; that development is beyond the scope of this book. We derive the asymptotic distribution of the roots and vectors when the population roots are all different (Theorem 13.5.1) and also when one root is multiple (Theorem 13.5.2). Since it can usually be assumed that the population roots are different unless there is information to the contrary, we summarize here Theorem 13.5.1.

As earlier, let the characteristic roots of $\boldsymbol{\Sigma}$ be $\lambda_1 > \cdots > \lambda_p$ and the corresponding characteristic vectors be $\boldsymbol{\beta}^{(1)}, \dots, \boldsymbol{\beta}^{(p)}$, normalized so $\boldsymbol{\beta}^{(i)\prime}\boldsymbol{\beta}^{(i)} = 1$ and satisfying $\beta_{1i} \geq 0$, $i = 1, \dots, p$. Let the roots and vectors of $\boldsymbol{S}$ be $l_1 > \cdots > l_p$ and $\boldsymbol{b}^{(1)}, \dots, \boldsymbol{b}^{(p)}$ normalized so $\boldsymbol{b}^{(i)\prime}\boldsymbol{b}^{(i)} = 1$ and satisfying $b_{1i} \geq 0$, $i = 1, \dots, p$. Let $d_i = \sqrt{n}(l_i - \lambda_i)$ and $\boldsymbol{g}^{(i)} = \sqrt{n}(\boldsymbol{b}^{(i)} - \boldsymbol{\beta}^{(i)})$, $i = 1, \dots, p$. Then in the limiting normal distribution the sets $d_1, \dots, d_p$ and $\boldsymbol{g}^{(1)}, \dots, \boldsymbol{g}^{(p)}$ are independent and $d_1, \dots, d_p$ are mutually independent. The element d_i has the limiting distribution $N(0, 2\lambda_i^2)$. The covariances of $\boldsymbol{g}^{(1)}, \dots, \boldsymbol{g}^{(p)}$ in the limiting distribution are

$$\mathscr{A}\mathscr{C}\left(\boldsymbol{g}^{(i)}\right) = \sum_{\substack{k=1 \\ k \neq i}}^{p} \frac{\lambda_i \lambda_k}{(\lambda_i - \lambda_k)^2} \boldsymbol{\beta}^{(k)} \boldsymbol{\beta}^{(k)\prime}, \tag{1}$$

$$\mathscr{A}\mathscr{C}\left(\boldsymbol{g}^{(i)}, \boldsymbol{g}^{(j)}\right) = -\frac{\lambda_i \lambda_j}{(\lambda_i - \lambda_j)^2} \boldsymbol{\beta}^{(j)} \boldsymbol{\beta}^{(i)\prime}, \qquad i \neq j. \tag{2}$$

In making inferences about a single ordered root, one treats l_i as approximately normal with mean λ_i and variance $2\lambda_i^2/n$. Since l_i is a consistent estimate of λ_i, the limiting distribution of

$$\sqrt{n}\,\frac{l_i - \lambda_i}{\sqrt{2}\, l_i} \tag{3}$$

is $N(0,1)$. A two-tailed test of the hypothesis $\lambda_i = \lambda_i^0$ has the (asymptotic) acceptance region

$$-z(\varepsilon) \le \sqrt{\frac{n}{2}}\,\frac{l_i - \lambda_i^0}{\lambda_i^0} \le z(\varepsilon), \tag{4}$$

where the value of the $N(0,1)$ distribution beyond $z(\varepsilon)$ is $\frac{1}{2}\varepsilon$. The interval (4) can be inverted to give a confidence interval for λ_i with confidence $1-\varepsilon$

$$\frac{l_i}{1 + \sqrt{2/n}\, z(\varepsilon)} \le \lambda_i \le \frac{l_i}{1 - \sqrt{2/n}\, z(\varepsilon)}. \tag{5}$$

Note that the confidence coefficient should be taken large enough so $\sqrt{2/n}\, z(\varepsilon) < 1$. Alternatively, one can use the fact that the limiting distribution of $\sqrt{n}\,(\log l_i - \log \lambda_i)$ is $N(0,2)$ by Theorem 4.2.3.

Inference about components of a vector $\boldsymbol{\beta}^{(i)}$ can be based on treating $\boldsymbol{b}^{(i)}$ as being approximately normal with mean $\boldsymbol{\beta}^{(i)}$ and (singular) covariance matrix $(1/n)$ times (1).

11.6.2. Confidence Region for a Characteristic Vector

We use the asymptotic distribution of the sample characteristic vectors to obtain a large-sample confidence region for the ith characteristic vector of $\boldsymbol{\Sigma}$ [Anderson (1963a)]. The covariance matrix (1) can be written

$$\mathbf{B}\boldsymbol{\Delta}_i^2\mathbf{B}' = \mathbf{B}_i^*\boldsymbol{\Delta}_i^{*2}\mathbf{B}_i^{*\prime}, \tag{6}$$

where $\boldsymbol{\Delta}_i$ is the $p \times p$ diagonal matrix with 0 as the ith diagonal element and $\sqrt{\lambda_i\lambda_j}/(\lambda_i - \lambda_j)$ as the jth diagonal element, $j \neq i$, $\boldsymbol{\Delta}_i^*$ is the $(p-1) \times (p-1)$ diagonal matrix obtained from $\boldsymbol{\Delta}_i$ by deleting the ith row and column, and $\mathbf{B}_i^*$ is the $p \times (p-1)$ matrix formed by deleting the ith column from $\mathbf{B}$. Then $\boldsymbol{h}^{(i)} = \boldsymbol{\Delta}_i^{*-1}\mathbf{B}_i^{*\prime}\sqrt{n}\,(\boldsymbol{b}^{(i)} - \boldsymbol{\beta}^{(i)})$ has a limiting normal distribution with mean $\mathbf{0}$ and covariance matrix

$$(7) \qquad \mathscr{C}(\boldsymbol{h}^{(i)}) = \boldsymbol{\Delta}_i^{*-1}\boldsymbol{\mathsf{B}}_i^{*\prime}(\boldsymbol{\mathsf{B}}_i^*\boldsymbol{\Delta}_i^{*2}\boldsymbol{\mathsf{B}}_i^{*\prime})\boldsymbol{\mathsf{B}}_i^*\boldsymbol{\Delta}_i^{*-1} = \boldsymbol{I}_{p-1},$$

and

$$(8) \qquad \boldsymbol{h}^{(i)\prime}\boldsymbol{h}^{(i)} = n(\boldsymbol{b}^{(i)} - \boldsymbol{\beta}^{(i)})'\boldsymbol{\mathsf{B}}_i^*\boldsymbol{\Delta}_i^{*-2}\boldsymbol{\mathsf{B}}_i^{*\prime}(\boldsymbol{b}^{(i)} - \boldsymbol{\beta}^{(i)})$$

has a limiting χ^2-distribution with $p-1$ degrees of freedom. The matrix of the quadratic form in $\sqrt{n}\,(\boldsymbol{b}^{(i)} - \boldsymbol{\beta}^{(i)})$ is

$$(9) \quad \boldsymbol{\mathsf{B}}_i^*\boldsymbol{\Delta}_i^{*-2}\boldsymbol{\mathsf{B}}_i^{*\prime} = \sum_{j=1}^{p} \boldsymbol{\beta}^{(j)}\left(\frac{\lambda_i}{\lambda_j} - 2 + \frac{\lambda_j}{\lambda_i}\right)\boldsymbol{\beta}^{(j)\prime} - \boldsymbol{\beta}^{(i)}\left(\frac{\lambda_i}{\lambda_i} - 2 + \frac{\lambda_i}{\lambda_i}\right)\boldsymbol{\beta}^{(i)\prime}$$

$$= \lambda_i\boldsymbol{\Sigma}^{-1} - 2\boldsymbol{I} + (1/\lambda_i)\boldsymbol{\Sigma}$$

because $\boldsymbol{\mathsf{B}}\boldsymbol{\Lambda}^{-1}\boldsymbol{\mathsf{B}}' = \boldsymbol{\Sigma}^{-1}$, $\boldsymbol{\mathsf{B}}\boldsymbol{\mathsf{B}}' = \boldsymbol{I}$, and $\boldsymbol{\mathsf{B}}\boldsymbol{\Lambda}\boldsymbol{\mathsf{B}}' = \boldsymbol{\Sigma}$. Then (8) is

$$(10) \quad n(\boldsymbol{b}^{(i)} - \boldsymbol{\beta}^{(i)})'(\lambda_i\boldsymbol{\Sigma}^{-1} - 2\boldsymbol{I} + (1/\lambda_i)\boldsymbol{\Sigma})(\boldsymbol{b}^{(i)} - \boldsymbol{\beta}^{(i)})$$

$$= n\boldsymbol{b}^{(i)\prime}(\lambda_i\boldsymbol{\Sigma}^{-1} - 2\boldsymbol{I} + (1/\lambda_i)\boldsymbol{\Sigma})\boldsymbol{b}^{(i)}$$

$$= n\left[\lambda_i\boldsymbol{b}^{(i)\prime}\boldsymbol{\Sigma}^{-1}\boldsymbol{b}^{(i)} + (1/\lambda_i)\boldsymbol{b}^{(i)\prime}\boldsymbol{\Sigma}\boldsymbol{b}^{(i)} - 2\right]$$

because $\boldsymbol{\beta}^{(i)\prime}$ is a characteristic vector of $\boldsymbol{\Sigma}$ with root λ_i and of $\boldsymbol{\Sigma}^{-1}$ with root $1/\lambda_i$. On the left-hand side of (10) we can replace $\boldsymbol{\Sigma}$ and λ_i by the consistent estimators $\boldsymbol{S}$ and l_i to obtain

$$(11) \quad n(\boldsymbol{b}^{(i)} - \boldsymbol{\beta}^{(i)})'\left[l_i\boldsymbol{S}^{-1} - 2\boldsymbol{I} + (1/l_i)\boldsymbol{S}\right](\boldsymbol{b}^{(i)} - \boldsymbol{\beta}^{(i)})$$

$$= n\left[l_i\boldsymbol{\beta}^{(i)\prime}\boldsymbol{S}^{-1}\boldsymbol{\beta}^{(i)} + (1/l_i)\boldsymbol{\beta}^{(i)\prime}\boldsymbol{S}\boldsymbol{\beta}^{(i)} - 2\right],$$

which has a limiting χ^2-distribution with $p-1$ degrees of freedom.

A confidence region for the ith characteristic vector of $\boldsymbol{\Sigma}$ with confidence $1-\varepsilon$ consists of the intersection of $\boldsymbol{\beta}^{(i)\prime}\boldsymbol{\beta}^{(i)} = 1$ and the set of $\boldsymbol{\beta}^{(i)}$ such that the right-hand side of (11) is less than $\chi^2_{p-1}(\varepsilon)$, where $\Pr\{\chi^2_{p-1} > \chi^2_{p-1}(\varepsilon)\} = \varepsilon$. Note that the matrix of the quadratic form (9) is positive semidefinite.

This approach also provides a test of the null hypothesis that the ith characteristic vector is a specified $\boldsymbol{\beta}_0^{(i)}$ ($\boldsymbol{\beta}_0^{(i)\prime}\boldsymbol{\beta}_0^{(i)} = 1$). The hypothesis is rejected if the right-hand side of (11) with $\boldsymbol{\beta}^{(i)}$ replaced by $\boldsymbol{\beta}_0^{(i)}$ exceeds $\chi^2_{p-1}(\varepsilon)$.

Mallows (1961) suggested a test of whether *some* characteristic vector of Σ is β_0. Let $\mathbf{B}_0$ be a $p \times (p-1)$ matrix such that $\beta_0' \mathbf{B}_0 = \mathbf{0}$. If the null hypothesis is true, $\beta_0' X$ and $\mathbf{B}_0' X$ are independent (because $\mathbf{B}_0$ is a nonsingular transform of the set of other characteristic vectors). The test is based on the multiple correlation between $\beta_0' X$ and $\mathbf{B}_0' X$. In principle, the test procedure can be inverted to obtain a confidence region. The usefulness of these procedures is limited by the fact that the hypothesized vector is not attached to a characteristic root; the interpretation depends on the root (e.g., largest versus smallest).

11.6.3. Exact Confidence Limits on the Characteristic Roots

We now consider a confidence interval for the entire set of characteristic roots of Σ, namely, $\lambda_1 \geq \cdots \geq \lambda_p$ [Anderson (1965a)]. We use the facts that $\beta^{(i)\prime}\Sigma\beta^{(i)} = \lambda_i$, $\beta^{(i)\prime}\beta^{(i)} = 1$, $i = 1, p$, $\beta^{(1)\prime}\Sigma\beta^{(p)} = 0 = \beta^{(1)\prime}\beta^{(p)}$. Then $\beta^{(1)\prime}X$ and $\beta^{(p)\prime}X$ are uncorrelated and have variances λ_1 and λ_p, respectively. Hence $n\beta^{(1)\prime}S\beta^{(1)}/\lambda_1$ and $n\beta^{(p)\prime}S\beta^{(p)}/\lambda_p$ are independently distributed as χ^2 with n degrees of freedom. Let l and u be two numbers such that

$$1 - \varepsilon = \Pr\{nl \leq \chi_n^2\}\Pr\{\chi_n^2 \leq nu\}. \tag{12}$$

Then

$$\begin{aligned} 1 - \varepsilon &= \Pr\left\{ l \leq \frac{\beta^{(1)\prime}S\beta^{(1)}}{\lambda_1}, \ \frac{\beta^{(p)\prime}S\beta^{(p)}}{\lambda_p} \leq u \right\} \\ &= \Pr\left\{ \frac{\beta^{(p)\prime}S\beta^{(p)}}{u} \leq \lambda_p, \ \lambda_1 \leq \frac{\beta^{(1)\prime}S\beta^{(1)}}{l} \right\} \\ &\leq \Pr\left\{ \min_{b'b=1} \frac{b'Sb}{u} \leq \lambda_p, \ \lambda_1 \leq \max_{b'b=1} \frac{b'Sb}{l} \right\} \\ &= \Pr\left\{ \frac{l_p}{u} \leq \lambda_p \leq \lambda_1 \leq \frac{l_1}{l} \right\}. \end{aligned} \tag{13}$$

THEOREM 11.6.1. *A confidence interval for the characteristic roots of* Σ *with confidence at least* $1 - \varepsilon$ *is*

$$l_p/u \leq \lambda_p \leq \lambda_1 \leq l_1/l, \tag{14}$$

where l *and* u *satisfy* (12).

A tighter inequality can lead to a better lower bound. The matrix $\boldsymbol{H} = n\boldsymbol{\beta}'\boldsymbol{S}\boldsymbol{\beta}$ has characteristic roots $nl_1, \dots, nl_p$ because $\boldsymbol{\beta}$ is orthogonal. We use the following lemma.

LEMMA 11.6.1. *For any positive definite matrix* $\boldsymbol{H}$

$$\mathrm{ch}_p(\boldsymbol{H}) \le \frac{1}{h^{ii}} \le \mathrm{ch}_1(\boldsymbol{H}), \qquad i = 1, \dots, p, \tag{15}$$

where $\boldsymbol{H}^{-1} = (h^{ij})$ *and* $\mathrm{ch}_p(\boldsymbol{H})$ *and* $\mathrm{ch}_1(\boldsymbol{H})$ *are the minimum and maximum characteristic roots of* $\boldsymbol{H}$, *respectively.*

PROOF. From Theorem A.2.4 in Appendix A we have $\mathrm{ch}_p(\boldsymbol{H}) \le h_{ii} \le \mathrm{ch}_1(\boldsymbol{H})$ and

$$\mathrm{ch}_p(\boldsymbol{H}^{-1}) \le h^{ii} \le \mathrm{ch}_1(\boldsymbol{H}^{-1}), \qquad i = 1, \dots, p. \tag{16}$$

Since $\mathrm{ch}_p(\boldsymbol{H}) = 1/\mathrm{ch}_1(\boldsymbol{H}^{-1})$ and $\mathrm{ch}_1(\boldsymbol{H}) = 1/\mathrm{ch}_p(\boldsymbol{H}^{-1})$, the lemma follows. Q.E.D.

The argument for Theorem 5.2.2 shows that $1/(\lambda_p h^{pp})$ is distributed as χ^2 with $n - p + 1$ degrees of freedom and Theorem 4.3.3 shows that h^{pp} is independent of h_{11}. Let l' and u' be two numbers such that

$$1 - \varepsilon = \Pr\{nl' \le \chi_n^2\}\Pr\{\chi_{n-p+1}^2 \le nu'\}. \tag{17}$$

Then

$$\begin{aligned} 1 - \varepsilon &= \Pr\left\{nl' \le \frac{h_{11}}{\lambda_1},\ \frac{1}{\lambda_p h^{pp}} \le nu'\right\} \\ &\le \Pr\left\{\frac{l_p}{u'} \le \lambda_p,\ \lambda_1 \le \frac{l_1}{l'}\right\} \end{aligned} \tag{18}$$

since $\mathrm{ch}_p(\boldsymbol{H}) = nl_p$ and $\mathrm{ch}_1(\boldsymbol{H}) = nl_1$.

THEOREM 11.6.2. *A confidence interval for the characteristic roots of* $\boldsymbol{\Sigma}$ *with confidence at least* $1 - \varepsilon$ *is*

$$\frac{l_p}{u'} \le \lambda_p \le \lambda_1 \le \frac{l_1}{l'}, \tag{19}$$

where l' *and* u' *satisfy* (17).

Anderson (1965a, 1965b) has shown that the above confidence bounds are optimal within the class of bounds

$$f(l_1, \dots, l_p) \le \lambda_p \le \lambda_1 \le g(l_1, \dots, l_p), \tag{20}$$

where f and g are homogeneous of degree 1 and are monotonically nondecreasing in each argument for fixed values of the others. If (20) holds with probability at least $1 - \varepsilon$, then a pair of numbers u' and l' can be found to satisfy (17) and

$$f(l_1, \dots, l_p) \le \frac{l_p}{u'}, \quad \frac{l_1}{l'} \le g(l_1, \dots, l_p). \tag{21}$$

The homogeneity condition means that the confidence bounds are multiplied by c^2 if the observed vectors are multiplied by c (which is a kind of scale invariance). The monotonicity conditions imply that an increase in the size of $\boldsymbol{S}$ results in an increase in the limits for $\boldsymbol{\Sigma}$ (which is a kind of consistency).

The confidence bounds given in (31) of Section 10.8 for the roots of $\boldsymbol{\Sigma}$ based on the distribution of the roots of $\boldsymbol{S}$ when $\boldsymbol{\Sigma} = \boldsymbol{I}$ are greater.

11.7. TESTING HYPOTHESES ABOUT THE CHARACTERISTIC ROOTS OF A COVARIANCE MATRIX

11.7.1. Testing a Hypothesis about the Sum of the Smallest Characteristic Roots

An investigator may raise the question whether the last $p - m$ principal components may be ignored, that is, whether the first m principal components furnish a good approximation to $\boldsymbol{X}$. He may want to do this if the sum of the variances of the last principal components is less than some specified amount, say γ. Consider the null hypothesis

$$H: \lambda_{m+1} + \cdots + \lambda_p \ge \gamma, \tag{1}$$

where γ is specified, against the alternative that the sum is less than γ. If the characteristic roots of $\boldsymbol{\Sigma}$ are different, it follows from Theorem 13.5.1 that

$$\sqrt{n}\left(\sum_{i=m+1}^{p} l_i - \sum_{i=m+1}^{p} \lambda_i \right) \tag{2}$$

has a limiting normal distribution with mean 0 and variance $2\Sigma_{i=m+1}^{p}\lambda_i^2$. The variance can be consistently estimated by $2\Sigma_{i=m+1}^{p}l_i^2$. Then a rejection region with (large-sample) significance level ε is

$$\sum_{i=m+1}^{p} l_i < \gamma - \frac{\sqrt{2\sum_{i=m+1}^{p} l_i^2}}{\sqrt{n}} z(2\varepsilon), \tag{3}$$

where $z(2\varepsilon)$ is the upper significance point of the standard normal distribution for significance level ε. The (large-sample) probability of rejection is ε if equality holds in (1) and is less than ε if inequality holds.

The investigator may alternatively want an upper confidence interval for $\Sigma_{i=m+1}^{p}\lambda_i$ with at least approximate confidence level $1-\varepsilon$. It is

$$\sum_{i=m+1}^{p} \lambda_i \leq \sum_{i=m+1}^{p} l_i + \frac{\sqrt{2\sum_{i=m+1}^{p} l_i^2}}{\sqrt{n}} z(2\varepsilon). \tag{4}$$

If the right-hand side is sufficiently small (in particular less than γ), the investigator has confidence that the sum of the variances of the $p-m$ smallest principal components is so small they can be neglected. Anderson (1963a) gave this analysis also in the case that $\lambda_{m+1} = \cdots = \lambda_p$.

11.7.2. Testing a Hypothesis about the Sum of the Smallest Characteristic Roots Relative to the Sum of All the Roots

The investigator may want to ignore the last $p-m$ principal components if their sum is small relative to the sum of all the roots (which is the trace of the covariance matrix). Consider the null hypothesis

$$H\colon f(\lambda) = \frac{\lambda_{m+1} + \cdots + \lambda_p}{\lambda_1 + \cdots + \lambda_p} \geq \delta, \tag{5}$$

where δ is specified, against the alternative that $f(\lambda) < \delta$. We use the fact that

$$\begin{aligned} \frac{\partial f(\lambda)}{\partial \lambda_i} &= -\frac{\lambda_{m+1} + \cdots + \lambda_p}{(\lambda_1 + \cdots + \lambda_p)^2}, & i &= 1,\dots,m, \\ \frac{\partial f(\lambda)}{\partial \lambda_i} &= \frac{\lambda_1 + \cdots + \lambda_m}{(\lambda_1 + \cdots + \lambda_p)^2}, & i &= m+1,\dots,p. \end{aligned} \tag{6}$$

Then the asymptotic variance of $f(\boldsymbol{l})$ is

$$2\left(\frac{\delta}{\operatorname{tr}\boldsymbol{\Sigma}}\right)^2(\lambda_1 + \cdots + \lambda_m^2) + 2\left(\frac{1-\delta}{\operatorname{tr}\boldsymbol{\Sigma}}\right)^2(\lambda_{m+1}^2 + \cdots + \lambda_p^2) \tag{7}$$

when equality holds in (5), by Theorem 4.2.3. The null hypothesis H is rejected if $\sqrt{n}\,[f(\boldsymbol{l}) - \delta]$ is less than the appropriate significance point of the standard normal distribution times the square root of (7) with λ's replaced by l's and $\operatorname{tr}\boldsymbol{\Sigma}$ by $\operatorname{tr}\boldsymbol{S}$. Alternatively one can construct a large-sample confidence region for $f(\boldsymbol{\lambda})$. A confidence region of approximate confidence $1-\varepsilon$ is $[z = z(2\varepsilon)]$

(8)

$$\frac{\sum_{i=m+1}^p \lambda_i}{\sum_{i=1}^p \lambda_i} \le \frac{\sum_{i=m+1}^p l_i}{\sum_{i=1}^p l_i} + z\frac{\left[2\left(\sum_{i=m+1}^p l_i\right)^2 \sum_{i=1}^m l_i^2 + 2\left(\sum_{i=1}^m l_i\right)^2 \sum_{i=m+1}^p l_i^2\right]^{\frac{1}{2}}}{\sqrt{n}\left(\sum_{i=1}^p l_i\right)^2}.$$

If the right-hand is sufficiently small, the investigator may be willing to let the first principal components represent the entire vector of measurements.

11.7.3. Testing Equality of the Smallest Roots

Suppose the observed $\boldsymbol{X} = \boldsymbol{V} + \boldsymbol{U} + \boldsymbol{\mu}$, where $\boldsymbol{V}$ and $\boldsymbol{U}$ are unobservable random vectors with means $\boldsymbol{0}$ and $\boldsymbol{\mu}$ is an unobservable vector of constants. If $\mathscr{E}\boldsymbol{U}\boldsymbol{U}' = \sigma^2\boldsymbol{I}$, $\boldsymbol{U}$ can be interpreted as composed of errors of measurement: uncorrelated components with equal variances. (It is assumed that all components of $\boldsymbol{X}$ are in the same units.) Then $\boldsymbol{V}$ can be interpreted as made up of the systematic parts and is supposed to lie in an m-dimensional space. Then $\mathscr{E}\boldsymbol{V}\boldsymbol{V}' = \boldsymbol{\Phi}$ is positive semidefinite of rank m. The observable covariance matrix $\boldsymbol{\Sigma} = \boldsymbol{\Phi} + \sigma^2\boldsymbol{I}$ has a characteristic root of σ^2 with multiplicity $p - m$ (Problem 4).

In this section we consider testing the null hypothesis that $\lambda_{m+1} = \cdots = \lambda_p$. That is equivalent to the null hypothesis that $\boldsymbol{\Sigma} = \boldsymbol{\Phi} + \sigma^2\boldsymbol{I}$, where $\boldsymbol{\Phi}$ is positive semidefinite of rank m. In Section 10.7, we saw that when $m = 0$, the likelihood ratio criterion was the $\frac{1}{2}pN$th power of the ratio of the geometric mean to the arithmetic mean of the sample roots. The analogous criterion here is the $\frac{1}{2}N$th power of

$$\frac{\prod_{i=m+1}^p l_i}{\left(\sum_{i=m+1}^p l_i\right)^{p-m}}(p-m)^{p-m}. \tag{9}$$

It is also the likelihood ratio criterion, but we shall not derive it. [See Anderson (1963a).] Let $\sqrt{n}\,(l_i - \lambda_{m+1}) = d_i$, $i = m+1, \dots, p$. The logarithm of (9) multiplied by $-n$ is asymptotically equivalent under the null hypothesis to

$$(10)\qquad -n\log\prod_{i=m+1}^{p} l_i + n(p-m)\log\frac{\sum_{i=m+1}^{p} l_i}{p-m}$$

$$= -n\sum_{i=m+1}^{p}\log\left(\lambda_{m+1} + n^{-\frac{1}{2}}d_i\right) + n(p-m)\log\frac{\sum_{i=m+1}^{p}\left(\lambda_{m+1} + n^{-\frac{1}{2}}d_i\right)}{p-m}$$

$$= n\left\{-\sum_{i=m+1}^{p}\log\left(1 + \frac{d_i}{\lambda_{m+1}n^{\frac{1}{2}}}\right) + (p-m)\log\left(1 + \frac{\sum_{i=m+1}^{p} d_i}{(p-m)\lambda_{m+1}n^{\frac{1}{2}}}\right)\right\}$$

$$= n\left\{-\sum_{i=m+1}^{p}\left[\frac{d_i}{\lambda_{m+1}n^{\frac{1}{2}}} - \frac{d_i^2}{2\lambda_{m+1}^2 n} + \cdots\right]\right.$$

$$\left. + (p-m)\left[\frac{\sum_{i=m+1}^{p} d_i}{(p-m)\lambda_{m+1}n^{\frac{1}{2}}} - \frac{\left(\sum_{i=m+1}^{p} d_i\right)^2}{2(p-m)^2\lambda_{m+1}^2 n} + \cdots\right]\right\}$$

$$= n\left\{\sum_{i=m+1}^{p}\frac{d_i^2}{2\lambda_{m+1}^2 n} - \cdots - \frac{\left(\sum_{i=m+1}^{p} d_i\right)^2}{2(p-m)\lambda_{m+1}^2 n} + \cdots\right\}$$

$$= \frac{1}{2\lambda_{m+1}^2}\left[\sum_{i=m+1}^{p} d_i^2 - \frac{1}{p-m}\left(\sum_{i=m+1}^{p} d_i\right)^2\right] + O_p(1).$$

It is shown in Section 13.5.2 that the limiting distribution of $d_{m+1}, \dots, d_p$ is the same as the distribution of the roots of a symmetric matrix $\boldsymbol{U}_{22} = (u_{ij})$, $i, j = m+1, \dots, p$, whose functionally independent elements are independent and normal with mean 0; an off-diagonal element u_{ij}, $i < j$, has variance λ_{m+1}^2 and a diagonal element u_{ii} has variance $2\lambda_{m+1}^2$. Then (10) has the limiting distribution of

$$(11)\qquad \left[1/(2\lambda_{m+1}^2)\right]\left[\operatorname{tr}\boldsymbol{U}_{22}^2 - 1/(p-m)(\operatorname{tr}\boldsymbol{U}_{22})^2\right]$$

$$= \left[1/(2\lambda_{m+1}^2)\right]\left(\operatorname{tr}\boldsymbol{U}_{22}\boldsymbol{U}_{22}' - 1/(p-m)(\operatorname{tr}\boldsymbol{U}_{22})^2\right]$$

$$= \frac{1}{2\lambda_{m+1}^2}\left[2\sum_{i<j} u_{ij}^2 + \sum_{i=m+1}^{p} u_{ii}^2 - \frac{1}{(p-m)}\left(\sum_{i=m+1}^{p} u_{ii}\right)^2\right].$$

Thus $\Sigma_{i<j} u_{ij}^2/\lambda_{m+1}^2$ is asymptotically χ^2 with $\frac{1}{2}(p-m)(p-m-1)$ degrees of freedom; $\frac{1}{2}[\Sigma_{i=m+1}^{p} u_{ii}^2 - (\Sigma_{i=m+1}^{p} u_{ii})^2/(p-m)]/\lambda_{m+1}^2$ is asymptotically χ^2 with $p-m-1$ degrees of freedom. Then (10) has a limiting χ^2-distribution with $\frac{1}{2}(p-m+2)(p-m-1)$ degrees of freedom. The hypothesis is rejected if the left-hand side of (10) is greater than the upper-tailed significance point of the χ^2-distribution. If the hypothesis is not rejected, the investigator may consider the last $p-m$ principal components to be composed entirely of error.

When the units of measurement are not all the same, the three hypotheses considered in Section 11.7 have questionable meaning. Corresponding hypotheses for the correlation matrix also have doubtful interpretation. Moreover, the last criterion does not have (usually) a χ^2-distribution. More discussion is given by Anderson (1963a).

The criterion (9) corresponds to the sphericity criterion of Section 10.7, and the number of degrees of freedom of the χ^2-distribution corresponds $[\frac{1}{2}(p-m)(p-m+1)-1]$.

PROBLEMS

1. (Sec. 11.2) Prove that the characteristic vectors of $\begin{pmatrix} 1 & \rho \\ \rho & 1 \end{pmatrix}$ are $\begin{pmatrix} 1/\sqrt{2} \\ 1/\sqrt{2} \end{pmatrix}$ and $\begin{pmatrix} 1/\sqrt{2} \\ -1/\sqrt{2} \end{pmatrix}$ corresponding to roots $1+\rho$ and $1-\rho$.

2. (Sec. 11.2) Verify that the proof of Theorem 11.2.1 yields a proof of Theorem A.2.1 of Appendix A for any real symmetric matrix.

3. (Sec. 11.2) Let $\boldsymbol{z} = \boldsymbol{y} + \boldsymbol{x}$, where $\mathscr{E}\boldsymbol{y} = \mathscr{E}\boldsymbol{x} = \boldsymbol{0}$, $\mathscr{E}\boldsymbol{y}\boldsymbol{y}' = \boldsymbol{\Phi}$, $\mathscr{E}\boldsymbol{x}\boldsymbol{x}' = \sigma^2\boldsymbol{I}$, $\mathscr{E}\boldsymbol{y}\boldsymbol{x}' = \boldsymbol{0}$. The p components of $\boldsymbol{y}$ can be called systematic parts and the components of $\boldsymbol{x}$ errors.

(*a*) Find the linear combination $\boldsymbol{\gamma}'\boldsymbol{z}$ of unit variance that has minimum error variance (i.e., $\boldsymbol{\gamma}'\boldsymbol{x}$ has minimum variance).

(*b*) Suppose $\phi_{ii} + \sigma^2 = 1$, $i = 1, \dots, p$. Find the linear function $\boldsymbol{\gamma}'\boldsymbol{z}$ of unit variance that maximizes the sum of squares of the correlations between z_i and $\boldsymbol{\gamma}'\boldsymbol{z}$, $i = 1, \dots, p$.

(*c*) Relate these results to principal components.

4. (Sec. 11.2) Let $\boldsymbol{\Sigma} = \boldsymbol{\Phi} + \sigma^2\boldsymbol{I}$, where $\boldsymbol{\Phi}$ is positive semidefinite of rank m. Prove that each characteristic vector of $\boldsymbol{\Phi}$ is a vector of $\boldsymbol{\Sigma}$ and each root of $\boldsymbol{\Sigma}$ is a root of $\boldsymbol{\Phi}$ plus σ^2.

5. (Sec. 11.2) Let the characteristic roots of $\boldsymbol{\Sigma}$ be $\lambda_1 \geq \lambda_2 \geq \cdots \geq \lambda_p \geq 0$.

(*a*) What is the form of $\boldsymbol{\Sigma}$ if $\lambda_1 = \lambda_2 = \cdots = \lambda_p > 0$? What is the shape of an ellipsoid of constant density?

(*b*) What is the form of $\boldsymbol{\Sigma}$ if $\lambda_1 > \lambda_2 = \cdots = \lambda_p > 0$? What is the shape of an ellipsoid of constant density?

(*c*) What is the form of $\boldsymbol{\Sigma}$ if $\lambda_1 = \cdots = \lambda_{p-1} > \lambda_p > 0$? What is the shape of the ellipsoid of constant density?

6. (Sec. 11.2) *Intraclass correlation*. Let

$$\boldsymbol{\Sigma} = \sigma^2\big[(1-\rho)\boldsymbol{I} + \rho\boldsymbol{\varepsilon}\boldsymbol{\varepsilon}'\big],$$

where $\boldsymbol{\varepsilon} = (1,\dots,1)'$. Show that for $\rho > 0$, the largest characteristic root is $\sigma^2[1 + (p-1)\rho]$ and the corresponding characteristic vector is $\boldsymbol{\varepsilon}$. Show that if $\boldsymbol{\varepsilon}'\boldsymbol{x} = 0$, then $\boldsymbol{x}$ is a characteristic vector corresponding to the root $\sigma^2(1-\rho)$. Show that the root $\sigma^2(1-\rho)$ has multiplicity $p-1$.

7. (Sec. 11.3) In the example, Section 9.6, consider the three pressing operations (x_2, x_4, x_5). Find the first principal component of this estimated covariance matrix. [*Hint*: Start with the vector $(1,1,1)$ and iterate.]

8. (Sec. 11.3) Prove directly the sample analogue of Theorem 11.2.1, where $\sum \boldsymbol{x}_\alpha = \boldsymbol{0}$, $\sum \boldsymbol{x}_\alpha \boldsymbol{x}_\alpha' = \boldsymbol{A}$.

9. (Sec. 11.3) Let l_1 and l_p be the largest and smallest characteristic roots of $\boldsymbol{S}$, respectively. Prove $\mathscr{E} l_1 \geq \lambda_1$ and $\mathscr{E} l_p \leq \lambda_p$.

10. (Sec. 11.3) Let $U_1 = \boldsymbol{\beta}^{(1)\prime}\boldsymbol{X}$ be the first population principal component with variance $\mathscr{V}(U_1) = \lambda_1$, and let $V_1 = \boldsymbol{b}^{(1)\prime}\boldsymbol{X}$ be the first sample principal component with (sample) variance l_1 (based on $\boldsymbol{S}$). Let $\boldsymbol{S}^*$ be the covariance matrix of a second (independent) sample. Show $\mathscr{E}\boldsymbol{b}^{(1)\prime}\boldsymbol{S}^*\boldsymbol{b}^{(1)} \leq \lambda_1$.

11. (Sec. 11.3) Suppose that $\sigma_{ij} > 0$ for every i, j $(\boldsymbol{\Sigma} = (\sigma_{ij}))$. Show that (a) coefficients of the first principal component are all of the same sign, and (b) coefficients of each other principal component cannot be all of the same sign.

12. (Sec. 11.4) Prove (5) and (6) when $\lambda_1 > \lambda_2$.

(*a*) Show $\boldsymbol{\Sigma}^i = \boldsymbol{\beta}\Lambda^i\boldsymbol{\beta}'$.

(*b*) Show

$$y_{(i)} = t_i \boldsymbol{\beta}\Lambda^i\boldsymbol{\beta}'\boldsymbol{x}_{(0)} = t_i\lambda_1^i\boldsymbol{\beta}\left(\frac{1}{\lambda_1}\Lambda\right)^i\boldsymbol{\beta}'\boldsymbol{x}_{(0)},$$

where $t_i = \prod_{j=0}^{i} s_j$ and $s_j = 1/\sqrt{\boldsymbol{x}_{(i)}'\boldsymbol{x}_{(i)}}$.

(*c*) Show

$$\lim_{i\to\infty}\left(\frac{1}{\lambda_1}\Lambda\right)^i = \boldsymbol{E}_{11},$$

where $\boldsymbol{E}_{11}$ has 1 in the upper left-hand position and 0's elsewhere.

(*d*) Show $\lim_{i \to \infty}(t_i \lambda_1^i)^2 = 1/(\beta^{(1)\prime} x_{(0)})^2$.
(*e*) Conclude the proof.

13. (Sec. 11.4) Let

$$\boldsymbol{\Sigma} = \begin{bmatrix} \sigma_{11} & \boldsymbol{\sigma}'_{(1)} \\ \boldsymbol{\sigma}_{(1)} & \boldsymbol{\Sigma}_{22} \end{bmatrix}, \qquad \boldsymbol{K} = \begin{bmatrix} 1 & \mathbf{0} \\ \mathbf{0} & \boldsymbol{H} \end{bmatrix},$$

where $\boldsymbol{H} = \boldsymbol{I}_{p-1} - 2\boldsymbol{\alpha}\boldsymbol{\alpha}'$ and $\boldsymbol{\alpha}$ has $p-1$ components. Show that $\boldsymbol{\alpha}$ can be chosen so that in

$$\boldsymbol{K}\boldsymbol{\Sigma}\boldsymbol{K} = \begin{bmatrix} \sigma_{11} & \boldsymbol{\sigma}'_{(1)}\boldsymbol{H} \\ \boldsymbol{H}\boldsymbol{\sigma}_{(1)} & \boldsymbol{H}\boldsymbol{\Sigma}_{22}\boldsymbol{H} \end{bmatrix}$$

$\boldsymbol{H}\boldsymbol{\sigma}_{(1)}$ has all 0 components except the first.

14. (Sec. 11.6) Show that

$$\log l_i - \sqrt{\frac{2}{n}}\, z(\varepsilon) < \log \lambda_i < \log l_i + \sqrt{\frac{2}{n}}\, z(\varepsilon)$$

is a confidence interval for $\log \lambda_i$ with approximate confidence $1 - \varepsilon$.

15. (Sec. 11.6) Prove that $u' < u$ if $l' = l$ and $p > 2$.

16. (Sec. 11.6) Prove that $u < u^*$ if $l = l^*$ and $p > 2$, where l^* and u^* are the l and u of Section 10.8.4.

17. The lengths, widths, and heights (in millimeters) of 24 male painted turtles [Jolicoeur and Mosimann (1960)] are given below. Find the (sample) principal components and their variances.

Case No.	Length	Width	Height	Case No.	Length	Width	Height
1	93	74	37	13	116	90	43
2	94	78	35	14	117	90	41
3	96	80	35	15	117	91	41
4	101	84	39	16	119	93	41
5	102	85	38	17	120	89	40
6	103	81	37	18	120	93	44
7	104	83	39	19	121	95	42
8	106	83	39	20	125	93	45
9	107	82	38	21	127	96	45
10	112	89	40	22	128	95	46
11	113	88	40	23	131	95	46
12	114	86	40	24	135	106	47

CHAPTER 12

Canonical Correlations and Canonical Variables

12.1. INTRODUCTION

In this section we consider two sets of variates with a joint distribution, and we analyze the correlations between the variables of one set and those of the other set. We find a new coordinate system in the space of each set of variates in such a way that the new coordinates display unambiguously the system of correlation. More precisely, we find linear combinations of variables in the sets that have maximum correlation; these linear combinations are the first coordinates in the new systems. Then a second linear combination in each set is sought such that the correlation between these is the maximum of correlations between such linear combinations as are uncorrelated with the first linear combinations. The procedure is continued until the two new coordinate systems are completely specified.

The statistical method outlined is of particular usefulness in exploratory studies. The investigator may have two large sets of variates and may wish to study the interrelations. If the two sets are very large, he may wish to consider only a few linear combinations of each set. Then he will want to study those linear combinations most highly correlated. For example, one set of variables may be measurements of physical characteristics, such as various lengths and breadths of skulls; the other variables may be measurements of mental characteristics, such as scores on intelligence tests. If the investigator is interested in relating these, he may find that the interrelation is almost completely described by the correlation between the first few canonical variates.

This theory was developed by Hotelling (1935), (1936).

In Section 12.2 the canonical correlations and variates in the population are defined; they imply a linear transformation to canonical form. Maximum likelihood estimators are sample analogs. Tests of independence and of the rank of a correlation matrix are developed on the basis of asymptotic theory in Section 12.4.

Another formulation of canonical correlations and variates is made in the case of one set being random and the other set consisting of nonstochastic variables; the expected values of the random variables are linear combinations of the nonstochastic variables (Section 12.6). This is the model of Section 8.2. One set of canonical variables consists of linear combinations of the random variables and the other set consists of the nonstochastic variables; the effect of the regression of a member of the first set on a member of the second is maximized. Linear functional relationships are studied in this framework.

Simultaneous equations models are studied in Section 12.7. Estimation of a single equation in this model is formally identical to estimation of a single linear functional relationship. The limited information maximum likelihood estimator and the two-stage least squares estimator are developed.

12.2. CANONICAL CORRELATIONS AND VARIATES IN THE POPULATION

Suppose the random vector $\boldsymbol{X}$ of p components has the covariance matrix Σ (which is assumed to be positive definite). Since we are only interested in variances and covariances in this chapter, we shall assume $\mathscr{E}\boldsymbol{X} = \boldsymbol{0}$ when treating the population. In developing the concepts and algebra we do not need to assume that $\boldsymbol{X}$ is normally distributed, though this latter assumption will be made to develop sampling theory.

We partition $\boldsymbol{X}$ into two subvectors of p_1 and p_2 components, respectively,

$$\boldsymbol{X} = \begin{pmatrix} \boldsymbol{X}^{(1)} \\ \boldsymbol{X}^{(2)} \end{pmatrix}. \tag{1}$$

For convenience we shall assume $p_1 \le p_2$. The covariance matrix is partitioned similarly into p_1 and p_2 rows and columns,

$$\Sigma = \begin{pmatrix} \Sigma_{11} & \Sigma_{12} \\ \Sigma_{21} & \Sigma_{22} \end{pmatrix}. \tag{2}$$

In the previous chapter we developed a rotation of coordinate axes to a new system in which the variance properties were clearly exhibited. Here we shall develop a transformation of the first p_1 coordinate axes and a transformation of the last p_2 coordinate axes to a new $(p_1 + p_2)$-system that will exhibit clearly the intercorrelations between $\boldsymbol{X}^{(1)}$ and $\boldsymbol{X}^{(2)}$.

Consider an arbitrary linear combination, $U = \boldsymbol{\alpha}'\boldsymbol{X}^{(1)}$, of the components of $\boldsymbol{X}^{(1)}$ and an arbitrary linear function, $V = \boldsymbol{\gamma}'\boldsymbol{X}^{(2)}$, of the components of $\boldsymbol{X}^{(2)}$. We first ask for the linear functions that have maximum correlation. Since the correlation of a multiple of U and a multiple of V is the same as the correlation of U and V, we can make an arbitrary normalization of $\boldsymbol{\alpha}$ and $\boldsymbol{\gamma}$. We therefore require $\boldsymbol{\alpha}$ and $\boldsymbol{\gamma}$ to be such that U and V have unit variance, that is,

$$(3) \qquad 1 = \mathscr{E}U^2 = \mathscr{E}\boldsymbol{\alpha}'\boldsymbol{X}^{(1)}\boldsymbol{X}^{(1)\prime}\boldsymbol{\alpha} = \boldsymbol{\alpha}'\boldsymbol{\Sigma}_{11}\boldsymbol{\alpha},$$

$$(4) \qquad 1 = \mathscr{E}V^2 = \mathscr{E}\boldsymbol{\gamma}'\boldsymbol{X}^{(2)}\boldsymbol{X}^{(2)\prime}\boldsymbol{\gamma} = \boldsymbol{\gamma}'\boldsymbol{\Sigma}_{22}\boldsymbol{\gamma}.$$

We note that $\mathscr{E}U = \mathscr{E}\boldsymbol{\alpha}'\boldsymbol{X}^{(1)} = \boldsymbol{\alpha}'\mathscr{E}\boldsymbol{X}^{(1)} = 0$ and similarly $\mathscr{E}V = 0$. Then the correlation between U and V is

$$(5) \qquad \mathscr{E}UV = \mathscr{E}\boldsymbol{\alpha}'\boldsymbol{X}^{(1)}\boldsymbol{X}^{(2)\prime}\boldsymbol{\gamma} = \boldsymbol{\alpha}'\boldsymbol{\Sigma}_{12}\boldsymbol{\gamma}.$$

Thus the algebraic problem is to find $\boldsymbol{\alpha}$ and $\boldsymbol{\gamma}$ to maximize (5) subject to (3) and (4).

Let

$$(6) \qquad \psi = \boldsymbol{\alpha}'\boldsymbol{\Sigma}_{12}\boldsymbol{\gamma} - \tfrac{1}{2}\lambda\left(\boldsymbol{\alpha}'\boldsymbol{\Sigma}_{11}\boldsymbol{\alpha} - 1\right) - \tfrac{1}{2}\mu\left(\boldsymbol{\gamma}'\boldsymbol{\Sigma}_{22}\boldsymbol{\gamma} - 1\right),$$

where λ and μ are Lagrange multipliers. We differentiate ψ with respect to the elements of $\boldsymbol{\alpha}$ and $\boldsymbol{\gamma}$. The vectors of derivatives set equal to zero are

$$(7) \qquad \frac{\partial\psi}{\partial\boldsymbol{\alpha}} = \boldsymbol{\Sigma}_{12}\boldsymbol{\gamma} - \lambda\boldsymbol{\Sigma}_{11}\boldsymbol{\alpha} = \mathbf{0},$$

$$(8) \qquad \frac{\partial\psi}{\partial\boldsymbol{\gamma}} = \boldsymbol{\Sigma}_{12}'\boldsymbol{\alpha} - \mu\boldsymbol{\Sigma}_{22}\boldsymbol{\gamma} = \mathbf{0}.$$

Multiplication of (7) on the left by $\boldsymbol{\alpha}'$ and (8) on the left by $\boldsymbol{\gamma}'$ gives

$$\boldsymbol{\alpha}'\boldsymbol{\Sigma}_{12}\boldsymbol{\gamma} - \lambda\boldsymbol{\alpha}'\boldsymbol{\Sigma}_{11}\boldsymbol{\alpha} = 0, \tag{9}$$

$$\boldsymbol{\gamma}'\boldsymbol{\Sigma}_{12}'\boldsymbol{\alpha} - \mu\boldsymbol{\gamma}'\boldsymbol{\Sigma}_{22}\boldsymbol{\gamma} = 0. \tag{10}$$

Since $\boldsymbol{\alpha}'\boldsymbol{\Sigma}_{11}\boldsymbol{\alpha} = 1$ and $\boldsymbol{\gamma}'\boldsymbol{\Sigma}_{22}\boldsymbol{\gamma} = 1$, this shows that $\lambda = \mu = \boldsymbol{\alpha}'\boldsymbol{\Sigma}_{12}\boldsymbol{\gamma}$. Thus (7) and (8) can be written as

$$-\lambda\boldsymbol{\Sigma}_{11}\boldsymbol{\alpha} + \boldsymbol{\Sigma}_{12}\boldsymbol{\gamma} = \mathbf{0}, \tag{11}$$

$$\boldsymbol{\Sigma}_{21}\boldsymbol{\alpha} - \lambda\boldsymbol{\Sigma}_{22}\boldsymbol{\gamma} = \mathbf{0}, \tag{12}$$

since $\boldsymbol{\Sigma}_{12}' = \boldsymbol{\Sigma}_{21}$. In one matrix equation this is

$$\begin{pmatrix} -\lambda\boldsymbol{\Sigma}_{11} & \boldsymbol{\Sigma}_{12} \\ \boldsymbol{\Sigma}_{21} & -\lambda\boldsymbol{\Sigma}_{22} \end{pmatrix}\begin{pmatrix} \boldsymbol{\alpha} \\ \boldsymbol{\gamma} \end{pmatrix} = \mathbf{0}. \tag{13}$$

In order that there be a nontrivial solution [which is necessary for a solution satisfying (3) and (4)], the matrix on the left must be singular; that is,

$$\begin{vmatrix} -\lambda\boldsymbol{\Sigma}_{11} & \boldsymbol{\Sigma}_{12} \\ \boldsymbol{\Sigma}_{21} & -\lambda\boldsymbol{\Sigma}_{22} \end{vmatrix} = 0. \tag{14}$$

The determinant on the left is a polynomial of degree p. To demonstrate this, consider a Laplace expansion by minors of the first p_1 columns. One term is $|-\lambda\boldsymbol{\Sigma}_{11}| \cdot |-\lambda\boldsymbol{\Sigma}_{22}| = (-\lambda)^{p_1+p_2}|\boldsymbol{\Sigma}_{11}| \cdot |\boldsymbol{\Sigma}_{22}|$. The other terms in the expansion are of lower degree in λ because one or more rows of each minor in the first p_1 columns does not contain λ. Since $\boldsymbol{\Sigma}$ is positive definite, $|\boldsymbol{\Sigma}_{11}| \cdot |\boldsymbol{\Sigma}_{22}| \neq 0$ (Corollary A.1.3 of Appendix A). This shows that (14) is a polynomial equation of degree p and has p roots, say $\lambda_1 \geq \lambda_2 \geq \cdots \geq \lambda_p$. ($\boldsymbol{\alpha}'$ and $\boldsymbol{\gamma}'$ complex conjugate in (9) and (10) prove λ real.)

From (9) we see that $\lambda = \boldsymbol{\alpha}'\boldsymbol{\Sigma}_{12}\boldsymbol{\gamma}$ is the correlation between $U = \boldsymbol{\alpha}'\boldsymbol{X}^{(1)}$ and $V = \boldsymbol{\gamma}'\boldsymbol{X}^{(2)}$ when $\boldsymbol{\alpha}$ and $\boldsymbol{\gamma}$ satisfy (13) for some value of λ. Since we want the maximum correlation, we take $\lambda = \lambda_1$. Let a solution to (13) for $\lambda = \lambda_1$ be $\boldsymbol{\alpha}^{(1)}, \boldsymbol{\gamma}^{(1)}$ and let $U_1 = \boldsymbol{\alpha}^{(1)\prime}\boldsymbol{X}^{(1)}$ and $V_1 = \boldsymbol{\gamma}^{(1)\prime}\boldsymbol{X}^{(2)}$. Then U_1 and V_1 are normalized linear combinations of $\boldsymbol{X}^{(1)}$ and $\boldsymbol{X}^{(2)}$, respectively, with maximum correlation.

We now consider finding a second linear combination of $\boldsymbol{X}^{(1)}$, say $U = \boldsymbol{\alpha}'\boldsymbol{X}^{(1)}$, and a second linear combination of $\boldsymbol{X}^{(2)}$, say $V = \boldsymbol{\gamma}'\boldsymbol{X}^{(2)}$, such that of all linear combinations uncorrelated with U_1 and V_1 these have maximum correlation. This procedure is continued. At the rth step we have obtained linear combinations $U_1 = \boldsymbol{\alpha}^{(1)\prime}\boldsymbol{X}^{(1)}$, $V_1 = \boldsymbol{\gamma}^{(1)\prime}\boldsymbol{X}^{(2)}, \ldots, U_r = \boldsymbol{\alpha}^{(r)\prime}\boldsymbol{X}^{(1)}$, $V_r = \boldsymbol{\gamma}^{(r)\prime}\boldsymbol{X}^{(2)}$ with corresponding correlations [roots of (14)] $\lambda^{(1)} = \lambda_1, \lambda^{(2)}, \ldots, \lambda^{(r)}$. We ask for a linear combination of $\boldsymbol{X}^{(1)}$, $U = \boldsymbol{\alpha}'\boldsymbol{X}^{(1)}$, and a linear combination of $\boldsymbol{X}^{(2)}$, $V = \boldsymbol{\gamma}'\boldsymbol{X}^{(2)}$, that of all linear combinations uncorrelated with $U_1, V_1, \ldots, U_r, V_r$ have maximum correlation. The condition that U be uncorrelated with U_i is

$$0 = \mathscr{E}UU_i = \mathscr{E}\boldsymbol{\alpha}'\boldsymbol{X}^{(1)}\boldsymbol{X}^{(1)\prime}\boldsymbol{\alpha}^{(i)} = \boldsymbol{\alpha}'\boldsymbol{\Sigma}_{11}\boldsymbol{\alpha}^{(i)}. \tag{15}$$

Then

$$\mathscr{E}UV_i = \alpha'\boldsymbol{\Sigma}_{12}\boldsymbol{\gamma}^{(i)} = \lambda^{(i)}\boldsymbol{\alpha}'\boldsymbol{\Sigma}_{11}\boldsymbol{\alpha}^{(i)} = 0. \tag{16}$$

The condition that V be uncorrelated with V_i is

$$0 = \mathscr{E}VV_i = \boldsymbol{\gamma}'\boldsymbol{\Sigma}_{22}\boldsymbol{\gamma}^{(i)}. \tag{17}$$

By the same argument we have

$$\mathscr{E}VU_i = \boldsymbol{\gamma}'\boldsymbol{\Sigma}_{21}\boldsymbol{\alpha}^{(i)} = 0. \tag{18}$$

We now maximize $\mathscr{E}U_{r+1}V_{r+1}$, choosing $\boldsymbol{\alpha}$ and $\boldsymbol{\gamma}$ to satisfy (3), (4), (15), and (17) for $i = 1, 2, \ldots, r$. Consider

$$\psi_{r+1} = \boldsymbol{\alpha}'\boldsymbol{\Sigma}_{12}\boldsymbol{\gamma} - \tfrac{1}{2}\lambda\left(\boldsymbol{\alpha}'\boldsymbol{\Sigma}_{11}\boldsymbol{\alpha} - 1\right) - \tfrac{1}{2}\mu\left(\boldsymbol{\gamma}'\boldsymbol{\Sigma}_{22}\boldsymbol{\gamma} - 1\right) + \sum_{i=1}^{r} \nu_i \boldsymbol{\alpha}'\boldsymbol{\Sigma}_{11}\boldsymbol{\alpha}^{(i)} + \sum_{i=1}^{r} \theta_i \boldsymbol{\gamma}'\boldsymbol{\Sigma}_{22}\boldsymbol{\gamma}^{(i)}, \tag{19}$$

where $\lambda, \mu, \nu_1, \ldots, \nu_r, \theta_1, \ldots, \theta_r$ are Lagrange multipliers. The vectors of partial derivatives of ψ_{r+1} with respect to the elements of $\boldsymbol{\alpha}$ and $\boldsymbol{\gamma}$ are set equal to zero, giving

$$\frac{\partial \psi_{r+1}}{\partial \boldsymbol{\alpha}} = \boldsymbol{\Sigma}_{12}\boldsymbol{\gamma} - \lambda\boldsymbol{\Sigma}_{11}\boldsymbol{\alpha} + \sum_{i=1}^{r} \nu_i \boldsymbol{\Sigma}_{11}\boldsymbol{\alpha}^{(i)} = \mathbf{0}, \tag{20}$$

$$\frac{\partial \psi_{r+1}}{\partial \boldsymbol{\gamma}} = \boldsymbol{\Sigma}_{21}\boldsymbol{\alpha} - \mu\boldsymbol{\Sigma}_{22}\boldsymbol{\gamma} + \sum_{i=1}^{r} \theta_i \boldsymbol{\Sigma}_{22}\boldsymbol{\gamma}^{(i)} = \mathbf{0}. \tag{21}$$

Multiplication of (20) on the left by $\boldsymbol{\alpha}^{(j)\prime}$ and (21) on the left by $\boldsymbol{\gamma}^{(j)\prime}$ gives

$$0 = \nu_j \boldsymbol{\alpha}^{(j)\prime} \boldsymbol{\Sigma}_{11} \boldsymbol{\alpha}^{(j)} = \nu_j, \tag{22}$$

$$0 = \theta_j \boldsymbol{\gamma}^{(j)\prime} \boldsymbol{\Sigma}_{22} \boldsymbol{\gamma}^{(j)} = \theta_j. \tag{23}$$

Thus Equations (20) and (21) are simply (11) and (12) or alternatively (13). We therefore take the largest λ_i, say, $\lambda^{(r+1)}$, such that there is a solution to (13) satisfying (3), (4), (15), and (17) for $i = 1, \ldots, r$. Let this solution be $\boldsymbol{\alpha}^{(r+1)}$, $\boldsymbol{\gamma}^{(r+1)}$, and let $U_{r+1} = \boldsymbol{\alpha}^{(r+1)\prime} \boldsymbol{X}^{(1)}$ and $V_{r+1} = \boldsymbol{\gamma}^{(r+1)\prime} \boldsymbol{X}^{(2)}$.

This procedure is continued step by step as long as successive solutions can be found which satisfy the conditions, namely, (13) for some λ_i, (3), (4), (15), and (17). Let m be the number of steps for which this can be done. Now we shall show that $m = p_1 (\leq p_2)$. Let $\mathbf{A} = (\boldsymbol{\alpha}^{(1)} \cdots \boldsymbol{\alpha}^{(m)})$, $\boldsymbol{\Gamma}_1 = (\boldsymbol{\gamma}^{(1)} \cdots \boldsymbol{\gamma}^{(m)})$ and

$$\boldsymbol{\Lambda} = \begin{pmatrix} \lambda^{(1)} & 0 & \cdots & 0 \\ 0 & \lambda^{(2)} & \cdots & 0 \\ \vdots & \vdots & & \vdots \\ 0 & 0 & \cdots & \lambda^{(m)} \end{pmatrix}. \tag{24}$$

Conditions (3) and (15) can be summarized as

$$\mathbf{A}' \boldsymbol{\Sigma}_{11} \mathbf{A} = \boldsymbol{I}. \tag{25}$$

Since $\boldsymbol{\Sigma}_{11}$ is of rank p_1 and $\boldsymbol{I}$ is of rank m, $m \leq p_1$. Now let us show that $m < p_1$ leads to a contradiction by showing that in this case there is another vector satisfying the conditions. Since $\mathbf{A}'\boldsymbol{\Sigma}_{11}$ is $m \times p_1$, there exists a $p_1 \times (p_1 - m)$ matrix $\boldsymbol{E}$ (of rank $p_1 - m$) such that $\mathbf{A}'\boldsymbol{\Sigma}_{11}\boldsymbol{E} = \mathbf{0}$. Similarly there is a $p_2 \times (p_2 - m)$ matrix $\boldsymbol{F}$ (of rank $p_2 - m$) such that $\boldsymbol{\Gamma}_1'\boldsymbol{\Sigma}_{22}\boldsymbol{F} = \mathbf{0}$. We also have $\boldsymbol{\Gamma}_1'\boldsymbol{\Sigma}_{21}\boldsymbol{E} = \boldsymbol{\Lambda}\mathbf{A}'\boldsymbol{\Sigma}_{11}\boldsymbol{E} = \mathbf{0}$ and $\mathbf{A}'\boldsymbol{\Sigma}_{12}\boldsymbol{F} = \boldsymbol{\Lambda}\boldsymbol{\Gamma}_1'\boldsymbol{\Sigma}_{22}\boldsymbol{F} = \mathbf{0}$. Since $\boldsymbol{E}$ is of rank $p_1 - m$, $\boldsymbol{E}'\boldsymbol{\Sigma}_{11}\boldsymbol{E}$ is nonsingular (if $m < p_1$), and similarly $\boldsymbol{F}'\boldsymbol{\Sigma}_{22}\boldsymbol{F}$ is nonsingular. Thus there is at least one root of

$$\begin{vmatrix} -\nu \boldsymbol{E}'\boldsymbol{\Sigma}_{11}\boldsymbol{E} & \boldsymbol{E}'\boldsymbol{\Sigma}_{12}\boldsymbol{F} \\ \boldsymbol{F}'\boldsymbol{\Sigma}_{21}\boldsymbol{E} & -\nu \boldsymbol{F}'\boldsymbol{\Sigma}_{22}\boldsymbol{F} \end{vmatrix} = 0, \tag{26}$$

because $|\boldsymbol{E}'\boldsymbol{\Sigma}_{11}\boldsymbol{E}| \cdot |\boldsymbol{F}'\boldsymbol{\Sigma}_{22}\boldsymbol{F}| \neq 0$. From the preceding algebra we see that there exist vectors $\boldsymbol{a}$ and $\boldsymbol{b}$ so that

$$\boldsymbol{E}'\boldsymbol{\Sigma}_{12}\boldsymbol{F}\boldsymbol{b} = \nu\boldsymbol{E}'\boldsymbol{\Sigma}_{11}\boldsymbol{E}\boldsymbol{a}, \tag{27}$$

$$\boldsymbol{F}'\boldsymbol{\Sigma}_{21}\boldsymbol{E}\boldsymbol{a} = \nu\boldsymbol{F}'\boldsymbol{\Sigma}_{22}\boldsymbol{F}\boldsymbol{b}. \tag{28}$$

Let $\boldsymbol{E}\boldsymbol{a} = \boldsymbol{g}$ and $\boldsymbol{F}\boldsymbol{b} = \boldsymbol{h}$. We now want to show that ν, $\boldsymbol{g}$, and $\boldsymbol{h}$ form a new solution $\lambda^{(m+1)}, \boldsymbol{\alpha}^{(m+1)}, \boldsymbol{\gamma}^{(m+1)}$. Let $\boldsymbol{\Sigma}_{11}^{-1}\boldsymbol{\Sigma}_{12}\boldsymbol{h} = \boldsymbol{k}$. Since $\mathrm{A}'\boldsymbol{\Sigma}_{11}\boldsymbol{k} = \mathrm{A}'\boldsymbol{\Sigma}_{12}\boldsymbol{F}\boldsymbol{b} = \boldsymbol{0}$, $\boldsymbol{k}$ is orthogonal to the rows of $\mathrm{A}'\boldsymbol{\Sigma}_{11}$ and therefore is a linear combination of the columns of $\boldsymbol{E}$, say $\boldsymbol{E}\boldsymbol{c}$. Thus the equation $\boldsymbol{\Sigma}_{12}\boldsymbol{h} = \boldsymbol{\Sigma}_{11}\boldsymbol{k}$ can be written

$$\boldsymbol{\Sigma}_{12}\boldsymbol{F}\boldsymbol{b} = \boldsymbol{\Sigma}_{11}\boldsymbol{E}\boldsymbol{c}. \tag{29}$$

Multiplication by $\boldsymbol{E}'$ on the left gives

$$\boldsymbol{E}'\boldsymbol{\Sigma}_{12}\boldsymbol{F}\boldsymbol{b} = \boldsymbol{E}'\boldsymbol{\Sigma}_{11}\boldsymbol{E}\boldsymbol{c}. \tag{30}$$

Since $\boldsymbol{E}'\boldsymbol{\Sigma}_{11}\boldsymbol{E}$ is nonsingular, comparison of (27) and (30) shows that $\boldsymbol{c} = \nu\boldsymbol{a}$, and therefore $\boldsymbol{k} = \nu\boldsymbol{g}$. Thus

$$\boldsymbol{\Sigma}_{12}\boldsymbol{h} = \nu\boldsymbol{\Sigma}_{11}\boldsymbol{g}. \tag{31}$$

In a similar fashion we show that

$$\boldsymbol{\Sigma}_{21}\boldsymbol{g} = \nu\boldsymbol{\Sigma}_{22}\boldsymbol{h}. \tag{32}$$

Therefore $\nu = \lambda^{(m+1)}$, $\boldsymbol{g} = \boldsymbol{\alpha}^{(m+1)}$, $\boldsymbol{h} = \boldsymbol{\gamma}^{(m+1)}$ is another solution. But this is contrary to the assumption that $\lambda^{(m)}, \boldsymbol{\alpha}^{(m)}, \boldsymbol{\gamma}^{(m)}$ was the last possible solution. Thus $m = p_1$.

The conditions on the λ's, $\boldsymbol{\alpha}$'s and $\boldsymbol{\gamma}$'s can be summarized as

$$\mathrm{A}'\boldsymbol{\Sigma}_{11}\mathrm{A} = \boldsymbol{I}, \tag{33}$$

$$\mathrm{A}'\boldsymbol{\Sigma}_{12}\boldsymbol{\Gamma}_1 = \boldsymbol{\Lambda}, \tag{34}$$

$$\boldsymbol{\Gamma}_1'\boldsymbol{\Sigma}_{22}\boldsymbol{\Gamma}_1 = \boldsymbol{I}. \tag{35}$$

Let $\boldsymbol{\Gamma}_2 = (\boldsymbol{\gamma}^{(p_1+1)} \cdots \boldsymbol{\gamma}^{(p_2)})$ be a $p_2 \times (p_2 - p_1)$ matrix satisfying

$$\boldsymbol{\Gamma}_2'\boldsymbol{\Sigma}_{22}\boldsymbol{\Gamma}_1 = \boldsymbol{0}, \tag{36}$$

$$\boldsymbol{\Gamma}_2'\boldsymbol{\Sigma}_{22}\boldsymbol{\Gamma}_2 = \boldsymbol{I}. \tag{37}$$

This matrix can be formed one column at a time; $\boldsymbol{\gamma}^{(p_1+1)}$ is a vector orthogonal to $\boldsymbol{\Sigma}_{22}\boldsymbol{\Gamma}_1$ and normalized so $\boldsymbol{\gamma}^{(p_1+1)\prime}\boldsymbol{\Sigma}_{22}\boldsymbol{\gamma}^{(p_1+1)} = 1$; $\boldsymbol{\gamma}^{(p_1+2)}$ is a vector orthogonal to $\boldsymbol{\Sigma}_{22}(\boldsymbol{\Gamma}_1\ \boldsymbol{\gamma}^{(p_1+1)})$ and normalized so $\boldsymbol{\gamma}^{(p_1+2)\prime}\boldsymbol{\Sigma}_{22}\boldsymbol{\gamma}^{(p_1+2)} = 1$, and so forth. Let $\boldsymbol{\Gamma} = (\boldsymbol{\Gamma}_1\ \boldsymbol{\Gamma}_2)$; this square matrix is nonsingular since $\boldsymbol{\Gamma}'\boldsymbol{\Sigma}_{22}\boldsymbol{\Gamma} = \boldsymbol{I}$. Consider the determinant

$$(38)\qquad \begin{vmatrix} \mathbf{A}' & \mathbf{0} \\ \mathbf{0} & \boldsymbol{\Gamma}_1' \\ \mathbf{0} & \boldsymbol{\Gamma}_2' \end{vmatrix} \cdot \begin{vmatrix} -\lambda\boldsymbol{\Sigma}_{11} & \boldsymbol{\Sigma}_{12} \\ \boldsymbol{\Sigma}_{21} & -\lambda\boldsymbol{\Sigma}_{22} \end{vmatrix} \cdot \begin{vmatrix} \mathbf{A} & \mathbf{0} & \mathbf{0} \\ \mathbf{0} & \boldsymbol{\Gamma}_1 & \boldsymbol{\Gamma}_2 \end{vmatrix}$$

$$= \begin{vmatrix} -\lambda\boldsymbol{I} & \boldsymbol{\Lambda} & \mathbf{0} \\ \boldsymbol{\Lambda} & -\lambda\boldsymbol{I} & \mathbf{0} \\ \mathbf{0} & \mathbf{0} & -\lambda\boldsymbol{I} \end{vmatrix}$$

$$= (-\lambda)^{p_2-p_1} \begin{vmatrix} -\lambda\boldsymbol{I} & \boldsymbol{\Lambda} \\ \boldsymbol{\Lambda} & -\lambda\boldsymbol{I} \end{vmatrix}$$

$$= (-\lambda)^{p_2-p_1}|-\lambda\boldsymbol{I}|\cdot|-\lambda\boldsymbol{I} - \boldsymbol{\Lambda}(-\lambda\boldsymbol{I})^{-1}\boldsymbol{\Lambda}|$$

$$= (-\lambda)^{p_2-p_1}|\lambda^2\boldsymbol{I} - \boldsymbol{\Lambda}^2|$$

$$= (-\lambda)^{p_2-p_1}\prod(\lambda^2 - \lambda^{(i)2}).$$

Except for a constant factor the above polynomial is

$$(39)\qquad \begin{vmatrix} -\lambda\boldsymbol{\Sigma}_{11} & \boldsymbol{\Sigma}_{12} \\ \boldsymbol{\Sigma}_{21} & -\lambda\boldsymbol{\Sigma}_{22} \end{vmatrix}.$$

Thus the roots of (14) are the roots of (38) set equal to zero, namely, $\lambda = \pm\lambda^{(i)}$, $i = 1,\dots,p_1$, and $\lambda = 0$ (of multiplicity $p_2 - p_1$). Thus $(\lambda_1,\dots,\lambda_p) = (\lambda_1,\dots,\lambda_{p_1},0,\dots,0,-\lambda_{p_1},\dots,-\lambda_1)$. The set $\{\lambda^{(i)2}\}$, $i = 1,\dots,p_1$, is the set $\{\lambda_i^2\}$, $i = 1,\dots,p_1$. To show that the set $\{\lambda^{(i)}\}$, $i = 1,\dots,p_1$, is the set $\{\lambda_i\}$, $i = 1,\dots,p_1$, we only need to show that $\lambda^{(i)}$ is nonnegative (and therefore is one of the λ_i, $i = 1,\dots,p_1$). We observe that

$$(40)\qquad \boldsymbol{\Sigma}_{12}\boldsymbol{\gamma}^{(r)} = -\lambda^{(r)}\boldsymbol{\Sigma}_{11}(-\boldsymbol{\alpha}^{(r)}),$$

$$(41)\qquad \boldsymbol{\Sigma}_{21}(-\boldsymbol{\alpha}^{(r)}) = -\lambda^{(r)}\boldsymbol{\Sigma}_{22}\boldsymbol{\gamma}^{(r)};$$

thus, if $\lambda^{(r)}, \boldsymbol{\alpha}^{(r)}, \boldsymbol{\gamma}^{(r)}$ is a solution, so is $-\lambda^{(r)}, -\boldsymbol{\alpha}^{(r)}, \boldsymbol{\gamma}^{(r)}$. If $\lambda^{(r)}$ were negative, then $-\lambda^{(r)}$ would be nonnegative and $-\lambda^{(r)} \geq \lambda^{(r)}$. But since $\lambda^{(r)}$ was to be maximum, we must have $\lambda^{(r)} \geq -\lambda^{(r)}$ and therefore $\lambda^{(r)} \geq 0$. Since the set $\{\lambda^{(i)}\}$ is the same as $\{\lambda_i\}$, $i = 1, \ldots, p_1$, we must have $\lambda^{(i)} = \lambda_i$.

Let

$$\boldsymbol{U} = \begin{pmatrix} U_1 \\ \vdots \\ U_{p_1} \end{pmatrix} = \mathbf{A}'\boldsymbol{X}^{(1)}, \tag{42}$$

$$\boldsymbol{V}^{(1)} = \begin{pmatrix} V_1 \\ \vdots \\ V_{p_1} \end{pmatrix} = \Gamma_1'\boldsymbol{X}^{(2)}, \tag{43}$$

$$\boldsymbol{V}^{(2)} = \begin{pmatrix} V_{p_1+1} \\ \vdots \\ V_{p_2} \end{pmatrix} = \Gamma_2'\boldsymbol{X}^{(2)}. \tag{44}$$

The components of $\boldsymbol{U}$ are one set of canonical variates and the components of $\boldsymbol{V} = \begin{pmatrix} \boldsymbol{V}^{(1)} \\ \boldsymbol{V}^{(2)} \end{pmatrix}$ are the other set. We have

$$\mathscr{E}\begin{pmatrix} \boldsymbol{U} \\ \boldsymbol{V}^{(1)} \\ \boldsymbol{V}^{(2)} \end{pmatrix}\left(\boldsymbol{U}' \ \boldsymbol{V}^{(1)\prime} \ \boldsymbol{V}^{(2)\prime}\right) = \begin{pmatrix} \mathbf{A}' & \mathbf{0} \\ \mathbf{0} & \Gamma_1' \\ \mathbf{0} & \Gamma_2' \end{pmatrix}\begin{pmatrix} \Sigma_{11} & \Sigma_{12} \\ \Sigma_{21} & \Sigma_{22} \end{pmatrix}\begin{pmatrix} \mathbf{A} & \mathbf{0} & \mathbf{0} \\ \mathbf{0} & \Gamma_1 & \Gamma_2 \end{pmatrix} \tag{45}$$

$$= \begin{pmatrix} \boldsymbol{I}_{p_1} & \Lambda & \mathbf{0} \\ \Lambda & \boldsymbol{I}_{p_1} & \mathbf{0} \\ \mathbf{0} & \mathbf{0} & \boldsymbol{I}_{p_2-p_1} \end{pmatrix},$$

where

$$\Lambda = \begin{pmatrix} \lambda_1 & 0 & \cdots & 0 \\ 0 & \lambda_2 & \cdots & 0 \\ \vdots & \vdots & & \vdots \\ 0 & 0 & \cdots & \lambda_{p_1} \end{pmatrix}. \tag{46}$$

DEFINITION 12.2.1. *Let $\boldsymbol{X} = (\boldsymbol{X}^{(1)\prime}\ \boldsymbol{X}^{(2)\prime})'$, where $\boldsymbol{X}^{(1)}$ has p_1 components and $\boldsymbol{X}^{(2)}$ has $p_2\ (= p - p_1 \geq p_1)$ components. The* rth *pair of canonical variates is the pair of linear combinations, $U_r = \boldsymbol{\alpha}^{(r)\prime}\boldsymbol{X}^{(1)}$ and $V_r = \boldsymbol{\gamma}^{(r)\prime}\boldsymbol{X}^{(2)}$, each of unit variance and uncorrelated with the first $r-1$ pairs of canonical variates and having maximum correlation. The correlation is the* rth *canonical correlation.*

THEOREM 12.2.1. *Let $\boldsymbol{X} = (\boldsymbol{X}^{(1)\prime}\ \boldsymbol{X}^{(2)\prime})'$ be a random vector with covariance matrix $\boldsymbol{\Sigma}$. The* rth *canonical correlation between $\boldsymbol{X}^{(1)}$ and $\boldsymbol{X}^{(2)}$ is the* rth *largest root of* (14). *The coefficients of $\boldsymbol{\alpha}^{(r)\prime}\boldsymbol{X}^{(1)}$ and $\boldsymbol{\gamma}^{(r)\prime}\boldsymbol{X}^{(2)}$ defining the* rth *pair of canonical variates satisfy* (13) *for $\lambda = \lambda_r$ and* (3) *and* (4).

We can now verify (without differentiation) that U_1, V_1 have maximum correlation. The linear combinations $\boldsymbol{a}'\boldsymbol{U} = (\boldsymbol{a}'\mathrm{A}')\boldsymbol{X}^{(1)}$ and $\boldsymbol{b}'\boldsymbol{V} = (\boldsymbol{b}'\Gamma')\boldsymbol{X}^{(2)}$ are normalized by $\boldsymbol{a}'\boldsymbol{a} = 1$ and $\boldsymbol{b}'\boldsymbol{b} = 1$. Since A and Γ are nonsingular, any vector $\boldsymbol{\alpha}$ can be written as $\mathrm{A}\boldsymbol{a}$ and any vector $\boldsymbol{\gamma}$ can be written as $\Gamma\boldsymbol{b}$ and hence any linear combinations $\boldsymbol{\alpha}'\boldsymbol{X}^{(1)}$ and $\boldsymbol{\gamma}'\boldsymbol{X}^{(2)}$ can be written as $\boldsymbol{a}'\boldsymbol{U}$ and $\boldsymbol{b}'\boldsymbol{V}$. The correlation between them is

$$\boldsymbol{a}'(\Lambda\ \boldsymbol{0})\boldsymbol{b} = \sum_{i=1}^{p_1} \lambda_i a_i b_i. \tag{47}$$

Let $\lambda_i a_i / \sqrt{\Sigma(\lambda_i a_i)^2} = c_i$. Then the maximum of $\boldsymbol{a}'(\Lambda\ \ \boldsymbol{0})\boldsymbol{b} = \sqrt{\Sigma(\lambda_i a_i)^2}\,\Sigma c_i b_i$ with respect to $\boldsymbol{b}$ is for $b_i = c_i$, since $\Sigma c_i b_i$ is the cosine of the angle between the vector $\boldsymbol{b}$ and $(c_1, \ldots, c_{p_1}, 0, \ldots, 0)$. Then (47) is

$$\sqrt{\Sigma\lambda_i^2 a_i^2} = \sqrt{\Sigma_2^{p_1}(\lambda_i^2 - \lambda_1^2)a_i^2 + \lambda_1^2}$$

and this is maximized by taking $a_i = 0$, $i = 2, \ldots, p_1$. Thus the maximized linear combinations are U_1 and V_1. In verifying that U_2 and V_2 form the second pair of canonical variates we note that lack of correlation between U_1 and a linear combination $\boldsymbol{a}'\boldsymbol{U}$ is $0 = \mathscr{E}U_1\boldsymbol{a}'\boldsymbol{U} = \mathscr{E}U_1\Sigma a_i U_i = a_1$ and lack of correlation between V_1 and $\boldsymbol{b}'\boldsymbol{V}$ is $0 = b_1$. The algebra used above gives the desired result with sums starting with $i = 2$.

We can derive a single matrix equation for $\boldsymbol{\alpha}$ or $\boldsymbol{\gamma}$. If we multiply (11) by λ and (12) by $\boldsymbol{\Sigma}_{22}^{-1}$, we have

$$\lambda\boldsymbol{\Sigma}_{12}\boldsymbol{\gamma} = \lambda^2\boldsymbol{\Sigma}_{11}\boldsymbol{\alpha}, \tag{48}$$

$$\boldsymbol{\Sigma}_{22}^{-1}\boldsymbol{\Sigma}_{21}\boldsymbol{\alpha} = \lambda\boldsymbol{\gamma}. \tag{49}$$

Substitution from (49) into (48) gives

$$\Sigma_{12}\Sigma_{22}^{-1}\Sigma_{21}\alpha = \lambda^2\Sigma_{11}\alpha \tag{50}$$

or

$$\left(\Sigma_{12}\Sigma_{22}^{-1}\Sigma_{21} - \lambda^2\Sigma_{11}\right)\alpha = \mathbf{0}. \tag{51}$$

The quantities $\lambda_1^2, \dots, \lambda_{p_1}^2$ satisfy

$$|\Sigma_{12}\Sigma_{22}^{-1}\Sigma_{21} - \nu\Sigma_{11}| = 0 \tag{52}$$

and $\alpha^{(1)}, \dots, \alpha^{(p_1)}$ satisfy (51) for $\lambda^2 = \lambda_1^2, \dots, \lambda_{p_1}^2$, respectively. The similar equations for $\gamma^{(1)}, \dots, \gamma^{(p_2)}$ occur when $\lambda^2 = \lambda_1^2, \dots, \lambda_{p_2}^2$ are substituted with

$$\left(\Sigma_{21}\Sigma_{11}^{-1}\Sigma_{12} - \lambda^2\Sigma_{22}\right)\gamma = \mathbf{0}. \tag{53}$$

THEOREM 12.2.2. *The canonical correlations are invariant with respect to transformations* $X^{(i)*} = C_i X^{(i)}$, *where* C_i *is nonsingular,* $i = 1, 2$, *and any function of* Σ *that is invariant is a function of the canonical correlations.*

PROOF. Equation (14) is transformed to

(54)

$$0 = \begin{vmatrix} -\lambda C_1\Sigma_{11}C_1' & C_1\Sigma_{12}C_2' \\ C_2\Sigma_{21}C_1' & -\lambda C_2\Sigma_{22}C_2' \end{vmatrix} = \begin{vmatrix} C_1 & \mathbf{0} \\ \mathbf{0} & C_2 \end{vmatrix} \cdot \begin{vmatrix} -\lambda\Sigma_{11} & \Sigma_{12} \\ \Sigma_{21} & -\lambda\Sigma_{22} \end{vmatrix} \cdot \begin{vmatrix} C_1' & \mathbf{0} \\ \mathbf{0} & C_2' \end{vmatrix}$$

and hence the roots are unchanged. Conversely let $f(\Sigma_{11}, \Sigma_{12}, \Sigma_{22})$ be a vector-valued function of Σ such that $f(C_1\Sigma_{11}C_1', C_1\Sigma_{12}C_2', C_2\Sigma_{22}C_2') = f(\Sigma_{11}, \Sigma_{12}, \Sigma_{22})$ for all nonsingular C_1 and C_2. If $C_1' = \mathbf{A}$ and $C_2 = \Gamma'$, then (54) is (38), which depends only on the canonical correlations. Then $f = f(I, (\Lambda, \mathbf{0}), I)$. Q.E.D.

We can make another interpretation of these developments in terms of prediction. Consider two random variables U and V with means 0 and variances σ_u^2 and σ_v^2 and correlation ρ. Consider approximating U by a multiple of V, say bV; then the mean square error of approximation is

$$\begin{aligned} \mathscr{E}(U - bV)^2 &= \sigma_u^2 - 2b\sigma_u\sigma_v\rho + b^2\sigma_v^2 \\ &= \sigma_u^2(1 - \rho^2) + (b\sigma_v - \rho\sigma_u)^2. \end{aligned} \tag{55}$$

This is minimized by taking $b = \sigma_u \rho / \sigma_v$. We can consider bV as a linear prediction of U from V; then $\sigma_u^2(1 - \rho^2)$ is the mean square error of prediction. The ratio of the mean square error of prediction to the variance of U is $\sigma_u^2(1 - \rho^2)/\sigma_u^2 = 1 - \rho^2$; the complement is a measure of the relative effect of V on U or the relative effectiveness of V in predicting U. Thus the greater ρ^2 or $|\rho|$ is, the more effective is V in predicting U.

Now consider the random vector $\boldsymbol{X}$ partitioned according to (1), and consider using a linear combination $V = \boldsymbol{\gamma}'\boldsymbol{X}^{(2)}$ to predict a linear combination $U = \boldsymbol{\alpha}'\boldsymbol{X}^{(1)}$. Then V predicts U best if the correlation between U and V is a maximum. Thus we can say that $\boldsymbol{\alpha}^{(1)\prime}\boldsymbol{X}^{(1)}$ is the linear combination of $\boldsymbol{X}^{(1)}$ that can be predicted best and $\boldsymbol{\gamma}^{(1)\prime}\boldsymbol{X}^{(2)}$ is the best predictor [Hotelling (1935)].

The mean square effect of V on U can be measured as

$$\mathscr{E}(bV)^2 = \rho^2 \frac{\sigma_u^2}{\sigma_v^2} \mathscr{E}V^2 = \rho^2 \sigma_u^2, \tag{56}$$

and the relative mean square effect can be measured by the ratio $\mathscr{E}(bV)^2/\mathscr{E}U^2 = \rho^2$. Thus maximum effect of a linear combination of $\boldsymbol{X}^{(2)}$ on a linear combination of $\boldsymbol{X}^{(1)}$ is made by $\boldsymbol{\gamma}^{(1)\prime}\boldsymbol{X}^{(2)}$ on $\boldsymbol{\alpha}^{(1)\prime}\boldsymbol{X}^{(1)}$.

It should be pointed out that in the special case of $p_1 = 1$, the one canonical correlation is the multiple correlation between $\boldsymbol{X}^{(1)} = X_1$ and $\boldsymbol{X}^{(2)}$.

The definition of canonical variates and correlations was made in terms of the covariance matrix $\boldsymbol{\Sigma} = \mathscr{E}(\boldsymbol{X} - \mathscr{E}\boldsymbol{X})(\boldsymbol{X} - \mathscr{E}\boldsymbol{X})'$. We could extend this treatment by starting with a normally distributed vector $\boldsymbol{Y}$ with $p + p_3$ components and define $\boldsymbol{X}$ as the vector having the conditional distribution of the first p components of $\boldsymbol{Y}$ given the value of the last p_3 components. This would mean treating $\boldsymbol{X}_\phi$ with mean $\mathscr{E}\boldsymbol{X}_\phi = \boldsymbol{\Theta}\boldsymbol{y}_\phi^{(3)}$; the elements of the covariance matrix would be the partial covariances of the first p elements of $\boldsymbol{Y}$.

The interpretation of canonical variates may be facilitated by considering the correlations between the canonical variates and the components of the original vectors [e.g., Darlington, Weinberg, and Wahlberg (1973)]. The covariance between the jth canonical variate U_j and X_i is

$$\mathscr{E}U_j X_i = \mathscr{E}\sum_{k=1}^{p_1} \alpha_k^{(j)} X_k X_i = \sum_{k=1}^{p_1} \alpha_k^{(j)} \sigma_{ki}. \tag{57}$$

Since the variance of U_j is 1, the correlation between U_j and X_i is

$$\operatorname{Corr}(U_j, X_i) = \frac{\sum_{k=1}^{p_1} \alpha_k^{(j)} \sigma_{ki}}{\sqrt{\sigma_{ii}}}. \tag{58}$$

An advantage of this measure is that it does not depend on the units of measurement of X_j. However, it is not a scalar multiple of the weight of X_i in U_j (namely, $\alpha_i^{(j)}$).

A special case is $\boldsymbol{\Sigma}_{11} = \boldsymbol{I}$, $\boldsymbol{\Sigma}_{22} = \boldsymbol{I}$. Then

$$\mathbf{A}'\mathbf{A} = \boldsymbol{I}, \quad \boldsymbol{\Gamma}'\boldsymbol{\Gamma} = \boldsymbol{I}, \quad \mathbf{A}'\boldsymbol{\Sigma}_{12}\boldsymbol{\Gamma} = (\boldsymbol{\Lambda}\ \mathbf{0}). \tag{59}$$

From these we obtain

$$\boldsymbol{\Sigma}_{12} = \mathbf{A}(\boldsymbol{\Lambda}\ \mathbf{0})\boldsymbol{\Gamma}', \tag{60}$$

where $\mathbf{A}$ and $\boldsymbol{\Gamma}$ are orthogonal and $\boldsymbol{\Lambda}$ is diagonal. This relationship is known as the *singular value decomposition* of $\boldsymbol{\Sigma}_{12}$. The elements of $\boldsymbol{\Lambda}$ are the square roots of the characteristic roots of $\boldsymbol{\Sigma}_{12}\boldsymbol{\Sigma}_{12}'$ and the columns of $\mathbf{A}$ are characteristic vectors. The diagonal elements of $\boldsymbol{\Lambda}$ are square roots of the (possibly nonzero) roots of $\boldsymbol{\Sigma}_{12}'\boldsymbol{\Sigma}_{12}$ and the columns of $\boldsymbol{\Gamma}$ are the characteristic vectors.

12.3. ESTIMATION OF CANONICAL CORRELATIONS AND VARIATES

12.3.1. Estimation

Let $\boldsymbol{x}_1, \dots, \boldsymbol{x}_N$ be N observations from $N(\boldsymbol{\mu}, \boldsymbol{\Sigma})$. Let $\boldsymbol{x}_\alpha$ be partitioned into two subvectors of p_1 and p_2 components, respectively,

$$\boldsymbol{x}_\alpha = \begin{pmatrix} \boldsymbol{x}_\alpha^{(1)} \\ \boldsymbol{x}_\alpha^{(2)} \end{pmatrix}, \qquad \alpha = 1, \dots, N. \tag{1}$$

The maximum likelihood estimator of $\boldsymbol{\Sigma}$ [partitioned as in (2) of Section 12.2] is

$$\hat{\boldsymbol{\Sigma}} = \begin{pmatrix} \hat{\boldsymbol{\Sigma}}_{11} & \hat{\boldsymbol{\Sigma}}_{12} \\ \hat{\boldsymbol{\Sigma}}_{21} & \hat{\boldsymbol{\Sigma}}_{22} \end{pmatrix} = \frac{1}{N} \sum_{\alpha=1}^{N} (\boldsymbol{x}_\alpha - \bar{\boldsymbol{x}})(\boldsymbol{x}_\alpha - \bar{\boldsymbol{x}})' \tag{2}$$

$$= \frac{1}{N} \begin{pmatrix} \sum\left(\boldsymbol{x}_\alpha^{(1)} - \bar{\boldsymbol{x}}^{(1)}\right)\left(\boldsymbol{x}_\alpha^{(1)} - \bar{\boldsymbol{x}}^{(1)}\right)' & \sum\left(\boldsymbol{x}_\alpha^{(1)} - \bar{\boldsymbol{x}}^{(1)}\right)\left(\boldsymbol{x}_\alpha^{(2)} - \bar{\boldsymbol{x}}^{(2)}\right)' \\ \sum\left(\boldsymbol{x}_\alpha^{(2)} - \bar{\boldsymbol{x}}^{(2)}\right)\left(\boldsymbol{x}_\alpha^{(1)} - \bar{\boldsymbol{x}}^{(1)}\right)' & \sum\left(\boldsymbol{x}_\alpha^{(2)} - \bar{\boldsymbol{x}}^{(2)}\right)\left(\boldsymbol{x}_\alpha^{(2)} - \bar{\boldsymbol{x}}^{(2)}\right)' \end{pmatrix}.$$

The maximum likelihood estimators of the canonical correlations Λ and the canonical variates defined by $\mathbf{A}$ and Γ involve applying the algebra of the previous section to $\hat{\Sigma}$. The matrices Λ, $\mathbf{A}$, and Γ_1 are uniquely defined if we assume the canonical correlations different and that the first nonzero element of each column of $\mathbf{A}$ is positive. The indeterminacy in Γ_2 allows multiplication on the right by a $(p_2 - p_1) \times (p_2 - p_1)$ orthogonal matrix; this indeterminacy can be removed by various types of requirements, for example, that the submatrix formed by the lower $p_2 - p_1$ rows be upper or lower triangular with positive diagonal elements. Application of Corollary 3.2.1 then shows that the maximum likelihood estimators of $\lambda_1, \ldots, \lambda_p$ are the roots of

$$\begin{vmatrix} -l\hat{\Sigma}_{11} & \hat{\Sigma}_{12} \\ \hat{\Sigma}_{21} & -l\hat{\Sigma}_{22} \end{vmatrix} = 0, \tag{3}$$

and the jth columns of $\hat{\mathbf{A}}$ and $\hat{\Gamma}_1$ satisfy

$$\begin{pmatrix} -l_j\hat{\Sigma}_{11} & \hat{\Sigma}_{12} \\ \hat{\Sigma}_{21} & -l_j\hat{\Sigma}_{22} \end{pmatrix} \begin{pmatrix} \hat{\alpha}^{(j)} \\ \hat{\gamma}^{(j)} \end{pmatrix} = \mathbf{0}, \tag{4}$$

$$\hat{\alpha}^{(j)\prime}\hat{\Sigma}_{11}\hat{\alpha}^{(j)} = 1, \tag{5}$$

$$\hat{\gamma}^{(j)\prime}\hat{\Sigma}_{22}\hat{\gamma}^{(j)} = 1. \tag{6}$$

$\hat{\Gamma}_2$ satisfies

$$\hat{\Gamma}_2'\hat{\Sigma}_{22}\hat{\Gamma}_1 = \mathbf{0}, \tag{7}$$

$$\hat{\Gamma}_2'\hat{\Sigma}_{22}\hat{\Gamma}_2 = \boldsymbol{I}. \tag{8}$$

When the other restrictions on Γ_2 are made, $\hat{\mathbf{A}}$, $\hat{\Gamma}$, and $\hat{\Lambda}$ are uniquely defined.

THEOREM 12.3.1. *Let $\boldsymbol{x}_1, \ldots, \boldsymbol{x}_N$ be N observations from $N(\boldsymbol{\mu}, \Sigma)$. Let Σ be partitioned into p_1 and p_2 $(p_1 \le p_2)$ rows and columns as in* (2) *in Section* 12.2 *and let $\boldsymbol{x}_\alpha$ be similarly partitioned as in* (1). *The maximum likelihood estimators of the canonical correlations are the roots of* (3) *where $\hat{\Sigma}_{ij}$ are defined by* (2). *The maximum likelihood estimators of the coefficients of the jth canonical components satisfy* (4), (5), *and* (6), $j = 1, \ldots, p_1$; *the remaining components satisfy* (7) *and* (8).

In the population the canonical correlations and canonical variates were found in terms of maximizing correlations of linear combinations of two sets of variates. The entire argument can be carried out in terms of the sample. Thus $\hat{\boldsymbol{\alpha}}^{(1)\prime}\boldsymbol{x}_\alpha^{(1)}$ and $\hat{\boldsymbol{\gamma}}^{(1)\prime}\boldsymbol{x}_\alpha^{(2)}$ have maximum sample correlation between any linear combinations of $\boldsymbol{x}_\alpha^{(1)}$ and $\boldsymbol{x}_\alpha^{(2)}$, and this correlation is l_1. Similarly, $\hat{\boldsymbol{\alpha}}^{(2)\prime}\boldsymbol{x}_\alpha^{(1)}$ and $\hat{\boldsymbol{\gamma}}^{(2)\prime}\boldsymbol{x}_\alpha^{(2)}$ have the second maximum sample correlation, and so forth.

It may also be observed that we could define the sample canonical variates and correlations in terms of $\boldsymbol{S}$, the unbiased estimator of $\boldsymbol{\Sigma}$. Then $\boldsymbol{a}^{(j)} = \sqrt{(N-1)/N}\,\hat{\boldsymbol{\alpha}}^{(j)}$, $\boldsymbol{c}^{(j)} = \sqrt{(N-1)/N}\,\hat{\boldsymbol{\gamma}}^{(j)}$, and l_j satisfy

$$\boldsymbol{S}_{12}\boldsymbol{c}^{(j)} = l_j \boldsymbol{S}_{11}\boldsymbol{a}^{(j)}, \tag{9}$$

$$\boldsymbol{S}_{21}\boldsymbol{a}^{(j)} = l_j \boldsymbol{S}_{22}\boldsymbol{c}^{(j)}, \tag{10}$$

$$\boldsymbol{a}^{(j)\prime}\boldsymbol{S}_{11}\boldsymbol{a}^{(j)} = 1, \tag{11}$$

$$\boldsymbol{c}^{(j)\prime}\boldsymbol{S}_{22}\boldsymbol{c}^{(j)} = 1. \tag{12}$$

We shall call the linear combinations $\boldsymbol{a}^{(j)\prime}\boldsymbol{x}_\alpha^{(1)}$ and $\boldsymbol{c}^{(j)\prime}\boldsymbol{x}_\alpha^{(2)}$ the sample canonical variates.

We can also derive the sample canonical variates from the sample correlation matrix,

$$\boldsymbol{R} = \left(\frac{\hat{\sigma}_{ij}}{\sqrt{\hat{\sigma}_{ii}}\sqrt{\hat{\sigma}_{jj}}}\right) = \left(\frac{s_{ij}}{\sqrt{s_{ii}s_{jj}}}\right) = (r_{ij}) = \begin{pmatrix} \boldsymbol{R}_{11} & \boldsymbol{R}_{12} \\ \boldsymbol{R}_{21} & \boldsymbol{R}_{22} \end{pmatrix}. \tag{13}$$

Let

$$\boldsymbol{S}_1 = \begin{pmatrix} \sqrt{s_{11}} & 0 & \cdots & 0 \\ 0 & \sqrt{s_{22}} & \cdots & 0 \\ \vdots & \vdots & & \vdots \\ 0 & 0 & \cdots & \sqrt{s_{p_1p_1}} \end{pmatrix}, \tag{14}$$

$$\boldsymbol{S}_2 = \begin{pmatrix} \sqrt{s_{p_1+1,p_1+1}} & 0 & \cdots & 0 \\ 0 & \sqrt{s_{p_1+2,p_1+2}} & \cdots & 0 \\ \vdots & \vdots & & \vdots \\ 0 & 0 & \cdots & \sqrt{s_{pp}} \end{pmatrix}. \tag{15}$$

Then we can write (9) through (12) as

$$R_{12}(S_2c^{(j)}) = l_j R_{11}(S_1 a^{(j)}), \tag{16}$$

$$R_{21}(S_1 a^{(j)}) = l_j R_{22}(S_2 c^{(j)}), \tag{17}$$

$$(S_1 a^{(j)})' R_{11}(S_1 a^{(j)}) = 1, \tag{18}$$

$$(S_2 c^{(j)})' R_{22}(S_2 c^{(j)}) = 1. \tag{19}$$

We can give these developments a geometric interpretation. The rows of the matrix $(x_1, \dots, x_N)$ can be interpreted as p vectors in an N-dimensional space and the rows of $(x_1 - \bar{x}, \dots, x_N - \bar{x})$ are the p vectors projected on the $(N-1)$-dimensional subspace orthogonal to the equiangular line. Denote these as $x_1^*, \dots, x_p^*$. Any vector u^* with components $\alpha'(x_1^{(1)} - \bar{x}^{(1)}, \dots, x_N^{(1)} - \bar{x}^{(1)}) = \alpha_1 x_1^* + \cdots + \alpha_{p_1} x_{p_1}^*$ is in the p_1-space spanned by $x_1^*, \dots, x_{p_1}^*$, and a vector v^* with components $\gamma'(x_1^{(2)} - \bar{x}^{(2)}, \dots, x_N^{(2)} - \bar{x}^{(2)}) = \gamma_1 x_{p_1+1}^* + \cdots + \gamma_{p_2} x_p^*$ is in the p_2-space spanned by $x_{p_1+1}^*, \dots, x_p^*$. The cosine of the angle between these two vectors is the correlation between $u_\alpha = \alpha' x_\alpha^{(1)}$ and $v_\alpha = \gamma' x_\alpha^{(2)}$, $\alpha = 1, \dots, N$. Finding α and γ to maximize the correlation is equivalent to finding the vectors in the p_1-space and the p_2-space such that the angle between them is least (i.e., has the greatest cosine). This gives the first canonical variates, and the first canonical correlation is the cosine of the angle. Similarly, the second canonical variates correspond to vectors orthogonal to the first canonical variates and with the angle minimized.

12.3.2. Computation

We shall discuss briefly computation in terms of the population quantities. Equations (50), (51), or (52) of Section 12.2 can be used. The computation of $\Sigma_{12}\Sigma_{22}^{-1}\Sigma_{21}$ can be accomplished by solving $\Sigma_{21} = \Sigma_{22}F$ for $\Sigma_{22}^{-1}\Sigma_{21}$ and then multiplying by Σ_{12}. If p_1 is sufficiently small, the determinant $|\Sigma_{12}\Sigma_{22}^{-1}\Sigma_{21} - \nu\Sigma_{11}|$ can be expanded into a polynomial in ν, and the polynomial equation may be solved for ν. The solutions are then inserted into (51) to arrive at the vectors α.

In many cases p_1 is too large for this procedure to be efficient. Then one can use an iterative procedure

$$\Sigma_{11}^{-1}\Sigma_{12}\Sigma_{22}^{-1}\Sigma_{21}\alpha(i) = \lambda^2(i+1)\alpha(i+1), \tag{20}$$

starting with an initial approximation $\boldsymbol{\alpha}(0)$; the vector $\boldsymbol{\alpha}(i+1)$ may be normalized by

$$\boldsymbol{\alpha}(i+1)'\boldsymbol{\Sigma}_{11}\boldsymbol{\alpha}(i+1) = 1. \tag{21}$$

Then $\lambda^2(i+1)$ converges to λ_1^2 and $\boldsymbol{\alpha}(i+1)$ converges to $\boldsymbol{\alpha}^{(1)}$ (if $\lambda_1 > \lambda_2$). This can be demonstrated in a fashion similar to that used for principal components using

$$\boldsymbol{\Sigma}_{11}^{-1}\boldsymbol{\Sigma}_{12}\boldsymbol{\Sigma}_{22}^{-1}\boldsymbol{\Sigma}_{21} = \mathbf{A}\boldsymbol{\Lambda}^2\mathbf{A}^{-1} \tag{22}$$

from (45) of Section 12.2. See Problem 9.

The right-hand side of (22) is $\sum_{i=1}^{p_1}\boldsymbol{\alpha}^{(i)}\lambda_i^2\tilde{\boldsymbol{\alpha}}^{(i)\prime}$, where $\tilde{\boldsymbol{\alpha}}^{(i)\prime}$ is the ith row of $\mathbf{A}^{-1}$. From the fact that $\mathbf{A}'\boldsymbol{\Sigma}_{11}\mathbf{A} = \boldsymbol{I}$, we find that $\mathbf{A}'\boldsymbol{\Sigma}_{11} = \mathbf{A}^{-1}$ and thus $\boldsymbol{\alpha}^{(i)\prime}\boldsymbol{\Sigma}_{11} = \tilde{\boldsymbol{\alpha}}^{(i)\prime}$. Now

$$\boldsymbol{\Sigma}_{11}^{-1}\boldsymbol{\Sigma}_{12}\boldsymbol{\Sigma}_{22}^{-1}\boldsymbol{\Sigma}_{21} - \lambda_1^2\boldsymbol{\alpha}^{(1)}\tilde{\boldsymbol{\alpha}}^{(1)\prime} = \sum_{i=2}^{p_1}\boldsymbol{\alpha}^{(i)}\lambda_i^2\tilde{\boldsymbol{\alpha}}^{(i)\prime} \tag{23}$$

$$= \mathbf{A}\begin{pmatrix} 0 & 0 & \cdots & 0 \\ 0 & \lambda_2^2 & \cdots & 0 \\ \vdots & \vdots & & \vdots \\ 0 & 0 & \cdots & \lambda_{p_1}^2 \end{pmatrix}\mathbf{A}^{-1}.$$

The maximum characteristic root of this matrix is λ_2^2. If we now use this matrix for iteration, we will obtain λ_2^2 and $\boldsymbol{\alpha}^{(2)}$. The procedure is continued to find as many λ_i^2 and $\boldsymbol{\alpha}^{(i)}$ as desired.

Given λ_i and $\boldsymbol{\alpha}^{(i)}$, we find $\boldsymbol{\gamma}^{(i)}$ from $\boldsymbol{\Sigma}_{21}\boldsymbol{\alpha}^{(i)} = \lambda_i\boldsymbol{\Sigma}_{22}\boldsymbol{\gamma}^{(i)}$ or $(1/\lambda_i)\boldsymbol{\Sigma}_{22}^{-1}\boldsymbol{\Sigma}_{21}\boldsymbol{\alpha}^{(i)} = \boldsymbol{\gamma}^{(i)}$. A check on the computations is provided by comparing $\boldsymbol{\Sigma}_{12}\boldsymbol{\gamma}^{(i)}$ and $\lambda_i\boldsymbol{\Sigma}_{11}\boldsymbol{\alpha}^{(i)}$.

For the sample we perform these calculations with $\hat{\boldsymbol{\Sigma}}_{ij}$ or $\boldsymbol{S}_{ij}$ substituted for $\boldsymbol{\Sigma}_{ij}$. It is often convenient to use $\boldsymbol{R}_{ij}$ in the computation (because $-1 < r_{ij} < 1$) to obtain $\boldsymbol{S}_1\boldsymbol{a}^{(j)}$ and $\boldsymbol{S}_2\boldsymbol{c}^{(j)}$; from these $\boldsymbol{a}^{(j)}$ and $\boldsymbol{c}^{(j)}$ are easy to compute.

Modern computational procedures are available for canonical correlations and variates similar to those sketched for principal components. Let

$$\boldsymbol{Z}_1 = \left(\boldsymbol{x}_1^{(1)} - \bar{\boldsymbol{x}}^{(1)}, \dots, \boldsymbol{x}_N^{(1)} - \bar{\boldsymbol{x}}^{(1)}\right), \tag{24}$$

$$(25) \qquad Z_2 = \left(x_1^{(2)} - \bar{x}^{(2)}, \dots, x_N^{(2)} - \bar{x}^{(2)} \right).$$

The QR decomposition of the transpose of these matrices (Section 11.4) is $Z_i' = Q_i R_i$, where $Q_i'Q_i = I_{p_i}$ and R_i is uppper triangular. Then $S_{ij} = Z_i Z_j' = R_i' Q_i' Q_j R_j$, $i, j = 1, 2$, and $S_{ii} = R_i' R_i$, $i = 1, 2$. The canonical correlations are the singular values of $Q_1'Q_2$ and the square roots of the characteristic roots of $(Q_1'Q_2)(Q_1'Q_2)'$ (by Theorem 12.2.2). Then the singular value decomposition of $Q_1'Q_2$ is $P(L\ O)T$, where P and T are orthogonal and L is diagonal. To effect the decomposition Householder transformations are applied to the left and right of $Q_1'Q_2$ to obtain an upper bidiagonal matrix, that is, a matrix with entries on the main diagonal and first superdiagonal. Givens matrices are used to reduce this matrix to a matrix that is diagonal to the degree of approximation required. For more detail see Kennedy and Gentle (1980), Sections 7.2 and 12.2, Chambers (1977), Björck and Golub (1973), and Golub and Luk (1976).

12.4. STATISTICAL INFERENCE

12.4.1. Tests of Independence and of Rank

In Chapter 9 we considered testing the null hypothesis that $X^{(1)}$ and $X^{(2)}$ are independent, which is equivalent to the null hypothesis that $\Sigma_{12} = \mathbf{0}$. Since $A'\Sigma_{12}\Gamma = (\Lambda\ \mathbf{0})$, it is seen that the hypothesis is equivalent to $\Lambda = \mathbf{0}$, that is, $\rho_1 = \cdots = \rho_{p_1} = 0$. The likelihood ratio criterion for testing this null hypothesis is the $N/2$ power of

$$(1) \qquad \frac{\begin{vmatrix} A_{11} & A_{12} \\ A_{21} & A_{22} \end{vmatrix}}{|A_{11}| \cdot |A_{22}|} = \frac{\begin{vmatrix} \hat{A}' & \mathbf{0} \\ \mathbf{0} & \hat{\Gamma}' \end{vmatrix}}{|\hat{A}'| \cdot |\hat{\Gamma}'|} \cdot \frac{\begin{vmatrix} A_{11} & A_{12} \\ A_{21} & A_{22} \end{vmatrix}}{|A_{11}| \cdot |A_{22}|} \cdot \frac{\begin{vmatrix} \hat{A} & \mathbf{0} \\ \mathbf{0} & \hat{\Gamma} \end{vmatrix}}{|\hat{A}| \cdot |\hat{\Gamma}|}$$

$$= \frac{\begin{vmatrix} I & \hat{\Lambda} & \mathbf{0} \\ \hat{\Lambda} & I & \mathbf{0} \\ \mathbf{0} & \mathbf{0} & I \end{vmatrix}}{|I| \cdot |I|} = \begin{vmatrix} I & \hat{\Lambda} \\ \hat{\Lambda} & I \end{vmatrix} = |I - \hat{\Lambda}^2| = \prod_{i=1}^{p_1} \left(1 - r_i^2\right),$$

where $r_1 = l_1 \geq \cdots \geq r_{p_1} = l_{p_1} \geq 0$ are the p_1 possibly nonzero sample canonical correlations. Under the null hypothesis, the limiting distribution of

Bartlett's modification of -2 times the logarithm of the likelihood ratio criterion, namely,

$$-\left[N - \tfrac{1}{2}(p+3)\right] \sum_{i=1}^{p_1} \log\left(1 - r_i^2\right) \tag{2}$$

is χ^2 with $p_1 p_2$ degrees of freedom. (See Section 9.4.) Note that it is approximately

$$N \sum_{i=1}^{p_1} r_i^2 = N \operatorname{tr} \boldsymbol{A}_{11}^{-1} \boldsymbol{A}_{12} \boldsymbol{A}_{21}^{-1} \boldsymbol{A}_{21}, \tag{3}$$

which is N times Nagao's criterion [(2) of Section 9.5].

If $\boldsymbol{\Sigma}_{12} \neq \boldsymbol{0}$, an interesting question is how many population canonical correlations are different from 0; that is, how many canonical variates are needed to explain the correlations between $\boldsymbol{X}^{(1)}$ and $\boldsymbol{X}^{(2)}$? The number of nonzero canonical correlations is equal to the rank of $\boldsymbol{\Sigma}_{12}$. The likelihood ratio criterion for testing the null hypothesis $H_k: \rho_{k+1} = \cdots = \rho_{p_1} = 0$, that is, that the rank of $\boldsymbol{\Sigma}_{12}$ is not greater than k, is $\prod_{i=k+1}^{p_1}(1 - r_i^2)^{\frac{1}{2}N}$ [Fujikoshi (1974)]. Under the null hypothesis

$$-\left[N - \tfrac{1}{2}(p+3)\right] \sum_{i=k+1}^{p_1} \log\left(1 - r_i^2\right) \tag{4}$$

has approximately the χ^2-distribution with $(p_1 - k)(p_2 - k)$ degrees of freedom. [Glynn and Muirhead (1978) suggest multiplying the sum in (4) by $N - k - \frac{1}{2}(p+3) + \sum_{i=1}^{k}(1/r_i^2)$; see also Lawley (1959).]

To determine the numbers of nonzero and zero population canonical correlations one can test that all the roots are 0. If that hypothesis is rejected, test that the $p_1 - 1$ smallest roots are 0, etc. Of course, these procedures are not statistically independent, even asymptotically. Alternatively, one could use a sequence of tests in the opposite direction: Test $\rho_{p_1} = 0$, then $\rho_{p_1 - 1} = \rho_{p_1} = 0$, and so on, until a hypothesis is rejected or until $\boldsymbol{\Sigma}_{12} = \boldsymbol{0}$ is accepted. Yet another procedure (which can only be carried out for small p_1) is to test $\rho_{p_1} = 0$, then $\rho_{p_1 - 1} = 0$ and so forth. In this procedure one would use r_j to test the hypothesis $\rho_j = 0$. The relevant asymptotic distribution will be discussed in Section 12.4.2.

12.4.2. Distributions of Canonical Correlations

The density of the canonical correlations is given in Section 13.4 for the case that $\mathbf{\Sigma}_{12} = \mathbf{0}$, that is, all the population correlations are 0. The density when some population correlations are different from 0 has been given by Constantine (1963) in terms of a hypergeometric function of two matrix arguments.

The large-sample theory is more manageable. Suppose the first k canonical correlations are positive, less than 1, and different, and suppose that $p_1 - k$ correlations are 0. Let

$$
\begin{aligned}
z_i &= \sqrt{N}\,\frac{r_i^2 - \rho_i^2}{2\rho_i(1-\rho_i^2)}, \qquad && i = 1,\dots,k, \\
z_i &= N r_i^2, && i = k+1,\dots,p_1. \qquad (5)
\end{aligned}
$$

Then in the limiting distribution $z_1,\dots,z_k$ and the set $z_{k+1},\dots,z_{p_1}$ are mutually independent, z_i has the limiting distribution $N(0,1)$, $i = 1,\dots,k$, and the density of the limiting distribution of $z_{k+1},\dots,z_{p_1}$ is

$$
\frac{\pi^{\frac{1}{2}(p_1-k)^2}\exp\left(-\frac{1}{2}\sum_{i=k+1}^{p_1} z_i\right)}{2^{\frac{1}{2}(p_1-k)(p_2-k)}\Gamma_{p_1-k}\left[\frac{1}{2}(p_1-k)\right]\Gamma_{p_1-k}\left[\frac{1}{2}(p_2-k)\right]} \times \prod_{i=k+1}^{p_1} z_i^{\frac{1}{2}(p_2-p_1-1)} \prod_{\substack{i,j=k+1\\ i<j}}^{p_1} (z_i - z_j). \qquad (6)
$$

This is the density of the characteristic roots of a $(p_1 - k)$-order matrix with distribution $W(\boldsymbol{I}_{p_1-k}, p_2-k)$; see Section 13.3. Note that the normalizing factor for the squared correlations corresponding to nonzero population correlations is $\sqrt{N}$ while the factor corresponding to zero population correlation is N. See Chapter 13.

In large samples, we treat r_i^2 as $N[\rho_i^2, (1/N)4\rho_i^2(1-\rho_i^2)^2]$ or r_i as $N[\rho_i, (1/N)(1-\rho_i^2)^2]$ (by Theorem 4.2.3) to obtain tests of ρ_i or confidence intervals for ρ_i. Lawley (1959) has shown that the transformation $z_i = \tanh^{-1}(r_i)$ [see Section 4.2.3] does not stabilize the variance and has a significant bias in estimating $\zeta_i = \tanh^{-1}(\rho_i)$.

12.5. AN EXAMPLE

In this section we consider a simple illustrative example. Rao [(1952), p. 245] gives some measurements on the first and second adult sons in a sample of 25 families. (These have been used in Problem 1 of Chapter 3 and Problem 41 of Chapter 4.) Let $x_{1\alpha}$ be the head length of the first son in the αth family, $x_{2\alpha}$ be the head breadth of the first son, $x_{3\alpha}$ be the head length of the second son, and $x_{4\alpha}$ be the head breadth of the second son. We shall investigate the relations between the measurements for the first son and for the second. Thus $\boldsymbol{x}_\alpha^{(1)\prime} = (x_{1\alpha}, x_{2\alpha})$ and $\boldsymbol{x}_\alpha^{(2)\prime} = (x_{3\alpha}, x_{4\alpha})$. The data can be summarized as†

$$\bar{\boldsymbol{x}}' = (185.72, 151.12, 183.84, 149.24),$$

(1)
$$\boldsymbol{S} = \frac{1}{24}\boldsymbol{A} = \begin{pmatrix} 95.2933 & 52.8683 & 69.6617 & 46.1117 \\ 52.8683 & 54.3600 & 51.3117 & 35.0533 \\ 69.6617 & 51.3117 & 100.8067 & 56.5400 \\ 46.1117 & 35.0533 & 56.5400 & 45.0233 \end{pmatrix}.$$

The matrix of correlations is

(2)
$$\boldsymbol{R} = \left(\begin{array}{cc|cc} 1.0000 & 0.7346 & 0.7108 & 0.7040 \\ 0.7346 & 1.0000 & 0.6932 & 0.7086 \\ \hline 0.7108 & 0.6932 & 1.0000 & 0.8392 \\ 0.7040 & 0.7086 & 0.8392 & 1.0000 \end{array}\right)$$

$$= \begin{pmatrix} \boldsymbol{R}_{11} & \boldsymbol{R}_{12} \\ \boldsymbol{R}_{21} & \boldsymbol{R}_{22} \end{pmatrix}.$$

All of the correlations are about 0.7 except for the correlation between the two measurements on second sons. In particular, $\boldsymbol{R}_{12}$ is nearly of rank one, and hence the second canonical correlation will be near zero.

We compute

(3)
$$\boldsymbol{R}_{22}^{-1}\boldsymbol{R}_{21} = \begin{pmatrix} 0.405{,}769 & 0.333{,}205 \\ 0.363{,}480 & 0.428{,}976 \end{pmatrix},$$

(4)
$$\boldsymbol{R}_{12}\boldsymbol{R}_{22}^{-1}\boldsymbol{R}_{21} = \begin{pmatrix} 0.544{,}311 & 0.538{,}841 \\ 0.538{,}841 & 0.534{,}950 \end{pmatrix}.$$

† Rao's computations are in error; his last "difference" is incorrect.

The determinantal equation is

$$(5) \qquad 0 = \begin{vmatrix} 0.544{,}311 - \nu & 0.538{,}841 - 0.7346\nu \\ 0.538{,}841 - 0.7346\nu & 0.534{,}950 - \nu \end{vmatrix}$$

$$= 0.460{,}363\nu^2 - 0.287{,}596\nu + 0.000{,}830.$$

The roots are 0.621,816 and 0.002,900; thus $l_1 = 0.788{,}553$ and $l_2 = 0.053{,}852$. Corresponding to these roots are the vectors

$$(6) \qquad \begin{aligned} S_1 a^{(1)} &= \begin{pmatrix} 0.552{,}166 \\ 0.521{,}548 \end{pmatrix}, \\ S_1 a^{(2)} &= \begin{pmatrix} 1.366{,}501 \\ -1.378{,}467 \end{pmatrix}, \end{aligned}$$

where S_1 is the diagonal matrix with diagonal elements $\sqrt{s_{11}} = 9.7618$, $\sqrt{s_{22}} = 7.3729$. We apply $(1/l_i)R_{22}^{-1}R_{21}$ to $S_1 a^{(i)}$ to obtain

$$(7) \qquad \begin{aligned} S_2 c^{(1)} &= \begin{pmatrix} 0.504{,}511 \\ 0.538{,}242 \end{pmatrix}, \\ S_2 c^{(2)} &= \begin{pmatrix} 1.767{,}281 \\ -1.757{,}288 \end{pmatrix}, \end{aligned}$$

where S_2 is the diagonal matrix with diagonal elements $\sqrt{s_{33}} = 10.0402$ and $\sqrt{s_{44}} = 6.7099$. We check these computations by calculating

$$(8) \qquad \begin{aligned} \frac{1}{l_1} R_{11}^{-1} R_{12}\left(S_2 c^{(1)}\right) &= \begin{pmatrix} 0.552{,}157 \\ 0.521{,}560 \end{pmatrix}, \\ \frac{1}{l_2} R_{11}^{-1} R_{12}\left(S_2 c^{(2)}\right) &= \begin{pmatrix} 1.365{,}151 \\ -1.376{,}741 \end{pmatrix}. \end{aligned}$$

The first vector in (8) corresponds closely to the first vector in (6); in fact, it is a slight improvement, for the computation is equivalent to an iteration on $S_1 a^{(1)}$. The second vector in (8) does not correspond as closely to the second vector in (6). One reason is that l_2 is correct to only four or five significant figures (as is $\nu_2 = l_2^2$) and thus the components of $S_2 c^{(2)}$ can be correct to only as many significant figures; secondly, the fact that $S_2 c^{(2)}$ corresponds to the smaller root means that the iteration decreases the accuracy instead of increasing it. Our final results are

$$
\begin{array}{lcc}
 & (1) & (2) \\
l_i = & 0.789, & 0.054, \\
\text{(9)} \quad a^{(i)} = & \begin{pmatrix} 0.0566 \\ 0.0707 \end{pmatrix}, & \begin{pmatrix} 0.1400 \\ -0.1870 \end{pmatrix}, \\
c^{(i)} = & \begin{pmatrix} 0.0502 \\ 0.0802 \end{pmatrix}, & \begin{pmatrix} 0.1760 \\ -0.2619 \end{pmatrix}.
\end{array}
$$

The larger of the two canonical correlations, 0.789, is higher than any of the individual correlations of a variable of the first set with a variable of the other. The second canonical correlation is very near zero. This means that to study the relation between two head dimensions of first sons and second sons we can confine our attention to the first canonical variates; the second canonical variates are correlated only slightly. The first canonical variate in each set is approximately proportional to the sum of the two measurements divided by their respective standard deviations; the second canonical variate in each set is approximately proportional to the difference of the two standardized measurements.

12.6. LINEARLY RELATED EXPECTED VALUES

12.6.1. Canonical Analysis of Regression Matrices

In this section we develop canonical correlations and variates for one stochastic vector and one nonstochastic vector. The expected value of the stochastic vector is a linear function of the nonstochastic vector (Chapter 8). We find new coordinate systems so that the expected value of each coordinate of the stochastic vector depends on only one coordinate of the nonstochastic vector; the coordinates of the stochastic vector are uncorrelated in the stochastic sense, and the coordinates of the nonstochastic vector are uncorrelated in the sample. The coordinates are ordered according to the effect sum of squares relative to the variance. The algebra is similar to that developed in Section 12.2.

One way of deriving the model is in terms of the conditional normal distribution. Suppose X partitioned as (1) in Section 12.2 has the normal distribution $N(\mu, \Sigma)$, where $\mu = (\mu^{(1)\prime}, \mu^{(2)\prime})'$ and Σ is partitioned as (2) in Section 12.2. Then the conditional distribution of $X^{(1)}$ given $X^{(2)} = x^{(2)}$ is $N(\mu^{(1)} + \beta(x^{(2)} - \mu^{(2)}),\ \Sigma_{11\cdot 2})$, where $\beta = \Sigma_{12}\Sigma_{22}^{-1}$ and $\Sigma_{11\cdot 2} = \Sigma_{11} - \Sigma_{12}\Sigma_{22}^{-1}\Sigma_{21}$, which we shall write as Ψ. Since we consider a set of random vectors $X_1^{(1)}, \dots, X_N^{(1)}$ with expected values depending on $x_1^{(2)}, \dots, x_N^{(2)}$, we can write the conditional expected value of $X_\phi^{(1)}$ as $\tau + \beta(x_\phi^{(2)} - \bar{x}^{(2)})$, where $\tau = \mu^{(1)} + \beta(\bar{x}^{(2)} - \mu^{(2)})$ can be considered as a parameter vector. We shall treat this model in its own right, not requiring that it be derived as a family of conditional distributions. This is the model of Section 8.2 with a slight change of notation.

Consider a linear combination of $X_\phi^{(1)}$, say $U_\phi = \alpha' X_\phi^{(1)}$. Then U_ϕ has variance $\alpha'\Psi\alpha$ and expected value

$$(1) \qquad \mathscr{E}U_\phi = \alpha'\tau + \alpha'\beta\left(x_\phi^{(2)} - \bar{x}^{(2)}\right).$$

The mean expected value is $(1/N)\sum_{\phi=1}^N \mathscr{E}U_\phi = \alpha'\tau$, and the "mean sum of squares due to $x^{(2)}$" is

$$(2) \qquad \frac{1}{n}\sum_{\phi=1}^{N}\left(\mathscr{E}U_\phi - \alpha'\tau\right)^2 = \frac{1}{n}\sum_{\phi=1}^{N}\alpha'\beta\left(x_\phi^{(2)} - \bar{x}^{(2)}\right)\left(x_\phi^{(2)} - \bar{x}^{(2)}\right)'\beta'\alpha$$

$$= \alpha'\beta S_{22}\beta'\alpha.$$

We can ask for the linear combination that maximizes the mean sum of squares relative to its variance; that is, the linear combination of dependent variables on which the independent variables have greatest effect. We want to maximize (2) subject to $\alpha'\Psi\alpha = 1$. That leads to the vector equation

$$(3) \qquad \left(\beta S_{22}\beta' - \kappa\Psi\right)\alpha = \mathbf{0}$$

for κ satisfying

$$(4) \qquad |\beta S_{22}\beta' - \kappa\Psi| = 0.$$

Multiplication of (3) on the left by α' shows that $\alpha'\beta S_{22}\beta'\alpha = \kappa$ for α and κ satisfying $\alpha'\Psi\alpha = 1$ and (3); to obtain the maximum we take the largest root of (4), say κ_1. Denote this vector by $\alpha^{(1)}$ and the corresponding random variable by $U_\phi^{(1)} = \alpha^{(1)\prime}X_\phi^{(1)}$. The expected value of this first canonical variable is $\mathscr{E}U_\phi^{(1)} = \alpha^{(1)\prime}[\beta(x_\phi^{(2)} - \bar{x}^{(2)}) + \tau]$. Let $\alpha^{(1)\prime}\beta = k\gamma^{(1)\prime}$, where k is determined so

$$(5)\qquad 1 = \frac{1}{n}\sum_{\phi=1}^{N}\left(\boldsymbol{\gamma}^{(1)\prime}\boldsymbol{x}_{\phi}^{(2)} - \frac{1}{N}\sum_{\eta=1}^{N}\boldsymbol{\gamma}^{(1)\prime}\boldsymbol{x}_{\eta}^{(2)}\right)^{2}$$

$$= \frac{1}{n}\sum_{\phi=1}^{N}\boldsymbol{\gamma}^{(1)\prime}\left(\boldsymbol{x}_{\phi}^{(2)} - \bar{\boldsymbol{x}}^{(2)}\right)\left(\boldsymbol{x}_{\phi}^{(2)} - \bar{\boldsymbol{x}}^{(2)}\right)'\boldsymbol{\gamma}^{(1)}$$

$$= \boldsymbol{\gamma}^{(1)\prime}\mathbf{S}_{22}\boldsymbol{\gamma}^{(1)}.$$

Then $k = \sqrt{\kappa_1}$. Let $v_{\phi}^{(1)} = \boldsymbol{\gamma}^{(1)\prime}(\boldsymbol{x}_{\phi}^{(2)} - \bar{\boldsymbol{x}}^{(2)})$. Then $\mathscr{E}U_{\phi}^{(1)} = \sqrt{\kappa_1}\, v_{\phi}^{(1)} + \boldsymbol{\alpha}^{(1)\prime}\boldsymbol{\tau}$.

Next let us obtain a linear combination $U_{\phi} = \boldsymbol{\alpha}'\boldsymbol{X}_{\phi}^{(1)}$ that has maximum effect sum of squares among all linear combinations with variance 1 and uncorrelated with $U_{\phi}^{(1)}$, that is, $0 = \mathscr{E}(U_{\phi} - \mathscr{E}U_{\phi})(U_{\phi}^{(1)} - \mathscr{E}U_{\phi}^{(1)})' = \boldsymbol{\alpha}'\boldsymbol{\Psi}\boldsymbol{\alpha}^{(1)}$. As in Section 12.2, we can set up this maximization problem with Lagrange multipliers and find that $\boldsymbol{\alpha}$ satisfies (3) for some κ satisfying (4) and $\boldsymbol{\alpha}'\boldsymbol{\Psi}\boldsymbol{\alpha} = 1$. The process is continued in a manner similar to that in Section 12.2. We summarize the results.

The jth canonical random variable is $U_{\phi}^{(j)} = \boldsymbol{\alpha}^{(j)\prime}\boldsymbol{X}_{\phi}^{(1)}$, where $\boldsymbol{\alpha}^{(j)}$ satisfies (3) for $\kappa = \kappa_j$ and $\boldsymbol{\alpha}^{(j)\prime}\boldsymbol{\Psi}\boldsymbol{\alpha}^{(j)} = 1$; $\kappa_1 \geq \kappa_2 \geq \cdots \geq \kappa_{p_1}$ are the roots of (4). We shall assume the rank of $\mathbf{B}$ is $p_1 \leq p_2$. (Then $\kappa_{p_1} > 0$.) $U_{\phi}^{(j)}$ has the largest effect sum of squares of linear combinations of unit variance and uncorrelated with $U_{\phi}^{(1)}, \ldots, U_{\phi}^{(j-1)}$. Let $\boldsymbol{\gamma}^{(j)} = (1/\sqrt{\kappa_j})\mathbf{B}'\boldsymbol{\alpha}^{(j)}$, $\nu_j = \boldsymbol{\alpha}^{(j)\prime}\boldsymbol{\tau}$ and $v_{\phi}^{(j)} = \boldsymbol{\gamma}^{(j)\prime}(\boldsymbol{x}_{\phi}^{(2)} - \bar{\boldsymbol{x}}^{(2)})$. Then

$$(6)\qquad \mathscr{E}U_{\phi}^{(j)} = \sqrt{\kappa_j}\, v_{\phi}^{(j)} + \nu_j,$$

$$(7)\qquad \frac{1}{n}\sum_{\phi=1}^{N}\left(v_{\phi}^{(j)} - \frac{1}{N}\sum_{\eta=1}^{N}v_{\eta}^{(j)}\right)^{2} = 1,$$

$$(8)\qquad \sum_{\phi=1}^{N}\left(v_{\phi}^{(j)} - \frac{1}{N}\sum_{\eta=1}^{N}v_{\eta}^{(j)}\right)\left(v_{\phi}^{(i)} - \frac{1}{N}\sum_{\eta=1}^{N}v_{\eta}^{(i)}\right) = 0, \qquad i \neq j.$$

If $p_2 > p_1$, then $\boldsymbol{\gamma}^{(p_1+1)}, \ldots, \boldsymbol{\gamma}^{(p_2)}$ can be chosen so $v_{\phi}^{(p_1+1)} = \boldsymbol{\gamma}^{(p_1+1)\prime}(\boldsymbol{x}_{\phi}^{(2)} - \bar{\boldsymbol{x}}^{(2)}), \ldots, v_{\phi}^{(p_2)} = \boldsymbol{\gamma}^{(p_2)\prime}(\boldsymbol{x}_{\phi}^{(2)} - \bar{\boldsymbol{x}}^{(2)})$ satisfy (7) and (8).

Let $\mathbf{A} = (\boldsymbol{\alpha}^{(1)} \cdots \boldsymbol{\alpha}^{(p_1)})$, $\boldsymbol{\Gamma}_1 = (\boldsymbol{\gamma}^{(1)} \cdots \boldsymbol{\gamma}^{(p_1)})$, $\boldsymbol{\Gamma}_2 = (\boldsymbol{\gamma}^{(p_1+1)} \cdots \boldsymbol{\gamma}^{(p_2)})$, $\boldsymbol{\Delta} = \operatorname{diag}(\delta_1, \ldots, \delta_{p_1}) = \operatorname{diag}(\sqrt{\kappa_1}, \ldots, \sqrt{\kappa_{p_1}})$, $\boldsymbol{U}_{\phi} = \mathbf{A}'\boldsymbol{X}_{\phi}^{(1)}$, $\boldsymbol{v}_{\phi}^{(1)} = \boldsymbol{\Gamma}_1'(\boldsymbol{x}_{\phi}^{(2)} - \bar{\boldsymbol{x}}^{(2)})$,

$v_\phi^{(2)} = \Gamma_2(x_\phi^{(2)} - \bar{x}^{(2)})$, and $v_\phi = (\nu_\phi^{(1)\prime}\ \nu_\phi^{(2)\prime})'$, $\phi = 1, \dots, N$. Then

$$\mathscr{E}(U_\phi - \mathscr{E}U_\phi)(U_\phi - \mathscr{E}U_\phi)' = \mathbf{A}'\Psi\mathbf{A} = I, \tag{9}$$

$$\mathscr{E}U_\phi = \Delta v_\phi^{(1)} + \nu, \qquad \phi = 1, \dots, N, \tag{10}$$

$$\frac{1}{n}\sum_{\phi=1}^{N}\left(v_\phi - \frac{1}{N}\sum_{\eta=1}^{N} v_\eta\right)\left(v_\phi - \frac{1}{N}\sum_{\eta=1}^{N} v_\eta\right)' = I. \tag{11}$$

The random canonical variates are uncorrelated and have variance 1. The expected value of each random canonical variate is a multiple of the corresponding nonstochastic canonical variate plus a constant. The nonstochastic canonical variates have sample variance 1 and are uncorrelated in the sample.

If $p_1 > p_2$, the maximum rank of $\boldsymbol{\beta}$ is p_2 and $\kappa_{p_2+1} = \cdots = \kappa_{p_1} = 0$. In that case we define $\mathbf{A}_1 = (\alpha^{(1)}, \dots, \alpha^{(p_2)})$ and $\mathbf{A}_2 = (\alpha^{(p_2+1)}, \dots, \alpha^{(p_1)})$, where $\alpha^{(1)}, \dots, \alpha^{(p_2)}$ (corresponding to positive κ's) are defined as before and $\alpha^{(p_2+1)}, \dots, \alpha^{(p_1)}$ are any vectors satisfying $\alpha^{(j)\prime}\Psi\alpha^{(j)} = 1$ and $\alpha^{(j)\prime}\Psi\alpha^{(i)} = 0$, $i \neq j$. Then $\mathscr{E}U_\phi^{(i)} = \delta_i v_\phi^{(i)} + \nu_i$, $i = 1, \dots, p_2$, and $\mathscr{E}U_\phi^{(i)} = \nu_i$, $i = p_2 + 1, \dots, p_1$.

In either case if the rank of $\boldsymbol{\beta}$ is $r \leq \min(p_1, p_2)$, there are r roots of (4) that are nonzero and hence $\mathscr{E}U_\phi^{(i)} = \delta_i v_\phi^{(i)} + \nu_i$ for $i = 1, \dots, r$.

12.6.2. Estimation

Let $x_1^{(1)}, \dots, x_N^{(1)}$ be a set of observations on $X_1^{(1)}, \dots, X_N^{(1)}$ with the probability structure developed in Section 12.6.1, and let $x_1^{(2)}, \dots, x_N^{(2)}$ be the set of corresponding independent variates. Then we can estimate τ, $\boldsymbol{\beta}$, and Ψ by

$$\hat{\tau} = \frac{1}{N}\sum_{\phi=1}^{N} x_\phi^{(1)} = \bar{x}^{(1)}, \tag{12}$$

$$\hat{\boldsymbol{\beta}} = A_{12}A_{22}^{-1} = S_{12}S_{22}^{-1}, \tag{13}$$

$$\begin{aligned}\hat{\Psi} &= \frac{1}{n}\sum_{\phi=1}^{N}\left[x_\phi^{(1)} - \bar{x}^{(1)} - \hat{\boldsymbol{\beta}}\left(x_\phi^{(2)} - \bar{x}^{(2)}\right)\right]\left[x_\phi^{(1)} - \bar{x}^{(1)} - \hat{\boldsymbol{\beta}}\left(x_\phi^{(2)} - \bar{x}^{(2)}\right)\right]' \\ &= \frac{1}{n}\left(A_{11} - A_{12}A_{22}^{-1}A_{21}\right) \\ &= S_{11} - S_{12}S_{22}^{-1}S_{21},\end{aligned} \tag{14}$$

where the $\boldsymbol{A}$'s and $\boldsymbol{S}$'s are defined as before. (It is convenient to divide by $n = N - 1$ instead of by N; the latter would yield maximum likelihood estimators.)

The sample analogues of (3) and (4) are

$$\mathbf{0} = \left(\hat{\boldsymbol{\beta}}\boldsymbol{S}_{22}\hat{\boldsymbol{\beta}}' - k\tilde{\boldsymbol{\Psi}}\right)\bar{\boldsymbol{a}} \tag{15}$$

$$= \left[\boldsymbol{S}_{12}\boldsymbol{S}_{22}^{-1}\boldsymbol{S}_{21} - k\left(\boldsymbol{S}_{11} - \boldsymbol{S}_{12}\boldsymbol{S}_{22}^{-1}\boldsymbol{S}_{21}\right)\right]\bar{\boldsymbol{a}},$$

$$0 = \left|\hat{\boldsymbol{\beta}}\boldsymbol{S}_{22}\hat{\boldsymbol{\beta}}' - k\tilde{\boldsymbol{\Psi}}\right| \tag{16}$$

$$= \left|\boldsymbol{S}_{12}\boldsymbol{S}_{22}^{-1}\boldsymbol{S}_{21} - k\left(\boldsymbol{S}_{11} - \boldsymbol{S}_{12}\boldsymbol{S}_{22}^{-1}\boldsymbol{S}_{21}\right)\right|.$$

The roots $k_1 \geq \cdots \geq k_{p_1}$ of (16) estimate the roots $\kappa_1 \geq \cdots \geq \kappa_{p_1}$ of (4) and the corresponding solutions $\bar{\boldsymbol{a}}^{(1)}, \dots, \bar{\boldsymbol{a}}^{(p_1)}$ of (15), normalized by $\bar{\boldsymbol{a}}^{(i)\prime}\tilde{\boldsymbol{\Psi}}\bar{\boldsymbol{a}}^{(i)} = 1$, estimate $\boldsymbol{\alpha}^{(1)}, \dots, \boldsymbol{\alpha}^{(p_1)}$. Then $\bar{\boldsymbol{c}}^{(j)} = (1/\sqrt{k_j})\hat{\boldsymbol{\beta}}'\bar{\boldsymbol{a}}^{(j)}$ estimate $\boldsymbol{\gamma}^{(j)}$, and $n_j = \bar{\boldsymbol{a}}^{(j)\prime}\bar{\boldsymbol{x}}^{(1)}$ estimate ν_j. The sample canonical variates are $\bar{\boldsymbol{a}}^{(j)\prime}\boldsymbol{x}_\phi^{(1)}$ and $\bar{\boldsymbol{c}}^{(j)\prime}(\boldsymbol{x}_\phi^{(2)} - \bar{\boldsymbol{x}}^{(2)})$, $j = 1, \dots, p_1$, $\phi = 1, \dots, N$. If $p_1 > p_2$, $p_1 - p_2$ more $\bar{\boldsymbol{a}}^{(j)}$'s can be defined satisfying $\bar{\boldsymbol{a}}^{(j)\prime}\tilde{\boldsymbol{\Psi}}\bar{\boldsymbol{a}}^{(j)} = 1$ and $\bar{\boldsymbol{a}}^{(j)\prime}\tilde{\boldsymbol{\Psi}}\bar{\boldsymbol{a}}^{(i)} = 0$, $i \neq j$.

12.6.3. Relations Between Canonical Variates

In Section 12.3, the roots $l_1 \geq \cdots \geq l_{p_1}$ were defined to satisfy

$$0 = \begin{vmatrix} -l\boldsymbol{S}_{11} & \boldsymbol{S}_{12} \\ \boldsymbol{S}_{21} & -l\boldsymbol{S}_{22} \end{vmatrix} = (-1)^{p_2} l^{p_2 - p_1} |\boldsymbol{S}_{22}|\,|\boldsymbol{S}_{12}\boldsymbol{S}_{22}^{-1}\boldsymbol{S}_{21} - l^2\boldsymbol{S}_{11}|. \tag{17}$$

Since (16) can be written

$$0 = |(1 + k)\boldsymbol{S}_{12}\boldsymbol{S}_{22}^{-1}\boldsymbol{S}_{21} - k\boldsymbol{S}_{11}|, \tag{18}$$

we see that $l_i^2 = k_i/(1 + k_i)$ and $k_i = l_i^2/(1 - l_i^2)$, $i = 1, \dots, p_1$. The vector $\boldsymbol{a}^{(i)}$ in Section 12.3 satisfies

$$\mathbf{0} = \left(\boldsymbol{S}_{12}\boldsymbol{S}_{22}^{-1}\boldsymbol{S}_{21} - l_i^2\boldsymbol{S}_{11}\right)\boldsymbol{a}^{(i)} \tag{19}$$

$$= \left(\boldsymbol{S}_{12}\boldsymbol{S}_{22}^{-1}\boldsymbol{S}_{21} - \frac{k_i}{1 + k_i}\boldsymbol{S}_{11}\right)\boldsymbol{a}^{(i)}$$

$$= \frac{1}{1 + k_i}\left[\boldsymbol{S}_{12}\boldsymbol{S}_{22}^{-1}\boldsymbol{S}_{21} - k_i\left(\boldsymbol{S}_{11} - \boldsymbol{S}_{12}\boldsymbol{S}_{22}^{-1}\boldsymbol{S}_{21}\right)\right]\boldsymbol{a}^{(i)},$$

which is equivalent to (15) for $k = k_i$. Comparison of the normalizations $\boldsymbol{a}^{(i)\prime}\boldsymbol{S}_{11}\boldsymbol{a}^{(i)} = 1$ and $\bar{\boldsymbol{a}}^{(i)\prime}(\boldsymbol{S}_{11} - \boldsymbol{S}_{12}\boldsymbol{S}_{22}^{-1}\boldsymbol{S}_{21})\bar{\boldsymbol{a}}^{(i)} = 1$ shows that $\bar{\boldsymbol{a}}^{(i)} = (1/\sqrt{1-l_i^2})\boldsymbol{a}^{(i)}$. Then $\bar{\boldsymbol{c}}^{(j)} = (1/\sqrt{k_j})\boldsymbol{S}_{22}^{-1}\boldsymbol{S}_{21}\bar{\boldsymbol{a}}^{(j)} = \boldsymbol{c}^{(j)}$.

We see that canonical variable analysis can be applied when the two vectors are jointly random and when one vector is random and the other is nonstochastic. The canonical variables defined by the two approaches are the same except for normalization. The measure of relationship between corresponding canonical variables can be the (canonical) correlation or it can be the ratio of "explained" to "unexplained" variance.

12.6.4. Testing Rank

The number of roots κ_j that are different from 0 is the rank of the regression matrix $\boldsymbol{\beta}$. It is the number of linear combinations of the regression variables that are needed to express the expected values of $\boldsymbol{X}_\phi^{(1)}$. We can ask whether the rank is k ($1 \le k \le p_1$ if $p_1 \le p_2$) against the alternative that the rank is greater than k. The hypothesis is

$$H_k: \kappa_{k+1} = \cdots = \kappa_{p_1} = 0. \tag{20}$$

The likelihood ratio criterion [Anderson (1951b)] is a power of

$$\prod_{i=k+1}^{p_1} (1+k_i)^{-1} = \prod_{i=k+1}^{p_1} (1-l_i^2). \tag{21}$$

Note that this is the same criterion as for the case of both vectors stochastic (Section 12.4). Then

$$-\left[N - \tfrac{1}{2}(p+3)\right] \sum_{i=k+1}^{p_1} \log(1-l_i^2) \tag{22}$$

has approximately the χ^2-distribution with $(p_1 - k)(p_2 - k)$ degrees of freedom.

The determination of the rank as any number between 0 and p_1 can be done as in Section 12.4.

12.6.5. Linear Functional Relationships

The study of Section 12.6 can be carried out in other terms. For example, the balanced one-way analysis of variance can be set up as

$$\boldsymbol{Y}_{\alpha j} = \boldsymbol{\nu}_\alpha + \boldsymbol{\mu} + \boldsymbol{U}_{\alpha j}, \quad \alpha = 1,\dots,m,\ j = 1,\dots,l, \tag{23}$$

where $\mathscr{E}U_\alpha = \mathbf{0}$, $\mathscr{E}U_\alpha U_\alpha' = \Psi$, $\sum_{\alpha=1}^m \boldsymbol{v}_\alpha = \mathbf{0}$, and

$$\Theta \boldsymbol{v}_\alpha = \mathbf{0}, \qquad \alpha = 1, \dots, m, \tag{24}$$

where Θ is $q \times p_1$ of rank q ($< p_1$). This is a special case of the model of Section 12.6.1 with $\phi = 1, \dots, N$, replaced by the pair of indices (α, j), $X_\phi^{(1)} = Y_{\alpha j}$, $\boldsymbol{\tau} = \boldsymbol{\mu}$, and $\boldsymbol{\beta}(x_\phi^{(2)} - \bar{x}^{(2)}) = \boldsymbol{v}_\alpha$ by use of dummy variables as in Section 8.8. The rank of $(\boldsymbol{v}_1, \dots, \boldsymbol{v}_m)$ is that of $\boldsymbol{\beta}$, namely, $r = p_1 - q$. There are q roots of (4) equal to 0 with

$$\boldsymbol{\beta} S_{22} \boldsymbol{\beta}' = l \sum_{\alpha=1}^m \boldsymbol{v}_\alpha \boldsymbol{v}_\alpha'. \tag{25}$$

The model (23) can be interpreted as repeated observations on $\boldsymbol{v}_\alpha + \boldsymbol{\mu}$ with error. The component equations of (24) are the linear functional relationships.

Let $\bar{y}_\alpha = (1/l)\sum_{j=1}^l y_{\alpha j}$ and $\bar{y} = (1/m)\sum_{\alpha=1}^m \bar{y}_\alpha$. The "sum of squares for effect" is

$$H = l \sum_{\alpha=1}^m (\bar{y}_\alpha - \bar{y})(\bar{y}_\alpha - \bar{y})' = n\hat{\boldsymbol{\beta}} S_{22} \hat{\boldsymbol{\beta}}' \tag{26}$$

with $m - 1$ degrees of freedom, and the "sum of squares for error" is

$$G = \sum_{\alpha=1}^m \sum_{j=1}^l (y_{\alpha j} - \bar{y}_\alpha)(y_{\alpha j} - \bar{y}_\alpha)' = n\tilde{\Psi} \tag{27}$$

with $m(l-1)$ degrees of freedom. The case $p_1 < p_2$ corresponds to $p_1 < l$. Then a maximum likelihood estimator of Θ is

$$\hat{\Theta} = (\bar{a}^{(r+1)}, \dots, \bar{a}^{(p_1)})', \tag{28}$$

the maximum likelihood estimators of $\boldsymbol{v}_\alpha$ are

$$\hat{\boldsymbol{v}}_\alpha = \tilde{\Psi}\hat{\Theta}'\hat{\Theta}(\bar{y}_\alpha - \bar{y}), \qquad \alpha = 1, \dots, n. \tag{29}$$

The estimator (28) can be multiplied by any nonsingular $q \times q$ matrix on the left to obtain another. For a fuller discussion, see Anderson (1984a) and Kendall and Stuart (1973).

12.7. SIMULTANEOUS EQUATIONS MODELS

12.7.1. The Model

Inference for structural equation models in econometrics is related to canonical correlations. The general model is

$$\boldsymbol{B}y_t + \Gamma z_t = \boldsymbol{u}_t, \qquad t = 1, \dots, T, \tag{1}$$

where $\boldsymbol{B}$ is $G \times G$ and Γ is $G \times K$. Here y_t is composed of G jointly dependent variables (endogenous), z_t is composed of K predetermined variables (exogenous and lagged dependent) which are treated as "independent" variables, and $\boldsymbol{u}_t$ consists of G unobservable random variables with

$$\mathscr{E}\boldsymbol{u}_t = \boldsymbol{0}, \qquad \mathscr{E}\boldsymbol{u}_t\boldsymbol{u}_t' = \Sigma. \tag{2}$$

We require $\boldsymbol{B}$ to be nonsingular. This model was initiated by Haavelmo (1944) and was developed by Koopmans, Marschak, Hurwicz, Anderson, Rubin, Leipnik, et al., 1944–1954, at the Cowles Commission for Research in Economics. Each component equation represents the behavior of some group (such as consumers or producers) and has economic meaning.

The set of structural equations (1) can be solved for y_t (because $\boldsymbol{B}$ is nonsingular):

$$y_t = \Pi z_t + v_t, \tag{3}$$

where

$$\Pi = -\boldsymbol{B}^{-1}\Gamma, \qquad v_t = \boldsymbol{B}^{-1}\boldsymbol{u}_t \tag{4}$$

with

$$\mathscr{E}v_t = \boldsymbol{0}, \qquad \mathscr{E}v_t v_t' = \boldsymbol{B}^{-1}\Sigma(\boldsymbol{B}')^{-1} = \Omega, \tag{5}$$

say. Equation (3) is called the *reduced form* of the model. It is a multivariate regression model. In principle, it is observable.

12.7.2. Identification by Specified Zeros

The structural equation (1) can be multiplied on the left by an arbitrary nonsingular matrix. To determine component equations that are economically meaningful, restrictions must be imposed. For example, in the case of demand and supply the equation describing demand may be distinguished by the fact that it includes consumer income and excludes cost of raw materials, which is in the supply equation. The exclusion of the latter amounts to specifying that its coefficient in the demand equation is 0.

We consider identification of a structural equation by specifying certain coefficients to be 0. It is convenient to treat the first equation. Suppose the variables are numbered so that the first G_1 jointly dependent variables are included in the first equation and the remaining $G_2 = G - G_1$ are not and the first K_1 predetermined variables are included and $K_2 = K - K_1$ are excluded. Then we can partition the coefficient matrices as

$$(\boldsymbol{B}\ \Gamma) = \begin{bmatrix} \boldsymbol{\beta}' & \mathbf{0} & \boldsymbol{\gamma}' & \mathbf{0} \\ - & - & - & - \end{bmatrix}, \tag{6}$$

where the vectors $\boldsymbol{\beta}$, $\mathbf{0}$, $\boldsymbol{\gamma}$, and $\mathbf{0}$ have G_1, G_2, K_1, and K_2 components, respectively. The reduced form is partitioned conformally into G_1 and G_2 sets of rows and K_1 and K_2 sets of columns:

$$\Pi = \begin{bmatrix} \Pi_{11} & \Pi_{12} \\ - & - \end{bmatrix}. \tag{7}$$

The relation between $\boldsymbol{B}$, Γ and Π can be expressed as

(8)

$$\begin{bmatrix} \boldsymbol{\gamma}' & \mathbf{0} \\ - & - \end{bmatrix} = \Gamma = -\boldsymbol{B}\Pi = -\begin{bmatrix} \boldsymbol{\beta}' & \mathbf{0} \\ - & - \end{bmatrix}\begin{bmatrix} \Pi_{11} & \Pi_{12} \\ - & - \end{bmatrix} = -\begin{bmatrix} \boldsymbol{\beta}'\Pi_{11} & \boldsymbol{\beta}'\Pi_{12} \\ - & - \end{bmatrix}.$$

The upper right-hand corner of (8) yields

$$\boldsymbol{\beta}'\Pi_{12} = \mathbf{0}. \tag{9}$$

To determine $\boldsymbol{\beta}$ $(G_1 \times 1)$ uniquely except for a constant of proportionality we need

$$\operatorname{rank}(\Pi_{12}) = G_1 - 1. \tag{10}$$

This implies

$$K_2 \geq G_1 - 1. \tag{11}$$

Addition of G_2 to (11) gives the *order condition*

$$G_2 + K_2 \geq G_1 + G_2 - 1 = G - 1. \tag{12}$$

The number of specified 0's in an identified equation must be at least equal to 1 less than the number of equations (or jointly dependent variables).

It can be shown that when $\boldsymbol{B}$ is nonsingular (10) holds if and only if the rank of the matrix consisting of the columns of $(\boldsymbol{B}\ \boldsymbol{\Gamma})$ with specified 0's in the first row is $G - 1$.

12.7.3. Estimation of the Reduced Form

The model (3) is a typical multivariate regression model. The observations are

$$\begin{pmatrix} y_1 \\ z_1 \end{pmatrix}, \ldots, \begin{pmatrix} y_T \\ z_T \end{pmatrix}. \tag{13}$$

The usual estimators of Π and Ω (Section 8.2) are

$$\boldsymbol{P} = \sum_{t=1}^{T} y_t z_t' \left(\sum_{t=1}^{T} z_t z_t' \right)^{-1}, \tag{14}$$

$$\hat{\Omega} = \frac{1}{T} \sum_{t=1}^{T} (y_t - \boldsymbol{P} z_t)(y_t - \boldsymbol{P} z_t)'. \tag{15}$$

These are maximum likelihood estimators if the v_t are normal.

If the z_t are exogenous (regardless of normality), then

$$\mathscr{E} \operatorname{vec} \boldsymbol{P} = \operatorname{vec} \Pi, \qquad \mathscr{C}(\operatorname{vec} \boldsymbol{P}) = \boldsymbol{A}^{-1} \otimes \Omega, \tag{16}$$

where

$$\boldsymbol{A} = \sum_{t=1}^{T} z_t z_t' \tag{17}$$

and $\operatorname{vec}(\boldsymbol{d}_1, \ldots, \boldsymbol{d}_m) = (\boldsymbol{d}_1', \ldots, \boldsymbol{d}_m')'$. If, furthermore, the v_t are normal, then $\boldsymbol{P}$ is normal and $T\hat{\Omega}$ has the Wishart distribution with covariance matrix Ω and $T - K$ degrees of freedom.

12.7.4. Estimation of the Coefficients of an Equation

First, consider the estimation of the vector of coefficients $\boldsymbol{\beta}$ when $K_2 = G_1 - 1$. Let

$$\boldsymbol{P} = \begin{pmatrix} \boldsymbol{P}_{11} & \boldsymbol{P}_{12} \\ \boldsymbol{P}_{21} & \boldsymbol{P}_{22} \end{pmatrix} \tag{18}$$

be partitioned as $\boldsymbol{\Pi}$. Then the probability is 1 that $\operatorname{rank}(\boldsymbol{P}_{12}) = G_1 - 1$ and the equation

$$\hat{\boldsymbol{\beta}}'\boldsymbol{P}_{12} = \mathbf{0} \tag{19}$$

has a nontrivial solution that is unique except for a constant of proportionality. This is the maximum likelihood estimator when the disturbance terms are normal.

If $K_2 \geq G_1$, then the probability is 1 that $\operatorname{rank}(\boldsymbol{P}_{12}) = G_1$ and (19) has only the trivial solution $\hat{\boldsymbol{\beta}} = \mathbf{0}$, which is unsatisfactory. To obtain a suitable estimator we find $\hat{\boldsymbol{\beta}}$ to minimize $\hat{\boldsymbol{\beta}}'\boldsymbol{P}_{12}$ in some sense relative to another function of $\hat{\boldsymbol{\beta}}'$.

Let z_t be partitioned into subvectors of K_1 and K_2 components:

$$z_t = \begin{pmatrix} z_t^{(1)} \\ z_t^{(2)} \end{pmatrix}, \tag{20}$$

$$\sum_{t=1}^{T} z_t z_t' = \boldsymbol{A} = \begin{pmatrix} \boldsymbol{A}_{11} & \boldsymbol{A}_{12} \\ \boldsymbol{A}_{21} & \boldsymbol{A}_{22} \end{pmatrix}, \tag{21}$$

$$\boldsymbol{A}_{22\cdot1} = \boldsymbol{A}_{22} - \boldsymbol{A}_{21}\boldsymbol{A}_{11}^{-1}\boldsymbol{A}_{12}. \tag{22}$$

Let y_t and $\boldsymbol{\Omega}$ be partitioned into G_1 and G_2 components:

$$y_t = \begin{pmatrix} y_t^{(1)} \\ y_t^{(2)} \end{pmatrix}, \tag{23}$$

$$\boldsymbol{\Omega} = \begin{pmatrix} \boldsymbol{\Omega}_{11} & \boldsymbol{\Omega}_{12} \\ \boldsymbol{\Omega}_{21} & \boldsymbol{\Omega}_{22} \end{pmatrix}. \tag{24}$$

Now set up the multivariate analysis of variance table for $y^{(1)}$.

Source	Sum of Squares
$z_t^{(1)}$	$\sum_{t=1}^{T} y_t^{(1)} z_t^{(1)\prime} A_{11}^{-1} z_t^{(1)} y_t^{(1)\prime}$
$z_t^{(2)} \perp z_t^{(1)}$	$P_{12} A_{22 \cdot 1} P_{12}'$
Error	$\sum_{t=1}^{T} (y_t^{(1)} - P_{11} z_t^{(1)} - P_{12} z_t^{(2)})(y_t^{(1)} - P_{11} z_t^{(1)} - P_{12} z_t^{(2)})'$
Total	$\sum_{t=1}^{T} y_t^{(1)} y_t^{(1)\prime}$

The first term in the analysis of variance table is the (vector) sum of squares of $y_t^{(1)}$ due to the effect of $z_t^{(1)}$. The second term is due to the effect of $z_t^{(2)}$ beyond the effect of $z_t^{(1)}$. The two add to $(PAP')_{11}$, which is the total effect of z_t, the predetermined variables.

We propose to find the vector $\hat{\beta}$ such that effect of $z_t^{(2)}$ on $\hat{\beta}' y_t^{(1)}$ beyond the effect of $z_t^{(1)}$ is minimized relative to the error sum of squares of $\hat{\beta}' y_t^{(1)}$. We minimize

$$\frac{\hat{\beta}'(P_{12} A_{22 \cdot 1} P_{12}')\hat{\beta}}{\hat{\beta}' \hat{\Omega}_{11} \hat{\beta}} = \frac{(\hat{\beta}' P_{12}) A_{22 \cdot 1} (\hat{\beta}' P_{12})'}{\hat{\beta}' \hat{\Omega}_{11} \hat{\beta}}. \tag{25}$$

This estimator has been called the *least variance ratio* estimator. Under normality and based only on the 0 restrictions on the coefficients of this single equation, the estimator is maximum likelihood and is known as the *limited information maximum likelihood* (LIML) estimator [Anderson and Rubin (1949)].

The algebra of minimizing (25) is to find the smallest root, say ν, of

$$|P_{12} A_{22 \cdot 1} P_{12}' - \lambda \hat{\Omega}_{11}| = 0 \tag{26}$$

and the corresponding vector satisfying

$$P_{12} A_{22 \cdot 1} P_{12}' \hat{\beta} = \nu \hat{\Omega}_{11} \hat{\beta}. \tag{27}$$

The vector is normalized according to some rule. A frequently used rule is to set one (nonzero) coefficient equal to 1, say the first, $\hat{\beta}_1 = 1$. If we write

(28) $$\boldsymbol{\beta} = \begin{pmatrix} 1 \\ \boldsymbol{\beta}^* \end{pmatrix}, \qquad \hat{\boldsymbol{\beta}} = \begin{pmatrix} 1 \\ \hat{\boldsymbol{\beta}}^* \end{pmatrix},$$

(29) $$\boldsymbol{\Pi}_{12} = \begin{pmatrix} \boldsymbol{\pi}_{12} \\ \boldsymbol{\Pi}_{12}^* \end{pmatrix}, \qquad \boldsymbol{P}_{12} = \begin{pmatrix} \boldsymbol{p}_{12} \\ \boldsymbol{P}_{12}^* \end{pmatrix},$$

(30) $$\hat{\boldsymbol{\Omega}}_{11} = \begin{pmatrix} \hat{\omega}_{11} & \hat{\boldsymbol{\omega}}_{(1)}' \\ \hat{\boldsymbol{\omega}}_{(1)} & \hat{\boldsymbol{\Omega}}_{11}^* \end{pmatrix},$$

then (27) can be replaced by the linear equation

(31) $$\left(\boldsymbol{P}_{12}^* \boldsymbol{A}_{22\cdot 1} \boldsymbol{P}_{12}^{*\prime} - \nu \hat{\boldsymbol{\Omega}}_{11}^*\right) \hat{\boldsymbol{\beta}}^* = -\left(\boldsymbol{P}_{12}^* \boldsymbol{A}_{22\cdot 1} \boldsymbol{p}_{12}' - \nu \hat{\boldsymbol{\omega}}_{(1)}\right).$$

The first component equation in (27) has been dropped because it is linearly dependent on the other equations [because ν is a root of (26)].

12.7.5. Relation to the Linear Functional Relationship

We now show that the model for the single linear functional relationship ($q = 1$) is identical to the model for structural equations in the special case that $G_2 = 0$, ($\boldsymbol{y}_t^{(1)} = \boldsymbol{y}_t$) and $z_t^{(1)} \equiv 1$ ($K_1 = 1$). Write the two models as

(32) $$\boldsymbol{X}_{\alpha j} = \boldsymbol{\mu} + \boldsymbol{v}_\alpha + \boldsymbol{U}_{\alpha j}, \qquad \alpha = 1, \dots, n, \quad j = 1, \dots, k,$$

where

(33) $$\sum_{\alpha=1}^{n} \boldsymbol{v}_\alpha = \boldsymbol{0},$$

and

(34) $$\boldsymbol{y}_t = \boldsymbol{\Pi}_1 + \boldsymbol{\Pi}_2 \boldsymbol{z}_t^{(2)} + \boldsymbol{v}_t, \qquad t = 1, \dots, T,$$

where $\boldsymbol{\Pi} = (\boldsymbol{\Pi}_1 \ \boldsymbol{\Pi}_2)$. The correspondence between the models is $p \leftrightarrow G = G_1$,

(35) $$\boldsymbol{X}_{\alpha j} \leftrightarrow \boldsymbol{y}_t, \qquad \boldsymbol{U}_{\alpha j} \leftrightarrow \boldsymbol{v}_t,$$

(36) $$(\alpha, j) \leftrightarrow t, \qquad nk \leftrightarrow T,$$

(37) $$\boldsymbol{\Psi} \leftrightarrow \boldsymbol{\Omega}, \qquad \boldsymbol{\mu} \leftrightarrow \boldsymbol{\Pi}_1.$$

We can write the model for the linear functional relationship with dummy variables. Define

(38) $$s_{\alpha j} = \begin{pmatrix} 0 \\ \vdots \\ 1 \\ \vdots \\ 0 \end{pmatrix} \leftarrow \alpha\text{th position}, \qquad \alpha = 1, \ldots, n-1,$$

(39) $$s_{nj} = \begin{pmatrix} -1 \\ \vdots \\ -1 \end{pmatrix}.$$

Then

(40) $$\boldsymbol{\mu} + \boldsymbol{\nu}_\alpha = (\boldsymbol{\mu}, \boldsymbol{\nu}_1, \ldots, \boldsymbol{\nu}_{n-1}) \begin{pmatrix} 1 \\ s_{\alpha j} \end{pmatrix}, \qquad \alpha = 1, \ldots, n,$$

where j may be suppressed. Note

(41) $$\boldsymbol{\nu}_n = -(\boldsymbol{\nu}_1 + \cdots + \boldsymbol{\nu}_{n-1}).$$

The correspondence is

(42) $$1 \leftrightarrow z_t^{(1)}, \qquad s_{\alpha j} \leftrightarrow z_t^{(2)},$$

(43) $$\boldsymbol{\mu} \leftrightarrow \Pi_1, \qquad (\boldsymbol{\nu}_1, \ldots, \boldsymbol{\nu}_{n-1}) \leftrightarrow \Pi_2,$$

(44) $$1 \leftrightarrow K_1, \qquad n-1 \leftrightarrow K_2,$$

(45) $$\boldsymbol{B}(\boldsymbol{\nu}_1, \ldots, \boldsymbol{\nu}_{n-1}) = \mathbf{0} \leftrightarrow \boldsymbol{\beta}'\Pi_2 = \mathbf{0}.$$

Let $\boldsymbol{P} = (\boldsymbol{P}_1 \ \boldsymbol{P}_2)$. In terms of the statistics we have the correspondence

(46) $$\hat{\boldsymbol{\mu}} = \bar{\boldsymbol{x}} \leftrightarrow \bar{\boldsymbol{y}},$$

(47) $$\hat{\boldsymbol{\nu}}_\alpha = \bar{\boldsymbol{x}}_\alpha - \bar{\boldsymbol{x}} \leftrightarrow \boldsymbol{P}_2.$$

Effect Matrix

(48) $$\boldsymbol{H} = k \sum_{\alpha=1}^{n} (\bar{\boldsymbol{x}}_\alpha - \bar{\boldsymbol{x}})(\bar{\boldsymbol{x}}_\alpha - \bar{\boldsymbol{x}})' \leftrightarrow \boldsymbol{P}_2 \boldsymbol{A}_{22\cdot 1} \boldsymbol{P}_2'$$

Error Matrix

(49)

$$\boldsymbol{G} = \sum_{\alpha=1}^{n} \sum_{j=1}^{k} (\boldsymbol{x}_{\alpha j} - \bar{\boldsymbol{x}}_\alpha)(\boldsymbol{x}_{\alpha j} - \bar{\boldsymbol{x}}_\alpha)' \leftrightarrow T\hat{\Omega} = \sum_{t=1}^{T} (\boldsymbol{y}_t - \boldsymbol{P}\boldsymbol{z}_t)(\boldsymbol{y}_t - \boldsymbol{P}\boldsymbol{z}_t)'.$$

Then the estimator $\hat{\boldsymbol{B}}$ of the linear functional relationship for $q = 1$ is identical to the LIML estimator [Anderson (1951b, 1976, 1984a)].

12.7.6. Asymptotic Theory as $T \to \infty$

The LIML estimator has an asymptotic normal distribution as the length of the observation series increases. We assume that $(1/T)\boldsymbol{A}$ approaches a positive definite limit. (If some predetermined variables are lagged endogenous variables, this limit must hold with probability 1.) We partition $\boldsymbol{\beta}$ and $\boldsymbol{\Pi}_{12}$ as in (28) and (29). Let

$$\sigma^2 = \boldsymbol{\beta}'\boldsymbol{\Omega}_{11}\boldsymbol{\beta} = \mathscr{V} u_{1t}. \tag{50}$$

Then

$$\frac{1}{\sigma}\left(\boldsymbol{\Pi}_{12}^{*}\boldsymbol{A}_{22\cdot 1}\boldsymbol{\Pi}_{12}^{*\prime}\right)^{\frac{1}{2}}\left(\hat{\boldsymbol{\beta}}^{*} - \boldsymbol{\beta}^{*}\right) \tag{51}$$

has a limiting normal distribution with mean $\boldsymbol{0}$ and covariance $\boldsymbol{I}$. We consider $\hat{\boldsymbol{\beta}}^{*}$ as approximately distributed according to

$$N\left[\boldsymbol{\beta}^{*}, \sigma^2\left(\boldsymbol{\Pi}_{12}^{*}\boldsymbol{A}_{22\cdot 1}\boldsymbol{\Pi}_{12}^{*\prime}\right)^{-1}\right] \tag{52}$$

[Anderson and Rubin (1950)]. Since $\boldsymbol{P}$ is a consistent estimator of $\boldsymbol{\Pi}$, in (52) $\boldsymbol{\Pi}_{12}^{*}$ can be replaced by $\boldsymbol{P}_{12}^{*}$.

Because of the correspondence between the LIML estimator and the MLE for the linear functional relationship as outlined in Section 12.7.5, this asymptotic theory can be translated for the latter. Suppose the single linear functional relationship is written as

$$0 = \boldsymbol{\beta}'\boldsymbol{v}_\alpha = \left(1\ \boldsymbol{\beta}^{*\prime}\right)\begin{pmatrix} v_{1\alpha} \\ \boldsymbol{v}_\alpha^{*} \end{pmatrix} = v_{1\alpha} + \boldsymbol{\beta}^{*\prime}\boldsymbol{v}_\alpha^{*}, \qquad \alpha = 1, \dots, n, \tag{53}$$

where

$$\boldsymbol{v}_\alpha = \begin{pmatrix} v_{1\alpha} \\ \boldsymbol{v}_\alpha^{*} \end{pmatrix}, \qquad \alpha = 1, \dots, n. \tag{54}$$

Let n ($\leftrightarrow K$) be fixed and let the number of replications $k \to \infty$ (corresponding to $T/K \to \infty$ for fixed K). Let $\sigma^2 = \boldsymbol{\beta}'\boldsymbol{\Psi}\boldsymbol{\beta}$.

Since $\Pi_{12}A_{22\cdot 1}\Pi_{12}'$ corresponds to $k\sum_{\alpha=1}^{n} v_\alpha^* v_\alpha^{*\prime}$, $\hat{\beta}^*$ here has the approximate distribution

$$N\left[\beta^*, \sigma^2\left(k\sum_{\alpha=1}^{n} v_\alpha^* v_\alpha^{*\prime}\right)^{-1}\right]. \tag{55}$$

12.7.7. Two-Stage Least Squares

The two-stage least squares (TSLS) estimator can be considered as a simplification of the LIML estimator. In (30) $\hat{\Omega}_{11}$ is a consistent estimator of Ω_{11}, ν is of probability order 1, and $P_{12}A_{22\cdot 1}P_{12}'$ is of probability order T. Thus the term $\nu\hat{\Omega}_{11}$ can be omitted without affecting the limiting distribution of $\sqrt{T}(\hat{\beta}^* - \beta^*)$. The TSLS estimator is then obtained from (31) as

$$\hat{\beta}_{\text{TSLS}}^{*\prime} = -\left(P_{12}^* A_{22\cdot 1} P_{12}^{*\prime}\right)^{-1} P_{12}^* A_{22\cdot 1} p_{12}'. \tag{56}$$

This estimator was suggested by Basmann (1957) and Theil (1961). It corresponds in the linear functional relationship setup to ordinary least squares on the first coordinate. If some other coefficient of β were set equal to one, the minimization would be in the direction of that coordinate.

Consider the general linear functional relationship when the error covariance matrix is unknown and there are replications. Constrain B to be

$$B = \left(I_m \; B^*\right). \tag{57}$$

Partition

$$v_\alpha = \begin{pmatrix} v_\alpha^{(1)} \\ v_\alpha^{(2)} \end{pmatrix}. \tag{58}$$

Then the least squares estimator of B^* is

(59)

$$\hat{B}_{\text{LS}}^* = -\sum_{\alpha=1}^{n}\left(\bar{x}_\alpha^{(1)} - \bar{x}^{(1)}\right)\left(\bar{x}_\alpha^{(2)} - \bar{x}^{(2)}\right)'\left[\sum_{\alpha=1}^{n}\left(\bar{x}_\alpha^{(2)} - \bar{x}^{(2)}\right)\left(\bar{x}_\alpha^{(2)} - \bar{x}^{(2)}\right)'\right]^{-1}.$$

For n fixed and $k \to \infty$,

$$\hat{B}_{\text{LS}}^* \underset{p}{\to} B^* \tag{60}$$

and

$$\sqrt{k}\,\mathrm{vec}\left(\hat{\boldsymbol{B}}^*_{\mathrm{LS}} - \boldsymbol{B}^*\right) \to N\left[\mathbf{0}, \left(\sum_{\alpha=1}^{n} \boldsymbol{v}_\alpha^{(2)} \boldsymbol{v}_\alpha^{(2)\prime}\right)^{-1} \otimes \boldsymbol{B}\boldsymbol{\Psi}\boldsymbol{B}'\right]. \tag{61}$$

[See Anderson (1984b).] It was shown by Anderson (1951c) that the q smallest sample roots are of such a probability order that the maximum likelihood estimator is asymptotically equivalent, that is, the limiting distribution of $\sqrt{k}\,\mathrm{vec}(\hat{\boldsymbol{B}}^*_{\mathrm{ML}} - \boldsymbol{B}^*)$ is the right-hand side of (61).

12.7.8. Other Asymptotic Theory

In terms of the linear functional relationship it may be more natural to consider $n \to \infty$ and k fixed. When $k = 1$ and the error covariance matrix is $\sigma^2 \boldsymbol{I}_p$, Gleser (1981) has given the asymptotic theory. For the simultaneous equations model, the corresponding conditions are $K_2 \to \infty$, $T \to \infty$, and K_2/T approaches a positive limit. Kunitomo (1980) has given an asymptotic expansion of the distribution in the case of $p = 2$ and $m = q = 1$.

When $n \to \infty$, the least squares estimator (i.e., minimizing the sum of squares of the residuals in one fixed direction) is not consistent; the LIML and TSLS estimators are not asymptotically equivalent.

12.7.9. Distributions of Estimators

Econometricians have studied intensively the distributions of TSLS and LIML estimator, particularly in the case of two endogenous variables.

Exact distributions have been given by Basmann (1961), (1963), Richardson (1968), Sawa (1969), Mariano and Sawa (1972), Phillips (1980) and Anderson and Sawa (1982). These have not been very informative because they are usually given in terms of infinite series the properties of which are unknown or irrelevant.

A more useful approach is by approximating the distributions. Asymptotic expansions of distributions have been made by Sargan and Mikhail (1971), Anderson and Sawa (1973), Anderson (1974), Kunitomo (1980), and others. Phillips (1982) has studied the Padé approach. See also Anderson (1977).

Tables of the distributions of the TSLS and LIML estimators in the case of two endogenous variables have been given by Anderson and Sawa (1977, 1979), Anderson, Kunitomo, and Sawa (1983a).

Anderson, Kunitomo, and Sawa (1983b) have graphed densities of the maximum likelihood estimator and the least squares estimator (minimizing in one direction) for the linear functional relationship (Section 12.6) for the case $p = 2$, $m = q = 1$, $\Psi = \sigma^2 \Psi_0$ and for various values of β, n, and

$$\delta^2 = \frac{1}{\sigma^2} \sum_{\alpha=1}^{n} (\mu_\alpha - \bar{\mu})^2. \tag{62}$$

PROBLEMS

1. (Sec. 12.2) Let $z_\alpha = z_{1\alpha} = 1$, $\alpha = 1, \dots, n$, and $\boldsymbol{\beta} = \beta$. Verify that $\alpha^{(1)} = \Sigma^{-1}\beta$. Relate this result to the discriminant function (Chapter 6).

2. (Sec. 12.2). Prove that the roots of (14) are real.

3. (Sec. 12.2) (*a*) Let $X' = (X^{(1)\prime} X^{(2)\prime})$, $\mathscr{E}X = \mathbf{0}$,

$$\mathscr{E}XX' = \begin{pmatrix} \Sigma_{11} & \Sigma_{12} \\ \Sigma_{21} & \Sigma_{22} \end{pmatrix},$$

$U = \alpha' X^{(1)}$, $V = \gamma' X^{(2)}$, $\mathscr{E}U^2 = 1 = \mathscr{E}V^2$, where α and γ are vectors. Show that choosing α and γ to maximize $\mathscr{E}UV$ is equivalent to choosing α and γ to minimize the generalized variance of $(U\ V)$.

(*b*) Let $X' = (X^{(1)\prime} X^{(2)\prime} X^{(3)\prime})$, $\mathscr{E}X = \mathbf{0}$,

$$\mathscr{E}XX' = \Sigma = \begin{pmatrix} \Sigma_{11} & \Sigma_{12} & \Sigma_{13} \\ \Sigma_{21} & \Sigma_{22} & \Sigma_{23} \\ \Sigma_{31} & \Sigma_{32} & \Sigma_{33} \end{pmatrix},$$

$U = \alpha' X^{(1)}$, $V = \gamma' X^{(2)}$, $W = \beta' X^{(3)}$, $\mathscr{E}U^2 = \mathscr{E}V^2 = \mathscr{E}W^2 = 1$. Consider finding α, γ, β to minimize the generalized variance of (U, V, W). Show that this minimum is invariant with respect to transformations $X^{*(i)} = A_i X^{(i)}$, $|A_i| \neq 0$.

(*c*) By using such transformations, transform Σ into the simplest possible form.

(*d*) In the case of $X^{(i)}$ consisting of two components, reduce the problem (of minimizing the generalized variance) to its simplest form.

(*e*) In this case give the derivative equations.

(*f*) Show that the minimum generalized variance is 1 if and only if $\Sigma_{12} = \mathbf{0}$, $\Sigma_{13} = \mathbf{0}$, $\Sigma_{23} = \mathbf{0}$. (*Note*: This extension of the notion of canonical variates does not yield to a "nice" explicit treatment.)

4. (Sec. 12.2) Let

$$X^{(1)} = AZ + Y^{(1)},$$

$$X^{(2)} = BZ + Y^{(2)},$$

where $Y^{(1)}, Y^{(2)}, Z$ are independent with mean zero and covariance matrices I with appropriate dimensionalities. Let $A = (a_1, \dots, a_k)$, $B = (b_1, \dots, b_k)$, and suppose that $A'A, B'B$ are diagonal with positive diagonal elements.

Show that the canonical variables for nonzero canonical correlations are proportional to $a_i'X^{(1)}, b_i'X^{(2)}$. Obtain the canonical correlation coefficients and appropriate normalizing coefficients for the canonical variables.

5. (Sec. 12.2) Let $\lambda_1 \geq \lambda_2 \geq \cdots \geq \lambda_q > 0$ be the positive roots of (14), where Σ_{11} and Σ_{22} are $q \times q$ nonsingular matrices.

(*a*) What is the rank of Σ_{12}?

(*b*) Write $\prod_{i=1}^q \lambda_i^2$ as the determinant of a rational function of Σ_{11}, Σ_{12}, Σ_{21}, and Σ_{22}. Justify your answer.

(*c*) If $\lambda_q = 1$, what is the rank of $\begin{pmatrix} \Sigma_{11} & \Sigma_{12} \\ \Sigma_{21} & \Sigma_{22} \end{pmatrix}$?

6. (Sec. 12.2) Let $\Sigma_{11} = (1-g)I_{p_1} + g\varepsilon_{p_1}\varepsilon_{p_1}'$, $\Sigma_{22} = (1-h)I_{p_2} + h\varepsilon_{p_2}\varepsilon_{p_2}'$, $\Sigma_{12} = k\varepsilon_{p_1}\varepsilon_{p_2}'$, where $-1/(p_1 - 1) < g < 1$, $-1/(p_2 - 1) < h < 1$, and k is suitably restricted. Find the canonical correlations and variates. What is the appropriate restriction on k?

7. (Sec. 12.3) Find the canonical correlations and canonical variates between the first two variables and the last three in Problem 42 of Chapter 4.

8. (Sec. 12.3) Prove directly the sample analogue of Theorem 12.2.1.

9. (Sec. 12.3) Prove that $\lambda_1^2(i+1) \to \lambda_1^2$ and $\alpha(i+1) \to \alpha^{(1)}$ if $\alpha(0)$ is such that $\alpha'(0)\Sigma_{11}\alpha^{(1)} \neq 0$. [*Hint*: Use $\Sigma_{11}^{-1}\Sigma_{12}\Sigma_{22}^{-1}\Sigma_{21} = \mathbf{A}\Lambda^2\mathbf{A}^{-1}$.]

10. (Sec. 12.6) Prove (9), (10), and (11).

11. Let $\lambda_1 \geq \lambda_2 \geq \cdots \geq \lambda_q$ be the roots of $|\Sigma_1 - \lambda\Sigma_2| = 0$, where Σ_1 and Σ_2 are $q \times q$ positive definite covariance matrices.

(*a*) What does $\lambda_1 = \lambda_q = 1$ imply about the relationship of Σ_1 and Σ_2?

(*b*) What does $\lambda_q > 1$ imply about the relationships of the ellipsoids $x'\Sigma_1^{-1}x = c$ and $x'\Sigma_2^{-1}x = c$?

(*c*) What does $\lambda_1 > 1$ and $\lambda_q < 1$ imply about the relationships of the ellipsoids $x'\Sigma_1^{-1}x = c$ and $x'\Sigma_2^{-1}x = c$?

12. (Sec. 12.4) For $q = 2$ express the criterion (2) of Section 9.5 in terms of canonical correlations.

13. Find the canonical correlations for the data in Problem 11 of Chapter 9.

CHAPTER 13

The Distributions of Characteristic Roots and Vectors

13.1. INTRODUCTION

In this chapter we find the distribution of the sample principal component vectors and their sample variances when all population variances are 1 (Section 13.3). We also find the distribution of the sample canonical correlations and one set of canonical vectors when the two sets of original variates are independent. This second distribution will be shown to be equivalent to the distribution of roots and vectors obtained in the next section. The distribution of the roots is particularly of interest because many invariant tests are functions of these roots. For example, invariant tests of the general linear hypothesis (Section 8.6) depend on the sample only through the roots of the determinantal equation

$$(1) \qquad \left|(\hat{\mathbf{B}}_{1\Omega} - \mathbf{B}_1^*)A_{11\cdot 2}(\hat{\mathbf{B}}_{1\Omega} - \mathbf{B}_1^*)' - lN\hat{\Sigma}_\Omega\right| = 0.$$

If the hypothesis is true, the roots have the distribution given in Theorem 13.2.2 or 13.2.3. Thus the significance level of any invariant test of the general linear hypothesis can be obtained from the distribution derived in the next section. If the test criterion is one of the ordered roots (e.g., the largest root), then the desired distribution is a marginal distribution of the joint distribution of roots.

The limiting distributions of the roots are obtained under fairly general conditions. These are needed to obtain other limiting distributions, such as the distribution of the criterion for testing that the smallest variance of principal components are equal.

13.2. THE CASE OF TWO WISHART MATRICES

13.2.1. The Transformation

Let us consider $\boldsymbol{A}^*$ and $\boldsymbol{B}^*$ $(p \times p)$ distributed independently according to $W(\boldsymbol{\Sigma}, m)$ and $W(\boldsymbol{\Sigma}, n)$ respectively $(m, n \geq p)$. We shall call the roots of

$$|\boldsymbol{A}^* - l\boldsymbol{B}^*| = 0 \tag{1}$$

the *characteristic roots of* $\boldsymbol{A}^*$ *in the metric of* $\boldsymbol{B}^*$ and the vectors satisfying

$$(\boldsymbol{A}^* - l\boldsymbol{B}^*)\boldsymbol{x}^* = \boldsymbol{0} \tag{2}$$

the *characteristic vectors of* $\boldsymbol{A}^*$ *in the metric of* $\boldsymbol{B}^*$. In this section we shall consider the distribution of these roots and vectors. Later it will be shown that the squares of canonical correlation coefficients have this distribution if the population canonical correlations are all zero.

First we shall transform $\boldsymbol{A}^*$ and $\boldsymbol{B}^*$ so that the distributions do not involve an arbitrary matrix $\boldsymbol{\Sigma}$. Let $\boldsymbol{C}$ be a matrix such that $\boldsymbol{C\Sigma C}' = \boldsymbol{I}$. Let

$$\begin{aligned} \boldsymbol{A} &= \boldsymbol{CA}^*\boldsymbol{C}', \\ \boldsymbol{B} &= \boldsymbol{CB}^*\boldsymbol{C}'. \end{aligned} \tag{3}$$

Then $\boldsymbol{A}$ and $\boldsymbol{B}$ are independently distributed according to $W(\boldsymbol{I}, m)$ and $W(\boldsymbol{I}, n)$ respectively (Section 7.3.3). Since

$$\begin{aligned} |\boldsymbol{A} - l\boldsymbol{B}| &= |\boldsymbol{CA}^*\boldsymbol{C}' - l\boldsymbol{CB}^*\boldsymbol{C}'| \\ &= |\boldsymbol{C}(\boldsymbol{A}^* - l\boldsymbol{B}^*)\boldsymbol{C}'| = |\boldsymbol{C}| \cdot |\boldsymbol{A}^* - l\boldsymbol{B}^*| \cdot |\boldsymbol{C}'|, \end{aligned}$$

the roots of (1) are the roots of

$$|\boldsymbol{A} - l\boldsymbol{B}| = 0. \tag{4}$$

The corresponding vectors satisfying

$$(\boldsymbol{A} - l\boldsymbol{B})\boldsymbol{x} = \boldsymbol{0} \tag{5}$$

satisfy

$$\begin{aligned} \mathbf{0} &= \boldsymbol{C}^{-1}(\boldsymbol{A} - l\boldsymbol{B})\boldsymbol{x} \\ &= \boldsymbol{C}^{-1}(\boldsymbol{C}\boldsymbol{A}^*\boldsymbol{C}' - l\,\boldsymbol{C}\boldsymbol{B}^*\boldsymbol{C}')\boldsymbol{x} \\ &= (\boldsymbol{A}^* - l\boldsymbol{B}^*)\boldsymbol{C}'\boldsymbol{x}. \end{aligned} \tag{6}$$

Thus the vectors $\boldsymbol{x}^*$ are the vectors $\boldsymbol{C}'\boldsymbol{x}$.

It will be convenient to consider the roots of

$$|\boldsymbol{A} - f(\boldsymbol{A} + \boldsymbol{B})| = 0 \tag{7}$$

and the vectors $\boldsymbol{y}$ satisfying

$$[\boldsymbol{A} - f(\boldsymbol{A} + \boldsymbol{B})]\,\boldsymbol{y} = \mathbf{0}. \tag{8}$$

The latter equation can be written

$$\begin{aligned} \mathbf{0} &= (\boldsymbol{A} - f\boldsymbol{A} - f\boldsymbol{B})\boldsymbol{y} \\ &= [(1-f)\boldsymbol{A} - f\boldsymbol{B}]\,\boldsymbol{y}. \end{aligned} \tag{9}$$

Since the probability that $f = 1$ is 0 (i.e., that $|-\boldsymbol{B}| = 0$), the above equation is

$$\left(\boldsymbol{A} - \frac{f}{1-f}\boldsymbol{B}\right)\boldsymbol{y} = \mathbf{0}. \tag{10}$$

Thus the roots of (4) are related to the roots of (7) by $l = f/(1-f)$ or $f = l/(1+l)$ and the vectors satisfying (5) are equal (or proportional) to those satisfying (8).

We now consider finding the distribution of the roots and vectors satisfying (7) and (8). Let the roots be ordered $f_1 > f_2 > \cdots > f_p > 0$ since the probability of two roots being equal is 0 [Okamoto (1973)]. Let

$$\boldsymbol{F} = \begin{pmatrix} f_1 & 0 & \cdots & 0 \\ 0 & f_2 & \cdots & 0 \\ \vdots & \vdots & & \vdots \\ 0 & 0 & \cdots & f_p \end{pmatrix}. \tag{11}$$

Suppose the corresponding vector solutions of (8) normalized by

$$y'(A+B)y = 1 \tag{12}$$

are $y_1, \ldots, y_p$. These vectors must satisfy

$$y_i'(A+B)y_j = 0, \qquad i \neq j, \tag{13}$$

because $y_i'Ay_j = f_j y_i'(A+B)y_j$ and $y_i'Ay_j = f_i y_i'(A+B)y_j$ and this can be only if (13) holds ($f_i \neq f_j$).

Let the $p \times p$ matrix Y be

$$Y = (y_1 \cdots y_p). \tag{14}$$

Equation (8) can be summarized as

$$AY = (A+B)YF, \tag{15}$$

and (12) and (13) give

$$Y'(A+B)Y = I. \tag{16}$$

From (15) we have

$$Y'AY = Y'(A+B)YF = F. \tag{17}$$

Multiplication of (16) and (17) on the left by $(Y')^{-1}$ and on the right by Y^{-1} gives

$$\begin{aligned} A + B &= (Y')^{-1}Y^{-1}, \\ A &= (Y')^{-1}FY^{-1}. \end{aligned} \tag{18}$$

Now let $Y^{-1} = E$. Then

$$\begin{aligned} A + B &= E'E, \\ A &= E'FE, \\ B &= E'(I - F)E. \end{aligned} \tag{19}$$

We now consider the joint distribution of E and F. From (19) we see that E and F define A and B uniquely. From (7) and (11) and the ordering $f_1 > \cdots > f_p$ we see that A and B define F uniquely. Equations (8) for $f = f_i$ and (12) define y_i uniquely except for multiplication by -1 (i.e., replacing y_i by $-y_i$). Since $YE = I$, this means that E is defined uniquely except that rows of $\boldsymbol{E}$ can be multiplied by -1. To remove this indeterminacy we require that $e_{i1} \geq 0$ (the probability that $e_{i1} = 0$ is 0). Thus $\boldsymbol{E}$ and $\boldsymbol{F}$ are uniquely defined in terms of $\boldsymbol{A}$ and $\boldsymbol{B}$.

13.2.2. The Jacobian

To find the density of $\boldsymbol{E}$ and $\boldsymbol{F}$ we substitute in the density of $\boldsymbol{A}$ and $\boldsymbol{B}$ according to (19) and multiply by the Jacobian of the transformation. We devote this subsection to finding the Jacobian

$$\left| \frac{\partial(\boldsymbol{A}, \boldsymbol{B})}{\partial(\boldsymbol{E}, \boldsymbol{F})} \right|. \tag{20}$$

Since the transformation from $\boldsymbol{A}$ and $\boldsymbol{B}$ to $\boldsymbol{A}$ and $\boldsymbol{G} = \boldsymbol{A} + \boldsymbol{B}$ has the Jacobian unity, we shall find

$$\left| \frac{\partial(\boldsymbol{A}, \boldsymbol{G})}{\partial(\boldsymbol{E}, \boldsymbol{F})} \right| = \left| \frac{\partial(\boldsymbol{A}, \boldsymbol{B})}{\partial(\boldsymbol{E}, \boldsymbol{F})} \right|. \tag{21}$$

First we notice that if $x_\alpha = f_\alpha(y_1, \ldots, y_n)$, $\alpha = 1, \ldots, n$, is a one-to-one transformation, the Jacobian is the determinant of the linear transformation

$$dx_\alpha = \sum_\beta \frac{\partial f_\alpha}{\partial y_\beta}\, dy_\beta, \tag{22}$$

where dx_α and dy_β are only formally differentials (i.e., we write these as a mnemonic device). If $f_\alpha(y_1, \ldots, y_n)$ is a polynomial, then $\partial f_\alpha / \partial y_\beta$ is the coefficient of y_β^* in the expansion of $f_\alpha(y_1 + y_1^*, \ldots, y_n + y_n^*)$ [in fact the coefficient in the expansion of $f_\alpha(y_1, \ldots, y_{\beta-1}, y_\beta + y_\beta^*, y_{\beta+1}, \ldots, y_n)$]. The elements of $\boldsymbol{A}$ and $\boldsymbol{G}$ are polynomials in $\boldsymbol{E}$ and $\boldsymbol{F}$. Thus the derivative of an element of $\boldsymbol{A}$ is the coefficient of an element of $\boldsymbol{E}^*$ or $\boldsymbol{F}^*$ in the expansion of $(\boldsymbol{E} + \boldsymbol{E}^*)'(\boldsymbol{F} + \boldsymbol{F}^*)(\boldsymbol{E} + \boldsymbol{E}^*)$ and the derivative of an element of $\boldsymbol{G}$ is the coefficient of an element of $\boldsymbol{E}^*$ or $\boldsymbol{F}^*$ in the expansion of $(\boldsymbol{E} + \boldsymbol{E}^*)'(\boldsymbol{E} + \boldsymbol{E}^*)$. Thus the Jacobian of the transformation from $\boldsymbol{A}$, $\boldsymbol{G}$ to $\boldsymbol{E}$, $\boldsymbol{F}$ is the determinant of the linear transformation

$$d\boldsymbol{A} = (d\boldsymbol{E})'\boldsymbol{F}\boldsymbol{E} + \boldsymbol{E}'(d\boldsymbol{F})\boldsymbol{E} + \boldsymbol{E}'\boldsymbol{F}(d\boldsymbol{E}), \tag{23}$$

$$d\boldsymbol{G} = (d\boldsymbol{E})'\boldsymbol{E} + \boldsymbol{E}'(d\boldsymbol{E}). \tag{24}$$

Since A and G (dA and dG) are symmetric, only the functionally independent component equations above are used.

Multiply (23) and (24) on the left by E'^{-1} and on the right by E^{-1} to obtain

$$E'^{-1}(dA)E^{-1} = E'^{-1}(dE)'F + dF + F(dE)E^{-1}, \tag{25}$$

$$E'^{-1}(dG)E^{-1} = E'^{-1}(dE)' + (dE)E^{-1}. \tag{26}$$

It should be kept in mind that (23) and (24) are now considered as a linear transformation without regard to how the equations were obtained.

Let

$$E'^{-1}(dA)E^{-1} = d\bar{A}, \tag{27}$$

$$E'^{-1}(dG)E^{-1} = d\bar{G}, \tag{28}$$

$$(dE)E^{-1} = dW. \tag{29}$$

Then

$$d\bar{A} = (dW)'F + dF + F(dW), \tag{30}$$

$$d\bar{G} = dW' + dW. \tag{31}$$

The linear transformation from dE, dF to dA, dG is considered as the linear transformation from dE, dF to dW, dF with determinant $|E^{-1}|^p = |E|^{-p}$ (because each row of dE is transformed by E^{-1}), the linear transformation from dW, dF to $d\bar{A}$, $d\bar{G}$, and the linear transformation from $d\bar{A}$, $d\bar{G}$ to $dA = E'(d\bar{A})E$, $dG = E'(d\bar{G})E$ with determinant $|E|^{p+1} \cdot |E|^{p+1}$ (from Section 7.3.3); and the determinant of the linear transformation from dE, dF to dA, dG is the product of the determinants of the three component transformations. The transformation (30), (31) is written in components as

$$\begin{aligned} d\bar{a}_{ii} &= df_i + 2f_i\,dw_{ii}, & \\ d\bar{a}_{ij} &= f_j\,dw_{ji} + f_i\,dw_{ij}, & i < j, \\ d\bar{g}_{ii} &= 2\,dw_{ii}, & \\ d\bar{g}_{ij} &= dw_{ji} + dw_{ij}, & i < j. \end{aligned} \tag{32}$$

The determinant is

$$
(33)\qquad
\begin{array}{r|cccc|}
 & df_i & dw_{ii} & dw_{ij}\ (i<j) & dw_{ij}\ (i>j) \\
d\bar{a}_{ii} & \boldsymbol{I} & 2\boldsymbol{F} & \boldsymbol{0} & \boldsymbol{0} \\
d\bar{g}_{ii} & \boldsymbol{0} & 2\boldsymbol{I} & \boldsymbol{0} & \boldsymbol{0} \\
d\bar{a}_{ij}\ (i<j) & \boldsymbol{0} & \boldsymbol{0} & \boldsymbol{M} & \boldsymbol{N} \\
d\bar{g}_{ij}\ (i<j) & \boldsymbol{0} & \boldsymbol{0} & \boldsymbol{I} & \boldsymbol{I}
\end{array}
$$

$$
= \begin{vmatrix} \boldsymbol{I} & 2\boldsymbol{F} \\ \boldsymbol{0} & 2\boldsymbol{I} \end{vmatrix} \cdot \begin{vmatrix} \boldsymbol{M} & \boldsymbol{N} \\ \boldsymbol{I} & \boldsymbol{I} \end{vmatrix} = 2^p |\boldsymbol{M} - \boldsymbol{N}|,
$$

where

(34)

$$
\boldsymbol{M} = \begin{array}{r}
 \\ da_{12} \\ \vdots \\ da_{1p} \\ \hline da_{23} \\ \vdots \\ da_{2p} \\ \hline \vdots \\ \hline da_{p-1,p}
\end{array}
\left[\begin{array}{ccc|ccc|c|c}
dw_{12} & \cdots & dw_{1p} & dw_{23} & \cdots & dw_{2p} & \cdots & dw_{p-1,p} \\
f_1 & \cdots & 0 & 0 & \cdots & 0 & & 0 \\
\vdots & & \vdots & \vdots & & \vdots & & \vdots \\
0 & \cdots & f_1 & 0 & \cdots & 0 & & 0 \\ \hline
0 & \cdots & 0 & f_2 & \cdots & 0 & & 0 \\
\vdots & & \vdots & \vdots & & \vdots & & \vdots \\
0 & \cdots & 0 & 0 & \cdots & f_2 & & 0 \\ \hline
 & & & & & & & \\ \hline
0 & \cdots & 0 & 0 & \cdots & 0 & & f_{p-1}
\end{array}\right]
$$

and

(35)

$$
\boldsymbol{N} = \begin{array}{r}
 \\ da_{12} \\ \vdots \\ da_{1p} \\ \hline da_{23} \\ \vdots \\ da_{2p} \\ \hline \vdots \\ \hline da_{p-1,p}
\end{array}
\left[\begin{array}{ccc|ccc|c|c}
dw_{21} & \cdots & dw_{p1} & dw_{32} & \cdots & dw_{p2} & \cdots & dw_{p,p-1} \\
f_2 & \cdots & 0 & 0 & \cdots & 0 & & 0 \\
\vdots & & \vdots & \vdots & & \vdots & & \vdots \\
0 & \cdots & f_p & 0 & \cdots & 0 & & 0 \\ \hline
0 & \cdots & 0 & f_3 & \cdots & 0 & & 0 \\
\vdots & & \vdots & \vdots & & \vdots & & \vdots \\
0 & \cdots & 0 & 0 & \cdots & f_p & & 0 \\ \hline
 & & & & & & & \\ \hline
0 & \cdots & 0 & 0 & \cdots & 0 & & f_p
\end{array}\right]
$$

Then

$$|\boldsymbol{M} - \boldsymbol{N}| = \prod_{i<j}(f_i - f_j). \tag{36}$$

The determinant of the linear transformation (23), (24) is

$$|\boldsymbol{E}|^{-p}|\boldsymbol{E}|^{p+1}|\boldsymbol{E}|^{p+1}2^p\prod_{i<j}(f_i - f_j) = 2^p|\boldsymbol{E}|^{p+2}\prod_{i<j}(f_i - f_j). \tag{37}$$

THEOREM 13.2.1. *The Jacobian of the transformation* (19) *is the absolute value of* (37).

13.2.3. The Joint Distribution of the Matrix E and the Roots

The joint density of $\boldsymbol{A}$ and $\boldsymbol{B}$ is

$$w(\boldsymbol{A}|\boldsymbol{I}, m)w(\boldsymbol{B}|\boldsymbol{I}, n) = C_1|\boldsymbol{A}|^{\frac{1}{2}(m-p-1)}|\boldsymbol{B}|^{\frac{1}{2}(n-p-1)}e^{-\frac{1}{2}\operatorname{tr}(\boldsymbol{A}+\boldsymbol{B})}, \tag{38}$$

where

$$C_1 = \left[2^{\frac{1}{2}p(n+m)}\Gamma_p\left(\tfrac{1}{2}n\right)\Gamma_p\left(\tfrac{1}{2}m\right)\right]^{-1}. \tag{39}$$

Therefore the joint density of $\boldsymbol{E}$ and $\boldsymbol{F}$ is

$$C_1|\boldsymbol{E}'\boldsymbol{F}\boldsymbol{E}|^{\frac{1}{2}(m-p-1)}|\boldsymbol{E}'(\boldsymbol{I}-\boldsymbol{F})\boldsymbol{E}|^{\frac{1}{2}(n-p-1)} \cdot e^{-\frac{1}{2}\operatorname{tr}\boldsymbol{E}'\boldsymbol{E}}2^p|\boldsymbol{E}'\boldsymbol{E}|^{\frac{1}{2}(p+2)}\prod_{i<j}(f_i - f_j). \tag{40}$$

Since $|\boldsymbol{E}'\boldsymbol{F}\boldsymbol{E}| = |\boldsymbol{E}'|\cdot|\boldsymbol{F}|\cdot|\boldsymbol{E}| = |\boldsymbol{F}|\cdot|\boldsymbol{E}'\boldsymbol{E}| = \prod_{i=1}^p f_i|\boldsymbol{E}'\boldsymbol{E}|$ and $|\boldsymbol{E}'(\boldsymbol{I}-\boldsymbol{F})\boldsymbol{E}| = |\boldsymbol{I}-\boldsymbol{F}|\cdot|\boldsymbol{E}'\boldsymbol{E}| = \prod_{i=1}^p(1-f_i)|\boldsymbol{E}'\boldsymbol{E}|$, the density of $\boldsymbol{E}$ and $\boldsymbol{F}$ is

$$2^pC_1|\boldsymbol{E}'\boldsymbol{E}|^{\frac{1}{2}(m+n-p)}e^{-\frac{1}{2}\operatorname{tr}\boldsymbol{E}'\boldsymbol{E}}\prod_{i=1}^p f_i^{\frac{1}{2}(m-p-1)}\prod_{i=1}^p(1-f_i)^{\frac{1}{2}(n-p-1)}\prod_{i<j}(f_i - f_j). \tag{41}$$

Clearly, $\boldsymbol{E}$ and $\boldsymbol{F}$ are statistically independent because the density factors into a function of $\boldsymbol{E}$ and a function of $\boldsymbol{F}$. To determine the marginal densities we have only to find the two normalizing constants (the product of which is 2^pC_1).

Let us evaluate

$$2^p \int |\boldsymbol{E}'\boldsymbol{E}|^{\frac{1}{2}(m+n-p)} e^{-\frac{1}{2}\operatorname{tr}\boldsymbol{E}'\boldsymbol{E}}\, d\boldsymbol{E}, \tag{42}$$

where the integration is $0 < e_{i1} < \infty$, $-\infty < e_{ij} < \infty$, $j \neq 1$. The value of (42) is unchanged if we let $-\infty < e_{i1} < \infty$ and multiply by 2^{-p}. Thus (42) is

$$(2\pi)^{\frac{1}{2}p^2} \int_{-\infty}^{\infty} \cdots \int_{-\infty}^{\infty} |\boldsymbol{E}'\boldsymbol{E}|^{\frac{1}{2}(m+n-p)} \left[\frac{1}{(2\pi)^{\frac{1}{2}p^2}} \exp\left(-\tfrac{1}{2} \sum_{i,j} e_{ij}^2 \right) \right] \prod de_{ij}. \tag{43}$$

Except for the constant $(2\pi)^{\frac{1}{2}p^2}$, (43) is a definition of the expectation of the $\frac{1}{2}(m+n-p)$ power of $|\boldsymbol{E}'\boldsymbol{E}|$ when the e_{ij} have as density the function within brackets. This expected value is the $\frac{1}{2}(m+n-p)$-th moment of the generalized variance $|\boldsymbol{E}'\boldsymbol{E}|$ when $\boldsymbol{E}'\boldsymbol{E}$ has the distribution $W(\boldsymbol{I}, p)$. (See Section 7.5.) Thus (43) is

$$(2\pi)^{\frac{1}{2}p^2} \frac{\Gamma_p\left[\frac{1}{2}(m+n)\right]}{\Gamma_p\left(\frac{1}{2}p\right)} 2^{\frac{1}{2}p(n+m-p)}. \tag{44}$$

Thus the density of $\boldsymbol{E}$ is

$$\frac{\Gamma_p\left(\frac{1}{2}p\right)}{2^{\frac{1}{2}p(m+n-2)} \pi^{\frac{1}{2}p^2} \Gamma_p\left[\frac{1}{2}(m+n)\right]} |\boldsymbol{E}'\boldsymbol{E}|^{\frac{1}{2}(m+n-p)} e^{-\frac{1}{2}\operatorname{tr}\boldsymbol{E}'\boldsymbol{E}}. \tag{45}$$

The density of f_i is (41) divided by (45); that is, the density of f_i is

$$C_2 \prod_{i=1}^{p} f_i^{\frac{1}{2}(m-p-1)} \prod_{i=1}^{p} (1-f_i)^{\frac{1}{2}(n-p-1)} \prod_{i<j} (f_i - f_j), \tag{46}$$

for $0 \le f_p \le \cdots \le f_1 \le 1$, where

$$C_2 = \frac{\pi^{\frac{1}{2}p^2} \Gamma_p\left[\frac{1}{2}(m+n)\right]}{\Gamma_p\left(\frac{1}{2}n\right) \Gamma_p\left(\frac{1}{2}m\right) \Gamma_p\left(\frac{1}{2}p\right)}. \tag{47}$$

The density of l_i is obtained from (46) by letting

$$f_i = \frac{l_i}{l_i + 1}; \tag{48}$$

we have

$$\frac{df_i}{dl_i} = \frac{1}{(l_i + 1)^2},$$

(49)
$$f_i - f_j = \frac{l_i - l_j}{(l_i + 1)(l_j + 1)},$$

$$1 - f_i = \frac{1}{l_i + 1}.$$

Thus the density of l_i is

(50)
$$C_2 \prod_{i=1}^{p} l_i^{\frac{1}{2}(m-p-1)} \prod_{i=1}^{p} (l_i + 1)^{-\frac{1}{2}(m+n)} \prod_{i<j} (l_i - l_j)$$

for $0 \le l_p \le \cdots \le l_1$.

THEOREM 13.2.2. *If $\boldsymbol{A}$ and $\boldsymbol{B}$ are distributed independently according to $W(\boldsymbol{\Sigma}, m)$ and $W(\boldsymbol{\Sigma}, n)$ respectively ($m \ge p, n \ge p$), the joint density of the roots of $|\boldsymbol{A} - l\boldsymbol{B}| = 0$ is* (50) *where C_2 is defined by* (47).

The joint density of $\boldsymbol{Y}$ can be found from (45) and the fact that the Jacobian is $|\boldsymbol{Y}|^{-2p}$. (See Theorem A.4.6 of Appendix A.)

13.2.4. The Distribution for A Singular

The matrix $\boldsymbol{A}$ above can be represented as $\boldsymbol{A} = \boldsymbol{W}_1\boldsymbol{W}_1'$, where the columns of $\boldsymbol{W}_1$ ($p \times m$) are independently distributed, each according to $N(\boldsymbol{0}, \boldsymbol{\Sigma})$. We now treat the case of $m < p$. If we let $\boldsymbol{B} + \boldsymbol{W}_1\boldsymbol{W}_1' = \boldsymbol{G} = \boldsymbol{CC}'$ and $\boldsymbol{W}_1 = \boldsymbol{CU}$, then the roots of

(51)
$$\begin{aligned} 0 &= |\boldsymbol{A} - f(\boldsymbol{A} + \boldsymbol{B})| = |\boldsymbol{W}_1\boldsymbol{W}_1' - f\boldsymbol{G}| \\ &= |\boldsymbol{CUU}'\boldsymbol{C}' - f\boldsymbol{CC}'| = |\boldsymbol{C}| \cdot |\boldsymbol{UU}' - f\boldsymbol{I}_p| \cdot |\boldsymbol{C}| \end{aligned}$$

are the roots of

(52)
$$|\boldsymbol{UU}' - f\boldsymbol{I}_p| = 0.$$

We shall show that the nonzero roots $f_1 > \cdots > f_m$ (these roots being distinct with probability 1) are the roots of

$$|\boldsymbol{U}'\boldsymbol{U} - f\boldsymbol{I}_m| = 0. \tag{53}$$

For each root $f \neq 0$ of (52) there is a vector $\boldsymbol{x}$ satisfying

$$\left(\boldsymbol{U}\boldsymbol{U}' - f\boldsymbol{I}_p\right)\boldsymbol{x} = \boldsymbol{0}. \tag{54}$$

Multiplication by $\boldsymbol{U}'$ on the left gives

$$\begin{aligned} \boldsymbol{0} &= \boldsymbol{U}'\left(\boldsymbol{U}\boldsymbol{U}' - f\boldsymbol{I}_p\right)\boldsymbol{x} \\ &= \left(\boldsymbol{U}'\boldsymbol{U} - f\boldsymbol{I}_m\right)(\boldsymbol{U}'\boldsymbol{x}). \end{aligned} \tag{55}$$

Thus $\boldsymbol{U}'\boldsymbol{x}$ is a characteristic vector of $\boldsymbol{U}\boldsymbol{U}'$ and f is the corresponding root.

It was shown in Section 8.4 that the density of $\boldsymbol{U} = \boldsymbol{U}_*'$ is (for $\boldsymbol{I}_p - \boldsymbol{U}\boldsymbol{U}'$ positive definite or $\boldsymbol{I}_m - \boldsymbol{U}_*\boldsymbol{U}_*'$ positive definite)

$$K|\boldsymbol{I}_p - \boldsymbol{U}\boldsymbol{U}'|^{\frac{1}{2}(n-p-1)} = K|\boldsymbol{I}_{p^*} - \boldsymbol{U}_*\boldsymbol{U}_*'|^{\frac{1}{2}(n^*-p^*-1)}, \tag{56}$$

where $p^* = m$, $n^* - p^* - 1 = n - p - 1$, and $m^* = p$. Thus $f_1, \dots, f_m$ must be distributed according to (46) with p replaced by m, m by p, and n by $n + m - p$, that is,

$$\frac{\pi^{\frac{1}{2}m^2}\Gamma_m\left[\frac{1}{2}(m+n)\right]}{\Gamma_m\left(\frac{1}{2}m\right)\Gamma_m\left[\frac{1}{2}(m+n-p)\right]\Gamma_m\left(\frac{1}{2}p\right)} \cdot \prod_{i=1}^{m}\left[f_i^{\frac{1}{2}(p-m-1)}(1-f_i)^{\frac{1}{2}(n-p-1)}\right]\prod_{i<j}(f_i - f_j). \tag{57}$$

THEOREM 13.2.3. *If $\boldsymbol{A}$ is distributed as $\boldsymbol{W}_1\boldsymbol{W}_1'$, where the m columns of $\boldsymbol{W}_1$ are independent, each distributed according to $N(\boldsymbol{0}, \boldsymbol{\Sigma})$, $m \leq p$, and $\boldsymbol{B}$ is independently distributed according to $W(\boldsymbol{\Sigma}, n)$, $n \geq p$, then the density of the nonzero roots of $|\boldsymbol{A} - f(\boldsymbol{A} + \boldsymbol{B})| = 0$ is given by* (57).

It is of interest to note that these distributions of roots were found independently and at about the same time by Fisher (1939), Girshick (1939), Hsu (1939a), Mood (1951), and Roy (1939). The development of the Jacobian in Section 13.2.2 is due mainly to Hsu [as reported by Deemer and Olkin (1951)].

13.3. THE CASE OF ONE NONSINGULAR WISHART MATRIX

In this section we shall find the distribution of the roots of

$$|\boldsymbol{A} - l\boldsymbol{I}| = 0, \tag{1}$$

where the matrix $\boldsymbol{A}$ has the distribution $W(\boldsymbol{I}, n)$. It will be observed that the variances of the principal components of a sample of $n + 1$ from $N(\boldsymbol{\mu}, \boldsymbol{I})$ are $1/n$ times the roots of (1). We shall find the following theorem useful:

THEOREM 13.3.1. *If the symmetric matrix $\boldsymbol{B}$ has a density of the form $g(l_1, \dots, l_p)$, where $l_1 > \cdots > l_p$ are the characteristic roots of $\boldsymbol{B}$, then the density of the roots is*

$$\frac{\pi^{\frac{1}{2}p^2} g(l_1, \dots, l_p) \prod_{i<j}(l_i - l_j)}{\Gamma_p(\frac{1}{2}p)}. \tag{2}$$

PROOF. From Theorem A.2.1 of Appendix A we know that there exists an orthogonal matrix $\boldsymbol{C}$ such that

$$\boldsymbol{B} = \boldsymbol{C}'\boldsymbol{L}\boldsymbol{C}, \tag{3}$$

where

$$\boldsymbol{L} = \begin{pmatrix} l_1 & 0 & \cdots & 0 \\ 0 & l_2 & \cdots & 0 \\ \vdots & \vdots & & \vdots \\ 0 & 0 & \cdots & l_p \end{pmatrix}. \tag{4}$$

If the l's are numbered in descending order of magnitude and if $c_{i1} \geq 0$, then (with probability 1) the transformation from $\boldsymbol{B}$ to $\boldsymbol{L}$ and $\boldsymbol{C}$ is unique. Let the matrix $\boldsymbol{C}$ be given the coordinates $c_1, \dots, c_{p(p-1)/2}$, and let the Jacobian of the transformation be $f(\boldsymbol{L}, \boldsymbol{C})$. Then the joint density of $\boldsymbol{L}$ and $\boldsymbol{C}$ is $g(l_1, \dots, l_p) f(\boldsymbol{L}, \boldsymbol{C})$. To prove the theorem we must show that

$$\int \cdots \int f(\boldsymbol{L}, \boldsymbol{C})\, dc_1 \dots dc_{p(p-1)/2} = \frac{\pi^{\frac{1}{2}p^2} \prod_{i<j}(l_i - l_j)}{\Gamma_p(\frac{1}{2}p)}. \tag{5}$$

We show this by taking a special case where $\boldsymbol{B} = \boldsymbol{U}\boldsymbol{U}'$ and $\boldsymbol{U}$ $(p \times m, m \geq p)$ has the density

$$\pi^{-\frac{1}{2}mp} \frac{\Gamma_p\left[\frac{1}{2}(m+n)\right]}{\Gamma_p\left(\frac{1}{2}n\right)} |\boldsymbol{I} - \boldsymbol{U}\boldsymbol{U}'|^{\frac{1}{2}(n-p-1)}. \tag{6}$$

Then by Lemma 13.3.1, which will be stated below, $\boldsymbol{B}$ has the density

$$\begin{aligned} \frac{\Gamma_p\left[\frac{1}{2}(m+n)\right]}{\Gamma_p\left(\frac{1}{2}m\right)\Gamma_p\left(\frac{1}{2}n\right)} |\boldsymbol{I} - \boldsymbol{B}|^{\frac{1}{2}(n-p-1)} |\boldsymbol{B}|^{\frac{1}{2}(m-p-1)} &= \frac{\Gamma_p\left[\frac{1}{2}(m+n)\right]}{\Gamma_p\left(\frac{1}{2}m\right)\Gamma_p\left(\frac{1}{2}n\right)} \prod_{i=1}^{p} (1 - l_i)^{\frac{1}{2}(n-p-1)} \prod_{i=1}^{p} l_i^{\frac{1}{2}(m-p-1)} \\ &= g^*(l_1, \ldots, l_p). \end{aligned} \tag{7}$$

The joint density of $\boldsymbol{L}$ and $\boldsymbol{C}$ is $f(\boldsymbol{L}, \boldsymbol{C}) g^*(l_1, \ldots, l_p)$. In the preceding section we proved that the marginal density of $\boldsymbol{L}$ is (50). Thus

$$\begin{aligned} \int \cdots \int g^*(l_1, \ldots, l_p) f(\boldsymbol{L}, \boldsymbol{C})\, d\boldsymbol{C} &= g^*(l_1, \ldots, l_p) \int \cdots \int f(\boldsymbol{L}, \boldsymbol{C})\, d\boldsymbol{C} \\ &= \frac{\pi^{\frac{1}{2}p^2} \prod (l_i - l_j)}{\Gamma_p\left(\frac{1}{2}p\right)} g^*(l_1, \ldots, l_p). \end{aligned} \tag{8}$$

This proves (5) and hence the theorem. Q.E.D.

The statement above (7) is based on the following lemma:

LEMMA 13.3.1. *If the density of* $\boldsymbol{Y}$ $(p \times m)$ *is* $f(\boldsymbol{Y}\boldsymbol{Y}')$, *then the density of* $\boldsymbol{B} = \boldsymbol{Y}\boldsymbol{Y}'$ *is*

$$\frac{|\boldsymbol{B}|^{\frac{1}{2}(m-p-1)} f(\boldsymbol{B}) \pi^{\frac{1}{2}pm}}{\Gamma_p\left(\frac{1}{2}m\right)}. \tag{9}$$

The proof of this, like that of Theorem 13.3.1, depends on exhibiting a special case; let $f(\boldsymbol{YY}') = (2\pi)^{-\frac{1}{2}pm} e^{-\frac{1}{2}\operatorname{tr} \boldsymbol{YY}'}$, then (9) is $w(\boldsymbol{B}|\boldsymbol{I}, m)$. Q.E.D.

Now let us find the density of the roots of (1). The density of $\boldsymbol{A}$ is

$$\frac{|\boldsymbol{A}|^{\frac{1}{2}(n-p-1)} e^{-\frac{1}{2}\operatorname{tr} \boldsymbol{A}}}{2^{\frac{1}{2}pn} \Gamma_p\left(\frac{1}{2}n\right)} = \frac{\prod_{i=1}^{p} l_i^{\frac{1}{2}(n-p-1)} \exp\left(-\frac{1}{2}\sum_{i=1}^{p} l_i\right)}{2^{\frac{1}{2}pn} \Gamma_p\left(\frac{1}{2}n\right)}. \tag{10}$$

Thus by the theorem we obtain as the density of the roots of $\boldsymbol{A}$

$$\frac{\pi^{\frac{1}{2}p^2} \prod_{i=1}^{p} l_i^{\frac{1}{2}(n-p-1)} \exp\left(-\frac{1}{2}\sum_{i=1}^{p} l_i\right) \prod_{i<j} (l_i - l_j)}{2^{\frac{1}{2}pn} \Gamma_p\left(\frac{1}{2}n\right) \Gamma_p\left(\frac{1}{2}p\right)}. \tag{11}$$

THEOREM 13.3.2. *If* $\boldsymbol{A}$ $(p \times p)$ *has the distribution* $\boldsymbol{W}(\boldsymbol{I}, n)$, *then the characteristic roots* $(l_1 \geq l_2 \geq \cdots \geq l_p \geq 0)$ *have the density* (11) *over the range where the density is not* 0.

COROLLARY 13.3.1. *Let* $v_1 \geq \cdots \geq v_p$ *be the sample variances of the sample principal components of a sample of size* $N = n + 1$ *from* $N(\boldsymbol{\mu}, \sigma^2 \boldsymbol{I})$. *Then* $(n/\sigma^2)v_i$ *are distributed with density* (11).

The characteristic vectors of $\boldsymbol{A}$ are uniquely defined (except for multiplication by -1) with probability 1 by

$$(\boldsymbol{A} - l\boldsymbol{I})\boldsymbol{y} = \boldsymbol{0}, \qquad \boldsymbol{y}'\boldsymbol{y} = 1, \tag{12}$$

since the roots are different with probability 1. Let the vectors with $y_{1i} \geq 0$ be

$$\boldsymbol{Y} = (\boldsymbol{y}_1, \ldots, \boldsymbol{y}_p). \tag{13}$$

Then

$$\boldsymbol{AY} = \boldsymbol{YL}. \tag{14}$$

From Section 11.2 we know that

$$\boldsymbol{Y}'\boldsymbol{Y} = \boldsymbol{I}. \tag{15}$$

Multiplication of (14) on the right by $\boldsymbol{Y}^{-1} = \boldsymbol{Y}'$ gives

$$A = YLY'. \tag{16}$$

Thus $\boldsymbol{Y}' = \boldsymbol{C}$, defined above.

Now let us consider the joint distribution of $\boldsymbol{L}$ and $\boldsymbol{C}$. The matrix $\boldsymbol{A}$ has the distribution of

$$\boldsymbol{A} = \sum_{\alpha=1}^{n} \boldsymbol{X}_\alpha \boldsymbol{X}_\alpha', \tag{17}$$

where the $\boldsymbol{X}_\alpha$ are independently distributed, each according to $N(\mathbf{0}, \boldsymbol{I})$. Let

$$\boldsymbol{X}_\alpha^* = \boldsymbol{Q}\boldsymbol{X}_\alpha, \tag{18}$$

where $\boldsymbol{Q}$ is any orthogonal matrix. Then the $\boldsymbol{X}_\alpha^*$ are independently distributed according to $N(\mathbf{0}, \boldsymbol{I})$ and

$$\boldsymbol{A}^* = \sum_{\alpha=1}^{n} \boldsymbol{X}_\alpha^* \boldsymbol{X}_\alpha^{*\prime} = \boldsymbol{Q}\boldsymbol{A}\boldsymbol{Q}' \tag{19}$$

is distributed according to $W(\boldsymbol{I}, n)$. The roots of $\boldsymbol{A}^*$ are the roots of $\boldsymbol{A}$; thus

$$\boldsymbol{A}^* = \boldsymbol{C}^{**\prime}\boldsymbol{L}\boldsymbol{C}^{**}, \tag{20}$$

$$\boldsymbol{C}^{**\prime}\boldsymbol{C}^{**} = \boldsymbol{I} \tag{21}$$

define $\boldsymbol{C}^{**}$ if we require $c_{i1}^{**} \geq 0$. Let

$$\boldsymbol{C}^* = \boldsymbol{C}\boldsymbol{Q}'. \tag{22}$$

Let

$$\boldsymbol{J}(\boldsymbol{C}^*) = \begin{pmatrix} \frac{c_{11}^*}{|c_{11}^*|} & 0 & \cdots & 0 \\ 0 & \frac{c_{21}^*}{|c_{21}^*|} & \cdots & 0 \\ \vdots & \vdots & & \vdots \\ 0 & 0 & \cdots & \frac{c_{p1}^*}{|c_{p1}^*|} \end{pmatrix}, \tag{23}$$

with $c_{i1}^*/|c_{i1}^*| = 1$ if $c_{i1}^* = 0$. Thus $J(C^*)$ is a diagonal matrix; the ith diagonal element is 1 if $c_{i1}^* \geq 0$ and is -1 if $c_{i1}^* < 0$. Thus

$$C^{**} = J(C^*)C^* = J(CQ')CQ'. \tag{24}$$

The distribution of C^{**} is the same as that of C. We now shall show that this fact defines the distribution of C.

DEFINITION 13.3.1. *If the random orthogonal matrix* E *of order* p *has a distribution such that* EQ' *has the same distribution for every orthogonal* Q, *the distribution of* E *is said to have the "Haar invariant" distribution* (*or normalized measure*).

The definition is possible because it has been proved that there is only one distribution with the required invariance property [Halmos (1950)]. It has also been shown that this distribution is the only one invariant under multiplication on the left by an orthogonal matrix (i.e., the distribution of QE is the same as that of E). From this it follows that the probability is $1/2^p$ that E be such that $e_{i1} \geq 0$. This can be seen as follows. Let $J_1, \dots, J_{2^p}$ be the 2^p diagonal matrices with elements $+1$ and -1. Since the distribution of J_iE is the same as that of E, the probability that $e_{i1} \geq 0$ is the same as the probability that the elements in the first column of J_iE are nonnegative. These events for $i = 1, \dots, 2^p$ are mutually exclusive and exhaustive (except for elements being 0 which have probability 0) and thus the probability of any one is $1/2^p$.

The conditional distribution of E given $e_{i1} \geq 0$ is 2^p times the Haar invariant distribution over this part of the space. We shall call it the conditional Haar invariant distribution.

LEMMA 13.3.2. *If the orthogonal matrix* E *has a distribution such that* $e_{i1} \geq 0$ *and if* $E^{**} = J(EQ')EQ'$ *has the same distribution for every orthogonal* Q, *then* E *has the conditional Haar invariant distribution.*

PROOF. Let the space V of orthogonal matrices be partitioned into the subspaces $V_1, \dots, V_{2^p}$ so that $J_iV_i = V_1$, say, where $J_1 = I$ and V_1 is the set for which $e_{i1} \geq 0$. Let μ_1 be the measure in V_1 defined by the distribution of E assumed in the lemma. The measure $\mu(W)$ of a (measurable) set W in V_i is defined as $(1/2^p)\mu_1(J_iW)$. Now we want to show that μ is the Haar invariant measure. Let W be any (measurable) set in V_1. The lemma assumes that

$2^p\mu(W) = \mu_1(W) = \Pr\{E \in W\} = \Pr\{E^{**} \in W\} = \Sigma\mu_1(J_i[WQ' \cap V_i]) = 2^p\mu(WQ')$. If U is any (measurable) set in V, then $U = \cup_{j=1}^{2^p}(U \cap V_j)$. Since $\mu(U \cap V_j) = (1/2^p)\mu_1[J_j(U \cap V_j)]$, by the above this is $\mu[(U \cap V_j)Q']$. Thus $\mu(U) = \mu(UQ')$. Thus μ is invariant and μ_1 is the conditional invariant distribution. Q.E.D.

From the lemma we see that the matrix C has the conditional Haar invariant distribution. Since the distribution of C conditional on L is the same, C and L are independent.

THEOREM 13.3.3. *If $C = Y'$, where $Y = (y_1, \ldots, y_p)$ are the normalized characteristic vectors of A with $y_{1i} \geq 0$ and where A is distributed according to $W(I, n)$, then C has the conditional Haar invariant distribution and C is distributed independently of the characteristic roots.*

From the preceding work we can generalize Theorem 13.3.1.

THEOREM 13.3.4. *If the symmetric matrix B has a density of the form $g(l_1, \ldots, l_p)$, where $l_1 > \cdots > l_p$ are the characteristic roots of B, then the joint density of the roots is* (2) *and the matrix of normalized characteristic vectors Y ($y_{1i} \geq 0$) is independently distributed according to the conditional Haar invariant distribution.*

PROOF. The density of QBQ', where $QQ' = I$, is the same as that of B (for the roots are invariant) and therefore the distribution of $J(Y'Q')Y'Q'$ is the same as that of Y'. Then Theorem 13.3.4 follows from Lemma 13.3.2. Q.E.D.

We shall give an application of this theorem to the case where $B = B'$ is normally distributed with the (functionally independent) components of B independent with means 0 and variances $\mathscr{E}b_{ii}^2 = 1$ and $\mathscr{E}b_{ij}^2 = \frac{1}{2}$ $(i < j)$.

THEOREM 13.3.5. *Let $B = B'$ have the density*

$$\pi^{-p(p+1)/4}2^{-\frac{1}{2}p}e^{-\frac{1}{2}\operatorname{tr} B^2}. \tag{25}$$

Then the characteristic roots $l_1 > \cdots > l_p$ of B have the density

$$2^{-\frac{1}{2}p}\pi^{p(p-1)/4}\Gamma_p^{-1}(\tfrac{1}{2}p)\exp\left(-\tfrac{1}{2}\sum_{i=1}^{p} l_i^2\right)\prod_{i<j}(l_i - l_j) \tag{26}$$

and the matrix $\mathbf{Y}$ of the normalized characteristic vectors ($y_{1i} \geq 0$) *is independently distributed according to the conditional Haar invariant distribution.*

PROOF. Since the characteristic roots of $\mathbf{B}^2$ are $l_1^2, \dots, l_p^2$ and $\operatorname{tr} \mathbf{B}^2 = \sum l_i^2$, the theorem follows directly. Q.E.D.

COROLLARY 13.3.2. *Let $n\mathbf{S}$ be distributed according to $W(\mathbf{I}, n)$, and define the diagonal matrix $\mathbf{L}$ and $\mathbf{B}$ by $\mathbf{S} = \mathbf{C}'\mathbf{L}\mathbf{C}$, $\mathbf{C}'\mathbf{C} = \mathbf{I}$, $l_1 > \cdots > l_p$, and $c_{i1} \geq 0$, $i = 1, \dots, p$. Then the density of the limiting distribution of $\sqrt{n}(\mathbf{L} - \mathbf{I}) = \mathbf{D}$ diagonal is* (26) *with l_i replaced by d_i, and the matrix $\mathbf{C}$ is independently distributed according to the conditional Haar measure.*

PROOF. The density of the limiting distribution of $\sqrt{n}(\mathbf{S} - \mathbf{I})$ is (25) and the diagonal elements of $\mathbf{D}$ are the characteristic roots of $\sqrt{n}(\mathbf{S} - \mathbf{I})$ and the columns of $\mathbf{C}'$ are the characteristic vectors. Q.E.D.

13.4. CANONICAL CORRELATIONS

The sample canonical correlations were shown in Section 12.3 to be the square roots of the roots of

(1) $$|\mathbf{A}_{12}\mathbf{A}_{22}^{-1}\mathbf{A}_{21} - f\mathbf{A}_{11}| = 0,$$

where

(2) $$\mathbf{A}_{ij} = \sum_{\alpha=1}^{N} \left(\mathbf{X}_\alpha^{(i)} - \overline{\mathbf{X}}^{(i)}\right)\left(\mathbf{X}_\alpha^{(j)} - \overline{\mathbf{X}}^{(j)}\right)', \qquad i, j = 1, 2,$$

and the distribution of

(3) $$\mathbf{X} = \begin{pmatrix} \mathbf{X}^{(1)} \\ \mathbf{X}^{(2)} \end{pmatrix}$$

is $N(\boldsymbol{\mu}, \boldsymbol{\Sigma})$, where

(4) $$\boldsymbol{\Sigma} = \begin{pmatrix} \boldsymbol{\Sigma}_{11} & \boldsymbol{\Sigma}_{12} \\ \boldsymbol{\Sigma}_{21} & \boldsymbol{\Sigma}_{22} \end{pmatrix}.$$

From Section 3.3 we know that the distribution of $\boldsymbol{A}_{ij}$ is the same as that of

(5) $$\boldsymbol{A}_{ij} = \sum_{\alpha=1}^{n} \boldsymbol{Y}_{\alpha}^{(i)} \boldsymbol{Y}_{\alpha}^{(j)\prime}, \qquad i,j = 1,2,$$

where $n = N - 1$ and

(6) $$\boldsymbol{Y} = \begin{pmatrix} \boldsymbol{Y}^{(1)} \\ \boldsymbol{Y}^{(2)} \end{pmatrix}$$

is distributed according to $N(\mathbf{0}, \boldsymbol{\Sigma})$. Let us assume that the dimensionality of $\boldsymbol{Y}^{(1)}$, say p_1, is not greater than the dimensionality of $\boldsymbol{Y}^{(2)}$, say p_2. Then there are p_1 nonzero roots of (1), say

(7) $$f_1 > f_2 > \cdots > f_{p_1}.$$

Now we shall find the distribution of $\{f_i\}$ when

(8) $$\boldsymbol{\Sigma}_{12} = \mathbf{0}.$$

For the moment assume $\{\boldsymbol{Y}_{\alpha}^{(2)}\}$ to be fixed. Then $\boldsymbol{A}_{22}$ is fixed, and

(9) $$\boldsymbol{B} = \boldsymbol{A}_{12} \boldsymbol{A}_{22}^{-1}$$

is the matrix of regression coefficients of $\boldsymbol{Y}^{(1)}$ on $\boldsymbol{Y}^{(2)}$. From Section 4.3 we know that

(10) $$\boldsymbol{A}_{11\cdot 2} = \sum_{\alpha=1}^{n} \left(\boldsymbol{Y}_{\alpha}^{(1)} - \boldsymbol{B}\boldsymbol{Y}_{\alpha}^{(2)}\right)\left(\boldsymbol{Y}_{\alpha}^{(1)} - \boldsymbol{B}\boldsymbol{Y}_{\alpha}^{(2)}\right)' = \boldsymbol{A}_{11} - \boldsymbol{B}\boldsymbol{A}_{22}\boldsymbol{B}'$$
$$= \boldsymbol{A}_{11} - \boldsymbol{A}_{12}\boldsymbol{A}_{22}^{-1}\boldsymbol{A}_{21}$$

and

(11) $$\boldsymbol{Q} = \boldsymbol{B}\boldsymbol{A}_{22}\boldsymbol{B}' = \boldsymbol{A}_{12}\boldsymbol{A}_{22}^{-1}\boldsymbol{A}_{21}$$

($\boldsymbol{\beta} = \mathbf{0}$) are independently distributed according to $W(\boldsymbol{\Sigma}_{11}, n - p_2)$ and $W(\boldsymbol{\Sigma}_{11}, p_2)$, respectively. In terms of $\boldsymbol{Q}$ Equation (1) defining f is

(12) $$|\boldsymbol{Q} - f(\boldsymbol{A}_{11\cdot 2} + \boldsymbol{Q})| = 0.$$

The distribution of f_i, $i = 1, \dots, p_1$, is the distribution of the nonzero roots of (12), and the density is given by (see Section 13.2)

$$(13)\quad \pi^{\frac{1}{2}p_1^2} \frac{\Gamma_{p_1}\left(\frac{1}{2}n\right)}{\Gamma_{p_1}\left[\frac{1}{2}(n-p_2)\right]\Gamma_p\left(\frac{1}{2}p_1\right)\Gamma_p\left(\frac{1}{2}p_2\right)}$$

$$\times \prod_{i=1}^{p_1} \left\{ f_i^{\frac{1}{2}(p_2-p_1-1)} (1-f_i)^{\frac{1}{2}(N-p_2-p_1-2)} \right\} \prod_{i<j}^{p_1} (f_i - f_j).$$

Since the conditional density (13) does not depend upon $\boldsymbol{Y}^{(2)}$, (13) is the unconditional density of the squares of the sample canonical correlation coefficients of the two sets $\boldsymbol{X}_\alpha^{(1)}$ and $\boldsymbol{X}_\alpha^{(2)}$, $\alpha = 1, \dots, N$. The density (13) also holds when the $\boldsymbol{X}^{(2)}$ are actually fixed variate vectors or have any distribution, so long as $\boldsymbol{X}^{(1)}$ and $\boldsymbol{X}^{(2)}$ are independently distributed and $\boldsymbol{X}^{(1)}$ has a multivariate normal distribution.

In the special case when $p_1 = 1$, $p_2 = p - 1$, (13) reduces to

$$(14)\quad \frac{\Gamma\left[\frac{1}{2}(N-1)\right]}{\Gamma\left[\frac{1}{2}(N-p)\right]\Gamma\left[\frac{1}{2}(p-1)\right]} f^{\frac{1}{2}(p-3)} (1-f)^{\frac{1}{2}(N-p-2)},$$

which is the density of the square of the sample multiple correlation coefficient between $\boldsymbol{X}^{(1)}$ ($p_1 = 1$) and $\boldsymbol{X}^{(2)}$ ($p_2 = p - 1$).

13.5. ASYMPTOTIC DISTRIBUTIONS IN THE CASE OF ONE WISHART MATRIX

13.5.1. All Population Roots Different

In Section 13.3 we found the density of the diagonal matrix $\boldsymbol{L}$ and the orthogonal matrix $\boldsymbol{B}$ defined by $\boldsymbol{S} = \boldsymbol{BLB}'$, $l_1 \geq \cdots \geq l_p$, and $b_{1i} \geq 0$, $i = 1, \dots, p$, when $n\boldsymbol{S}$ is distributed according to $W(\boldsymbol{I}, n)$. In this section we find the asymptotic distribution of $\boldsymbol{L}$ and $\boldsymbol{B}$ when $n\boldsymbol{S}$ is distributed according to $W(\boldsymbol{\Sigma}, n)$ and the characteristic roots of $\boldsymbol{\Sigma}$ are different. (Corollary 13.3.2 gave the asymptotic distribution when $\boldsymbol{\Sigma} = \boldsymbol{I}$.)

THEOREM 13.5.1. *Suppose $n\boldsymbol{S}$ has the distribution $W(\boldsymbol{\Sigma}, n)$. Define diagonal $\boldsymbol{\Lambda}$ and $\boldsymbol{L}$ and orthogonal $\boldsymbol{\beta}$ and $\boldsymbol{B}$ by*

$$(1)\quad \boldsymbol{\Sigma} = \boldsymbol{\beta\Lambda\beta}', \qquad \boldsymbol{S} = \boldsymbol{BLB}',$$

$\lambda_1 > \lambda_2 > \cdots > \lambda_p$, $l_1 \geq l_2 \geq \cdots \geq l_p$, $\beta_{1i} \geq 0$, $b_{1i} \geq 0$, $i = 1, \ldots, p$. *Define* $\boldsymbol{G} = \sqrt{n}(\boldsymbol{B} - \boldsymbol{\beta})$ *and diagonal* $\boldsymbol{D} = \sqrt{n}(\boldsymbol{L} - \boldsymbol{\Lambda})$. *Then the limiting distribution of* $\boldsymbol{D}$ *and* $\boldsymbol{G}$ *is normal with* $\boldsymbol{D}$ *and* $\boldsymbol{G}$ *independent and the diagonal elements of* $\boldsymbol{D}$ *are independent. The diagonal element* d_i *has the limiting distribution* $N(0, 2\lambda_i^2)$. *The covariance matrix of* $\boldsymbol{g}_i$ *in the limiting distribution of* $\boldsymbol{G} = (\boldsymbol{g}_1, \ldots, \boldsymbol{g}_p)$ *is*

$$\mathscr{A}\mathscr{C}(\boldsymbol{g}_i) = \sum_{\substack{k=1 \\ k \neq i}}^{p} \frac{\lambda_i \lambda_k}{(\lambda_i - \lambda_k)^2} \boldsymbol{\beta}_k \boldsymbol{\beta}_k', \tag{2}$$

where $\boldsymbol{\beta} = (\boldsymbol{\beta}_1, \ldots, \boldsymbol{\beta}_p)$. *The covariance matrix of* $\boldsymbol{g}_i$ *and* $\boldsymbol{g}_j$ *in the limiting distribution is*

$$\mathscr{A}\mathscr{C}(\boldsymbol{g}_i, \boldsymbol{g}_j) = -\frac{\lambda_i \lambda_j}{(\lambda_i - \lambda_j)^2} \boldsymbol{\beta}_j \boldsymbol{\beta}_i', \qquad i \neq j. \tag{3}$$

PROOF. The matrix $n\boldsymbol{T} = n\boldsymbol{\beta}'\boldsymbol{S}\boldsymbol{\beta}$ is distributed according to $W(\boldsymbol{\Lambda}, n)$. Let

$$\boldsymbol{T} = \boldsymbol{Y}\boldsymbol{L}\boldsymbol{Y}', \tag{4}$$

where $\boldsymbol{Y}$ is orthogonal. In order that (4) determine $\boldsymbol{Y}$ uniquely, we require $y_{ii} \geq 0$. Let $\sqrt{n}(\boldsymbol{T} - \boldsymbol{\Lambda}) = \boldsymbol{U}$ and $\sqrt{n}(\boldsymbol{Y} - \boldsymbol{I}) = \boldsymbol{W}$. Then (4) can be written

$$\boldsymbol{\Lambda} + \frac{1}{\sqrt{n}}\boldsymbol{U} = \left(\boldsymbol{I} + \frac{1}{\sqrt{n}}\boldsymbol{W}\right)\left(\boldsymbol{\Lambda} + \frac{1}{\sqrt{n}}\boldsymbol{D}\right)\left(\boldsymbol{I} + \frac{1}{\sqrt{n}}\boldsymbol{W}\right)', \tag{5}$$

which is equivalent to

$$\boldsymbol{U} = \boldsymbol{W}\boldsymbol{\Lambda} + \boldsymbol{D} + \boldsymbol{\Lambda}\boldsymbol{W}' + \frac{1}{\sqrt{n}}(\boldsymbol{W}\boldsymbol{D} + \boldsymbol{W}\boldsymbol{\Lambda}\boldsymbol{W}' + \boldsymbol{D}\boldsymbol{W}') + \frac{1}{n}\boldsymbol{W}\boldsymbol{D}\boldsymbol{W}'. \tag{6}$$

From $\boldsymbol{I} = \boldsymbol{Y}\boldsymbol{Y}' = (\boldsymbol{I} + (1/\sqrt{n})\boldsymbol{W})(\boldsymbol{I} + (1/\sqrt{n})\boldsymbol{W}')$, we have

$$\boldsymbol{0} = \boldsymbol{W} + \boldsymbol{W}' + \frac{1}{\sqrt{n}}\boldsymbol{W}\boldsymbol{W}'. \tag{7}$$

We shall proceed heuristically and justify the method later. If we neglect terms of order $1/\sqrt{n}$ and $1/n$ in (6) and (7), we obtain

$$\boldsymbol{U} = \boldsymbol{W}\boldsymbol{\Lambda} + \boldsymbol{D} + \boldsymbol{\Lambda}\boldsymbol{W}', \tag{8}$$

$$\boldsymbol{0} = \boldsymbol{W} + \boldsymbol{W}'. \tag{9}$$

When we substitute $\boldsymbol{W}' = -\boldsymbol{W}$ from (9) into (8) and write the result in components, we obtain $w_{ii} = 0$,

$$(10) \qquad d_i = u_{ii}, \qquad i = 1, \dots, p,$$

$$(11) \qquad w_{ij} = \frac{u_{ij}}{\lambda_j - \lambda_i}, \qquad i \neq j,\ i, j = 1, \dots, p.$$

(Note $w_{ij} = -w_{ji}$.) From Theorem 3.4.4 we know that in the limiting normal distribution of $\boldsymbol{U}$ the functionally independent elements are statistically independent with means 0 and variances $\mathscr{AV}(u_{ii}) = 2\lambda_i^2$ and $\mathscr{AV}(u_{ij}) = \lambda_i \lambda_j$, $i \neq j$. Then the limiting distribution of $\boldsymbol{D}$ and $\boldsymbol{W}$ is normal and $d_1, \dots, d_p, w_{12}, w_{13}, \dots, w_{p-1,p}$ are independent with means 0 and variances $\mathscr{AV}(d_i) = 2\lambda_i^2$, $i = 1, \dots, p$, and $\mathscr{AV}(w_{ij}) = \lambda_i \lambda_j / (\lambda_j - \lambda_i)^2$, $j = i+1, \dots, p$, $i = 1, \dots, p - 1$. Each column of B is $\pm$ the corresponding column of $\boldsymbol{\beta Y}$; since $\boldsymbol{Y} \xrightarrow{p} \boldsymbol{I}$, $\boldsymbol{\beta Y} \xrightarrow{p} \boldsymbol{\beta}$ and with arbitrary high probability each column of $\boldsymbol{B}$ is nearly identical to the corresponding column of $\boldsymbol{\beta Y}$. Then $\boldsymbol{G} = \sqrt{n}(\boldsymbol{B} - \boldsymbol{\beta})$ has the limiting distribution of $\boldsymbol{\beta}\sqrt{n}(\boldsymbol{Y} - \boldsymbol{I}) = \boldsymbol{\beta W}$. The asymptotic variances and covariances follow.

Now we justify the limiting distribution of $\boldsymbol{D}$ and $\boldsymbol{W}$. The equations $\boldsymbol{T} = \boldsymbol{YLY}'$ and $\boldsymbol{I} = \boldsymbol{YY}'$ and conditions $l_1 > \cdots > l_p$, $y_{ii} > 0$, $i = 1, \dots, p$, define a $1-1$ transformation of $\boldsymbol{T}$ to $\boldsymbol{Y}$, $\boldsymbol{L}$ except for a set of measure 0. The transformation from $\boldsymbol{Y}$, $\boldsymbol{L}$ to $\boldsymbol{T}$ is continuously differentiable. The inverse is continuously differentiable in a neighborhood of $\boldsymbol{Y} = \boldsymbol{I}$ and $\boldsymbol{L} = \Lambda$ since the equations (8) and (9) can be solved uniquely. Hence $\boldsymbol{Y}$, $\boldsymbol{L}$ as a function of $\boldsymbol{T}$ satisfies the conditions of Theorem 4.2.3. Q.E.D.

13.5.2. One Root of Higher Multiplicity

In Section 11.7.3 we used the asymptotic distribution of the q smallest sample roots when the q smallest population roots are equal. We shall now derive that distribution.

Let

$$(12) \qquad \Lambda = \begin{pmatrix} \Lambda_1 & \boldsymbol{0} \\ \boldsymbol{0} & \lambda^* \boldsymbol{I}_q \end{pmatrix},$$

where the diagonal elements of the diagonal matrix Λ_1 are different and are larger than λ^* (> 0). Let $\boldsymbol{U} = \sqrt{n}(\boldsymbol{T} - \Lambda)$, $\boldsymbol{D} = \sqrt{n}(\boldsymbol{L} - \Lambda)$, and $\boldsymbol{Y}$ be partitioned similarly:

$$U = \begin{pmatrix} U_{11} & U_{12} \\ U_{21} & U_{22} \end{pmatrix}, \quad D = \begin{pmatrix} D_1 & 0 \\ 0 & D_2 \end{pmatrix}, \quad Y = \begin{pmatrix} Y_{11} & Y_{12} \\ Y_{21} & Y_{22} \end{pmatrix}. \tag{13}$$

Define $W_{11} = \sqrt{n}\,(Y_{11} - I)$, $W_{12} = \sqrt{n}\,Y_{12}$, and $W_{21} = \sqrt{n}\,Y_{21}$. Then (5) can be written

$$\begin{aligned}
(14)\quad & \begin{pmatrix} \Lambda_1 & 0 \\ 0 & \lambda^* I_q \end{pmatrix} + \frac{1}{\sqrt{n}} \begin{pmatrix} U_{11} & U_{12} \\ U_{21} & U_{22} \end{pmatrix} \\
&= \left[\begin{pmatrix} I_{p-q} & 0 \\ 0 & Y_{22} \end{pmatrix} + \frac{1}{\sqrt{n}} \begin{pmatrix} W_{11} & W_{12} \\ W_{21} & 0 \end{pmatrix} \right] \\
&\quad \times \left[\begin{pmatrix} \Lambda_1 & 0 \\ 0 & \lambda^* I_q \end{pmatrix} + \frac{1}{\sqrt{n}} \begin{pmatrix} D_1 & 0 \\ 0 & D_2 \end{pmatrix} \right] \\
&\quad \times \left[\begin{pmatrix} I_{p-q} & 0 \\ 0 & Y_{22}' \end{pmatrix} + \frac{1}{\sqrt{n}} \begin{pmatrix} W_{11}' & W_{21}' \\ W_{12}' & 0 \end{pmatrix} \right] \\
&= \begin{pmatrix} \Lambda_1 & 0 \\ 0 & \lambda^* Y_{22} Y_{22}' \end{pmatrix} + \frac{1}{\sqrt{n}} \left[\begin{pmatrix} D_1 & 0 \\ 0 & Y_{22} D_2 Y_{22}' \end{pmatrix} \right. \\
&\quad \left. + \begin{pmatrix} W_{11}\Lambda_1 & \lambda^* W_{12} Y_{22}' \\ W_{21}\Lambda_1 & 0 \end{pmatrix} + \begin{pmatrix} \Lambda_1 W_{11}' & \Lambda_1 W_{21}' \\ \lambda^* Y_{22} W_{12}' & 0 \end{pmatrix} \right] \\
&\quad + \frac{1}{n} M,
\end{aligned}$$

where the submatrices of M are sums of products of Y_{22}, Λ_1, $\lambda^* I_q$, D_k, W_{kl}, and $1/\sqrt{n}$. The orthogonality of Y implies

$$\begin{aligned}
(15)\quad I_p &= YY' \\
&= \begin{pmatrix} I_{p-q} & 0 \\ 0 & Y_{22} Y_{22}' \end{pmatrix} + \frac{1}{\sqrt{n}} \left[\begin{pmatrix} W_{11} & W_{12} Y_{22}' \\ W_{21} & 0 \end{pmatrix} + \begin{pmatrix} W_{11}' & W_{21}' \\ Y_{22} W_{12}' & 0 \end{pmatrix} \right] \\
&\quad + \frac{1}{n} N,
\end{aligned}$$

where the submatrices of N are sums of products of W_{kl}. From (15) we see that

(16)
$$\boldsymbol{Y}_{22}\boldsymbol{Y}_{22}' = \boldsymbol{I}_q - \frac{1}{n}\boldsymbol{N}_{22}.$$

When this is substituted into the lower right-hand corner of (14), we obtain

(17)
$$\boldsymbol{U}_{22} = \boldsymbol{Y}_{22}\boldsymbol{D}_2\boldsymbol{Y}_{22}' + \frac{1}{\sqrt{n}}\left(\boldsymbol{M}_{22} - \lambda^*\boldsymbol{N}_{22}\right).$$

The limiting distribution of $(1/\lambda^*)\boldsymbol{U}_{22}$ has density (25) of Section 13.3 with p replaced by $p - q$. Then the limiting distribution of $\boldsymbol{D}_2$ and $\boldsymbol{Y}_{22}$ is the distribution of $\boldsymbol{D}_2^*$ and $\boldsymbol{Y}_{22}^*$ defined by $\boldsymbol{U}_{22}^* = \boldsymbol{Y}_{22}^*\boldsymbol{D}_2^*\boldsymbol{Y}_{22}^{*\prime}$, where $(1/\lambda^*)\boldsymbol{U}_{22}^*$ has the density (25) of Section 13.3.

To justify the preceding derivation we note that $\boldsymbol{D}_2$ and $\boldsymbol{Y}_{22}$ are functions of $\boldsymbol{U}$ depending on n that converge to the solution of $\boldsymbol{U}_{22}^* = \boldsymbol{Y}_{22}^*\boldsymbol{D}_2^*\boldsymbol{Y}_{22}^{*\prime}$. We can use the following theorem given by Anderson (1963a) and due to Rubin.

THEOREM 13.5.2. *Let $F_n(\boldsymbol{u})$ be the cumulative distribution function of a random matrix $\boldsymbol{U}_n$. Let $\boldsymbol{V}_n$ be a matrix-valued function of $\boldsymbol{U}_n$, $\boldsymbol{V}_n = \boldsymbol{f}_n(\boldsymbol{u}_n)$, and let $G_n(\boldsymbol{v})$ be the (induced) distribution of $\boldsymbol{V}_n$. Suppose $F_n(\boldsymbol{u}) \to F(\boldsymbol{u})$ in every continuity point of $F(\boldsymbol{u})$, and suppose for every continuity point $\boldsymbol{u}$ of $\boldsymbol{f}(\boldsymbol{u})$, $\boldsymbol{f}_n(\boldsymbol{u}_n) \to \boldsymbol{f}(\boldsymbol{u})$ when $\boldsymbol{u}_n \to \boldsymbol{u}$. Let $G(\boldsymbol{v})$ be the distribution of the random matrix $\boldsymbol{V} = \boldsymbol{f}(\boldsymbol{U})$, where $\boldsymbol{U}$ has the distribution $F(\boldsymbol{u})$. If the probability of the set of discontinuities of $\boldsymbol{f}(\boldsymbol{u})$ according to $F(\boldsymbol{u})$ is 0, then*

(18)
$$\lim_{n\to\infty} G_n(\boldsymbol{v}) = G(\boldsymbol{v})$$

in every continuity point of $G(\boldsymbol{v})$.

The details of verifying that $\boldsymbol{U}(n)$ and

(19)
$$\left(\boldsymbol{D}_2(n), \boldsymbol{Y}_{22}(n)\right) = \boldsymbol{f}_n(\boldsymbol{U}(n))$$

satisfy the conditions of the theorem have been given by Anderson (1963a).

13.6. ASYMPTOTIC DISTRIBUTIONS IN THE CASE OF TWO WISHART MATRICES

In Section 13.2 we studied the distributions of the roots $l_1 \geq l_2 \geq \cdots \geq l_p$ of

(1)
$$|\boldsymbol{S}^* - l\boldsymbol{T}^*| = 0$$

and the vectors satisfying

$$(\boldsymbol{S}^* - l\boldsymbol{T}^*)\boldsymbol{x}^* = \boldsymbol{0} \tag{2}$$

and $\boldsymbol{x}^{*\prime}\boldsymbol{T}^*\boldsymbol{x}^* = 1$ when $\boldsymbol{A}^* = m\boldsymbol{S}^*$ and $\boldsymbol{B}^* = n\boldsymbol{T}^*$ are distributed independently according to $W(\boldsymbol{\Sigma}, m)$ and $W(\boldsymbol{\Sigma}, n)$, respectively. In this section we study the asymptotic distributions of the roots and vectors as $n \to \infty$ when $\boldsymbol{A}^*$ and $\boldsymbol{B}^*$ are distributed independently according to $W(\boldsymbol{\Phi}, m)$ and $W(\boldsymbol{\Sigma}, n)$, respectively, and $m/n \to \eta > 0$. We shall assume that the roots of

$$|\boldsymbol{\Phi} - \lambda\boldsymbol{\Sigma}| = 0 \tag{3}$$

are distinct. (In Section 13.2 $\lambda_1 = \cdots = \lambda_p = 1$.)

THEOREM 13.6.1. *Let $m\boldsymbol{S}^*$ and $n\boldsymbol{T}^*$ be independently distributed according to $W(\boldsymbol{\Phi}, m)$ and $W(\boldsymbol{\Sigma}, n)$, respectively. Let $\lambda_1 > \lambda_2 > \cdots > \lambda_p$ (> 0) be the roots of* (3) *and let $\boldsymbol{\Lambda}$ be the diagonal matrix with the roots as diagonal elements in descending order; let $\boldsymbol{\gamma}_1, \dots, \boldsymbol{\gamma}_p$ be the solutions to*

$$(\boldsymbol{\Phi} - \lambda_i\boldsymbol{\Sigma})\boldsymbol{\gamma} = \boldsymbol{0}, \qquad i = 1, \dots, p, \tag{4}$$

$\boldsymbol{\gamma}'\boldsymbol{\Sigma}\boldsymbol{\gamma} = 1$, and $\gamma_{1i} \geq 0$, and let $\boldsymbol{\Gamma} = (\boldsymbol{\gamma}_1, \dots, \boldsymbol{\gamma}_p)$. Let $l_1 \geq \cdots \geq l_p$ (> 0) be the roots of (1), *and let $\boldsymbol{L}$ be the diagonal matrix with the roots as diagonal elements in descending order; let $\boldsymbol{x}_1^*, \dots, \boldsymbol{x}_p^*$ be the solutions to* (2) *for $l = l_i$, $i = 1, \dots, p$, $\boldsymbol{x}^{*\prime}\boldsymbol{T}^*\boldsymbol{x}^* = 1$, and $x_{1i}^* > 0$, and let $\boldsymbol{X}^* = (\boldsymbol{x}_1^*, \dots, \boldsymbol{x}_p^*)$. Define $\boldsymbol{Z}^* = \sqrt{n}(\boldsymbol{X}^* - \boldsymbol{\Gamma})$ and diagonal $\boldsymbol{D} = \sqrt{n}(\boldsymbol{L} - \boldsymbol{\Lambda})$. Then the limiting distribution of $\boldsymbol{D}$ and $\boldsymbol{Z}^*$ is normal with means $\boldsymbol{0}$ as $n \to \infty$, $m \to \infty$, and $m/n \to \eta$ (> 0). The asymptotic variances and covariances that are not 0 are*

$$\mathscr{AV}(d_i) = 2\frac{\lambda_i^2(1+\eta)}{\eta}, \tag{5}$$

$$\mathscr{AC}(\boldsymbol{z}_i^*) = \sum_{\substack{k=1\\k\neq i}}^{p} \frac{\lambda_i(\lambda_k + \eta\lambda_i)}{\eta(\lambda_k - \lambda_i)^2}\boldsymbol{\gamma}_k\boldsymbol{\gamma}_k' + \frac{1}{2}\boldsymbol{\gamma}_i\boldsymbol{\gamma}_i', \tag{6}$$

$$\mathscr{AC}(d_i, \boldsymbol{z}_i^*) = \lambda_i\boldsymbol{\gamma}_i, \tag{7}$$

$$\mathscr{AC}(\boldsymbol{z}_i^*, \boldsymbol{z}_j^*) = -\frac{\lambda_i\lambda_j(1+\eta)}{\eta(\lambda_j - \lambda_i)^2}\boldsymbol{\gamma}_j\boldsymbol{\gamma}_i', \qquad i \neq j. \tag{8}$$

PROOF. Let

$$S = \Gamma' S^* \Gamma, \qquad T = \Gamma' T^* \Gamma. \tag{9}$$

Then mS and nT are distributed independently according to $W(\Lambda, m)$ and $W(I, n)$, respectively (Section 7.3.3). Then $l_1, \ldots, l_p$ are the roots of

$$|S - lT| = 0. \tag{10}$$

Let $x_1, \ldots, x_p$ be the solutions to

$$(S - l_i T)x = 0, \qquad i = 1, \ldots, p, \tag{11}$$

and $x'Tx = 1$, and let $X = (x_1, \ldots, x_p)$. Then $x_i^* = \Gamma x_i$ and $X^* = \Gamma X$ except possibly for multiplication of columns of X (or X^*) by -1. If $Z = \sqrt{n}(X - I)$, then $Z^* = \Gamma Z$ (except possibly for multiplication of columns by -1).

We shall now find the limiting distribution of D and Z. Let $\sqrt{n}(S - \Lambda) = U$ and $\sqrt{n}(T - I) = V$. Then U and V have independent limiting normal distributions with means 0. The functionally independent elements of U and V are statistically independent in the limiting distribution. The variances are $\mathscr{E}u_{ii}^2 = 2(n/m)\lambda_i^2 \to 2\lambda_i^2/\eta$, $\mathscr{E}u_{ij}^2 = (n/m)\lambda_i\lambda_j \to \lambda_i\lambda_j/\eta$, $i \neq j$, $\mathscr{E}v_{ii}^2 = 2$, $\mathscr{E}v_{ij}^2 = 1$, $i \neq j$.

From the definition of L and X we have $SX = TXL$, $X'TX = I$, and $X'SX = L$. If we let $X^{-1} = G$, we obtain

$$S = G'LG, \qquad T = G'G. \tag{12}$$

We require $g_{ii} > 0$, $i = 1, \ldots, p$. Since $S \xrightarrow{p} \Lambda$ and $T \xrightarrow{p} I$, $L \xrightarrow{p} \Lambda$, and $G \xrightarrow{p} I$. Let $\sqrt{n}(G - I) = H$. Then we write (12) as

$$\Lambda + \frac{1}{\sqrt{n}}U = \left(I + \frac{1}{\sqrt{n}}H'\right)\left(\Lambda + \frac{1}{\sqrt{n}}D\right)\left(I + \frac{1}{\sqrt{n}}H\right), \tag{13}$$

$$I + \frac{1}{\sqrt{n}}V = \left(I + \frac{1}{\sqrt{n}}H'\right)\left(I + \frac{1}{\sqrt{n}}H\right). \tag{14}$$

These can be rewritten

$$U = D + \Lambda H + H'\Lambda + \frac{1}{\sqrt{n}}(DH + H'D + H'\Lambda H) + \frac{1}{n}H'DH, \tag{15}$$

$$V = H + H' + \frac{1}{\sqrt{n}}H'H. \tag{16}$$

If we neglect the terms of order $1/\sqrt{n}$ and $1/n$ (as in Section 13.5), we can write

$$U = D + \Lambda H + H'\Lambda, \tag{17}$$

$$V = H + H', \tag{18}$$

$$U - V\Lambda = D + \Lambda H - H\Lambda. \tag{19}$$

The diagonal elements of (18) and the components of (19) are

$$v_{ii} = 2h_{ii}, \tag{20}$$

$$u_{ii} - \lambda_i v_{ii} = d_i, \tag{21}$$

$$u_{ij} - v_{ij}\lambda_j = (\lambda_i - \lambda_j)h_{ij}, \qquad i \neq j. \tag{22}$$

The limiting distribution of $\boldsymbol{H}$ and $\boldsymbol{D}$ is normal with means 0. The pairs (h_{ij}, h_{ji}) of off-diagonal elements of $\boldsymbol{H}$ are independent with variances

$$\mathscr{AV}(h_{ij}) = \frac{\lambda_j(\lambda_i + \eta\lambda_j)}{\eta(\lambda_i - \lambda_j)^2}, \qquad i \neq j, \tag{23}$$

and covariances

$$\mathscr{AC}(h_{ij}, h_{ji}) = -\frac{\lambda_i\lambda_j(1 + \eta)}{\eta(\lambda_i - \lambda_j)^2}, \qquad i \neq j. \tag{24}$$

The pairs (d_i, h_{ii}) of diagonal elements of $\boldsymbol{D}$ and $\boldsymbol{H}$ are independent with variances (5),

$$\mathscr{AV}(h_{ii}) = \frac{1}{2}, \tag{25}$$

and covariance

$$\mathscr{AC}(d_i, h_{ii}) = -\lambda_i. \tag{26}$$

The diagonal elements of $\boldsymbol{D}$ and $\boldsymbol{H}$ are independent of the off-diagonal elements of $\boldsymbol{H}$.

That the limiting distribution of $\boldsymbol{D}$ and $\boldsymbol{H}$ is normal is justified by Theorem 4.2.3. $\boldsymbol{S}$ and $\boldsymbol{T}$ are polynomials in $\boldsymbol{L}$ and $\boldsymbol{G}$ and their derivatives are polynomials and hence continuous. Since Equations (12) with auxiliary conditions can be solved uniquely for $\boldsymbol{L}$ and $\boldsymbol{G}$, the inverse function is also continuously differentiable at $\boldsymbol{L} = \boldsymbol{\Lambda}$ and $\boldsymbol{G} = \boldsymbol{I}$. By Theorem 4.2.3, $\boldsymbol{D} = \sqrt{n}(\boldsymbol{L} - \boldsymbol{\Lambda})$ and $\boldsymbol{H} = \sqrt{n}(\boldsymbol{G} - \boldsymbol{I})$ have a limiting normal distribution. In turn $\boldsymbol{X} = \boldsymbol{G}^{-1}$ is continuously differentiable at $\boldsymbol{G} = \boldsymbol{I}$ and $\boldsymbol{Z} = \sqrt{n}(\boldsymbol{X} - \boldsymbol{I}) = \sqrt{n}(\boldsymbol{G}^{-1} - \boldsymbol{I})$ has the limiting distribution of $-\boldsymbol{H}$. (Expand $\sqrt{n}[(\boldsymbol{I} + (1/\sqrt{n})\boldsymbol{H})^{-1} - \boldsymbol{I}]$.) Since $\boldsymbol{G} \xrightarrow{p} \boldsymbol{I}$, $\boldsymbol{X} \xrightarrow{p} \boldsymbol{I}$ and $x_{ii} > 0$, $i = 1, \dots, p$, with probability approaching 1. Then $\boldsymbol{Z}^* = \sqrt{n}(\boldsymbol{X}^* - \boldsymbol{\Gamma})$ has the limiting distribution of $\boldsymbol{\Gamma Z}$. (Since $\boldsymbol{X} \xrightarrow{p} \boldsymbol{I}$, $\boldsymbol{X}^* = \boldsymbol{\Gamma X} \xrightarrow{p} \boldsymbol{\Gamma}$ and $x_{1i} > 0$, $i = 1, \dots, p$, with probability approaching 1.) The asymptotic variances and covariances (6) to (8) are obtained from (23) to (26). Q.E.D.

PROBLEMS

1. (Sec. 13.2) Prove Theorem 13.2.1 for $p = 2$ by calculating the Jacobian directly.

2. (Sec. 13.2) Prove Theorem 13.3.2 for $p = 2$ directly by representing the orthogonal matrix $\boldsymbol{C}$ in terms of the cosine and sine of an angle.

3. (Sec. 13.2) Consider the distribution of the roots of $|\boldsymbol{A} - l\boldsymbol{B}| = 0$ when $\boldsymbol{A}$ and $\boldsymbol{B}$ are of order two and are distributed according to $W(\boldsymbol{\Sigma}, m)$ and $W(\boldsymbol{\Sigma}, n)$, respectively.

(a) Find the distribution of the largest root.
(b) Find the distribution of the smallest root.
(c) Find the distribution of the sum of the roots.

4. (Sec. 13.2) Prove that the Jacobian $|\partial(\boldsymbol{G}, \boldsymbol{A})/\partial(\boldsymbol{E}, \boldsymbol{F})|$ is $\prod(f_i - f_j)$ times a function of $\boldsymbol{E}$ by showing that the Jacobian vanishes for $f_i = f_j$ and that its degree in f_i is the same as that of $\prod(f_i - f_j)$.

5. (Sec. 13.3) Give the Haar invariant distribution explicitly for the 2×2 orthogonal matrix represented in terms of the cosine and sine of an angle.

6. (Sec. 13.3) Let $\boldsymbol{A}$ and $\boldsymbol{B}$ be distributed according to $W(\boldsymbol{\Sigma}, m)$ and $W(\boldsymbol{\Sigma}, n)$ respectively. Let $l_1 > \cdots > l_p$ be the roots of $|\boldsymbol{A} - l\boldsymbol{B}| = 0$ and $m_1 > \cdots > m_p$ be the roots of $|\boldsymbol{A} - m\boldsymbol{\Sigma}| = 0$. Find the distribution of the m's from that of the l's by letting $n \to \infty$.

7. (Sec. 13.3) Prove Lemma 13.3.1 in as much detail as Theorem 13.3.1.

8. Let $\boldsymbol{A}$ be distributed according to $W(\boldsymbol{\Sigma}, n)$. In case of $p = 2$ find the distribution of the characteristic roots of $\boldsymbol{A}$. [*Hint*: Transform so that $\boldsymbol{\Sigma}$ goes into a diagonal matrix.]

9. From the result in Problem 6 find the distribution of the sphericity criterion (when the null hypothesis is not true).

10. (Sec. 13.3) Show that $\boldsymbol{X}$ $(p \times n)$ has the density $f_X(\boldsymbol{X}'\boldsymbol{X})$ if and only if $\boldsymbol{T}$ has the density

$$\frac{2^p \pi^{pn/2}}{\Gamma_p(n/2)} \sum_{i=1}^{p} t_{ii}^{n-i} f_X(\boldsymbol{T}\boldsymbol{T}'),$$

where $\boldsymbol{T}$ is the lower triangular matrix with positive diagonal elements such that $\boldsymbol{T}\boldsymbol{T}' = \boldsymbol{X}'\boldsymbol{X}$. Srivastava and Khatri (1979). [*Hint*: Compare Lemma 13.3.1 with Corollary 7.2.1.]

11. (Sec. 13.5.2) In the case that the covariance matrix is (12) find the limiting distribution of $\boldsymbol{D}_1$, $\boldsymbol{W}_{11}$, $\boldsymbol{W}_{12}$, and $\boldsymbol{W}_{21}$.

CHAPTER 14

Factor Analysis

14.1. INTRODUCTION

Factor analysis is based on a model in which the observed vector is partitioned into an unobserved "systematic" part and an unobserved "error" part. The components of the error vector are considered as uncorrelated or independent, while the systematic part is taken as a linear combination of a relatively small number of unobserved factor variables. The analysis separates the effects of the factors, which are of basic interest, from the errors. From another point of view the analysis gives a description or explanation of the interdependence of a set of variables in terms of the factors without regard to the observed variability. This approach is to be compared with principal component analysis which describes or "explains" the *variability* observed. Factor analysis was developed originally for the analysis of scores on mental tests; however, the methods are useful in a much wider range of situations, for example, analyzing sets of tests of attitudes, sets of physical measurements, and sets of economic quantities. When a battery of tests is given to a group of individuals, it is observed that the score of an individual on a given test is more related to his scores on other tests than to the scores of other individuals on the other tests; that is, usually the scores for any particular individual are interrelated to some degree. This interrelation is "explained" by considering a test score of an individual as made up of a part which is peculiar to this particular test (called error) and a part which is a function of more fundamental quantities called "scores of primary abilities" or "factor scores." Since they enter several test scores, it is their effect that connects the various test scores. Roughly the idea is simply that a person who is "more intelligent" in some respects will do better on many tests than someone who is less intelligent.

The model for factor analysis is defined and discussed in Section 14.2. Maximum likelihood estimators of the parameters are derived in the case that the factor scores and errors are normally distributed, and a test that the model fits is developed. The large-sample distribution theory is given for the estimators and test criterion (Section 14.3). Maximum likelihood estimators for fixed factors do not exist, but alternative estimation procedures are suggested (Section 14.4). Some aspects of interpretation are treated in Section 14.5. The maximum likelihood estimators are derived when the factors are normal and identification is effected by specified zero loadings. Finally the estimation of factor scores is considered. Anderson (1984a) discusses the relationship of factor analysis to principal components and linear functional and structural relationships.

14.2. THE MODEL

14.2.1. Definition of the Model

Let the observable vector $\boldsymbol{X}$ be written as

$$\boldsymbol{X} = \boldsymbol{\Lambda} \boldsymbol{f} + \boldsymbol{U} + \boldsymbol{\mu}, \tag{1}$$

where $\boldsymbol{X}$, $\boldsymbol{U}$, and $\boldsymbol{\mu}$ are column vectors of p components, $\boldsymbol{f}$ is a column vector of m $(\leq p)$ components, and $\boldsymbol{\Lambda}$ is a $p \times m$ matrix. We assume that $\boldsymbol{U}$ is distributed independently of $\boldsymbol{f}$ and with mean $\mathscr{E}\boldsymbol{U} = \boldsymbol{0}$ and covariance matrix $\mathscr{E}\boldsymbol{U}\boldsymbol{U}' = \boldsymbol{\Psi}$, which is diagonal. The vector $\boldsymbol{f}$ will be treated alternatively as a random vector and as a vector of parameters that varies from observation to observation.

In terms of mental tests each component of $\boldsymbol{X}$ is a score on a test or battery of tests. The corresponding component of $\boldsymbol{\mu}$ is the average score of this test in the population. The components of $\boldsymbol{f}$ are the scores of the mental factors; linear combinations of these enter into the test scores. The coefficients of these linear combinations are the elements of $\boldsymbol{\Lambda}$, and these are called *factor loadings*. Sometimes the elements of $\boldsymbol{f}$ are called *common* factors because they are common to several different tests; in the first presentation of this kind of model [Spearman (1904)] $\boldsymbol{f}$ consisted of one component and was termed the *general* factor. A component of $\boldsymbol{U}$ is the part of the test score not "explained" by the common factors. This is considered as made up of the error of measurement in the test plus a "specific factor," this specific factor having to do only with this particular test. Since in our model (with one set of observations on each individual) we cannot distinguish between these two components of the coordinate of $\boldsymbol{U}$, we shall simply term the element of $\boldsymbol{U}$ as the error of measurement.

The specification of a given component of $\boldsymbol{X}$ is similar to that in regression theory (or analysis of variance) in that it is a linear combination of other variables. Here, however, $\boldsymbol{f}$, which plays the role of the independent variable, is not observed.

We can distinguish between two kinds of models. In one we consider the vector $\boldsymbol{f}$ to be a random vector, and in the other we consider $\boldsymbol{f}$ to be a vector of nonrandom quantities that varies from one individual to another. In the second case, it is more accurate to write $\boldsymbol{X}_\alpha = \Lambda \boldsymbol{f}_\alpha + \boldsymbol{U} + \boldsymbol{\mu}$. The nonrandom factor score vector may seem a better description of the systematic part, but it poses problems of inference because the likelihood function may not have a maximum. In principle, the model with random factors is appropriate when different samples consist of different individuals; the nonrandom factor model is suitable when the specific individuals involved and not just the structure are of interest.

When $\boldsymbol{f}$ is taken as random, we assume $\mathscr{E}\boldsymbol{f} = \boldsymbol{0}$. (Otherwise, $\mathscr{E}\boldsymbol{X} = \Lambda\mathscr{E}\boldsymbol{f} + \boldsymbol{\mu}$, and $\boldsymbol{\mu}$ can be redefined to absorb $\Lambda\mathscr{E}\boldsymbol{f}$.) Let $\mathscr{E}\boldsymbol{f}\boldsymbol{f}' = \boldsymbol{\Phi}$. Our analysis will be made in terms of first and second moments. Usually, we shall consider $\boldsymbol{f}$ and $\boldsymbol{U}$ to have normal distributions. If $\boldsymbol{f}$ is not random, then $\boldsymbol{f} = \boldsymbol{f}_\alpha$ for the αth individual. Then we shall assume usually $(1/N)\sum_{\alpha=1}^N \boldsymbol{f}_\alpha = \boldsymbol{0}$ and $(1/N)\sum_{\alpha=1}^N \boldsymbol{f}_\alpha \boldsymbol{f}_\alpha' = \boldsymbol{\Phi}$.

There is a fundamental indeterminacy in this model. Let $\boldsymbol{f} = \boldsymbol{C}\boldsymbol{f}^*$ ($\boldsymbol{f}^* = \boldsymbol{C}^{-1}\boldsymbol{f}$) and $\Lambda^* = \Lambda\boldsymbol{C}$, where $\boldsymbol{C}$ is a nonsingular $m \times m$ matrix. Then (1) can be written as

$$\boldsymbol{X} = \Lambda^*\boldsymbol{f}^* + \boldsymbol{U} + \boldsymbol{\mu}. \tag{2}$$

When $\boldsymbol{f}$ is random, $\mathscr{E}\boldsymbol{f}^*\boldsymbol{f}^{*\prime} = \boldsymbol{C}^{-1}\boldsymbol{\Phi}(\boldsymbol{C}^{-1})' = \boldsymbol{\Phi}^*$; when $\boldsymbol{f}$ is nonrandom, $(1/N)\sum_{\alpha=1}^N \boldsymbol{f}_\alpha^*\boldsymbol{f}_\alpha^{*\prime} = \boldsymbol{\Phi}^*$. The model with Λ and $\boldsymbol{f}$ is equivalent to the model with Λ^* and $\boldsymbol{f}^*$; that is, by observing $\boldsymbol{X}$ we cannot distinguish between these two models.

Some of the indeterminacy in the model can be eliminated by requiring that $\mathscr{E}\boldsymbol{f}\boldsymbol{f}' = \boldsymbol{I}$ if $\boldsymbol{f}$ is random, or $\sum_{\alpha=1}^N \boldsymbol{f}_\alpha \boldsymbol{f}_\alpha' = N\boldsymbol{I}$ if $\boldsymbol{f}$ is not random. In this case the factors are said to be *orthogonal*; if $\boldsymbol{\Phi}$ is not diagonal, the factors are said to be *oblique*. When we assume $\mathbf{\Phi} = \mathbf{I}$, then $\mathscr{E} \mathbf{f}^* \mathbf{f}^{*\prime} = \mathbf{C}^{-1}(\mathbf{C}^{-1})' = \mathbf{I}$ ($\mathbf{I} = \mathbf{CC}'$). The indeterminacy is equivalent to multiplication by an orthogonal matrix; this is called the problem of *rotation*. Requiring that $\mathbf{\Phi}$ be diagonal means that the components of $\mathbf{f}$ are independently distributed when $\mathbf{f}$ is assumed normal. This has an appeal to psychologists because one idea of common mental factors is (by definition) that they are independent or uncorrelated quantities.

A crucial assumption is that the components of $\mathbf{U}$ are uncorrelated. Our viewpoint is that the errors of observation and the specific factors are by definition uncorrelated. That is, the interrelationships of the test scores are caused by the common factors, and that is what we want to investigate. There is another point of view on factor analysis that is fundamentally quite different; that is, that the common factors are supposed to explain or account for as much of the variance of the test scores as possible. To follow this point of view, we should use a different model.

A geometric picture helps the intuition. Consider a p-dimensional space. The columns of $\mathbf{\Lambda}$ can be considered as m vectors in this space. They span some m-dimensional subspace; in fact, they can be considered as coordinate axes in the m-dimensional space, and $\mathbf{f}$ can be considered as coordinates of a point in that space referred to this particular axis-system. This subspace is called the *factor space*. Multiplying $\mathbf{\Lambda}$ on the right by a matrix corresponds to taking a new set of coordinate axes in the factor space.

If the factors are random, the covariance matrix of the observed $\mathbf{X}$ is

$$(3) \quad \mathbf{\Sigma} = \mathscr{E}(\mathbf{X} - \boldsymbol{\mu})(\mathbf{X} - \boldsymbol{\mu})' = \mathscr{E}(\mathbf{\Lambda f} + \mathbf{U})(\mathbf{\Lambda f} + \mathbf{U})' = \mathbf{\Lambda \Phi \Lambda}' + \mathbf{\Psi}.$$

If the factors are orthogonal ($\mathscr{E}\mathbf{ff}' = \mathbf{I}$), then (3) is

$$(4) \qquad \mathbf{\Sigma} = \mathbf{\Lambda\Lambda}' + \mathbf{\Psi}.$$

If $\mathbf{f}$ and $\mathbf{U}$ are normal, all the information about the structure comes from (3) [or (4)] and $\mathscr{E}\mathbf{X} = \boldsymbol{\mu}$.

14.2.2. Identification

Given a covariance matrix $\mathbf{\Sigma}$ and a number m of factors, we can ask whether there exist a triplet $\mathbf{\Lambda}$, $\mathbf{\Phi}$ positive definite, and $\mathbf{\Psi}$ positive definite and diagonal to satisfy (3); if so, is the triplet unique? Since any triplet can be transformed into an equivalent structure $\mathbf{\Lambda C}$, $\mathbf{C}^{-1}\mathbf{\Phi C}'^{-1}$, and $\mathbf{\Psi}$, we can put m^2 independent conditions on $\mathbf{\Lambda}$ and $\mathbf{\Phi}$ to rule out this indeterminacy. The number of components in the observable $\boldsymbol{\Sigma}$ and the number of conditions (for uniqueness) is $\frac{1}{2}p(p+1)+m^2$; the number of parameters in $\boldsymbol{\Lambda}$, $\boldsymbol{\Phi}$ and $\boldsymbol{\Psi}$ is pm, $\frac{1}{2}m(m+1)$, and p, respectively. If the excess of observed quantities and conditions over number of parameters, namely, $\frac{1}{2}[(p-m)^2-p-m]$ is positive we can expect a problem of existence but anticipation of uniqueness if a set of parameters exists. If the excess is negative, we can expect existence but possibly not uniqueness; if the excess is 0, we can hope for both existence and uniqueness (or at least a finite number of solutions). The question of existence of a solution is whether there exists a diagonal matrix $\boldsymbol{\Psi}$ with nonnegative diagonal entries such that $\boldsymbol{\Sigma}-\boldsymbol{\Psi}$ is positive semidefinite of rank m. Anderson and Rubin (1956) include most of the known results on this problem.

If a solution exists and is unique, the model is said to be *identified*. As noted above, some m^2 conditions have to be put on $\boldsymbol{\Lambda}$ and $\boldsymbol{\Phi}$ to eliminate a transformation $\boldsymbol{\Lambda}^*=\boldsymbol{\Lambda C}$ and $\boldsymbol{\Phi}^*=\boldsymbol{C}^{-1}\boldsymbol{\Phi C}'^{-1}$. We have referred above to the condition $\boldsymbol{\Phi}=\boldsymbol{I}$, which forces a transformation $\boldsymbol{C}$ to be orthogonal. [There are $\frac{1}{2}m(m+1)$ component equations in $\boldsymbol{\Phi}=\boldsymbol{I}$.] For some purposes, it is convenient to add the restrictions that

$$\boldsymbol{\Gamma}=\boldsymbol{\Lambda}'\boldsymbol{\Psi}^{-1}\boldsymbol{\Lambda} \tag{5}$$

is diagonal. If the diagonal elements of $\boldsymbol{\Gamma}$ are ordered and different ($\gamma_{11}>\gamma_{22}>\cdots>\gamma_{mm}$), $\boldsymbol{\Lambda}$ is uniquely determined. Alternative conditions are that the first m rows of $\boldsymbol{\Lambda}$ form a lower triangular matrix. A generalization of this condition is to require that the first m rows of $\boldsymbol{B\Lambda}$ form a lower triangular matrix, where $\boldsymbol{B}$ is given in advance. (This condition is implied by the so-called centroid method.)

Simple Structure. These are conditions proposed by Thurstone (1947, p. 335) for choosing a matrix out of the class $\boldsymbol{\Lambda C}$ that will have particular psychological meaning. If $\lambda_{i\alpha}=0$, then the αth factor does not enter into the ith test. The general idea of "simple structure" is that many tests should not depend on all the factors when the factors have real psychological meaning. This suggests that given a $\boldsymbol{\Lambda}$ one should consider all rotations, that is, all matrices $\boldsymbol{\Lambda C}$, where $\boldsymbol{C}$ is orthogonal, and choose the one giving most 0 coefficients. This matrix can be considered as giving the simplest structure and presumably the one with most meaningful psychological interpretation. It should be remembered that the psychologist can construct his or her tests so that they depend on the assumed factors in different ways.

The positions of the 0's are not chosen in advance, but rotations $\boldsymbol{C}$ are tried until a Λ is found satisfying these conditions. It is not clear that these conditions effect identification. Reiersøl (1950) has modified Thurstone's conditions so that there is only one rotation that satisfies the conditions, thus effecting identification.

Zero Elements in Specified Positions. Here we consider a set of conditions that requires of the investigator more a priori information. He or she must know that some particular tests do not depend on some specific factors. In this case, the conditions are that $\lambda_{i\alpha} = 0$ for specified pairs (i, α); that is, that the αth factor does not affect the ith test score. In this case we do not assume that $\mathscr{E}\boldsymbol{ff}' = \boldsymbol{I}$. These conditions are similar to some used in econometric models. The coefficients of the αth column are identified except for multiplication by a scale factor if (a) there are at least $m-1$ zero elements in that column and if (b) the rank of $\Lambda^{(\alpha)}$ is $m-1$, where $\Lambda^{(\alpha)}$ is the matrix composed of the rows containing the assigned 0's in the αth column with those assigned 0's deleted (i.e., the αth column deleted). (See Problem 1.) The multiplication of a column by a scale constant can be eliminated by a normalization, such as $\phi_{\alpha\alpha} = 1$ or $\lambda_{i\alpha} = 1$ for *some* i for *each* α. If $\phi_{\alpha\alpha} = 1$, $\alpha = 1, \dots, m$, then $\boldsymbol{\Phi}$ is a correlation matrix.

It will be seen that there are m normalizations and a minimum of $m(m-1)$ zero conditions. This is equal to the number of elements of $\boldsymbol{C}$. If there are more than $m-1$ zero elements specified in one or more columns of Λ, then there may be more conditions than are required to take out the indeterminacy in $\Lambda\boldsymbol{C}$; in this case the conditions may restrict $\Lambda\boldsymbol{\Phi}\Lambda'$.

As an example, consider the model

$$\boldsymbol{X} = \boldsymbol{\mu} + \begin{bmatrix} 1 & 0 \\ \lambda_{21} & 0 \\ \lambda_{31} & \lambda_{32} \\ 0 & \lambda_{42} \\ 0 & 1 \end{bmatrix} \begin{bmatrix} v \\ a \end{bmatrix} + \boldsymbol{U} \tag{6}$$

$$= \boldsymbol{\mu} + \begin{bmatrix} v \\ \lambda_{21} v \\ \lambda_{31} v + \lambda_{32} a \\ \lambda_{42} a \\ a \end{bmatrix} + \boldsymbol{U},$$

for the scores on five tests, where v and a are measures of verbal and arithmetic ability. The first two tests are specified to depend only on verbal ability while the last two tests depend only on arithmetic ability. The normalizations put verbal ability into the scale of the first test and arithmetic ability into the scale of the fifth test.

Koopmans and Reiersøl (1950), Anderson and Rubin (1956), and Howe (1955) suggested the use of preassigned 0's for identification and developed maximum likelihood estimation under normality for this case. [See also Lawley (1958).] Jöreskog (1969) called factor analysis under these identification conditions "confirmatory" factor analysis; with arbitrary conditions or with "rotation to simple structure," it has been called "exploratory" factor analysis.

Other Conditions. A convenient set of conditions is to require the upper square submatrix of Λ to be the identity. This assumes that the upper square matrix without this condition is nonsingular. In fact, if $\Lambda^* = (\Lambda_1^{*\prime}, \Lambda_2^{*\prime})'$ is an arbitrary $p \times m$ matrix with Λ_1^* square and nonsingular, then $\Lambda = \Lambda^* \Lambda_1^{*-1} = (\boldsymbol{I}_m, \Lambda_2')'$ satisfies the condition. (This specification of the leading $m \times m$ submatrix of Λ as $\boldsymbol{I}_m$ is a convenient identification condition and does not imply any substantive meaning.)

14.2.3. Units of Measurement

We have considered factor analysis methods applied to covariance matrices. In many cases the unit of measurement of each component of $\boldsymbol{X}$ is arbitrary. For instance, in psychological tests the unit of scoring has no intrinsic meaning.

Changing the units of measurement means multiplying each component of $\boldsymbol{X}$ by a constant; these constants are not necessarily equal. When a given test score is multiplied by a constant, the factor loadings for the test are multiplied by the same constant and the error variance is multiplied by square of the constant. Suppose $\boldsymbol{D}\boldsymbol{X} = \boldsymbol{X}^*$, where $\boldsymbol{D}$ is a diagonal matrix with positive diagonal elements. Then (1) becomes

$$\boldsymbol{X}^* = \Lambda^* \boldsymbol{f} + \boldsymbol{U}^* + \mu^*, \tag{7}$$

where $\mu^* = \mathscr{E}\boldsymbol{X}^* = \boldsymbol{D}\mu$, $\Lambda^* = \boldsymbol{D}\Lambda$, and $\boldsymbol{U}^* = \boldsymbol{D}\boldsymbol{U}$ has covariance matrix $\boldsymbol{\Psi}^* = \boldsymbol{D}\boldsymbol{\Psi}\boldsymbol{D}$. Then

$$\mathscr{E}(\boldsymbol{X}^* - \mu^*)(\boldsymbol{X}^* - \mu^*)' = \Lambda^* \boldsymbol{\Phi} \Lambda^{*\prime} + \boldsymbol{\Psi}^* = \boldsymbol{\Sigma}^*, \tag{8}$$

where $\Sigma^* = D\Sigma D$. Note that if the identification conditions are $\Phi = I$ and $\Lambda'\Psi^{-1}\Lambda$ diagonal, then Λ^* satisfies the latter condition. If Λ is identified by specified 0's and the normalization is by $\phi_{\alpha\alpha} = 1$, $\alpha = 1, \dots, m$ (i.e., Φ is a correlation matrix), then $\Lambda^* = D\Lambda$ is similarly identified. (If the normalization is $\lambda_{i\alpha} = 1$ for specified i for each α, each column of $D\Lambda$ has to be renormalized.)

A particular diagonal matrix D consists of the reciprocals of the observable standard deviations, $d_{ii} = 1/\sqrt{\sigma_{ii}}$. Then $\Sigma^* = D\Sigma D$ is the correlation matrix.

We shall see later that the maximum likelihood estimators with identification conditions Γ diagonal or specified 0's transform in the above fashion; that is, the transformation $x_\alpha^* = Dx_\alpha$, $\alpha = 1, \dots, N$, induces $\hat{\Lambda}^* = D\hat{\Lambda}$ and $\hat{\Psi}^* = D\hat{\Psi}D$.

14.3. MAXIMUM LIKELIHOOD ESTIMATORS FOR RANDOM ORTHOGONAL FACTORS

14.3.1. Maximum Likelihood Estimators

In this section we find the maximum likelihood estimators of the parameters when the observations are normally distributed, that is, the factor scores and errors are normal [Lawley (1940)]. Then $\Sigma = \Lambda\Phi\Lambda' + \Psi$. We impose conditions on Λ and Φ to make them just identified. These do not restrict $\Lambda\Phi\Lambda'$; it is a positive definite matrix of rank m. For convenience we suppose that $\Phi = I$ (i.e., the factors are orthogonal or uncorrelated) and that $\Gamma = \Lambda'\Psi^{-1}\Lambda$ is diagonal. Then the likelihood depends on the mean μ and $\Sigma = \Lambda\Lambda' + \Psi$. The maximum likelihood estimators of Λ and Φ under some other conditions effecting just identification [e.g., $\Lambda = (I_m, \Lambda_2')'$] are transformations of the maximum likelihood estimators of Λ under the preceding conditions. If $x_1, \dots, x_N$ are a set of N observations on X, the likelihood function for this sample is

$$(1) \qquad L = (2\pi)^{-\frac{1}{2}pN} |\Sigma|^{-\frac{1}{2}N} \exp\left[-\frac{1}{2} \sum_{\alpha=1}^{N} (x_\alpha - \mu)'\Sigma^{-1}(x_\alpha - \mu) \right].$$

The maximum likelihood estimator of the mean μ is $\hat{\mu} = \bar{x} = (1/N)\sum_{\alpha=1}^{N} x_\alpha$. Let

$$(2) \qquad A = \sum_{\alpha=1}^{N} (x_\alpha - \bar{x})(x_\alpha - \bar{x})'.$$

Next we shall maximize the logarithm of (1) with $\boldsymbol{\mu}$ replaced by $\hat{\boldsymbol{\mu}}$; this is[†]

$$(3) \qquad -\tfrac{1}{2}pN\log 2\pi - \tfrac{1}{2}N\log|\boldsymbol{\Sigma}| - \tfrac{1}{2}\operatorname{tr}\boldsymbol{A}\boldsymbol{\Sigma}^{-1}.$$

(This is the logarithm of the "concentrated likelihood.") From $\boldsymbol{\Sigma}\boldsymbol{\Sigma}^{-1} = \boldsymbol{I}$, we obtain for any parameter θ

$$(4) \qquad \frac{\partial\boldsymbol{\Sigma}^{-1}}{\partial\theta} = -\boldsymbol{\Sigma}^{-1}\frac{\partial\boldsymbol{\Sigma}}{\partial\theta}\boldsymbol{\Sigma}^{-1}.$$

Then the partial derivative of (3) with regard to ψ_{ii}, a diagonal element of $\boldsymbol{\Psi}$, is $-N/2$ times

$$(5) \qquad \sigma^{ii} - \sum_{k,j=0}^{p} c_{kj}\sigma^{ji}\sigma^{ik},$$

where $\boldsymbol{\Sigma}^{-1} = (\sigma^{ij})$ and $(c_{ij}) = \boldsymbol{C} = (1/N)\boldsymbol{A}$. In matrix notation, (5) set equal to 0 yields

$$(6) \qquad \operatorname{diag}\boldsymbol{\Sigma}^{-1} = \operatorname{diag}\boldsymbol{\Sigma}^{-1}\boldsymbol{C}\boldsymbol{\Sigma}^{-1},$$

where $\operatorname{diag}\boldsymbol{H}$ indicates the diagonal terms of the matrix $\boldsymbol{H}$. Equivalently $\operatorname{diag}\boldsymbol{\Sigma}^{-1}(\boldsymbol{\Sigma} - \boldsymbol{C})\boldsymbol{\Sigma}^{-1} = \operatorname{diag}\boldsymbol{0}$. The derivative of (3) with respect to $\lambda_{k\tau}$ is $-N$ times

$$(7) \qquad \sum_{j=1}^{p}\sigma^{kj}\lambda_{j\tau} - \sum_{h,g,j=1}^{p}\sigma^{kh}c_{hg}\sigma^{gj}\lambda_{j\tau}, \quad k = 1,\ldots,p,\ \tau = 1,\ldots,m.$$

In matrix notation (7) set equal to 0 yields

$$(8) \qquad \boldsymbol{\Sigma}^{-1}\boldsymbol{\Lambda} = \boldsymbol{\Sigma}^{-1}\boldsymbol{C}\boldsymbol{\Sigma}^{-1}\boldsymbol{\Lambda}.$$

We have

$$(9) \qquad \boldsymbol{\Sigma}\boldsymbol{\Psi}^{-1}\boldsymbol{\Lambda} = (\boldsymbol{\Lambda}\boldsymbol{\Lambda}' + \boldsymbol{\Psi})\boldsymbol{\Psi}^{-1}\boldsymbol{\Lambda} = \boldsymbol{\Lambda}\boldsymbol{\Gamma} + \boldsymbol{\Lambda} = \boldsymbol{\Lambda}(\boldsymbol{\Gamma} + \boldsymbol{I}).$$

From this we obtain $\boldsymbol{\Psi}^{-1}\boldsymbol{\Lambda}(\boldsymbol{\Gamma} + \boldsymbol{I})^{-1} = \boldsymbol{\Sigma}^{-1}\boldsymbol{\Lambda}$. Multiply (8) by $\boldsymbol{\Sigma}$ and use the

[†] We could add the restriction that the off-diagonal elements of $\boldsymbol{\Lambda}'\boldsymbol{\Psi}^{-1}\boldsymbol{\Lambda}$ are 0 with Lagrange multipliers, but the Lagrange multipliers become 0 when the derivatives are set equal to 0. The restrictions do not affect the maximum.

above to obtain

$$\Lambda(\Gamma + I) = C\Psi^{-1}\Lambda \tag{10}$$

or

$$(C - \Psi)\Psi^{-1}\Lambda = \Lambda\Gamma. \tag{11}$$

Next we want to show that $\Sigma^{-1} - \Sigma^{-1}C\Sigma^{-1} = \Sigma^{-1}(\Sigma - C)\Sigma^{-1}$ is $\Psi^{-1}(\Sigma - C)\Psi^{-1}$ when (8) holds. Multiply the latter by Σ on the left and on the right to obtain

(12)

$$\begin{aligned}\Sigma\Psi^{-1}(\Sigma - C)\Psi^{-1}\Sigma &= (\Lambda\Lambda' + \Psi)\Psi^{-1}(\Psi + \Lambda\Lambda' - C)\Psi^{-1}(\Lambda\Lambda' + \Psi)\\ &= \Psi + \Lambda\Lambda' - C\end{aligned}$$

because

$$\begin{aligned}\Lambda\Lambda'\Psi^{-1}(\Psi + \Lambda\Lambda' - C) &= \Lambda\Lambda' + \Lambda\Gamma\Lambda' - \Lambda\Lambda'\Psi^{-1}C\\ &= \Lambda\left[(I + \Gamma)\Lambda' - \Lambda'\Psi^{-1}C\right]\\ &= \mathbf{0}\end{aligned} \tag{13}$$

by virtue of (10). Thus

$$\Sigma^{-1}(\Sigma - C)\Sigma^{-1} = \Psi^{-1}(\Sigma - C)\Psi^{-1}. \tag{14}$$

Then (6) is equivalent to $\operatorname{diag}\Psi^{-1}(\Sigma - C)\Psi^{-1} = \operatorname{diag}\mathbf{0}$. Since Ψ is diagonal, this equation is equivalent to

$$\operatorname{diag}(\Lambda\Lambda' + \Psi) = \operatorname{diag} C. \tag{15}$$

The equations for the estimators $\hat{\Lambda}$ and $\hat{\Psi}$ are (10), (15) and $\Lambda'\Psi^{-1}\Lambda = \text{diag}$.

We can multiply (11) on the left by $\Psi^{-\frac{1}{2}}$ to obtain

$$\Psi^{-\frac{1}{2}}(C - \Psi)\Psi^{-\frac{1}{2}}(\Psi^{-\frac{1}{2}}\Lambda) = (\Psi^{-\frac{1}{2}}\Lambda)\Gamma, \tag{16}$$

which shows that the columns of $\Psi^{-\frac{1}{2}}\Lambda$ are characteristic vectors of $\Psi^{-\frac{1}{2}}(C - \Psi)\Psi^{-\frac{1}{2}} = \Psi^{-\frac{1}{2}}C\Psi^{-\frac{1}{2}} - I$ and the corresponding diagonal elements of Γ are the characteristic roots. [In fact, the characteristic vectors of $\boldsymbol{\Psi}^{-\frac{1}{2}}\boldsymbol{C}\boldsymbol{\Psi}^{-\frac{1}{2}} - \boldsymbol{I}$ are the characteristic vectors of $\boldsymbol{\Psi}^{-\frac{1}{2}}\boldsymbol{C}\boldsymbol{\Psi}^{-\frac{1}{2}}$ because $(\boldsymbol{\Psi}^{-\frac{1}{2}}\boldsymbol{C}\boldsymbol{\Psi}^{-\frac{1}{2}} - \boldsymbol{I})\boldsymbol{x} = \gamma\boldsymbol{x}$ is equivalent to $\boldsymbol{\Psi}^{-\frac{1}{2}}\boldsymbol{C}\boldsymbol{\Psi}^{-\frac{1}{2}}\boldsymbol{x} = (1+\gamma)\boldsymbol{x}$.] The vectors are normalized by $(\boldsymbol{\Psi}^{-\frac{1}{2}}\Lambda)'(\boldsymbol{\Psi}^{-\frac{1}{2}}\Lambda) = \Lambda'\boldsymbol{\Psi}^{-1}\Lambda = \Gamma$. The characteristic roots are chosen to maximize the likelihood. To evaluate the maximized likelihood function we calculate

$$\begin{aligned}(17)\qquad \operatorname{tr}\boldsymbol{C}\hat{\boldsymbol{\Sigma}}^{-1} &= \operatorname{tr}\boldsymbol{C}\hat{\boldsymbol{\Sigma}}^{-1}(\hat{\boldsymbol{\Sigma}} - \hat{\Lambda}\hat{\Lambda}')\hat{\boldsymbol{\Psi}}^{-1}\\ &= \operatorname{tr}\left[\boldsymbol{C}\hat{\boldsymbol{\Psi}}^{-1} - (\boldsymbol{C}\hat{\boldsymbol{\Sigma}}^{-1}\hat{\Lambda})\hat{\Lambda}'\hat{\boldsymbol{\Psi}}^{-1}\right]\\ &= \operatorname{tr}\left[\boldsymbol{C}\hat{\boldsymbol{\Psi}}^{-1} - \hat{\Lambda}\hat{\Lambda}'\hat{\boldsymbol{\Psi}}^{-1}\right]\\ &= \operatorname{tr}\left[(\hat{\Lambda}\hat{\Lambda}' + \hat{\boldsymbol{\Psi}})\hat{\boldsymbol{\Psi}}^{-1} - \hat{\Lambda}\hat{\Lambda}'\hat{\boldsymbol{\Psi}}^{-1}\right]\\ &= p.\end{aligned}$$

The third equality follows from (8) multiplied on the left by $\hat{\boldsymbol{\Sigma}}$; the fourth equality follows from (15) and the fact that $\hat{\boldsymbol{\Psi}}$ is diagonal. Next we find

$$\begin{aligned}(18)\qquad |\hat{\boldsymbol{\Sigma}}| &= |\hat{\boldsymbol{\Psi}}^{\frac{1}{2}}|\cdot|\hat{\boldsymbol{\Psi}}^{-\frac{1}{2}}\hat{\Lambda}\hat{\Lambda}'\hat{\boldsymbol{\Psi}}^{-\frac{1}{2}} + \boldsymbol{I}_p|\cdot|\hat{\boldsymbol{\Psi}}^{\frac{1}{2}}|\\ &= |\hat{\boldsymbol{\Psi}}|\cdot|\hat{\Lambda}'\hat{\boldsymbol{\Psi}}^{-\frac{1}{2}}\hat{\boldsymbol{\Psi}}^{-\frac{1}{2}}\hat{\Lambda} + \boldsymbol{I}_m|\\ &= |\hat{\boldsymbol{\Psi}}|\cdot|\hat{\Gamma} + \boldsymbol{I}_m|\\ &= \prod_{i=1}^{p}\hat{\psi}_{ii}\prod_{j=1}^{m}(\hat{\gamma}_j + 1).\end{aligned}$$

The second equality is $|\boldsymbol{U}\boldsymbol{U}' + \boldsymbol{I}_p| = |\boldsymbol{U}'\boldsymbol{U} + \boldsymbol{I}_m|$ for $\boldsymbol{U}$ $(p \times m)$, which is proved as in (14) of Section 8.4. From the fact that the characteristic roots of $\boldsymbol{\Psi}^{-\frac{1}{2}}(\boldsymbol{C} - \boldsymbol{\Psi})\boldsymbol{\Psi}^{-\frac{1}{2}}$ are the roots $\gamma_1 > \gamma_2 > \cdots > \gamma_p$ of $0 = |\boldsymbol{C} - \boldsymbol{\Psi} - \gamma\boldsymbol{\Psi}| = |\boldsymbol{C} - (1+\gamma)\boldsymbol{\Psi}|$

$$(19)\qquad \frac{|\boldsymbol{C}|}{|\hat{\boldsymbol{\Psi}}|} = \prod_{i=1}^{p}(1 + \hat{\gamma}_i).$$

[Note that the roots $1 + \gamma_i$ of $\boldsymbol{\Psi}^{-\frac{1}{2}}\boldsymbol{C}\boldsymbol{\Psi}^{-\frac{1}{2}}$ are positive; the roots γ_i of $\boldsymbol{\Psi}^{-\frac{1}{2}}(\boldsymbol{C} - \boldsymbol{\Psi})\boldsymbol{\Psi}^{-\frac{1}{2}}$ are not necessarily positive; usually some will be negative.]

Then

$$|\hat{\boldsymbol{\Sigma}}| = \frac{|\boldsymbol{C}|\prod_{j\in S}(1+\hat{\gamma}_j)}{\prod_{i=1}^{p}(1+\hat{\gamma}_i)} = \frac{|\boldsymbol{C}|}{\prod_{j\notin S}(1+\hat{\gamma}_j)}, \tag{20}$$

where S is the set of indices corresponding to the roots in $\hat{\boldsymbol{\Gamma}}$. The logarithm of the maximized likelihood function is

$$-\tfrac{1}{2}pN\log 2\pi - \tfrac{1}{2}N\log|\boldsymbol{C}| - \tfrac{1}{2}N\sum_{j\notin S}\log(1+\hat{\gamma}_j) - \tfrac{1}{2}Np. \tag{21}$$

The largest roots $\hat{\gamma}_1 > \cdots > \hat{\gamma}_m$ should be selected for diagonal elements of $\hat{\boldsymbol{\Gamma}}$. Then $S = (1,\ldots,m)$. The logarithm of the concentrated likelihood (3) is a function of $\boldsymbol{\Sigma} = \boldsymbol{\Lambda}\boldsymbol{\Lambda}' + \boldsymbol{\Psi}$. This matrix is positive definite for every $\boldsymbol{\Lambda}$ and every diagonal $\boldsymbol{\Psi}$ that is positive definite; it is also positive definite for some diagonal $\boldsymbol{\Psi}$'s that are not positive definite. Hence there is not necessarily a relative maximum for $\boldsymbol{\Psi}$ positive definite. The concentrated likelihood function may increase as one or more diagonal elements of $\boldsymbol{\Psi}$ approaches 0. In that case the derivative equations may not be satisfied for $\boldsymbol{\Psi}$ positive definite.

The equations for the estimators (11) and (15) can be written as polynomial equations (multiplying (11) by $|\boldsymbol{\Psi}|$), but cannot be solved directly. There are various iterative procedures for finding a maximum of the likelihood function, including steepest descent, Newton–Raphson, scoring (using the information matrix), and Fletcher and Powell. [See Lawley and Maxwell (1971), Appendix II, for a discussion.]

Since there may not be a relative maximum in the region for which $\psi_{ii} > 0$, $i = 1,\ldots,p$, an iterative procedure may define a sequence of values of $\hat{\boldsymbol{\Lambda}}$ and $\hat{\boldsymbol{\Psi}}$ that includes $\hat{\psi}_{ii} < 0$ for some indices i. Such negative values are inadmissible because ψ_{ii} is interpreted as the variance of an error. One may impose the condition that $\psi_{ii} \geq 0$, $i = 1,\ldots,p$. Then the maximum may occur on the boundary (and not all of the derivative equations will be satisfied). For some indices i the estimated variance of the error is 0; that is, some test scores are exactly linear combinations of factor scores. If the identification conditions $\boldsymbol{\Phi} = \boldsymbol{I}$ and $\boldsymbol{\Lambda}'\boldsymbol{\Psi}^{-1}\boldsymbol{\Lambda}$ diagonal are dropped, we can find a coordinate system for the factors such that the test scores with 0 error variance can be interpreted as (transformed) factor scores. That interpretation does not seem useful. [See Lawley and Maxwell (1971) for further discussion.]

An alternative to requiring ψ_{ii} to be positive is to require ψ_{ii} to be bounded away from 0. A possibility is $\psi_{ii} \geq \varepsilon\sigma_{ii}$ for some small ε, such as 0.005. Of course, the value of ε is arbitrary; increasing ε will decrease the value of the maximum if the maximum is not in the interior of the restricted region and the derivative equations will not all be satisfied.

The nature of the concentrated likelihood is such that more than one relative maximum may be possible. Which maximum an iterative procedure approaches will depend on the initial values. Rubin and Thayer (1982) have given an example of three sets of estimates from three different initial estimates using the EM algorithm.

The EM (expectation-maximization) algorithm is a possible computational device for maximum likelihood estimation [Dempster, Laird, and Rubin (1977) and Rubin and Thayer (1982)]. The idea is to treat the unobservable $\boldsymbol{f}$'s as missing data. Under the assumption that $\boldsymbol{f}$ and $\boldsymbol{U}$ have a joint normal distribution, the sufficient statistics are the means and covariances of the $\boldsymbol{X}$'s and $\boldsymbol{f}$'s. The E step of the algorithm is to obtain the expectation of the covariances on the basis of trial values of the parameters. The M step is to maximize the likelihood function on the basis of these covariances; this step provides updated values of the parameters. The steps alternate, and the procedure usually converges to the maximum likelihood estimators. (See Problem 3.)

As noted in Section 14.3, the structure is equivariant and the factor scores are invariant under changes in the units of measurement of the observed variables $\boldsymbol{X} \to \boldsymbol{D}\boldsymbol{X}$, where $\boldsymbol{D}$ is a diagonal matrix with positive diagonal elements and Λ is identified by $\Lambda'\Psi^{-1}\Lambda$ is diagonal. If we let $\boldsymbol{D}\Lambda = \Lambda^*$, $\boldsymbol{D}\Psi\boldsymbol{D} = \Psi^*$, and $\boldsymbol{D}\boldsymbol{C}\boldsymbol{D} = \boldsymbol{C}^*$, then the logarithm of the likelihood function is a constant plus a constant times

$$(22)\quad -\log|\Psi^* + \Lambda^*\Lambda^{*\prime}| - \operatorname{tr}\boldsymbol{C}^*(\Psi^* + \Lambda^*\Lambda^{*\prime})^{-1}$$

$$= -\log|\Psi + \Lambda\Lambda'| - \operatorname{tr}\boldsymbol{C}(\Psi + \Lambda\Lambda')^{-1} - 2\log|\boldsymbol{D}|.$$

The maximum likelihood estimators of Λ^* and Ψ^* are $\hat{\Lambda}^* = \boldsymbol{D}\hat{\Lambda}$, $\hat{\Psi}^* = \boldsymbol{D}\hat{\Psi}\boldsymbol{D}$, and $\hat{\Lambda}^{*\prime}\hat{\Psi}^{*-1}\hat{\Lambda}^* = \hat{\Lambda}\hat{\Psi}^{-1}\hat{\Lambda}$ is diagonal. That is, the estimated factor loadings and error variances are merely changed by the units of measurement.

It is often convenient to use $d_{ii} = 1/\sqrt{c_{ii}}$ so $\boldsymbol{D}\boldsymbol{C}\boldsymbol{D} = (r_{ij})$ is made up of the sample correlation coefficients. The analysis is independent of the units of measurements. This fact is related to the fact that psychological test scores do not have natural units.

The fact that the factors do not depend on the location and scale factors is one reason for considering factor analysis as an analysis of interdependence.

It is convenient to give some rules of thumb for initial estimates of the communalities, $\sum_{j=1}^{m}\hat{\lambda}_{ij}^2 = 1 - \hat{\psi}_{ii}$, in terms of observed correlations. One rule is to use the $R^2_{i\cdot 1,\ldots,i-1,i+1,\ldots,p}$. Another is to use $\max_{h \neq i}|r_{ih}|$.

14.3.2. Test of the Hypothesis That the Model Fits

We shall derive the likelihood ratio test that the model fits; that is, that for a specified m the covariance matrix can be written as $\boldsymbol{\Sigma} = \boldsymbol{\Psi} + \boldsymbol{\Lambda}\boldsymbol{\Lambda}'$ for some diagonal positive definite $\boldsymbol{\Psi}$ and some $p \times m$ matrix $\boldsymbol{\Lambda}$. The likelihood ratio criterion is

$$(23) \qquad \frac{\max_{\boldsymbol{\mu},\boldsymbol{\Lambda},\boldsymbol{\Psi}} L(\boldsymbol{\mu}, \boldsymbol{\Psi} + \boldsymbol{\Lambda}\boldsymbol{\Lambda}')}{\max_{\boldsymbol{\mu},\boldsymbol{\Sigma}} L(\boldsymbol{\mu}, \boldsymbol{\Sigma})} = \frac{|\boldsymbol{C}|^{\frac{1}{2}N}}{|\hat{\boldsymbol{\Psi}} + \hat{\boldsymbol{\Lambda}}\hat{\boldsymbol{\Lambda}}|^{\frac{1}{2}N}} = \prod_{j=m+1}^{p} (1 + \hat{\gamma}_j)^{\frac{1}{2}N}$$

because the unrestricted maximum likelihood estimator of $\boldsymbol{\Sigma}$ is $\boldsymbol{C}$, $\operatorname{tr}\boldsymbol{C}(\hat{\boldsymbol{\Psi}} + \hat{\boldsymbol{\Lambda}}\hat{\boldsymbol{\Lambda}})^{-1} = p$ by (17), and $|\boldsymbol{C}|/|\hat{\boldsymbol{\Sigma}}|$ from (20). The null hypothesis is rejected if (23) is too small. We can use -2 times the logarithm of the likelihood ratio criterion:

$$(24) \qquad -N \sum_{j=m+1}^{p} \log(1 + \hat{\gamma}_j)$$

and reject the null hypothesis if (24) is too large.

If the regularity conditions for $\hat{\boldsymbol{\Psi}}$ and $\hat{\boldsymbol{\Lambda}}$ to be asymptotically normally distributed hold, the limiting distribution of (24) under the null hypothesis is χ^2 with degrees of freedom $\frac{1}{2}[(p-m)^2 - p - m]$, which is the number of elements of $\boldsymbol{\Sigma}$ plus the number of identifying restrictions minus the number of parameters in $\boldsymbol{\Psi}$ and $\boldsymbol{\Lambda}$. (See Section 14.2.2.) Bartlett (1950) has suggested replacing N by[†] $N - (2p+11)/6 - 2m/3$.

[†]This factor is heuristic. If $m = 0$, the factor from Chapter 9 is $N - (2p+11)/6$; Bartlett suggested replacing N and p by $N - m$ and $p - m$, respectively.

From (15) and the fact that $\hat{\gamma}_1, \ldots, \hat{\gamma}_p$ are the characteristic roots of $\hat{\Psi}^{-\frac{1}{2}}(C - \hat{\Psi})\hat{\Psi}^{-\frac{1}{2}}$ we have

$$
\begin{aligned}
(25) \qquad 0 &= \operatorname{tr} \hat{\Psi}^{-\frac{1}{2}}(C - \hat{\Psi} - \hat{\Lambda}\hat{\Lambda}')\hat{\Psi}^{-\frac{1}{2}} \\
&= \operatorname{tr} \hat{\Psi}^{-\frac{1}{2}}(C - \hat{\Psi})\hat{\Psi}^{-\frac{1}{2}} - \operatorname{tr} \hat{\Psi}^{-\frac{1}{2}}\hat{\Lambda}\hat{\Lambda}'\hat{\Psi}^{-\frac{1}{2}} \\
&= \operatorname{tr} \hat{\Psi}^{-\frac{1}{2}}(C - \hat{\Psi})\hat{\Psi}^{-\frac{1}{2}} - \operatorname{tr} \hat{\Gamma} \\
&= \sum_{i=1}^{p} \hat{\gamma}_i - \sum_{i=1}^{m} \hat{\gamma}_i = \sum_{i=m+1}^{p} \hat{\gamma}_i .
\end{aligned}
$$

If $|\hat{\gamma}_j| < 1$ for $j = m + 1, \ldots, p$, we can expand (24) as

$$
(26) \qquad -N \sum_{j=m+1}^{p} \left(\hat{\gamma}_j - \tfrac{1}{2}\hat{\gamma}_j^2 + \tfrac{1}{3}\hat{\gamma}_j^3 - \cdots \right) = \tfrac{1}{2}N \sum_{j=m+1}^{p} \left(\hat{\gamma}_j^2 - \tfrac{2}{3}\hat{\gamma}_j^3 + \cdots \right).
$$

The criterion is approximately $\frac{1}{2}N \sum_{j=m+1}^{p} \hat{\gamma}_j^2$. The estimators $\hat{\Psi}$ and $\hat{\Lambda}$ are found so that $C - \hat{\Psi} - \hat{\Lambda}\hat{\Lambda}'$ is small in a statistical sense or, equivalently, so $C - \hat{\Psi}$ is approximately of rank m. Then the smallest $p - m$ roots of $\hat{\Psi}^{-\frac{1}{2}}(C - \hat{\Psi})\hat{\Psi}^{-\frac{1}{2}}$ should be near 0. The criterion measures the deviations of these roots from 0. Since $\hat{\gamma}_{m+1}, \ldots, \hat{\gamma}_p$ are the nonzero roots of $\hat{\Psi}^{-\frac{1}{2}}(C - \hat{\Sigma})\hat{\Psi}^{-\frac{1}{2}}$, we see that

$$
\begin{aligned}
(27) \qquad \frac{1}{2} \sum_{j=m+1}^{p} \hat{\gamma}_j^2 &= \tfrac{1}{2} \operatorname{tr} \left[\hat{\Psi}^{-\frac{1}{2}}(C - \hat{\Sigma})\hat{\Psi}^{-\frac{1}{2}} \right]^2 \\
&= \tfrac{1}{2} \operatorname{tr} \hat{\Psi}^{-1}(C - \hat{\Sigma})\hat{\Psi}^{-1}(C - \hat{\Sigma}) \\
&= \sum_{i<j} \frac{(c_{ij} - \hat{\sigma}_{ij})^2}{\hat{\psi}_{ii}\hat{\psi}_{jj}}
\end{aligned}
$$

because the diagonal elements of $C - \hat{\Sigma}$ are 0.

In many situations the investigator does not know a value of m to hypothesize. He or she wants to determine the smallest number of factors such that the model is consistent with the data. It is customary to test successive values of m. The investigator starts with a test that the number of factors is a specified m_0 (possibly 0 or 1). If that hypothesis is rejected, one proceeds to test that the number is $m_0 + 1$. One continues in that fashion until a hypothesis is accepted or until $\frac{1}{2}[(p-m)^2 - p - m] \le 0$. In the last event one concludes that no nontrivial factor model fits. Unfortunately, the probabilities of errors under this procedure are unknown, even asymptotically.

14.3.3. Asymptotic Distributions of the Estimators

The maximum likelihood estimators $\hat{\mathbf{\Lambda}}$ and $\hat{\mathbf{\Psi}}$ maximize the average concentrated log likelihood functions $L^*(\mathbf{C}, \mathbf{\Lambda}^*, \mathbf{\Psi}^*)$ given by (3) divided by N for $\mathbf{\Sigma}^* = \mathbf{\Psi}^* + \mathbf{\Lambda}^*\mathbf{\Lambda}^{*\prime}$, subject to $\mathbf{\Lambda}^{*\prime}\mathbf{\Psi}^{*-1}\mathbf{\Lambda}^*$ being diagonal. If $\mathbf{C}$ is a consistent estimator of $\mathbf{\Sigma}$ (the "true" covariance matrix), then $L^*(\mathbf{C}, \mathbf{\Lambda}^*, \mathbf{\Psi}^*) \to L^*(\mathbf{\Psi} + \mathbf{\Lambda}\mathbf{\Lambda}', \mathbf{\Lambda}^*, \mathbf{\Psi}^*)$ uniformly in probability in a neighborhood of $\mathbf{\Lambda}$, $\mathbf{\Psi}$ and $L^*(\mathbf{\Psi} + \mathbf{\Lambda}\mathbf{\Lambda}', \mathbf{\Lambda}^*, \mathbf{\Psi}^*)$ has a unique maximum at $\mathbf{\Psi}^* = \mathbf{\Psi}$ and $\mathbf{\Lambda}^* = \mathbf{\Lambda}$. Because the function is continuous, the $\mathbf{\Lambda}^*, \mathbf{\Psi}^*$ that maximize $L^*(\mathbf{C}, \mathbf{\Lambda}^*, \mathbf{\Psi}^*)$ must converge stochastically to $\mathbf{\Lambda}, \mathbf{\Psi}$.

THEOREM 14.3.1. *If $\mathbf{\Lambda}$ and $\mathbf{\Psi}$ are identified by $\mathbf{\Lambda}'\mathbf{\Psi}^{-1}\mathbf{\Lambda}$ being diagonal and the diagonal elements are different and ordered and if $\mathbf{C} \xrightarrow{p} \mathbf{\Psi} + \mathbf{\Lambda}\mathbf{\Lambda}'$, then $\hat{\mathbf{\Psi}} \xrightarrow{p} \mathbf{\Psi}$ and $\hat{\mathbf{\Lambda}} \xrightarrow{p} \mathbf{\Lambda}$.*

A sufficient condition for $\mathbf{C} \xrightarrow{p} \mathbf{\Sigma}$ is that $(\mathbf{f}' \ \mathbf{U}')'$ have a distribution with finite second-order moments.

The estimators $\hat{\mathbf{\Lambda}}$ and $\hat{\mathbf{\Psi}}$ are the solutions to Equations (10), (15), and $\mathbf{\Lambda}'\mathbf{\Psi}^{-1}\mathbf{\Lambda} = \text{diag}$. These equations are polynomial equations. The derivatives of $\hat{\mathbf{\Lambda}}$ and $\hat{\mathbf{\Psi}}$ as functions of $\mathbf{C}$ are continuous unless they become infinite. Anderson and Rubin (1956) investigated conditions for the derivative to be finite and proved the following theorem:

THEOREM 14.3.2. *Let*

$$(\theta_{ij}) = \mathbf{\Theta} = \mathbf{\Psi} - \mathbf{\Lambda}(\mathbf{\Lambda}'\mathbf{\Psi}^{-1}\mathbf{\Lambda})^{-1}\mathbf{\Lambda}'. \tag{28}$$

If (θ_{ij}^2) is nonsingular, if $\mathbf{\Lambda}$ and $\mathbf{\Psi}$ are identified by the condition that $\mathbf{\Lambda}'\mathbf{\Psi}^{-1}\mathbf{\Lambda}$ is diagonal and the diagonal elements are different and ordered, if $\mathbf{C} \xrightarrow{p} \mathbf{\Psi} + \mathbf{\Lambda}\mathbf{\Lambda}'$, and if $\sqrt{N}(\mathbf{C} - \mathbf{\Sigma})$ has a limiting normal distribution, then $\sqrt{N}(\hat{\mathbf{\Lambda}} - \mathbf{\Lambda})$ and $\sqrt{N}(\hat{\mathbf{\Psi}} - \mathbf{\Psi})$ have a limiting normal distribution.

For example, $\sqrt{N}(\boldsymbol{C} - \boldsymbol{\Sigma})$ will have a limiting distribution if $(\boldsymbol{f}' \; \boldsymbol{U}')'$ has a distribution with finite fourth moments.

The covariance matrix of the limiting distribution of $\sqrt{N}(\hat{\boldsymbol{\Lambda}} - \boldsymbol{\Lambda})$ and $\sqrt{N}(\hat{\boldsymbol{\Psi}} - \boldsymbol{\Psi})$ is too complicated to derive or even present here. Lawley (1953) found covariances for $\sqrt{N}(\hat{\boldsymbol{\Lambda}} - \boldsymbol{\Lambda})$ appropriate for $\boldsymbol{\Psi}$ known, and Lawley (1967) extended his work to the case of $\boldsymbol{\Psi}$ estimated. [See also Lawley and Maxwell (1971).] Jennrich and Thayer (1973) corrected an error in his work.

The covariance of $\sqrt{N}(\hat{\psi}_{ii} - \psi_{ii})$ and $\sqrt{N}(\hat{\psi}_{jj} - \psi_{jj})$ in the limiting distribution is

$$2\psi_{ii}^2\psi_{jj}^2\xi^{ij}, \qquad i, j = 1, \ldots, p, \tag{29}$$

where $(\xi^{ij}) = \boldsymbol{\Xi}^{-1}$ and $\boldsymbol{\Xi} = (\theta_{ij}^2)$. The other covariances are too involved to give here.

While the asymptotic covariances are too complicated to give insight into the sampling variability, they can be programmed for computation. In that case the parameters are replaced by their consistent estimators.

14.3.4. Minimum-Distance Methods

An alternative to maximum likelihood is generalized least squares. The estimators are the values of $\boldsymbol{\Psi}$ and $\boldsymbol{\Lambda}$ that minimize

$$\operatorname{tr}(\boldsymbol{C} - \boldsymbol{\Sigma})\boldsymbol{H}(\boldsymbol{C} - \boldsymbol{\Sigma})\boldsymbol{H}, \tag{30}$$

where $\boldsymbol{\Sigma} = \boldsymbol{\Psi} + \boldsymbol{\Lambda}\boldsymbol{\Lambda}'$ and $\boldsymbol{H} = \boldsymbol{\Sigma}^{-1}$ or some consistent estimator of $\boldsymbol{\Sigma}^{-1}$. When $\boldsymbol{H} = \boldsymbol{\Sigma}^{-1}$, the objective function is of the form

$$[\boldsymbol{c} - \boldsymbol{\sigma}(\boldsymbol{\Psi}, \boldsymbol{\Lambda})]'[\operatorname{cov}\boldsymbol{c}]^{-1}[\boldsymbol{c} - \boldsymbol{\sigma}(\boldsymbol{\Psi}, \boldsymbol{\Lambda})], \tag{31}$$

where $\boldsymbol{c}$ represents the elements of $\boldsymbol{C}$ arranged in a vector, $\boldsymbol{\sigma}(\boldsymbol{\Psi}, \boldsymbol{\Lambda})$ is $\boldsymbol{\Psi} + \boldsymbol{\Lambda}\boldsymbol{\Lambda}'$ arranged in a corresponding vector, and $\operatorname{cov}\boldsymbol{c}$ is the covariance matrix of $\boldsymbol{c}$ under normality [Anderson (1973a)]. Jöreskog and Goldberger (1972) use $\boldsymbol{C}^{-1}$ for $\boldsymbol{H}$ and minimize

$$\operatorname{tr}(\boldsymbol{C} - \boldsymbol{\Sigma})\boldsymbol{C}^{-1}(\boldsymbol{C} - \boldsymbol{\Sigma})\boldsymbol{C}^{-1} = \operatorname{tr}(\boldsymbol{I} - \boldsymbol{\Sigma}\boldsymbol{C}^{-1})^2. \tag{32}$$

The matrix of derivatives with respect to the elements of $\boldsymbol{\Lambda}$ set equal to 0 form the matrix equation

$$C^{-1}(C - \Sigma)C^{-1}\Lambda = \mathbf{0}. \tag{33}$$

This can be rewritten as

$$\Lambda = \Sigma C^{-1}\Lambda. \tag{34}$$

Multiplication on the left by $\Sigma^{-1}C\Sigma^{-1}$ yields (8), which leads to (10). This estimator of Λ given Ψ is the same as the maximum likelihood estimator except for normalization of columns. The equation obtained by setting the derivatives of (32) with respect to Ψ equal to 0 is

$$\operatorname{diag} C^{-1}\left[(\Psi + \Lambda\Lambda') - C\right]C^{-1} = \operatorname{diag}\mathbf{0}. \tag{35}$$

An alternative is to minimize

$$\tfrac{1}{2}\operatorname{tr}\left\{(\Psi + \Lambda\Lambda')^{-1}\left[C - (\Psi + \Lambda\Lambda')\right]\right\}^2. \tag{36}$$

This leads to (8) or (10) and

$$\operatorname{diag}\Sigma^{-1}C\Sigma^{-1}(C - \Sigma)\Sigma^{-1} = \operatorname{diag}\mathbf{0}. \tag{37}$$

Browne (1974) has shown that the GLS estimator of Ψ has the same asymptotic distribution as the maximum likelihood estimator. Dahm and Fuller (1981) have shown that if $\operatorname{cov} c$ in (31) is replaced by a matrix converging to $\operatorname{cov} c$ and Ψ, Λ, and Φ depend on some parameters, then the asymptotic distributions are the same as for maximum likelihood.

14.3.5. Relation to Principal Component Analysis

What is the relation to principal component analysis proposed by Hotelling (1933)? As explained in Chapter 11, the vector of sample principal components is the orthogonal transformation $B'X$, where the columns of B are the characteristic vectors of C normalized by $B'B = I$. Then

$$C = BTB' = \sum_{i=1}^{p} b_i t_i b_i', \tag{38}$$

where T is the diagonal matrix with diagonal elements $t_1, \ldots, t_p$, the characteristic roots of C. If $t_{m+1}, \ldots, t_p$ are small, C can be approximated by

$$B_1 T_1 B_i' = \sum_{i=1}^{m} b_i t_i b_i', \tag{39}$$

where T_1 is the diagonal matrix with diagonal elements $t_1, \ldots, t_m$, and X is approximated by

$$B_1 B_1' X = \sum_{i=1}^{m} (b_i' X) b_i. \tag{40}$$

Then the sample covariance of the difference between X and the approximation (40) is the sample covariance of

$$X - B_1 B_1' X = B_2 B_2' X, \tag{41}$$

which is $B_2 T_2 B_2' = \sum_{i=m+1}^{p} b_i t_i b_i'$ and the sum of the variances of the components is $\sum_{i=m+1}^{p} t_i$. Here T_2 is the diagonal matrix with $t_{m+1}, \ldots, t_p$ as diagonal elements.

This analysis is in terms of some common unit of measurement. The first m components "explain" a large proportion of the "variance," $\operatorname{tr} C$. When the units of measurement are not the same (e.g., when the units are arbitrary), it is customary to standardize each measurement to (sample) variance 1. However, then the principal components do not have the interpretation in terms of variance.

Another difference between principal component analysis and factor analysis is that the former does not separate the error from the systematic part. This fault is easily remedied, however. Thomson (1934) proposed the following estimation procedure for the factor analysis model. A diagonal matrix Ψ is subtracted from C and the principal component analysis is carried out on $C - \Psi$. However, Ψ is determined so $C - \Psi$ is close to rank m. The equations are

$$(C - \Psi)\Lambda = \Lambda L, \tag{42}$$

$$\operatorname{diag}(\Psi + \Lambda\Lambda) = \operatorname{diag} C, \tag{43}$$

$$\Lambda'\Lambda = L \text{ diagonal}. \tag{44}$$

The last equation is a normalization and takes out the indeterminacy in Λ. This method allows for the error terms, but still depends on the units of measurement. The estimators are consistent but not (asymptotically) efficient in the usual factor analysis model.

14.3.6. The Centroid Method

Before the availability of high-speed computers, the centroid method was used almost exclusively because of its computational ease. For the sake of history we give a sketch of the method. Let $\boldsymbol{R}^*$ be the correlation reduced matrix, that is, the matrix consisting of r_{ij}, $i \neq j$, and $1 - \hat{\psi}^*_{ii}$, where $\hat{\psi}^*_{ii}$ is an initial estimate of the error variance in standard deviation units. Thomson's principal components approach is first to find the m characteristic vectors of $\boldsymbol{R}_0 = \boldsymbol{R}^*$ corresponding to the m largest characteristic roots. As indicated in Chapter 11, one computational method involves starting with an initial estimate of the first vector, say $\boldsymbol{x}^{(0)}$, calculating $\boldsymbol{x}^{(1)} = \boldsymbol{R}_0 \boldsymbol{x}^{(0)}$, and iterating. At the rth step $\boldsymbol{x}^{(r)}$ is approximately $\gamma_1 \boldsymbol{x}^{(r-1)}$, where γ_1 is the largest root and $\boldsymbol{x}^{(r)\prime}\boldsymbol{x}^{(r)} \sim \gamma_1^2 \boldsymbol{x}^{(r-1)\prime}\boldsymbol{x}^{(r-1)}$. Then $\boldsymbol{y}_1 = \boldsymbol{x}^{(r)} / \sqrt{\gamma_1 \boldsymbol{x}^{(r-1)\prime}\boldsymbol{x}^{(r-1)}}$ is approximately the first characteristic vector normalized so $\boldsymbol{y}_1' \boldsymbol{y}_1 = \gamma_1$. To obtain the second vector apply the same procedure to $\boldsymbol{R}_1 = \boldsymbol{R}^* - \boldsymbol{y}_1 \boldsymbol{y}_1'$.

The centroid method can be considered as a very rough approximation to the principal component approach. With psychological tests the correlation matrix usually consists of positive entries, and the first characteristic vector has all positive components, often of about the same value. The centroid method uses $\boldsymbol{\varepsilon} = (1, \dots, 1)'$ as the "initial estimate of the first vector." Then $\boldsymbol{R}^*\boldsymbol{\varepsilon} = \boldsymbol{x}^{(1)}$ is the first iterate and should be an approximation to the first characteristic vector. An approximation to the first characteristic root is $\boldsymbol{\varepsilon}'\boldsymbol{R}^*\boldsymbol{\varepsilon}/\boldsymbol{\varepsilon}'\boldsymbol{\varepsilon}$. Then $\boldsymbol{y}_1 = \boldsymbol{x}^{(1)} / \sqrt{\boldsymbol{\varepsilon}'\boldsymbol{R}^*\boldsymbol{\varepsilon}}$ is an approximation to the first characteristic vector of $\boldsymbol{R}^*$ normalized to have length squared γ_1. The operations can be carried out on an adding machine or desk calculator because $\boldsymbol{R}^*\boldsymbol{\varepsilon}$ amounts to adding across rows and $\boldsymbol{\varepsilon}'\boldsymbol{R}^*\boldsymbol{\varepsilon}$ is the sum of those row totals.

The second characteristic vector is orthogonal to the first. A vector orthogonal to $\boldsymbol{\varepsilon}$ is $\boldsymbol{\varepsilon}^*$ consisting of $p/2$ 1's and $p/2$ -1's. Then $\boldsymbol{R}_1\boldsymbol{\varepsilon}^* = \boldsymbol{x}_2$ is an approximation to the second characteristic vector and $\boldsymbol{\varepsilon}^{*\prime}\boldsymbol{R}_1\boldsymbol{\varepsilon}^*\boldsymbol{\varepsilon}^{*\prime}\boldsymbol{\varepsilon}^*$ approximates the second characteristic root. These operations involve changing signs of entries of $\boldsymbol{R}_1$ and adding. The positions of the -1's in $\boldsymbol{\varepsilon}^*$ are selected to maximize $\boldsymbol{\varepsilon}^{*\prime}\boldsymbol{R}_1\boldsymbol{\varepsilon}^*$. The procedure can be continued.

14.4. ESTIMATION FOR FIXED FACTORS

Let $\boldsymbol{x}_\alpha = (x_{1\alpha}, \dots, x_{p\alpha})'$ be an observation on $\boldsymbol{X}_\alpha$ given by

$$\boldsymbol{X}_\alpha = \boldsymbol{\Lambda} \boldsymbol{f}_\alpha + \boldsymbol{\mu} + \boldsymbol{U}_\alpha \tag{1}$$

with $\boldsymbol{f}_\alpha$ being a nonstochastic vector (an incidental parameter), $\alpha = 1, \dots, N$, satisfying $\sum_{\alpha=1}^{N} \boldsymbol{f}_\alpha = \mathbf{0}$. The likelihood function is

$$(2) \quad L = \frac{1}{\left[(2\pi)^p \prod_{i=1}^{p} \psi_{ii}\right]^{N/2}} \prod_{i=1}^{p} \exp\left\{ -\frac{1}{2} \sum_{\alpha=1}^{N} \frac{\left(x_{i\alpha} - \mu_i - \sum_{j=1}^{m} \lambda_{ij} f_{j\alpha}\right)^2}{\psi_{ii}} \right\}.$$

This likelihood function does not have a maximum. To show this fact, let $\mu_1 = 0$, $\lambda_{11} = 1$, $\lambda_{1j} = 0$, $j \neq 1$, $f_{1\alpha} = x_{1\alpha}$. Then $x_{1\alpha} - \mu_1 - \sum_{j=1}^{m} \lambda_{ij} f_{j\alpha} = 0$ and ψ_{11} does not appear in the exponent but appears only in the constant. As $\psi_{11} \to 0$, $L \to \infty$. Thus the likelihood does not have a maximum, and the maximum likelihood estimators do not exist [Anderson and Rubin (1956)]. Lawley (1941) set the partial derivatives of the likelihood equal to 0, but Solari (1969) showed that the solution is only a stationary value, not a maximum.

Since maximum likelihood estimators do not exist in the case of fixed factors, what estimation methods can be used? One possibility is to use the maximum likelihood method appropriate for random factors. It was stated by Anderson and Rubin (1956) and proved by Fuller, Pantula, and Amemiya (1982) in the case of identification by 0's that the asymptotic normal distribution of the maximum likelihood estimators for the random case is the same as for fixed factors.

The sample covariance matrix under normality has the noncentral Wishart distribution [Anderson (1946a)] depending on $\boldsymbol{\Psi}$, $\boldsymbol{\Lambda\Phi\Lambda}'$, and $N-1$. Anderson and Rubin (1956) proposed maximizing this likelihood function. However, one of the equations is difficult to solve. Again the estimators are asymptotically equivalent to the maximum likelihood estimators for the random factor case.

14.5. FACTOR INTERPRETATION AND TRANSFORMATION

14.5.1. Interpretation

The identification restrictions of $\boldsymbol{\Lambda}'\boldsymbol{\Psi}^{-1}\boldsymbol{\Lambda}$ diagonal or the first m rows of $\boldsymbol{\Lambda}$ being $\boldsymbol{I}_m$ may be convenient for computing the maximum likelihood estimators, but the components of the factor score vector may not have any intrinsic meaning. We saw in Section 14.2 that 0 coefficients may give meaning to a factor by the fact that this factor does not affect certain tests. Similarly, large factor loadings may help in interpreting a factor. The coefficient of verbal ability, for example, should be large on tests that look like they are verbal.

In psychology each variable or factor usually has a natural positive direction: more answers right on a test and more of the ability represented by the factor. It is usually expected that more ability leads to higher performance; that is, the factor loading should be positive if it is not 0. Therefore, roughly speaking, for the sake of interpretation, one may look for factor loadings that are either 0 or positive and large.

14.5.2. Transformations

The maximum likelihood estimators on the basis of some arbitrary identification conditions including $\boldsymbol{\Phi} = \boldsymbol{I}$ are $\hat{\boldsymbol{\Lambda}}$ and $\hat{\boldsymbol{\Psi}}$. We consider transformations

$$(1) \qquad \hat{\boldsymbol{\Lambda}}^* = \hat{\boldsymbol{\Lambda}} \boldsymbol{P}, \qquad \hat{\boldsymbol{\Phi}}^* = \boldsymbol{P}^{-1}(\boldsymbol{P}^{-1})' = (\boldsymbol{P}'\boldsymbol{P})^{-1}.$$

If the factors are to be orthogonal, then $\boldsymbol{\Phi}^* = \boldsymbol{I}$ and $\boldsymbol{P}$ is orthogonal. If the factors are permitted to be oblique, $\boldsymbol{P}$ can be an arbitrary nonsingular matrix and $\hat{\boldsymbol{\Phi}}^*$ an arbitrary positive definite matrix.

The rows of $\hat{\boldsymbol{\Lambda}}$ can be plotted in an m-dimensional space. Figure 14.1 is a plot of the rows of a 5×2 matrix $\hat{\boldsymbol{\Lambda}}$. The coordinates refer to factors and the

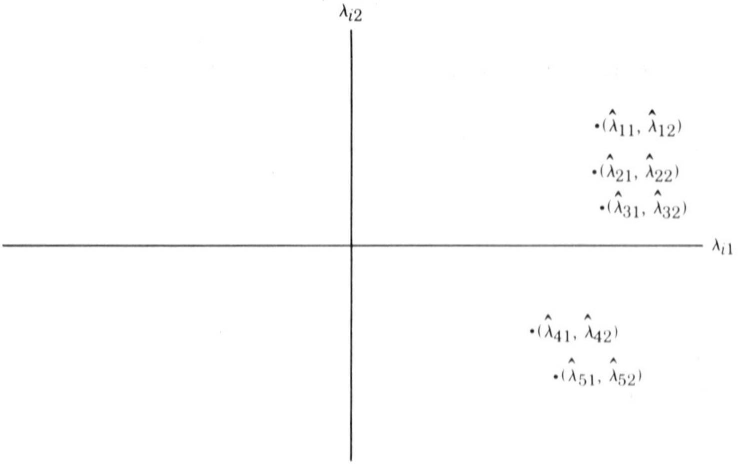

Figure 14.1. Rows of $\hat{\boldsymbol{\Lambda}}$.

points refer to tests. If $\boldsymbol{\Phi}^*$ is required to be $\boldsymbol{I}_m$, we are seeking a rotation of coordinate axes in this space. In the example that is graphed, a rotation of 45° would put all of the points into the positive quadrant, that is, $\lambda^*_{ij} \geq 0$. One of the new coordinates would be large for each of the first three points and small for the other two points, and the other coordinate would be small for the first three and large for the last two. The first factor is representative of what is common to the first three tests and the second factor represents what is common to the last two tests.

If $m > 2$, a general rotation can be approximated manually by a sequence of two-dimensional rotations.

If $\boldsymbol{\Phi}^*$ is not required to be $\boldsymbol{I}_m$, the transformation $\boldsymbol{P}$ is simply nonsingular. If the normalization of the jth column of Λ is $\lambda_{i(j),j} = 1$, then

$$1 = \hat{\lambda}^*_{i(j),j} = \sum_{k=1}^{m} \hat{\lambda}_{i(j),k} p_{kj}; \tag{2}$$

each column of $\boldsymbol{P}$ satisfies such a constraint. If the normalization is $\phi_{jj} = 1$, then

$$1 = \phi^*_{jj} = \sum_{k=1}^{m} (p^{jk})^2, \tag{3}$$

where $(p^{jk}) = \boldsymbol{P}^{-1}$.

Of the various computational procedures that are based on optimizing an objective function, we describe the varimax method proposed by Kaiser (1958) to be carried out on pairs of factors. Horst (1965), Chapter 18, extended the method to be done on all factors simultaneously. A modified criterion is

$$\sum_{j=1}^{m} \sum_{i=1}^{p} \left(\lambda^{*2}_{ij} - \frac{\sum_{h=1}^{p} \lambda^{*2}_{hj}}{p} \right)^2 = \sum_{j=1}^{m} \left[\sum_{i=1}^{p} \lambda^{*4}_{ij} - \frac{\left(\sum_{h=1}^{p} \lambda^{*2}_{hj} \right)^2}{p} \right], \tag{4}$$

which is proportional to the sum of the column variances of the squares of the transformed factor loadings. The orthogonal matrix $\boldsymbol{P}$ is selected so as to maximize (4). The procedure tends to maximize the scatter of λ^{*2}_{ij} within columns. Since $\lambda^{*2}_{ij} \geq 0$, there is a tendency to obtain some large loadings and some near 0. Kaiser's original criterion was (4) with λ^{*2}_{ij} replaced by $\lambda^{*2}_{ij} / \sum_{h=1}^{m} \lambda^{*2}_{ih}$.

Lawley and Maxwell (1971) describe other criteria. One of them is a measure of similarity to a predetermined $p \times m$ matrix of 1's and 0's.

14.5.3. Orthogonal Versus Oblique Factors

In the case of "orthogonal" factors the components are uncorrelated in the population or in the sample according to whether the factors are considered random or fixed. The idea of uncorrelated factor scores has appeal. Some psychologists claim that the orthogonality of the factor scores is essential if one is to consider the factor scores more basic than the test scores. Considerable debate has gone on among psychologists concerning this point. On the other side, Thurstone (1947), page vii, has said "it seems just as unnecessary to require that mental traits shall be uncorrelated in the general population as to require that height and weight be uncorrelated in the general population."

As we have seen, given a pair of matrices $\boldsymbol{\Lambda}, \boldsymbol{\Phi}$, equivalent pairs are given by $\boldsymbol{\Lambda P}$, $\boldsymbol{P}^{-1}\boldsymbol{\Phi P}'^{-1}$ for nonsingular $\boldsymbol{P}$'s. The pair may be selected (i.e., the $\boldsymbol{P}$ given $\boldsymbol{\Lambda}, \boldsymbol{\Phi}$) as the one with the most meaningful interpretation in terms of the subject matter of the tests. The idea of simple structure is that with 0 factor loadings in certain patterns the component factor scores can be given meaning regardless of the moment matrix. Permitting $\boldsymbol{\Phi}$ to be an arbitrary positive definite matrix allows more 0's in $\boldsymbol{\Lambda}$.

Another consideration in selecting transformations or identification conditions is autonomy, or permanence, or invariance with regard to certain changes. For example, what happens if a selection of the constituents of a population is made? In case of intelligence tests, suppose a selection is made, such as college admittees out of high school seniors, that can be assumed to involve the primary abilities. One can envisage that the relation between unobserved factor scores $\boldsymbol{f}$ and observed test scores $\boldsymbol{x}$ is unaffected by the selection, that is, that the matrix of factor loadings $\boldsymbol{\Lambda}$ is unchanged. The variance of the errors (and specific factors), the diagonal elements of $\boldsymbol{\Psi}$, may also be considered as unchanged by the selection because the errors are uncorrelated with the factors (primary abilities).

Suppose there is a "true" model, $\boldsymbol{\Lambda}, \boldsymbol{\Phi}, \boldsymbol{\Psi}$, and the investigator applies identification conditions that permit him to discover it. Next, suppose there is a selection that results in a new population of factor scores so that their covariance matrix is $\boldsymbol{\Phi}^*$. When the investigator analyzes the new observed covariance matrix $\boldsymbol{\Psi} + \boldsymbol{\Lambda \Phi}^*\boldsymbol{\Lambda}'$ will he find $\boldsymbol{\Lambda}$ again? If part of the identification conditions are that the factor moment matrix is $\boldsymbol{I}$, then he will obtain a different factor loading matrix. On the other hand, if the identification conditions are entirely on the factor loadings (specified 0's and 1's), the factor loading matrix from the analysis is the same as before.

The same consideration is relevant in comparing two populations. It may be reasonable to consider that $\Psi_1 = \Psi_2$, $\Lambda_1 = \Lambda_2$, but $\Phi_1 \neq \Phi_2$. To test the hypothesis that $\Phi_1 = \Phi_2$, one wants to use identification conditions that agree with $\Lambda_1 = \Lambda_2$ (rather than $\Lambda_1 = \Lambda_2 C$). The conditions should be on the factor loadings.

What happens if more tests are added (or deleted)? In addition to observing $X = \Lambda f + \mu + U$, suppose one observes $X^* = \Lambda^* f + \mu^* + U^*$, where U^* is uncorrelated with U. Since the common factors f are unchanged, Φ is unchanged. However, the (arbitrary) condition $\Lambda'\Psi^{-1}\Lambda$ being diagonal is changed; use of this type of condition would lead to a rotation of $(\Lambda' \; \Lambda^{*\prime})$.

14.6. ESTIMATION FOR IDENTIFICATION BY SPECIFIED ZEROS

We now consider estimation of Λ, Ψ, and Φ when Φ is unrestricted and Λ is identified by specified 0's and 1's. We assume that each column of Λ has at least $m-1$ 0's in specified positions and that the submatrix consisting of the rows of Λ containing the 0's specified for a given column is of rank $m-1$. (See Section 14.2.2.) We further assume that each column of Λ has 1 in a specified position or, alternatively, that the diagonal element of Φ corresponding to that column is 1. Then the model is identified.

The likelihood function is given by (1) of Section 14.3. The derivatives of the likelihood function set equal to 0 are

$$\operatorname{diag} \Sigma^{-1}\left[C - (\Psi + \Lambda\Phi\Lambda')\right]\Sigma^{-1} = \operatorname{diag} \mathbf{0}, \tag{1}$$

$$\Lambda'\Sigma^{-1}\left[C - (\Psi + \Lambda\Phi\Lambda')\right]\Sigma^{-1}\Lambda = \mathbf{0} \tag{2}$$

for positions in Φ that are not specified, and

$$\Sigma^{-1}\left[C - (\Psi + \Lambda\Phi\Lambda')\right]\Sigma^{-1}\Lambda = \mathbf{0} \tag{3}$$

for positions in Λ not specified, where

$$\Sigma = \Psi + \Lambda\Phi\Lambda. \tag{4}$$

These equations cannot be simplified as in Section 14.3.1 because (3) holds only for unspecified positions in Λ, and hence one cannot multiply by Σ on the left. [See Howe (1955), Anderson and Rubin (1956), and Lawley (1958).]

These equations are not useful for computation. The likelihood function, however, can be maximized numerically.

As noted before, a change in units of measurements, $\boldsymbol{X}^* = \boldsymbol{DX}$ results in a corresponding change in the parameters Λ and Ψ if identification is by 0 in specified positions of Λ and normalization is by $\phi_{jj} = 1$, $j = 1, \ldots, m$. It is readily verified that the derivative equations (1), (2), (3), and (4) are changed in a corresponding manner.

14.7. ESTIMATION OF FACTOR SCORES

It is frequently of interest to estimate the factor scores of the individuals in the group being studied. In the model with nonstochastic factors the factor scores are incidental parameters that characterize the individuals. As we have seen (Section 14.4), the maximum likelihood estimators of the parameters $(\Psi, \Lambda, \mu, f_1, \ldots, f_N)$ do not exist. We shall therefore study the estimation of the factor scores on the basis that the structural parameters (Ψ, Λ, μ) are known.

When f_α is considered as an incidental parameter, $x_\alpha - \mu$ is an observation from a distribution with mean Λf_α and covariance matrix Ψ. The weighted least squares estimator of f_α is

$$\text{(1)} \qquad \begin{aligned} \hat{f}_\alpha &= (\Lambda'\Psi^{-1}\Lambda)^{-1}\Lambda'\Psi^{-1}(x_\alpha - \mu) \\ &= \Gamma^{-1}\Lambda'\Psi^{-1}(x_\alpha - \mu), \end{aligned}$$

where $\Gamma = \Lambda'\Psi^{-1}\Lambda$ (not necessarily diagonal). This estimator is unbiased and its covariance matrix is

$$\text{(2)} \qquad \mathscr{E}(\hat{f}_\alpha - f_\alpha)(\hat{f}_\alpha - f_\alpha)' = (\Lambda'\Psi^{-1}\Lambda)^{-1} = \Gamma^{-1}$$

by the usual generalized least squares theory [Bartlett (1937b, 1938)]. It is the minimum variance unbiased linear estimator of f_α. If x_α is normal, the estimator is also maximum likelihood.

When f_α is considered random [Thomson (1951)], we suppose X_α and f_α have a joint normal distribution with mean vector $(\mu', \mathbf{0}')'$ and covariance matrix

$$\text{(3)} \qquad \mathscr{C}\begin{pmatrix} X \\ f \end{pmatrix} = \begin{pmatrix} \Psi + \Lambda\Phi\Lambda' & \Lambda\Phi \\ \Phi\Lambda' & \Phi \end{pmatrix}.$$

Then the regression of $\boldsymbol{f}$ on $\boldsymbol{X}$ (Section 2.5) is

$$\mathscr{E}(\boldsymbol{f}|\boldsymbol{X}) = \boldsymbol{\Phi\Lambda}'(\boldsymbol{\Psi} + \boldsymbol{\Lambda\Phi\Lambda}')^{-1}(\boldsymbol{x} - \boldsymbol{\mu}) \tag{4}$$

$$= \boldsymbol{\Phi}(\boldsymbol{\Phi} + \boldsymbol{\Phi\Gamma\Phi})^{-1}\boldsymbol{\Phi\Lambda}'\boldsymbol{\Psi}^{-1}(\boldsymbol{x} - \boldsymbol{\mu}).$$

The estimator or predictor of $\boldsymbol{f}_\alpha$ is

$$\hat{\boldsymbol{f}}_\alpha^* = \boldsymbol{\Phi}(\boldsymbol{\Phi} + \boldsymbol{\Phi\Gamma\Phi})^{-1}\boldsymbol{\Phi\Lambda}'\boldsymbol{\Psi}^{-1}(\boldsymbol{x}_\alpha - \boldsymbol{\mu}). \tag{5}$$

If $\boldsymbol{\Phi} = \boldsymbol{I}$, the predictor is

$$\hat{\boldsymbol{f}}_\alpha^* = (\boldsymbol{I} + \boldsymbol{\Gamma})^{-1}\boldsymbol{\Lambda}'\boldsymbol{\Psi}^{-1}(\boldsymbol{x}_\alpha - \boldsymbol{\mu}). \tag{6}$$

When $\boldsymbol{\Gamma}$ is also diagonal, the jth element of (6) is $\gamma_j/(1 + \gamma_j)$ times the jth element of (1). In the conditional distribution of $\boldsymbol{x}_\alpha$ given $\boldsymbol{f}_\alpha$ (for $\boldsymbol{\Phi} = \boldsymbol{I}$)

$$\mathscr{E}(\hat{\boldsymbol{f}}_\alpha^*|\boldsymbol{f}_\alpha) = (\boldsymbol{I} + \boldsymbol{\Gamma})^{-1}\boldsymbol{\Gamma}\boldsymbol{f}_\alpha, \tag{7}$$

$$\mathscr{C}(\hat{\boldsymbol{f}}_\alpha^*|\boldsymbol{f}_\alpha) = (\boldsymbol{I} + \boldsymbol{\Gamma})^{-1}\boldsymbol{\Gamma}(\boldsymbol{I} + \boldsymbol{\Gamma})^{-1}, \tag{8}$$

$$\mathscr{E}\left[(\hat{\boldsymbol{f}}_\alpha^* - \boldsymbol{f}_\alpha)(\hat{\boldsymbol{f}}_\alpha^* - \boldsymbol{f}_\alpha)'|\boldsymbol{f}_\alpha\right] = (\boldsymbol{I} + \boldsymbol{\Gamma})^{-1}(\boldsymbol{\Gamma} + \boldsymbol{f}_\alpha\boldsymbol{f}_\alpha')(\boldsymbol{I} + \boldsymbol{\Gamma})^{-1}, \tag{9}$$

$$\mathscr{E}(\hat{\boldsymbol{f}}_\alpha^* - \boldsymbol{f}_\alpha)(\hat{\boldsymbol{f}}_\alpha^* - \boldsymbol{f}_\alpha)' = (\boldsymbol{I} + \boldsymbol{\Gamma})^{-1}. \tag{10}$$

This last matrix describing the mean square error is smaller than (2) describing the unbiased estimator. The estimator (5) or (6) is a Bayes estimator and is appropriate when $\boldsymbol{f}_\alpha$ is treated as random.

PROBLEMS

1. (Sec. 14.2) *Identification by 0's.* Let

$$\boldsymbol{\Lambda} = \begin{pmatrix} \boldsymbol{0} & \boldsymbol{\Lambda}^{(1)} \\ \boldsymbol{\lambda}_{(1)} & \boldsymbol{\Lambda}_{(1)} \end{pmatrix}, \qquad \boldsymbol{C} = \begin{pmatrix} c_{11} & \boldsymbol{c}_{12} \\ \boldsymbol{c}_{21} & \boldsymbol{C}_{22} \end{pmatrix},$$

where $\boldsymbol{C}$ is nonsingular. Show that

$$\boldsymbol{\Lambda C} = \begin{pmatrix} \boldsymbol{0} & \boldsymbol{\Lambda}^{*(1)} \\ \boldsymbol{\lambda}_{(1)}^* & \boldsymbol{\Lambda}_{(1)}^* \end{pmatrix}$$

implies

$$C = \begin{pmatrix} c_{11} & c_{12} \\ 0 & C_{22} \end{pmatrix}$$

if and only if $\Lambda^{(1)}$ is of rank $m - 1$.

2. (Sec. 14.3) For $p = 3$, $m = 1$, and $\Lambda = \lambda$, prove $|\theta_{ij}^2| = \prod_{i=1}^3 (\lambda_i^2/\psi_{ii})$.

3. (Sec. 14.3) *The EM algorithm.*

(*a*) If f and U are normal and f and X were observed, show that the likelihood function based on $(x_1, f_1), \dots, (x_N, f_N)$ would be

$$\prod_{\alpha=1}^{N} \left\{ \frac{1}{(2\pi)^{\frac{1}{2}p} \prod_{i=1}^{p} \psi_{ii}} \exp\left[-\frac{1}{2} \sum_{i=1}^{p} \frac{\left(x_{i\alpha} - \mu_i - \sum_{j=1}^{m} \lambda_{ij} f_{j\alpha}\right)^2}{\psi_{ii}} \right] \right.$$

$$\left. \times \frac{1}{(2\pi)^{\frac{1}{2}m} |\Phi|^{\frac{1}{2}}} \exp\left[-\tfrac{1}{2} f_\alpha' \Phi^{-1} f_\alpha \right] \right\}.$$

(*b*) Show that when the factor scores are included as data the sufficient set of statistics is $\bar{x}$, $\bar{f}$, $C_{xx} = C$,

$$C_{xf} = \frac{1}{N} \sum_{\alpha=1}^{N} (x_\alpha - \bar{x})(f_\alpha - \bar{f})',$$

$$C_{ff} = \frac{1}{N} \sum_{\alpha=1}^{N} (f_\alpha - \bar{f})(f_\alpha - \bar{f})'.$$

(*c*) Show that the conditional expectations of the covariances in (*b*) given $X = (x_1, \dots, x_N)$, Λ, Φ, and Ψ are

$$C_{xx}^* = \mathscr{E}(C_{xx} | X, \Lambda, \Phi, \Psi) = C_{xx},$$

$$C_{xf}^* = \mathscr{E}(C_{xf} | X, \Lambda, \Phi, \Psi) = C_{xx}(\Psi + \Lambda\Phi\Lambda')^{-1}\Lambda\Phi,$$

$$C_{ff}^* = \mathscr{E}(C_{ff} | X, \Lambda, \Phi, \Psi) = \Phi\Lambda'(\Psi + \Lambda\Phi\Lambda')^{-1} C_{xx} (\Psi + \Lambda\Phi\Lambda')^{-1}\Lambda\Phi$$

$$+ \Phi - \Phi\Lambda'(\Psi + \Lambda\Phi\Lambda')^{-1}\Lambda\Phi.$$

(*d*) Show that the maximum likelihood estimators of Λ and Ψ given $\Phi = I$ are

$$\hat{\Lambda} = C_{xf}^* C_{ff}^{*-1},$$

$$\hat{\Psi} = C_{xx}^* - C_{xf}^* C_{ff}^{*-1} C_{xf}^{*\prime}.$$

APPENDIX A

Matrix Theory

A.1. DEFINITION OF A MATRIX AND OPERATIONS ON MATRICES

In this appendix we summarize some of the well-known definitions and theorems of matrix algebra. A number of results that are not always contained in books on matrix algebra are proved here.

An $m \times n$ matrix $\boldsymbol{A}$ is a rectangular array of real numbers

$$\text{(1)} \qquad \boldsymbol{A} = \begin{pmatrix} a_{11} & a_{12} & \cdots & a_{1n} \\ a_{21} & a_{22} & \cdots & a_{2n} \\ \vdots & \vdots & & \vdots \\ a_{m1} & a_{m2} & \cdots & a_{mn} \end{pmatrix},$$

which may be abbreviated (a_{ij}), $i = 1, 2, \ldots, m$, $j = 1, 2, \ldots, n$. Capital boldface letters will be used to denote matrices whose elements are the corresponding lower-case letters with appropriate subscripts. The sum of two matrices $\boldsymbol{A}$ and $\boldsymbol{B}$ of the same numbers of rows and columns, respectively, is defined by

$$\text{(2)} \qquad \boldsymbol{A} + \boldsymbol{B} = (a_{ij}) + (b_{ij}) = (a_{ij} + b_{ij}).$$

The product of a matrix by a real number λ is defined by

$$\text{(3)} \qquad \lambda \boldsymbol{A} = \boldsymbol{A} \lambda = (\lambda a_{ij}).$$

It can be verified that these operations have the algebraic properties

(4) $$\boldsymbol{A}+\boldsymbol{B}=\boldsymbol{B}+\boldsymbol{A},$$

(5) $$(\boldsymbol{A}+\boldsymbol{B})+\boldsymbol{C}=\boldsymbol{A}+(\boldsymbol{B}+\boldsymbol{C}),$$

(6) $$\boldsymbol{A}+(-1)\boldsymbol{A}=(0),$$

(7) $$(\lambda+\mu)\boldsymbol{A}=\lambda\boldsymbol{A}+\mu\boldsymbol{A},$$

(8) $$\lambda(\boldsymbol{A}+\boldsymbol{B})=\lambda\boldsymbol{A}+\lambda\boldsymbol{B},$$

(9) $$\lambda(\mu\boldsymbol{A})=(\lambda\mu)\boldsymbol{A}.$$

The matrix (0) with all elements 0 is denoted as $\mathbf{0}$. The operation $\boldsymbol{A}+(-1\boldsymbol{B})$ is denoted as $\boldsymbol{A}-\boldsymbol{B}$.

If $\boldsymbol{A}$ has the same number of columns as $\boldsymbol{B}$ has rows, that is, $\boldsymbol{A}=(a_{ij})$, $i=1,\ldots,l$, $j=1,\ldots,m$, $\boldsymbol{B}=(b_{jk})$, $j=1,\ldots,m$, $k=1,\ldots,n$, then $\boldsymbol{A}$ and $\boldsymbol{B}$ can be multiplied according to the rule

(10) $$\boldsymbol{AB}=(a_{ij})(b_{jk})=\left(\sum_{j=1}^{m}a_{ij}b_{jk}\right),\qquad i=1,\ldots,l,\ k=1,\ldots,n;$$

that is, $\boldsymbol{AB}$ is a matrix with l rows and n columns, the element in the ith row and kth column being $\sum_{j=1}^{m}a_{ij}b_{jk}$. The matrix product has the properties

(11) $$(\boldsymbol{AB})\boldsymbol{C}=\boldsymbol{A}(\boldsymbol{BC}),$$

(12) $$\boldsymbol{A}(\boldsymbol{B}+\boldsymbol{C})=\boldsymbol{AB}+\boldsymbol{AC},$$

(13) $$(\boldsymbol{A}+\boldsymbol{B})\boldsymbol{C}=\boldsymbol{AC}+\boldsymbol{BC}.$$

The relationships (11)–(13) hold provided one side is meaningful (i.e., the numbers of rows and columns are such that the operations can be performed) since it follows then that the other side is also meaningful. Because of (11) we can write

(14) $$(\boldsymbol{AB})\boldsymbol{C}=\boldsymbol{A}(\boldsymbol{BC})=\boldsymbol{ABC}.$$

The product $\boldsymbol{BA}$ may be meaningless even if $\boldsymbol{AB}$ is meaningful, and even when both are meaningful they are not necessarily equal.

The *transpose* of the $l\times m$ matrix $\boldsymbol{A}=(a_{ij})$ is defined to be the $m\times l$ matrix $\boldsymbol{A}'$ which has in the jth row and ith column the element that $\boldsymbol{A}$ has in the ith row and jth column. The operation of transposition has the properties

$$(A')' = A, \tag{15}$$

$$(A + B)' = A' + B', \tag{16}$$

$$(AB)' = B'A', \tag{17}$$

again with the restriction which is understood throughout this book that at least one side is meaningful.

A vector x with m components can be treated as a matrix with m rows and one column. Therefore, the above operations hold for vectors.

We shall now be concerned with square matrices of the same size, which can be added and multiplied at will. The number of rows and columns will be taken to be p. A is called *symmetric* if $A = A'$. A particular matrix of considerable interest is the *identity* matrix

$$I = \begin{pmatrix} 1 & 0 & 0 & \cdots & 0 \\ 0 & 1 & 0 & \cdots & 0 \\ 0 & 0 & 1 & \cdots & 0 \\ \vdots & \vdots & \vdots & & \vdots \\ 0 & 0 & 0 & \cdots & 1 \end{pmatrix} = (\delta_{ij}), \tag{18}$$

where δ_{ij}, the Kronecker delta, is defined by

$$\delta_{ij} = \begin{cases} 1, & i = j, \\ 0, & i \neq j. \end{cases} \tag{19}$$

The identity matrix satisfies

$$IA = AI = A. \tag{20}$$

We shall write the identity as I_p when we wish to emphasize that it is of order p. Associated with any square matrix A is the determinant $|A|$, defined by

$$|A| = \sum (-1)^{f(j_1, \ldots, j_p)} \prod_{i=1}^{p} a_{ij_i}, \tag{21}$$

where the summation is taken over all permutations $(j_1, \ldots, j_p)$ of the set of integers $(1, \ldots, p)$, and $f(j_1, \ldots, j_p)$ is the number of transpositions required to change $(1, \ldots, p)$ into $(j_1, \ldots, j_p)$. A transposition consists of interchanging two numbers, and it can be shown that, although one can transform $(1, \ldots, p)$ into $(j_1, \ldots, j_p)$ by transpositions in many different ways, the number of transpositions required is always even or always odd so that $(-1)^{f(j_1, \ldots, j_p)}$ is consistently defined. Then

$$|AB| = |A| \cdot |B|. \tag{22}$$

Also

$$|A| = |A'|. \tag{23}$$

A submatrix of A is a rectangular array obtained from A by deleting rows and columns. A *minor* is the determinant of a square submatrix of A. The minor of an element a_{ij} is the determinant of the submatrix of a square matrix A obtained by deleting the ith row and jth column. The *cofactor* of a_{ij}, say A_{ij}, is $(-1)^{i+j}$ times the minor of a_{ij}. It can be shown that

$$|A| = \sum_{j=1}^{p} a_{ij} A_{ij} = \sum_{j=1}^{p} a_{jk} A_{jk}. \tag{24}$$

If $|A| \neq 0$, there exists a unique matrix B such that $AB = I$. B is called the inverse of A and is denoted by A^{-1}. Let a^{hk} be the element of A^{-1} in the hth row and kth column. Then

$$a^{hk} = \frac{A_{kh}}{|A|}. \tag{25}$$

The operation of taking the inverse satisfies

$$(AC)^{-1} = C^{-1}A^{-1}, \tag{26}$$

since

$$(AC)(C^{-1}A^{-1}) = A(CC^{-1})A^{-1} = AIA^{-1} = AA^{-1} = I. \tag{27}$$

Also $I^{-1} = I$ and $A^{-1}A = I$. Furthermore, since the transposition of (27) gives $(A^{-1})'A' = I$ we have $(A^{-1})' = (A')^{-1}$.

A matrix whose determinant is not zero is called *nonsingular*. If $|A| \neq 0$, then the only solution to

$$Az = 0 \tag{28}$$

is the trivial one of $z = \mathbf{0}$ [by multiplication of (28) on the left by A^{-1}]. If $|A| = 0$, there is at least one nontrivial solution (that is, $z \neq \mathbf{0}$). Thus an equivalent definition of A being nonsingular is that (28) have only the trivial solution.

A set of vectors $z_1, \ldots, z_r$ is said to be *linearly independent* if there exists no set of scalars $c_1, \ldots, c_r$, not all zero, such that $\sum_{i=1}^{r} c_i z_i = \mathbf{0}$. A $q \times p$ matrix D is said to be of *rank* r if the maximum number of linearly independent columns is r. Then every minor of order $r+1$ must be zero (from the remarks in the preceding paragraph applied to the relevant $r+1$ order square matrix) and at least one minor of order r must be nonzero. Conversely, if there is at least one minor of order r that is nonzero, there is at least one set of r columns (or rows) which is linearly independent. If all minors of order $r+1$ are zero, there cannot be any set of $r+1$ columns (or rows) that are linearly independent, for such linear independence would imply a nonzero minor of order $r+1$, but this contradicts the assumption. Thus rank r is equivalently defined by the maximum number of linearly independent rows, by the maximum number of linearly independent columns, or the maximum order of nonzero minors.

We now consider the quadratic form

$$x'Ax = \sum_{i,j=1}^{p} a_{ij} x_i x_j, \tag{29}$$

where $x' = (x_1, \ldots, x_p)$ and $A = (a_{ij})$ is a symmetric matrix. This matrix A and the quadratic form are called *positive semidefinite* if $x'Ax \geq 0$ for all x. If $x'Ax > 0$ for all $x \neq \mathbf{0}$, A and the quadratic form are called *positive definite*. In this book *positive definite* implies the matrix is symmetric.

THEOREM A.1.1. *If C with p rows and columns is positive definite, and if B with p rows and q columns, $q \leq p$, is of rank q, then $B'CB$ is positive definite.*

PROOF. Given a vector $y \neq \mathbf{0}$, let $x = By$. Since B is of rank q, $By = x \neq \mathbf{0}$. Then

$$\begin{aligned} y'(B'CB)y &= (By)'C(By) \\ &= x'Cx > 0. \end{aligned} \tag{30}$$

The proof is completed by observing that $\boldsymbol{B}'\boldsymbol{C}\boldsymbol{B}$ is symmetric. As a converse, we observe that $\boldsymbol{B}'\boldsymbol{C}\boldsymbol{B}$ is positive definite only if $\boldsymbol{B}$ is of rank q, for otherwise there exists $\boldsymbol{y} \neq \boldsymbol{0}$ such that $\boldsymbol{B}\boldsymbol{y} = \boldsymbol{0}$. Q.E.D.

COROLLARY A.1.1. *If $\boldsymbol{C}$ is positive definite and $\boldsymbol{B}$ is nonsingular, then $\boldsymbol{B}'\boldsymbol{C}\boldsymbol{B}$ is positive definite.*

COROLLARY A.1.2. *If $\boldsymbol{C}$ is positive definite, then $\boldsymbol{C}^{-1}$ is positive definite.*

PROOF. $\boldsymbol{C}$ must be nonsingular for if $\boldsymbol{C}\boldsymbol{x} = \boldsymbol{0}$ for $\boldsymbol{x} \neq \boldsymbol{0}$, then $\boldsymbol{x}'\boldsymbol{C}\boldsymbol{x} = \boldsymbol{0}$ for this $\boldsymbol{x}$, but this is contrary to the assumption that $\boldsymbol{C}$ is positive definite. Let $\boldsymbol{B}$ in Theorem A.1.1 be $\boldsymbol{C}^{-1}$. Then $\boldsymbol{B}'\boldsymbol{C}\boldsymbol{B} = (\boldsymbol{C}^{-1})'\boldsymbol{C}\boldsymbol{C}^{-1} = (\boldsymbol{C}^{-1})'$. Transposing $\boldsymbol{C}\boldsymbol{C}^{-1} = \boldsymbol{I}$, we have $(\boldsymbol{C}^{-1})'\boldsymbol{C}' = (\boldsymbol{C}^{-1})'\boldsymbol{C} = \boldsymbol{I}$. Thus $\boldsymbol{C}^{-1} = (\boldsymbol{C}^{-1})'$. Q.E.D.

COROLLARY A.1.3. *The $q \times q$ matrix formed by deleting $p - q$ rows of a positive definite matrix $\boldsymbol{C}$ and the corresponding $p - q$ columns of $\boldsymbol{C}$ is positive definite.*

PROOF. This follows from Theorem A.1.1 by forming $\boldsymbol{B}$ by taking the $p \times p$ identity matrix and deleting the columns corresponding to those deleted from $\boldsymbol{C}$. Q.E.D.

The *trace* of a square matrix $\boldsymbol{A}$ is defined as $\operatorname{tr}\boldsymbol{A} = \sum_{i=1}^{p} a_{ii}$. The following properties are verified directly:

$$\operatorname{tr}(\boldsymbol{A} + \boldsymbol{B}) = \operatorname{tr}\boldsymbol{A} + \operatorname{tr}\boldsymbol{B}, \tag{31}$$

$$\operatorname{tr}\boldsymbol{A}\boldsymbol{B} = \operatorname{tr}\boldsymbol{B}\boldsymbol{A}. \tag{32}$$

A square matrix $\boldsymbol{A}$ is said to be diagonal if $a_{ij} = 0$, $i \neq j$. Then $|\boldsymbol{A}| = \prod_{i=1}^{p} a_{ii}$, for in (24) $|\boldsymbol{A}| = a_{11}A_{11}$, and in turn A_{11} is evaluated similarly.

A square matrix $\boldsymbol{A}$ is said to be triangular if $a_{ij} = 0$ for $i > j$ or alternatively for $i < j$. If $a_{ij} = 0$ for $i > j$, the matrix is *upper* triangular, and, if $a_{ij} = 0$ for $i < j$, the matrix is *lower* triangular. The product of two upper triangular matrices is upper triangular, for the i, jth term $(i > j)$ of $\boldsymbol{A}\boldsymbol{B}$ is $\sum_{k=1}^{p} a_{ik}b_{kj} = 0$ since $a_{ik} = 0$ for $k < i$ and $b_{kj} = 0$ for $k > j$. Similarly, the product of two lower triangular matrices is lower triangular. The determinant of a triangular matrix is the product of the diagonal elements. The inverse of a nonsingular triangular matrix is triangular in the same way.

THEOREM A.1.2. *If $\boldsymbol{A}$ is nonsingular, there exists a nonsingular lower triangular matrix $\boldsymbol{F}$ such that $\boldsymbol{FA} = \boldsymbol{A}^*$ is nonsingular upper triangular.*

PROOF. Let $\boldsymbol{A} = \boldsymbol{A}_1$. Define recursively $\boldsymbol{A}_g = (a_{ij}^{(g)}) = \boldsymbol{F}_{g-1}\boldsymbol{A}_{g-1}$, $g = 2, \dots, p$, where $\boldsymbol{F}_{g-1} = (f_{ij}^{(g-1)})$ has elements

$$f_{jj}^{(g-1)} = 1, \qquad j = 1, \dots, p, \tag{33}$$

$$f_{i,g-1}^{(g-1)} = -\frac{a_{i,g-1}^{(g-1)}}{a_{g-1,g-1}^{(g-1)}}, \qquad i = g, \dots, p, \tag{34}$$

$$f_{ij}^{(g-1)} = 0, \qquad \text{otherwise.} \tag{35}$$

Then

$$a_{ij}^{(g)} = 0, \qquad i = j+1, \dots, p, \; j = 1, \dots, g-1, \tag{36}$$

$$a_{ij}^{(g)} = a_{ij}^{(g-1)}, \qquad i = 1, \dots, g-1, \; j = 1, \dots, p, \tag{37}$$

$$a_{ij}^{(g)} = a_{ij}^{(g-1)} + f_{i,g-1}^{(g-1)} a_{g-1,j}^{(g-1)} = a_{ij}^{(g-1)} - \frac{a_{i,g-1}^{(g-1)} a_{g-1,j}^{(g-1)}}{a_{g-1,g-1}^{(g-1)}}, \qquad i, j = g, \dots, p. \tag{38}$$

Note that $\boldsymbol{F} = \boldsymbol{F}_{p-1}, \dots, \boldsymbol{F}_1$ is lower triangular and the elements of $\boldsymbol{A}_g$ in the first $g-1$ columns below the diagonal are 0; in particular $\boldsymbol{A}^* = \boldsymbol{FA}$ is upper triangular. From $|\boldsymbol{A}| \neq 0$ and $|\boldsymbol{F}_{g-1}| = 1$, we have $|\boldsymbol{A}_{g-1}| \neq 0$. Hence $a_{11}^{(1)}, \dots, a_{g-2,g-2}^{(g-2)}$ are different from 0 and the last $p-g$ columns of $\boldsymbol{A}_{g-1}$ can be numbered so $a_{g-1,g-1}^{(g-1)} \neq 0$; then $f_{i,g-1}^{(g-1)}$ is well-defined. Q.E.D.

The equation $\boldsymbol{FA} = \boldsymbol{A}^*$ can be solved to obtain $\boldsymbol{A} = \boldsymbol{LR}$, where $\boldsymbol{R} = \boldsymbol{A}^*$ is upper triangular and $\boldsymbol{L} = \boldsymbol{F}^{-1}$ is lower triangular and has 1's on the main diagonal (because $\boldsymbol{F}$ is lower triangular and has 1's on the main diagonal). This is known as the *LR decomposition.*

COROLLARY A.1.4. *If $\boldsymbol{A}$ is positive definite, there exists a lower triangular nonsingular matrix $\boldsymbol{F}$ such that $\boldsymbol{FAF}'$ is diagonal and positive definite.*

PROOF. From Theorem A.1.2, there exists a lower triangular nonsingular matrix $\boldsymbol{F}$ such that $\boldsymbol{FA}$ is upper triangular and nonsingular. Then $\boldsymbol{FAF}'$ is upper triangular and symmetric; hence it is diagonal. Q.E.D.

COROLLARY A.1.5. *The determinant of a positive definite matrix $\boldsymbol{A}$ is positive.*

PROOF. From the construction of $\boldsymbol{FAF}'$

$$\boldsymbol{FAF}' = \begin{bmatrix} a_{11}^{(1)} & 0 & 0 & \cdots & 0 \\ 0 & a_{22}^{(2)} & 0 & \cdots & 0 \\ 0 & 0 & a_{33}^{(3)} & \cdots & 0 \\ \vdots & \vdots & \vdots & & \vdots \\ 0 & 0 & 0 & \cdots & a_{pp}^{(p)} \end{bmatrix} \tag{39}$$

is positive definite and hence $a_{gg}^{(g)} > 0$, $g = 1, \ldots, p$, and $0 < |\boldsymbol{FAF}'| = |\boldsymbol{F}| \cdot |\boldsymbol{A}| \cdot |\boldsymbol{F}| = |\boldsymbol{A}|$. Q.E.D.

COROLLARY A.1.6. *If $\boldsymbol{A}$ is positive definite, there exists a lower triangular matrix $\boldsymbol{G}$ such that $\boldsymbol{GAG}' = \boldsymbol{I}$.*

PROOF. Let $\boldsymbol{FAF}' = \boldsymbol{D}^2$, and let $\boldsymbol{D}$ be the diagonal matrix whose diagonal elements are the positive square roots of the diagonal elements of $\boldsymbol{D}^2$. Then $\boldsymbol{G} = \boldsymbol{D}^{-1}\boldsymbol{F}$ serves the purpose. Q.E.D.

COROLLARY A.1.7. (Cholesky decomposition) *If $\boldsymbol{A}$ is positive definite, there exists a unique lower triangular matrix $\boldsymbol{T}$ ($t_{ij} = 0$, $i < j$) with positive diagonal elements such that $\boldsymbol{A} = \boldsymbol{TT}'$.*

PROOF. From Corollary A.1.6, $\boldsymbol{A} = \boldsymbol{G}^{-1}(\boldsymbol{G}')^{-1}$, where $\boldsymbol{G}$ is lower triangular. Then $\boldsymbol{T} = \boldsymbol{G}^{-1}$ is lower triangular. Q.E.D.

In effect this theorem was proved in Section 7.2 for $\boldsymbol{A} = \boldsymbol{VV}'$.

A.2. CHARACTERISTIC ROOTS AND VECTORS

The *characteristic roots* of a square matrix $\boldsymbol{B}$ are defined as the roots of the *characteristic equation*

$$(1) \qquad |\boldsymbol{B} - \lambda \boldsymbol{I}| = 0.$$

For example, with $\boldsymbol{B} = \begin{pmatrix} 5 & 2 \\ 2 & 5 \end{pmatrix}$,

$$(2) \quad |\boldsymbol{B} - \lambda \boldsymbol{I}| = \begin{vmatrix} 5 - \lambda & 2 \\ 2 & 5 - \lambda \end{vmatrix} = 25 - 4 - 10\lambda + \lambda^2 = \lambda^2 - 10\lambda + 21.$$

The degree of the equation (1) is the order of the matrix $\boldsymbol{B}$ and the constant term is $|\boldsymbol{B}|$.

A matrix $\boldsymbol{C}$ is said to be orthogonal if $\boldsymbol{C}'\boldsymbol{C} = \boldsymbol{I}$; it follows that $\boldsymbol{C}\boldsymbol{C}' = \boldsymbol{I}$. Let the vectors $\boldsymbol{x}' = (x_1, \dots, x_p)$ and $\boldsymbol{y}' = (y_1, \dots, y_p)$ represent two points in a p-dimensional Euclidean space. The distance squared between them is $D(\boldsymbol{x}, \boldsymbol{y}) = (\boldsymbol{x} - \boldsymbol{y})'(\boldsymbol{x} - \boldsymbol{y})$. The transformation $\boldsymbol{z} = \boldsymbol{C}\boldsymbol{x}$ can be thought of as a change of coordinate axes in the p-dimensional space. If $\boldsymbol{C}$ is orthogonal, the transformation is distance-preserving, for

$$(3) \quad \begin{aligned} D(\boldsymbol{C}\boldsymbol{x}, \boldsymbol{C}\boldsymbol{y}) &= (\boldsymbol{C}\boldsymbol{y} - \boldsymbol{C}\boldsymbol{x})'(\boldsymbol{C}\boldsymbol{y} - \boldsymbol{C}\boldsymbol{x}) \\ &= (\boldsymbol{y} - \boldsymbol{x})'\boldsymbol{C}'\boldsymbol{C}(\boldsymbol{y} - \boldsymbol{x}) = (\boldsymbol{y} - \boldsymbol{x})'(\boldsymbol{y} - \boldsymbol{x}) = D(\boldsymbol{x}, \boldsymbol{y}). \end{aligned}$$

Since the angles of a triangle are determined by the lengths of its sides, the transformation $\boldsymbol{z} = \boldsymbol{C}\boldsymbol{x}$ also preserves angles. It consists of a rotation together with a possible reflection of one or more axes. We shall denote $\sqrt{\boldsymbol{x}'\boldsymbol{x}}$ by $\|\boldsymbol{x}\|$.

THEOREM A.2.1. *Given any symmetric matrix* $\boldsymbol{B}$, *there exists an orthogonal matrix* $\boldsymbol{C}$ *such that*

$$(4) \qquad \boldsymbol{C}'\boldsymbol{B}\boldsymbol{C} = \boldsymbol{D} = \begin{pmatrix} d_1 & 0 & \cdots & 0 \\ 0 & d_2 & \cdots & 0 \\ \vdots & \vdots & & \vdots \\ 0 & 0 & \cdots & d_p \end{pmatrix}.$$

If $\boldsymbol{B}$ *is positive semidefinite,* $d_i \geq 0$, $i = 1, \dots, p$; *if* $\boldsymbol{B}$ *is positive definite,* $d_i > 0$, $i = 1, \dots, p$.

The proof is given in the discussion of principal components in Section 11.2 for the case of $\boldsymbol{B}$ positive semidefinite and holds for $\boldsymbol{B}$ symmetric. The characteristic equation (1) under transformation by $\boldsymbol{C}$ becomes

$$
\begin{aligned}
(5) \qquad 0 &= |\boldsymbol{C}'| \cdot |\boldsymbol{B} - \lambda \boldsymbol{I}| \cdot |\boldsymbol{C}| = |\boldsymbol{C}'(\boldsymbol{B} - \lambda \boldsymbol{I})\boldsymbol{C}| \\
&= |\boldsymbol{C}'\boldsymbol{B}\boldsymbol{C} - \lambda \boldsymbol{I}| = |\boldsymbol{D} - \lambda \boldsymbol{I}| \\
&= \begin{vmatrix} d_1 - \lambda & 0 & \cdots & 0 \\ 0 & d_2 - \lambda & \cdots & 0 \\ \vdots & \vdots & & \vdots \\ 0 & 0 & \cdots & d_p - \lambda \end{vmatrix} = \prod_{i=1}^{p} (d_i - \lambda).
\end{aligned}
$$

Thus the characteristic roots of $\boldsymbol{B}$ are the diagonal elements of the transformed matrix $\boldsymbol{D}$.

If λ_i is a characteristic root of $\boldsymbol{B}$, then a vector $\boldsymbol{x}_i$ not identically $\boldsymbol{0}$ satisfying

$$(6) \qquad (\boldsymbol{B} - \lambda_i \boldsymbol{I})\boldsymbol{x}_i = \boldsymbol{0}$$

is called a *characteristic vector* of the matrix $\boldsymbol{B}$ corresponding to the characteristic root λ_i. Any scalar multiple of $\boldsymbol{x}_i$ is also a characteristic vector. When $\boldsymbol{B}$ is symmetric, $\boldsymbol{x}_i'(\boldsymbol{B} - \lambda_i \boldsymbol{I}) = \boldsymbol{0}$. If the roots are distinct, $\boldsymbol{x}_j'\boldsymbol{B}\boldsymbol{x}_i = 0$ and $\boldsymbol{x}_j'\boldsymbol{x}_i = 0$, $i \neq j$. Let $\boldsymbol{c}_i = (1/\|\boldsymbol{x}_i\|)\boldsymbol{x}_i$ be the ith normalized characteristic vector, and let $\boldsymbol{C} = (\boldsymbol{c}_1, \ldots, \boldsymbol{c}_p)$. Then $\boldsymbol{C}'\boldsymbol{C} = \boldsymbol{I}$ and $\boldsymbol{B}\boldsymbol{C} = \boldsymbol{C}\boldsymbol{D}$. These lead to (4). If a characteristic root has multiplicity m, then a set of m corresponding characteristic vectors can be replaced by m linearly independent linear combinations of them. The vectors can be chosen to satisfy (6) and $\boldsymbol{x}_j'\boldsymbol{x}_i = 0$ and $\boldsymbol{x}_j'\boldsymbol{B}\boldsymbol{x}_i = 0$, $i \neq j$.

A characteristic vector lies in the direction of the principal axis (see Chapter 11). The characteristic roots of $\boldsymbol{B}$ are proportional to the squares of the reciprocals of the lengths of the principal axes of the ellipsoid

$$(7) \qquad \boldsymbol{x}'\boldsymbol{B}\boldsymbol{x} = 1$$

since this becomes under the rotation $\boldsymbol{y} = \boldsymbol{C}\boldsymbol{x}$

$$(8) \qquad 1 = \boldsymbol{y}'\boldsymbol{D}\boldsymbol{y} = \sum_{i=1}^{p} d_i y_i^2 .$$

For a pair of matrices $\boldsymbol{A}$ (nonsingular) and $\boldsymbol{B}$ we shall also consider equations of the form

$$|\boldsymbol{B} - \lambda \boldsymbol{A}| = 0. \tag{9}$$

The roots of such equations are of interest because of their invariance under certain transformations. In fact, for nonsingular $\boldsymbol{C}$, the roots of

$$|\boldsymbol{C}'\boldsymbol{B}\boldsymbol{C} - \lambda(\boldsymbol{C}'\boldsymbol{A}\boldsymbol{C})| = 0 \tag{10}$$

are the same as those of (9) since

$$|\boldsymbol{C}'\boldsymbol{B}\boldsymbol{C} - \lambda \boldsymbol{C}'\boldsymbol{A}\boldsymbol{C}| = |\boldsymbol{C}'(\boldsymbol{B} - \lambda\boldsymbol{A})\boldsymbol{C}| = |\boldsymbol{C}'| \cdot |\boldsymbol{B} - \lambda\boldsymbol{A}| \cdot |\boldsymbol{C}| \tag{11}$$

and $|\boldsymbol{C}'| = |\boldsymbol{C}| \neq 0$.

By Corollary A.1.6 we have that if $\boldsymbol{A}$ is positive definite there is a matrix $\boldsymbol{E}$ such that $\boldsymbol{E}'\boldsymbol{A}\boldsymbol{E} = \boldsymbol{I}$. Let $\boldsymbol{E}'\boldsymbol{B}\boldsymbol{E} = \boldsymbol{B}^*$. From Theorem A.2.1 we deduce that there exists an orthogonal matrix $\boldsymbol{C}$ such that $\boldsymbol{C}'\boldsymbol{B}^*\boldsymbol{C} = \boldsymbol{D}$, where $\boldsymbol{D}$ is diagonal. Defining $\boldsymbol{E}\boldsymbol{C}$ as $\boldsymbol{F}$, we have the following theorem:

THEOREM A.2.2. *Given $\boldsymbol{B}$ positive semidefinite and $\boldsymbol{A}$ positive definite, there exists a nonsingular matrix $\boldsymbol{F}$ such that*

$$\boldsymbol{F}'\boldsymbol{B}\boldsymbol{F} = \begin{pmatrix} \lambda_1 & 0 & \cdots & 0 \\ 0 & \lambda_2 & \cdots & 0 \\ \vdots & \vdots & & \vdots \\ 0 & 0 & \cdots & \lambda_p \end{pmatrix}, \tag{12}$$

$$\boldsymbol{F}'\boldsymbol{A}\boldsymbol{F} = \boldsymbol{I}, \tag{13}$$

where $\lambda_1 \geq \cdots \geq \lambda_p$ (≥ 0) are the roots of (9). If $\boldsymbol{B}$ is positive definite, $\lambda_i > 0$, $i = 1, \ldots, p$.

Corresponding to each root λ_i there is a vector $\boldsymbol{x}_i$ satisfying

$$(\boldsymbol{B} - \lambda_i \boldsymbol{A})\boldsymbol{x}_i = \boldsymbol{0} \tag{14}$$

and $\boldsymbol{x}_i'\boldsymbol{A}\boldsymbol{x}_i = 1$. If the roots are distinct $\boldsymbol{x}_j'\boldsymbol{B}\boldsymbol{x}_i = 0$ and $\boldsymbol{x}_j'\boldsymbol{A}\boldsymbol{x}_i = 0$, $i \neq j$. Then $\boldsymbol{F} = (\boldsymbol{x}_1, \ldots, \boldsymbol{x}_p)$. If a root has multiplicity m, then a set of m linearly independent $\boldsymbol{x}_i$'s can be replaced by m linearly independent combinations of them. The vectors can be chosen to satisfy (14) and $\boldsymbol{x}_j'\boldsymbol{B}\boldsymbol{x}_i = 0$ and $\boldsymbol{x}_j'\boldsymbol{A}\boldsymbol{x}_i = 0$, $i \neq j$.

THEOREM A.2.3. (The Singular Value Decomposition) *Given an* $n \times p$ *matrix*, $\boldsymbol{X}$, $n \geq p$, *there exists an* $n \times n$ *orthogonal matrix* $\boldsymbol{P}$, *a* $p \times p$ *orthogonal matrix* $\boldsymbol{Q}$, *and an* $n \times p$ *matrix* $\boldsymbol{D}$ *consisting of a* $p \times p$ *diagonal positive semidefinite matrix and an* $(n-p) \times p$ *zero matrix such that*

$$\boldsymbol{X} = \boldsymbol{PDQ}. \tag{15}$$

PROOF. From Theorem A.2.1, there exists a $p \times p$ orthogonal matrix $\boldsymbol{Q}$ and a diagonal matrix $\boldsymbol{E}$ such that

$$\boldsymbol{QX'XQ'} = \begin{pmatrix} \boldsymbol{E}_1 & \boldsymbol{0} \\ \boldsymbol{0} & \boldsymbol{0} \end{pmatrix}, \tag{16}$$

where $\boldsymbol{E}_1$ is diagonal and positive definite. Let $\boldsymbol{XQ'} = \boldsymbol{Y} = (\boldsymbol{Y}_1\ \boldsymbol{Y}_2)$, where the number of columns of $\boldsymbol{Y}_1$ is the order of $\boldsymbol{E}_1$. Then $\boldsymbol{Y}_2'\boldsymbol{Y}_2 = \boldsymbol{0}$, and hence $\boldsymbol{Y}_2 = \boldsymbol{0}$. Let $\boldsymbol{P}_1 = \boldsymbol{Y}_1\boldsymbol{E}_1^{-\frac{1}{2}}$. Then $\boldsymbol{P}_1'\boldsymbol{P}_1 = \boldsymbol{I}$. An $n \times n$ orthogonal matrix $\boldsymbol{P} = (\boldsymbol{P}_1\ \boldsymbol{P}_2)$ satisfying the theorem is obtained by adjoining $\boldsymbol{P}_2$ to make $\boldsymbol{P}$ orthogonal. Then the upper left-hand corner of $\boldsymbol{D}$ is $\boldsymbol{E}_1^{\frac{1}{2}}$ and the rest of $\boldsymbol{D}$ consists of zeros. Q.E.D.

THEOREM A.2.4. *Let* $\boldsymbol{A}$ *be positive definite and* $\boldsymbol{B}$ *be positive semidefinite. Then*

$$\lambda_p \leq \frac{\boldsymbol{x}'\boldsymbol{Bx}}{\boldsymbol{x}'\boldsymbol{x}} \leq \lambda_1, \tag{17}$$

where λ_1 *and* λ_p *are the largest and smallest roots of* (1), *and*

$$\lambda_p \leq \frac{\boldsymbol{x}'\boldsymbol{Bx}}{\boldsymbol{x}'\boldsymbol{Ax}} \leq \lambda_1, \tag{18}$$

where λ_1 *and* λ_p *are the largest and smallest roots of* (9).

PROOF. The inequalities (17) were essentially proved in Section 11.2, and can also be derived from (4). The inequalities (18) follow from Theorem A.2.2. Q.E.D.

A.3. PARTITIONED VECTORS AND MATRICES

Consider the matrix $\boldsymbol{A}$ defined by (1) of Section A.1. Let

$$
(1)\qquad
\begin{aligned}
\boldsymbol{A}_{11} &= (a_{ij}), && i = 1,\dots,p;\ j = 1,\dots,q,\\
\boldsymbol{A}_{12} &= (a_{ij}), && i = 1,\dots,p;\ j = q+1,\dots,n,\\
\boldsymbol{A}_{21} &= (a_{ij}), && i = p+1,\dots,m;\ j = 1,\dots,q,\\
\boldsymbol{A}_{22} &= (a_{ij}), && i = p+1,\dots,m;\ j = q+1,\dots,n.
\end{aligned}
$$

Then we can write

$$
(2)\qquad \boldsymbol{A} = \begin{pmatrix} \boldsymbol{A}_{11} & \boldsymbol{A}_{12} \\ \boldsymbol{A}_{21} & \boldsymbol{A}_{22} \end{pmatrix}.
$$

We say that $\boldsymbol{A}$ has been partitioned into submatrices $\boldsymbol{A}_{ij}$. Let $\boldsymbol{B}$ $(m \times n)$ be partitioned similarly into submatrices $\boldsymbol{B}_{ij}$, $i, j = 1, 2$. Then

$$
(3)\qquad \boldsymbol{A} + \boldsymbol{B} = \begin{pmatrix} \boldsymbol{A}_{11} + \boldsymbol{B}_{11} & \boldsymbol{A}_{12} + \boldsymbol{B}_{12} \\ \boldsymbol{A}_{21} + \boldsymbol{B}_{21} & \boldsymbol{A}_{22} + \boldsymbol{B}_{22} \end{pmatrix}.
$$

Now partition $\boldsymbol{C}$ $(n \times r)$ as

$$
(4)\qquad \boldsymbol{C} = \begin{pmatrix} \boldsymbol{C}_{11} & \boldsymbol{C}_{12} \\ \boldsymbol{C}_{21} & \boldsymbol{C}_{22} \end{pmatrix},
$$

where $\boldsymbol{C}_{11}$ and $\boldsymbol{C}_{12}$ have q rows and $\boldsymbol{C}_{11}$ and $\boldsymbol{C}_{21}$ have s columns. Then

$$
\begin{aligned}
(5)\qquad \boldsymbol{AC} &= \begin{pmatrix} \boldsymbol{A}_{11} & \boldsymbol{A}_{12} \\ \boldsymbol{A}_{21} & \boldsymbol{A}_{22} \end{pmatrix} \begin{pmatrix} \boldsymbol{C}_{11} & \boldsymbol{C}_{12} \\ \boldsymbol{C}_{21} & \boldsymbol{C}_{22} \end{pmatrix} \\
&= \begin{pmatrix} \boldsymbol{A}_{11}\boldsymbol{C}_{11} + \boldsymbol{A}_{12}\boldsymbol{C}_{21} & \boldsymbol{A}_{11}\boldsymbol{C}_{12} + \boldsymbol{A}_{12}\boldsymbol{C}_{22} \\ \boldsymbol{A}_{21}\boldsymbol{C}_{11} + \boldsymbol{A}_{22}\boldsymbol{C}_{21} & \boldsymbol{A}_{21}\boldsymbol{C}_{12} + \boldsymbol{A}_{22}\boldsymbol{C}_{22} \end{pmatrix}.
\end{aligned}
$$

To verify this, consider an element in the first p rows and first s columns of

$\boldsymbol{AC}$. The i, jth element is

$$\sum_{k=1}^{n} a_{ik}c_{kj}, \qquad i \le p,\ j \le s. \tag{6}$$

This sum can be written

$$\sum_{k=1}^{q} a_{ik}c_{kj} + \sum_{k=q+1}^{n} a_{ik}c_{kj}. \tag{7}$$

The first sum is the i, jth element of $\boldsymbol{A}_{11}\boldsymbol{C}_{11}$, the second sum is the i, jth element of $\boldsymbol{A}_{12}\boldsymbol{C}_{21}$, and therefore the entire sum (6) is the i, jth element of $\boldsymbol{A}_{11}\boldsymbol{C}_{11} + \boldsymbol{A}_{12}\boldsymbol{C}_{21}$. In a similar fashion we can verify that the other submatrices of $\boldsymbol{AC}$ can be written as in (5).

We note in passing that if $\boldsymbol{A}$ is partitioned as in (2), then the transpose of $\boldsymbol{A}$ can be written

$$\boldsymbol{A}' = \begin{pmatrix} \boldsymbol{A}'_{11} & \boldsymbol{A}'_{21} \\ \boldsymbol{A}'_{12} & \boldsymbol{A}'_{22} \end{pmatrix}. \tag{8}$$

If $\boldsymbol{A}_{12} = \boldsymbol{0}$ and $\boldsymbol{A}_{21} = \boldsymbol{0}$, then for $\boldsymbol{A}$ positive definite and $\boldsymbol{A}_{11}$ square

$$\boldsymbol{A}^{-1} = \begin{pmatrix} \boldsymbol{A}_{11}^{-1} & \boldsymbol{0} \\ \boldsymbol{0} & \boldsymbol{A}_{22}^{-1} \end{pmatrix}. \tag{9}$$

The matrix on the right exists because $\boldsymbol{A}_{11}$ and $\boldsymbol{A}_{22}$ are nonsingular. That the right-hand matrix is the inverse of $\boldsymbol{A}$ is verified by multiplication

$$\begin{pmatrix} \boldsymbol{A}_{11} & \boldsymbol{0} \\ \boldsymbol{0} & \boldsymbol{A}_{22} \end{pmatrix} \begin{pmatrix} \boldsymbol{A}_{11}^{-1} & \boldsymbol{0} \\ \boldsymbol{0} & \boldsymbol{A}_{22}^{-1} \end{pmatrix} = \begin{pmatrix} \boldsymbol{I} & \boldsymbol{0} \\ \boldsymbol{0} & \boldsymbol{I} \end{pmatrix}, \tag{10}$$

which is a partitioned form of $\boldsymbol{I}_p$.

We also note that

$$\begin{vmatrix} \boldsymbol{A}_{11} & \boldsymbol{0} \\ \boldsymbol{0} & \boldsymbol{A}_{22} \end{vmatrix} = \begin{vmatrix} \boldsymbol{A}_{11} & \boldsymbol{0} \\ \boldsymbol{0} & \boldsymbol{I} \end{vmatrix} \cdot \begin{vmatrix} \boldsymbol{I} & \boldsymbol{0} \\ \boldsymbol{0} & \boldsymbol{A}_{22} \end{vmatrix} = |\boldsymbol{A}_{11}| \cdot |\boldsymbol{A}_{22}|. \tag{11}$$

The evaluation of the first determinant in the middle is made by expanding according to minors of the last row; the only nonzero element in the sum is the last which is 1 times a determinant of the same form with $\boldsymbol{I}$ of order one less. The procedure is repeated until $|\boldsymbol{A}_{11}|$ is the minor. Similarly,

$$(12) \qquad \begin{vmatrix} \boldsymbol{A}_{11} & \boldsymbol{A}_{12} \\ \boldsymbol{0} & \boldsymbol{A}_{22} \end{vmatrix} = \begin{vmatrix} \boldsymbol{I} & \boldsymbol{0} \\ \boldsymbol{0} & \boldsymbol{A}_{22} \end{vmatrix} \cdot \begin{vmatrix} \boldsymbol{A}_{11} & \boldsymbol{A}_{12} \\ \boldsymbol{0} & \boldsymbol{I} \end{vmatrix} = |\boldsymbol{A}_{11}| \cdot |\boldsymbol{A}_{22}|.$$

A useful fact is that if $\boldsymbol{A}_1$ of q rows and p columns is of rank q, there exists a matrix $\boldsymbol{A}_2$ of $p-q$ rows and p columns such that

$$(13) \qquad \boldsymbol{A} = \begin{pmatrix} \boldsymbol{A}_1 \\ \boldsymbol{A}_2 \end{pmatrix}$$

is nonsingular. This statement is verified by numbering the columns of $\boldsymbol{A}$ so that $\boldsymbol{A}_{11}$ consisting of the first q columns of $\boldsymbol{A}_1$ is nonsingular (at least one $q \times q$ minor of $\boldsymbol{A}_1$ is different from zero) and then taking $\boldsymbol{A}_2$ as $(\boldsymbol{0}\ \boldsymbol{I})$; then

$$(14) \qquad |\boldsymbol{A}| = \begin{vmatrix} \boldsymbol{A}_{11} & \boldsymbol{A}_{12} \\ \boldsymbol{0} & \boldsymbol{I} \end{vmatrix} = |\boldsymbol{A}_{11}|,$$

which is not equal to zero.

THEOREM A.3.1. *Let the square matrix* $\boldsymbol{A}$ *be partitioned as in* (2) *so that* $\boldsymbol{A}_{22}$ *is square. If* $\boldsymbol{A}_{22}$ *is nonsingular, let*

$$(15) \qquad \boldsymbol{B} = \begin{pmatrix} \boldsymbol{I} & -\boldsymbol{A}_{12}\boldsymbol{A}_{22}^{-1} \\ \boldsymbol{0} & \boldsymbol{I} \end{pmatrix}, \qquad \boldsymbol{C} = \begin{pmatrix} \boldsymbol{I} & \boldsymbol{0} \\ -\boldsymbol{A}_{22}^{-1}\boldsymbol{A}_{21} & \boldsymbol{I} \end{pmatrix}.$$

Then

$$(16) \qquad \boldsymbol{BA} = \begin{pmatrix} \boldsymbol{A}_{11} - \boldsymbol{A}_{12}\boldsymbol{A}_{22}^{-1}\boldsymbol{A}_{21} & \boldsymbol{0} \\ \boldsymbol{A}_{21} & \boldsymbol{A}_{22} \end{pmatrix}, \qquad \boldsymbol{AC} = \begin{pmatrix} \boldsymbol{A}_{11} - \boldsymbol{A}_{12}\boldsymbol{A}_{22}^{-1}\boldsymbol{A}_{21} & \boldsymbol{A}_{12} \\ \boldsymbol{0} & \boldsymbol{A}_{22} \end{pmatrix},$$

$$(17) \qquad \boldsymbol{BAC} = \begin{pmatrix} \boldsymbol{A}_{11} - \boldsymbol{A}_{12}\boldsymbol{A}_{22}^{-1}\boldsymbol{A}_{21} & \boldsymbol{0} \\ \boldsymbol{0} & \boldsymbol{A}_{22} \end{pmatrix}.$$

If $\boldsymbol{A}$ *is symmetric,* $\boldsymbol{C} = \boldsymbol{B}'$.

THEOREM A.3.2. *Let the square matrix* $\boldsymbol{A}$ *be partitioned as in* (2) *so that* $\boldsymbol{A}_{22}$ *is square. If* $\boldsymbol{A}_{22}$ *is nonsingular,*

$$|\boldsymbol{A}| = |\boldsymbol{A}_{11} - \boldsymbol{A}_{12}\boldsymbol{A}_{22}^{-1}\boldsymbol{A}_{21}| \cdot |\boldsymbol{A}_{22}|. \tag{18}$$

PROOF. Equation (18) follows from (16) because $|\boldsymbol{B}| = 1$. Q.E.D.

COROLLARY A.3.1. *For* $\boldsymbol{C}$ *nonsingular*

$$|\boldsymbol{C} + \boldsymbol{y}\boldsymbol{y}'| = |\boldsymbol{C}|(1 + \boldsymbol{y}'\boldsymbol{C}^{-1}\boldsymbol{y}). \tag{19}$$

PROOF. By Theorem A.3.2, the left-hand side of (19) is

$$\begin{vmatrix} 1 & -\boldsymbol{y}' \\ \boldsymbol{y} & \boldsymbol{C} \end{vmatrix} = \begin{vmatrix} \boldsymbol{C} & \boldsymbol{y} \\ -\boldsymbol{y}' & 1 \end{vmatrix}, \tag{20}$$

which is the right-hand side of (19) by the same theorem. Q.E.D.

THEOREM A.3.3. *Let the nonsingular matrix* $\boldsymbol{A}$ *be partitioned as in* (2) *so that* $\boldsymbol{A}_{22}$ *is square. If* $\boldsymbol{A}_{22}$ *is nonsingular, let* $\boldsymbol{A}_{11\cdot 2} = \boldsymbol{A}_{11} - \boldsymbol{A}_{12}\boldsymbol{A}_{22}^{-1}\boldsymbol{A}_{21}$. *Then*

$$\boldsymbol{A}^{-1} = \begin{pmatrix} \boldsymbol{A}_{11\cdot 2}^{-1} & -\boldsymbol{A}_{11\cdot 2}^{-1}\boldsymbol{A}_{12}\boldsymbol{A}_{22}^{-1} \\ -\boldsymbol{A}_{22}^{-1}\boldsymbol{A}_{21}\boldsymbol{A}_{11\cdot 2}^{-1} & \boldsymbol{A}_{22}^{-1}\boldsymbol{A}_{21}\boldsymbol{A}_{11\cdot 2}^{-1}\boldsymbol{A}_{12}\boldsymbol{A}_{22}^{-1} + \boldsymbol{A}_{22}^{-1} \end{pmatrix}. \tag{21}$$

PROOF. From Theorem A.3.1,

$$\boldsymbol{A} = \boldsymbol{B}^{-1}\begin{pmatrix} \boldsymbol{A}_{11\cdot 2} & \boldsymbol{0} \\ \boldsymbol{0} & \boldsymbol{A}_{22} \end{pmatrix}\boldsymbol{C}^{-1}. \tag{22}$$

Hence

$$\begin{aligned} \boldsymbol{A}^{-1} &= \boldsymbol{C}\begin{pmatrix} \boldsymbol{A}_{11\cdot 2} & \boldsymbol{0} \\ \boldsymbol{0} & \boldsymbol{A}_{22} \end{pmatrix}^{-1}\boldsymbol{B} \\ &= \begin{pmatrix} \boldsymbol{I} & \boldsymbol{0} \\ -\boldsymbol{A}_{22}^{-1}\boldsymbol{A}_{21} & \boldsymbol{I} \end{pmatrix}\begin{pmatrix} \boldsymbol{A}_{11\cdot 2}^{-1} & \boldsymbol{0} \\ \boldsymbol{0} & \boldsymbol{A}_{22}^{-1} \end{pmatrix}\begin{pmatrix} \boldsymbol{I} & -\boldsymbol{A}_{12}\boldsymbol{A}_{22} \\ \boldsymbol{0} & \boldsymbol{I} \end{pmatrix}. \end{aligned} \tag{23}$$

Multiplication gives the desired result. Q.E.D.

COROLLARY A.3.2. *If* $x' = (x^{(1)\prime}\ x^{(2)\prime})$, *then*

(24)

$$x'A^{-1}x = \left(x^{(1)} - A_{12}A_{22}^{-1}x^{(2)}\right)'A_{11\cdot 2}^{-1}\left(x^{(1)} - A_{12}A_{22}^{-1}x^{(2)}\right) + x^{(2)\prime}A_{22}^{-1}x^{(2)}.$$

PROOF. From the theorem

(25)

$$\begin{aligned} x'A^{-1}x &= x^{(1)\prime}A_{11\cdot 2}^{-1}x^{(1)} - x^{(1)\prime}A_{11\cdot 2}^{-1}A_{12}A_{22}^{-1}x^{(2)} \\ &\quad - x^{(2)\prime}A_{22}^{-1}A_{21}A_{11\cdot 2}^{-1}x^{(1)} + x^{(2)\prime}\left(A_{22}^{-1}A_{21}A_{11\cdot 2}^{-1}A_{12}A_{22}^{-1} + A_{22}^{-1}\right)x^{(2)}, \end{aligned}$$

which is equal to the right-hand side of (24). Q.E.D.

THEOREM A.3.4. *Let the nonsingular matrix* A *be partitioned as in* (2) *so that* A_{22} *is square. If* A_{22} *is nonsingular,*

$$(26)\quad \left(A_{22} - A_{21}A_{11}^{-1}A_{12}\right)^{-1} = A_{22}^{-1}A_{21}(A_{11} - A_{12}A_{22}^{-1}A_{21})^{-1}A_{12}A_{22}^{-1} + A_{22}^{-1}.$$

PROOF. The lower right-hand corner of A^{-1} is the right-hand side of (26) by Theorem A.3.3 and is also the left-hand side of (26) by interchange of 1 and 2. Q.E.D.

THEOREM A.3.5. *Let* U *be* $p \times m$. *The conditions that* $I_p - UU'$, $I_m - U'U$, *and*

$$(27)\qquad \begin{pmatrix} I_p & U \\ U' & I_m \end{pmatrix}$$

are positive definite are equivalent.

PROOF. We have

$$\begin{aligned} (28)\quad (v'\ w')\begin{pmatrix} I_p & U \\ U' & I_m \end{pmatrix}\begin{pmatrix} v \\ w \end{pmatrix} &= v'v + v'Uw + w'U'v + w'w \\ &= v'\left(I_m - UU'\right)v + (U'v + w)'(U'v + w). \end{aligned}$$

The second term on the right-hand side is nonnegative; the first term is positive for all $\boldsymbol{v} \neq \mathbf{0}$ if and only if $\boldsymbol{I}_m - \boldsymbol{U}'\boldsymbol{U}$ is positive definite. Reversing the roles of $\boldsymbol{v}$ and $\boldsymbol{w}$ shows that (27) is positive definite if and only if $\boldsymbol{I}_p - \boldsymbol{U}\boldsymbol{U}'$ is positive definite. Q.E.D.

A.4. SOME MISCELLANEOUS RESULTS

THEOREM A.4.1. *Let $\boldsymbol{C}$ be $p \times p$, positive semidefinite, and of rank r ($\leq p$). Then there is a nonsingular matrix $\boldsymbol{A}$ such that*

$$\boldsymbol{ACA}' = \begin{pmatrix} \boldsymbol{I}_r & \mathbf{0} \\ \mathbf{0} & \mathbf{0} \end{pmatrix}. \tag{1}$$

PROOF. Since $\boldsymbol{C}$ is of rank r, there is a $(p-r) \times p$ matrix $\boldsymbol{A}_2$ such that

$$\boldsymbol{A}_2\boldsymbol{C} = \mathbf{0}. \tag{2}$$

Choose $\boldsymbol{B}$ $(r \times p)$ such that

$$\begin{pmatrix} \boldsymbol{B} \\ \boldsymbol{A}_2 \end{pmatrix} \tag{3}$$

is nonsingular. Then

$$\begin{pmatrix} \boldsymbol{B} \\ \boldsymbol{A}_2 \end{pmatrix} \boldsymbol{C}(\boldsymbol{B}' \ \boldsymbol{A}_2') = \begin{pmatrix} \boldsymbol{BC} \\ \mathbf{0} \end{pmatrix} (\boldsymbol{B}' \ \boldsymbol{A}_2') = \begin{pmatrix} \boldsymbol{BCB}' & \mathbf{0} \\ \mathbf{0} & \mathbf{0} \end{pmatrix}. \tag{4}$$

This matrix is of rank r and therefore $\boldsymbol{BCB}'$ is nonsingular. By Corollary A.1.6 there is a nonsingular matrix $\boldsymbol{D}$ such that $\boldsymbol{D}(\boldsymbol{BCB}')\boldsymbol{D}' = \boldsymbol{I}$. Then

$$\boldsymbol{A} = \begin{pmatrix} \boldsymbol{DB} \\ \boldsymbol{A}_2 \end{pmatrix} = \begin{pmatrix} \boldsymbol{D} & \mathbf{0} \\ \mathbf{0} & \boldsymbol{I} \end{pmatrix} \begin{pmatrix} \boldsymbol{B} \\ \boldsymbol{A}_2 \end{pmatrix} \tag{5}$$

is a nonsingular matrix such that (1) holds. Q.E.D.

LEMMA A.4.1. *If $\boldsymbol{E}$ is $p \times p$, symmetric, and nonsingular, there is a nonsingular matrix $\boldsymbol{F}$ such that*

$$\boldsymbol{FEF}' = \begin{pmatrix} \boldsymbol{I} & \mathbf{0} \\ \mathbf{0} & -\boldsymbol{I} \end{pmatrix}, \tag{6}$$

where the order of $\boldsymbol{I}$ is the number of positive characteristic roots of $\boldsymbol{E}$ and the order of $-\boldsymbol{I}$ is the number of negative characteristic roots of $\boldsymbol{E}$.

PROOF. From Theorem A.2.1 we know there is an orthogonal matrix $\boldsymbol{G}$ such that

$$(7) \qquad \boldsymbol{GEG}' = \begin{pmatrix} h_1 & 0 & \cdots & 0 \\ 0 & h_2 & \cdots & 0 \\ \vdots & \vdots & & \vdots \\ 0 & 0 & \cdots & h_p \end{pmatrix},$$

where $h_1 \geq \cdots \geq h_q > 0 > h_{q+1} \geq \cdots \geq h_p$ are the characteristic roots of $\boldsymbol{E}$. Let

$$(8) \quad \boldsymbol{K} = \begin{pmatrix} 1/\sqrt{h_1} & \cdots & 0 & 0 & \cdots & 0 \\ \vdots & & \vdots & \vdots & & \vdots \\ 0 & \cdots & 1/\sqrt{h_q} & 0 & \cdots & 0 \\ 0 & \cdots & 0 & 1/\sqrt{-h_{q+1}} & \cdots & 0 \\ \vdots & & \vdots & \vdots & & \vdots \\ 0 & \cdots & 0 & 0 & \cdots & 1/\sqrt{-h_p} \end{pmatrix}.$$

Then

$$(9) \qquad \boldsymbol{KGEG'K}' = (\boldsymbol{KG})\boldsymbol{E}(\boldsymbol{KG})' = \begin{pmatrix} \boldsymbol{I} & \boldsymbol{0} \\ \boldsymbol{0} & -\boldsymbol{I} \end{pmatrix}. \quad \text{Q.E.D.}$$

COROLLARY A.4.1. *Let $\boldsymbol{C}$ be $p \times p$, symmetric, and of rank r $(\leq p)$. Then there is a nonsingular matrix $\boldsymbol{A}$ such that*

$$(10) \qquad \boldsymbol{ACA}' = \begin{pmatrix} \boldsymbol{I} & \boldsymbol{0} & \boldsymbol{0} \\ \boldsymbol{0} & -\boldsymbol{I} & \boldsymbol{0} \\ \boldsymbol{0} & \boldsymbol{0} & \boldsymbol{0} \end{pmatrix},$$

where the order of $\boldsymbol{I}$ is the number of positive characteristic roots of $\boldsymbol{C}$ and the order of $-\boldsymbol{I}$ is the number of negative characteristic roots, the sum of the orders being r.

PROOF. The proof is the same as that of Theorem A.4.1 except that Lemma A.4.1 is used instead of Corollary A.1.6.

LEMMA A.4.2. *Let $\boldsymbol{A}$ be $n \times m$ $(n > m)$ such that*

$$\boldsymbol{A}'\boldsymbol{A} = \boldsymbol{I}. \tag{11}$$

There exists an $n \times (n-m)$ matrix $\boldsymbol{B}$ such that $(\boldsymbol{A}\ \boldsymbol{B})$ is orthogonal.

PROOF. Since $\boldsymbol{A}$ is of rank m there exists an $n \times (n-m)$ matrix $\boldsymbol{C}$ such that $(\boldsymbol{A}\ \boldsymbol{C})$ is nonsingular. Take $\boldsymbol{D}$ as $\boldsymbol{C} - \boldsymbol{A}\boldsymbol{A}'\boldsymbol{C}$; then $\boldsymbol{D}'\boldsymbol{A} = \boldsymbol{0}$. Let $\boldsymbol{E}(n-m) \times (n-m)$ be such that $\boldsymbol{E}'\boldsymbol{D}'\boldsymbol{D}\boldsymbol{E} = \boldsymbol{I}$. Then $\boldsymbol{B}$ can be taken as $\boldsymbol{D}\boldsymbol{E}$. Q.E.D.

LEMMA A.4.3. *Let $\boldsymbol{x}$ be a vector of n components. Then there exists an orthogonal matrix $\boldsymbol{O}$ such that*

$$\boldsymbol{O}\boldsymbol{x} = \begin{pmatrix} c \\ 0 \\ \vdots \\ 0 \end{pmatrix}, \tag{12}$$

where $c = \sqrt{\boldsymbol{x}'\boldsymbol{x}}$.

PROOF. Let the first row of $\boldsymbol{O}$ be $(1/c)\boldsymbol{x}'$. The other rows may be chosen in any way to make the matrix orthogonal. Q.E.D.

LEMMA A.4.4. *Let $\boldsymbol{B} = (b_{ij})$ be a $p \times p$ matrix. Then*

$$\frac{\partial|\boldsymbol{B}|}{\partial b_{ij}} = B_{ij}, \qquad i, j = 1, \dots, p. \tag{13}$$

PROOF. The expansion of $|\boldsymbol{B}|$ by elements of the ith row is

$$|\boldsymbol{B}| = \sum_{h=1}^{p} b_{ih} B_{ih}. \tag{14}$$

Since B_{ih} does not contain b_{ij}, the lemma follows. Q.E.D.

LEMMA A.4.5. *Let $b_{ij} = \beta_{ij}(c_1, \dots, c_n)$ be the i, jth element of a $p \times p$ matrix $\boldsymbol{B}$. Then*

$$\frac{\partial|\boldsymbol{B}|}{\partial c_g} = \sum_{i,h=1}^{p} \frac{\partial|\boldsymbol{B}|}{\partial b_{ih}} \cdot \frac{\partial \beta_{ih}(c_1, \dots, c_n)}{\partial c_g} = \sum_{i,h=1}^{p} B_{ih} \frac{\partial \beta_{ih}(c_1, \dots, c_n)}{\partial c_g}. \tag{15}$$

THEOREM A.4.2. *If* $A = A'$,

$$\frac{\partial|A|}{\partial a_{ii}} = A_{ii}, \tag{16}$$

$$\frac{\partial|A|}{\partial a_{ij}} = 2A_{ij}, \qquad i \neq j. \tag{17}$$

PROOF. Equation (16) follows from the expansion of $|A|$ according to elements of the ith row. To prove (17) let $b_{ij} = b_{ji} = a_{ij}$, $i, j = 1, \ldots, p$; $i \leq j$. Then by Lemma A.4.5,

$$\frac{\partial|B|}{\partial a_{ij}} = B_{ij} + B_{ji}. \tag{18}$$

Since $|A| = |B|$ and $B_{ij} = B_{ji} = A_{ij} = A_{ji}$, (17) follows. Q.E.D.

THEOREM A.4.3.

$$\frac{\partial}{\partial x}(x'Ax) = 2Ax, \tag{19}$$

where $\partial/\partial x$ denotes taking partial derivatives with respect to each component of x and arranging the partial derivatives in a column.

PROOF. Let h be a column vector of as many components as x. Then

$$\begin{aligned}(x+h)'A(x+h) &= x'Ax + h'Ax + x'Ah + h'Ah \\ &= x'Ax + 2h'Ax + h'Ah.\end{aligned} \tag{20}$$

The partial derivative vector is the vector multiplying h' in the second term on the right. Q.E.D.

DEFINITION A.4.1. *Let $A = (a_{ij})$ be a $p \times m$ matrix and $B = (b_{\alpha\beta})$ be a $q \times n$ matrix. The $pm \times qn$ matrix with $a_{ij}b_{\alpha\beta}$ as the element in the i, αth row and the j, βth column is called the Kronecker or direct product of A and B and is denoted by $A \otimes B$; that is,*

$$A \otimes B = \begin{pmatrix} a_{11}B & a_{12}B & \cdots & a_{1m}B \\ a_{21}B & a_{22}B & \cdots & a_{2m}B \\ \vdots & \vdots & & \vdots \\ a_{p1}B & a_{p2}B & \cdots & a_{pm}B \end{pmatrix}. \tag{21}$$

Some properties are the following when the orders of matrices permit the indicated operations:

$$(A \otimes B)(C \otimes D) = (AC) \otimes (BD), \tag{22}$$

$$(A \otimes B)^{-1} = A^{-1} \otimes B^{-1}. \tag{23}$$

THEOREM A.4.4. *Let the i characteristic root of A $(p \times p)$ be λ_i and the vector be $(x_{1i}, \dots, x_{pi})'$ and let the αth root of B $(q \times q)$ be ν_α and the vector be y_α, $\alpha = 1, \dots, q$. Then the i, αth root of $A \otimes B$ is $\lambda_i \nu_\alpha$ and the vector is $x_i \otimes y_\alpha = (x_{1i} y_\alpha', \dots, x_{pi} y_\alpha')'$, $i = 1, \dots, p$, $\alpha = 1, \dots, q$.*

PROOF.

$$(A \otimes B)(x_i \otimes y_\alpha) = \begin{pmatrix} a_{11}B & \cdots & a_{1p}B \\ \vdots & & \vdots \\ a_{p1}B & \cdots & a_{pp}B \end{pmatrix} \begin{pmatrix} x_{1i} y_\alpha \\ \vdots \\ x_{pi} y_\alpha \end{pmatrix} \tag{24}$$

$$= \begin{pmatrix} \sum_j a_{1j} x_{ji} B y_\alpha \\ \vdots \\ \sum_j a_{pj} x_{ji} B y_\alpha \end{pmatrix}$$

$$= \begin{pmatrix} \lambda_i x_{1i} B y_\alpha \\ \vdots \\ \lambda_i x_{pi} B y_\alpha \end{pmatrix} = \lambda_i \nu_\alpha \begin{pmatrix} x_{1i} y_\alpha \\ \vdots \\ x_{pi} y_\alpha \end{pmatrix}. \qquad \text{Q.E.D.}$$

THEOREM A.4.5.

$$|A \otimes B| = |A|^q |B|^p. \tag{25}$$

PROOF. The determinant of any matrix is the product of its roots; therefore

$$|A \otimes B| = \prod_{i=1}^{p} \prod_{\alpha=1}^{q} \lambda_i \nu_\alpha = \left(\prod_{i=1}^{p} \lambda_i \right)^q \left(\prod_{\alpha=1}^{q} \nu_\alpha \right)^p. \tag{26}$$

Q.E.D.

THEOREM A.4.6. *The Jacobian of the transformation* $\boldsymbol{E} = \boldsymbol{Y}^{-1}$ (*from* $\boldsymbol{E}$ *to* $\boldsymbol{Y}$) *is* $|\boldsymbol{Y}|^{-2p}$, *where* p *is the order of* $\boldsymbol{E}$ *and* $\boldsymbol{Y}$.

PROOF. From $\boldsymbol{EY} = \boldsymbol{I}$, we have

$$\left(\frac{\partial}{\partial\theta}\boldsymbol{E}\right)\boldsymbol{Y} + \boldsymbol{E}\left(\frac{\partial}{\partial\theta}\boldsymbol{Y}\right) = \boldsymbol{0}, \tag{27}$$

where

$$\left(\frac{\partial}{\partial\theta}\boldsymbol{E}\right) = \begin{pmatrix} \frac{\partial e_{11}}{\partial\theta} & \cdots & \frac{\partial e_{1p}}{\partial\theta} \\ \vdots & & \vdots \\ \frac{\partial e_{p1}}{\partial\theta} & \cdots & \frac{\partial e_{pp}}{\partial\theta} \end{pmatrix}. \tag{28}$$

Then

$$\left(\frac{\partial}{\partial\theta}\boldsymbol{E}\right) = -\boldsymbol{E}\left(\frac{\partial}{\partial\theta}\boldsymbol{Y}\right)\boldsymbol{E} = -\boldsymbol{Y}^{-1}\left(\frac{\partial}{\partial\theta}\boldsymbol{Y}\right)\boldsymbol{Y}^{-1}. \tag{29}$$

If $\theta = y_{\alpha\beta}$, then

$$\left(\frac{\partial}{\partial y_{\alpha\beta}}\boldsymbol{E}\right) = -\boldsymbol{E}\varepsilon_{\alpha\beta}\boldsymbol{E} = -\boldsymbol{e}_{\cdot\alpha}\boldsymbol{e}_{\beta\cdot}, \tag{30}$$

where $\varepsilon_{\alpha\beta}$ is a $p \times p$ matrix with all elements 0 except the element in the αth row and βth column which is 1 and $\boldsymbol{e}_{\cdot\alpha}$ is the αth column of $\boldsymbol{E}$ and $\boldsymbol{e}_{\beta\cdot}$ is the βth row. Thus $\partial e_{ij}/\partial y_{\alpha\beta} = -e_{i\alpha}e_{\beta j}$. Then the Jacobian is the determinant of a $p^2 \times p^2$ matrix

$$\operatorname{mod}\left|\frac{\partial e_{ij}}{\partial y_{\alpha\beta}}\right| = |e_{i\alpha}e_{\beta j}| = |\boldsymbol{E} \otimes \boldsymbol{E}'| = |\boldsymbol{E}|^p|\boldsymbol{E}'|^p = |\boldsymbol{E}|^{2p} = |\boldsymbol{Y}|^{-2p}. \tag{31}$$

Q.E.D.

THEOREM A.4.7. *Let* $\boldsymbol{A}$ *and* $\boldsymbol{B}$ *be symmetric matrices with characteristic roots* $a_1 \ge a_2 \ge \cdots \ge a_p$ *and* $b_1 \ge b_2 \ge \cdots \ge b_p$, *respectively, and let* $\boldsymbol{H}$ *be a* $p \times p$ *orthogonal matrix. Then*

$$\max_{\boldsymbol{H}} \operatorname{tr} \boldsymbol{HAH}'\boldsymbol{B} = \sum_{j=1}^{p} a_j b_j, \qquad \min_{\boldsymbol{H}} \operatorname{tr} \boldsymbol{HA}'\boldsymbol{H}'\boldsymbol{B} = \sum_{j=1}^{p} a_j b_{p+1-j}. \tag{32}$$

Proof. Let $\boldsymbol{A} = \boldsymbol{H}_a \boldsymbol{D}_a \boldsymbol{H}_a'$ and $\boldsymbol{B} = \boldsymbol{H}_b \boldsymbol{D}_b \boldsymbol{H}_b'$, where $\boldsymbol{H}_a$ and $\boldsymbol{H}_b$ are orthogonal and $\boldsymbol{D}_a$ and $\boldsymbol{D}_b$ are diagonal with diagonal elements $a_1, \ldots, a_p$ and $b_1, \ldots, b_p$ respectively. Then

$$
\begin{aligned}
(33) \qquad \max_{\boldsymbol{H}^*} \operatorname{tr} \boldsymbol{H}^* \boldsymbol{A} \boldsymbol{H}^{*\prime} \boldsymbol{B} &= \max_{\boldsymbol{H}^*} \operatorname{tr} \boldsymbol{H}^* \boldsymbol{H}_a \boldsymbol{D}_a \boldsymbol{H}_a' \boldsymbol{H}^{*\prime} \boldsymbol{H}_b \boldsymbol{D}_b \boldsymbol{H}_b' \\
&= \max_{\boldsymbol{H}^*} \operatorname{tr} \boldsymbol{H}_b' \boldsymbol{H}^* \boldsymbol{H}_a \boldsymbol{D}_a \left(\boldsymbol{H}_b' \boldsymbol{H}^* \boldsymbol{H}_a \right)' \boldsymbol{D}_b \\
&= \max_{\boldsymbol{H}} \operatorname{tr} \boldsymbol{H} \boldsymbol{D}_a \boldsymbol{H}' \boldsymbol{D}_b .
\end{aligned}
$$

We have

$$
\begin{aligned}
(34) \quad \operatorname{tr} \boldsymbol{H} \boldsymbol{D}_a \boldsymbol{H}' \boldsymbol{D}_b &= \sum_{i=1}^{p} \left(\boldsymbol{H} \boldsymbol{D}_a \boldsymbol{H}' \right)_{ii} b_i \\
&= \sum_{i=1}^{p-1} \sum_{j=1}^{i} \left(\boldsymbol{H} \boldsymbol{D}_a \boldsymbol{H}' \right)_{jj} (b_i - b_{i+1}) + b_p \sum_{j=1}^{p} \left(\boldsymbol{H} \boldsymbol{D}_a \boldsymbol{H}' \right)_{jj} \\
&\le \sum_{i=1}^{p-1} \sum_{j=1}^{i} a_j (b_i - b_{i+1}) + b_p \sum_{j=1}^{p} a_j \\
&= \sum_{i=1}^{p} a_i b_i
\end{aligned}
$$

by Lemma A.4.6 below. The minimum in (32) is treated as the negative of the maximum with $\boldsymbol{B}$ replaced by $-\boldsymbol{B}$ [von Neumann (1937)]. Q.E.D.

Lemma A.4.6. *Let* $\boldsymbol{P} = (p_{ij})$ *be a doubly stochastic matrix* ($p_{ij} \ge 0, \sum_{i=1}^{p} p_{ij} = 1, \sum_{j=1}^{p} p_{ij} = 1$). *Let* $y_1 \ge y_2 \ge \cdots \ge y_p$. *Then*

$$
(35) \qquad \sum_{i=1}^{k} y_i \ge \sum_{i=1}^{k} \sum_{j=1}^{n} p_{ij} y_j , \qquad k = 1, \ldots, p.
$$

Proof.

$$
(36) \qquad \sum_{i=1}^{k} \sum_{j=1}^{p} p_{ij} y_j = \sum_{j=1}^{p} g_j y_j ,
$$

where $g_j = \sum_{i=1}^k p_{ij}$, $j = 1, \ldots, p$ $(0 \le g_j \le 1, \sum_{j=1}^p g_j = k)$. Then

$$\sum_{j=1}^p g_j y_j - \sum_{j=1}^k y_j = -\sum_{j=1}^k y_j + y_k\left(k - \sum_{j=1}^p g_j\right) + \sum_{j=1}^p g_j y_j \tag{37}$$

$$= \sum_{j=1}^k (y_j - y_k)(g_j - 1) + \sum_{j=k+1}^p (y_j - y_k) g_j$$

$$\le 0. \quad \text{Q.E.D.}$$

COROLLARY A.4.2. *Let $\boldsymbol{A}$ be a symmetric matrix with characteristic roots $a_1 \ge a_2 \ge \cdots \ge a_p$. Then*

$$\max_{\boldsymbol{R}'\boldsymbol{R} = \boldsymbol{I}_k} \operatorname{tr} \boldsymbol{R}'\boldsymbol{A}\boldsymbol{R} = \sum_{i=1}^k a_i. \tag{38}$$

PROOF. In Theorem A.4.7 let

$$\boldsymbol{B} = \begin{pmatrix} \boldsymbol{I}_k & \boldsymbol{0} \\ \boldsymbol{0} & \boldsymbol{0} \end{pmatrix}. \quad \text{Q.E.D.} \tag{39}$$

A.5. GRAM–SCHMIDT ORTHOGONALIZATION AND THE SOLUTION OF LINEAR EQUATIONS

A.5.1. Gram–Schmidt Orthogonalization

The derivation of the Wishart density in Section 7.2 included the Gram–Schmidt orthogonalization of a set of vectors; we shall review that development here. Consider the p linearly independent n-dimensional vectors $\boldsymbol{v}_1, \ldots, \boldsymbol{v}_p$ $(p \le n)$. Define $\boldsymbol{w}_1 = \boldsymbol{v}_1$,

$$\boldsymbol{w}_i = \boldsymbol{v}_i - \sum_{j=1}^{i-1} \frac{\boldsymbol{v}_i'\boldsymbol{w}_j}{\|\boldsymbol{w}_j\|^2} \boldsymbol{w}_j, \qquad i = 2, \ldots, p. \tag{1}$$

Then $\boldsymbol{w}_i \ne \boldsymbol{0}$, $i = 1, \ldots, p$, because $\boldsymbol{v}_1, \ldots, \boldsymbol{v}_p$ are linearly independent, and

$$\boldsymbol{w}_i'\boldsymbol{w}_j = 0, \qquad i \ne j, \tag{2}$$

as was proved by induction in Section 7.2. Let $\boldsymbol{u}_i = (1/\|\boldsymbol{w}_i\|)\boldsymbol{w}_i$, $i = 1, \ldots, p$. Then $\boldsymbol{u}_1, \ldots, \boldsymbol{u}_p$ are *orthonormal*; that is, they are orthogonal and of unit length. Let

$$U = (\boldsymbol{u}_1, \ldots, \boldsymbol{u}_p). \tag{3}$$

Then $\boldsymbol{U}'\boldsymbol{U} = \boldsymbol{I}$. Define $t_{ii} = \|\boldsymbol{w}_i\|$ (> 0),

$$t_{ij} = \frac{\boldsymbol{v}_i'\boldsymbol{w}_j}{\|\boldsymbol{w}_j\|} = \boldsymbol{v}_i'\boldsymbol{u}_j, \qquad j = 1, \ldots, i-1,\ i = 2, \ldots, p, \tag{4}$$

and $t_{ij} = 0$, $j = i+1, \ldots, p$, $i = 1, \ldots, p-1$. Then $\boldsymbol{T} = (t_{ij})$ is a lower triangular matrix. We can write (1) as

$$\boldsymbol{v}_i = \|\boldsymbol{w}_i\|\boldsymbol{u}_i + \sum_{j=1}^{i-1} (\boldsymbol{v}_i'\boldsymbol{u}_j)\boldsymbol{u}_j = \sum_{j=1}^{i} t_{ij}\boldsymbol{u}_j, \qquad i = 1, \ldots, p, \tag{5}$$

that is,

$$\boldsymbol{V} = (\boldsymbol{v}_1, \ldots, \boldsymbol{v}_p) = \boldsymbol{U}\boldsymbol{T}'. \tag{6}$$

Then

$$\boldsymbol{A} = \boldsymbol{V}'\boldsymbol{V} = \boldsymbol{T}\boldsymbol{U}'\boldsymbol{U}\boldsymbol{T}' = \boldsymbol{T}\boldsymbol{T}' \tag{7}$$

as shown in Section 7.2. Note if $\boldsymbol{V}$ is square, we have decomposed an arbitrary nonsingular matrix into the product of an orthogonal matrix and an upper triangular matrix with positive diagonal elements; this is sometimes known as the *QR decomposition*. The matrices $\boldsymbol{U}$ and $\boldsymbol{T}$ in (6) are unique.

These operations can be done in a different order. Let $\boldsymbol{V} = (\boldsymbol{v}_1^{(0)}, \ldots, \boldsymbol{v}_p^{(0)})$. For $k = 1, \ldots, p-1$ define

$$t_{kk} = \|\boldsymbol{v}_k^{(k-1)}\|, \qquad \boldsymbol{u}_k = \frac{1}{t_{kk}}\boldsymbol{v}_k^{(k-1)}, \tag{8}$$

$$t_{jk} = \boldsymbol{v}_j^{(k-1)\prime}\boldsymbol{u}_k, \qquad j = k+1, \ldots, p, \tag{9}$$

$$\boldsymbol{v}_j^{(k)} = \boldsymbol{v}_j^{(k-1)} - t_{jk}\boldsymbol{u}_k, \qquad j = k+1, \ldots, p. \tag{10}$$

Finally $t_{pp} = \|\boldsymbol{v}_p^{(p-1)}\|$ and $\boldsymbol{u}_p = (1/t_{pp})\boldsymbol{v}_p^{(p-1)}$. It can easily be verified that the same orthonormal vectors $\boldsymbol{u}_1, \ldots, \boldsymbol{u}_p$ and the same triangular matrix (t_{ij}) are given by the two procedures.

The numbering of the columns of $\boldsymbol{V}$ is arbitrary. For numerical stability it is usually best at any given stage to select the largest of $\|\boldsymbol{v}_j^{(k-1)}\|$ to call t_{kk}.

Instead of constructing $\boldsymbol{w}_i$ as orthogonal to $\boldsymbol{w}_1, \ldots, \boldsymbol{w}_{i-1}$, we can construct it as orthogonal to $\boldsymbol{v}_1, \ldots, \boldsymbol{v}_{i-1}$. Let $\boldsymbol{w}_1 = \boldsymbol{v}_1$ and define

$$\boldsymbol{w}_i = \boldsymbol{v}_i + \sum_{j=1}^{i-1} f_{ij} \boldsymbol{v}_j \tag{11}$$

such that

$$\begin{aligned} 0 = \boldsymbol{v}_h' \boldsymbol{w}_i &= \boldsymbol{v}_h' \boldsymbol{v}_i + \sum_{j=1}^{i-1} f_{ij} \boldsymbol{v}_h' \boldsymbol{v}_j \\ &= a_{hi} + \sum_{j=1}^{i-1} a_{hj} f_{ij}, \qquad h = 1, \ldots, i-1. \end{aligned} \tag{12}$$

Let $\boldsymbol{F} = (f_{ij})$, where $f_{ii} = 1$ and $f_{ij} = 0$, $i < j$. Then

$$\boldsymbol{W} = (\boldsymbol{w}_1, \ldots, \boldsymbol{w}_p) = \boldsymbol{V}\boldsymbol{F}'. \tag{13}$$

Let $\boldsymbol{D}_t$ be the diagonal matrix with $\|\boldsymbol{w}_j\| = t_{jj}$ as the jth diagonal element. Then $\boldsymbol{U} = \boldsymbol{W}\boldsymbol{D}_t^{-1} = \boldsymbol{V}\boldsymbol{F}'\boldsymbol{D}_t^{-1}$. Comparison with $\boldsymbol{V} = \boldsymbol{U}\boldsymbol{T}'$ shows that $\boldsymbol{F} = \boldsymbol{D}\boldsymbol{T}^{-1}$. Since $\boldsymbol{A} = \boldsymbol{T}\boldsymbol{T}'$, we see that $\boldsymbol{F}\boldsymbol{A} = \boldsymbol{D}\boldsymbol{T}'$ is upper triangular. Hence $\boldsymbol{F}$ is the matrix defined in Theorem A.1.2.

There are other methods of accomplishing the *QR* decomposition that may be computationally more efficient or more stable. A *Householder matrix* has the form $\boldsymbol{H} = \boldsymbol{I}_n - 2\boldsymbol{\alpha}\boldsymbol{\alpha}'$, where $\boldsymbol{\alpha}'\boldsymbol{\alpha} = 1$, and is orthogonal and symmetric. Such a matrix $\boldsymbol{H}_1$ (i.e., a vector $\boldsymbol{\alpha}$) can be selected so that the first column of $\boldsymbol{H}_1\boldsymbol{V}$ has 0's in all positions except the first, which is positive. The next matrix has the form

$$\boldsymbol{H}_2 = \begin{pmatrix} 1 & \boldsymbol{0} \\ \boldsymbol{0} & \boldsymbol{I}_{n-1} \end{pmatrix} - 2\begin{pmatrix} 0 \\ \boldsymbol{\alpha} \end{pmatrix}(0\ \boldsymbol{\alpha}') = \begin{pmatrix} 1 & \boldsymbol{0} \\ \boldsymbol{0} & \boldsymbol{I}_{n-1} - \boldsymbol{\alpha}\boldsymbol{\alpha}' \end{pmatrix}. \tag{14}$$

The $(n-1)$-component vector $\boldsymbol{\alpha}$ is chosen so that the second column of $\boldsymbol{H}_1\boldsymbol{V}$ has all 0's except the first two components, the second being positive. This process is continued until

$$H_{p-1} \cdots H_2 H_1 V = \begin{bmatrix} T' \\ 0 \end{bmatrix}, \tag{15}$$

where T' is upper triangular and $\mathbf{0}$ is $(n-p) \times p$. Let

$$H' = H_1 \cdots H_{p-1} = (H^{(1)}\; H^{(2)}), \tag{16}$$

where $H^{(1)}$ has p columns. Then from (15) we obtain

$$V = H^{(1)} T'. \tag{17}$$

Since the decomposition is unique, $H^{(1)} = U$.

Another procedure uses *Givens matrices*. A Givens matrix G_{ij} is I except for the elements $g_{ii} = \cos\theta = g_{jj}$ and $g_{ij} = \sin\theta = -g_{ji}$, $i \neq j$. It is orthogonal. Multiplication of V on the left by such a matrix leaves all rows unchanged except the ith and jth; θ can be chosen so that the i, jth element of $G_{ij}V$ is 0. Givens matrices $G_{21}, \ldots, G_{n1}$ can be chosen in turn so $G_{n1} \cdots G_{21}V$ has all 0's in the first column except the first element, which is positive. Next $G_{32}, \ldots, G_{n2}$ can be selected in turn so that when they are applied the resulting matrix has 0's in the second column except for the first two elements. Let

$$G' = G'_{21} \cdots G'_{n1} G'_{32} \cdots G'_{n,p-1} = (G^{(1)}\; G^{(2)}). \tag{18}$$

Then we obtain

$$V = G' \begin{bmatrix} T' \\ 0 \end{bmatrix} = G^{(1)} T', \tag{19}$$

and $G^{(1)} = U$.

A.5.2. Solution of Linear Equations

In the computation of regression coefficients and other statistics, we need to solve linear equations

$$Ax = y, \tag{20}$$

where A is $p \times p$ and positive definite. One method of solution is Gaussian elimination of variables or pivotal condensation. In the proof of Theorem A.1.2 was the construction of a lower triangular matrix $\boldsymbol{F}$ with diagonal elements 1 such that $\boldsymbol{FA} = \boldsymbol{A}^*$ is upper triangular. If $\boldsymbol{Fy} = \boldsymbol{y}^*$, then the equation is

$$\boldsymbol{A}^*\boldsymbol{x} = \boldsymbol{y}^*. \tag{21}$$

In coordinates this is

$$\sum_{j=i}^{p} a_{ij}^* x_j = y_i^*. \tag{22}$$

Let $a_{ij}^{**} = a_{ij}^*/a_{ii}^*$, $y_i^{**} = y_i^*/a_{ii}^*$, $j = i, i+1, \ldots, p$, $i = 1, \ldots, p$. Then

$$x_i = y_i^{**} - \sum_{j=i+1}^{p} a_{ij}^{**} x_j; \tag{23}$$

these equations are to be solved successively for $x_p, x_{p-1}, \ldots, x_1$. The calculation of $\boldsymbol{FA} = \boldsymbol{A}^*$ is known as the forward solution, and the solution of (17) as the backward solution.

Since $\boldsymbol{FAF}' = \boldsymbol{A}^*\boldsymbol{F}' = \boldsymbol{D}^2$ diagonal, (23) is $\boldsymbol{A}^{**}\boldsymbol{x} = \boldsymbol{y}^{**}$, where $\boldsymbol{A}^{**} = \boldsymbol{D}^{-2}\boldsymbol{A}^*$ and $\boldsymbol{y}^{**} = \boldsymbol{D}^{-2}\boldsymbol{y}^*$. Solving this equation is

$$\boldsymbol{x} = \boldsymbol{A}^{**-1}\boldsymbol{y}^{**} = \boldsymbol{F}'\boldsymbol{y}^{**}. \tag{24}$$

The computation is

$$\boldsymbol{x} = \boldsymbol{F}_1' \cdots \boldsymbol{F}_{p-1}'\boldsymbol{D}^{-2}\boldsymbol{F}_{p-1} \cdots \boldsymbol{F}_1 \boldsymbol{y}. \tag{25}$$

The multiplier of $\boldsymbol{y}$ in (25) indicates a sequence of row operations which yields $\boldsymbol{A}^{-1}$.

The operations of the forward solution transform $\boldsymbol{A}$ to the upper triangular matrix $\boldsymbol{A}^*$. As seen in Section A.5.1, the triangularization of a matrix can be done by a sequence of Householder transformations or by a sequence of Givens transformations.

From $\boldsymbol{FA} = \boldsymbol{A}^*$, we obtain

$$|\boldsymbol{A}| = \prod_{i=1}^{p} a_{ii}^{(i)}, \tag{26}$$

which is the product of the diagonal elements of $\boldsymbol{A}^*$, resulting from the forward solution. We also have

$$(27) \qquad \boldsymbol{y}'\boldsymbol{A}^{-1}\boldsymbol{y} = (\boldsymbol{F}\boldsymbol{y})'\boldsymbol{D}^{-2}(\boldsymbol{F}\boldsymbol{y}) = \boldsymbol{y}^{*\prime}\boldsymbol{D}^{-2}\boldsymbol{y}^{*} = \boldsymbol{y}^{*\prime}\boldsymbol{y}^{**}.$$

The forward solution gives a computation for the quadratic form which occurs in T^2 and other statistics.

APPENDIX B

TABLE 1

WILKS' LIKELIHOOD CRITERION: FACTORS $C(p, m, M)$ TO ADJUST TO χ^2_{pm} WHERE $M = n - p + 1$

5% Significance Level

$M \backslash m$	$p = 3$ 2	4	6	8	10	12	14	16
1	1.295	1.422	1.535	1.632	1.716	1.791	1.857	1.916
2	1.109	1.174	1.241	1.302	1.359	1.410	1.458	1.501
3	1.058	1.099	1.145	1.190	1.232	1.272	1.309	1.344
4	1.036	1.065	1.099	1.133	1.167	1.199	1.229	1.258
5	1.025	1.046	1.072	1.100	1.127	1.154	1.179	1.204
6	1.018	1.035	1.056	1.078	1.101	1.123	1.145	1.167
7	1.014	1.027	1.044	1.063	1.082	1.101	1.121	1.139
8	1.011	1.022	1.036	1.052	1.068	1.085	1.102	1.119
9	1.009	1.018	1.030	1.043	1.058	1.073	1.088	1.102
10	1.007	1.015	1.025	1.037	1.050	1.063	1.076	1.089
12	1.005	1.011	1.019	1.028	1.038	1.048	1.059	1.070
15	1.003	1.008	1.013	1.020	1.027	1.035	1.043	1.052
20	1.002	1.004	1.008	1.012	1.017	1.022	1.028	1.034
30	1.001	1.002	1.004	1.006	1.009	1.011	1.015	1.018
60	1.000	1.001	1.001	1.002	1.002	1.003	1.004	1.006
∞	1.000	1.000	1.000	1.000	1.000	1.000	1.000	1.000
χ^2_{pm}	12.5916	21.0261	28.8693	36.4150	43.7730	50.9985	58.1240	65.1708

TABLE 1 (*Continued*)

$M \setminus m$	18	20	22	$p = 4$ 2	4	6	8	10
1	1.971	2.021	2.067	1.407	1.451	1.517	1.583	1.644
2	1.542	1.580	1.616	1.161	1.194	1.240	1.286	1.331
3	1.377	1.408	1.438	1.089	1.114	1.148	1.183	1.218
4	1.286	1.313	1.338	1.057	1.076	1.102	1.130	1.159
5	1.228	1.251	1.273	1.040	1.055	1.076	1.099	1.122
6	1.188	1.208	1.227	1.030	1.042	1.059	1.078	1.097
7	1.158	1.176	1.193	1.023	1.033	1.047	1.063	1.080
8	1.135	1.151	1.167	1.018	1.027	1.038	1.052	1.067
9	1.117	1.132	1.147	1.015	1.022	1.032	1.044	1.057
10	1.103	1.116	1.129	1.012	1.018	1.027	1.038	1.049
12	1.081	1.092	1.103	1.009	1.014	1.020	1.029	1.038
15	1.060	1.069	1.078	1.006	1.009	1.014	1.020	1.027
20	1.040	1.046	1.052	1.003	1.006	1.009	1.013	1.017
30	1.021	1.025	1.029	1.002	1.003	1.004	1.006	1.009
60	1.007	1.008	1.009	1.000	1.001	1.001	1.002	1.003
∞	1.000	1.000	1.000	1.000	1.000	1.000	1.000	1.000
χ^2_{pm}	72.1532	79.0819	85.9649	15.5073	26.2962	36.4150	46.1943	55.7585

$M \setminus m$	12	14	16	18	20	$p = 5$ 2	4	6
1	1.700	1.751	1.799	1.843	1.884	1.503	1.483	1.514
2	1.373	1.413	1.450	1.485	1.518	1.209	1.216	1.245
3	1.252	1.284	1.314	1.343	1.371	1.120	1.130	1.154
4	1.186	1.213	1.239	1.264	1.288	1.079	1.089	1.108
5	1.145	1.168	1.190	1.212	1.233	1.056	1.065	1.081
6	1.118	1.137	1.157	1.176	1.194	1.042	1.050	1.063
7	1.097	1.115	1.132	1.149	1.165	1.033	1.040	1.051
8	1.082	1.097	1.113	1.128	1.143	1.026	1.032	1.042
9	1.070	1.084	1.098	1.111	1.125	1.022	1.027	1.035
10	1.061	1.073	1.086	1.098	1.110	1.018	1.023	1.030
12	1.047	1.058	1.068	1.078	1.088	1.013	1.017	1.023
15	1.035	1.042	1.050	1.058	1.066	1.009	1.011	1.016
20	1.022	1.027	1.033	1.039	1.045	1.005	1.007	1.010
30	1.011	1.014	1.018	1.021	1.024	1.002	1.003	1.005
60	1.003	1.004	1.005	1.007	1.008	1.001	1.001	1.001
∞	1.000	1.000	1.000	1.000	1.000	1.000	1.000	1.000
χ^2_{pm}	65.1708	74.4683	83.6753	92.8083	101.879	18.3070	31.4104	43.7730

TABLE 1 (*Continued*)

						$p = 6$		
$M \setminus m$	8	10	12	14	16	2	6	8
1	1.556	1.600	1.643	1.683	1.722	1.587	1.520	1.543
2	1.280	1.315	1.350	1.383	1.415	1.254	1.255	1.279
3	1.182	1.211	1.240	1.267	1.294	1.150	1.163	1.184
4	1.131	1.155	1.179	1.203	1.226	1.100	1.116	1.134
5	1.100	1.120	1.141	1.161	1.181	1.072	1.088	1.103
6	1.079	1.097	1.114	1.132	1.150	1.055	1.069	1.082
7	1.065	1.080	1.095	1.111	1.127	1.043	1.056	1.068
8	1.054	1.067	1.081	1.095	1.109	1.035	1.046	1.057
9	1.046	1.057	1.070	1.082	1.095	1.029	1.039	1.048
10	1.039	1.050	1.061	1.072	1.083	1.024	1.034	1.042
12	1.030	1.038	1.047	1.057	1.066	1.018	1.025	1.032
15	1.021	1.028	1.034	1.042	1.049	1.012	1.018	1.023
20	1.013	1.018	1.022	1.027	1.033	1.007	1.011	1.014
30	1.007	1.009	1.012	1.014	1.018	1.003	1.006	1.007
60	1.002	1.003	1.004	1.004	1.006	1.001	1.002	1.002
∞	1.000	1.000	1.000	1.000	1.000	1.000	1.000	1.000
χ^2_{pm}	55.7585	67.5048	79.0819	90.5312	101.879	21.0261	50.9985	65.1708

			$p = 7$					$p = 8$
$M \setminus m$	10	12	2	4	6	8	10	2
1	1.573	1.605	1.662	1.550	1.530	1.538	1.557	1.729
2	1.307	1.335	1.297	1.263	1.266	1.282	1.303	1.336
3	1.208	1.232	1.178	1.165	1.173	1.189	1.208	1.206
4	1.154	1.175	1.121	1.116	1.124	1.139	1.155	1.142
5	1.120	1.138	1.089	1.087	1.095	1.108	1.122	1.105
6	1.097	1.113	1.068	1.068	1.075	1.086	1.099	1.081
7	1.081	1.095	1.054	1.055	1.062	1.071	1.083	1.065
8	1.068	1.081	1.044	1.045	1.051	1.060	1.070	1.053
9	1.059	1.070	1.036	1.038	1.043	1.051	1.060	1.044
10	1.051	1.061	1.031	1.032	1.037	1.044	1.053	1.038
12	1.040	1.048	1.023	1.024	1.029	1.034	1.042	1.028
15	1.029	1.035	1.016	1.017	1.020	1.024	1.031	1.019
20	1.018	1.023	1.010	1.011	1.013	1.016	1.019	1.012
30	1.010	1.012	1.005	1.005	1.006	1.008	1.010	1.006
60	1.003	1.004	1.001	1.001	1.002	1.002	1.003	1.001
∞	1.000	1.000	1.000	1.000	1.000	1.000	1.000	1.000
χ^2_{pm}	79.0819	92.8083	23.6848	41.3371	58.1240	74.4683	90.5312	26.2962

TABLE 1 (*Continued*)

$M \backslash m$	8	$p = 9$ 2	4	6	$p = 10$ 2
1	1.538	1.791	1.614	1.558	1.847
2	1.288	1.373	1.309	1.293	1.408
3	1.195	1.232	1.201	1.196	1.257
4	1.144	1.162	1.144	1.144	1.182
5	1.113	1.121	1.110	1.112	1.137
6	1.091	1.094	1.088	1.090	1.107
7	1.076	1.076	1.071	1.074	1.087
8	1.064	1.062	1.060	1.062	1.072
9	1.055	1.052	1.050	1.053	1.061
10	1.048	1.045	1.043	1.046	1.052
12	1.038	1.034	1.033	1.035	1.039
15	1.027	1.023	1.023	1.025	1.028
20	1.017	1.014	1.015	1.016	1.017
30	1.009	1.007	1.007	1.008	1.009
60	1.003	1.002	1.002	1.002	1.002
∞	1.000	1.000	1.000	1.000	1.000
χ^2_{pm}	83.6753	28.8693	50.9985	72.1532	31.4104

1% Significance Level

$M \backslash m$	$p = 3$ 2	4	6	8	10	12	14	16
1	1.356	1.514	1.649	1.763	1.862	1.949	2.026	2.095
2	1.131	1.207	1.282	1.350	1.413	1.470	1.523	1.571
3	1.070	1.116	1.167	1.216	1.262	1.306	1.346	1.384
4	1.043	1.076	1.113	1.150	1.187	1.221	1.254	1.285
5	1.030	1.054	1.082	1.112	1.141	1.170	1.198	1.224
6	1.022	1.040	1.063	1.087	1.112	1.136	1.159	1.182
7	1.016	1.031	1.050	1.070	1.091	1.111	1.132	1.152
8	1.013	1.025	1.041	1.058	1.075	1.093	1.111	1.129
9	1.010	1.021	1.034	1.048	1.064	1.080	1.095	1.111
10	1.009	1.017	1.028	1.041	1.055	1.069	1.082	1.097
12	1.006	1.012	1.021	1.031	1.042	1.053	1.064	1.076
15	1.004	1.009	1.014	1.021	1.030	1.038	1.047	1.056
20	1.002	1.005	1.009	1.013	1.019	1.024	1.030	1.036
30	1.001	1.002	1.004	1.007	1.009	1.012	1.016	1.019
60	1.000	1.001	1.001	1.002	1.003	1.004	1.005	1.006
∞	1.000	1.000	1.000	1.000	1.000	1.000	1.000	1.000
χ^2_{pm}	16.8119	26.2170	34.8053	42.9798	50.8922	58.6192	66.2062	73.6826

TABLE 1 (*Continued*)

$M \setminus m$	18	20	22	$p = 4$ 2	4	6	8	10
1	2.158	2.216	2.269	1.490	1.550	1.628	1.704	1.774
2	1.616	1.657	1.696	1.192	1.229	1.279	1.330	1.379
3	1.420	1.453	1.485	1.106	1.132	1.168	1.207	1.244
4	1.315	1.344	1.371	1.068	1.088	1.115	1.146	1.176
5	1.249	1.274	1.297	1.047	1.063	1.085	1.109	1.134
6	1.204	1.226	1.246	1.035	1.048	1.066	1.086	1.107
7	1.171	1.190	1.209	1.027	1.037	1.052	1.070	1.088
8	1.146	1.163	1.180	1.021	1.030	1.043	1.058	1.073
9	1.127	1.142	1.157	1.017	1.025	1.036	1.048	1.062
10	1.111	1.125	1.139	1.014	1.021	1.030	1.041	1.054
12	1.087	1.099	1.110	1.010	1.015	1.023	1.031	1.041
15	1.065	1.074	1.083	1.007	1.010	1.016	1.022	1.029
20	1.043	1.049	1.056	1.004	1.006	1.010	1.014	1.019
30	1.023	1.027	1.031	1.002	1.003	1.005	1.007	1.009
60	1.007	1.009	1.010	1.000	1.001	1.001	1.002	1.003
∞	1.000	1.000	1.000	1.000	1.000	1.000	1.000	1.000
χ^2_{pm}	81.0688	88.3794	95.6257	20.0902	31.9999	42.9798	53.4858	63.6907

$M \setminus m$	12	14	16	18	20	$p = 5$ 2	4	6
1	1.838	1.896	1.949	1.999	2.045	1.606	1.589	1.625
2	1.424	1.467	1.507	1.545	1.580	1.248	1.253	1.284
3	1.280	1.314	1.347	1.378	1.408	1.141	1.150	1.175
4	1.205	1.234	1.261	1.287	1.313	1.092	1.101	1.121
5	1.159	1.183	1.207	1.230	1.252	1.065	1.074	1.090
6	1.128	1.149	1.169	1.189	1.208	1.049	1.056	1.070
7	1.106	1.124	1.142	1.160	1.177	1.038	1.044	1.056
8	1.089	1.105	1.121	1.137	1.153	1.031	1.036	1.046
9	1.076	1.091	1.105	1.119	1.133	1.025	1.030	1.039
10	1.066	1.079	1.092	1.105	1.118	1.021	1.025	1.033
12	1.051	1.062	1.073	1.083	1.094	1.015	1.019	1.025
15	1.037	1.045	1.053	1.062	1.071	1.010	1.013	1.017
20	1.024	1.029	1.035	1.041	1.047	1.006	1.008	1.011
30	1.012	1.015	1.019	1.022	1.026	1.003	1.004	1.005
60	1.004	1.005	1.006	1.007	1.008	1.001	1.001	1.001
∞	1.000	1.000	1.000	1.000	1.000	1.000	1.000	1.000
χ^2_{pm}	73.6826	83.5134	93.2168	102.816	112.329	23.2093	37.5662	50.8922

TABLE 1 (*Continued*)

$M \setminus m$	8	10	12	14	16	$p = 6$ 2	6	8
1	1.672	1.721	1.768	1.813	1.855	1.707	1.631	1.656
2	1.321	1.359	1.396	1.431	1.465	1.300	1.294	1.319
3	1.204	1.235	1.265	1.294	1.323	1.175	1.183	1.205
4	1.145	1.171	1.196	1.221	1.245	1.116	1.129	1.148
5	1.110	1.131	1.153	1.174	1.196	1.084	1.097	1.113
6	1.087	1.105	1.124	1.143	1.161	1.063	1.076	1.090
7	1.071	1.087	1.103	1.119	1.136	1.050	1.061	1.074
8	1.059	1.073	1.087	1.102	1.116	1.040	1.051	1.062
9	1.050	1.062	1.075	1.088	1.101	1.033	1.043	1.052
10	1.043	1.054	1.065	1.077	1.089	1.028	1.037	1.045
12	1.033	1.041	1.051	1.060	1.070	1.021	1.028	1.035
15	1.023	1.030	1.037	1.044	1.052	1.014	1.020	1.024
20	1.015	1.019	1.024	1.029	1.034	1.008	1.012	1.015
30	1.007	1.010	1.012	1.015	1.019	1.004	1.006	1.008
60	1.002	1.003	1.004	1.005	1.006	1.001	1.002	1.002
∞	1.000	1.000	1.000	1.000	1.000	1.000	1.000	1.000
χ^2_{pm}	63.6907	76.1539	88.3794	100.425	112.329	26.2170	58.6192	73.6826

$M \setminus m$	10	12	$p = 7$ 2	4	6	8	10	$p = 8$ 2
1	1.687	1.722	1.797	1.667	1.642	1.648	1.666	1.879
2	1.348	1.378	1.348	1.305	1.306	1.321	1.342	1.394
3	1.230	1.255	1.207	1.188	1.194	1.210	1.229	1.238
4	1.169	1.191	1.140	1.130	1.138	1.152	1.169	1.163
5	1.131	1.150	1.102	1.097	1.105	1.117	1.132	1.120
6	1.106	1.122	1.078	1.076	1.083	1.094	1.107	1.092
7	1.087	1.102	1.062	1.061	1.067	1.077	1.089	1.074
8	1.074	1.086	1.050	1.050	1.056	1.065	1.075	1.060
9	1.063	1.075	1.042	1.042	1.047	1.055	1.065	1.050
10	1.055	1.065	1.035	1.036	1.041	1.048	1.056	1.043
12	1.042	1.051	1.026	1.027	1.031	1.037	1.044	1.032
15	1.030	1.037	1.018	1.019	1.022	1.025	1.032	1.022
20	1.019	1.024	1.011	1.012	1.014	1.017	1.020	1.013
30	1.010	1.013	1.005	1.006	1.007	1.009	1.011	1.007
60	1.003	1.004	1.001	1.002	1.002	1.003	1.003	1.002
∞	1.000	1.000	1.000	1.000	1.000	1.000	1.000	1.000
χ^2_{pm}	88.3794	102.816	29.1412	48.2782	66.2062	83.5134	100.425	31.9999

TABLE 1 (*Continued*)

	$p = 9$				$p = 10$
$M \setminus m$	8	2	4	6	2
1	1.646	1.953	1.740	1.671	2.021
2	1.326	1.436	1.355	1.333	1.476
3	1.215	1.267	1.226	1.218	1.296
4	1.158	1.185	1.161	1.158	1.207
5	1.123	1.138	1.122	1.122	1.155
6	1.099	1.107	1.096	1.098	1.121
7	1.082	1.086	1.078	1.080	1.098
8	1.069	1.070	1.065	1.067	1.081
9	1.059	1.059	1.055	1.058	1.068
10	1.051	1.050	1.047	1.050	1.058
12	1.040	1.038	1.036	1.037	1.044
15	1.028	1.026	1.026	1.027	1.031
20	1.018	1.016	1.016	1.017	1.019
30	1.009	1.008	1.008	1.009	1.010
60	1.003	1.002	1.002	1.003	1.003
∞	1.000	1.000	1.000	1.000	1.000
χ^2_{pm}	93.2168	34.8053	58.6192	81.0688	37.5662

TABLE 2

TABLES OF SIGNIFICANCE POINTS FOR THE LAWLEY–HOTELLING TRACE TEST

$$\Pr\left\{\frac{n}{m} W \geq x_{\alpha}\right\} = \alpha$$

$p = 2$, 5% Significance Level

$n \backslash m$	2	3	4	5	6	8	10	12	15
2	9.859*	10.659*	11.098*	11.373*	11.562*	11.804*	11.952*	12.052*	12.153*
3	58.428	58.915	59.161	59.308	59.407	59.531	59.606	59.655	59.705
4	23.999	23.312	22.918	22.663	22.484	22.250	22.104	22.003	21.901
5	15.639	14.864	14.422	14.135	13.934	13.670	13.504	13.391	13.275
6	12.175	11.411	10.975	10.691	10.491	10.228	10.063	9.949	9.832
7	10.334	9.594	9.169	8.893	8.697	8.440	8.277	8.164	8.048
8	9.207	8.488	8.075	7.805	7.614	7.361	7.201	7.090	6.975
10	7.909	7.224	6.829	6.570	6.386	6.141	5.984	5.875	5.761
12	7.190	6.528	6.146	5.894	5.715	5.474	5.320	5.212	5.100
14	6.735	6.090	5.717	5.470	5.294	5.057	4.905	4.798	4.686
18	6.193	5.571	5.209	4.970	4.798	4.566	4.416	4.309	4.198
20	6.019	5.405	5.047	4.810	4.640	4.410	4.260	4.154	4.042
25	5.724	5.124	4.774	4.542	4.374	4.147	3.998	3.892	3.780
30	5.540	4.949	4.604	4.374	4.209	3.983	3.835	3.729	3.617
35	5.414	4.829	4.488	4.260	4.096	3.872	3.724	3.618	3.505
40	5.322	4.742	4.404	4.178	4.014	3.791	3.643	3.538	3.425
50	5.198	4.625	4.290	4.066	3.904	3.682	3.535	3.429	3.315
60	5.118	4.549	4.217	3.994	3.833	3.611	3.465	3.359	3.245
70	5.062	4.496	4.165	3.944	3.783	3.562	3.416	3.310	3.196
80	5.020	4.457	4.127	3.907	3.747	3.526	3.380	3.274	3.159
100	4.963	4.403	4.075	3.856	3.696	3.476	3.330	3.224	3.109
200	4.851	4.298	3.974	3.757	3.598	3.380	3.234	3.127	3.012
∞	4.744	4.197	3.877	3.661	3.504	3.287	3.141	3.035	2.918

*Multiply by 10^2

TABLE 2 (*Continued*)

$p = 2$, 1% Significance Level

$n \backslash m$	2	3	4	5	6	8	10	12	15
2	2.467†	2.667†	2.776†	2.844†	2.891†	2.952†	2.989†	3.014†	3.039†
3	2.985*	2.990*	2.992*	2.994*	2.995*	2.996*	2.997*	2.997*	2.998*
4	74.275	71.026	69.244	68.116	67.337	66.332	65.712	65.290	64.862
5	38.295	35.567	34.070	33.121	32.465	31.615	31.088	30.729	30.364
6	26.118	23.794	22.517	21.706	21.143	20.413	19.958	19.648	19.332
7	20.388	18.326	17.191	16.469	15.967	15.313	14.905	14.626	14.341
8	17.152	15.268	14.229	13.567	13.106	12.504	12.127	11.868	11.603
10	13.701	12.038	11.120	10.531	10.121	9.582	9.243	9.010	8.769
12	11.920	10.388	9.541	8.996	8.615	8.113	7.796	7.577	7.351
14	10.844	9.399	8.597	8.082	7.720	7.242	6.939	6.729	6.511
18	9.617	8.278	7.533	7.053	6.714	6.265	5.979	5.780	5.572
20	9.236	7.932	7.206	6.736	6.406	5.966	5.685	5.489	5.284
25	8.604	7.360	6.666	6.217	5.899	5.476	5.204	5.013	4.813
30	8.219	7.013	6.339	5.903	5.593	5.180	4.914	4.726	4.529
35	7.959	6.780	6.120	5.692	5.389	4.982	4.720	4.535	4.339
40	7.773	6.613	5.964	5.542	5.243	4.841	4.582	4.398	4.204
50	7.523	6.389	5.754	5.341	5.048	4.653	4.397	4.216	4.023
60	7.363	6.247	5.621	5.214	4.924	4.534	4.280	4.100	3.908
70	7.252	6.148	5.529	5.125	4.838	4.451	4.199	4.020	3.829
80	7.171	6.075	5.461	5.061	4.775	4.391	4.140	3.961	3.770
100	7.059	5.976	5.369	4.972	4.690	4.308	4.059	3.881	3.691
200	6.843	5.785	5.191	4.803	4.525	4.150	3.903	3.727	3.538
∞	6.638	5.604	5.023	4.642	4.369	4.000	3.757	3.582	3.393

†Multiply by 10^4. *Multiply by 10^2.

TABLE 2 (*Continued*)

$p = 3$, 5% Significance Level

$n \backslash m$	3	4	5	6	8	10	12	15	20
3	25.930*	26.996*	27.665*	28.125*	28.712*	29.073*	29.316*	29.561*	29.809*
4	1.188*	1.193*	1.196*	1.198*	1.200*	1.202*	1.203*	1.204*	1.205*
5	42.474	41.764	41.305	40.983	40.562	40.300	40.120	39.937	39.750
6	25.456	24.715	24.235	23.899	23.458	23.182	22.992	22.799	22.600
7	18.752	18.056	17.605	17.288	16.870	16.608	16.427	16.241	16.051
8	15.308	14.657	14.233	13.934	13.540	13.290	13.118	12.941	12.758
10	11.893	11.306	10.921	10.649	10.287	10.057	9.897	9.732	9.560
12	10.229	9.682	9.323	9.068	8.727	8.509	8.357	8.198	8.033
14	9.255	8.736	8.394	8.149	7.822	7.612	7.465	7.311	7.150
16	8.618	8.118	7.788	7.553	7.236	7.031	6.887	6.736	6.577
18	8.170	7.685	7.364	7.135	6.825	6.624	6.483	6.334	6.177
20	7.838	7.365	7.051	6.826	6.522	6.325	6.185	6.038	5.882
25	7.294	6.841	6.539	6.323	6.029	5.837	5.700	5.556	5.401
30	6.965	6.524	6.231	6.020	5.732	5.543	5.409	5.265	5.112
35	6.745	6.313	6.025	5.818	5.534	5.348	5.214	5.072	4.919
40	6.588	6.162	5.878	5.673	5.393	5.208	5.076	4.934	4.781
50	6.377	5.961	5.682	5.481	5.205	5.022	4.891	4.750	4.597
60	6.243	5.832	5.558	5.359	5.086	4.904	4.774	4.633	4.480
70	6.150	5.744	5.471	5.274	5.003	4.823	4.693	4.553	4.399
80	6.082	5.679	5.408	5.212	4.943	4.763	4.634	4.493	4.339
100	5.989	5.590	5.322	5.128	4.860	4.682	4.552	4.413	4.258
200	5.810	5.419	5.156	4.965	4.702	4.525	4.397	4.257	4.102
∞	5.640	5.256	4.999	4.812	4.552	4.377	4.250	4.110	3.954

*Multiply by 10^2

TABLE 2 (*Continued*)

$n \setminus m$	3	4	5	6	8	10	12	15	20
			$p = 3$, 1% Significance Level						
3	6.484†	6.750†	6.917†	7.031†	7.178†	7.267†	7.328†	7.389†	7.451†
4	5.990*	5.995*	5.998*	6.000*	6.002*	6.003*	6.005*	6.006*	6.007*
5	1.274*	1.242*	1.222*	1.208*	1.190*	1.179*	1.172*	1.164*	1.156*
6	59.507	57.032	55.462	54.377	52.973	52.102	51.509	50.906	50.292
7	37.994	35.993	34.721	33.840	32.695	31.984	31.498	31.002	30.496
8	28.308	26.599	25.511	24.755	23.771	23.157	22.737	22.308	21.868
10	19.737	18.355	17.471	16.855	16.050	15.544	15.197	14.840	14.472
12	15.973	14.765	13.990	13.448	12.737	12.288	11.978	11.659	11.328
14	13.905	12.803	12.096	11.599	10.945	10.530	10.243	9.946	9.638
16	12.610	11.581	10.918	10.452	9.836	9.444	9.172	8.890	8.596
18	11.729	10.751	10.120	9.676	9.087	8.712	8.450	8.178	7.893
20	11.091	10.152	9.545	9.117	8.549	8.186	7.932	7.668	7.390
25	10.075	9.201	8.634	8.233	7.699	7.356	7.115	6.803	6.596
30	9.479	8.644	8.102	7.718	7.205	6.874	6.641	6.395	6.135
35	9.087	8.280	7.755	7.382	6.883	6.560	6.332	6.091	5.834
40	8.811	8.023	7.511	7.146	6.656	6.339	6.115	5.877	5.623
50	8.448	7.686	7.189	6.836	6.360	6.050	5.831	5.597	5.346
60	8.220	7.474	6.988	6.642	6.174	5.870	5.653	5.422	5.172
70	8.063	7.329	6.850	6.509	6.047	5.746	5.531	5.302	5.053
80	7.948	7.224	6.750	6.412	5.955	5.656	5.443	5.215	4.967
100	7.793	7.081	6.614	6.281	5.830	5.534	5.323	5.096	4.850
200	7.498	6.808	6.356	6.032	5.593	5.304	5.096	4.873	4.627
∞	7.222	6.554	6.116	5.801	5.373	5.089	4.885	4.664	4.419

†Multiply by 10^4. *Multiply by 10^2.

TABLE 2 (*Continued*)

$n \setminus m$	4	5	6	8	10	12	15	20	25
				$p = 4$, 5% Significance Level					
4	49.964*	51.204*	52.054*	53.142*	53.808*	54.258*	54.71*	55.17*	55.46*
5	1.996*	2.001*	2.005*	2.009*	2.011*	2.013*	2.015*	2.016*	2.017*
6	65.715	64.999	64.497	63.841	63.432	63.151	62.866	62.573	62.396
7	37.343	36.629	36.129	35.474	35.064	34.782	34.495	34.200	34.019
8	26.516	25.868	25.413	24.814	24.437	24.178	23.912	23.639	23.471
10	17.875	17.326	16.938	16.424	16.098	15.872	15.640	15.399	15.250
12	14.338	13.848	13.500	13.037	12.741	12.535	12.321	12.099	11.961
14	12.455	12.002	11.680	11.248	10.972	10.778	10.577	10.366	10.234
16	11.295	10.868	10.563	10.154	9.890	9.705	9.512	9.309	9.181
18	10.512	10.104	9.812	9.419	9.165	8.986	8.798	8.600	8.475
20	9.950	9.556	9.274	8.893	8.645	8.471	8.287	8.093	7.970
25	9.059	8.688	8.422	8.062	7.826	7.659	7.482	7.293	7.173
30	8.538	8.182	7.927	7.578	7.350	7.188	7.015	6.829	6.710
35	8.197	7.852	7.603	7.263	7.040	6.880	6.710	6.526	6.408
40	7.957	7.619	7.375	7.041	6.821	6.664	6.495	6.313	6.195
50	7.640	7.313	7.075	6.750	6.535	6.380	6.214	6.033	5.916
60	7.442	7.120	6.887	6.568	6.356	6.203	6.038	5.858	5.740
70	7.305	6.988	6.758	6.443	6.232	6.081	5.917	5.738	5.620
80	7.206	6.892	6.665	6.351	6.143	5.992	5.829	5.650	5.532
100	7.071	6.762	6.537	6.228	6.021	5.872	5.710	5.531	5.413
200	6.814	6.514	6.295	5.993	5.791	5.644	5.484	5.305	5.186
∞	6.574	6.282	6.069	5.774	5.576	5.431	5.272	5.094	4.974

*Multiply by 10^2

TABLE 2 (*Continued*)

$n \setminus m$	4	5	6	8	10	12	15	20	25
			$p = 4$, 1% Significance Level						
4	12.491†	12.800†	13.012†	13.283†	13.449†	13.561†	13.67†	13.79†	13.87†
5	9.999*	10.004*	10.008*	10.012*	10.014*	10.016*	10.018*	10.02*	10.02*
6	1.938*	1.906*	1.885*	1.857*	1.840*	1.828*	1.816*	1.804*	1.797*
7	85.053	82.731	81.125	79.047	77.759	76.882	75.989	75.082	74.522
8	51.991	50.178	48.921	47.290	46.276	45.583	44.877	44.156	43.715
10	29.789	28.478	27.566	26.376	25.632	25.121	24.597	24.060	23.731
12	21.965	20.889	20.138	19.154	18.534	18.108	17.668	17.215	16.936
14	18.142	17.199	16.539	15.670	15.121	14.742	14.349	13.943	13.691
16	15.916	15.059	14.457	13.662	13.157	12.807	12.444	12.066	11.831
18	14.473	13.674	13.112	12.368	11.894	11.564	11.221	10.863	10.639
20	13.466	12.710	12.177	11.470	11.018	10.703	10.374	10.030	9.814
25	11.924	11.237	10.751	10.103	9.687	9.395	9.089	8.766	8.562
30	11.055	10.409	9.951	9.338	8.943	8.665	8.372	8.060	7.863
35	10.499	9.880	9.440	8.851	8.470	8.200	7.915	7.611	7.418
40	10.114	9.514	9.087	8.514	8.142	7.879	7.600	7.301	7.110
50	9.614	9.040	8.631	8.079	7.720	7.465	7.194	6.902	6.713
60	9.305	8.747	8.319	7.811	7.460	7.210	6.943	6.655	6.468
70	9.095	8.549	8.158	7.630	7.284	7.037	6.774	6.488	6.301
80	8.944	8.405	8.020	7.498	7.157	6.912	6.651	6.367	6.181
100	8.739	8.211	7.833	7.321	6.985	6.744	6.486	6.204	6.019
200	8.354	7.848	7.484	6.990	6.664	6.429	6.176	5.898	5.714
∞	8.000	7.513	7.163	6.686	6.369	6.140	5.892	5.616	5.432

†Multiply by 10^4. *Multiply by 10^2.

TABLE 2 (*Continued*)

$n \backslash m$	5	6	8	10	12	15	20	25	40
			$p = 5$,	5% Significance Level					
5	81.991*	83.352*	85.093*	86.160†	86.88†	—	—	—	—
6	3.009*	3.014*	3.020*	3.024†	3.027†	3.029†	3.032†	—	—
7	93.762	93.042	92.102	91.515	91.113	90.705	90.29	90.04	—
8	51.339	50.646	49.739	49.170	48.780	48.382	47.973	47.723	47.35
10	27.667	27.115	26.387	25.927	25.610	25.284	24.947	24.740	24.422
12	20.169	19.701	19.079	18.683	18.409	18.124	17.830	17.647	17.365
14	16.643	16.224	15.666	15.309	15.059	14.800	14.530	14.361	14.100
16	14.624	14.239	13.722	13.389	13.157	12.914	12.659	12.499	12.250
18	13.326	12.963	12.476	12.161	11.939	11.708	11.463	11.310	11.068
20	12.424	12.078	11.612	11.310	11.097	10.874	10.637	10.488	10.252
25	11.046	10.728	10.297	10.016	9.817	9.606	9.381	9.239	9.010
30	10.270	9.969	9.559	9.291	9.099	8.896	8.679	8.539	8.314
35	9.774	9.484	9.088	8.828	8.642	8.444	8.230	8.093	7.869
40	9.429	9.147	8.761	8.507	8.325	8.130	7.919	7.783	7.561
50	8.982	8.711	8.339	8.092	7.915	7.725	7.518	7.383	7.161
60	8.706	8.441	8.077	7.836	7.662	7.474	7.269	7.135	6.912
70	8.517	8.257	7.899	7.661	7.489	7.304	7.100	6.967	6.743
80	8.381	8.124	7.770	7.535	7.365	7.181	6.978	6.845	6.621
100	8.197	7.945	7.597	7.365	7.197	7.014	6.813	6.680	6.455
200	7.850	7.607	7.271	7.045	6.881	6.702	6.503	6.370	6.142
∞	7.531	7.295	6.970	6.750	6.590	6.414	6.217	6.084	5.850

†Multiply by 10^4. *Multiply by 10^2.

TABLE 2 (*Continued*)

$n \backslash m$	5	6	8	10	12	15	20	25	40
			$p = 5$, 1% Significance Level						
5	20.495*	20.834*	21.267*	21.53*	—	—	—	—	—
6	15.014*	15.019*	15.025*	15.029*	15.033*	15.03*	15.06*	—	—
7	2.735*	2.704*	2.665*	2.640*	2.623*	2.606*	2.590*	2.579*	—
8	1.150*	1.128*	1.099*	1.081*	1.069*	1.057*	1.044*	1.036*	—
10	48.048	46.670	44.877	43.758	42.992	42.210	41.408	40.921	—
12	31.108	30.065	28.701	27.846	27.257	26.653	26.031	25.648	25.06
14	24.016	23.145	22.001	21.279	20.781	20.268	19.736	19.408	18.90
16	20.240	19.472	18.459	17.817	17.373	16.913	16.435	16.138	15.678
18	17.929	17.228	16.302	15.713	15.304	14.878	14.435	14.159	13.727
20	16.380	15.727	14.862	14.310	13.925	13.525	13.105	12.843	12.431
25	14.107	13.529	12.759	12.265	11.918	11.555	11.172	10.930	10.547
30	12.880	12.345	11.629	11.167	10.842	10.500	10.136	9.906	9.538
35	12.115	11.607	10.926	10.486	10.174	9.845	9.494	9.271	8.911
40	11.593	11.105	10.448	10.022	9.720	9.401	9.058	8.839	8.484
50	10.928	10.465	9.841	9.434	9.144	8.836	8.504	8.290	7.940
60	10.523	10.076	9.471	9.076	8.794	8.493	8.167	7.956	7.609
70	10.251	9.814	9.223	8.835	8.559	8.263	7.941	7.732	7.386
80	10.055	9.626	9.045	8.663	8.390	8.097	7.779	7.571	7.225
100	9.793	9.374	8.806	8.432	8.164	7.876	7.561	7.355	7.009
200	9.306	8.907	8.363	8.004	7.745	7.465	7.157	6.953	6.606
∞	8.863	8.482	7.961	7.615	7.365	7.093	6.790	6.588	6.236

*Multiply by 10^2

TABLE 2 (*Continued*)

$p = 6$, 5% Significance Level

$n \backslash m$	6	8	10	12	15	20	25	30	35
10	45.722	44.677	44.019	43.567	43.103	42.626	42.334	42.136	41.993
12	28.959	28.121	27.590	27.223	26.843	26.451	26.209	26.044	25.925
14	22.321	21.600	21.141	20.821	20.489	20.144	19.929	19.783	19.677
16	18.858	18.210	17.795	17.505	17.202	16.886	16.688	16.553	16.455
18	16.755	16.157	15.772	15.501	15.218	14.921	14.735	14.607	14.513
20	15.351	14.788	14.424	14.168	13.899	13.615	13.436	13.313	13.223
25	13.293	12.786	12.456	12.222	11.975	11.711	11.544	11.428	11.343
30	12.180	11.705	11.395	11.173	10.939	10.687	10.526	10.414	10.331
35	11.484	11.031	10.733	10.520	10.293	10.049	9.892	9.782	9.700
40	11.009	10.571	10.282	10.075	9.853	9.614	9.460	9.351	9.270
50	10.402	9.983	9.706	9.507	9.293	9.060	8.908	8.801	8.721
60	10.031	9.625	9.355	9.160	8.951	8.721	8.572	8.465	8.385
70	9.781	9.383	9.118	8.927	8.720	8.494	8.345	8.239	8.159
80	9.601	9.209	8.948	8.759	8.555	8.330	8.182	8.076	7.996
100	9.360	8.976	8.720	8.534	8.333	8.110	7.963	7.857	7.777
200	8.910	8.542	8.295	8.115	7.919	7.701	7.555	7.449	7.369
500	8.659	8.300	8.059	7.882	7.689	7.473	7.328	7.222	7.140
1000	8.579	8.223	7.983	7.808	7.616	7.400	7.255	7.149	7.067
∞	8.500	8.146	7.908	7.734	7.543	7.328	7.183	7.077	6.994

TABLE 2 (*Continued*)

			$p = 6$,	1% Significance Level					
$n \backslash m$	6	8	10	12	15	20	25	30	35
10	86.397	83.565	81.804	80.602	79.376	78.124	77.360	76.845	76.474
12	46.027	44.103	42.899	42.073	41.227	40.359	39.826	39.466	39.206
14	32.433	30.918	29.966	29.309	28.634	27.936	27.507	27.215	27.004
16	25.977	24.689	23.875	23.311	22.729	22.126	21.753	21.498	21.314
18	22.292	21.146	20.418	19.913	19.389	18.844	18.505	18.273	18.105
20	19.935	18.886	18.217	17.752	17.267	16.761	16.445	16.229	16.071
25	16.642	15.737	15.156	14.749	14.324	13.875	13.592	13.397	13.254
30	14.944	14.118	13.586	13.211	12.816	12.398	12.133	11.949	11.814
35	13.913	13.138	12.635	12.281	11.906	11.506	11.252	11.074	10.943
40	13.223	12.482	12.000	11.659	11.298	10.911	10.663	10.490	10.361
50	12.358	11.661	11.206	10.882	10.538	10.167	9.927	9.759	9.633
60	11.839	11.169	10.730	10.417	10.083	9.721	9.486	9.320	9.196
70	11.493	10.841	10.413	10.107	9.779	9.424	9.192	9.028	8.905
80	11.246	10.607	10.187	9.886	9.563	9.212	8.983	8.819	8.697
100	10.917	10.295	9.886	9.592	9.276	8.930	8.703	8.541	8.419
200	10.312	9.723	9.333	9.052	8.748	8.412	8.190	8.030	7.908
500	9.980	9.409	9.030	8.755	8.458	8.128	7.907	7.747	7.625
1000	9.874	9.308	8.933	8.661	8.365	8.037	7.817	7.657	7.534
∞	9.770	9.210	8.838	8.568	8.274	7.948	7.728	7.568	7.446

TABLE 2 (*Continued*)

$n \backslash m$	8	10	12	15	20	25	30	35
	$p = 7$,		5% Significance Level					
10	85.040	84.082	83.426	82.755	82.068	81.648	81.364	81.159
12	42.850	42.126	41.627	41.113	40.583	40.257	40.037	39.877
14	29.968	29.373	28.961	28.534	28.091	27.817	27.631	27.495
16	24.038	23.519	23.158	22.781	22.389	22.145	21.978	21.857
18	20.692	20.222	19.893	19.549	19.189	18.964	18.809	18.696
20	18.561	18.125	17.819	17.498	17.159	16.947	16.800	16.694
25	15.587	15.202	14.930	14.642	14.337	14.143	14.009	13.911
30	14.049	13.693	13.440	13.172	12.884	12.701	12.573	12.478
35	13.113	12.776	12.535	12.278	12.002	11.825	11.700	11.608
40	12.485	12.160	11.927	11.679	11.411	11.237	11.115	11.025
50	11.695	11.386	11.165	10.927	10.668	10.500	10.381	10.292
60	11.219	10.921	10.706	10.475	10.221	10.056	9.938	9.850
70	10.901	10.610	10.400	10.173	9.923	9.760	9.643	9.555
80	10.674	10.388	10.181	9.957	9.710	9.548	9.432	9.344
100	10.371	10.091	9.889	9.669	9.426	9.265	9.150	9.062
200	9.812	9.545	9.350	9.138	8.902	8.744	8.629	8.542
500	9.504	9.244	9.054	8.846	8.613	8.456	8.342	8.254
1000	9.405	9.148	8.959	8.753	8.521	8.365	8.250	8.162
∞	9.308	9.053	8.866	8.661	8.431	8.275	8.160	8.072

TABLE 2 (*Continued*)

$n \backslash m$	8	10	12	15	20	25	30	35
	$p = 7$, 1% Significance Level							
10	185.93	182.94	180.90	178.83	176.73	175.44	174.57	173.92
12	71.731	69.978	68.779	67.552	66.296	65.528	65.010	64.636
14	44.255	42.978	42.099	41.197	40.269	39.698	39.311	39.032
16	33.097	32.057	31.339	30.599	29.834	29.361	29.039	28.806
18	27.273	26.374	25.750	25.105	24.435	24.019	23.735	23.529
20	23.757	22.949	22.388	21.804	21.195	20.816	20.556	20.367
25	19.117	18.440	17.965	17.469	16.947	16.619	16.392	16.227
30	16.848	16.239	15.810	15.360	14.882	14.580	14.370	14.216
35	15.512	14.945	14.544	14.121	13.670	13.383	13.183	13.036
40	14.634	14.095	13.713	13.309	12.876	12.599	12.405	12.262
50	13.553	13.049	12.691	12.310	11.899	11.634	11.448	11.309
60	12.914	12.432	12.088	11.720	11.323	11.065	10.882	10.746
70	12.492	12.024	11.690	11.332	10.942	10.689	10.509	10.374
80	12.193	11.736	11.408	11.056	10.673	10.422	10.244	10.110
100	11.797	11.353	11.034	10.691	10.316	10.070	9.894	9.761
200	11.077	10.658	10.356	10.028	9.667	9.427	9.254	9.123
500	10.685	10.230	9.987	9.668	9.314	9.078	8.906	8.774
1000	10.561	10.160	9.869	9.553	9.202	8.966	8.795	8.663
∞	10.439	10.043	9.755	9.441	9.092	8.857	8.686	8.555

TABLE 2 (*Continued*)

$p = 8$, 5% Significance Level

$n \backslash m$	8	10	12	15	20	25	30	35
14	42.516	41.737	41.198	40.641	40.066	39.711	39.470	39.296
16	31.894	31.242	30.788	30.318	29.829	29.525	29.318	29.167
18	26.421	25.847	25.446	25.028	24.591	24.319	24.132	23.996
20	23.127	22.605	22.239	21.856	21.454	21.201	21.028	20.902
25	18.770	18.324	18.009	17.677	17.325	17.102	16.947	16.834
30	16.626	16.221	15.934	15.629	15.303	15.095	14.950	14.843
35	15.356	14.977	14.707	14.418	14.109	13.910	13.771	13.668
40	14.518	14.156	13.898	13.621	13.322	13.129	12.994	12.893
50	13.482	13.142	12.898	12.636	12.351	12.165	12.034	11.936
60	12.866	12.540	12.305	12.051	11.774	11.593	11.465	11.368
70	12.459	12.142	11.912	11.665	11.393	11.215	11.088	10.992
80	12.169	11.858	11.634	11.390	11.122	10.946	10.820	10.725
100	11.785	11.483	11.264	11.026	10.763	10.590	10.465	10.370
200	11.084	10.798	10.589	10.362	10.108	9.939	9.816	9.722
500	10.701	10.423	10.221	9.999	9.751	9.584	9.461	9.367
1000	10.579	10.304	10.104	9.884	9.637	9.470	9.348	9.254
∞	10.459	10.188	9.989	9.771	9.526	9.360	9.238	9.144

$p = 8$, 1% Significance Level

$n \backslash m$	8	10	12	15	20	25	30	35
14	65.793	64.035	62.828	61.592	60.323	59.545	59.019	58.639
16	44.977	43.633	42.707	41.754	40.771	40.164	39.753	39.456
18	35.265	34.146	33.373	32.573	31.745	31.232	30.882	30.629
20	29.786	28.808	28.129	27.425	26.691	26.235	25.924	25.697
25	23.001	22.212	21.661	21.085	20.480	20.100	19.838	19.647
30	19.867	19.173	18.686	18.173	17.631	17.288	17.051	16.876
35	18.077	17.440	16.991	16.516	16.011	15.690	15.466	15.301
40	16.924	16.324	15.900	15.451	14.970	14.662	14.447	14.288
50	15.528	14.975	14.582	14.163	13.711	13.420	13.216	13.063
60	14.715	14.190	13.815	13.414	12.980	12.698	12.499	12.351
70	14.184	13.677	13.313	12.925	12.502	12.226	12.031	11.885
80	13.810	13.315	12.960	12.580	12.165	11.894	11.701	11.556
100	13.317	12.839	12.496	12.127	11.722	11.457	11.267	11.124
200	12.429	11.983	11.660	11.311	10.925	10.669	10.484	10.343
500	11.951	11.521	11.210	10.871	10.495	10.244	10.061	9.921
1000	11.800	11.375	11.067	10.732	10.359	10.109	9.927	9.787
∞	11.652	11.233	10.928	10.597	10.227	9.978	9.796	9.656

TABLE 2 (*Continued*)

$n \backslash m$	10	12	15	20	25	30	35
		$p = 10$,	5% Significance Level				
14	98.999	98.013	97.002	95.963	95.326	94.9	94.6
16	58.554	57.814	57.050	56.260	55.772	55.44	55.20
18	43.061	42.454	41.824	41.169	40.762	40.485	40.284
20	35.146	34.620	34.071	33.497	33.140	32.895	32.716
25	26.080	25.660	25.219	24.753	24.458	24.255	24.107
30	22.140	21.773	21.384	20.970	20.706	20.523	20.388
35	19.955	19.618	19.260	18.876	18.630	18.458	18.331
40	18.569	18.252	17.914	17.550	17.316	17.151	17.029
50	16.913	16.622	16.309	15.969	15.748	15.592	15.476
60	15.960	15.684	15.385	15.059	14.847	14.695	14.582
70	15.341	15.074	14.786	14.469	14.261	14.113	14.002
80	14.907	14.647	14.365	14.055	13.851	13.705	13.595
100	14.338	14.087	13.814	13.513	13.313	13.170	13.061
200	13.319	13.085	12.828	12.542	12.351	12.212	12.106
500	12.774	12.548	12.301	12.023	11.836	11.699	11.594
1000	12.602	12.379	12.134	11.859	11.674	11.538	11.432
∞	12.434	12.214	11.972	11.700	11.515	11.380	11.275

$n \backslash m$	10	12	15	20	25	30	35
		$p = 10$,	1% Significance Level				
14	180.90	178.28	175.62	172.91	171.24	170	—
16	89.068	87.414	85.270	83.980	82.91	82.2	81.7
18	59.564	58.328	57.055	55.742	54.933	54.384	53.990
20	45.963	44.951	43.905	42.821	42.150	41.693	41.362
25	31.774	31.029	30.253	29.440	28.932	28.583	28.328
30	26.115	25.489	24.832	24.139	23.701	23.399	23.177
35	23.116	22.556	21.966	21.338	20.939	20.663	20.459
40	21.267	20.749	20.201	19.615	19.241	18.980	18.787
50	19.114	18.646	18.148	17.611	17.266	17.023	16.842
60	17.901	17.462	16.992	16.484	16.154	15.922	15.748
70	17.124	16.703	16.252	15.762	15.443	15.216	15.046
80	16.583	16.175	15.738	15.260	14.948	14.726	14.559
100	15.881	15.490	15.069	14.608	14.305	14.088	13.925
200	14.641	14.280	13.889	13.457	13.169	12.962	12.803
500	13.986	13.641	13.266	12.848	12.569	12.366	12.210
1000	13.780	13.441	13.070	12.658	12.381	12.179	12.023
∞	13.581	13.246	12.881	12.472	12.198	11.997	11.842

TABLE 3

Tables of Significance Points for the Bartlett–Nanda–Pillai Trace Test

$$\Pr\left\{\frac{n+m}{m}V \geq x_\alpha\right\} = \alpha$$

p=2

α	n \ m	1	2	3	4	5	6	7	8	9	10	15	20
.05	13	5.499	4.250	3.730	3.430	3.229	3.082	2.970	2.881	2.808	2.747	2.545	2.431
	15	5.567	4.310	3.782	3.476	3.271	3.122	3.008	2.917	2.842	2.779	2.572	2.455
	19	5.659	4.396	3.858	3.546	3.336	3.183	3.066	2.972	2.895	2.831	2.616	2.493
	23	5.718	4.453	3.911	3.595	3.383	3.228	3.109	3.013	2.935	2.869	2.650	2.524
	27	5.759	4.495	3.950	3.632	3.418	3.261	3.141	3.045	2.966	2.899	2.677	2.548
	33	5.801	4.539	3.992	3.672	3.456	3.299	3.178	3.081	3.001	2.934	2.709	2.578
	43	5.845	4.586	4.037	3.716	3.499	3.341	3.219	3.122	3.041	2.974	2.746	2.613
	63	5.891	4.635	4.086	3.764	3.547	3.389	3.266	3.169	3.088	3.020	2.791	2.657
	83	5.914	4.661	4.112	3.790	3.573	3.415	3.293	3.195	3.114	3.046	2.818	2.683
	123	5.938	4.688	4.139	3.818	3.601	3.443	3.321	3.223	3.143	3.075	2.847	2.713
	243	5.962	4.715	4.168	3.846	3.630	3.472	3.351	3.254	3.174	3.106	2.880	2.748
	∞	5.991	4.744	4.197	3.877	3.661	3.504	3.384	3.287	3.208	3.141	2.918	2.788
.01	13	7.499	5.409	4.570	4.094	3.780	3.555	3.383	3.248	3.138	3.047	2.751	2.587
	15	7.710	5.539	4.671	4.180	3.857	3.625	3.448	3.309	3.196	3.101	2.795	2.625
	19	8.007	5.732	4.824	4.312	3.976	3.734	3.550	3.405	3.287	3.188	2.867	2.686
	23	8.206	5.868	4.935	4.409	4.064	3.815	3.627	3.478	3.356	3.255	2.923	2.735
	27	8.349	5.970	5.019	4.483	4.131	3.878	3.686	3.534	3.410	3.307	2.968	2.775
	33	8.500	6.080	5.111	4.566	4.207	3.950	3.754	3.600	3.473	3.368	3.021	2.823
	43	8.660	6.201	5.214	4.659	4.294	4.032	3.833	3.675	3.547	3.439	3.085	2.881
	63	8.831	6.333	5.329	4.764	4.393	4.127	3.925	3.765	3.634	3.525	3.163	2.955
	83	8.920	6.404	5.392	4.823	4.449	4.181	3.977	3.815	3.684	3.574	3.210	3.000
	123	9.012	6.478	5.459	4.885	4.508	4.238	4.033	3.871	3.739	3.628	3.263	3.052
	243	9.108	6.556	5.529	4.951	4.572	4.301	4.095	3.932	3.800	3.689	3.323	3.113
	∞	9.210	6.638	5.604	5.023	4.642	4.369	4.163	4.000	3.867	3.757	3.393	3.185

p=3

α	n \ m	1	2	3	4	5	6	7	8	9	10	15	20
.05	14	6.989	5.595	5.019	4.684	4.458	4.293	4.165	4.063	3.979	3.908	3.672	3.537
	16	7.095	5.673	5.082	4.738	4.507	4.338	4.207	4.103	4.017	3.944	3.702	3.563
	20	7.243	5.787	5.177	4.822	4.583	4.409	4.274	4.166	4.077	4.002	3.751	3.606
	24	7.341	5.866	5.245	4.883	4.639	4.461	4.323	4.213	4.122	4.046	3.790	3.640
	28	7.410	5.925	5.295	4.929	4.682	4.501	4.362	4.250	4.158	4.081	3.821	3.668
	34	7.482	5.987	5.351	4.980	4.730	4.547	4.406	4.293	4.200	4.121	3.857	3.702
	44	7.559	6.055	5.412	5.037	4.784	4.599	4.457	4.342	4.248	4.169	3.901	3.743
	64	7.639	6.129	5.480	5.101	4.846	4.660	4.516	4.400	4.305	4.225	3.955	3.795
	84	7.681	6.168	5.517	5.137	4.880	4.693	4.549	4.433	4.338	4.257	3.986	3.826
	124	7.724	6.209	5.556	5.174	4.917	4.730	4.585	4.469	4.374	4.293	4.022	3.862
	244	7.768	6.251	5.597	5.214	4.957	4.769	4.624	4.508	4.413	4.333	4.063	3.904
	∞	7.815	6.296	5.640	5.257	4.999	4.812	4.667	4.552	4.457	4.377	4.110	3.954
.01	14	8.971	6.855	5.970	5.457	5.112	4.862	4.669	4.516	4.390	4.285	3.939	3.743
	16	9.245	7.006	6.083	5.551	5.195	4.937	4.738	4.581	4.451	4.343	3.986	3.783
	20	9.639	7.236	6.258	5.698	5.326	5.056	4.849	4.684	4.549	4.436	4.063	3.850
	24	9.910	7.403	6.387	5.808	5.424	5.146	4.933	4.764	4.625	4.509	4.124	3.903
	28	10.106	7.528	6.486	5.893	5.501	5.217	4.999	4.827	4.685	4.567	4.174	3.948
	34	10.317	7.667	6.598	5.990	5.588	5.298	5.076	4.900	4.756	4.635	4.233	4.001
	44	10.545	7.821	6.724	6.101	5.690	5.393	5.167	4.986	4.839	4.715	4.305	4.067
	64	10.790	7.994	6.867	6.230	5.809	5.505	5.274	5.090	4.939	4.813	4.394	4.151
	84	10.920	8.088	6.947	6.301	5.876	5.569	5.335	5.150	4.998	4.871	4.448	4.203
	124	11.056	8.188	7.032	6.379	5.948	5.639	5.403	5.216	5.063	4.935	4.510	4.263
	244	11.196	8.294	7.124	6.463	6.028	5.716	5.478	5.290	5.136	5.007	4.581	4.334
	∞	11.345	8.406	7.222	6.554	6.116	5.801	5.562	5.372	5.218	5.089	4.664	4.419

TABLE 3 (*Continued*)

$p=4$

α	$n \backslash m$	1	2	3	4	5	6	7	8	9	10	15	20
.05	15	8.331	6.859	6.245	5.885	5.642	5.462	5.323	5.212	5.119	5.041	4.779	4.627
	17	8.472	6.952	6.318	5.947	5.696	5.512	5.369	5.255	5.160	5.080	4.811	4.654
	21	8.671	7.091	6.429	6.043	5.782	5.591	5.443	5.324	5.225	5.143	4.864	4.701
	25	8.805	7.190	6.510	6.114	5.846	5.650	5.498	5.377	5.276	5.191	4.906	4.738
	29	8.901	7.263	6.571	6.168	5.896	5.696	5.542	5.418	5.316	5.230	4.939	4.768
	35	9.004	7.343	6.640	6.229	5.952	5.749	5.593	5.467	5.363	5.275	4.980	4.805
	45	9.113	7.431	6.716	6.298	6.017	5.811	5.652	5.524	5.418	5.329	5.029	4.851
	65	9.229	7.528	6.802	6.378	6.092	5.883	5.721	5.592	5.485	5.395	5.090	4.910
	85	9.291	7.580	6.849	6.421	6.134	5.923	5.761	5.631	5.523	5.432	5.127	4.945
	125	9.354	7.635	6.899	6.469	6.179	5.968	5.804	5.674	5.566	5.475	5.168	4.987
	245	9.419	7.693	6.952	6.519	6.228	6.016	5.852	5.721	5.613	5.522	5.216	5.035
	∞	9.488	7.754	7.009	6.574	6.282	6.069	5.905	5.774	5.667	5.576	5.272	5.094
.01	15	10.293	8.188	7.276	6.737	6.373	6.105	5.898	5.731	5.594	5.479	5.095	4.874
	17	10.619	8.360	7.401	6.840	6.462	6.184	5.971	5.799	5.658	5.539	5.144	4.916
	21	11.095	8.625	7.598	7.003	6.604	6.313	6.089	5.909	5.762	5.638	5.225	4.987
	25	11.428	8.818	7.744	7.126	6.712	6.411	6.180	5.995	5.844	5.716	5.290	5.044
	29	11.672	8.966	7.858	7.222	6.798	6.490	6.253	6.064	5.909	5.779	5.344	5.091
	35	11.938	9.131	7.987	7.332	6.897	6.581	6.338	6.145	5.986	5.853	5.408	5.149
	45	12.228	9.318	8.135	7.460	7.012	6.688	6.439	6.241	6.079	5.942	5.487	5.221
	65	12.545	9.529	8.306	7.610	7.149	6.816	6.561	6.358	6.192	6.052	5.586	5.314
	85	12.715	9.645	8.402	7.695	7.227	6.890	6.632	6.426	6.258	6.117	5.646	5.371
	125	12.893	9.769	8.505	7.787	7.313	6.971	6.710	6.503	6.333	6.190	5.715	5.439
	245	13.080	9.902	8.617	7.889	7.408	7.062	6.798	6.588	6.417	6.273	5.796	5.519
	∞	13.277	10.045	8.739	8.000	7.513	7.163	6.897	6.686	6.513	6.369	5.892	5.616

$p=5$

α	$n \backslash m$	1	2	3	4	5	6	7	8	9	10	15	20
.05	16	9.589	8.071	7.430	7.052	6.795	6.605	6.457	6.338	6.239	6.155	5.873	5.706
	18	9.761	8.179	7.512	7.120	6.854	6.659	6.506	6.384	6.282	6.196	5.906	5.735
	22	10.007	8.340	7.639	7.228	6.949	6.745	6.586	6.458	6.352	6.263	5.961	5.784
	26	10.176	8.457	7.732	7.308	7.021	6.810	6.647	6.516	6.407	6.316	6.006	5.823
	30	10.298	8.544	7.803	7.370	7.077	6.862	6.696	6.562	6.451	6.358	6.042	5.856
	36	10.429	8.641	7.883	7.440	7.141	6.922	6.752	6.616	6.503	6.408	6.086	5.896
	46	10.571	8.748	7.974	7.521	7.216	6.992	6.819	6.680	6.565	6.468	6.140	5.945
	66	10.724	8.868	8.077	7.615	7.303	7.075	6.899	6.757	6.640	6.541	6.208	6.009
	86	10.805	8.933	8.134	7.667	7.353	7.122	6.944	6.802	6.684	6.584	6.248	6.049
	126	10.890	9.002	8.195	7.724	7.407	7.174	6.995	6.851	6.733	6.633	6.295	6.095
	246	10.978	9.076	8.261	7.786	7.466	7.232	7.052	6.907	6.788	6.688	6.350	6.150
	∞	11.071	9.154	8.332	7.853	7.531	7.296	7.115	6.970	6.851	6.750	6.414	6.217
.01	16	11.534	9.451	8.521	7.966	7.587	7.306	7.088	6.912	6.767	6.644	6.230	5.989
	18	11.902	9.642	8.658	8.077	7.682	7.391	7.165	6.983	6.833	6.707	6.281	6.033
	22	12.449	9.939	8.876	8.255	7.835	7.528	7.291	7.100	6.943	6.810	6.366	6.106
	26	12.837	10.159	9.040	8.390	7.954	7.635	7.389	7.192	7.030	6.893	6.434	6.166
	30	13.125	10.328	9.168	8.497	8.048	7.720	7.468	7.266	7.100	6.960	6.491	6.216
	36	13.442	10.518	9.314	8.621	8.158	7.820	7.561	7.354	7.183	7.040	6.559	6.277
	46	13.790	10.735	9.483	8.765	8.287	7.939	7.673	7.460	7.284	7.137	6.644	6.355
	66	14.176	10.984	9.681	8.936	8.442	8.083	7.808	7.589	7.409	7.258	6.752	6.455
	86	14.385	11.122	9.793	9.034	8.531	8.167	7.888	7.666	7.483	7.330	6.818	6.518
	126	14.606	11.270	9.914	9.142	8.630	8.260	7.977	7.752	7.567	7.412	6.895	6.592
	246	14.839	11.431	10.047	9.260	8.740	8.364	8.077	7.849	7.663	7.506	6.985	6.681
	∞	15.086	11.605	10.193	9.392	8.863	8.482	8.192	7.961	7.773	7.615	7.093	6.790

TABLE 3 (*Continued*)

p=6

α	n \ m	1	2	3	4	5	6	7	8	9	10	15	20
.05	17	10.794	9.247	8.585	8.193	7.926	7.728	7.573	7.448	7.344	7.256	6.956	6.778
	19	10.993	9.367	8.676	8.268	7.990	7.785	7.625	7.496	7.389	7.299	6.990	6.808
	23	11.282	9.550	8.817	8.386	8.093	7.878	7.711	7.576	7.464	7.369	7.048	6.858
	27	11.483	9.684	8.922	8.475	8.172	7.950	7.777	7.638	7.523	7.425	7.095	6.899
	31	11.630	9.784	9.003	8.545	8.234	8.007	7.830	7.688	7.570	7.471	7.134	6.934
	37	11.790	9.897	9.094	8.624	8.306	8.073	7.892	7.747	7.627	7.525	7.181	6.976
	47	11.964	10.024	9.199	8.716	8.390	8.151	7.966	7.817	7.694	7.590	7.239	7.029
	67	12.154	10.166	9.319	8.824	8.490	8.245	8.056	7.903	7.778	7.672	7.312	7.099
	87	12.255	10.245	9.387	8.885	8.547	8.299	8.108	7.954	7.827	7.720	7.357	7.142
	127	12.362	10.328	9.459	8.951	8.609	8.359	8.165	8.010	7.882	7.774	7.409	7.193
	247	12.474	10.417	9.538	9.024	8.678	8.425	8.230	8.074	7.945	7.836	7.470	7.254
	∞	12.592	10.513	9.623	9.104	8.755	8.500	8.303	8.146	8.017	7.908	7.543	7.328
.01	17	12.722	10.664	9.724	9.157	8.767	8.478	8.252	8.069	7.917	7.788	7.351	7.093
	19	13.126	10.874	9.873	9.277	8.869	8.567	8.332	8.143	7.986	7.853	7.403	7.137
	23	13.736	11.202	10.111	9.469	9.034	8.714	8.465	8.266	8.100	7.961	7.490	7.213
	27	14.173	11.446	10.292	9.617	9.162	8.828	8.570	8.363	8.192	8.048	7.561	7.275
	31	14.501	11.635	10.433	9.734	9.264	8.921	8.655	8.442	8.267	8.119	7.621	7.328
	37	14.865	11.850	10.596	9.871	9.384	9.030	8.756	8.537	8.356	8.204	7.693	7.392
	47	15.270	12.097	10.787	10.032	9.527	9.160	8.878	8.652	8.466	8.309	7.783	7.474
	67	15.723	12.382	11.011	10.224	9.700	9.319	9.027	8.794	8.602	8.440	7.899	7.581
	87	15.970	12.542	11.138	10.335	9.800	9.413	9.115	8.878	8.683	8.520	7.971	7.649
	127	16.233	12.716	11.278	10.457	9.912	9.517	9.215	8.974	8.776	8.610	8.055	7.729
	247	16.513	12.903	11.432	10.593	10.037	9.635	9.328	9.084	8.883	8.715	8.154	7.827
	∞	16.812	13.108	11.602	10.745	10.178	9.770	9.458	9.210	9.008	8.838	8.274	7.948

p=7

α	n \ m	1	2	3	4	5	6	7	8	9	10	15	20
.05	18	11.961	10.396	9.719	9.316	9.040	8.835	8.675	8.545	8.437	8.345	8.031	7.843
	20	12.184	10.528	9.817	9.396	9.109	8.896	8.730	8.596	8.484	8.390	8.067	7.874
	24	12.513	10.731	9.972	9.525	9.220	8.996	8.821	8.680	8.563	8.464	8.127	7.926
	28	12.744	10.880	10.088	9.622	9.306	9.073	8.892	8.746	8.626	8.523	8.176	7.969
	32	12.915	10.994	10.178	9.699	9.374	9.135	8.950	8.800	8.676	8.572	8.216	8.005
	38	13.102	11.123	10.281	9.787	9.453	9.208	9.017	8.864	8.737	8.630	8.266	8.049
	48	13.308	11.267	10.399	9.890	9.547	9.294	9.098	8.941	8.811	8.701	8.328	8.106
	68	13.534	11.433	10.537	10.012	9.658	9.398	9.197	9.036	8.902	8.789	8.407	8.180
	88	13.657	11.524	10.614	10.082	9.722	9.459	9.255	9.092	8.956	8.842	8.456	8.226
	128	13.786	11.623	10.698	10.158	9.793	9.526	9.320	9.155	9.018	8.903	8.513	8.281
	248	13.923	11.728	10.790	10.242	9.872	9.602	9.394	9.226	9.089	8.972	8.580	8.348
	∞	14.067	11.842	10.890	10.334	9.960	9.687	9.477	9.309	9.170	9.053	8.661	8.431
.01	18	13.874	11.841	10.895	10.321	9.923	9.627	9.395	9.206	9.049	8.915	8.460	8.188
	20	14.310	12.069	11.056	10.448	10.031	9.721	9.479	9.283	9.121	8.982	8.512	8.233
	24	14.974	12.426	11.314	10.655	10.207	9.876	9.619	9.412	9.240	9.095	8.602	8.310
	28	15.456	12.694	11.510	10.815	10.344	9.999	9.731	9.515	9.337	9.186	8.676	8.374
	32	15.822	12.902	11.665	10.943	10.455	10.098	9.821	9.599	9.416	9.261	8.738	8.429
	38	16.230	13.141	11.845	11.092	10.586	10.215	9.930	9.700	9.511	9.351	8.814	8.496
	48	16.688	13.416	12.056	11.269	10.742	10.357	10.061	9.824	9.628	9.463	8.909	8.582
	68	17.206	13.737	12.306	11.482	10.932	10.532	10.224	9.978	9.776	9.605	9.033	8.696
	88	17.491	13.919	12.449	11.605	11.043	10.635	10.321	10.070	9.865	9.691	9.110	8.768
	128	17.796	14.116	12.607	11.743	11.168	10.751	10.431	10.176	9.967	9.790	9.201	8.854
	248	18.124	14.333	12.782	11.897	11.308	10.883	10.557	10.298	10.085	9.906	9.309	8.960
	∞	18.475	14.571	12.977	12.070	11.468	11.034	10.703	10.439	10.223	10.043	9.441	9.092

TABLE 3 (*Continued*)

$p=8$

α	$n \backslash m$	1	2	3	4	5	6	7	8	9	10	15	20
.05	19	13.101	11.524	10.835	10.423	10.141	9.930	9.766	9.632	9.521	9.426	9.100	8.904
	21	13.346	11.667	10.941	10.509	10.214	9.995	9.824	9.685	9.570	9.472	9.136	8.935
	25	13.710	11.889	11.109	10.647	10.333	10.101	9.920	9.774	9.652	9.550	9.198	8.988
	29	13.970	12.054	11.235	10.753	10.425	10.184	9.996	9.844	9.718	9.612	9.249	9.032
	33	14.163	12.180	11.334	10.837	10.499	10.250	10.057	9.902	9.772	9.663	9.292	9.070
	39	14.377	12.323	11.448	10.934	10.585	10.329	10.130	9.970	9.837	9.725	9.344	9.116
	49	14.614	12.487	11.580	11.048	10.688	10.423	10.218	10.053	9.917	9.801	9.409	9.176
	69	14.877	12.674	11.734	11.183	10.811	10.538	10.326	10.156	10.016	9.897	9.494	9.254
	89	15.021	12.779	11.822	11.261	10.883	10.605	10.390	10.218	10.075	9.955	9.547	9.304
	129	15.173	12.892	11.918	11.347	10.962	10.680	10.462	10.288	10.143	10.021	9.609	9.363
	249	15.335	13.015	12.023	11.442	11.051	10.765	10.544	10.367	10.221	10.098	9.682	9.436
	∞	15.507	13.148	12.138	11.549	11.152	10.862	10.638	10.459	10.312	10.188	9.771	9.526
.01	19	14.999	12.992	12.043	11.463	11.060	10.758	10.521	10.328	10.167	10.030	9.558	9.275
	21	15.463	13.235	12.215	11.598	11.174	10.857	10.610	10.409	10.241	10.099	9.612	9.321
	25	16.177	13.620	12.491	11.819	11.360	11.021	10.757	10.543	10.366	10.216	9.704	9.399
	29	16.700	13.910	12.703	11.991	11.507	11.151	10.874	10.651	10.467	10.310	9.780	9.465
	33	17.100	14.137	12.871	12.128	11.626	11.256	10.970	10.740	10.550	10.389	9.844	9.521
	39	17.549	14.398	13.067	12.290	11.766	11.383	11.086	10.848	10.651	10.485	9.924	9.591
	49	18.058	14.702	13.297	12.482	11.935	11.536	11.227	10.980	10.776	10.604	10.024	9.681
	69	18.640	15.058	13.573	12.716	12.143	11.725	11.403	11.146	10.934	10.756	10.155	9.801
	89	18.962	15.261	13.733	12.853	12.265	11.838	11.509	11.247	11.030	10.849	10.238	9.877
	129	19.310	15.484	13.909	13.005	12.403	11.965	11.630	11.362	11.142	10.956	10.335	9.970
	249	19.684	15.729	14.106	13.177	12.559	12.112	11.769	11.495	11.271	11.083	10.453	10.083
	∞	20.090	16.000	14.327	13.371	12.738	12.280	11.930	11.652	11.424	11.233	10.597	10.227

$p=10$

α	$n \backslash m$	1	2	3	4	5	6	7	8	9	10	15	20
.05	21	15.322	13.733	13.027	12.603	12.311	12.093	11.922	11.782	11.666	11.566	11.222	11.013
	23	15.604	13.897	13.147	12.700	12.393	12.164	11.985	11.840	11.719	11.616	11.260	11.045
	27	16.033	14.154	13.340	12.857	12.526	12.282	12.091	11.937	11.808	11.699	11.325	11.100
	31	16.344	14.347	13.487	12.978	12.631	12.375	12.176	12.015	11.881	11.768	11.380	11.146
	35	16.580	14.497	13.603	13.075	12.716	12.451	12.245	12.079	11.941	11.824	11.426	11.186
	41	16.843	14.669	13.737	13.188	12.816	12.541	12.328	12.156	12.014	11.893	11.482	11.236
	51	17.140	14.868	13.895	13.323	12.936	12.651	12.430	12.251	12.104	11.979	11.555	11.301
	71	17.476	15.100	14.083	13.486	13.083	12.786	12.556	12.371	12.218	12.089	11.650	11.388
	91	17.662	15.231	14.191	13.581	13.169	12.866	12.632	12.443	12.288	12.156	11.710	11.443
	131	17.861	15.375	14.310	13.687	13.266	12.957	12.718	12.526	12.368	12.234	11.781	11.511
	251	18.076	15.532	14.443	13.806	13.376	13.061	12.818	12.622	12.461	12.325	11.867	11.594
	∞	18.307	15.705	14.591	13.940	13.501	13.180	12.933	12.735	12.572	12.434	11.972	11.700
.01	21	17.197	15.234	14.284	13.698	13.288	12.980	12.736	12.537	12.371	12.228	11.733	11.432
	23	17.707	15.507	14.476	13.849	13.413	13.088	12.832	12.624	12.449	12.301	11.789	11.478
	27	18.505	15.941	14.788	14.096	13.621	13.268	12.993	12.769	12.584	12.426	11.885	11.559
	31	19.101	16.273	15.029	14.290	13.786	13.413	13.123	12.888	12.693	12.528	11.965	11.628
	35	19.562	16.535	15.222	14.447	13.920	13.531	13.230	12.986	12.785	12.614	12.034	11.687
	41	20.088	16.839	15.448	14.632	14.080	13.674	13.359	13.106	12.897	12.720	12.119	11.761
	51	20.692	17.196	15.718	14.855	14.274	13.848	13.519	13.255	13.037	12.852	12.229	11.858
	71	21.394	17.623	16.045	15.130	14.516	14.067	13.722	13.445	13.216	13.024	12.374	11.989
	91	21.790	17.868	16.236	15.292	14.660	14.199	13.845	13.561	13.327	13.130	12.466	12.074
	131	22.221	18.141	16.450	15.475	14.824	14.350	13.986	13.695	13.456	13.254	12.577	12.177
	251	22.692	18.444	16.690	15.683	15.012	14.525	14.152	13.853	13.608	13.402	12.712	12.307
	∞	23.209	18.783	16.964	15.923	15.231	14.730	14.346	14.041	13.791	13.581	12.881	12.472

TABLE 4

Tables of Significance Points for the Roy Maximum Root Test

$$\Pr\left\{\frac{m+n}{m}R \geq x_\alpha\right\} = \alpha$$

$p=2$

α	n \ m	1	2	3	4	5	6	7	8	9	10	15	20
.05	13	5.499	3.736	3.011	2.605	2.342	2.157	2.018	1.910	1.823	1.752	1.527	1.407
	15	5.567	3.807	3.078	2.668	2.401	2.211	2.069	1.959	1.869	1.796	1.562	1.436
	19	5.659	3.905	3.173	2.759	2.487	2.293	2.148	2.033	1.940	1.864	1.618	1.484
	23	5.718	3.971	3.239	2.822	2.548	2.352	2.204	2.087	1.993	1.915	1.661	1.521
	27	5.759	4.018	3.286	2.868	2.593	2.396	2.247	2.129	2.033	1.954	1.696	1.552
	33	5.801	4.068	3.336	2.918	2.643	2.445	2.294	2.175	2.079	1.998	1.736	1.588
	43	5.845	4.120	3.391	2.973	2.697	2.498	2.347	2.228	2.131	2.049	1.783	1.631
	63	5.891	4.176	3.449	3.032	2.757	2.558	2.407	2.288	2.190	2.109	1.840	1.686
	83	5.914	4.205	3.480	3.064	2.789	2.591	2.440	2.321	2.223	2.142	1.873	1.718
	123	5.938	4.235	3.512	3.097	2.823	2.626	2.476	2.356	2.259	2.178	1.909	1.755
	243	5.962	4.265	3.545	3.132	2.859	2.663	2.513	2.395	2.298	2.217	1.951	1.797
	∞	5.991	4.297	3.580	3.169	2.897	2.702	2.554	2.436	2.340	2.261	1.998	1.847
.01	13	7.499	4.675	3.610	3.040	2.681	2.432	2.249	2.109	1.997	1.907	1.625	1.478
	15	7.710	4.834	3.742	3.154	2.782	2.523	2.333	2.186	2.069	1.973	1.676	1.519
	19	8.007	5.064	3.937	3.325	2.936	2.664	2.463	2.307	2.182	2.080	1.758	1.587
	23	8.206	5.223	4.074	3.448	3.048	2.768	2.559	2.397	2.268	2.161	1.823	1.641
	27	8.349	5.339	4.176	3.540	3.133	2.847	2.634	2.468	2.335	2.225	1.876	1.686
	33	8.500	5.465	4.287	3.642	3.228	2.936	2.718	2.548	2.412	2.299	1.938	1.740
	43	8.660	5.600	4.409	3.755	3.334	3.037	2.815	2.641	2.501	2.386	2.013	1.807
	63	8.831	5.747	4.543	3.881	3.454	3.153	2.926	2.749	2.607	2.488	2.105	1.891
	83	8.920	5.825	4.616	3.950	3.520	3.217	2.989	2.810	2.666	2.547	2.160	1.943
	123	9.012	5.906	4.692	4.022	3.591	3.285	3.056	2.877	2.732	2.612	2.222	2.002
	243	9.108	5.991	4.772	4.100	3.666	3.360	3.130	2.950	2.804	2.683	2.292	2.072
	∞	9.210	6.080	4.856	4.182	3.747	3.440	3.209	3.029	2.884	2.763	2.373	2.154

$p=3$

α	n \ m	1	2	3	4	5	6	7	8	9	10	15	20
.05	14	6.989	4.517	3.544	3.010	2.669	2.430	2.254	2.117	2.008	1.919	1.639	1.491
	16	7.095	4.617	3.634	3.092	2.745	2.501	2.319	2.178	2.065	1.973	1.682	1.526
	20	7.243	4.760	3.767	3.215	2.859	2.608	2.420	2.274	2.156	2.059	1.751	1.585
	24	7.341	4.858	3.859	3.302	2.942	2.686	2.495	2.345	2.224	2.124	1.805	1.631
	28	7.410	4.929	3.927	3.367	3.004	2.746	2.552	2.400	2.277	2.176	1.849	1.669
	34	7.482	5.004	4.001	3.439	3.073	2.812	2.616	2.462	2.338	2.234	1.901	1.715
	44	7.559	5.086	4.081	3.517	3.149	2.887	2.689	2.534	2.408	2.303	1.962	1.771
	64	7.639	5.173	4.169	3.604	3.235	2.972	2.773	2.616	2.489	2.383	2.038	1.842
	84	7.681	5.220	4.216	3.651	3.282	3.019	2.820	2.663	2.535	2.429	2.082	1.885
	124	7.724	5.268	4.265	3.701	3.332	3.069	2.870	2.713	2.586	2.479	2.132	1.934
	244	7.768	5.318	4.317	3.754	3.386	3.123	2.924	2.768	2.641	2.535	2.189	1.991
	∞	7.815	5.370	4.371	3.810	3.443	3.181	2.983	2.828	2.701	2.596	2.253	2.059
.01	14	8.971	5.416	4.106	3.412	2.978	2.680	2.462	2.295	2.163	2.055	1.724	1.552
	16	9.245	5.613	4.265	3.548	3.098	2.787	2.559	2.384	2.245	2.132	1.782	1.598
	20	9.639	5.905	4.507	3.757	3.284	2.956	2.714	2.527	2.378	2.257	1.877	1.676
	24	9.910	6.111	4.681	3.910	3.422	3.082	2.831	2.636	2.481	2.354	1.954	1.740
	28	10.106	6.264	4.811	4.026	3.528	3.180	2.922	2.722	2.562	2.431	2.016	1.792
	34	10.317	6.431	4.955	4.156	3.647	3.292	3.027	2.821	2.657	2.521	2.091	1.857
	44	10.545	6.614	5.116	4.303	3.784	3.420	3.148	2.938	2.768	2.628	2.182	1.937
	64	10.790	6.815	5.296	4.469	3.940	3.568	3.291	3.075	2.901	2.757	2.295	2.040
	84	10.920	6.923	5.393	4.560	4.027	3.652	3.372	3.153	2.977	2.832	2.363	2.103
	124	11.056	7.037	5.497	4.658	4.120	3.742	3.460	3.239	3.062	2.915	2.441	2.177
	244	11.196	7.157	5.608	4.763	4.221	3.841	3.556	3.334	3.155	3.008	2.530	2.264
	∞	11.345	7.284	5.725	4.875	4.331	3.948	3.663	3.440	3.260	3.112	2.634	2.369

TABLE 4 (*Continued*)

$p=4$

α	$n \backslash m$	1	2	3	4	5	6	7	8	9	10	15	20
	15	8.331	5.211	4.013	3.365	2.955	2.670	2.459	2.297	2.168	2.063	1.736	1.564
	17	8.472	5.336	4.123	3.464	3.045	2.752	2.536	2.368	2.235	2.126	1.784	1.603
	21	8.671	5.517	4.287	3.613	3.182	2.880	2.655	2.481	2.341	2.227	1.864	1.670
	25	8.805	5.644	4.403	3.721	3.283	2.975	2.745	2.566	2.422	2.304	1.928	1.724
	29	8.901	5.736	4.490	3.802	3.360	3.048	2.814	2.632	2.486	2.365	1.979	1.769
.05	35	9.004	5.837	4.585	3.893	3.446	3.130	2.893	2.708	2.559	2.436	2.040	1.822
	45	9.113	5.946	4.690	3.993	3.543	3.224	2.984	2.796	2.645	2.520	2.114	1.889
	65	9.229	6.065	4.806	4.106	3.653	3.332	3.090	2.900	2.746	2.619	2.206	1.974
	85	9.291	6.128	4.869	4.168	3.714	3.392	3.149	2.959	2.804	2.677	2.261	2.026
	125	9.354	6.195	4.935	4.234	3.779	3.457	3.214	3.023	2.868	2.740	2.322	2.087
	245	9.419	6.265	5.005	4.304	3.850	3.527	3.284	3.093	2.939	2.811	2.393	2.157
	∞	9.488	6.338	5.080	4.380	3.926	3.603	3.361	3.171	3.017	2.890	2.475	2.242
	15	10.293	6.080	4.549	3.744	3.244	2.901	2.651	2.461	2.310	2.188	1.812	1.618
	17	10.619	6.308	4.731	3.898	3.378	3.021	2.760	2.559	2.401	2.273	1.875	1.668
	21	11.095	6.650	5.010	4.137	3.589	3.211	2.933	2.720	2.550	2.412	1.981	1.754
	25	11.428	6.896	5.213	4.315	3.748	3.356	3.067	2.844	2.666	2.521	2.066	1.825
	29	11.672	7.080	5.368	4.451	3.872	3.469	3.172	2.942	2.759	2.609	2.137	1.884
.01	35	11.938	7.284	5.542	4.606	4.013	3.600	3.294	3.057	2.868	2.713	2.221	1.956
	45	12.228	7.510	5.737	4.782	4.175	3.752	3.437	3.193	2.998	2.837	2.326	2.048
	65	12.545	7.762	5.958	4.984	4.364	3.930	3.607	3.356	3.155	2.989	2.458	2.166
	85	12.715	7.899	6.080	5.096	4.470	4.031	3.704	3.450	3.246	3.078	2.538	2.240
	125	12.893	8.045	6.210	5.218	4.585	4.142	3.812	3.555	3.348	3.178	2.630	2.326
	245	13.080	8.199	6.350	5.349	4.710	4.263	3.930	3.671	3.462	3.290	2.737	2.430
	∞	13.277	8.363	6.500	5.491	4.848	4.397	4.062	3.801	3.591	3.418	2.863	2.555

$p=5$

α	$n \backslash m$	1	2	3	4	5	6	7	8	9	10	15	20
	16	9.589	5.856	4.448	3.694	3.218	2.890	2.648	2.463	2.316	2.196	1.824	1.630
	18	9.761	6.002	4.575	3.806	3.320	2.982	2.734	2.542	2.390	2.265	1.878	1.674
	22	10.007	6.218	4.765	3.978	3.477	3.128	2.869	2.669	2.509	2.378	1.967	1.747
	26	10.176	6.370	4.903	4.104	3.593	3.237	2.971	2.765	2.601	2.465	2.037	1.807
	30	10.298	6.483	5.006	4.200	3.683	3.321	3.051	2.842	2.674	2.535	2.095	1.857
.05	36	10.429	6.606	5.121	4.307	3.785	3.418	3.144	2.930	2.759	2.617	2.165	1.918
	46	10.571	6.742	5.249	4.428	3.901	3.529	3.251	3.034	2.859	2.714	2.250	1.994
	66	10.724	6.892	5.392	4.566	4.034	3.658	3.377	3.156	2.979	2.832	2.357	2.092
	86	10.805	6.973	5.471	4.643	4.108	3.731	3.448	3.227	3.048	2.900	2.421	2.152
	126	10.890	7.058	5.554	4.724	4.189	3.810	3.527	3.304	3.125	2.976	2.494	2.223
	246	10.978	7.148	5.643	4.812	4.276	3.897	3.613	3.390	3.210	3.061	2.578	2.306
	∞	11.071	7.244	5.738	4.907	4.371	3.992	3.708	3.485	3.306	3.157	2.676	2.407
	16	11.534	6.701	4.963	4.055	3.492	3.108	2.829	2.616	2.448	2.312	1.895	1.680
	18	11.902	6.954	5.163	4.223	3.638	3.238	2.945	2.722	2.546	2.403	1.962	1.733
	22	12.449	7.340	5.473	4.487	3.870	3.446	3.135	2.896	2.707	2.554	2.076	1.825
	26	12.837	7.620	5.703	4.685	4.047	3.606	3.282	3.033	2.835	2.673	2.169	1.902
	30	13.125	7.832	5.880	4.840	4.186	3.733	3.400	3.142	2.938	2.770	2.246	1.966
.01	36	13.442	8.069	6.079	5.016	4.346	3.881	3.537	3.271	3.060	2.886	2.339	2.046
	46	13.790	8.335	6.306	5.219	4.532	4.053	3.699	3.425	3.206	3.026	2.456	2.147
	66	14.176	8.635	6.566	5.454	4.750	4.259	3.894	3.611	3.385	3.198	2.604	2.279
	86	14.385	8.800	6.711	5.587	4.874	4.377	4.007	3.720	3.490	3.301	2.695	2.362
	126	14.606	8.977	6.867	5.731	5.010	4.506	4.132	3.841	3.608	3.416	2.800	2.461
	246	14.839	9.165	7.036	5.888	5.159	4.650	4.272	3.978	3.741	3.547	2.924	2.579
	∞	15.086	9.367	7.218	6.060	5.324	4.810	4.428	4.132	3.893	3.697	3.070	2.724

$p=6$

α	n \ m	1	2	3	4	5	6	7	8	9	10	15	20
.05	17	10.794	6.470	4.861	4.005	3.468	3.098	2.827	2.620	2.455	2.322	1.908	1.693
	19	10.993	6.634	5.001	4.128	3.579	3.199	2.920	2.705	2.535	2.396	1.965	1.740
	23	11.282	6.880	5.216	4.320	3.753	3.359	3.068	2.844	2.665	2.519	2.062	1.819
	27	11.483	7.056	5.372	4.462	3.884	3.481	3.182	2.951	2.767	2.615	2.139	1.884
	31	11.630	7.188	5.491	4.571	3.985	3.576	3.272	3.036	2.848	2.693	2.203	1.939
	37	11.790	7.334	5.625	4.695	4.101	3.686	3.376	3.136	2.943	2.785	2.280	2.006
	47	11.964	7.495	5.774	4.836	4.235	3.813	3.498	3.253	3.057	2.894	2.375	2.090
	67	12.154	7.675	5.944	4.997	4.390	3.963	3.643	3.394	3.194	3.028	2.495	2.200
	87	12.255	7.774	6.038	5.088	4.477	4.048	3.727	3.476	3.274	3.107	2.568	2.268
	127	12.362	7.878	6.138	5.185	4.573	4.141	3.818	3.566	3.363	3.195	2.652	2.348
	247	12.474	7.989	6.246	5.291	4.676	4.244	3.920	3.667	3.463	3.294	2.749	2.444
	∞	12.592	8.107	6.362	5.405	4.790	4.357	4.033	3.780	3.576	3.408	2.864	2.561
.01	17	12.722	7.296	5.360	4.352	3.730	3.306	2.998	2.764	2.580	2.431	1.974	1.739
	19	13.126	7.570	5.574	4.531	3.885	3.444	3.122	2.877	2.683	2.526	2.044	1.795
	23	13.736	7.992	5.912	4.817	4.135	3.667	3.325	3.063	2.855	2.687	2.165	1.892
	27	14.173	8.303	6.164	5.034	4.328	3.841	3.484	3.210	2.993	2.816	2.264	1.974
	31	14.501	8.541	6.360	5.204	4.480	3.980	3.612	3.329	3.104	2.921	2.347	2.043
	37	14.865	8.808	6.583	5.400	4.657	4.142	3.763	3.470	3.237	3.047	2.448	2.128
	47	15.270	9.112	6.839	5.628	4.864	4.334	3.943	3.640	3.399	3.201	2.576	2.238
	67	15.723	9.458	7.136	5.895	5.111	4.565	4.161	3.848	3.598	3.393	2.739	2.383
	87	15.970	9.650	7.303	6.047	5.252	4.699	4.289	3.971	3.716	3.507	2.840	2.475
	127	16.233	9.856	7.484	6.213	5.408	4.847	4.431	4.108	3.850	3.637	2.958	2.584
	247	16.513	10.079	7.682	6.395	5.580	5.012	4.591	4.264	4.002	3.786	3.097	2.717
	∞	16.812	10.319	7.897	6.596	5.772	5.198	4.772	4.442	4.177	3.959	3.264	2.882

$p=7$

α	n \ m	1	2	3	4	5	6	7	8	9	10	15	20
.05	18	11.961	7.063	5.258	4.304	3.708	3.298	2.999	2.770	2.589	2.442	1.989	1.753
	20	12.184	7.243	5.411	4.437	3.827	3.406	3.098	2.861	2.674	2.522	2.049	1.802
	24	12.513	7.516	5.647	4.647	4.016	3.580	3.258	3.011	2.814	2.653	2.151	1.887
	28	12.744	7.714	5.821	4.803	4.160	3.713	3.382	3.127	2.924	2.757	2.235	1.956
	32	12.915	7.863	5.954	4.925	4.272	3.817	3.481	3.220	3.012	2.842	2.304	2.015
	38	13.102	8.030	6.104	5.063	4.401	3.939	3.596	3.330	3.117	2.942	2.388	2.088
	48	13.308	8.216	6.275	5.222	4.551	4.081	3.732	3.461	3.243	3.063	2.492	2.180
	68	13.534	8.426	6.471	5.407	4.727	4.251	3.895	3.619	3.396	3.213	2.625	2.300
	88	13.657	8.541	6.579	5.511	4.827	4.348	3.990	3.711	3.486	3.301	2.706	2.376
	128	13.786	8.665	6.697	5.624	4.937	4.455	4.095	3.814	3.588	3.401	2.800	2.465
	248	13.923	8.797	6.824	5.747	5.058	4.573	4.211	3.929	3.702	3.514	2.910	2.573
	∞	14.067	8.938	6.961	5.882	5.191	4.705	4.343	4.060	3.833	3.645	3.041	2.705
.01	18	13.874	7.872	5.744	4.640	3.960	3.498	3.163	2.908	2.707	2.545	2.050	1.797
	20	14.310	8.164	5.971	4.829	4.124	3.642	3.292	3.026	2.816	2.646	2.124	1.855
	24	14.974	8.619	6.332	5.133	4.389	3.879	3.507	3.222	2.997	2.814	2.250	1.956
	28	15.456	8.957	6.605	5.367	4.595	4.065	3.676	3.379	3.143	2.951	2.355	2.042
	32	15.822	9.218	6.818	5.551	4.759	4.214	3.814	3.506	3.262	3.063	2.443	2.115
	38	16.230	9.514	7.063	5.765	4.952	4.390	3.977	3.659	3.406	3.199	2.551	2.206
	48	16.688	9.852	7.346	6.015	5.179	4.600	4.173	3.843	3.581	3.366	2.688	2.324
	68	17.206	10.243	7.679	6.313	5.452	4.855	4.413	4.072	3.799	3.575	2.866	2.481
	88	17.491	10.461	7.867	6.483	5.610	5.003	4.554	4.207	3.929	3.701	2.976	2.581
	128	17.796	10.697	8.073	6.670	5.785	5.169	4.713	4.360	4.078	3.846	3.106	2.700
	248	18.124	10.954	8.298	6.878	5.980	5.356	4.894	4.535	4.248	4.012	3.260	2.847
	∞	18.475	11.233	8.546	7.108	6.200	5.567	5.099	4.736	4.446	4.207	3.447	3.030

TABLE 4 (*Continued*)

$p=8$

α	$n \backslash m$	1	2	3	4	5	6	7	8	9	10	15	20
.05	19	13.101	7.640	5.645	4.594	3.941	3.493	3.166	2.916	2.719	2.559	2.067	1.812
	21	13.346	7.834	5.808	4.737	4.067	3.607	3.270	3.012	2.808	2.643	2.130	1.863
	25	13.710	8.132	6.063	4.962	4.270	3.792	3.441	3.171	2.956	2.782	2.238	1.952
	29	13.970	8.350	6.253	5.131	4.425	3.935	3.574	3.295	3.074	2.893	2.326	2.025
	33	14.163	8.515	6.399	5.264	4.547	4.049	3.680	3.396	3.169	2.984	2.400	2.088
	39	14.377	8.701	6.566	5.416	4.688	4.181	3.806	3.515	3.283	3.092	2.490	2.165
	49	14.614	8.912	6.757	5.593	4.854	4.338	3.955	3.658	3.420	3.224	2.603	2.264
	69	14.877	9.151	6.977	5.800	5.050	4.526	4.136	3.832	3.589	3.388	2.747	2.395
	89	15.021	9.283	7.101	5.917	5.163	4.634	4.241	3.935	3.689	3.486	2.836	2.478
	129	15.173	9.426	7.235	6.046	5.287	4.755	4.358	4.050	3.802	3.597	2.940	2.576
	249	15.335	9.579	7.381	6.187	5.424	4.889	4.491	4.180	3.931	3.725	3.063	2.695
	∞	15.507	9.745	7.541	6.342	5.577	5.040	4.640	4.329	4.078	3.872	3.210	2.843
.01	19	14.999	8.435	6.119	4.921	4.185	3.685	3.323	3.048	2.832	2.658	2.125	1.853
	21	15.463	8.743	6.357	5.119	4.355	3.836	3.458	3.171	2.944	2.762	2.201	1.913
	25	16.177	9.226	6.739	5.439	4.634	4.084	3.682	3.376	3.134	2.938	2.332	2.018
	29	16.700	9.589	7.030	5.687	4.853	4.280	3.861	3.541	3.287	3.081	2.442	2.107
	33	17.100	9.871	7.259	5.885	5.028	4.439	4.007	3.676	3.414	3.200	2.535	2.184
	39	17.549	10.194	7.524	6.115	5.234	4.627	4.181	3.839	3.566	3.344	2.649	2.280
	49	18.058	10.565	7.833	6.387	5.480	4.854	4.392	4.037	3.754	3.523	2.795	2.405
	69	18.640	10.998	8.199	6.713	5.778	5.131	4.653	4.284	3.990	3.749	2.986	2.573
	89	18.962	11.242	8.408	6.901	5.952	5.294	4.808	4.432	4.132	3.886	3.105	2.680
	129	19.310	11.508	8.638	7.109	6.146	5.478	4.983	4.601	4.295	4.044	3.246	2.810
	249	19.684	11.798	8.891	7.341	6.364	5.685	5.183	4.794	4.483	4.228	3.415	2.970
	∞	20.090	12.117	9.173	7.601	6.610	5.922	5.413	5.019	4.703	4.445	3.622	3.171

$p=10$

α	$n \backslash m$	1	2	3	4	5	6	7	8	9	10	15	20
.05	21	15.322	8.761	6.395	5.158	4.392	3.869	3.489	3.199	2.970	2.785	2.217	1.925
	23	15.604	8.979	6.577	5.315	4.531	3.994	3.602	3.303	3.067	2.875	2.285	1.980
	27	16.033	9.320	6.864	5.566	4.756	4.199	3.790	3.477	3.229	3.028	2.403	2.075
	31	16.344	9.573	7.082	5.759	4.931	4.359	3.939	3.616	3.360	3.151	2.500	2.156
	35	16.580	9.769	7.252	5.912	5.071	4.488	4.060	3.730	3.467	3.253	2.581	2.225
	41	16.843	9.992	7.448	6.090	5.234	4.641	4.203	3.865	3.596	3.376	2.682	2.311
	51	17.140	10.247	7.676	6.299	5.429	4.824	4.376	4.030	3.754	3.527	2.810	2.423
	71	17.476	10.543	7.944	6.548	5.663	5.047	4.590	4.235	3.952	3.719	2.977	2.572
	91	17.662	10.709	8.097	6.691	5.800	5.178	4.716	4.358	4.070	3.834	3.081	2.668
	131	17.861	10.890	8.265	6.850	5.952	5.324	4.858	4.496	4.206	3.967	3.204	2.783
	251	18.076	11.087	8.450	7.027	6.122	5.490	5.021	4.656	4.363	4.122	3.350	2.925
	∞	18.307	11.303	8.654	7.224	6.315	5.679	5.207	4.840	4.546	4.304	3.529	3.102
.01	21	17.197	9.534	6.851	5.470	4.624	4.051	3.636	3.322	3.075	2.877	2.271	1.962
	23	17.707	9.867	7.107	5.682	4.806	4.211	3.779	3.452	3.194	2.987	2.351	2.025
	27	18.505	10.399	7.523	6.029	5.107	4.478	4.021	3.672	3.397	3.175	2.491	2.137
	31	19.101	10.805	7.846	6.302	5.346	4.693	4.216	3.851	3.564	3.330	2.608	2.233
	35	19.562	11.125	8.103	6.522	5.541	4.868	4.376	4.000	3.702	3.460	2.709	2.315
	41	20.088	11.495	8.405	6.782	5.772	5.078	4.570	4.180	3.871	3.619	2.835	2.420
	51	20.692	11.928	8.761	7.093	6.052	5.335	4.808	4.403	4.081	3.819	2.996	2.558
	71	21.394	12.441	9.190	7.473	6.397	5.654	5.107	4.686	4.350	4.076	3.211	2.745
	91	21.790	12.735	9.439	7.695	6.601	5.845	5.287	4.857	4.515	4.234	3.347	2.867
	131	22.221	13.059	9.716	7.944	6.832	6.062	5.494	5.055	4.705	4.418	3.510	3.016
	251	22.692	13.417	10.025	8.225	7.094	6.310	5.732	5.285	4.928	4.636	3.707	3.201
	∞	23.209	13.816	10.373	8.545	7.395	6.598	6.010	5.556	5.193	4.895	3.952	3.438

TABLE 5

SIGNIFICANCE POINTS FOR THE MODIFIED LIKELIHOOD RATIO TEST OF EQUALITY OF COVARIANCE MATRICES BASED ON EQUAL SAMPLE SIZES

$\Pr\{-2\log\lambda^* \geq x\} = 0.05$

$n_g \backslash q$	2	3	4	5	6	7	8	9	10
					$p = 2$				
3	12.18	18.70	24.55	30.09	35.45	40.68	45.81	50.87	55.87
4	10.70	16.65	22.00	27.07	31.97	36.76	41.45	46.07	50.64
5	9.97	15.63	20.73	25.56	30.23	34.79	39.26	43.67	48.02
6	9.53	15.02	19.97	24.66	29.19	33.61	37.95	42.22	46.45
7	9.24	14.62	19.46	24.05	28.49	32.82	37.07	41.26	45.40
8	9.04	14.33	19.10	23.62	27.99	32.26	36.45	40.57	44.65
9	8.88	14.11	18.83	23.30	27.62	31.84	35.98	40.06	44.08
10	8.76	13.94	18.61	23.05	27.33	31.51	35.61	36.65	43.64
					$p = 3$				
5	19.2	30.5	41.0	51.0	60.7	70.3	79.7	89.0	98.3
6	17.57	28.24	38.06	47.49	56.68	65.69	74.58	83.37	92.09
7	16.59	26.84	36.29	45.37	54.21	62.89	71.45	79.91	88.29
8	15.93	25.90	35.10	43.93	52.54	60.99	69.33	77.56	85.72
9	15.46	25.22	34.24	42.90	51.34	59.62	67.79	75.86	83.86
10	15.11	24.71	33.59	42.11	50.42	58.58	66.62	74.57	82.45
11	14.83	24.31	33.08	41.50	49.71	57.76	65.71	73.56	81.35
12	14.61	23.99	32.67	41.01	49.13	57.11	64.97	72.75	80.46
13	14.43	23.73	32.33	40.60	48.66	56.57	64.37	72.08	79.72
					$p = 4$				
6	30.07	48.63	65.91	82.6	98.9	115.0	131.0	—	—
7	27.31	44.69	60.90	76.56	91.89	107.0	121.9	137.0	152.0
8	25.61	42.24	57.77	72.78	87.46	101.9	116.2	130.4	144.6
9	24.46	40.56	55.62	70.17	84.42	98.45	112.3	126.1	139.8
10	23.62	39.34	54.05	68.27	82.19	95.91	109.5	122.9	136.3
11	22.98	38.41	52.85	66.81	80.49	93.95	107.3	120.5	133.6
12	22.48	37.67	51.90	65.66	79.14	92.41	105.5	118.5	131.5
13	22.08	37.08	51.13	64.73	78.04	91.16	104.1	117.0	129.7
14	21.75	36.59	50.50	63.96	77.14	90.12	103.0	115.7	128.3
15	21.47	36.17	49.97	63.31	76.38	89.25	102.0	114.6	127.1

TABLE 5 (*Continued*)

$n_g \backslash q$	2	3	4	5	6	7	$n_g \backslash q$	2	3	4	5
			$p = 5$						$p = 6$		
8	39.29	65.15	89.46	113.0	—	—	10	49.95	84.43	117.0	—
9	36.70	61.40	84.63	107.2	129.3	151.5					
10	34.92	58.79	81.25	103.1	124.5	145.7	11	47.43	80.69	112.2	142.9
							12	45.56	77.90	108.6	138.4
11	33.62	56.86	78.76	100.0	120.9	141.6	13	44.11	75.74	105.7	135.0
12	32.62	55.37	76.83	97.68	118.2	138.4	14	42.96	74.01	103.5	132.2
13	31.83	54.19	75.30	95.81	116.0	135.9	15	42.03	72.59	101.6	129.9
14	31.19	53.24	74.06	94.29	114.2	133.8					
15	30.66	52.44	73.02	93.03	112.7	132.1	16	41.25	71.41	100.1	128.0
							17	40.59	70.41	98.75	126.4
16	30.21	51.77	72.14	91.95	111.4	130.6	18	40.02	69.55	97.63	125.0
							19	39.53	68.80	96.64	123.8
							20	39.11	68.14	95.78	122.7

TABLE 6

CORRECTION FACTORS FOR SIGNIFICANCE POINTS FOR THE SPHERICITY TEST

	5% Significance Level					
$n \backslash p$	3	4	5	6	7	8
4	1.217					
5	1.074	1.322				
6	1.038	1.122	1.383			
7	1.023	1.066	1.155	1.420		
8	1.015	1.041	1.088	1.180	1.442	
9	1.011	1.029	1.057	1.098	1.199	1.455
10	1.008	1.021	1.040	1.071	1.121	1.214
12	1.005	1.013	1.023	1.039	1.060	1.093
14	1.004	1.008	1.015	1.024	1.037	1.054
16	1.003	1.006	1.011	1.017	1.025	1.035
18	1.002	1.005	1.008	1.012	1.018	1.025
20	1.002	1.004	1.006	1.010	1.014	1.019
24	1.001	1.002	1.004	1.006	1.009	1.012
28	1.001	1.002	1.003	1.004	1.006	1.008
34	1.000	1.001	1.002	1.003	1.004	1.005
42	1.000	1.001	1.001	1.002	1.002	1.003
50	1.000	1.000	1.001	1.001	1.002	1.002
100	1.000	1.000	1.000	1.000	1.000	1.000
χ^2	11.0705	16.9190	23.6848	31.4104	40.1133	49.8018

TABLE 6 (*Continued*)

$n\backslash p$	3	4	5	6	7	8
		1% Significance Level				
4	1.266					
5	1.091	1.396				
6	1.046	1.148	1.471			
7	1.028	1.079	1.186	1.511		
8	1.019	1.049	1.103	1.213	1.542	
9	1.013	1.034	1.067	1.123	1.234	1.556
10	1.010	1.025	1.047	1.081	1.138	1.250
12	1.006	1.015	1.027	1.044	1.068	1.104
14	1.004	1.010	1.018	1.028	1.041	1.060
16	1.003	1.007	1.012	1.019	1.028	1.039
18	1.002	1.005	1.009	1.014	1.020	1.028
20	1.002	1.004	1.007	1.011	1.015	1.021
24	1.001	1.003	1.005	1.007	1.010	1.013
28	1.001	1.002	1.003	1.005	1.007	1.009
34	1.001	1.001	1.002	1.003	1.004	1.006
42	1.000	1.001	1.001	1.002	1.003	1.003
50	1.000	1.001	1.001	1.001	1.002	1.002
100	1.000	1.000	1.000	1.000	1.000	1.001
χ^2	15.0863	21.6660	29.1412	37.5662	46.9629	57.3421

TABLE 7†

Significance Points for the Modified Likelihood Ratio Test $\Sigma = \Sigma_0$

$\Pr\{-2\log\lambda_1^* \geq x\} = 0.05$

n	5%	1%
$p = 2$		
2	13.50	19.95
3	10.64	15.56
4	9.69	14.13
5	9.22	13.42
6	8.94	13.00
7	8.75	12.73
8	8.62	12.53
9	8.52	12.38
10	8.44	12.26
$p = 4$		
7	25.8	30.8
8	24.06	29.33
9	23.00	28.36
10	22.28	27.66
11	21.75	27.13
12	21.35	26.71
13	21.03	26.38
14	20.77	26.10
15	20.56	25.87
$p = 3$		
4	18.8	25.6
5	16.82	22.68
6	15.81	21.23
7	15.19	20.36
8	14.77	19.78
9	14.47	19.36
10	14.24	19.04
11	14.06	18.80
12	13.92	18.61
13	13.80	18.45
14	13.70	18.31
15	13.62	18.20
$p = 5$		
9	32.5	40.0
10	31.4	38.6
11	30.55	37.51
12	29.92	36.72
13	29.42	36.09
14	29.02	35.57
15	28.68	35.15
16	28.40	34.79
17	28.15	34.49
18	27.94	34.23
19	27.76	34.00
20	27.60	33.79
$p = 6$		
12	40.9	49.0
13	40.0	47.8
14	39.3	47.0
15	38.7	46.2
16	38.22	45.65
17	37.81	45.13
18	37.45	44.70
19	37.14	44.32
20	36.87	43.99
21	36.63	43.69
22	36.41	43.43
24	36.05	42.99
26	35.75	42.63
28	35.49	42.32
30	35.28	42.07
$p = 7$		
18	48.6	56.9
19	48.2	56.3
20	47.7	55.8
21	47.34	55.36
22	47.00	54.96
24	46.43	54.28
26	45.97	53.73
28	45.58	53.27
30	45.25	52.88
32	44.97	52.55
34	44.73	52.27
$p = 8$		
24	58.4	67.1
26	57.7	66.3
28	57.09	65.68
30	56.61	65.12
32	56.20	64.64
34	55.84	64.23
36	55.54	63.87
38	55.26	63.55
40	55.03	63.28
$p = 9$		
28	70.1	79.6
30	69.4	78.8
32	68.8	78.17
34	68.34	77.60
36	(67.91)	(77.08)
38	(67.53)	(76.65)
40	67.21	76.29
45	66.54	75.51
50	66.02	74.92
55	65.61	74.44
60	65.28	74.06
$p = 10$		
34	(82.3)	(92.4)
36	81.7	91.8
38	81.2	91.2
40	80.7	90.7
45	79.83	89.63
50	79.13	88.83
55	78.57	88.20
60	78.13	87.68
65	77.75	87.26
70	77.44	86.89
75	77.18	86.59

†Entries in parentheses have been interpolated or extrapolated into Korin's table. p = number of variates; N = number of observations; $n = N - 1$. $\lambda_1^* = n\log|\Sigma_0| - np - n\log|S| + n\,\mathrm{tr}(S\Sigma_0^{-1})$, where S is the sample covariance matrix.

References

At the end of each reference in brackets is a list of sections in which that reference is used.

Abramowitz, Milton, and Irene Stegun (1972), *Handbook of Mathematical Functions with Formulas, Graphs, and Mathematical Tables*, National Bureau of Standards. U.S. Government Printing Office, Washington, D.C. [8.5]

Abruzzi, Adam (1950), *Experimental Procedures and Criteria for Estimating and Evaluating Industrial Productivity*, doctoral dissertation, Columbia University Library. [9.7, 9.P]

Adrian, Robert (1808), Research concerning the probabilities of the errors which happen in making observations, etc., *The Analyst or Mathematical Museum*, **1**, 93–109. [1.2]

Aitken, A. C. (1937), Studies in practical mathematics, II. The evaluation of the latent roots and latent vectors of a matrix, *Proceedings of the Royal Society of Edinburgh*, **57**, 269–305. [11.4]

Anderson, R. L., and T. A. Bancroft (1952), *Statistical Theory in Research*, McGraw-Hill, New York. [8.P]

Anderson, T. W. (1946a), The non-central Wishart distribution and certain problems of multivariate statistics, *Annals of Mathematical Statistics*, **17**, 409–431. (Correction, **35** (1964), 923–924.) [14.4]

Anderson, T. W. (1946b), Analysis of multivariate variance, unpublished. [10.6]

Anderson, T. W. (1951a), Classification by multivariate analysis, *Psychometrika*, **16**, 31–50. [6.5, 6.9]

Anderson, T. W. (1951b), Estimating linear restrictions on regression coefficients for multivariate normal distributions, *Annals of Mathematical Statistics*, **22**, 327–351. (Correction, *Annals of Statistics*, **8** (1980), 1400.) [12.6, 12.7]

Anderson, T. W. (1951c), The asymptotic distribution of certain characteristic roots and vectors, *Proceedings of the Second Berkeley Symposium on Mathematical Statistics and Probability* (Jerzy Neyman, ed.), University of California, Berkeley, 105–130. [12.7]

Anderson, T. W. (1955a), Some statistical problems in relating experimental data to predicting performance of a production process, *Journal of the American Statistical Association*, **50**, 163–177. [8.P]

Anderson, T. W. (1955b), The integral of a symmetric unimodal function over a symmetric convex set and some probability inequalities, *Proceedings of the American Mathematical Society*, **6**, 170–176. [8.10]

Anderson, T. W. (1963a), Asymptotic theory for principal component analysis, *Annals of Mathematical Statistics*, **34**, 122–148. [11.3, 11.6, 11.7, 13.5]

Anderson, T. W. (1963b), A test for equality of means when covariance matrices are unequal, *Annals of Mathematical Statistics*, **34**, 671–672. [5.P]

Anderson, T. W. (1965a), Some optimum confidence bounds for roots of determinantal equations, *Annals of Mathematical Statistics*, **36**, 468–488. [11.6]

Anderson, T. W. (1965b), Some properties of confidence regions and tests of parameters in multivariate distributions (with discussion), *Proceedings of the IBM Scientific Computing Symposium in Statistics, October 21–23, 1963*, IBM Data Processing Division, White Plains, New York, 15–28. [11.6]

Anderson, T. W. (1969), Statistical inference for covariance matrices with linear structure, *Multivariate Analysis II* (P. R. Krishnaiah, ed.), Academic, New York, 55–66.

Anderson, T. W. (1973a), Asymptotically efficient estimation of covariance matrices with linear structure, *Annals of Statistics*, **1**, 135–141. [14.3]

Anderson, T. W. (1973b), An asymptotic expansion of the distribution of the Studentized classification statistic *W*, *Annals of Statistics*, **1**, 964–972. [6.6]

Anderson, T. W. (1973c), Asymptotic evaluation of the probabilities of misclassification by linear discriminant functions, *Discriminant Analysis and Applications* (T. Cacoullos, ed.), Academic, New York, 17–35. [6.6]

Anderson, T. W. (1974), An asymptotic expansion of the distribution of the limited information maximum likelihood estimate of a coefficient in a simultaneous equation system, *Journal of the American Statistical Association*, **69**, 565–573. (Correction, **71** (1976), 1010.) [12.7]

Anderson, T. W. (1976), Estimation of linear functional relationships: approximate distributions and connections with simultaneous equations in econometrics (with discussion), *Journal of the Royal Statistical Society B*, **38**, 1–36. [12.7]

Anderson, T. W. (1977), Asymptotic expansions of the distributions of estimates in simultaneous equations for alternative parameter sequences, *Econometrica*, **45**, 509–518. [12.7]

Anderson, T. W. (1984a), Estimating linear statistical relationships, *Annals of Statistics*, **12**, 1–45. [10.6, 12.6, 12.7, 14.1]

Anderson, T. W. (1984b), Asymptotic distribution of an estimator of linear functional relationships, unpublished. [12.7]

Anderson, T. W., and R. R. Bahadur (1962), Classification into two multivariate normal distributions with different covariance matrices, *Annals of Mathematical Statistics*, **33**, 420–431. [6.10]

Anderson, T. W., Somesh Das Gupta, and George P. H. Styan (1972), *A Bibliography of Multivariate Statistical Analysis*, Oliver & Boyd, Edinburgh. (Reprinted by Robert E. Krieger, Malabar, Florida, 1977.) [Preface]

Anderson, T. W., Naoto Kunitomo, and Takamitsu Sawa (1983a), Evaluation of the distribution function of the limited information maximum likelihood estimator, *Econometrica*, **50**, 1009–1027. [12.7]

Anderson, T. W., Naoto Kunitomo, and Takamitsu Sawa (1983b), Comparison of the densities of the TSLS and LIMLK estimators, *Global Econometrics, Essays in Honor of Lawrence R. Klein* (F. Gerard Adams and Bert Hickman, eds.), MIT, Cambridge, MA, 103–124. [12.7]

Anderson, T. W., and Ingram Olkin (1979), Maximum likelihood estimation of the parameters of a multivariate normal distribution, Technical Report No. 38, ONR Contract N00014-75-C-0442, Department of Statistics, Stanford University.

Anderson, T. W., and Herman Rubin (1949), Estimation of the parameters of a single equation in a complete system of stochastic equations, *Annals of Mathematical Statistics*, **20**, 46–63. (Reprinted in *Readings in Econometric Theory* (J. Malcolm Dowling and Fred R. Glahe, eds.), Colorado Associated University, 1970, 358–375.) [12.7]

Anderson, T. W., and Herman Rubin (1950), The asymptotic properties of estimates of the parameters of a single equation in a complete system of stochastic equations, *Annals of Mathematical Statistics*, **21**, 570–582. (Reprinted in *Readings in Econometric Theory* (J. Malcolm Dowling and Fred R. Glahe, eds.), Colorado Associated University, 1970, 376–388.) [12.7]

Anderson, T. W., and Herman Rubin (1956), Statistical inference in factor analysis, *Proceedings of the Third Berkeley Symposium on Mathematical Statistics and Probability* (Jerzy Neyman, ed.), Vol. V, University of California, Berkeley and Los Angeles, 111–150. [14.2, 14.3, 14.4, 14.6]

Anderson, T. W., and Takamitsu Sawa (1973), Distributions of estimates of coefficients of a single equation in a simultaneous system and their asymptotic expansions, *Econometrica*, **41**, 683–714. [12.7]

Anderson, T. W., and Takamitsu Sawa (1977), Tables of the distribution of the maximum likelihood estimate of the slope coefficient and approximations, Technical Report No. 234, Economics Series, Institute for Mathematical Studies in the Social Sciences, Stanford University, April 1977. [12.7]

Anderson, T. W., and Takamitsu Sawa (1979), Evaluation of the distribution function of the two-stage least squares estimate, *Econometrica*, **47**, 163–182. [12.7]

Anderson, T. W., and Takamitsu Sawa (1982), Exact and approximate distributions of the maximum likelihood estimator of a slope coefficient, *Journal of the Royal Statistical Society B*, **44**, 52–62. [12.7]

Anderson, T. W., and George P. H. Styan (1982), Cochran's theorem, rank additivity and tripotent matrices, *Statistics and Probability: Essays in Honor of C. R. Rao* (G. Kallianpur, P. R. Krishnaiah, and J. K. Ghosh, eds.), North-Holland, Amsterdam, 1–23. [7.4]

Anderson, T. W., and Akimichi Takemura (1982), A new proof of admissibility of tests in multivariate analysis, *Journal of Multivariate Analysis*, **12**, 457–468. [8.10]

Barnard, M. M. (1935), The secular variations of skull characters in four series of Egyptian skulls, *Annals of Eugenics*, **6**, 352–371. [8.8]

Barnes, E. W. (1899), The theory of the gamma function, *Messenger of Mathematics*, **29**, 64–129. [8.5]

Bartlett, M. S. (1934), The vector representation of a sample, *Proceedings of the Cambridge Philosophical Society*, **30**, 327–340. [8.3]

Bartlett, M. S. (1937a), Properties of sufficiency and statistical tests, *Proceedings of the Royal Society of London A*, **160**, 268–282. [10.2]

Bartlett, M. S. (1937b), The statistical conception of mental factors, *British Journal of Psychology*, **28**, 97–104. [14.7]

Bartlett, M. S. (1938), Further aspects of the theory of multiple regression, *Proceedings of the Cambridge Philosophical Society*, **34**, 33–40. [14.7]

Bartlett, M. S. (1939), A note on tests of significance in multivariate analysis, *Proceedings of the Cambridge Philosophical Society*, **35**, 180–185. [7.2, 8.6]

Bartlett, M. S. (1947), Multivariate analysis, *Journal of the Royal Statistical Society, Supplement*, **9**, 176–197. [8.8]

Bartlett, M. S. (1950), Tests of significance in factor analysis, *British Journal of Psychology (Statistics Section)*, **3**, 77–85. [14.3]

Basmann, R. L. (1957), A generalized classical method of linear estimation of coefficients in a structural equation, *Econometrica*, **25**, 77–83. [12.7]

Basmann, R. L. (1961), A note on the exact finite sample frequency functions of generalized classical linear estimators in two leading overidentified cases, *Journal of the American Statistical Association*, **56**, 619–636. [12.7]

Basmann, R. L. (1963), A note on the exact finite sample frequency functions of generalized classical linear estimators in a leading three-equation case, *Journal of the American Statistical Association*, **58**, 161–171. [12.7]

Bennett, B. M. (1951), Note on a solution of the generalized Behrens–Fisher problem, *Annals of the Institute of Statistical Mathematics*, **2**, 87–90. [5.5]

Berger, J. O. (1975), Minimax estimation of location vectors for a wide class of densities, *Annals of Statistics*, **3**, 1318–1328. [3.5]

Berger, J. O. (1976), Admissibility results for generalized Bayes estimators of coordinates of a location vector, *Annals of Statistics*, **4**, 334–356. [3.5]

Berger, J. O. (1980a), A robust generalized Bayes estimator and confidence region for a multivariate normal mean, *Annals of Statistics*, **8**, 716–761. [3.5]

Berger, J. O. (1980b), *Statistical Decision Theory, Foundations, Concepts, and Methods*,

Springer-Verlag, New York. [3.4, 6.2, 6.7]

Berndt, Ernst R., and N. Eugene Savin (1977), Conflict among criteria for testing hypotheses in the multivariate linear regression model, *Econometrica*, **45**, 1263–1277. [8.6]

Bhattacharya, P. K. (1966), Estimating the mean of a multivariate normal population with general quadratic loss function, *Annals of Mathematical Statistics*, **37**, 1819–1824. [3.5]

Birnbaum, Allan (1955), Characterizations of complete classes of tests of some multiparametric hypotheses, with applications to likelihood ratio tests, *Annals of Mathematical Statistics*, **26**, 21–36. [5.6, 8.10]

Björck, A., and G. Golub (1973), Numerical methods for computing angles between linear subspaces, *Mathematics of Computation*, **27**, 579–594. [12.3]

Blackwell, David, and M. A. Girshick (1954), *Theory of Games and Statistical Decisions*, John Wiley & Sons, New York. [6.2, 6.7]

Bonnesen, T., and W. Fenchel (1948), *Theorie der Konvexen Körper*, Chelsea, New York. [8.10]

Bose, R. C. (1936a), On the exact distribution and moment-coefficients of the D^2-statistic, *Sankhyā*, **2**, 143–154. [3.3]

Bose, R. C. (1936b), A note on the distribution of differences in mean values of two samples drawn from two multivariate normally distributed populations, and the definition of the D^2-statistic, *Sankhyā*, **2**, 379–384. [3.3]

Bose, R. C., and S. N. Roy (1938), The distribution of the studentised D^2-statistic, *Sankhyā*, **4**, 19–38. [5.4]

Bowker, A. H. (1960), A representation of Hotelling's T^2 and Anderson's classification statistic W in terms of simple statistics, *Contributions to Probability and Statistics* (I. Olkin, S. G. Ghurye, W. Hoeffding, W. G. Madow and H. B. Mann, eds.), Stanford University, Stanford, California, 142–149. [5.2]

Bowker, A. H., and R. Sitgreaves (1961), An asymptotic expansion for the distribution function of the W-classification statistic, *Studies in Item Analysis and Prediction* (Herbert Solomon, ed.), Stanford University, Stanford, California, 293–310. [6.6]

Box, G. E. P. (1949), A general distribution theory for a class of likelihood criteria, *Biometrika*, **36**, 317–346. [8.5, 9.4, 10.4, 10.5]

Bravais, Auguste (1846), Analyse mathématique sur les probabilités des erreurs de situation d'un point, *Mémoires Présentés par Divers Savants à l'Académie Royale des Sciences de l'Institut de France*, **9**, 255–332. [1.2]

Brown, G. W. (1939), On the power of the L_1 test for equality of several variances, *Annals of Mathematical Statistics*, **10**, 119–128. [10.2]

Browne, M. W. (1974), Generalized least squares estimates in the analysis of covariance structures, *South African Statistical Journal*, **8**, 1–24. [14.3]

Chambers, John M. (1977), *Computational Methods for Data Analysis*, John Wiley & Sons, New York. [12.3]

Chan, Tony F., Gene H. Golub, and Randall J. LeVeque (1981), Algorithms for computing the sample variance: analysis and recommendations, unpublished. [3.2]

Chernoff, Herman (1956), Large sample theory: parametric case, *Annals of Mathematical Statistics*, **27**, 1–22. [8.6]

Chu, S. Sylvia, and K. C. S. Pillai (1979), Power comparisons of two-sided tests of equality of two covariance matrices based on six criteria, *Annals of the Institute of Statistical Mathematics*, **31**, 185–205. [10.6]

Chung, Kai Lai (1974), *A Course in Probability Theory*, 2nd ed., Academic, New York. [2.2]

Clemm, D. S., P. R. Krishnaiah, and V. B. Waikar (1973), Tables for the extreme roots of the Wishart matrix, *Journal of Statistical Computation and Simulation*, **2**, 65–92. [10.8]

Clunies-Ross, C. W., and R. H. Riffenburgh (1960), Geometry and linear discrimination, *Biometrika*, **47**, 185–189. [6.10]

Cochran, W. G. (1934), The distribution of quadratic forms in a normal system, with applications to the analysis of covariance, *Proceedings of the Cambridge Philosophical Society*, **30**, 178–191. [7.4]

Constantine, A. G. (1963), Some non-central distribution problems in multivariate analysis, *Annals of Mathematical Statistics*, **34**, 1270–1285. [12.4]

Constantine, A. G. (1966), The distribution of Hotelling's generalised T_0^2, *Annals of Mathematical Statistics*, **37**, 215–225. [8.6]

Consul, P. C. (1966), On the exact distributions of the likelihood ratio criteria for testing linear hypotheses about regression coefficients, *Annals of Mathematical Statistics*, **37**, 1319–1330. [8.4]

Consul, P. C. (1967a), On the exact distributions of likelihood ratio criteria for testing independence of sets of variates under the null hypothesis, *Annals of Mathematical Statistics*, **38**, 1160–1169. [9.3]

Consul, P. C. (1967b), On the exact distributions of the criterion W for testing sphericity in a p-variate normal distribution, *Annals of Mathematical Statistics*, **38**, 1170–1174. [10.7]

Courant, R., and D. Hilbert (1953), *Methods of Mathematical Physics*, Interscience, New York. [9.10]

Cramér, H. (1946), *Mathematical Methods of Statistics*, Princeton University, Princeton. [2.6, 3.2, 3.4, 6.5, 7.2]

Dahm, P. Fred, and Wayne A. Fuller (1981), Generalized least squares estimation of the functional multivariate linear errors in variables model, unpublished. [14.3]

Daly, J. F. (1940), On the unbiased character of likelihood-ratio tests for independence in normal systems, *Annals of Mathematical Statistics*, **11**, 1–32. [9.2]

Daniels, H. E., and M. G. Kendall (1958), Short proof of Miss Harley's theorem on the correlation coefficient, *Biometrika*, **45**, 571–572. [4.2]

Darlington, R. B., S. L. Weinberg, and H. J. Walberg (1973), Canonical variate analysis and related techniques, *Review of Educational Research*, **43**, 433–454. [12.2]

Das Gupta, Somesh (1965), Optimum classification rules for classification into two multivariate normal populations, *Annals of Mathematical Statistics*, **36**, 1174–1184. [6.6]

Das Gupta, S., T. W. Anderson, and G. S. Mudholkar (1964), Monotonicity of the power functions of some tests of the multivariate linear hypothesis, *Annals of Mathematical Statistics*, **35**, 200–205. [8.10]

David, F. N. (1937), A note on unbiased limits for the correlation coefficient, *Biometrika*, **29**, 157–160. [4.2]

David, F. N. (1938), *Tables of the Ordinates and Probability Integral of the Distribution of the Correlation Coefficient in Small Samples*, Cambridge University, Cambridge. [4.2]

Davis, A. W. (1968), A system of linear differential equations for the distribution of Hotelling's generalized T_0^2, *Annals of Mathematical Statistics*, **39**, 815–832. [8.6]

Davis, A. W. (1970a), Exact distributions of Hotelling's generalized T_0^2, *Biometrika*, **57**, 187–191. [Preface, 8.6]

Davis, A. W. (1970b), Further applications of a differential equation for Hotelling's generalized T_0^2, *Annals of the Institute of Statistical Mathematics*, **22**, 77–87. [Preface, 8.6]

Davis, A. W. (1971), Percentile approximations for a class of likelihood ratio criteria, *Biometrika*, **58**, 349–356. [10.8, 10.9]

Davis, A. W. (1972a), On the marginal distributions of the latent roots of the multivariate beta matrix, *Annals of Mathematical Statistics*, **43**, 1664–1670. [8.6]

Davis, A. W. (1972b), On the distributions of the latent roots and traces of certain random matrices, *Journal of Multivariate Analysis*, **2**, 189–200. [8.6]

Davis, A. W. (1980), Further tabulation of Hotelling's generalized T_0^2, *Communications in Statistics*, **B9**, 321–336. [Preface, 8.6]

Davis, A. W., and J. B. F. Field (1971), Tables of some multivariate test criteria, Technical Report No. 32, Division of Mathematical Statistics, C.S.I.R.O., Canberra, Australia. [10.8]

Davis, Harold T. (1933), *Tables of the Higher Mathematical Functions*, Vol. I, Principia Press, Bloomington, Indiana. [8.5]

Davis, Harold T. (1935), *Tables of the Higher Mathematical Functions*, Vol. II, Principia Press, Bloomington, Indiana. [8.5]

Deemer, Walter L., and Ingram Olkin (1951), The Jacobians of certain matrix transformations useful in multivariate analysis. Based on lectures of P. L. Hsu at the University of North Carolina, 1947, *Biometrika*, **38**, 345–367. [13.2]

De Groot, Morris H. (1970), *Optimal Statistical Decisions*, McGraw-Hill, New York. [3.4, 6.2, 6.7]

Dempster, A. P., N. M. Laird, and D. B. Rubin (1977), Maximum likelihood from incomplete data via the EM algorithm (with discussion), *Journal of the Royal Statistical Society B*, **39**, 1–38. [14.3]

Diaconis, Persi, and Bradley Efron (1983), Computer-intensive methods in statistics,

Scientific American, **248**, 116–130. [4.3]

Eaton, M. L., and M. D. Perlman (1974), A monotonicity property of the power functions of some invariant tests for MANOVA, *Annals of Statistics*, **2**, 1022–1028. [8.10]

Efron, Bradley (1982), *The Jackknife, the Bootstrap, and Other Resampling Plans*, Society for Industrial and Applied Mathematics, Philadelphia. [4.2]

Efron, Bradley, and Carl Morris (1977), Stein's paradox in statistics, *Scientific American*, **236**, 119–127. [3.5]

Elfving, G. (1947), A simple method of deducing certain distributions connected with multivariate sampling, *Skandinavisk Aktuarietidskrift*, **30**, 56–74. [7.2]

Erdélyi, A. (1954), *Tables of Integral Transforms*, Vol. I. Based, in part, on notes left by Harry Bateman. McGraw-Hill, New York. [4.2]

Ferguson, Thomas Shelburne (1967), *Mathematical Statistics: a Decision Theoretic Approach*, Academic, New York. [3.4, 6.2, 6.7]

Fisher, R. A. (1915), Frequency distribution of the values of the correlation coefficient in samples from an indefinitely large population, *Biometrika*, **10**, 507–521. [4.2, 7.2]

Fisher, R. A. (1921), On the "probable error" of a coefficient of correlation deduced from a small sample, *Metron*, **1**, Part 4, 3–32. [4.2]

Fisher, R. A. (1924), The distribution of the partial correlation coefficient, *Metron*, **3**, 329–332. [4.3]

Fisher, R. A. (1928), The general sampling distribution of the multiple correlation coefficient, *Proceedings of the Royal Society of London, A*, **121**, 654–673. [4.4]

Fisher, R. A. (1936), The use of multiple measurements in taxonomic problems, *Annals of Eugenics*, **7**, 179–188. [5.3, 6.5, 11.5]

Fisher, R.A. (1939), The sampling distribution of some statistics obtained from non-linear equations, *Annals of Eugenics*, **9**, 238–249. [13.2]

Fisher, R. A. (1947a), *The Design of Experiments* (4th ed.), Oliver and Boyd, Edinburgh. [8.9]

Fisher, R. A. (1947b), The analysis of covariance method for the relation between a part and the whole, *Biometrics*, **3**, 65–68. [3.P]

Fisher, R. A., and F. Yates (1942), *Statistical Tables for Biological, Agricultural and Medical Research* (2nd ed.), Oliver and Boyd, Edinburgh. [4.2]

Fog, David (1948), The geometrical method in the theory of sampling, *Biometrika*, **35**, 46–54. [7.2]

Foster, F. G. (1957), Upper percentage points of the generalized beta distribution, II, *Biometrika*, **44**, 441–453. [8.6]

Foster, F. G. (1958), Upper percentage points of the generalized beta distribution, III, *Biometrika*, **45**, 492–503. [8.6]

Foster, F. G., and D. H. Rees (1957), Upper percentage points of the generalized beta distribution, I, *Biometrika*, **44**, 237–247. [8.6]

Frisch, R. (1929), Correlation and scatter in statistical variables, *Nordic Statistical*

Journal, **8**, 36–102. [7.5]

Fujikoshi, Y. (1973), Monotonicity of the power functions of some tests in general MANOVA models, *Annals of Statistics*, **1**, 388–391. [8.6]

Fujikoshi, Y. (1974), The likelihood ratio tests for the dimensionality of regression coefficients, *Journal of Multivariate Analysis*, **4**, 327–340. [12.4]

Fujikoshi, Y., and M. Kanazawa (1976), The ML classification statistic in covariate discriminant analysis and its asymptotic expansions, *Essays in Probability and Statistics*, 305–320. [6.6]

Fuller, Wayne A., Sastry, G. Pantula, and Yasuo Amemiya (1982), The covariance matrix of estimators for the factor model, unpublished. [14.4]

Gabriel, K. R. (1969), Simultaneous test procedures—Some theory of multiple comparisons, *Annals of Mathematical Statistics*, **40**, 224–250. [8.7]

Gajjar, A. V. (1967), Limiting distributions of certain transformations of multiple correlation coefficients, *Metron*, **26**, 189–193. [4.4]

Galton, Francis (1889), *Natural Inheritance*, MacMillan, London. [1.2, 2.5]

Gauss, K. F. (1823), *Theory of the Combination of Observations*, Göttingen. [1.2]

Giri, N. (1977), *Multivariate Statistical Inference*, Academic, New York. [7.2, 7.7]

Giri, N., and J. Kiefer (1964), Local and asymptotic minimax properties of multivariate tests, *Annals of Mathematical Statistics*, **35**, 21–35. [5.6]

Giri, N., J. Kiefer, and C. Stein (1963), Minimax character of Hotelling's T^2 test in the simplest case, *Annals of Mathematical Statistics*, **34**, 1524–1535. [5.6]

Girshick, M. A. (1939), On the sampling theory of roots of determinantal equations, *Annals of Mathematical Statistics*, **10**, 203–224. [13.2]

Gleser, Leon Jay (1981), Estimation in a multivariate "errors in variables" regression model: Large sample results, *Annals of Statistics*, **9**, 24–44. [12.7]

Glynn, W. J., and R. J. Muirhead (1978), Inference in canonical correlation analysis, *Journal of Multivariate Analysis*, **8**, 468–478. [12.4]

Golub, Gene H., and Franklin T. Luk (1976), Singular value decomposition: applications and computations, unpublished. [12.3]

Grubbs, F. E. (1954), Tables of 1% and 5% probability levels of Hotelling's generalized T^2 statistics, Technical Note No. 926, Ballistic Research Laboratory, Aberdeen Proving Ground, Maryland. [8.6]

Gupta, Shanti S. (1963), Bibliography on the multivariate normal integrals and related topics, *Annals of Mathematical Statistics*, **34**, 829–838. [2.3]

Gurland, John (1968), A relatively simple form of the distribution of the multiple correlation coefficient, *Journal of the Royal Statistical Society B*, **30**, 276–283. [4.4]

Gurland, J., and R. Milton (1970), Further consideration of the distribution of the multiple correlation coefficient, *Journal of the Royal Statistical Society B*, **32**, 381–394. [4.4]

Haavelmo, T. (1944), The probability approach in econometrics, *Econometrica*, **12**, Supplement, 1–118. [12.7]

Haff, L. R. (1980), Empirical Bayes estimation of the multivariate normal covariance matrix, *Annals of Statistics*, **8**, 586–597. [7.8]

Halmos, P. R. (1956), *Measure Theory*, D. van Nostrand, New York. [13.3]

Harley, B. I. (1956), Some properties of an angular transformation for the correlation coefficient, *Biometrika*, **43**, 219–224. [4.2]

Harley, B. I. (1957), Further properties of an angular transformation of the correlation coefficient, *Biometrika*, **44**, 273–275. [4.2]

Harris, Bernard, and Andrew P. Soms (1980), The use of the tetrachoric series for evaluating multivariate normal probabilities, *Journal of Multivariate Analysis*, **10**, 252–267. [2.3]

Hayakawa, Takesi (1967), On the distribution of the maximum latent root of a positive definitive symmetric random matrix, *Annals of the Institute of Statistical Mathematics*, **19**, 1–17. [8.6]

Heck, D. L. (1960), Charts of some upper percentage points of the distribution of the largest characteristic root, *Annals of Mathematical Statistics*, **31**, 625–642. [8.6]

Hickman, W. Braddock (1953), *The Volume of Corporate Bond Financing Since 1900*, Princeton University, Princeton, 82–90. [10.7]

Hoel, Paul G. (1937), A significance test for component analysis, *Annals of Mathematical Statistics*, **8**, 149–158. [7.5]

Hooker, R. H. (1907), The correlation of the weather and crops, *Journal of the Royal Statistical Society*, **70**, 1–42. [4.2]

Horst, Paul (1965), *Factor Analysis of Data Matrices*, Holt, Rinehart, and Winston, New York. [14.5]

Hotelling, Harold (1931), The generalization of Student's ratio, *Annals of Mathematical Statistics*, **2**, 360–378. [5.1, 5.P]

Hotelling, Harold (1933), Analysis of a complex of statistical variables into principal components, *Journal of Educational Psychology*, **24**, 417–441, 498–520. [11.1, 14.3]

Hotelling, Harold (1935), The most predictable criterion, *Journal of Educational Psychology*, **26**, 139–142. [12.2]

Hotelling, Harold (1936), Relations between two sets of variates, *Biometrika*, **28**, 321–377. [12.1]

Hotelling, Harold (1947), Multivariate quality control, illustrated by the air testing of sample bombsights, *Techniques of Statistical Analysis* (C. Eisenhart, M. Hastay, and W. A. Wallis, eds.), McGraw-Hill, New York, 111–184. [8.6]

Hotelling, Harold (1951), A generalized T test and measure of multivariate dispersion, *Proceedings of the Second Berkeley Symposium on Mathematical Statistics and Probability* (Jerzy Neyman, ed.), University of California, Los Angeles and Berkeley, 23–41. [8.6, 10.7]

Hotelling, Harold (1953), New light on the correlation coefficient and its transforms (with discussion), *Journal of the Royal Statistical Society B*, **15**, 193–232. [4.2]

Howe, W. G. (1955), *Some Contributions to Factor Analysis*, U.S. Atomic Energy Commission Report, Oak Ridge National Laboratory, Oak Ridge, Tennessee. [14.2, 14.6]

Hsu, P. L. (1938), Notes on Hotelling's generalized *T*, *Annals of Mathematical Statistics*, **9**, 231–243. [5.4]

Hsu, P. L. (1939a), On the distribution of the roots of certain determinantal equations, *Annals of Eugenics*, **9**, 250–258. [13.2]

Hsu, P. L. (1939b), A new proof of the joint product moment distribution, *Proceedings of the Cambridge Philosophical Society*, **35**, 336–338. [7.2]

Hsu, P. L. (1945), On the power functions for the E^2-test and the T^2-test, *Annals of Mathematical Statistics*, **16**, 278–286. [5.6]

Hudson, M. (1974), Empirical Bayes estimation, Technical Report No. 58, NSF contract GP 30711X-2 Department of Statistics, Stanford University. [3.5]

Immer, F. R., H. D. Hayes, and LeRoy Powers (1934), Statistical determination of barley varietal adaptation, *Journal of the American Society of Agronomy*, **26**, 403–407. [8.9]

Ingham, A. E. (1933), An integral which occurs in statistics, *Proceedings of the Cambridge Philosophical Society*, **29**, 271–276. [7.2]

Ito, K. (1956), Asymptotic formulae for the distribution of Hotelling's generalized T_0^2 statistic, *Annals of Mathematical Statistics*, **27**, 1091–1105. [8.6]

Ito, K. (1960), Asymptotic formulae for the distribution of Hotelling's generalized T_0^2 statistic, II, *Annals of Mathematical Statistics*, **31**, 1148–1153. [8.6]

Izenman, Alan Julian (1980), Assessing dimensionality in multivariate regression, *Analysis of Variance*, *Handbook of Statistics*, Vol. 1 (P. R. Krishnaiah, ed.), North-Holland, Amsterdam, 571–591. [3.P]

James, A. T. (1954), Normal multivariate analysis and the orthogonal group, *Annals of Mathematical Statistics*, **25**, 40–75. [7.2]

James, A. T. (1964), Distributions of matrix variates and latent roots derived from normal samples, *Annals of Mathematical Statistics*, **35**, 475–501. [8.6]

James, W., and C. Stein (1961), Estimation with quadratic loss, *Proceedings of the Fourth Berkeley Symposium on Mathematical Statistics and Probability* (Jerzy Neyman, ed.), Vol. I, 361–379. University of California, Berkeley. [3.5, 7.8]

Japanese Standards Association (1972), *Statistical Tables and Formulas with Computer Applications*. [Preface]

Jennrich, Robert I., and Dorothy T. Thayer (1973), A note on Lawley's formulas for standard errors in maximum likelihood factor analysis, *Psychometrika*, **38**, 571–580. [14.3]

Jolicoeur, Pierre, and J. E. Mosimann (1960), Size and shape variation in the painted turtle, a principal component analysis, *Growth* **24**, 339–354. (Also in *Benchmark Papers in Systematic and Evolutionary Biology* (E. H. Bryant and W. R. Atchley, eds.), **2** (1975), 86–101. [11.P]

Jöreskog, K. G. (1969), A general approach to confirmatory maximum likelihood factor analysis, *Psychometrika*, **34**, 183–202. [14.2]

Jöreskog, K. G., and Arthur S. Goldberger (1972), Factor analysis by generalized least squares, *Psychometrika*, **37**, 243–260. [14.3]

Kaiser, Henry F. (1958), The varimax criterion for analytic rotation in factor analysis, *Psychometrika*, **23**, 187–200. [14.5]

Kanazawa, M. (1979), The asymptotic cut-off point and comparison of error probabilities in covariate discriminant analysis, *Journal of the Japan Statistical Society*, **9**, 7–17. [6.6]

Kelley, T. L. (1928), *Crossroads in the Mind of Man*, Stanford University, Stanford. [4.P, 9.P]

Kendall, M. G., and Alan Stuart (1973), *The Advanced Theory of Statistics* (3rd ed.) Vol. 2, Charles Griffin, London. [12.6]

Kennedy, William J., Jr., and James E. Gentle (1980), *Statistical Computing*, Marcel Dekker, New York. [12.3]

Khatri, C. G. (1963), Joint estimation of the parameters of multivariate normal populations, *Journal of Indian Statistical Association*, **1**, 125–133. [7.2]

Khatri, C. G. (1966), A note on a large sample distribution of a transformed multiple correlation coefficient, *Annals of the Institute of Statistical Mathematics*, **18**, 375–380. [4.4]

Khatri, C. G. (1968), A note on exact moments of arc sine correlation coefficient with the help of characteristic function, *Annals of the Institute of Statistical Mathematics*, **20**, 143–149. [4.2]

Khatri, C. G. (1972), On the exact finite series distribution of the smallest or the largest root of matrices in three situations, *Journal of Multivariate Analysis*, **2**, 201–207. [8.6]

Khatri, C. G., and K. C. Sreedharan Pillai (1966), On the moments of the trace of a matrix and approximations to its non-central distribution, *Annals of Mathematical Statistics*, **37**, 1312–1318. [8.6]

Khatri, C. G., and K. C. S. Pillai (1968), On the non-central distributions of two test criteria in multivariate analysis of variance, *Annals of Mathematical Statistics*, **39**, 215–226. [8.6]

Khatri, C. G., and K. V. Ramachandran (1958), Certain multivariate distribution problems, I (Wishart's distribution), *Journal of the Maharaja Sayajairo, University of Baroda*, **7**, 79–82. [7.2]

Kiefer, J. (1957), Invariance, minimax sequential estimation, and continuous time processes, *Annals of Mathematical Statistics*, **28**, 573–601. [7.8]

Kiefer, J. (1966), Multivariate optimality results, *Multivariate Analysis*, (Parachuyi R. Krishnaiah, ed.), Academic, New York, 255–274. [7.8]

Kiefer, J., and R. Schwartz (1965), Admissible Bayes character of T^2-, R^2-, and other fully invariant tests for classical multivariate normal problems, *Annals of Mathematical Statistics*, **36**, 747–770. [5.P, 9.9, 10.10]

Klotz, Jerome, and Joseph Putter (1969), Maximum likelihood estimation of the multivariate covariance components for the balanced one-way layout. *Annals of Mathematical Statistics*, **40**, 1100–1105. [10.6]

Kolmogorov, A. (1950), *Foundations of the Theory of Probability*, Chelsea, New York. [2.2]

Konishi, Sadanori (1978a), An approximation to the distribution of the sample correlation coefficient, *Biometrika*, **65**, 654–656. [4.2]

Konishi, Sadanori (1978b), Asymptotic expansions for the distributions of statistics based on a correlation matrix, *Canadian Journal of Statistics*, **6**, 49–56. [4.2]

Konishi, Sadanori (1979), Asymptotic expansions for the distributions of functions of a correlation matrix, *Journal of Multivariate Analysis*, **9**, 259–266. [4.2]

Koopmans, T. C., and Olav Reiersøl (1950), The identification of structural characteristics, *Annals of Mathematical Statistics*, **21**, 165–181. [14.2]

Korin, B. P. (1968), On the distribution of a statistic used for testing a covariance matrix, *Biometrika*, **55**, 171–178. [10.8]

Korin, B. P. (1969), On testing the equality of k covariance matrices, *Biometrika*, **56**, 216–218. [10.5]

Kramer, K. H. (1963), Tables for constructing confidence limits on the multiple correlation coefficient, *Journal of the American Statistical Association*, **58**, 1082–1085. [4.4]

Krishnaiah, P. R. (1978), Some recent developments on real multivariate distributions, *Developments in Statistics* (P. R. Krishnaiah, ed.), Vol. 1, Academic, New York, 135–169. [8.6]

Krishnaiah, P. R. (1980), Computations of some multivariate distributions, *Analysis of Variance*, *Handbook of Statistics*, Vol. 1 (P. R. Krishnaiah, ed.), North-Holland, Amsterdam, 745–971.

Krishnaiah, P. R., and T. C. Chang (1972), On the exact distributions of the traces of $S_1(S_1 + S_2)^{-1}$ and $S_1S_2^{-1}$, *Sankhyā*, *A*, **34**, 153–160. [8.6]

Krishnaiah, P. R., and F. J. Schuurmann (1974), On the evaluation of some distributions that arise in simultaneous tests for the equality of the latent roots of the covariance matrix, *Journal of Multivariate Analysis*, **4**, 265–282. [10.7]

Kshirsagar, A. M. (1959), Bartlett decomposition and Wishart distribution, *Annals of Mathematical Statistics*, **30**, 239–241. [7.2]

Kudo, H. (1955), On minimax invariant estimates of the transformation parameter, *Natural Science Report*, **6**, 31–73, Ochanomizu University, Tokyo, Japan. [7.8]

Kunitomo, Naoto (1980), Asymptotic expansions of the distributions of estimators in a linear functional relationship and simultaneous equations, *Journal of the American Statistical Association*, **75**, 693–700. [12.7]

Lachenbruch, P. A., and M. R. Mickey (1968), Estimation of error rates in discriminant analysis, *Technometrics*, **10**, 1–11. [6.6]

Laplace, P. S. (1811), Mémoire sur les intégrales définies et leur application aux probabilités, *Mémoires de l'Institut Impérial de France*, *Année 1810*, 279–347. [1.2]

Lawley, D. N. (1938), A generalization of Fisher's z test, *Biometrika*, **30**, 180–187. [8.6]

Lawley, D. N. (1940), The estimation of factor loadings by the method of maximum likelihood, *Proceedings of the Royal Society of Edinburgh*, *Sec. A*, **60**, 64–82. [14.3]

Lawley, D. N. (1941), Further investigations in factor estimation, *Proceedings of the Royal Society of Edinburgh*, *Sec. A*, **61**, 176–185. [14.4]

Lawley, D. N. (1953), A modified method of estimation in factor analysis and some large sample results, *Uppsala Symposium on Psychological Factor Analysis, 17–19 March 1953*, Uppsala, Almqvist and Wiksell, 35–42. [14.3]

Lawley, D. N. (1958), Estimation in factor analysis under various initial assumptions, *British Journal of Statistical Psychology*, **11**, 1–12. [14.2, 14.6]

Lawley, D. N. (1959), Tests of significance in canonical analysis, *Biometrika*, **46**, 59–66. [12.4]

Lawley, D. N. (1967), Some new results in maximum likelihood factor analysis, *Proceedings of the Royal Society of Edinburgh*, *Sec. A*, **87**, 256–264. [14.3]

Lawley, D. N., and A. E. Maxwell (1971), *Factor Analysis as a Statistical Method* (2nd ed.), American Elsevier, New York. [14.3, 14.5]

Lee, Y. S. (1971a), Asymptotic formulae for the distribution of a multivariate test statistic: power comparisons of certain multivariate tests, *Biometrika*, **58**, 647–651. [8.6]

Lee, Y. S. (1971b), Some results on the sampling distribution of the multiple correlation coefficient, *Journal of the Royal Statistical Society B*, **33**, 117–130. [4.4]

Lee, Y. S. (1972), Tables of upper percentage points of the multiple correlation coefficient, *Biometrika*, **59**, 175–189. [4.4]

Lehmann, E. L. (1959), *Testing Statistical Hypotheses*, John Wiley & Sons, New York. [4.2, 5.6]

Lehmer, Emma (1944), Inverse tables of probabilities of errors of the second kind, *Annals of Mathematical Statistics*, **15**, 388–398. [5.4]

Loève, M. (1977), *Probability Theory I*, (4th ed.), Springer-Verlag, New York. [2.2]

Loève, M. (1978), *Probability Theory II*, (4th ed.), Springer-Verlag, New York. [2.2]

Madow, W. G. (1938), Contributions to the theory of multivariate statistical analysis, *Transactions of the American Mathematical Society*, **44**, 454–495. [7.2]

Mahalanobis, P. C. (1930), On tests and measures of group divergence, *Journal and Proceedings of the Asiatic Society of Bengal*, **26**, 541–588. [3.3]

Mahalanobis, P. C., R. C. Bose, and S. N. Roy (1937), Normalisation of statistical variates and the use of rectangular co-ordinates in the theory of sampling distributions, *Sankhyā*, **3**, 1–40. [7.2]

Mallows, C. L. (1961), Latent vectors of random symmetric matrices, *Biometrika*, **48**, 133–149. [11.6]

Mariano, Roberto S., and Takamitsu Sawa (1972), The exact finite-sample distribution of the limited-information maximum likelihood estimator in the case of two included endogenous variables, *Journal of the American Statistical Association*, **67**, 159–163. [12.7]

Marshall, A. W., and I. Olkin (1979), *Inequalities: Theory of Majorization and Its Applications*, Academic, New York. [8.10]

Mathai, A. M. (1971), On the distribution of the likelihood ratio criterion for testing linear hypotheses on regression coefficients, *Annals of the Institute of Statistical*

Mathematics, **23**, 181–197. [8.4]

Mathai, A. M. (1972), The exact distributions of three multivariate statistics associated with Wilks' concept of generalized variance, *Sankhyā A*, **34**, 161–170. [7.5]

Mathai, A. M., and R. S. Katiyar (1979), Exact percentage points for testing independence, *Biometrika*, **66**, 353–356. [9.3]

Mathai, A. M., and P. N. Rathie (1980), The exact non-null distribution for testing equality of covariance matrices, *Sankhyā A*, **42**, 78–87. [10.4]

Mathai, A. M., and R. K. Saxena (1973), *Generalized Hypergeometric Functions with Applications in Statistics and Physical Sciences*, Lecture Notes No. 348, Springer-Verlag, New York. [9.3]

Mauchly, J.W. (1940), Significance test for sphericity of a normal n-variate distribution, *Annals of Mathematical Statistics*, **11**, 204–209. [10.7]

Mauldon, J. G. (1955), Pivotal quantities for Wishart's and related distributions, and a paradox in fiducial theory, *Journal of the Royal Statistical Society B*, **17**, 79–85. [7.2]

McLachlan, G. J. (1973), An asymptotic expansion of the expectation of the estimated error rate in discriminant analysis, *Australian Journal of Statistics*, **15**, 210–214. [6.6]

McLachlan, G. J. (1974a), An asymptotic unbiased technique for estimating the error rates in discriminant analysis, *Biometrics*, **30**, 239–249. [6.6]

McLachlan, G. J. (1974b), Estimation of the errors of misclassification on the criterion of asymptotic mean square error, *Technometrics*, **16**, 255–260. [6.6]

McLachlan, G. J. (1974c), The asymptotic distributions of the conditional error rate and risk in discriminant analysis, *Biometrika*, **61**, 131–135. [6.6]

McLachlan, G. J. (1977), Constrained sample discrimination with the studentized classification statistic W, *Communications in Statistics-Theory and Methods*, **A6**, 575–583. [6.6]

Memon, A. Z., and M. Okamoto (1971), Asymptotic expansion of the distribution of the Z statistic in discriminant analysis, *Journal of Multivariate Analysis*, **1**, 294–307. [6.6]

Mijares, T. A. (1964), *Percentage Points of the Sum $V_1^{(s)}$ of s Roots ($s = 1 - 50$)*, The Statistical Center, University of the Philippines, Manila. [8.6]

Mikhail, N. N. (1965), A comparison of tests of the Wilks–Lawley hypothesis in multivariate analysis, *Biometrika*, **52**, 149–156. [8.6]

Mises, R. von (1945), On the classification of observation data into distinct groups, *Annals of Mathematical Statistics*, **16**, 68–73. [6.8]

Mood, A. M. (1951), On the distribution of the characteristic roots of normal second-moment matrices, *Annals of Mathematical Statistics*, **22**, 266–273. [13.2]

Morris, Blair, and Ingram Olkin (1964), Some estimation and testing problems for factor analysis models, unpublished. [10.6]

Mudholkar, G. S. (1966), On confidence bounds associated with multivariate analysis of variance and non-independence between two sets of variates, *Annals of Mathematical Statistics*, **37**, 1736–1746. [8.7]

Mudholkar, Govind S., and Madhusudan C. Trivedi (1980), A normal approximation for the distribution of the likelihood ratio statistic in multivariate analysis of variance, *Biometrika*, **67**, 485–488. [8.5]

Mudholkar, Govind S., and Madhusudan C. Trivedi (1981), A normal approximation for the multivariate likelihood ratio statistics, *Statistical Distributions in Scientific Work* (C. Taillie, et al., eds.), Vol. 5, 219–230, D. Reidel Publishing. [8.5]

Muirhead, R. J. (1970), Asymptotic distributions of some multivariate tests, *Annals of Mathematical Statistics*, **41**, 1002–1010. [8.6]

Muirhead, Robb J. (1982), *Aspects of Multivariate Statistical Theory*, John Wiley and Sons, New York. [7.7]

Nagao, Hisao (1973a), On some test criteria for covariance matrix, *Annals of Statistics*, **1**, 700–709. [9.5, 10.2, 10.7, 10.8]

Nagao, Hisao (1973b), Asymptotic expansions of the distributions of Bartlett's test and sphericity test under the local alternatives, *Annals of the Institute of Statistical Mathematics*, **25**, 407–422. [10.5, 10.6]

Nagao, Hisao (1973c), Nonnull distributions of two test criteria for independence under local alternatives, *Journal of Multivariate Analysis*, **3**, 435–444. [9.4]

Nagarsenker, B. N., and K. C. S. Pillai (1972), *The Distribution of the Sphericity Test Criterion*, ARL 72-0154, Aerospace Research Laboratories. [Preface]

Nagarsenker, B. N., and K. C. S. Pillai (1973a), The distribution of the sphericity test criterion, *Journal of Multivariate Analysis*, **3**, 226–235. [10.7]

Nagarsenker, B. N., and K. C. S. Pillai (1973b), Distribution of the likelihood ratio criterion for testing a hypothesis specifying a covariance matrix, *Biometrika*, **60**, 359–364. [10.8]

Nagarsenker, B. N., and K. C. S. Pillai (1974), Distribution of the likelihood ratio criterion for testing $\Sigma = \Sigma_0$, $\mu = \mu_0$, *Journal of Multivariate Analysis*, **4**, 114–122. [10.9]

Nanda, D. N. (1948), Distribution of a root of a determinantal equation, *Annals of Mathematical Statistics*, **19**, 47–57. [8.6]

Nanda, D. N. (1950), Distribution of the sum of roots of a determinantal equation under a certain condition, *Annals of Mathematical Statistics*, **21**, 432–439. [8.6]

Nanda, D. N. (1951), Probability distribution tables of the larger root of a determinantal equation with two roots, *Journal of the Indian Society of Agricultural Statistics*, **3**, 175–177. [8.6]

Narain, R. D. (1948), A new approach to sampling distributions of the multivariate normal theory, I, *Journal of the Indian Society of Agricultural Statistics*, **1**, 59–69. [7.2]

Narain, R. D. (1950), On the completely unbiassed character of tests of independence in multivariate normal systems, *Annals of Mathematical Statistics*, **21**, 293–298. [9.2]

National Bureau of Standards, United States (1959), *Tables of the Bivariate Normal Distribution Function and Related Functions*, U.S. Government Printing Office,

Washington, D.C. [2.3]

von Neumann, J. (1937), Some matrix-inequalities and metrization of matric-space, *Tomsk University Review*, **1**, 286–300. Reprinted in *John von Neuman Collected Works* (A. H. Taub, ed.), **4** (1962), Pergamon, New York, 205–219. [A.4]

Neveu, Jacques (1965), *Mathematical Foundations of the Calculus of Probability*, Holden-Day, San Francisco. [2.2]

Ogawa, J. (1953), On the sampling distributions of classical statistics in multivariate analysis, *Osaka Mathematics Journal*, **5**, 13–52. [7.2]

Okamoto, Masashi (1963), An asymptotic expansion for the distribution of the linear discriminant function, *Annals of Mathematical Statistics*, **34**, 1286–1301. (Correction, **39** (1968), 1358–1359.) [6.6]

Okamoto, Masashi (1973), Distinctness of the eigenvalues of a quadratic form in a multivariate sample, *Annals of Statistics*, **1**, 763–765. [13.2]

Olkin, Ingram, and John W. Pratt (1958), Unbiased estimation of certain correlation coefficients, *Annals of Mathematical Statistics*, **29**, 201–211. [4.2, 4.4]

Olkin, Ingram, and S. N. Roy (1954), On multivariate distribution theory, *Annals of Mathematical Statistics*, **25**, 329–339. [7.2]

Olson, C. L. (1974), Comparative robustness of six tests in multivariate analysis of variance, *Journal of the American Statistical Association*, **69**, 894–908. [8.6]

Pearson, E. S., and H. O. Hartley (1972), *Biometrika Tables for Statisticians*, Vol. II, Cambridge (England), Published for the Biometrika Trustees at the University Press. [Preface], [8.4]

Pearson, E. S., and S. S. Wilks (1933), Methods of statistical analysis appropriate for *k* samples of two variables, *Biometrika*, **25**, 353–378. [10.5]

Pearson, K. (1896), Mathematical contributions to the theory of evolution—III. Regression, heredity and panmixia, *Philosophical Transactions of the Royal Society of London, Series A*, **187**, 253–318. [2.5, 3.2]

Pearson, K. (1900), On the criterion that a given system of deviations from the probable in the case of a correlated system of variables is such that it can be reasonably supposed to have arisen from random sampling, *Philosophical Magazine*, **50** (fifth series), 157–175. [3.3]

Pearson, K. (1901), On lines and planes of closest fit to systems of points in space, *Philosophical Magazine*, **2** (sixth series), 559–572. [11.2]

Pearson, K. (1930), *Tables for Statisticians and Biometricians*, Part I (3rd ed.), Cambridge University, Cambridge. [2.3]

Pearson, K. (1931), *Tables for Statisticians and Biometricians*, Part II, Cambridge University, Cambridge. [2.3, 6.8]

Perlman, M. D. (1980), Unbiasedness of the likelihood ratio tests for equality of several covariance matrices and equality of several multivariate normal populations, *Annals of Statistics*, **8**, 247–263. [10.2, 10.3]

Perlman, M. D., and I. Olkin (1980), Unbiasedness of invariant tests for MANOVA and other multivariate problems, *Annals of Statistics*, **8**, 1326–1341. [8.10]

Phillips, P. C. B. (1980), The exact distribution of instrumental variable estimators in an equation containing $n + 1$ endogenous variables, *Econometrica*, **48**, 861–878. [12.7]

Phillips, P. C. B. (1982), A new approach to small sample theory, unpublished, Cowles Foundation for Research in Economics, Yale University. [12.7]

Pillai, K. C. S. (1954), On some distribution problems in multivariate analysis, Mimeo Series No. 54, Institute of Statistics, University of North Carolina, Chapel Hill, North Carolina. [8.6]

Pillai, K. C. S. (1955), Some new test criteria in multivariate analysis, *Annals of Mathematical Statistics*, **26**, 117–121. [8.6]

Pillai, K. C. S. (1956), On the distribution of the largest or the smallest root of a matrix in multivariate analysis, *Biometrika*, **43**, 122–127. [8.6]

Pillai, K. C. S. (1960), *Statistical Tables for Tests of Multivariate Hypotheses*, Statistical Center, University of the Philippines, Manila. [8.6]

Pillai, K. C. S. (1964), On the moments of elementary symmetric functions of the roots of two matrices, *Annals of Mathematical Statistics*, **35**, 1704–1712. [8.6]

Pillai, K. C. S. (1965), On the distribution of the largest characteristic root of a matrix in multivariate analysis, *Biometrika*, **52**, 405–414. [8.6]

Pillai, K. C. S. (1967), Upper percentage points of the largest root of a matrix in multivariate analysis, *Biometrika*, **54**, 189–194. [8.6]

Pillai, K. C. S., and A. K. Gupta (1969), On the exact distribution of Wilks's criterion, *Biometrika*, **56**, 109–118. [8.4]

Pillai, K. C. S., and Y. S. Hsu (1979), Exact robustness studies of the test of independence based on four multivariate criteria and their distribution problems under violations, *Annals of the Institute of Statistical Mathematics*, **31**, Part A, 85–101. [8.6]

Pillai, K. C. S., and K. Jayachandran (1967), Power comparisons of tests of two multivariate hypotheses based on four criteria, *Biometrika*, **54**, 195–210. [8.6]

Pillai, K. C. S., and K. Jayachandran (1970), On the exact distribution of Pillai's $V^{(s)}$ criterion, *Journal of the American Statistical Association*, **65**, 447–454. [8.6]

Pillai, K. C. S., and T. A. Mijares (1959), On the moments of the trace of a matrix and approximations to its distribution, *Annals of Mathematical Statistics*, **30**, 1135–1140. [8.6]

Pillai, K. C. S., and B. N. Nagarsenker (1971), On the distribution of the sphericity test criterion in classical and complex normal populations having unknown covariance matrices, *Annals of Mathematical Statistics*, **42**, 764–767. [10.7]

Pillai, K. C. S., and P. Samson, Jr. (1959), On Hotelling's generalization of T^2, *Biometrika*, **46**, 160–168. [8.6]

Pillai, K. C. S., and T. Sugiyama (1969), Non-central distributions of the largest latent roots of three matrices in multivariate analysis, *Annals of the Institute of Statistical*

Mathematics, **21**, 321–327. [8.6]

Pillai, K. C. S., and D. L. Young (1971), On the exact distribution of Hotelling's generalized T_0^2, *Journal of Multivariate Analysis*, **1**, 90–107. [8.6]

Plana, G. A. A. (1813), Mémoire sur divers problèmes de probabilité, *Mémoires de l'Académie Impériale de Turin, pour les Années 1811–1812*, **20**, 355–408. [1.2]

Pólya, G. (1949), Remarks on computing the probability integral in one and two dimensions, *Proceedings of the Berkeley Symposium on Mathematical Statistics and Probability* (J. Neyman, ed.), 63–78. [2.P]

Rao, C. R. (1948a), The utilization of multiple measurements in problems of biological classification, *Journal of the Royal Statistical Society B*, **10**, 159–193. [6.9]

Rao, C. R. (1948b), Tests of significance in multivariate analysis, *Biometrika*, **35**, 58–79. [5.3]

Rao, C. Radhakrishna (1951), An asymptotic expansion of the distribution of Wilks's criterion, *Bulletin of the International Statistical Institute*, **33**, Part 2, 177–180. [8.5]

Rao, C. R. (1952), *Advanced Statistical Methods in Biometric Research*, John Wiley & Sons, New York. [12.5]

Rao, C. R. (1973), *Linear Statistical Inference and Its Applications* (2nd ed.), John Wiley & Sons, New York. [4.2]

Rasch, G. (1948), A functional equation for Wishart's distribution, *Annals of Mathematical Statistics*, **19**, 262–266. [7.2]

Reiersøl, Olav (1950), On the identifiability of parameters in Thurstone's multiple factor analysis, *Psychometrika*, **15**, 121–149. [14.2]

Richardson, D. H. (1968), The exact distribution of a structural coefficient estimator, *Journal of the American Statistical Association*, **63**, 1214–1226. [12.7]

Rothenberg, Thomas J. (1977), Edgeworth expansions for multivariate test statistics, IP-255, Center for Research in Management Science, University of California, Berkeley. [8.6]

Roy, S. N. (1939), p-statistics or some generalisations in analysis of variance appropriate to multivariate problems, *Sankhyā*, **4**, 381–396. [13.2]

Roy, S.N. (1945), The individual sampling distribution of the maximum, the minimum and any intermediate of the p-statistics on the null-hypothesis, *Sankhyā*, **7**, 133–158. [8.6]

Roy, S. N. (1953), On a heuristic method of test construction and its use in multivariate analysis, *Annals of Mathematical Statistics*, **24**, 220–238. [8.6, 10.6]

Roy, S. N. (1957), *Some Aspects of Multivariate Analysis*, John Wiley & Sons, New York. [8.6, 10.6, 10.8]

Ruben, Harold (1966), Some new results on the distribution of the sample correlation coefficient, *Journal of the Royal Statistical Society B*, **28**, 513–525. [4.2]

Rubin, Donald B., and Dorothy T. Thayer (1982), EM algorithms for ML factor analysis, *Psychometrika*, **47**, 69–76. [14.3]

Šalaevskiĭ, O. V. (1968), The minimax character of Hotelling's T^2 test (Russian),

Doklady Akademii Nauk SSSR, **180**, 1048–1050. [5.6]

Šalaevskiĭ (Shalaevskii), O. V. (1971), Minimax character of Hotelling's T^2 test. I. *Investigations in Classical Problems of Probability Theory and Mathematical Statistics*, V. M. Kalinin and O. V. Shalaevskii (Seminar in Mathematics, V. I. Steklov Institute, Leningrad, Vol. 13), Consultants Bureau, New York. [5.6]

Sargan, J. D., and W. M. Mikhail (1971), A general approximation to the distribution of instrumental variables estimates, *Econometrica*, **39**, 131–169. [12.7]

Sawa, Takamitsu (1969), The exact sampling distribution of ordinary least squares and two-stage least squares estimators, *Journal of the American Statistical Association*, **64**, 923–937. [12.7]

Schatzoff, M. (1966a), Exact distributions of Wilks's likelihood ratio criterion, *Biometrika*, **53**, 347–358. [8.4]

Schatzoff, M. (1966b), Sensitivity comparisons among tests of the general linear hypothesis, *Journal of the American Statistical Association*, **61**, 415–435. [8.6]

Scheffé, Henry (1942), On the ratio of the variances of two normal populations, *Annals of Mathematical Statistics*, **13**, 371–388. [10.2]

Scheffé, Henry (1943), On solutions of the Behrens–Fisher problem, based on the t-distribution, *Annals of Mathematical Statistics*, **14**, 35–44. [5.5]

Schuurmann, F. J., P. R. Krishnaiah, and A. K. Chattopadhyay (1975), Exact percentage points of the distribution of the trace of a multivariate beta matrix, *Journal of Statistical Computation and Simulation*, **3**, 331–343. [8.6]

Schuurmann, F. J., and V. B. Waikar (1973), Tables for the power function of Roy's two-sided test for testing hypothesis $\Sigma = I$ in the bivariate case, *Communications in Statistics*, **1**, 271–280. [10.8]

Schuurmann, F. J., V. B. Waikar, and P. R. Krishnaiah (1975), Percentage points of the joint distribution of the extreme roots of the random matrix $S_1(S_1 + S_2)^{-1}$, *Journal of Statistical Computation and Simulation*, **2**, 17–38. [10.6]

Schwartz, R. (1967), Admissible tests in multivariate analysis of variance, *Annals of Mathematical Statistics*, **38**, 698–710. [8.10]

Serfling, Robert J. (1980), *Approximation Theorems of Mathematical Statistics*, John Wiley & Sons, New York. [4.2]

Simaika, J. B. (1941), On an optimum property of two important statistical tests, *Biometrika*, **32**, 70–80. [5.6]

Siotani, Minoru (1980), Asymptotic approximations to the conditional distributions of the classification statistic Z and its studentized form Z^*, *Tamkang Journal of Mathematics*, **11**, 19–32. [6.6]

Siotani, M., and R. H. Wang (1975), Further expansion formulae for error rates and comparison of the W- and the Z-procedures in discriminant analysis, Technical Report No. 33, Department of Statistics, Kansas State University, Manhattan, Kansas. [6.6]

Siotani, M., and R. H. Wang (1977), *Asymptotic Expansions for Error Rates and Comparison of the W-Procedure and the Z-Procedure in Discriminant Analysis*, North-

Holland, Amsterdam. [6.6]

Sitgreaves, Rosedith (1952), On the distribution of two random matrices used in classification procedures, *Annals of Mathematical Statistics*, **23**, 263–270. [6.5]

Solari, M. E. (1969), The "maximum likelihood solution" of the problem of estimating a linear functional relationship, *Journal of the Royal Statistical Society B*, **31**, 372–375. [14.4]

Spearman, Charles (1904), "General-intelligence," objectively determined and measured, *American Journal of Psychology*, **15**, 201–293. [14.2]

Srivastava, M. S., and C. G. Khatri (1979), *An Introduction to Multivariate Statistics*, North-Holland, New York. [10.9, 13.P]

Stein, C. (1956a), The admissibility of Hotelling's T^2-test, *Annals of Mathematical Statistics*, **27**, 616–623. [5.6]

Stein, C. (1956b), Inadmissibility of the usual estimator for the mean of a multivariate normal distribution, *Proceedings of the Third Berkeley Symposium on Mathematical and Statistical Probability* (Jerzy Neyman, ed.), Vol. I, 197–206, University of California, Berkeley. [3.5]

Stein, C. (1974), Estimation of the parameters of a multivariate normal distribution. I. Estimation of the means, Technical Report No. 63 NSF Contract GP 30711X-2, Department of Statistics, Stanford University. [3.5]

Student (W. S. Gosset) (1908), The probable error of a mean, *Biometrika*, **6**, 1–25. [3.2]

Subrahmaniam, Kocherlota, and Kathleen Subrahmaniam (1973), *Multivariate Analysis: A Selected and Abstracted Bibliography, 1957–1972*, Marcel Dekker, New York. [Preface]

Sugiura, Nariaki, and Hisao Nagao (1968), Unbiasedness of some test criteria for the equality of one or two covariance matrices, *Annals of Mathematical Statistics*, **39**, 1686–1692. [10.8]

Sugiyama, T. (1967), Distribution of the largest latent root and the smallest latent root of the generalized B statistic and F statistic in multivariate analysis, *Annals of Mathematical Statistics*, **38**, 1152–1159. [8.6]

Sugiyama, T., and K. Fukutomi (1966), On the distribution of the extreme characteristic roots of the matrices in multivariate analysis, *Reports of Statistical Application Research, Union of Japanese Scientists and Engineers*, **13**. [8.6]

Sverdrup, Erling (1947), Derivation of the Wishart distribution of the second order sample moments by straightforward integration of a multiple integral, *Skandinavisk Aktuarietidskrift*, **30**, 151–166. [7.2]

Tang, P. C. (1938), The power function of the analysis of variance tests with tables and illustrations of their use, *Statistical Research Memoirs*, **2**, 126–157. [5.4]

Theil, H. (assisted by J. S. Cramer, H. Moerman, and A. Russchen) (1961), *Economic Forecasts and Policy*, (2nd rev. ed.), North-Holland, Amsterdam. Contributions to Economic Analysis No. XV, (first published 1958). [12.7]

Thomson, Godfrey H. (1934), Hotelling's method modified to give Spearman's "g," *Journal of Educational Psychology*, **25**, 366–374. [14.3]

Thomson, Godfrey H. (1951), *The Factorial Analysis of Human Ability* (5th ed.), University of London, London. [14.7]

Thurstone, L. L., (1947), *Multiple-Factor Analysis*, University of Chicago, Chicago. [14.2, 14.5]

Tukey, J. W. (1949), Dyadic anova, an analysis of variance for vectors, *Human Biology*, **21**, 65–110. [8.9]

Wald, A. (1943), Tests of statistical hypotheses concerning several parameters when the number of observations is large, *Transactions of the American Mathematical Society*, **54**, 426–482. [4.2]

Wald, A. (1944), On a statistical problem arising in the classification of an individual into one of two groups, *Annals of Mathematical Statistics*, **15**, 145–162. [6.4, 6.5]

Wald, A. (1950), *Statistical Decision Functions*, John Wiley & Sons, New York. [6.2, 6.7, 8.10]

Wald, A., and R. J. Brookner (1941), On the distribution of Wilks' statistic for testing the independence of several groups of variates, *Annals of Mathematical Statistics*, **12**, 137–152. [8.4, 9.3]

Walker, Helen M. (1931), *Studies in the History of Statistical Method*, Williams and Wilkins, Baltimore. [1.1]

Welch, P. D., and R. S. Wimpress (1961), Two multivariate statistical computer programs and their application to the vowel recognition problem, *Journal of the Acoustical Society of America*, **33**, 426–434. [6.10]

Whittaker, E. T., and G. N. Watson (1943), *A Course of Modern Analysis*, Cambridge University, Cambridge. [8.5]

Widder, David Vernon (1941), *The Laplace Transform*, Princeton University, Princeton, New Jersey. [4.2]

Wijsman, Robert A. (1979), Constructing all smallest simultaneous confidence sets in a given class, with applications to MANOVA, *Annals of Statistics*, **7**, 1003–1018. [8.7]

Wijsman, Robert A. (1980), Smallest simultaneous confidence sets with applications in multivariate analysis, *Multivariate Analysis*, **V**, 483–498. [8.7]

Wilkinson, James Hardy (1965), *The Algebraic Eigenvalue Problem*, Clarendon, Oxford. [11.4]

Wilkinson, J. H., and C. Reinsch (1971), *Linear Algebra*, Springer-Verlag, New York. [11.4]

Wilks, S. S. (1932), Certain generalizations in the analysis of variance, *Biometrika*, **24**, 471–494. [7.5, 8.3, 10.4]

Wilks, S. S. (1934), Moment-generating operators for determinants of product moments in samples from a normal system, *Annals of Mathematics*, **35**, 312–340. [8.3]

Wilks, S. S. (1935), On the independence of k sets of normally distributed statistical variables, *Econometrica*, **3**, 309–326. [8.4, 9.3, 9.P]

Wishart, John (1928), The generalised product moment distribution in samples from a normal multivariate population, *Biometrika*, **20A**, 32–52. [7.2]

Wishart, John (1948), Proofs of the distribution law of the second order moment statistics, *Biometrika*, **35**, 55–57. [7.2]

Wishart, John, and M. S. Bartlett (1933), The generalised product moment distribution in a normal system, *Proceedings of the Cambridge Philosophical Society*, **29**, 260–270. [7.2]

Woltz, W. G., W. A. Reid, and W. E. Colwell (1948), Sugar and nicotine in cured bright tobacco as related to mineral element composition, *Proceedings of the Soil Sciences Society of America*, **13**, 385–387. [8.P]

Yamauti, Ziro (1977), *Concise Statistical Tables*, Japanese Standards Association. [Preface]

Yates, F., and W. G. Cochran (1938), The analysis of groups of experiments, *Journal of Agricultural Science*, **28**, 556. [8.9]

Yule, G. U. (1897a), On the significance of Bravais' formulae for regression & c., in the case of skew correlation, *Proceedings of the Royal Society of London*, **60**, 477–489. [2.5]

Yule, G. U. (1897b), On the theory of correlation, *Journal of the Royal Statistical Society*, **60**, 812–854. [2.5]

Zehna, P. W. (1966), Invariance of maximum likelihood estimators, *Annals of Mathematical Statistics*, **37**, 744. [3.2]

Index

Absolutely continuous distribution, 8
Additivity of Wishart matrices, 254
Admissible, definition of, 83, 199, 226
Admissible procedures, 226
Admissible test, definition of, 184
  Stein's theorem for, 185
Almost invariant test, 183
Analysis of variance, random effects model, 425
  likelihood ratio test of, 426
  see also Multivariate analysis of variance
Anderson, Edgar, Iris data of, 99
A posteriori density, 84
  of $\boldsymbol{\mu}$, 84
  of $\boldsymbol{\mu}$ and $\boldsymbol{\Sigma}$, 270
  of μ, given $\bar{x}$ and S, 272
  of $\boldsymbol{\Sigma}$, 270
Asymptotic distribution of a function, 121
Asymptotic expansions of distributions of likelihood ratio criteria, 316
  of gamma function, 312

Barley yields in two years, 346
Bartlett decomposition, 251
Bartlett-Nanda-Pillai trace criterion, *see* Linear hypothesis
Bayes estimator of covariance matrix, 272
Bayes estimator of mean vector, 84
Bayes procedure, 83
  extended, 85
Bayes risk, 83
Bernoulli polynomials, 312
Best estimator of covariance matrix, invariant with respect to triangular linear transformations, 276, 278
  proportional to sample covariance matrix, 274
Best linear predictor, 39, 491
  and predictand, 491
  see also Canonical correlations and variates
Beta distribution, 164
Bhattacharya's estimator of the mean, 95
Bivariate normal density, 22, 37
  distribution, 22, 37
    computation of, 24
Bootstrap method, 125

$\mathcal{C}$, 18
Canonical analysis of regression coefficients, 502
  sample, 505
Canonical correlations and variates, 480, 489
  asymptotic distribution of sample correlations, 499
  computation of, 495
  distribution of sample, 540
  invariance of, 490
  maximum likelihood estimators of, 495
  sample, 494
  testing number of nonzero correlations, 498
  use in testing hypotheses of rank of covariance matrix, 498
Central limit theorem, multivariate, 81
Characteristic function, 43
  continuity theorem for, 48
  inversion of, 48
  of the multivariate normal distribution, 45
Characteristic roots and vectors, 587, 588
  asymptotic distributions of, 540, 544
  distribution of roots of a symmetric matrix, 537
  distribution of roots of Wishart matrix, 534

of Wishart matrix in the metric of another, 522
asymptotic distribution of, 545
distribution of, 530
see also Principal components
Chi-squared distribution, 280
noncentral, 76
Cholesky decomposition, 586
Classification into normal populations:
Bayes:
into one of several, 229
into one of two, 205
discriminant function, 207
sample, 209
example, 231
invariance of procedures, 216
likelihood criterion for, 215
unequal covariance matrices, 235
linear, for unequal covariance matrices, 235
admissible, 239
maximum likelihood rule, 216
minimax:
one of several, 230
one of two, 207
one of several, 229
one of two, 205
W-statistic, 209
asymptotic distribution of, 212
asymptotic expansion of misclassification probabilities, 218
Z-statistic, 216
asymptotic expansion of misclassification probabilities, 222
see also Classification procedures
Classification procedures:
admissible, 199, 226
into several populations, 227
into two populations, 202, 203
a priori probabilities, 197
Bayes, 83, 198
and admissible, 202, 203, 228
into several populations, 225
into two populations, 205
complete class of, 199, 226
essentially, 199
minimal, 199
costs of misclassification, definition of, 197
expected loss from misclassification, 198, 225
minimax, 199
for two populations, 204
probability of misclassification, 198, 217
see also Classification into normal populations
Cochran's theorem, 258
Coefficient of alienation, 397
Complete class of procedures, 83
essentially, 83
minimal, 83
Completeness, definition of, 78
of sample mean and covariance matrix, 78
Complex normal distribution, 58
characteristic function of, 58
linear transformation in, 58
maximum likelihood estimators for, 100
Complex Wishart distribution, 281
Components of variance, 425
Concentration ellipsoid, 52, 79
Conditional density, 12
normal, 36
Conditional probability, 11
Conjugate family of distributions, 268
Consistency, definition of, 80
Contours of constant density, 23
Correlation coefficient:
canonical, *see* Canonical correlations and variates
confidence interval for, 116
by use of Fisher's z, 124
distribution of sample, asymptotic, 122
bootstrap method for, 125
tabulation of, 113
when population is not zero, 113
when population is zero, 108
distribution of set of sample, 268
Fisher's z, 122
geometric interpretation of sample, 66
invariance of population, 22
invariance of sample, 99
likelihood ratio test, 118
maximum likelihood estimator of, 65
as measure of association, 23
moments of, 151
monotone likelihood ratio of, 150
multiple, *see* Multiple correlation coefficient
partial, *see* Partial correlation coefficient
in the population (simple, Pearson, product-moment, total), 21
sample (Pearson), 65, 103
test of equality of two, 124
test by hypothesis about, 114

by Fisher's z, 123
power of, 115
test that it is zero, 108
unbiased estimator of, 119
Cosine of angle between two vectors, 65. *See also* Correlation coefficient
Covariance, 18
Covariance matrix, 18
asymptotic distribution of sample, 81
Bayes estimator of, 272
characterization of distribution of sample, 71
confidence bounds for quadratic form in, 439
consistency of sample as estimator of population, 81
distribution of sample, 249
estimation, *see* Best estimator of
geometrical interpretation of sample, 66
with linear structure, 101
maximum likelihood estimator of, 63
computation of, 63
when the mean vector is known, 99
of normal distribution, 21
sample, 71
singular, 33
tests of hypotheses, *see* Testing that a covariance matrix is a given matrix; Testing that a covariance matrix is proportional to a given matrix; Testing that a covariance matrix and mean vector are equal to a given matrix and vector; Testing equality of covariance matrices; Testing equality of covariance matrices and mean vectors; Testing independence of sets of variates
unbiased estimator of, 71
Cramer-Rao lower bound, 80
Critical function, 182
Cumulants, 50
of a multivariate normal distribution, 50
Cumulative distribution function (cdf), 7

Decision procedure, 82
Degenerate normal distribution, 31, 33
Density, 7
conditional, 12
normal, 36
marginal, 9
normal, 29, 31
multivariate normal, 21
Determinant, 581
derivative of, 598
symmetric matrix, 599
Dirichlet distribution, 284
Discriminant function, *see* Classification into normal populations
Distance, 587
Distance between two populations, 75
Distribution, *see* Canonical correlations; Characteristic roots; Correlation coefficient; Covariance matrix; Cumulative distribution function; Density; Generalized variance; Mean vector; Multiple correlation coefficient; Multivariate normal density; Multivariate t-distribution; Non central chi-squared distribution; Non central F-distribution; Non central T^2 distribution; Partial correlation coefficient; Principal components; Regression coefficients; T-test and independence of sets of variates; Wishart distribution
Distribution of matrix of sums of squares and cross-products, *see* Wishart distribution
Duplication formula for gamma function, 76, 112, 303

$\mathcal{E}$, 9
Efficiency of vector estimate, definition of, 79
Ellipsoid of concentration of vector estimate, 52, 79
Ellipsoid of constant density, 34
Equiangular line, 66
Exp, 16
Expected value of complex-valued function, 44
Exponential family of distributions, 185
Extended region, 351

Factor analysis, 550
centroid method, 569
communalities, 563
confirmatory, 556
EM algorithm, 562, 577
exploratory, 556
general factor, 551
identification of structure in, 553
by specified zeros, 553, 576

Factor analysis (*Continued*)
loadings, 551
maximum likelihood estimators, 559
asymptotic distribution of, 565
in case of identification by zeros, 574
nonexistence for fixed factors, 570
for random factors, 559
minimum distance methods, 566
model, 551
oblique factors, 553
orthogonal factors, 552
principal component analysis, relation to, 567
rotation of factors, 553
scores, estimation of, 575
simple structure, 554
space of factors, 553
transformations, 571
tests of fit for, 563
units of measurement, 556
varimax criterion, modified, 572
Factorization theorem, 77
Fisher's z, 122
asymptotic distribution of, 123
moments of, 123
see also Correlation coefficient; Partial correlation coefficient

$\Gamma_p(t)$, 251
Gamma function, multivariate, 251
Generalized analysis of variance, *see* Multivariate analysis of variance; Regression coefficients and function
Generalized T^2, *see* T^2-test and statistic
Generalized variance, 259
asymptotic distribution of sample, 266
distribution of sample, 264
geometric interpretation of sample in N dimensions, 262
in p dimensions, 263
invariance of, 457
moments of sample, 264
sample, 260
General linear hypothesis, *see* Linear hypothesis, testing of; Regression coefficients and function
Gram-Schmidt orthogonalization, 245, 603

Haar invariant distribution of orthogonal matrices, 536
Hadamard's inequality, 54
Head lengths and breadths of brothers, 97
Hotelling's T^2, *see* T^2 -test and statistic
Hypergeometric function, 113

Incomplete beta function, 324
Independence, 10
mutual, 11
of normal variables, 28
of sample mean vector and sample covariance matrix, 71
tests of, *see* Correlation coefficient; Multiple correlation coefficient; Testing independence of sets of variates
Information matrix, 80
Integral of a symmetric unimodal function over a symmetric convex set, 363
Intraclass correlation, 478
Invariance, *see* Classification into normal populations; Correlation coefficient; Generalized variance; Linear hypothesis; Multiple correlation coefficient; Partial correlation coefficient; T^2-test; Testing that a covariance matrix is a given matrix; Testing that a covariance matrix is proportional to a given matrix; Testing equality of covariance matrices; Testing equality of covariance matrices and mean vectors; Testing independence of sets of variates
Inverted Wishart distribution, 268
Iris, four measurements on, 98, 168

Jacobian, 13
James-Stein estimator, 86
for arbitrary known covariance matrix, 93
average mean squared error of, 90

Kronecker delta, 69
Kronecker product of matrices, 599
characteristic roots of, 600
determinant of, 600

Latin square, 372
Lawley-Hotelling trace criterion, *see* Linear hypothesis
Likelihood, induced, 64
Likelihood function for sample from multivariate normal distribution, 60

Likelihood loss function for covariance matrix, 273
Likelihood ratio test, definition of, 117. *See also* Correlation coefficient; Linear hypothesis; Mean vector; Multiple correlation coefficient; Regression coefficients; T^2-test; Testing that a covariance matrix is given matrix; Testing that a covariance matrix is proportional to given matrix; Testing that a covariance matrix and mean vector are equal to a given matrix and vector; Testing equality of covariance matrices; Testing equality of covariance matrices and mean vectors; Testing independence of sets of variates
Linear combinations of normal variables, distribution of, 25
Linear equations, solution of, 606
  by Gaussian elimination, 606
Linear functional relationship, 507
  relation to simultaneous equations, 514
Linear hypothesis, testing of:
  admissibility of, 349
    necessary condition for, 361
  Bartlett-Nanda-Pillai trace criterion, 326
    admissibility of, 374
    asymptotic expansion of distribution of, 328
    as Bayes procedure, 373
    table of significance points of, 630
    tabulation of power of, 328
  canonical form of, 296
  comparison of powers, 330
  invariance of criteria, 322
  Lawley-Hotelling trace criterion, 323
    admissibility of, 374
    asymptotic expansion of distribution of, 325
    monotonicity of power function of, 367
    table of significance points of, 616
    tabulation of, 326
  likelihood ratio criterion, 293
    admissibility of, 373
    asymptotic expansion of distribution of, 317
    as Bayes procedure, 373
    distributions of, 301, 304
    F-approximation to distribution of, 321
    geometric interpretation of, 296
    moments of, 303
    monotonicity of power function of, 367
    normal approximation to distribution of, 318
    table of significance points, 609
    tabulation of distribution of, 308
    Wilks' Λ, 293
  monotonicity of power function of an invariant test, 361
  Roy's maximum root criterion, 329
    distribution for $p = 2$, 329
    monotonicity of power function of, 367
    table of significance points, 634
    tabulation of distribution of, 329
  step-down test, 309
  see also Regression coefficients and function
Linearly independent vectors, 583
Linear transformation of a normal vector, 25, 31, 33
Loss, 82
LR decomposition, 585

Mahalanobis distance, 75, 206
  sample, 219
Majorization, 352
  weak, 352
Marginal density, 9
  distribution, 9
  normal, 29, 31
Mathematical expectation, 9
Matrix, 579
  bidiagonal, upper, 497
  characteristic roots and vectors of, *see* Characteristic roots and vectors
  cofactor in, 582
  convexity, 356
  definition of, 579
  diagonalization of symmetric, 587
  doubly stochastic, 602
  Givens, 464, 606
  Householder, 464, 605
  identity, 581
  inverse, 582
  minor of, 582
  nonsingular, 582
  operations with, 579
  positive definite, 583
  positive semidefinite, 583
  rank of, 583
  symmetric, 581

Matrix (*Continued*)
trace of, 584
transpose, 580
triangular, 584
tridiagonal, 464
Matrix of sums of squares and cross-products of deviations from the means, 61
Maximum likelihood estimators, *see* Canonical correlations and variates; Correlation coefficient; Covariance matrix; Mean vector; Multiple correlation coefficient; Partial correlation coefficient; Principal components; Regression coefficients; Variance
Maximum likelihood estimator of function of parameters, 64
Maximum of the likelihood function, 62
Maximum of variance of linear combinations, 456. *See also* Principal components
Mean vector, 18
asymptotic normality of sample, 81
completeness of sample as an estimator of population, 78
confidence region for difference of two:
when common covariance matrix is known, 74
when covariance matrix is unknown, 165
consistency of sample as estimate of population, 81
distribution of sample, 71
efficiency of sample, 80
improved estimator when covariance matrix is unknown, 173
maximum likelihood estimator of, 63
sample, 60
simultaneous confidence regions for linear functions of, 167
testing equality of, in several distributions, 194
testing equality of two when common covariance matrix is known, 74
tests of hypothesis about:
when covariance matrix is known, 73
when covariance matrix is unknown, *see* T-test
see also James-Stein estimator
Minimax, 85
Missing observations, maximum likelihood estimators, 154
Modulus, 13
Moments, 10, 49
factoring of, 11
from marginal distributions, 10
of normal distributions, 18, 49
Monotone region, 351
in majorization, 352
Multiple correlation coefficient:
adjusted, 143
distribution of sample:
conditional, 144
when population correlation is not zero, 146
when population correlation is zero, 140
geometric interpretation of sample, 137
invariance of population, 54
invariance of sample, 152
likelihood ratio test that is zero, 141
as maximum correlation between one variable and linear combination of other variables, 40
maximum likelihood estimator of, 137
moments of sample, 147, 151
optimal properties of, 148
population, 40
sample, 135
tabulation of distribution of, 148
unbiased estimator of, 147
Multivariate analysis of variance (MANOVA), 342
Latin square, 372
one-way, 342
two-way, 342
see also Linear hypothesis, testing of
Multivariate beta distribution, 372
Multivariate gamma function, 251
Multivariate normal density, 21
distribution, 21
computation of, 24
Multivariate t-distribution, 273, 283

$n(x|\mu,\Sigma)$, 21
$N(\mu,\Sigma)$, 21
Neyman-Pearson Fundamental Lemma, 241
Noncentral chi-squared distribution, 76
Noncentral F-distribution, 174
tables of, 174
Noncentral T^2-distribution, 174

Orthonormal vectors, 604

Parallelotope, 260
  volume of, 260
Partial correlation coefficient:
  computational formulas for, 41, 43
  confidence intervals for, 133
  distribution of sample, 133
  geometric interpretation of sample, 127
  invariance of population, 56
  invariance of sample, 152
  maximum likelihood estimator of, 127
  in the population, 37
  recursion formula for, 43
  sample, 127
  tests about, 133
Partial covariance, 36
  estimator of, 126
Partial variance, 36
Partioning of a matrix, 591
  addition of, 591
  of a covariance matrix, 27
  determinant of, 594
  inverse of, 594
  multiplication of, 591
Partioning of a vector, 591
  of a mean vector, 26
  of a random vector, 26
Pearson correlation coefficient, *see* Correlation coefficient
Plane of closest fit, 459
Polar coordinates, 279
Positive definite matrix, 583
Positive part of a function, 91
  of the James-Stein estimator, 92
Positive semidefinite matrix, 583
Precision matrix, 268
  unbiased estimator of, 270
Principal axes of ellipsoids of constant density, 458. *See also* Principal components
Principal components, 451
  asymptotic distribution of sample, 468, 540
  computation of, 462
  confidence region for, 470, 472
  distribution of sample, 534, 537
  maximum likelihood estimator of, 460
  population, 456
  testing hypotheses about, 473, 474, 475
Probability element, 8
Product-moment correlation coefficient, *see* Correlation coefficient

QL algorithm, 464
QR algorithm, 464
  decomposition, 604
Quadratic form, 583
Quadratic loss function for covariance matrix, 273

r, 65
$\mathcal{R}$ (real part), 251
Random matrix, 17
  expected value of, 18
Random vector, 17
Randomized test, definition of, 182
Rectangular coordinates, 251
  distribution of, 249, 251
Regression coefficients and function, 36
  confidence regions for, 33
  distribution of sample, 291
  geometric interpretation of sample, 127
  maximum likelihood estimator of, 289
  partial correlation, connection with, 54
  sample, 289
  simultaneous confidence intervals for, 337, 338
  testing hypotheses of rank of, 507
  testing they are zero, in case of one dependent variable, 141
Residuals from regression, 39
Risk function, 83

Simple correlation coefficient, *see* Correlation coefficient
Simultaneous equations, 509
  estimation of coefficients, 512
    Least Variance Ratio (LVR), 513
    Limited Information Maximum Likelihood (LIML), 513
    Two Stage Least Squares (TSLS), 517
  identification by zeros, 510
  reduced form, 509
    estimation of, 511
Singular normal distribution, 31, 33
Singular value decomposition, 492, 590
Spherical normal distribution, 23
Sphericity test, *see* Testing that a covariance matrix is proportional to a given matrix
Standardized variable, 23
Stochastic convergence, 101
  of a sequence of random matrices, 101

Sufficiency, definition of, 77
Sufficiency of sample mean vector and covariance matrix, 77
Surface area of unit sphere, 280
Surfaces of constant density, 23
Symmetric matrix, 581

T^2-statistic, 163. *See also* T^2-test and statistic
T^2-test and statistic, 159
  admissibility of, 188
  as Bayes procedure, 190
  distribution of statistic, 163
  geometric interpretation of statistic, 163
  invariance of, 163
  as likelihood ratio test of mean vector, 163
  limiting distribution of, 163
  noncentral distribution of statistic, 174
  optimal properties of, 181
  power of, 174
    tables of, 174
  for testing equality of means when covariance matrices are different, 175
  for testing equality of two mean vectors when covariance matrix is unknown, 167
  for testing symmetry in mean vector, 170
  as uniformly most powerful invariant test of mean vector, 181, 182
Testing that a covariance matrix is a given matrix, 434
  invariant tests of, 438
  likelihood ratio criterion for, 434
  modified likelihood ratio criterion for, 434
    asymptotic expansion of distribution of, 438
    moments of, 436
    table of significance points, 641
  Nagao's criterion, 438
Testing that a covariance matrix is proportional to a given matrix, 427
  invariant tests, 432
  likelihood ratio criterion for, 429
    admissibility, 449
    asymptotic expansion of distribution of, 432
    moments of, 430
    table of significance points, 639
  Nagao's criterion, 432
Testing that a covariance matrix and mean vector are equal to a given matrix and vector, 440
  likelihood ratio criterion for, 441
    asymptotic expansion of distribution of, 442
    moments of, 441
Testing equality of covariance matrices, 405
  invariant tests, 423
  likelihood ratio criterion for, 406
    invariance of, 408
  modified likelihood ratio criterion for, 406
    admissibility of, 445, 446
    asymptotic expansion of distribution of, 420
    distribution of, 414
    moments of, 416
    table of significance points, 638
  Nagao's criterion for, 408
Testing equality of covariance matrices and mean vectors, 408
  likelihood ratio criterion for, 409
    asymptotic expansion of distribution of, 421
    distribution of, 414
    moments of, 416
    unbiasedness of, 410
Testing independence of sets of variates, 376
  and canonical correlations, 498
  likelihood ratio criterion for, 379
    admissibility of, 397
    asymptotic expansion of distribution of, 386
    distribution of, 383
    invariance of, 381
    moments of, 384
    monotonicity of power function of, 402
    unbiasedness of, 381
  Nagao's test, 388
    asymptotic expansion of distribution of, 388
  stepdown tests, 389
Testing rank of regression matrices, 507
Tests of hypotheses, *see* Correlation coefficient; Generalized analysis of variance; Linear hypothesis; Mean vector; Multiple correlation coefficient; Partial correlation coefficient; Regression coefficients; T^2-test and statistic
Tetrachoric functions, 24
Total correlation coefficient, *see* Correlation coefficient

Trace of a matrix, 584
Transformation of variables, 13

Unbiased estimator, definition of, 71
Unbiased test, definition of, 362
Uniform distribution over ellipsoid, 14

Variance, 18
  generalized, *see* Generalized variance
  maximum likelihood estimator of, 64

$w(\boldsymbol{A}\mid\Sigma,n)$, 249
$W(\Sigma,n)$, 249
$w^{-1}(\boldsymbol{B}\mid\Psi,m)$, 268
$W^{-1}(\Psi,m)$, 268
Wishart distribution, 249
  characteristic function of, 254
  geometric interpretation of, 249
  marginal distributions of, 255
  noncentral, 570
  for $p = 2$, 111

z, *see* Fisher's z
Zonal polynomials, 468

Applied Probability and Statistics (Continued)

DRAPER and SMITH • Applied Regression Analysis, *Second Edition*
DUNN • Basic Statistics: A Primer for the Biomedical Sciences, *Second Edition*
DUNN and CLARK • Applied Statistics: Analysis of Variance and Regression
ELANDT-JOHNSON and JOHNSON • Survival Models and Data Analysis
FLEISS • Statistical Methods for Rates and Proportions, *Second Edition*
FOX • Linear Statistical Models and Related Methods
FRANKEN, KÖNIG, ARNDT, and SCHMIDT • Queues and Point Processes
GALAMBOS • The Asymptotic Theory of Extreme Order Statistics
GIBBONS, OLKIN, and SOBEL • Selecting and Ordering Populations: A New Statistical Methodology
GNANADESIKAN • Methods for Statistical Data Analysis of Multivariate Observations
GOLDBERGER • Econometric Theory
GOLDSTEIN and DILLON • Discrete Discriminant Analysis
GREENBERG and WEBSTER • Advanced Econometrics: A Bridge to the Literature
GROSS and CLARK • Survival Distributions: Reliability Applications in the Biomedical Sciences
GROSS and HARRIS • Fundamentals of Queueing Theory
GUPTA and PANCHAPAKESAN • Multiple Decision Procedures: Theory and Methodology of Selecting and Ranking Populations
GUTTMAN, WILKS, and HUNTER • Introductory Engineering Statistics, *Third Edition*
HAHN and SHAPIRO • Statistical Models in Engineering
HALD • Statistical Tables and Formulas
HALD • Statistical Theory with Engineering Applications
HAND • Discrimination and Classification
HILDEBRAND, LAING, and ROSENTHAL • Prediction Analysis of Cross Classifications
HOAGLIN, MOSTELLER, and TUKEY • Understanding Robust and Exploratory Data Analysis
HOEL • Elementary Statistics, *Fourth Edition*
HOEL and JESSEN • Basic Statistics for Business and Economics, *Third Edition*
HOGG and KLUGMAN • Loss Distributions
HOLLANDER and WOLFE • Nonparametric Statistical Methods
IMAN and CONOVER • Modern Business Statistics
JAGERS • Branching Processes with Biological Applications
JESSEN • Statistical Survey Techniques
JOHNSON and KOTZ • Distributions in Statistics
Discrete Distributions
Continuous Univariate Distributions—1
Continuous Univariate Distributions—2
Continuous Multivariate Distributions
JOHNSON and KOTZ • Urn Models and Their Application: An Approach to Modern Discrete Probability Theory
JOHNSON and LEONE • Statistics and Experimental Design in Engineering and the Physical Sciences, Volumes I and II, *Second Edition*
JUDGE, HILL, GRIFFITHS, LÜTKEPOHL and LEE • Introduction to the Theory and Practice of Econometrics
JUDGE, GRIFFITHS, HILL and LEE • The Theory and Practice of Econometrics
KALBFLEISCH and PRENTICE • The Statistical Analysis of Failure Time Data